# Transportation Engineering

## Volume II: Railways, Airports, Docks and Harbours, Bridges and Tunnels


**C Venkatramaiah**
Formerly, Professor of Civil Engineering
S V University College of Engineering
Tirupati

**Universities Press**

Transportation Engineering Volume II: Railways, Airports, Docks and Harbours, Bridges and Tunnels

UNIVERSITIES PRESS (INDIA) PRIVATE LIMITED

*Registered office*
3-6-747/1/A & 3-6-754/1, Himayatnagar, Hyderabad 500 029, Telangana, India
info@universitiespress.com; www.universitiespress.com

*Distributed by*
Orient Blackswan Private Limited

*Registered office*
3-6-752 Himayatnagar, Hyderabad 500 029, Telangana, India

*Other offices*
Bengaluru / Bhopal / Chennai / Guwahati / Hyderabad / Jaipur / Kolkata / Lucknow
Mumbai / New Delhi / Noida / Patna / Vijayawada


First published 2016
Reprinted 2018

ISBN: 978-81-7371-998-1



*Typeset in* Palatino 10/12 by
MacroTex Solutions, Chennai 600 091, Tamil Nadu, India

501433

*Printed at*
Graphica Printers & Binders, Hyderabad 500 013

*Published by*
Universities Press (India) Private Limited
3-6-747/1/A & 3-6-754/1, Himayatnagar, Hyderabad 500 029, Telangana, India

*To*
*my sons-in-law,*

*Dr G V Marutheswar*
*Professor*
*Department of EEE*
*S V University College of Engineering*
*Tirupati*

*and*

*Mr M Kishore Kumar*
*Distinguished Technologist*
*Hewlett–Packard Enterprise*
*Bengaluru*

# Contents

# Preface

This book is complementary to the author's *Transportation Engineering, Volume I: Highway Engineering*. It has been designed in such a way that, along with the first volume, it effectively covers the syllabi of Transportation Engineering prescribed by several universities across the country.

The book consists of five parts:

Part I: Railway Engineering
Part II: Airport Engineering
Part III: Dock and Harbour Engineering
Part IV: Bridge Engineering
Part V: Tunnel Engineering

While the first three parts, along with Volume I deal at length with the primary modes of transportation, the fourth and fifth parts are essential links to these modes. Bridges serve the purpose of overcoming different kinds of obstructions such as water bodies and valleys in order to maintain continuity of road and railway routes, and tunnels help to reduce the length of the route. Tunnels may also serve the function of transporting water, sewage, oil and other fluids.

The weightage attached to Railway Engineering in the syllabus of Transportation Engineering – I/II is much more than that associated with Airports, Harbours, Bridges and Tunnels. Bearing this fact in mind, the treatment of these topics in this volume has been made commensurate with the relative emphasis placed on them.

The salient features of this book are:

1. The S.I. system of units is adopted primarily, while the MKS System is used where considered necessary for compatibility with data from the original research.
2. The latest IRS, IRC and IS Codes, including the special publications, and the practices recommended by them are adopted throughout.
3. The sequence of topics and sub-topics has been laid out in logical order.
4. Emphasis is laid on current practices used in India. (Practices in other countries are given to the minimum extent, where considered necessary, so as to avoid confusion to the student.)
5. Adequate number of numerical examples is included to help the student in understanding the concepts and their application.
6. The comprehensive summary given at the end of each chapter is intended to help the student in a quick-review of the topics on the eve of an examination.
7. Adequate number of questions and practice problems are given to enable the student understand the concepts and their application.
8. The answers to numerical problems, chapter-wise, are given in Appendix-I for the convenience of the student.
9. Symbols and nomenclature adopted are such that they are consistent to a large extent. They are given in Appendix-II.
10. There are several abbreviations and acronyms in widespread use in this subject. These are listed in Appendix-III.
11. A comprehensive list of References, arranged in alphabetical order, is provided in Appendix-IV; the relevant IRS codes, IRC codes and Special Publications are listed separately.
12. An exhaustive index of key words is provided at the end of the book.

13. Lecture slides and solutions to chapter-end problems are provided as online resources. These can be accessed at
    http://www.universitiespress.com/cvenkatramaiah/transportationengineeringvol2

References have been made to Indian Railway Standard Codes and Manuals, relating to different aspects of Railway Engineering, such as railway track and its components, including railway bridges. The use of relevant material from these codes and manuals in the book for the student's benefit is gratefully acknowledged. The permission granted by IRC and the RDSO for citing their Codes and Special Publications is thankfully appreciated. The reader is advised to refer to the latest available versions of the Codes and Special Publications.

The author also acknowledges the permission granted by OECD for reproducing excerpts of their material on "Safety in Tunnels" and by PIARC for including the salient provisions of "Operation and Maintenance of Tunnels" from the Road Tunnels Manual and Reports 1 and 2 of OECD and PIARC on the transport of dangerous goods through road tunnels.

All sources have been duly acknowledged to the best of the author's knowledge and belief. Omissions, if any, are only inadvertent and not deliberate, and they will be taken care of at the earliest opportunity if pointed out. The author hopes his effort would be appreciated by the academic community— students and teachers alike. While all efforts have been made to make this work error-free, it is possible that inadvertent lapses may have crept in. The author will be grateful if the readers could point them out.

## ACKNOWLEDGEMENTS

The author acknowledges the authors of several reference books and papers from which he gained a significant quantum of subject knowledge.

The author wishes to express his grateful thanks to his wife, Smt. Lakshmi Suseela, for her unstinted cooperation, and for putting up with inconveniences during the preparation of the manuscript; his daughters, Smt. Sarada and Smt. Usha Padmini, for their appreciation, co-operation and support; and to his sons-in-law, Prof. G V Marutheswar and Mr Kishore Kumar for their invaluable support in providing the several codes that have been cited across the book.

The author thanks all his students who have always recognised and appreciated his efforts in teaching them.

The author expresses his thanks to his publishers, Universities Press (India) Pvt. Ltd for bringing out this book in a relatively short time. In particular, he is grateful to

- Mr Madhu Reddy, Director, Universities Press, and Mr Thomas Mathew Rajesh, Vice-president, for their excellent support and for encouraging the author to take up this project placing confidence in his capabilities.
- Ms Sreelatha Menon, Managing Editor, for her appreciation and co-operation.
- Mr Kallol Das, for providing the syllabi and question papers of various universities.
- Mr Ramesh M R, for his copy-editing efforts to enhance clarity of the text, and for his excellent co-operation in maintaining high quality in production.

**C Venkatramaiah**

# PART I

## Railway Engineering

Chapter 1

# INTRODUCTION AND HISTORICAL ASPECTS

**CHAPTER OBJECTIVES**

This chapter enables the student to:

- Learn the general history and development of railways in India as well as in other countries
- Understand the evolution of railways in India during the British period, and its subsequent expansion by the Government of India during the post-Independence period
- Gain insight into railway administration and the role of railways in national development.

## 1.1 INTRODUCTION

One of the most important modes of transportation is the Railway. With the development of civilisation, mankind developed various modes of transport for carrying people and goods from one place to another. Of these modes, the Railway network of any country, especially of a developing country like India, has come to be considered as the most important asset in achieving all-round progress — economic, social, cultural, industrial, commercial and political.

The various phases of the construction and development of railways are:

- Planning a railway route as part of network
- Engineering survey and alignment of the route
- Selection of the various components of the railway track
- Geometric design aspects of the railway
- Construction of the railway track along with accessories such as stations, station yards, different tracks and control facilities like signals and passenger amenities
- Operation of the railway with passenger trains, goods trains and so on with the necessary rolling stock such as coaches, wagons and engines or locomotives
- Maintenance of all these facilities.

*Railway Engineering* is concerned with the successful implementation of all these phases.

## 1.2 HISTORICAL ASPECTS OF RAILWAYS

It is pertinent to study the historical aspects of railways in general, and trace the development of the Indian Railways in particular, before proceeding to consider the various phases of development constituting the subject of Railway Engineering.

### History of Railways

The development of railways is closely linked to that of civilisation. In olden times, humans used animal power for pulling loaded carts. Gradually, efforts were made to replace animal power with mechanical power.

The idea of pulling a wheeled vehicle placed on timber baulks by horses led to the principle of a rail track. Iron plates were fixed to prolong the life of timber baulks; the term 'plate laying' for laying a new track originated from this.

Railways originated in England; but the first efforts to develop a steam engine were made by a French man, Nicholas Joseph Cugnot, in 1769. Richard Trevithick, a British Engineer, designed and built a steam locomotive in 1804. But these were meant for traction on roads only.

George Stephenson, a British engineer, is considered to have built the first steam locomotive engine for traction on rails. The first public railway in the world to operate passenger and freight traffic was between Darlington and Stockton in England on 27 September 1825. The first train pulled several wagons and hundreds of people at a speed of over 20 km/h. Stephenson's improved locomotive engine, the 'Rocket', hauled the first passenger train between Liverpool and Manchester on 15 September 1830.

Simultaneously, the USA and European countries other than England started constructing and operating their own railway routes. The first railway line in the USA was between Mohawk and Hudson near New York in 1833 and that in Germany was between Nuremberg and Furth (Nürnberg and Fürth in German) in the year 1835.

### Development of Railways in India

During the British rule, Lord Dalhousie, has been credited with the introduction of railways in India, and putting up proposals for several important routes. The first idea of a railway line between Madras (now Chennai) and Bangalore (now Bengaluru) was mooted in 1831, but it took several years to materialise.

The following table gives the more important developments in India Railways in chronological order:

| Year | Important Developments |
|---|---|
| 1844 | East Indian Railway company was formed by R. M. Stephenson. |
| 1850 | Construction of a railway line started between Bombay and Thane. |
| 1853 | The first railway line between Bombay (now Mumbai) to Thane became operational for passenger traffic on April 16. The 34-km stretch was covered by the first train in 1 hr 15 min. |
| 1854 | The 39-km long railway line from Howrah to Hooghly became operational on August 15. |
| 1856 | The line between Vyasarpadi and Wallajah Road (100 km), built by the Madras Railway Company, became operational. |

Cont'd

**Table Cont'd**

| Year | Important Developments |
|---|---|
| 1866 | Calcutta (now Kolkata), British India's capital was linked to Delhi, Bombay and Amritsar. |
| 1850–'68 | Railway lines were constructed by private companies under the Early Guarantee System. In this system, the Government guaranteed a minimum percentage of return, but retained the right to purchase these routes after 25 to 50 years. |
| 1855–'60 | Companies such as Madras State Railway (MSR), East India Railway (EIR), Great Indian Peninsula Railway (GIP), and Bombay, Baroda and Central India Railway (BB & CIR) were formed. |
| 1869–'81 | The railways developed very fast as state-owned railways or company-managed ones under the new guarantee system with government control. |
| 1873 | The first metre gauge line from Delhi to Farrukh Nagar (61 km) was opened. |
| 1881 | Darjeeling Himalayan Railway (Narrow gauge: 2 ft), the first hill railway was opened. |
| 1881–'97 | Several new companies were given contracts, prominent among them being the Madras Southern Maharatta Railway (MSMR) and the Bengal Nagpur Railway (BNR). |
| 1887 | Victoria Terminus (now Chhatrapati Shivaji Terminus) railway station opened in Bombay (now Mumbai). |
| 1903 | Kalka to Shimla narrow gauge line (96 km) was opened to traffic. |
| 1905 | Railway board established with one president and two members. |
| 1920 | 'Indian Railway Inquiry committee' was appointed with Sir William Acworth as the Chairman to study the Indian Railways. |
| 1924 | Railway finances separated from general finances; nationalisation of railways started. |
| | Railway board was reconstituted with the Chief commissioner as President and two members. |
| 1925 | Local train system on the electrified section – Bombay (now Mumbai) to Kurla became operational. |
| 1931 | Electric trains started on the double-line section from Madras Beach to Tambaram. |
| 1936 | Air conditioning was introduced in passenger coaches. |
| 1937 | Burma (now Myanmar) got separated and 3200 km of railway routes taken out of Indian Railways. |
| 1939–'42 | During World War-II, 26 branch lines were closed; railway workshops were used for the manufacture of defence materials. |
| 1947 | India gained independence; railway lines got divided between India and Pakistan. |
| 1947–'51 | Regrouping of the existing railways was done and six railway zones formed: Central Railway (CR), Northern Railway (NR), Eastern Railway (ER), Southern Railway (SR), Western Railway (WR), and North-Eastern Railway (NER). |
| 1950 | Chittaranjan Locomotive Works Started production of steam locomotives. |
| 1952 | Integral Coach Factory (ICF) was set up in Perambur, Madras (now Chennai) to produce welded steel, lightweight integral coaches. |
| 1954 | The position of Chief Commissioner for railways was renamed as Chairman of the Railway Board. |

Cont'd

**Table Cont'd**

| Year | Important Developments |
|---|---|
| 1957 | Research, Design and Standards Organisation (RD&SO) was set up in Lucknow, merging the various standards committees. Railway Protection Force (RPF) was also set up in the same year |
| 1961 | Diesel Locomotive Works was set up in Varanasi; Chittaranjan Locomotive works started manufacturing electric locomotives during the same year. |
| 1969 | Divisional system was adopted uniformly on Indian Railways, New Delhi–Howrah Rajdhani Express running at 120 km/h was introduced during the same year. |
| 1974 | Rail India Technical and Economic Services (RITES) started. |
| 1976 | Indian Railway Construction Company (IRCON) was formed |
| 1984 | First metro rail was introduced in Kolkata. |
| 1985 | Computerised passenger reservation system was introduced. |
| 1987 | Rail Coach Factory (RCF) was established in Kapurthala. |
| 1988 | Container corporation of India (CONCOR) was established. First Shatabdi Express was run from New Delhi and Thansi. |
| 1998 | Konkan Railway System became operational on 26 January. Fairy Queen, the oldest working steam locomotive in the world was given the Guinness Records Certificate. |
| 1999 | Darjeeling Himalayan Railway declared World Heritage Site by UNESCO, Centenary celebrations of Nilgiri Mountain Railway. The largest route relay interlocking system in the world at Delhi railway station was awarded Guinness Records Certificate. |
| 2002 | Jan Shatabdi trains were introduced. |
| | Two more railway zones – North Western Railway (Jaipur) and East Central Railway (Hazipur) – came into being on 1st October. |
| 2003 | With effect from 1st April, Indian Railways has 16 Zones and 59 Divisions. |
| 2008 | Garib Rath Express between Secunderabad and Visakhapatnam was inaugurated on 24 October. |
| | Kalka – Shimla railway declared as World Heritage site by UNESCO. |
| 2009 | The document Railway Vision 2020 (2011–2020) was placed before the parliament. |
| 2010 | First regular passenger service in the Kashmir valley was opened to traffic. |

The funds allocated during the various Five-year Plans, starting from 1951–'56, facilitated the development and expansion including modernisation of the Indian Railways.

## Organisation of Indian Railways

The construction and operation of railways in India are in accordance with the *Indian Railway Act*, 1890, amended from time to time, the latest amendment having been enacted in 2003. The executive authority of administering railways is vested with the *Government of India*. A separate *Ministry of Railways* takes care of this through the railway board.

Indian Railways is the largest public undertaking of the Government of India in terms of assets, finances and number of employees.

*Railway Board*: The responsibility of management of Indian Railways is that of the Railway Board, under overall supervision of the Ministry of Railways. All aspects such as the regulation, construction, maintenance and operation are taken care of by the railway board; this consists of a chairman, a financial commissioner, and five more members. The chairman is the ex-officio principal secretary and reports to the Minister for Railways; he is responsible for dealing with technical and administrative matters.

The financial commissioner is empowered to take decisions on all financial matters.

The members are each in charge, separately, of staff, traffic, civil engineering, mechanical engineering and electrical engineering.

The railway board is further assisted by a number of additional members and executive directors in the different directorates, but without the power to take policy decisions.

**Research, Designs and Standards Organisation (RD&SO)**

This is located in Lucknow, and consists of a team of specialists from different fields headed by a director general. RD&SO is the technical consultant to the railway board, the zonal railways and production units as also to public and private sector undertakings with regard to the design and standardisation of railway equipment. RD&SO's quality management system has been awarded ISO 9001: 2000 certification.

**Zonal Railways**

The entire railway network was originally divided into six zones, later into nine, and presently into sixteen zones. These are:

1. Central (CR)
2. Eastern (ER)
3. East Central (ECR)
4. East Coast (ECOR)
5. Northern (NR)
6. North Central (NCR)
7. North Eastern (NER)
8. North Frontier (NFR OR NEFR)
9. North Western (NWR)
10. Southern (SR)
11. South Central (SCR)
12. South Eastern (SER)
13. South East Central (SECR)
14. South Western (SWR)
15. Western (WR)
16. West Central (WCR)

In addition, the *Kolkata Metro* has been awarded Zonal Status, and the *Konkan Railway System* has also been considered that way for administrative purposes.

The zonal railways take care of all railway matters in their respective zones. Each zone is headed by a general manager, who is assisted by an additional general manager and a deputy general manager, along with several senior/chief managers or officers for public relations, personnel, finance and accounts, operations, commercial matters, security, stores, medical wing and construction administration. The various engineering wings are headed by chief engineers of the particular discipline.

**Production units:** There are six production units – the Chittaranjan Locomotive Works at Chittaranjan (diesel locomotives), Integral Coach Factory at Chennai (coaches), Diesel Component Works at Patiala (diesel components), Rail Coach Factory at Kapurthala (coaches) and Wheel and Axle Plant at Bengaluru (wheels and axles).

### Central Organisation for Modernisation of Workshops (COFMOW)

Established in 1979 as an agency to implement the various programmes of modernisation of railway workshops, this was necessitated by the fact that most of the workshops were over a hundred years old and in need of reconstruction. This organisation is responsible for the selection, procurement, and induction of the latest workshop technologies and modern machinery and plant and functions with financial assistance from the World Bank.

### Commission of Railway Safety

This came into being, in its present form, in 1961. The Commission is headed by a Chief Commissioner, and presided by nine commissioners. Its headquarters is in Lucknow while the other offices are located at Delhi, Mumbai, Kolkata, Chennai, Bengaluru and Secunderabad. The commission advises the railway board on all matters of safety in construction and operation of railways.

In addition to the units described above, there are several undertakings under the Ministry of Railways, of which the significant ones shall now be considered.

**Rail India Technical and Economic Services Ltd (RITES):** This provides consultancy services on all aspects of railways. It is a multi-disciplinary consultancy organisation in the field of transportation covering all modes.

RITES has provided their services in over 60 countries in South-east Asia, the UK, the USA, the Middle-east, Europe, Africa and Latin America.

**Indian Railways Construction Company Ltd (IRCON):** Established in 1976 as a specialised agency for the construction of major railway projects not only in India but also in other countries, IRCON has diversified its activities to other related sectors such as highways, expressways, bridges, flyovers, airports, and mass rapid transit systems, apart from IT and telecommunication services in several countries across the world.

**Dedicated Freight Corridor Corporation of India Ltd (DFCCIL):** This is for planning, construction, maintenance, and operation of a dedicated freight corridor, exclusively for running freight trains at a speed of 100 km/h to cover about 3,300 route kilometres in the Eastern and Western Corridors, funded by Japan and the World Bank, respectively.

**Indian Railway Finance Corporation Ltd (IRFC):** This was set up in 1986 for raising funds for the development of Indian Railways – through the issue of bonds, loans from financial institutions and external borrowings.

**Container Corporation of India Ltd (CONCOR):** This was set up in 1989 for developing multimodal transport to support the country's trade by carrying domestic cargo traffic in ISO containers, adopting the latest technology. It has a nation-wide network of 56 container terminals and provides logistic services for international and domestic cargo industry in India.

**Kokan Railway Corporation Ltd (KRCL):** This is the first railway project in the country to be built on build, operate and transfer (BOT) principle. Four states – Maharashtra, Goa, Karnataka and Kerala – participated along with the Ministry of Railways. The 760-km line took 8 years to be completed and opened in 1998. The salient features of Kor :an railways are presented in Table 1.1.

**Indian Railways Catering and Tourism Corporation Ltd (IRCTC):** This is a new corporation meant to provide catering services on trains, boost rail tourism, popularise the internet ticketing

**Table 1.1** Salient features of Konkan railway

| **Route length** | **760 km** |
|---|---|
| (Mumbai to Mangalore along the west coast) | |
| **Track** | |
| Gauge | Broad Gauge (1676 mm) |
| Ruling gradient: | 1: 150 (Compensated 0.67 per cent) |
| Rails: | Welded rails |
| Sleepers | Pre-stressed concrete (more block) |
| Number of curves: | 320 |
| **Earthwork** | |
| Maximum height of an embankment: | 25 m |
| Deepest cutting | 28 m |
| Total earth work | 88.77 million cu. m |
| **Stations** | |
| Total number of stations | 59 |
| **Bridges** | |
| Major bridges | 179 (Lineal waterway 21.50 km) |
| Minor bridges | 1919 (Lineal waterway 5.73 km) |
| **Road crossings** | |
| Road overbridges and underbridges: | 300 |
| Longest bridge | Sharavati bridge near Honnavar: 2065.8 m |
| Tallest viaduct | Panval Nadi: 64 m high |
| (420-m long PSC box girder) | |
| **Tunnels** | 92 (Covering a total length of 83 km) |
| Longest tunnel | Karbude tunnel: 6.5 km |
| **Speed** | |
| Maximum speed | 100 km/h |
| Average speed | 36 to 45 km/h |
| **Anti-collision device (ACD)** | Patented device to prevent collisions (Raksha Kavach). |
| **Sky bus metro** | Eco-friendly and economic mass rapid transportation device. |
| **Advantage** | Significant reduction in distance from Mangalore to Mumbai and Delhi. |

system, and make packaged drinking water (Rail Neer) readily available to the passenger. This is another successful venture of Indian Railways.

**RailTel Corporation of India Ltd (RailTel):** RailTel was set up in September, 2000, with an objective to build a nation-wide network of telecom multimedia by creating an optic fibre cable (OFC) network.

RailTel has modernised train control and communication system by providing OFC network along the railway track including high band-width communication facilities at over 3775 stations, and an efficient railway STD service.

RailTel has also set up cyber cafes for high-speed internet browsing, e-mail, audio chatting, and video conferencing.

**Centre for Railway Information System (CRIS):** This has been established to develop and provide the design and information of all computer services on the railways relating to freight operation, passenger reservation, train enquiry and track management.

**Rail Land Development Authority (RLDA):** This was set up in 2005 through an amendment of the Indian Railways Act, 1989, for the development of vacant railway land for commercial use with a view to generate additional revenue for Indian Railways.

**Indian Railway Welfare Organisation (IRWO):** This was set up in 1989 for meeting the housing needs of railway employees in service, as also those who have retired. The headquarters is in New Delhi. This organisation builds houses for railway employees on 'no profit–no loss basis'.

Several housing colonies have been constructed and many more are in progress in various places across the country.

**Centralised Training Institutes (CTI's) for railway employees:** The following are the six CTI's:

1. Railway staff college, Vadodara.
2. Indian Railways Institute of Civil Engineering, Pune
3. Indian Railways Institute of Signal and Telecommunication Engineering, Secunderabad.
4. Indian Railways Institute of Electrical Engineering, Nasik.
5. Indian Railways Institute of Mechanical Engineering, Jamalpur.
6. Indian Railways Institute of Transport Management, Lucknow.

While the first provides training in general management, the others provide railway officers with specialised technical training in their respective areas.

There are over 200 training centres for non-gazetted employees of railways. Training has been made mandatory in certain categories dealing with safety.

## 1.3 ROLE OF RAILWAYS IN TRANSPORTATION

Transportation, or movement of persons and goods from one place to another, is rightly regarded as an index of social, cultural, economic and commercial progress of any country. Movement of raw materials, distribution of manufactured goods and marketing of agricultural produce require fast and efficient means of transportation, which is the base of trade and commerce. Human kind has devised various modes of transport across land, water and air.

**Land transport:** Highways and Railways
**Water transport:** Transport on canals, rivers, seas, oceans, and other water bodies.
**Air transport:** Airways

The above modes exclude miscellaneous modes such as pipelines, conveyers, cableways and ropeways, which serve special needs under specific circumstances.

Railways fall under the category of modes with just one degree of freedom with restraint in the lateral and vertical directions.

Highways provide two degrees of freedom since vehicles are restrained only vertically.

Aircraft and submarines can move or be turned in any direction; thus they have three degrees of freedom. Railways have a special role in land transport in view of the capability to transport heavy and large volumes of goods and large number of persons over long distances with safety and comfort.

Railways play a significant role in transportation in view of their distinct advantages, especially in long distance coverage, facilitating the progress of the nation on several fronts.

Specifically, the Indian Railway System is the second largest in the world, and has received priority in the matter of development of transportation.

## Role of Indian Railways in National Development

With a glorious heritage of over 160 years of service, Indian Railways have been playing a significant role in the development of the nation on several fronts – specifically, economic, political and social. Being the largest network and public sector undertaking, it is practically the lifeline of the nation.

**Economic front:** The following are the contributions of Indian Railways on the economic front:

- Industrial growth due to quick and efficient transportation of raw materials
- Fast and efficient distribution of the finished products
- Boost in commerce and trade, resulting in higher standard of living
- Large scale and commercial farming
- Reduced cost of transportation of goods over long distances
- Easy mobility of labour for industry
- Mobility of people resulting in uniform population density
- Price stabilisation through movement of commodities.
- Quick transport of essential commodities during famines and war
- Energy-efficiency in operation
- Efficient land use
- Environment-friendliness in operation.

**Political front:** The effect of the railway network on the political front can be listed as follows:

- National integration, unity and sense of national pride
- Logistic support for national defence during war.
- Administrative convenience for the Central and State Governments through fast and efficient transport.
- Maintenance of internal security, law and order.
- Facilitation of mass migration of people at times of need.

**Social front:** The following are the desirable effects:

- Revolutionary changes in social values, beliefs and way of living
- Mixing of people resulting in the emergence of new social values based on reason, logic and scientific approach to human relations.
- People of different religions and languages using the railways foster brotherhood and unity.
- Broadening of the social outlook of people in far-flung and remote regions of the country.
- Boosting pilgrimage to places of religious significance
- Encouraging tourism – both domestic and foreign
- Relatively safe and convenient mode of travel for all categories of persons.

## Important Statistics of Indian Railways

Important statistics of certain aspects of Indian Railways are given below to give an idea of the magnitude of this prestigious public sector undertaking of the Government of India:

Number of *employees*: Over 13.1 lakh (Nearly 25% are engineering employees)
Number of *route kilometres*: 65,436
Number of *route km electrified*: 21,614 (20KV-AC)
*Running track* km (including double lines): Over 88,000
*Total track kilometres*: 1,15,000
Number of *railway stations*: 7,172
Number of *major bridges* (length more than 500 m): 48
Total number of *all bridges*: Over 1.33 lakh
Number of *locomotives*: 10499 (43 steam, 5633 diesel, 4823 electric)
*Freight wagons*: 2,39,281
*Passenger coaches*: 62,924
Number of *trains run daily*: 12,617 (passenger); 7,421 (freight)
Number of *passengers carried* during 2013–'14: 842.5 crore (Over 2.3 crore daily, of which nearly 50% were suburban passengers).
*Freight carried* during 2013–'14: Over 105 crore tonnes.
*Total revenues* during 2013–'14: Over 1,44,000 crore (₹94,000 crore from passenger traffic)
*Net income* during 2013–'14: ₹15,700 crore.
*Net social service obligation cost* during 2013–'14: ₹20,000 crore

(Social service cost includes freight of essential commodities at concessional rates; concessional rates for traffic bound for drought-hit and flood-affected areas and non-remunerative lines operated purely in public interest)

## Long-term Plan of Indian Railways – Vision 2020

A long-term plan of Indian Railways, put down as *Vision 2020*, was presented to the Parliament in December 2009 to cover the period 2011–'20. The vision document laid down the plan of action as stated below: "Indian Railways shall provide efficient, affordable, customer-focused and environmentally sustainable integrated transport solutions. It shall be a vehicle of inclusive growth, connecting regions, communities, ports and centres of industry, commerce, tourism and pilgrimage across the country. The reach and access of its services will be continuously expanded and improved by its integrated team of committed, empowered and satisfied employees and by the use of cutting edge technology."

Vision 2020 will address the following four *strategic national goals*:

- Inclusive development, both geographically and socially
- Strengthening national integration
- Large-scale generation of productive employment
- Environmental sustainability.

### Primary Goals of Vision 2020

1. Aiming for a higher growth trajectory.
2. Expansion of railway network
3. Creation of capacity
4. Zero-tolerance for accidents

5. Reducing the Indian Railways' carbon foot print (to make it more environment-friendly.)
6. Improving quality of service to passengers.
7. High-speed rail travel and high-speed corridors.
8. Improving technological excellence.
9. Multiplication of lines to have separate dedicated freight corridors in a few major trunk routes.

A few targets in numbers are as follows:

| | |
|---|---|
| Doubling | 12,000 km |
| New lines | 25,000 km |
| Electrification | 14,000 km |
| Wagons | 2,90,000 (approx.) |
| Diesel locomotives | 5,400 (approx.) |
| Electric locomotives | 4,300 (approx.) |
| Passenger coaches | 51,000 (approx.) |
| World-class stations | 50 stations |

## Comparison of Railway and Highway Transportation

Table 1.2 gives a comparison of railway and highway transport in certain important aspects:

**Table 1.2** Railways versus Highways

| S. No | Parameter | Railway | Highway |
|---|---|---|---|
| 1 | Right of way | More | Less |
| 2 | Cost | High | Comparatively less |
| 3 | Right of access | Restricted to trains only | Open to all vehicles |
| 4 | Tractive resistance | Low; reduces haulage costs | High; almost five times that of a train on rails |
| 5 | Flexibility | Railways can be used only between fixed points | More flexibility; can provide door-to-door service |
| 6 | Suitability | Heavy loads and large numbers over long distances | Light loads and smaller numbers over shorter distances. |
| 7 | Gradients and curves | Flatter | Can be relatively sharper |
| 8 | Operational controls | More controls like signalling and interlocking | Relatively less – except traffic signals at intersections |
| 9 | Employment potential | Very high | Relatively less |
| 10 | Power requirements | High | Relatively less |
| 11 | Safety | Relatively safe | High accident-rate; relatively less safe |
| 12 | Environmental pollution | Relatively less | Relatively high due to automobiles |

## 1.4 RECENT DEVELOPMENTS IN RAILWAYS

The recent trends/developments in railways around the world have the following objectives:

- Improving efficiency and safety
- Automation in railway operation and control
- High speeds – 120 to 250 km/h
- Developing multi-purpose vehicles and linking railways with other modes of transportation for flexibility, efficiency and economy.

The more important developments in a few countries are:

1. **The USA:** Tracked air-cushioned vehicles (TACV) for high speeds of over 500 km/h. Full automation in the operation and control of trains.
2. **Japan:** Magnetic levitation (MAGLEV) vehicles based on the principle of magnetic repulsion for speeds of up to 450 km/h.
3. **France:** TACV for speeds up to 400 km/h.
4. **Germany:** MAGLEV for speeds up to 450 km/h. Double-purpose vehicles for use on a road or a railway track for speeds up to 100 km/h.
5. **The U.K.:** High-speed rolling stock with full automation on conventional track for speeds up to 400 km/h.
6. **Russia:** Refrigerated trains for transport of perishable goods and food material. Modernised tracks for speeds up to 200 km/h.
7. **Canada:** Crewless freight trains and inter-modal transport with combined tickets for rail and road transport.
8. **Italy:** Automation in train control to achieve speeds up to 250 km/h. Better safety provisions at level crossings.

### Recent Developments in Indian Railways

- Modernisation and up gradation of track standards
- Use of welded rails, concrete sleepers and thicker ballast for better load capacity
- Conversion of shorter gauges to broad gauge for achieving uniformity of gauge
- Switching over to electric and diesel traction
- Automation in signalling and train control
- Adoption of modern methods and devices for track maintenance
- Computerisation of customer services.
- Air-conditioning of coaches and catering services
- Container services for freight traffic
- Super-fast trains such as Shatabdi and Rajdhani expresses
- Long-distance superfast trains such as Him Sagar Express between Jammu Tawi and Kanyakumari.
- Luxury trains such as Palace on Wheels to promote tourism
- World-class stations
- Metro Railways such as Kokata Metro, Delhi Metro, Bengaluru Metro and Hyderabad Metro (under construction)
- Adoption of public–private partnership (PPP) models such as build–operate–transfer (BOT) for big railway projects (Example: Konkan Railway, built on BOT–basis)

## SUMMARY

- One of the most important modes of transportation is the Railway. Successful implementation of the various phases of development and operation of a railway constitutes 'Railway Engineering'.
- Railways originated in England. George Stephenson built the first steam locomotive for railway traction; the first train ran from Darlington to Stockton in 1825.
- The first train in India ran from Bori Bunder in Mumbai to Thane; the 34 km journey took one hour and fifteen minutes. This was in 1853.
- There had been steady growth of railways in India except during the World Wars.
- There has been a spurt of growth in Indian Railways after independence, especially from the generous funding received during the successive 5-year plans.
- The responsibility of all aspects of railways is vested in the Government of India – the Ministry of Railways through the empowered Railway Board.
- There are presently 16 railway zones and several divisions for administrative convenience.
- There are six production units for the manufacture of locomotives, coaches and wagons.
- The Research, Design and Standards Organisation (RD & SO) in Lucknow is a premiere institution for research and standards in Railways.
- Railways play a special role in land transport.
- The role of Railways in national development is very significant on the economic, political and social fronts.
- Indian Railways is the largest public sector undertaking with about 66,000 route kilometres and over 13.1 lakh employees.
- Railways have the inherent advantage of bulk transport over long distances.
- Recent developments for high speeds are TACV and MAGLEV.
- The most important objective of improvement and expansion of Railways is automation in operation and control of trains consistent with safety.

## QUESTIONS

1. What are the important characteristics of railways?
2. Give a brief comparison of railways and highways.
3. Enumerate the advantages of railways.
4. Write a brief note on the development of Indian Railways.
5. Write a note on the role of Indian Railway in national development.
6. List the different zones of Indian Railways and their headquarters.
7. Write a brief note on high-speed trains.
8. What are salient features of the Konkan Railway?
9. Write brief notes on:

   (a) Containerisation of freight transport.

   (b) Route kilometerage and running kilometerage

   (c) Automatic train operation

10. Write a short note on:

    (a) Railway board

    (b) Manufacturing units of Indian Railways

    (c) BOT railway projects

    (d) Social cost of Indian Railways

11. Write a note on the organisational structure of Indian Railways.
12. Write a note on 'Indian Railways – Vision 2020'.
13. Explain the role of rail transportation in the development of the country. (VTU)
14. Write about the 'Research, Design and Standards Organisation (RDSO)' of the Railways. Where is it located? (KIIT, Odisha)

Chapter 2

# RAILWAY PLANNING, ALIGNMENT AND SURVEYS

**CHAPTER OBJECTIVES**

This chapter focuses on:

- The planning aspects of a railway network
- The factors to be considered for choosing a good alignment of a railway route between any two points,
- The various surveys required to achieve a good alignment and transfer it on to the ground,
- The conventional and modern methods of surveying using contemporary instruments and techniques.

## 2.1 RAILWAY NETWORK PLANNING

In planning a railway network for a country, the basic principle to be followed is: 'Proceed from the whole to the part'.

This means that broad planning of the railway routes is required to connect important places in the country and it has to be first carried out. This results in a broad network.

In the next stage, railway lines connecting the intermediate places to bigger places in the network have to be planned. This results in a small inner network within the main framework.

In the final stage, feeder lines to these main and secondary railway routes are planned; this results in tertiary networks connecting smaller places.

The tertiary lines are fed by other modes of land transport such as roads for providing access to the various railway stations along the different railway routes.

It must be remembered that the final alignment based on engineering surveys must be decided upon before planning the railway routes in the secondary network, and later in the tertiary network in the appropriate sequence.

Piece-meal construction of railway routes without evolving a broad plan, often called the 'master plan', results in an irregular network, which is not desirable.

## 2.2 RAILWAY TRACK ALIGNMENT

Determining or fixing the centre line of a proposed railway line on the ground is referred to as the *alignment*.

Alignment in the horizontal plane, or horizontal alignment, refers to the direction of the railway track in plan, including the horizontal curves at locations where there is change of direction.

Alignment in the vertical plane, or vertical alignment, refers to the direction of the line in the vertical plane, including the vertical curves at locations where there is change of gradient.

### Importance of Choosing a Good Alignment

When a new railway line is planned, sufficient care has to be exercised in choosing a good alignment fulfilling the basic as well as the specific requirements of the particular case. The importance of choosing a good railway alignment at the planning stage cannot be overstated, for the following reasons:

1. Railway projects involve large capital outlays, and change of an alignment after construction is neither easy nor desirable.
2. Major structures such as bridges, which are expensive, will be involved in a railway route and they cannot be abandoned, except under extra-ordinary circumstances.
3. Land acquisition for a modified alignment is also expensive in view of land appreciation after the construction of a railway track.
4. Even if a route proves to be unremunerative, railways' social obligation may require it to operate the line in public interest.

### Basic Requirements of an Ideal Alignment

The following are the basic requirements of an ideal alignment of a railway line:

#### Purpose of the New Route

- Strategic consideration: As in the case of border areas of the country, for defence purposes.
- Connecting centres of trade/industry: For augmenting the transportation facilities between such centres.
- Developing backward regions: As in the case of north-eastern states.
- Shortening of distances on existing railway routes: An exemplary instance is the Konkan railway line, which fulfils the twin purpose of shortening the distances between Mangalore and cities like Mumbai and Delhi, and the development of the west-coast region between Maharashtra and Kerala in the vicinity of Western Ghats.

#### Integrated Development

The new route must fit into the overall scheme or Master Plan of the region for integrated development of the network.

### Economic Considerations

The following are the aspects which need to be considered under this head:

- Shortest route: The route length should be as small as possible to keep the overall cost low.
- Economic cost: The initial cost of construction and cost of maintenance put together must be as low as practicable. Earthwork and cross-drainage works affect the construction cost, while steep gradients and sharp curves affect maintenance cost.
- Operation cost: This is also affected by sharp curves and steep gradients, which may be avoided, if possible.

### Safety and Comfort

Maximum safety and comfort to the passengers, an important criterion, can be fulfilled by appropriate design of horizontal curves with transitions, and easy vertical curves at change of gradients, as also cant on horizontal curves.

### Aesthetic Considerations

A train journey should be visually pleasing; this can be achieved if the alignment passes through places of natural and scenic beauty.

## Choosing a Good Alignment

It is not possible to align a railway route such that it is the shortest or the most direct route between the terminal points, since such a route will involve practical difficulties in a rolling or mountainous terrain. In view of this, the following points must be taken care of while choosing a good alignment:

### Obligatory Points

These are also called 'controlling points' and are those through which the alignment must pass for strategic, political, commercial, or technical reasons. The following are such points which fall in one or more of these categories:

- Important places: Of strategic, political, or commercial importance
- Existing saddles and passes: In crossing a hilly terrain, an alignment should pass through existing saddles and passes so as to avoid high embankments and deep cuts.
- Major bridge sites: Suitable bridge sites which make the cost of construction of major bridges low become obligatory points.
- Tunnel sites: When a tunnel is economical compared to a deep cut, suitable tunnel sites are obligatory points.

### Geometric Design Standards

A train journey should be made in safety and comfort. Adoption of transitions and smooth vertical curves addresses these concerns.

### Road Crossings

Suitable points for perpendicular crossing are obligatory for simple design and avoiding accidents.

### Geological Considerations

Good and stable soil formation is the best for an alignment to pass.

### Areas Prone to Flooding

The alignment must avoid zones of possible flooding.

### Monuments of Importance

The alignment must avoid monuments of importance.

### Traffic Considerations

The alignment must be such that it attracts maximum traffic so as to earn good revenues from passengers as well as freight.

### Materials and Labour

Places where the required materials for construction and labour are available can be chosen for alignment.

### Economic Considerations

The alignment must be favourable for making the costs of construction and maintenance a minimum; a cost-effective alignment giving maximum returns would be ideal.

### Terrain and Topography of the Area

This is an important criterion in the selection of a good alignment for a new railway route. On plain terrain, the alignment is simple and easy; obligatory points control it and the gradients and curves can be easy and flat.

But on mountainous terrain, the length of the route has to be necessarily increased in order to keep the gradients within the ruling values.

The following are some of the common methods for 'developing' an alignment:

### Zigzag Line Method

The alignment is made to traverse in a zigzag line; it follows a convenient side slope approximately at right-angles to the general direction of the alignment, and it is made to turn about 180° in a horse-shoe pattern to gain elevation.

A typical schematic of such a development is shown in Fig. 2.1.

### Switch-back Method

This method involves a reversal of direction by means of a switch, twice in opposite directions as shown in Fig. 2.2.

The train has to necessarily stop at the switch, which is usually located in a station yard.

Referring to the above figure, P and Q are two switches and $P_1$ and $Q_1$ are buffer stops.

A train coming from S will stop at $Q_1$ and move back to line QP. It stops at $P_1$ again and then traces the line along PR.

This method facilitates considerable gain in elevation in the case of steep side slopes.

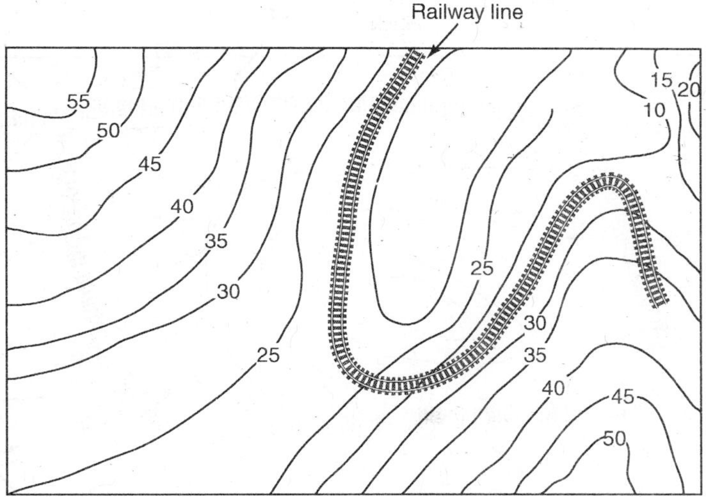


FIG. 2.1 *Zigzag line alignment*

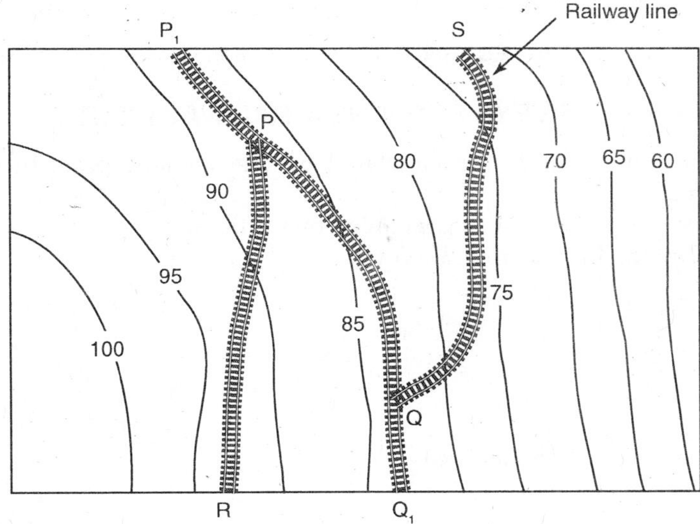


FIG. 2.2 *Switch-back alignment*

### Spiral or Complete Loop Method

This method is convenient for use in a narrow valley where a viaduct has been constructed at a considerable height across the valley, as shown in Fig. 2.3.

In such a situation as this, a complete loop or spiral alignment is constructed such that the track crosses the same point again at a height through a tunnel or a flyover.

Thus it can be seen that the topography of the area and water bodies at the location may require an alignment that deviates from a smooth straight and has to be developed in a manner appropriate to the nature of the terrain at the particular point.

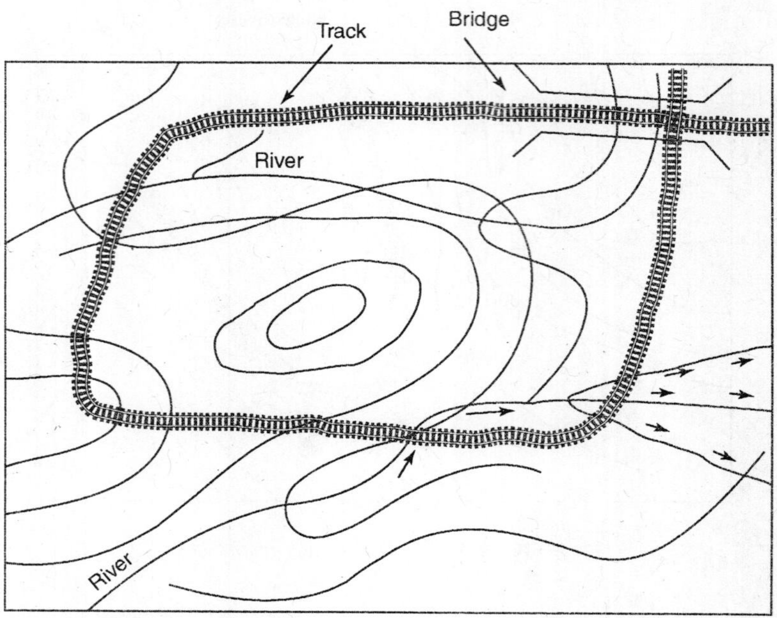


**FIG. 2.3** *Spiral or complete loop alignment*

## 2.3 ENGINEERING SURVEYS FOR A NEW RAILWAY ROUTE

Railway alignment and location are facilitated by a systematic procedure for conducting 'Engineering Surveys'.

These are either preceded by or accompanied by traffic survey.

The following are the Engineering Surveys:

1. Study of topographic maps
2. Reconnaissance survey
3. Preliminary survey
4. Final location survey
5. Soil survey
6. Electrification survey (where needed).

### Traffic Survey

This comprises a detailed study of the traffic conditions in the area of the proposed route. The objectives are to identify

- The most promising route with respect to traffic and economic viability
- The traffic potential – existing and projected for the design period – for the proposed railway route
- The standard of the railway track needed to satisfy the anticipated traffic needs.

Traffic survey is usually conducted along with the initial phases of the engineering surveys in order to assess the technical feasibility and the economics of alternative proposals or routes. The two teams conducting traffic and engineering surveys should work with close co-ordination for obtaining reliable data.

The following aspects are to be covered in the collection of information during traffic survey:

- Character of human resources, or the people
- Population growth rate and related information
- Social mobility of people in the area
- Details of existing trade and commerce
- Existing industries and anticipated industrial growth; expected freight from their output
- Agricultural produce and expected freight for distribution as well as export
- Natural resources in the area and potential for exploitation in the near future
- Places of tourist interest and expected tourist traffic, if any
- Existing transportation facilities in the area.

Based on the information collected on all these aspects, the results of a traffic survey have to be put in the form of a detailed report to give an idea of the traffic potential and anticipated growth of both passenger traffic and freight traffic. This would form the basis for judging both the need and economic viability of the proposed railway route.

## Engineering Surveys

The following engineering surveys are now considered:

### Study of Topographic Maps

Topographic maps are available with the Survey of India, from where they can be obtained. These are contour maps with 15 to 30 m contour intervals, showing the topographic features such as rivers, valleys, ridges, and hills. Depending upon the elevations of the terminal points as well as the obligatory points, and bearing in mind the ruling gradients and other related geometric standards, two or three alternative routes may be chosen, after a careful study of the topographic maps. Further study of these alternative routes has to be made as part of the other surveys.

### Reconnaissance Survey

The objective of reconnaissance survey is to physically examine the possible alignments observed during the study of the topographic maps. This is usually carried out using simple surveying instruments.

Details of certain features, not available from the map study, are collected during this phase. Some of the details that could be collected from this survey are:

1. Approximate gradients and radii of horizontal curves necessary.
2. Controlling points on the alignments.
3. Obstructions from permanent structures not shown in the maps.
4. Existing transportation facilities – railways as well as highways.
5. Ponds, lakes, valleys, ridges, hillocks, and so on.
6. Information relating to cross-drainage works such as culverts, causeways and bridges, required along each of the possible routes.
7. Geological features and information relating to the strata along the routes.
8. Stability of slopes in the case of hilly terrain.
9. Sources of construction materials – borrow areas for earth materials and ballast materials.
10. Approximate locations of railway stations.

11. Climate factors, hydrologic information such as water-table location, water sources, maximum flood level of streams and rivers.
12. Availability of labour, power and water supply along the route.

A report on the results of reconnaissance survey will be very convenient.

**Preliminary Survey**

The objectives of a preliminary survey are:

(i) To compare the proposed routes chosen during reconnaissance for a good alignment
(ii) To conduct detailed surveys on the chosen alternative routes (accurate field work is to be carried out)
(iii) To estimate the quantities of earthwork and other materials to facilitate the preparation of detailed and abstract estimates of the project cost
(iv) To choose the best alignment from all the considerations relevant to the project.

The detailed surveys needed are:

**Traverse survey:** Open traverses are run with the aid of a theodolite and tapes. The lengths of each of the lines and the deflection angles where there is a change in direction are also measured accurately.

(Curves are not considered at this stage)

**Levelling:** Longitudinal section along the proposed route and cross-sections at intervals of 30 m to 100 m along the route are to be taken, depending upon whether the terrain is plain or rolling.

**Contouring:** This is done in the vicinity of the route, covering at least 250 m on either side of the proposed centre line. Tacheometry may be used for this purpose. Benchmarks are connected to GTS benchmarks and are established at intervals of at least 250 m.

**Additional details:** Drainage, cross-drainage works, hydrologic data, soil data, existing features like buildings, lakes, rivers, power lines, and geological details, are also collected more accurately than during reconnaissance.

Based on all these studies, the final location of the selected route is made on paper, which has to be translated on to the ground in the next stage of location survey.

**Final Location Survey**

The location of the chosen route is first marked accurately on the plan of the area; this is called *paper location*

Next, the chosen location of the route for the railway line has to be demarcated on to the ground. This includes pegging the centre line, establishing bench marks, and determining of the reduced levels of the pegged out stations, including the critical locations of change in slope of the natural ground.

**Pegging the centre line:** The centre line of the final route is established by fixing pegs on the ground, all the angles being measured accurately with the aid of a theodolite.

The spacing of pegs depends upon the nature of the terrain – 50 m for plain terrain and 20 m for hilly terrain. The pegs must be referenced with respect to at least three reference points in the vicinity, so that they may be re-established when necessary.

Cross-sections should be taken at 50 to 100 m intervals in plain terrain, 50 m in rolling terrain, and 20 m in hilly terrain.

Bench marks should be established by precise levelling.

The following data are also obtained:

- Right of way
- Land need for geometric features
- Materials, equipment and labour
- Land acquisition costs
- Data for components
- Cost calculations.

### Soil Survey

The nature and extent of the soil strata in and around the route have to be ascertained. The preliminary ideas of the soils and rocks, and other geological details obtained from the reconnaissance and preliminary surveys have to be followed up by a soil survey which is required to identify and classify the soils for use in the design and constructions of the subgrade and embankments.

Undisturbed soil samples have to be obtained at intervals of at least 1 km along the route in order to ascertain the characteristics of strength, settlement and bearing capacity from appropriate laboratory and field tests.

### Survey for Electrification

This is necessary only if there is a proposal to carry out electrification of the railway track.

**Survey on foot:** This survey is recommended to be conducted by going on foot along the chosen route with the following objectives:

- Checking the track geometrics at intervals of 200 m.
- Information of signalling details proposed.
- Checking the position of cabins and other buildings.
- Information on overhead structures along the line.
- Information on highway crossings.
- Details of major bridges and tunnels.

Thus, all details that can have bearing on electrification are sought to be collected.

## Conventional Methods of Surveying

Certain devices and instruments are used – some for approximate surveys and the others for more accurate surveys while conducting engineering surveys for a new railway route. These are summarised below:

| S. No. | Equipment | Purpose |
|---|---|---|
| | | **A. Reconnaissance survey** |
| 1. | Aneroid barometer | To determine the relative heights or altitudes of places |
| 2. | Abney level | To determine gradients (up to an accuracy of 10 minutes of arc) |
| 3. | Prismatic compass | To determine bearings of lines of the route |
| 4. | A pair of binoculars | To observe distant objects |

Cont'd

Table Cont'd

| S. No. | Equipment | Purpose |
|---|---|---|
| 5. | Pedometer | To determine approximate distances |
| | | **B. Preliminary survey** |
| 1. | Dumpy level | To determine reduced levels for longitudinal and cross sections |
| 2. | Theodolite | To conduct traversing |
| 3. | Tacheometer | To plot main features |
| 4. | Prismatic compass | To determine the magnetic bearing of lines |
| 5. | Plane table | To plot interior details |
| | | **C. Location survey** |
| 1. | Dumpy level and precise level | To determine reduced levels |
| 2. | Steel tape | To measure distances accurately |
| 3. | Theodolite | To determine angles accurately |

## Modern Methods of Conducting Surveys

In hilly terrain, modern surveying techniques are far more convenient for preliminary surveys with a view to choosing the best alignment of a railway line from among a few alternatives.

The common methods in this category are:

1. Electronic distance measurement (EDM)
2. Total station (TS)
3. Geographic information system (GIS)
4. Global positioning system (GPS)
5. Aerial photogrammetry
6. Remote sensing (Satellite imagery)
7. Digital terrain modelling (DTM)

### Electronic Distance Measurement (EDM)

Electronic distance meter works on the principle of measuring the time taken by electromagnetic waves to travel the distance between an observed station and the EDM device; knowing the velocity of these waves, the distance is calculated and it is made to appear on the screen of the device through calibration. A typical EDM device can measure distances up to 10 km to an accuracy of one centimetre.

### Total Station (TS)

This instrument works on the principle of EDM; it is a combination of a theodolite and an EDM device. In effect, it can be used for accurate measurement of distances as well as angles, the latter to an accuracy of 1″. Measurements are automatically recorded and displayed both in the horizontal and vertical planes. This tends to reduce human errors in the measurements.

### Geographic Information System (GIS)

The importance of GIS lies in its capacity to integrate different kinds of data in storing, updating, retrieving, and displaying the data. Buffering and network analysis and techniques of GIS, are useful in identifying the best alignment.

### Global Positioning System (GPS)

A GPS device is capable of fixing the co-ordinates of any point on the globe making use of data from a set of satellites circling the earth at an altitude of over 10,000 km. Distances of the order of thousands of kilometres can be measured in any weather and at any time of the day.

### Aerial Photogrammetry

Photographs are taken by aerial cameras from aircraft flying at known altitudes, of the ground below; these are collated into mosaics and distances and spot elevations (and contour maps therefrom) are obtained with the aid of stereo-plotters.

Interpretation of aerial photographs requires skill and experience.

### Remote Sensing (Satellite Imagery)

Satellite images of object on earth, taken from Indian as well as foreign satellites circling the earth at an altitude of 600 to 900 km, are used with remote sensing techniques for preparing maps of the earth's surface and features like soils, crops, and other geological features.

Satellite imagery from ISRO is updated every month; these images along with topographic maps can provide a good understanding of ground conditions.

High resolution satellite imagery is very valuable in planning routes through urban areas, where the details of individual building are also provided.

These techniques need skill interpretation and use.

### Digital Terrain Modelling (DTM)

Complete details of the chosen route for the railway line can be obtained from contour maps in conjunction with digital terrain modelling. The most economical alignment may be defined with the aid of a computer into which the data and model are input.

All other details for the preliminary survey may be got with aid of the contour maps, plots from photogrammetry and other software using a computer.

### Merits and Demerits of Modern Methods of Surveying

#### Merits

- Rapidity of surveying
- Elimination of human errors in preparing a comprehensive and up-to-date map of the area
- Updating old maps without the need for field work
- Estimating the correct project cost
- Accelerated construction programme

#### Demerits

- Interpretation of data from aerial photographs and remote sensing from satellite imagery require skill through specialised training.
- Both aerial photography and satellite imagery require formalities like governmental and defence clearances.
- EDM, total station, GIS, GPS, and stereo-plotting equipment are expensive.
- Subsoil conditions cannot be ascertained accurately.

On the whole, for large projects, the use of modern methods is easily justified.

## Engineering Drawings

The following engineering drawings must be enclosed with the project report:

(i) General map of the locality:
All the existing ground features and the proposed alignment have to be shown.
Scale: 1 cm = 20 km (1:2000000)

(ii) Index maps of the site:
These show the general topography of the area.
Scale: 1 cm = 2.5 km (1:250000)

(iii) Index plan and section:
Scales: 1 cm = 500 m (Horizontal)
1 cm = 12.5 m (Vertical)

(iv) Longitudinal section and cross-sections along the route:
Scales: 1 cm = 50 m (Horizontal)
1 cm = 5 m (Vertical)

(v) Detailed plans of bridges with spans greater than 9 m.

(vi) Station yards plans:
Scale: 1 cm = 50 m

(vii) Detailed drawings of bridges and buildings:
Scale: 1 cm = 1 m

(viii) Plans of road crossings:
Scale: 1 cm = 50 m

(ix) Details of all important features such as rivers, roads and canals, lying within 300 m on either side of the centre line of the proposed alignment.

## Project Report

A comprehensive project report must be prepared and submitted to the competent authority along with all the engineering drawings.

A project report must contain the following:

- Introduction and background
- Chosen alignment with all the obligatory points, gradients and curves
- Comparative details of alternative routes investigated
- Railway track details proposed
- Standards of construction
- Sourcing of materials and labour
- Detailed economic studies establishing viability
- Plant and equipment for construction
- Conclusions
- Final recommendations, and phasing, if any
- Estimated cost of the project

## 2.4 ENVIRONMENTAL IMPACT ASSESSMENT (EIA) OF RAILWAY PROJECTS

In the recent past, there has been significant awareness globally on the environmental degradation being caused human activities in the construction and industrial sector. The need for protection of the natural environment for posterity has been well recognised by advanced countries like the USA, as also by developing countries like India. As a consequence of widespread debate and discussion in several international forums, several countries have made *Environmental Impact Assessment (EIA)* mandatory before constructing any new project; the transportation sector is no exception.

Details about EIA are available on the Web site, www.envfor.nic.in and – UNEP 2002. (For highway projects, this is covered by IRC:104–1989)

The conditions that together shape and characterise the environment in a certain area are:

- The landscape, ground and geology, water, topography, climate and vegetation
- Ecosystems, plants, animals, and habitats
- Development and land use
- Historical development and traces of such development in the landscape.
- The environmental situation, air quality, noise, pollution and other disturbances.

A condition in the environment that can be measured, for example, air quality, water quality, or sound quality, can be monitored and assessed; it can then be compared with standards, base values and limits to conclude whether it is acceptable. Environmental impact means a change in environmental quality that can be measured or recorded.

Environmental impact analysis deals with positive and negative effects of the project and cost-effective preventive measures for any possible damage—soil erosion, slides, submergence due to floods, loss of vegetation, loss of forest cover, harm to wild life, and adverse impact on ecological balance.

The Government of India has, as in several other countries, made it mandatory that an 'environmental impact study and assessment' be made for the execution of major infrastructure projects, highway and railway projects included.

A few projects, for which environmental impact assessment has been made along with suitable recommendations for mitigating the expected damage, if any, are cited below from within India and abroad:

- Konkan Railway Project, KRCL, India
- Jaipur Metro Railway, JMRC, India.
- Hajia Railway Project, China
- Hochi Minh City Metro Rail Project, Vietnam
- Hong Kong Express Rail Link, Hong Kong
- European Union Projects (in accordance with the Directive of the Court of Justice).

These are all recent examples.

In conclusion, environment impact assessment studies have become a mandatory and integral part of any major infrastructure project in any sector, including the transport sector.

## SUMMARY

- Railway network planning needs to be made using an integrated approach. The basic policy of proceeding from the whole to the part is to be followed.
- Alignment of a new railway route involves that in the horizontal plane, with curves at changes in direction, and in the vertical plane, with vertical curves at changes of gradient. It is important to choose good alignment.
- The basic requirements of an ideal alignment must take into consideration the purpose of the new route, the need to fit the alignment into the overall scheme of the master plan for the area, and the economic viability of the new line.

- In choosing a good alignment, obligatory or control points, geometric design standards for safety and comfort, perpendicular river crossings, traffic requirements, economic considerations, and finally terrain characteristics are the governing characteristics.
- Traffic survey and engineering surveys such as map study of the area, reconnaissance and preliminary and location surveys are essential, while soil survey and electrification survey may be done, if required.
- Conventional methods of surveying involve the use of Abney level, clinometer and prismatic compass besides pedometer for approximate measurements; for accurate measurements, the plane table, dumpy level, precise level, the Theodolite and tacheometer are used.
- Modern methods of surveying involve the use of electronic distance measuring devices including the total station, the geographic information system, global positioning system, aerial photogrammetry, remote sensing techniques (satellite imagery), and digital terrain modelling. These methods are justified in the case of large projects.
- A detailed project report along with the engineering drawings is necessary for the competent authority to approve the project proposal.
- Environmental impact assessment studies and recommendations to mitigate the possible adverse effects have become mandatory for any new infrastructure project including new railway routes.
  The Konkan Railway project and the Jaipur Metro Railway Project are two recent examples where such studies have been conducted.

## QUESTIONS

1. Explain the possible reasons for the construction of a new railway line.
2. What is the meaning of track alignment? List the basic requirements of a good alignment.
3. Explain the factors which govern the alignment of a new railway route.
4. Write brief notes on:
   (a) Obligatory points
   (b) Topography of the country
5. Explain the following with appropriate sketches:
   (a) Zigzag development
   (b) Switch-back development
6. What are the various surveys conducted to choose an alignment for a new railway route? Explain the scope of each.
7. What are objectives of a reconnaissance survey? List the different items of information gathered in it.
8. List the instruments used during reconnaissance, preliminary and location surveys, stating the function of each.
9. What are the various engineering drawings to accompany a project report? Give the scales recommended for each.
10. List the items to be included in a project report.

11. Write brief notes on:
    (a) Factors influencing a proposed alignment.
    (b) Role of traffic survey in the alignment of a railway line.
    (c) Pegging the centre line of an alignment.
12. Write short notes on:
    (a) Preliminary survey
    (b) Location survey
    (c) Railway electrification survey.
13. What factors influence the alignment of a railway track? Explain. (JNTU-H)
14. Briefly explain the surveys conducted for track alignment. (WB University of Technology)

Chapter 3

# RAILWAY TRACK AND ITS COMPONENTS

## CHAPTER OBJECTIVES

This chapter discusses the railway track and its components in detail. It covers the following topics:

- The various gauges, the uni-gauge policy of Indian Railways, and the classification of the various railway routes
- Details about the standard rails, their manufacture and important properties; rail joints and welded rails, and defects which occur during operation
- Sleepers, the various materials and the designs used, as also the spacing and sleeper density; rail fastenings and fittings for connecting rail to rail and rail to sleepers
- Ballast, its functions, and the various materials used including the specifications
- The formation, soil subgrade and the strength and stability, including drainage aspects
- The concepts of tractive effort and hauling capacity of a locomotive and the governing factors
- Standard specifications for track structure for various routes on Indian Railways
- Concepts regarding the analysis of stresses in the track and its components.

## 3.1 RAILWAY TRACK

*Railway track* is the railroad on which trains run; it consists of two rails, set parallel to each other at a fixed distance between them and fastened to sleepers, which are in turn embedded in a layer of ballast. The entire structure is supported on a prepared *formation* of earth material, called the *subgrade*.

Each of the components of a railway track has its own functions to perform, to enable the track to transmit the loads from the trains running over it to the ground below in a safe manner.

In the early days of the development of railways, a temporary track used to be laid by the side of the proposed track purely for the purpose of carrying the materials necessary for construction. This temporary track used to be dismantled subsequently. The final track is referred as the *permanent way* to differentiate it from the temporary one used and dismantled.

Also, in the early days prior to the development of railways in England, temporary tracks were laid and used to transport material from collieries. With the advent of railways, the railway track built to allow trains to run over it came to be known as the *permanent way*, a popular designation for a railway track.

### Requirements of a Good Permanent Way

A sound permanent way is an important prerequisite for safety and comfort of railway passengers. The following requirements must be fulfilled for this:

- The gauge of the railway track should be uniform and accurate, the tolerance being very little.
- The alignment should be straight and should be free of kinks, except in the case of horizontal curves for which transitions should be provided at the ends.
- The longitudinal gradient should be gentle and uniform; at locations of change of gradient necessitated by the terrain conditions, a suitable vertical curve must be introduced.
- The levels of the rails should be equal laterally, except on horizontal curves, where appropriate superelevation should be provided to counteract the ill-effects of the centrifugal force.
- The track should be sufficiently elastic and resilient to absorb the dynamic forces or loads from running trains.
- The lateral strength of the track should be adequate to maintain stability under lateral stresses due to different reasons including temperature changes in welded rails.
- Replacement and renewal of the various components of the railway track should be easy.
- Provision of a good drainage system for the track is also an important requirement.
- The track structure should be such that the initial cost and maintenance should be as low as possible.

## 3.2 GAUGE OF A RAILWAY

**Definition:** According to the Indian Railway practice, *gauge* of a railway is defined as the minimum clear horizontal distance between the inner faces of the rails (Fig. 3.1). (The inner faces are also called the 'running' faces.)

In countries such as the USA, Japan and a few European countries, gauge is defined as the clear distance between the inner faces of the rails, measured at a vertical distance ranging from 14 to 16 mm below the top of the rail.

### Gauges on Indian Railways

On the recommendation of W. Simms, Consulting Engineer to the Government of India, the East India Company, which introduced railways and built the early routes in India, adopted

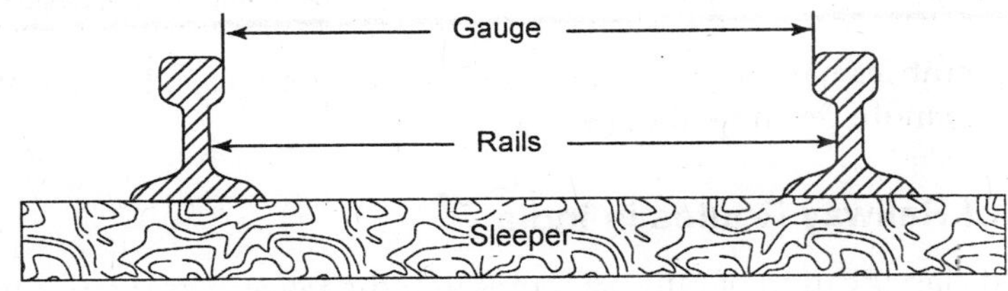


**FIG. 3.1** *Gauge of a railway*

a gauge of 1.676 m (the equivalent of 5ft. 6 in. in the FPS units prevailing at that time.) This is referred to as the *Broad Gauge* (BG) or *Indian Standard Gauge*.

In due course, the Government of India decided to build cheaper railways on economic considerations. This resulted in the introduction of the *Metre Gauge*, measuring exactly 1 metre (the equivalent of 3 ft. $3\frac{3}{8}$ in.).

After some more years, two more gauges shorter than the Metre Gauge were introduced: Narrow Gauges (NG) measuring 762 mm (2 ft. 6 in.) and 610 mm. (2 ft.). These gauges have been used for the construction of mountain railways, for railways in thinly populated areas with low traffic, and feeder railways to mines, factories and ports.

The total route kilometres of these gauges on Indian Railways are given below:

**Route lengths of various gauges on Indian Railways as on 31.03.2014**

| Gauge | Width (mm) | Route-km (approx.) | Percent of total |
|---|---|---|---|
| Broad gauge (BG) | 1676 | 60000 | 90.91 |
| Metre gauge (MG) | 1000 | 4000 | 6.06 |
| Narrow gauge (NG) | 762 | 2000 | 3.03 |
| | 610 | ... | ... |
| | Total | 66 000 | 100 |

**Note:** The Standard Gauge of 1435 mm had been used for the Calcutta Tramway, inaugurated in the year 1902. This is India's only tram system, the oldest in Asia. Delhi Metro also used this gauge for a stretch.

### Gauges on Railways in Other Countries

A gauge of 1.525 m (5 ft.) was adopted originally in England, when the flanges of the wheels were outside the rails. But the wheels with inside flanges came to be preferred from the point of view of lateral stability and guiding the wheels along the rails in a better way. With the top width of the rail fixed at 45 mm at that time, the gauge became $(1.525 - 2 \times 0.045)$ m or 1.435 m. This became the standard gauge later on, not only in England, but in most European countries.

The following is a summary of gauges adopted in different countries in the world:

| Type of gauge | Gauge | Countries |
|---|---|---|
| Standard gauge (SG) | 1435 | Canada, China, England, Turkey, the USA |
| Broad gauge (BG) | 1676 | Argentina, Brazil, India, Pakistan, Sri Lanka |
| Broad gauge | 1525 | Russia, Finland |
| Cape gauge | 1067 | Africa, Australia, Japan, New Zealand |
| Metre gauge (MG) | 1000 | Argentina, France, India, Switzerland |

Various other countries have adopted 23 other different gauges, while the above gauges are considered to be the more prominent ones.

## Classification of Railway Routes in India

The railway board classification of railway routes in India takes into account the traffic carried and the maximum permissible speed, besides the importance of the route.

One broad classification is given below:

**Trunk Routes**

Permissible speed: 120 to 160 km/h for BG and 100 to 75 for M. G.

**Broad Gauge**

1. Delhi – Mughalsarai – Howrah
2. Delhi – Kota – Mumbai
3. Delhi – Jhansi – Nagpur – Chennai
4. Howrah – Bagpur – Mumbai
5. Mumbai – Guntakal – Chennai
6. Howrah – Vijayawada – Chennai

**Metre Gauge**

1. Lucknow – Gorakhpur – Guwahati
2. Delhi – Jaipur – Ahmedabad
3. Chennai – Madurai – Thiruvananthapuram

**Main Lines**

**BG:** 10 Gross Million Tonnes (G.M.T.) per annum or more of traffic. Permissible speed: 100 km/h. to 120 km/h.

**MG:** 2.5 G.M.T. per annum or more of traffic. Permissible speed: 75 km/h

**Branch Lines**

**BG:** Less than 10 G.M.T. per annum of traffic. Permissible speed: less than 100 km/h

**MG:** Less than 2.5 G.M.T. per annum of traffic. Permissible speed: less than 75 km/h

Specifications and track standards for each of these categories differ.

Another classification purely based on the *maximum speed* on the railway line is as follows:

Based on this criterion, the Indian Railways have classified the *broad gauge routes* into five different classes:

Class- A, Class-B, Class-C, Class-D, and Class-E

Since 1999, two more classes have been added: Class-D Special between classes-C and D; and, Class-E Special between classes-D and E.

## Class-A lines

Sanctioned speed: 160 km/h. The following fall in this category:

1. New Delhi–Howrah (Rajdhani route)
2. New Delhi–Bombay central (Frontier Mail route)
3. New Delhi–Chennai central (Grand Trunk route)
4. Ratnagiri–Sawantwadi (Konkan Railway)

A few more are contemplated for upgradation for even higher speeds. Two examples are:

5. Mumbai central–Ahmedabad
6. Bangalore–Chennai central

**Class-B lines**

Sanctioned speed: 130 km/h

There are 25 routes in this category. A few of them are:

1. Kalyan–Pune–Wadi–Secunderabad–Kazipet
2. Vadodara–Ahmedabad
3. Mathura–Ratlam
4. Ahmedabad–Ajmer–Jaipur–Delhi
5. Kharagpur–Visakhapatnam–Vijayawada
6. Allahabad–Jabalpur–Itarsi–Bhusawal

**Class-C Lines**

The criterion is not speed, but all suburban sections fall in this category. There are 18 of them. All suburban sections originating from Mumbai CST, Sealdah–Howrah and Chennai Central are included.

**D-Special Class (1999)**

Sanctioned speed: 100 km/h. These are routes with high traffic density or high expected growth of traffic. There are 25 of them. The following fall in this category:

1. Gundur–Renigunta
2. Guntakal–Hospet
3. Lucknow–Varanasi
4. Lucknow–Kanpur

**Class-D Lines**

Sanctioned speed: Up to 100 km/h. There are 13 of them. The following are a few lines in this category:

1. Guntur–Guntakal
2. Vijayawada-Bhimavaram– Nidadavolu
3. Salem–Byappanahalli
4. Secunderabad–Dronachalam

**E-Special Class (1999)**

Sanctioned speed: Less than 100 km/hr. These are routes with high traffic density or high expected growth of traffic. There are 38 of them, which are branch lines. Two examples of this class are:

1. Dornakal–Bhadrachalam Road–Singareni Collieries
2. Kirandul–Koraput

**Class-E Lines**

All other sections and branch lines with speeds less than 100 km/h.

**Metre gauge** routes have been classified primarily into three groups-Q-class, R-class and S-class. The R-class is further sub-divided into three sub-groups: R-1, R-2 and R-3 in decreasing order of the traffic carried.

### Q-Class (1999)

Traffic density is more than 2.5 gross million tonnes per year. Maximum permissible speed is more than 75 km/hr. There are 18 of them. A few examples are:

1. Delhi–Jaipur (Permitted speed: 105 km/h. for the Pink City express)
2. Villupuram–Thanjavur–Tiruchirapally
3. Dindigul–Madurai
4. Chennai Beach–Villupuram

### R-Class

Sanctioned speed: upto 75 km/h.

**R-1:** Traffic density more than 5 G.M.T. per year. (Slated for gauge conversion soon.) Example: Secunderabad–Guntakal

**R-2:** Traffic density from 2.5 GMT to 5 GMT per year. Example: Guntakal–Hospet

**R-3:** Traffic density from 1.5 GMT to 2.5 GMT per year. Example: Madurai–Rameswaram

### S-Class

All other remaining metre gauge routes for which the sanctioned speed is less than 75 km/h and traffic density less than 1.5 G1MT per year make up the S-class.

## Choice of Gauge

Gauge choice has now lost its relevance and is not as significant as it had been in the early stages of development of railways in the various countries of the world. The choice now is very limited since new railway lines in any country are constructed to adhere to the standard gauge that had already been adopted.

However, the following discussion on the factors influencing the choice of gauge has more academic value than practical value.

### Economic Considerations

The following are the consequences of adopting a wider gauge:

(i) Increase in cost for land acquisition, earthwork for the formation, and track components such as ballast, sleepers and rails.
(ii) Marginal increase in cost of cross-drainage works such as culverts and bridges.
(iii) The cost of rolling stock such as coaches and wagons is almost independent of the gauge for the same volume of traffic and freight.
(iv) The cost of structures such as buildings, stations, platforms, signalling, and control devices is also not much dependent on the gauge.

Thus, it may be observed that there is only a marginal increase in cost if a wider gauge is used.

### Traffic Considerations

The size of wagons, the speed, and the hauling capacity of the engine have bearing on the volume of traffic. The following points are relevant in the *present* context:

(i) Larger wagons and coaches increase the traffic, theoretically speaking.

(ii) Greater speeds and better compatibility with stability and safety features are possible with wider gauge.
(iii) Signalling equipment and operation control costs are practically independent of the gauge.

### Nature of the Terrain

Steeper gradients and sharper curves, which are inevitable in hilly terrain, are more suited for a narrow gauge than for a wide gauge.

## Uniformity of Gauge

A railway network with uniform gauge contributes significantly to a smooth, speedy and efficient operation of trains from the point of view of both passengers and freight. Irrespective of the marginal advantage of a wider gauge or the limitations of a narrower gauge, a single, judiciously selected gauge should be adopted for the entire country. This is because the overall advantage of uniformity of gauge outweighs other considerations.

### Problems Posed by a Multi-gauge System

The value of uniformity of gauge has been widely recognized in many countries of the world.

However, India has been using multiple gauges because of historical reasons and facing a host of problems created by it. The British, who introduced railways to India during their rule in less than three decades after it was introduced in their own country, had a purely commercial approach, with no social obligation towards their subjects. Thus, they built some important lines using broad gauge, but preferred the metre gauge and other narrower gauges in the less developed regions of the country with a view to decreasing their financial commitment, although marginally. Thus, India holds the dubious distinction of having railway lines with the world's broadest (BG-1676 mm) as well as narrowest (NG-610 mm) gauges. There is the broad gauge (Indian Standard Gauge), the metre gauge and the two narrow gauges (762 mm and 610 mm), besides the world standard gauge of 1435 mm in small pockets for certain special purposes! This multiplicity of gauges is really mind-boggling, a 'gift' from the British.

Of course, after attaining independence, the Government of India began the process of rectifying this anomaly, though at a slow pace.

### Disadvantages of the Multi-gauge System

There are numerous demerits of change of gauge, or *break of gauge*:

**Inconvenience to passengers**: Change of trains, mid-journey, causes inconvenience to passengers in several ways:

(i) Missing connecting trains in case of delays.
(ii) Strain associated with climbing stairs to access foot-over bridges.
(iii) Transporting luggage from one platform to another.
(iv) Getting accommodation in the connecting trains.
(v) Uncertainties in reaching the destination.

**Difficulty in trans-shipment of goods:** The problems that arise are:

(i) Damage to goods
(ii) Theft or misplacement of goods

(iii) Delay in receipt of goods at the destination.
(iv) Deployment of more staff and labour during trans-shipment of goods.

**Hindrance to fast movement of passengers and goods:** Change of gauge causes loss of time during trans-shipment and becomes a major hindrance to fast movement.

**Additional facilities for different gauges:** Additional facilities are required at stations and station yards to deal with tracks of different gauges.

**Unbalanced economic growth:** Multiplicity of gauges and the trans-shipment difficulties that are caused lead to unbalanced economic growth of areas with railways of different gauges.

**Rolling stock:** Rolling stock relating to different gauge leads to inefficient use of both due to break of gauge.

**Gauge conversion:** As development takes place and passenger and freight traffic increase in areas in which railways of narrower gauges exist, gauge conversion projects become a necessity leading to avoidable expenditure and strain on the national exchequer. In view of all these problems, the development of railways to a certain standard gauge throughout the country has been well recognised.

## Unigauge Policy of Indian Railways

*Project Unigauge* is an ongoing exercise by the Indian Railways to standardise most of the rail gauges in India to 1676 mm Broad Gauge. A broad gauge network allows bigger trains, higher speeds and greater stability. In India, which has a high population density, broad gauge provides comfortable travel and is considered to be economically feasible in the long run.

As a result of Project Unigauge, the share of the Broad Gauge in total route-km has been steadily rising and that of the Metre Gauge has been declining; the present values are over 90% for the BG and about 6% for the MG.

Prior to the starting of Project Unigauge in 1991, some sporadic conversion of MG tracks to BG, such as Bangalore–Ooty–Guntakal and Pune–Miraj had been done. But the pace has significantly increased in the wake of Project Unigauge.

One metre gauge and four narrow gauge **heritage railway routes** in hilly regions will not be converted to broad gauge.

These are:

- Nilgiri Mountain Railway (Tamil Nadu): MG (1000 mm)- 46 km
- Kalka–Shimla Railway (Himachal Pradesh): NG (762 mm)-94 km
- Kangra Valley Railway (Himachal Pradesh): NG (762 km)-164 km
- Darjeeling Himalayan Railway (West Bengal): NG (610 mm)-79 km
- Matheran Hill Railway (Maharashtra): NG (610 mm)-20 km

Urban Mass Rapid Transport systems (Metro Rail) built or under construction in six large cities in India use width 1676 mm Broad Gauge and 1435 mm Standard Gauge. These Metro Railways are independent corporations and do not come under the jurisdiction of Indian Railways. Project Unigauge does not, therefore, apply to them.

Indian Railways Vision 2020 has, as one of its goals, complete conversion of tracks to the broad gauge, except, of course, the heritage railway lines.

This goal is expected to be reached by the year 2020, and India will have mostly uniform gauge of the 1676 mm Broad Gauge for the vast majority of its railway lines.

The *benefits of the unigauge policy* are:

- Elimination of transport bottlenecks and fast movement of passengers and goods.
- Elimination of trans-shipment delays, damage to goods and inconvenience to passengers.
- Availability of alternative routes in times of need
- Maximum and the best usage of rolling stock
- Maximum utilisation of track and reduction in operating costs.
- Reduction of the need for additional facilities at places where lines of different gauges merge.
- Uniformly good transport infrastructure the various regions of the country.
- Balanced economy growth for all regions.

There are two more gauges, which are not actually gauges, but represent only limiting dimensions—either for loading of freight wagons or for fixed structures such as bridges and tunnels. These are the *Loading Gauge* and the *Construction Gauge*.

**Loading Gauge**

This represents the maximum height and width to which rolling stock such as locomotives and coaches and wagons can be built or loaded, as shown in Fig. 3.2.

| Gauge | H(mm) | B(mm) | These dimensions are fixed based on the maximum dimensions of the rolling stock |
|---|---|---|---|
| BG | 4725 | 3660 | |
| MG | 3430 | 2745 | |

This template-like structure consists of a vertical post with a horizontal arm from which a steel arc is suspended as shown. This structure is meant to ensure that the top and the sides of the load will clear all structures such as bridges and tunnels along the route.

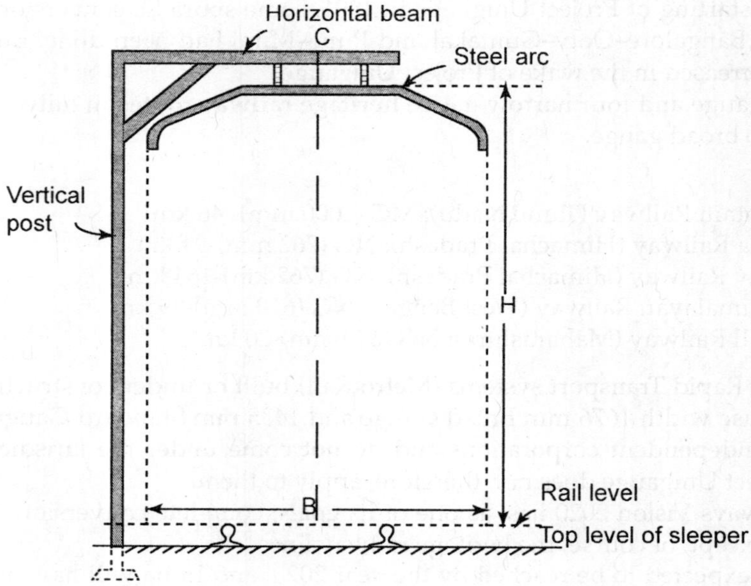


**FIG. 3.2** *Loading gauge*

### Construction Gauge

The *construction gauge* is fixed by adding the necessary clearances to the loading gauge so that the trains can move safely without any fouling or infringement. The fixed structures such as bridges, tunnels and platform sheds near stations are all built in accordance with the construction gauge so that the extreme edges at the top and the sides remain clear of the loading gauge.

The loading gauge and the construction gauge have bearing on the carrying capacity of the train. The trend in other countries is to make these gauges larger than those fixed by the British. This has led to larger locomotives in such countries when compared with those of England.

## 3.3 TRACK STRUCTURE

As discussed in Section 3.1, the railway track or permanent way consists of a combination of rails fixed to sleepers, which are embedded in ballast laid on a prepared formation, the whole set resting on a subgrade of earth material.

This set-up is shown schematically in a typical cross-section of the permanent way (Fig. 3.3)

**Rails:** These are the top-most component of a railway track. They are manufactured from special rail steel in standard lengths (depending upon the gauge) and are laid horizontally or at a chosen longitudinal gradient, parallel to each other, and separated by the constant distance, or the track gauge.

The joints between different lengths are properly treated and connected by appropriate fittings.

Rails are fastened to sleepers placed below them at intervals such that the gauge and levels are maintained.

**Sleepers:** Sleepers support the rails and are laid transversely at proper spacing. They are made to rest on ballast, suitably packed and boxed with it. They may be of wood, steel, reinforced concrete, or pre-stressed concrete.

**Rail fastenings:** These are meant to fix one rail length to the adjacent ones, as also to fix the rails to the sleepers below at predetermined spacing.

These are invariably made of steel.

**Ballast:** This is to hold the sleepers in position. The sleepers are boxed with ballast to provide them lateral support. There should be adequate depth of ballast below the sleepers; it is called

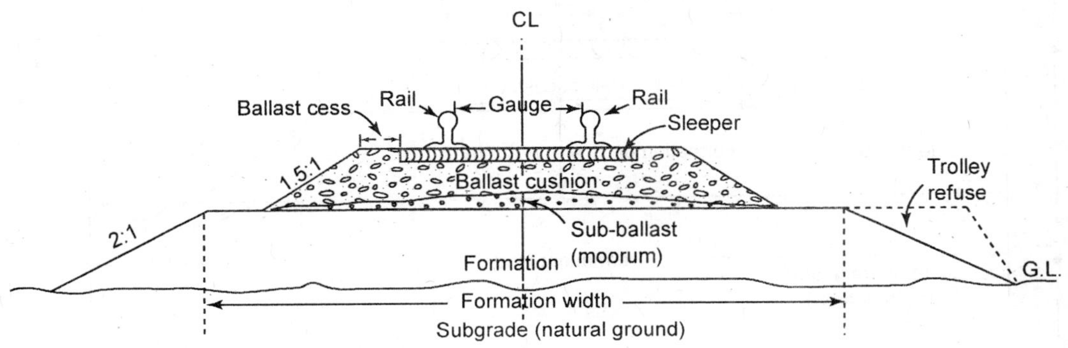


**FIG. 3.3** *A schematic of track structure in cross-section*

ballast cushion. Similarly, the top width of ballast and the base width are such that additional quantity is packed for providing lateral stability to the track.

**Formation:** It is the prepared earth embankment over which the ballast material is laid, and packed. It is compacted well for stability.

**Subgrade:** The general ground over which the formation rests is called the subgrade; it consists of earth material or rock, if available at the desired level. Load from the trains and the track are ultimately transmitted by the track components to the subgrade, which should have adequate bearing capacity.

## Typical Cross-sections of Railway Track

A railway track may be laid on an embankment or in a cutting, depending upon the relative elevations of the natural ground and formation level or track level at the location. It may be set in broad gauge or metre gauge, and the track may be a single line or a double line.

The minimum ballast cushion is 20 cm. Side drains are provided on both sides of the embankment. However, in case there is adequate cross slope across the natural ground, one side drain on the higher side is sufficient.

Typical cross-sections of a railway track on embankment and cutting are shown for single line as well as double line sections in Figs. 3.4 to 3.7.

These cross-sections are for straight sections of the railway track.

In the case of curved sections, since cant or superelevation is required to be given at the top levels of the rails with the outer level being at a higher than the inner, the ballast section will be given the necessary modifications while the formation level is kept horizontal.

Thus, the ballast section will be of a larger depth on the outer side of the horizontal curve than on the inner.

## Electrified Tracks

Indian Railways has a significant length of electrified broad gauge tracks on trunk routes and main lines.

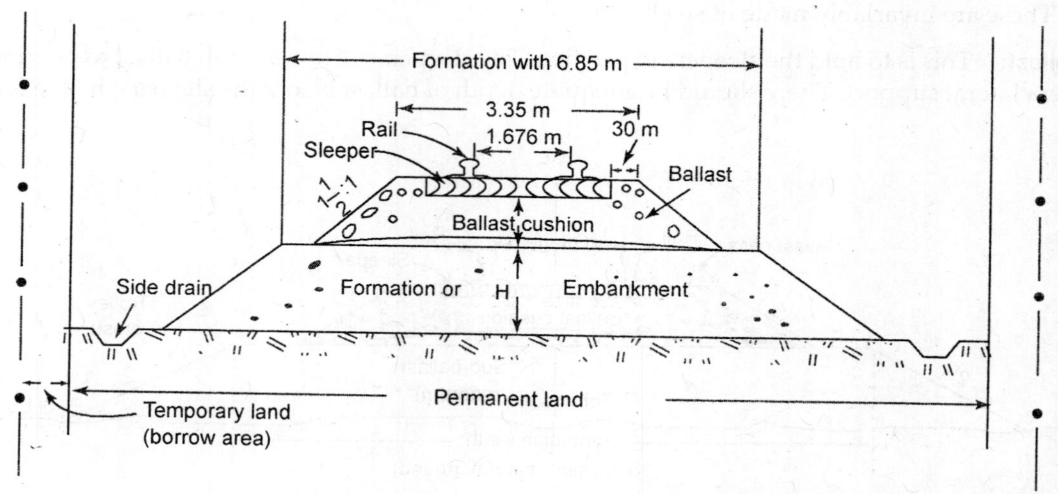


**FIG. 3.4** *Single-line broad gauge track (embankment)*

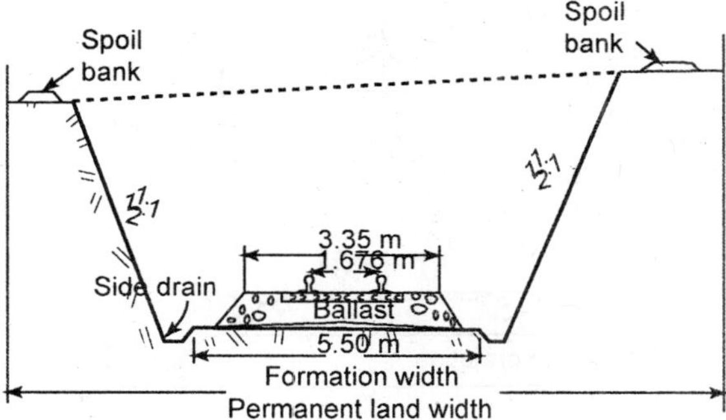


**FIG. 3.5** *Single-line broad gauge track (cutting)*

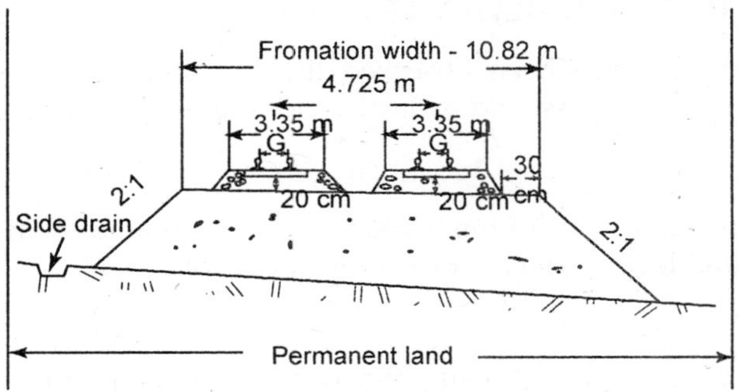


**FIG. 3.6** *Double-line broad gauge track (embankment)*

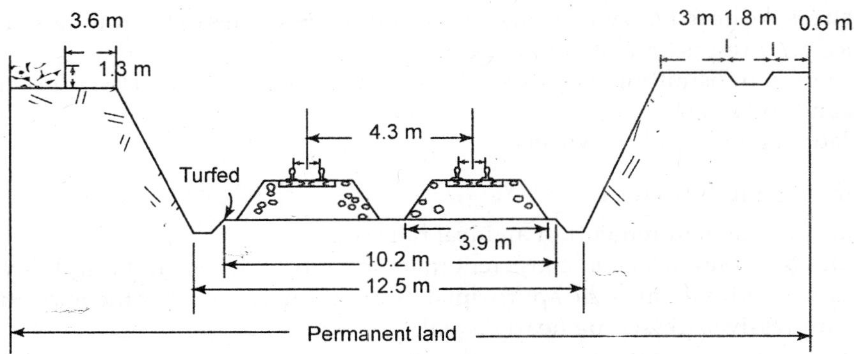


**FIG. 3.7** *Double-line broad gauge track (cutting)*

Traction poles support the necessary accessories for electric traction to complete the electric circuit through the electric locomotive and the rails. These poles are erected on the formation with concrete bases.

A typical cross-section of a track in an electrified section is shown schematically in Fig. 3.7 (a).

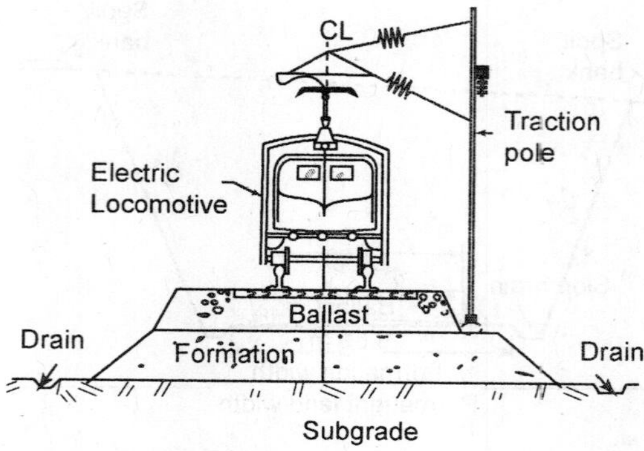


**FIG. 3.7(a)** *Electrified section of broad gauge track (cross-section)*

## 3.4 COMPONENTS OF A RAILWAY TRACK

In the previous section, it has been observed that the railway track consists of various components such as rails, sleepers, and ballast, besides other components.

### Rails

Rails are the topmost components of the track; two rails, laid parallel to each other, provide a smooth, continuous and level surface for the movements of trains.

#### Functions of Rails

Rails are expected to perform the following functions in a railway track:

- Provide a smooth and hard surface for the movement of trains
- Provide very little friction between the steel wheel and steel rail, compared to that between a pneumatic or radial tyre and a roadway
- Bear the stresses due to heavy vertical loads, through the wheels of the rolling stock, braking forces, thermal stresses and lateral forces
- Act as girders in transmitting the loads to the sleepers in such a way as to reduce the stresses in ballast and the subgrade
- Serve as lateral guides for the wheels.

#### Requirements of a Rail Section

The following are the requirements of an ideal rail section:

(i) It should be economical and consistent in its strength, stiffness and durability
(ii) The centre of gravity must be approximately at mid-height so that the maximum tensile and compressive stresses are nearly equal.
(iii) Relative distribution of the material in the three main parts—the head, the web and the foot must be balanced.
(iv) The height should be adequate to provide enough vertical stiffness for transmitting loads to the sleepers.
(v) The widths of head and foot must be sufficient to provide good lateral stiffness
(vi) The wearing surface of the head must be sufficiently hard and the depth of the head must also be adequate to minimise the vertical wear
(vii) The thickness of the web must be adequate to provide flexural rigidity in the horizontal plane

(viii). The width of the foot must be sufficient to prevent overturning, especially on horizontal curves

(ix) The fishing angles, or the shape of the bottom of the head and the top of the foot, must be properly designed for efficient transmission of loads at rail joints

(x) The fillets and their radii must be designed properly so as to prevent concentration of stresses

(xi) The rail section must be such that the rail specimen can withstand stresses caused by impact loads from the moving trains.

(xii) The material and manufacturing process of the rail must be such that all the functions expected of it are performed to the maximum extent.

These are really stringent conditions and are ideal; in practice, however, some degree of departure is tolerated.

## Types of Rails

Historically speaking, rails of wood, cast iron, and wrought iron were used first; however, after the advent of structural steel in general, and rail steel in particular; rails are being manufactured from rail steel.

The first rails used were the *double-headed (DH) type*, made of an I- or dumb-bell shaped section. The idea was to reverse the top and bottom when the top got worn out and prolong the life. However, it was observed that the bottom table got dented and became rough because of continuous contact with the chairs, making reuse impossible. This led to the development of *bull-headed rail (BH)*, with more metal in the head to withstand the wear and tear in a better manner. This type had been in use in England for a long time. The major drawback of this type was the requirement of chairs for fixing it to sleepers.

The *flat-footed (FF) rail*, also called the *Vignole rail* after its inventor, was developed with the significant advantage that it could be fastened to the sleepers directly with the aid of spikes. This section is considered to be an economical design, with greater strength and lateral stability when compared to a bull-headed rail for a given cross-sectional area.

This has become the most popular type with over 90 per cent of the world's railways using it. The Indian Railways have standardised the flat-footed rail for adoption on their railway tracks.

Figures 3.8 to 3.10 show the shapes of sections of the double–headed, bull-headed and flat-footed rails, respectively:

**Standard rail sections:** The rail is specified by its weight per unit length kg/m (lbs/yard in the FPS units).

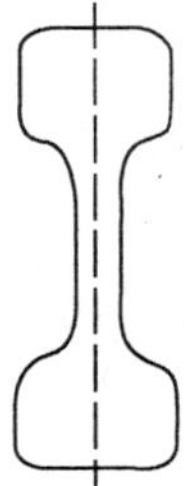

**FIG. 3.8** *Double-headed rail*

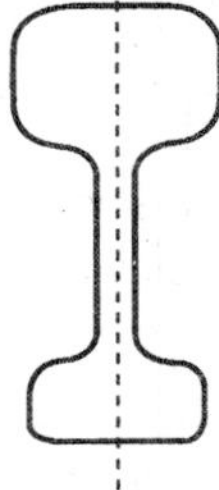

**FIG. 3.9** *Bull-headed rail*

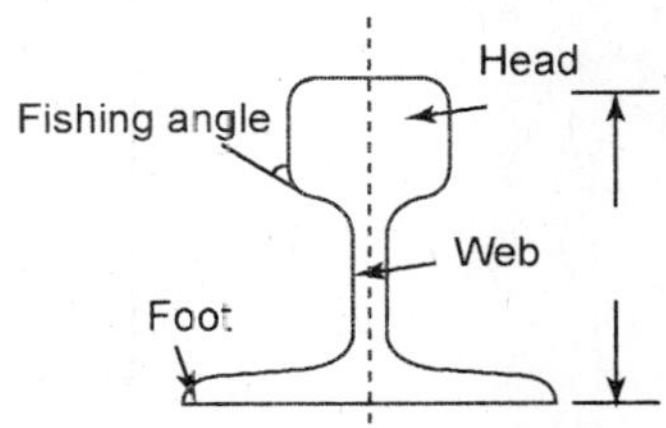


**FIG. 3.10** *Flat-footed rail*

The weight and section of a rail is governed by:

- Maximum axle load
- Maximum allowable speed
- Type and spacing of sleepers
- Ballast cushion.

A thumb rule governing the maximum axle load, $W_a$ and the weight in kg/m ($w$) is given below:

$$W_a = 560 \cdot w \tag{3.1}$$

The standard rail sections prescribed by the Indian Railways are given in Table 3.1:

**Table 3.1** Standard rail sections on Indian Railways

| Gauge | Rail section | Authority |
|---|---|---|
| Broad gauge | 60 kg/m | UIC |
| | 52 kg/m | IRS |
| | 90 lb/yd | RBS |
| Metre gauge | 90 lb/yd | RBS |
| | 75 lb/yd | RBS |
| | 60 lb/yd | RBS |
| Narrow gauge | 50 lb/yd | RBS |

UIC : International Union of Railways
IRS : Indian Railway Standard
RBS : Revised British Standard

Detailed dimensions of standard rail sections are shown in Fig. 3.11:

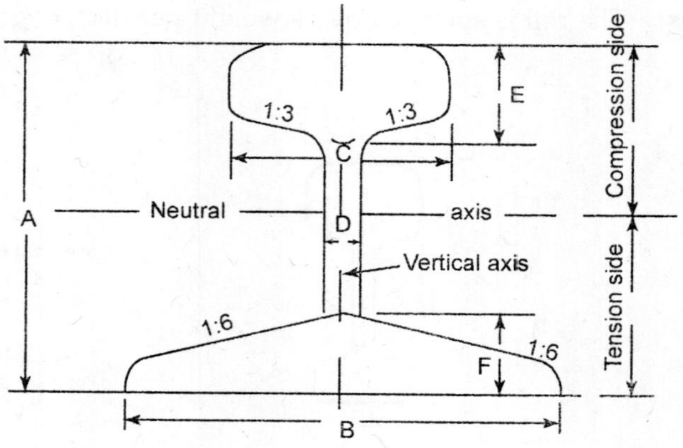

**FIG. 3.11** *Standard flat-footed rail section*

Detailed dimensions of the standard rail sections used by Indian Railways are given below:

**Details of Standard Rail Sections Used on Indian Railways**

| Rail section | Wt./m (kg) | Area of section (mm²) | Dimensions (mm) | | | | | |
|---|---|---|---|---|---|---|---|---|
| | | | A | B | C | D | E | F |
| 50 R | 24.80 | 3168 | 104.8 | 100.0 | 52.4 | 9.9 | 32.9 | 15.1 |
| 60 R | 29.76 | 3800 | 114.3 | 109.5 | 57.2 | 11.1 | 35.7 | 16.7 |
| 75 R | 37.13 | 4737 | 128.6 | 122.5 | 61.9 | 13.1 | 39.7 | 18.7 |
| 90 R | 44.61 | 5895 | 142.9 | 136.5 | 66.7 | 13.9 | 43.7 | 20.6 |
| 52 KG (IRS) | 51.89 | 6615 | 156.0 | 136.0 | 67.0 | 15.5 | 51.0 | 29.0 |
| 60 KG (UIC) | 60.34 | 7686 | 172.0 | 150.0 | 74.3 | 16.5 | 51.0 | 31.5 |

The 90 R-rail section is considered to be adequate only for a traffic density of about 10 GMT (gross million tonnes per km per annum), speeds of up to 100 km/h, axle loads up to main line standard, and a service life of about 20 years. In order to overcome these limitations, the Indian Railways designed a heavier section of 52 kg/m in 1959 for heavier and faster traffic. This section is used on all BG main lines for speeds up to 130 km/h and traffic density up to 25 GMT.

Furthermore, a new design for the 60 kg-UIC rail section has been evolved by the Indian Railways to meet the requirements of traffic for speeds of up to 160 km/h and a traffic density of about 35 GMT.

The need to enhance the service life of rails has arisen with the advent of concrete sleepers that have a service life of 50 years.

A 52-kg IRS flat-footed rail section is shown in Fig. 3.12.

## Manufacture of Rails and Rail Steel

Steel for the manufacture of rails is obtained from the open hearth or duplex process. The process uses a basic oxygen or electric arc furnace initially. The rail manufacturing process consists of continuous casting of blooms, rolling of rails including controlled cooling, and finishing including eddy current and ultrasonic testing.

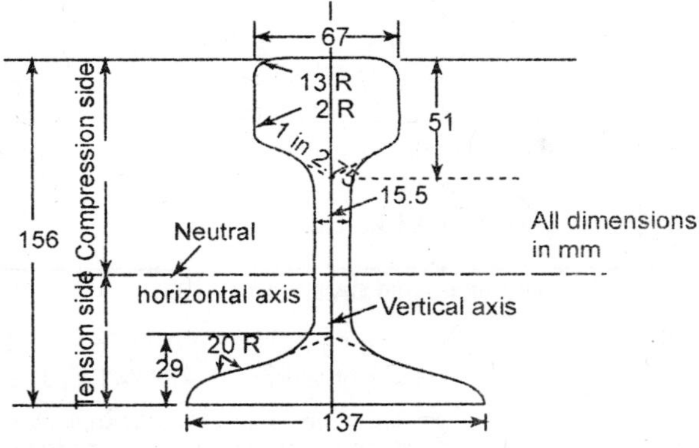


**FIG. 3.12** *IRS-52 kg Rail Section (FF)*

**Rail steel:** The desirable chemical composition of rail steel is as follows:

| | |
|---|---|
| Carbon | 0.6 to 0.8% |
| Manganese | 0.8 to 1.3% |
| Silicon | 0.3 to 0.5% |
| Sulphur | 0.04% (max.) |
| Phosphorus | 0.04% (max.) |
| Aluminium | 0.02% (max.) |
| Liquid hydrogen | 3% |

The desirable value of ultimate tensile strength (UTS) is 880 MPa ($\approx$ 90 kg/mm$^2$).

For special locations such as points and crossings and sharp curves, medium manganese steel, an alloy steel, is preferred (Manganese content: 1.10 to 1.40%).

## Rail Specification –T-12-2009 of Indian Railways

Rail specification of Indian Railways is revised from time to time based on traffic trends and track behaviour, and the latest research into these and related aspects. T-12-2009 is the latest specification in this series, which covers prime rails of UTS-880 MPa as also special classes of rail steel such as vanadium (VN) and corrosion-resistant steel such as copper molybdenum (CM), and nickel chromium copper (NC).

The quality of steel used for rails is determined by the manufacturing process, chemical composition, acceptance tests and qualifying criteria. Tolerances in sectional dimensions of rails, length of rail, end straightness, in ultrasonic testing of rails, and surface defects are also specified.

The following are the *acceptance tests*:

1. Chemical analysis
2. Tension test
3. Sulphur print
4. Hardness test
5. Falling weight or tap test
6. Hydrogen content

End-hardened rails and head-hardened rails are used to minimise wear at the ends and at the head, respectively.

### Branding of Rails

Every rail has a brand on its web, which is repeated at intervals of 3 metres or less. IRS-T-12-2009 specifies that brand marks should be rolled on one side of the web, with letters of at least 20 mm height.

A typical *brand mark* appears somewhat like this:

| **IRS-52 kg-880-SAIL VIII 14→OB** | |
|---|---|
| IRS-52 kg | IRS rail section of weight 52 kg/m |
| 880 | Grade of rail section – 880 MPa (UTS) |
| SAIL | Steel Authority of India Ltd (Manufacturer) |

Cont'd

**Table Cont'd**

| | |
|---|---|
| VIII 14 | August 2014 (Month and year of manufacture) |
| → | shows the directions of the top of the ingot |
| OB | Open hearth basic (process of manufacture) |

### Length of Rails

The considerations in deciding the length of individual rails are as follows:

1. Convenience of manufacturing and facilities at the steel plant
2. Facilities for transporting long rails, especially on curves in the longest wagons available
3. Convenience in lifting and handling
4. Expansion gap at the rail joints. (This is not considered as a constraint now, since it is recognised that the restraint on the rail caused by the fastenings of the rail to the sleepers counteracts the tendency for expansion due to an increase in temperature to a considerable extent.)
5. Large internal thermal stresses in long rails.

It is well recognised that the longer is the rail, the less would be the number of joints per km and fittings at joints, as also the maintenance costs; further, the riding comfort is more. Consistent with the constraints, it, is therefore better to use longer rails. (Of course, a rail length should be more than the rigid wheel base or the distance between consecutive axles.)

Considering the above factors, Indian Railways has standardised the rail length as 13 m (previously 42 ft) for Broad Gauge and 12 m (previously 39 ft) for Metre Gauge and Narrow Gauge tracks. (In the USA and Europe, a rail length of up to 30 m has been used).

In view of the advantages of longer rails, Indian Railways is planning to use 39 m/78 m rails in the near future, as these are being produced by SAIL and Bhilai steel plant. (It is also planned to manufacture 130 m long rails for use on high speed tracks.)

## Rail Joints

In the early days, rails were rolled in short lengths due to constraints in rolling and transportation. These rails had to be connected at their ends to the adjacent rails, thus requiring what is known as 'rail joints'. The length of rails was also supposed to be governed by the expansion gap at these joints, which was considered to increase with increase in rail length. (Although later, it was understood that this is not a constraint in using longer rails owing to the restraint from rail fastenings — rail to rail and rail to sleepers.)

A rail joint has become an inevitable feature of a railway track; it is considered to be a necessary evil as it creates a lot of maintenance problems.

### Disadvantages of a Rail Joint

A rail joint is the weakest link in a railway track. At the joint, there is a break in continuity of the rail in both the horizontal and vertical planes because of the expansion gap and imperfections in the levels of rail heads due to wear. When the wheels of the train negotiate the gap at the joint, a severe jolt occurs, which loosens the ballast underneath and also the fittings at the joint. This causes heavy wear of the track components and poses maintenance problems.

The following are the *ill-effects* of a rail joint:

- Additional maintenance for the joint and related components to maintain the gauge and levels

- Decrease in the durability of rails, sleepers, and fastenings due to the impact of moving loads on the joint
- Higher chances of rail failure due to battering of rail ends
- Noise caused when the trains negotiate the joints causes noise pollutions and discomfort.
- Reduction in riding comfort due to the wear and tear of track components.
- Increase in fuel consumption due to the extra effort required while negotiating the joints.
- Greater susceptibility to sabotage through the removal of fish plates and rails.

### Requirements of an Ideal Rail Joint

An ideal rail joint is one which provides the same strength and stiffness as the parent rail. This demand is of a tall order and can never be met.

However, the characteristics of an ideal rail joint are the following:

- It should have the same strength and stiffness as the parent rail.
- It should hold both the rail ends in the correct location and levels to provide the maximum possible conformity. This prevents wheel jumping at the joint.
- It should provide adequate gap between rail ends to take care of the expansion or contraction due to temperature changes.
- It should provide for the normal rail-end wear without disturbing the alignment and gauge.
- It should provide sufficient elasticity and resistance to longitudinal forces.
- It should provide for easy replacement of rails as and when needed.
- The overall cost, that is, the initial cost and maintenance cost of the rail joint should be minimum.

As has already been pointed out it is very difficult to design a rail joint which meets all these requirements.

### Types of Rail Joints

Rail joints may be classified on the basis of the positions of sleepers near the joint:

1. Supported joint
2. Suspended joint
3. Bridge joints

**1. Supported joint:** In this type, the rail ends are supported directly on the sleeper, thereby avoiding cantilever action (Fig. 3.13):

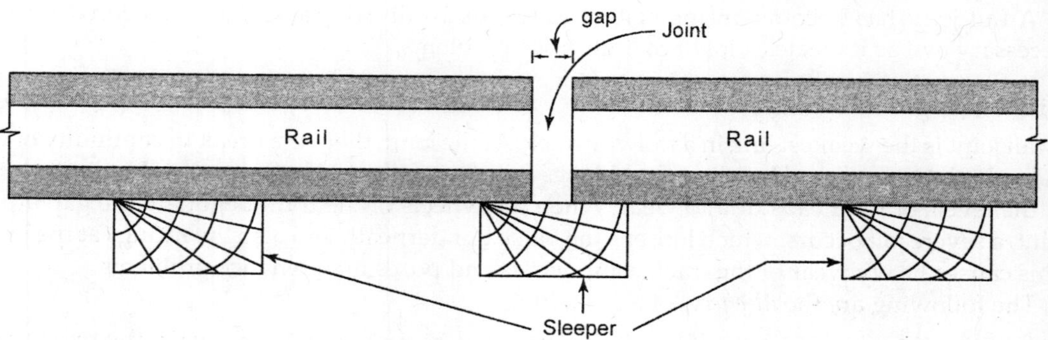


**FIG. 3.13** *Supported rail joint*

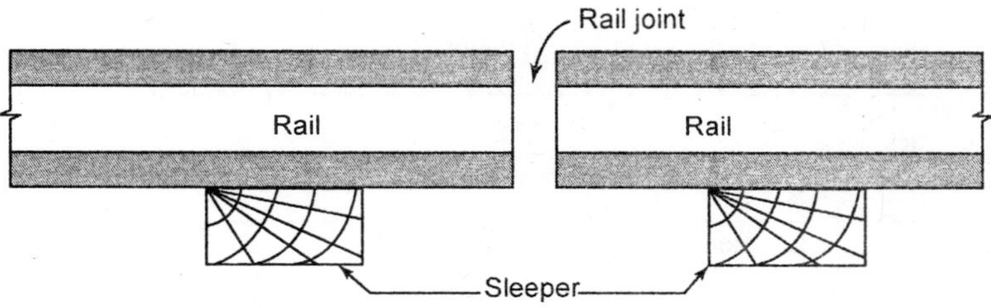


**FIG. 3.14** *Suspended rail joint*

This was expected to reduce the wear and tear of the rails; however, in practice, the support tends to raise the rail ends slightly, resulting in a hard run on the joint. Further, the supporting sleeper undergoes wear and tear involving more maintenance.

**2. Suspended joint:** In this type, the rail ends are suspended, some portion of the rail being cantilevered at the joint from the adjacent sleepers. Because of this cantilever action, the packing under the sleepers adjacent to the joints becomes loose under the impact of the moving train loads. However, this type of joint is popular not only in India but also in several other countries.

A schematic of this type of joint is shown in Fig. 3.14.

**3. Bridge joint:** Bridge joint is similar to the suspended joint except that, in the former, the adjacent sleepers on either side of the joint are connected by a *metal flat* or a corrugated plate, called a *bridge plate*. These are shown schematically in Figs. 3.15 and 3.16.

Bridge joints are not generally used on Indian Railways owing to the complexity of design and uncertainty of structural action.

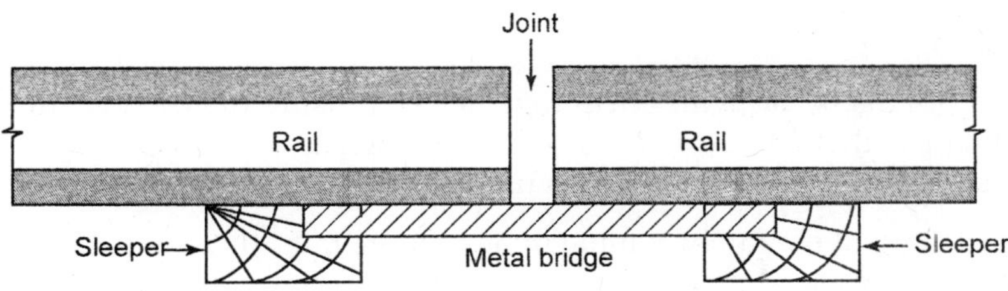


**FIG. 3.15** *Bridge joint with metal flat*

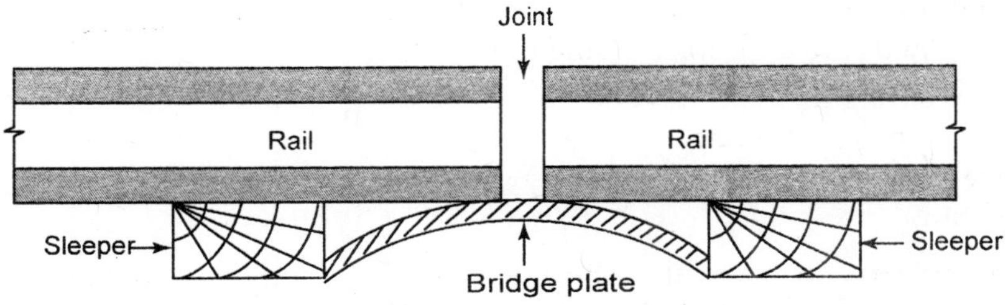


**FIG. 3.16** *Bridge joint with bridge plate*

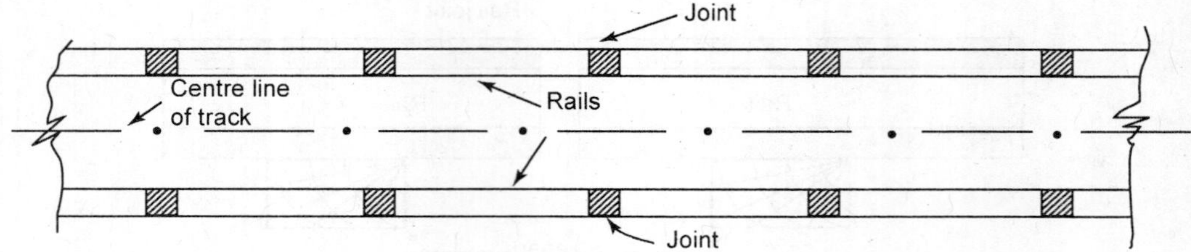


**FIG. 3.17** *Square joints*

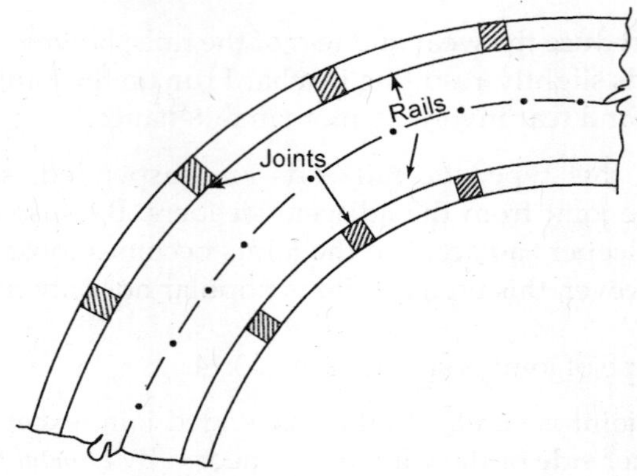


**FIG. 3.18** *Staggered rail joints*

Based on the relative position of the rail joints along the railway track, they may be classified as *square joints* and *staggered joints*. These are shown in Figs. 3.17 and 3.18, respectively.

## Welding of Rails

Joining the ends of rails by bringing them together and fusing them under the influence of intense heat with or without an external material is known as *welding of rails*.

The purposes for which welding of rails is used could be one of the following.

- Increasing the length of the rail by joining three or more rails, thereby reducing the number of joints and fastenings, leading to economy and better riding comfort.
- Repairing worn-out rails to make them fit for re-use.
- To build up worn-out points and rail ends on sharp curves.

### Methods of Welding Rails

The following methods are used for welding rails:

1. Gas pressure welding
2. Electric arc or metal arc welding
3. Flash butt welding
4. Thermit welding

**1. Gas pressure welding:** In this method, heat is produced by combining oxygen and acetylene gases. The rail ends to be welded are brought together and the burner through which the gases are ignited releases a flame at a temperature of 1200°C.

The metal melts, resulting in fusion and welding together of the ends. The process is completed under a load of 20 t and a temperature of 850°C, and allowed to cool naturally.

The disadvantage is limited output and irregular availability of gas.

**2. Electric arc or metal arc welding:** In this method, heat is generated by passing an electric current across a gap between two conductors or metal electrodes brought close together to produce an arc of electric current between them. This produces a lot of heat, which causes the rail ends to fuse or get welded.

There are different techniques of operating this method. Indian Railways has started using it recently on a trial basis.

**3. Flash butt welding:** Heat is generated in this method by passing a very high electric current of the order of 35,000 A between the interfaces of the rail ends which are brought very close together. The rail ends are preheated using an oxy-acetylene torch a predetermined number of times (15 for 52 kg-rails) before they are welded. Considerable heat is generated; the rail ends are passed to and fro by the welding machine until the temperature rises to the fusion limit of 1000°C to 1500°C. The rail ends are pressed together with a force of about 37 t as the final flashing joins the two rail ends together. The total time needed is about 3 min. and the loss in rail length is about 20 mm for each weld.

The method can be adopted most economically and efficiently only in a workshop or a big plant. High-quality welded joints are produced by the flash butt welding method, which has become the standard method of welding on Indian Railways.

There are flash butt welding plants at at least eight places in different railways zones. Even the most modern automatic flash butt welding machines have been procured by Indian Railways.

**4. Thermit welding:** This method, developed by Gold Schmidt of Germany, is the only form of site welding adopted universally. The principle of the method involves an exothermic chemical reaction that takes place when a mixture of finely divided aluminium and iron oxide (called 'thermit mixture) is ignited resulting in the formation of aluminium oxide and iron:

$$Fe_2O_3 + 2\,Al \rightarrow 2\,Al_2O_3 + 2\,Fe^+ + \text{(Heat energy)} \tag{3.2}$$

(159 g of iron oxide + 548 of aluminium give 102 g of aluminium oxide, 112 g of iron, and 182 kCal of heat energy; it takes just 15 to 25 seconds to achieve a temperature of about 2450°C)

The released iron is in the molten state and welds the ends of the rails; the lighter aluminium oxide floats on top as the slag).

The conventional welding takes about 45 minutes; to overcome this, *shot preheating* or *SKV method* has been developed, which takes only about 15 minutes; however, this requires almost double the amount of mixture but much less petrol. Hence, this is used for on-site welding on busy routes. But the weld is considered to be weak in fatigue compared to a flash butt weld.

### Development of Welded Rails

Indian Railways has been increasingly using welded rails to enhance the potential of the track to carry trains at greater speeds consistent and better riding comfort, with the bonus of reduced maintenance costs.

Rails were originally welded into panels of the three to five rails; these were called *short-welded panels (SWP)* or *short-welded rails (SWR)*, and could expand or contract throughout their length. In due course, a new concept of *locking up stresses* in the rail, which resisted any changes in the rail dimensions, has been developed. This has led to *long-welded rails (LWR)* of lengths of 1 km or more; in fact, this has further led to the development of *continuous welded rail (CWR)*, which continues even through station yards and points and crossings, deriving maximum advantage of the welded track.

**Theory of long-welded rails:** The expansion of a rail is a function of its length, the change in temperature, and the coefficient of linear expansion of the rail material. A free rail would undergo this change in length; but the rail is fastened to sleepers, which in turn are embedded in, ballast. This causes a restraint on the expansion, which gives rise to locked-up internal stresses' in the rail material. This thermal force is given by the equation

$$P = E \cdot A \cdot \alpha \cdot t \tag{3.3}$$

where P = Force in tonnes

E = Modulus of elasticity of rail metal in $t/cm^2$ (2,150)

A = Cross-sectional area of rail in $cm^2$ (66.15 $cm^2$ for 52-kg rail)

$\alpha$ = Coefficient of linear expansion ($11.52 \times 10^{-6}/°C$)

t = Change in temperature in °C. This leads to $P = 31.5$ At (Here, $P$ is in kg if $A$ is in kg/m)

For a 52-kg rail, $P$ works out to 1.638t/°C of temperature change.

## Breathing Length of Long Welded Rails

In the case of long welded rails, thermal expansion takes place at the end of rails owing to inadequacy of the restraint due to rail fastenings and the ballast. The resistance offered by the structure will be adequate to prevent thermal expansion of a certain length of the central portion of the LWR. This is called the *non-breathing length* and the remaining length, distributed in two equal halves at either end of the LWR, is known as the *breathing length* of the LWR. The change in length of the end portions is given by

$$\delta l = \frac{1}{2} l \cdot \alpha \cdot t \tag{3.4}$$

where l is the breathing length of the rail (For Indian Railways, $l$ is about 100 m for BG tracks either end).

There are some prohibited locations for LWR. These are:

- Locations of heavy wear, corrugation or corrosion
- Locations where the formation is weak.
- Locations where the soil is susceptible to pumping, needing frequent screening of ballast.
- Locations susceptible to flooding, breaches and subsidence.

**Switch expansion joint (SEJ):** This is a device installed at the end of an LWR to allow for expansion of the breathing length.

## Short-welded Rails (SWR)

On Indian Railways, short-welded rails are restricted to three rail lengths in order to limit the thermal forces to an acceptable value.

## Continuous Welded Rails (CWR)

As the switch expansion joints used at the end of LWRs have been a source of weakness, continuous welded rails for several kilometres on end have been laid on European Railways;

this is coming into vogue even on Indian Railways, the SEJs being provided at station yards to facilitate quick attention and maintenance whenever required.

## Rail Wear and Other Defects

During the continued operation of a railway track with trains for passenger and goods traffic, defects are bound to develop in the rails. The primary defects are of wear of different parts of the rail, including the battering of rail ends. In addition, miscellaneous defects also occur. Further, rails are also affected by the phenomenon of *creep*. All these lead to rail failure, ultimately necessitating the renewal of rails.

### Wear of Rails

The friction between the rail and the wheel during the passage of moving loads in the form of loaded trains causes rail wear, especially at the rail head.

Wear and tear of the rail head is caused not only by the impact of moving loads, but also by the forces of acceleration, deceleration and braking, abrasion between the rail and the wheel, the presence of materials such as sand, and climatic and weather effects such as temperature, rain, and, such other allied factors.

**Types of wear:** A rail may undergo wear and tear in the following positions:

- Top of the rail head (vertical wear)
- Sides of the rail-head (lateral wear)
- Ends of the rail (rail-end batter)

Consequently, loss of metal and hence loss of weight of the rail occurs. When this loss exceeds the limit of tolerance, renewal of rail is called for. Wear of rails occurs prominently at some special locations such as on sharps curves (centrifugal effect), on steep gradients (extra tractive force applied by the engine), in tunnels (due to weather effects), and on approaches to railway stations (acceleration and deceleration effects).

Measurement of wear of rails is now possible by the use of special instruments designed to measure the rail profile and automatically record it on graph paper.

The following are the *methods of reducing rail wear*:

- Better maintenance of ballast section, alignment and gauge.
- Use of heavier and stronger (with greater UTS) rails with greater wear resistance.
- Reduction of the number of rail joints by welding of rail lengths.
- Lubricating the gauge faces of the outer rail on horizontal curves.
- Uses of bearing plates in the case of wood sleepers.
- Providing check rails on sharp curves
- Interchanging the inner and outer rails on horizontal curves, to extend the life of the rail and service life of the track.

When the extent of rail wear exceeds the tolerance limit and causes stresses greater than permissible values, the worn out rails are to be replaced by new rails.

**Rail-end batter:** The pounding action of wheel loads on rail joints batters the ends of rails there in course of time. This causes wear at the contact surfaces between the rails and the sleepers, and the ballast in the vicinity of the joint gets loosened; the rail fastenings at the joint also become loose and all these effects lead to further batter of the rail ends.

Quantitatively, rail-end batter is defined as the difference between the rail at the end and at a point 30 cm away from the end. If the batter is 3 mm or more, the rail ends can be cropped and the rail re-used.

Lateral wear is considerable on sharp curves, where the provisions of rail lubricators tend to reduce the wear significantly. There are different types of mechanical devices that can be attached to wheels to provide such lubrication.

## Other Defects in Rails

A few other types of defects that occur in the rails on continuous use are given below:

### Corrugation of Rails

Corrugation results in minute depressions of different sizes and shapes on the rail surface at irregular intervals. The factors leading to the formation of corrugation in rails are given below:

- High nitrogen content in the rails
- Steep gradients
- High axle loads and high speeds
- Spots where acceleration and deceleration of trains take place
- Spots of weak formation
- Long tunnels
- Electrified tracks
- Presence of sand in the vicinity of the track
- Humidity as in coastal areas.

Although a few theories have been advanced for the reasons for corrugation on rails, none of them has been considered satisfactory. However, experience has shown that the above factors could be possible reasons.

Rail corrugation is an undesirable feature as it causes rapid oscillation of rails, with the usual consequences of loosening the fastenings and ballast.

This has one other unpleasant consequence – the production of a 'roaring sound' when trains pass, possibly due to locking of air in the corrugation. This phenomenon, called 'roaring of rails' is very unpleasant and creates excessive noise, causing inconvenience to passengers.

Corrugation can be removed by grinding the rail head by a fraction of a millimetre; however, no method of rail grinding has been standardised by Indian Railways.

But German railways use a rail grinding train which moves at 30 km/h and grinds both rails simultaneously.

### Corrosion or Rusting of Rails

The main cause of corrosion is the effect of acid gases dissolved in the moisture film coating the rail. These gases abound in industrial and coastal areas. Incidence of corrosion is high near ash-pits, water columns, wet tunnels and passenger yards receiving discharge of night soil.

Disused rails corrode more than those subjected to regular fast traffic.

The practice of Indian Railways for protection of rails against corrosion is to apply a coat of zinc chromate or red oxide primer followed by three coats of bituminous emulsion. The surface has to be thoroughly de-scaled and cleaned before application of these coats.

### Hogging of Rails

*Hogging* means bending in the vertical plane, usually downwards, at the ends of a rail length. Poor maintenance of the rail joint and/or weak formation leads to hogging of rails. Track quality deteriorates, leading to rough riding.

Hogged rails may be rectified by what is called *cropping;* this involves the cutting off the hogged ends and drilling fresh holes for the fish plates. The rails are then pulled back equal to the length cut off.

Using a defogging machine to straighten the end is another option, but it may not be very efficient. *Measured shovel packing* during the maintenance of ballast is considered to be a good remedial measure.

### Scabbing of Rails

This occurs due to falling of patches or chunks of metal from the rail table (top portion), and is generally seen as an elliptical depression, the surface revealing a progressive fracture with numerous cracks around it.

### Wheel Burns

These are caused by the slipping of the driving wheel of locomotives on the rail surface; the heat generated by this creates a depression on the rail table. Wheel burns occur on steep gradients or at places of frequent braking. They affect the riding quality, as also the strength.

Slipping occurs when the circumferential movement of the wheel is more than the forward movement of the axle; when the reverse happens, skidding occurs. Both could cause defects in rail head.

### Shelling

This is the progressive horizontal separation of metal that occurs on the gauge face at the upper corner. High bearing pressure on a small contact area, producing heavy internal shear stresses, tends to cause shelling.

### Kinks in Rails

*Kinks* are formed when the rail ends near a joint move slightly out of position.

Loose packing of ballast at rail joints, gauge and alignment defects, and uneven wear of rail heads on either side could lead to kinks. They cause loss of riding comfort and could lead to severe damage if they are not attended to in time. Obviously, proper maintenance at rail joints and on sharp curves can prevent kinks in rails.

### Buckling of Rails

*Buckling* occurs when the free expansion of the rail due to change in temperature is prevented, resulting in serious change in position and the alignment.

Buckling may sometimes lead to serious derailments; hence, adequate preventive as well as remedial measures are necessary to control and eliminate buckling of rails.

### Rail Failure

A rail is considered to have failed if it has to be removed from the track on account of serious defects. Usually, fatigue cracks are the origin of a majority of rail failures.

The *forms of rail failure* may be:

- Crushed heads
- Split heads
- Horizontal or transverse fissures in the head
- Metal flow in the head
- Rail breakage
- Split web.

The main *causes of rail failure* could be one or more of the following:

(i) Inherent manufacturing defects in rails
(ii) Rolling stock defects
(iii) Abnormal traffic effects
(iv) Excessive corrosion of rails
(v) Poor maintenance of rail joints
(vi) Improper maintenance of track
(vii) Delayed renewal of track
(viii) Defects in welding of rails
(ix) Derailments of trains or train accidents.

## Detection of Flaws in Rails

*Flaw*, here, means a defect, which will ultimately lead to breaking or fracture of the rail. Therefore, detection of flaws and timely remedial actions is imperative for safe running of trains.

Given below are the major methods of detecting of flaws in rails:

- Detailed visual examination of the rail ends with a magnifying glass after opening up the rail joint.
- Using rail flaw detection equipment.

### Ultrasonic Rail Flaw Detector (USFD)

This non-destructive method of testing rails has been in use on Indian Railways for detecting flaws in rails.

Different techniques of ultrasonic testing have been evolved, using different types of equipment. Some of these are:

- Frequency modulation
- Pulse echo
- Transmission
- Resonance
- Acoustic range.

The first two techniques are in use on Indian Railways.

### Kraut Kramer Multi-probe Rail Testing Trolley

This is the most common type of equipment used. Since it is a hand trolley, the progress is slow — only two to three kilometres a day. The defects noticed are classified as follows:

1. IMR defects: Serious flaws that call for immediate removal of the rail.

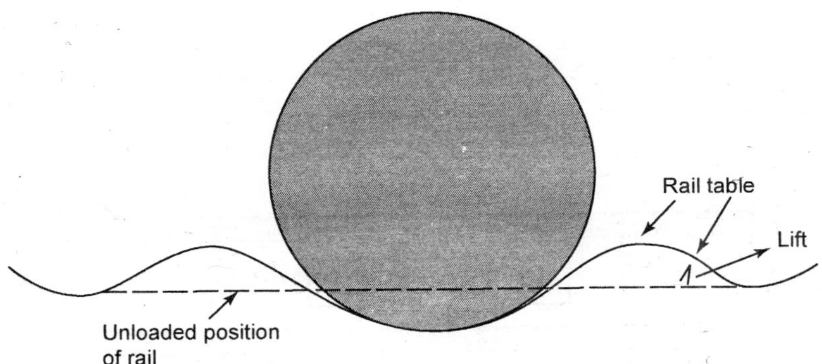


**FIG. 3.19** *Wave motion theory of creep*

2. REM defects: Flaws warranting early removal.
3. OBS defects: Lese serious defects, which need to be kept under observation.

*Ultrasonic Rail Flaw Testing Car*, which can cover 100 to 200 km per day has been used by German Railways.

## Creep of Rails

*Creep* is defined as the longitudinal movement of rails with respect to the sleepers. The tendency of movement of the rail will be in the direction of dominant traffic. Creep is a common phenomenon in a railway track, but its magnitude varies from place to place. The magnitude of creep may be several centimetres in a month at certain locations, but may be practically negligible at certain other places.

### Theories Relating to Creep

Certain theories have been put forward to explain the phenomenon of creep, but none of them has been found to be satisfactory. However, the following three theories are the more important ones:

**(A) Wave motion theory:** According to this theory, moving loads induce wave motion in the resilient track, which causes a deflection in the rail under the load. The portion of the rail directly under the wheel gets slightly depressed due to the wheel load; this leads to the formation of a wavy surface, as shown in Fig. 3.19.

As the wheels of the train move forward, the depressions under the wheels also move forward while the previously depressed portions spring back to the original level. As the train proceeds, the induced wave motion moves the rail forward. The motion of the wheels on the waves exert an ironing effect on the rail, which, in turn, causes a longitudinal movement of the rail; this leads to the creep of the rail.

**(B) Percussion theory:** According to this theory, creep is caused due to the impact of the wheels on the rail end ahead of a joint. As the wheel leaves the trailing rail at the joint, the rail gets pushed forward longitudinally in the direction of the traffic. The continuous movements of several wheels over the joint push the landing or facing rail forward, which results in creep (Fig. 3.20).

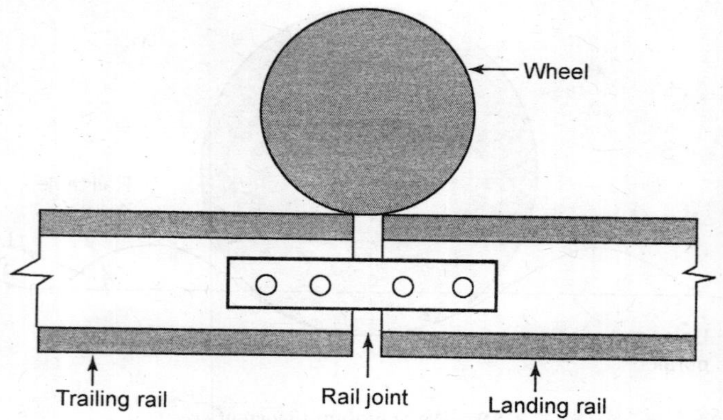


**FIG. 3.20** *Percussion theory of creep*

**(C) Drag theory:** According to this theory, the backward thrust of the driving wheels of the locomotive pushes the rail backward, while the thrust of the other wheels and those of wagons/coaches pushes the rail in the direction of movement of the train. This results in the longitudinal movement of the rail in the direction of traffic, which causes creep.

**Factors Causing Creep**

The following are the main factors causing rail creep:

**(i) Ironing effect of the moving wheels:** The moving wheels cause ironing effect while on the waves formed in the rail under the wheel loads, thereby making the rail move in the direction of traffic. This results in rail creep.

**(ii) Acceleration and deceleration of trains:** Acceleration while starting to move and deceleration while stopping tend to cause creeps. Acceleration causes backward thrust through the wheels, tending to push the rail backward; but the deceleration, required to stop a train needs the application of brakes, and the braking effect tends to push the rail forward. Thus, these actions cause creep in one direction or the other.

**(iii) Imbalance in traffic:** In a single-line section, the volume of traffic in the two directions is variable; creep in such cases develops in the direction in which the traffic predominates.

In a double-line section, traffic on each of the two tracks is only in one direction. Therefore, creep in each track occurs in the direction of traffic.

**(iv) Poor maintenance of track:** The following maintenance lapses also cause creep:

- Inadequate ballast
- Poorly maintained rail joints
- Poor maintenance of fastenings of rails to sleepers
- Light rails for heavy traffic
- Improper expansion gap at rail joint
- Weak formation and lack of drainage.

**(v) Temperature changes:** Expansion of rail during hot weather conditions tends to cause creep of the rail.

### Effects of Creep:

The following are the general effects of rail creep:

(i) Sleepers move out of their position due to rail creep and become out of square; that is, their perpendicularity to the centre line of the track is disturbed. This can affect the gauge as well as the alignment.

(ii) Expansion gaps at rail joints get disturbed; the gaps get reduced at certain joints and widened at certain others. Joints get jammed, resulting in stresses in the rail and rail fastenings. This became a major problem at switch expansion joints in the case of long welded rails.

(iii) Rail renewal, when called for, presents problems as the new rail length may not match the rail to be removed in terms of its length, because of creep effects.

(iv) Points and crossings get distorted due to creep, and maintenance of gauge and alignment becomes a problem.

(v) Interlocking mechanism of the points and crossings also gets disturbed by rail creep and require quick attention to ensure safety of trains.

(vi) Excessive creep can even lead to buckling of the track; this is another reason to lay emphasis on efficient maintenance of rail joints and other track components.

(vii) Certain miscellaneous effects such as breakage of bolts and formation of kinks in the alignment also tend to occur because of rail creep.

### Measurement of Creep

A device called *creep indicator* is used to measure the creep of a rail. It consists of two creep posts, driven one on either side of the track. Generally rail pieces are used to serve as creep posts, which are driven at about 1 km intervals. The top level of the posts is kept at the same level as that of the sleeper or that of the rail.

A chisel mark is made at the side of the bottom flange of the rail on either side of the track. A string is stretched between the two creep posts. As creep occurs, the rail moves longitudinally along with the chisel mark. The distance between the chisel mark and the string, measured parallel to the track, is the amount of creep.

Creep is measured at quarterly intervals on Indian Railways, and a creep register is maintained with all the necessary particulars for every prescribed location along the track.

The maximum prescribed limit for creep is 150 mm. Also not more than six consecutive rails should be jammed in a single track at any location.

Creep is not allowed near approaches to points and crossings.

### Locations Susceptible to Creep

The following are the locations which are more susceptible to creep:

- Valleys in long and steep gradients
- Approaches to bridges
- Approaches to points and crossings
- Approaches to level crossings

- Steep gradients and sharp curves
- Junction points of track with wood sleepers and steel sleepers.

**Adjustment of Creep**

When the creep exceeds the prescribed limit of 150 mm, it should be adjusted as soon as possible by pulling back the rails. This should be done by following the required precautions.

The steps in this procedure are given below:

(i) Rail joints and expansion gaps are inspected along the route to know the current status.
(ii) The amount of creep to be adjusted and the correct expansion gap are to be decided in advance.
(iii) The fastenings at the joint are loosened/removed to the extent required.
(iv) The rails are pulled back one by one with a rope attached to a hook. This should be done carefully without disturbing the rail joints unduly.

The procedure is laborious and slow and can be taken up only for isolated lengths of the track.

To increase the pace, five rail lengths may be tackled at one time. Instead of using a rope, a cut rail piece of about 5 m length is used to give blows to the rails to be adjusted.

For extensive work, a 'creep adjuster' is used.

**Measures to Reduce Creep**

The most important measure to reduce creep of rails is to use what are called *creep anchors*. The function of creep anchors is to fit the rail smugly against the sleepers.

A variety of creep anchors have been tried, and the *fair T* and *fair V* anchors have been standardised for use on Indian Railways. The *fair V* anchor, which is more popular, is shown in Fig. 3.21. At least 8 anchors must be provided per rail length for effective prevention of creep of rail.

The following measures, in addition to creep anchors, may be adopted to minimise creep:

- Good maintenance of track – proper packing of sleepers with well compacted shoulder ballast
- Regular inspection of rail joints and expansion gaps for jamming, so that remedial measures may be taken in time
- Use of anti-creep bearing plates which should be provided on wood sleepers.
- Use of trough type steel sleepers, which are known to provide excellent grip to the rails, thereby preventing creep.

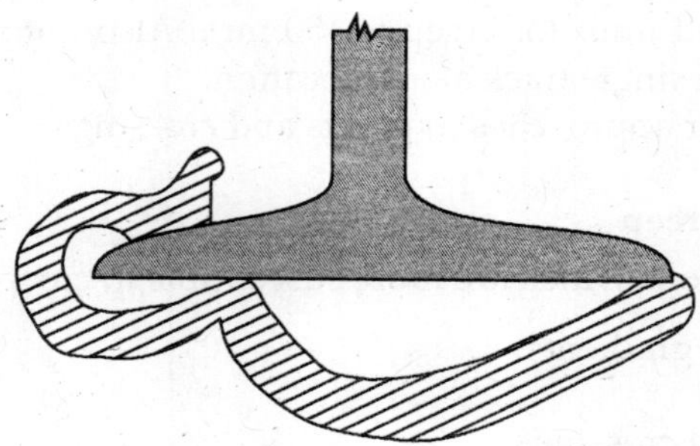

**FIG. 3.21** *Fair V anchor*

## Renewal of Rails

The need for renewal of rails arises for any of the following reasons:

- Rail fractures or rail failures, which could be sudden
- Excessive wear of rails beyond the limits of correction or repair
- Shortened service life due to special reasons such as accidents and derailments
- Maintenance of the track falling below the prescribed standards
- Periodic renewal as part of the approved maintenance plan of the Railways.

### Further Criteria for Renewal

(i) More than 6% loss in weight for 52 kg/m rails on broad gauge.
(ii) Corrosion beyond 1.5 mm in the web and the foot of the rail.
(iii) More than 8 mm of vertical wear for 52 kg/m rail on BG.
(iv) Lateral wear of rail of more than 8 mm to 10 mm on BG.

## SLEEPERS

*Sleepers* are track components placed below rails; they are meant to support the rails to which they are fastened and transmit the loads to the ballast and subgrade below.

Historically, sleepers consisting of stone slabs or timber joists were placed longitudinally below the rails, but were soon rejected because of rough and nosy riding. These were replaced by transverse sleepers or cross-sleepers, which are universally adopted; they are placed in a direction perpendicular to the centre-line of the track at specified intervals, which could vary at special locations along the track. (They have been known as *Ties* in American terminology.)

### Function of Sleepers

The important functions of sleepers as follows:

- To hold the rails to the correct gauge and alignment
- To transfer the load uniformly from the rails to a larger area of ballast
- To act as an elastic medium between the rails and the ballast to absorb impact of the moving loads of the trains.
- To provide longitudinal and lateral stability to the track.
- To provide the means to rectify the geometry of the track.
- To provide a firm and even support to the rails.

### Requirements of Sleepers

In order to perform these functions, an ideal sleeper should fulfil the following requirements:

(i) It should have sufficient bearing area so that the ballast underneath it is not crushed.
(ii) It should be such that it is possible to adjust and maintain the gauge correctly.
(iii) The designs of the sleeper and the fastenings should be such that fixing and removing the rails is relatively easy.
(iv) The weight of the sleepers should not be too high so that handling is easy.

(v) The material and design of the sleeper should be such that it is sufficiently elastic in order to absorb the vibrations caused by the dynamic loads from trains.
(vi) The design should be such that it should facilitate track circuiting.
(vii) The material and design of the sleeper should be such that it does not get damaged during transport and handling.
(viii) The sleeper should possess anti-theft and anti-sabotage properties.
(ix) The economic cost of the sleepers, or the sum of the initial cost and maintenance cost, should be minimum.

## Classification of Sleepers Based on the Material

Based on the material used, sleepers may be classified as follows:

1. Wooden (or timber) sleepers
2. Steel sleepers
3. Cast-iron sleepers
4. Reinforced concrete sleepers
5. Pre-stressed concrete sleepers

## Wooden (or Timber) Sleepers

These are considered to be the ideal type of sleepers and have been extensively used.

The size of the sleeper is chosen such that it is sufficient to provide the necessary bearing area and flexural strength as a beam. While the length and width govern the former, the latter depends on depth of the sleepers and is related to stiffness. The standard sizes of wood sleepers used on Indian Railways are given below:

**Sizes and bearing areas of wooden sleepers**

| Gauge | Size (cm) | Bearing area per sleeper ($m^2$) |
|---|---|---|
| BG | 275 × 25 × 13 | 0.4650* |
| MG | 180 × 20 × 11.5 | 0.3098 |
| NG | 150 × 18 × 11.5 | 0.2090 |

*The bearing length under each rail seat is 92 cm giving an area of 2325 $cm^2$ under each rail seat, and the total bearing area is twice this. (The bearing length is more because of the hard packing of ballast under the rail seat.)

For bridges and point and crossings, thicker sections are used—25 cm × (15 cm or 18 cm).

### Specifications for Timber Sleeper

- Sound and seasoned wood; as far as possible, sap-wood.
- Free from defects such as knots, shakes and cracks.
- Free from splitting and warping.
- Made preferably of 'pole' or 'boxed heart' type, with the pith at the centre of the cross section
- Should be straight, salon parallel at the top and bottom faces with ends cut square; slab and quarter type cross-sections may also be used (Fig. 3.22).

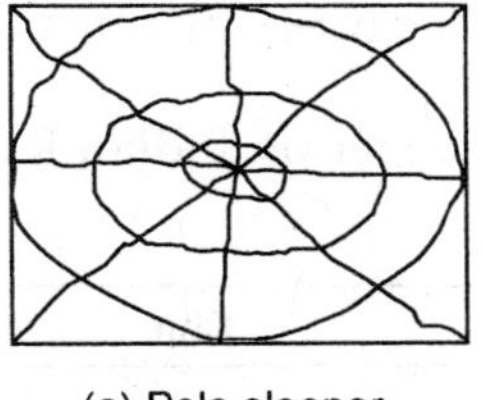

(a) Pole sleeper

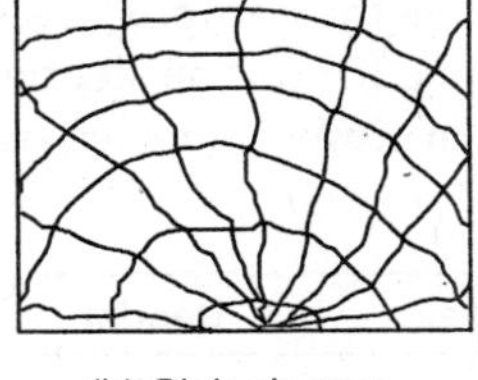

(b) Slab sleeper

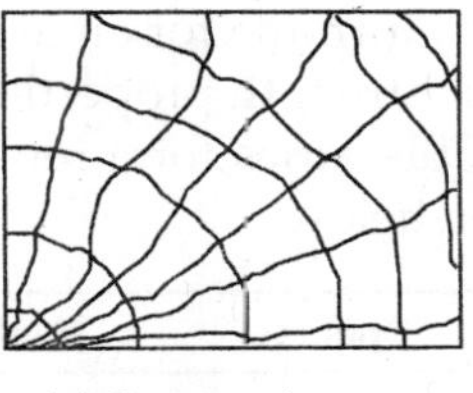

(c) Quarter sleeper

**FIG. 3.22** *Cross-sections of timber sleeper with respect to the location of pith*

### Type of Wood for Sleepers

There are two broad categories of wood for sleepers:

(i) Durable wood e.g., teak, sal, rosewood, vengai
(ii) Non-durable wood e.g., chir, fir, deodar, jamun, kail

Sleepers from durable wood do not require any treatment other than seasoning and may be laid directly on the track (U)*. (Sal wood is not usually treated.) Sleepers from non-durable wood require treatment before being laid on the track (T)[1].

The primary service life of sleepers made from durable wood is about 19 years and 31 years for BG and MG respectively. If non-durable type of wood is used, the service life, in turn, is about 12.5 years and 15.5 years for the non-durable type (T)*.

## Composite Sleeper Index

The composite sleeper index (**CSI**) evolved from a combination of the properties of strength and hardness and is used to assess the suitability of a particular timber for use as a sleeper based on its mechanical strength.

CSI is given by the formula:

$$CSI = \frac{(S + 10H)}{20} \tag{3.5}$$

Here, $S$ is the index of the general strength for both green and dry timber at 12% moisture content; $H$ is the index for general hardness for both green and dry timber at 12% moisture content. The minimum CSI values prescribed by Indian Railways are given below:

### Sleeper Location and Corresponding CSI

| Sleeper location | CSI (min) |
|---|---|
| Normal track | 783 |
| Crossings | 1352 |
| Bridges | 1455 |

The Forest Research Institute (FRI), Dehradun, has published strength coefficients of various timber species as percentage of that of teak, a well-known all-purpose timber in India. As the strength of timber varies in different directions and under different types of stress, the FRI has

[1] U: Untreated sleepers are those from naturally durable species of wood.
T: Treated sleepers are those made of all other types of wood.

evolved a formula for an index, known as the composite sleeper index (CSI), by considering a few basic strength properties of relevance for use as sleeper.

The CSI-values for a few well-known timber species used for sleeper on Indian Railways are given below:

| Species | Sal | Teak | Deodar | Fir | Chir |
|---|---|---|---|---|---|
| CSI | 112 | 82 | 63 | 58 | 54 |

Indian Railways have prescribed the following minimum values of CSI for sleepers from any timber depending on their location on the track.

| | |
|---|---|
| Track sleepers | 54 |
| Crossing sleepers | 94 |
| Bridge sleepers | 102 |

Bearing plates are recommended on all sleepers, treated or untreated, with a CSI-value of 82 or less.

## Seasoning of Timber

*Seasoning* means drying up of the sap or the juices contained in a living tree, before using the timber; unless this is done, the sleeper tends to twist, warp and decay.

Seasoning may be done using artificial techniques of controlled heating, or natural or *air seasoning*, in which the timber is exposed to air circulation by an appropriate stacking method, preferably under cover.

The following are the stacking methods used:

(i) One-in-nine method
(ii) Open crib method
(iii) Close crib method

### One-in-nine Method

In this method, triangular gaps are left in stacking for air circulation (Fig. 3.23).

This method is suitable for all conifers including chir, fir and deodar.

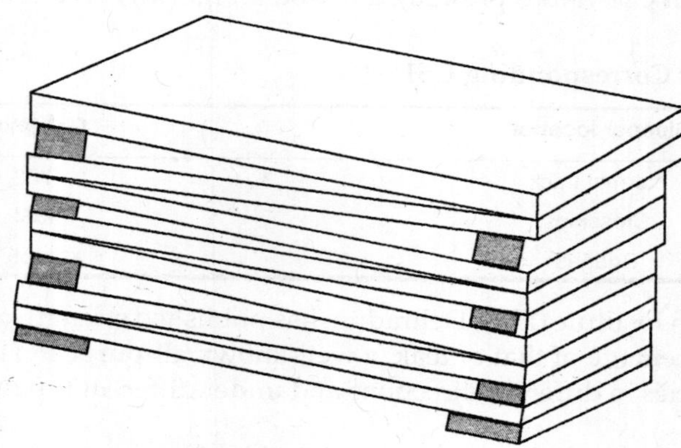

**FIG. 3.23** *One-in-nine method of stacking*

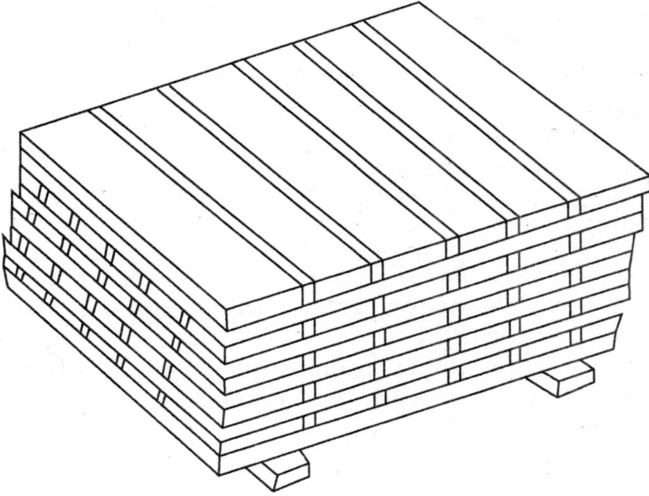

**FIG. 3.24** *Open-crib method of stacking*

### Open-crib Method

This method is the same as one-in-nine method in purpose and effect, but is space-saving. The space between the sleepers can be adjusted to control the rate of drying. (Fig. 3.24).

### Close-crib method

This method is suitable for refractory type of timber, which is very hard and develop cracks if controlled drying is not done. Sal is an example for this kind.

### Treatment of sleepers

This consists of filling up the cells of wood with a preservative solution such as creosote oil (creosoting), Zinc chloride salt solution (Burnetising), or Mercuric dichloride salt solution (Kyanising). The preservative is forced into the wood under pressure using one of the following methods of impregnation:

(i) Full cell – Bethel process
(ii) Empty cell – Rueping process
(iii) Empty cell – Lowry process

The life of the sleepers can be increased by 30% to 50% by treating sleepers.

Before subjecting sleepers to treatment, they are left to age for a few months. During this period, chemicals are sprayed to prevent possible decay. The moisture should not more than 25% when the sleepers are subjected to treatment.

### Advantages of wooden sleepers

- Cheap and obtained from nature.
- Good shock-absorbing quality.
- Ease of handling
- Convenient for track circuiting.
- Suited to modern maintenance methods

- Suitable to use without stone ballast on bridges and ash-pits
- Ease of correction of alignment.

**Disadvantages**

- Less durability than other materials
- Difficult to maintain gauge and alignment
- Susceptible to fire haggard.
- Negligible scrap values.

India Railways has prohibited the use of new wooden sleepers not only for technical and economic reasons, but also due to environmental concerns in order to avoid deforestation and depletion of precious natural resources.

Consequent to this policy, the sleeper treatment plants set up by the Indian Railways in Punjab, Assam, UP and Kerala are currently not working.

## Steel Sleepers

Steel sleepers have been developed by Indian Railways owing to the shortage of wooden sleepers. *Steel channel sleepers* have been developed particularly for use on girder bridges with plate girder or open web girders. Economic considerations also have been responsible for the use of steel sleepers.

### Steel Trough (ST) Sleepers

A sleeper of this type essentially consists of a rolled steel plate of about 2 mm thickness pressed into a trough shape and the rail seat canted to 1 in 20. The ends of the section are flattened out in the shape of a spade which helps retain the ballast at the ends (Fig. 3.25).

The section at the rail seat is made flatter and shorter than that at the middle of the sleeper. Three alternative types have been designed by varying the procedure of connecting the rail to the sleeper:

(a) ST sleeper with pressed-up lugs and keys
(b) ST sleeper with loose jaws inserted into holes and tightened by keys.
(c) ST sleeper with clips and bolts.

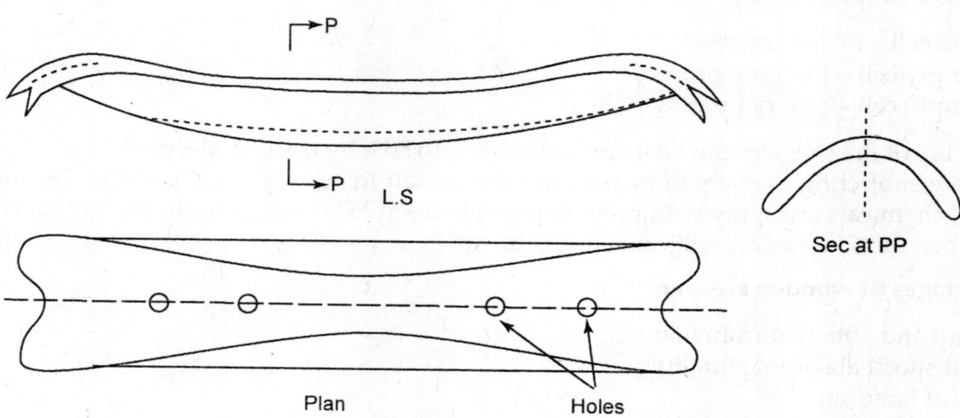


**FIG. 3.25** *A schematic view of steel trough sleeper*

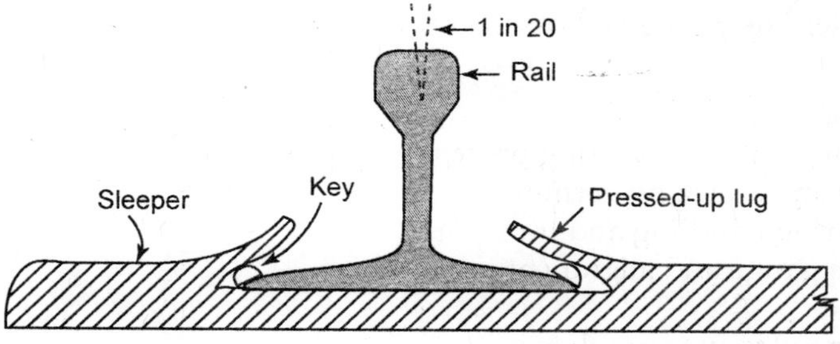


**FIG. 3.26** *ST sleeper with pressed-up lugs*

**(a) ST sleeper with pressed-up lugs and keys:** The lugs or jaws are pressed out of the plate self to accommodate the foot of the rail and the key, as shown in Fig. 3.26.

The rail is held in position by means of two standard keys as shown. The type presents maintenance problems as the keys get loosened due to the impact loads and the vibrations caused.

**(b) ST sleeper with loose jaws inserted into holes:** In this type, two holes are punched into either side of the plate to accommodate loose jaws, which are specially designed. The rails are held in position by means of two standard keys driven into the loose jaws, as shown in Fig. 3.27.

**(c) ST sleeper with clips and bolts:** In this type, bolts are inserted into clips which hold the rail to the sleeper at either end and tightened, as shown in Fig. 3.28.

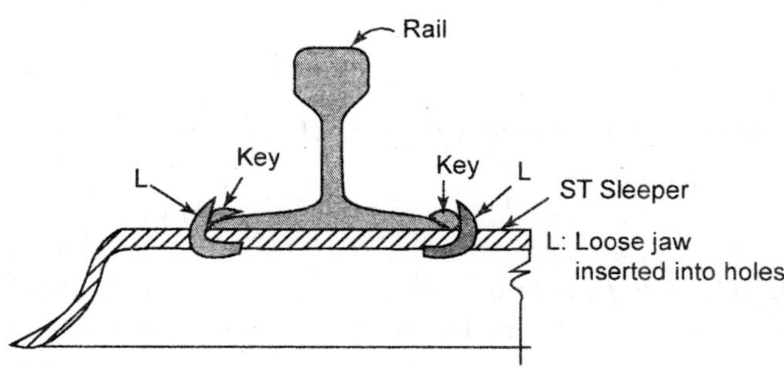


**FIG. 3.27** *Sleeper with loose jaws inserted into holes*

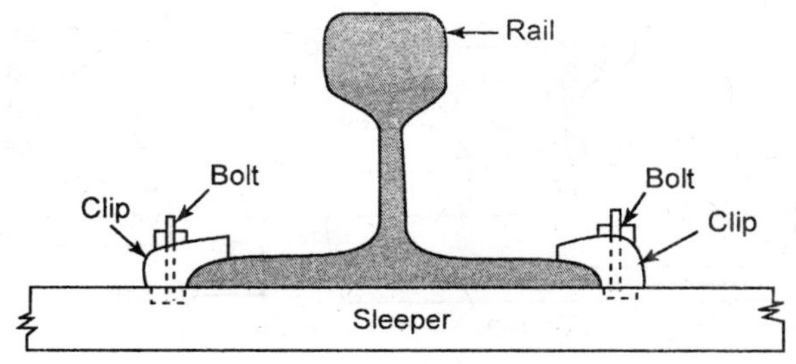


**FIG. 3.28** *ST sleeper with clips and bolts*

**Advantages of Steel Sleepers Over Wooden Sleepers**

- More durability
- High lateral rigidity
- Standard and simple manufacturing process
- Free from decay and attack by vermin
- Less damage during handling and transport
- Easy to maintain gauge and alignment.
- High scrap value
- Relatively less maintenance problems
- Not susceptible to fire hazard.

**Disadvantages**

- Liable to corrosion
- Liable to cracks at rail seats
- Unsuitable for track-circuited sections
- Liable to become centre-bound because of slopes at either end toward the centre
- Design is rail-specific, i.e., different types of rails need different designs of the sleepers and connections to the rails.

## Cast-iron Sleepers

Cast-iron (CI) sleepers have been used in India to the extent of about 8 per cent of the track.

The following types have been used:

(a) Pot or bowl Sleepers
(b) Plate sleepers
(c) Box sleepers
(d) CST-9 sleepers.

Of these, box type have become obsolete. Even pot sleepers are not being procured any more.

**Pot or Bowl Sleepers**

This type consists of two hollow pots or bowls of circular or elliptical shape placed inverted. The two pots are connected by a tie-rod by means of cotters and gibs, which permit the adjustment of gauge up to ±3 mm. The rail seat at the top of the pot, which is given a cant of 1 in 20, allows the rail to the placed on it and held in position by means of a key (Fig. 3.29).

[Total weight: 114kg; this type cannot be used on curves sharper than 4° on B.G.]

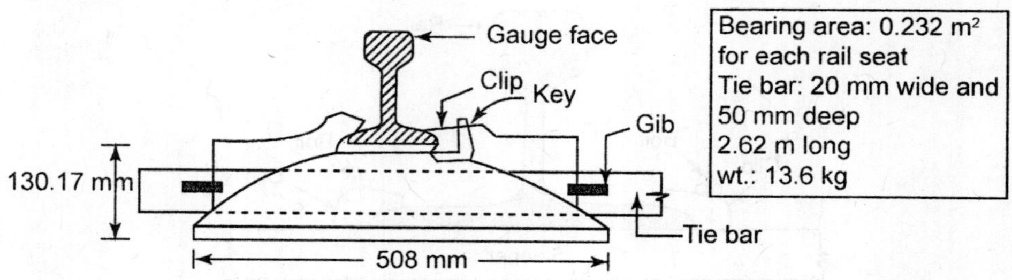


**FIG. 3.29** *CI Pot sleeper*

### Plate Sleeper

This type consists of rectangular plates, 865 mm × 305 mm, the 305 mm-dimension parallel to the rails, with projecting ribs under the plates. The ribs provide lateral stability. The bars are used to hold the plates in position; once again, gibs and cotters, distance pieces, or keys are used as fasteners for firm grip.

The effective bearing area per sleeper is 0.464 m$^2$ on BG. Both pot sleepers and plate sleepers can be used with free-float (FF) rails as well as with bull-head (BH)-rails, but need to be cast differently. The rail is made with a cant of 1 in 20.

Even this type is superseded by an improved type — the CST-9 sleepers.

### Box Sleepers

This is, for all practical purposes, obsolete.

### CST-9 Sleepers

The CST-9 sleeper is a standard one, which is being used extensively on Indian Railways. The designation, CST-9, stands for 'Central Standard Trial-9', the ninth of the series produced by the Central Standard Office of the Indian Railways.

This sleeper is a combination of pot, plate, and box type of sleepers. It consists of two triangular inverted pots on either side of the rail seat, a central plate with a projected keel, and a box on top of the plate. The two plates are connected by a tie bar by means of four cotters. The rails are held to the sleepers by two-way keys provided at each rail seat on the side of the gauge face. The gauge can be adjusted by ± 5mm by altering the relative position of the four cotters.

The CST-9 sleeper is shown in Fig. 3.30:

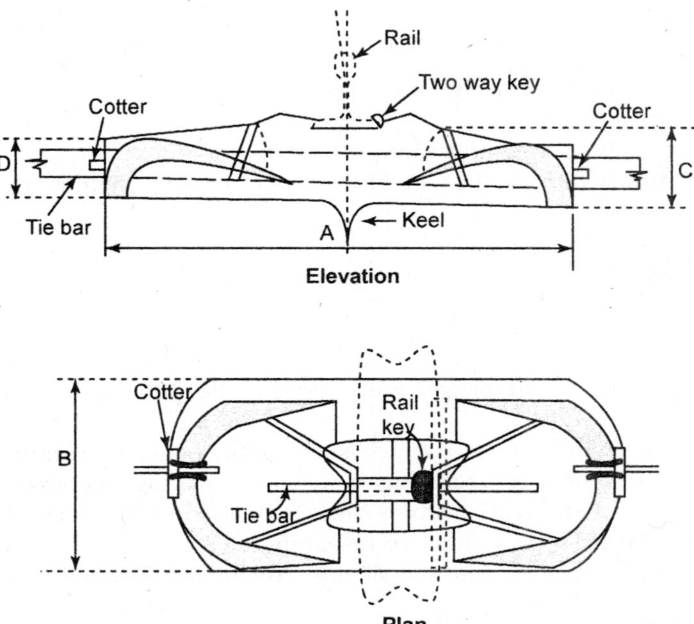


**FIG. 3.30** *CST-9 Sleeper*

The dimensions of the CST-9 sleeper for the 52-kg rail on broad gauge are:

A = 800 mm; B = 330 mm; C = 140 mm; D = 89 mm.

The weight is 43.55 kg.

The weight of the whole assembly along with fastening is 102 kg.

The effective bearing area provided is almost equal to that provided by a standard BG-wooden sleeper. A design with reverse jaws is also available as an anti-sabotage measure; at least three such sleepers are provided per rail.

Although this sleeper is the most popular one, it has certain *disadvantages*, also as given below:

- Unsuitable for the breathing lengths of long welded tracks because of its rigidity.
- Rail seat wear causes the keys to become loose.
- Due to less metal under rail seat, the vibrations are directly transmitted to the ballast. Hence, the ballast requires frequent attention.
- Limited lateral strength for use in long welded rails
- Not suitable for mechanical maintenance operations.

**Advantages of Cast-iron Sleepers**

The following are the advantages of cast-iron sleeper steel trough sleepers:

- Easy to manufacture
- High scrap value
- Less corrosion
- Less probability of cracking at rail seat.

**Disadvantages**

- Less lateral stability
- Unsuitable for track-circuiting
- Gauge maintenance is difficult as tie be bars get bent
- Susceptible to breakage
- Not suitable for mechanical maintenance procedures.

In spite of these disadvantages, the CI CST-9 sleeper is one of the most popular sleepers on Indian Railways.

## Concrete Sleepers

Concrete sleepers have evolved mainly due to economic considerations, as also due to changing track patterns for high speeds and heavy traffic. Heavier rail sections and long welded rails necessitated a heavier and better type of sleepers to be used on the modern railway track.

After the advent of cement concrete, which is a versatile material, Monnier suggested even in 1877 that reinforced cement concrete may be used for making sleepers for railway tracks. The design was improved and RCC sleepers came to be used in Austria and Italy.

After the Second World War, extensive research in France, Germany and the U.K. led to the development and use of concrete sleepers on railway tracks in several European countries.

The Germans adopted a beam type sleeper and considered it as a rigid continuous beam which is supported on a firm bed. But the French engineers designed the sleepers as two separate blocks connected by a tie-bar and resting on a resilient bed of ballast. The former design is based on static loading while the latter allows for a slight differential settlement of ballast. The latter theory of beams on elastic foundations is complex; necessitating an empirical basis. (The through type had a tendency for cracking on the tension side while the block-and-tie type were satisfactory.) In India, the shortage of wood for sleepers led to the adoption of steel and cast-iron sleepers.

The need to use a significant quantity of pig iron for sleepers, which could otherwise be used for industries, coupled with the inadequacy of metal sleepers for higher speeds and long welded rails have caused cement concrete sleepers, with their larger weight and greater strengths in the longitudinal and lateral directions to become the preferred type on Indian Railways.

Concrete sleepers have been designed for the following conditions:

### B.G. Sleeper

(i) Vertical load of 15 t at the rail seat
(ii) Vertical load of 13 t and a lateral load of 7 t towards the outside of one rail only
(iii) Vertical load of 13 t at rail seat plus a reaction at the Centre of sleeper equal to half the load under the rail seat.
(iv) Bending moment: 1.33 tm at the rail seat and 0.52 tm at the centre of sleeper

A section of the composite or block-and-type sleeper is shown in Fig. 3.31 for one block.

### Advantages of Concrete Sleepers

- More strength and stability owing to their weight
- Suited to long welded rails by virtue of their lateral strength and resistance to buckling
- Suited to track-circuiting, as concrete is a poor conductor of electricity
- Easy to maintain gauge, alignment and cross-level as elastic fastenings are used
- Packing ballast is better retained
- Mechanical maintenance methods can be easily applied
- No fire hazards
- No damage from pests or corrosion under normal conditions
- High durability – lasts for 40 to 50 years; rail and sleeper renewals can be matched to advantage.
- Suited for mass production in plants.

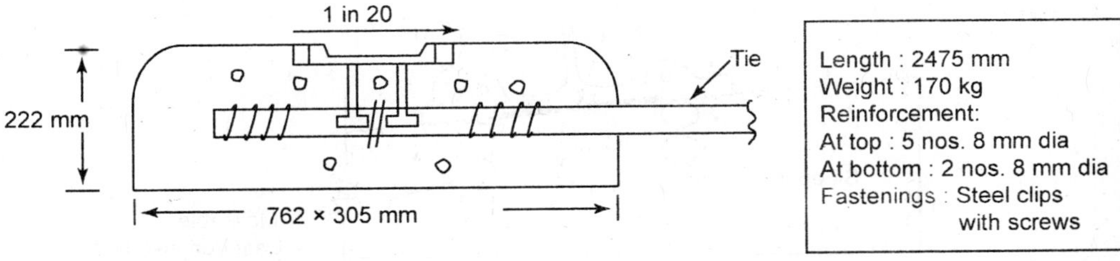


**FIG. 3.31** *R.C. Sleeper – Block and Tie type*

**Disadvantages**

- Cannot be handled without mechanical equipment in view of their large weight
- Not suitable for beater type of packing ballast
- Heavy damage in case of derailment
- Very little scrap value

**Mono-block vs. Two-block Sleepers**

- Mono-block sleepers give better stability to the track.
- Mono-block sleepers have longer working life; the tie-bar in a two-block sleeper is a source of weakness and may be subjected to corrosion.
- Mono-block sleepers are likely to become centre-bound.
- During derailments the tie-bar of two-block sleepers gets severely damaged.
- Alignment and gauge are more easily affected with two-block sleepers, requiring frequent attention.

## Pre-stressed Concrete (PSC) Sleepers

After the advent of pre-stressed concrete for use in the manufacture of sleepers on railway tracks in European countries such as France and Germany, pre-stressed concrete has been extensively used for sleepers on Indian Railways; more than 90 per cent of the total number of sleepers is of this type now.

Details of this type of sleeper are given below.

**Mono-block Pre-stressed Concrete Sleepers of the Pre-tension Type**

A mono-block pre-stressed concrete sleeper of the pre-tension type using Pandrol lips is shown in Fig. 3.32.

With an overall length of 2750 mm, the concrete sleeper weighs 270 kg nearly. A cant of 1 in 20 is provided at the surface for a distance of 175 mm on either side of the centre line of the rail for the rail filling. The sleeper is pre-stressed with 18 high-tensile steel strands of 3 mm × 3 mm diameter and 126 mm-diameter mild steel links.

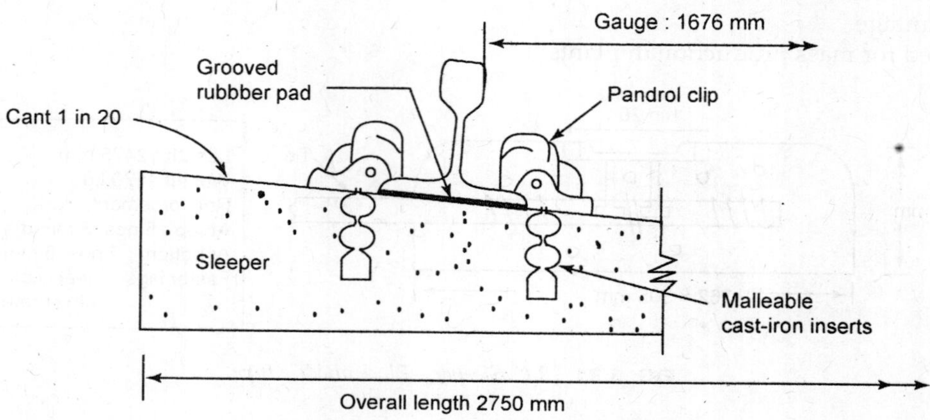


**FIG. 3.32** *Mono-block pre-stressed concrete sleeper*

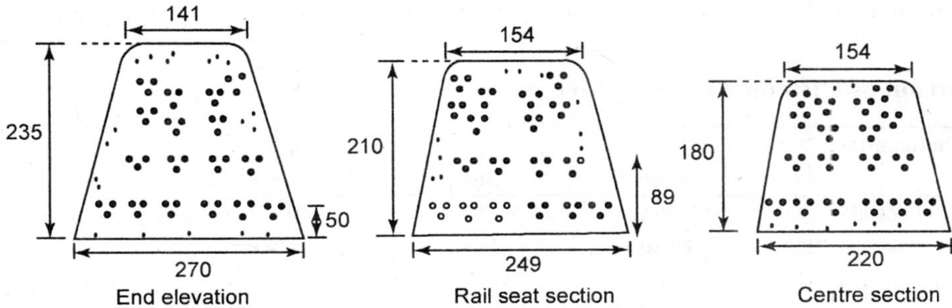


**FIG. 3.33** *Important dimensions of typical section of PSC sleeper*

The initial pre-stress is 100 kg/cm²; the 28-day crushing strength of the concrete is not less than 525 kg/cm².

The rail rests on a grooved rubber pad of size 130 mm square, with the grooves parallel to the axis of the rail. The fastenings for a 52kg-rail are Pandrol clips, which are held in malleable cast-iron inserts as shown in Fig. 3.32.

The important dimensions of the sleeper at the end, at rail seat section, and at the centre section are shown in Fig. 3.33.

Each strand is to be tensioned with an initial force of 2730 kg.

### Mono-block Post-tension Type of Pre-stressed Concrete Sleepers for B.G.

The first factory to manufacture sleepers of this type was set up in Allahabad, with German collaboration in 1981. The salient features of the post-tension type of PSC sleepers are:

| | |
|---|---|
| Length | 2750 m |
| Width at centre | 160 mm at top |
| Depth at centre | 200 mm at bottom |
| Weight | 295 kg |
| Initial pre-stressing force | 37 t |
| Final pre-stressing force | 31 t |
| Minimum 28-day strength of concrete | 550 kg/cm² |
| Minimum strength at the time of applying pre-stress | 450 kg/cm² |

The use of post-tensioned PSC sleepers has not been successful on Indian Railways and their manufacture has since been stopped.

### Special Sleepers

For special locations like turnouts, sleepers with special design are manufactured and used.

Although certain disadvantages such as heavy damage during derailments, high initial cost and high standard of maintenance besides the need for mechanical handling exist, concrete and pre-stressed concrete sleepers have been used extensively on Indian Railways.

A comparison of the different types of sleepers in respect of their important aspects is given below:

**Comparison of Different Types of Sleepers**

| Characteristic | Type of Sleeper | | | |
|---|---|---|---|---|
| | Wood | Steel | CI | RCC and PSC |
| Service life | 12–15 yrs | 40–50 yrs | 40–50 yrs | 50–60 yrs |
| Weight of sleeper (BG) | 83 kg | 79 kg | 79 kg | 270 kg |
| Maintenance | Manual or mechanised | Manual or mechanised | Manual | Mechanised only |
| Maintenance cost | High | Medium | Medium | Low |
| Gauge adjustment | Difficult | Easy | Easy | Not possible |
| Track elasticity | Good | Good | Good | Best |
| Creep | Excessive | Less | Less | Minimum |
| Track circuiting | Best | Difficult; insulating pads are necessary | Difficult; insulating pads are necessary | Easy |
| Suitability for fastening | Suitable for conventional fastening (CF) and elastic fastening (EF) | Suitable for CF and EF | Suitable for CF only. | Suitable for EF only. |
| Handling | Manual handling; no damage to sleepers while handling | Manual handling; no damage to sleeper while handling. | Manual handling; liable to break by rough handling | No manual handling; gets damaged by rough handling |
| Damage by white ants and Corrosion | Can be damaged by white ants. | Can be damaged by Corrosion. | Can be damaged by corrosion. | No damage by while ants or by Corrosion |
| Scrap value | Low | Higher than Wooden sleepers. | High | None |

**Usage of Different Types of Sleepers**

It is estimated that on BG track of Indian Railways, about 94% of RCC and PSC sleepers, 4% of CI sleepers, 1.5% of steel sleepers and less than 0.5% of wooden sleepers are in use.

## Sleeper Density

Sleeper density is defined as the number of sleepers per rail length. It is defined as $(N + x)$, where $N$ denotes the length of the rail in metres, and $x$ is a number adopted based on the axle load, speed, type and section of the rails, type and strength of the sleepers, nature of ballast, depth of the ballast cushion, and the strength of the formation.

On a BG track on which 13 m rails are used, if a sleeper density of $(N + 6)$ is adopted, 19 sleepers are used per rail length.

In fact, sleeper density varies from $(N + 4)$ to $(N + 6)$ for BG track in India. In advanced countries such as the U.S.A., France and Germany, the sleeper density is of the order of $(N + 8)$ to $(N + 11)$.

Another way of specifying the sleeper density is by the number of sleepers per kilometre length of track. If the sleeper density is $(N + 6)$ for 13 m rail lengths, the numbers sleepers per km will be around 1463. This kind of specification is more relevant is the case of long welded rail tracks.

Indian Railways, as also Railways in many other countries, are now using this specification.

### Spacing of Sleepers

Spacing of sleepers is dependent on the sleeper density. An important point to be remembered is that, at special locations along the track such as rail joints, points and crossings and major structures, sleepers are spaced more closely than on the other normal zones of the track.

Even from this aspect, it is considered better to specify the number of sleepers per kilometre length of track rather than sleeper density. Near rail joints, the sleepers adjacent to the joint are called 'joint sleepers'; on BG track, the centre-to-centre of these is 300 mm to 380 mm for wooden and metal sleepers respectively. The sleepers on either side of the joint sleepers are called 'first shoulder sleeper'; they are placed at 610 mm from the joint sleepers.

On either side of the first shoulder sleepers are the second shoulder sleepers; they are placed 700 to 720 mm from the first shoulder sleeper (This is for a sleeper density of $N + 4$). From the next set onwards, the sleepers are referred to as the intermediate sleepers, which are spaced at 830 to 840 mm for a sleeper density of $N + 4$. [All these values are for fish-plated tracks.] For welded tracks, for a sleeper density of 1660 per kilometre, centre–to-centre spacing of sleepers will be 600 mm (LWR tracks). It is now well recognised that for high speeds and heavy loads, the sleeper spacing must be as low as possible. (However, for facilitating maintenance operations of the ballast, a certain minimum spacing of the order of 60 cm is required and this is considered as a limitation.)

## Rail Fastenings and Track Fittings

The function of fastenings and fittings in a railway track is to hold the rails in the correct position with respect to the alignment and gauge to ensure smooth and safe running of trains. These may be required to join one rail to the adjacent ones, at also to fix them to the sleepers. The fittings and fastenings have to be designed so that they serve the purpose for which they are intended.

The fittings and fastenings commonly used in Indian Railway are listed below:

| | |
|---|---|
| Joining rail to rail | Fish plates, Junction fish plates, fish bolts and nuts. |
| Joining rail to wooden sleepers | Dog spikes, round spikes, screw spikes, fang bolts; bearing plates. |
| Joining sleepers to bridge girders | Hook bolts or dog Bolts |
| Joining rail to steel trough sleepers | Loose jaws and keys. |
| Joining rail to cast-iron sleepers | Cotters and tie-bars. |
| Joining B. H. rail to sleepers | Chairs. |
| Elastic fastenings | Pandrol clip, spring steel clip |
| Other fitting | Malleable cast-iron inserts, rubber pads. |

### Important Requirements of a Rail Fastening

- Has to absorb shocks and vibrations from impact loads
- Has to maintain gauge and alignment correctly
- Should be easy to fix and adjust, with minimum number of parts

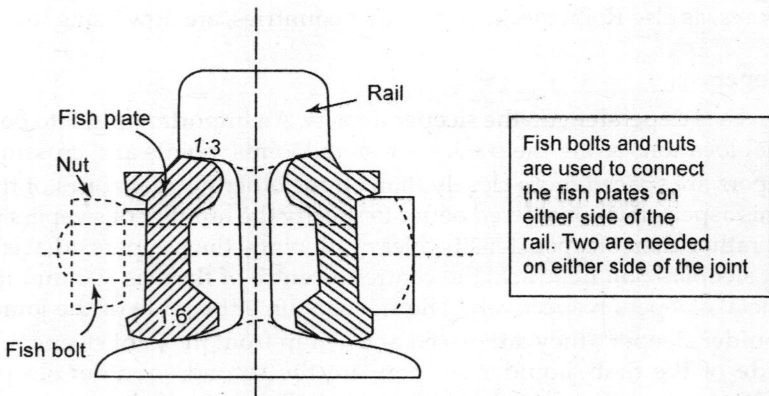


**FIG. 3.34** *Fish Plates and fish bolts*

- Should not affect the rail and the sleeper adversely
- Should be so designed to prevent sabotage
- Should have the minimum overall cost including maintenance.

### Fish Plates

The function of a fish plate in a rail joint is to hold the two rails together in both horizontal and vertical planes. The name 'fish plates' is said to have derived from the shape of its section resembling a fish. The fish plates snugly fit the underside of the flange and top side of the base of the rail (Fig. 3.34).

Fish plates are made of special steel with the following specification:

| | |
|---|---|
| Carbon | 0.30 to 0.42% |
| Manganese | 0.60% (max.) |
| Silicon | 0.15% (max.) |
| Sulphur | 0.06% (max.) |
| Phosphorus | 0.06% (max.) |
| Tensile strength | 5600 kg/cm$^2$ |
| Elongation | 20% (min.) |

The section area of the two fish plates together is almost the same as that of the rail section, so that the strength is nearly the same; but since they fall short in height or depth, their strength in vertical bending is less than that of the rail. The fishing angles of the fish plates match those of the rail both at the top and the bottom for a snug fit. For a 52-kg rail, the length of fish plate is 610 mm, the diameter of the fish bolt holes is 27 mm, and weight per pair is 28.71 kg.

### Junction Fish Plates

*Junction fish plates* or *combination fish plates* are used to connect rails of different sections. They are made to cover the rail section at either end adequately up to the centre where the rail section changes. Another feature is the elimination of the expansion gap in order to increase the strength. Although there is a difference in the depths of junction fish plates used, the top table for the two rail sections is set to be common (Fig. 3.35).

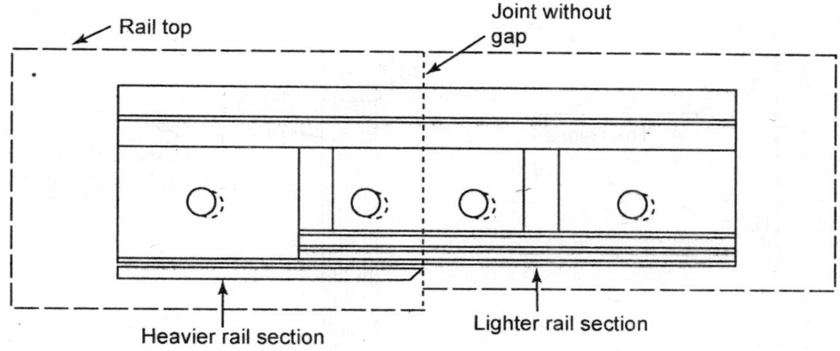


**FIG. 3.35** *Junction fish plates*

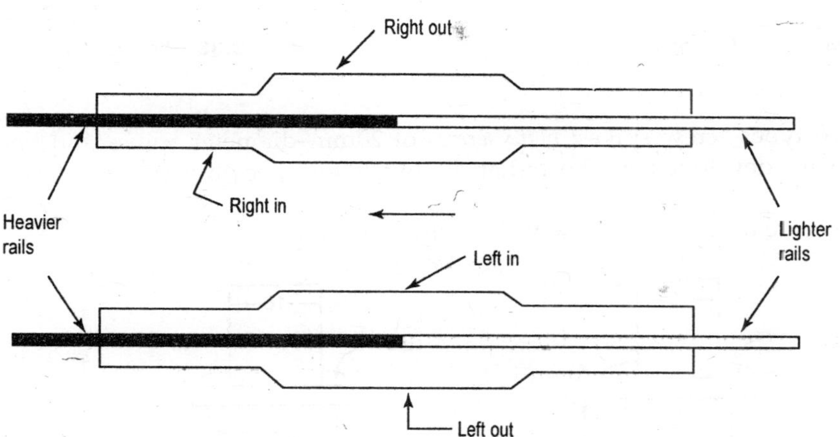


**FIG. 3.36** *Identification of junction fish plates*

A uniform system of marking is used to identify junction plates—right in, right out, left in, and left out, based on their position with the direction from the lighter rail to the heavier rail (Fig. 3.36).

### Dog Spikes

These are simple fastenings to fix rails to wooden sleepers. The dog spikes get their name because the head of these spikes appear similar to a dog's ear (Fig. 3.37).

The number of dog spikes required is 2 (one on either side staggered) on straight track, 3 (two outside and one inside) on curved track, and 4 (two each on outside and inside) on joint sleepers and bridges. The section of the spike is 16 mm square, and the length varies from 120 mm to 160 mm for BG, depending on their location.

### Round Spikes

These are used in conjunction with anti-creep bearing plates, as also for fixing switches to wooden sleepers. They have a circular section of diameter 18 mm (Fig. 3.38).

Round spikes have become obsolete.

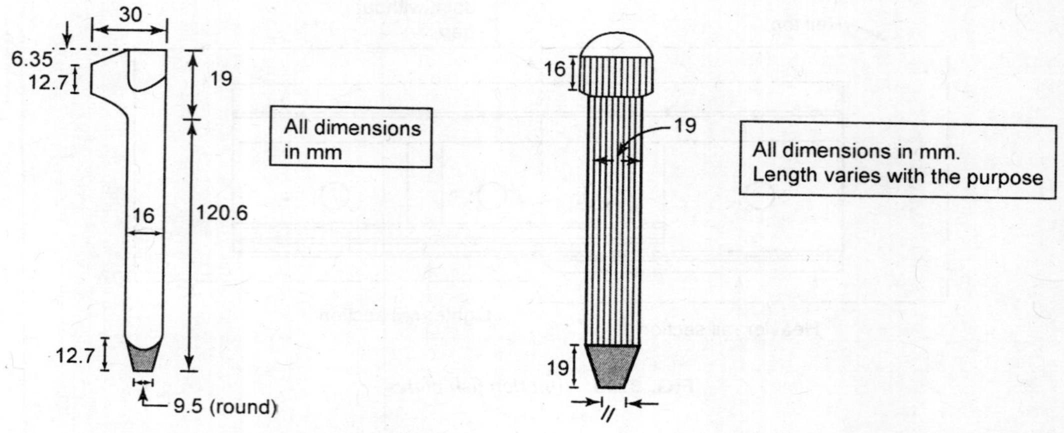


**FIG. 3.37** *Dog Spike*

**FIG. 3.38** *Round Spikes*

### Screw Spikes

There are two types screw spikes: *plate screws* of 20mm–diameter spikes and *rail screws* of 22 mm-diameter developed by the Indian Railways for specific uses. These are shown in Fig. 3.39.

Plate screws are used to fix rails with the help of bearing plates.

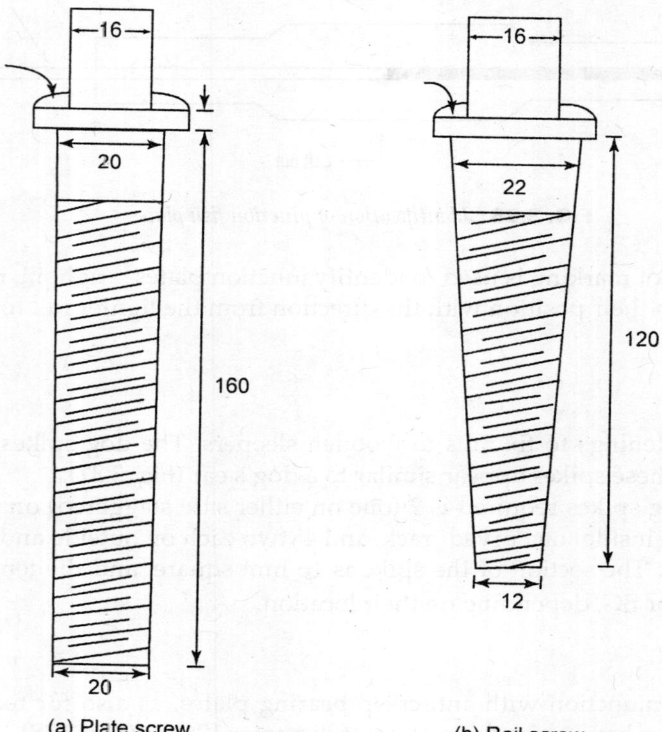


**FIG. 3.39** *Screw spikes*

Screw spikes are used to fix the rails directly to the sleepers with or without bearing plates. They are also used on bridges and on platform tracks. These spikes conserve the life of wooden sleeper; so they should be preferred to dog spikes and round spikes.

### Fang Bolts

These are employed under the switches for fixing slide chairs to the sleepers; gauge is preserved better by using these bolts (Fig. 3.40):

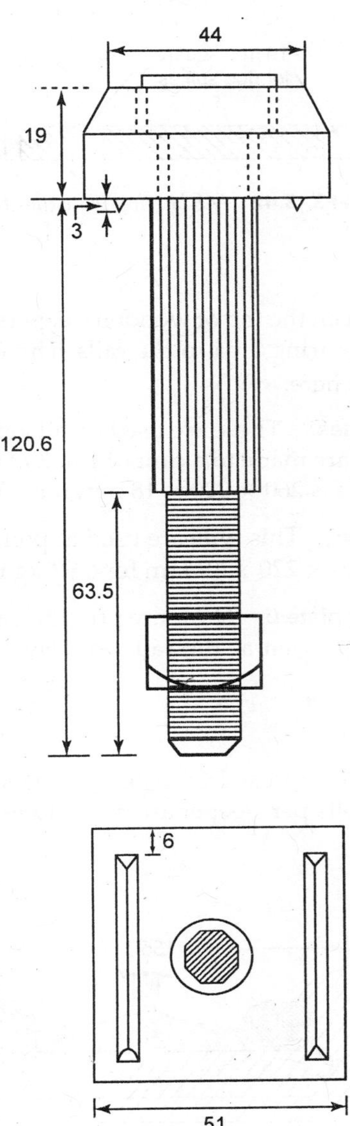


**FIG. 3.40** *Fang bolt*

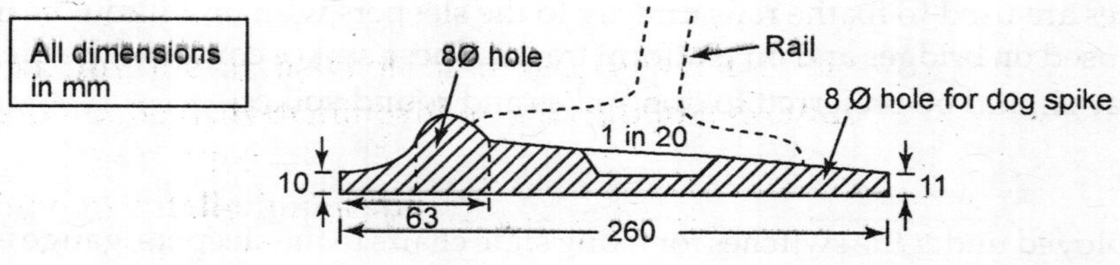


**FIG. 3.41** *M.S. bearing Plate*

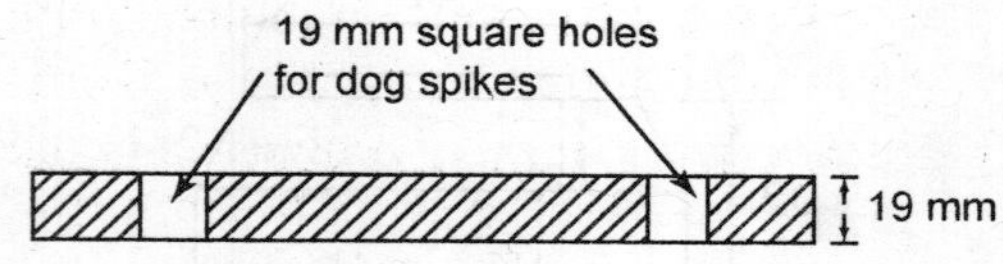


**FIG. 3.42** *MS bearing plate (flat type)*

### Bearing Plates

These are plates inserted between the top of wooden sleepers and the bottom of the rails. Their purpose is to provide better bearing area to the rails. Three types of bearing plates used on Indian Railways are described here.

**(i) Mild steel canted bearing plates:** These are used on all joints and curves to provide a larger bearing area to the rails. They are made to a cant of 1 in 20 and with a groove in the middle to prevent rocking. The usual size is 260 × 220 × 18 mm for a 52 kg-rail (Fig. 3.41).

**(ii) MS bearing plates of flat type:** This type are used at points and crossings of a turnout. No cant is provided. The size is 260 × 220 × 19 mm for a 52-kg rail (Fig. 3.42).

**(iii) Cast-iron anti-creep bearing plates:** These are provided at locations susceptible to develop rail creep. A cant of 1 in 20 is given at the rail seat. The size 285 × 205 m for B.G. tracks (Fig. 3.43).

### Hook Bolts

*Hook bolts,* also called *dog bolts,* are used to fasten sleepers to the top flange of a girder on bridges. A minimum of two bolts per sleeper are used along with bearing plates.

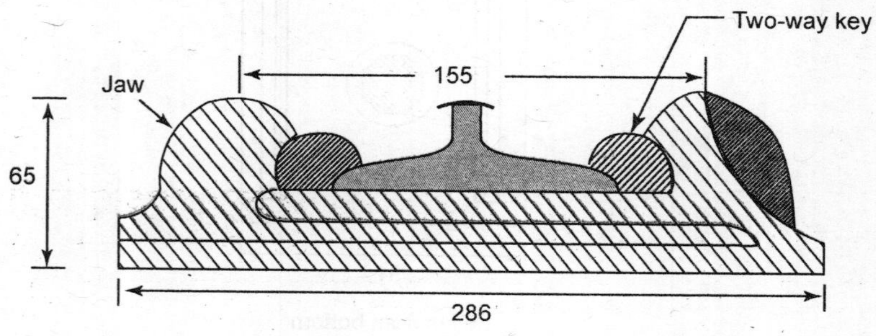


**FIG. 3.43** *C.I. anti-creep bearing plates*

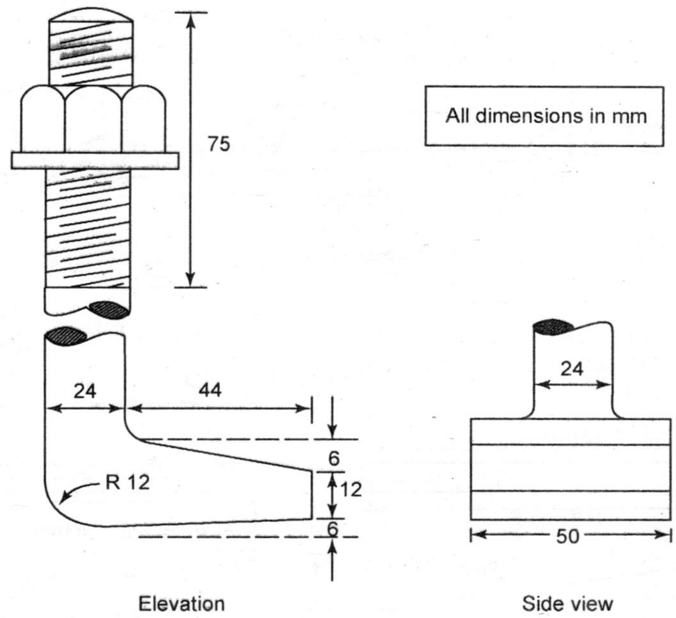


**FIG. 3.44** *Hook bolt*

There are two types: straight lip type and sloping lip type; the latter is shown in Fig. 3.44.

### Loose Jaws

Loose jaws, along with key are used to hold the rails to steel trough sleepers. They are manufactured using spring steel; each jaw weighs 288 g (Fig. 3.45).

### Two-way Keys

Two-way keys have universal use for fixing steel trough sleepers, and cast-iron sleepers of the pot type and the CST-9 type to the rail. A two-way taper is provided at both ends; as such, the key can be driven in either direction. A special rolled section is used in the manufacture of these keys. The length is about 190 mm with a taper of 1 in 32 for use on BG tracks. An

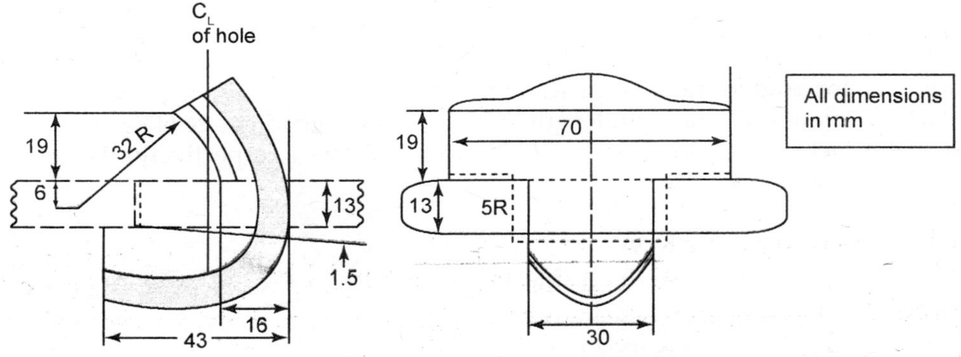


**FIG. 3.45** *Spring steel loose jaw*

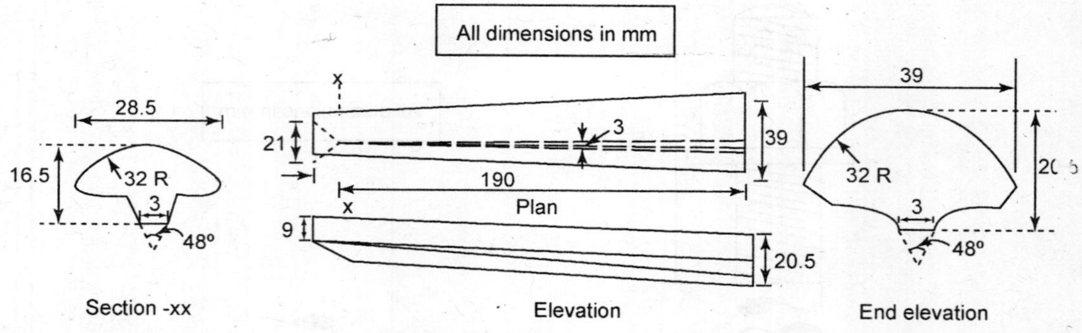


**FIG. 3.46** *Two-way keys*

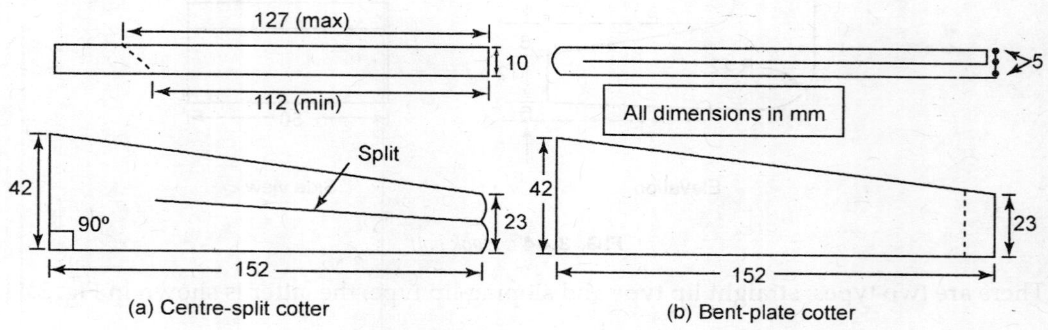


**FIG. 3.47** *MS Cotters*

adjustment ± 3 mm is possible in gauge by altering the extent to which the key is driven in (Fig. 3.46).

### Cotters

These are used for fixing tie-bars to cast-iron sleepers. Four different types of cotters, based on the method of their splitting, are used on Indian Railways. Two of them – centre-split cotter and bent-plate cotter are shown in Fig. 3.47.

The other types are side-spilt and solid end-spilt cotters.

### MS Tie Bars

These are used for holding the two plates of CI-CST-9 sleeper together. The normal section of a tie-bar is 50 mm × 12 mm with length of 2720 mm for broad gauge (The corresponding dimensions for metre gauge are 45 mm × 10 mm, and 1870 mm, respectively). The tie-bar for broad gauge is shown in Fig. 3.48.

### Cast-iron Chairs With Spring Keys

Bull-headed rails are supported on cast-iron chairs and held in position by spring keys. These chairs are fixed to the sleepers by means of spikes.

A CI-chair with spring keys is shown in Fig. 3.49.

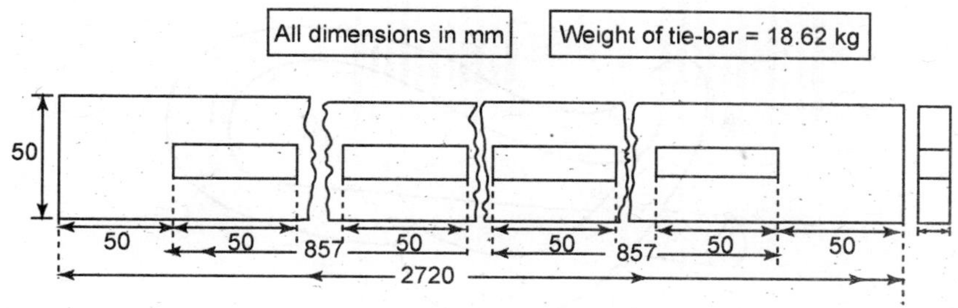


**FIG. 3.48** *MS tie-bar for BG*

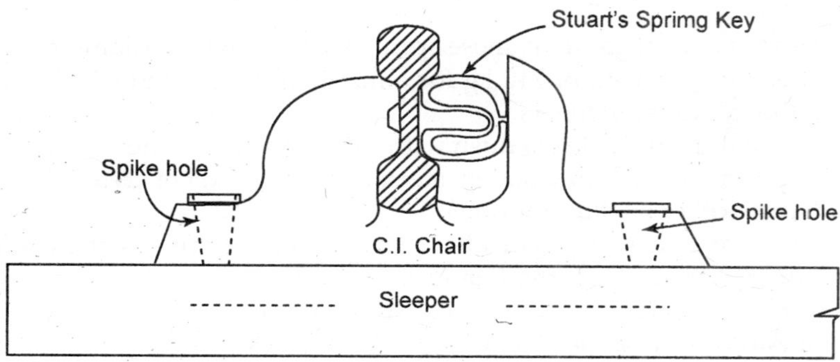


**FIG. 3.49** *Cast-iron Chair with spring key*

## Elastic Fastenings

Rigid fastenings tend to become loose under the impact loads from moving trains even at moderate speeds of the order of 100 km/h. Track parameters such as gauge and alignment get affected beyond tolerance limits frequently, requiring attention of the maintenance engineer more often than desirable.

In the wake of high speeds and heavy loads of trains in the modern era, there came a necessity to develop a fastening which maintains the track parameters and dampens the vibrations from the dynamic forces of moving trains. This resulted in the development of elastic fastenings.

The requirements of an elastic fastening are practically the same as those of any rail fastening already discussed; but in addition, it should be efficient in absorbing the shocks and vibrations caused by the dynamic forces from trains moving at high speed carrying heavy loads. In effect, besides ensuring the correctness of the track geometry and other parameters, elastic fastenings should lead to minimum maintenance.

Another important feature is that they should be universally applicable for use with wooden, steel, cast-iron, and concrete sleepers.

An elastic fastening is usually in the form of clips. Various types of clips have been developed.

## Pandrol Clip

This is also known as an *elastic rail clip*, and is a standard type of elastic fastening used on Indian Railways. It requires very little attention towards its maintenance and is a *fit-and-forget*

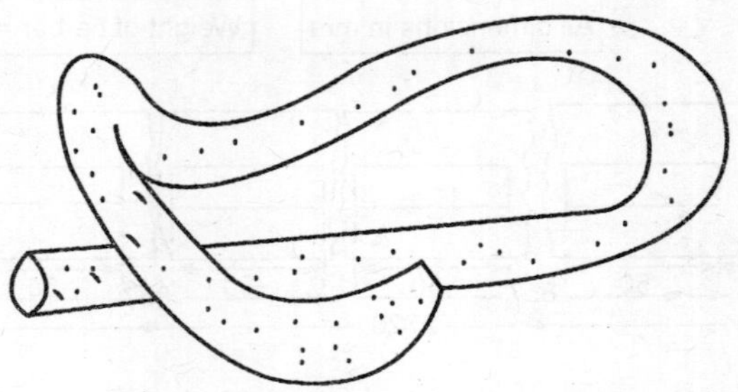

**FIG. 3.50** *Pandrol clip (general view)*

type. This clip is made of a silico-manganese spring steel bar with a diameter of 20.6 mm and is heat treated. It exerts a tie load of 710 kg for a nominal deflection of 11.4 mm. This load is adequate to grip the rail to the sleepers.

It is driven such that the outer leg is flush with the outer face of the CI insert.

The Pandrol clip can be fixed on wooden, steel, cast-iron, and concrete sleepers along with a base plate and a few other ancillary fittings.

In the case of concrete sleepers, malleable cast-iron inserts are punched directly into the sleepers during their manufacture, the Pandrol clip is fixed in the holes of the CI insert. To increase the elasticity, a 4.5 mm-thick grooved rubber pad is provided under the rail seat. A general view of a Pandrol clip is shown in Fig. 3.50.

The *disadvantages* of this clip are:

- Lack of facility for gauge adjustment
- Small contact area at the rail foot causing indentation.
- Easily removable with a hammer, with possible room for sabotage.
- During service it may get caught inside the CI insert.

### Spring Steel Clip

Spring steel clip, being tried with pre-stressed concrete sleepers on Indian Railways, is shown in Fig. 3.51.

The rail rests on a grooved rubber pad and is held in the correct position by a pair of spring clips. The clip is pressed by means of a nut tightened on a 22 mm-bolt, which is inserted

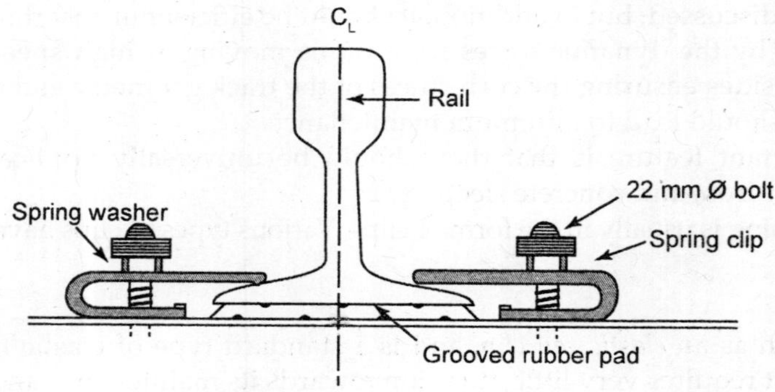


**FIG. 3.51** *Spring steel clip*

from the underside of the sleepers. The nut is tightened to a torque of 13.5 kg.cm, to obtain a resisting force of 1 tonne per pair of clips.

### Malleable Cast-iron Inserts

These are directly fixed to the concrete sleepers during manufacture.

Two types of inserts are being used on Indian Railways:

(i) Stem-type inserts for use in pre-tension pre-stressed concrete sleeper; the weight of this insert is 1.6 kg.
(ii) Gate-type inserts for use in post-tension pre-stressed concrete sleepers; the weight of this insert is 1.7 kg.

### Rubber Pads

A rubber pad is an integral part of an elastic fastening. It is provided between the rails and the sleepers.

The functions of a rubber pad are:

- Absorbing shock
- Dampening vibrations
- Resisting lateral movement of rail
- Providing electrical insulation between the rails
- Preventing abrasion of the button surface of the rail.

Grooved rubber pads of 4.5 mm thickness made of special quality rubber are used by Indian Railways. The function of the grooves is to limit the lateral expansion of rubber under the pressure of dynamic loads; they also aid in the uniform distribution of load on the sleepers. A schematic of the rubber pad is shown in Fig. 3.52:

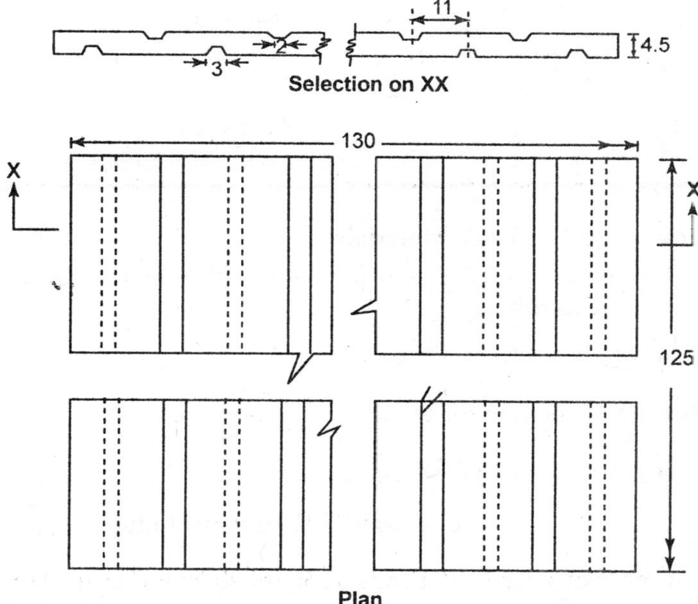


**FIG. 3.52** *Grooved rubber pad*

While 4.5 mm-thick rubber pads have given a service life of only six to seven years, 6 mm-thick rubber pads have been shown to give a service life of 15 to 20 years.

## Number of Fastenings

The number of various track fittings and fastenings required per sleeper for different types of sleepers are given below.

### Number of Fastenings per Sleeper

| S. No. | Type of sleeper | Type of fastening | Number |
|---|---|---|---|
| 1. | Wooden | Dog spikes or screw-spikes | 8 |
| | | CI bearing plates | 2 |
| | | Keys for CI bearing Plates | 4 |
| | | Plates screws | 8 |
| | | Pandrol clips | 4 |
| | | Rubber pads | 2 |
| 2. | Steel trough | Loose jaws | 4 |
| | | Keys | 4 |
| | | Pandrol clip | 4 |
| 3. | CST-9 | Plates | 2 |
| | | Tie-bar | 1 |
| | | Cotters | 4 |
| | | Keys | 4 |
| | | Pandrol clips | 4 |
| | | Rubber pads | 2 |
| 4. | Concrete (RCC/PSC) | Pandrol clips | 4 |
| | | MCI inserts | 4 |
| | | Rubber pads | 2 |
| | | Nylon insulating liners for clips | 4 |

### Calculation of Requirement of Track Materials

The calculation of the quantity of different items of track material required per kilometre length of BG track is made as follows.

**1. Rails:** The standard rails are 13 m- long for BG.

$$\text{Number of rails per km for BG} = \frac{1000}{13} \times 2 = 154.$$

$$\text{Weight of 52 kg-rail per km} = 52 \times 154$$

$$= 104104 \text{ kg} = 104 \text{ tonnes}$$

**2. Sleepers:** The number of sleepers depends upon the sleeper density. For a sleeper density of (N + 6):

$$\text{Number of sleepers per rail} = (13 + 6) = 19$$

$$\text{Number of sleeper per km} = \frac{1000}{3} \times 19 = 1463$$

**3. Fittings and fastenings**

**For a wooden sleeper track:**

$$\text{Number of fish plates per km} = 2 \times (\text{number of rail per km})$$
$$= 2 \times 154 = 308$$
$$\text{Number of fish bolt} = 4 \times 154 = 616$$
$$\text{Number of bearing plates} = 1463 \times 2 = 2926$$
$$\text{Number of spikes} = 1463 \times 4 = 5852$$

**For a concrete sleepers track:**

$$\text{Number of elastic rail clips} = 1463 \times 4 = 5852$$
$$\text{Number of grooved rubber pads} = 1463 \times 2 = 2926$$
$$\text{Number of nylon insulating liners for clips} = 1463 \times 4 = 5852$$

**Note:**

- Fastenings for other type of sleepers can also be calculated based on their requirements shown earlier.

The requirements of different items may be calculated for a MG track also in a similar manner

## Certain Specific Aspects Related to Rails and Sleepers

Under this head, the following specific aspects will be considered:

(a) Coning of wheels
(b) Tilting of rails
(c) Adzing of sleepers.

### (a) Coning of wheels

The tread of the wheels of a railway vehicle (a locomotive, a coach, or a wagon) is not made horizontal, but sloped in the form of the frustum of a cone; this is to enable the vehicle to move smoothly on the track both on straights and on horizontal curves.

On straight reaches, the circumference of the treads of both the wheels is equal, as the wheels are aligned centrally on a straight and level surface.

However, on a horizontal curve, the outer wheel has to negotiate more distance than the inner one. The vehicle also tends to move in the outward radial direction due to the centrifugal force acting on it while negotiating the curve. To counteract this, the circumference of the tread of the outer wheel is made greater than of the inner wheel; this is achieved by making the wheel tread slope downwards from inside to outside of the track. Coning of the wheel helps this to be realised (Fig. 3.53).

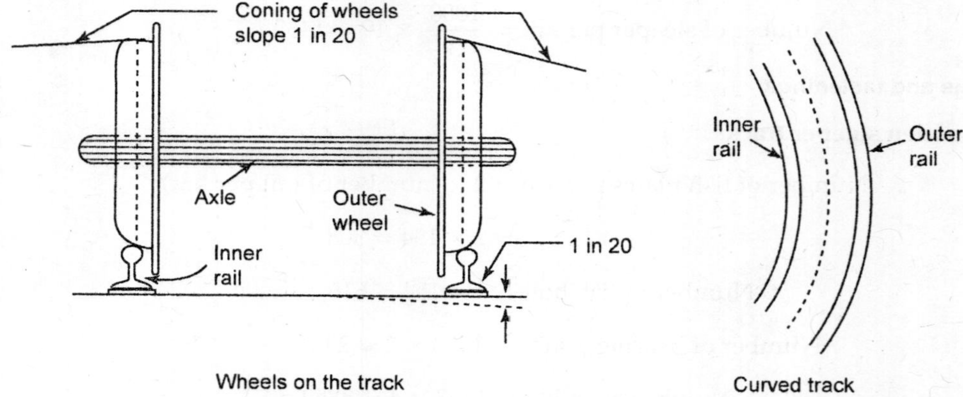


**FIG. 3.53** *Coning of Wheels*

Because of the extra distance to be traversed by the outer wheel on the horizontal curve of a rail track, the outer wheel tends to slide as it is connected to the inner wheel by the axle.

Coning of the wheels prevents the flange of the wheel touching the side of the rail, thus preventing jerks and an uncomfortable ride. This also reduces the wear and tear of the wheel flange. The slip *s* is given by

$$s = \frac{2\pi\theta}{360}.G \tag{3.6}$$

where $\theta$ is the central angle fixed by the rigid base and G is the gauge.

The slip is nearly 0.029 m per degree of the curve.

### (b) Tilting of Rails

Rails are tilted inward at an angle of 1 in 20 to match the slope of coning of the wheels of a railway vehicle. This is intended to reduce the wear and tear of the rails as well as of the tread of the wheels.

There is heavy wear of the rails as the pressure from the wheel acts near the inner face of the rail. Further, eccentric loading of rails causes lateral bending stresses. This uneven loading may cause damage to the sleepers too.

To counteract these effects, rails are tilted at a slope of 1 in 20, which is also the slope of the wheel cone. The tilting of rails may be achieved in two ways: by 'adzing' the wooden sleeper, or by providing canted bearing plates to suit this slope.

### (c) Adzing of Wooden Sleepers

For making the rails to be tilted inwards at a cant of 1 in 20, wooden sleepers have to be cut to this slope at the rail seat before laying them in position to take on the rails.

This procedure of cutting the wooden sleeper at a slope of 1 in 20 (to match the coning of the wheels) so that the rails are tilted accordingly is known as 'adzing'. It has to be done with great care since any variation of this slope from sleeper of sleeper causes a rough ride.

The adzed surface of a wooden sleeper is treated with coal tar or creosote oil to ensure protection of the surface.

It must be noted that adzing of a wooden sleeper is done only when a bearing plate is not used. In case a bearing plate is used, irrespective of the type of sleeper, the top surface of the bearing plate is made to a slope of 1 in 20 to achieve the desired purpose.

## Check Rails and Guard Rails

These are lengths of rail laid parallel to the track. They are usually laid at a fixed distance away from the running rail and fastened to it with an appropriate fastening such as a check block or a cast-iron distance piece, along with bolts.

*Check rails* are used at the following locations:

(i) Points and crossings (for guiding the wheel through the narrow clearance at the nose of the crossing.

(ii) Sharp curves (for preventing the wheel from causing excessive wear of the outer rail.)

(iii) Bridge approaches on curves (as a safeguard against derailment.)

*Guard rails* are used at the following locations:

(i) Level crossings (for providing clearance to the running wheels — the gap between the extended portions of the guard rails beyond the roadway should be filled with ballast for contiguity with the road surface.)

(ii) Girder bridges (including pre-stressed concrete bridges without deck slab and an open floor).

(iii) Major ballasted deck bridges (for preventing derailment and consequent damage.)

High embankments and deep cuttings (for preventing derailments and consequent severe damages.)

[In (ii) to (iv), guide rails prevent the derailment of vehicle from the railway track bed.]

Guard rails on bridges are fastened to the sleepers independent of the running rails.

The top of the guard rails is kept at the same level as the head of the running rail; if kept lower, it should not be more than 25 mm below the running rail. More of this has been discussed under Points and Crossings in Chapter 5

## Ballast

*Ballast* is a layer of selected granular material placed below and around the sleepers for transmitting the loads from above to the formation. Its primary function is to provide longitudinal and lateral stability to the railway track and to maintain adequate drainage to it.

### Function of Ballast

Ballast is expected to serve the following functions:

- It provides a hard bed for the sleepers
- It holds the sleepers in position during the movement of trains.
- It helps distribute the load from the sleepers on to a large area of the formation.
- It absorbs the shock and vibrations from the train loads and provides resilience to the track to enhance the riding comfort.
- It provides resistance to the track against longitudinal and lateral forces for stability.
- It provides effective drainage to the track.
- It facilitates the easy maintenance of the level and alignment of the track.

- It protects the formation against exposure to rain and wind.
- It inhibits the growth of vegetation in the vicinity of the track.

**Requirements of a Good Ballast Material**

In order to perform the function expected of the ballast, a good ballast material should satisfy the following requirements:

(i) It should be tough and resist both attrition and abrasion under the action of train loads.
(ii) It should be sufficiently hard or possess adequate crushing strength.
(iii) It should be durable and should not get disintegrated even in adverse weather conditions.
(iv) It should allow good drainage of rain water.
(v) It should be non-porous so that it does not absorb water.
(vi) In should consist of cubical and angular particles so that it retains its position longitudinally and laterally under traffic in all zones of the track, by virtue of its internal friction.
(vii) It should be easily workable with the standard implements used for maintenance.
(viii) It should not affect the sleepers and rails by any kind of chemical action under prolonged exposure and contact.
(xi) It should be cheap, economical and easily available.

It is difficult to obtain a material which satisfies all these requirements; but there are materials which fulfil many of the important ones.

**Different Materials for ballast**

The following materials are used for ballast on Indian Railways:

**(i) Broken stone:** This is the most commonly used ballast material on Indian Railways. This is obtained from hard stones such as granite, traps and quartzite. Good quality hard stone is used as ballast for high-speed trunk and main routes.

**(ii) Gravel:** Naturally available gravel from river and stream beds makes a good ballast material, but it is not preferred for high-speed tracks. It is used only in special circumstances.

**(iii) Moorum:** This is the material obtained from the decomposition of laterite. It is normally used as the initial ballast of a new track.

**(iv) Kankar:** It is derived from lime and is soft; this is also used as an initial ballast for new tracks, or for MG and NG section with light traffic in the event of non-availability of a better material.

**(v) Sand:** Coarse sand ballast is used under cast-iron pot sleepers, and also with wooden and steel trough sleepers in sections with low traffic. It has good drainage properties and provides a relatively silent running. But, it easily gets blown away by the wind because of its low unit weight and causes excessive rail wear. It also corrodes the moving parts of the rolling stock or railway vehicles. If it is covered with a layer of brick pieces, if serves its functions well.

**(vi) Cinder or coal ash:** This material is commonly used as ballast in station yards and sidings. It is also used as initial ballast for new track, in view of its easy availability and low cost. However, it should not be used with steel sleepers and fittings, because of its corrosive action.

**(vii) Brickbats (broken brick):** Over-burnt bricks, broken into small pieces, may be used as a ballast material as a temporary measure in the case of non-availability of other materials. It gets easily powdered and makes the track dusty. Constant abrasion with this dust deposited on the top of the rails tends to cause corrugations. Brick ballast, however, provides good drainage.

**(viii) Selected earth** Selected earth material may also the used as ballast for tracks of sidings, or as a temporary ballast on a new formation. Hardened clay and disintegrated rock are suitable for this purpose.

Each of these materials has its own merits and demerits for use as a ballast material. A comparison of some of the more important of these ballast materials is presented in the Table 3.2.

**Table 3.2** Comparison of different ballast materials

| S. No. | Ballast material | Advantages | Disadvantages |
|---|---|---|---|
| 1 | Broken stone | Hard and durable when got from hard rocks | High initial cost |
| | | Stable and gives resilience to the track | Angular shape may injure wooden sleepers |
| | | God drainage properties | Difficult to procure |
| | | Economic in the long run | |
| 2 | Sand | Good packing material for CI pot sleeper | Causes excessive wear of rails and parts of rolling stock |
| | | Good drainage properties | Blown away easily. |
| | | Noiseless track | Poor retentivity of packing |
| | | Relatively cheap | Difficult to maintain. |
| 3 | Moorum | Prevents seepage of water | Very soft and gets powdered easily |
| | | Cheap, if available locally | Maintenance is difficult |
| | | Provides aesthetic appearance | Average quality of track |
| 4 | Cinder or coal ash | Easy availability on railways | Harmful for steel sleepers. |
| | | Good drainage | Corrodes steel sleepers and rail bottom |
| | | Very cheap | Soft and easily pulverized |
| | | | Difficult to maintain |

## Size of Ballast Material

The size of ballast material has been universally adopted to be 50 mm for all types of sleepers on Indian Railways.

Special locations such as points and crossings are subject to greater impact loads; this requires ballast with greater frictional resistance, better amenability to packing, finer degree of adjustments and improved compaction. It has been considered that a size of 25 mm for ballast particles satisfies these requirements.

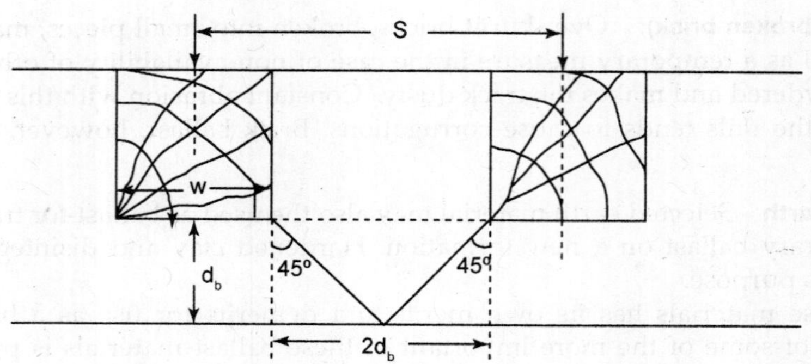


**FIG. 3.54** *Depth of ballast cushion*

Hence, ballast of 25 mm size is used at such locations, as well as on main lines and trunk routes on Indian Railways. The specification provides for graded ballast of 25 mm to 65 mm, the maximum portion being in the ronge of 40 mm to 50 mm size.

### Design of Ballast Section

The standard ballast profile is a trapezoidal section with a certain top width, side slopes of 1:1 to $1\frac{1}{2}$ :1 (Horizontal: Vertical), and a ballast cushion of a designed depth. The most crucial of all these is the ballast cushion underneath the sleepers, which determines the extent of load distribution on to the formation, the load-carrying capacity of the railway track and the speeds which it can sustain.

### Depth of Ballast Cushion

The load from the sleepers is transferred to the formation through ballast. Though the contours of equal pressure are bulb-shaped, the load dispersion may be assumed, for the sake of simplicity, to be at 45º to the vertical. For even distribution of load to the formation, the depth of ballast cushion should be such that there is no overlap of the dispersion lines.

Based on this, the depth of ballast can be determined as shown in Fig. 3.54.

Let w be the width of sleeper, $d_b$ the depth of ballast cushion, and S, the centre-to-centre spacing of sleepers.

Form the figure, the following relation is obvious:

$$S = w + 2d_b \tag{3.7}$$

Or

$$d_b = \frac{(S - w)}{2} \tag{3.8}$$

In accordance with this, the minimum ballast cushion prescribed by Indian Railways is 150 to 200 mm. But, on main lines and trunk routes, it ranges from 250 to 300 mm.

### Ballast Profile

For a normal track with fish-plated rail joints, the standard ballast profile for broad gauge is shown in Fig. 3.55.

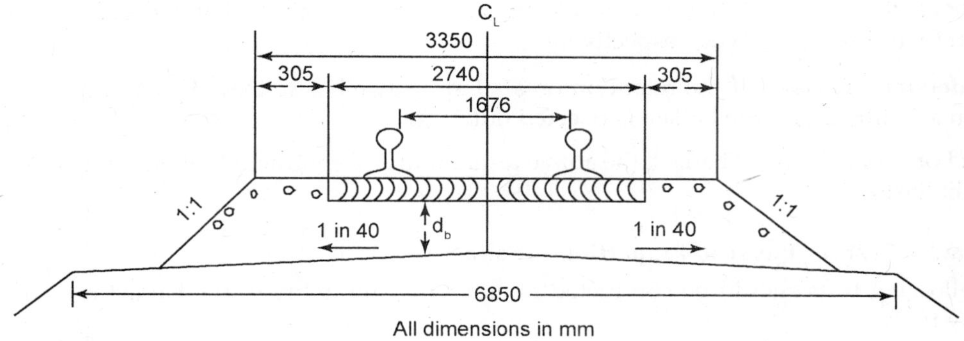


**FIG. 3.55** *Standard ballast profile for BG*

## Quantity of Ballast

The quantity of ballast required per unit length of the track is dependent primarily on the ballast cushion, top width and side slopes of the ballast profile, the nature of the horizontal alignment at the location – straight or curved, and the type of sleepers.

For straight reaches on trunk routes of BG with a ballast cushion of 300 mm, the quantity of ballast required per metre length is around 1.588 $m^3$; on main lines with a cushion of 250 mm, it is 1.375 $m^3$, and on others with a cushion of 200 mm, it is around 1.167 $m^3$.

On curved reaches, it ranges from 1.634$m^3$ to 1.202 $m^3$. These are the guidelines prescribed by Indian Railways.

## Specifications for Ballast

The following specifications are followed for stone ballast for all types of sleepers on normal tracks on Indian Railways.

**(a) General qualities:** The following general qualities are required for the stone ballast: It should be hard, durable, angular in shape, and free from organic matter. Generally, it should be made of cubical particles and should not be flaky.

In general, machine-crushed ballast should be used for all BG routes; hand-crushed ballast may be used for MG and NG tracks.

**(b) Physical properties:** The following are to be satisfied when tested in accordance with IS: 2386 (IV) (as given in Vol. I: Highway Engineering):

| | |
|---|---|
| Abrasion value | 30% (max.) |
| Impact value | 20% (max.) |
| Water absorption | 1% (max.) |

**(c) Size and gradation:** The following requirements should be satisfied:

| | |
|---|---|
| % retained on 65 mm square mesh sieve | 5% (max.) |
| % retained on 40 mm square mesh sieve | 40 to 60% |
| % retained on 20 mm square mesh sieve | ≮ 98% |

**(d) Oversized ballast:** A maximum of 10% and 70 % may be permitted with respect to retention on 65 mm and 40 mm sieves, respectively.

**(e) Undersized ballast:** If the retention on a 40 mm sieve is less than 40% and it is less than 98% on a 20 mm sieve, the ballast is rejected outright.

(**Note:** For every 100 $m^3$ of ballast, the above tests should be conducted for ensuring the quality of the ballast.)

### Laboratory Test for Physical Properties of Ballast

The following tests should be conducted to judge the suitability of the ballast material for a railway track:

(i) Aggregate abrasion value
(ii) Aggregate impact value
(iii) Flakiness index (for shape)
(iv) Specific gravity and water absorption

(**Note:** These tests have already been dealt with in detail in Vol. I – Highway Engineering, in accordance with the relevant Indian Standards code of practice.)

### Renewal of Ballast

On prolonged and repeated application of wheel loads on the track components, the ballast material gets crushed to smaller sizes than originally laid. This increase in finer sizes leads to the filling of voids in the aggregate and consequent prevention of free flow of water through the ballast. This affects the drainage significantly and should be avoided.

To make up the loss of ballast due to the crushing under repeated loading, and other reasons like that due to rain and wind, the ballast has to be renewed from time to time, as part of the track maintenance programme.

The fine material is removed by passing the ballast on wire-mesh screens of appropriate size and the loss in the ballast material is compensated, and re-laid on the track to the standard profile.

### Measurement of Ballast

This is done either in stacks or in wagons, depending upon the purpose and quality required.

### Ballast-less track

The primary disadvantage of the traditional track structure is the high cost of maintenance, especially in the restoration of track geometry by renewal of the ballast profile from time to time. This can be overcome by using a 'ballast-less' track. In its simplest form, this consists of a continuous slab of concrete with the rails supported on a resilient pad placed over it.

The variations include continuous reinforced concrete slab and precast pre-stressed concrete units laid on a base.

The main disadvantage of ballast-less track is its high initial lost; if it is to be adopted in an existing ballasted track, long closure of the track would be required. The overall economic cost over the design life is very attractive owing to the minimisation of the maintenance cost.

Ballast-less track is usually considered for new high-speed tracks or special locations like tunnels with maintenance difficulties. Long service life, high lateral strength, absence of track turbulence common with ballasted tracks at high speeds, besides very little maintenance needs, are advantages of a ballast-less track.

In India, the maintenance needs of ballasted tracks are high during monsoon months. In addition, there is a risk of environmental hazard created by dust during the operation of high-speed trains on ballasted tracks, apart from the need for frequent cleaning to remove ballast contamination. All these factors that undermine the use of ballast can be mitigated by the use of ballast-less tracks.

Most of the high-speed lines in countries such as Japan, South Korea, Taiwan, and China are ballast-less tracks.

The new Beijing–Shanghai high-speed line is a ballast-less track.

In India, ballast-less track should be a preferred new technology to be used on all the proposed high-speed routes.

## Subgrade and Formation

*Subgrade* is the naturally occurring earth material or soil which is prepared to receive the ballast and the railway track above it. The prepared flat surface that is ready to receive the ballast is called the *formation*. This is an important component of the track, since it supports the entire track structure.

The *functions of the formation* are:

- To provide a smooth and level bed for laying the track above
- To bear the load transmitted to it through the ballast from the moving trains and distribute it to a large area of the soil below, such that the pressure transmitted to it is less than its safe bearing capacity
- To facilitate drainage
- To provide stability to the track on the whole.

### EarthWork for Formation

To align the track according to the permissible gradients and to avoid frequent changes of gradient, it becomes necessary for the level of the formation to be above the natural ground level or below it at different locations along the alignment.

If the desired formation level is above the natural ground level, a raised bank called an *embankment* is constructed above the natural ground. If the desired level is below the natural ground level, a cut or an excavation is made and a formation is prepared to the required level. This is called a *cutting*. In the former case, the track is said to be *in embankment* and in the latter, it is said to be *in cutting*.

In extreme cases where the height of the embankment is excessive, the track is carried over a bridge or a viaduct; where the depth of cutting would be excessive, a tunnel is made through the ground and the track carried through it.

The height of the formation depends upon the ground level and the gradients adopted. The side slopes depend upon the shearing strength of the soil and stability considerations. The width of the formation depends upon the number of tracks to be laid, the gauge, the sleeper dimensions, the ballast profile and such other factors.

The recommended widths of formation for the broad gauge single- and double-tracks have already been given in the previous Section. In general, the width of formation required is more when the track is in cutting than when it is in embankment.

The side slopes of both the embankment and the cutting depend upon the factors governing slope stability – the angle of internal friction and the shearing strength of the soil. The stability of an earth slope is investigated by the *slip circle method* or the *friction circle method*. (These have been covered in some detail in Chapter 6 of Transportation Engineering, Vol. I: Highway Engineering. A more detailed analysis can be found in any standard book on geotechnical engineering.) In practice, a slope of 2:1 (Horizontal : Vertical) may be required for an embankment, and $1\frac{1}{2}$ :1 or 1:1 for a cutting; an even steeper slope such as 0.5:1 may be adopted in stiff soil or rock in cutting.

**Balancing of Earthwork**

When a new railway line is planned, soil surveys and soil exploration must be undertaken at appropriate intervals along the route. An effort must be made to align the route in such a wa1y that the amount of soil excavated from the cuttings is sufficient to form the necessary embankments. If this type of balancing of earthwork can be achieved, the trouble of disposing of surplus excavated material or of obtaining material for embankments from other sources will be avoided.

If the excavated material is not sufficient to form the embankments, it should be obtained from borrow pits located by the side of the alignment. In low-lying country, the track must be on high embankment to avoid flooding during monsoon periods.

**Guidelines for Execution of Earthwork**

The following guidelines are prescribed by Indian Railways for the execution of earthwork:

**Soil Surveys and Soil Exploration**

These are to be carried out as part of the preliminary and final location surveys for any railway project, as has already been mentioned in the previous chapter.

Exploratory borings and soil sampling should be undertaken along the alignment at intervals of 500 m or less depending upon the expected soil strata, as also at the proposed locations for borrow pits.

**Classification of Soils**

If the track-bed or formation is laid on natural ground, i.e., either on level ground or in cutting, the control of water is an important factor with respect to the subgrade material.

The broad categories of material encountered are:

| Soils | |
|---|---|
| Cohesion-less soils | Gravel, sand |
| Cohesive soils | Clay, fine silt |
| Organic soils | Organic soils, peat |
| **Rocks** | |
| Igneous rocks | Granite, basalt or trap |
| Sedimentary rocks | Sandstone, limestone |
| Metamorphic rocks | Slate, marble |

Organic soils and peat should never be used for forming embankments; cuttings also should be avoided in such soils, as far as possible.

**Design of Formation**

An appropriate design and construction procedure for the formation is essential for trouble-free service from a railway track. The necessary conditions to be satisfied are:

- Embankment/cutting should be structurally, or geometrically, stable.
- Settlement in the soil mass supporting the subgrade should be within permissible limits.
- Bearing capacity failure, or shear failure, should be avoided.

The water content at which a soil is compacted has a significant effect of the properties of the compacted soil. The design parameters should be such that the formation is safe even under the most adverse conditions.

**Compaction of Earthwork**

The method of compaction should satisfy the conditions of economy and efficiency to achieve the maximum dry density for the compacted fill. The *methods of compaction* available are:

- Compaction in layers by using rollers — suitable for sandy and silty soils with moderate cohesion; vibratory rollers permit greater layer thickness (IS: 2720-PtVIII-1983).
- Compaction by sheepsfoot rollers, effective in breaking for clods and filling large void spaces— suitable for cohesive soils such as clayey sands (SC), clayey gravels (GC) and silt-sand and silt-gravel mixtures with significant clay fraction.
- Vibratory compaction is the most effective one for cohesion Bess soils (IS: 2720-PtXIV-1983); layer thickness should not exceed 60 cm. Poorly graded sands with Uniformity Coefficient less than 2 should not be used, to safeguard it against liquefaction under dynamic loads.

**General Points**

Long-term stability is vital in the case of cuttings.

1. Drainage of the formation is important.
2. Side slopes of embankments and cuttings should be protected by special measures, such as 'turfing', if necessary, to prevent erosion.
3. The borrow pits should be located sufficiently away from the toe of the embankment to prevent base failures.
4. Embankments in cohesive soils need special treatment for ensuring a stable formation.
5. High-fill construction in swampy and marshy lands needs special investigation.

**Unstable Formations and Treatment Methods**

Unstable formations cause the following problems:

- Increased maintenance due to variations in track levels, especially during extreme hot, cold, or wet seasons.
- Loss of ballast which sinks into weak formation.
- Slope failure through slips and consequent disruption of traffic.

**Defects and Failures in Formation**

Defects and failures arise in formations due to several reasons:

(i) Unstable embankments and cuttings result in excessive creep deformations in the track.
(ii) Swelling and shrinkage of soil in the formation result in heaving and unequal settlement of the formation, disturbing track levels and alignment.
(iii) Bearing capacity failures due to inadequate side slopes or cess width of the formation, or the ballast cushion.
(iv) Miscellaneous ways of losing formation soil – through ants, rodents, erosion through seepage, and so on.

The following precautions may be taken to avoid or at least minimise formation failure.

- Settlements under train loads should be less than 10 mm for the normal traffic loads expected
- Settlement is an important aspect which helps in checking the stability of a formation, evaluating the effectiveness of a treatment, and combining the behaviour of two subgrade materials.
- Any indication of instability of a formation shall be attended to as early as possible to:
  - eliminate the probability of disruption to traffic or speed restrictions;
  - maintain the quality of the track;
  - avoid untimely renewals of track components; and,
  - minimise track maintenance.
- For any modernisation of the track for high-speed traffic, a stable formation is an essential prerequisite.

**Treatment Methods for Formation Soils**

One or more of the following methods may be employed depending on the specific need:

(a) Provision of a 30 cm to 100 cm deep blanket of coarse grained material
(b) Grouting of formation with cement mortar or lime slurry under pressure (Grouting may serve only for a few years requiring attention again.)
(c) Provision of sub-surface drainage to stabilise the base, i.e., ballast and lankest
(d) Vertical sand drains
(e) Moorum blanket
(f) Use of geo-textiles as cut-offs.

The last three methods are used to tackle foundation problems but they may be used in this context also, if they are considered necessary at a particular location.

To summarise, the main causes of the defects in formations and the remedial measures found effective in each case are given in Table 3.3:

**Table 3.3** Causes and remedial measures for formation defects

| S. No. | Nature of defect | Cause | Action/Treatment |
|---|---|---|---|
| 1. | Variation in track level | (a) Deficient track structure<br>(b) Unstable foundation | Improve drainage<br>Replenish ballast after screening<br>Track components to be checked<br>Investigate the causes and evolve a treatment plan. |
| 2. | Unstable embankment or cutting | (a) Inadequate safety of slopes<br>(b) Inadequate cess width<br>(c) Improper subsurface drainage | Proper slopes and drainage to be provided.<br>Provision of sand layers. |
| 3. | Inadequate track foundations | Bearing capacity failures due to<br>(a) Inadequate ballast and blanket<br>(b) Inadequate cess | Provide<br>Necessary thickness for the blanket.<br>Proper cess width. |

Cont'd

Table 3.3 Cont'd

| S. No. | Nature of defect | Cause | Action/Treatment |
|---|---|---|---|
| 4. | Acute defects in summer season | Swelling and shrinkage of soils. | 1 m blanket for full cess width, and on top of slopes also, if necessary<br><br>Wider cess to keep the zone of moisture changes away from the soil mass supporting the track. |
| 5. | Loss of soil | Termites, ants, rats and porcupines. | Find solution from local agricultural and forest authorities.<br><br>Grouting of holes with clay. |

## RAILWAY TRACK DRAINAGE

Drainage of a railway track involves the interception, collection, and disposal of water from, upon, or under the track. It is accomplished by the adoption of a proper surface and sub-surface drainage system.

### Need for Track Drainage

Improper drainage adversely affects track maintenance. Excess water in the formation and/or ballast affects the stability of the embankment by significantly reducing the bearing power of the soil.

The following are the ill-effects of lack of proper track drainage:

- *Settlement*—differential settlement of portions of the embankment, variation in longitudinal and cross-levels affecting the safety and riding quality of the track.
- *Reduction in bearing capacity*—a maximum of 50 per cent reduction can occur for the soil under submerged condition; this can lead to instability and failure.
- *Embankment failure*—this is the consequence of reduction in bearing capacity due to the presence of excess water, which can lead to slip on the slopes.
- *Formation of ballast pockets*—the effect of impact loads from running trains results in ballast pockets, where ballast is pushed into the formation.
- *Shrinkage and swelling effects*—these occur in soils with poor bearing capacity in the presence of excess water; for example, in black cotton soils. This can lead to loss of ballast and uneven settlement.
- *Clogging of ballast*—the slush formed in the formation gets pushed up into the ballast, badly affecting its performance. This requires screening and renewal of ballast, with increased maintenance.

#### Sources of Water in the Track and Formation

- *Gravity water*: Rainwater moves into the subgrade by gravity. Provision of cross-fall, side drains and lowering of the cess are some measures to reduce this.

- *Capillary water*: Water can rise into the subgrade by capillary action from below. Provision of a pervious layer in the embankment can act as a capillary break.
- *Water from adjacent areas*: If water is present in adjacent low-lying areas, it can seep into the subgrade. This can be reduced by providing paved catch-water drains and inverted filters and underground drains.
- *Hygroscopic moisture from the atmosphere*—This is relatively small and has very little effect on the total moisture content of the soil.

## Requirements of a Good Drainage System

The following are the requirements of a good drainage system:

(i) Surface water from rain or from adjacent areas should not be allowed to percolate into the formation of the track.
(ii) Side drains should be properly designed so that the surface water is carried away.
(iii) The erosion of banks and their side slopes should be prevented by appropriate surface drainage measures.
(iv) The formation and subgrade should be of coarse-grained soils, so that providing surface drainage becomes relatively easy.
(v) The subsurface drainage arrangements should be designed such that water retained in the formation is collected and led away.
(vi) Longitudinal drains should be designed such that the water is finally led to an appropriate outfall, from where it can be drained off.
(vii) Special arrangements have to be made for drainage in waterlogged areas.
(viii) The drains and other arrangements require to be inspected regularly in order to clear them periodically for proper functioning of the system.

### Track Drainage Systems

Track drainage consists of two distinctive phases:

(a) Surface drainage
(b) Sub-surface drainage.

### Surface Drainage

Rain water and water from adjacent areas should be drained off by making provisions for well-designed and effective surface drains.

Surface drainage in mid-sections between railway stations consists of the following features:

- *Side drains*: These should be provided along the track in cutting, where cess level is not above the ground level. All drains must have adequate longitudinal gradient to allow free flow of water.
- *Lining of drains*: This is mandatory on trunk routes, and desirable on others, unless the drain is in rock strata.
- *Surplus ballast*: The shoulders of the track should be cleared of surplus ballast, as it retards drainage and encourages growth of vegetation.
- *Level crossings*: Adequate openings should be provided under level crossings in order to allow the water in the side drains in cuttings to flowout.

**Surface drainage in station yards**

The following are the guidelines:

- *Open surface drains*: To make inspection and cleaning convenient, the surface drains should be left open.
- *Drains between tracks*: Track drains should be saucer shaped, to collect water from both sides.
- *Drain should not be above cess level*: If it is inevitable, weep-holes are to be provided.
- *Network of longitudinal and cross-drains*: These are required to drain off storm water effectively.

**Sub-surface drainage:** There are various sources through which water reaches the sub-surface such as percolation of rainwater, seepage from adjacent areas and water bodies, and capillary water.

Various forms of sub-surface drainage systems are:

(i) *Inverted filter*: This is a blanket of cohesionless graded material, which acts as a capillary break. It can be made of gravel, coarse sand, fine sand, and silt from top to bottom and is provided underneath the ballast.

(ii) *Pairing of catch water drains*: This is an effective method of preventing entry of water into the subgrade.

(iii) *Installation of sand piles*: Sand piles of 20 cm diameter at close intervals will drain off water through them, preventing stagnation of water in the subgrade.

(iv) *Provision of perforated pipes*: Perforated pipes of 30 cm diameter at appropriate places will drain off water pockets in the subgrade.

(v) *Cement mortar grouting*: Cement mortar grouting under pressure will inhibit the collection and movement of water in the subgrade.

(vi) *Provision of a blanket under ballast*, which is also known as *sub-ballast*.

### Blanket or Sub-ballast

A blanket consists of an intervening layer of free draining material in the body of the embankment just underneath the ballast cushion. The blanket serves two functions:

- It prevents the entry of rainwater into the formation soil by draining it away to the sides.
- It minimises the puncturing of stone ballast into the formation and prevents loss of ballast.

The blanket should cover the entire formation width from the shoulders. The depth should be about 30 cm in clay soil. If the formation soil is weak, a thicker layer of 60 cm may be necessary. The material should be properly graded. The plastic fines should not be more than 5 per cent. Non-plastic fines can be allowed up to 12%

The uniformity coefficient should be between 4 and 7, or above 7. The coefficient of curvature should be between 1 and 3.

The following soils do not need a blanket:

- Rocky beds.
- Well-graded gravel (GW)
- Well-graded sand (SW)
- Soils conforming to the specifications of the blanket material.

The *minimum thickness of blanket* depends upon the nature of the subgrade soil — it ranges from 45 cm for poorly graded sand and gravel to 100 cm for silts and clays with plasticity.

The blanket or sub-ballast layer is given side slopes of 1 in 40 from the centre of the width to either side for draining off water.

## 3.5 TRACTION, TRACTIVE EFFORT AND HAULING CAPACITY

### Traction on Railways

The motive power needed for traction on railways is provided by the *locomotive*, or the railway engine. The locomotive is one of the three items of *rolling stock* on railways:

The other two are *coaches* and *wagons*.

Based on the energy source for drawing power, locomotives are of three types:

1. Steam locomotives
2. Diesel locomotives
3. Electric locomotives.

**(1) Steam locomotives:** These locomotives derive the motive power from steam generated in a boiler; the thermal energy of steam is converted into the mechanical energy needed for traction.

**(2) Diesel locomotives:** These derive the motive power from an internal combustion engine, called the 'diesel engine', and use high-speed diesel oil as fuel.

**(3) Electric locomotives:** These locomotives derive the power from an electric conductor running along the track; in other words, electrical energy is converted into mechanical energy needed for motion.

### Types of Traction

The three types of locomotives provide three types of traction on Indian Railways:

(a) Steam traction
(b) Diesel traction
(c) Electric traction.

Diesel and electric locomotives are more efficient than steam locomotives, with greater hauling capacity and better capability to carry heavy loads at high speeds. In view of this, diesel and electric traction have been gradually replacing steam traction; the fact that, out of a total of 10,499 locomotives, only 43 are steam, 5,633 are diesel and 4,823 are electric, on Indian Railways, substantiates this.

#### Steam Locomotives

A steam locomotive is identified by the arrangement of its wheels. It has two types of wheels: the driving wheels and the idle wheels. The driving wheels are coupled together and are directly connected with the pistons of the engine cylinder. With increase in diameter of the driving wheels, the tractive power increases, but the speed decreases. The hauling capacity is dependent on the number of driving wheels. The idle wheels merely serve the purpose of distributing the load of the locomotive on the track. Usually, idle wheels will be distributed ahead of the driving wheels and behind them. The three digit code such as 4-6-2 for the locomotive indicates four idle wheels in the front, six driving wheels, and two idle wheels after the driving wheels. (The number of axles will be half the number of the wheels.) A tender or a mini-wagon stores fuel and is coupled with the main locomotive.

### Diesel Locomotives

A four-stroke diesel engine is used for these. Depending upon the type of transmission of the power generated to the driving wheels—mechanical, hydraulic, or electric–diesel locomotives can be classified as the conventional type, the diesel–hydraulic type, or the diesel–electric type; the last type have become very popular on Indian Railways.

These are being manufactures at Diesel Loco Works, Varanasi.

### Electric Locomotives

Electric motors, which draw power from an overhead distribution system through pantographs mounted on the locomotives, are used to run electric locomotives. Different systems—1500 V dc, 750 V dc, 25 kV ac single phase and ac three-phase—are used.

While the best motors for electric traction are dc series motors, the most economical way to transmit power is to directly transmit ac power at industrial frequency. Indian Railways has locomotives which use the best of both systems—locomotives with dc series motors that take power at 25 kV ac single phase at a frequency of 50 Hz; the voltage is stepped down to 1500 V or 750 V by a transformer present in the locomotive, converted to dc by silicon rectifiers, and this power is fed to the dc series motors inside the locomotives.

The best of these locomotives are being produced at Chittaranjan Loco Works.

## Classification of Locomotives

According to the standard classification of Indian Railways, steam locomotives are designated by a two-letter code; the first letter indicates the gauge: X (or W) for BG and Y for MG, and the second letter indicates the service: A (or L) for Light Passenger, B (or P) for Standard Passenger, C for Heavy Passenger, D (or G) for Standard Goods, E for Heavy Goods, and M for Mixed Goods and Passenger.

For diesel and electric locomotives, one more alphabet—D, A, or C, is used to indicate diesel, ac electric or dc electric engines, respectively. Thus, XB indicates a Standard Passenger locomotive for BG, YC a Heavy Passenger locomotive for MG, XE a Heavy Goods locomotive for BG, and WAG a BG ac electric locomotive to haul a standard goods train. The wheel configuration for passenger locomotives is generally 4–6–2, and that for goods locomotives is 2–8–2.

The heaviest axle load on BG is 22.9 tonnes, and that on MG is 16.2 tonnes.

## Rolling Stock

*Rolling stock* includes, besides locomotives, passenger coaches, goods wagons, electric multiple units (EMUs), diesel multiple units (DMUs), and special wagons, such as BOX wagons.

### Coaches

Conventional coaches are for different classes such as I- and II-Sleeper, AC-Two tier, AC-Three tier, and AC-First.

EMUs are part of suburban trains. The three basic structural designs are:

(i) Integral coaches (ICF, Perambur, Chennai), built of light metal.

(ii) Integral coaches (BEML, Bengaluru), built of light metal.
(iii) Non-integral coaches (Indian Railways Standard Design-IRS), built with wooden body.

The coaches from ICF and BEML are the latest and are popular.

### Goods Wagons

These are primarily meant for goods traffic. They may be closed or open type—the latter type are more in number. Special BOX wagons are meant to carry coal and other bulk traffic.

(BOX—indicates bogie wagon of the open and high sided type.)

Bogie wagons, supported on more than four wheels, are now the most popular type on Indian Railways.

## Brake Systems

Brakes in the locomotive are applied to stop a moving train.

The two basic systems are:

(i) Compressed air brakes.
(ii) Vacuum brakes.

The air brake system gets a sluggish response from a long train; further, if the train gets divided into two parts for any reason, the rear portion does not get the power to operate the brakes, which can result in an accident.

However, the air brake system is considered superior in ensuring uniform application of the brakes along the entire train length, quicker process of deceleration, gradual retardation of the wagons, and a fast release of brakes throughout the train length to facilitate a quicker restart.

But, the vacuum brake system is used mostly on Indian Railways. Sometimes, a combination of the two systems could be used for locomotives hauling high-speed trains.

## Resistances to the Motion of a Train

Several resistances are encountered by a train in motion on a railway track. These are listed below:

(a) Friction during movement between the internal parts of locomotives and between the various parts of the coaches and wagons, as also the friction between the rails and the wheels of the rolling stock. (This force is independent of the speed of the train.)
(b) Resistance due to the wave action produced by the impact of the wheels on the rail, as also due to the track irregularities such as minute corrugations on the rails; and, the resistance due to the unevenness in the longitudinal and cross-levels. (This force is proportional to the speed of the train, and the weight of the train.)
(c) Atmospheric resistances to a train moving at a great speed. (This force is proportional to the weight of the train and to the square of speed of the train.)
(d) Resistances due to track geometry—Vertical alignment with a rising gradient and horizontal alignment with a curve.
(e) Resistances during the operation of a train—on starting and due to acceleration. These are usually encountered at stations. Resistance during starting depends on the weight of the locomotive, as also that of trailing vehicles. Resistance due to acceleration is dependent on the total weight of the train and the acceleration. (These are also dependent on weight of the train, and, on the gradient and degree of curvature.)

A train has to develop enough power to overcome all these resistances and keep itself in uniform motion at a constant velocity.

### Resistance due to Friction

This resistance is independent of speed. The sources are:

- **Journal friction:** This depends on the type of bearing, the nature of the lubricant, the condition and temperature of the bearing. In the case of roller bearings, this resistance varies from 0.5 to 1.0 kg per tonne of load.
- **Internal resistance:** This resistance is produced due to the movement of various parts of the locomotives, the coaches and the wagons.
- **Rolling resistance:** This is due to the interaction between the rail and the wheel during the motion of the train.

The total resistance due to friction, $R_f$, is given by

$$R_f = 0.0016W \tag{3.9}$$

where $W$ is the weight of the train in tonnes.
It is seen that this is independent of the speed of the train.

### Resistance due to Wave Action ($R_{wa}$)

When a train moves at a constant speed, a certain amount of resistance develops due to the wave action in the rail and due to track irregularities in longitudinal and cross-levels. These resistances vary with the speed of the train. Precise calculation of this component has not been possible, but the following formula has been suggested based on experience:

$$R_{wa} = 0.00008WV \tag{3.10}$$

where $W$ = Weight of the train in tonnes

$V$ = Speed of the train in km/h.

($R_{wa}$ will be obtained in tonnes.)

### Atmospheric Resistance due to Wind

Wind resistance consists of side resistance, head resistance and tail resistance. The exact magnitude of resistance depends upon the size and shape of the vehicle, its speed, as also the velocity and direction of the wind. The exposed area of the vehicle and the wind's velocity and direction govern the wind resistance to the motion of the train (Fig. 3.56):

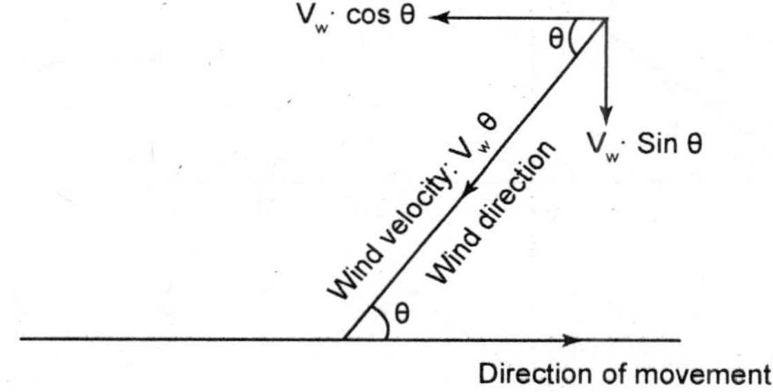


**FIG. 3.56** *Atmospheric resistance due to wind*

Wind pressure on the vehicle is the maximum when it acts at **60°** to the direction of motion of the train. Atmospheric wind resistance, $R_{at}$, may be got from the following empirical formula:

$$R_{at} = 0.000017\ AV^2$$

where $A$ is the area of the vehicle exposed to the wind (m²),
and $V$ is the wind velocity in km/h.

However, the relative velocity of the wind with respect to the vehicle is more relevant since the resistance actually varies with the square of the velocity.

The following empirical formula evolved on the basis of research studies is used:

$$R_{at} = 0.000\,0006\ WV^2 \tag{3.11}$$

where $W$ is the weight of the train in tonnes and $V$ the velocity of the train in km/h. ($R_{at}$ is obtained in tonnes.)

### Resistances Due to Track Geometry

Resistances due to track geometry include resistances developed due to the gradient in vertical alignment and the curves in horizontal alignment.

**(i) Resistance due to gradient:** To negotiate a rising gradient, a train requires extra effort in order to overcome gravity.

Assuming that a wheel load of W moves on a rising gradient AB as shown in Fig. 3.57, the forces acting on the wheel are:

- Wheel load $W$, acting downward

The normal component, $W_n$, acts on the rail, perpendicular to AB.

The component parallel to the gradient is to be counteracted by the resistance due to gradient, $R_g$.

ΔPQR is the triangle of forces. ΔABC and ΔPQR are similar.

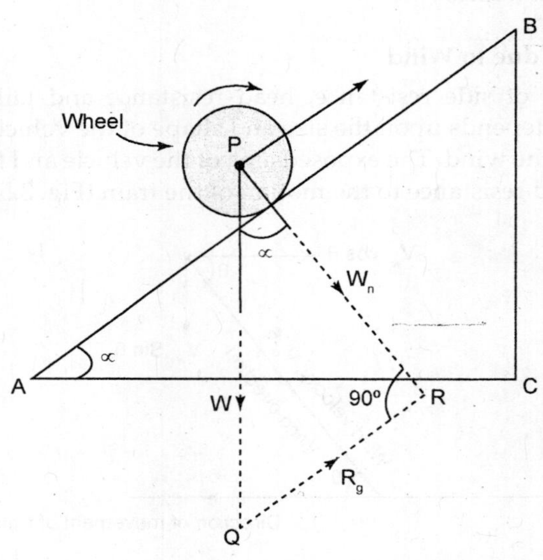


**FIG. 3.57** *Resistance due to gradient*

$$R_g = W.\sin\text{–} = W.\frac{BC}{AB}$$

For flat gradients common in railways,
AB is approximately equal to AC.

$$\therefore \quad R_g = W.\frac{BC}{AC} = W.\tan,$$

$$\therefore \quad R_g = \frac{W \times g}{100}, \; g \text{ being the gradient in \%.}$$

Hence, the resistance due to gradient,

$$R_g = W.\frac{g}{100} \tag{3.12}$$

(If $W$ is in tonnes, $R_g$ is also obtained in tonnes).

[Note: When the train moves down a falling gradient, gravity aids the motion.]

**(ii) Resistance due to curvature of track:** This is also called *curve resistance*. When a train negotiates a horizontal curve, resistance is caused for the following reasons:

- The rigid wheel base of the vehicle does not allow it to adapt to the curved track; the frame takes up a tangential position as shown in Fig. 3.58. The flange of the outer wheel of the leading axle rubs against the inner face of the outer rail, thus triggering resistance to motion.
- Longitudinal slip may also occur because of the tendency of forward motion of the wheels on a curved track. The outer flange of the trailing axle remains clear and tends to derail. The position will be acute with longer wheel base on sharp curves.
- Transverse slip, which increases the friction between the wheel flange and the rails, could be another source of resistance on curves.
- Inadequate superelevation, or cant, increases the pressure on the outer rail; similarly, excess cant puts more pressure on the inner rails. These effects also could cause resistance.

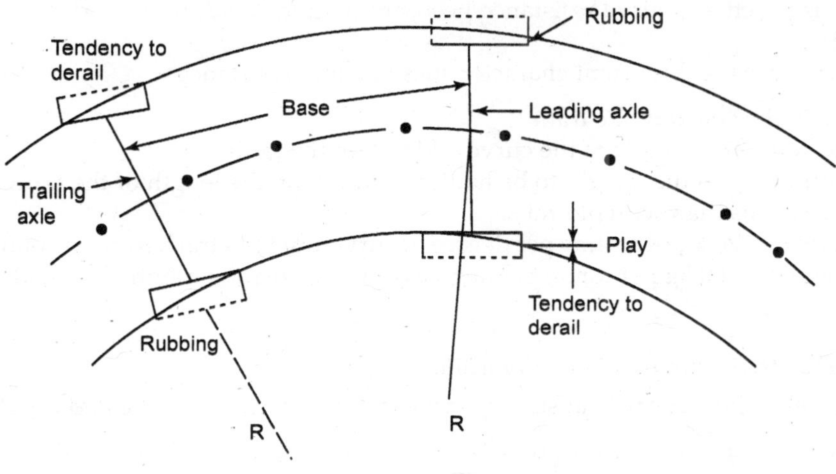


**FIG. 3.58** *Resistance due to curvature*

- Poor track maintenance—bad alignment, worn-out rails and improper levels can also increase the curve resistance.

Curve resistance can be determined from the following equation:

Curve resistance, $$R_c = C.\frac{FG}{R} \tag{3.13}$$

where $F$ = force of sliding friction

$G$ = gauge of the track,

$R$ = mean radius of the curve, and,

$C$ = a constant, which depends upon several factors.

$F$ = the force of sliding friction, depends upon the coefficient of sliding friction.

Eq. (3.13) is derived from the work done by sliding friction in negotiating the extra distance to be covered by the outer wheels than that by the inner wheels for the same central angle.

The following empirical formulae are used for determining the curve resistance:

For BG, $$R_c = 0.0004\ W{\cdot}D \tag{3.14}$$

For MG, $$R_c = 0.0003\ W{\cdot}D \tag{3.15}$$

In these equations,
$W$ is the weight of the train in tonnes, and
$D$ is the degree of the curve.

**Grade compensation on curves:** Curve resistance is often compensated by a reduction in the gradient. The effect of curve resistance is thus translated in terms of resistance due to gradient.

The compensation recommended is 0.04 percent on BG and 0.03 percent on MG for every degree of curvature.

(Thus the compensation needed may be got by multiplying by the degree of the curve.)

When this is practised, curve resistance need not be calculated; only grade resistance has to be applied.

The following are the important characteristics of curve resistance

- Increases with the speed of the train
- Depends on the central angle of the curve and not on the radius
- Depends primarily on the weight to be hauled, and not on the length of the train.
- Less with new rails than with old rails.
- Can be as low as 0.20 kg per tonne per degree of curvature for a track in good condition, and may be as high as 1.00 kg per tonne per degree of curvature for a rough track with worn-out rails.

### Resistances Due to Starting and Acceleration

These resistances occur generally at stations, where trains start, accelerate and decelerate.

**Resistance on starting ($R_s$):** This is given by

$$R_s = 0.15\ W_l + 0.005\ W_v \tag{3.16}$$

where $W_l$ = Weight of locomotive (tonnes), and

$W_v$ = Weight of trailing vehicles (tonnes).

**Resistance due to acceleration:** This is given by

$$R_{ac} = 0.028\,a\,W \tag{3.17}$$

where $W$ = total weight of the train = $(W_l + W_v)$ in tonnes,

and $a$ = acceleration, (Km/h/s).

The various resistances are summarised in Table 3.4:

**Table 3.4** Resistances to the motion of a train

| S. No. | Nature of resistance | Value of resistance | Remarks |
|---|---|---|---|
| 1. | Friction ($R_f$) | $0.0016\,W$ | Independent of speed. |
| 2. | Wave action, track irregularities, etc. ($R_{wa}$) | $0.00008\,WV$ | Dependent on speed. |
| 3. | Atmospheric, due to wind ($R_{at}$) | $0.000\,0006\,WV^2$ | Depend on the square of speed. |
| 4. | Gradient ($R_g$) | $W.\frac{g}{100}$ | Dependent on slope. |
| 5. | Curvature of track ($R_c$) | BG: $0.0004\,WD$<br>MG: $0.0003\,WD$ | Dependent on degree of curve. |
| 6. | Starting ($R_s$) | $0.15\,W_l + 0.005\,W_v$ | Dependent on weight. |
| 7. | Acceleration ($R_{ac}$) | $0.028\,aW$ | Dependent on weight and acceleration. |

## Tractive Effort of a Locomotive

The tractive effort of a locomotive is the force it can generate for hauling a train. This should be sufficient to haul a train at the maximum permissible speed.

Tractive effort is, in general, equal to or a little greater than the hauling capacity of the locomotive. The assessment of the tractive effort depends upon the type of locomotive.

### Steam Locomotive

The tractive effort ($T_e$) can be calculated, in this case, by equating the total power generated by the steam engine to the work done by the driving wheels.

Work done by a two-cylinder steam engine = $2p \times A \times 2L$

where $p$ = Difference in steam pressure between the two sides of the cylinder,

$A$ = Area of the piston, and

$L$ = Length of the stroke.

In terms of $d$, the diameter of the piston, this will be

$$\text{Work done} = 2p \times \left(\frac{\pi d^2}{4}\right) \times 2L = \pi P d^2 L \tag{3.18}$$

Work done in one revolution of the driving wheel of the locomotive

$$= \text{Tractive effort} \times \text{Circumference of the wheel}$$

$$= T_e \times \pi D \tag{3.19}$$

Equating (3.18) and (3.19),

$$\pi p d^2 \cdot L = T_e \cdot \pi D$$

$$\Rightarrow \quad T_e = \frac{p \cdot d^2 \cdot L}{D} \tag{3.20}$$

It is obvious that the tractive effort decreases with an increase in the diameter of the driving wheel of the locomotive.

**Diesel Locomotive**

Tractive effort of a diesel locomotive can be got from the empirical formula:

$$T_e = \frac{308 \times \text{RHP}}{V} \text{ (kg)} \tag{3.21}$$

where RHP is the rated horse-power of the engine, and $V$ is the velocity in km/h.

**Electric Locomotive**

The tractive effort of an electric locomotive varies inversely with a certain power of the speed.

The empirical formulae used are:

For an ac electric locomotive:

$$T_e = \frac{a}{V^5} \tag{3.22}$$

For a dc electric locomotive:

$$T_e = \frac{a}{V^3} \tag{3.23}$$

$a$ here is a constant which depends upon certain characteristics of the locomotive.

## Hauling Power of a Locomotive

The hauling power of a locomotive depends upon the weight exerted on the deriving wheels and the friction between the wheel and the rail.

The coefficient of friction depends upon the condition of the rail surface and the speed of the locomotive. It has an inverse relationship with the velocity and direct relationship with the roughness of the rail surface.

For high speeds it is about 0.1 and for low speeds, it is about 0.2.

Wetness, dryness, and the degree of roughness of the rail surface affect the friction.

$$\text{Hauling Power} = \begin{pmatrix}\text{No. of pairs}\\ \text{of driving wheels}\end{pmatrix} \times \begin{pmatrix}\text{Weight exerted on}\\ \text{each driving axle}\end{pmatrix} \times (\text{Coefficient of friction}) \quad (3.24)$$

**Comparison of the Types of Traction**

The important characteristics of the three types of traction are given in the Table 3.5:

**Table 3.5** A comparison of steam, diesel and electric traction

| Characteristics | Steam traction | Diesel traction | Electric traction |
|---|---|---|---|
| | **I Design characteristics** | | |
| 1. Design | Simple design of engine, but bulky and heavy. | Design is not simple, but the weight of the engine is less | Design is complicated but relatively low weight |
| 2. Tractive effort | Low, as torque is not uniform. | Higher as torque is uniform. | Even greater tractive effort may be obtained. |
| 3. Ratio of HP and weight | 75 kg per HP. | 45 kg per HP. | 25 kg per HP. |
| 4. Thermal efficiency | About 7% | About 25% | About 90% |
| 5. Reversal | Requires a turn table for reversal. | Reversal not required. Only driver and guard have to interchange positions. | Reversing is not required. |
| 6. Life of locomotive | About 40 years. | About 30 years. | Over 30 years. |
| | **II Operational characteristics** | | |
| 1. Smoke and fire | Both smoke and fire present. | No fire and little smoke. | No smoke and no fire. |
| 2. Promptness of service | Takes time for raising enough steam. | Ready service. | Ready service. |
| 3. Monthly kilometerage | 3,500 km | 9,00 km | 10,000 km |
| 4. Speed | Only low speeds possible. | Higher speeds possible. | Very high speeds, even on steep grades because of high overload capacity. |
| 5. Rate of acceleration. | Very low | Better | Best |
| 6. Haulage | Only a limited number of coaches. | Greater number of coaches. | Large number of coaches. |
| 7. Track condition | Damage to track due to hammer blow action. | No damage to track. | Smooth run; no damage to track. |
| 8. Repairs and renewals | Frequent | Relatively less | Minimum |
| 9. Future prospects | Steam locos have been phased out on Indian Railways. | Bright prospects; more diesel locos are being produced. | Best prospects; increasing number of electric locos are being produced. |

## 3.6 SPECIFICATIONS FOR TRACKS ON INDIAN RAILWAYS

A railway track consists of several components such as rails, sleepers, track fittings and fastenings, as well as ballast.

The Railway Board provides the necessary specifications for the various components of BG and MG tracks of Indian Railways.

### Track Structure on Indian Railways

The following track structure has been recommended for Indian Railways:

**Broad Gauge Routes**

| S. No. | Components | Traffic density GMT | Specification (A, B, C routes) | Remarks |
|---|---|---|---|---|
| 1. | Rails | More than 20 | 60 kg | 90 UTS |
| | | 10–20 | 60 kg | 52/90 for D&E- routes. |
| | | 5–10 | 60 kg or 52/90 | 60 kg for A-routes |
| | | Less than 5 | 52/90 | Second hand for E-routes. |
| 2. | (a) Sleepers | ... | Concrete (RCC/PSC) or Steel | Concrete, steel and CST-9 for D & E-routes. |
| | (b) Sleeper density (Number/km) | More than 20 | 1660 | - |
| | | 10–20 | 1660 | 1540 for D & E-routes |
| | | Less than 10 | 1540 | 1660 for A-routes |
| | | Loop lines | 1340 | - |
| 3. | Fastenings | ... | Elastic | Conventional also for C, D and E-routes. |
| 4. | Ballast Cushion | ... | 300 mm | A-route |
| | | | 250mm | B and C-routes |
| | | | 200 mm | D-routes |

**Metre Gauge Routes**

| S. No. | Components | Traffic density GMT | Specification | Remarks |
|---|---|---|---|---|
| 1. | Rails | More than 5 | 90 R new | – |
| | | 2.5 to 5 | 90 R (SH)/75 R (new) | – |
| | | 1.5 to 2.5 | 50 kg SH/90 R (SH) | 90 R (SH) to 75 R (SH) for S-routes |
| 2. | (a) Sleepers | " | Concrete steel CST–9 | For all Routes |
| | (b) Sleepers density | " | (M + 7) | (M + 3) for S-routes |
| 3. | Ballast cushion | " | 300 mm/200mm | For R-routes |
| | | | 200 mm/ 150 mm | For S-routes |

## General Features of Indian Railways

Indian Railways is the second largest railway system (after Russian Railways) in the world, wholly owned by the Government of India.

### Track

It consist of rails, sleepers, rail fastenings and track fittings, ballast, and the formation. This is the most important and major asset of Indian Railways.

### Locomotives

Indian Railways have a total of 10,499 locomotives, of which only 43 are steam locomotives and the rest are diesel and electric locomotives.

In about five decades the number of steam locomotives has been phased out from over 10,800 to the present insignificant number. The operational efficiency in both passenger and freight traffic has therefore, improved significantly.

### Electrification and Electric Traction

Electric traction was first introduced in 1925 on a small section of the Bombay area (now Mumbai). This was on 1500 V dc system. Until 1957, electric traction was confined only to the suburban sections of Bombay and Madras.

Electrification of Howrah–Burdwan suburban section of Calcutta was completed in 1958. Thereafter, electrification continued to cover the trunk routes, connecting Calcutta, Delhi, Bombay and Madras, in a phased manner. The system was 25 kV single phase AC.

Presently, over 21,614 km of routes have been electrified, constituting about 32 percent of the total rout-kilometerage.

Electric locomotives based on the 25 kV single-phase AC system are being produced at the Chittaranjan loco works.

### Diesel Traction

Diesel traction was started in a small way prior to the Second World War with the introduction of diesel shunting engines. From 1956, diesel traction was extended to high-density routes for speedy transportation of freight and passenger traffic.

Diesel traction based on new generation technology, with the 4000 HP, three-phase AC-diesel locomotives, has been possible with the diesel locomotives with state-of-the-art features being produced at the Diesel Locomotive Works, Varanasi. Passenger locomotives have a speed potential of 130 km/h.

### Rolling Stock

Rolling stock of Indian Railways is being modernised by producing new designs of fuel-efficient locomotives, high-speed coaches, and high-speed bogies for freight traffic. The bogie wagons – BCN, BTPN, and BOXN type – have a higher payload and speed potential for optimal utilisation of track capacity.

### Railway Safety

Indian Railways has been striving to improve the standard of railway safety. This has been possible through the Railway Safety Fund, new technologies like anti-collision devices, continuous track circuiting and such other means, modernisation of infrastructure—track, rolling stock and signalling equipment, and preventive maintenance.

**Some Interesting Facts of Indian Railways**

1. Only-rack railway: The Nilgiri Mountain Railway.
2. Fastest train: Shatabdi Express from New Delhi to Bhopal. (150 km/h)
3. Slowest train: Nilgiri Railway train from Mettupalayam to Udhagamandalam (10.24 km/h).
4. Oldest working locomotive: The Fair Queen, the Steam locomotive built in England in 1855 (service between Delhi cantt. and Alwar)
5. Railway Station at the highest altitude: Ghoom on the Darjeeling Himalayan Railway (2,258 m)
6. Section with maximum bridges and tunnels: Kalka–Shimla route (102 tunnels and 809 bridges).
7. Longest platform: Kharagpur (833 m)
8. Longest tunnel: Kharbude tunnel on Konkan Railway between Mumbai and Goa (645 m)
9. Longest bridge: Delhi-on-Sone bridge over some river near Sasaram on the Kolkata–Delhi main line (3.06 km long).
10. Tallest viaduct: Panvel Nadi Viaduct on Konkan Railway (64 m high)
11. Largest railway yard: Mughalsarai (near Varanasi)
12. UNESCO World Heritage Sites:

    The Darjeeling Himalayan Railway from

    (a) New Jalpaiguri to Darjeeling.
    (b) Kalka–Shimla Railway

13. Longest passenger train: The New Delhi—Prayag Raj Express with 24/26 coaches.
14. The world's tallest railway bridge is under construction in the Kashmir valley on the Chenab river, as a part of the Udhampur–Srinagar–Baramulla railway project of the Indian Railways. It is expected to reach a height of 359 m.

Of the total length of 1,315 m, the main arch span of the bridge will be 480 m. The deep Chenab river valley being prone to high wind pressure, the bridge is designed to withstand very high wind velocities.

The special blast-proof steel being used for the bridge is 63 mm-thick, and the concrete pillars are designed to make the bridge terrorism-proof. The location is in the Seismic Zone V of the Himalayas, and is therefore designed to be earthquake resistant.

The bridge is expected to be completed by March 2016 and will have a life span of 120 years.

## 3.7 STRESSES IN RAILWAY TRACK

A railway track is designed to transfer the forces produced by rolling wheel loads on the rail head, in a safe way; this takes place successively through rail, fastenings, sleeper, ballast, and ultimately to the formation. The track is not firmly embedded in the ground unlike the other civil engineering structures; in view of this, engineers consider a track to be floating on the foundation. Because of this peculiar feature, the theoretical analysis of track stresses is considered to be rather approximate. The variables involved are numerous and a number of assumptions have to be made to obtain the values of stresses.

But a usable analytical approach is certainly valuable in the comparison and validation of experimental data.

In the Indian Railways, the following parameters are considered for assessing the strength of a track for speed clearance and such other purposes:

(i) Stresses in the rail
(ii) Rail-wheel contact stresses
(iii) Dynamic overload due to unsuspended masses
(iv) Fish plate and bolt hole stresses
(v) Sleeper stresses
(vi) Ballast stresses
(vii) Pressures in the formation or subgrade.

### Stresses in the Rail

The basis for computation of stresses in the rail is the elastic theory, which assumes that the track support is elastic, that is, the bed consisting of the sleepers, the ballast and the formation taken together is considered to be elastic.

The path of the uniformly loaded wheels will be horizontal, and the track will be depressed on the approach of the wheel and recovers as soon as the wheel has passed. As a consequence, the rail bends in a continuous curve between the wheels and not between the sleepers. Due to continuity of the track, there will be both downward and upward bending, with a point of contra-flexure in between these zones.

### Relief of Stress

The flexural effects of adjoining wheels overlap. The net bending moment in the rail at a particular point is the difference between the bending moments at that point caused by the consecutive wheel loads. Consequently, the bending moment caused by one wheel at any point in the rail is reduced by the reverse bending moment of the wheel loads on its either side. This reduction in bending moment reduces the longitudinal bending stress in the rail section.

This is known as *relief of stress*.

## Forces Acting on the Track

The following are the various forces acting on the track:

(a) Vertical loads—static loads, dynamic effects including the effect of speed, the hammer blow effect including the steam effect, the inertia of reciprocating parts, and so on.
(b) Longitudinal forces due to tractive effort, braking forces, thermal forces, etc.
(c) Lateral forces due to eccentric vertical loading, shunting separations, and the movement of live load.
(d) Stresses due to wheel and rail contact.
(e) Stresses due to surface defects of rails.

### Virtual Wheel Load or Talbot Load

The actual static wheel load has to be modified to allow for the dynamic or impact effect caused by the factors mentioned in (a) above. The effect of adjoining wheels is also to be considered. Therefore, the instantaneous load applied on the rail is much higher than the static wheel load.

The net wheel load obtained after modifications is called the *virtual wheel load*, or *Talbot load*, after the famous American track engineer, Prof. A. N. Talbot, who did pioneering work in the subject of track stresses.

This net wheel load is used to calculate the stresses in the rail.

### Track Modulus

The elasticity of the track is characterised by what is called the *track modulus (U)*. This is similar to the modulus of elasticity, which is an index of the resistance to deformation of a material.

Track Modulus is defined as the *load per unit length of the rail required to produce a unit depression of the track*. This factor is used in the determination of the track stresses.

From extensive research conducted to determine the track modulus, it has been observed that the depression immediately below the load is not directly proportional to the load in the entire load range. Due to gaps between the rail and sleeper, between sleeper and ballast, and the voids in the ballast, the track depression is disproportionately higher in the initial stages of loading. These voids and gaps get closed under a certain initial load, and thereafter, further depression per unit load is smaller and is in truly elastic range. It is found that an initial load of 4 tonnes for BG and 3 tonnes for MG causes higher depressions, beyond which the truly elastic range commences.

Thus, it is found that a single value of track modulus cannot completely define the load-depression behaviour of a track. A linear load-depression relationship is assumed in the initial stage, which yields a track-modulus value, called the *Initial Track Modulus* ($U_i$); the linear relationship after this in the truly elastic range yields another value, called the *Elastic Track Modulus* ($U_e$). This double-moduli approach, considered to be more realistic, is used in calculating track stresses on Indian Railways. The track modulus varies with the gauge, weight of rail section, the type of sleepers and sleeper density, and the ballast cushion.

The track modulus values recommended by Indian Railways is given in Table 3.6

**Table 3.6** Track modulus values

| Gauge | Track Standard | Initial values $U_i$ (kg/cm²) | Elastic values $U_e$ (kg/cm²) |
|---|---|---|---|
| BG | 90 R rails, (N + 3) sleeper density and 200 mm- ballast cushion. | 75 | 300 |
| BG | 52-kg rails, (N + 6) Sleeper density and 250 mm—ballast cushion | 120 | 380 |
| MG | 50 R, 60 R, 75 R rails, (N + 3) Sleeper density, and 200 mm — ballast cushion | 50 | 250 |
| BG LWR track | 52 kg/60kg rails, concrete Sleepers, 250 mm — ballast cushion | 183.7 | 419.1 |

**Note:** For other sleeper densities, the track modulus values are taken to be inversely proportional to the intermediate sleeper spacing.

### Dynamic Augment of Vertical Loads – Speed Factor

Because of vertical impact due to speed and track irregularities, and the rail vibrations, the instantaneous dynamic wheel load is much more than the static wheel load. The dynamic load is obtained by increasing the static load, by an incremental amount given by the *speed factor* (or *the impact factor*).

The following formula was used on Indian Railways till 1965 for the speed factor (SF):

$$SF = \frac{V}{18.2\sqrt{U}} \tag{3.25}$$

where $V$ = Speed in km/h,

and $U$ = Track modulus in kg/cm².

Since 1966, the *German formula* given by Schram has been adopted, which is given below:

- For speeds up to 100 km/h:

$$SF = \frac{V^2}{30,000} \tag{3.26a}$$

- For speeds above 100 km/h:

$$SF = \frac{4V^2}{10^5} - \frac{1.5V^3}{10^7} \tag{3.26b}$$

($V$ is the speed in km/h)

[For example, at a speed of 100 km/h and $U$-value of 100 kg/cm², the Indian formula gives a speed factor of 55%, where as the German formula gives a value of 33%.]

Graphs showing the speed factor for different speeds for various types of rolling stock have been plotted.

These are used for the design of track components for the maximum loads that are expected on a given route for the conditions of operation.

### Effect of the Leading Wheel – Lift of Rail

When a wheel load is applied on a rail, it causes certain depression, and a wave-like effect is created with the passage of the train. As explained earlier under *Rail Creep*, the portion of the rail immediately under the wheel gets depressed due to the wheel load, and the portion ahead of the wheel gets lifted in a wavy formation. The analysis of the stresses in the rail is based on the assumption that the rail is supported on a number of sleepers elastically. Due to this elasticity, it follows that negative reactions and negative bending moments can develop at these supports. However, in practice, a small lift in the rail is always possible due to loose fastenings; this lift in the rail will increase the bending stress in the rail by about 10 per cent.

The effect of this lift will be negligible if the wheels of the vehicle are closely spaced; the lift, in this case, is suppressed by the leading wheel.

In view of this, the effect of the lift in the rail on bending stress is considered only when the lead axle is at a distance greater than six times the distance of the point of contra-flexure from the load.

### Hammer Blow Effect

The revolving masses in the driving and coupled wheels of a steam locomotive, such as crank pins and connecting rods, are completely balanced by counter weights placed diametrically opposite to them. But the reciprocating masses of the piston, piston rod and cross head, by virtue of inertia and oscillatory movement, produce alternating forces, and tend to cause the locomotive to oscillate sideways. To reduce this tendency, a weight is placed at the opposite side of the crank. While the horizontal component of the centrifugal force of this added weight produces the desired effect, the vertical component causes variation of the wheel pressure on the rail. This is called the *hammer blow*.

This effect occurs only in the case of steam locomotives. In the course of rotation of the wheel, this effect gets reversed in sign and effect.

### Steam Effect

In a steam locomotive, steam pressure acts on the piston and is transmitted to the driving wheels through the connecting rod. The vertical component of the force on crank pins and connecting rod is at an angle to the piston rod.

This causes what is known as the *steam effect*. This effect does not synchronise with the hammer blow effect due to overbalance, and adds to it only during some part of the revolution of the crank shaft.

### Inertia of Reciprocating Masses

The reciprocating masses, due to their inertia and acceleration, alter the forces on the piston and in the connecting rod during the revolution of the wheel.

The maximum combined force of the hammer blow, the steam effect, and inertia of each driving wheel, and the hammer blow effect of the coupled wheels do not act simultaneously due to the phase-difference in the angular positions of the counterweights in the wheels.

Curves may be plotted for these three effects for one complete revolution of the wheel; the maximum combined effect is got by summing the values from the three curves taking the sign of each force duly into consideration.

### Longitudinal Bending Stresses in the Rail

These stresses occur due to vertical loads on the rail. Determination of these stresses is based on the assumption that the rail is a long beam continuously supported by an elastic foundation. The bending moment diagram for one half of the rail between consecutive supports is shown schematically in Fig. 3.59:

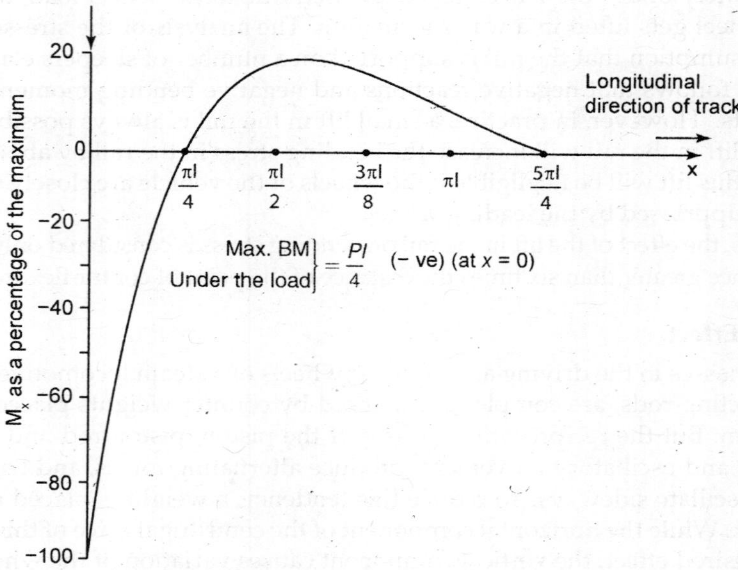


**FIG. 3.59** *Bending moment in the rail due to wheel load*

The bending moment, $M$, is given by

$$M = 0.25Pe^{\frac{-X}{1}}\left(\sin\frac{x}{1} - \cos\frac{x}{1}\right) \tag{3.27}$$

where $P$ = wheel load,

$EI$ = flexural rigidity of the rail section,

$l = \left(\frac{EI}{U}\right)^{\frac{1}{4}}$, the characteristic length,

$U$ = Track modulus, and

$x$ = distance of the point from the load.

From Eq. (3.27), the bending moment is zero at points where $x = \frac{\pi l}{4}, \frac{5\pi l}{4}, \ldots,$

and maximum at points $x = 0, \frac{\pi l}{2}, \frac{3\pi l}{2}, \ldots$

Now, the procedure for the determination of the stresses in the rail can be summarised as follows:

Let $X_{i,e}$ = the distance from the load to the point of contra-flexure (cm) (*i*: for initial load, *e*: for elastic range)

$M_o$ = the bending moment immediately under an isolated wheel load P (cm·t)

$f_c$ = compressive stress in the rail head (t/cm²)

$f_t$ = tensile stress in the rail foot (t/cm²)

$P$ = wheel load (t)

$U_{i,e}$ = track modulus kg/cm² (i: initial load range, e: elastic range)

$I_{xx}$ = vertical moment of inertia in rail section (cm⁴)

$Z_c$ = Section modulus of rail in compression (cm³)

$Z_t$ = Section modulus of rail tension (cm³)

$d$ = depression (or deflection) of track (cm)

The relevant formulae are:

$$X_{i,e} = 42.33\left(\frac{I_{XX}}{U_{i,e}}\right) \text{cm} \tag{3.28}$$

$$M_o = 0.319\, P \times X_{i,e} \text{ cm·t} \tag{3.29}$$

$$f_c = \frac{M_0}{Z_c} \text{ t/cm}^2 \tag{3.30}$$

$$f_t = \frac{M_o}{Z_t} \text{ t/cm}^2 \tag{3.31}$$

$$d = \frac{9.25P}{\left(I_{xx}.U_{i,e}^{3}\right)^{\frac{1}{4}}} \text{ cm} \tag{3.32}$$

The permissible value of bending stress due to a vertical load is 23.5 kg/mm² (or 2.35 t/cm²). The relief factor for bending moments and that for depressions due to the adjacent wheel loads may be obtained from the charts shown in Fig. 3.60.

The procedure for the determination of bending stresses under vertical bending due to the wheel loads and that for the use of the above chart will be illustrated by means of a numerical example in the next section.

### Stresses Due to the Eccentricity of Vertical Load

If the vertical load, $P$, acts at an eccentricity of $e$ on the head of the rail, it produces bending in the vertical plane and twisting of the rail section. The torque is taken to act at the middle of the rail length when held at the ends.

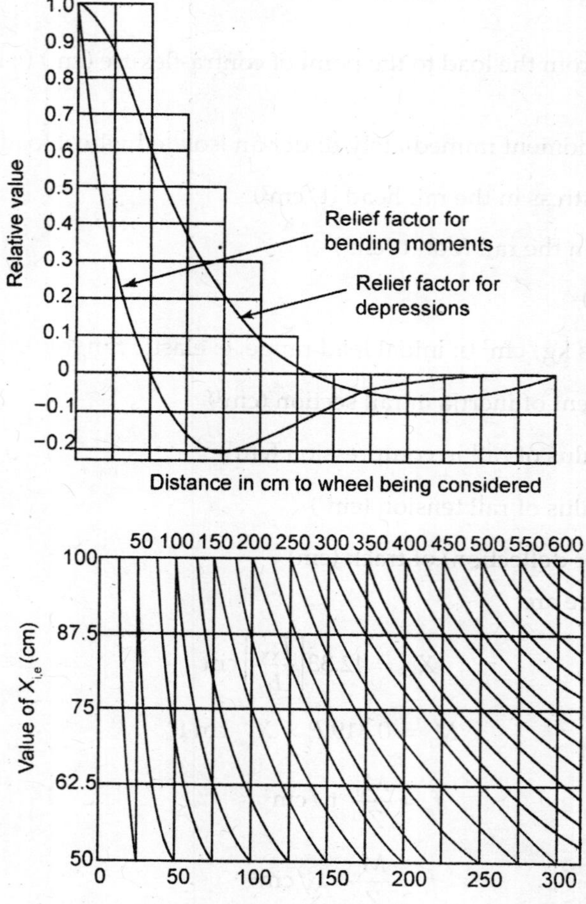


**FIG. 3.60** *Relief factors for bending moments and depressions in rail*

This causes additional bending of head and foot of the rail. These bending stresses can be calculated by assuming that the maximum bending moment is half of the torque.

### Longitudinal Forces on the Rail

Longitudinal stresses are developed in the rail due to tractive effort of the locomotive and its braking force. An approximation of these forces is given below:

(i) **Tractive effort** (for alternating current)—30 to 40% of the weight of the locomotive.
(ii) **Braking force**—15 to 20% of the weight of the locomotive plus 10–15% of trailing load.

**Temperature stresses:** These are also longitudinal stresses. Tensile stresses are induced in winter due to contraction while compressive stresses are produced in summer due to expansion. The extreme values can be 10.75 kg/mm$^2$ in winter and 9.5 kg/mm$^2$ in summer.

### Stresses Due to Surface Defects

A flat on the wheel or a low spot in the rail causes additional stress in the rail section. Empirical studies indicate an additional deflection of about 1.5 times the depth of the flat or low spot occurs at a critical speed of about 30 km/h. An additional bending moment is caused due to this, and results in additional stresses in the rail section.

### Rail Wheel Contact Stress

The maximum contact shear stress ($F_{max}$) at the contact point between the wheel and the rail is given by the empirical formula

$$F_{max}\ (\text{kg/mm}^2) = 4.13\sqrt{\frac{P}{R}} \qquad (3.33)$$

where $P$ = static wheel load (kg) ( + 1000 kg on curves),

and $R$ = radius of the worn-out wheel. (mm). The maximum permissible value is 21.6 kg/mm$^2$.

### Resultant Stress

The resultant stress from all the forces acting on the rail has to be checked against the maximum permissible stress in the rail steel.

### Stresses in Sleepers

Sleepers are subjected to several forces—static and dynamic. Wheel loads are transmitted to the sleepers from both the rails, forces from the fastenings, design, dimensions and strength of sleeper material.

The distribution of pressure under the sleeper is difficult to determine; the pattern depends on the sleeper characteristics as well as on the firmness of packing under the sleeper. As the ballast yields under the load, pressure distribution under the sleeper varies with the standards of maintenance.

The two extreme conditions based on the packing of ballast are:

**End-bound sleeper:** The ballast under the sleeper and the ends of the sleeper are well compacted and hard-packed, for a newly compacted one. Under these conditions, the

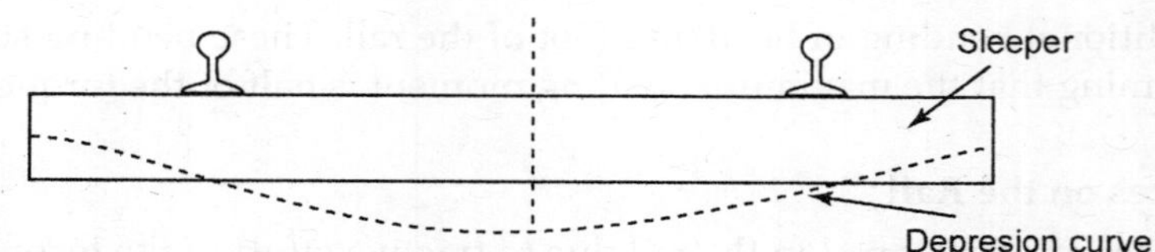


**FIG. 3.61** *End-bound sleeper*

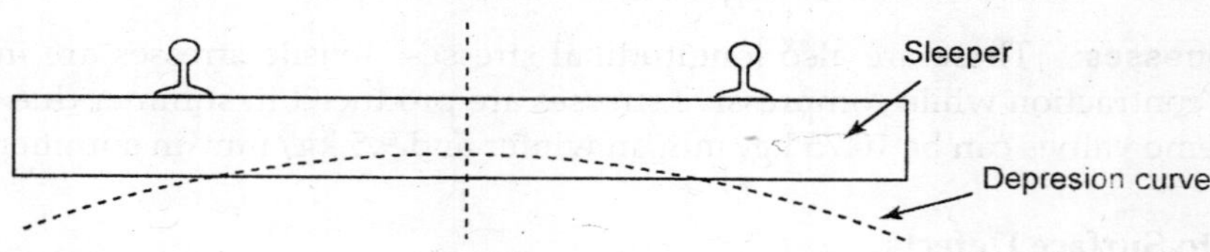


**FIG. 3.62** *Centre-bound sleeper*

deflection of the sleeper under the action of wheel loads is greater at the centre than at the ends; this is called an *end-bound sleeper* and can be represented as shown in Fig. 3.61.

**Centre-bound sleeper:** As trains pass over the track, the packing of ballast under the sleeper, especially under the rail seat, tends to become loose because of the hammering action of the moving loads. Loose packing may add to this effect, and packing at the middle becoming a harder than that under the rail seat. This results in greater depression at the ends than at the middle. This is called a *centre-bound sleeper* (Fig. 3.62).

Based on the elastic theory, the maximum load on the rail seat, $P_s$, is given by

$$P_s = \frac{PS}{Zl} \tag{3.34}$$

where $P$ = Wheel load,

$S$ = Sleeper spacing,

$Z$ = Modulus of section for the rail, and

$l$ = Characteristic length $\left[=\left(\frac{EI}{U}\right)^{\frac{1}{4}}\right]$

The maximum load on the rail seat is 30 to 50% of the dynamic wheel load, depending upon the packing under the sleeper besides various other factors.

Provision of bearing plates under the rail seats is considered to reduce the stresses in the sleeper.

## Stresses on Ballast

The load from the rail is transmitted to the sleeper, from which in turn, it is transferred to the ballast. The efficiency of transmission depends on the elastic characteristics of the sleeper including its shape and size, and the ballast cushion and degree of compaction under the sleeper.

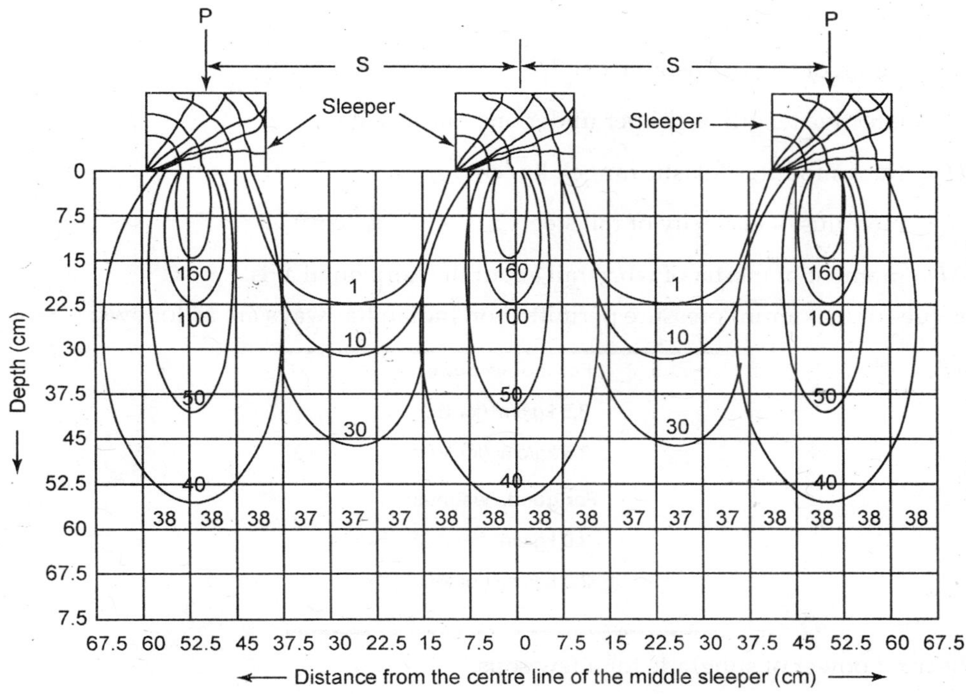


**FIG. 3.63** *Stresses in the ballast (After Prof. Talbot)*

Prof. A. N. Talbot has analysed the pressure distribution under the ballast under the sleeper and has found that the pressure distribution would be bulb-shaped (Fig. 3.63).

The important conclusions from this figure are:

(i) The stress in the sleeper is maximum at the centre of its width; it decreases from the centre towards the ends.

(ii) The pressure under the sleeper is uniform at a depth approximately equal to the spacing between the sleepers.

## Pressure on Formation or Subgrade

The loads from the trains and the track structure are finally transmitted to the subgrade, which has to carry it without shear failure or undue settlement.

The spacing between the sleepers, the ballast characteristics such as the size, cushion and compaction under the sleeper, and the nature of the subgrade—all these factors play a significant role in the distribution of pressure on the subgrade.

The maximum pressure, $p_f$ (max.), on the formation is calculated from the following formula:

$$p_f\,(\text{max.}) = \frac{2PS}{\pi DL}\left(\frac{U_e}{64EI}\right)^{\frac{1}{4}} \tag{3.35}$$

where $P$ = live wheel load, tonnes

$S$ = sleeper spacing, cm

$D$ = depth of ballast cushion, cm

$L$ = effective length of sleeper under one rail seat, cm

$U_e$ = track modulus (elastic range), kg/cm$^2$.

$E$ = modulus of elasticity of rail steel ($2.11 \times 10^6$ kg/cm$^2$)

$I$ = moment of inertia of worn rail about the horizontal axis.

The values of maximum pressure permitted on Indian Railways are as follows:

| |
|---|
| **For motive power:** |
| 3.5 kg/cm$^2$ (for BG) |
| 2.5 kg/cm$^2$ (for MG) |
| **For goods wagons:** |
| 3.0 kg/cm$^2$ (for BG) |
| 2.3 kg/cm$^2$ (for MG) |

The bearing power of subgrade for a few soils:

| | |
|---|---|
| Soft clay/Loose sand | 1.10 kg/cm$^2$ (approx.) |
| Dry clay/Dense sand | 1.150 kg/cm$^2$ |
| Compacted soil | Around 3 kg/cm$^2$ |

## 3.8 NUMERICAL EXAMPLES

### EXAMPLE 3.1

A train with a total weight of 1200 tonnes ascends a gradient of 1 in 250. Determine the resistance due to gradient encountered by the train.

**Solution**

Grade resistance, $$R_g = \frac{W \times g\%}{100}$$

$$W = 1200 \text{ t}$$

$$g = \frac{1}{250} = 0.4\%$$

$$\therefore \quad R_g = \frac{1200 \times 1}{250}$$

$$= \underline{4.80 \text{ t}}$$

## EXAMPLE 3.2

Determine the curve resistance when a train when a train of weight 1200 t negotiates a 6°-curve on a BG tracks.

### Solution

Curve resistance, $R_c = 0.0004 \times WD$ (BG)

$W = 1200$ t

$D = 6$

$\therefore$ $R_c = 0.0004 \times 1200 \times 6 = \underline{2.88 \text{ t}}$

## EXAMPLE 3.3

What should be the gradient for a BG track, if the grade resistance together with the curve resistance due to a 4° curve is equal to the resistance due to the ruling gradient of 1 in 200?

### Solution

Let $W$ be the weight of train and the required gradient $g$ %.

Curve resistance for a 4°-curve on BG $= 0.0004 \times 4 \times W$

$$\text{Grade resistance for a grade of } g\% = \frac{Wg}{100}$$

$$\text{Grade resistance for the ruling gradient} = \frac{W}{200}$$

Then,
$$\frac{W}{200} = \frac{Wg}{100} + 0.0016\,W$$

$$g = \frac{100}{W}(0.0050\,W - 0.0016\,W) = 0.34\%$$

or
$$g = \frac{0.34}{100} = \frac{1}{300} \text{ or } \underline{1 \text{ in } 300}.$$

## EXAMPLE 3.4

A locomotive on a MG track has three pairs of driving wheels, each carrying 22.5 tonnes. Determine the maximum load it can haul on a level track with a curvature of 3° at 75 km/h.

### Solution

$$V = 75 \text{ km/h}$$

Assuming the coefficient of friction as 1/6, the hauling capacity of the locomotive in this case is given by

$$\frac{1}{6} \times 3 \times 22.5 \text{ t} = 11.25 \text{ t}$$

The train resistances are (in terms of the weight of train, $W$):

$$R_f = 0.0016\ W;$$

$$R_{wa} = 0.00008\ W \times 75 = 0.0060\ W$$

$$R_{at} = 0.0000006 \times WV^2 = 0.0000006 \times 75 \times 75\ W = 0.0033W$$

$$R_c = 0.0003\ \text{WD (for MG)} = 0.0003 \times 3\ W = 0.0009\ W$$

$$\text{Total train resistance} = (0.0016 + 0.0060 + 0.0033 + 0.0009)\ W = 0.0118\ W$$

Equating this to the hauling capacity of 11.25 t,

$$0.018\ W = 11.25$$

$$\Rightarrow \quad W = \frac{11.25}{0.0118}\ t = 953\ t$$

The maximum load it can haul, under these conditions, is 950 tonnes.

## EXAMPLE 3.5

Determine the steepest gradient which a train of 18 wagons and a locomotive can negotiate, with the following data: weight of each wagon = 18 tonnes; weight of locomotive = 150 tonnes; tractive effort of the locomotive = 3 kg/tonne and that of the wagons = 2.5 kg/tone. The speed of the train is 50 km/h

### Solution

$$V = 50\ \text{km/h}$$

(i) Total rolling resistance is the sum of that of the locomotive and of the wagons.

$\therefore$ Rolling resistance ($R_f$) = (150 × 3 + 18 × 18 × 2.5) kg = 1260 kg

$\therefore$ $R_f = 1.26$ t

(ii) Total train resistance = $R_{wa} + R_f + R_{at} + R_g$

Weight of train, $W = 150 + 18 \times 18 = 574$ t

$\therefore$ Total train resistance

$$= 0.00008 \times 574 \times 50 + 1.26 + 0.0000006 \times 574 \times 50^2 + \frac{574 \times g}{100}, \ g \text{ being percent grade.}$$

Equating this to the tractive effort of 18 tonnes,

$$18 = 4.417 + 5.74\ g$$

$$\therefore \quad g = 2.366\% \approx 1 \text{ in } 42$$

The steepest gradient it can negotiate is therefore 1 in 42.

## EXAMPLE 3.6

Calculate the maximum permissible train load that can be hauled by a locomotive with four pairs of driving wheels with an axle load of 20 tonnes each on a BG track with a ruling gradient of 1 in 200 and a maximum curvature of 3°, running at a speed of 60 km/h. Assume the coefficient of friction as 0.2.

**Solution**

$$V = 60 \text{ km/h}$$

(i) Hauling capacity of the locomotive

$$= 4 \times 20 \times 0.2 = 16 \text{ tonnes.}$$

(ii) Total train resistance

$$= R_f + R_{wa} + R_{at} + R_g + R_c$$

$$= 0.0016\ W + 0.0008 \times 60\ W + 0.0000006 \times 60^2\ W + W \times \frac{1}{200} + 0.0004 \times 3 \times W$$

$$= 0.01476\ W$$

Equating the hauling capacity to the total train resistance,

$$0.01476\ W = 16$$

$$\Rightarrow \quad W = \frac{16}{0.01476} = 1084 \text{ t}$$

$\therefore$ The maximum permissible train load may be taken as <u>1080 tonnes</u>.

## EXAMPLE 3.7

(i) Calculate the maximum permissible load that a BG locomotive with three pairs of driving wheels with an axle load of 20 tonnes each can pull on a straight level track at a speed of 75 km/h.

(ii) Also determine the reduction in speed if it has to negotiate a rising gradient of 1 in 250.

(iii) What would be the further reduction in speed if the train has to negotiate a 3° curve on this gradient? Assume the friction coefficient as 0.2.

**Solution**

$$f = 0.2 \text{ and } V = 75 \text{ km/h}$$

(i) Hauling capacity = 3 × 20 × 0.2 = 12 tonnes.

Total train resistance on a straight level track at a speed of 75 km/h is given by

$$= R_f + R_{wa} + B_{at}$$

$$= 0.0016\ W + 0.00008 \times 75\ W + 0.0000006 \times 75^2\ W$$

$$= 0.010975\ W$$

Equating this to the hauling capacity and solving for $W$,

$$W = \frac{12}{0.01975} = \underline{1090 \text{ tonnes}}.$$

(ii) On a rising gradient, the additional resistance is

$$W \times \frac{1}{250} = 0.004\ W$$

$\therefore$ $$12 = W(0.0016 + 0.00008\,V + 0.0000006\,V^2 + 0.004)$$

$$= 1090\,(0.0016 + 0.00008\,V + 0.0000006\,V^2 + 0.004)$$

Solving this equation for $V$, $V = 49$ km/h

$\therefore$ Reduction in speed = (75–49) = 26 km/h

(iii) On a curve of 3° on a rising gradient of 1 in 250, curve resistance $= 0.0004 \times 3\,W = 0.0012\,W$

$\therefore$ $$12 = W(0.0016 + 0.00008\,V + 0.0000006\,V^2 + 0.004 + 0.0012)$$

$$= 1090\,(0.0068 + 0.00008\,V + 0.0000006\,V^2)$$

Solving for $V$, $V = 40$ km/h.

$\therefore$ Further reduction in speed = (49–40) = 9 km/h

## EXAMPLE 3.8

Calculate the minimum theoretical length of a LWR beyond which the central portion of the rail would not be subjected to any thermal expansion, given:

Rail section: 52 kg/m
Cross-sectional area: 65 cm²
Coefficient of thermal expansion of rail steel = $1.15 \times 10^{-5}$ /°C
Rise in temperature after laying the track = 30°C
Modulus of elasticity of rail steel = $20 \times 10^5$ kg/cm²
Sleeper spacing = 65 cm
Average restraining force per sleeper = 300 kg.

**Solution**

$$\alpha = 1.15 \times 10^{-5}\ /°\text{C} \qquad E = 20 \times 10\ \text{kg/cm}^2$$

$$t = 30°\text{C} \qquad A = 65\ \text{cm}^2$$

Thermal force, $$F = \alpha \cdot t \cdot A \cdot E$$

$$= 1.15 \times 10^{-5} \times 30 \times 65 \times 20 \times 10^5\ \text{kg} = 44,850\ \text{kg}$$

Resistance offered per sleeper = 300 kg

Number of sleepers required to restrain a force of 44,850 kg is $\dfrac{44,850}{300} \approx 150$

Breathing length at each end for a sleeper spacing of 0.65 m is 150 × 0.65 = 97.5 m =100 m (approx.) at either end.

Total breathing length on both sides (presuming from portion at the middle) = 100 × 2 = 200 m.

Therefore, the minimum theoretical length of LWR is 200 m. This will prevent creep.

## EXAMPLE 3.9

Calculate the bending stresses in the rail due to a wagon mooring at a speed of 80 km/h on a 52kg-rail, with the following data:

Wheel load = 10 t; Distance to the adjacent wheel = 150 cm

Initial track modulus, $U_i = 120$ kg/cm$^2$

Elastic track modulus, $U_e = 420$ kg/cm$^2$

Moment of inertia of worn out rail section, $I_{xx} = 1924$ cm$^4$

Section modulus in compression, $Z_c = 240$ cm$^3$

Section modulus in tension, $Z_t = 256$ cm$^3$

**Solution**

(i) Calculate $X_i$, the distance of the point of contra-flexure from the load:

$$X_{i,e} = 42.33\left(\frac{I_{xx}}{U_i}\right)^{1/4}$$

For the initial load

$$X_i = 42.33\left(\frac{1924}{120}\right)^{\frac{1}{4}} = \underline{85.10 \text{ cm}}$$

For the remaining load in the elastic range:

$$X_e = 42.33\left(\frac{1924}{420}\right)^{\frac{1}{4}} = \underline{61.93 \text{ cm}}$$

(ii) Speed factor (or impact factor or dynamic augmentation):

$$\text{The speed factor, SF} = \frac{V^2}{30{,}000} = \frac{80^2}{30{,}000} = 0.2133 \text{ (for } V_x < 100 \text{ km/h)}$$

The dynamic wheel load = 10 × 1.2133 = 12.133 t

(iii) Effect of adjacent wheel:

Bending moment under the wheel load ($P$) is given by eq. 3.29:

$$M_o = 0.319\, P \times X_{i,e}$$

The relief factors for bending moment from Fig. 3.60 are given below:

For a distance of 150 cm:

Relief factor for $X_i = 85.10$ cm is 0.20, and

Relief factor for $X_e = 61.93$ cm is 0.18 (both negative)

Therefore, the effect of adjacent wheel will be as given below:

For an initial load of 4t; 4 × (–0.20) = –0.80 t

For the remaining load: 8.133 × (–0.18) = –1.46 t

(iv) Determination of the virtual wheel load or Talbot load:

Talbot load for an initial load of 4t

$$4 - 0.80 = 3.20 \text{ t}$$

Talbot load for the remaining load of 8.133 t:

$$8.133 - 1.160 = 6.673 \text{ t}$$

$$6X_i = 510.6 \text{ cm} \quad 6X_c = 372 \text{ cm}$$

Since the distance to the adjacent wheel is less than $6X_i/6X_e$, the effect of the leading axle need not be considered.

$$\text{Maximum Talbot load} = (3.200 + 6.673) \text{ t}$$

$$T_1 = \underline{9.873 \text{ t}}$$

(v) Bending moment in the rail

$$M_o = 0.319 \times T_{li} \times X_i + 0.319 \times T_{le} \times X_e$$

$$= 0.319 \times 3.20 \times 85.10 + 0.319 \times 6.673 \times 61.93$$

$$= 86.87 + 131.83 = \underline{218.70 \text{ cm·t}}$$

(vi) Stress in the rail head (compressions)

$$= \frac{M_o}{Z_c} = \frac{218.70}{2.56} = \underline{0.911 \text{ t/cm}^2}$$

Stress in the rail foot (tensile)

$$= \frac{M_o}{Z_t} = \frac{218.70}{2.56} = \underline{0.854 \text{ t/cm}^2}$$

## SUMMARY

- Railway track consists of two parallel rails, supported on cross-sleepers and embedded in ballast material, which is supported on a formation or subgrade of soil. This is also called the Permanent way in contrast to temporary lines laid for transport of materials.
- The horizontal distance between the inner or running faces of the rails is called the Gauge of the railway. There are three gauges – the broad gauge (BG), the metre Gauge (MG), and the narrow Gauges on Indian Railways. The broad gauge (1.676 m) and the metre gauge (1.000 m) are the more popular gauges, with the latter being gradually converted to BG. A standard gauge of 1.435 m is common in several countries.
- Railway routes are classified as Trunk routes, Main lines and Branch lines based on their importance and traffic. Based on speed, they are classified as A, B, C, D, E, D spl and E spl on BG and Q, R, $R_1$, $R_2$, $R_3$, and S on MG.
- Choice of gauge is based on economic, traffic and terrain considerations. For historical reasons, India has a multi-gauge system, which involves a lot of difficulty in trans-shipment of passengers and goods. Hence, Indian Railways have decided to adopt the uni-gauge policy, which is being implemented in stages. Consequently, over 90% of the route kilometerage is on broad gauge now, and the remaining portions are expected to be completed soon, except for hill routes.
- The track structure of rails and sleeper is supported on a cushion of ballast to absorb the impact loads from the wheels of trains, and transmit and distribute them to the formation or soil subgrade such that the pressure does not exceed the safe bearing power.
- The most popular rail section is the flat-footed rails called the Vignole rail; this is the standard section on Indian Railways-52 kg/m or 90 lb/yd (90R) sections for BG, and 75 R and 60R sections for MG.

- Rails are manufactured from special rail steel, obtained from the open-hearth furnace or Duplex process. The ultimate tensile strength should be a minimum of 880 MPa.
- The length of rails is governed by manufacturing, transporting and handling convenience, besides expansion gap considerations at the ends. The standard length is 13 m for BG and 12 m for MG. Rail joints are usually of the fish-plated type. A joint being a weak point in the track, it is has to be designed well – the three types are supported type (on the sleeper), suspended type (cantilevering from the sleepers at either end), and the bridge type (supported by a bridge plate to the sleepers at either end). Rail joints are staggered on horizontal curves.
- Rail joints cause discomfort to the passengers and also are weak points in the track, structurally speaking. In order to avoid, or at least minimise the discomfort, a few rail lengths are welded together (Short-welded rail – SWR). Significant lengths of track are now being welded on Indian Railways. Long-welded rails (LWR) of a few kilometres (along with switch expansion joints at the ends), and even continuous welded rails (CWR) for several kilometres are being used.
  In the case of LWR'S small portions at either end, which are free to expand, are called the 'breathing lengths.'
- Several types of defects occur in the rails during operation – rail wear, end-batter, corrugations (which cause unpleasant sound when trains run, the rails being then called 'roaring rails'), corrosion, and kinks. These lead to rail failure beyond a point. Rail creep is another important defect, which depends on many factors. Rail creep causes many ill-effects including the jamming of expansion gaps and buckling, if this happens beyond a limit.
  Adjustment of creep should be done in the initial stages by pulling back the rails to control the damage. A few measures such as the provision of creep anchors tend to reduce rail creep. Good maintenance of the track also controls it.
  Certain criteria for rail wear (such as 6% loss of weight and others) are used to indicate the need for rail renewal.
- The main function of sleepers is to provide proper support to the rails and elastically reduce the impact of the dynamic wheel load and transmit them to the bed of ballast in which they are embedded. The common types are of wood or timber, steel, cast iron, and concrete (RCC or PSC). Timber or wooden sleepers need seasoning and also treatment, in some cases with creosote oil pressure grouting, for increasing their durability.
  Steel trough type sleepers are also used; cast-iron sleepers of the pot, plate or box type have been used, the latest being CST-9 type.
  Concrete sleepers have better life and provide stronger track by virtue of their weight. Reinforced cement concrete sleepers, as well as pre-stressed concrete sleepers have become very popular on Indian Railways. Precast pre-tensioned type pre-stressed concrete sleepers are the most popular ones (mono-block type).
- Sleeper density is expressed as $(N + x)$ where $N$ is the number of metres in rail length; $(N + 4)$ on BG means $(13 + 4)$ or 17 sleepers per rail length. The more the sleeper density, the better for track stability.
  This is also expressed as the number of sleepers per kilometre length of the track.
  Based on this, the spacing of sleepers is decided. This need not be uniform, being lesser at special locations like rail joints, horizontal curves (and) points and crossings.
- Rail fastenings and track fittings are required to connect rail to rail, rail to sleepers, and other such purposes.
  Fish-plates, spikes, bolts, loose jaws and keys, chairs, and elastic fastenings such as Pandrol clips, malleable CI inserts, bearing plates and grooved rubber pads are some of the more important fittings and fastenings.

- Coning of wheels, tilting of rails, and adzing of sleepers are popular measures to increase stability of operation in running the vehicles.
- Ballast holds the sleepers in position, provides a firm bed, facilities drainage of track, absorbs shocks and distributes the loads on to larger area of formation for providing stability. Broken stone, moorum, kankar, and even sand can be used as permanent ballast materials; coal ash or cinder, broken brick and selected earth material may be used as temporary ballast for new tracks or for sidings and loop lines.
  Ballast cushion below the sleepers contributes to the strength of the track and boxing at their edges gives lateral rigidity.
  Specifications are prescribed and tests for suitability recommended for a material to be used as ballast. Of late, ballast-less tracks have been designed and used on bridge decks and tunnels, which is the trend for modern railway tracks.
- The formation or soil subgrade is meant to bear the pressure transmitted by the ballast from above and distribute it to the natural soil below such that neither shear failure nor excessive settlement occurs.
  If the subgrade soil is weak, it can be strengthened by treatment; several methods are available, for example, sand piles, lime columns, cement grouting, and so on, to improve drainage and increase the strength.
- Railway track drainage is an important and integral part of track design and construction. Surface drainage (example: side drains) and sub-surface drainage (example: previous blankets and inverted fillers) are the two components in this.
- Tractive effort is the power generated by the locomotive to pull the train; it depends on the engine characteristics and type of traction – steam, diesel, or electric.
  Hauling capacity depends upon the number of driving axles and the weight imposed on them, and the coefficient of friction. Tractive effort is equal to or a little greater than the hauling capacity. Rolling stock comprises the locomotives, coaches and wagons.
- A train encounters several kinds of resistances to its motion. Some of the important ones are: friction or rolling resistance (independent of the speed), speed resistance due to wave action (varies as the weight of the train), gradient resistance on a rising grade, curve resistance on a horizontal curve (varies as the weight of train and the degree of curvature), and inertia resistance during starting and acceleration (depend on the weight).
  Grade compensation or reduction in gradient on horizontal curves is practised to nullify the effect of curve resistance.
  The hauling power will be either equal or a little less than the tractive effort of the locomotive. Diesel and electric locomotives has greater tractive power than steam locomotives; the latter are being phased out with their number reduced to a negligible percentage of the total stock of locomotives.
- Indian Railways have prescribed standard specifications for the track structure on broad gauge and metre gauge tracks on trunk routes, main and branch lines of different classes; these cover the various components such as the rails, sleepers, fastenings, and ballast.
- Stresses in the railway track and its components are difficult to be analysed due to the peculiar nature of the track being supported continuously on an elastic bed. Reasonably good approximations may be got based on this theory and on the assumption of double-modulus—initial track modulus up to a certain initial load (around 4 t for BG) below which depressions are disproportionately high, modulus being low, and the elastic track modulus beyond this load with smaller depressions. The initial value is much smaller than the elastic value.

A speed or impact factor is used to augment the wheel load to take into account the dynamic nature of loads. Bending moments (considering stress relief due to leading axle, if applicable), and bending stresses in the rail head (compression) and the rail foot (tension) may be calculated. Stresses under sleepers depend upon the nature of packing of ballast below it. The sleeper may be end-bound or centre-bound. Pressure on formation may be calculated based on wheel load, sleeper spacing, track modulus and the flexural rigidity of rail section, which should be less than the safe bearing power of the soil.

## QUESTIONS AND PROBLEMS

1. Why is a railway track called the Permanent Way? What are the ideal requirements of a Permanent Way?
2. State the basic functions of the various components of a railway track, and mark the components on a typical cross-section of the track.
3. Define the 'gauge' of a railway. List the gauges adopted in India, stating the reasons for multiplicity of gauges.
4. Mention the disadvantages of non-uniformity of gauges in a railway network.
5. Explain the uni-gauge policy of Indian Railways and the advantages claimed. What are the indications of the progress of the uni-gauge policy?
6. Draw typical cross-sections of a railway track for the following conditions:

   BG single-line in embankment

   BG double-line in cutting.

   BG single-line in embankment in an electrified section.
7. What are the considerations in the choice of gauge for a new railway line?
8. Write brief notes on:

   Coning of wheels and tilting of rails

   Adzing of sleepers

   Rail steel.
9. What are functions and requirements of rails in a railway track?
10. What are the considerations in the determination of the rail section and length of rail?
11. Classify the different types of defects in rails, and briefly explain them.
12. Write short notes on:

    Roaring rails

    Short-welded rails

    Long-welded rails.
13. Sketch and explain a fish-plates rail joint.
14. Describe the types of rail joints with sketches

15. Write short notes on:

    Dog spike and screw spike

    Bearing plate

    Pandrol chip

16. Give the specification for the track structure of a BG Trunk route in India.
17. What are the functions of sleepers in a railway track?
18. What are the advantages of timber sleepers? How are wooden sleepers treated to enhance their service life?
19. Write short notes on:

    Seasoning wooden sleeper

    Steel trough sleeper

    CST-9 sleeper.

20. Explain 'sleeper density'. What sleeper density is recommended for a broad gauge track with fast traffic?
21. What are the requirements of a good ballast material? Describe a good material for permanent ballasting.
22. What is screening of ballast? When should the ballast be renewed?
23. Write brief notes on:

    Ballast-less track

    Mono-block pre-tensioned pre-stressed concrete sleepers

    Drainage of a railway track

    Loading gauge and construction gauge.

24. What is the purpose of 'formation' in a railway track?

    How is a weak formation material improved? Explain any one method.

25. (a) What are the different types of railway traction used on Indian Railways?

    (b) What are the components of rolling stock?

26. (a) Differentiate between the tractive effort and the hauling capacity of a locomotive.

    (b) Explain the resistance due to friction encountered by a moving train.

27. Write brief notes on:

    Curve resistance

    Grade compensation

    Atmosphere resistance due to wind.

28. Explain the double-module theory for determination of the stresses in a railway track.
29. Write short notes on:

    Dynamic effect of wheel loads and speed factor

Relief of stress due to leading

Stresses in the rail head axle and rail foot

30. Write brief notes on:

    End-bound and centre-bound sleepers

    Stresses in ballast

    Pressure on the sub grade.

31. A train of total weight 900 tonnes negotiates an 8º curve on a MG line. What is the curve resistance encountered by the train?

32. What would be the gradient on a BG track when the grade resistance together with the curve resistance due to a curve of 3º curvature equals the ruling gradient of 1 in 150?

33. A locomotive has a hauling capacity of 1200 kg. What load can it pull on a straight level track at a speed of 96 km/h?

    What should be the reduction in speed it if has to pull this load on a rising grade of 1 in 180?

34. A locomotive has four pairs of driving wheels with an axle load of 22.5 tonnes. How much load can it pull on a straight level track at a speed of 80 km/h?

    If the train has to ascend a grade of 1 in 160 with this load, how much speed reduction is necessary?

35. Determine the steepest gradient on a straight track, which a train with 20 wagons can negotiate with the following data:

    Weight of wagon = 20 tonnes

    Rolling resistance of wagon = 2.5 kg/tonne

    Speed of the train = 60 km/h

    Weight of locomotive = 120 tonnes

    Tractive of effect of the locomotive = 12 tonnes

    Rolling resistance of the locomotive = 3.5 kg/tonne

36. Calculate the compensated gradient on a BG track with a 4º Curvature on a ruling gradient of 1 in 250

37. Calculate the minimum theoretical length of a LWR beyond which the central portion of the rail would not be subjected to thermal expansion, given the following:

    Cross-sectional area of the rail = 66 $cm^2$

    Coefficient of thermal expansion of rail

    Steel = $1.1 \times 10^{-3}$/°C

    Temperature variation after laying the track = 22°C

    Modulus of elasticity of rail steel = $2.2 \times 10^6$ kg/$cm^2$

    Sleeper spacing = 72cm

    Average restraining force per sleeper per rail =300 kg.

38. The wheel load of a vehicle is 11 tonnes with an adjacent wheel at 220 cm carrying the same load. Track modulus = 75 kg/cm$^2$. The moment of inertia of the rail section = 1650 cm$^4$

    Determine the bending moment under the wheel load.

    Determine the stresses in the rail head and rail foot if the modulus of section in compression and tension are respectively 220 cm$^3$ and 242 cm$^3$.

39. (a) Illustrate with sketches the various fittings for

    Wooden sleepers

    C.I. Sleepers

    Elastic fastenings

    (b) Discuses the merits and demerits of each. (JNTU-H)

40. Give a brief description of various gauges adopted in our country and bring out the importance of having a uniform gauge throughout. (JUTH-H)

41. Write short notes on:

    (a) Spikes

    (b) Blocks

    (c) Chairs and keys

    (d) Fish plates and fish bolts. (JUTH-H)

42. (a) Write about the importance of the following in a railway track:

    Size of ballast

    Quality of ballast

    Depth of ballast

    Screening of ballast. (JNTU-H)

    (b) Discuss the rail fastenings used in Indian Railway system. (JNTU-H)

43. (a) Discuss the factors on which sleeper density depend and how it is expressed for fish plated track and for welded track. (JUTH-H)

    (b) An MG track has a sleeper density of (M + 5). If the rail length is 13 m, what is the number of sleepers under one rail length? (JUTH-H)

44. Enumerate the factors to be considered in the design of fish plates. Give suitable sketches of a rail joint and indicate how such design considerations are implemented in practice. (JUTH-H)

45. With the help of a typical diagram, explain the components of a permanent way and also discuss about their functions in the railway track. (JUTH-H)

46. Explain the various theories used to explain the creep of rails with neat sketches. How do we prevent creep? (JUTH-H)

47. (a) Discuss the factors which govern the choice of gauge of a railway.

    (b) Define creep. What are the causes and effects of creep? (JUTH-H)

48. (a) What are the requirements of a good ballast? Mention the different types of ballast used in permanent way.

(b) Calculate the quantity of materials required for the construction of BG track of length 19.500 m with the rail section of 52 kg/m and standard length of 13 m. Take sleeper density as $(n + 4)$.

(c) Determine the maximum train load that can be handled by a locomotive having four pairs of driving wheels of an axle load of 28 tonnes each. On a straight track, the train runs at a speed of 90 km/h. Also determine the reduction of speed if the train has to run on an upward gradient of 1 in 150. If the train moves on this upward gradient with 3°-curve, what would be the reduction of speed?

$$\mu = \frac{1}{6}$$

49. (a) Define a permanent way and mention its components.

(b) What are the different types of rails? What are the advantages of using FF rails?

(c) The sleeper density of a BG track is $(n + 6)$. If the track is laid with welded rails 26 m long, what is the number of sleepers required on the rail length? (Kerala Univ. & UT Tech. Univ).

50. What is meant by wear of rails? How do you classify the Wear? Discuss the various causes of wear and suggest suitable measures to reduce the wear of wails. (Kerala Univ.)

51. What should be the length of track (i) to overcome the temperature stress, (ii) to prevent creep for equilibrium $A = 80$ cm$^2$, $\alpha = 1.5 \times 10^{-5}$ / °C, $E = 25 \times 10^5$ kg/cm$^2$, and rise in temperature = 30°C. Assume 500 kg/m as resistance (Kerala Univ.)

52. (a) Write a short note on sleepers.

(b) Explain loading gauge and construction gauge.

(c) Discuss adzing of sleepers.

(d) Why are heavy rails preferred to light rails?

(e) Discuss the use of ballast less track. (Punjab Tech. Univ.)

53. (a) Explain and draw the neat sketches of rail sections.

(b) Discuss the advantages of pre-stressed concrete sleepers over R.C.C. sleepers. Draw a stretch of mono-block pre-stressed concrete sleeper. (Punjab Tech. Univ.)

54. (a) Draw a typical cross-section of a permanent way on embankment. Describe the requirements of an ideal permanent way.

(b) Discuss any one theory of creep of rails.

(c) What are the requirements of fish plates? Discuss the reasons for failure of fish plates. (UP Tech. Univ.)

55. (a) What is the minimum depth of a ballast section? Explain also the screening of ballast.

(b) Give any four laboratory tests for getting the physical properties of the ballast. Also discuss the impact value test for ballast. (UP Tech. Univ.)

56. (a) Differentiate between hauling capacity and tractive effort of a locomotive.

(b) Calculate the maximum permissible train load that can be pulled by a locomotive with four pairs of driving wheels with an axle load of 28.42 t each on a BG track with a ruling gradient of 1 in 200 and a maximum curvature of 3°, travelling at a speed of 48.3 km/h. Take the coefficient of friction as 0.2 (UP Tech. Univ.)

57. Using a sleeper density of $(n + 6)$, find out how many sleepers would be required for constructing 640 metres-long BG track (WB Tech. Univ. and Mumbai Univ.)

58. (a) A locomotive on BG track has three pairs of driving wheels, each carrying 20 tonnes. What maximum load can it pull on a level track with curvature 2° at 50 km/h? (WB Tech. Univ.)

(b) What would be expression for sleeper density if there are 22 sleepers in a rail length of 19 m? (WB Tech. Univ.)

59. What is the maximum permissible train load for a locomotive with four pairs of driving wheels carrying an axle load of 24 t each? The speed is 80 km/h on a straight level BG track. What is the reduction in speed if the train has to negotiate a rising gradient of 1 in 200? Assume $\mu = \frac{1}{6}$. (WB Tech. Univ.) )

Chapter 4

# GEOMETRIC DESIGN OF RAILWAY TRACK

**CHAPTER OBJECTIVES**

This chapter enables the student to:

- Understand the various geometric elements relating to the horizontal and vertical alignments of a railway track including the gradients, the various types of horizontal curves – simple, compound, reverse and transition curves
- Comprehend the concepts of equilibrium cant, cant excess, cant deficiency and negative cant; determine the length of transition required, and differentiate between the types of transition curves used on railways.
- Discern the concepts of extra widening of gauge and extra lateral clearances on horizontal curves; appreciate the criteria for safe speeds; and, the type and design features of vertical curves appropriate to the railway track.

## 4.1 INTRODUCTION

Geometric design of a railway track includes all parameters which determine the geometry of the track. As with highways, this aspect is crucial to ensure safe and efficient operation of a railway network.

Primarily, these parameters are related to alignment of the railway track—both in the horizontal and vertical planes.

The general and broad treatment of alignment has already been covered in Chapter 2. In this chapter, particular aspects such as gradients in the vertical alignment, including vertical curves, and the various types of curves in the horizontal alignment, including the superelevation, will be studied in detail.

## 4.2 NECESSITY FOR GEOMETRIC DESIGN

The following considerations point to the need for proper geometric design of a railway track:

1. Ensuring efficient and safe operation of trains
2. Achieving the maximum permissible design speeds in different sections of the track

3. Carrying the design axle loads and transmitting them safely to the formation through the track components such as rails, sleepers and ballast
4. Preventing derailments and accidents, thereby maximising railway safety
5. Minimising maintenance needs for the track
6. Presenting an aesthetic appearance of the track.

## 4.3 GRADIENT

A *gradient* or *grade* becomes necessary in the vertical alignment of a railway track when it cannot be kept horizontal. This may happen in the following situations:

- The stations or places connected are at different elevations.
- The terrain is rolling, mountainous, or hilly.
- The track is aligned to follow the natural ground as far as possible in order to minimise the cost of earthwork

With respect to the direction of movement of the train, a rising gradient is one in which the track gains elevation, and a falling gradient is one in which the track loses elevation in the forward direction (*up* or *down*, respectively).

A gradient is quantitatively represented by the distance in which a rise or fall of one unit of elevation occurs; sometimes, it is also represented by the rise or fall in one hundred units of distance, or as per cent (1 in $n$ or $g$ percent).

The distance travelled may be measured along the track or the corresponding horizontal projection. Both are supposed to yield the same value of the gradient, since the gradients adopted in Railways are relatively very flat. (This amounts to taking $\sin \theta$ as being approximately equal to $\tan \theta$, where $\theta$ is the angle made by the track with the horizontal.)

### Types of Gradients Used on the Railways

The following are the types of gradients used:

1. Ruling gradient
2. Momentum gradient
3. Pusher or helper gradient
4. Gradients in station yards.

#### Ruling Gradient

The ruling gradient is the steepest gradient which exists in a particular section. This determines the maximum load which can be hauled by a locomotive on that section; the locomotive should have adequate power to haul the train load over this gradient at the maximum permissible speed.

Owing to large variations in the topography of the country, in the traffic, and in the locomotives used, Indian Railways does not specify a fixed ruling gradient, in general.

However, the following ruling gradients are adopted

1 in 150 to 1 in 250 on plain terrain, and

1 in 100 to 1 in 150 on hilly terrain.

(These are applicable when only one locomotive pulls the train.)

The idea of a ruling gradient is that no other gradient can be steeper than this in the particular section.

### Momentum Gradient

A gradient steeper than the ruling gradient can be used at certain locations such as a valley, where a rising gradient follows a falling gradient. At such locations, a train running on the falling grade acquires good momentum and kinetic energy, which will be sufficient for it to negotiate a rising grade steeper than the ruling value.

However, there should be no obstacles such as signals in this zone, since they may halt the train or reduce its speed.

### Pusher or Helper Gradient

In hilly terrain, gradients steeper than the ruling gradient may have to be adopted to reduce the length of the route and the overall cost. In such a case, the hauling power of one locomotive is not adequate to pull the train, and an extra engine is needed to push the train up the grade; such a gradient is known as a 'pusher' gradient or a 'helper' gradient. The extra engine used is called a pusher engine or a banking engine.

Examples of such gradients are found in mountain railways such as the Darjeeling Himalayan Railway and the Nilgiris Mountain Railway, besides the Konkan Railway in the Western Ghats.

### Gradients in Station Yards

The gradients in station yards have to be necessarily flat in order to prevent standing vehicles from rolling away from the yard due to gravity, as also to reduce the additional starting resistance.

However, flatter gradients such as 1 in 1000 are adopted in station yards to ensure good drainage; the maximum gradient in station yards on Indian Railways is recommended as 1 in 400.

### Grade Compensation on Horizontal Curves

It has already been seen in Chapter 3 that the train encounters additional resistance to its motion on horizontal curves. If the curve is situated on a rising gradient which is the ruling value, the train cannot pull the maximum load on this section. In order to avoid such a contingency, the gradients on horizontal curves are compensated to the following extent on Indian Railways, so that these values are lower than the ruling gradients:

- On BG: 0.04 percent per degree curvature, or 70/R, whichever is smaller.
- On MG: 0.03 percent per degree curvature, or 52.5/R, whichever is smaller.
- On NG: 0.02 percent per degree curvature, or 35/R, whichever is smaller.

(R is the radius of the curve in metres.)

[Note: In view of the relevance of this concept and the values, they are repeated here.

## 4.4 CIRCULAR CURVES

Curves are introduced in the horizontal alignment when there is a need to change in the direction of the track, either to connect obligatory points or to avoid undesirable locations in the terrain. A curve facilitates a gradual change in direction in the horizontal plane.

Curves are either circular or non-circular (or transitional).

## Types of Circular Curves

The basic types of circular curves are:

1. Simple curves
2. Compound curves
3. Reverse curves

### Simple Curves

A 'simple' curve is a single circular arc connecting two straights or tangents.

### Compound Curves

A compound curve consists of two circular arcs of different radii, curving in the same direction. It can be visualised as connecting three straights, the middle one being the common tangent for both the arcs.

### Reverse Curves

A 'reverse' curve has two circular arcs of the same or different radii, curving in opposite directions. This too can be visualised as connecting three straights. The middle one is again the common tangent for both the arcs.

These three types are shown in Fig. 4.1.

## Degree of a Curve

A curve may be defined or specified by its radius, or by its 'degree'. The degree of a curve ($D$) is defined as the angle subtended at the centre by a standard arc length of 30.5 m (the equivalent of 100 ft).

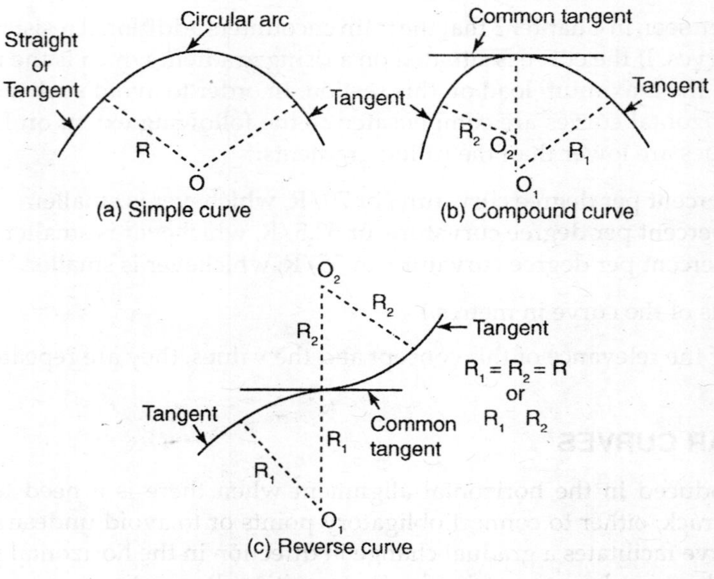

(a) Simple curve

(b) Compound curve

(c) Reverse curve

**FIG. 4.1** *Types of circular curves*

If $R$ is the radius of the curve in metres,

$$D = \frac{360}{2\pi R} \times 30.5 = \frac{1750}{R} \text{ (approx.)} \tag{4.1}$$

Also,

$$D = \frac{5730}{R} \quad \text{(when } R \text{ is in feet)}$$

In case where the radius is very large as in the Railways, the arc length and the corresponding chord length are almost equal. Thus the standard length may be taken conveniently as the chord itself.

(**Note**: Degree and radius are inversely related.)

### Elements of a Simple Curve

The elements of a simple curve are shown in Fig. 4.2.

The following relationships follow from the geometry of the figure and the properties of the circle:

$$T = R\tan\frac{\phi}{2} \tag{4.2}$$

$$L_c = 2R\,\text{Sin}\frac{\phi}{2} \tag{4.3}$$

$$L_a = R\left(\text{Sec}\frac{\phi}{2} - 1\right) \tag{4.4}$$

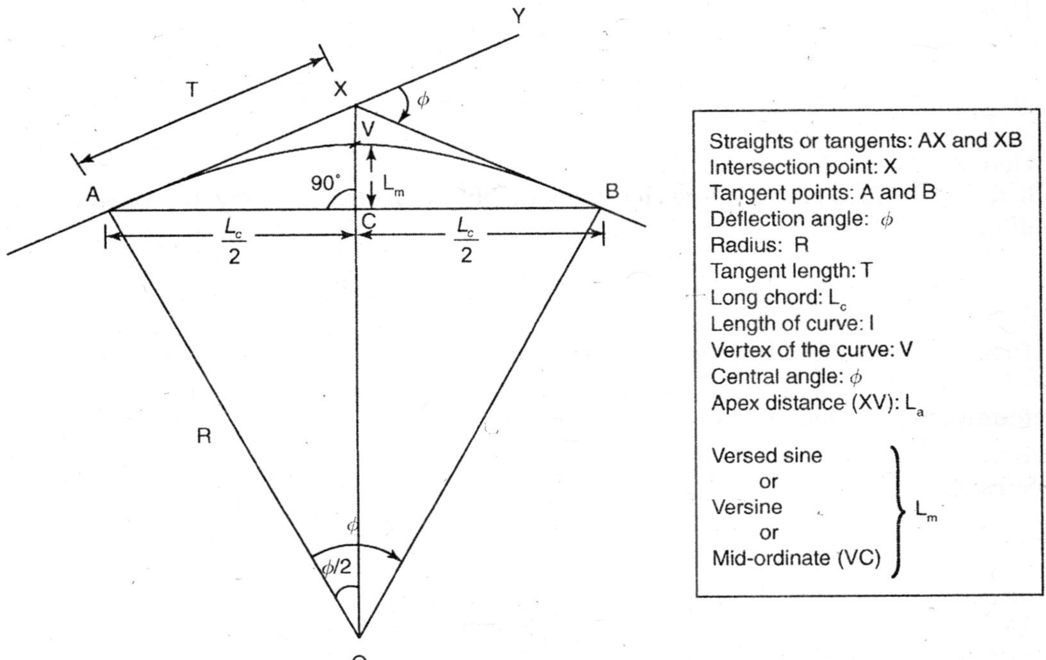


**FIG. 4.2** *Elements of a simple curve*

$$L_m = R\left(1-\text{Cos}\frac{\phi}{2}\right) = R \text{ versine}\frac{\phi}{2} \quad (4.5)$$

$$l = R\cdot\phi\,(\phi \text{ in radians}) \quad (4.6)$$

$$l = \frac{\pi R\phi}{180}(\phi \text{ in degrees}) \quad (4.7)$$

$$l = \frac{\phi}{D}(\phi \text{ in degrees}) \quad (4.8)$$

[This relation (Eq. 4.8) is exact for the arc definition and approximate for the chord definition of the degree of the curve; but the difference is insignificant for large radii, as those used in the Railways.]

**Relationship Between Radius and Versine of a Curve**

Referring to Fig. 4.2, versine of the curve, $L_m$, (or mid-ordinate, $Vc$) may be obtained from the property of a circle

$$L_m\cdot(2R-L_m) = \frac{L_c}{2}\times\frac{L_c}{2}$$

$$2RL_m - L_m^2 = \frac{L_c^2}{4}$$

Since $L_m$ is very small, $L_m^2$ can be neglected when compared to $2R\cdot L_m$

Then
$$2R\cdot L_m = \frac{L_c^2}{4}$$

or
$$L_m = \frac{L_c^2}{8R} \quad (4.9)$$

Here, $L_m$, $L_c$ and $R$ are in the same units.

If $R$ is the radius and $L_c$ is the chord length, both in metres, and say, the versine $L_m$ is in centimetres.

$$L_m = \frac{100\times L_c^2}{8R} = 12.5\frac{L_c^2}{R}\text{ cm} \quad (4.10)$$

Thus, when the chord length and the corresponding versine are known, $R$ can be calculated.

**Determination of Degree of Curve in the Field**

A chord length of 11.8 m is convenient for this purpose.

Substituting this for $L_c$ in Eq. (4.10),

$$L_m = \frac{12.5\times11.8\times11.8}{R}$$

or
$$R = \frac{1750}{L_m}(\text{approx.})$$

Substituting this in Eq. (4.1) for the degree of the curve

$$D = L_m \text{(approx.) in centimetres} \tag{4.11}$$

Thus, the versine of the curve in centimetres on a 11.8 m chord is its Degree.

### Maximum Degree of a Curve

The maximum permissible degree of a curve depends on several factors including the gauge, wheel base of the vehicle, maximum permissible superelevation, and allied factors such as the location along the track.

The maximum permissible values of the degree of the curve used on Indian Railways are:
BG:10 (Corresponding minimum radius = 175 m)
MG:16 (Corresponding minimum radius = 109 m)
NG:40 (Corresponding minimum radius = 44 m)
(On turnouts, the maximum values of degree are somewhat less than these.)

### Setting Out Circular Curves

There are various methods for setting out points on the ground on circular curves or 'pegging out'. The following methods can be used, depending on the type of tools available for the purpose.

**If only chain and tape are available:**

(i) Offsets from long chord (or verse sine)
(ii) Successive bisection of arcs (or quartering of verse sine)
(iii) Offsets from tangents
(iv) Chord deflections or deflection distances. (This is the best method as the work is done in close proximity to the curve)

**If a theodolite is available:**

(i) Rankine's method tangential angles (accurate)
(ii) Two-theodolite method using tangential angles (chaining can be avoided)
(iii) Tacheometric method using tangential angles (A single tacheometer is adequate, since distances are obtained using the principle of tacheometry)

(The details of these methods are out of scope of the present treatment, and they are available in any standard book on 'Surveying').

## Compound Curves

As already defined, a compound curve consists of two or more circular arcs of different radii, curving in the same direction.

Compound curves may be required when space restrictions preclude a single circular curve, and when there are property boundaries which should not be violated.

The geometry of a compound curve is shown in Fig. 4.3.

The following are the relevant equations:

$$\phi = \phi_1 + \phi_2 \tag{4.12}$$

$$T_1 \sin\phi = R_1 \text{ Versine } \phi + (R_2 - R_1)\text{Versine } \phi_2 \tag{4.13}$$

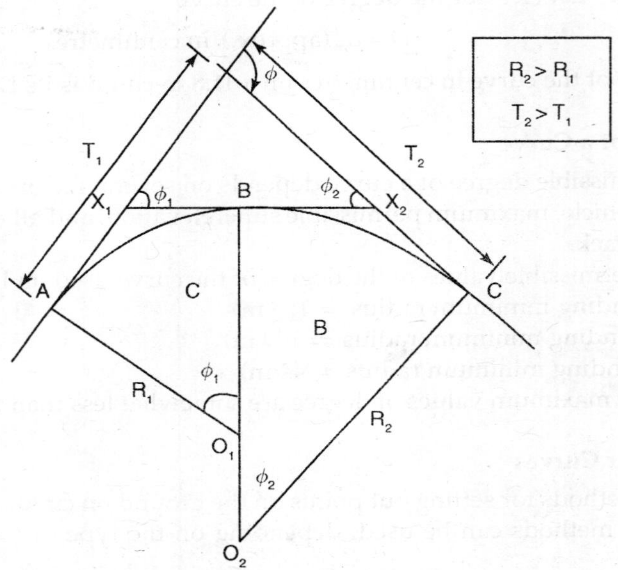


**FIG. 4.3** *Geometry of a compound curve*

$$T_2 \sin\phi = R_2 \text{ Versine}\phi - (R_2 - R_1) \text{ Versine } \phi_1 \tag{4.14}$$

If four of the seven elements of the compound curve are known, the remaining three can be obtained by solving Eqs. (4.12) to (4.14) simultaneously.

[The derivation of these equations is beyond the scope of the present treatment.]

## Reverse Curves

Here, the two circular arcs curve in opposite directions with a common tangent at the junction, and they connect the two end straights. The junction point is the point of reverse curvature or contraflexure. These curves can be used when the deflection angle is very small or even zero (parallel-end tangents), the latter being a special case. They are usually avoided on main lines, but may be used in sidings and yards.

The geometric principles of reverse curves are similar to those for compound curves. The formulae for compound curves, with a change of sign in some of the terms, are applicable to reverse curves.

Other factors introducing variations in reverse curves are that the radii of the two arcs need not be unequal, and the intersection point of the end straights may fall on either side of the common tangent, and also on either side of the end tangent points. Furthermore, reverse curves can be introduced between parallel and also non-parallel end straights.

### Reverse Curves Between Non-parallel End Tangents

Four different cases arise; only one of the cases is shown in Fig. 4.4.

The general equation applicable to all four cases is

$$\phi = \phi_1 \sim \phi_2 \tag{4.15}$$

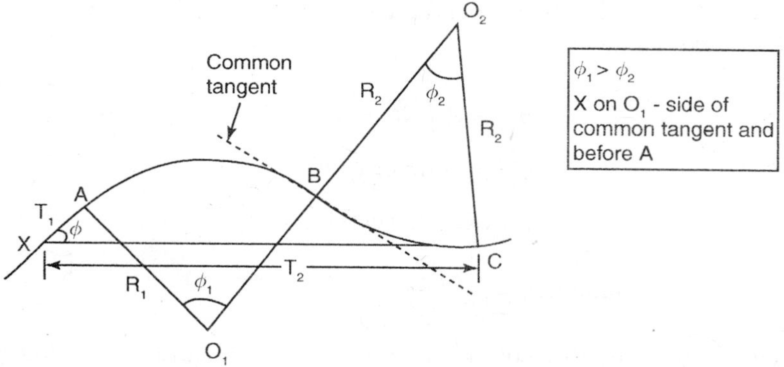


**FIG. 4.4** *Reverse curve between non-parallel tangents*

The other relationships can be derived in exactly the same manner as for a compound curve. The equations for the case shown are given below:

$$T_1 \sin\phi = (R_1 + R_2)\text{ Versine } \phi_2 - R_1 \text{ Versine } \phi \tag{4.16}$$

$$T_2 \sin\phi = (R_1 + R_2)\text{ Versine } \phi_1 - R_2 \text{ Versine } \phi \tag{4.17}$$

Once again, if any four of the seven elements
$\phi$, $\phi_1$, $\phi_2$, $T_1$, $T_2$, $R_1$ and $R_2$ – are known, the other three can be solved from Eqs. (4.15) to (4.17).
The other three cases are:

- $\phi_1 > \phi_2$ and X on $O_1$ -side of common tangent and beyond A.
- $\phi_1 < \phi_2$ and X on $O_2$ -side of common tangent and before C.
- $\phi_1 < \phi_2$ and X on $O_2$ -side of common tangent and beyond C.

The equations for these three cases are similar to Eqns. (4.16) and (4.17), except for change of sign of some of the terms. The first and the fourth cases lead to the same equations.

### Reverse Curve Between Parallel Tangents

In this case, $\phi = 0$ and $T_1$ and $T_2$ are infinitely large.

This case is of considerable practical importance since it is used for cross-over between parallel tracks in railways. Figure 4.5 shows a reverse curve between parallel tangents:

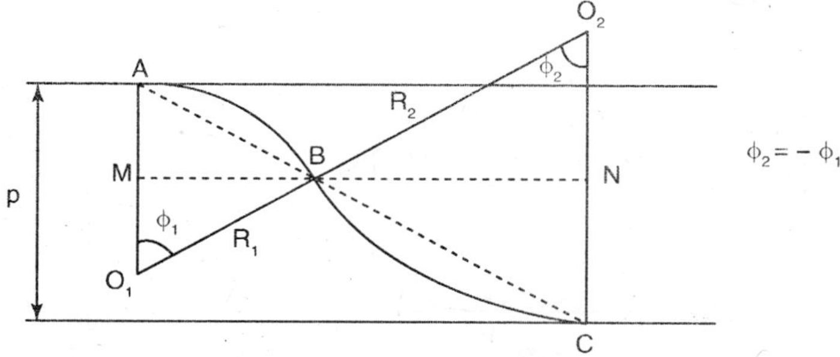


**FIG. 4.5** *Reverse curve between parallel tangents*

Referring to Fig. 4.5, let $p$ be the perpendicular distance between the parallel tangents. Let MBN be parallel to the tangents.

Then

$$p=\overline{AM}+\overline{CN}=R_1 \text{ versine}\,\phi_1+R_2 \text{ versine}\,\phi_2$$

Since $$\phi_1=\phi_2\text{, numerically,}$$

$$p=(R_1+R_2)\text{ versine}\,\phi_1$$

or $$\text{versine}\,\phi_1=\frac{p}{(R_1+R_2)} \tag{4.18}$$

Using this equation, if any three of the quantities – $p$, $R_1$, $R_2$, and $\phi_1$ – are known, the fourth can be calculated.

Usually, $p$, $R_1$ and $R_2$ will be known or fixed.

It can be observed from the geometry of the figure that A, B and C are always collinear.

Now, $$\overline{AB}=2R_1\sin\frac{\phi_1}{2}$$

and $$\overline{BC}=2R_2\sin\frac{\phi_1}{2}$$

$\therefore$ $$\overline{AC}=2(R_1+R_2)\sin\frac{\phi_1}{2}=\frac{2p}{AC}(R_1+R_2)$$

or $$\overline{AC}=\sqrt{2p(R_1+R_2)} \tag{4.19}$$

Since the curve is relatively short, this chord can be used to set it out. If the two radii are equal, we get an even more special case, B lying midway between the parallel tangents and AB and BC being equal.

**Reverse Curve of Three Arcs From a Single Tangent**

Figure 4.6 shows such a reverse curve of three arcs of equal radius, leaving a straight at A and joining it back at E. If $O_1N$ is drawn perpendicular to AB, we have $\overline{AN}=\frac{1}{4}\cdot\overline{AF}$.

Triangle $AO_1N$ and AMF are similar, and so it can be shown that

$$\overline{AF}=2\sqrt{Rp}$$

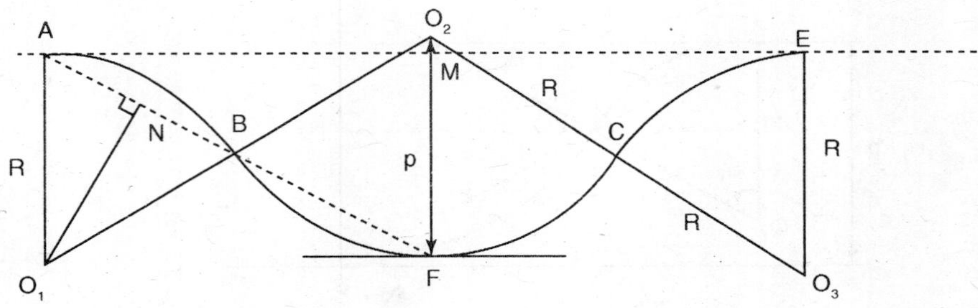


**FIG. 4.6** *Reverse curve of three arcs of equal radius*

and $$\overline{AM}^2 = 4Rp - p^2$$

or $$\overline{AM} = \overline{EM} = \sqrt{p(4R - p)} \tag{4.20}$$

If the mid-point M of the deviation is known, its beginning A and end E can be located. Sometimes, such an arrangement is used for temporary diversions on highways. In railways, this is modified slightly with the middle portion being a straight and is used between two parallel tracks, and is called a gauntleted track.

This and the other arrangements of reverse curves are used for different kinds of turnouts from main tracks for diverting trains from one track to another.

The details of these arrangements will be studied in the next chapter.

## 4.5 SUPERELEVATION OR CANT

According to Newton's first law of motion, a body moving in a straight line continues to do so unless it is forced to change its direction by an external force. When a railway vehicle is forced to negotiate a curved path along a horizontal curve in the alignment of the track, it is subjected to a constant radial acceleration at every point tangential to the curve, against its tendency to move along a straight direction. Consequently, it is subjected to a force given by the product of its mass and this radial acceleration; this force which is in the radial direction at any point along the path is called the *centrifugal force*, directed outward, or away from the centre of the circular path. As the vehicle moves at a constant speed along the curve, the dynamic equilibrium is maintained by a force, equal and opposite to the centrifugal force, which is known as the *centripetal force*.

This is shown schematically in Fig. 4.7, along with the weight of the vehicle ($W$) and the centrifugal force ($F$), both acting at the centre of gravity of the vehicle, the former vertically and the latter horizontally:

The radial acceleration, $$a = \frac{v^2}{R} (\text{m/s}^2)$$

where $v$ is the constant velocity (m/s)
and $R$ is the radius of the horizontal curve (m)

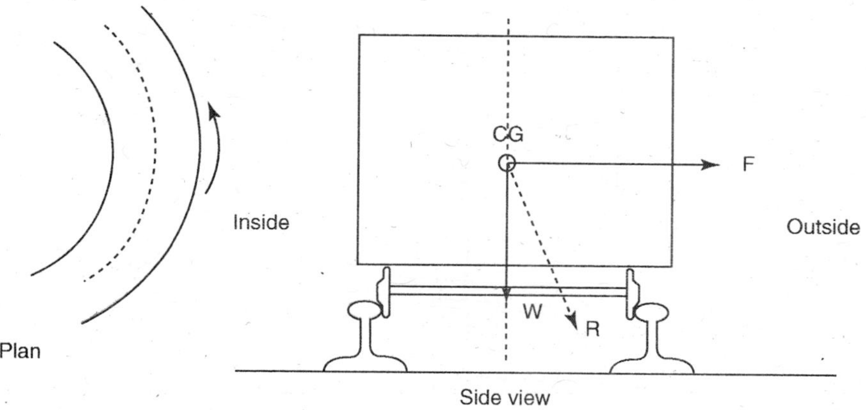


**FIG. 4.7** *Centrifugal effect for motion along a curve*

Centrifugal force, $$F = \frac{M \cdot v^2}{R}$$

where $M$ is the mass of the vehicle.

$$\therefore \quad F = \frac{Wv^2}{gR} \left(\text{since } M = \frac{W}{g}\right) \tag{4.21}$$

The resultant, $R$, of the forces $F$ and $W$ is inclined to the vertical outward, thus causing more pressure to be exerted on the outer rail than on the inner. This irregular or unequal stressing of the rails causes greater wear of the outer rail than the inner, and affects the other track fittings and components.

Besides this, the tendency of the centrifugal force is to cause discomfort to the passengers in the vehicle unless it is counteracted appropriately. The method adopted for counteracting the centrifugal force is to make the plane of the top of the rails normal to the resultant of the centrifugal force and the weight to generate a counteracting gravity component. This is achieved by raising the outer rail of the curve with respect to the inner as shown in Fig. 4.8. This is known as providing *superelevation*.

From the figure,

$$\tan\alpha = \frac{F}{W} = \frac{v^2}{gR} \quad \text{from Eq. (4.21)}$$

This is called the 'rate of superelevation', $e$; also, this is known as the 'centrifugal ratio'.

The amount of superelevation, $E$, is given by

$$E = G \cdot e = G \cdot \tan\alpha = \frac{Gv^2}{gR} \tag{4.22}$$

If $V$ is the speed in km/h, $R$ is the radius in metres, this can be shown to be

$$E = \frac{G \cdot V^2}{127R} \tag{4.23}$$

If $G$ is expressed in millimetres, the superelevation, $E$, is obtained in millimetres.

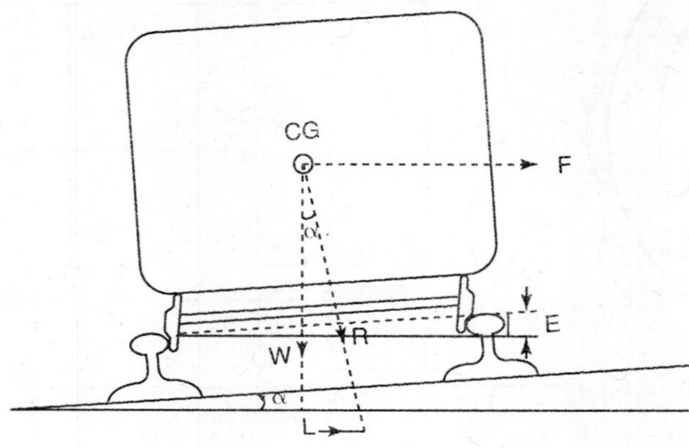


**FIG. 4.8** *Counteracting centrifugal effect by superelevation*

Here $G$, the dynamic gauge, should be taken as the gauge plus the width of the rail head; this is only for the purpose of computing the amount of superelevation.

For the Indian Railways, $G$ is taken as 1750 mm for the Broad Gauge and 1058 mm for the Metre Gauge.

**Functions of Superelevation**

The important functions of superelevation are:

- Bringing about an even distributions of pressure on both rails
- Reducing the wear of rails and of the rolling stock
- Counteracting the effect of lateral force
- Minimising the discomfort to the passengers caused by the effect of centrifugal force.

The term, *cant*, is employed in Railways to refer to the *superelevation* very often.

**Equilibrium Superelevation or Cant**

The superelevation or cant calculated from Eq. (4.23) for a particular speed, $V$, is called the *equilibrium superelevation* or cant for this particular speed. This is because the load on both the rails will be equal if this calculated superelevation is provided, and the vehicle runs at the speed $V$; this ensures perfect equilibrium for the vehicle.

**Equilibrium Speed**

It can be observed that the superelevation or cant varies with the square of the speed. In a section where goods, passenger and express trains run at different speeds, the cant provided for any particular speed will be in excess for slow trains and less than required for fast trains. Excess cant causes extra pressure on the inner rail, and less cant causes more pressure on the outer rail as also discomfort to the passengers. Improper cant can cause disturbance to track geometry, and also cause a tendency to overturn slow vehicles like empty goods wagons.

Therefore, a judicious compromise should be arrived at in deciding the speed for which cant is to be provided. This speed is called the *equilibrium speed* for the section. The equilibrium cant provided for this speed will lead to *cant excess* for trains moving at lower speeds and *cant deficiency* for trains running at higher speeds than this.

The speed that can actually be attained in any given section based on track geometry by both fast and slow trains will be chosen as the equilibrium speed.

**Determination of Equilibrium Speed**

The *earlier practice* of prescribing the equilibrium speed on a level track under average conditions was as follows:

(i) Where the maximum sanctioned speed of the section was more than 50 km/h, 75 percent of this maximum speed was taken as the value of the equilibrium speed, subject to a choice between minimum speed of 50 km/h and the safe speed of the curve, whichever is less.

(ii) Where the maximum speed sanctioned was 50 km/h or less, the maximum sanctioned speed or the safe speed of the curve, whichever was less, was taken as the equilibrium speed.

**Revised method:** The earlier practice has been recently revised by Indian Railways. According to this revised practice, the following factors shall be taken into consideration:

- Maximum permissible speed which can actually be achieved both by fast trains and by goods trains
- Gradients in the section
- Composition of both slow and fast trains
- Number of stoppages
- Speed restrictions – permanent and temporary.

Once the equilibrium speed is determined, the amount of superelevation to be provided is calculated as usual from the formula:

$$\underset{\text{(in mm)}}{E} = \frac{G.V^2}{127R} = \frac{13.78V^2}{R} \quad \text{(for BG, with a dynamic gauge of 1750 mm)} \qquad (4.24)$$

and

$$\underset{\text{(in mm)}}{E} = \frac{8.33V^2}{R} \quad \text{(for MG, with a dynamic gauge of 1058 mm)} \qquad (4.25)$$

### Maximum Values of Superelevation or Cant

The limits of maximum superelevation or cant on Indian Railways are given below:

(i) *Broad gauge*: Group A, B and C routes: 165 mm
(This may be increased to 185 for A- routes in future).
Group D and E routes: 140 mm

(ii) *Metre gauge* All routes: 90 mm
(This may be increased to 100 mm on high-speed routes.)

(iii) Narrow gauge (762 mm): 65 mm
(This may be increased to 75 mm with special permission)

### Cant Deficiency

*Cant deficiency* is the difference between the equilibrium cant necessary for the maximum permissible speed on the curve and the actual cant provided.

Cant deficiency has to be limited because:

- Higher cant deficiency causes discomfort to passengers because of the unbalanced centrifugal force.
- Higher cant deficiency causes greater lateral forces and rail wear due to unbalanced centrifugal force.

The maximum permissible values of cant deficiency on Indian Railways are as follows:

Broad gauge: 75 mm (For speeds of over 100 km/h on Group A and B routes with special permission)

Metre gauge: 50 mm for all routes.

### Cant Excess

The difference between the equilibrium cant required for a lower speed than the maximum permissible speed and the actual cant provided for the latter is called *cant excess*.

Excess cant causes greater pressure on the inner rail, and hence greater wear. Therefore, it should be limited. The maximum cant excess values permitted on Indian Railways are:

Broad gauge: 75 mm

Metre gauge: 65 mm

Booked speeds of goods trains should be taken into account for calculating the cant excess for a particular section.

### Negative Superelevation or Cant

When a main line is on a curve and a branch line taking off from it also lies on a curve of contrary flexure, the provision of cant or superelevation for both the lines in the correct manner becomes impossible for a short distance near the take-off point. This is shown in a schematic way in Fig. 4.9.

Referring to Fig. 4.9, the outer rail of the main line curve, AB, must be higher than the corresponding inner rail, CD. However, for the branch line, CF must be higher than AE, or point C must be higher than A, contradictory to the requirement of the main line.

In such a case, the branch line curve will have a negative cant for a short distance from the point of take-off. This also follows from the requirement of the canted curve of the outer rail and the turnout sleeper being in the same inclined plane. In order to ensure that the norms laid down for maximum cant and cant deficiency are not violated, speeds of trains on both the main line and the branch line (especially the latter) must be necessarily restricted.

The negative cant for the branch line and the reduction in speed over the main line can be calculated as follows:

(i) The equilibrium cant for the branch line curve is first calculated using the formula

$$E = \frac{G.V^2}{127R}$$

(ii) The equilibrium cant $E$ is reduced by the permissible cant deficiency $E_d$, to obtain the resultant cant, denoted as $E_b$:

$$E_b = E - E_d$$

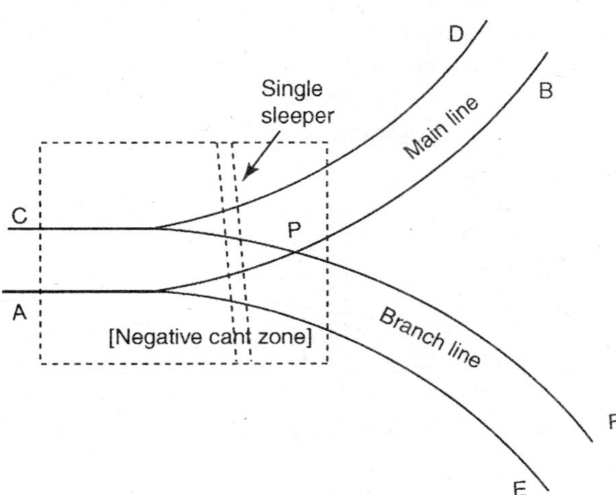


**FIG. 4.9** *Branch line curved track taking off a main line curved track – gives rise to negative cant*

The value of $E_d$ is generally higher than that of $E$, and therefore, $E_b$ is normally negative. The branch line thus has a negative cant of $E_b$.

(iii) The maximum permissible speed on the main line, which has the positive superelevation of $E_b$, is now calculated by adding the allowable cant deficiency $E_d$ to it ($E_b + E_d$), i.e., the speed corresponding to this cant. The safe speed for the curve is also calculated, and the lower of these two values is taken as the maximum permissible speed on the main line.

## 4.6 TRANSITION CURVES

A *transition curve* consists of a non-circular arc which facilitates a gradual change of curvature between a straight and a circular curve, or between the two branches of a compound curve or those of a reverse curve. It is common practice to provide transition curves in railways, as in highways.

Abrupt curvature changes not only cause discomfort to passengers, but also puts unequal pressure on the outer and inner rails of a track as the vehicle negotiates the curve. Transition curves prevent this.

Transition curves also serve as a medium for the gradual introduction of superelevation or cant while passing from a straight on to a circular curve; similarly, they allow gradual elimination of cant when moving back to the next straight.

The change of curvature can be made continuous, and as gradual as desired, by the introduction of a suitable transition curve. In fact, the effect of a sudden change of curvature is much more serious in the case of reverse curves at the point of reverse curvature.

Transition curves are required at both ends of a circular curve, which, coupled with end transitions, is referred to as the *combined* or *composite* curve.

### Requirements of an Ideal Transition Curve

When a transition curve is introduced between a straight and a circular curve, the requirements of an ideal transition are as follows:

(i) The transition should be tangential to the straight; that is, its curvature should be zero at its origin on the straight.

(ii) The curvature should increase along the transition at the same rate as the cant increases.

(iii) The full value of the superelevation required for the circular curve should be attained at the junction of the transition curve and the circular curve.

(iv) The transition curve should join the circular curve tangentially; that is, its curvature should be equal to that of the circular curve at the junction.

It is universal practice to the increase the cant at a uniform rate. To accomplish this, the curvature of the transition at any point is proportional to the distance of that point from the origin or the starting point.

### Length of Transition Curve

In railways, the length of a transition curve is determined based on accepted norms for permissible rate of change of cant, cant deficiency, and radial acceleration.

The corresponding formulae for the length of a transition on Indian Railways are given below:

a. $$L = 0.008\, E_a \cdot V_m \tag{4.26}$$

b. $$L = 0.008\, E_d \cdot V_m \tag{4.27}$$

c. $$L = 0.72\, E_a \qquad (4.28)$$

Here,

$L$ = Length of transition in metres,

$V_m$ = Maximum permissible speed in km/h,

$E_a$ = Actual cant on the curve in mm, and

$E_d$ = Cant deficiency in mm.

The largest value obtained from the above three equations shall be adopted as the desirable length of transition.

Equations (4.26) and (4.27) are based on rate of gain of cant and cant deficiency of 35 mm per second as a vehicle travels along the transition. Equation (4.28) is based on the maximum cant gradient (or spatial rate) of 1.4 mm per metre or 1 in 720.

On high-speed routes, transition length may be determined considering possible future speeds of 160 km/h for Group A lines and 130 km/h for Group–B lines.

In exceptional cases where space is not available for providing transitions in accordance with these norms, the length may be reduced to a maximum of 2/3 of the desirable length calculated from Eqns. (4.26) and (4.27), and to 0.36 $E_a$, and the largest of the three values shall be adopted.

This is based on the criteria that the time-rate of gain of cant and cant deficiency will not exceed 55 mm per second, and the maximum cant gradient (space-rate) will be limited to 2.8 mm per metre or 1 in 360.

This relaxation applies to broad gauge only.

## Types of Transition Curves

The following are the types of transition curves used:

### Euler's Spiral or Clothoid

This is an ideal transition curve, for which the equation is

$$\phi = \frac{l^2}{2RL} \qquad (4.29)$$

where

$\phi$ = Angle made by the tangent to the transition curve with respect to the x-axis at any point on the curve,

$R$ = Radius of the circular curve.

$L$ = Length of the transition, and

$l$ = Length of transition up to the point considered.

Although this satisfies the requirements of a transition curve perfectly, it is not considered to be convenient for calculations needed to set out the curve in the field.

### Cubic spiral

The equation to the cubic spiral is

$$y = \frac{l^2}{6RL} \qquad (4.30)$$

where $y$ = y-co-ordinate of any point on the spiral, and $R$, $L$ and $l$ are as defined earlier.

This is also a good transition curve, but it is also not preferred for the same reason as for the Euler spiral.

**Bernoulli's Lemniscate**

In this curve, the radius decreases or the curvature increases as the length increases; also, rate of decrease of radius decreases with increase of length.

The polar equation is $$\rho = k\sqrt{\sin 2\alpha} \tag{4.31}$$

where

$\rho$ = Length of polar ray ($\rho$ = for $\alpha = 0$ and $90^\circ$),
$\alpha$ = Polar angle, and
$k$ = A constant, which is maximum for $\alpha = 45^\circ$.

This is convenient for setting out the curve by the method of tangential angles. This is also a symmetrical curve which can be used even for large deflection angles,

The lemniscate is used in modern highway practice, but it is not used in railways.

**Cubic Parabola**

This is also known as 'Froude's transition curve' and is obtained as an approximation of Euler spiral.

The equation is

$$y = \frac{x^3}{6RL} \tag{4.32}$$

Here, $x$ and $y$ are the Cartesian co-ordinates of a point on the curve with the starting point as the origin. This is very convenient for the computations needed to set out the curve in the field.

This is the most popular transition curve used on the Indian Railways; both the curvature and cant increase at a linear rate. The increase of cant is made such that the inner rail continues to be at the same level and the outer rail is raised linearly along the transition so that the full cant is attained at the start of the circular curve.

These curves are superimposed and shown in Fig. 4.10:

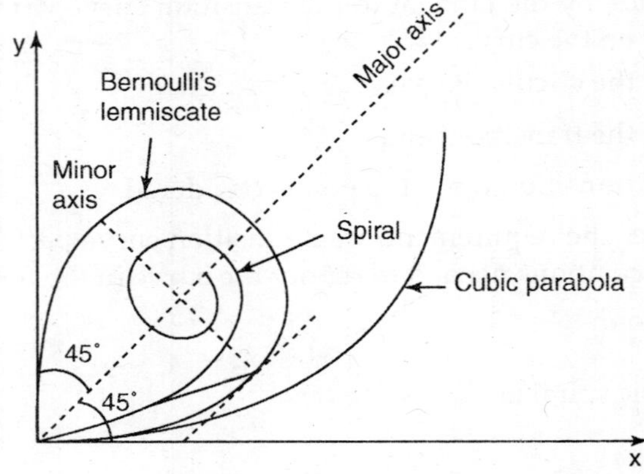


**FIG. 4.10** *Lemniscate, spiral and cubic parabola*

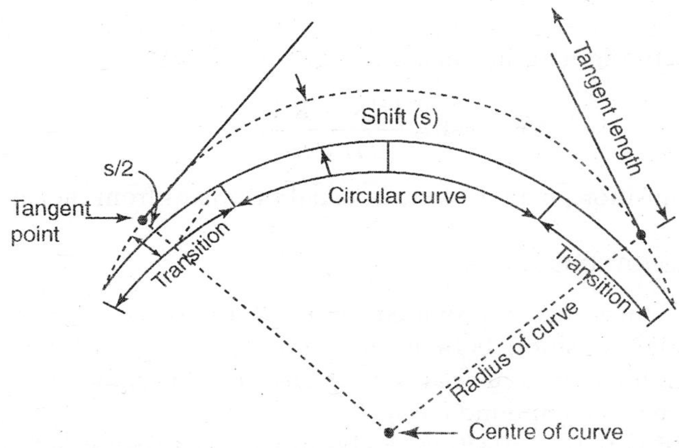


**FIG. 4.11** *Shift of circular curve due to transition*

## Shift of Circular Curve due to Transition

Introducing transitions at the ends of a simple circular curve necessitates shifting the circular curve from its original position while preserving its radius, in order to accommodate the transition. Alternatively, the circular curve can be located in its position without transitions, if the two tangents are shifted outwards from the curve to an extent that would permit introduction of the transitions. Of these two approaches, the first, in which the main curve is shifted to accommodate the transition is the common practice. This is shown in Fig. 4.11.

From the geometry of the combined curve and the properties of the cubic parabolic transition curve, it can be shown that the *shift*, *s*, is given by

$$s = \frac{L^2}{24R} \tag{4.33}$$

It can also be shown that at the mid-length of the transition, the magnitude of the shift of the curve is half the total shift. Thus, the line denoting the shift and the transition length bisect each other.

The transition curves can be set out by offsets from tangents or by Rankine's method of tangential angles.

Modified tangent length, $T' = (R+s)\tan\frac{\phi}{2} + \frac{L}{2}$ (for a cubic parabolic transition) (4.34)

## Transitions for Compound Curves

A common transition curve may be provided between the two circular arcs of a compound curve. The length of transition, in this case, can be got from the formula

$$L = 0.008\,(E_{a1} - E_{a2}).\,V_m \tag{4.35}$$

$$L = 0.008\,(E_{d1} - E_{d2}).\,V_m \tag{4.36}$$

The greater value from these two equations is taken into consideration.

Here, $E_{a1}$ and $E_{d1}$ are the cant and cant deficiency respectively for the first arc and $E_{a2}$ and $E_{d2}$ are the corresponding values for the second.

$V_m$ is the maximum permissible speed in km/h.

$L$ is the length of the transition in metres and the transition curves at the beginning and end of the compound curve are dealt with as usual with the data for the first arc and with those of the second arc, respectively. All formulae relating to the transition can be applied by substituting $R_1R_2/(R_2-R_1)$ for $R$; the shift at the junctions is given by

$$s_j = \frac{L^2(R_2 - R_1)}{24 R_1 R_2} \tag{4.37}$$

Setting out the transitions may be done as usual by offsets from the tangents.

### Transitions for Reverse Curves

A common transition curve may be provided between the two circular arcs of reverse curvature. The total length of the transition, both at the beginning and end of the reverse curve and at the junction between the two circular arcs of reverse curvature, may be obtained in the same manner as in the case of a compound curve.

All formulae relating to the transition can be applied by substituting $R_1R_2/(R_2+R_1)$ for $R$; the shift at the junction, $s_j$, is given by

$$s_j = \frac{L^2(R_2 R_1)}{24 \cdot R_1 R_2} \tag{4.38}$$

This follows from the fact that, at the point of reverse curvature, the distance between the shifted circular arcs is given by $(s_1 + s_2)$ in this case while it is $(s_1 \sim s_2)$ in the case of a compound curve.

It has been stipulated that for high-speed Group-A and Group-B routes, a minimum straight length of 50 m should be kept between the two circular arcs constituting a reverse curve. (In the case of a high-speed MG route, this is 30 m.)

High speeds of the order of 100 to 130 km/h should not be permitted on the reverse curves unless the above stipulation is satisfied, or the junction transitions are suitably extended.

Reverse curves are therefore considered inappropriate on high-speed routes unless they are unavoidable from terrain or other constraints.

## 4.7 EXTRA CLEARANCES ON CURVES

Certain minimum lateral clearances have been prescribed to ensure that vehicles travel with the same degree of safety all over the system of a particular gauge. On horizontal curves, the centre line of the vehicle moves away from the centre line of the track, projecting inwards at the middle of the length, and outwards at the ends. (This happens due to the rigidity of the frame of the vehicle.)

This necessitates extra clearance for structures; the one at the middle of the length of the vehicle is called *overthrow* and that at the ends is called *end-throw*.

The position of a bogie on a curve is shown schematically in Fig. 4.12.

Let $R$ be the radius of the curve, $L$ the overall length of the vehicle, and $C$ the centre-to-centre distance of the bogies.

From the geometry of the circle,

Overthrow (inside), $$x_0 = \frac{C^2}{8R} \tag{4.39}$$

End throw (outside), $$x_2 \frac{L^2}{8R} - \frac{C^2}{8R} = \frac{(L^2 - C^2)}{8R} \tag{4.40}$$

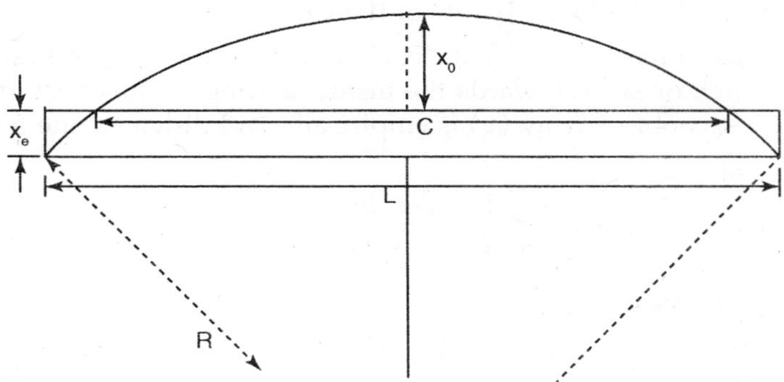


**FIG. 4.12** *Extra lateral clearances on curves*

These are the extra lateral clearances at the centre and at the ends, respectively, on the inside and on the outside of the curve.

**Effect of leaning of vehicle due to superelevation**

On account of superelevation, vehicles tend to lean towards the inside of the curve as shown in Fig. 4.13.

The extra clearance needed due to the effect of leaning of the vehicle on curves towards the inside of the curve, $x_l$, is

$$\underset{(\mathrm{mm})}{x_l} = \frac{h.E}{G} \tag{4.41}$$

where,

$h$ = Height of the vehicle (mm)

$E$ = Superelevation provided (mm), and

$G$ = Dynamic gauge (mm)

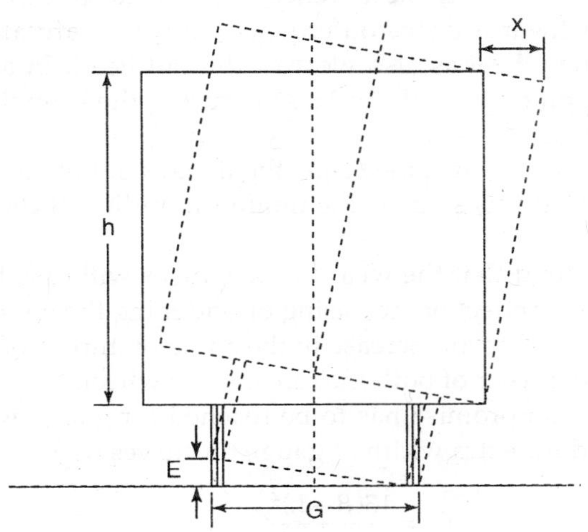


**FIG. 4.13** *Effect of leaning of vehicle*

**Effect of Lurch or Sway Towards the Inside of the Curve**

Unbalanced centrifugal forces due to cant deficiency cause the vehicle suspension to react, causing additional lurch or sway towards the inside, adding to the effect of lean. The extra clearance needed on account of sway ($x_s$) is empirically laid down as one-fourth of that due to learning:

$$x_s = \frac{1}{4}.x_l = \frac{1}{4}.\frac{hE}{G} \tag{4.42}$$

**Lateral Clearance on Curves**

The resultant clearances required are

On the inside:
$$ec_i = x_0 + x_l + x_s = \left(\frac{C^2}{8R} + \frac{h.E}{G} + \frac{1}{4}.\frac{hE}{G}\right) \tag{4.43}$$

On the outside:
$$ex_0 = x_e = \left(\frac{L^2 - c^2}{8R}\right) \tag{4.44}$$

(**Note**: Extra clearances are required between adjacent curved tracks and for platforms; however, these are outside the scope of the present discussion. In this case of platforms, the calculated clearance is reduced by 51 mm on the inside and 25 mm on the outside to prevent excessive gap from the footboard.)

## 4.8 WIDENING OF GAUGE ON CURVES

A railway vehicle normally assumes the central position a straight track, with the flanges of both the wheels staying clear of the rails. The moment it enters a curved portion of the track, the flange of the outer wheel of the leading axle continues to travel straight until it rubs against the rail. The outer wheel travels a longer distance than the inner one due to coning of the wheels, but the rigidity of the wheel base causes the trailing axle to occupy a different position. In order to make up for this difference in the distance travelled by the wheels, the inner wheel must slip backwards or the outer wheel slide forward along the rail. This causes wear of the flanges of the wheel as also wear of the rail head. In addition to slipping and sliding along the rail, slipping across the rail also occurs to the keep the front axle of the bogie always perpendicular to the line of motion.

If the curve becomes sharper, at some stage, the flanges of both axles of the bogie begin to touch the rails; the radius at this stage is the minimum radius of the curve to accommodate the bogie.

It can be easily understood that the wear of the flanges will ease the passage of the bogie along the curve, as it has the effect of increasing or widening the gauge. However, any excess of gauge widening could lead to an increase in the angle of impact of the wheel flange on the rail, leading again to excess wear of both the rail and the wheel.

Therefore, a judicious compromise has to be reached for gauge widening on curves. The following formula is used for extra width of gauge on curves ($G_e$):

$$G_e = \frac{13(B+L)^2}{R} \tag{4.45}$$

where

$B$ = Wheel base of the vehicle in metres

$R$ = Radius of the curve in metres

$L$ = Lap of the flange in metres

$L$ is given by the equation

$$L = 0.02\sqrt{h^2 + Dh} \quad (L \text{ in metres}) \tag{4.46}$$

where

$h$ = Depth of flange below the top of the rail (cm)

and D = diameter of the wheel of the vehicle (cm).

The following standards are prescribed for gauge tolerances on Indian Railways:

## Gauge Widening and Tolerances

| Gauge of railway | Radius (m) | Gauge tolerance |
|---|---|---|
| 1. BG (1,676mm) | (a) Straight track – including curves of radius 400m or more. | 3 mm tight, i.e., – 3mm (Gauge : 1673 mm) |
| | (b) Curves of radius less than 400 mm | 5 mm slack, i.e., + 5mm<br>(Gauge: 1681 mm) |
| 2. MG (1,000 m) | (a) Straight track – including curves of 300 m radius or more. | Zero tolerance (Gauge: 1000 mm) |
| | (b) 200 to 300 m radius | Up to 5mm slack (+ 5mm)<br>(Gauge: 1005 mm |
| | (c) Less than 200 m radius | Up to 15 mm slack (+5mm))<br>(Gauge: 1015 mm) |

## Check Rails on Curves

Rails laid on curves are subjected to grinding, slipping and sliding actions under moving vehicles, and need frequent renewals. Worn-out outer rails show the signs of grinding on the inside rail head. Inner rail head is often spread due to transverse slipping (Fig. 4.14):

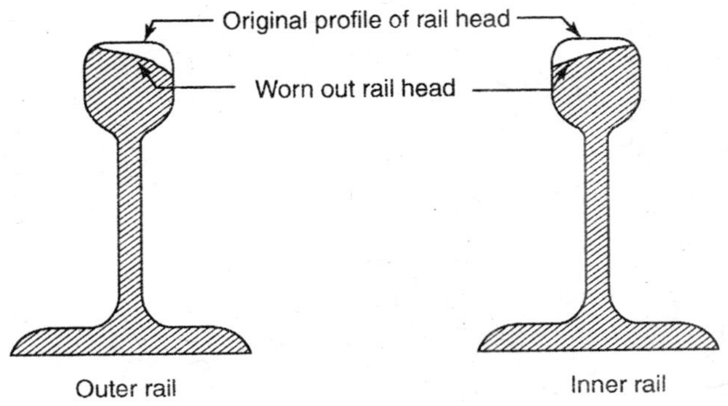


**FIG. 4.14** *Rail head wear on curves*

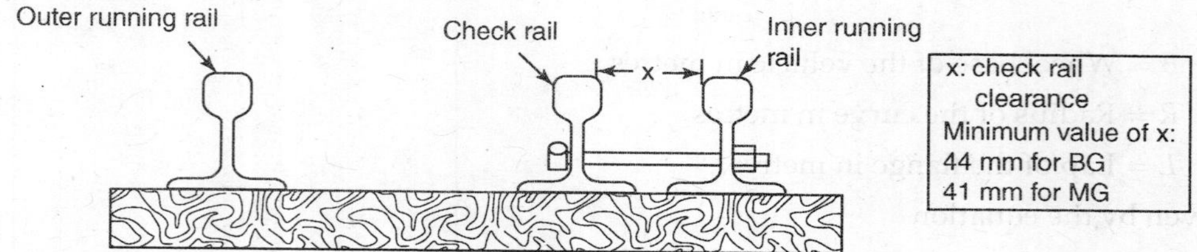


**FIG. 4.15** *Schematic of check rail*

The pressure of the flange of the leading wheel against the rail head leads to the risk of the wheel climbing over the rail, besides wearing the rail head. To minimise such a risk, a 'check rail' is fixed inside the inner rail, parallel to it (Fig. 4.15).

Functions of check rails:

- Reduction of lateral wear on outer rail
- Prevention of the outer wheel flange from mounting the rail
- Prevention of the possibility of derailment of the vehicle.

## 4.9 SAFE SPEED ON CURVES

*Safe speed* refers to a speed which prevents overturning and/or derailment of a railway vehicle with a reasonable margin of safety.

The old procedure of determining safe speed on Indian Railways was based on the empirical formula given by Martin. This formula involves the square root of the radius of the curve only. As this formula is no longer used, the details are not given here.

The new formulae that are presently applied on the Indian Railways are based on the rate of increase of cant and cant deficiency, which is considered to be more rational.

These formulae are presented below:

### Fully Transitioned Curves

The maximum permissible speed for transitioned curves is obtained from the following formulae:

(a) Broad gauge:

$$V = 0.27\sqrt{(E_a + E_d).R} \tag{4.47}$$

(b) Metre gauge:

$$V = 0.347\sqrt{(E_a + E_d).R} \tag{4.48}$$

$E_a$ = Actual cant (mm)

$E_d$ = Cant deficiency (mm)

$R$ = Radius of curve (mm)

$V$ = Maximum permissible speed (km/h)

These formulae are based on the basic formula for cant

$$E = \frac{G.V^2}{127R} \tag{4.49}$$

($G$ is the dynamic gauge, which is 1750 mm for BG and 1058 mm for MG)

(c) Narrow gauge (762 mm):

$$V = 3.65\sqrt{R-6} \text{ (Subject to a maximum of 50 km/h)} \quad (4.50)$$

This is based on the old Martin formula.

## Non-transitioned Curves

### With Cant on Virtual Transitions

The determination of maximum permissible speed on curves without transition is based on the concept of 'virtual transitions'. The linear velocity of a vehicle moving with uniform velocity on a straight track begins to change into angular velocity as soon as the first bogie of the train crosses the tangent point; this change continues until the last bogie crosses the tangent point. The full value of the angular velocity is attained at this stage. The change in motion from a straight to a curved condition takes place over the shortest distance between the bogie centres, which is considered as the virtual transition'. Normally, this distance is 14.8 m on B G., 13.7 m on MG, and 10.3 m on NG, commencing at half this distance before the tangent point. (This means that half the virtual transition is on the straight and the remaining half on the curve).

The deficiency of cant is considered to be gained in the length of virtual transition, and the cant also gained in a similar manner; the cant gradient in any case not being steeper than 1 in 360 for BG and 1 in 720 for MG and NG.

The safe speed should be got on the basis of the cant that can be provided based on these criteria, increased by the permissible cant deficiency.

### Non-transitioned Curves With no Cant

The safe speed is calculated on the basis of cant deficiency that can be permitted on the curve, considering the rate at which this is gained or lost over the virtual transition.

### Curves Laid With Inadequate Transition

In this case, the safe speed is arrived at on the basis of the actual cant deficiency, which can be provided taking into consideration the gradient for limiting cant deficiency.

(This becomes important for high-speed routes, when the length of transition curve cannot be increased easily.)

In addition to these, the *maximum sanctioned speed* on the section shall also be considered, and never exceeded.

**Maximum sanctioned speed:** This is the maximum permissible speed authorised by the Commissioner of Railway Safety. It is determined after an analysis of the condition of the track, the type of locomotive and of the rolling stock, the efficiency of signalling and interlocking, and such other allied factors. This shall never be exceeded in the particular section.

### Simplified Method of Calculation of Permissible Cant and Speed

This simplified procedure is applicable to most cases except for very flat curves.

The following are the steps in the procedure:

1. The cant for maximum speed of the section is calculated using $E = GV^2/127R$.

2. The cant for the slowest goods train in the section is calculated similarly. The permissible cant excess is added to this.
3. The cant for the equilibrium speed of the section (if decided) is also calculated.
4. The minimum of the three values obtained above becomes the permissible cant, $E_a$.
5. To the permissible cant, $E_a$, the permissible cant deficiency is added, and the maximum permissible speed for this sum is obtained using the Eqn. (4.47) for BG.

The above considerations in the determination of safe speed on curves can be better understood by going through the numerical examples worked out in Section 4.11.

## 4.10 VERTICAL CURVES

An angle in the vertical plane is formed when two different gradients meet. This angle is smoothened out by providing a vertical curve in order to prevent an uncomfortable ride.

Vertical curves are of two types:

(a) Summit curves and (b) Sag curves

The curvature should be introduced gradually. Unlike in the case of horizontal curves, there is no 'ideal' transition for vertical curves. But the effect of change of gradient can be made relatively small if the radius of curvature is large enough at the beginning.

The *minimum radii* recommended on Indian Railways are given below:

| | | |
|---|---|---|
| Broad gauge | Group –A routes | 4000 m |
| | Group –B routes | 3000 m |
| | Other routes | 2500m |
| Metre gauge | All routes | 2500m |

The curve which is generally used is the simple parabola, for which the rate of change of gradient is constant along its length, unlike that for a circular arc. Theoretically speaking, the cubic parabola is superior. The rate of change of gradient varies uniformly with distance in this case; this means that the curvature is introduced gradually. However, the simple parabola is preferred since its computations are relatively easy and it adequately fulfils the required conditions.

In fact, when the radius of curvature is large (say, more than ten times the length), there is practically no difference between the shape of a circle and that of a parabola.

### Summit Curves

Summit curves are formed in the following situations:

(i) An up-grade followed by a down-grade
(ii) An up-grade followed by a flatter up-grade
(iii) A down-grade followed by a steeper down-grade

These are shown in Fig. 4.16.

### Sag Curves

Sag curves are formed in the following situations:

(i) A down-grade followed by an up-grade.
(ii) A down-grade followed by a flatter down grade

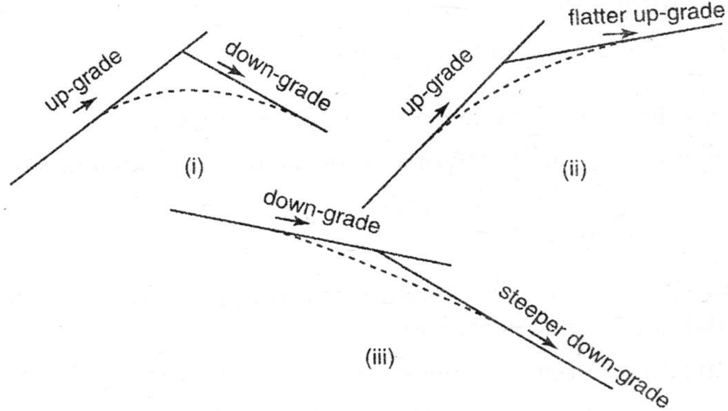


**FIG. 4.16** *Summit curves*

(iii) An up-grade followed by a steeper up-grade

These are shown in Fig. 4.17.

## Length of a Vertical Curve

The length of a vertical curve depends upon the algebraic difference of the gradients and the type of curve formed – summit or sag.

### Old Practice of Determination of the Length of a Vertical Curve

Hitherto, the length of a vertical curve was determined based on the rate of change of gradients permissible; this was taken as 0.1 per cent between successive chords of 30.5 (100 ft) length for summits, whereas it was taken as 0.05 per cent between successive chords of the same 30.5 m length.

Therefore the length of the vertical curve, $L$, in metres was given by

$$L = \left(\frac{G}{r_g}\right) \times 30.5 \tag{4.51}$$

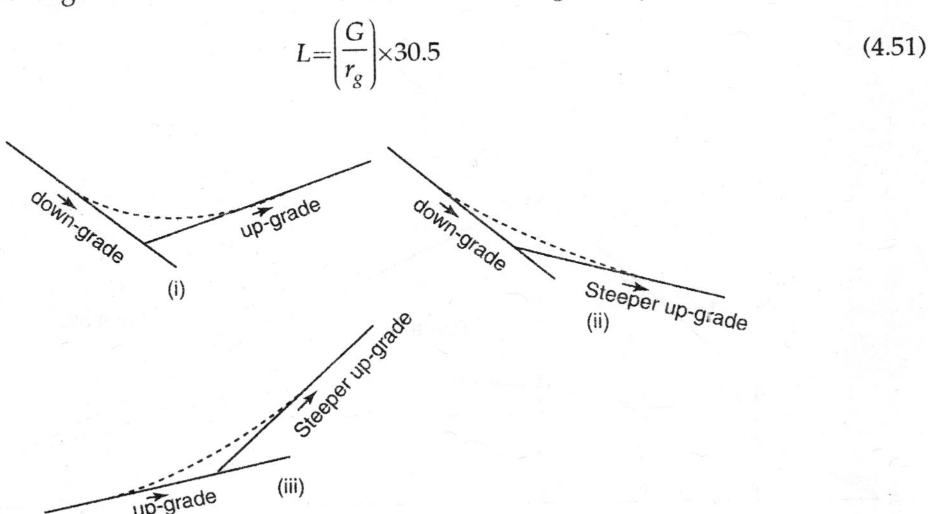


**FIG. 4.17** *Sag curves*

where $G$ = Algebraic difference of gradients, percent,

and, $r_g$ = Rate of change of gradients in percent.

**New method of determination of the length of a vertical curve**

According to the new method, the length of a vertical curve is calculated as

$$L = R \cdot G \tag{4.52}$$

where,

$R$ = Radius of the vertical curve (minimum recommended by Indian Railways for the particular gauge and route) and

$G$ = Algebraic difference of gradients, expressed in radians.

($R$ and $L$ are in the same units – metres). The length calculated in this manner is smaller than that obtained by the old method, but it is considered to be reasonable and adequate.

**Setting out Vertical Curves**

Two methods are popular for setting out vertical curves:

(i) Tangent corrections
(ii) Chord gradients

Of these, the method of tangent corrections is simpler and more convenient for setting out in the field. The following are the steps followed (Fig. 4.18).

(Up-grade is considered positive and downgrade negative.)

1. The length of the vertical curve is first calculated. The changes and reduced levels (RLs) of the intersection, and those of the tangent points, are calculated.
2. Tangent corrections, $e_t$, are then calculated from the equation

$$y = c.x^2 \tag{4.53}$$

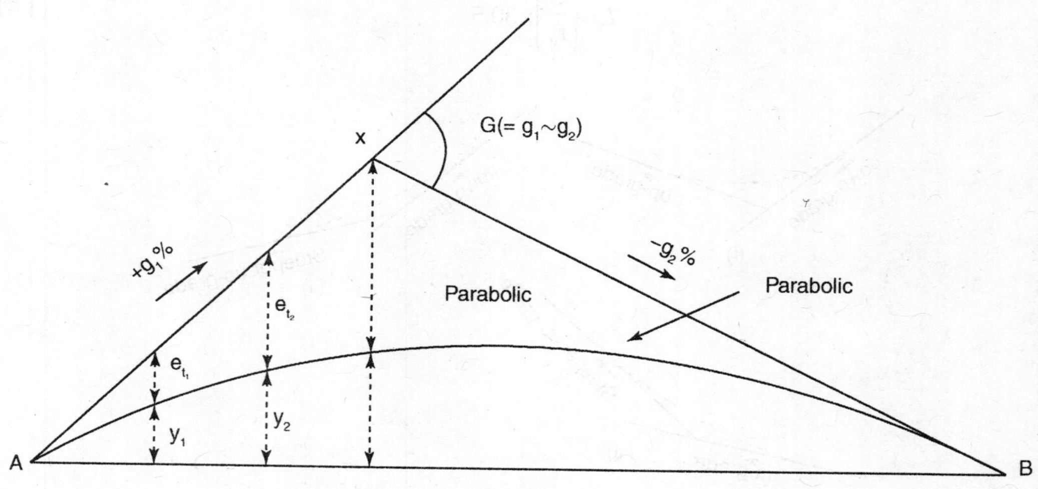


**FIG. 4.18** *Tangent corrections method*

where $$c = \frac{G}{4n} \text{ or } \frac{G}{400}n \text{ (if } G \text{ is in per cent)} \tag{4.54}$$

Here,

$y$ = Vertical ordinate

$x$ = Horizontal distance from the tangent point

$G$ = Algebraic difference of gradients ($g_1 \sim g_2$),

$n$ = Number of chords for half the length of the curve.

3. The reduced levels of the stations on the curve are obtained by algebraically adding the tangent corrections to those of the points on the tangent.

   These corrections indicate the depth of cut or height of fill to fix points on the vertical curve.

**Recommendations About Vertical Curves**

The following recommendations about vertical curves are made by Indian Railways:

**Vertical Curves**

(i) Vertical curves must not be provided in the transition portion of horizontal curves, as it is difficult to maintain track geometry.
(ii) Turnouts must not take off at locations where vertical curves are provided, because the maintenance of the turnouts poses difficulties.
(iii) Vertical curves must not be provided over deck bridges without ballast.
(iv) In a rolling profile, sags and summits should not be too close to each other, as it would lead to problems in train operation.
(v) Vertical curves need not be provided if the algebraic difference of grades is less than 0.4 per cent.

**Summit curves**

(vi) Summit vertical curves must be avoided when the track is on a horizontal circular curve especially when the track is with CWR or LWR, as the resistance to buckling of the track decreases.
(vii) Summit curves must be avoided at the centre of long tunnels, as it could affect ventilation.

**Sag curves**

(viii) Sag curve should not be located in a cutting or a tunnel as drainage may pose problems.
(ix) Sag curves must be maintained well as they are accident-prone; the wagons at the lowest point have a tendency to bunch up and get lifted off the track since the front wagons get slowed down, moving against gravity over the up-grade.

## 4.11 NUMERICAL EXAMPLES

### EXAMPLE 4.1

What gradient should be adopted on a section with a ruling gradient of 1 in 150 on a BG line to accommodate a 2°–curve?

**Solution**

$$\text{Ruling gradient} = 1/150 = 0.67\%$$

Compensation for a 2° curve on BG line = 0.04 × 2 = 0.08%

∴ Compensated gradient = (0.67 – 0.08) % = 0.59%

This comes to 1 in 170

## EXAMPLE 4.2

What is the equilibrium cant on a MG line on a 4° curve for an average speed of 75 km/h? What is the maximum permissible speed allowing the maximum cant deficiency? Assume the curve to be fully transitioned.

**Solution**

Radius, $$R = \frac{1750}{4} \text{ m} = 437.5 \text{ m}$$

For a fully transitioned curve, equilibrium cant for a speed of 75 km/h

$$E = \frac{8.33V^2}{R} = \frac{8.33 \times 75^2}{437.5} \text{ mm}$$

$$= 107 \text{ mm}$$

But, maximum cant permitted is only 90 mm.

Maximum cant deficiency = 50 mm

Cant for calculating speed = 90 + 50 = 140mm

∴ $$140 = \frac{8.33V_m^2}{437.5}$$

or $$V_m = \sqrt{\frac{140 \times 437.5}{8.33}} \text{ km/h}$$

$$= 85.75 \text{ km/h}$$

∴ Maximum permissible speed = 85 km/h

## EXAMPLE 4.3

Determine the equilibrium speed for which superelevation is to be maintained for the following split-up of the trains and their speeds in the section:

12 trains at 50 km/h

8 trains at 60 km/h

4 trains at 70 km/h

1 train at 90 km/h

The track is on BG and the main line curve is a 2°- curve. Determine the superelevation to be provided allowing cant deficiency of 75mm, and the speed for this section.

**Solution**

$$\text{Equilibrium speed} = \text{Weighted average speed}$$
$$= (12 \times 50 + 8 \times 60 + 4 \times 70 + 1 \times 90)/25$$
$$= 58 \text{ km/h}$$

Equilibrium cant for this peed

$$E = \frac{13.78V^2}{R} = \frac{13.78 \times 58^2}{875}$$
$$= 53 \text{ mm}$$

$$\left(\text{R for a } 2°\text{- curve} = \frac{1750}{2} \text{ m} = 875 \text{ m}\right)$$

Allowing the maximum permissible cant deficiency of 75 mm, the cant for which the speed is to be obtained is (53+ 75) mm = 128 mm

$$\therefore \quad 128 = \frac{13.75 \times V_m^{\ 2}}{875}$$

or
$$V_m = \sqrt{\frac{128 \times 875}{13.78}} = 90 \text{km/h}$$

Maximum permissible speed for this section = <u>90 km/h</u>

## EXAMPLE 4.4

A turnout takes off as a 4°– curve with contrary flexure from a BG main line on a 2°- curve. The speed on the branch line is restricted to 35 km/h. Determine the maximum permissible speed on the main line.

**Solution**

Cant on the branch line

$$E = \frac{13.78V^2}{R} = \frac{13.78 \times 35^2}{437.5} \text{ mm}$$
$$= 38.6 \text{ or } 40 \text{ mm}$$

(R for a 4°-curve is $\frac{1750}{4}$ m or 437.5 m)

Negative cant for the branch line

$$E_b = E - E_d = (40 - 75) = -35 \text{ mm}$$

since the permissible cant deficiency is 75 mm

This becomes the positive cant for the main line.

$$E_a = +35 \text{ mm}$$

Adding the permissible cant deficiency of 75 mm to this, we get the theoretical cant for which the permissible speed ($V_m$) on the main line is to be calculated.

$\therefore$ $E$ for the main line = (35 + 75) = 110 mm

Since this is less than the maximum permissible cant for BG, which is 140 mm,

$$E = 110 \text{ mm}$$

$$110 = \frac{13.78 \times V_m^2}{R} = \frac{13.78 \times V_m^2}{875}$$

$$\left(R = \frac{1750}{2} = 875 \text{ m for the main line}\right)$$

$$\therefore \quad V_m = \sqrt{\frac{110 \times 875}{13.78}} = 83.6 \text{km/h}$$

or <u>80 km/h</u>

## EXAMPLE 4.5

Calculate the superelevation, maximum permissible speed, and the length of transition for a 3°-curve on a BG section with a maximum sanctioned speed of 90 km/h. Assume the equilibrium speed to be 75 km/h and the booked speed of the goods train to be 50 km/h.

**Solution**

$$\text{Radius of curve} = \frac{1750}{D} = \frac{1750}{3}$$
$$= 583.3 \text{ m}$$

Equilibrium superelevation for 75 km/h

$$= \frac{GV^2}{127R} = \frac{1750 \times 75^2}{127 \times 583.3}$$
$$= 133 \text{ mm}$$

Equilibrium superelevation for maximum sanctioned

$$\text{speed of 90 km/h} = \frac{1750 \times 90^2}{127 \times 583.3} = 191 \text{mm}$$

$$\text{Cant deficiency} = 191 - 133 = 58 \text{ mm } (< 100 \text{ mm, so OK.})$$

$$\text{Actual cant} = 191 - 58 = 133 \text{ mm } (< 165 \text{ mm, so OK.})$$

Equilibrium cant for a goods train speed of 50 km/h

$$= \frac{1750 \times 50^2}{127 \times 583.3} = 59 \text{mm}$$

$$\text{Cant excess} = \text{Actual cant} - 59 \text{ mm } (133 - 59)$$
$$= 74 \text{ mm } (< 75 \text{ mm, so OK})$$

So the actual cant of 135 mm should be provided (rounding off to the higher multiple of 5mm).

Safe speed or speed potential $= 0.27\sqrt{(135 + 100) \times 583.3} = 99.6 \text{km/h}$, or 100 km/h (approx.)

The maximum permissible speed is the lower value of <u>90 km/h</u>. Length of transition is the maximum of the following:

(i) Rate of change of cant of 35 mm/s

$$L = 0.008\,(E_a \cdot V_m) = 0.008 \times 135 \times 90 = 97.2\text{ m}$$

(ii) Rate of change of cant deficiency of 35mm/s: $L = 0.008\,(E_d \cdot V_m) = 0.008 \times 135 \times 90 = 72\text{ m}$

(iii) Cant gradient of 1 in 720:

$$L = 0.72 \times 135\text{ m} = 97.2\text{ mm}$$

$\therefore$ Length of transition = 98 mm (or 100 m if space permits).

## EXAMPLE 4.6

Calculate the maximum permissible speed on a 1°-curve on a high-speed BG route with maximum sanctioned speed of 120 km/h. The superelevation provided is 55 mm and the length of transition is 55 m, which cannot be increased due to the proximity of the yard.

**Solution**

$$\text{Radius of the curve} = \frac{1750}{D} = \frac{1750}{1}$$

$$= 1750\text{ m}$$

Safe speed on the curve (theoretical value)

$$V = 0.27\sqrt{(E_a + E_d)\cdot R} = 0.27\sqrt{(55+100)\cdot 1750}$$

$$= 140.6\text{ km/h}$$

Speed based on transition length:

(i) Rate of change of cant (max.55 mm/s):

$$V_m = \frac{198L}{E_d} = \frac{198\times 55}{55}$$

$$= 198\text{ km/h}$$

(ii) Rate of change of cant deficiency (max. 55 mm/s):

$$V_m = \frac{198L}{E_d} = \frac{198\times 55}{100}$$

$$= 108.9\text{ km/h}$$

The smaller value is 108 km/h

(iii) Cant gradient (not more than 1 in 720):

$$1 \div \frac{55\times 1000}{55} = \text{or 1in1000 (less than 1 in 720, so O.K.)}$$

The maximum permissible speed in the section would be the least of the sanctioned speed of 120 km/h and the theoretical safe speed of 140 km/h.

However, if the cant deficiency can be made less than the maximum of 100 mm, the maximum permissible speed over the circular curve can be increased.

Hence, the maximum permissible speed over the curve can be kept at 120 km/h and that over the restricted transition at 108 km/h.

## EXAMPLE 4.7

Two high-level platforms are to be provided on the inside as well as the outside of a 3°-curve on a BG track with a superelevation of 150 mm. Determine the extra clearances required for these platforms, both on the inside and outside of the curve. Assume the length of the bogie as 21 m, centre-to-centre bogie distance, 14 m, and height of platform as 840 mm.

### Solution

$$\text{Radius of the curve} = \frac{1750}{3} = \frac{1750}{3} = 583.3 \text{ m}$$

Extra clearance required on the inside of the curve:

$$ec_i = \text{Overthrow} + \text{Lean} + \text{Sway} - 51$$

$$= \frac{C^2}{8R} + \frac{h \cdot E}{G} + \frac{1}{4} \cdot \frac{hE}{G} - 51$$

$$L = 21000 \text{ mm};$$

$$C = 14000 \text{ mm};$$

$$E = 150 \text{ mm};$$

$$G = 1750 \text{ m};$$

$$h = 840 \text{ mm}.$$

$$\therefore \quad ec_i = \frac{(14000)^2}{8 \times 583300} + \frac{840 \times 150}{1750} \times \frac{5}{4} - 51$$

$$= 81 \text{ mm or } 80 \text{ mm (approx.)}$$

Extra clearance required on the outside of the curve:

$$ec_o = x_e - 25 = \left( \frac{L^2 - C^2}{8R} \right)$$

$$= \frac{(1000)^2 (21^2 - 14^2)}{8 \times 583300} 25$$

$$= 27.5 \text{ mm.}$$

$$= \underline{30 \text{ mm (approx.)}}$$

## EXAMPLE 4.8

The wheel base of a vehicle on a BG track is 6 m. The wheel diameter is 1500 mm and the flanges project 30 mm below the top of the rail. Radius of the curve is 175 m. Determine the extra width of gauge required.

### Solution

$$\text{Lap of flange, } L = 0.02\sqrt{h^2 + Dh}$$

$$h = 3 \text{ cm}, D = 150 \text{ cm}$$

$$L = 0.02\sqrt{3^2 + 150 \times 3}\,\text{m} = 0.43\,\text{m}$$

$$\text{Extra width of gauge, } G_e = \frac{13(B+L)^2}{R}$$

Wheel base, $B = 6$ m; $R = 175$ m

$$G_e = \frac{13(6+0.43)^2}{175}\,\text{cm} = 3\,\text{cm}$$

$$= 30 \text{ mm}$$

In Railway practice, an extra width of 10 mm is considered adequate for curves with a radius less than 400 m.

## EXAMPLE 4.9

Calculate the length of the vertical curve between a rising gradient of 1.5 % and a falling gradient of 1%. Compare the values obtained from the earlier practice and from the new method.

**Summit curve:**

Change of gradient, $G = 1.5 - (-1) = 2.5$ %

Rate of change of gradient for summit curves is not to exceed 0.1% on a 30.5 m chord.

∴ Length of the vertical curve, $L = \frac{2.5}{0.1} \times 30.5 = 762.5$ m

According to the new method,

$$L = R.\,G$$

$R = 4000$ m for BG high speed routes.

$$\therefore \quad L = \frac{4000 \times 2.5}{100} = \underline{100 \text{ m}}.$$

This is considered to be realistic and easy to set in the field.

## EXAMPLE 4.10

For the vertical curve of Example 4.9, the intersection point has a chainage of 1000 m and its RL is 100 m. Calculate the RL and the chainage of various points on the curve in order to set out the vertical curve at 10 m intervals.

### Solution

The chainage of intersection of the gradients is given as 1000 m.

∴ Chainage of the first tangent point or

$$\text{Beginning of the curve} = 1000 - \frac{100}{2} = 950 \text{ m}$$

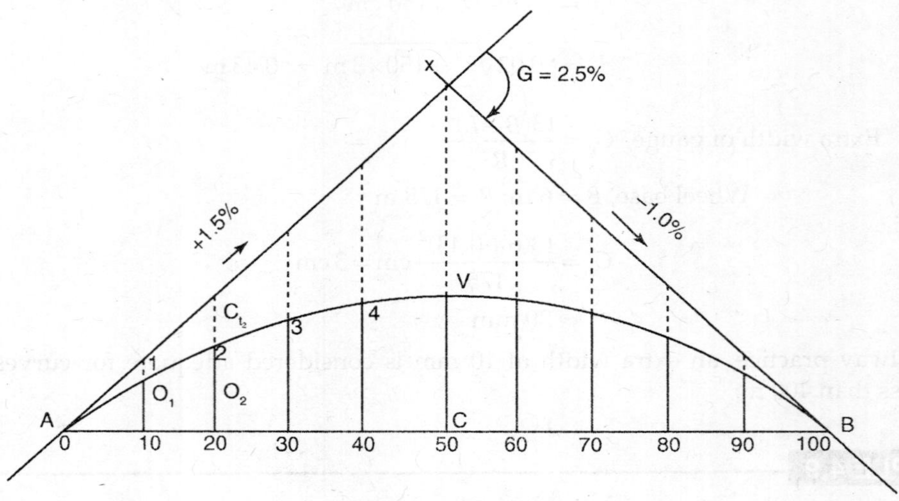


**FIG 4.19** *Vertical summit curve (Example 4.10)*

Chainage of the end of the curve = 950 + 100 = 1050 m

$$\text{RL of } A = 100 - \frac{1.50}{100} \times 50$$

$$= 99.25 \text{ m}$$

$$\text{RL of } B = 100 - \frac{1}{100} \times 50$$

$$= 99.50 \text{ m}$$

Increase in RL for 100 m = 99.50 – 99.25 = 0.25 m

Increase in RL for 10 m = 0.025 m

The vertical curve is shown in Fig. 4.19.

There are two approaches to obtain the RL's of the point on the vertical curve.

In the first, the RL's of points on the first gradient can be got by adding the increase in level for each chord until the RL of X is got. The tangent corrections for each point on the gradient can be computed using $y = cx^2$, and subtracted from the RL's of points on the grade, to obtain the RL's of the respective points on the curve. A similar procedure is used to obtain the RL's for the other half of the curve; the only difference is in obtaining the RL's of the points on the falling grade using the gradient for it.

In the second approach, the offsets $O_1$, $O_2$, ... can be got from the formula

$$O_{1,2} \ldots = \frac{x(L-x)}{2R}$$

*R* here is 4,000 m for BG as recommended by Indian Railways. These are added to the RL's of the respective chord points between A and B, obtained proportionate to the difference in the RL's of A and B to gives the RL's of the points on the vertical curve.

The results obtained from the second approach are shown in the following table.

**Offsets from long chord method**

| Chainage (m) | RL of point (m) | Offset (m) | RL of points on the curve (m) |
|---|---|---|---|
| A 950 | 99.2500 | – | 99.25000 |
| 960 | 99.2750 | $\frac{10\times 90}{2\times 4000}=0.1125$ | 99.3875 |
| 970 | 99.3000 | $\frac{20\times 80}{2\times 4000}=0.2000$ | 99.5000 |
| 980 | 99.3250 | $\frac{30\times 70}{2\times 4000}=0.2625$ | 99.5875 |
| 990 | 99.3500 | $\frac{40\times 60}{2\times 4000}=0.3000$ | 99.6500 |
| V 1000 | 99.3750 | $\frac{50\times 50}{2\times 4000}=0.3125$ | 99.6875 |
| 1010 | 99.4000 | $\frac{60\times 40}{2\times 4000}=0.3000$ | 99.7000 |
| 1020 | 99.4250 | $\frac{70\times 30}{2\times 4000}=0.2625$ | 99.6875 |
| 1030 | 99.4500 | $\frac{80\times 20}{2\times 4000}=0.2000$ | 99.6500 |
| 1040 | 99.4750 | $\frac{90\times 10}{2\times 4000}=0.1125$ | 99.5875 |
| B 1050 | 99.5000 | – | 99.5000 |

The points on the curve can be accessed by making the cuts necessary to reach the particular elevation, and pegged out.

The method of tangent corrections is also illustrated below:

$$\text{RL of C (the mid-point of AB)} = \frac{1}{2}(99250 + 99.500)$$

$$= 99.3750$$

$$\text{RL of the vertex, V, of the curve} = \tfrac{1}{2}(\text{RL of } X + \text{RL of C})$$

$$= \tfrac{1}{2}(100.000 + 99.375) = 99.6875$$

$$\therefore \quad \overline{CV} = 99.6875 - 99.3750$$

$$= 0.3125m = \overline{XV}$$

This is the tangent correction for the vertex, V. Tangent corrections for the intermediate points.

1, 2, 3, 4, ... can be got as follows:

$$c_{t_1} = \left(\frac{10}{50}\right)^2 \times 0.3125 = 0.0125\text{m}$$

$$c_{t_2} = \left(\frac{20}{50}\right)^2 \times 0.3125 = 0.0500\text{m}$$

$$c_{t_3} = \left(\frac{30}{50}\right)^2 \times 0.3125 = 0.1125\text{m}$$

$$c_{t_4} = \left(\frac{40}{50}\right)^2 \times 0.3125 = 0.2000\text{m}$$

The elevations of the corresponding points on the gradient AX can be got easily from the known value of $x$ and the gradient 1.5%.

The RL's of the points 1, 2, 3, 4, ... on the curve are obtained by subtracting the tangent corrections from the grade elevations.

For the other half of the curve the tangent corrections of points starting from the other end of the curve are the same as those for the first half.

Only the grade elevations differ as the other gradient is 1 % and they may be easily got. The RL's of the points on the curve can be obtained by subtracting the tangent corrections from the corresponding grade elevations.

The results obtained from this approach are tabulated in a similar way, as below:

**Tangent corrections method**

| Chainage (m) | Grade elevation (m) | Tangent correction (m) | Curve elevation (m) |
|---|---|---|---|
| A 950 | 99.2500 | – | 99.2500 |
| 960 | 99.400 | 0.0125 | 99.3875 |
| 970 | 99.5500 | 0.0500 | 99.5000 |
| 980 | 99.7000 | 0.1125 | 99.5875 |
| 990 | 99.8500 | 0.2000 | 99.6500 |
| V 1000 | 100.0000 | 0.3125 | 99.6875 |
| 1010 | 99.9000 | 0.2000 | 99.7000 |
| 1020 | 99.8000 | 0.1125 | 99.6875 |
| 1030 | 99.7000 | 0.0500 | 99.6500 |
| 1040 | 99.6000 | 0.0125 | 99.5875 |
| B 1050 | 99.5000 | – | 99.5000 |

(It is observed that the results are precisely the same as got earlier) The other main method is the 'chord gradients' method.

A *chord gradient* is the difference in elevation between consecutive points on the vertical curve. Equations may be derived for chord gradients and applied consecutively, starting from either end of the curve, to obtain the RL's of the intermediate points on the vertical curve. It can be shown that this method also yields exactly the same results.

## SUMMARY

- Geometric design of a railway track includes all parameters relating to the geometry – primarily those related to the horizontal alignment and the vertical alignment.
  Proper design of these parameters is necessary for ensuring efficient and safe operation of trains to carry loads at the maximum permissible speed. Appropriate design also minimises maintenance needs.
- Gradient in the vertical alignment becomes necessary when the track cannot be kept horizontal due to terrain or other considerations.
  The types of gradients used in railways are the ruling (or maximum permissible) gradient, momentum gradient (used in valleys to utilise the momentum gained on the falling grade to go up the rising grade), pusher or helper gradient (used in hill railways), needing an extra engine, and the flatter gradients used in station yards.
- The gradient has to be reduced or compensated with respect to the ruling value on horizontal curves in order to keep the effect of curve resistance to traction nullified. This is called *grade compensation* on horizontal curves.
- The curves in the horizontal alignment are classified as simple, compound and reverse curves. A simple curve consists of a single circular arc to achieve the desired deviation or deflection of direction; a compound curve achieves it by means of two circular arcs of different radii, curving in the same direction; and, a reverse curve consists of two circular arcs of the same radius or different radii curving in opposite directions.
- A circular curve can be specified either by its radius or its 'degree'. The degree of the curve is defined as the angle subtended at the centre by a standard arc length (30.5 m, the equivalent of 100 ft); for this, the degree is given as 1750/R (approx.)
  For curves with very large radius, the arc length and corresponding chord length do not differ much; thus, it is immaterial whether standard length is that of the chord or the arc, especially in railways where the curves used are very flat, i.e., with relatively large radii.
  The degree and radius are inversely related to each other. Different methods of setting out using offsets and angles are available.
- The mid-ordinate of a simple circular curve is also called the versine of the curve. This has practical use in the maintenance of curves in railways.
  Another interesting fact regarding versine is that when the versine is measured on 11.8 m chord, its value in centimetres equals the degree of the curve. The maximum degree (or minimum radii) values to be used on BG and MG tracks are prescribed by Indian Railways.
- The elements of compound curves and reverse curves can be calculated from the geometry of these curves; using these calculations, the curves may be set out using either offsets or tangential angles. Reverse curves between parallel tangents are used popularly for cross-over between parallel tracks in railways.
- Superelevation or cant is provided on horizontal curves to counteract the centrifugal force and its ill-effects; this consists of raising the outer rail above the inner. Superelevation causes the pressure on the outer and the inner rails to be equalised; further, rail head wear of the outer rail is reduced and discomfort to the passengers is minimised. The sum of the gauge and the width of rail head, called 'dynamic gauge', is used in the calculations for cant.
- The cant calculated for a particular speed from the standard equation $E = \frac{GV^2}{127R}$ is called the equilibrium cant for that speed.
  Since several types of trains with varying speeds run in a section of a railway line, a particular compromise value has to be chosen to provide the cant on the curve, this speed

is called the 'equilibrium speed'. It is the weighted average of the speeds of all the trains which operate in the section.

Maximum values of cant are prescribed by Indian Railways for BG and MG routes.

- The extent by which the actual cant provided falls short of that required for a higher value of speed of a fast train is called the 'cant deficiency'. On the other hand, the actual cant exceeds that required for slow trains, and this difference is known as the 'cant excess'.

  Maximum permissible values of cant deficiency and cant excess are specified by Indian Railways.

  When a branch line takes off in a curved manner from a curved main line of reverse curvature, a peculiar situation arises – the outer rail of the branch line becomes the inner one of the main line for a short distance. Because of the cant on the main line, a negative cant occurs for a short length of the branch line. Limits are specified for negative cant that can be permitted on Indian Railways.
- A 'transition curve' is a non-circular arc which permits gradual increase of curvature between a straight portion and a circular curve. This helps to reduce passenger discomfort that arises as a result of the centrifugal force acting on the curve.

  A transition curve serves another important function as the medium of gradual introduction of cant so that the full value of cant required on the circular curve is attained at the junction.
- The length of transition curve required is calculated based on the following criteria:

  (a) Time-rate of change of cant

  (b) Time rate of change of cant deficiency

  (c) Space-rate of change of cant.

  The maximum of these is to be adopted.
- The types of transition curves are Euler's spiral, cubic spiral, Bernoulli's lemniscate, and cubic parabola; the last of these is popularly used in railways because of the simplicity of setting out calculations.

  The circular curves get shifted due to the introduction of transitions.
- Transition curves are required to be introduced at the junctions of circular arcs in compound curves and reverse curves.
- Extra lateral clearances are required on curves due to the rigid wheel base of long vehicles –the effect of curvature causes over-throw (inside) and end-throw (outside). Besides this, the effect of leaning due to cant, and the sway effect also require additional clearances. The calculated values of total clearances should be reduced by 51 mm on the inside and 25 mm on the outside in the case of platforms.
- Extra widening of gauge on curves is provided to reduce rail-head wear; however, this is limited to 10 mm on BG to prevent accidents. Check rails are also provided on curves to reduce the lateral wear of rail-head, and also to prevent derailment.
- Safe speed on curves is determined based on the following criteria:

  (a) For transitioned curves – the equilibrium speed calculated for the sum of cant and maximum cant deficiency.

  (b) Non-transitioned curves

  with cant on virtual transition (14.8 m on BG)

  with no cant-based on permitted cant deficiency

  (c) Curves with inadequate transition –based on limiting cant deficiency gradient.

  The minimum of these, subject to the maximum permissible sanctioned speed for the section based on several criteria, is taken as the safe speed on the curves in the section.
- Vertical curves are introduced when the change of gradient is more than 0.4 per cent. They may be summit curves or sag curves. Usually a parabolic arc is used for vertical curves;

the minimum radius is prescribed as 4000 m for BG high-speed routes. Length of a vertical curve is determined based on the maximum permissible rate of change of gradient – 0.1 percent for summits and 0.05 per cent for sags. This is supposed to yield very high lengths of transition, and so, a new method has been recommended by Indian Railways. This involves the product of the minimum radius and algebraic difference of the gradients, and it gives reasonable length for vertical curves.

Vertical curves are set out by the method of tangent corrections or by that of chord gradients; the former is the more popular method.

Indian Railways has made recommendation regarding the use of vertical curves and their avoidance at special locations such as transitions, turnouts, bridges, on horizontal curves, and in long tunnels.

## QUESTIONS AND PROBLEMS

1. Enumerate the parameters which are included in the geometric design of a railway track.
2. Explain the necessary features of a good geometric design of a railway track.
3. Define the 'gradient' of a railway track and the reasons for providing it.
4. Write short notes on:
   (i) Momentum gradient
   (ii) Ruling gradient
5. Write short notes on:
   (i) Pusher gradient
   (ii) Gradients in station yards
6. Explain grade compensation on horizontal curves and the norms for it on Indian Railways.
7. Explain the 'degree' of a simple circular curve and derive the relationship between degree and radius.
8. Enumerate the various methods of setting out a simple circular curve.
9. What is 'versine' of a curve? How is it useful in the determination of the degree of a curve in the field?
10. What is 'superelevation' and what are its functions?
11. Differentiate between 'equilibrium cant' and 'equilibrium speed' on a railway.
12. Write short notes on:
    (i) Cant deficiency
    (ii) Cant excess
    (iii) Negative cant
13. What are the maximum permissible values of cant and cant deficiency prescribed on Indian Railways?
14. Differentiate between compound curves and reverse curves. What are the circumstances which warrant their use on railways?

15. Differentiate between 'gauge' and 'dynamic gauge' of a railway?
16. What is a 'transition curve' and what are its functions?
17. What are the criteria used in fixing the length of transition curves in railways? Explain with the standards prescribed in this regard.
18. Enumerate the types of transition curves. Which one of these is preferred on Indian Railways and why?
19. Explain 'shift' when dealing with transition curves.
20. What are the locations in compound curves and reverse curves where transitions need be introduced?
21. Write brief notes on:

    (i) Extra lateral clearances on curves

    (ii) Gauge widening on curves.
22. What is 'safe speed on curves? Explain briefly how this is evaluated on Indian Railways.
23. What is the maximum sanctioned speed on a section of a railway route? How is it determined by the railway authorities?
24. What are 'vertical curves' and when are they needed? What is the limit of change of gradient below which vertical curves are not provided on Indian Railways?
25. Why is a parabolic arc preferred to a circular arc for vertical curves?
26. What are the methods of determining the length of vertical curves on Indian Railways? State the earlier practice, as also the new method.
27. Explain briefly the methods of setting out vertical curves. Which method is popularly used on Indian Railways?
28. What is the compensated gradient on a section with a ruling gradient of 1 in 180 on a BG route to accommodate a 3°-curve?
29. What is the equilibrium cant on a 3°-curve on a BG route for an equilibrium speed of 80 km/h? What is the permissible speed allowing the maximum cant deficiency, assuming the curve to be fully transitioned?
30. Determine the equilibrium speed for a section of a BG high-speed route with the following data of the number of trains running on it and their speeds:

    4 trains at 110 km/h

    8 trains at 80 km/h

    10 trains at 60 km/h

    8 trains at 50 km/h
31. A turnout takes off as a 5°-curve with contrary flexure from a BG main line on a 2°-curve. The speed on the branch line is restricted to 30 km/h. Determine the maximum permissible speed on the main line.
32. The actual cant provided on a transitioned curve of a BG route is 120 mm. Determine the length of the transition required if the rate of change of cant/cant deficiency is not to

exceed 35 mm/s and the cant gradient is not to exceed 1 in 720. Assume the sanctioned maximum speed on the section as 100 km/h.

33. Determine the extra lateral clearances required on the inside and outside of a 2°-curve of a BG track with a superelevation of 120 mm. Assume the following:

    Height of the vehicle = 3.35 m

    Over-all length of the bogie = 20 m

    Centre- to-centre bogie distance = 15 m

34. Determine the extra gauge-widening required with the following data:

    Wheel base of vehicle = 6.6 m

    Wheel diameter = 1800 mm

    Flange projection below top of rail = 30 mm

    Radius of the curve = 180 m

35. Calculate the length of a vertical curve for a summit where two grades of + 2% and – 1% meet. Compare the values obtained from the rate of change of gradient criterion and from the new method adopted by the Indian Railways.

36. (a) Write an explanatory rote on cant deficiency.

    (b) On a BG track routes involving high speeds, a transition curve of 95 m has been laid and a superelevation of 75 mm has been provided. The degree of the curve is 1° and the maximum sanctioned speed for the section is 170 km/h. Determine the maximum permissible speed.

37. (a) Explain negative superelevation and state the circumstance in which it occurs.

    (b) Write a short note on cant deficiency. What are the permissible values for different gauges in India?

    (c) On a BG track with a 4°-curve, if the speed is to be restricted to 68 km/h, what superelevation should be given after allowing for the permissible cant deficiency? (JNTU-H)

38. What should be the equilibrium cant on a MG track 6° curve for an average speed of 60 km/h? Also, find out the permissible speed allowing the maximum cant deficiency. (JNTU-H)

39. A 5°-curve diverges from a 3°-main curve in the reverse direction in the layout of a BG yard. If the speed on the branch line is restricted to 35 km/h, determine the restricted speed on the main line. (VTU)

40. Determine the equilibrium cant on a 2-degree curve on a broad gauge, if 16 trains, 10 trains, 8 trains, 4 trains, and 2 trains are running at speeds of 50, 60, 70, 80, and 100 km/h, respectively. (VTU)

41. In a layout in a BG yard, an 8°-curve diverges from a 5°-main curve in a reverse direction. If the maximum permissible speed on the main curve is 65 km/h, determine the restricted speed on the diverging curve. (VTU)

42. A 6°-curve branches off from a 6°-main curve in an opposite direction in the layout of a BG yard. If the speed on the branch line is limited to 35.5 km/h, determine the restricted speed on the main line. (Punjab Technical University)

43. Calculate the superelevation and the maximum permissible speed for a 2°-BG transitioned curve on a high-speed route with a maximum sanctioned speed of 110 km/h. The speed for calculating the equilibrium superelevation, as decided by the Chief Engineer, is 80 km/h and the booked speed of goods trains is 50 km/h (UP Tech uny.)
44. Determine the length of transition curve for MG 4°-curve, having a cant of 8 cm. The maximum permissible speed on the curve is 60 km/h. What is the shift of the curve due to the transition? (Kalinga Tech. Uny., Odisha)
45. A sag curve joins a downgrade of 1% followed by an upgrade of 0.6%. Determine the length of the parabolic vertical curve if the allowable rate of change of gradient is 0.2% for a chain length of 30 m. (Kalinga Tech. Uny., Odisha)

Chapter 5

# POINTS AND CROSSINGS

**CHAPTER OBJECTIVES**

This chapter elucidates:

- The need and functions of points and crossings in railways to transfer a train from one track to another
- The design aspects of the components of switches constituting points and crossings
- The methods of design of a turnout and level crossings, their types and design aspects.

## 5.1 INTRODUCTION

Special arrangements are needed to transfer railway vehicles from one track to another. The tracks may be parallel to each other, diverging from each other or converging towards each other. Such arrangement used to change the course of locomotives and carriages from one track to another are called 'points and crossings'; these are necessary because the wheels of railway vehicles are provided with inside flanges, and consequently, which are used for navigating from one track to another.

The points, also called 'switches', help in diverting the vehicle, and the 'crossings' are provided with gaps in the rails to help the flanged wheels roll over them in the desired direction.

One complete set of points and crossings constitute what is known as a 'turnout'.

(Turnings on railway lines are thus in contrast to those on highways where no special arrangement is needed for a vehicle to move from one road to another which takes off or meets the first.)

## 5.2 TURNOUTS

As mentioned earlier, a turnout consists of one set of switches and crossing. In fact, the simplest arrangement of points and crossing is on a turnout taking off from a straight track.

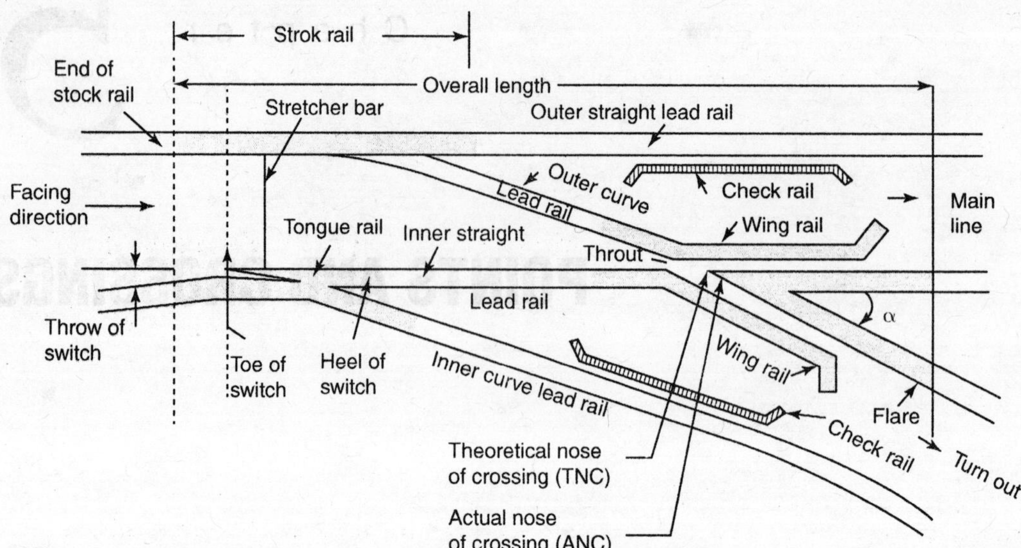


**FIG. 5.1** *Schematic plan view of a right-hand turnout*

### Left-hand and Right-hand Turnouts

A turnout is designated as a right-hand turnout or a left-hand turnout, depending on the direction in which it diverts the traffic—to the right or to the left.

### Elements of a Turnout

A schematic plan view of a right-hand turnout is shown in Fig. 5.1.

### Main Parts of a Turnout

**Set of switches:** A pair of stock rails, a pair of tongue rails, a pair of heel blocks, slide chairs, two or more stretcher bars, and a gauge tie plate.

**Crossing:** A nose consisting of a point rail and a splice rail, two wing rails, and two check rails.

**Lead rails:** Four sets of lead rails.

A schematic line diagram of a left-hand turnout is shown in Fig. 5.2.

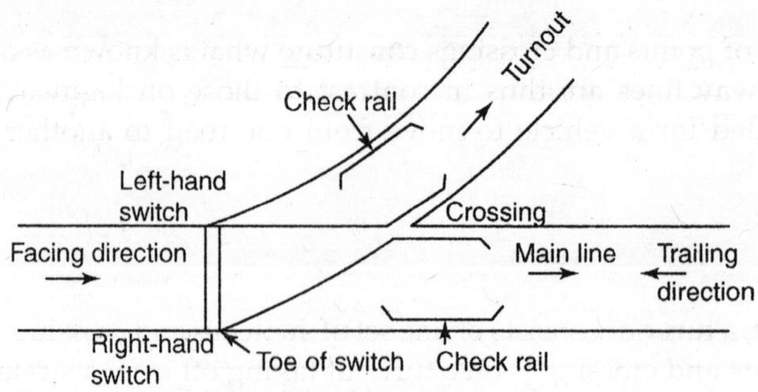


**FIG. 5.2** *Schematic line diagram of a left-hand turnout*

When standing at the toe of a switch, the direction towards the crossing is called the *facing direction*; the direction opposite to this is called the *trailing direction*. When the vehicle moves in the facing direction, it traverses the point or switch first, from its thin end to the thick end, and then negotiates the crossing; the reverse would be the case when it moves in the trailing direction.

Thus, the important parts of a turnout are the following:

**Stock rail:** It is the running rail against which a tongue rail operates.

**Tongue rail:** It is a tapered rail, attached to a running rail at its thicker end, made movable through a hinge at the heel of the switch.

It is made of high-carbon steel or manganese steel so that it resists wear at its sharp end.

**Points or switch:** A switch consists of tongue and stock rails with necessary fittings and connections.

**Crossing:** This a device introduced at the junction where two rails cross each other to permit the wheel flange of a vehicle to pass from one track to another.

**Check rails:** These are introduced at the opposite side of the crossing on both the tracks – the main and the turnout.

## 5.3 POINTS OR SWITCHES

The details of a set of points or switches are shown in Fig. 5.3.

The important parts of a pair of points or switches are:

(i) A pair of stock rails (AB and CD), of medium-manganese steel
(ii) A pair of tongue rails (PQ and RS), also called 'switch rails', of medium-manganese steel to withstand wear. The end of the tongue rail is machined to a very thin section for a snug

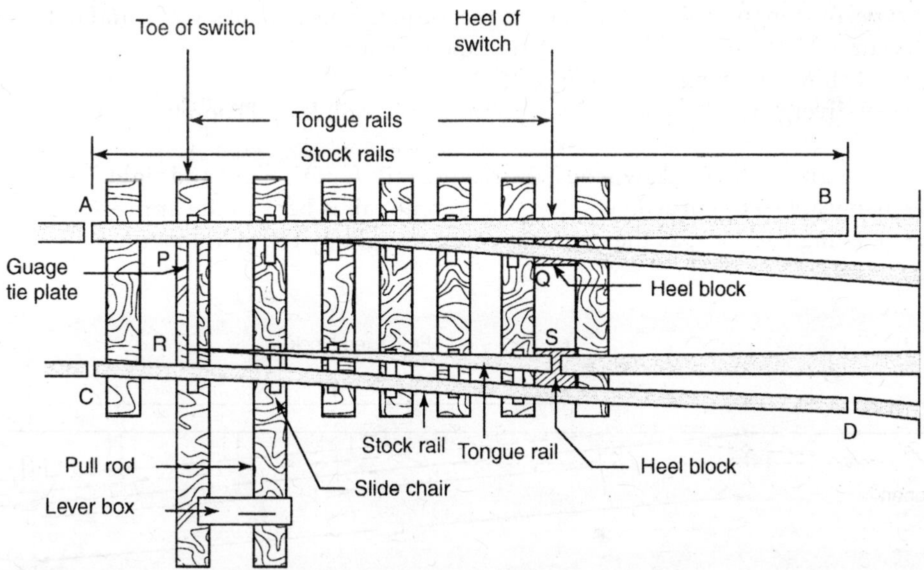


**FIG. 5.3** *Details of a switch*

fit with the stock rail; this tapered thin end is the 'toe' and the other thicker end is the 'heel'.

(iii) Two more stretcher bars connecting the tongue rails very near the toe, in order to hold them at a fixed distance from each other.

(iv) A pair of heel blocks to hold the heel of the tongue rails at the standard clearance from the stock rails.

(v) A number of slide chairs to support the tongue rail and enable its movement towards or away from the stock rail.

(vi) A gauge tie plate to fix and ensure correct gauge at the points or switches.

## Types of Switches

There are two types of switches—*stud switch* and *split switch.*

In the stud type, no tongue rail is provided; instead, a small length of the track is moved from one side to the other. As this type of switch is no longer used on Indian Railways, further details are not given.

Split switches have come into vogue and are commonly used on Indian Railways. A split switch consists of a pair of stock rails and a pair of tongue rails.

*Split switches* are again classified into two types:

i Fixed heel or spring type switches
ii Loose heel type switches

### Fixed Heel Switch

In this type of design, the tongue rails are rigidly connected to stock rails at the heel, by means of distance blocks, bolts and nuts. The movement of the blades of the tongue rails is effected by springing from the first fixed block.

Again, in the fixed heel switch, there are two types—*straight switch* and *curved switch;* in the straight type, the tongue rail is straight from the toe to the heel of the switch, and in the curved type, it is curved to the same radius as that of the turnout.

Figure 5.4 shows a schematic of a fixed heel switch.

($d$ = Heel divergence; $L_{ta}$ = Actual length of tongue rail; $L_{tt}$ = Theoretical length of tongue rail)

In both straight and curved switches, the tongue rails for left-hand and right-hand turnouts are interchangeable. For curved switches, the tongue rail is bent to the appropriate radius at the time of laying the track, as they are fabricated straight by the manufacturers.

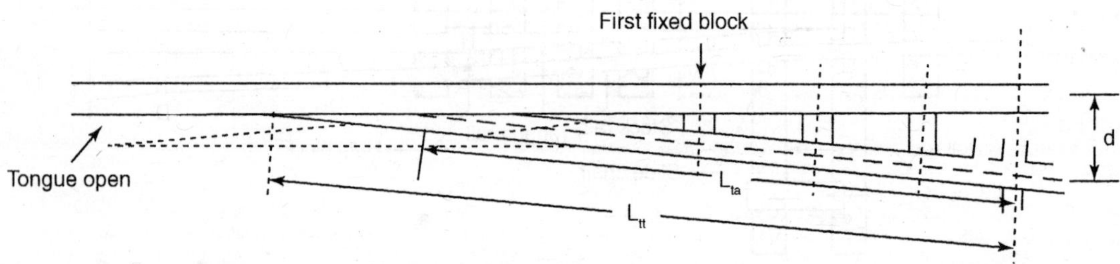


**FIG. 5.4** *Schematic of a fixed heel switch*

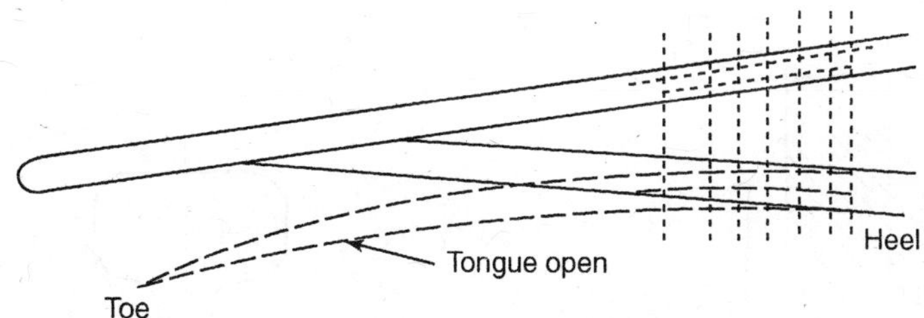


**FIG. 5.5** *Schematic of a loose-heel switch*

### Loose-heel Switch

In this type of design, the tongue rails pivot about heel joints, held in position by heel blocks and fish plates. The fishplate holding the tongue rail is bent suitably to permit free movement of the latter, and the fish bolts in this are kept loose. A schematic of a loose-heel switch is shown in Fig. 5.5.

In view of the relatively sharper change of curvature and consequent knocking of the wheels of vehicles, loose heel switches are not recommended for new layouts.

The following factors are to be considered in the determination of the length of tongue and stock rails:

(a) The tongue rail should be longer than the greatest distance between adjacent wheels of four-wheeled vehicles, whether they are of the same or adjacent vehicles; otherwise, the switch tends to rise and gape at the toe, each time a wheel load bears on the heel of the switch.
(b) Since the switch forms a virtual bend in the track, the switch angle should be as small as possible. In straight switches, longer switches achieve this.
(c) The stock rail should be sufficiently long, so as to keep the joint away from the tongue rail. In view of these considerations, the lengths of straight switches are 6400 mm and 4725 mm on BG for 1 in 12 and 1 in 8.5 crossings, respectively.

## Design of a Switch

The toe portion of a switch may be designed in two different ways; based on this, switches may be classified as — the *ordinary* or *undercut switch* and the *overriding switch.*

### Undercut Switch

In this type, the foot of the stock rail is planed to accommodate the tongue rail, as shown in Fig. 5.6.

### Overriding Switch

In this type, the foot of the tongue rail is planed to override that of the stock rail, the full section of which is maintained.

Overriding switches are standardised on Indian Railways.

The *advantages of overriding switch* are:

- Since the full section is maintained for the stock rail, the switch does not become weak.
- The planing is confined only to the tongue rail, and hence, it is economical.
- Although the tongue rail has a relatively thin edge of 6 mm thickness, it is supported by the stock rail for the whole of the weakened portion of the length; therefore, the combined strength of the two rails is greater than that of the tongue rail in the undercut switch.

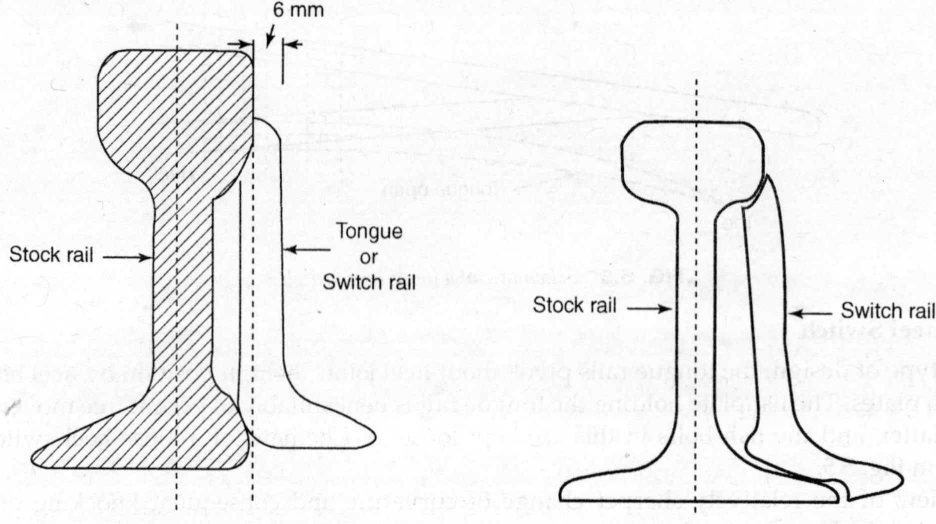


**FIG. 5.6** *Undercut switch*

**FIG. 5.7** *Overriding switch*

Figure 5.7 shows an overriding switch.

In the overriding switch the head of the tongue rail is kept higher than that of the stock rail by 6 mm, for the planed part of its length. This reduces the amount of undercut in the foot of the tongue rail. The difference in height is adjusted by placing special machined bearing plates behind the heel.

The only important disadvantage of an overriding switch is its relatively thin blade tip which is liable to damage under lateral forces imposed by the train loads.

**Important Definitions Relating to Switches**

**(i) Gauge tie plates** These are made out of steel flats (12 mm thickness for BG and 10 mm for MG), provided over the sleepers directly under the toe of the switches, and under the nose of the crossing to maintain exact gauge at these important points.

**(ii) Heel divergence:** This is the shortest distance between the gauge faces of the stock rail and the tongue rail at the heel of the switch and is made up of the flange-way clearance and the width of tongue rail head at the heel.

**(iii) Lead rails:** Ordinary or plain rails that are laid between the switches and the crossing assemblies.

**(iv) Left-hand or right-hand rails:** Stock rails, tongue rails and other fittings are designated left-hand or right-hand depending on their location when viewed at in the facing direction of the points.

**(v) Slide chairs:** These are provided under the tongue rails to provide smooth lateral movement of the rails.

**(vi) Stepped plates:** These are used to adjust the difference in height of the tongue rail and the stock rail at the heel of the switch.

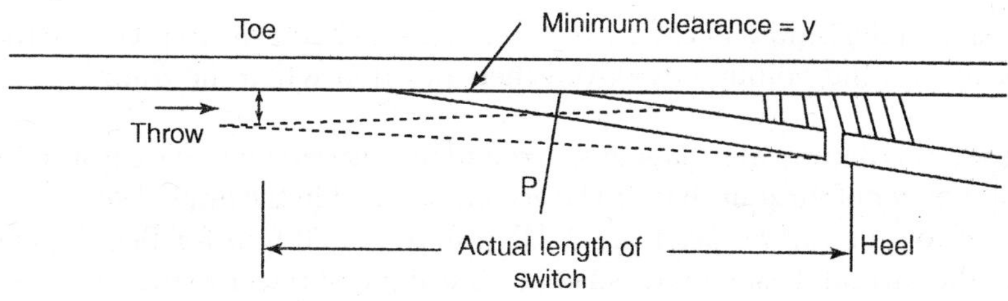


**FIG. 5.8** *Throw of switch*

**(vii) Stretcher bars:** Made of spring steel flats, these are rigidly fixed to the tongue rails by brackets; their purpose is to maintain the tongue rails at the correct distance apart.

**(viii) Switch angle:** This is the angle between the gauge faces of the stock rail and the tongue rail at its toe, when the switch is in the closed position. It is dependent upon the heel divergence and the length of the tongue rail.

**(ix) Switch length:** It is the free length of tongue rail from the toe to the heel of a switch.

**(x) Theoretical toe of switch:** It is the point of intersection of the gauge faces of tongue rail at the toe when extended, and the stock rail.

**(xi) Throw of switch:** It is the distance through which a tongue rail moves at its toe from its closed position to open position. This is fixed from the consideration of maximum flange-way clearance between the tongue rail and the stock rail at the point where the former attains the full head width and is closest to the stock rail in the open position (point P in Fig. 5.8).

Throw of switch is fixed as 115 mm for BG and 100 mm for MG, so as to get a minimum flange-way clearance of 44 mm and 41 mm respectively, at P.

**(xii) Flange-way clearance:** This is the distance between the adjoining faces of the running rail and the check rail/wing rail at the nose of the crossing. It is meant to provide free passage to wheel flanges. The values fixed for BG and MG are 44 mm (max. 48 mm) and 41 mm (max. 44 mm), respectively.

## Design Aspects of Tongue Rails

### Straight and Curved Tongue Rails

Tongue rails may be either straight or curved; straight-tongue rails are manufactured easily and can be used for right-hand as well as left-hand turnouts. The disadvantage of straight-tongue rails is that the trains get jolted due to abrupt change in the alignment.

Curved tongue rails are bent to the curvature of the turnout from the toe to the heel of the switch; they allow for smooth turning of trains, but their use is restricted to the specific curvature for which they are made.

### Length of Tongue Rails

The length of tongue rail from the heel to the toe varies with the gauge and the switch angle. For a fixed value of heel divergence, the longer the tongue rail, the smaller is the switch angle and consequently, the smoother is the entry to the switch. Also, the length of the tongue rail should be more than the rigid wheel base of a four-wheeled vehicle, to preclude the possibility of derailment in case the points move from their position when the train is running on the switch.

Longer tongue rails occupy more layout space, and this places some space restrictions in station yards, where a number of turnouts have to be accommodated in the limited space available.

The length of tongue rail ranges from 4,725 mm to 11,150 mm for BG, depending on the designation of the turnout based on its size, i.e., the angle of the crossing.

### Relation Between the Heel Divergence and Switch Angle

The relation between the heel divergence and the switch angle may be obtained from the following formulae:

$$\sin\beta = \frac{d}{L_{ta}} \text{ (ignoring the thickness of the tongue rail at toe)} \quad (5.1)$$

where $\beta$ = switch angle,

$d$ = heel divergence, and,

$L_{ta}$ = actual length of tongue rail (along its centre–line)

If the thickness of the tongue rail is considered,

$$\sin\beta = \frac{(d-t)}{L_{ta}} \quad (5.2)$$

where $t$ = thickness of the tongue rail at toe. The theoretical length of the tongue rail ($L_{tt}$) is its length measured along its inner face; this is somewhat greater than its actual length.

$$L_{tt} = \frac{d}{\sin\beta} \quad (5.3)$$

If the difference between the theoretical length and actual length of the tongue rail is $x$,

$$x = L_{tt} - L_{ta} = \frac{t}{\sin\beta} \quad (5.4)$$

(from Eqs. 5.2 and 5.3)

[These relationships can be easily derived from the geometry of Fig. 5.4, and are not given here.]

## 5.4 CROSSINGS

A *crossing* is a device introduced at the point where two rails cross each other so as to permit the flanges of the wheels of a vehicle to pass from one track to another. A gap is provided from the throw to the nose of the crossing, over which the wheel glides or jumps. As the wheel negotiates this gap, the other wheel is guided properly with the aid of a check rail; this prevents the possibility of the wheel flange striking the nose.

Figure 5.9 shows a crossing along with the component parts.

The *components of a crossing* are:

(i) Point rail and splice rail

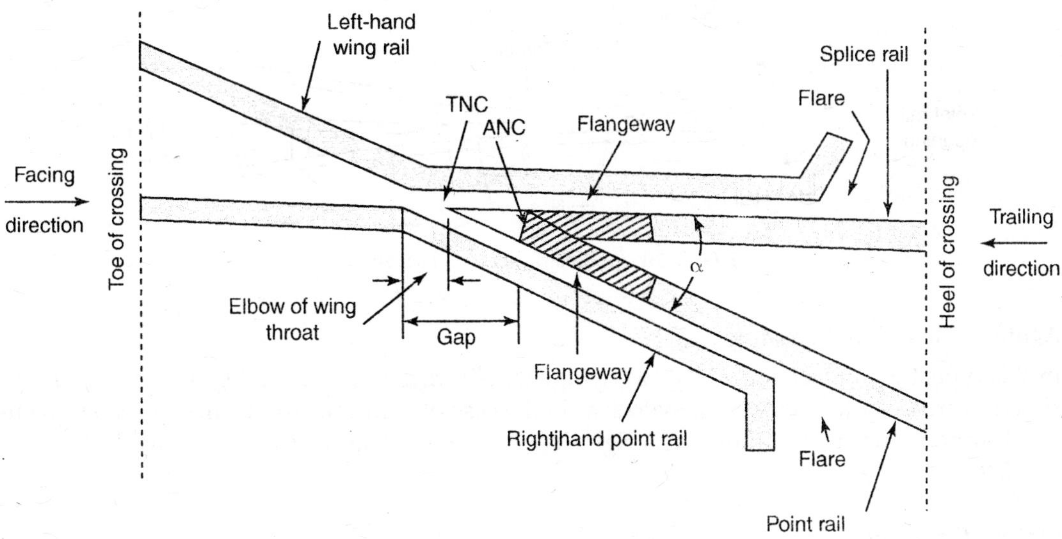


**FIG. 5.9** *A crossing and its components*

(ii) Wing rails
(iii) Check rails.

### Point Rail and Splice Rail

Point rail and splice rail are machined to form a nose. The point rail ends at the nose, and the splice rail joins it a little behind the nose. Although theoretically, the point rail should end in a point or in a knife edge, it is actually made to end in a blunt nose (thickness 6 mm) so that the fine end is not broken off under wheel loads.

The toe of the blunt nose is called the actual nose of crossing (ANC), and the theoretical point where the gauge faces of the point and splice rails meet is called the theoretical nose of crossing (TNC). The V-joint is planed to a depth of 6 mm at the nose and made to run out in 89 mm to prevent a wheel running in the facing direction from hitting the nose.

### Wing Rails

There are two wing rails – the right-hand and left-hand wing rails – which converge to form a throat, and diverge again on either side of the nose. These rails are flared at the ends to facilitate the entry and exit of the flanged-wheel in the gap.

### Check Rails

A pair of check rails guide the wheel flanges and provide a path for them, thereby preventing them from moving sideways; otherwise, it may result in the wheel hitting the nose of the crossing as it moves in the facing direction.

## Types of Crossings

Based on the geometry, crossings are classified as *acute-angled* or *Vee* crossings and *Obtuse* crossings.

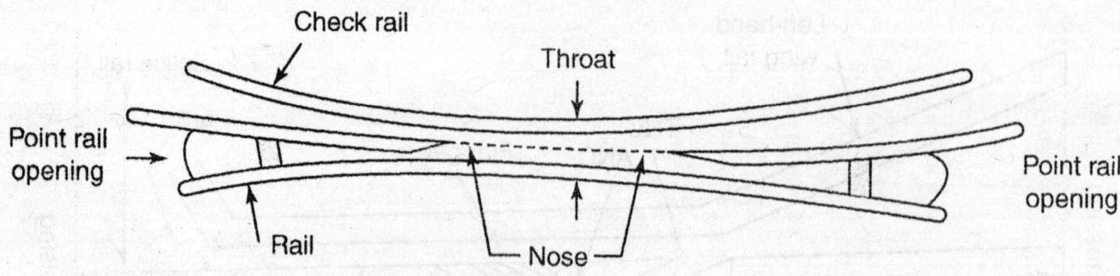


**FIG. 5.10** *Schematic of an obtuse crossing*

### Acute-angle or Vee Crossing

In this type, the intersection of the two gauge faces forms an acute angle; this occurs when a right rail crosses a left rail, or vice-versa, as in the case of turn-outs (in the first case, it is a left-hand turnout and in the other, it is a right-hand turnout). This appears as shown in Figs. 5.1 and 5.9.

### Obtuse Crossing

In this type, the intersection of the two gauge faces forms an obtuse angle; this occurs when the right rail (or the left rail) of one track crosses the right rail (or the left rail) of another track (at an angle other than 90°). This happens in the case of what is called a 'diamond crossing'. An obtuse crossing is shown in Fig. 5.10.

This is also called a *double crossing* since it has two point nails and two noses as shown.

Based on the method of manufacture, crossings can be classified as (i) Built-up crossing, (ii) Cast steel crossing, (iii) Combined built-up and cast steel crossing, and (iv) Spring crossing or movable crossing.

**(i) Built-up crossing:** In this type, the Vee-section consisting of point and splice rails, and the two wing rails are assembled together by means of bolts and nuts and distance blocks or pieces to form the crossing. This is commonly used on Indian Railways because of the advantage of low initial cost; however, this type lacks rigidity and requires frequent checking of bolts, which may get loosened, or broken under fast and heavy traffic. On the other hand, this type is also advantageous since each constituent may be repaired or replaced separately.

**(ii) Cast steel crossing:** This is a one-piece type of crossing made of cast steel; it is comparatively rigid and has high initial cost. The advantage is that it requires very little maintenance.

However, these crossings pose a number of maintenance problems in the event of their needing repair.

Recently, cast manganese steel (CMS) crossings have been have been developed by Indian Railways; these are stronger and offer more resistance to wear; hence, they have longer life. In fact, their average life is about four times that of a built-up crossing, which typically lasts for less than two years.

Another advantage of CMS crossing is the absence of bolts and other components that tend to get loosened under fast and heavy traffic. Therefore, CMS crossing are used for Group-A routes where the traffic density is more than 20 GMT.

**(iii) Combined built-up and cast-steel crossing:** This is a combination of the built-up and the cast-steel types; it consists of a cast steel nose finished to ordinary rail faces to form the two

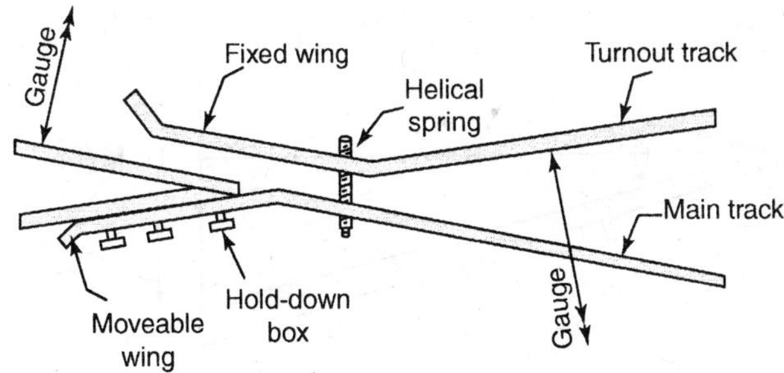


**FIG. 5.11** *Spring or movable crossing*

legs of the crossing. This allows easy repair of worn-out wing rails by welding, but the nose is liable to sudden fracture.

**(iv) Spring or movable crossing:** This is shown in Fig. 5.11.

In this type of crossing, one wing rail is movable and is held against the Vee of the crossing with a strong helical spring, while the other wing rail is kept fixed. When a vehicle passes on the main track, the movable wing rail is snug with the crossing and the vehicle; therefore, the vehicle does not negotiate any gap at the crossing. In case the vehicle has to pass over the turnout track, the movable wing is forced out by the wheel flanges, thereby enabling it to negotiate a gap of the same order as that in a normal turnout.

This type of crossing is considered especially useful when high-speed traffic passes over the main track, while low-speed traffic runs over the turnout traffic.

## Number and Angle of Crossing

A crossing is designated either by the angle which the gauge faces make with each other, or by the number of the crossing, which is more common. There are three methods of relating the angle of crossing, $\alpha$, and the crossing number, $N$, as given below:

### Centre Line Method

In this method, $N$ is measured along the centre line of the crossing (Fig. 5.12).

$$\cot\frac{\alpha}{2} = \frac{N}{(1/2)}$$

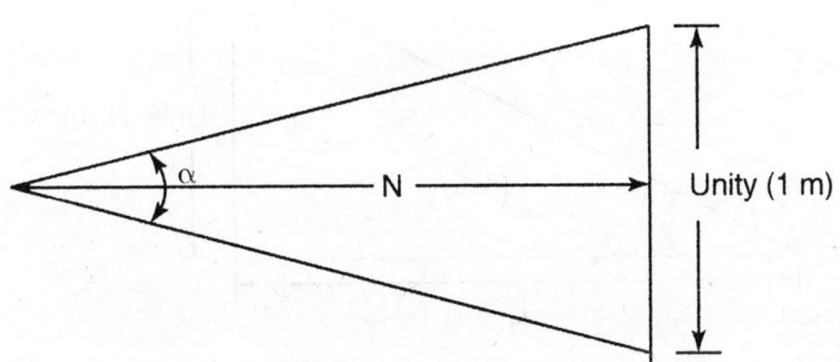


**FIG. 5.12** *Centre line method for N*

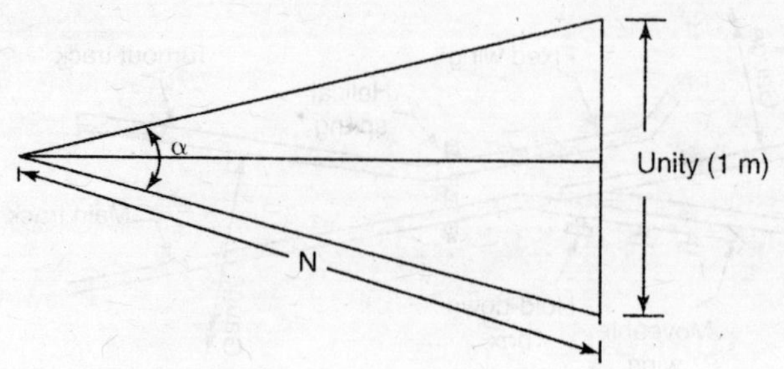


FIG. 5.13 *Isosceles triangle method for N*

or $$N = \frac{1}{2} \cot \frac{\alpha}{2} \tag{5.5}$$

(This method is used in the UK and the USA).

**Isosceles Triangle Method**

In this method, $N$ is taken as one of the equal sides of an isosceles triangle (Fig. 5.13).

$$\sin \frac{\alpha}{2} = \frac{1/2}{N} = \frac{1}{2N}$$

or $$\operatorname{cosec} \frac{\alpha}{2} = 2N$$

or $$N = \frac{1}{2} \operatorname{cosec} \frac{\alpha}{2} \tag{5.6}$$

**Right Angle Method**

In this method, $N$ is measured along the base of right-angled triangle (Fig. 5.14).

$$\cot \alpha = \frac{N}{1} = N$$

or $$N = \cot \alpha \tag{5.7}$$

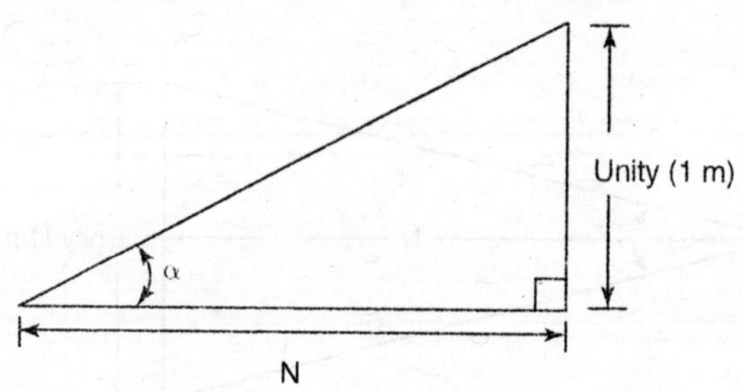


FIG. 5.14 *Right angle method for N*

Also known as Cole's method, this method is used on Indian Railways.

*N* being the cotangent of the angle formed by the two gauge faces, gives the smallest angle for the same value of *N*.

To determine the number of a crossing at the site of a railway track, the point where the offset gauge face of the turnout track is 1 m is marked. The distance of this point, in metres, from the theoretical nose of the crossing gives the crossing number, *N*.

## 5.5 DESIGN OF A TURNOUT

The simplest turnout is one which takes off from a straight main track, and consists of a set of points or switches and a crossing.

There are two methods for the design of such a turnout:

(a) *Cole's method*; and
(b) *IRS method*

The following terms are relevant to the design of turnouts:

**(i) Curve lead (*CL*):** This is the distance from the tangent point (*T*) to the theoretical nose of crossing (TNC), measured along the length of the main track.

**(ii) Switch lead (*SL*):** This is the distance from the tangent point (*T*) to the heel of the switch (*L*), measured along the length of the main track.

**(iii) Heel divergence (*d*):** This is the distance between the main line and the turnout line at the heel.

**(iv) Lead of crossing (*L*):** This is the difference between the curve lead and switch lead:

Lead of crossing (or crossing lead), $L = (CL - SL)$

**(v) Radius of turnout (*R*):** It is actually the radius of the centre-line of the turnout curve ($R_0$) plus half the gauge ($G$).

or
$$R = R_0 + \frac{1}{2}G$$

As the radius of the curve is very large,

$R \approx R_0$, for all practical purposes.

**(vi) Angle of crossing ($\alpha$):** It is the angle between the main line and the tangent of the turnout line.

### Cole's Method

This is a method used to design a turnout taking off from a straight track. The nomenclature is shown in Fig. 5.15.

The curvature begins at a point on the straight main track ahead of the toe of the switch (at the theoretical toe) and ends at the theoretical nose of the crossing. The offset of the curve at the heel of the switch (perpendicular to the main track) is the 'heel divergence'. If the tongue rail and the wing rail up to the TNC were curved, there would be no kinks in this layout; but, more commonly, they are straight. This results in kinks at three points – actual toe of the

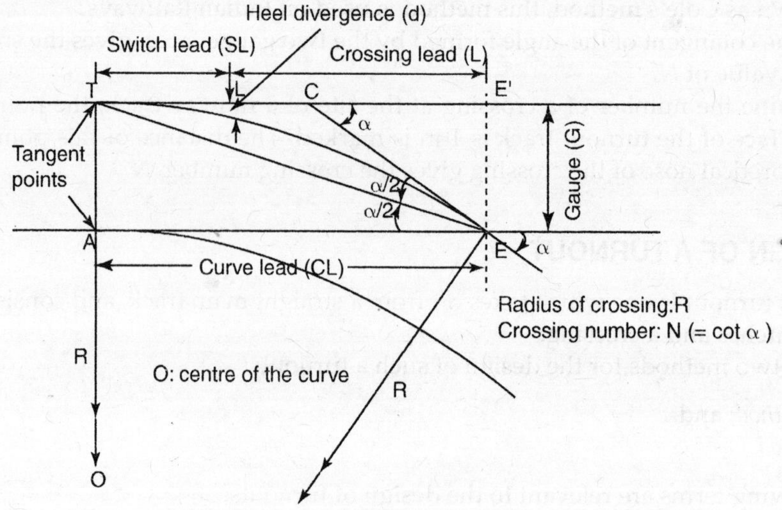


**FIG. 5.15** *Design of turnout (Cole's method)*

switch, heel of the switch and at the first distance block of the crossing. The calculations for the design are as follows:

**Curve Lead (*CL*)**

In $\Delta$ ATE, AT = G and $\angle$AET = $\alpha/2$.

$$\tan\tfrac{\alpha}{2} = \tfrac{AT}{AE} = \tfrac{G}{CL}$$

or
$$CL = G.\cot\tfrac{\alpha}{2} \quad (5.8)$$

Also,
$$\begin{aligned} \text{CL} &= \text{E'C} + \text{CT} \\ &= \text{E'C} + \text{CE} \\ &= \text{G} \cot\alpha + \text{G} \operatorname{cosec}\alpha \end{aligned}$$

$$\text{CL} = \text{GN} + \text{G}\sqrt{1+N^2} \quad (5.9)$$

or
$$\text{CL} = 2\text{GN (approx)} \quad (5.10)$$

**Switch Lead (*SL*)**

Heel divergence (*d*) is the offset for the length of tangent, TL.

From the properties of a circle,

$$\text{SL} \times \text{SL} = d\,(2R - d)$$

or
$$\text{SL} = \sqrt{2Rd - d^2} \quad (5.11)$$

**Crossing Lead (*L*)**

$$L = \text{CL} - \text{SL}$$

or
$$L = G\cot\tfrac{\alpha}{2} - \sqrt{2Rd - d^2} \quad (5.12)$$

or $$L = 2GN - \sqrt{2Rd - d^2} \text{ (approx)} \tag{5.13}$$

**Radius of Turnout (*R*)**

In Δ AOE,

$$OE = OT = R$$

$$OA = R - G$$

$$OE^2 = OA^2 + AE^2$$

$$OE^2 = (R - G)^2 + (CL)^2$$

or $$R^2 = (R - G)^2 + (GN + G\sqrt{1+N^2})^2$$

$$= R^2 - 2RG + G^2 + G^2N^2 + G^2(1+N^2) + 2G^2N\sqrt{1+N^2}$$

$$2RG = 2G^2(1+N^2) + 2G^2N\sqrt{1+N^2}$$

or $$R = G(1+N^2) + GN\sqrt{1+N^2} \tag{5.14}$$

or $$R = 1.5\,G + 2GN^2 \text{ (approx)} \tag{5.15}$$

## IRS Method

The layout of turnout for the IRS method of design is shown in Fig. 5.16.
This is the standard layout used by Indian Railways.

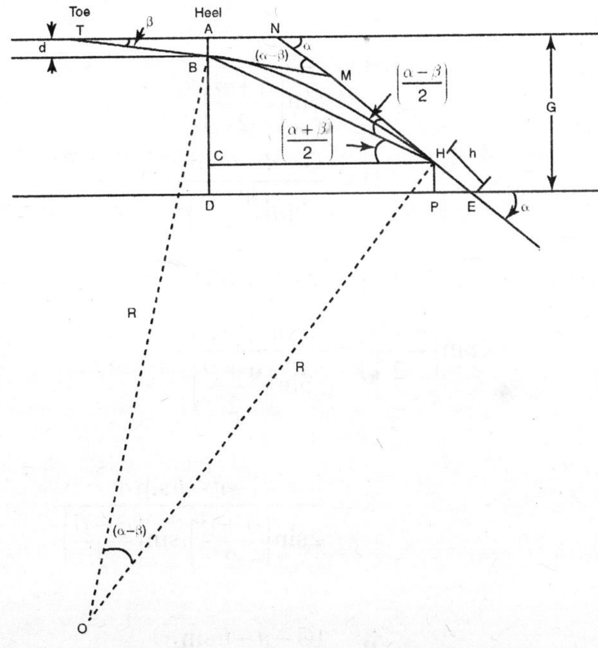


**FIG. 5.16** *Design of turnout (IRS method)*

In this layout, the curve begins from the heel of the switch and ends at the toe of the crossing (at the centre of the first distance block). The crossing is straight and no kink is felt at this point. The only kink occurs at the toe of the switch.

The calculations in this method are somewhat involved, and therefore, it is used only when precision is required.

**Crossing Lead (L)**

In Δ BMH,

$$BM = MH$$

$$\angle MHB = \angle MBH \frac{(\alpha - \beta)}{2}$$

$$\therefore \quad BHC = \frac{(\alpha + \beta)}{2} \text{ (from the geometry of the figure)}$$

$$BC = AD - (AB + CD) = G - (d + h \sin \alpha)$$

$$\therefore \quad L = (G - d - h\sin\alpha) \cot \frac{(\alpha + \beta)}{2} + h\cos\alpha \tag{5.16}$$

**Radius of Curve (R)**

In Δ OBH,

$$B\hat{O}H = \alpha - \beta$$

$$BH = 2R \operatorname{Sin}\left(\frac{\alpha - \beta}{2}\right) \tag{5.17}$$

In Δ BHC,

$$BH = \frac{BC}{\operatorname{Sin}\left(\frac{\alpha + \beta}{2}\right)}$$

$$BH = \frac{G - d - h\operatorname{Sin}\alpha}{\sin\left(\frac{\alpha + \beta}{2}\right)} \tag{5.18}$$

Equating the two expressions for BH (Eqs. 5.17 and 5.18),

$$2R\operatorname{Sin}\left(\frac{\alpha - \beta}{2}\right) = \frac{G - d - h\operatorname{Sin}\alpha}{\operatorname{Sin}\left(\frac{\alpha + \beta}{2}\right)}$$

or

$$R = \frac{G - d - h\sin\alpha}{2\sin\left(\frac{\alpha + \beta}{2}\right).\sin\left(\frac{\alpha - \beta}{2}\right)}$$

or

$$R = \frac{(G - d - h\sin\alpha)}{(\cos\beta - \cos\alpha)} \tag{5.19}$$

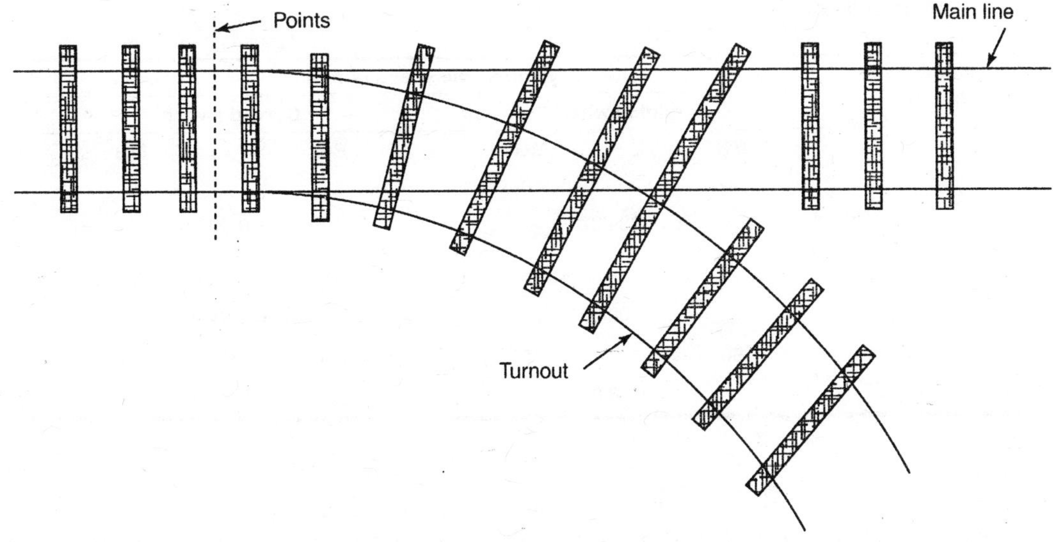


**FIG. 5.17** *Sleepers at points and crossings*

## Raised Check Rails for Obtuse Crossings

In order to provide a guided pathway in the throat portion of a BG obtuse diamond crossing, the check rails are raised by welding a 25 mm-thick MS plate.

## Sleepers at Points and Crossings

Sleepers are normally laid perpendicular to the track. At points and crossings, sleepers have to cater to the needs of both the main line and the turnout portion. Hence *longer sleepers* are used for some length of the track as shown in Fig. 5.17.

If the sleepers of one track lie in between those of another track, they are called *interlaced sleepers*.

Standard turnouts on Indian Railways:

The standard turnouts used on Indian Railways are given below:

1 in $8\frac{1}{2}$ (crossing number, $N = 8.5$) for goods trains

$$\left.\begin{array}{l} 1 \text{ in } 12 (N = 12) \\ 1 \text{ in } 16 (N = 16) \end{array}\right\} \text{for passenger trains}$$

$$\left.\begin{array}{l} 1 \text{ in } 20 (N = 20) \\ 1 \text{ in } 24 (N = 24) \end{array}\right\} \text{for high speed tracks}$$

## Speeds on Turnouts

The speed potential of a turnout track is dependent on the lateral guiding forces, which, in turn, depend on the switch entry angles.

The following permissible speeds are recommended on BG and MG turnouts of different sizes, with straight switches as well as curved switches.

**Permissible Speeds on Turnouts**

| Designation of turnout | Permissible speed (km/h) | | | |
|---|---|---|---|---|
| | Straight switch | | Curved switch | |
| | BG | MG | BG | MG |
| 1 in $8\frac{1}{2}$ | 10 | 10 | 25 | 15 |
| ($\alpha$ = 642'35') | $\beta$ = 1°34'27" | | $\beta$ = 0°27'35" | |
| 1 in 12 | 15 | 15 | 40 | 25 |
| ($\alpha$ = 4°45'49") | $\beta$ = 1°08'00" | | $\beta$ = 0°24'27" | |
| 1 in 16 | 50 | 30 | 50 | 30 |
| ($\alpha$ = 3°34'35") | $\beta$ = 1°08'00" | | $\beta$ = 0°24'27" | |

**Notes:**

(i) For symmetrical splits, relatively greater speeds are permitted.
(ii) For thick web tongue rails and for improved design on PSC sleepers, allowed speeds are higher.
(iii) Switch angles for curved switches are lower than the corresponding values for straight switches.
(iv) Turn-in curves shall not be sharper than turn-out curves.

**Main Factors Responsible for Low Speeds on Turnouts**

- Sudden change in the direction upon entry on to the switch.
- Absence of transition from curved lead to the straight crossing.
- Absence of superelevation over the turnout curve.
  (To counteract this, flat bearing plates are provided over the sleepers under the rails.)
- Gaps in the gauge face and the running table at the crossing.
- Variation in cross level due to raised switch rails.

The trend in Indian Railways is to improve the turnout design by overcoming the above limitations in order to permit higher speeds on turnouts, as in advanced countries.

The Delhi Metro is an example of this trend.

## 5.6 TRACK JUNCTIONS AND LAYOUTS

Track junctions consist of several sets of points and crossings. The objective is to enable railway vehicles to cross from one track to another. There can be several types of track junctions with specific track layouts depending on the requirements of traffic in and around station yards.

The commonly used layouts are the following:

(i) Turnout from a curved main track
(ii) Symmetrical split
(iii) Three-throw switch
(iv Double turnout
(v) Crossover between parallel tracks
(vi) Diamond crossing

(vii) Diamond crossing with single slip
(viii) Diamond crossing with double slip
(ix) Square Crossing
(x) Scissors crossover
(xi) Gauntleted track
(xii) Gathering line
(xiii) Triangle
(xiv) Double junctions

## Turnout From a Curved Main Track

A turnout can take off from a curved main track. Two cases arise – turnout of a *similar flexure* (to that of the main track) or a turnout of *contrary flexure.*

Figure 5.18 shows a schematic of a turnout of similar flexure, which continues to run in the same direction as the main line curve even after taking off from it.

The degree of the turnout curve will be higher than that of the mainline curve.

Let

$D_S$ = Degree of the outer rail of the turnout curve when it takes off the main track
$D_m$ = Degree of the inner rail of the main track on which the crossing lies,
$D_t$ = Degree of the outer rail of the turnout curve on which the crossing lies, and
$R_s$, $R_m$ and $R_t$ = The corresponding radii which are inversely related to the degrees.

(For smooth entry on to the turnout, the radius of its initial portion is larger than that of the later portion towards the crossing; that is, $D_s < D_t$.)

The following relations hold:

$$D_t = (D_s + D_m) \qquad (5.20)$$

$$R_t = \frac{R_m . R_s}{(R_m + R_s)} \qquad (5.21)$$

For a turnout of contrary flexure, which takes off towards the direction opposite to that of the main line curve, the corresponding relations, with the same notation, are:

$$D_t = (D_s - D_m) \qquad (5.22)$$

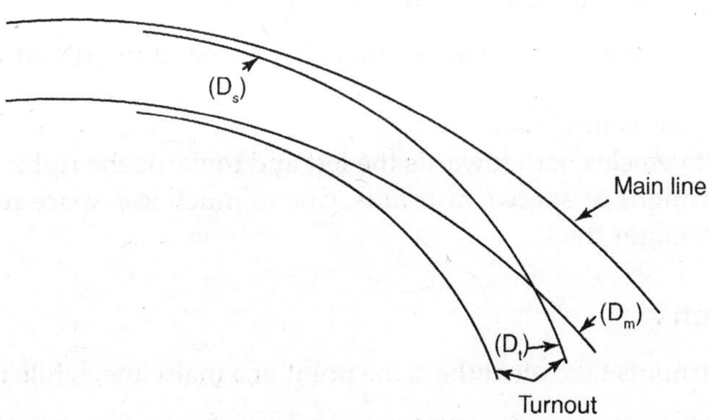


**FIG. 5.18** *Schematic of a turnout of similar flexure*

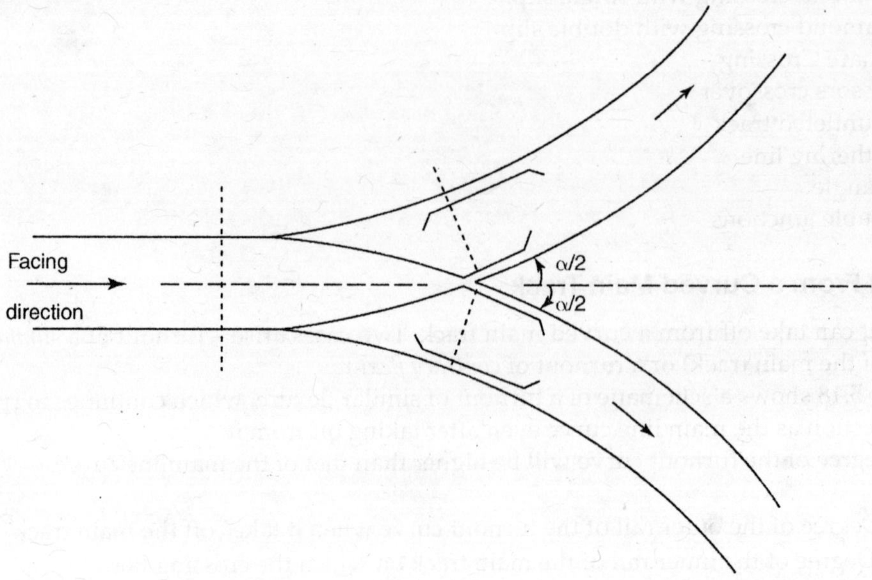


**FIG. 5.19** *Schematic of a symmetrical split*

$$R_t = \frac{R_m . R_s}{(R_m - R_s)} \tag{5.23}$$

The only difference here, is that $D_m$ is the degree of the *outer rail* of the main track on which the crossing lies.

## Symmetrical Split

This is a special case of turnout of contrary flexure in which the radii of the two curves are the same. In other words, looking in the facing direction, a straight track splits up in opposite directions with equal radii.

Figure 5.19 shows a schematic of a symmetrical split.

The salient features of a symmetrical split are:

(i) A pair of points or switches, one acute-angled crossing, four curved lead rails, and two check rails
(ii) Symmetry about the centre line
(iii) Facility to divert vehicles both towards the left and towards the right; and,
(iv) Suitability in locations of space constraints, due to much less space requirements than a turnout from a straight track.

## Three-throw Switch

In this layout, two turnouts take off at the same point of a main line, while the latter continues along a straight line.

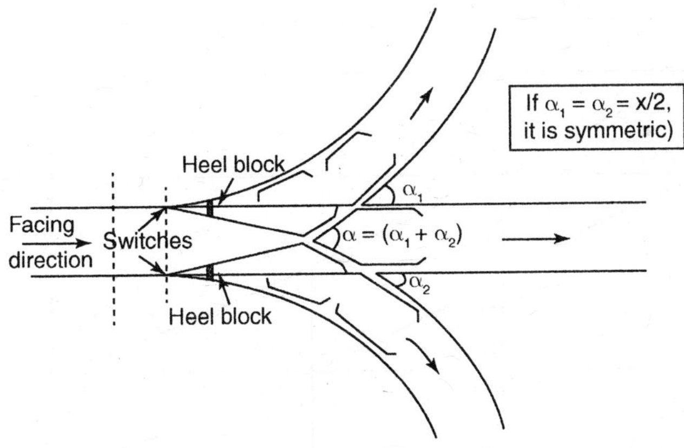


**FIG. 5.20** *Schematic of three-throw switch of contrary flexure*

A three-throw switch can be of contrary flexure for the turnouts, or similar flexure; Fig. 5.20 shows the former type:

This layout has two switches and each switch has two tongue rails placed on either side. A combined heel block is provided to support both the tongue rails. Movement of a vehicle is possible either straight, to the right, or to the left. Three-throw switches are no longer used in view of the possibility of derailment at double switches, especially at high speeds.

## Double Turnout

Also called a *tandem*, this layout is an improvement over a three-throw switch, because the turnouts are staggered and take off from the main line at two different places. This overcomes the defect of the three-throw switch as the switches are separated by adequate distance.

A double turnout can be of contrary flexure when the two turnouts take off on either side of the main track, or of similar flexure if they take off on to the same side of it.

Figure 5.21 shows a double turnout of contrary flexure:

Double turnouts are used at locations with space constraints, where traffic is heavy.

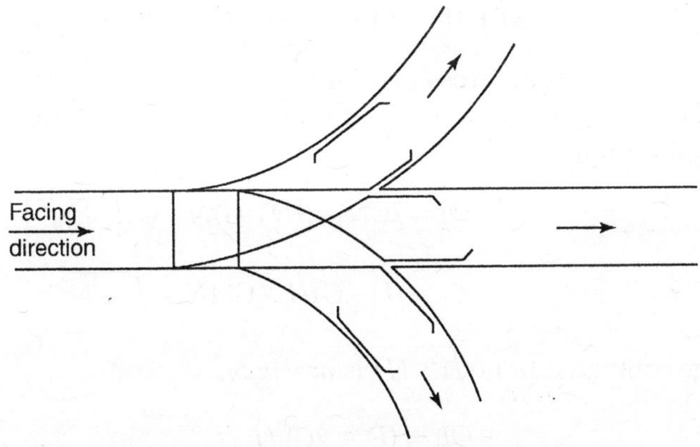


**FIG. 5.21** *Schematic of a double turnout of contrary flexure*

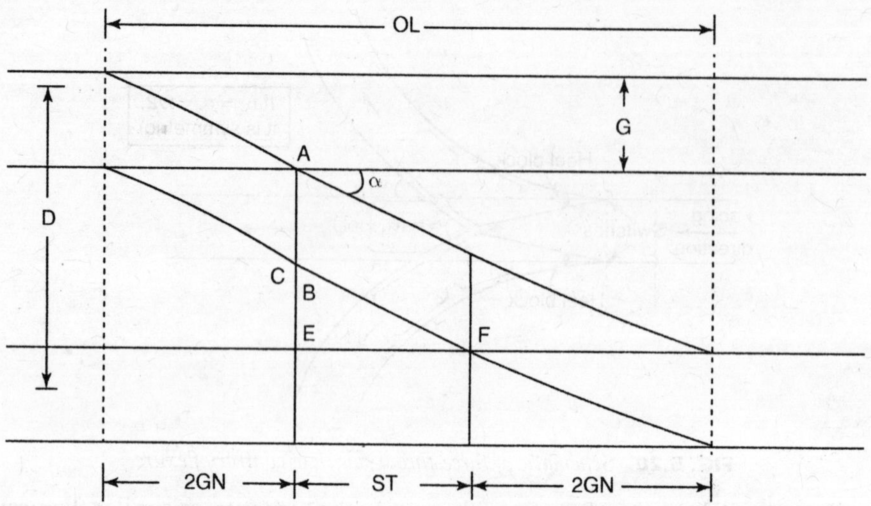


**FIG. 5.22** *Schematic of a crossover between parallel tracks*

## Crossover Between Parallel Tracks

A crossover between parallel tracks consists of one turnout from each of them, of reverse curvature, which join between the two parallel tracks. There may be a straight portion between the two turnouts, which is more common. This is shown schematically in Fig. 5.22.

Such a crossover can be designed either by Cole's method or by IRS method.

### Cole's Method

In this layout, two parallel tracks at a centre-to-centre distance of $D$ between them are connected by a crossover with a small length of straight portion between the two theoretical noses of crossing.

The length of the straight portion parallel to the tracks, $ST$, is given by

$$ST = (D - G)N - G\sqrt{1+N^2} \quad (5.24)$$

Here $G$ is the gauge and $N$ is the crossing number.

The overall length, $OL$, of the crossover is the sum of the curve leads of the two turnouts and length of straight portion, $ST$.

$$OL = 2GN + \{(D-G)N - G\sqrt{1+N^2}\} + 2GN$$

$$= (D-G)N + G(4N - \sqrt{1+N^2}) \quad (5.25)$$

Since $N^2$ is very large compared to 1, $\sqrt{1+N^2}$ is nearly $N$.

Therefore, $OL = (D + 2G)N$

Also, $OL = 2GN + ST + 2GN$

or $OL = 4GN + ST \quad (5.26)$

**IRS Method**

The distance from the TNC, measured along the straight track, is given by:

$$ST = (D - G - G \sec \alpha) \cot \alpha \qquad (5.27)$$

or

$$ST = D \cot \alpha - G \cot \frac{\alpha}{2} \qquad (5.28)$$

Here, $ST$ is the distance from one TNC to the other, along the straight track; $D$ is the centre-to-centre distance between the two parallel tracks, $G$ is the gauge, and $\alpha$ is the crossing angle.

Similarly, the distance between the TNC's *along the crossover* is given by

$$\overline{CF} = (D - G - G \sec\alpha) \operatorname{cosec}\alpha + G \tan\alpha \qquad (5.29)$$

## Diamond Crossing

A *diamond crossing* is provided when two tracks cross each other; the tracks may be of the same gauge or of different gauges. Typically, it consists of two acute crossings and two obtuse crossings, as shown in Fig. 5.23.

The acute crossings are those at A and C, and the obtuse crossings are those at B and D. ABCD is a rhombus.

The various lengths needed for setting out are obtained as follows:

$$EB = DF = AE. \cot \alpha = GN \qquad (5.30)$$

$$AB = BC = G \operatorname{cosec} \alpha \qquad (5.31)$$

$$\text{Diagonal } AC = G \operatorname{cosec} \frac{\alpha}{2} \qquad (5.32)$$

$$\text{Diagonal } BD = G \sec \frac{\alpha}{2} \qquad (5.33)$$

The gaps at B and D increase with decreasing $\alpha$. This increases the risk of derailment, especially for vehicles with smaller wheels.

Therefore, the flattest diamond crossing permitted on Indian Railways is 1 in 8.5.

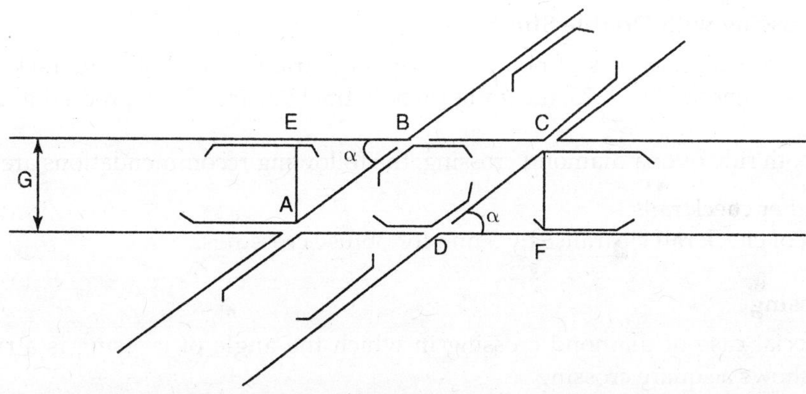


**FIG. 5.23** *Schematic of a diamond crossing*

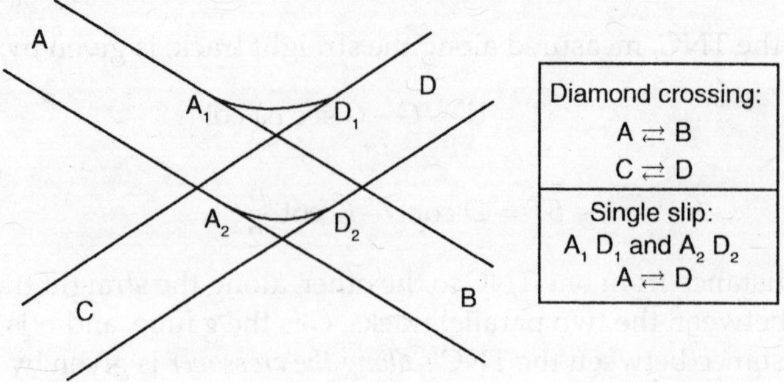


**FIG. 5.24** *Schematic of a diamond crossing with single slip*

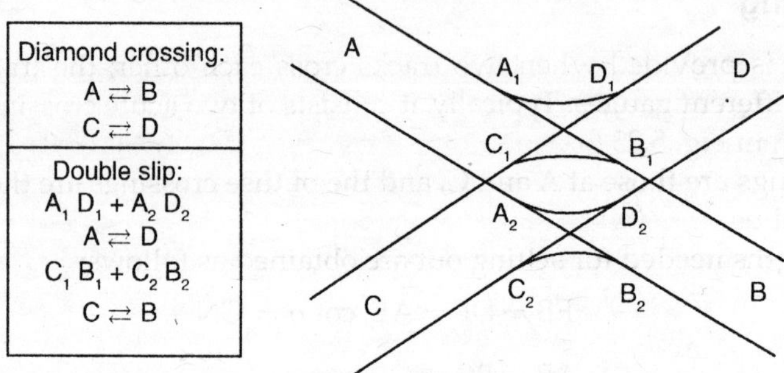


**FIG. 5.25** *Diamond crossing with double slip*

### Diamond Crossing with Single Slip

With a simple diamond crossing, the trains cannot change from one track to another; an arrangement which includes slips can provide this facility.

There are two types of slips – the single slip and the double slip.

With a single slip, trains from only one direction can change tracks; this is shown in Fig. 5.24. The train on track A can change to track D, and vice versa; but the train on track C can proceed only to track D.

### Diamond Crossing with Double Slip

In this case, there are four sets of points, which allow trains from both the tracks to change tracks. With reference to Fig. 5.25, the trains on both tracks A and C can proceed to either track B or track D:

For a smooth ride over a diamond crossing, the following recommendations are made:

- 25 mm higher check rails.
- Reduction of check rail clearance by 3 mm for obtuse crossings.

### Square Crossing

This is a special case of diamond crossing in which the angle of crossing is a right angle. Figure 5.26 shows a square crossing.

A square crossing is not preferred because of the disadvantage of rough riding due to the simultaneous negotiation of gaps on either side.

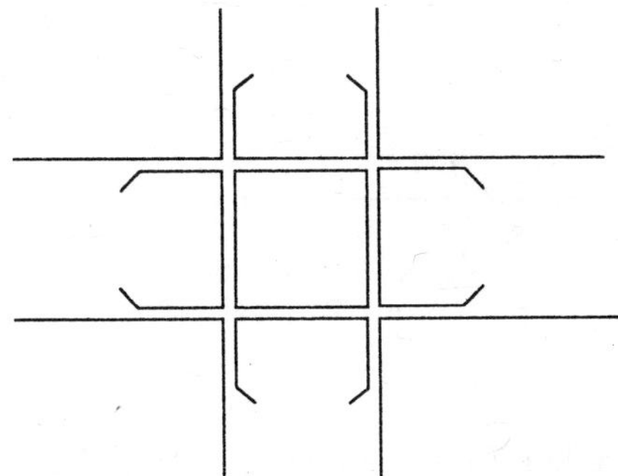

FIG. 5.26 *Schematic of a square crossing*

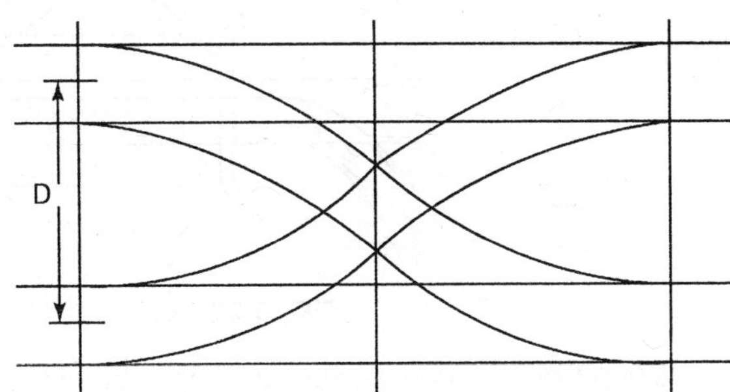


FIG. 5.27 *Schematic of a Scissors Crossover*

## Scissors Crossover

A scissors crossover is formed when two crossovers overlap exactly opposite to each other; it consists of four turnouts and a diamond crossing, as shown in Fig. 5.27.

This consists of four pairs of switches, six acute crossings, two obtuse crossings and check rails. A layout of this type permits transferring of a vehicle from one track to another and vice versa, both in the forward direction and the backward direction. The same functions can be performed by two crossovers following one after the other; but the advantage of the scissors crossover is the saving of space. This is preferred in busy passenger yards when two trains are dealt with on a single long platform.

The best arrangement for such a cross-over is that in which the acute crossings of the diamonds are far away from the main-line tracks, as it helps to provide a smoother riding experience than otherwise. The centre-to-centre distance of the parallel tracks must be adequate to facilitate this.

## Gauntleted Track

It is an arrangement in which two tracks of the same or different gauges are run together for a certain length.

It is useful in two situations:

(i) As a temporary diversion on a double-line section to allow one of the tracks to shift and run along with the other for a certain length, when the former is under repair (Fig. 5.28); and,

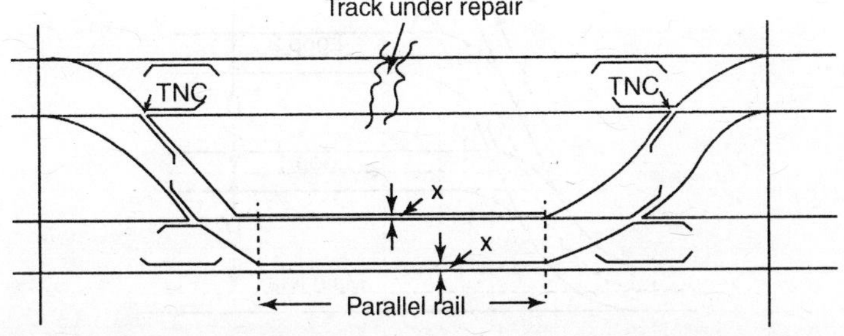


FIG. 5.28 *Gauntleted track as a temporary diversion*

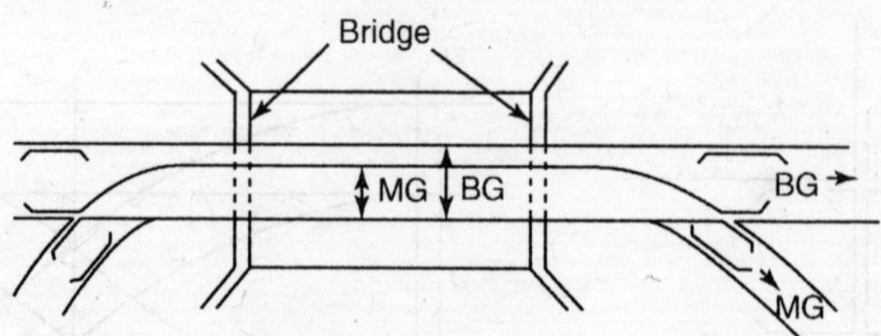


**FIG. 5.29** *Gauntleted track for mixed gauge*

(ii) When two tracks of different gauges have to run together on the same bridge (Fig. 5.29).

Salient features of a gauntleted track are:

- Two sets of crossings without switches.
- Two tracks laid on the same set of sleepers.
- Useful when part of a double-line section is under repair.
- Economical to use for mixed gauge, say BG and MG, for a short stretch.

## Gathering Line

Also called a *ladder track*, this is a track where a number of parallel tracks gather or merge on; alternatively, several parallel tracks can branch off from a gathering line.

There are two patterns of a gathering line:

(i) Gathering line at crossing angle (ladder track)
(ii) Gathering line at limiting angle
(Herring bone grid or balloon layout)

### Gathering Line at Crossing Angle (Ladder Track)

In this layout, the angle of inclination of the gathering line is the same as that of the turnout. Since the angle of the ladder track is equal to the angle of crossing, the two tracks intersect at the theoretical nose of crossing, and no curve is introduced at the turnout to connect the parallel tracks.

Figure 5.30 shows this layout.

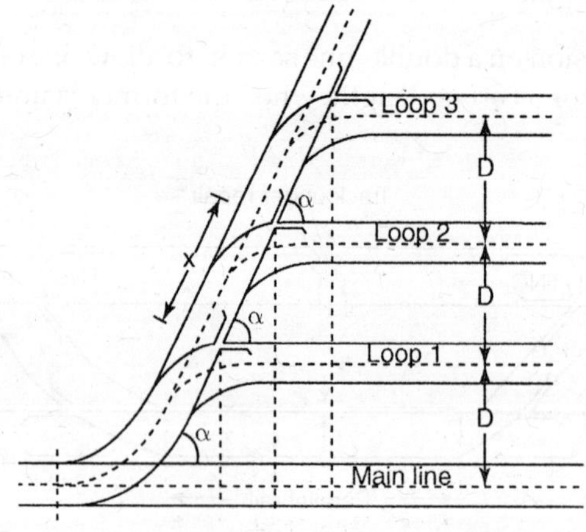


**FIG 5.30** *Gathering line at crossing angle (Ladder track)*

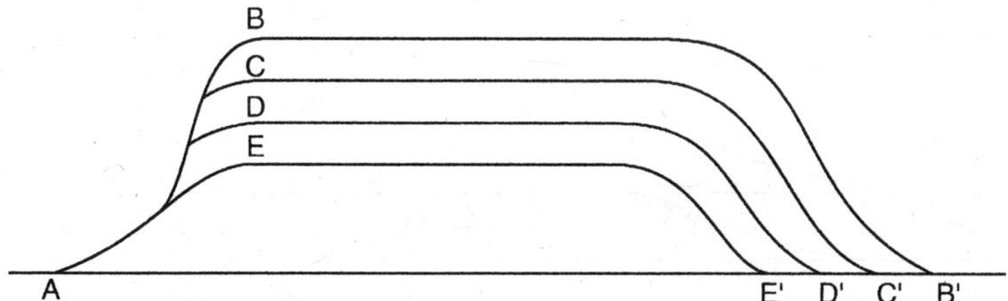


**FIG. 5.31** *A schematic of a gathering line at limiting angle (Herring bone grid)*

### Gathering Line at Limiting Angle (Herring Bone Grid)

In this layout, the angle of the gathering line is greater than the crossing angle and a curve follows the back leg of the crossing which again is followed by the stock joint of the next turnout. This results in saving of space.

This is also known as a 'Herring bone grid'

A schematic of this arrangement is shown in Fig. 5.31.

The limiting angle, $\theta$, of the gathering line is defined by

$$\text{Sin}\theta = \frac{\text{Space between two adjacent parallel tracks}}{\text{Overall length of turnout}}$$

or

$$\text{Sin}\,\theta = \frac{D}{x} \tag{5.34}$$

(Refer Fig 5.30 for $D$ and $x$)

## Triangle

A triangle is provided in big yards for changing the direction of an engine.

A *turntable* is also used for the same purpose, but it is an expensive and cumbersome device, although it occupies much less space.

A triangle consists of one symmetrical split and two turnouts along with lead rails, check rails, etc. It is shown in Fig. 5.32.

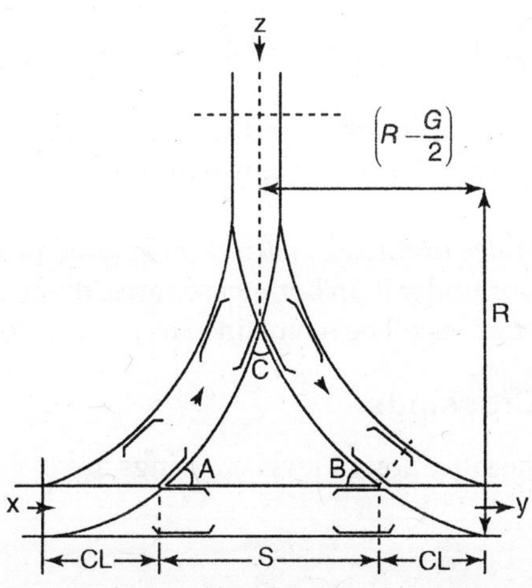


**FIG. 5.32** *Schematic of a Triangle*

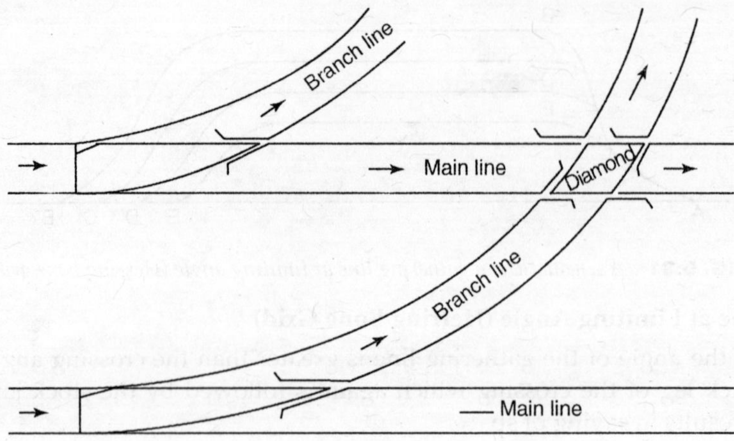


**FIG. 5.33** *Schematic of a typical double junction*

*To change the direction of an engine* standing at X, it is first taken to Z, then to Y, and then back to X. The direction of the engine then gets reversed.

The need to change the direction of an engine arises only in the case of a steam locomotive; it is not required for diesel and electric locomotive, as they can be operated conveniently from both sides. With the phasing out of steam locomotives on Indian Railways, the triangle and the turntable are almost redundant.

### Double Junctions

These are formed when there are two or more main lines from which branch lines take off in the same direction.

The layout of a double junction consists of a number of turnouts with one or more diamond crossings depending upon the number parallel main lines.

A typical double junction is shown in Fig. 5.33.

The main lines may be either straight or curved; the branch lines may be either single or double lines. Double junctions commonly occur in congested yards in and around stations.

## 5.7 LEVEL CROSSINGS

Level crossings are provided to allow road traffic to cross the railway track at the points of intersection of a highway with a railway line. Since the crossing is at the same level, it is called a *level crossing*.

Alternatively, a road *overbridge* or a road *underbridge* may be provided so that the road traffic passes over the railway track or under it. In both these cases, the necessary clearance between the road surface and the railway track shall be maintained as prescribed by the railway authorities.

### Classification of Level Crossings

The factors governing the classification of level crossings are:

- Class of the road
- Volume of road traffic

- Number of trains passing the crossing
- Sight distance or visibility conditions.

The classification is given below:

(i) Special class
(ii) A – class
(iii) B – class
(iv) C – class
(v) D – class

The primary criterion is the *Train Vehicle Units* (TVU's), a combination of rail traffic or road traffic. The TVU is calculated as the product of the number of trains running over the section in a 24-hour period and the number of vehicles passing the point on the roadway during the same period. For this purpose, the weighted average number of vehicles or passenger car units (PCU's) may be used.

Based on this criterion, the different classes of level crossings are given in Table 5.1:

**Table 5.1** Criteria for classification of level crossings

| Class | Criteria |
|---|---|
| Special class | TVU's greater than 50,000 |
| A-class | TVU's between 30,000 and 50,000. Line capacity utilisation > 80% (on single line) and road vehicles > 1000 |
| $B_1$-class | TVU's between 30,000 and 25,000. Road vehicles > 750 |
| $B_2$-class | TVU's between 25,000 and 20,000. Road vehicles > 750 |
| C-class | All other crossings for roads not covered above |
| D-class | For cattle crossing |

**Notes:**

1. The normal position of gates will be open to road traffic when TVU is greater than 25,000 and gates are interlocked.

   If TVU is less than 25,000, the gates will normally be closed to road traffic.
2. The following facilities will be provided in the order of priority: (a) Telephones (b) Lifting barriers (c) Interlocking of manned crossing gates.

**Special class:** These are busiest in terms of road traffic as on national highways. Normally, the gates are open to road traffic, and they are closed when a train has to pass. The gates of the level crossing are interlocked with signals. Special class level crossings are manned round the clock by gatemen in eight-hour shifts.

**A-class:** These are also busy in terms of road traffic as they are on important roads; the gates are normally open to road traffic in this case also. The only difference is that only two gatemen work in 12-hour shifts, as these crossings are not as busy as those of special class.

**$B_1$- and $B_2$- classes:** These level crossings are relatively less busy; they are usually on metalled or WBM roads. The gates are normally closed to road traffic, but they can be kept open if the gates are interlocked with signals. Two gatemen, working in 12-hour shifts, man them.

**C-class:** These are provided on unmetalled roads; some of them are unmanned owing to low volume of road traffic.

**D-class:** These level crossings are meant only for cattle and pedestrians. They are invariably left unguarded.

There are over 40,000 level crossings on Indian Railways, of which about 5,000 fall under special, A- and B-classes, about 30,000 fall under C-class, and the remaining fall under D-class.

(**Note:** These are only approximate figures, which are likely to change when the level crossings are upgraded based on changing traffic needs, or even eliminated at a few places by construction of over bridges or underbridges.)

## Square or Skew Level Crossings

Level crossings should preferably cross the railway line at right angles. In case right angles are rendered impossible due to the road approaches, the angle of crossing should not be less than 45°. Such crossings are called *skew* crossings, while 90°-crossings are *square* crossings. Even at skew crossings, the gate posts and the gates shall be fixed square to the road. A typical layout of a square level crossing is shown in Fig. 5.34.

### Dimensions of Level Crossings

Typical dimensions of square level crossings of A- and B- classes are shown in Table 5.2:

**Table 5.2** Typical dimensions of A- and B-class level crossings

| Item | A-class (on Class-I road) | B-class (on Class-II rod) |
|---|---|---|
| Width of gates (max.) | 9 m | 7.5 m |
| Length of guard rails (min.) | 11 m | 9.5 m |
| Distance of gate post from centre-line of track (max.) | 3 m | 3 m |
| Distance of cabin from centre-line of track (max.) | 6m | 6 m |

Cont'd

**Table 5.2 Cont'd**

| Item | A-class (on Class-I road) | B-class (on Class-II rod) |
|---|---|---|
| Width of metalled road between gates (min.) | Same as width of gate | Some as width of gate |
| Gradient | | |
| (a) between gates | Level | Level |
| (b) beyond gates | Level up to a distance of 8 m; a grade of 1 in 30 beyond that. | Level upto a distance of 8 m; a grade of 1 in 20 beyond that. |
| Length of road outside the gates (min.) | 22.5 m | 15 m |
| Fencing on lines parallel to the track | 15 m | 15 m |
| Provision of lights at the gates for the benefit of road users | Red when gate is closed, and white when it is open to road traffic | Same as for A-class |

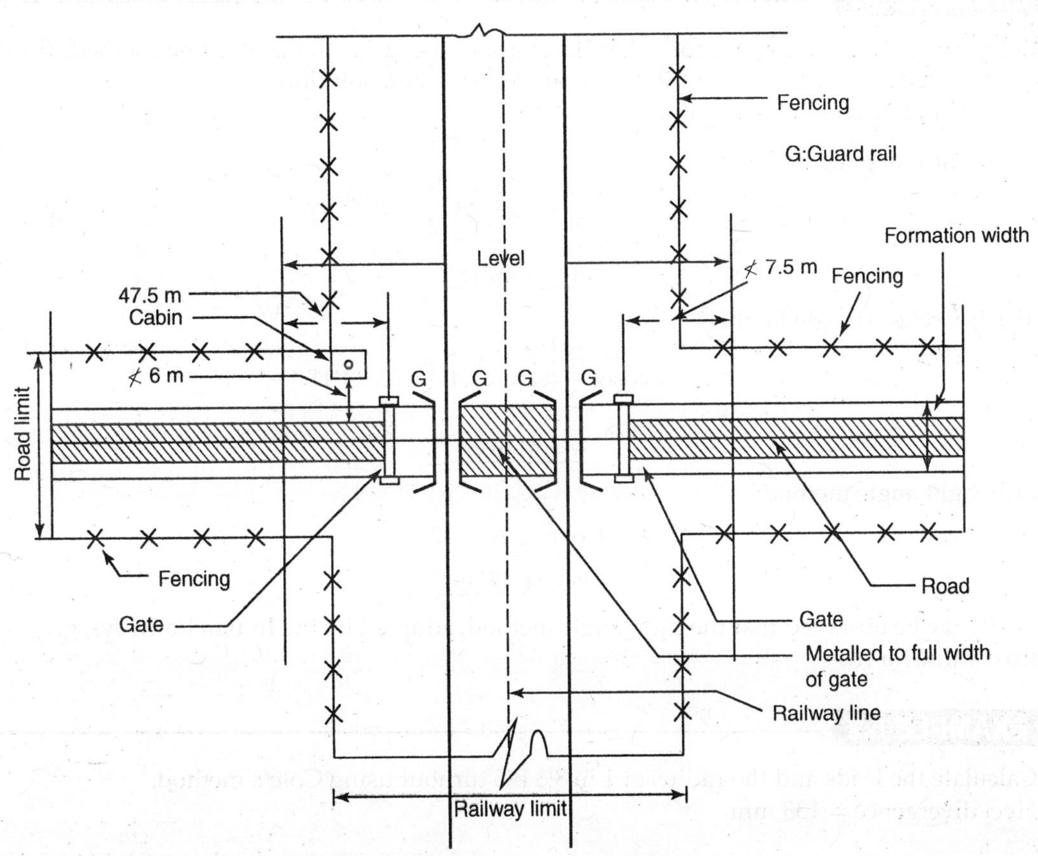


**FIG. 5.34** *A typical layout of a square level crossing*

### Track Structure at Level Crossings

Following are the important aspects of the track structure at level crossings:

(i) Concrete sleepers of special design should preferably be used at level crossings. If this is not feasible, wooden sleepers of U-category should be used.
(ii) If wooden sleepers of U-category are used, bearing plates should be used; four spikes should be provided at each rail seat.
(iii) Rail joints should preferably be avoided in running rails as well as on check rails within the level crossing, with 3 mm clearance on either side.
(iv) There should be no combination rail joint (at changes of rail sections) on the level crossing proper and its approaches up to a distance of 50 m.
(v) If the level crossing is on an SWR stretch, fish-plated joints should be avoided on the level crossing and also within 6 m distance from the end of the level crossing.
(vi) A level crossing should not fall within the breathing length of an LWR stretch.

## 5.8 NUMERICAL EXAMPLES

### EXAMPLE 5.1

Calculate the crossing angle for a 1 in 12 crossing using: (i) the centre line method, (ii) the isosceles triangle method, and (iii) the right angle method solution:

For a 1 in 12 crossing (N = 12):

(i) Centre-line method:

$$\cot\frac{\alpha}{2} = 2N = 24 \quad \frac{\alpha}{2} = 2^\circ 23'09''$$

$$\therefore \quad \underline{\alpha = 4^\circ 46'18''}$$

(ii) Isosceles triangle method:

$$\text{cosec}\frac{\alpha}{2} = 2N = 24 \quad \frac{\alpha}{2} = 2^\circ 23'17''$$

$$\therefore \quad \underline{\alpha = 4^\circ 46'34''}$$

(iii) Right angle method:

$$\text{Cot}\ \alpha = N = 12$$

$$\therefore \quad \underline{\alpha = 4^\circ 45'49''}$$

(It may be observed that the right angle method, adopted by the Indian Railways, gives the least value for $\alpha$).

### EXAMPLE 5.2

Calculate the leads and the radius of 1 in 8½ BG turnout using Cole's method.
Heel divergence = 133 mm.

**Solution**

$$G = 1.676 \text{ m};$$

$$d = 133 \text{ mm} = 0.133 \text{ m}$$

$$\alpha = 6^\circ 42'35''$$

$$N = 8.5$$

$$\text{Curve lead (CL)} = GN + G\sqrt{1+N^2}$$

$$= 1.676 \times 8.5 + 1.676\sqrt{1+(8.5)^2}$$

$$= 28.590 \text{ m}$$

$$\approx \underline{28.6 \text{ m}}$$

$$\text{Radius of turnout (R)} = 1.5\,G + 2GN^2$$

$$= 1.5 \times 1.676 + 2 \times 1.676 \times (8.5)^2$$

$$= 244.696 \text{ m} \approx \underline{244.7 \text{ m}}$$

$$\text{Switch lead (SL)} = \sqrt{2Rd - d^2} = \sqrt{2 \times 244.7 \times 0.133 - (0.133)^2}$$

$$= 8.067 \text{ m}$$

$$\approx \underline{8.1 \text{ m}}$$

$$\text{Crossing lead (L)} = CL - SL = (28.6 - 8.1)\text{m}$$

$$= \underline{20.5 \text{ m}}$$

## EXAMPLE 5.3

Calculate the lead and radius of a 1 in 8½ turnout with straight switches, using the IRS method. Take the heel divergence as 130 mm and the length of the straight portion of the turnout from the tangent point of crossing to the TNC ($h$ in Fig. 5.16) as 860 mm. $\alpha = 6^\circ 42'35''$ and $\beta = 1^\circ 34'27''$.

### Solution

$G = 1676$ mm; $d = 130$ mm; $h = 860$ mm

$$\alpha = 6^\circ 42'35''$$

$$\beta = 1^\circ 34'27''$$

$$\text{Crossing lead } (L) = (G - d - h \sin \alpha) \cot \frac{(\alpha+\beta)}{2} + h\cos\alpha$$

$$= (1676 - 130 - 860 \times 0.1168) \times 13.8089 + 860 \times 0.993$$

$$= 20{,}815.46 \text{ or } 20{,}820 \text{ mm (approx)}$$

$$\text{Radius } (R) = \frac{G - d - h\sin\alpha}{2\sin\left(\frac{\alpha+\beta}{2}\right).\sin\left(\frac{\alpha-\beta}{2}\right)}$$

$$\frac{(1676-130-860\times 0.1168)}{2\times 0.7223\times 0.0448}=\underline{22{,}336\ \text{mm}}$$

## EXAMPLE 5.4

A turnout takes off a straight BG track with a 1 in 12 crossing. Determine the lead and radius of the turnout with the following data: heel divergence (d) = 130 mm; crossing angle ($\alpha$) = 4°45′49″; switch angle ($\beta$) = 1°08′00″; and, straight length between the TNC and the tangent point of crossing (h) = 1.40 m.

**Solution**

$$\alpha = 4^\circ 45'49''$$

$$\beta = 1^\circ 08'00''$$

$$G = 1.676\ \text{m}$$

$$d = 0.130\ \text{m}$$

$$N = 12$$

$$h = 1.400\ \text{m}$$

$$\text{Radius } (R) = \left(\frac{G-d-h\sin\alpha}{\cos\beta-\cos\alpha}\right)$$

$$=\left(\frac{1.676-0.130-1.400\times \sin 4^\circ 45'49''}{\cos 1^\circ 08'00''-\cos 4^\circ 45'49''}\right)=438.762\ \text{m (say 440 m)}$$

$$\text{Crossing lead } (L) = (h \text{ Cos } \alpha + G - d - h \text{ Sin } \alpha) \cot \cot\left(\frac{\alpha+\beta}{2}\right)$$

$$= 1.400 \cos 4^\circ 45'49'' + (1.676 - 0.130 - 1.400 \sin 4^\circ 45'49'') \times \cot 2^\circ 56'54''$$

$$= 1.400 \times 0.9965 + 1.42974 \times 19.415 = \underline{29.154\ \text{m}}$$

## EXAMPLE 5.5

A 1 in 8½ crossover is proposed to be laid between two BG parallel tracks with a centre-to-centre distance of 5.5 m. Determine the length of the straight portion of the track and the overall length of the crossover. Use Cole's method.

**Solution**

$$\text{Straight track (ST)} = (D-G)\ N - G\sqrt{1+N^2}$$

$$D = 5.5\ \text{m}\ N = 8.5\ G = 1.676\ \text{m}$$

$$\therefore \quad \text{ST} = (5.500-1.676)\times 8.5 - 1.676\sqrt{1+(8.5)^2}$$

$$= 18.16\ \text{m}$$

$$\text{Overall length (OL)} = \text{ST} + 4\text{GN}$$

$$= 18.16 + 4 \times 1.676 \times 8.5 = \underline{75.15 \text{ m}}$$

## EXAMPLE 5.6

A 1 in 12 crossover is laid between two parallel BG tracks placed at a centre-to-centre distance of 5.5 m Calculate the straight length, overall length, radius of the turnout curve and the lead distance. Take heel divergence as 130 mm.

**Solution**

$$D = 5.5 \text{ m}; G = 1.676 \text{ m}; N = 12$$

$$\text{Straight length (ST)} = (5.500 - 1.676) \times 12 - 1.676\sqrt{1+144}$$

$$= 45.888 - 20.180 = 25.71 \text{ m}$$

$$\text{Overall length (OL)} = \text{ST} + 4\text{GN}$$

$$= 25.71 + 4 \times 1.676 \times 12$$

$$= \underline{106.16 \text{ m}}$$

$$\text{Radius of the turnout curve } (R) = 1.5\,G + 2\,GN^2$$

$$= 1.500 \times 1.676 + 2 \times 1.676 \times 12^2 = 485 \text{ m}$$

$$\text{Lead of crossing } (L) = (\text{Curve lead} - \text{Switch lead})$$

$$= 2GN - \sqrt{2Rd - d^2}$$

$$= 2 \times 1.676 \times 12 - \sqrt{2 \times 485 \times 0.130 - (0.130)^2}$$

$$= 40.20 - 11.23 \approx \underline{29 \text{ m}}$$

## EXAMPLE 5.7

A 1 in 12 crossover of IRS type is laid between two BG parallel tracks with their centres 5.5 m apart. Calculate the straight length and the distance from TNC to TNC along the crossover.

**Solution**

$$G = 1.676 \text{ m};$$

$$N = 12;$$

$$D = 5.5 \text{ m};$$

$$\alpha = 4^\circ 45' 49''$$

$$\text{Straight length (ST)} = D \operatorname{Cot} \alpha - G \operatorname{Cot} \alpha/2$$

$$= 5.5 \times 12 - 1.676 \times 24.04$$

or $\quad ST = 25.7$ m

Distance from TNC to TNC along the crossover,

$$\overline{CF} = (D - G - G\sec\alpha)\operatorname{cosec}\alpha + G\tan\alpha$$

$$= (5.500 - 1.676 - 1.682) \times 12.04 + 1.676 \times (1/12)$$

$$= 25.93 \text{ m} \approx \underline{26 \text{ m}}$$

## EXAMPLE 5.8

Two BG tracks cross each other at 1 in 12. Calculate the important dimensions of the diamond crossing.

**Solution**

Referring to Fig.5.23,

$$G = 1.676 \text{ m};$$

$$N = 12$$

$$\alpha = 4^\circ\, 45'49'' \; (= \cot^{-1}12)$$

The important dimensions are:

$$\overline{EB} = \underline{1.676} \times 12 = 20.102 \approx \underline{20.10 \text{ m}}$$

$$\overline{AB} = BC = G \operatorname{cosec}\alpha$$

$$= 1.676 \times \operatorname{cosec} 4^\circ 45'49''$$

$$= 20.184 \text{ m} \approx \underline{20.20 \text{ m}}$$

$$\overline{AC} = G \operatorname{cosec}\frac{\alpha}{2} = 1.676 \times \operatorname{cosec} 2^\circ 22'54.''5$$

$$= 40.329 \text{ m} \approx \underline{40.33 \text{ m}}$$

$$\overline{BD} = G \sec\frac{\alpha}{2} 1.676 \times \sec 2^\circ 22'54.5 = \underline{1.678 \text{ m}}$$

## EXAMPLE 5.9

Determine the angle of switch if the heel divergence is equal to 120 mm and the actual length of the tongue rail is 4560 mm. Neglect the thickness of the tongue rail at the toe.

**Solution**

$$d = 120 \text{ mm};$$

$$L_{ta} = 4560 \text{ mm}$$

$$\operatorname{Sin}\beta = \frac{d}{L_{ta}} = \frac{120}{4560}$$

$$=0.0263158$$

$\therefore$ Angle of switch, $\underline{\beta = 1°30'28''}$

## EXAMPLE 5.10

A switch has a tongue rail of actual length 4.500 m. If the heel divergence is 120 mm and the thickness of the tongue at toe is 6.3 mm, determine the switch angle, theoretical length of the tongue rail, and the difference between the theoretical and the actual lengths of the tongue rail.

**Solution**

$$L_{ta} = 4500 \text{ mm}$$

$$d = 120 \text{ mm}$$

$$t = 6.3 \text{ mm}$$

$$\text{Sin } \beta = \frac{(d-t)}{L_{ta}} = \frac{(120-6.3)}{4500}$$

$$= \frac{113.7}{4500} = 0.0252667$$

$\therefore$

$$\underline{\beta = 1°26'52''}$$

$$L_{tt} = \frac{d}{\sin\beta} = \frac{120}{0.0252667} \text{ mm}$$

$$= \underline{4750 \text{ mm}}$$

This is the theoretical length of the tongue rail.

The difference between the theoretical and actual lengths of the tongue rails

$$= 4750 - 4500 = \underline{250 \text{ mm}}.$$

## SUMMARY

- Points and crossings are special arrangements needed to move a train from one track to another; the need for this arises owing to the inside flanges of wheels requiring gaps for the movement.

  Points or switches divert a vehicle while crossings help the flanged wheels to roll over the gaps in the desired direction.
- A turnout consists of one set of switches and a crossing; the simplest layout is a turnout taking off from a straight track. A turnout is designated as a right-hand or a left-hand turnout depending on the direction in which it diverts the traffic, as viewed in the facing direction. (Looking towards the crossing, standing at the points or switches gives the facing direction; the reverse is the trailing direction.)

  The main parts of a turnout are a pair of switches or points, a crossing and four lead rails.
- The important components of points or switches are a pair of stock rails, a pair of tongue (or switch) rails (of medium-manganese steel) tapered towards the toe, stretcher bars connecting the tongue rails at the toe, a pair of heel blocks, slide chairs, and a gauge tie-plate.

- A split type switch is the more commonly used one; it may be fixed heel or spring type (which, again, may be straight or curved) or the loose heel type. The loose-heel type is not preferred for new layouts.
  Based on the design, a switch may be of the undercut type or of the overriding type; in the former, both the stock rail and the tongue rail need to be cut and planed for a snug fit at closure of the switch, while in the latter, the planing is confined only to the tongue rail. The overriding switch is more advantageous and is therefore preferred.
  Switch angle, throw of switch and heel divergence are important design aspects of switches.
- A crossing is a device introduced at the point where two rails cross each other so as to permit the flanges of the wheels to pass from one track to another. As the wheel negotiates the gap at the nose of the crossing, the other wheel is guided by means of a check rail.
  The main components of a crossing are the point rail, the splice rail, check rails, and wing rails.
- Based on the geometry, crossings are either 'acute-angled' or 'obtuse-angled'. In the acute angled or V-crossing, the gauge faces of the point and splice rails meet at an acute angle; this happens when the right rail of one track crosses the left rail of another, or vice-versa.
  In case the gauge faces meet at an obtuse angle, it is called an obtuse crossing; this occurs when the same side or similar rails of two tracks cross each other. This happens in the case of a layout called a 'diamond crossing', at the obtuse corners of the rhombus.
  Based on the method of manufacture, crossings may be of the built-up type, cast steel type, or a combination of these two types. Cast manganese steel crossings (CMS), developed recently by Indian Railways, are considered to be stronger.
  Spring or movable type of crossing with a strong steel helical spring is considered especially useful for taking high-speed traffic on the main track, with low-speed traffic running on the turnout track.
  A crossing is designated either by the angle ($\alpha$) between the gauge faces of point and splice rails, or by the number ($N$) of the crossing, which is more common.
  Three methods are available to relate $\alpha$ and $N$ – the centre line method ($N = \frac{1}{2}\cot\frac{\alpha}{2}$), the isosceles triangle method $\left(N = \frac{1}{2}\operatorname{cosec}\frac{\alpha}{2}\right)$, and the right angle method ($N = \cot\alpha$); the last one, which gives the smallest angle for a given crossing number, is the method adopted by Indian Railways.
- The design of a turnout consists in determining the curve lead, the switch lead, and the crossing lead (all measured parallel to the main straight track).
  Cole's method of design is relatively simple; the curvature is taken to start from a point on the straight main track ahead of the toe of the switch and end at the theoretical nose of crossing.
  The offset of the curve at the heel of the switch (taken perpendicular to the main track) is called the *heel divergence*. This arrangement results in kinks at the actual toe of the switch, heel of the switch and at the first distance block of the crossing.
  In the IRS method, which is the standard layout used on Indian Railways, the curve begins at the heel of the switch and ends at the toe of the crossing (at the centre of the first distance block). The only kink occurs at the toe of the switch. The calculations are a little involved, but the layout is superior.
  Sleepers are laid at points and crossings to cover both the tracks for some distance.
  The standard turnouts used on Indian Railways are: 1 in 8.5 ($N = 8.5$), 1 in 12 and 1 in 16; the first one is used for goods traffic tracks and the other two for passenger traffic tracks. Recently, 1 in 20 and 1 in 24 crossings have been developed for use on high-speed tracks.

The permissible speeds on turnouts are much lower than those on the tracks. These are dependent on switch entry angles, and are lower for curved switches than for straight ones. Hence, the permissible speeds are greater for the former.
Also, there are several other factors that are responsible for the low speeds on turnouts; but the latest trend on Indian Railways is to improve the switch design in order to permit higher speeds on turnouts as in advanced countries.

- Track junctions and layouts involve several sets of points and crossings and are mostly used in and around station yards to fulfil specific needs.
  The more important layouts are the symmetrical split, the three-throw switch, the double turnout, crossover between parallel tracks, the diamond crossing, diamond crossings with single and double slips, scissors crossover, gathering line, and the double junctions. Each one serves a specific purpose, with limitations and advantages of its own.
- Level crossings are provided when a railway line and a roadway cross each other at the same level.
  Level crossings are classified based on the class of road, volume of road traffic, number of trains passing the point, and visibility conditions, as special class, A-, $B_1$-, $B_2$-,C-, and D-class level crossings.
  The criterion is Train Vehicle Units (TVU), a combination of road and rail traffic index, defined as the product of the number of trains running in a 24-hour period and the number of road vehicles passing the point in the same period (for the latter, a weighted average number of the various types of vehicles, such as the PCU's or Passenger Car Units is considered).
  As the traffic increases, the level crossings may be replaced by road overbridges or road underbridges, avoiding traffic conflicts between the railway and the roadway.

## QUESTIONS AND PROBLEMS

1. Explain the need for points and crossings in railways.
2. Draw a neat sketch of a left-hand turnout taking off from a BG main track which is straight, and explain briefly the parts. (VTU and JNTU-H)
3. Differentiate a fixed-heel switch from a loose-heel switch.
4. What is meant by crossing angle and crossing number in railways? How are they related?
5. Explain the methods of measuring the crossing number.

   Which is the method adopted by Indian Railways and what is its special feature?
6. Explain the types of crossings based on the method of manufacture.
7. Explain the basis of turnout design by Cole's method and IRS method with appropriate sketches.
8. What are the standard sizes of turnouts used on Indian Railways? Explain the situations under which each is suitable.
9. Explain the factors which govern the permissible speeds on turnouts.
10. Differentiate between turnouts from a curved main track with similar flexure and that with contrary flexure. Give neat sketches.
11. Differentiate between a symmetrical split and a three-throw switch with neat sketches.

12. How is a double turnout superior to a three-throw switch? Explain with neat sketches.
13. Explain how a crossover between parallel tracks is designed when a straight portion is introduced between the turnouts.

    How are the straight length and the overall length (parallel to the straight tracks) determined? Give neat sketches.
14. Explain a diamond crossing and give details of where it occurs. How is it improved by the introduction of slips? Give neat sketches.
15. Explain how a gauntleted track is useful in running mixed gauge tracks together for a short stretch.
16. Differentiate between a gathering line at crossing angle and that at the limiting angle (herring bone grid). Give neat sketches.
17. What is the purpose of a triangle layout? Explain its operation with a sketch.
18. When are double junctions formed? Explain with a neat sketch.
19. What is a level crossing? How are level crossings classified?
20. Explain the salient features of a square level crossing with a neat sketch.
21. List the important aspects track structure at level crossings.
22. Write brief notes on:

    (i) Heel divergence

    (ii) Switch angle

    (iii) Actual nose of crossing
23. Write brief critical notes on:

    (i) Theoretical nose of crossing

    (ii) Flangeway clearance

    (iii) Special steel for points and crossings
24. Distinguish between:

    (i) Right-hand turnout and left-hand turnout

    (ii) Undercut switch and overriding switch

    (iii) Crossing angle and crossing number.
25. Distinguish between:

    (i) Facing points and trailing points

    (ii) Acute crossing and obtuse crossing

    (iii) Single slip and double slip.
26. A switch has the following dimensions:

    Actual length of the tongue rail = 4.8 m

    Heel divergence = 135 mm

Thickness at the toe of the tongue rail = 6 mm

Determine the angle of the switch and the theoretical length of the tongue rail.

27. Determine the crossing angle for a crossing number of 20 by

    (i) Centre line method,

    (ii) Isosceles triangle method, and by

    (iii) Right angle method.

    What is your observation and which value would you recommend for use on Indian railways?

28. What is the necessity of points and crossings on railways? (Kerala University)

29. What is meant by a crossing? What are the essential requirements of a good crossing?

    Discuss the various types of crossings in use on Indian Railways. (Kerala University)

30. Sketch and explain the operation of a right-hand turnout. (Kerala University)

31. (a) Distinguish between switch lead and curve lead.

    (b) Explain the constituents of crossings on railways.

    (c) Calculate the necessary leads to set out a 1 in 8.5 turnout, taking off from a straight BG track with its curve starting from the toe of the switch and passing through the theoretical nose of crossing. Given heel divergence ($d$) = 11.4 cm. (Punjab Tech. University.)

32. (a) What is the difference between a tongue rail and a stock rail?

    (b) Write short notes on:

    (i) Level crossing

    (ii) Types of switches

    (iii) Throw of switch

    (iv) Interlaced sleepers

    (Kalinga Institute of Technology, Odisha)

33. What is a turnout? Explain, with sketches, left-hand and right-hand turnouts. (Mumbai University)

Chapter 6

# STATIONS AND YARDS – LAYOUT AND EQUIPMENT

**CHAPTER OBJECTIVES**

This chapter sets out to:

- Define the purpose of a railway station, criteria for site selection, and the various facilities required at railway stations.
- Explain the classification of stations based on their functions – block stations, halts, flag stations, junctions and terminus stations; the layouts of different types of stations, station yards, and the various yards at stations to fulfil different functions for passengers and goods traffic; the use of marshalling yards and their operation
- Enumerate the different types of equipment needed at stations and station yards in order that their functions are fulfilled.

## 6.1 INTRODUCTION

The need for railway stations is obvious – as trains are meant for carrying passengers and goods traffic, stations are needed along the railway route for passengers to entrain and alight and goods to be booked and carried to their destinations.

For operating passenger and goods traffic, several types of yards are required in and around railway stations. Appropriate facilities, equipment and machinery are necessarily provided for this purpose.

## 6.2 RAILWAY STATIONS

The *primary purpose* of a railway station is to provide facilities to book passenger and goods traffic, and to operate the trains.

### Functions of a Railway Station

A railway station is provided to perform one or more of the following functions:

(i) Entraining and detraining of passengers

(ii) Loading and unloading of goods
(iii) Controlling the movement of trains
(iv) Facilitating the crossing and passing operations of trains in a single-line section, where traffic is entertained in both directions.
(v) Providing facilities for locomotives to refuel with water and coal, or diesel oil, as the case may be.
(vi) Providing facilities for attaching or detaching wagons of goods trains.
(vii) Facilitating change of engines and/or crew.
(viii) Providing facilities for sorting wagons/coaches to form new trains.
(ix) Providing basic amenities such as food and water for passengers.
(x) Providing shelter to passengers in times of emergencies like accidents and natural calamities such as cyclones and floods, which disrupt rail traffic.

## Site Selection for a Railway Station

The following factors need to be considered in the selection of a site for a railway station.

**(a) Adequate land:** Availability of adequate land for the station buildings and auxiliary works, and the station yards, including provision for future expansion.

**(b) Nature of the terrain:** Fairly level ground in plain terrain with good drainage arrangement is preferable; this is to take care of the desirable and maximum gradients of 1 in 1000 and 1 in 400 specified by Indian railways. In the case of rolling terrain, stations must preferably be located on summits and not in valleys for easy operation.

**(c) Alignment:** Stations must be selected on the straight portion of the alignment of the track since clear visibility of signals is possible only with such an alignment. Operation on curves poses certain problems.

**(d) Accessibility:** The station should be easily accessible, with proximity to towns and villages. Approach roads from nearby places improve the accessibility and convenience to passengers.

**(e) Basic amenities:** Availability of adequate water supply for passengers and for operational needs is an important criterion. Power supply must be ensured. Finally, the relevant authorities of civil administration should be consulted before deciding upon the site for the station.

## Facilities Required at Railway Stations

A well-designed railway station in well-maintained surroundings adds to the aesthetics, especially if the main building is attractive with pleasing architecture. Service to the passengers and other users is the main consideration and hence, the best amenities should be provided. The structure should cope not only with the existing demands but also anticipated demands in the design life of 20 to 25 years, with provision for expansion.

The following is a broad classification of the facilities required at stations:

**(a) Passenger requirements:** Waiting rooms and retiring rooms, big waiting halls with seating arrangements; enquiry, booking and reservation offices; platforms and platform shelters; refreshment rooms, canteens, cafeteria, tea stalls, book stalls; bathrooms and toilets, drinking water supply clocks and public telephones; announcement systems, train indications of arrivals and departures including platform numbers, CCTV, foot-over bridges, escalators/elevators at big stations, subways, transport facilities to and from the station and so on.

**(b) Traffic requirements:** Station buildings, station master's and ASM's offices, railway protection force, station equipment for control of trains, signals and signal cabins, reception and departure lines, sidings to deal with trains to be stationed before despatch, and so on.

**(c) Rolling stock requirements:** Locomotive sheds, workshops, watering and fuelling arrangements, turn-table/triangle for reversing steam locomotives, inspections or examination pits, ash pits, and so on.

**(d) Staff requirements:** Rest houses and rooms for officers and staff; running rooms for guards; rooms for Ticket collectors and TTE's ; running rooms for locomotive drivers; staff canteens; staff quarters, and so on.

## Classification of Railway Stations

The main considerations in classifying stations are:

(a) Operational
(b) Functional
(c) Financial

### Operation Considerations

In accordance with the rules evolved by the Indian Railways, stations are broadly classified as 'Block stations' and 'Non-block stations'. Block stations are further classified as A-class, B-class, and C-class stations.

Non-block stations are classified as D-class or flag stations.

### Functional Considerations

This classification is based on the functions which a station is required to perform. Accordingly, stations may be halt/flag stations, wayside stations, junction stations and terminus stations.

### Financial Considerations

Depending upon passenger/goods earnings and allied considerations, stations are classified as A-, B-, C-, D-, E-, and F-, stations. This classification is an indicator of the density of passenger traffic.

In summary, the following are the desirable criteria for stations:

- Least cost
- Least number of signals
- Flexibility in shunting operations
- Line capacity and maximum utilisation
- Faster movement of trains
- Increasing passengers and goods earnings.

### Block Stations

A station at which the driver of a train has to obtain an 'authority to proceed' to the next block section is known as a 'block station'. This restriction applies to a system of operation in which the railway line is divided into what are known as 'Block Sections' at the ends of which block stations will be located.

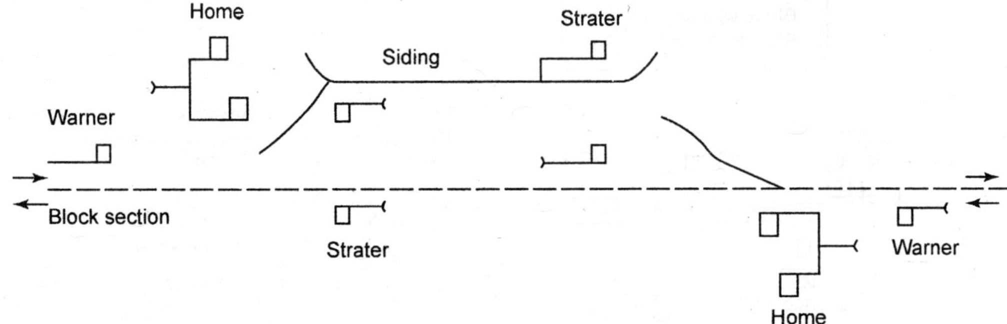


**FIG. 6.1** *Schematic of class-A station on a single-line section*

In fact, the permission to approach and the permission to proceed are granted to trains only at block stations. Permission to approach is granted to a train coming from the previous block only after ensuring that the line on to which it is received is clear, and the appropriate points are set and locked.

Typically, a block section is 5 to 10 km long and block stations are provided at either end. This system ensures that a 'space interval' exists between running trains, thus preventing accidents.

**A-class stations:** These are normally provided on double-line sections. A 'line clear' signal is necessary for receiving a train approaching from the previous block. The signal is given only after the correct points are set and locked; also, no shunting operations are permitted after the line clear has been granted.

A schematic of a class-A station on a single-line section is shown in Fig. 6.1 while that on a double-line section is shown in Fig. 6.2.

A-class stations are suitable for sections in which fast trains run.

The *signals required* are (as the train approaches a station):

*Warner*: This is located at a warning distance from the home signal; it indicates whether the section beyond is clear or not.

*Home*: This is the first stop signal.

*Starter*: Placed at an adequate distance from the home signal, it marks the point up to which the line should be clear for the train to be given permission to approach.

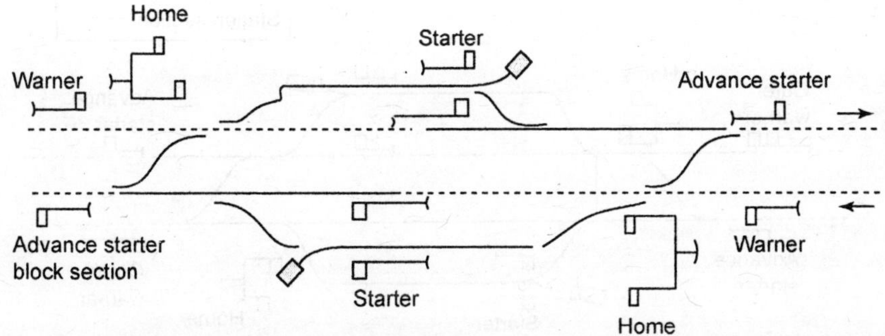


**FIG. 6.2** *Schematic of A-class station on a double-line section*

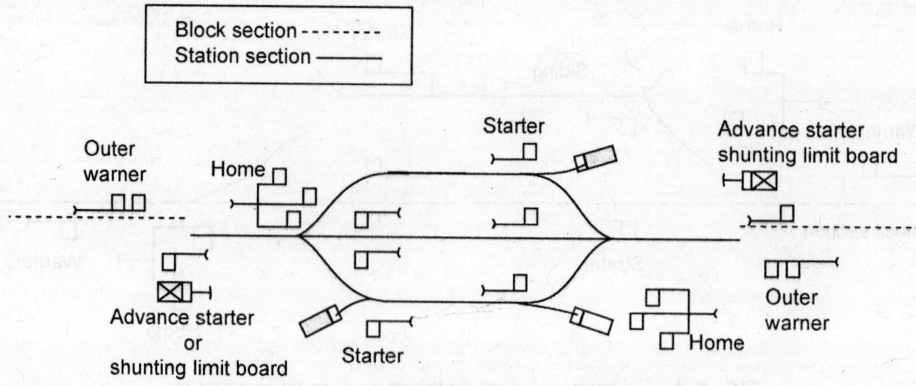


**FIG. 6.3** *Schematic of a B-class station on single-line section*

*Advance starter*: This is optional and is meant to allow the driver to further increase the speed of the train.

### Advantages of the Layout

- Use of fewer signals.
- Enhanced safety due to the provision of a warner before the home signal.
- Trains stop within the station limits normally.

### Disadvantages

- No shunting operations are permitted once line clear has been given.
- A line has to be kept clear up to the starter signal for receiving a train, reducing the flexibility of operation.

**B-class stations:** These are provided on single-line as well as double-line sections and are the most common type on Indian Railways due to the greater flexibility of working.

A schematic of a B-class station on a single-line and double-line section are shown in Figs. 6.3 and 6.4 respectively.

The line has to be clear up to an adequate distance beyond the outer signal before 'permission to approach' can be given to a train.

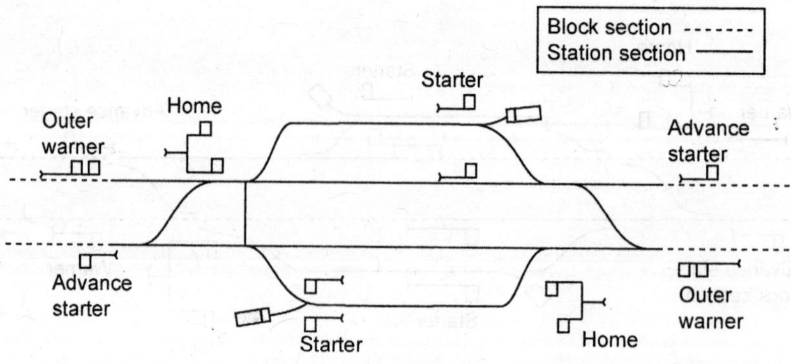


**FIG. 6.4** *Schematic of a B-class station on double-line sections*

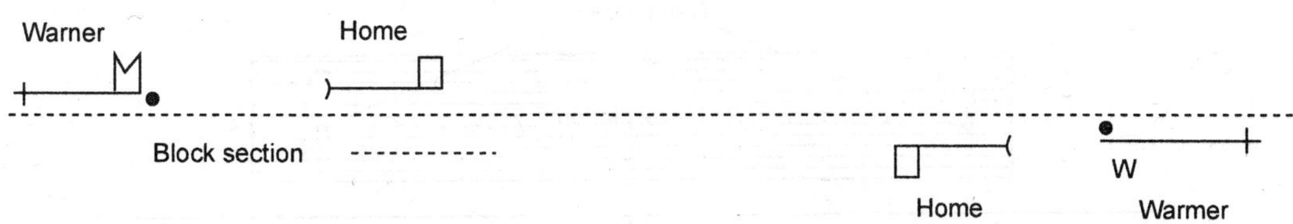


**FIG. 6.5** *Schematic of a C-class station on a single line section*

The minimum signals required are:

*Outer*: The first stop signal; it can be below the warner also.

*Home*: Protects the facing points and is placed at an adequate distance from the outer signal.

*Starter*: This is also provided on a double line section.

*Warner*: Provided on the outer arm post, fast traffic can be allowed while permitting shunting operations even when clear signal is given.

**C-class station:** These stations serve only as the ends of block sections, without booking passengers. Basically, it divides a long block section into smaller blocks. Normally, trains do not stop at these stations.

Figures 6.5 and 6.6 show a schematic of a C-class station on a single-line section and double-line section respectively.

The minimum signals are:

*Warner*: Placed at an adequate warning distance ahead of the home signal, it indicates whether the line ahead is clear or not.

*Home*: It is the first stop signal.

C-class stations increase the line capacity by achieving faster movement of trains.

**Non-block or D-class stations:** Located between two block stations, they do not form the boundary of any block section. No signals are provided at such stations.

A D-class station without any siding is called a *flag station*.

Sometimes, a siding may be provided through a crossover. This can be operated only with the help of a key, which can be released with the aid of a ball token. A D-class station with an outlying siding is called a *DK station*.

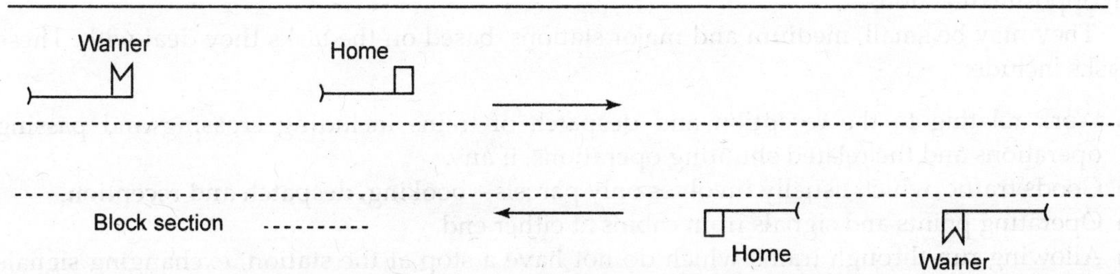


**FIG. 6.6** *Schematic of a C-class station on a double-line section*

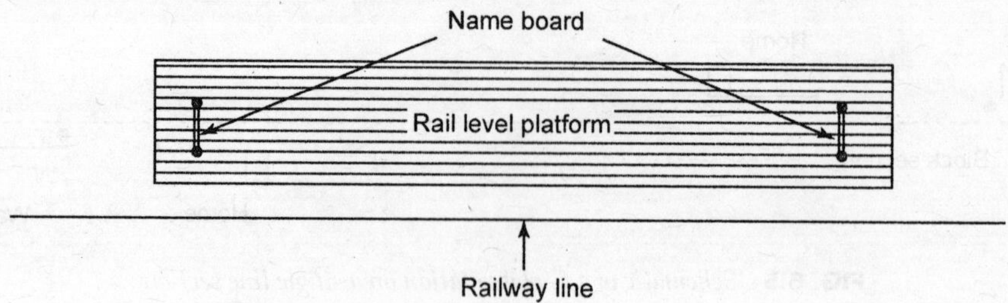


**FIG. 6.7** *Typical layout of a halt station*

**Functional classification of stations**

The layout varies in size and facilities for operation based on the type and volume of traffic, and the proximity to cities and industrial areas.

The following are the categories:

(i) Halts
(ii) Flag stations
(iii) Wayside stations
(iv) Junction stations
(v) Terminus stations

**Halts:** A halt is the simplest station where trains can stop. Only a rail-level platform is provided with a name board at either end. A small waiting shed, if provided, can also serve as a booking office.

No yard or station building is provided.

Selected passenger trains stop for a minute or two for entraining and detraining passengers.

A typical layout of a halt station is shown in Fig. 6.7.

**Flag stations:** These are more important than halt stations, with the provision of a station building and staff. A control is provided for communication.

A small waiting hall, booking office, platforms with benches, and drinking water are also provided.

Sometimes, a siding is also provided for stabling wagons booked for that station.

**Wayside stations:** Also called *crossing stations*, these have the facilities for controlling the movement of trains on block sections. The name 'crossing station' is derived from the fact that they were originally conceived for single-line sections to facilitate crossing of trains running in opposite directions.

They may be small, medium and major stations, based on the tasks they deal with. These tasks include:

- Work relating to the reception and despatch of trains including crossing and passing operations and the related shunting operations, if any.
- Goods traffic, which usually involves only parcels – booking, despatch and reception.
- Operating points and signals from cabins at either end.
- Allowing run-through trains which do not have a stop at the station, exchanging signals with the driver and guard of the train.

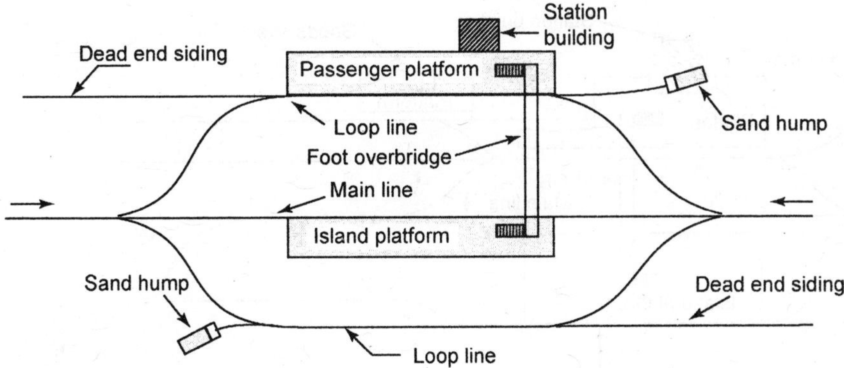


**FIG. 6.8** *A typical layout of a wayside station on a single-line section*

A typical layout of a wayside station is shown in Fig 6.8.

The salient features of way-side stations are: two loop lines with dead-end sidings, two platforms – one near the station and the other, an island platform which can deal with two stopping trains and a foot-over bridge for passengers to move from one platform with the other.

Figure 6.9 shows a layout of a crossing station on a double-line section with a common loop line:

The salient features are:

- In addition to the two main lines for up and down traffic, one common loop is provided to receive trains from either direction.
- Two platforms are provided – one main platform beside the station building and one island platform which can deal with traffic on one mainline and on the common loop.
- A foot over-bridge that connects the two platforms.
- Two emergency crossovers are provided on either side of the station so that only single-line operations can be made in case of an emergency.

Alternatively, two loop lines – one for traffic in each direction – may be provided parallel and adjacent to the main lines with platforms on either side of the lines, with a foot over-bridge connecting them. Emergency crossovers are provided for both the main lines on either side.

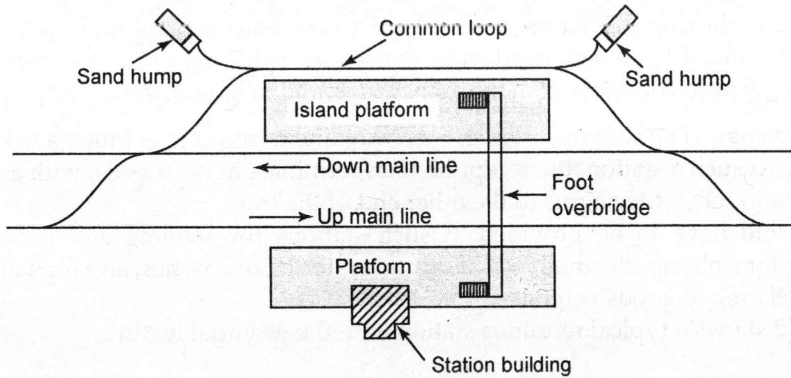


**FIG. 6.9** *Layout of a crossing station on a double-line section with a common loop*

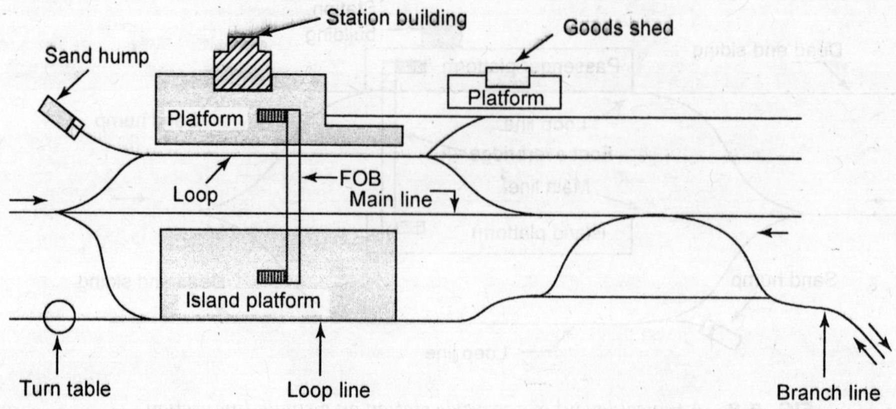


**FIG. 6.10** *A typical junction station with single main and a single branch line*

**Junction stations:** When three or more lines coming from different directions meet, a junction station is provided at such a point. Usually, at junctions, trains arrive from branch lines and return to their originating stations or proceed to other stations. Figure 6.10 shows a typical layout of a junction station with a single main line and a single branch line, which is the simplest:

The essential features of junction stations are:

- Two platforms – one for main line and the other, an island platform.
  Main line as well as branch line trains can be received on the main platform.
- A foot over-bridge connecting the two platforms
- A goods siding and a good shed to deal with goods traffic.
- A turntable for reversing the direction of an engine, if necessary.
- Emergency crossovers on either side to switch to a single-line operation in case of an emergency.

A bigger junction station is one where two main lines and two branch lines meet, with traffic from four different directions.

Figure 6.11 shows a typical junction station with two main lines and two branch lines with the essential features:

In addition to the essential features mentioned above, junction stations also have a number of island platforms, sidings with dead ends, crossovers including a scissors crossover to cater to the additional needs of trains from the branch lines and to handle the expected goods traffic.

**Terminus Stations:** The station at which a railway line terminates is called a *terminus station* or a *terminal*. At such a station, the reception lines terminate at dead ends, with a provision to turn around and move the engine to the other end of the train.

Facilities will have to be provided at such stations for stabling and servicing of the locomotives, for stabling, cleaning, washing, and watering of coaches, and for stabling and all other tasks relating to goods wagons.

Figure 6.12 shows a typical terminus station with the essential features:

### Classification Based on Financial Considerations

The six categories are A, B, C, D, E, and F, as given below:

A: Main line stations with high annual earnings; divisional and zonal headquarters

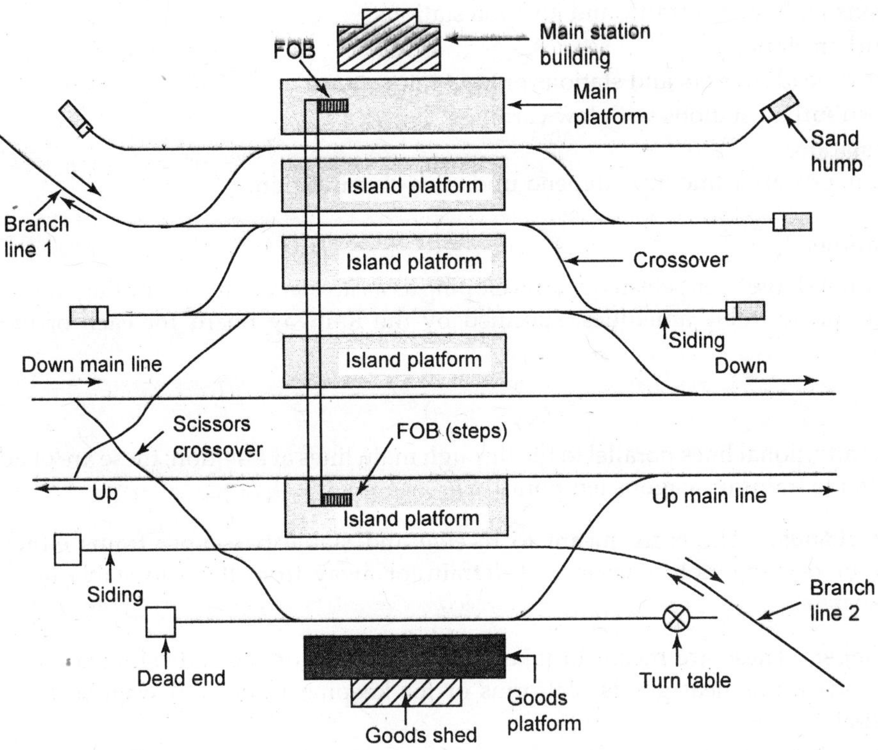


**FIG. 6.11** *A typical junction station with two main lines and two branch lines*

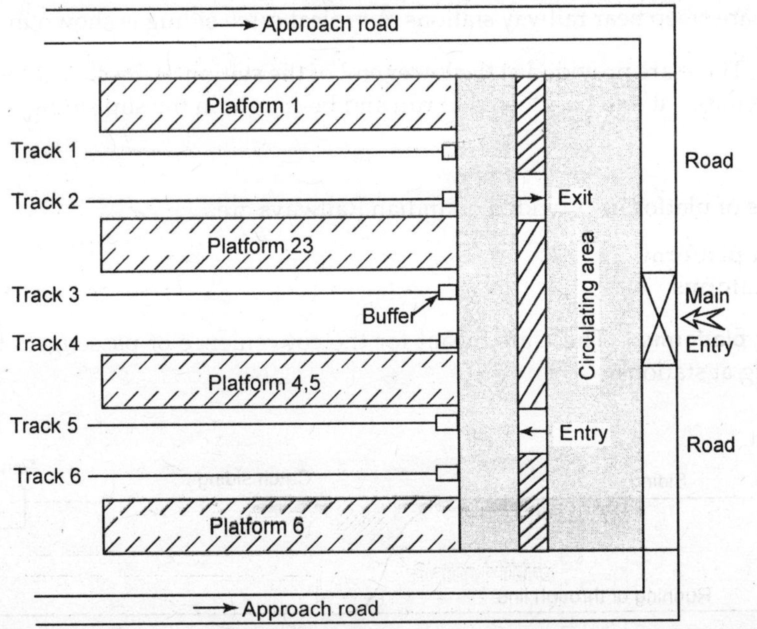


**FIG. 6.12** *A typical terminus station*

B: Stations with tourist traffic and junction stations
C: Suburban stations
D: District headquarters and stations of local importance
E: Non-suburban stations with low earnings
F: Halt stations
Minimum essential amenities depend upon this classification.

### Model Stations

Certain stations have been selected for developing as *model stations* by upgrading the amenities to the level of desirable amenities, specified by the Railway Board for each of the above categories.

### Sidings

Sidings are additional lines parallel to the through main lines at a station; these are of adequate length to stable trains or wagons temporarily.

**Passenger sidings:** These are meant to receive and stable passenger trains either at the platforms for passengers to entrain and detrain, or away from them to stable temporarily before despatch.

**Goods sidings:** These are meant to take care of goods trains. Goods sidings are used for stabling goods trains near goods platforms or for keeping trains and wagons temporarily before despatch.

**Catch sidings:** These sidings are meant to stop runaway wagons which move along the track in an unauthorised manner; at the end of the tracks, buffer stops or sand humps are provided to enable the wagon or train to stop safely; these are required in hilly terrains where the gradients are steep near railway stations. A typical catch siding is shown in Fig. 6.13.

**Slip sidings:** These are provided at the lower end of the station, If, by chance, a vehicle enters the station premises, it will be allowed to run and be caught in the slip siding.

### Platforms

The two types of platforms provided on Indian Railways are:

(a) Passenger platforms
(b) Goods platforms

**(a) Passenger platforms:** These are meant for the convenience of passengers for entraining and detraining at stations:

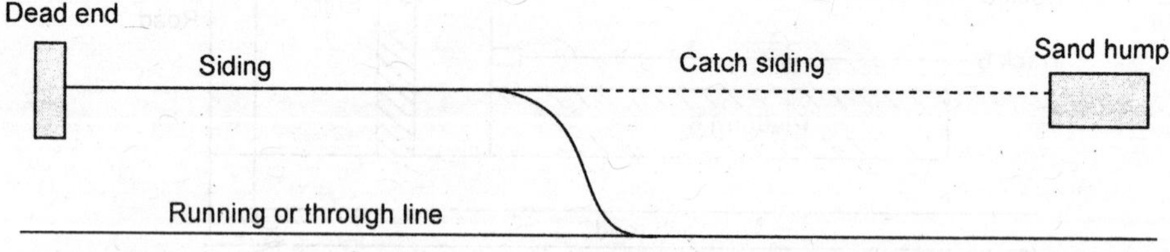


**FIG. 6.13** *A typical catch siding*

These are again classified as

(i) Rail-level platforms
(ii) Low-level platforms
(iii) High-level platforms

**(i) *Rail-level platforms:*** These are provided at unimportant way-side stations at rail-level.

**(ii) *Low-level platforms:*** These are provided at less important main line stations. The height is 455 mm.

**(iii) *High-level platforms:*** These are provided at all important main line stations. Their height is 760 mm to 840 mm on BG and 305 mm to 405 mm on MG.

***Length:*** Platforms should be of sufficient length to accommodate the longest train on the station. It may be 180 m to 450 m.

***Width:*** Minimum platform width prescribed is 3.66 m

***Ramp:*** A ramp of 1 in 6 is provided at the ends.

***Shelter:*** A minimum length of 60 m has to be provided with shelter.

***Station name boards:*** There has to be at least two name boards – one at each end, perpendicular to the track. The name should be in Hindi, English and the regional language. Minimum height to the underside is 1.8 m.

Water supply and toilet facilities have to be provided.

**(b) Goods platforms:** These are provided for loading and unloading of goods and parcels.

***Height:*** 1.07 m for BG and 0.69 m for MG.

***Length:*** Minimum 60 m.

***Width:*** Minimum 3.1 m

**Paving:** Bituminous carpet or concrete

***Other facilities:*** Weighing facilities, mobile cranes, or fixed overhead gantry cranes for handling heavy/bulky goods; adequate storage accommodation.

Figure 6.14 shows a typical section of a platform:

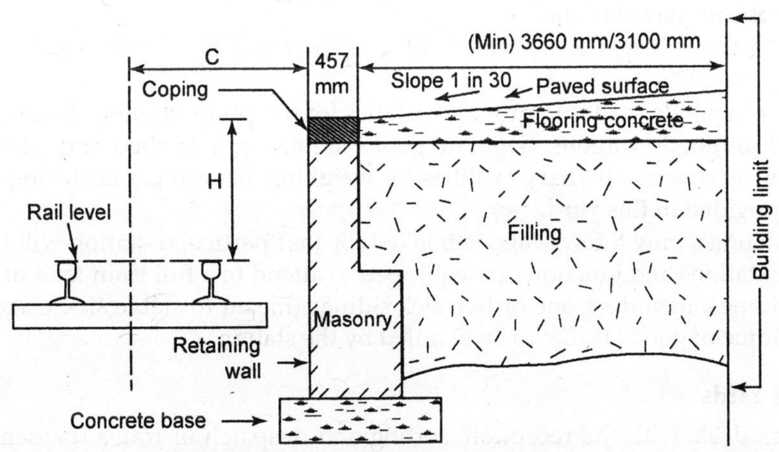


**FIG. 6.14** *A typical section of a platform*

Clearance C: 1.676 m for BG and 1.350 m for MG.

Dock platform: Sometimes, a branch line may be extended up to the main platform for a little distance for the convenience in transhipment of passengers; such an extended portion is called the *Dock platform*.

## 6.3 STATION YARDS

A yard within the station premises is a system of tracks laid to serve the specific purpose of dealing with railway passenger as well as goods traffic..These purposes include receiving and dispatching of trains, besides stabling, sorting, marshalling, arranging new trains, and all other related activities. Yard activities are governed by rules and regulations prescribed by the railway board.

### Types of Yards

The following are types of station yards:

1. Passenger or coaching yards
2. Goods yards
3. Marshalling yards
4. Locomotive yards
5. line yards

Each of these yards performs a set of specific functions related to the passenger traffic as well as goods traffic.

#### Passenger or Coaching Yard

The main function of this yard is to deal with the reception and despatch of passenger trains. The station building and the platforms also are considered to be part of the coaching yard.

In addition, depending upon the volume of traffic, facilities for cleaning, washing, and watering of coaches, examination of coaches, and maintenance of AC units in the case of air-conditioned coaches are also provided. Additional lines provided for this purpose, called *passenger sidings*, form part of a passenger yard. The number of such sidings is large at junction and terminus stations, while a limited number of passenger sidings are provided at most stations except halt/flag stations.

#### Goods Yard

The function of a goods yard is to provide facilities for reception, stabling, loading, unloading, and despatch of goods wagons. Separate goods sidings, goods shed and goods platforms, along with the necessary auxiliary facilities for weighing, handling and storing of goods and parcels are provided in this yard.

In small stations, only a few wagons booked for that particular station will be dealt with, while bigger stations and junctions are equipped to attend to a full train-load of wagons. The number of sidings, including one or two sick sidings (meant to stable sick wagons) depends upon the volume of goods traffic to be handled by the station.

#### Marshalling Yards

A yard which deals with the reception, sorting and despatch of trains (passenger or goods trains) is known as a *marshalling yard*. In view of its functions and importance, marshalling yards are discussed in detail a little later.

· **Locomotive Yards**

The primary function of a locomotive yard is to house locomotives. Facilities for replenishing water and fuel, as well as for examining, servicing, and repairing coaches are provided. These loco-yards also have workshops and stores for ease of maintenance. In addition, they contain a number of facilities so that adequate number of locomotives can be simultaneously stabled, examined, serviced, repaired, and hauled out for use. Approach tracks, exit tracks, different pits like examination pits and ash pits, turn table for reversing steam locomotives, workshops and stores, and such other facilities form part of a locomotive yard. The layout can be rectangular or semicircular in shape, and adequate space must be available.

**Sick Line Yard**

Whenever a wagon or a coach becomes defective, it is said to be 'sick' and is taken to one of the sick lines in the 'sick line yards'. Such sick wagons or coaches are examined and repaired in this yard, and all facilities needed for this purpose, including a small office, store, and workshop, are provided in the sick line yard. The number of lines depends upon the requirements and importance of the station. Junction stations and terminus stations are necessarily provided with sick line yards.

## Marshalling Yard

This is the yard where goods trains are received, sorted out, and new trains are formed according to their destinations, with wagons arranged in station order, and properly marshalled and despatched in the appropriate routes. Some of the wagons may be empty for further booking at different stations while others may be loaded at the originating station itself.

Marshalling operations are carried out so that the wagons can be conveniently detached without the need for much shunting at wayside stations.

A marshalling yard consists of sub-yards–one each for the functions of reception of trains, sorting of wagons/trains, and departure of formed trains.

These are:

(a) Reception yard or reception lines
(b) Sorting yard or sorting lines
(c) Departure yard or departure lines.

Figure 6.15 shows one type of layout for a marshalling yard:

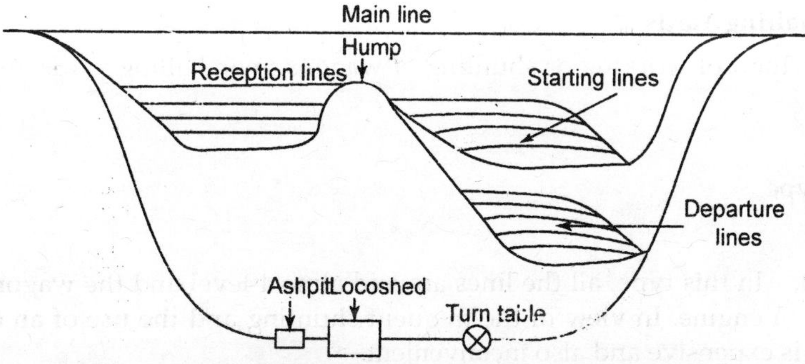


**FIG. 6.15** *Typical layout of a marshalling yard*

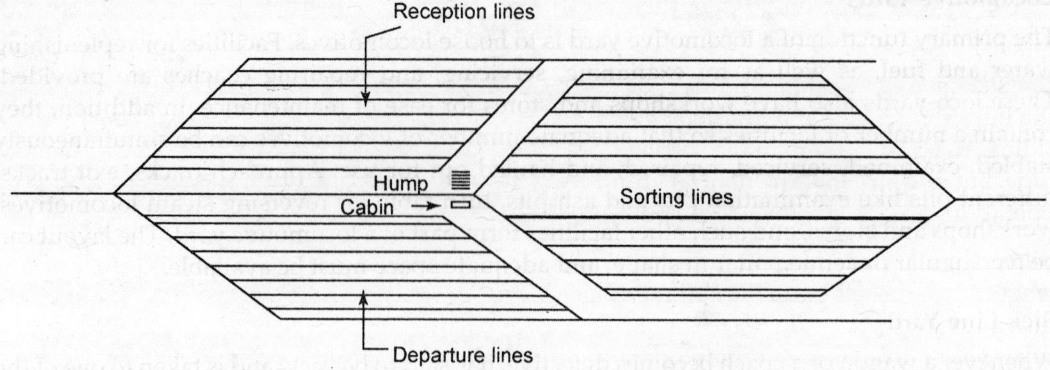


**FIG. 6.16** *Layout of a marshalling yard (with a central hump)*

Figure 6.16 shows another type of marshalling yard layout with a central hump.

### Design Principles

The following principles are to be borne in mind in the design of a marshalling yard:

(i) It should be designed in such a way that there is minimum detention of wagons in the yard so that sorting of wagons is done as fast as possible.

(ii) Necessary facilities such as a long shunting neck, a correctly placed hump, appropriate braking arrangement, weigh bridge, ash pits, and water column should be provided.

(iii) Through traffic on the main lines should be expeditiously dealt with without interference from shunting operations.

(iv) Unidirectional movement of wagons is desirable.

(v) Conflicting movements of wagons/engines should be avoided.

(vi) The lead distances for the movement of wagons/engines should be as short as possible.

(vii) Marshalling yards should be provided with good lighting for night-working.

(viii) Adequate space should be provided for future expansion.

[**Note:** In the case of double-line sections, separate marshalling yards are normally provided for up- and down-traffic, far behind the respective main lines.]

### Types of Marshalling Yards

Based on the method of working or shunting of wagons, marshalling yards are classified as follows:

- Flat type
- Gravitation type
- Hump type

**Flat type of yard:** In this type, all the lines are laid almost level and the wagons are moved with the aid of an engine. In view of the frequent shunting and the use of an engine for all operations, this is expensive and also inconvenient.

This type is used only when there is severe limitation on space or when the number of wagons to be dealt with is low.

**Gravitation type of yard:** In this type, the natural ground being favourable, some portion of the lines can be laid at a gradient. When tracks are laid in this way, the wagons move to the siding assigned to them in the sorting lines by the action of gravity. This is the least expensive type, but it is rare that the natural ground is so favourable; therefore, sometimes the shunting operations are assisted by engine power besides gravity.

**Hump type of yard:** In this type, an artificial hump is created by means of earthwork. The wagons are pushed up to the summit of the hump with an engine and let go, so that they slide down under gravity and reach the sidings assigned to each one of them. In a way, it is similar to gravitation yards as shunting operations are carried out under the influence of gravity.

The gradients recommended for the pushing stretch and the downward gradients beyond the hump are decide based on the topography, type of wagons, spacing between successive groups of wagons and test runs, and the braking method employed to stop the wagons.

**Speed regulation of wagons:** The speed is regulated to keep them stable in the sorting lines without any damage. Two methods are available for speed regulation:

- Mechanical method.
- Non-mechanical method.

*Mechanical method:* In this method, wagons are slowed down automatically with the help of *retarders*, which are in the shape of bars fixed on either side of the track. These bars are operated electrically or electromechanically to press against the sides of the moving wheels, which retards the motion and make the wagons stop at the appropriate place.

This is also called *automatic braking*, and is extensively used in advanced countries.

*Non-mechanical method:* In this, the wagon speed is controlled with the aid of hand brakes or skids. A shunting helper runs alongside the rolling wagon and applies a hand brake to it at an appropriate place, making the wagon to stop.

In another approach, skids are placed on the track, which get dragged by the rolling wagon; the friction developed retards the wagon and makes it stop at the desired location.

The *gradients recommended in hump yards* are:

**Mechanical braking**

Rising gradient up to the hump: 1 in 50 to 1 in 125
Top of hump: Level
First falling grade after hump: 1 in 17 to 1 in 20
Next falling grades: 1 in 200 to level

**Non-mechanical braking**

Rising gradient: 1 in 50 to 1 in 100
First falling grade: 1 in 25 to 1 in 35
Next falling grades: 1 in 80 to 1 in 200.
Falling gradient of the sidings: 1 in 400 to 1 in 600

[**Note:** These are only recommended values, and there can be deviations based on factors relevant to the site and other operating conditions.]

### Design of the Components of a Marshalling Yard

**Reception yard:** The number of lines depends upon the number of trains expected and their frequency. One reception line serves three to four trains normally. The length of siding is 700 m for BG and 650 m for MG.

**Shunting neck:** The length should be more than that of the longest train expected.

**Hump:** The design criteria are safe and efficient handling of wagons. Average height is 2.5 m to 3 m for an ordinary hump, while it is 3.5 to 6 m for a mechanised operation.

**Sorting yard:** The number of lines required depends upon the number of destinations for which the trains are to be assembled. The length of a sorting line is kept 15 to 20 % more than that of a normal train. The layout can be of the ladder type or the balloon type. The ladder type of layout has 'ladder track' (discussed in Chapter 5) as its main feature; otherwise, it is called the 'ballon' type. The difference lies in HOW the various parallel tracks for sorting are connected/laid out for sorting operations. Figures 6.15 and 6.16 show the typical layout of a marshalling yard. The speed control may be mechanical or non-mechanical.

**Departure yard:** The number of lines depends on the number of trains to be despatched and on the frequency of their departure. The need for the departure yard is obviated if assembled trains can be sent out straight from the sorting yard; this saves a lot of space, but problems could arise if the departures of trains are delayed for operational reasons.

[**Note:** The recent trend of carrying goods traffic as train-loads from point to point, avoiding the loading of piecemeal wagons, has considerably reduced the need and importance of marshalling yards for dealing with goods trains.]

## 6.4 EQUIPMENT AT STATIONS AND YARDS

The following are the *equipment* required at *stations*:

**For passengers:** Clocks, public telephones, CCTV for security and TV for entertainment, escalators/elevators at big and important stations for the convenience of the elderly.

**For traffic and staff:** Train control equipment, flags and lamps, and other auxiliary equipment.

**Common to all:** Public address systems, water supply, coolers for drinking water, first-aid facilities.

The following are the various *equipment* and facilities needed in different yards:

1. Cranes
2. Weigh bridge
3. Loading ramps
4. Ash pits or pans
5. Drop pits and examination pits
6. Water columns
7. Turntable
8. Traverser
9. Washing platforms
10. Buffer stops
11. Scotch blocks
12. Derailing switch
13. Sand humps

14. Fouling marks
15. Loading gauge and construction gauge

(**Note:** The last mentioned items have already been covered in Chapter 3.)

### Cranes

Cranes are provided in goods sheds for loading and unloading heavy or bulky items, such as timber logs, cars, trucks, and heavy machinery.

The following types of cranes are commonly used in goods yards near goods platforms:

**Fixed jib crane:** This is fixed at a convenient location on the goods platforms for loading and unloading heavy goods from wagons.

**Mobile crane:** Mounted on a truck or a wagon, it can be moved anywhere on the platform.

**Overhead gantry crane:** This consists of two horizontal steel girders, supported by a number of vertical steel columns at suitable intervals along their length. A travelling platform which is fitted with hoisting equipment is fixed between the girders. It can move to and fro on the girders. Wagons or trucks are brought under the gantry for loading and unloading goods.

### Weighbridge

A weigh bridge is used to weigh a loaded wagon to obtain the weight of the goods contained in it. A short length of track supported on beams located in a pit under it is made to rest on knife edges attached to levers.

When a wagon is brought on to the weigh bridge, its weight can be read by a pointer on a graduated disc located in an adjacent structure.

A weigh bridge is usually provided on a siding, and not on a through track.

### End-loading Ramps

End-loading ramps are provided to facilitate the unloading of the wagons at their rear end. Such ramps are mainly used for unloading cars, motor-bikes, and such mobile vehicles, which can be driven away on to an approach road via the ramp after getting unloaded.

A typical end-loading ramp is shown in Fig. 6.17:

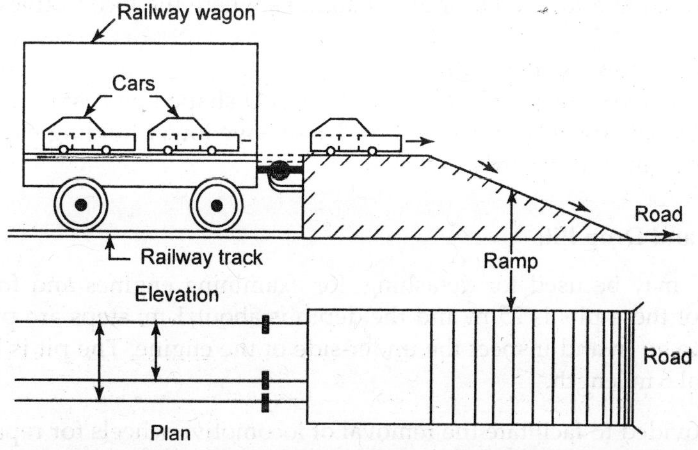


**FIG. 6.17** *End–loading ramp*

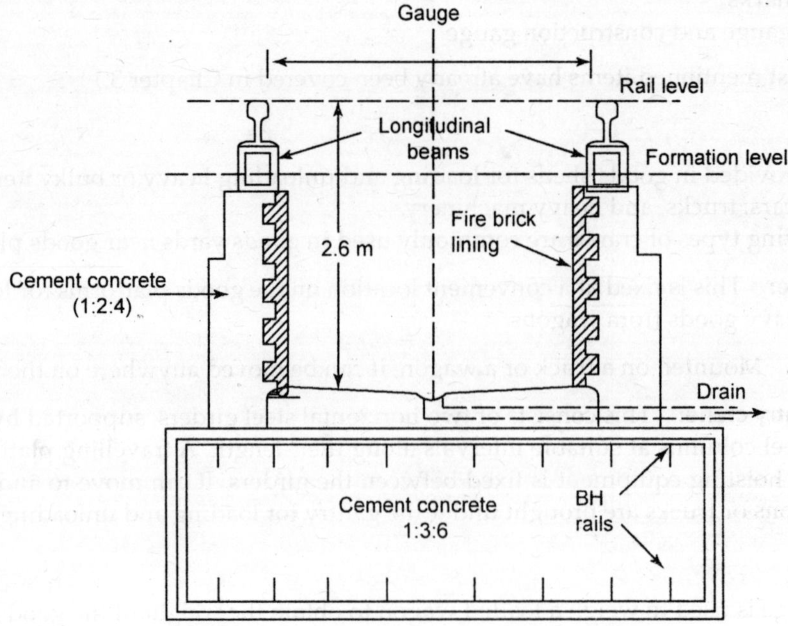


**FIG. 6.18** *A typical cross-section of an ash-pit*

The *essential features* are:

- A siding with a dead end, buffer stop, and a platform with a ramp.
- Platforms at a height of 1.30 m for BG and 0.86 m for MG.
- A small gap between the buffer stop and the ramp platform to minimise damage to the latter; this gap is covered by the hinged plates of the wagon while unloading.

### Ash Pits and Ash Pans

Ash pits, or de-ashing pits are meant to collect the ashes falling from locomotives. Usually, such pits are provided in a loco-shed or at big stations where the locomotives collect water, while de-ashing.

A typical ash pit is shown in Fig. 6.18.

*Ash pans* also serve the same purpose; they consist of U-shaped precast reinforced concrete units placed side by side for retention of ashes. These have lower holding capacity than ash pits and are provided in station yards.

### Examination Pits and Drop Pits

**Examination pits:** may be used for de-ashing, for examining engines and for repairs. The minimum length of these pits is 25 m and the depth is about 1 m; steps are provided at the ends for the staff to go in and inspect the under-side of the engine. The pit is lined with fire bricks at the central 6 m length.

**Drop pits:** are provided to facilitate the removal of locomotive wheels for repair or renewal. These pits are at right angles to the track.

Mobile jacks are used to remove the wheels of the engine.

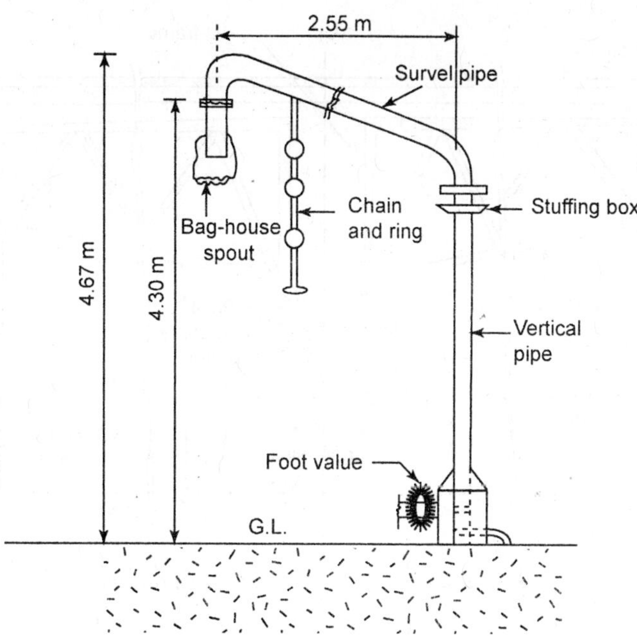


**FIG. 6.19** *Schematic of a water column*

## Water Columns

These are meant to supply water to the locomotives, and are provided both in station yards and in locomotive yards.

Figure 6.19 shows a schematic of a water column.

## Turntable

A turntable is a device used for reversing or changing the direction of a locomotive. It is usually provided at junctions, terminus stations, locomotive yards and marshalling yards.

It consists of a track supported on two parallel girders, which are braced together and supported or suspended from a central pivot. Flangeless wheels are attached to the ends of girders, and these roll over the rails (called 'race rails') along the circumference of the pit. The turntable is installed in a circular masonry pit. The girders supporting the track move around the race rail, and can be fixed in any desired position with the aid of locking bolts.

The base of the pit slopes towards the centre to drain rainwater into a sump, from which it can be drained off.

Two or more tracks radiate outwards from the edge of the circular pit; they can be aligned and connected temporarily to the track on the turntable. The diameter should be adequate accommodate the longest engine. The normal sizes on Indian Railways are:

BG: 30.5 m, 22.9 m, and 19.8 m
MG: 19.8 m and 15.75 m.

A vertical section of a turntable is shown in Fig. 6.20.

The steps in reversing the direction of an engine are:

(i) The turntable is revolved either manually or electrically on the pivot and is brought in line with the track on which the engine is made to stand. This position is then locked with the locking bolts.

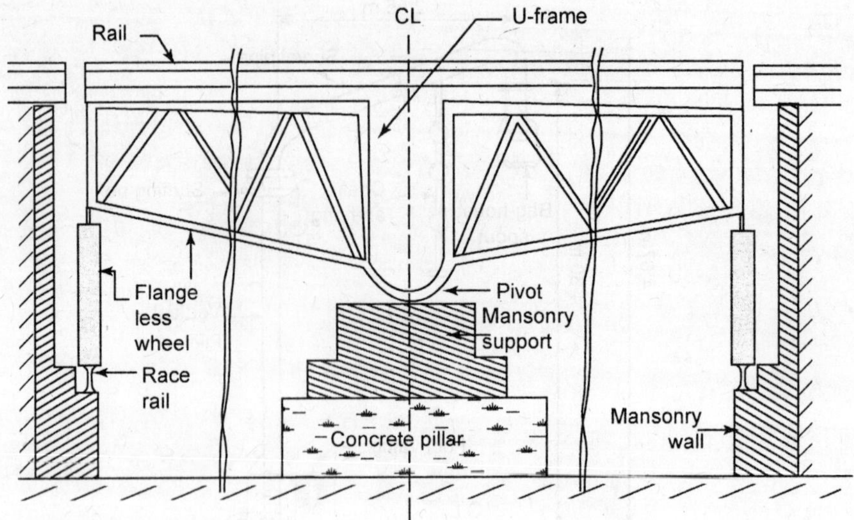


**FIG. 6.20** *Typical cross-section of a turntable*

(ii) The engine is brought on to the track on the turntable and the locking bolts are removed.
(iii) The turntable is now rotated until its track aligns with the track on to which the engine has to be sent.
(iv) This position is now locked with the bolts, and the engine is shunted on to the required track.

Since steam locomotives have been practically phased out on Indian Railways, all equipment needed to deal with steam locomotives, such as ash-pits, water column, and turntables have lost their relevance, and are now of historical importance only.

Triangle-shaped tracks, called *triangle*, are meant to serve the same purpose as that of a turntable, has also lost its relevance.

**Traverser**

This is a device for transferring vehicles from one track to a parallel track without the use of a turnout or a crossover. It consists of a platform with a track that is mounted on rollers, which can traverse to and fro and can be made to fall in line with the track on either side.

This is an expensive arrangement and is provided in large workshops so that a coach or a locomotive may be shifted from one shop to another for a series of checks and repairs. Space limitation for providing a crossover may be overcome in this manner.

Figure 6.21 shows a traverser. The operation of a traverser is obvious – the railway vehicle stands on one of the parallel tracks, say A. The traverser track is aligned first with Track A and the vehicle is brought on to the traverser. Then the traverser is moved on rollers in line with Track B on to which the vehicle has to be transferred. The vehicle is then shifted to Track B.

**Washing Platforms**

Junctions and terminus stations are provided with carriage washing platforms on two or three long sidings which can accommodate a full length of train.

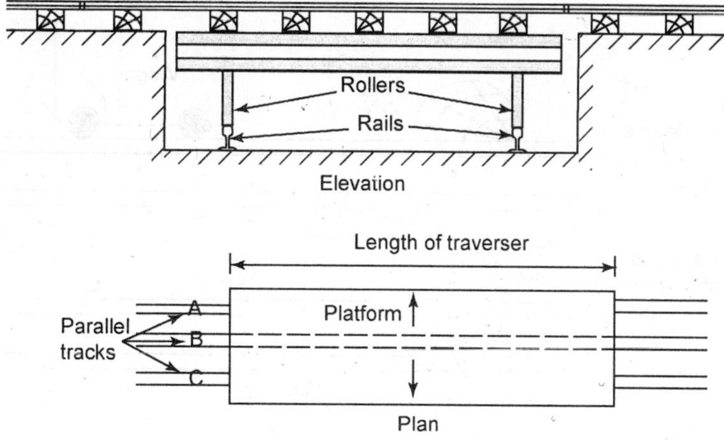


**FIG. 6.21** *Schematic of a traverser*

The salient features are:

- The height equals that of carriage floor; the width is 610 mm; the platforms are of cement concrete
- A number of hydrants to supply water under pressure
- Positioned between two tracks so that two trains can be washed simultaneously
- The tracks are supported on masonry structures with adequate drainage arrangements.

### Buffer Stop

A *buffer stop* is provided at the dead end of a siding to ensure that vehicles stop at it without going off the track. It is a kind of barrier across the track which prevents vehicles from going beyond a particular end point.

Figure 6.22 shows a side view of a buffer stop:

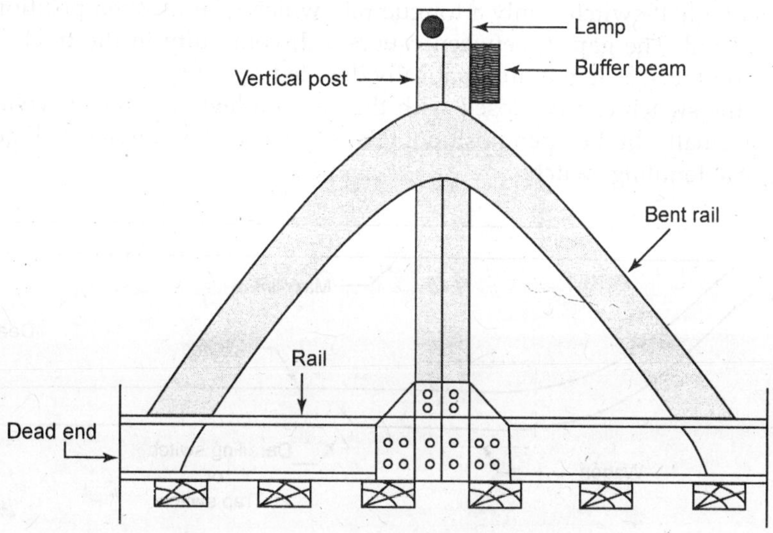


**FIG. 6.22** *Buffer stop*

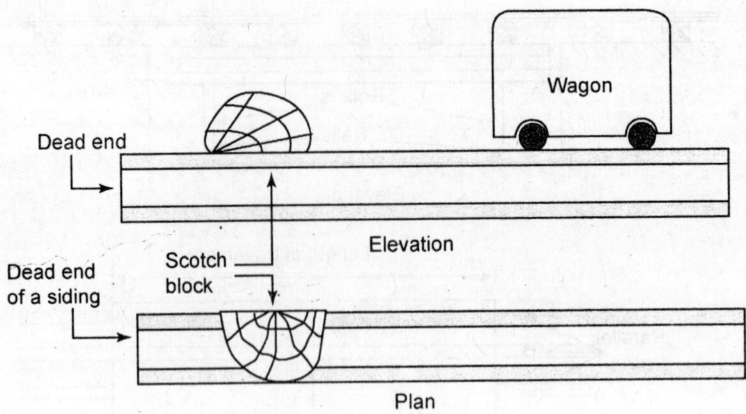


**FIG. 6.23** *Scotch block*

The *essential features* of a buffer stop are:

- It should be strong enough to take the impact of a moving vehicle.
- It should have a buffer disc with a cross-sleeper, which is painted red, with a red lamp at the middle for night-working.
- The track should be straight for some distance before the buffer stop.
- Visibility from a long distance is another good feature.

### Scotch Blocks

These are wooden blocks placed on the rail and held in place with the aid of a device to form an obstruction, so that a vehicle is not allowed to go beyond it.

Figure 6.23 shows a scotch block.

### Derailing Switch

This consists of a half-switch—only a tongue rail, which is in its open position, facing away from the stock rail. The gap in between causes a discontinuity in the track; thus a vehicle cannot go beyond this, as it gets automatically derailed.

If desired, the switch can be closed with the aid of a lever to allow a vehicle to traverse beyond this normally. In the open position, this is also called a *trap switch*. Figure 6.24 shows a schematic of a derailing switch:

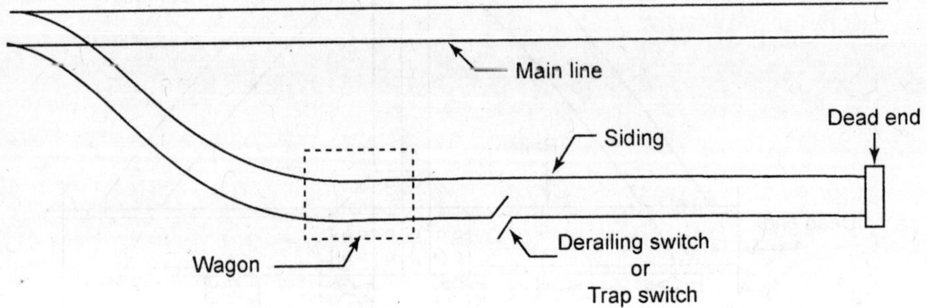


**FIG. 6.24** *Schematic of a derailing switch*

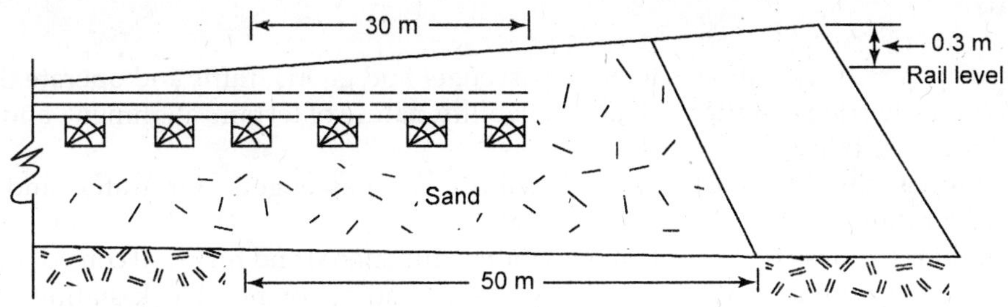


**FIG. 6.25** *A typical section of a sand hump*

### Sand Hump

This is considered to be the best method to isolate and stop a moving vehicle without causing any damage to it. It is provided on a loop line in order to stop an incoming train which tries to overshoot. It consists of simply a mound of sand of a specified cross-section, which covers the track at the end of a dead-end siding; it is laid on a rising gradient to aid the resistance of the sand in bringing the vehicle to a stop.

Figure 6.25 shows a typical section of a sand hump.

### Fouling Mark

This is provided between two converging tracks at the particular point beyond which two vehicles should not be allowed simultaneously to move along the two tracks. The minimum centre-to-centre distance of the tracks stipulated for this purpose is 4,265 mm for BG and 3,660 mm for MG lines.

Figure 6.26 shows a fouling mark.

The *salient features* of a fouling mark are:

- It consists of a stone or concrete block or an old sleeper painted white.
- It should be visible from a distance; the letters FM are painted in black.
- The top of the fouling mark should extend up to the top of the ballast section.
- It should be fixed perpendicular to the track firmly into the ground.

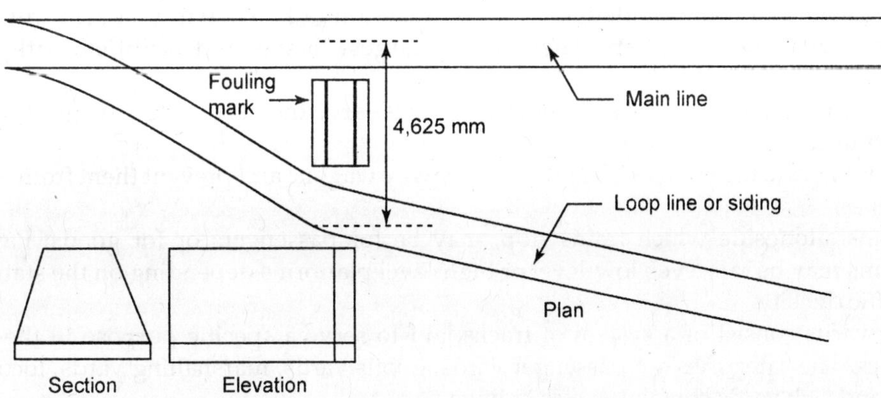


**FIG. 6.26** *Fouling mark*

## SUMMARY

- The purpose of a railway station is to book passenger and goods traffic and operate trains.
- The criteria for selection of a site are land availability, nature of terrain, alignment, and more importantly, accessibility.
- Railway stations should have adequate facilities for passengers, for traffic and train operation, for rolling stock, and for railway staff.
- Railway stations are classified based on operational, functional and financial considerations. From operational criteria, stations are either block stations or non-block stations. Block stations are classified as A-class, B-class and C-class stations, while non-block stations are graded as D-class or Flag stations.
  From functional criteria, stations are classified as Halts, Wayside stations, Junctions and Terminus stations.
  From financial or earnings criteria, stations are classified as A-, B-, C-, D-, E-, and F- stations
- A block station is that at which a train driver should receive the authority to proceed and enter the next block section of the railway line. This is relevant to the Absolute block system of train operation in which only one train is allowed to be in one block section at any time, the railway line being divided into a convenient number of block sections. Thus, a space interval is enforced between trains for safety.
- Block stations are divided into A-, B-, and C- class stations depending upon the traffic and importance.
  The least important stations are non-block stations or Flag stations or D-class stations, which do not mark the ends of a block section.
- Halts are the simplest as they are merely stations where a few trains stop for two minutes for the convenience of passengers.
  Flag stations are a little more important than halts; station building and staff, and sometimes a siding, are provided.
  Wayside stations or crossing stations have facilities to control the movement of trains. They facilitate trains to cross each other on a single-line section. Reception and despatch of trains and goods traffic, train control by points and signals, and such other tasks are carried out at such stations. Loop lines and sidings are also provided depending upon traffic density.
  Junction stations are those at which one or more branch lines meet a main line; consequently, passenger as well as goods traffic will be significant. In view of this, adequate facilities for different sections of users and administration of railways are provided.
  Terminus stations are provided at big terminuses on railways where several routes join or from where they originate. They are the largest in size and facilities, with several platforms.
- Sidings or lines parallel to the through main lines serve the needs of passenger trains and goods trains.
  Catch sidings are meant to catch and stop runaway wagons and prevent them from entering main lines.
- Platforms, alongside which trains stop, may be for passengers or for goods. Passenger platforms may be rail-level, low-level or high-level platforms depending on the station and the traffic density.
- Station yards consist of a system of tracks laid to serve a specific purpose in the station premises. The categories are passenger yards, goods yards, marshalling yards, locomotive yards, and sick yards containing sick sidings.

- Marshalling yards are those where goods trains are received, sorted, new trains are formed according to their destinations with wagons arranged in station order, and despatched to their respective destinations.

  Marshalling yards consist of sub-yards—reception lines, sorting lines and despatch lines arranged in a convenient manner with minimum leads for shunting operations. Based on the method of working, the types are flat, gravitation, or the hump type, the last one being the most common one.

  In the hump type of yard, a hump is created at a convenient location with respect to the various lines; wagons/trains are pushed up to the hump by engine power, and allowed to roll down into the appropriate lines.

  Speed regulation of wagons may be automatic or mechanical; in the latter method, retarders, operated electrically or electromechanically, press against the sides of the moving wheels and bring the vehicles to stop.

  The alternative method is to apply handbrakes manually by staff running alongside the moving wagons, or by placing skids on the track to obstruct motion, to bring the vehicles to stop by their frictional resistance.
- Several kinds of equipment are necessary in stations and station yards. Equipment at stations include clocks, public telephones, TV and CCTV, elevators/escalators, train control equipment, flags and lamps, public address systems, water coolers for drinking water, and first-aid facilities.

  Equipment in station yards include cranes, weighbridge, end-loading ramps, water-columns, buffer stops, shunting neck, loading gauge and construction gauge for dealing with goods traffic; ash pits, ash pans, examination pits, water columns, turntable, and traverser in and around locomotive sheds. In addition, there are also scotch blocks, derailing switches and sand humps in dead-end sidings to deal with runaway wagons and prevent damage to other railway vehicles and tracks.

## QUESTIONS

1. (a) What is the purpose of a railway station?

   (b) What are the considerations in selecting a site for a railway station?
2. What are the facilities required at a railway station for

   (i) the passengers and the public,

   (ii) the traffic and operations, and (iii) the railway staff?
3. Describe briefly the considerations in classifying railway stations on Indian Railways.
4. Discuss briefly A-class, B-class and C-class stations and the minimum facilities required for each.
5. Sketch a schematic plan of an A-class station on a single-line section, showing the minimum signals necessary for operation
6. Sketch a typical layout of a wayside station on a single-line section, with minimum goods traffic.
7. What are the various sidings needed in and around railway stations and what are their specific functions?

8. Differentiate between:
   (a) Passenger platform and goods platform
   (b) Main platform and island platform
   (c) Main platform and dock platform
9. Differential between:
   (a) Flag station and block station
   (b) Wayside station and junction station
   (c) Junction station and terminus station.
10. What are the various types of yards in and around stations?
11. What are the facilities required for goods yards with heavy goods traffic?
12. What is a locomotive yard and what are the minimum facilities needed at such a yard?
13. Explain briefly 'sick line yards'.
14. What is a 'marshalling yard' and what are its functions?
15. What are the various lines which have to be provided in marshalling yards?
16. Sketch a typical layout of a marshalling yard, incorporating the essential features.
17. Enumerate the methods of working of a marshalling yard.
18. Describe the features of working of a hump-type marshalling yard.
19. What are the methods adopted for speed regulation of wagons in marshalling yards?
20. Differentiate between skids and retarders.
21. Write brief notes on:
   (a) Water column
   (b) Ash pit
   (c) Sand-hump
22. Write brief critical notes on:
   (a) Turntable
   (b) Traverser
   (c) Buffer stop
23. Write short notes on:
   (a) Cranes
   (b) Scotch blocks
   (c) Derailing switch
24. Write short notes on:
   (a) Shunting neck
   (b) Fouling mark
   (c) Washing platform

25. What is a marshalling yard? Explain the functions of a marshalling yard with a sketch. (VTU)
26. Write short notes on any two of the following:
    (a) Turntable
    (b) Water column
    (c) Buffer stop (VTU)
27. (a) What are the functions of a railway station?
    (b) What are the gradients recommended in station yards? (Kerala University)
28. Draw a typical layout of a locomotive yard and mention its chief requirements. (Punjab Tech. University)
29. Briefly describe the various factors that influence the selection of site for a railway station.

    Sketch A-class and C-class railway stations and explain their features. (UP Tech. University)
30. What is the purpose of providing marshalling yards? What are the points to be considered in their design? Also mention the main features of marshalling yards. (UP Tech. University)
31. (a) Differentiate between a station and a yard.
    (b) How are railway stations classified from the operational point of view?
    (c) Give the various classes of railway stations and describe each type in detail. (Kalinga Institute of Technology, Odisha)
32. Classify the types of railway stations. (West Bengal University of Technology)
33. (a) Draw a neat sketch of the layout of a marshalling yard.
    (b) What are the different types of station yards?
    (c) Explain in brief the working of the marshalling yards stating three important functions of these yards. (Mumbai University)
34. Explain the different types of yards in a station. (Mumbai University)
35. Explain, in brief, the working of a marshalling yard. (Mumbai University)

Chapter 7

# SIGNALLING, INTERLOCKING, AND TRAIN CONTROL

**CHAPTER OBJECTIVES**

This chapter describes:

- The mechanism of signalling and interlocking and the process by which it helps to ensure safety in the movement of trains
- The operation, function and location for classifying signals – two-aspect and three-aspect semaphore signals and their indications, mechanical and electrical signalling systems which operate both signals and points for reception and despatch of trains on different tracks, and the use of coloured light signals in sections of electric traction along with track circuiting.
- The setting of points in such a way as to avoid conflict and ensure safety; electrical and electronic systems of interlocking
- Methods of train control and operation such as centralised train control, route relay interlocking, and train protection and warning systems that enhance railway safety.

## 7.1 INTRODUCTION

The purpose of signalling and interlocking is to control the movement of trains for safety and efficiency in operation. Interlocking of signals, points, block instruments used for operation and other associated equipment are meant specifically to ensure safe running of trains. Maximum track utilisation can be realised only with an efficient and fool-proof system of signalling and interlocking.

Signalling can also be viewed as a medium of communication between the station master or the traffic controller in an automatic control system on the one hand, and the loco pilot on the other. (Locomotive drivers are now called *loco pilots* in accordance with the railway terminology approved by the Railway Board.)

Signalling, in olden days, was rather primitive and used policemen on horseback to ensure that the track was clear. Signalling was introduced by British railways for the first time in 1842. The procedure of interlocking points and the corresponding signals to ensure safety in train operation was developed later in 1867.

Absolute block system of train control has been the most widely used method. *Route relay interlocking, centralised train control,* and such other modern signalling systems are of recent origin.

## 7.2 SIGNALLING

*Signalling* consists of the use of various types of signals for the operation and control of trains.

### Objectives of Signalling

(i) Regulating train movement for safe running at maximum permissible speeds, leading to maximum utilisation of the track.
(ii) Maintaining a safe distance between trains running on the same track.
(iii) Facilitating crossing and passing operations of trains running in opposite directions and in the same direction, respectively, on a single-line section.
(iv) Regulating the arrival into and departure from station yards, of the several trains running in the section.
(v) Facilitating safe and efficient shunting operations in station yards while dealing with both passenger traffic and goods traffic.
(vi) Ensuring the safety of both railway traffic and road traffic at level crossings.
(vii) Giving appropriate indications during maintenance operation and repair of track.

A well-designed system of signals and signalling is certain to achieve the above objectives, when supported by the appropriate methods of interlocking of points and signals.

### Classification of Signals

Railway signals may be classified based on different criteria; they are:

#### Operation

(a) Audible or detonating signals
(b) Visible signals

(i) Hand signals
(ii) Fixed signals

#### Function

(a) Stop signals
(b) Permission or caution signals
(c) Shunting signals

#### Location

(a) Reception signals: Outer, router, home
(b) Departure signals: Starter, advance starter

These are all stop signals.

**Construction**

(a) Semaphore signals

- (i) Lower quadrant system (two-aspect)
- (ii) Upper quadrant system (three-aspect)

(b) Coloured light signals

Two-aspect, three-aspect, and four-aspect types.

**Special Purposes**

- (i) Repeater signal
- (ii) Co-acting signal
- (iii) Router signal
- (iv) Calling-on signal
- (v) Home and warner combination signal
- (vi) Point-indicator signal
- (vii) Temporary signals
  - → Caution indicator
  - → Speed indicator
  - → Stop indicator
  - → Shunting limit indicator
  - → Termination (of restrictions) indicator

**Audible or detonating signals:** Detonators and fog signals are used in foggy weather when other signals do not serve the purpose due to poor visibility. Their sound attracts the attention of loco pilots as the detonators, fixed to the rail by means of chips, explode as a train passes. These are fixed about 90 m before a signal, thereby alerting the driver to look out for the signal indication ahead.

**Visible signals:** These signals are visible from a distance, and draw the attention of loco pilots in view of their strategically planned positions alongside the track.

**Hand signals:** The basic forms of hand signals are flags fixed to wooden handles; they are either red or green, the former indicating 'STOP' and the latter, 'PROCEED'. These are used during day time.

At night, hand-held lamps, with movable red and green slides are used; the indications are the same as those for the flags. If a yellow side is also there, it may be used to indicate 'PROCEED WITH CAUTION'.

Even bare hands may be used for indicating the appropriate signals during day time if even flags are not available.

Hand signals may also be used to give the appropriate indication to an approaching loco pilot during maintenance and track repairs, as also during the execution of engineering works. The signals may be used by station masters, train guards, cabinmen, gangmen, pointsmen, or any other authorised person.

**Fixed Signals**

A fixed signal is one which is firmly fixed to the ground by the side of the track, and therefore, they are at pre-planned locations, serving different purposes.

Functionally speaking, fixed signals may be stop signals, permission or caution signals, or shunting signals.

**Stop signals:** These are usually *semaphore* type signals; they are designed in such a way that, in the event of failure of the equipment, they will always be in the *stop* position. This *fail-safe* mechanism is the characteristic feature of semaphore-type fixed signals.

**Semaphore signals:** The origin of *semaphore* is from Greek—*Sema* means 'sign' and *phor* means to 'bear'. A semaphore signal consists of a movable arm pivoted to a vertical post through a horizontal pin. A typical semaphore signal is shown in Fig. 7.1.

The arm of the signal is painted red with a vertical white stripe on the side facing the drivers; the other side is painted white with a black vertical stripe.

The signal consists of a spectacle arm, a pivot or a spindle, a counter weight spring stop, a crank and a crank rod, and a chain and pulley arrangement for the wire from the cabin. The whole assembly is housed on top of a tubular or a lattice post.

A steel ladder is also provided to reach the arm and the lamp for any adjustment. For visibility of the signal position at night a lamp is fixed at the top of the post. The spectacle containing green and red glasses moves along with the arm; the glasses are arranged such that the driver sees the red light when the arm is horizontal. When the arm is lowered the green light is visible.

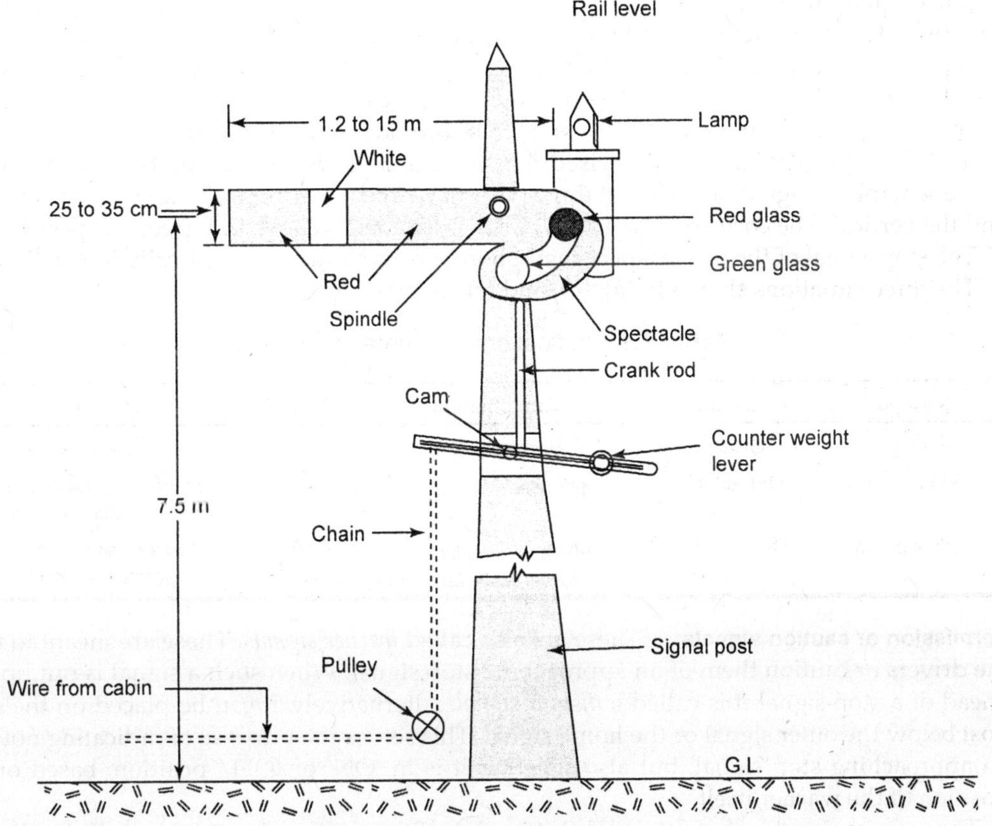


**FIG. 7.1** *A typical semaphore Signal*

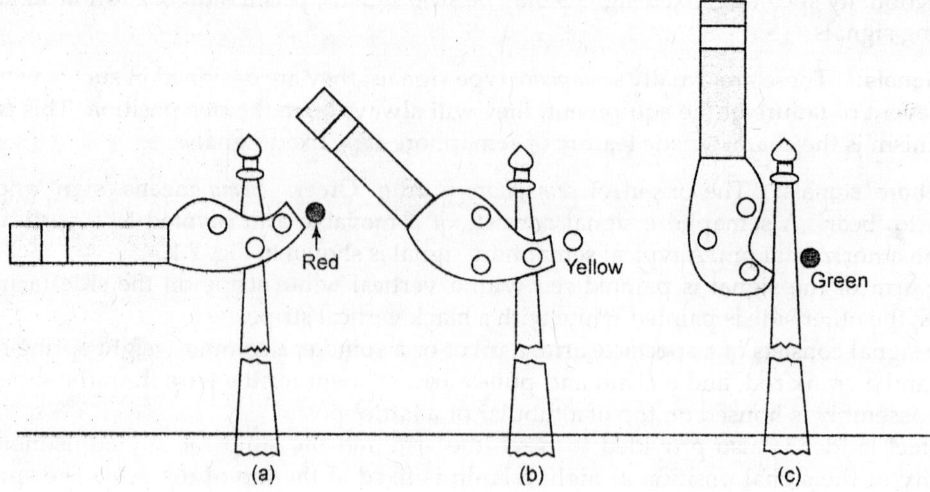


**FIG. 7.2** *Schematic of a three-aspect semaphore signal*

The semaphore signal with a red square-ended arm is used as a stop signal; it is used as a warner signal with a red fish-tailed end of the arm.

Lower quadrant semaphore signals move only in the fourth quadrant of a circle and have only two colour aspects—red and green. The horizontal position shows red while an inclined lower position at about 45° below the horizontal shows green. If lowered further to a vertical position, the signal merges with the post and cannot be easily discerned.

To overcome this difficulty and to be able to provide additional information, *multi-aspect upper quadrant signals* have been devised. The three-aspect system is common; in this, the arm of the semaphore signal can be set at three positions—the horizontal, 45° above the horizontal, and the vertical. The corresponding colour aspects are red, yellow and green, respectively.

The stop signal of the three-aspect semaphore type is shown schematically in Fig. 7.2.

The three situations shown in (a), (b), and (c) are given in Table 7.1:

**Table 7.1** Indications on semaphore signal

| Situation | Indication | Arm position | Colour | Meaning |
|---|---|---|---|---|
| Situation (a) | ON-stop | Horizontal arm | Red | Stop dead |
| Situation (b) | OFF-caution | 45° above horizontal | Yellow | Proceed with caution and check for the next signal |
| Situation (c) | OFF-proceed | 90° above horizontal (vertically up) | Green | Proceed at maximum permitted speed. |

**Permission or caution signals:** These are also called *warner signals*. These are meant to warn the drivers or caution them of an approaching stop signal. When such a signal is put up well ahead of a stop signal it is called a *distant signal*. Alternatively, it can be placed on the same post below the outer signal or the home signal. This serves the purpose of indicating not only an approaching stop signal, but also whether it is in 'ON' or 'OFF' position, based on the position of the warner itself.

The warner signal arm is *fish-tailed* at its end, and the white strip on it is also fish-tailed, so as to differentiate it from a stop signal.

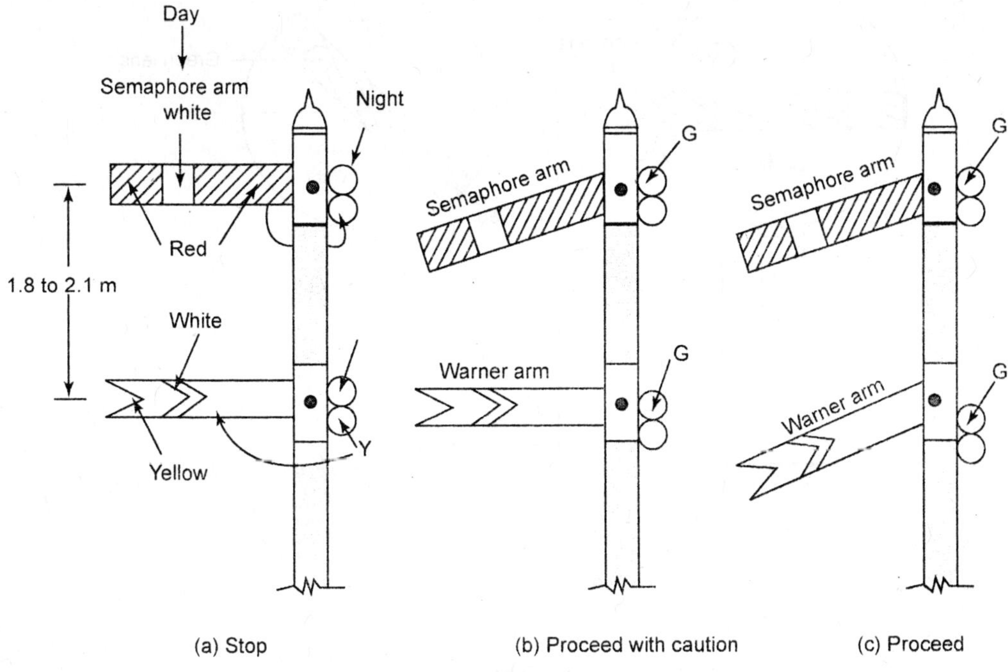


**FIG. 7.3** *Warner below an outer signal*

Figure 7.3 shows a warner signal fixed below an outer signal in the three possible positions: The corresponding indications are given in Table 7.2:

**Table 7.2** Indications on caution signals

| Position | Day indication on semaphore signal | Night indication on semaphore signal | Meaning |
|---|---|---|---|
| (a) Stop | Both stop and warner horizontal | Red and Yellow | Stop |
| (b) Proceed with caution | Stop signal inclined at 45° downwards and warner horizontal | Green and Green | Proceed WITH CAUTION and check the next signal |
| (c) Proceed | Stop signal inclined at 45° downwards as also warner signal (both inclined.) | Green and Green | PROCEED at the maximum permitted speed. |

**Shunting signals:** These are dwarf signals and are used for regulating the shunting of vehicles in station yards. These may be placed on an independent short post or a regular signal post below the semaphore signals. In areas where semaphore signals are used, shunt signals are of the disc type.

A typical disc type shunting signal is shown in Fig. 7.4.

A shunting signal consists of a circular disc with a red band on a white background. The disc revolves around a pivot. Two holes are provided – one for red and the other for green lamps for indication during night time.

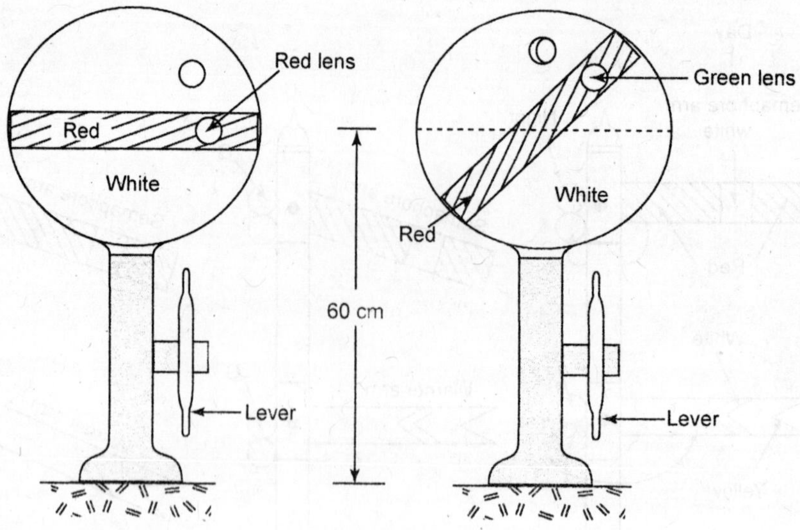


**FIG. 7.4** *Disc type shunt signals*

The horizontal red band and red light indicate danger or stop, with no permission for shunting. Inclined red band and green light indicate that the signal is off and that shunting is permitted.

### Coloured Light Signals

In this system of signalling, coloured lights are used to indicate the track condition to the loco pilots during both day-time and night-time. Power beams of light are emitted to catch their attention, especially during the day. The light from an electric lamp (12 V–33 W) is passed through a combination of lenses and is focused as a parallel beam of light; this is protected by special lenses and hoods. Therefore, it can be distinctly seen even in bright sunlight. these lights are fixed on a vertical post at about the level of the driver's eye. Only one aspect is displayed at a time.

Coloured light signals are used in automatic signalling sections, suburban sections, and routes carrying high-density fast traffic. Such signs can be two-aspect (green and red), three-aspect (green, yellow and red), or four-aspect (green, double yellow and red). Green indicates 'proceed', yellow 'proceed with caution', and red indicates stop.

A typical schematic of a three-aspect coloured light signal is shown in Fig. 7.5.

The signal indications are as usual. In a four-aspect signal, yellow, green, yellow and red in that order appear from top to bottom. Two yellow lights displayed together means 'pass the signal at full speed, but pass the next at cautious speed.'

In the case of coloured light signals provided in sections where automatic signalling exists, the signal is always green or shows the 'proceed' indication. As soon as the train enters a section, the signal changes automatically to red by the passage of the train itself. Then, the signal turns to yellow indicating to the driver to proceed with caution; finally, when the train enters the next block section, the signal turns green, indicating to the next approaching loco pilot that they can proceed at full permissible speed.

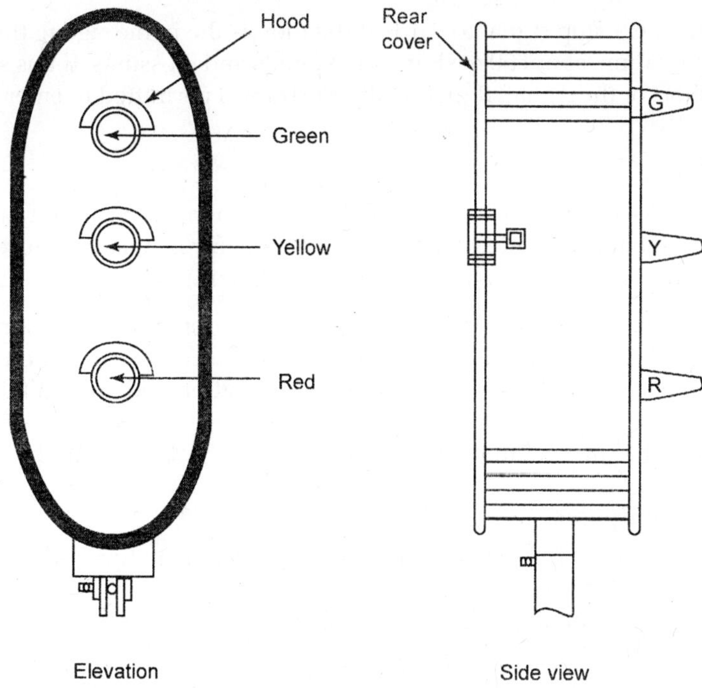


**FIG. 7.5** *Schematic of a three-aspect coloured light signal*

Thus, each aspect of the signal gives information not only about the signal itself, but also about condition of the track ahead or of the next section. This enables a loco pilot to proceed confidently even at full permissible speed.

**Shunting signals with coloured lights:** These may be provided at the top of a separate short post, or on a main signal post at a lower level.

When it is provided on an independent post, two white lights in a horizontal line indicate the 'On' or danger position, while the two white lights appearing on a line inclined at 45° to the horizontal indicate 'off' or permission for shunting.

An arrow at the top of the signal indicates the line to which the signal belongs.

**Stop signals with reference to location in a station:** The stop signals provided in and around a station are the outer and the home signal for the reception of an approaching train, and the starter and the advance starter for its departure from the station.

### Reception Signals

**Outer signal:** This is the first stop signal of a train near a station at the point of entry into the station limit from a block section. It is usually provided at a distance of 400 m from the next signal, that is, the home signal. It usually has a warner signal below the stop signal. If the outer signal is 'on', the train has to stop at least 9 m ahead of it, and can proceed with caution if the warner below is 'off'. If both are 'off', the driver can proceed confidently, expecting the next signal–home–also to be 'off'. There is no need to reduce the speed in this case.

**Home signal:** The next stop signal towards the station is the home signal; this is provided at the entrance to the station, about 50 m short of the points and crossings. If this signal is 'on', the train must stop short of the signal. If this is 'off', the train is permitted to enter with caution.

### Departure Signals

**Starter signal:** This is a stop signal and marks the limit up to which a train can occupy a line. A separate starter signal is provided for each line. This controls the departure of a train from the station. Once this signal is 'off', the train is free to depart.

**Advance starter signal:** This is the last stop signal for the departure of a train from a station. This is provided beyond the outermost points, marking the end of the station limits.

The block section starts at the advance starter of one station and ends at the outer signal of the next station.

A train can leave the station limits only when the advance starter is in the 'off' position.

### Signals for Special Purposes

Certain signals serve special functions, sometimes temporary in nature, as in the case of signals that indicate track maintenance and repairs.

Some of the more important special-purpose signals are discussed here.

**Repeater signal:** When a signal is not likely to be visible to the loco pilot from an adequate distance either due to a sharp curve before the signal, or for any other reason, a repeater signal is provided at a suitable position at the rear of main signal. A circular plate on which the letter 'R' is painted is fixed below the signal.

The semaphore arm is painted yellow with a black vertical band.

**Co-acting signal:** If a signal is not likely to be visible to a driver due to some obstruction such as an overbridge, another signal is provided on the same post, which acts in unison with the original one. This is called a 'co-acting signal.'

**Router (or a routing) signal:** When the signals relating to both main and branch lines are provided on the same post, they are called 'routing' signals. The signal meant for the particular line on to which a train is to be received will be kept in the 'off' position, giving permission for the approaching train to proceed on to the line.

If the train is to be received on to a loop line, the speed has to be reduced to 15 km/h, but if it is being received on to the main line, the speed can be a little higher.

Conventionally, the signals relating to the main line will be kept at a higher level than those for a loop line.

**Calling-on signal:** This consists of a shorter arm fixed on a home signal post below the main semaphore arm, as shown in Fig. 7.6.

When the main home signal is 'on' and the calling-on signal is 'off', the train is to proceed with caution until the next stop signal; that is, the calling-on signal is used to 'call' the train which is waiting beyond the home signal.

This is particularly useful when the main signal fails and is therefore it is always 'on', until it is rectified.

**Home and warner combination signal:** This is similar to a warner signal below a home signal on the same vertical post, as shown in Fig. 7.3.

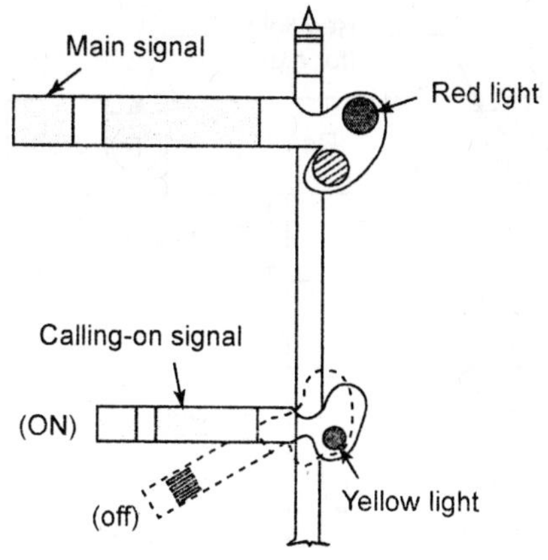

**FIG. 7.6** *Calling-on signal*

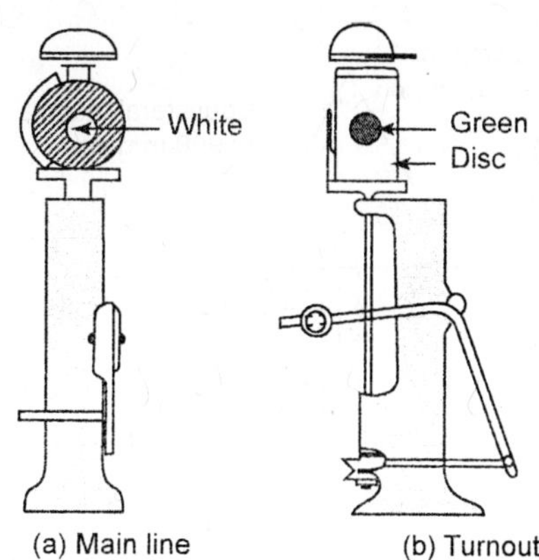


**FIG. 7.7** *Point indicator signals*

**Point indicator signals:** These are provided to indicate whether points have been set for the main line or for the turn-out and consist of an open box with two white circular discs on opposite sides of the box and green bands on the remaining sides. The box is made to rotate automatically about a vertical axis when the points are set.

The white disc indicates that the points are set for the main line; when the points are set for the turnout, the green bands are visible to the loco pilot. At night, white light indicates main line setting and green light indicates turnout setting. Figure 7.7 shows point indicator signals:

### Temporary Signals

These signals are used temporarily—such as when the track is under maintenance and repairs, when shunting operations are under progress, or when caution indication is to be given for an obstruction. Some of these signals are explained here.

**Caution indicator:** The driver is advised to reduce the speed and run with caution by means of this signal, shown in Fig. 7.8:

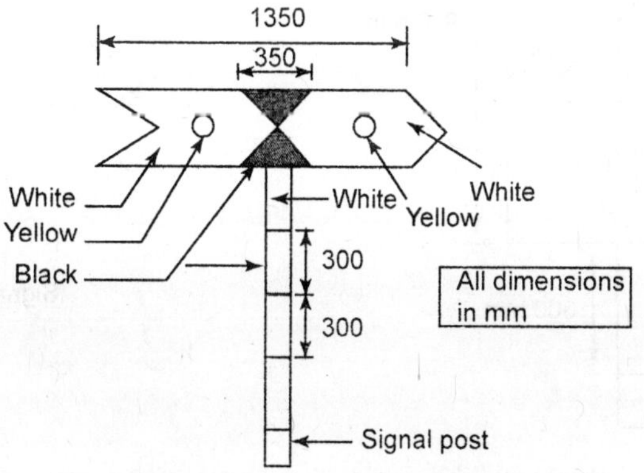


**FIG. 7.8** *Schematic of a caution indicator*

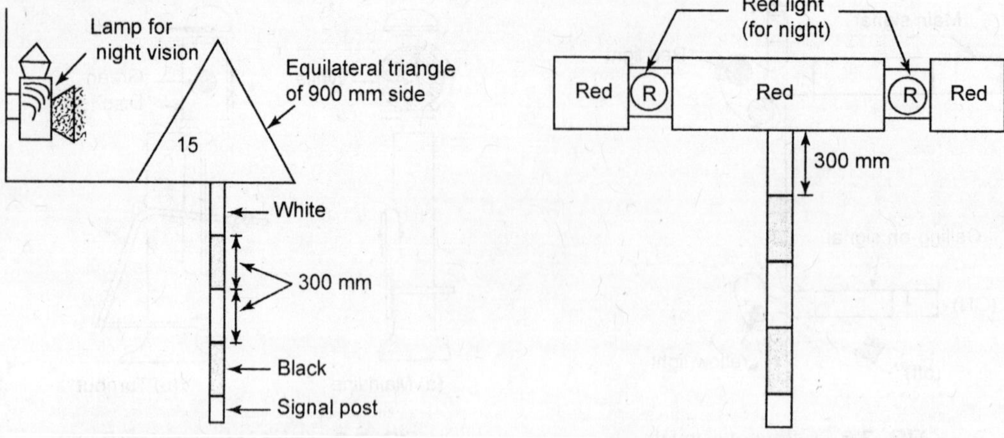


**FIG. 7.9** *Schematic of a speed indicator*

**FIG. 7.10** *Schematic of stop indicator*

**Speed indicator:** This signal indicates to the driver the speed to be maintained; a schematic view of this signal is shown in Fig. 7.9.

**Stop indicator:** This signal indicates that the driver has to stop the train at once; a schematic view of this signal is shown in Fig. 7.10.

**Shunting limit indicator:** This marks the limit of shunting operations, and is shown in Fig. 7.11.

**Termination signal:** This indicates the termination of restrictions and permits the driver to resume the normal speed. Figure 7.12 shows a schematic view of this signal:

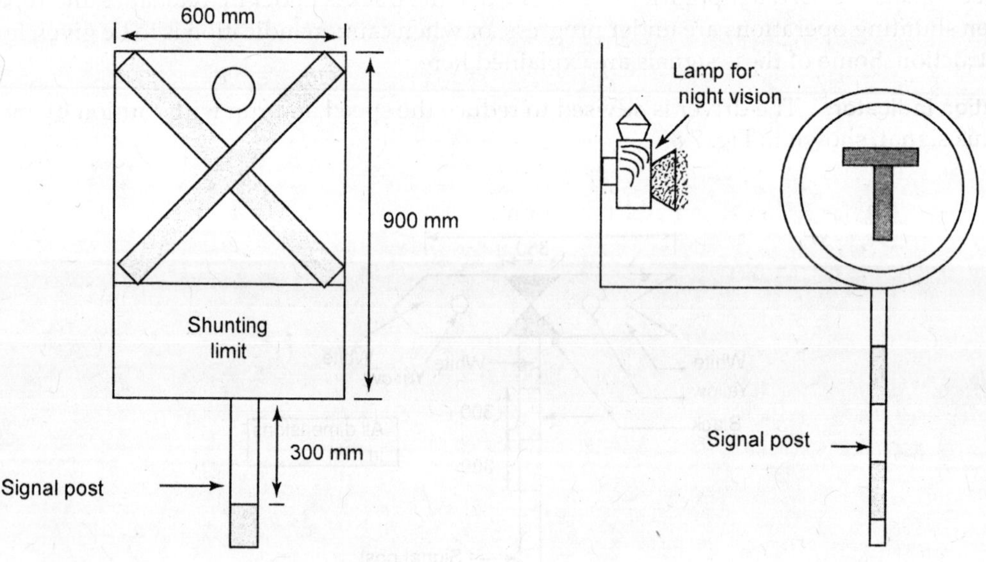


**FIG. 7.11** *Shunting limit indicator*

**FIG. 7.12** *Schematic of a Termination signal*

## 7.3 SIGNALLING SYSTEMS

Signalling systems may be broadly classified as follows:

1. Mechanical system
2. Electrical system

The primary *components* in each of these systems are:

(i) Signals and points-operated units
(ii) Transmission system
(iii) Levers and press buttons-operating units
(iv) Monitoring units
(v) Interlocking system (to be considered in detail in the next Section.)

### Mechanical System of Signalling

Signals and points are the operated units, and both the signalling system and the interlocking system depend for their operation on a mechanical system.

**Signals:** Semaphore signals are those which are operated using a mechanical system. The signals may be of the lower quadrant type or of the upper quadrant type. In the first of these, the lower left-hand quadrant of a circle is used to display an indication to the loco pilot. This idea was probably developed because of 'KEEP LEFT' rules for driving on highways in the UK and in India. The arm of the signal is normally kept horizontal (ON position, indicating STOP–danger ahead), which is lowered (OFF position, indicating PROCEED–line clear) when an approaching train has to be given permission to come into the section ahead. A third position is not possible due to design and visibility problems. In the upper quadrant type, three aspects are possible-horizontal arm, arm inclined at 45°, and arm vertical, the middle position indicating PROCEED WITH CAUTION and the third one indicating PROCEED; these are for stop signals, and similar indications are given for caution or warner signals. (These have been discussed in detail in the previous section.)

**Points:** Points or switches are set mechanically and locks and stretcher bars are used to keep them in any particular position. A solid steel rod of 33 mm diameter runs from the lever in the cabin and is connected to the points by means of cranks and compensators. This system is restricted to a particular distance from the cabin, because of transmission losses.

**Point locks:** A point lock ensures that each set of points is set correctly to the desired position. It is provided between the two tongue rails at the toe of the switches. It consists of a plunger which moves in the casing of the facing point lock. The plunger rod is connected to the signal cabin through a lock bar. There are also stretcher blades, one each for the tongue rails. There are two notches in each blade, which move inside the plunger casing along with the tongue rails. When the points are set for a particular line, the notch in the stretcher blade rests in its proper position and the plunger rod enters the notch and locks the switch in the required position to prevent the movement of points when a train passes over them. It gets unlocked only when the next operation of the points is carried out.

**Detectors:** The functions of a detector for points are:

(a) Detecting any defect in the connection between the points and the corresponding lever.
(b) Ensuring the actuation of the correct signal corresponding to the points.

A detector can be of the mechanical type or of the electrical type. In the mechanical type, the point is held in position by virtue of its design itself. Thus, strictly speaking, it is not a separate device.

A detector consists of one slide for points and a set of slides for signals, housed in a detector box. The signal slides are set perpendicular to the point slides and no vertical movement is possible between them. The signal slide has only one notch, but the point slide has as many notches as there are signals relevant to this set of points.

A particular signal can be lowered only when the notch in that particular signal slide coincides and engages with the corresponding notch in the point slide. If the points are correctly set for the main line, the notch in the point slide comes to rest opposite the notch of the main line signal slide. The main line signal lever can then be pulled and the corresponding signal lowered. Thus, the point slide moves and its notch rests in its correct position only if the points are properly set as intended.

The slide (or linear) type of mechanical detector is shown in Fig. 7.13.

This type of slide detector is used for single-wire signalling. For double-wire signalling, a double-wheel detector, which is a rotary type detector, is used. This rotates in a vertical plane and serves the functions expected of a detector efficiently in the double-wire system of signalling.

**Lock bar:** This is provided so as to make it impossible to change the points when a train is passing over it. It is an angle-section of length greater than that of the longest wheel base of a vehicle. To hold the lock bar in place, revolving clips are provided on the inner face of a rail. The length of this bar is 12.8 m for BG and 12.2 m for MG.

The design is such that when the lever in the cabin is pulled, the lock bar rises slightly above the rail level and then comes down. In the event of a vehicle being positioned at the same place, the flange of the wheel obstructs the lock bar which cannot, therefore, rise above the rail level; this precludes the operation of the points at that time.

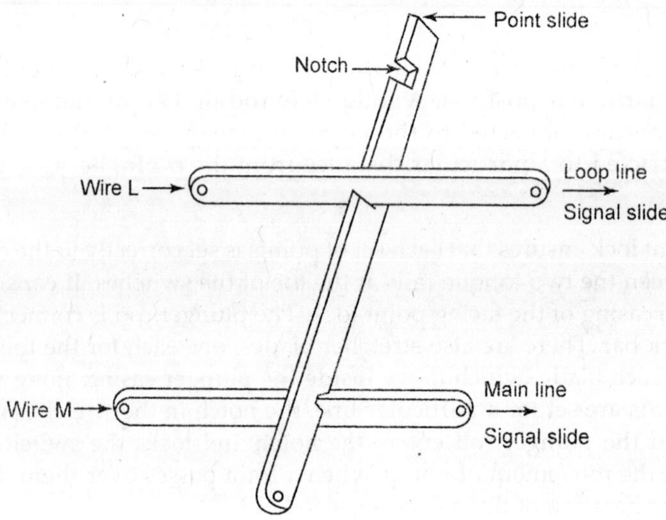


**FIG. 7.13** *Mechanical detector (for single-wire signalling)*

### Transmission Systems

The act of pulling a lever in a cabin is transmitted to the operated unit like a signal or a set of points by means of a transmission system.

There are two types of transmission systems – the single-wire system and the double-wire system. The single-wire system, which was the first to be used, is being superseded by the double-wire system, in view of the latter's advantages over the design of the former.

While only one wire connects the lever and the signal in the single-wire system, a loop of two wires running parallel to each other is wrapped over a drum-lever in the double wire system; this system works on the principle of pull and push arrangement.

### Single-wire System

The following equipment are needed in this system of transmission:

(a) Lever frame
(b) Signal transmission wire
(c) Cabin wire adjuster
(d) Signal and its component parts

*(a) Lever frame*: The function of this is to operate the single-wire system; in addition, it actuates the interlocking system provided to ensure safety. The lever frame may be of the direct locking type or the catch handle type. The signals, points, locks and the lock bars are operated by pulling the appropriate levers.

*(b) Signal transmission wire*: The wire is of galvanised steel of size 4.00 mm to 3.25 mm; it is hauled over a horizontal or vertical wheel by pulling the lever. When the lever is pulled back to its normal position, the counter-weight connected to the signal arm brings the wire back to its original position.

*(c) Cabin wire adjuster*: This helps in adjusting the variation of the length of the wire due to temperature changes; unless this is done, inadequate lowering or drooping of the signal arm may occur, leading to ambiguity.

*(d) Signal and its component parts*: These, along with the fail-safe system involving counterweights have already been dealt with under semaphore signals (Fig. 7.1)

**Disadvantages of the single-wire system:** Drooping and incorrect lowering of the signal occurs if the wire becomes slack for any reason. This conveys ambiguous or misleading information to the loco pilot and can lead to accidents.

There can be several reasons for the wire to become slack: Sagging due to self-weight, change of length due to temperature variations, elastic stretching due to the tension in the wire when the lever is pulled, entangling of the wire accidentally into the pulley stake and the wheels, and finally, breakage of the wire. In view of these limitations, the range of single-wire transmission is only about 950 m. Actually solid rods of 30 to 33 mm diameter are used to connect the levers to the points; the rods move on roller guides fixed at 2 m intervals and a crank is used at every change of direction.

**Temperature Compensator for the rods:** This is provided to neutralise the effect of temperature changes in the rods. Figure 7.14 shows a schematic of the compensator:

A temperature compensator consists of a pair of cranks—one acute and the other obtuse—connected by a link and is designed to absorb the change in length due to temperature.

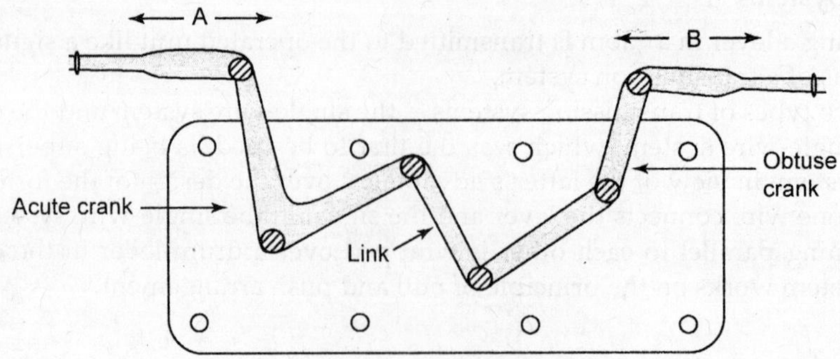


FIG. 7.14 *Temperature Compensator*

variation. The compensator is usually at the middle of the length of the rod up to a length of 36.5 m. If the length is more, compensators can be placed at quarter points.

**Double-wire System**

In this system, the effort is transmitted with the aid of two wires from the lever to the signals and points. Each wire is of solid galvanised steel of size 4 mm to 3.25 mm, attached to pulley stakes and driven firmly into the ground. The two wires are so connected as to form a loop between the lever and the signal. when the lever is operated, the wire gets pulled, and when it is brought back, the wire gets pushed. This push–push mechanism causes the drum to rotate in one direction when the lever is pulled and in the other direction when it is restored. The rotary motion of the drum gets converted to linear motion by means of cams and cranks, which finally actuates the signal. Double-wire mechanism is schematically shown in Fig. 7.15.

In between the drums, the wire passes over wheels, pulleys and pulley-stakes, double-wire compensators, and cranks. (All these details are not shown here.)

**Advantage of the double-wire system:** The advantages of the double-wire system are:

- Operation is easier than with the single-wire system.
- Better suited to multi-aspect signalling.
- No drooping of wires, resulting in increased safety.

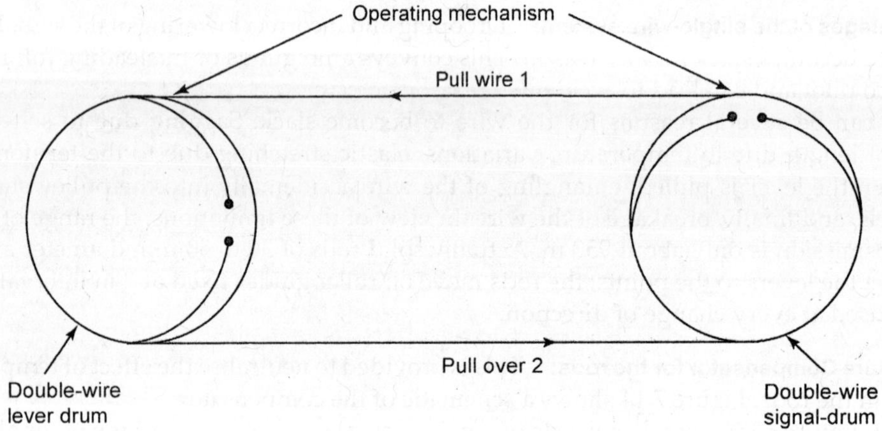


FIG. 7.15 *Schematic of double-wire mechanism*

- Range of operation is greater than that in the single-wire system.
- Operating cost is less than that of the single-wire system.
- Less possibility of external interference.
- Signal return to the 'on' position needs pulling back of the lever, which makes it specifically safe.
- The working of the double-wire compensator is automatic without need for manual use of the cabin wire adjuster.
- This requires only one cabin as against two in the single-wire system.
- Not easily affected by weather conditions.
- Requires less maintenance.
- Double-wire compensators protect the system against wire breakage.

In the case of double-wire system, a rotary type double-wheel detector does the function of ensuring the correct setting of the points and locks. Double-wheel detectors are used as monitoring units on the Indian Railways.

## Electrical System of Signalling

This system is becoming increasingly popular on Indian Railways with the expansion of several major electrification projects. In this system, electrical energy is used for displaying signal aspects. The transmission of power is done electrically, and the signals and points are operated using push buttons. The monitoring of the system as well as the interlocking is done electrically.

In the case of electronic or solid state signalling system, the operation of the signals in done electrically, but the interlocking is done electronically.

The electrical signalling system consists of either coloured light signals, or semaphore signals, with their arms operated by electric motors. Points are operated by converting the rotary movement of the electrical machines fastened to sleepers near the points into a linear push or pull. The point machines are operated with a 24 V dc supply, or by a 110 V dc supply. The operation of this machine involves first unlocking, bringing the point from normal to reverse or vice-versa, and then locking the point once again. The operating time is only 3 to 5 seconds.

The medium of transmission is either an overhead line or an underground copper cable. The use of overhead lines is limited; in big yards, underground cables are preferred.

Push buttons or rotary switches are used for operating the equipment. The complete layout of the yard is represented on a monitoring console with all details of tracks, points, signals, and so on.

### Monitoring System for Electrical Signalling

A monitoring system mainly consists of point detectors, track circuits and axle counters.

**Electrical point Detector:** This type of detector detects the points and ensures that they are set properly. The working principle is the same 'slide system', as in the case of the mechanical system. The slides in the electrical system are so adjusted that a gap of 3 mm is left between the tongue rail and the stock rail; this ensures that there is no contact between them and precludes the possibility of the signal being turned off at any time.

**Track Circuiting:** A 'track circuit' is formed along the running rails and connected to the signal and cabin. The purpose of the track circuit is to indicate the presence, or otherwise, of a railway vehicle on the track, thereby showing whether the line is clear or not.

For setting up a track circuit, the ends of the rails forming the track circuit have to be isolated by insulating the rail joints. When the rails are laid on wooden sleepers, they are electrically

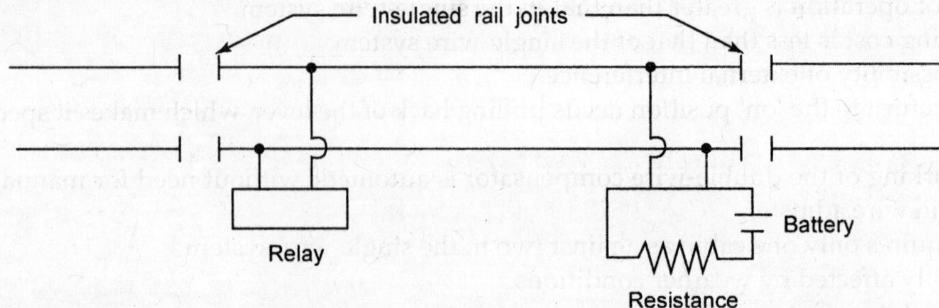


**FIG. 7.16** *principle of track-circuiting*

insulated from each other; in view of the scarcity of wood, concrete sleepers are now being extensively used also serve the purpose of insulating the rail joints.

The ends of the rails on one side are connected to a battery through resistances, while on the other side, the ends of the rail are connected to a relay, which is energised by the electrical energy from the battery when the track is free or unoccupied by a railway vehicle. This electrified track is the 'primary circuit'.

As soon as the track is occupied by a railway vehicle, the two rails are short-circuited by the wheel and axle of the vehicle, the major portion of the current flowing through them. This deprives the relay of adequate energy feed from the battery, thus de-energising it. This breaks the 'secondary circuit' from the relay connected with the signals, thus ensuring that the signals concerned are automatically set to indicate danger.

The basic principle of track circuiting is shown in Fig. 7.16.

A schematic of a dc track circuit is shown in Fig. 7.17.

**Insulated joints:** Track-circuited sections are electrically insulated from the rest of the track by insulated joints on either side of these sections.

The most common types of insulated joints are:

**(a) *Conventional insulated joints*:** In this type, the standard fishplates are skimmed on the fishing planes to accommodate channel type insulator of nylon/fibre glass between the rail and the fishplates. Ferrules or bushes are used to insulate fish bolts and end-posts between the rail ends.

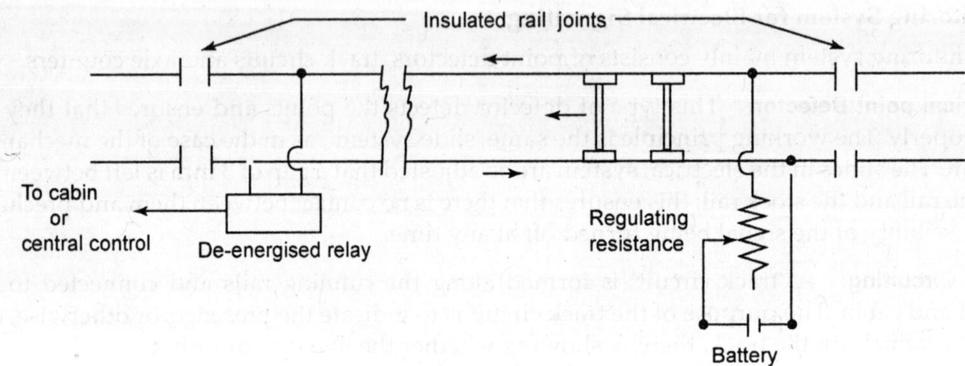


**FIG. 7.17** *Schematic of a track circuit (dc)*

An important feature of these joints is that there is no allowance for expansion of the rail. The joints require special care with regard to their maintenance; they should not butt against an SWR, but have at least one rail length on either side and it should be box-anchored.

**(b)** ***Glued insulated joints*****:** These joints are made of web-fitting type of fishplates glued to the rails using a high polymer adhesive. High-tensile steel bolts are used to ensure that the rail and fishplates act monolithically under high frequency vibrations. Although glue provides sufficient insulation, separate fibre glass insulating liners are used for additional safety margin. Suitably designed fibre glass end-posts are used between the rail-ends. Fish bolts are covered with fibre glass insulating sleeves.

Once the glue is fully set, no relative movement is possible between the rail and the other components; thus, the rail joint behaves like a welded joint for all practical purposes.

Great care should be exercised in the preparation of glued joints; the guidelines issued should be scrupulously followed.

**Axle counters:** An essential requirement of a track circuit is the insulation of two consecutive rails from each other. Sleepers are excellent insulators for rail joints. Owing to the shortage of timber for use as sleepers and also due to the prohibitive cost of concrete sleepers, a device known as 'axle counter' has come to be used as a substitute for track-circuiting to detect the presence of a railway vehicle on a track.

The principle lies in installing a pair of rail inductors at either end of a block section of the track for counting the axles. As soon as a vehicle enters the section from one end, the number of axles entering are automatically counted. Similarly, when the vehicle leaves the section, the axles are counted once again at the other end. If the count of axles is the same at both the ends it indicates that the section is clear or unoccupied. If the number of axles counted at the exit end is less than that at the entry end of the section, it indicates that the section is occupied by a train or other railway vehicle, and therefore, the line is not clear.

Axle counters are progressively used on Indian Railways in sections where trains are operated on the Absolute Block System.

## 7.4 INTERLOCKING

### Purpose and Definition

*Interlocking* is a system used to ensure the safety of trains. With the installation of several sets of points and their corresponding signals and the introduction of high-speed traffic, the need to eliminate human error has become imperative in order that accidents involving the loss of life and property are completely avoided.

Interlocking is defined as a technique by which it is ensured that before a signal is brought to 'Off' position, giving 'green signal' or 'line clear' indication to a train, the points which control the line are properly set, locked, and held in that position until the entire line is traversed by the train, and at the same time, all points and signals, the operation of which would lead to conflicting movements, are locked against such movements.

The signal system and the interlocking system are so designed that the failure of any device or equipment automatically brings the signal to the 'On' position, indicating danger, thus ensuring safety.

## Essential Principle of Interlocking

All equipment and devices must be so designed and interlocked to comply with the following stipulations:

(i) It should not be possible to turn off a signal unless all the relevant points for the line on which the train is to be received are correctly set, the facing points are locked, and all interlocked level crossings in the vicinity are closed to road traffic.

(ii) The line should be fully isolated before the signal is turned off, precluding the possibility of any other vehicle entering this line.

(iii) Once the signal is turned off, it should not be possible to make any adjustment in the points on the route, including those on the isolated line.
Similarly, no interlocked gates should be released and opened to road traffic until the signal is brought to the 'On' position.

(iv) It should not be possible to turn off any two signals at the same time as it can lead to conflicting indications or movements.

(v) The points should also be interlocked to avoid conflicting movements.

## Standards of Interlocking

On Indian Railways, the standards of signalling and interlocking have been classified into the following four categories, depending upon the speed of through trains running on the main lines:

| | |
|---|---|
| Up to 50 km/h | Standard-I |
| Up to 100 km/h | Standard-II |
| Up to 130 km/h | Standard-III |
| Up to 160 km/h | Standard-IV |

### Signalling

The types of signalling equipment to be provided at different stations and other requirements to be fulfilled for each of these categories differ. Two-aspect signalling is still permitted only on Standard-I, and starter signals are not mandatory. In Standards II, III and IV, the signalling has to be of the multi-aspect type, and starter signals are mandatory.

### Points

The method of interlocking of points is the key-locking in standard-I, which is an indirect method of interlocking between signals and points. In Standard-II, a plunger type facing point lock is used; but in Standards III and IV, the points are operated from a central location, and the interlocking between points and signals is direct.

### Isolation of Lines

In standard-I, isolation of a line for which points and signals are set, is optional. In Standards II, III and IV, isolation of the main line from the adjacent lines is mandatory.

## Methods of Interlocking

The following are the methods of interlocking used on Indian Railways:

1. Key system of interlocking
2. Mechanical system of interlocking
3. Electrical system of interlocking
4. Electronic system of interlocking

### Key System of Interlocking

This is the simplest method of interlocking, still used in small stations on branch lines on Indian Railways. This method is sometimes said to be an 'indirect method of interlocking'.

The principle can be understood by means of a simple example:

Let there be a station with a main line and a loop line; the point can be set for the main line or the branch line.

(a) The point has two keys, the first is Key-A, which can be taken out when the point is set and locked for the main line.
    Similarly, Key-B can be taken out when the point is set and locked for the loop line. At any particular time, either Key-A or Key-B can be taken out, depending upon whether the route is set for the main line or for the loop line.
(b) The lever frame operating the signals is provided with two levers. The lever relating to the main line signal can be operated only by Key-A. Similarly, the signal for the loop line can be operated only by Key-B.
(c) If the train is to be received on the main line, the points for the main line are set and locked, releasing the Key-A. This key is used to operate the main line signal lever, thus placing it in the 'off' position. Since Key-A cannot be used to operate the loop line signal lever, the possibility of the point and the signal being contradictory or conflicting is automatically eliminated. Similar logic applies to the point and signal for the loop line with respect to Key-B.

### Mechanical System of Interlocking

Mechanical interlocking on lever frames is an improved method compared to the key system. This method involves the use of plungers and tie bars. The plungers are made of hardened steel, 300 mm × 16 mm and have notches in them.

The tappets and tie bars are placed at right angles to the plungers and are provided with suitably shaped C.I. or steel pieces that fit exactly in the notches of the tappets, which are normally of section-38 mm × 16 mm.

The whole arrangement is housed in a locking trough. Each lever is attached to the plunger, which has notches of shape suitable to accommodate the locking tappets.

When a lever is pulled, the plunger to which it is connected moves; the tappet accommodated in the notch of the plunger is pushed out at right angles to the direction of movement of the plunger due to wedge action. The motion is transmitted to all the other tappets that are connected to this tappet through a tie bar. Due to this motion, the other tappets either get pushed into or out of the respective notches of the other plungers depending upon the interlocking provided. In case the other tappet is free but slips into the notch of the other plunger, it locks the lever connected to this plunger. Hence, the other lever also gets locked in that position, and therefore, it cannot be operated.

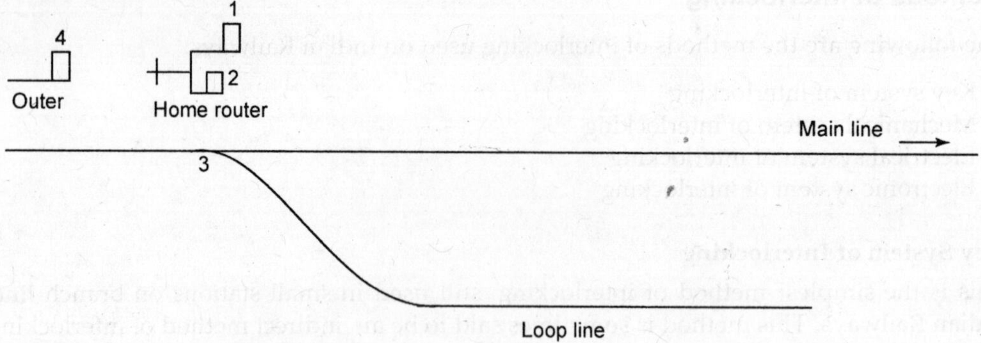


**FIG. 7.18** *Main and loop with signals and points*

However, if the tappet was earlier inside the notch of the plunger, thereby locking the lever and is now pushed out of the notch, the other lever becomes free to be operated.

This type of arrangement sets a predetermined sequence for pulling the levers, which actuates the interlocking.

To illustrate this method, consider an example of one main line and one loop line, along with the corresponding points and signals (Fig. 7.18). The procedure for interlocking will be as follows:

The levers for signals and points shall satisfy the following:

(a) The standard setting of the points is for the main line. The main line home signal shall be interlocked with facing points such that when this signal is 'off', the points cannot be altered. With reference to the figure, this means that '1 locks 3'.
The position of tappets and lock for this is shown schematically in Fig. 7.19.

(b) The signal for the loop line shall be interlocked with the facing points so that it cannot be 'off' unless the points are set for the loop; further, it cannot be 'off' when the points are set for the main line.

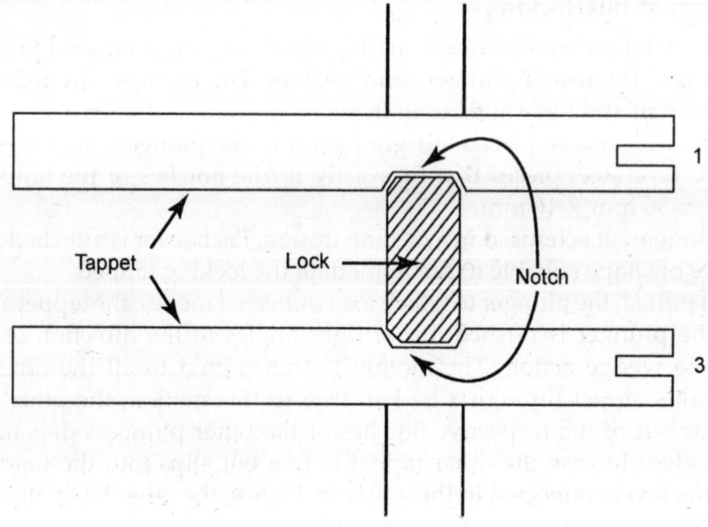


**FIG. 7.19** *Lever 1 locks Lever 3*

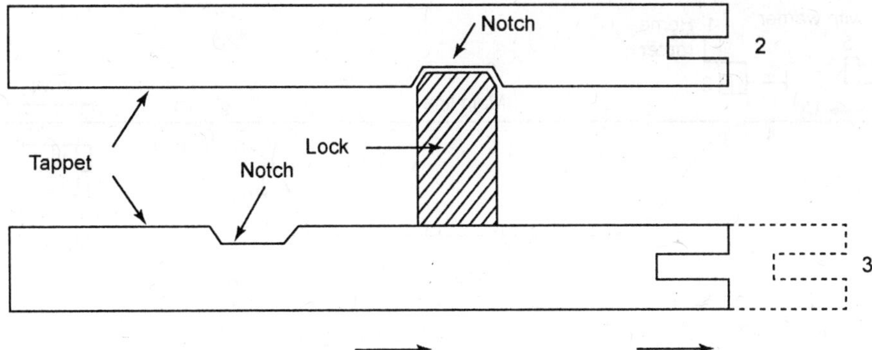


**FIG. 7.20** *Lever 3 releases Lever 2 or Lever 2 back-locks Lever 3*

This means that '3 releases 2' or '2 back-locks 3'. The position of tappets and lock for this is shown schematically in Fig. 7.20.

(c) The outer signal shall be interlocked with the routing signal such that it cannot be 'off' until one of the signals on the router is 'off'. This means that '1 or 2 releases 4' or '4 back-locks 1 or 2'. The position of tappets and locks for this is shown schematically in Fig. 7.21. Once the interlocking relations among the various levers are established, these are put down conveniently in the form of an interlocking table as given below:

**Interlocking Table**

| Lever for | Lever no. | Releases | Locks | Back-locks |
|---|---|---|---|---|
| Home router main (signal) | 1 | 4 | 3 | – |
| Home router loop (signal) | 2 | 4 | – | 3 |
| Points | 3 | 2 | 1 | – |
| Outer (signal) | 4 | – | – | 1/2 |

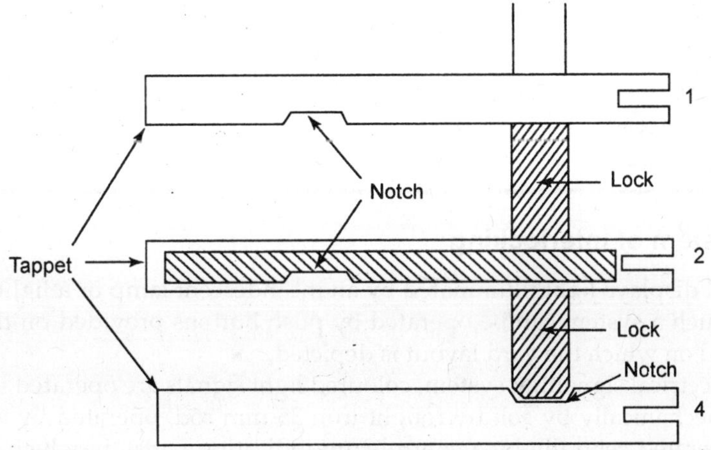


**FIG. 7.21** *Lever 1 or 2 releases Lever 4 or Lever 4 back-locks Lever 1 or 2*

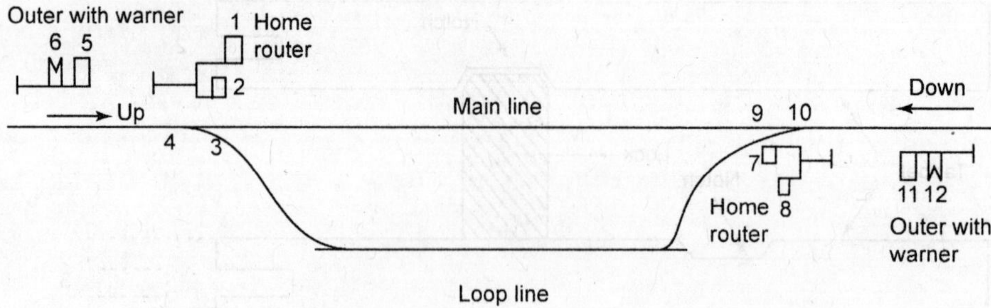


**FIG.7.22** *Mechanical interlocking for points and signals for a two-line station.*

**Mechanical interlocking of points and signals of a two-line railway station:** The principle of mechanical interlocking can be extended to any numbers of points and signals. A relatively simple example of interlocking of points and signals of a two-line railway station may be considered (Fig. 7.22).

The signals, points and point locks are shown, numbered separately for the Up and Down directions. The corresponding interlocking table is given here.

**Interlocking Table**

| Lever function | Lever no. | Releases | Locks | Back locks |
|---|---|---|---|---|
| Signal up home main | 1 | 5,6 | 3, 7, 8 | 4 |
| Signal up home loop | 2 | 5 | 7, 8 | 3/4 |
| Points facing up main | 3 | 2 | 1, 12 | – |
| Points lock for 3 | 4 | 1/2 | 3, 12 | – |
| Signal up outer | 5 | 6 | – | 1/2 |
| Signal up warner | 6 | – | 9, 10 | 1/5 |
| Signal down home main | 7 | 11/12 | 1, 2, 9 | 10 |
| Signal down home loop | 8 | 11 | 1, 2 | 9/10 |
| Points facing down main | 9 | 8 | 7, 6 | – |
| Point lock for 9 | 10 | 7/8 | 6, 9 | – |
| Signal down outer | 11 | 12 | – | 7/8 |
| Signal down warner | 12 | – | 3, 4 | 7/11 |

## Electrical System of Interlocking

Since the signal displays light illuminated by an incandescent lamp or a light emitting diode (LED) signal, such a system can be operated by push buttons provided on the control-cum-indication panel on which the yard layout is depicted.

Under the electrical signalling system, coloured light signals are operated by levers; points are operated mechanically by solid wrought iron 33 mm rod, operated by an electric point machine; or operated centrally by a control-cum-indication panel, in which case, points are operated by an electric point machine in conjunction with coloured light signals.

Using levers to operate electric signals is a hybrid procedure. It is used a temporary measure to suit 25 kV ac traction, set to be converted later to operation by control-cum-indication panel.

### Electrical Interlocking of Points and Signals of a Two-line Railway Station

Electric interlocking is achieved through electric switches or 'relays'. Relays use the principle of electromagnetism—a soft iron core wrapped inside a wire coil turns into an electromagnet when electric current is passed through the wire. An armature attached to this electromagnet makes contact when it is attracted to the magnet, and breaks contact when it is not attracted to it.

The entire system is housed in a glass or a metal box, and is known as the 'Relay,' whose function is to close or break the circuit as required by the interlocking operation.

## Electronic System of Interlocking

In electrical interlocking, relays are used to translate the interlocking relations in the selection table into logic circuits.

In electronic interlocking, the logic circuits are converted into Boolean equations, which are transformed into a programme loaded on the processor of a computer.

The layout of the whole yard may be loaded on the computer, and the train operation may be carried out using a mouse and the Visual Display Unit (VDU); or, a control-cum-indication panel may be used. Both these methods may also be used, and the control may be transferred by means of a switch provided on the control-cum-indication panel.

Finally, the relays are energised by the system after evaluating the interlocking relations, which transmit the appropriate commands to the points and to the signal.

When a VDU and mouse is used, the particular portion of the signal and the interlocking plan is selected, and the signal required is taken 'off' by a mouse click on the signal profile on the VDU. The system works inside, following the same principles as in the case of the electrically interlocked system, but the requirements are analysed electronically and the relays are energised to transmit the final command to operate the points and bring the signal concerned to its 'off' position.

## Slotting of Signals in Interlocking

*Slotting* is used in interlocking so that a signal is controlled by more than one source—from two or more cabins for enhancing safety in operation.

The two *principles* on which a slotted signal works are:

- The signal cannot be placed in its 'off' position until the controls from all the remote sources are operated
- The signal can be placed in its 'on' danger position by any of the remote sources of operation.

There are two methods of slotting of signals:

(i) Mechanical system
(ii) Electrical system.

The *mechanical system* is shown schematically in Fig. 7.23.

A mild steel flat connects the three levers as shown. It is fixed to the first lever and free with respect to the second and third levers. Lever 2 is connected to the wire coming from Cabin-I and Lever 3 is connected to the signal wire coming from Cabin-II.

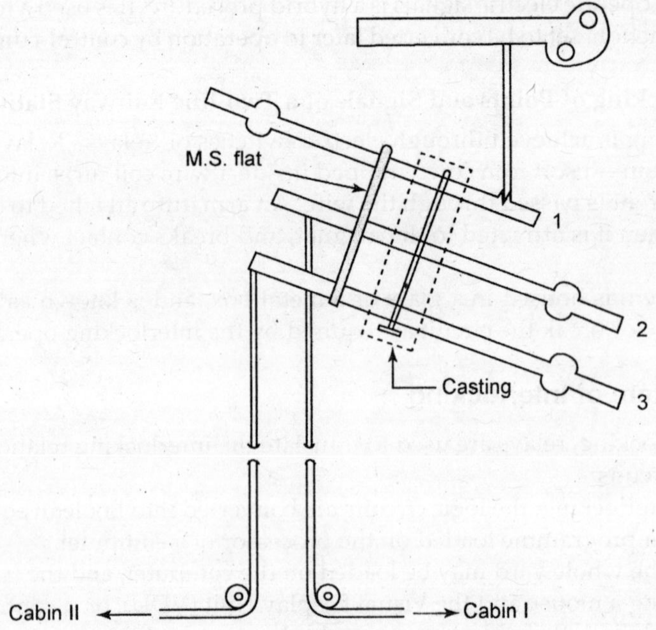


**FIG. 7.23** *Slotting of signals (mechanical system)*

The mechanical arrangement is made in such a way that Lever 1 cannot be operated until Levers 2 and 3 are pulled. This ensures that the signal cannot be brought to its 'off' position until this requirement of the mechanism is fulfilled.

In an electrical system, all the movements associated with the mechanical system are carried out electrically and its working principle is practically the same as that of the mechanical system.

## 7.5 TRAIN CONTROL SYSTEMS

There are several systems for controlling the movement of trains. These can be broadly classified into two systems—time interval system and space interval system.

The following methods are adopted for controlling the movement of trains on Indian Railways:

1. One-engine-only system
2. Following train system (time interval principle)
3. Pilot guard system
4. Train staff and ticket system
5. Absolute block system
6. Automatic block system (based on automatic signalling)
7. Train control through modern signalling installations
   (a) Panel interlocking
   (b) Route relay interlocking
   (c) Solid state interlocking

(d) Centralised train control (CTC)
(e) Train protection and warning system (TPWS)
(f) Automatic warning system (AWS)

The first four methods are very restrictive. They are practically obsolete and are now of historical interest only.

The fifth method is the one that is currently used on most stations in Indian Railways. The sixth method is adopted where the line is provided with continuous track circuiting in conjunction with electric traction.

The methods listed under train control through modern signalling installations are the most modern and involve recent techniques using advanced technologies.

### One-engine-only System

In this system, only one train is permitted to be in a section at one time. A loco pilot is authorised to run the train in any section by means of a 'wooden staff' or a 'token'. Since this cannot be at two places at the same time, the safety of the trains is ensured.

As there is only one engine and one train that work to and fro on the same section, this system is possible only on short branch lines with limited traffic. If more than one train has to be despatched in the same direction, this system cannot be used.

### Following Train System

This system works on the principle of time interval; trains follow each other after a specific time interval. Trains sent after the first one is permitted to run only at a speed of 25 km/h. If the time interval is adequate, safely is ensured to a limited extent.

This system may be used in emergencies and in short stretches of double-lines.

### Pilot Guard System

In this system, one person, known as the 'pilot guard', accompanies the train, riding on the foot plate of the engine, as the authority for the train to proceed. The red uniform and a name badge worn by the pilot guard is the identification for the person. Even in this system, trains can follow each other after a fixed time interval. But normally the pilot guard returns to the first station with another train in the reverse direction. This may be possible only on short single-line sections in the event of communication failure.

### Train Staff and Ticket System

This is similar to the pilot guard system. A wooden staff or a ticket serves as the authority to proceed. There is only one such staff for a section, which is kept at one of the two stations of that section. The staff is interlocked with a ticket box at the station, in which there are printed tickets, and they cannot be taken out so long as the box is locked. Only the station that has the staff can despatch a train.

If only one train is to be despatched, the staff is given to the loco-pilot as the authority to proceed. If more than one train are to be sent, the tickets are used as the authority to proceed for all the trains except the last, for which the staff is used.

The speed of trains is restricted to 25 km/h and the time interval between consecutive trains shall not be less than 15 minutes.

A similar system is followed by the other station also. Safety is ensured since only one ticket can be issued at any given time.

**Absolute Block System**

In this system, the entire route is divided into what are called *block sections*. A block section is the line between two block stations (with facilities for controlling trains). *Block instruments* installed at each one of these stations are connected electrically. A token kept in a block instrument can be taken out with the consent of or understanding between the two station masters at either end of the section on a single-line section. On a double-line section, bringing the departure or starter signal to its 'off' position is the authority to proceed since there is no token system.

A train is not sent from a station until and unless the previous train has arrived at the next station, ensuring that only one train is there at one time in any section. The procedure in this system is known as the *lock and block* method, the instruments used being known a *block instruments*.

**Block instruments:** Each station has two block instruments, one for each station on its either side. The instruments of two adjacent stations are electrically inter-connected.

These instruments can be operated by the station masters of the stations at either end of the section. A round metal ball, called a *token*, is taken as the authority to proceed into the next block section. On a single-line section, this token is contained inside the block instrument.

**Single-line token instruments:** The token can be taken out of the block instrument of the station from which a train is to be despatched, by turning a handle of the instrument towards the end labelled *Train going to side*. This can be done only with the cooperation of the station master of the station on the other side of the section, who has to turn the handle of his/her instrument towards the end labelled *Train coming from side*.

Once this is done at both ends, the handles of both the instruments are locked in their respective positions; they can be normalised only after the train arrives at the next station and the token given by the driver is inserted into the block instrument at that station. The procedure of locking and releasing of the block instruments is called 'the lock and block' procedure. The block instrument for a single-line is shown in Fig. 7.24.

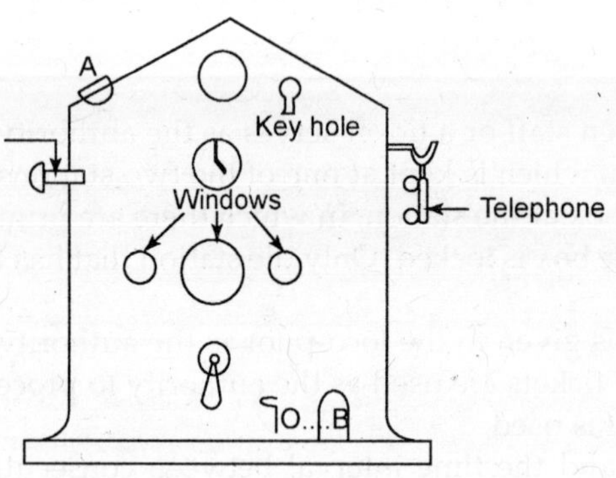


**FIG. 7.24** *Schematic of a block instrument (single-line)*

**Tokenless block instruments:** In the *token* method, the operations come to a halt if, by chance, the token is misplaced, causing a lot of delay and inconvenience. To overcome this difficultly, tokenless block instruments have been developed.

The principle of the block system is followed, but without using a token. The last signal which permits the train to enter the block section—the advance starter—is interlocked with the block instrument in such a way that it is not possible to turn this signal off unless the block instruments have got the line clear command.

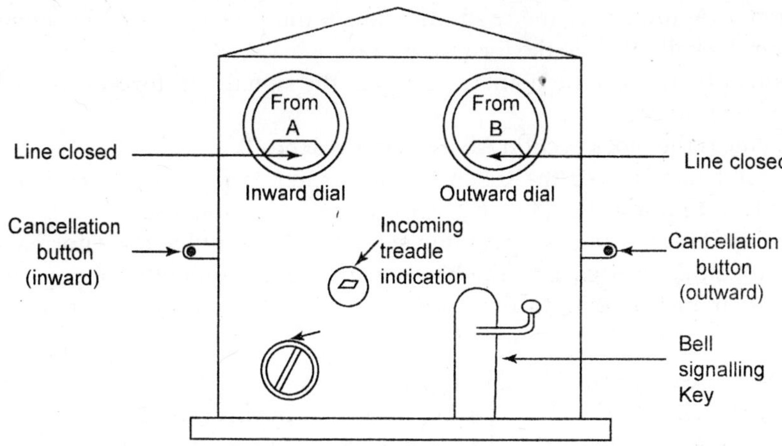


**FIG. 7.25** *Block instrument for a double-line section*

**Double-line block instruments:** In a double-line section, the traffic being uni-directional, one train on each line can be permitted to remain in a block section.

The block instrument in this case consists of a commutator handle and two indicator needles placed vertically, one above the other. The station master turns the block instrument commutator to the *line clear side*. This causes the electrical circuit to make contact so that the advance starter of the station from which a train is to be despatched can be placed in its 'off' position. The block instruments for a double-line section are shown schematically in Fig.7.25.

*The essential features of the absolute block system are:*

- No train is allowed to leave a station unless permission has been received in advance from the next block station.
- On a single line, *permission to approach* is not given unless the line is clear for the entire block section into which the train is to be despatched.
- When two trains are running in the same direction on the same track, the second train should not be given permission to approach till the entire length of the first train is within the limits of the home signal, all the signals behind the first train are 'on', and the line is clear.

### Working Procedure of the Absolute Block System

This may be explained with reference to a block section, AB, shown in Fig. 7.26.

(i) The station master at A seeks 'line clear' from his/her counterpart at B over the telephone.

(ii) The person at B ensures that the line is clear, agrees to receive the train from B and grants 'line clear'. Both simultaneously operate their block instruments enabling A to extract the token.

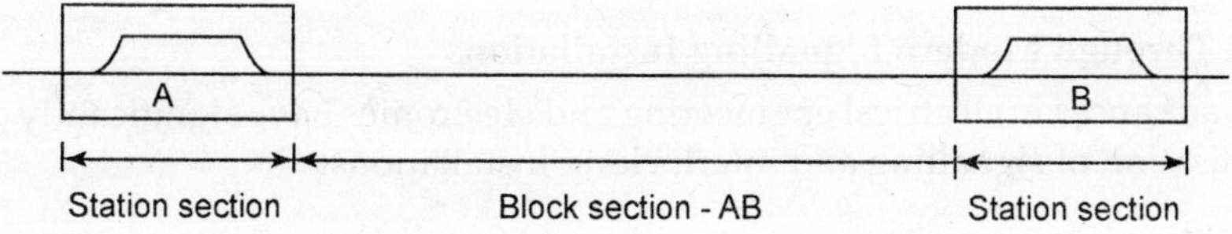


**FIG.7.26** *Working procedure of absolute block system*

(iii) The person at A authorises the cabin-man to set the points, places the signal 'off' and hands over the ball token to the loco-pilot.
(iv) The person at B also sets the points and places the signal 'off' for the line on which the train is to be received.
(v) The train enters the block section AB and later reaches B.
(vi) The loco-pilot hands over the token to the person at B, who places it in the block instrument, and informs the person at A of the arrival of the train.
(vii) The points at Station B and the reception signals are restored to their normal positions.
(viii) Once the token is inserted into the block instrument at B, both the persons at A and B restore their instruments to the normal 'line closed' position, thereby making the instruments ready for the next operation.

The same procedure is repeated by the station masters at B and C, the next block station, in order to allow the train into the block section BC. Since the system ensures space interval between consecutive trains, it is absolutely safe. Most stations on Indian Railways adopt this system.

### Automatic Block System (Automatic Signalling)

The absolute block system ensures safety but it is time-consuming as a train has to wait until the previous train clears the entire block section. Thus, the track capacity cannot be utilised fully.

In the automatic block system, which requires *continuous track circuiting*, more trains can be accommodated in the section by dividing it into smaller automatic block sections. The essential features of this system are:

(i) The line should be provided with continuous track circuiting.
(ii) The line is divided into a series of automatic block signalling sections, entry into them being controlled by a stop signal.
(iii) These stop signals shall be controlled by the track circuits, fulfilling the following conditions:
- The signal shall not be 'off' unless the line is clear in advance, not only up to the next stop signal but also for an adequate distance beyond it.
- As soon as s train passes the signal, the signal should automatically assume its 'On' position.

#### Advantages of automatic signalling

(a) Elimination of human error, leading to greater safety.
(b) Increased efficiency leading to greater track utilisation or **track capacity.**[1]
(c) No block instruments are needed.
(d) In case of power failure, the signals automatically assume the 'On' position.
(e) It is facilitated by electric traction leading to modernisation and high speeds.
(f) Less number of personnel required for operation of signals, leading to economy.

### Train Control Through Modern Signalling Installations

Technological advances in electrical engineering and electronics have significantly contributed to the modernisation of signalling and interlocking installations.

[1] *Track capacity* or *line capacity* means the maximum number of trains that can be operated per day on a given length of track or railway line, or on a given section or route. The corresponding number for a yard is the *terminal capacity*.

A few of these are presented here.

**(a) Panel interlocking** All points and signals are operated electrically from a central cabin with the aid of switches mounted on a panel, on which the yard layout also is depicted. Relays are used for electrical interlocking of points and signals.

Thus, multi-cabin operations can be eliminated and with the removal of inter-cabin control and signal slotting, a lot of time is saved. Greater realisation of line capacity is possible by a smaller number of personnel. There are around 2800 panel interlocking installations on Indian Railways.

**(b) Route relay interlocking:** This is an improvement over panel interlocking. In panel interlocking, several switches relating to points and signals have to be operated. However, route relay interlocking needs the use of only a pair of switches for automatic operation of all the tasks; the desired route for the train is set automatically by the operation of all points on the route, as also the required signal.

During this operation, it is ascertained that the line is clear and that there are no conflicting movement of the trains running on the section. Of course, an essential requirement is that the entire track in the yard has to be track-circuited.

All conditions and indications of all signals are shown on a panel with the diagram of the yard. Once 'line clear' is given for a certain route, the relevant portion of the diagram gets illuminated with white lights. The moment the train comes on, the lights turn red, and when the train has cleared, the lights go off automatically.

Route relay interlocking is useful in busy yards with high traffic density; complete safety is ensured by track-circuiting of the entire yard.

There are about 250 route relay interlocked yards on Indian Railways.

**(c) Solid state interlocking:** Railways in advanced countries have introduced micro-processor-based solid state interlocking system on a large scale. The advantages are less space requirement, less power needs, easy installation and maintenance, and greater reliability.

Indian Railways have introduced the first solid state interlocking system at Srirangam, a temple city in Tamilnadu. Solid state interlocking systems are being used progressively on Indian Railways.

**(d) Centralised Train Control (CTC):** In this system, the operation of all points and signals is centralised at one place for the whole section. A single person, called the CTC operator, controls all the operations. Thus, a CTC operator does the work of the station masters of several stations of the section and operates all the points and signals through remote control. Usually, one CTC panel is provided to control all stations up to about 120 km on either side. The operation is carried out with the aid of knobs. The panel indicates the status of the various tracks—occupied or clear.

In a CTC system, panel interlocking is provided at all stations, which are connected electrically to the CTC. Thus, there can be interaction between the CTC and the panels at the stations, and vice-versa.

The *advantages of CTC* are:

- Maximum realisation of line capacity or track capacity.
- Considerable reduction of skilled personnel.
- Easy detection of defects in the track, leading to quick and effective maintenance.

**(e) Train protection and warning system (TPWS):** This system was made mandatory in the UK in 1999 for adoption on all passenger train lines. When a signal is fitted with TPWS, the brakes of a train are automatically applied if the train approaches a signal in 'on' condition too fast. It does so for any violation of speed restrictions as also at buffer stops fitted with TPWS.

Thus, the risk of derailment and train collisions is drastically reduced.

The Railway Safety Review Committee has recommended its use for improving safety on Railways.

TPWS is a spot-transmission-based control system to be used in conjunction with the existing signalling system.

Movement commands are generated track-side and transmitted to the train. TPWS provides a continuous speed supervision system. The system comprises a track-side subsystem and a computer-based system evaluator with an indication panel and a tachometer installed in the locomotive. A sealed microprocessor chip mounted on the sleepers acts as an interface between the loco computer and the interlocking system. If the loco-pilot does not react to the commands or warnings, brakes are automatically applied, and if necessary, even emergency brakes are applied.

**(f) Automatic Warning System (AWS):** This provides audio-visual warning to the loco-pilots and prevents them from going beyond a signal in its danger position. This is also called *auxiliary warning system*.

Electromagnets fixed on the track convey the signal aspect to the loco-pilot cabin. The message is transmitted electronically by an antenna at the bottom of the locomotive. If the driver fails to control the train according to the signal aspect, the brakes are applied automatically. This system has been developed and is in use in advanced countries such as Japan, Germany, Switzerland, and the USA.

AWS of Siemens' design has been in use the Central and Western zones of Indian Railways—for example, in Mumbai suburban sections.

There are two types of AWS, depending upon the manner in which the information is transferred to the loco-cabin from the track:

(i) *Intermittent type*: The updating of information about the track condition and signals is done only at the signals. The loco-pilot, therefore, is authorised to run the train based on the available data until further updating.

(ii) *Continuous type*: The system updates the information continuously, uploading any change immediately. This includes changes in speed and changes in the route ahead.

**Anti-collision Device (ACD):** As mentioned in the first chapter, an anti-collision device has been developed indigenously, and installed on the Konkan Railway.

This works based on the global positioning system (GPS), a microprocessor-based central processing unit, and a digital modem for communication with other ACD's.

There are two types of ACD's—mobile type for the locomotives and stationary type for stations and level crossings. Exchange of information between the ACD's is possible within 3 km. Based on the ACD interaction, the loco-ACD decides to apply the brakes if necessary.

## SUMMARY

- The purpose of signalling and interlocking is to control the movement of trains for ensuring safety and achieving efficiency of operation. A well-designed system of signalling, supported by an appropriate method of interlocking achieves the objectives to a large extent.
- Signals are classified based on different criteria such as operation, function, location, construction, and special purposes. Audible or detonating signals are used in foggy weather when visibility is impaired. Visible signals can be hand signals or fixed signals, the former being used for limited purpose. Fixed signals are normally of the semaphore type except in the case of those meant to guide shunting operations. Stop signals and warner signals or caution signals are the two signal categories based on their function. With reference to location in and around a station yard, the stop signals used for reception of a train are the outer, home and router signals and, the starter and the advance starter signals are for the departure of a train. With reference to construction, the signals are of the semaphore type (2-aspect lower quadrant or the 3-aspect upper quadrant) or the coloured light signals (2-, 3-, or the 4-aspect types). Repeater, router, calling-on, point indicator, and temporary signals such as speed, stop-shunting limit, and termination (of restrictions) indications are all for special purposes.
- Signalling systems are of the mechanical or electrical type. Signals and points are the operated units. Points need point locks, lock bars, and detectors (mechanical or electrical type) for their operation. Signalling may be of the single-wire system or the double-wire loop system, the latter having several advantages over the former.
  Temporary compensators are used to absorb the changes in length of the rodding due to temperature changes.
- Electrical signalling system involves the use of coloured light signals (both for day and night indication); it is used in sections of electric traction.
- The monitoring systems for electrical signalling needs track circuiting via the rails as conductors.
- Interlocking is a system used to ensure the safety of trains. It is defined as a technique by which it is ensured that before a signal is turned 'off', giving 'green signal' or 'line clear' indication to a train, the points which control the line are properly set, locked, and held in that position until the entire line is traversed by the train.
  The signal and interlocking systems are so designed that the failure of any device or equipment automatically brings the signal to the 'ON' position, indicating danger, thereby ensuring safety. The essential principles of interlocking are all so formulated as to ensure safety. There are four speed-based interlocking standards on Indian Railways-I to IV, with increasing speed. The methods of interlocking are the key system (indirect method which is practically obsolete), and the mechanical, electrical and electronic systems. The mechanical method using tappets and locks is the most commonly used one. Interlocking tables are prepared to indicate the locking, releasing and back-locking operations achieved by the pulling of the levers for various points and signals.
- While electrical interlocking uses relays, in electronic interlocking the programme is loaded on to a computer. Slotting of signals (operated mechanically or electrically) is required to coordinate inter-cabin control when the signals are controlled by more than one operator.
- The earlier methods of train control and operation such as the following train method and the pilot guard method are now obsolete. The most common system is the Absolute Block system, in which a line divided into block sections and only one train is allowed in any section at a given time. Block instruments, with or without ball tokens are used. Automatic

Block system or Automatic Signalling involves the actuation of signals by the wheels of the train, and needs continuous track circuiting; it is thus convenient in sections where electric traction is used.

- Modern methods of signalling systems are panel interlocking, route relay interlocking, solid state (or electronic) interlocking, centralised train control coupled with panel interlocking, connected electrically to the CTC, Train Protection and Warning system, and Automatic Warning System; in the last two methods, commands generated trackside are transmitted to the loco-cabin. The indigenous anti-collision device developed by the Konkan railway is becoming an invaluable tool for enhancing railway safety.

## QUESTIONS

1. What are the objectives of signalling?
2. Why are signals necessary in railways?
3. (a) Differentiate between hand signals and fixed signals.

   (b) Draw a neat sketch of a two-aspect semaphore signal and explain its working.
4. Explain the various signals used in station yards, depending upon their location, mentioning the purpose of each.
5. What are the various criteria for the classification of signals? Give one example for each.
6. Write short notes on:

   (a) Home signal

   (b) Advance starter signal

   (c) Warner signal.
7. Write short notes on:

   (a) Router signal

   (b) Calling on signal

   (c) Shunting signals
8. Write short notes on:

   (a) Caution signal

   (b) Speed signal

   (c) Termination of restriction signal
9. Explain the difference between two-aspect and three-aspect semaphore signals with sketches.
10. Explain the salient features of electrical system of signalling.
11. Write on explanatory note on track circuiting.
12. What are the important components of the mechanical system of signalling?
13. Define 'interlocking' and state its primary purpose.
14. What are the essential principles of interlocking?

15. What are categories of interlocking standards on Indian Railways?
16. Explain, with a neat sketch, the mechanical interlocking of points and signals in a two-line railway station with the minimum required signals. Prepare an interlocking table for this system.
17. Explain the features of the Absolute Block system of train operation on a single-line section.
18. Explain the salient features of automatic signalling system.
19. Write short notes on:
    (a) Tappets and locks
    (b) Time-interval system of train operation
    (c) Slotting of signals in interlocking.
20. Write a note on 'Panel Interlocking'.
21. Explain the system of 'Route Relay Interlocking'.
22. Write short notes on:
    (a) Detector
    (b) Temperature compensator.
23. Differentiate between single-wire signalling and double-wire signalling.
24. Bring out the advantages of double-wire loop system of signalling compared to the single-wire system.
25. How does the electronic system of interlocking differ from the electrical system of interlocking?
26. Explain briefly the working of a semaphore signal, with the help of a neat sketch. (VTU)
27. What are the objects of signalling? (Kerala University)
28. Describe the working of 'Absolute Block System' of signalling. (Kerala University)
29. Explain the semaphore signal and its working. (Punjab Tech University)
30. What are the different systems of controlling the movement of trains? Briefly explain the absolute block system of working for single-line and double-line stations. (UP Tech. University)
31. Write short notes on:
    (i) Semaphore signal
    (ii) Shunting signal (West Bengal Tech. University).
32. Write short notes on:
    (i) Semaphore signal
    (ii) Warner signal (Mumbai University).
33. Classify the various types of signals. Explain the different types of signals with sketches. (Mumbai University)

34. State the various methods for controlling train movements. Explain absolute block system with sketches. (Mumbai University)
35. Explain the working of 'interlocking of signals' in railways. (Mumbai University)
36. Explain the essential principles of interlocking. (Kalinga Institute of Technology)
37. Describe the interlocking system of a single line station. (Kalinga Institute of Technology)
38. Give the classification of signalling based on functional characteristics. Describe the working principle of semaphore signalling with a neat sketch. (Kalinga Institute of Technology)

Chapter 8

# TRACK CONSTRUCTION AND MAINTENANCE

**CHAPTER OBJECTIVES**

This chapter enables the student to understand:

- The methods of track construction – both conventional labour-oriented methods and modern mechanised methods involving portable track equipment for different tasks such as sleeper-laying, rail-handling, levelling and alignment of the track, as well as heavy tamping and track-laying machinery
- The various track maintenance operations for gauge, levels, ballast section, rails and rail joints, sleepers, and track drainage structures; conventional manual maintenance methods involving the use of simple tools; annual maintenance cycle for the track components; track testing and recording cars to identify defects; modern methods of track maintenance using off-track and on-track tampers; measured shovel packing (MSP) and directed track maintenance (DTM); manual and mechanised methods of track renewal; and diversions for facilitating repairs and construction.

## 8.1 INTRODUCTION

A new railway line in an area is required for:

- Development of backward areas
- Connecting existing and potential trade centres and industrial areas
- Providing short-cuts between places already connected by railway lines
- Political considerations
- Strategic and political considerations near border areas.

*Doubling* projects may be planned to cope with increase in traffic, adding another line on the route.

*Gauge conversion* projects may be taken up to cater to the needs of traffic and as part of the Unigauge policy.

In all these cases, construction of a railway track has to be carried out, which involves several steps. There are different methods of railway track construction—conventional and modern.

Maintenance of a railway track under operation is an obvious need owing to the dynamic effects of railway traffic–passengers as well as goods trains. For this, there are conventional methods involving manual operations and modern methods involving mechanised and automated procedures.

A track may need upgrading or rehabilitation after prolonged use. For this, standards have been evolved. When the wear and tear of track components exceeds the standard limits of tolerance, the track is re-laid or renewed using manual or mechanised methods.

## 8.2 TRACK CONSTRUCTION

The operation of laying track components such as the sleepers and rails on a prepared formation and connecting them appropriately for the construction of a railway track is known as *plate laying*. Ballasting is done later. The point from which the laying is commenced is called the *base* and the point up to which the laying has progressed is called the *rail head*.

### Construction of a New Railway Line

The main steps involved in the construction of a new railway line are the following:

(a) Land acquisition
(b) Earth work
(c) Construction of cross-drainage works such as bridges
(d) Plate laying or track-linking
(e) Ballasting
(f) Opening to traffic

#### Land Acquisition

Work relating to land acquisition should be taken well in advance in order to complete all formalities and overcome legal hurdles, if any. Land acquisition needs co-ordination with the state government concerned in accordance with the procedure laid down in the relevant Act. Giving statutory notice and paying the requisite compensation to the owners are part of the task. The area acquired should be adequate for the formation, berms and borrow pits with provision for future expansion such as doubling. A width of 25 m is considered sufficient; extra width is required for station yards, level crossings, and bridge approaches. For small stations, an area of 150 m × 1000 m is adequate for the yard.

#### Earthwork for Formation

Depending upon the desired formation level and the natural ground level at a particular point, the formation may be laid in an embankment or a cutting. Normally, formation laid in an embankment is preferred as it presents few drainage problems. The height of the embankment depends on the high flood level in the area due to heavy rainfall during monsoon, with a reasonable free board.

The standard widths for BG and MG are given below for track in embankment and in cutting, both for single-line and double-line:

Broad gauge

| | Single-line | Double-line |
|---|---|---|
| Embankment | 6.85 m | 12.155 m |
| Cutting | 6.25 m | 11.155 m |

Metre gauge

| | | |
|---|---|---|
| Embankment | 5.85 m | 9.81 m |
| Cutting | 5.25 m | 9.21 m |

(For double-line, the spacing between the centres of the tracks is 4.725 m for BG and 3.96 m for MG.) The points to be borne in mind with regard to earthwork have already been given in Chapter 3.

### Construction of Cross-drainage Works–Bridges and Culverts

Depending upon the topography of the area and the waterway to be crossed, culverts may be of RCC Hume pipes or slabs, and bridges may be of RCC slabs, plate girders, steel trusses or of PSC girders.

Since bridges are major structures, they are built to accommodate a double line where only a single line is being laid at the time, in order to allow for future expansions. They must be designed to bear the load of the heaviest locomotive likely to run on the section.

### Plate Laying or Track Linking

When the formation is ready, plate laying or track linking is taken up. This consists of laying the track components such as sleepers, rails and rail fastenings. The following methods are used for plate-laying:

**Tram-line method or side method:** In this method, a temporary track, known as a 'tram-line', is laid by the side of the proposed railway line. This tram-line is used to transport track materials to the site.

This temporary track may be left as such, and may be used for doubling, if it is planned for execution in the near future.This method is considered useful in flat terrain, where laying a tram-line is relatively easy.

A modification of this method is the *side method*, in which a service road is built parallel to the track to be constructed, and is used to transport track material and girders for bridges using heavy trucks. Once again, only a flat terrain is considered suitable for adopting this method.

**American method:** In this method, rails and sleepers are assembled in a depot, and these assembled panels are transported to the site by means wagons along with the necessary cranes. The panels are unloaded by means of cranes at the rail head, linked to the panels already laid, and placed in their final position.

This method, first used in the USA, is adopted in developed countries, particularly where RCC and PSC sleepers are used, which are very heavy and cannot be easily handled using manual method.

**Telescopic method:** This is the most widely used method on Indian Railways. In this method, all the required materials—the rails, the sleeper, rail fastenings and fittings—are procured in advance and stored in a base depot.

The material is then transported to the rail head and unloaded. The track materials are then linked and packed; the rail head is thus advanced. The track already laid is used to transport the track materials for further work by means of wagons. The work of track linking is continued so that the work proceeds until the entire track is laid.

## Track Linking Procedure

The conventional method of track linking using manual labour involves the following steps:

Transporting the materials to the rail head

Linking of track

Packing of track

Ballasting of track

**Transporting the materials to the rail head:** The materials are procured and unloaded with the help of *material gangs* (labour meant for loading materials at the base depot into either dip lorries or into material trains) at the rail head. The first base depot lies at the starting point of the line. After the work has progressed for about 5 km, a subsidiary depot is established at the new rail head, and a portion of the track material from the first base depot is transferred to the next base depot. All the operations of loading at the depot and unloading at the rail head are carried out with the help of the material gangs.

**Linking of track:** After the track material is unloaded at the rail head, the tracks are linked with the help of *linking gangs*. Labourers of the linking gang actively engage in operations related to the linking process.

The procedure of linking involves the following steps.

The sleepers are laid roughly at the desired spacing based on the sleeper density decided upon. The centres of the sleepers are made to be at the centre line of the alignment, defined by means of a stretched string.

Rails are carried by the gangmen using rail tongs and laid on the formation near their final position. A dozen or more men are required to carry a rail (which weighs nearly 500 kg). This can be obviated by using a device, known as *Anderson rail carrier*, shown in Fig. 8.1.

The rails have markings to indicate the position of sleepers, at which the latter may be placed.

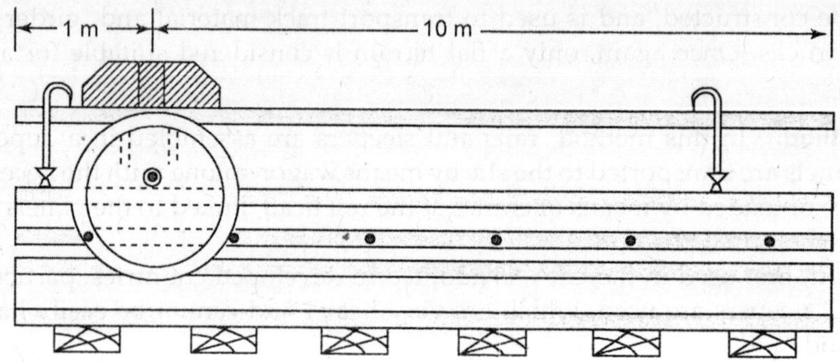


**FIG. 8.1** *Anderson rail carrier*

Fish plates and bolts are kept near the rail joints Fittings for sleepers are kept near the ends of the latter.

Depending upon the type of sleeper, appropriate fittings are used to fix the rails to the sleepers. Bearing plates are provided where necessary.

The gaps at the rail joints are provided with aid of steel liners or 'shims' of appropriate thickness.

The linking gangs work under the supervision of an engineer and carry out their instructions.

**Packing of track:** This is done thoroughly with the aid of beaters by the *packing-in gangs.* The points to be taken care of are the gradient, curvature where the track is on a curve and the cross levels which should be uniform or with superelevation depending upon whether the track is straight or curved.

No hollow spaces should be left after packing.

**Ballasting of track:** In the case of a new track, ballasting is done after at least two monsoon periods in order to allow the formation to settle completely.

A *ballast train* is used for this purpose, the ballast being unloaded through hoppers on to the track, and spread by a plough attached to one of the wagons. Alternatively, the ballast stored on the cess of the formation may be placed on the track manually. The ballast must be thoroughly packed under the track with the designed section and cushion.

**Requirements of track materials:** All the materials required for track construction such as rails, sleepers, and rail fastenings and track fittings are calculated for 1 km length of the track.

Since two rails are needed for the track, the weight of the rails required per km length can be easily calculated knowing the weight of the rail per metre. The number of sleepers per kilometre length of the track can be calculated from the sleeper density fixed for the section.

The number of fittings and fastenings may be calculated, depending on the type of sleepers used. The nature of the fittings and their number including those at rail joints can be calculated. This has been dealt with in adequate detail in Chapter 3.

*Note*: The work relating to doubling and gauge conversion is similar, except for a few differences.

### Modern Methods of Construction of Railway Lines

Modern methods involve the use of:

sophisticated surveying techniques
modern geotechnology for the preparation of the subgrade and the formation; and
modern and sophisticated track machines for track laying and linking.

**Modern surveying techniques:** These include satellite imageries, aerial photographs and digital terrain modelling. They are employed in deciding the initial alignment as well as the final location for the new railway line. Modern surveying instruments and techniques are effective in expediting the work, increasing the precision and reliability and reducing the cost.

**Modern geotechnology for subgrade preparation:** Significant advancements have been made in the field of geotechnology and geotechnical engineering in the recent past. These can be successfully used in the design of subgrade for a railway line to provide a stable formation.

When the natural soil is not good, some of the techniques adopted for stabilising it are:

(a) Removal of soft soil for a limited depth and replacement with suitable soil.
(b) Preloading for acceleration of consolidation settlement

(c) Provision of vertical sand-drains
(d) Provision of inverted filters
(e) Provision of stone columns in thick clay soils
(f) Cement grouting
(g) Vibro-compaction
(h) Use of geotextiles.

**Note:** Modern surveying techniques been covered in Chapter 2 and the use of geo-technology for subgrade preparation has been discussed in Chapter 3 in more detail.

**Mechanised Track Construction**

Most of the track construction in India has been carried out by means of simple road-based equipment, employing manual labour. The final tamping has been done by *on-track tamping machines.* Manual handling of heavy track materials such as concrete sleepers and long rails is hazardous and may also damage the track components.

Two methods of mechanised track construction are briefly described here:

**Using Simple Portable Track Equipment**

The brief procedure is given below:

A layer of ballast, 100mm thick, is placed on the prepared subgrade. Road-based equipment is used for its transport. Spreading may be done by a paver-type of equipment.

Rails or rail panels of 2 to 3 rail lengths are brought to the site on road-trailers, and gently placed on the ballast with the help of truck-mounted cranes (to a gauge of 3400 mm). The rails are joined together using fish plates and fish bolts with appropriate expansion gaps, in which liners or shims are inserted temporarily.

A sleeper-laying machine (*Robel's Unit-PA 1-20ES*) is used for the transport and placement of concrete sleepers in position. This machine moves on the auxiliary rails already laid on 3400 mm gauge. Sleepers are loaded on to this machine with the aid of a truck-mounted crane. They are stacked at suitable locations along the line, from the manufacturing plant in rail/road vehicles. The sleeper-laying machine weighs about 15 t and progresses at a rate of about 150 m/hr.

When a certain portion of the track is covered in this way, rails are placed on the sleepers with the help of *Robel's Rail Threader* (wt. 1.5 t), or a lighter *Robel's Rail Mover* (wt. 115 kg). Next, the track is lifted, levelled and aligned with the help of Robel's Machine (No. HGR 47.230, wt. 1.5 t).

Then, the first round of tamping is done by means of Plasser Tamping Machine –Minima 2 (Wt. 7.5 t). This progresses at a rate of 100 to 120 m/hr.

Additional ballast is spread using rail-mounted dumpers. Further lifting and levelling is done by Robel's machine. Then, a second round of tamping of track is carried out with *Plasser Tamping Machine.*

Welding of rails into long rail lengths may be carried out by flash butt or alumino-thermit welding. A third round of tamping may be given if necessary. Ballast-profiling machine is used for obtaining the desired profile.

A self-loading tamping machine – Plasser 08-275 ZW (Wt. 30 t) – is used for the final round of tamping. The track may now be thrown open for operating traffic.

IRCON (International), a public sector company, has been using this method in their railway construction projects.

### Using a Track-laying Machine

Plasser and Theurer SVM 1000 S, a high-capacity track laying machine can be used.

The salient features of this machine are:

- Automatic placing of sleepers at the specified spacing
- Carrying and placing the rails ahead of the machine.

This machine can be conveniently transported to the site on standard railway flat-bed wagons.

The front of the machine runs on a crawler track end the rear end is supported on a wagon.

Sleepers are transported to the laying unit of the machine by conveyor belts, and aligned correctly.

New rails are threaded on to the base plates by means of rail guide clamps; these are transported in front of the laying-machine with the aid of guiding rollers and rail conveyor units are mounted on the machine.

A sound-proof cabin in front of the laying unit houses the control units. Another cabin is located in the area where the rails are on the base plates.

The laying-machine is powered by a diesel engine and operates hydraulically. In addition to the sleeper laying unit, the machine also comprises the following:

- Gantry crane
- Drive unit
- Rail pulling and conveying device
- Tractor for pulling long rails in an area where the track is still to be laid. The tractor is equipped with rail clamps.

The sequence of operations is given below:

(i) A ballast layer of about 150mm is placed over the prepared subgrade.

(ii) The track-laying train with integrated rail supply unit is brought to the site. The rail-pulling unit pulls the rails from the rail wagons and moves them across the track-laying machine in the front. The rails are carried on a tractor and placed on the ballast.

(iii) The machine lays the concrete sleepers at regular spacing, and the rails are brought into position on these sleepers. (A portal crane picks up the sleepers from the wagons and places them on conveyor belts, thus feeding the track-laying machine continuously with the concrete sleepers.)

(iv) Elastic fastenings are fitted to the rails using simple tools and tightened.

(v) The track is tamped with on-track tamping machines.

(vi) More layers of ballast in thicknesses of 100 mm each, are placed, tamped and further consolidated.

(vii) The rails can be continuously welded, de-stressed after proper consolidation, and thrown open to traffic.

### Advantages of the Track-laying Machine

The following are the advantages of using a track-laying machine:

- Continuous action of track-laying
- Handling of long rails and easy unloading
- Automatic laying of sleepers at selected spacing
- No locomotive is needed as the machine is diesel-powered

- Easily transported on standard flat-bed wagons
- Efficient handling of rails and sleepers
- Flexibility of operation of the various units
- Easy setting up and closing down of the machine
- Speedy execution and high track quality.

## 8.3 TRACK MAINTENANCE – CONVENTIONAL METHOD

Unlike most civil engineering structures which are supported by solid foundations, railway track rests on a floating foundation of ballast; further, it is subjected to heavy and high-speed traffic and the consequent impact loads.

Hence, the railway track needs regular maintenance in order that it fulfils its functions effectively. Repeated movement of loads from the trains cause attrition, loss of ballast and wear and tear of other track components such as rails, sleepers and track fastenings. Rains, floods, winds and temperature changes increase the maintenance needs of the track structure and its components.

**Necessity of Track Maintenance**

The railway track needs regular maintenance so that trains run at permissible speeds with safety and comfort to passengers. Maintenance needs arise because of the following:

Packing of ballast under the sleepers becomes loose, thereby disturbing the track geometry under the constant pounding received from train loads. This affects the safety of the track.

Track fittings and fastenings become loose due to the vibrations under dynamic train loads and also cause wear of the track components.

In addition, weathering effects of the atmosphere and the forces of nature such as rain and temperature variations cause wear and tear of the track components.

**Advantages of Regular Track Maintenance**

A well-maintained track provides comfort and safety to trains and the passengers.

Quick attention to small details such as loosened bolts, keys and spikes, and other fittings prevents the loss of material.

Regular track maintenance results in the reduction of operating costs.

Regular maintenance reduces the frequency of track renewals, which entail heavy expenditure.

Good maintenance enhances not only the life of the track but also that of the rolling stock.

**Important Items of Track Maintenance**

1. Gauge
2. Longitudinal and cross levels, including cant on curves.
3. Horizontal alignment – straight or curved
4. Vertical alignment – level or gradient
5. Packing, recoupment or renewal, and deep screening of ballast sections
6. Rails and rail joints
7. Squaring and packing-repacking of joint sleepers.
8. Track drainage structures, slopes of embankments and cuttings, catch water drains and cross-drainage structures.

A systematic programme is drawn up for regular track maintenance; each major activity is taken up at a certain specified periodicity in a cyclic manner. This ensures that all essential operations of regular maintenance are taken care of sequentially and efficiently.

### Maintenance Equipment and Tools

The equipment and tools commonly used for track maintenance are given in Table 8.1 along with the purpose of each item:

**Table 8.1** Equipment and tools for track maintenance

| S. No | Equipment /Tool | Purpose |
|---|---|---|
| 1. | Rail gauge | To check the gauge |
| 2. | Gauge-cum-level | To check the gauge and also cross level |
| 3. | Cant board | To check the cant or superelevation |
| 4. | Mallet or wood hammer | To check the packing of sleepers |
| 5. | Canne-a-boule (Wooden mallet) | To assess the voids under sleepers |
| 6. | T-Square | To check the squareness of sleepers |
| 7. | Sleeper tong | To carry sleepers |
| 8. | Rail tong | To lift and carry rails |
| 9. | Beater or pick-axe | To pack ballast under the sleeper |
| 10. | Crowbar | To lift the track for surfacing and to correct track alignment. |
| 11. | Jim crow | To bend or break the rails. |
| 12. | Spiking hammer | To drive spikes |
| 13. | Keying hammer | To drive keys |
| 14. | Spanner | To tighten bolts |
| 15. | Wire claw or ballast rake | To draw out ballast for screening and packing |
| 16. | Phowrah (shovel) | To cut earth or pull out ballast |
| 17. | Box spanner | To drive rail screws or plate screws. |

Some of these items are shown in Fig. 8.2.

## Maintenance Cycle

Normally, an annual maintenance cycle is drawn up and followed on Indian Railways as part of regular and systematic maintenance of the track. This consists of the following operations:

Through packing
Overhauling
Picking up slacks

### Through Packing

Detailed guidelines have been laid down in the Permanent Way Manual for the work of through packing. The following operations constitute through packing and are carried out in sequence.

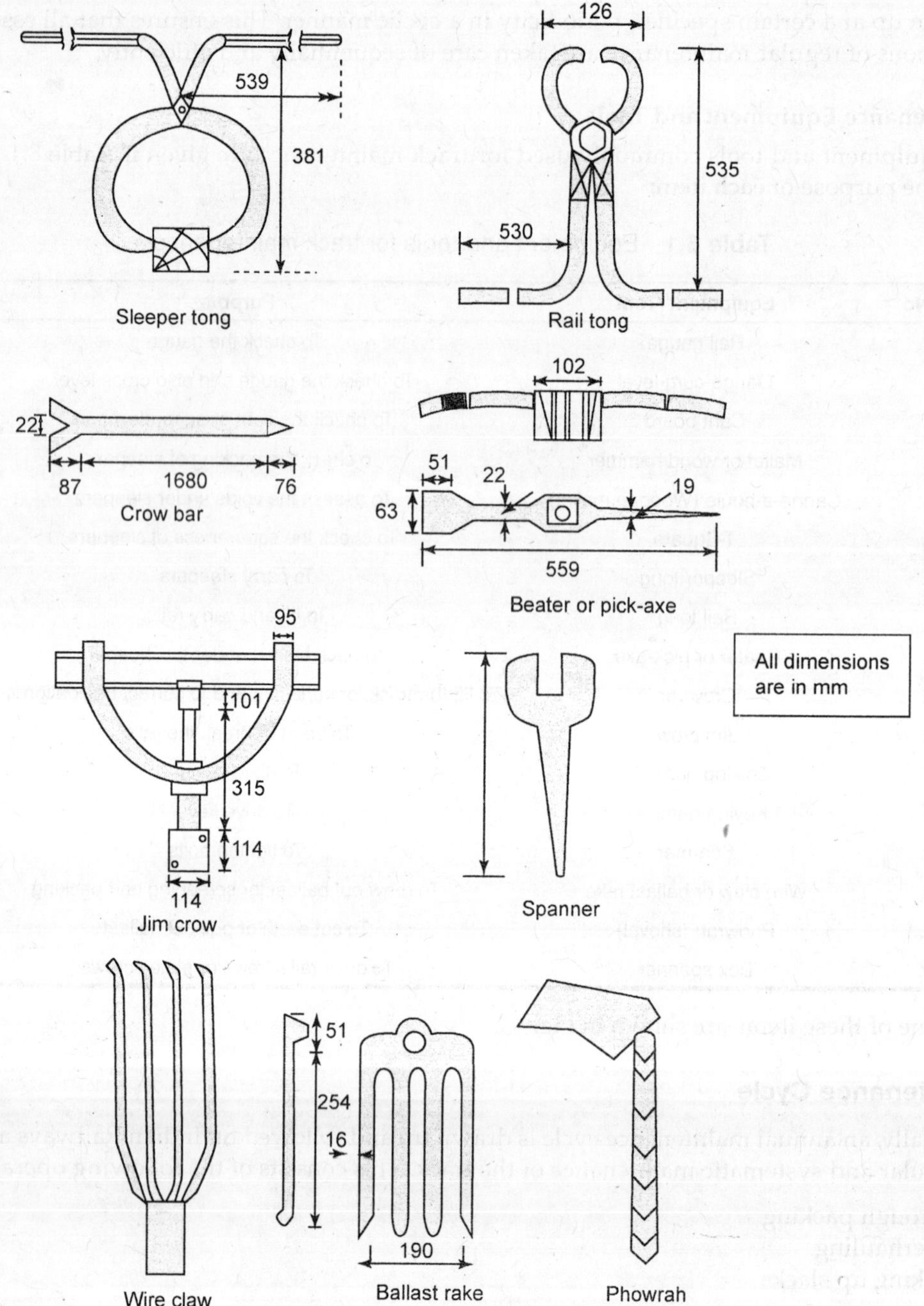


**FIG. 8.2** *Track maintenance tools*

**(a) Opening of rail road:** The ballast on either side of the rail seats is dug out and opened to a depth of 50 mm below the bottom of the sleeper with the aid of a wire claw; this should extend from the end of sleepers on the outside to 450mm (BG) inside of the rail seat (350mm for MG), without disturbing the core under the sleepers. The ballast should be drawn by phowrahs/ shovels outward and inward; the ridge formed at the centre should not project more than 50mm above the rail level. The length opened should be such that it can be repacked before the end of the day.

**(b) Examination of rails, sleepers and fastenings:** Rails should be examined at the underside for corrosion, end-cracking, wear of the head; fish-plates at rail joints should be checked for wear and fish bolts for their tightness. Wire brushes are used to clean the rust and Jim crow is used for correction of kinks in rails.

Sleepers should be inspected for their condition, especially at the rail seats, and should be checked for split and decay if they are wooden sleepers.

Fastenings and fittings should be checked for their condition and tightness/grip, and replaced if necessary.

**(c) Squaring of sleepers:** When sleepers get out of square, the gauge gets disturbed and kinks occur. A reference rail, called the sighting rail, is first checked for the correct positions of the sleepers and marked with chalk; a T-square is used to make the corresponding marks on the other rail. The cores of sleepers which are out of square are picked with beaters, the fastenings are loosened, and the sleepers are squared. This should be done by placing crowbars firmly against the sleeper and pushing it to its correct position. Then the gauge is checked, fastenings tightened and the cores repacked.

**(d) Correcting the alignment:** The alignment of the track is checked visually from a distance of about 60 m during the day when visibility is good. Small deviations in the alignment are corrected by slewing the track after loosening the cores at the ends and drawing out sufficient ballast. Crowbars are planted deep into the ballast at about 30° from the vertical for slewing, for which at least six workmen are required.

**(e) Gauging:** This involves the correction and preservation of the gauge; it is done after ensuring that the sleepers are truly square. One of the rails is taken as the base rail, and the other rail is adjusted to the correct gauge.

The gauge adjustment depends upon the type of sleeper. It is done by re-spiking at the correct location in the case of wooden sleeper, by adjusting the keys in the case of steel trough sleepers, and the cotters for CST-9 sleepers. Gauging should not disturb the alignment; the latter may be rechecked, if necessary.

Tolerances have been specified for the gauge separately for BG and MG lines on straights and on curves. These have been already been dealt with in Chapter 3.

Preservation of gauge is necessary for comfortable riding and for reducing wear of the track components as well as of the rolling stock.

**(f) Packing of sleepers:** Each sleeper is packed firmly and uniformly to ensure that the rails are at the correct levels – longitudinally and laterally. On curves, the correct cant has to be ensured.

The ballast under the sleeper bed is properly packed—by four men, two at each rail, on either side of the sleepers. In the case of wooden and steel sleepers, it should be ensured that the sleepers are not centre-bound, by making a trough in the ballast section at the centre. After packing two rail lengths, the rail at the other side should be checked for proper cross level.

At the end, slight distortions in alignment, if any, should be corrected. Beaters or pick axes are used for packing.

**(g) Repacking of joint sleepers:** The 'joint' and 'shoulder' sleepers should be repacked before boxing at the ends, and the cross levels checked.

**(h) Boxing of ballast section and dressing:** The final operation is through packing. The ballast section is properly boxed and dressed with the help of a special template. The cess is also dressed similarly and its level is properly adjusted for ensuring drainage; high cess level affects drainage while low cess level causes ballast spread and wastage.
*Note*: Through packing is not done in the rainy season.

### Disadvantages of Beater Packing

(i) It is strenuous for the gangmen.
(ii) Quality of packing is not consistent; it varies with the gangman.
(iii) Pulverisation of ballast pieces occurs under beater strokes; ballast life and drainage are affected.
(iv) Sleepers – wood sleepers and even concrete sleepers – are likely to be damaged by the beaters.
(v) Assessment of labour required and its output is difficult.
(vi) It does not meet the standards required for heavy- and high-density fast traffic.

## Overhauling

Overhauling of track consists of the following sequence of operations:

**(a) Shallow screening and making up of ballast:** The crib ballast between sleepers is opened up to a depth of 50 to 75 mm at the bottom of the sleepers, sloping from the centre to the end. The ballast in the shoulders is opened to full depth. The removed ballast is screened and put back, without disturbing the packing under the sleepers. Consecutive spaces between sleepers should not be opened at the same time.

Screening in alternate panels should be completed in one rail length so that several rail lengths are not stripped of ballast simultaneously.

If there are drains across the track, they should first be cleaned and filled with boulders to prevent the packing from working out and forming slacks.

**(b) Through packing of track:** All the steps under through packing, described earlier, should be carried out again to the extent necessary.

**(c) Making of cess:** If cess is higher than necessary, it should be cut. Similarly, it has to be made up to the required level wherever it is low. A cess template may be used for the purpose of checking the section.

Overhauling should be preferably completed before summer. The process of overhauling should be done in a section at least once in 3 to 4 years, depending upon a variety of factors such as the age of the track, speed and volume of traffic, the type of traction and the amount of rainfall in the section.

### Picking up Slacks

*Slacks* are points in the track where the running of railway vehicles is faulty; such locations require more frequent attention than that provided by packing or overhauling. Slacks usually occur at the following locations:

(a) Rail joints with excessive gaps
(b) Locations of poor drainage

(c) Places where the formation is unstable
(d) Places where ballast conditions are poor
(e) Approaches to level crossings, bridges, curves, points and crossings, ash pits and water columns
(f) Near platform lines and gauntleted tracks where good packing cannot be achieved.

The need for picking up slacks is determined by visual examination and track recording. The sleepers needing attention are marked with chalk, and are attended to as in through packing. When a joint has to be picked up, at least three sleepers on either side are packed.

If the joint is low, the fish plate is slightly loosened, and the rail ends are lifted; then the joint is thoroughly packed and the fishplate tightened again.

Normally, one or two days in a week are allotted for picking up slacks.

Sometimes, only the following segments are packed:

(i) Joint sleepers and two sleepers on either side
(ii) Intermediate sleepers where needed
(iii) A few sleepers near approaches to level crossings and bridges.

Points and crossings need attention throughout the year.

Separate charts are maintained – one for the track in a block section and the other for tracks in the station yards – for planning the maintenance operations.

## Deep Screening of Ballast

*Deep screening* of the ballast is done to ensure that a clean ballast cushion of the required depth is available; this is necessary for providing proper drainage and elasticity for the track. Otherwise, track geometry gets disturbed, affecting the performance of the track.

Speed restrictions have to be imposed in the section where deep screening is done. Adequate quantity of ballast, the necessary tools and equipment, and the labour force should be arranged before commencing the operation of deep screening, which should not be done during monsoon.

The procedure consists of the following steps:

(i) A group of four sleepers are tackled one at a time. Each of the sleepers, 1 to 4, is tackled individually (Fig. 8.3).
(ii) The ballast from Spaces A and B is removed fully and the track is supported on wooden blocks.
(iii) The ballast under Sleeper 1 is removed, screened, and packed again under the same sleeper.
(iv) The wooden block is removed from Space A.
(v) The ballast in Space C is removed, screened and placed in Space A, adding extra quantity if required.

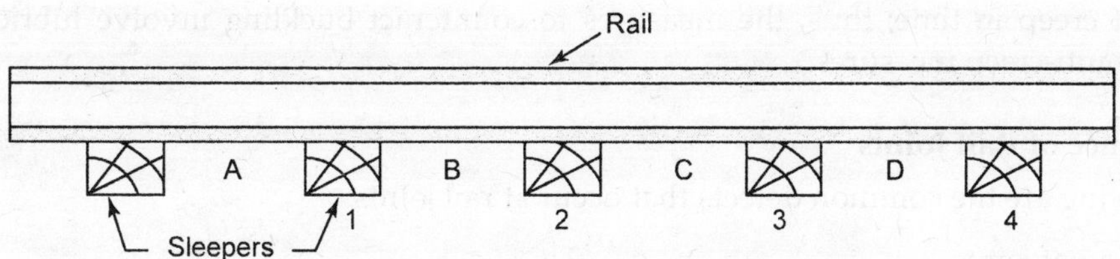


**FIG. 8.3** *Deep screening of ballast*

(vi) The wooden block removed from Space A is placed in Space C. Sleeper 2 is provided with wooden blocks on either side to support the track.
(vii) The ballast under Sleeper 2 is removed, screened, replaced and packed.
(viii) The ballast from Space D is removed, screened and placed in Space B, adding extra ballast if necessary.
(ix) The wooden block is removed from Space B and placed in Space D to support the track.

This procedure is repeated for the next set of sleepers, until the work planned for the day is completed. It may be observed that when ballast is removed from under one sleeper, there is support from at least four sleepers to the track. The track is lifted, if required, not more than 50 mm at one time.

Deep screening is done under a speed restriction of 15 km/h. This speed can be restored to the section speed in stages, in a total period of 21 days for manual packing, and in 10 days for machine packing. During this period, picking up slacks and packing is also carried out in phases.

The cross level and longitudinal level are also checked and restored before the section speed is restored or allowed after the deep screening operation. (Speed-restriction boards should be put up at the appropriate places.)

It is considered desirable to commence deep screening in a direction opposite to the traffic in a double-line section.

Deep screening is usually carried out prior to track renewal/ sleeper renewal, and before the introduction of machine maintenance, or converting the track into long welded rails (LWR).

### Maintenance of Rails

The aspects to be taken care of are the rail surface, rail creep and buckling of track.

**(a) Rail surface:** The surface of rails is susceptible to corrosion, corrugation, side wear, head wear, and so on. Each defect should be dealt with due care at the proper time for the upkeep of the rails.

**(b) Rail creep:** It is the tendency of the rails to expand gradually in the direction of heavier traffic. Rail creep causes distortion of gauge, loosening of joints and breakage of fastenings and fittings. It also causes sleepers to become out-of-square and buckling of the traffic, in the extreme case.

The phenomenon of creep, its causes, effects and the precautions that could be taken to reduce creep or prevent it to the extent possible have been studied in detail under 'rails' in Chapter 3.

As part of the maintenance operations, rail creep should be periodically measured and corrected as and when it exceeds the prescribed tolerance limits.

Provision of anchors to arrest creep and provision of anti-creep fastenings are considered to be effective measures in this regard.

**(c) Buckling of track:** This tendency is more in hot weather, when the compressive forces exceed the lateral or longitudinal resistance of the track. Unless this is corrected in time, accidents are likely to occur. The symptoms are kinks and absence of gaps between the rails at noon during hot weather. The reasons could be non-lubrication of joints and failure to counteract creep in time; thus, the measures to counteract buckling involve lubrication of joints and anti-creep measures.

### Maintenance of Rail Joints

The following are the common defects that occur at rail joints:

(a) Hogging of joints
(b) High joints

(c) Blowing joints
(d) Pumping joints

**(a) Hogging of joints:** This is a phenomenon in which the rail ends get depressed due to loose or faulty fastenings and poor maintenance of joints. The running quality of the track gets affected by hogging.

This can be remedied by tightening the fittings, over-packing the joint sleepers, cropping the hogged portion of the rail and by dehogging the rail or removing the bends by a dehogging machine.

**(b) High joints:** This condition results in an uncomfortable ride, similar to a camel-back ride and may be caused by changes in track structure such as the use of wooden sleepers at the joints in a metal sleeper track (metal sleepers settle more than wooden sleepers), sinking of intermediate sleepers, or over-packing of joint sleepers .

The remedy is to lift and pack metal sleepers, without disturbing the wooden sleepers.

**(c) Blowing joints:** Such joints blow out fine dust when a train passes, and they are covered with a fine layer of dust. Loose packing of joint sleepers, unclean ballast, and battered rail ends and other surface defects of rails will result in *blowing joints.*

The obvious remedy is to clean the ballast and to carry out through packing of joint sleepers, tightening of fittings, improve rail ends, adjust for rail creep, and reduce the expansion gaps as may be necessary.

**(d) Pumping joints:** In rainy season, blowing joints become *pumping joints*, which pump out mud and slush from the mud pockets formed below the joints.

All the remedies which are useful for blowing joints are applicable for pumping joints too; in addition, improvement in drainage yields good results.

### Lubrication of Rail Joints

This is an important part of systematic or regular track maintenance. The purpose of lubrication is to allow for free expansion or contraction of the rail and to reduce wear and tear on the fishing planes of rails and the fish plates.

The lubricant is a mixture of plumbago (dry graphite) and kerosene oil in the ratio 3:2 for the fish plates. Black oil (or reclaimed oil) is used for the lubrication of fish bolts and nuts.

The surfaces should be thoroughly cleaned with wire brushes and jute before applying the lubricant.

Lubrication is carried out once a year during winter; it should not be undertaken during summer or rainy season.

Joints should be opened one at a time; only one fish plate should be taken out at a time. At no time should there be less than one fish plate and three fish bolts connecting the two rails. This ensures safety even if slow moving traffic is permitted before the completion of the job.

### Raising and Lowering of the Track

To keep a good riding surface, lifting (or raising) or lowering of the track becomes necessary under certain circumstances.

**Lifting of track:** This becomes necessary when the track undergoes regrading due to remodelling of the yards, construction of a bridge or creating approaches to level crossings. It may also be necessitated by the occurrence of sag due to the puncturing of ballast into yielding formation.

The points to be observed in the lifting of the track are:

(i) Pegs should be fixed at the correct level at suitable intervals alongside the track.
(ii) Lifting should not exceed 75 mm at a time. Heavy lifting, if necessary, should be carried out in stages, under speed restriction and with the protection of traffic by means of appropriate engineering signals.
(iii) In the case of a single line, lifting should start from the downhill end and be carried up the grade. In the case of a double-line it should be carried out in the direction opposite to that of the traffic. The easement gradient for the passage of trains should not exceed 1 in 500.
(iv) If the track is to be lifted under bridges and in tunnels, standard clearances should be ensured.
(v) In the case of horizontal curves, the correct cant should be ensured.
(vi) The track is finally ballasted, through packed and boxed and the cess brought to the correct level.

**Lowering of track:** This should be resorted to only under inevitable circumstances under suitable speed restriction and protection of engineering signals. The steps are given below:

(i) Trenches should be made across the track at every 30 m to the final level.
(ii) The spaces between the sleepers in cleared, the core beneath is broken and levelled into the space between the sleepers. The operation is repeated until the desired level is reached.
(iii) The track is then ballasted, through packed and boxed, the cess being cut to the proper level.
(iv) Lowering, as in the case of lifting, should be restricted to a maximum of 75 mm at a time, and the grade for passage of trains should not exceed 1 in 500.
(v) As opposed to lifting, lowering should be carried out in the direction of the falling grade.

Lifting or lowering of the track should be carried out under the supervision of a competent official such as a Permanent Way Inspector (PWI).

### Maintenance of Sleepers

This consists primarily of squaring the sleepers, picking up slacks and packing under the sleepers.

**Squaring the sleepers:** This is important for maintaining the correct gauge without any kinks in the rails. The sleepers should be set such that they are exactly at right angles to the track. The procedure for this has already been studied under through packing.

**Picking up slacks:** The gang mate (the leader of the gang) and the gang men should stand on the cess on either side, about one rail length apart and observe the sleepers for any slack/ movement during the passage of a train. Loose sleepers should be chalk-marked.

**Packing under the sleepers:** Ballast should be firmly and uniformly packed under the rail seats.

**Defective sleepers:** Wood sleepers are durable in general, but they are likely to get affected by respiking during the maintenance of fittings. When it is considered that it has been affected sufficiently, it should be replaced with a new one. Concrete and pre-stressed concrete (PSC) sleepers need little attention. Steel sleepers should be checked for corrosion.

### Maintenance of Level Crossings

The maintenance of level crossings involves the following:

**(a) Tree and bush clearing:** This is necessary for ensuring good visibility of the roadway from the railway and vice-versa.

**(b) Overhauling of track at level crossing:** Usually, level crossings should be overhauled at least once a year, or more frequently if required. However, if the track is on concrete sleepers, it should be overhauled with each cycle of machine packing. At any rate, railway tracks should be reconditioned at least once in two years. During overhauling, the condition of rails, fastenings, rail joints and sleepers should be examined. Wooden sleepers should be coated with tar.

Track components in contact with the road should be thoroughly cleaned with wire brush. Flangeway clearances, cross-level, gauge and alignment should be checked and corrected, if necessary and the track packed thoroughly.

**(c) Maintenance of check rails:** Check rails are removed for tamping, overhauling, destressing or renewal and refixed as quickly as possible; a speed restriction of 30 km/h is imposed if the check rail is not immediately fixed.

**(d) Maintenance of whistle boards and stop boards:** The 'whistle' boards and 'stop' boards provided at the approaches to level crossings should be maintained properly.

**(e) Height gauges:** These are located at a minimum distance of 8 m from the gate posts; these should be maintained to see that the vehicles on the road do not exceed the height of the gauge. Monthly inspection by the Permanent Way Inspector is mandatory for all level crossings.

### Maintenance in Track-circuited Sections

The following precautions are required to be taken in the maintenance of track-circuited sections:

(i) No tool or metal object should be placed across or touching the two rails, since this may cause short-circuiting.
(ii) All gauges, levels, trolleys and any other vehicles on track-circuited lengths should be insulated.
(iii) During the maintenance operations, care should be taken to see that no track-circuit fittings, such as rail bonding wires or lead wires are damaged. (Signalling staff may be associated with the maintenance work, if necessary).
(iv) Proper drainage should be ensured so that there is no accumulation of water touching the two rails as this can cause the failure of the track-circuit.
(v) The ballast should be at least 50 mm below the foot of the rail, since wet ballast during rainy season can cause circuit-failure.

(Insulated joints are used to electrically insulate track-circuited sections from the adjoining lengths of the track.)

### Track Maintenance in Electrified Sections

In electrified sections, overhead conductors carry high-voltage electric currents for feeding the traction motors of electric locomotives, the return current flows through the rail partially or fully. To ensure continuity in the electrical circuit and proper earthing in case of leakage of current, traction bonds such as longitudinal rail bonds and cross bonds (in DC tractions) are attached to the rails.

The following precautions are to be taken during maintenance work in electrified routes, so as to avoid disruption of traffic and to ensure personal safety:

(i) No damage should be caused to any of the traction bonds. During track renewal, replacement of fractured rails and removal of fish plates, temporary metallic jumpers should be provided for maintaining the continuity of the track.

(ii) Special care needs to be taken during deep screening of ballast and track-relaying to maintain the distance between the track components, electric masts and OHE wires.
(iii) Permanent Way tools shall be used along with gloves to ensure the safety of workers.
(iv) No work should be carried out within 2 m of the live parts unless the power is shut down.
(v) Long metallic bodies like new or released rails lying along the track should be earthed to take care of induced voltage
(vi) For similar reasons of the danger of induced voltages, maintenance or other railway staff are advised to be away from the track when an electric train is within 250 m of their location.

## Programme of Track Maintenance

An annual programme of regular track maintenance by the conventional manual methods, as recommended by the Indian Railways is given below:

**Annual Programme of Conventional System of Track Maintenance**

| Period | Details of maintenance work |
|---|---|
| 1. Post-monsoon: About six months after the monsoon. | (a) Attention to run-down length in the entire gang beat to restore the section to good shape. |
| | (b) One cycle of through packing/directed track maintenance end-to-end of the gang length, including overhauling of the nominated sections. |
| | (c) Normally, 4–5 days per week are allotted for through packing/overhauling and the remaining days for picking up slacks, attention to level crossings, points and crossings and bridge approaches. lubrication of rail joints, gap adjustment and realignment of curves are to be done during this period. |
| 2. Pre-monsoon: About two months prior to the onset of monsoon. | Normally, 2–4 days in a week are allotted to clearing of side drains and catch-water drains, clearing waterways, cess repairs, and picking up slacks. Normal systematic maintenance may be carried out during the other days. |
| 3. Monsoon period: About four months | Attention to the track as required; this consists primarily of picking up slacks and attention to drains and waterways. During abnormally heavy rains, patrolling of the track by gangs should be carried out in addition to regular monsoon patrolling. |

**Note:** (i) Scattered or casual renewals, creep adjustment, and earth-work repairs are done as required.
(ii) Special instructions are to be followed, as laid down for the maintenance of LWR/CWR.

**Maintenance Planning:** Every PWI must prepare a maintenance plan of the section one month in advance. This should include inspection of points and crossings, curves, deep screening, creep adjustment, and such other items. Adequate material, tools, and labour are to be arranged in advance.

## Protection of Trains and Track During Engineering Work

There are three categories of engineering work:

**(a)** *Routine maintenance work* such as through packing, picking up slacks, and overhauling. This does not involve any danger to traffic; hence, no speed restriction or any kind of signals is required.

**(b)** *Short duration work* such as casual renewal of rails and sleepers, lubrication of rail joints and adjustment of creep.

Hand signals, banner flags and fog signals are required to be used at specified distance. Temporary speed restriction is enforced until the work is completed.

**(c)** *Long duration work* such as track relaying, bridge construction, and track renewal, may require diversions extending over weeks.

Continuous restriction of speed is to be in force for the entire period of the work.

[1]Temporary engineering signals in the form of caution, stop, speed and termination indicators are at specified distances to provide protection to trains. All concerned will be advised in advance about the work and the restrictions.

The types of signals/indicators used along with their placement are shown for two situations in Figs. 8.4 and 8.5.

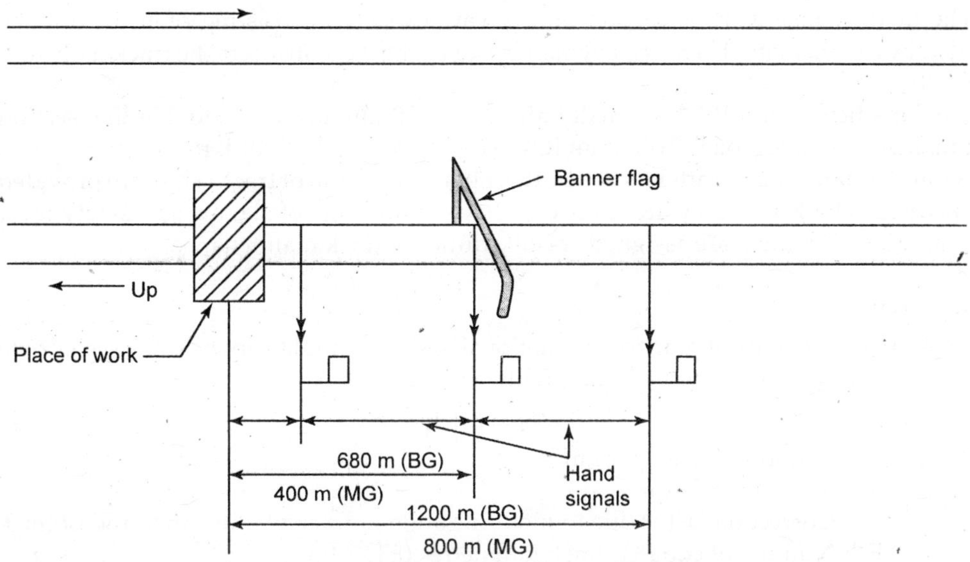


**FIG. 8.4** *Protection of track and trains by speed restriction for engineering works of short duration*

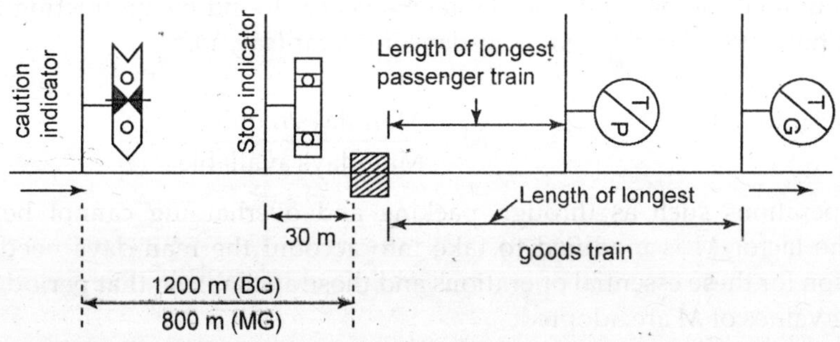


**FIG. 8.5** *Protection of track and trains for engineering works of long duration*

[1]These signals have been dealt with in Chapter 7.

## Organisation Structure

Track maintenance on Indian Railways is well-organised through a systematic organisation structure.

The responsibility of maintaining the track is entrusted to the Civil Engineering Wing, headed by the Principal Chief Engineer of each zonal railway. Other functional chief engineers for track, for bridges, for planning and design, and one called Chief engineer-General assist them. The chief engineers are assisted by deputy chief engineers and executive engineers.

The direct responsibility of the maintenance of all civil engineering works including the track lies with the Divisions. Each division has Divisional Superintending Engineers, Divisional Engineers, Assistant Engineers, Permanent Way Inspectors (PWI's), and Inspector of Works (IOW's).

The maintenance work is carried out by various *gangs*, consisting of 10 to 20 gangmen, headed by a gang mate. There is a keyman in each gang, who inspects the track daily to ensure its safety.

The jurisdiction of a PWI is usually around 60–70 km and for a double-line section, it is 50 km; that of an additional PWI or an IOW is 30–40 km on BG trunk routes.

A railway gang usually attends to a beat of about 6 to 10 km of track. They are provided with the necessary tools and they are trained to have knowledge of signals and safety measures. The mate and gangmen are assigned specific duties of track maintenance.

### Gang Strength

For ensuring uniformity of practice on a national basis, a formula known as the *special committee formula* has been evolved by the Indian Railways:

$$N = MKE \tag{8.1}$$

where N = Number of men per km

M = Manpower factor

K = Correction factor due to modernisation of track maintenance and of the track

E = Number of equated track kilometres (ETKM).

**Manpower factor:** This is the basic requirement of gangmen per kilometre; it is the weightage for the actual man-days required to carry out normal track maintenance in a year with respect to the actual number of man-days available.

Let the total manpower needed be $X$.

The actual number of man-days available in a year is found by subtracting the number of Sundays and holidays from the total man-days in a year (say, $Y$).

Then

$$M = \frac{\text{Man-days required}}{\text{Man-days available}} = \frac{X}{Y} \tag{8.2}$$

Certain operations such as through packing and overhauling cannot be done during monsoon. The factor $M$ is modified to take into account the man-days needed in the dry working season for these essential operations and those available in that period. On this basis, the following values of $M$ are adopted:

$$M \text{ (for BG)} = 1.47$$

$$M \text{ (for MG)} = 1.21$$

**Correction factor (K):** This factor depends on the track section and type of maintenance. Values of $K$ adopted for conventional manual method of maintenance are given below:

| S. No. | Type of track section | Factor $K$ |
|---|---|---|
| 1. | Fish-plated track, including running layout, loops and sidings. | 1.00 |
| 2. | SWP | 0.95 |
| 3. | LWR on metal and wooden sleepers | 0.85 |

**Note:** (i) Man-power requirement for Measured Shovel-Packing (MSP) and Directed Track Maintenance (DTM) is considered to be the same as for conventional manual maintenance.

(ii) Man-power requirement for mechanical track maintenance with on-track tampers is somewhat less than that for the conventional manual maintenance. This is reflected in the reduced values of $K$ for this type of maintenance.

[These aspects will be studied in the next Section.]

**Equated track kilometre (ETKM):** In view of the diverse conditions of track such as different gauges, traffic densities, formation, alignment and rainfall, a concept called **equated track kilometre** has been evolved by which all types of tracks are converted to **equivalent standard track,** the formula for which is :

$$E = L.\ U(1 + A + B + C) \tag{8.3}$$

where $E$ = Equated track kilometre
$L$ = Running track kilometre
$U$ = Traffic density factor
$A$ = Formation factor
$B$ = Alignment factor
$C$ = Rainfall factor

**Traffic density factor (U):** This depends on the traffic density and the nature of the section. For broad gauge, $U$ ranges from 1.4 to 0.8 with decreasing traffic density.

For metre gauge, $U$ ranges from 1.2 to 0.6.

*Note*: In double or multiple lines, $U$ may be assessed for the average traffic density of the lines and applied to all the lines. Wherever the speeds exceed 105 km/h, $U$-values should be increased by 0.1.

**Formation factor (A):** This depends on the nature of the soil, and is given below:

Stable soil $A = 0$

Ordinary unstable soil $A = 0.1$

Shrinkable soil/black cotton soil $A = 0.2$

**Alignment factor (B):** This factor is applicable to curves of 1.5º or more for BG and 3º or more for MG.

($A$ and $B$ should be assessed in proportion to the length falling under bad soil conditions/curves compared to the total length.)

**Rainfall factor (C):** This is dependent on rainfall. When the rainfall is less than 200 cm/year $C = 0$. For rainfall of 200 cm/year $C = 0.1$. For every additional 10 cm of annual rainfall, an increase of 0.01 is applied for $C$, subject to a maximum of 0.2 for the factor.

[A slightly different approach is used for calculating $E$ for yards, track connections and layouts, which are adopted on Indian Railways.]

**Modified Formula for Gang Strength**

The modified formula which is currently used is:

$$N = 0.95\ \text{MKE} \tag{8.4}$$

The following modifications are recommended to this formula to take care of increased GMT, higher speeds, heavier axle loads and the use of BOXN wagons:

(i) For traffic density more than 20 GMT per annum on BG sections, the following values of $U$ are recommended:

| Traffic density | 20–25 GMT | 25–35 GMT | 35–45 GMT | 45–55 GMT |
|---|---|---|---|---|
| $U$ | 1.4 | 1.5 | 1.6 | 1.7 |

(ii) For BG sections where five or more loaded trains consisting of BOXN or equivalent wagons are operated per day, $U$ may be further increased by 0.1.

(iii) A further addition of 0.1 may be made for $U$ both for BG and MG where the speeds are above 110 km/h and 75 km/h, respectively.

*Note*: The equated track kilometres shall be worked out kilometre-wise for each section on each gauge separately and added.

**Normal strength of a gang:** To the number of gangmen obtained from the formula, keymen and gang mate shall be added, with an element of leave reserve.

Gang length does not normally exceed 6 km for main lines and 8 km for branch lines. The total number of gangmen is limited to about 18, excluding keymen (one for each line) and the gang mate, who heads the gang.

**Track patrolling:** Patrolling of railway tracks is done to ensure safety of the track and of the trains. This involves to and fro movement of the patrolman along the track to watch out for any defect that could affect the safety.

The types of patrolling in vogue on Indian Railways are:

**(a)** ***Keyman's daily patrol***: This is done on foot—one rail in one direction and the other rail in the reverse direction, reaching the starting point. The keys and other fittings are tightened.

(b) ***Gauge patrol during heavy rain or storm***: This is carried out by the gang, headed by the mate; any unusual occurrences are observed and evaluated for necessary action.

**(c)** ***Night patrolling during monsoon***: This is done according to a planned programme to detect any damage to the track such as settlements or breaches caused by flood waters.

These are in addition to the *trolley inspection*, which is normally done once a month by the Assistant Engineer (AEN) in charge of the section. During monsoon patrolling, a train, push trolley, or motor trolley may be used.

The PWI is expected to inspect the entire section with the help of a push trolley at least once in a week. All important features such as bridges, curves, points and crossings, and level crossings are inspected in a planned manner by the Assistant Engineer as well as the PWI.

### Standards for Work Load of PWI's and AEN's

A study of the work load of PWI's and AEN's of open line was conducted by the Railway Board in order to quantify it for purposes of comparison. A new terminology—Integrated Track Kilometre (ITKM), was introduced for quantifying the work related to not only track, but also bridges and buildings; this new unit was intended to serve the purpose of comparing work load and laying yardsticks.

**Integrated Track Kilometres (ITKM):** This is a composite unit taking into account the track, bridges and buildings. The equated track kilometre on account of these features is found separately for each, and summed up to obtain the ITKM.

**ETKM for track:** Running track kilometres and weighted lengths of sidings are added to get the ETKM for track. Ten sets of points and crossings are considered equivalent to one track kilometre. One set of points and crossings is considered to be equivalent to 1 turnout, 1 diamond, 2 cross-overs, and 5 scissors.

**ETKM for buildings:** Based on equivalent cost of maintenance, 1500 $m^2$ of plinth area is considered as one equated track kilometre. The total ETKM for buildings can then be found.

**ETKM for bridges:** This is based on waterway.

100 m of minor bridges per track = 2 ETKM

100 m of major bridges per track = 3 ETKM

Now the ITKM can be found by adding these three ETKM's.

**Work load of PWI:** This varies from 110 to 170 ITKM on BG (single line, double line, or multiple lines)

**Work load of AEN:** 425 to 550 ITKM

(Further details are found in the guidelines issued by the Railway Board.)

## Track Recording

Ordinary track inspection methods are purely subjective and do not provide a uniform yardstick for comparison. This led to the development of objective methods of assessment of the quality of track by *track recording* with the aid of *track recorders*. These are devices which provide a continuous record of various track parameters from which a reliable assessment of the running quality of the track may be obtained.

### Track Recording Cars

These are broadly classified as the *mechanical type* and the *electronic type*.

The following are the various track-recording cars used:

1. Track recording trolley
2. Hallade track recorder
3. Oscillograph car
4. Portable accelerometer
5. Portable oscillations monitoring system – OMS-2000
6. Amsler's track recording car
7. Plasser and Theurer's modern track recording cars

**Mechanical Track-recording Car:**

This has two bogies – one with two axles and the other with three axles. The measuring bogie is the three-axled one with a wheel base of 3.6 m and an axle load of 7 tonnes. The central axle has flangeless cylindrical wheels, the other two having normal flanged wheels with 1 in 20 taper or coning.

A continuous record of unevenness (of left and right rails) gauge, twist, curvature/alignment (of left and right rails), and speed is obtained while running on the track at about 70 to 80 km/h.

**Principles of recording track parameters:** The following are the principles/basis for recording track parameters:

(i) *Unevenness*: It is measured over two selected chord lengths of 3.6 m and 9.6 m. This is the deviation in the vertical plane (depression/rise) of the middle of the measuring bogie with reference to the average position of the two outer axles (Scale 1:1).

(ii) *Gauge*: The changes in gauge are picked up by spring-loaded gauge feelers in contact with the gauge faces at 14–17 mm below the rail table. It is recorded to a scale of 1:1.

(iii) *Twist*: This is calculated from dynamic cross level, or the difference in the vertical profile of the left and right rails on two different chords of 3.6 m and 4.8 m. (The gyroscope or an inclinometer measures twist more precisely (Scale 1:1).

(iv) *Curvature/Alignment*: Curvature of the left and right rails are measured over two selected chord lengths of 3.6 m and 7.2 m as their verse sines in mm (Scale 1:1).

**Electronic Track-recording Car**

The *Amsler track-recording car*, used on Indian Railways was first imported from Switzerland, and later developed indigenously.

The construction features of this car are the same as those of the mechanical type of cars, except that the sensing elements of the various track parameters are transmitted electronically and recorded graphically to provide a continuous record of the parameters. The mid-chord processor facilitates the conversion of the versines measured to a common base of 7.2 m chord. Linear Voltage Differential Transducers (LVDT) pick up the signals and facilitate the recording on a continuous basis in a graphical form.

**Frequency of track-recording:** Track recording cars are run regularly according to the following frequencies:

BG High-speed routes (More than 130 km/h) ... 2 months
High-speed routes (More than 110 km/h) ... 3 months
A & B routes ... 4 months
C & D routes ... 6 months
E routes ... 12 months

**Analysis of track-recording charts:** This is done to arrive at the nature of attention required in each stretch—immediate attention, early picking up of slacks, and so on. A critical study of these charts help to reveal the trends of track deterioration and needs for track improvement. The extent of irregularities in unevenness, gauge, twist, and alignment for classifying the track into A, B, C and D categories has been laid down, and it serves as an indication of performance of the permanent way officials in maintaining the track geometry. A *combined track recording* *index (CTR)* has also been evolved based on unevenness, gauge, twist and alignment. A value of 80 for this index is considered *outstanding* and below 50 is considered *poor* in maintenance and riding quality. Statistical concepts such as *standard deviation* are used to measure track irregularities and set standards.

A *microprocessor-based track monitoring system* is capable of operation up to a speed of 150 km/h. *Plasser and Theurer's Modern Track-recording Cars* are the most sophisticated equipment produced and used in advanced countries for modern track maintenance. (Plasser and Theurer is an Austrian firm.)

## 8.4 TRACK MAINTENANCE – MODERN METHODS

Modern mechanised or automated methods of track maintenance involve the use of modern track equipment and machines. The following methods fall in this category:

- Mechanical maintenance with aid of track machines
- Measured shovel packing (MSP)
- Directed track maintenance (DTM)

The emphasis in these methods is on the identification and rectification of defects, quality control and the use of modern techniques of maintenance.

Although the age-old 'beater packing' using manual labour has stood the test of time, recent technological advancements in the field of railways and traction have necessitated the switch-over to modern mechanised methods of track maintenance.

In these methods, the maintenance staff needs less physical strength and more mechanical skill. Further, these methods meet the requirements of track maintenance for higher speeds and heavier axle loads. In addition, modern methods overcome the limitations of beater packing apart from providing the following *advantages*:

- It helps to maintain longitudinal and lateral alignment of LWR track laid on concrete sleepers to close tolerances. The time required for maintenance is considerably reduced, with minimal disruption of traffic.
- Track geometry can be preserved for a relatively longer duration.
- Maintenance of high speed routes is easily carried out.

### Tamping Machines

Tamping machines are broadly classified as *off-track* tamping machines and *on-track* tamping machines.

#### Off-track Tamping Machines

These are hand-tools driven by compressed air, petrol engine or electricity. Tools of the first category are pneumatic in nature requiring an air-compressor. The second are self-contained with a petrol engine and the third type require an electric generator. Jacks are used to lift the track; the tamping work is carried out in between train movements, thus requiring no restrictions on traffic.

The following are the *requirements* for off-track tampers:

(i) Relatively light tampers are easy to operate.
(ii) Less noisy tampers are safe and comfortable to use.
(iii) The tools must be reliable, requiring minimum servicing.

(iv) Electric tools must meet safety requirements to prevent the possibility of electric shocks.
(v) The output must be good.
(vi) The track produced after tamping should be stable for a long time.

### Off-track Tampers Used on Indian Railways

**Pionjar tampers:** These are driven by an in-built petrol engine. Each tool weighs 27 kg and gives a high power stroke, resulting in better tamping. The output can be about 100 sleepers per day with a set of two tools. (ATLAS-make comes in this category.)

**Kango tampers:** These work with the help of an electric motor drive and pneumatic piston through coupling. Both impact and vibration are combined in the stroke delivering high energy for tamping. Sets of 4 to 8 tools are driven with power supply to half the number. Even one or two tools can be used where portability is important. Each tool weighs about 18 kg. A four-tool set can give an output of 64 sleepers per day.

**Bosch tampers:** These are electrically driven by two 5 kVA generators. Four tampers form a set, each tool weighing about 10 kg.

The *advantages* of this type of tampers are:

- Ease of operation.
- Fully damped stroke
- Variable compaction facility

With each set, an output of 120 sleepers per day can be achieved. (HILTI-make comes in this category.)

### Disadvantages of Off-track Tampers

- Necessity to lift the track with the help of jacks and opening up the ballast to the bottom of the sleepers
- Considerable amount of manual work
- Need for transporting tracks to the work spot.
- Unsuitability for use on a heavy track structure with concrete sleepers and long welded rails

In view of the above, off-track tampers have limited use of packing isolated sleepers.

### On-track Tamping Machines

These are self-propelled machines and the tamping is carried out automatically through controls in the operator's cabin. On-track tampers can be of the light type or the heavy type.

**Light on-track tampers:** The weight of a unit is only 1 to 1.2 tonnes. There are two tamping units mounted on a common motorised trolley. Lifting and levelling the track has to be done manually in advance. The output can be 150 sleepers per day. These are only slightly more advantageous than off-track tampers, and therefore their use has been discontinued on Indian Railways.

**Heavy on-track tampers:** These are heavy-weighing about 30 tonnes or more. The functions of the initial models were limited to tamping alone, but the latest models can perform lifting, levelling, tamping, and lining of track in an automatic manner. Once they are mounted on the track, some time is needed to dismount them, in view of their weight.

Traffic should be blocked for a sufficiently long time for economic working of on-track tampers.

**On-track Tamping Machines on Indian Railways**

Plasser and Theurer of Austria supplied their first on-track tamping machine (VKR-05) to Indian Railways in 1966. Later, a factory was set up in Faridabad and the machine components were progressively indigenised. *Plassermatic 06-16 SLC on-track tamping machines* made in this factory could lift, level, tamp and align the track automatically.

Improved versions of this machine are being used on Indian Railways. Vibratory squeezing of ballast under the sleepers brings about a high degree of efficiency of tamping by adjusting the tamping pressure. The machine consists of independent units for tamping, track lifting, levelling and lining operations; they afford great flexibility since only one or two operations may be carried out depending on the need. Efficient packing and correction of errors in longitudinal and cross-levels and in alignment is achieved, along with efficient packing under the sleepers. Traffic blocks of 4 to 5 hours in two spells will be necessary for economy in operation. The output could be around 600 sleepers per hour. The SLC-type machines have now been phased out and replaced with improved versions such as *Universal Main Liner 06-16*, *Universal Tamper (UT)*, and *Duo-matic 08-32 on-track tamping machine*. In the last machine, 32 tamping tools are used to pack two sleepers at a time.

Some more machines for special uses are *Points and crossings-tamping machines*, *Ballast cleaning machines*, *Universal ballast distribution and grading machine*, *Dynamic track stabiliser*, *Multipurpose tamping machine*, and *Formation rehabilitation machine*.

**Saving Time Through Mechanised Track Maintenance**

Mechanised track maintenance brings about excellent quality for the track; especially where concrete sleepers are laid and LWR/CWR are used. However, significant traffic blocks extending to about 4 to 5 hours are inherent for economy in operation of heavy track maintenance machines. This necessitates good planning to derive maximum advantages of mechanised track maintenance.

About 200 tamping machines and 70 ballast cleaners are in operation on Indian Railways, in addition to a significant number of dynamic track stabilisers.

In view of the expansion of Indian Railways in a significant way, the need for further manufacture and use of modern equipment for track maintenance cannot be overemphasised.

**Small Track Machines**

A large number of small track machines are being used on the Indian Railways for conventional manual maintenance and casual or unplanned repair jobs.

**Measured Shovel Packing (MSP)**

This is an improved form of manual packing, with a scientific approach, but without the use of sophisticated equipment. This was first introduced in England in the year 1910, and was further improved upon in France in 1927. Later on, the method became popular in many advanced countries.

The basic principle of the method is to leave the well-compacted portion of the ballast bed undisturbed as far as possible, and eliminate the minor unevenness in levels due to the voids created by filling the underside of sleepers with measured quantities of stone chips of prescribed sizes. Therefore, MSP is not used immediately after track renewal or deep screening or when the lift required is more than 30 mm.

Indian Railways recommends this method for adoption for the following jobs:

(a) Through packing of wooden sleeper track
(b) Packing of wooden sleepers at the joints in a metal sleeper track
(c) Through packing of points and crossings with wooden and concrete sleepers
(d) Dehogging of rails
(e) Packing of LWR sections with wooden or concrete sleepers

(Mechanised maintenance with on-track tampers is preferred for LWR sections since any degree of lack of care in MSP can endanger the safety of the track.)

**Equipment needed for MSP**

The following equipment are needed for MSP:

(i) Canne-a-Boule
(ii) Dansometer
(iii) Fleximeter
(iv) Gauge-cum-level
(v) Viseur and Mire
(vi) Non-infringing track jack
(vii) Packing Shovels
(viii) Dozing shovel and measuring can
(ix) Plain shovels

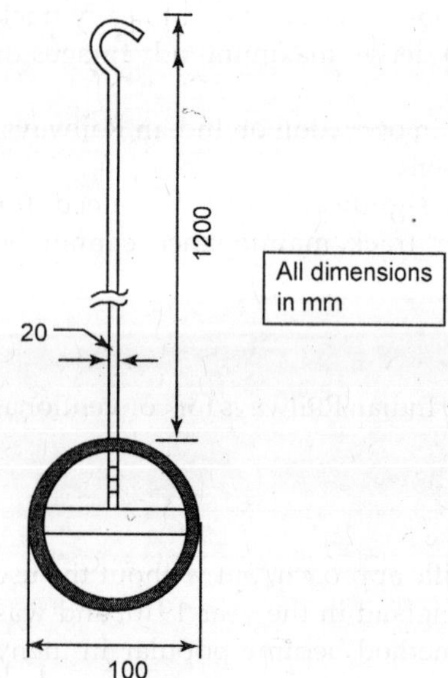


**FIG. 8.6** *Canne-a-Boule*

**(i) Canne-a-Boule:** This is used to determine the extent of voids under the sleepers. It consists of an iron ball of 100 mm diameter with a mild steel rod handle of 20 mm diameter and 1200 mm length (Fig. 8.6).

**(ii) Dansometer:** This is used to measure the depression of a sleeper when a train passes over it. The dancing rod which rests on the sleeper is supported on a tripod, the three legs of which rest on the ballast (Fig. 8.7).

The extent to which the friction sleeve moves down determines the voids under the sleeper.

**(iii) Fleximeter:** This instrument is used to measure the total depression of rail under load. If the rail-to-sleeper fastenings are tight, the results given by the fleximeter and dansometer should be the same; otherwise, their difference indicates the extent of looseness of the fastenings (Fig. 8.8).

**(iv) Gauge-cum-level:** This measures the gauge and the cross-levels to an accuracy of 1 mm. The cross-level is

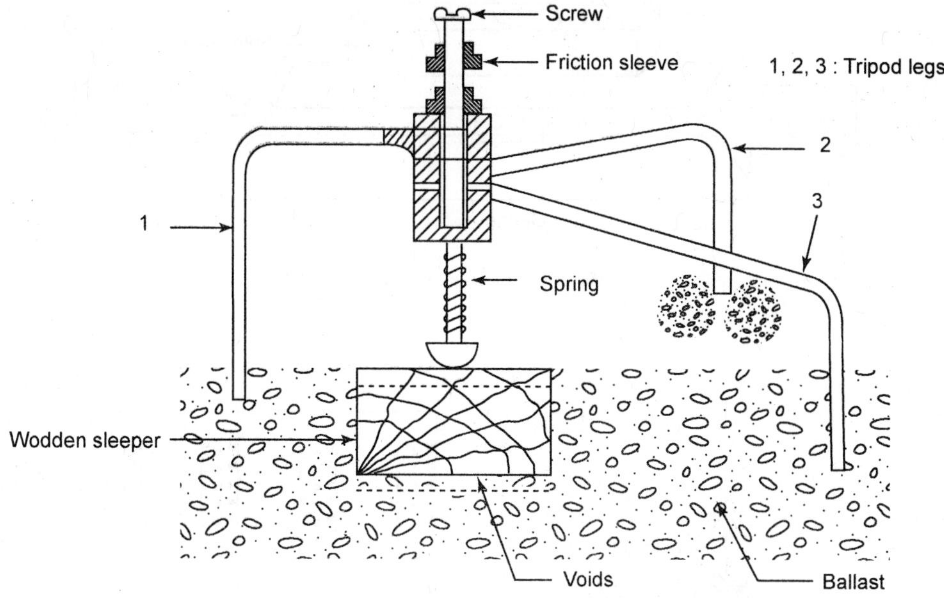


**FIG. 8.7** *Dansometer*

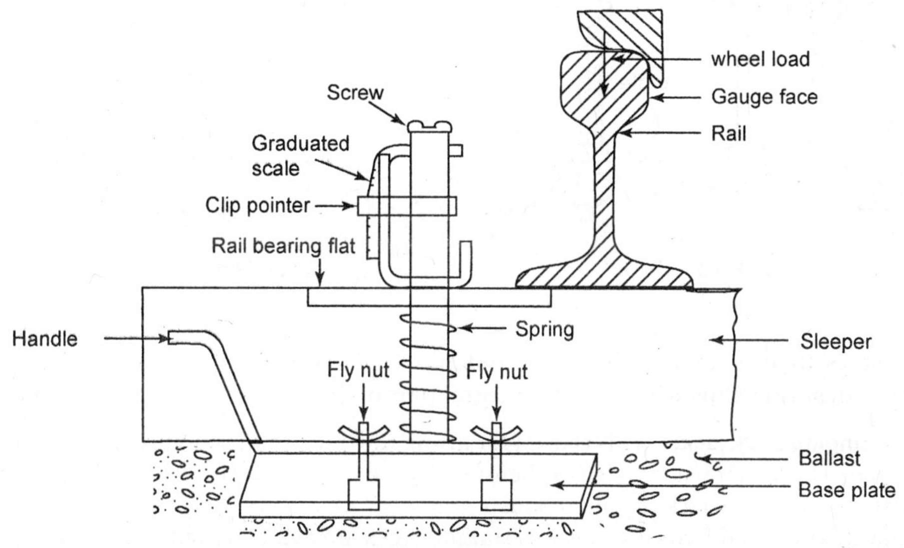


**FIG. 8.8** *Fleximeter*

measured with the help of a sensitive spirit level, 200 mm long, with a sensitivity up to an angle of 2′30″ (Fig. 8.9).

**(v) Viseur and Mire:** These are analogous to the levelling instrument and staff used to measure the unevenness of the rail top. By fixing a collimation line and moving the 'mire' from sleeper to sleeper, the lift required for any particular sleeper to make an even rail top is determined.

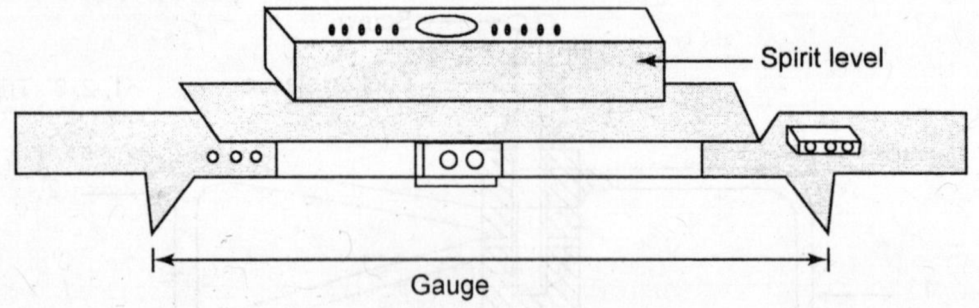


FIG. 8.9 *Gauge-cum-level*

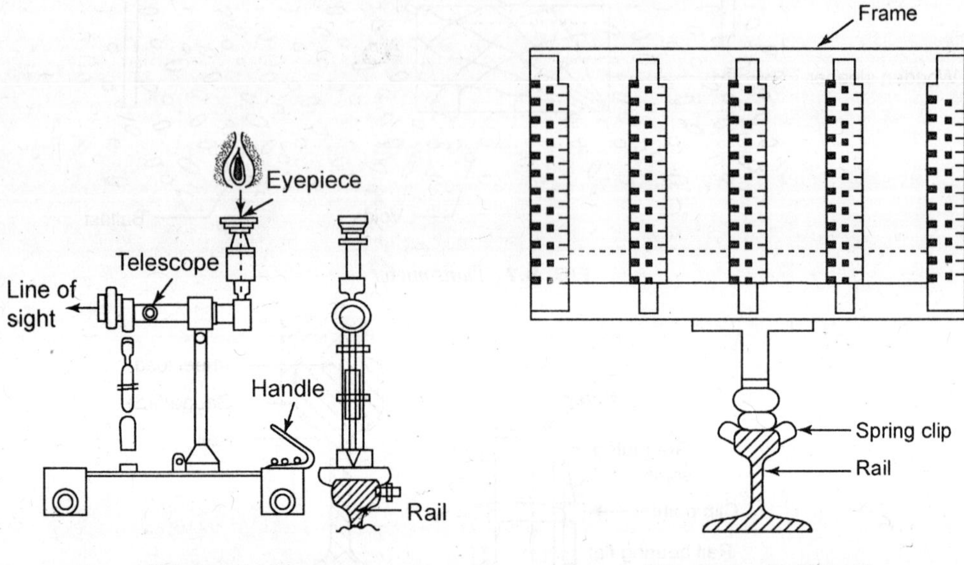


FIG. 8.10 *Viseur*

FIG. 8.11 *Staff (mire)*

Packing plates are used to raise the viseur and mire to the required level; these are fixed to the rail head with spring clips which are an integral part of the instrument (Fig. 8. 10 and 8.11).

**(vi) Non-infringing track jack:** This is used for lifting the track so that stone chips can be spread over the bed. The jack must be released before the arrival of the train. There is a quick-release mechanism to lower the jack instantaneously.

Both mechanical and hydraulic non-infringing jacks are available with capacities from 5 to 10 tonnes, with a lift of 75–100 mm.

The mechanical type is shown schematically in Fig. 8.12.

**(vii) Packing Shovel:** This is used for distributing stone chips on the sleeper bed. The throw of the blade is 100 mm for BG and 85 mm for MG (Fig. 8.13).

**(viii) Dozing shovel and measuring can:** These are used to measure and provide the required quantity of stone chips into the packing shovel (Fig. 8.14).

**(ix) Plain shovel:** This is used for spreading stone chips over sleeper beds at places where packing shovels cannot be used for want of space (Fig. 8.15).

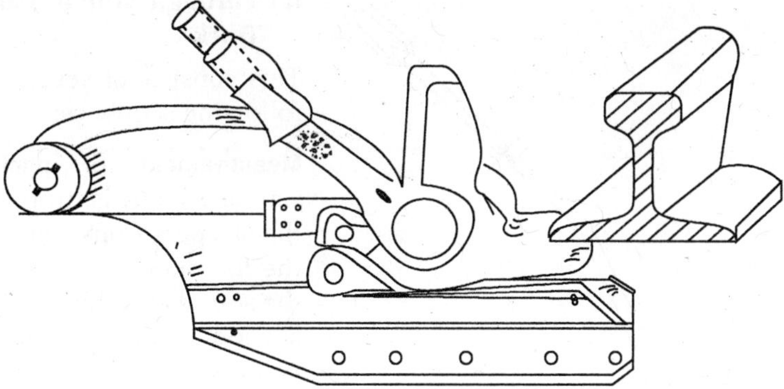

**FIG. 8.12** *Non-infringing jack (mechanical type)*

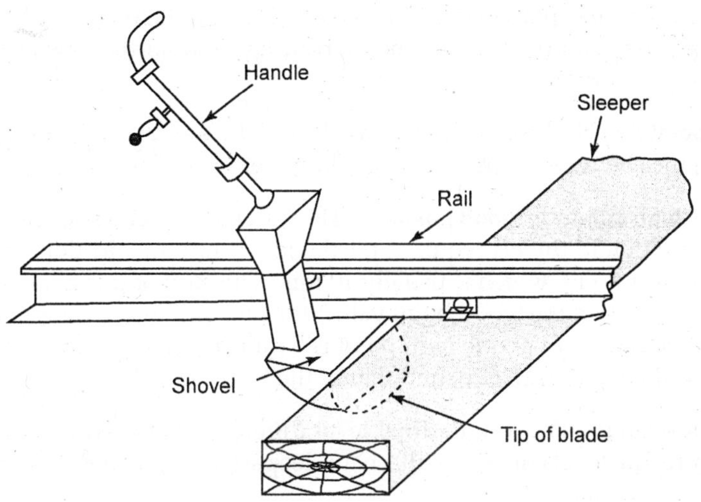


**FIG. 8.13** *Packing Shovel*

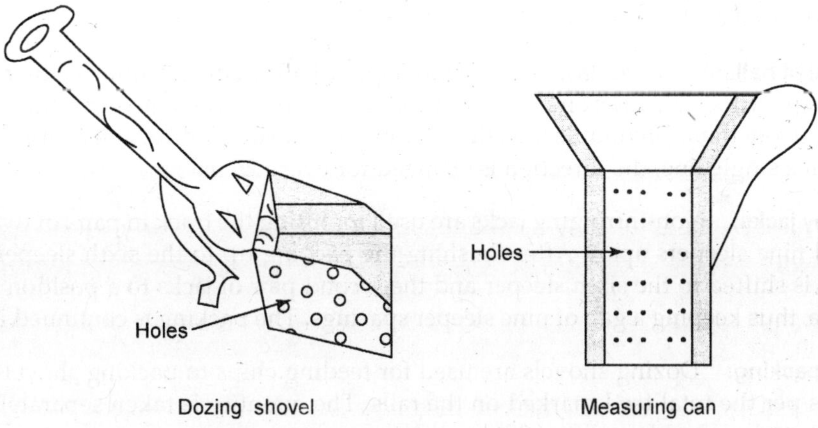


**FIG. 8.14** *Dozing shovel and measuring can*

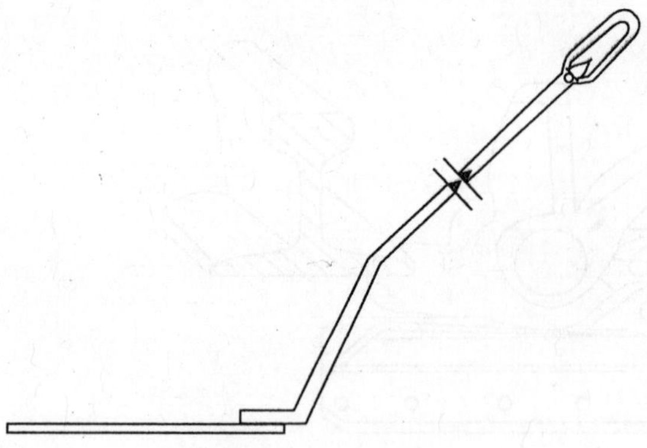

**FIG. 8.15** *Plain Shovel*

**(a) Through MSP of Wooden Sleeper Track**

This consists of several operations in the following sequence:

**Measurement of packing voids:** The Canne-a-boule is dropped at either end of the sleepers from a height of 400 mm and the looseness of the sleeper is judged by the sound and the rebound of the ball. For a good rebound and healthy sound, the value is taken as zero, and it is increased depending on the reduction of rebound and the dullness of sound. These values are marked at the ends of sleepers with a chalk. At a few selected places, dansometer readings are taken after the passage of a train; these are compared with the Canne-a-boule readings and corrections are applied to them; if necessary.

**Fixing 'high points':** High points are located on both the rails by sighting along the gauge faces; the distance between high points should not exceed 25 m.

**Conversion of 'high points' to 'good points':** The difference in cross-levels at all the high points is determined to obtain the lift required for the lower rail. In addition, a general lift is given even to the highest point, which is usually 10 mm. The sum of this general rise and the cross-level correction is called the 'good point value' for each rail for each high point. (For example, if the cross-level difference at one high point is 5 mm and the general rise is 10 mm, the good point value for the higher rail is 10 and that of the lower rail 15.)

**Longitudinal levelling:** The longitudinal level difference between two high points is got by Viseur and mire; the levels on every alternate sleeper are taken and those on the intermediate sleepers are interpolated.

**Total feed – Mark definitive:** The total feed is the sum of the longitudinal level correction and the packing void. This is also known as *mark definitive,* and this figure is marked on the inner side of the foot of the rail.

**Removal of ballast:** Wire claws are used to remove ballast for 250 mm on either side of the rail for a width of 150–200 mm; and up to the bottom of the sleeper. Only one side of the sleeper is opened at any time. On double line, the opening is done in the direction of approaching trains, while on a single line, the direction is changed for every 25 m.

**Lifting by jacks:** Non-infringing jacks are used for lifting the track in pairs of two, one for each rail and nine sleepers apart. After finishing the packing up to the sixth sleeper, the first pair of jacks is shifted to the sixth sleeper and the second pair of jacks to a position six sleepers in advance, thus keeping a gap of nine sleeper spacings. The packing is continued in this manner.

**Shovel packing:** Dozing shovels are used for feeding chips to packing shovels, the quantity being as per the total feed marked on the rails. The quantity is taken separately for each rail seat, fed, and packed for a lift of 20 mm lift at one time.

**Providing ramps:** At the end of the day's work, a ramp or run out of 1 mm per sleeper is given to the track. The balance quantity is fed in continuation at the beginning of next day's work.

**Alignment:** Track alignment can also be checked by viseur and mire and corrected.

**Checking:** After passage of at least two trains, the cross-levels and longitudinal level are checked and corrected if necessary.

### (b) MSP of Wooden Sleepers at Joints:

The following are the steps in the procedure

**Total lift for the rails:** This is the sum of the lift based on the void content at the joint sleepers and the longitudinal lift due to sag in the track. For the sighting rail, the total lift is the longitudinal lift plus that for the voids. Similarly, for the other rail, the total lift is the sum of the longitudinal lift, the lift for the voids in this rail, and the correction for cross-levels.

**Lifting:** The joint sleepers are then lifted by jacks.

**Packing:** Packing shovels are used for this purpose.

**Alignment:** The variations due to beater packing for intermediate sleepers and MSP at joint sleepers are determined by viseur and mire and corrected.

### (c) MSP of Points and Crossings

This has been known to yield good results for wooden and concrete sleepers, although it may have to be repeated in the case of steel sleepers. The steps in the operation are as follows:

**Looseness or dance of sleepers:** This is measured for the main line and the turnout side, as well as for the lead curve, separately. However, on the switch portion, the dance need not be measured separately up to the point where the track divergence is less than 400 mm. Near the crossing, it may be measured on both sides and the average considered.

**High Points:** These are located in the usual way on the main line; but if any point is located on the switch portion, it should be located on the stock rail.

**Cross-level correction at high points:** This is done in the same way as in through MSP, converting the high points to good points. (On switch portion, good points are located on stock rail.)

**Longitudinal dips:** Viseur and mire are used to measure longitudinal dips as in through MSP.

**Total feed or Mark definitive:** The total feed is got by the addition of packing voids and the longitudinal dip. In the switch and crossing portions where the two rails are close to each other, the combined feed or additional dosage is worked out for the space between the two rails proportionately.

On the turnout side, the lift of the rails is got by taking into account the tilt given to the sleeper on account of the lift given to the rails on the main line. These values are obtained proportionate to the distance of the turnout rail from the main line rails and the gauge. The values thus obtained are added to the dance values for looseness to arrive at the total feed. (The sleeper is considered as a rigid body for this purpose.)

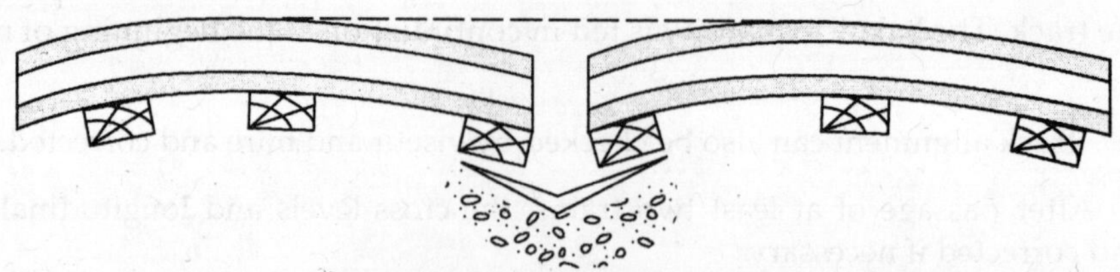

**FIG. 8.16** *Dehogging of rail ends*

**Shovel Packing:** Ballast is removed completely after lifting the track on jacks, and the calculated total feeds of ballast pieces are packed by shovels.

**Alignment:** This is checked after the passage of trains, and corrected if necessary.

**Traffic block:** At least a block of 1½ hours is necessary for MSP of a turnout.

### (d) Dehogging of Rails

Dehogging of rail ends can be done with the aid of MSP (Fig. 8.16).

(i) The dip at the joint is measured on a 1.5 m chord ($a$).
(ii) The dance of the joint sleepers is, say, $d$.
(iii) The total feed for the joint sleeper is given by $a + d + a'$ when $a' = a$ (subject to a maximum of 5 mm).
(iv) Shovel packing is done for the joint sleepers, without touching the adjacent sleepers.
(v) After two days, the adjacent sleepers are packed without disturbing the joint sleepers.
(vi) The dip is measured again, and the dehogging operation repeated, if necessary.

### Advantages of MSP

- The track can be maintained to better tolerances.
- It gives more life to the track than beater packing.
- MSP enhances the life of wooden sleepers as it causes much less damage to the sleepers' underside than beater packing.
- Centre-bound sleepers are avoided with MSP.
- Less strenuous and more scientific than beater packing.

### Disadvantages of MSP

- It is applicable only to flat-bottom sleepers.
- Special-sized stone chips are required; hence, it is costly.
- It can be done only on a considerable ballast bed.
- Instruments used are delicate and need careful handling.
- The calculations for the feeds of material in different situations are involved, calling for high degree of skill and care in implementation.

### MSP Versus On-track Tamping

Experience with on-track tampers has shown that the track geometry deteriorates rapidly under traffic, as early as in just one week. It has also been observed that the track reverts to the condition before tamping; this phenomenon is referred to 'ballast memory'. Tamping machines thus fail to erase this ballast memory.

On the other hand, track geometry is better retained by MSP with a general lift of 40 mm, reducing the frequency of maintenance needs to almost one-fourth when compared to on-track tamping.

However, on-track tamping causes much less damage to the ballast than MSP; thus, the resilience of the track and drainage are better retained by on-track tamping.

In view of the speed of work, tamping machines were preferred initially, but because of the advantage of MSP retaining track geometry better interest in the use of MSP got revived although it is slower.

### Modifications of MSP

A major drawback of MSP is its labour-intensive procedure. Track measurement, opening, feeding, packing and closing of ballast are, for the most part, carried out manually.

British Railways has evolved the following methods to overcome this disadvantage:

**(a) Semi-mechanised method:** In this, a portable compressor is used to blow the stone chips underneath the sleeper through a hollow metal tube. This obviates the need for opening of the ballast for feeding the chips.

However, the measurement of track geometry, calculation of feeds and feeding of stone chips in the metal tube is done manually. Also, the compressor has to be moved from one place to another manually.

**(b) Fully mechanised method:** A fully mechanised system for MSP involves the use of an equipment-PBI-84, a three-bogie vehicle weighing about 80 t, which gives a pneumatic ballast injection. It has its own ballast hopper and can carry 10 t of stone, sufficient for a normal working shift. The output can be around 400 m of track per hour. An on-board microprocessor measures the track geometry and calculates the feed for each sleeper; the machine lifts, aligns the track, and blows the requisite quantity of stone chips—all these operations carried out automatically.

*On Indian Railways,* a stone-blower may provide a good economic solution for the maintenance of branch lines deficient in ballast which are laid on steel/CST-9 sleepers, as tamping machines cannot be deployed here efficiently.

## Directed Track Maintenance (DTM)

Conventional through packing is based on the concept that loosening and repacking of ballast under rail sleepers according to a predetermined cycle scheduled in a year automatically improves the elasticity and the riding quality of the track. Since this requires each and every sleeper to be attended to, this method is bound to involve a certain amount of unproductive work irrespective of the packing condition and track geometry. In spite of this, the system of through packing, overhauling and picking up slacks has wide acceptance as the track quality is reasonably good.

Recent advancements in track technology such as the use of long welded and continuous welded rails (LWR and CWR), laying down of track tolerances for riding comfort, and improved track monitoring for an objective assessment of track condition have influenced the system of track maintenance.

The system of Directed Maintenance of Track is one such development; this essentially consists of improved supervision of track with emphasis on proper identification of defects, and undertaking the work of removal of these defects. This leads to better standard of track maintenance because sleepers and ballast that are well consolidated and in good conditions are not disturbed. Further, it also leads to greater economy.

*Directed Track Maintenance (DTM)* is similar to conventional maintenance except that in this system, the emphasis is on need-based maintenance rather than on routine cycle from one end to the other.

### Periodic Inspection and Need-based Maintenance

The following are the elements in this system:

(i) Expansion gap adjustment
(ii) De-stressing of LWR/CWR
(iii) Creep adjustment
(iv) Realignment of curves
(v) Cleaning of drains and drainage arrangements

The steps involved in the operations are:

- Location of defects by recording of track geometry
- Identification of the spots with defects needing immediate attention
- Recording of defects in an inspection book
- Rectification of defects
- Quality check by the DTM-supervisor, and recording of the work done.

**Location of defects:** It is desirable to have a continuous record of track geometry and riding characteristics such as unevenness, gauge, cross-level, twist and alignment.

Track recording car is used of this purpose and based on an analysis of the defects recorded, stretches needing immediate attention, those which need regular attention, and those which should be left without any disturbance are identified.

Index tolerances have been fixed for the various parameters of track geometry and defects are identified based on when they exceed the tolerance limit.

**Rectification of defects:** Adequate attention is given as needed to activities such as through packing, picking up slacks and through MSP at joints at the specific locations identified. The procedures are the same as in the conventional system.

**Quality check:** This is carried out by the DTM-Supervisor and the results are recorded in the Unit Inspection Book.

### Instruments Needed for DTM

(i) Nylon cord, 25 m long, for checking the alignment and unevenness
(ii) Stepped gauge (with an accuracy of 1 mm)
(iii) Gauge-cum-level
(iv) Canne-a-boule
(v) Steel scale, 15 cm long (reading to 1 mm)
(vi) A straight edge, 1 m long.
(vii) Track Recording-cum-curve corrector (TRCC)

The beat length is usually 6 to 8 km, and the beat strength is 15 to 20 gangmen. All items of work such as lubrication of fish plates, and patrolling during monsoon and heavy rains have to be carried out on a planned and regular basis.

**Future of DTM on Indian Railways:** DTM is considered to be the most rational and economical method of track maintenance. Hence, it is recommended for use on A, B and C routes. DTM is a kind of forerunner of the *track management system* in so far as it optimises track maintenance inputs in a need-based manner.

### Modern Track Management System

In directed maintenance of track, only that portion of the track where irregularities in track geometry exceed prescribed tolerances or index values is attended to for rectification. The remaining part of the track and the sleepers are left untouched. However, the process of decision-making is somewhat subjective, in spite of the index values of tolerances in track geometry.

*Modern Track Management System* deals with the track structure in totality. All inputs in track construction, monitoring, maintenance, and renewals where needed are analysed to arrive at the best cost-effective solution to provide a desired level of service. A logical extension of track maintenance is *track renewal*, which has to be justified as an economic alternative based on scientific analysis. This is possible only by quantification and attaching appropriate price tags to certain elements such as transit delays, inconvenience and discomfort to passengers, and safety hazards. Thus, direct and indirect costs are to be worked out for cost–benefit analysis.

### Basic Requirements of Track Management System

**(a) Sound track structure:** Strong formation, good drainage, and well-designed track components.

**(b) Data base:** All updated information about the track features—formation bearing power, ballast, depth and condition, drainage system, sleepers, track fittings, rails, alignment features like gradient and curves, LWR/CWR, rail joints, speeds, axle loads, and GMT-details is needed.

**(c) Objective track evaluation system:** Track-recording car or ultrasonic rail testing car is used to record all features of track geometry.

**(d) Fixation of threshold values for track parameters:** This needs to be done for each category of lines.

**(e) Track maintenance planning cell:** This should be equipped with computers to store, retrieve and analyse all the track data in order to fix the optimum programme of maintenance.

**(f) Track degradation models:** These models are developed for future planning of maintenance and renewal work. Upgrading track geometry and track structure to meet specific needs will be made possible by proper study of these models.

### Track Management System on Indian Railways

In view of the potential of this system in the optimal use of resources, this is being adopted by the India Railways in a phased manner. For this purpose, a special cell was created in the Centre for Railway Information System (CRIS).

The following are the *salient features*:

- Database
- Track monitoring system
- Threshold values
- Computer hardware and software

The output will be in the form of programme sheets.

**Advantages of TMS**

- Optimisation of material and labour inputs
- Optimum use of on-track machines
- Timely planning for replacement of track components and track renewal

## 8.5 TRACK STANDARDS AND REHABILITATION

There are wide variations in traffic density and speeds on the railway lines in India. For economy in construction and maintenance, it is necessary that the standards of the track should be commensurate with their needs. Track standards consist of guidelines for the type, strength and quality of the track components such as rails, sleepers, track fitting and fastenings, ballast, switches and crossings for different speeds and traffic densities.

**Considerations in the Choice of Track Structure**

- Traffic density
- Axle load
- Wheel diameter
- Permissible speeds
- Traffic mix pattern
- Track maintenance method
- Desired comfort level for passengers.

**Choice of rails:** The parameters for rails are—

- Type of rail
- Weight per metre
- UTS
- Hardness
- Length of rail from the rolling mill
- Method of welding for welded rails.

**Quantification of Advantages of Heavier and Stronger Rails**

- Life time GMT is proportional to $(UTS)^2$, $(wt/m)^3$, and $\dfrac{1}{(\text{Axle load})^3}$
- LWR/CWR reduces fuel cost by 5%
- Heavier rails reduce soil pressure, resulting in reduction of track maintenance cost.

**Choice of sleepers:** The materials used are wood, cast-iron, steel, RCC and PSC. The choice depends upon:

- Ease of maintenance (flat-bottom sleepers are good in this respect.)
- Traffic density
- Subgrade strength
- Maintenance method (concrete sleepers are better suited to mechanised system of track maintenance)
- Technical considerations (LWR is better maintained with concrete sleepers).
- Environmental considerations (Wood sleepers should be avoided to prevent deforestation.)

**Sleeper density:** This is important for track strength. The factors governing sleeper density are:

- Axle load
- Traffic density
- Strength of rail
- Type of sleeper
- Ballast cushion
- Nature and strength of formation
- Concrete sleepers are preferred for turnouts.

**Switches and crossings:** The factors governing the type of switches and crossings are:

- Speed on the straight and on the turnout
- Volume of traffic
- Tonnage of traffic

Cast Manganese Steel (CMS) crossings are preferred for high traffic density.

**Subgrade stresses:** Higher values of sleeper density and ballast cushion reduce subgrade stresses significantly.

Track structure for different routes on BG and on MG has been discussed in Chapter 3.

### Track Structures on Dedicated Freight Corridors (DFC's)

*Dedicated freight corridors (DFC's)* have been introduced on Indian Railways for fast movement of goods. These are lines designed to meet the specific requirements of specialised rolling stock moving at their designated speeds. These corridors can carry high amount of goods traffic without interference from slower passenger traffic; the added advantage is augmentation of passenger services without interference of goods traffic on routes earmarked for the former. The following are the parameters for *construction and track structure of dedicated freight corridors*:

| | |
|---|---|
| Gauge | 1676 mm |
| Maximum curvature | 2.5° |
| Curve compensation | 0.04% per degree of curvature |
| Ruling gradient | 1 in 200 (compensated) |
| Vertical clearance | 7.1 m |
| Centre-to-centre distance between DFC-tracks | 5.5 m |
| Centre-to-centre distance between existing track and DFC track | 6.0 m |
| Rails | 60 kg UIC/68 kg, 90 UTS |
| Sleepers | PSC, 1660 nos./km for main line and 1549 nos./km for loop line sidings |
| Ballast cushion | 300 mm |
| Points and crossings | 60 kg rails, 1 in 12 curved Switches (CMS) |

With an axle load of 32.5 t and a trailing load of 15,000 tonnes, DFC's come under the category of Heavy Haul Operation.

### Track Rehabilitation

In countries like India, resource constraint calls for a degraded track to be operated even if at a restricted speed, although the track is overdue for renewal. Optimal solutions may be got by proper appreciation of the basic concepts of track technology in such cases, consistent with safety of operation.

Safety can be ensured if continuity of the two rails, maintenance of gauge within acceptable tolerances and maintenance of vertical and horizontal alignments are ensured.

### General Run-down of the Track

This means that disturbance has occurred to the track geometry to a significant extent. Deterioration in cross and longitudinal levels occurs even with the passage of a few trains.

The remedy is often in ensuring proper drainage of track and a clean ballast cushion. Good sleeper density is another remedy.

**Defects in the rails:** These can be remedied by cropping rail-ends, changing them end to end, interchanging left and right rails, shifting worn-out rails from curves to straight sections, dehogging, and finally, imposing speed restrictions.

**Over-aged and worn-out sleepers:** The remedies depend on the material and type of sleepers. Shifting the rail seat and re-spiking it with the aid of wooden plugs can be the remedies for worn-out wooden sleepers.

Similarly, the remedies appropriate to steel and cast-iron sleepers can be the use of steel pads or plywood pads of appropriate size and of oversize keys and cotters, depending upon the particular type of defect observed. Deep screening and provision of clean ballast cushion helps the restoration of concrete sleepers.

### Upgrading of Track

While *rehabilitation* refers to the use of simple remedies to enhance the life of the track for its continued use for operating traffic even at restricted speeds, pending further and more expensive rectification measures such as track renewal, *upgrading* of a track refers to the improvement of the structures of an existing track to cater to the needs of faster and heavier traffic, or for any other special purpose.

Upgrading usually involves the replacement of rails by heavier and stronger ones, increasing the sleeper density, changing the sleepers to RCC or PSC sleepers, and increasing the ballast cushion.

(*Track renewal* is dealt with in the next section, *upgrading* is considered in the next chapter.)

## 8.6 TRACK RENEWAL

The impact of wheel loads from running trains causes heavy wear and tear of the track components. Even in the absence of wear and tear, the track may be unable to cope with the increasing quantum of traffic. In either of these cases, *track renewals* are carried out periodically to ensure the safety and efficiency of the track.

Track renewal consists of replacement of the primary track components—rails and sleepers, either separately or together. If the renewal is done using new material, it is said to be *primary renewal*; if it is done using released serviceable material, it is called *secondary renewal*, which is used only in the case of branch lines and other relatively unimportant sections.

Rails which have undergone wear may be reconditioned at a *rail re-profiling plant*; this involves straightening, ultrasonic testing, cutting off the defective portion, and re-profiling of the rail-head and gauge face.

Cast iron (CST-9) sleepers are reconditioned to restore the rail seat and the jaws to their original condition.

Such reconditioned rails and sleepers, and sleepers released during track upgrading which are serviceable are used for secondary renewals.

## Classification of Track Renewals

Track renewals may be classified as follows:

(a) Complete track renewal
(b) Through rail renewal
(c) Through sleeper renewal
(d) Casual renewals

**(a) Complete track renewal (CTR):** This involves the renewal of all the track components for a particular stretch. Recoupment of ballast and provision of the designed ballast cushion with clean ballast are also done along with *complete track renewal*.

**(b) Through rail renewal (TTR):** This consists of renewal of all the rails in a stretch.

**(c) Through sleeper renewal (TSR):** This entails the renewal of all sleepers in a given stretch. As in the case of complete track renewal, recoupment of ballast and provision of the full cushion with clean material are also done with TSR.

**(d) Casual renewals:** This implies the replacement of unserviceable rails, sleepers and fastenings with identical sections of serviceable or new track components. Such renewals are carried out in isolated locations and not in continuous stretches.

The aforesaid renewals may be considered as part of normal maintenance operations. All these categories may be primary or secondary as the case may be, depending on the need.

## Criteria for Track Renewal

These criteria are given below separately for the main track components—rails and sleepers.

### Criteria for Rail Renewal

In general, rails are renewed on service life-cum-condition basis. Since fatigue is the primary cause of rail failure, rails are often removed from the track to prevent fatigue failures long before they are worn out to an unserviceable stage.

In the absence of a reliable measure of cumulative fatigue damage before crack formation, the criterion for rail renewal is set at a certain GMT of traffic, or a predetermined amount of wear.

The following are the criteria for renewal of rails:

(i) Incidence of rail fracture
(ii) Rail wear consisting of

- Loss of section
- Wear due to corrosion

- Vertical wear
- Lateral wear

(iii) Maintenance of track to the prescribed standards
(iv) Shortened service life
(v) Plan-based renewal

**(i) Incidence of rail fracture:** A spate of rail fractures on a particular length indicates the necessity for rail renewal. Ultrasonic testing should be done, results analysed and rail renewal proposed only if the fractures are high—4% in a year or 10% overall. If the failure is at a joint, end-cropping with or without welding may be useful.

**(ii) Rail wear:** This may include any of the following:

*Loss of section*: Loss in weight should not exceed 6% for 52 kg rails and 5% for 90 R rails on BG; if the loss is more, it calls for rail renewal.

*Wear due to corrosion*: Corrosion is caused prominently in coastal areas. The criterion for rail renewal is corrosion of more than 1.5 mm in the web and the foot of the rail.

*Vertical wear*: Vertical wear causing a depth reduction of the rail head with the risk of wheel flanges grazing the fish plates is considered to be the limit calling for renewal. The limits are 13 mm for 60 kg rail, 8 mm for 52 kg rail and 6 mm for 90 R rail on BG.

*Lateral wear*: This is measured at a point 13–15 mm below the rail table. The limits of lateral wear beyond which renewal is called for are given below for BG:

*Straight portion*:
A and B routes ...6 mm
C and D routes ...8 mm

*Curves*:
A and B routes...8 mm
C and D routes... 10 mm

**(iii) Maintenance of track of the prescribed standards** Rail renewals may become necessary due to poor running quality of the track in spite of extra maintenance, high cost of maintenance for ensuring safety, and poor condition of the rail due to hogging, battering, wheel burn and corrugation.

**(iv) Shortened service life:** Renewal is planned when the rail has carried the following minimum traffic (on BG):

For 60 kg/m rail (72 UTS)....550 GMT
For 52 kg/m rail (72 UTS)....350 GMT
Since the rails for MG are lighter, the minimum traffic for renewal is lesser.

**(v) Plan-based renewals:** These are aimed at modernising the track structure on selected routes in the shortest possible time. It may sometimes involve premature renewals.

### Criteria for Sleeper Renewal

The presence of about 30 to 35% of unserviceable sleepers in the track justifies through sleeper renewal in any stretch. Of the released sleepers, those which are serviceable are used for casual renewals elsewhere.

**Wooden sleepers:** These become unserviceable when they are split and broken with deeply cut rail seats. Deterioration due to insect attack or respiking calls for renewal.

**Steel sleepers:** These become unserviceable due to excessive corrosion and when the lugs/rail seats get cracked. Their life is considered to be 40 to 50 years.

**Cast iron sleepers:** These become unserviceable when the tie bar and the rail seat get excessively corroded, when the plates get cracked, or when the lugs/rail seats get excessively worn out.

The average life of a CI sleeper is about 40 to 50 years. Inability to hold the gauge or to retain the packing of the ballast can be considered to be the criteria for loss of serviceability.

**Concrete sleepers:** RCC sleepers and PSC sleepers are stronger than the other types owing to their weight, and are more durable. They may need renewal only in the case of upgradation of the track for higher speeds and heavier loads.

## Execution of Track Renewal or Track Relaying

This may be done either manually or by a mechanised method.

### Preliminary Work for Track Renewal

(i) Unloading track materials
(ii) Ballasting the section
(iii) Fixing level and centre-line
(iv) Arranging traffic blocks.

**(i) Unloading track materials:** Track materials such as rails, sleepers and fastenings are to be brought and unloaded at the location where they are to be laid. Fastenings/fittings being small are likely to be lost; so they are stored in depots near the site. Rail dolly is used for easy transportation of the rails.

**(ii) Ballasting the section:** Ballast is brought in ballast trains, deep screened in advance, and unloaded at the site at particular locations on either side of the track.

**(iii) Fixing level and centre line:** Centre-line and level pegs are provided at 50 m intervals.

**(iv) Arranging traffic blocks:** This should be finalised in consultation with all concerned regarding the operation.

### Manual Method of Track Renewal

When the preliminary work is completed, the actual job of track renewal is undertaken. The following are the steps in the process:

(a) A traffic block is set up for two to four hours to facilitate the work of relaying.
(b) The old or existing track is dismantled and the material stacked on the side opposite to the new material.
(c) Ballast should be dressed and levelled to the correct profile.
(d) The sleepers are laid in position, and the rails are linked using proper expansion liner.
(e) Packing is done correctly, levelled, and the track aligned.

(About 100 m of track may be laid in one hour by 20 gangmen.)

### Mechanised Track Renewal

Modern tracks are laid using heavy concrete sleepers, which cannot be easily handled by using manual labour. Rough handling may also damage the sleepers since they are quite brittle. Mechanised track renewal/relaying results in fast progress of work and helps to maintain a good quality track, and is now preferred on Indian Railways for all important routes.

The method may be semi-mechanised using prefabricated panels, in which case the main job of relaying and linking is done with the help of *Plasser Quick Relaying System (PQRS)*; it may be fully mechanised using *Track Relaying Trains (TRT)*.

### Semi-mechanised Method using Prefabricated Panels

Mechanical re-laying is mostly carried out with the help of PQRS equipment manufactured by an Austrian company, plasser and Theurer, in a factory in Faridabad, India.

This equipment consists of two self-propelled portal cranes, with a capacity of 5 to 9 t, which run on an auxiliary track with a gauge of 3.4 m. Rails are fitted to concrete sleepers with fastenings to make prefabricated panels in a base workshop. These panels are loaded in special BFR-wagons and transported to the site along with portal cranes. (BFR stands for: B – bogie wagon, F – flat car, and R – rapid or forced discharge). After the old track is dismantled and the preparatory work is completed, the new panels are laid with the aid of portal cranes. A fully equipped PQRS-rake along with the portal cranes is deployed for the track-relaying.

The work of dismantling old track and relaying with new prefabricated panels is carried out taking up two panels at a time; the panels taken out are loaded on empty BFR-wagons in the PQRS-rake.

The preparation of the ballast bed is done manually. The process is repeated a number of times with two panels at a time, the new panels being connected or linked-up with track fittings and fastenings. A crane of 9 t capacity is needed to lift a 13 m-rail panel with concrete sleepers. The output in this semi-mechanised method can be around 200 m per hour, or 800 m in a four-hour block. This method considered suited to Indian conditions.

### Fully Mechanised Method with Track Renewal Trains

Track renewal trains of modern design can carry out track-relaying automatically with minimum manual work. Indian Railways has purchased a few TRT's from Tamper Corporation of the USA.

A TRT is designed to perform all the operations of replacement of rails and sleepers simultaneously and continuously. Removal of old rails and sleepers, stocking, levelling and compacting of ballast bed, placing new sleepers in their correct positions, laying of new rails and removal of released rails are all done sequentially. There are separate units for dealing with rails, sleepers and ballast in the main vehicle. A gantry of 5 t capacity and a set of BFR wagons also of the same capacity form part of a TRT. The main vehicle weighs about 110 t and is 45 m long. A maximum rate of progress of 720 m per hour can be achieved using this fully mechanised method with a TRT. Speed restrictions are imposed during track renewals in a planned manner.

### Diversions

*Diversions* are set up at such locations in a track where there is some kind of obstruction or where a construction work is in progress.

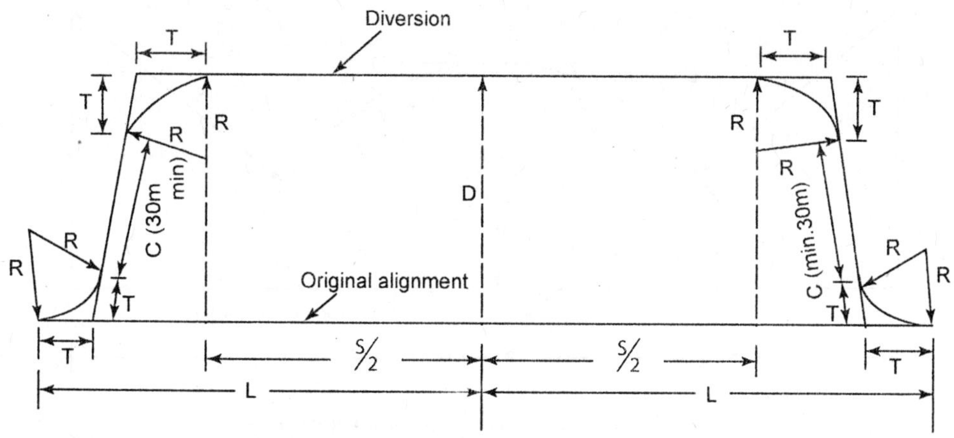


**FIG. 8.17** *A typical semi-permanent diversions*

Diversions may be temporary for not more than 10 days, or may be semi-permanent as in the case of reconstruction of lines, bridges and such major engineering work needing a diversion for a much longer duration. At a temporary diversion, a train must stop and proceed at low speed; at a semi-permanent diversion, trains may proceed non-stop at reduced speed. The plan of a typical semi-permanent diversion is shown in Fig. 8.17.

The formulae used for setting out this diversion are:

$$L = \sqrt{C^2 + 4RD - D^2} + \frac{S}{2} \tag{8.5}$$

$$T = \frac{RD}{\left(L - \frac{S}{2} + C\right)} \tag{8.6}$$

Here,

L = Length of half the diversion measured along the original alignment

C = Straight length between the circular arcs of reverse curvature connecting the original alignment to the diversion

R = Radius of the circular arcs

D = Maximum distances of the diversion from the original alignment

S = Straight portion of the diversion parallel to the original alignment

T = Tangent length of the circular arcs.

**Note:** The radius should not be less than 450 m for BG. The gradient should not be steeper than 1 in 100.

If the whole diversion consists of two sets of reverse curves without any straight portions, it appears as shown in Fig. 8.18.

For such a layout,

$$L = \sqrt{D(4R - D)} \tag{8.7}$$

(**Note:** This layout is generally not preferred because of too many reverse curves which restrict the speed)

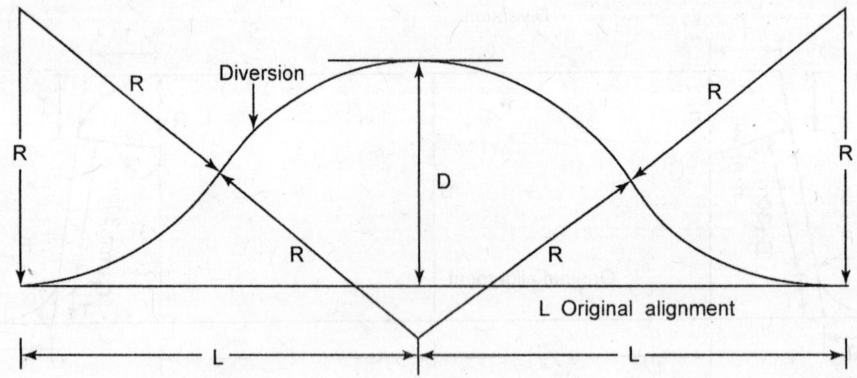


**FIG. 8.18** *Diversions without straight portions*

## SUMMARY

- 'Plate laying' means laying out sleepers and rails on a prepared formation and connecting them correctly to produce the railway track, which is ballasted.
- The methods of plate laying are tram line method, American method and telescopic method; the last one is popularly used on Indian Railways. These are manual methods. In the telescopic method, the track already constructed is used for transporting materials to the plate laying site. Labour is divided into three groups—material gangs, linking-in gangs and packing gangs; their specific duties are handling and placing the materials where needed, linking the sleepers and rails by means of fastenings, and placing and packing of ballast.
- Modern method of track construction involves the use of modern surveying methods such as satellite imagery and aerial photographs, modern geo-technology for subgrade/formation preparation and mechanised construction using machines. Mechanised construction may involve the use of portable track equipment for different tasks; Robel's units for sleeper-laying, Robel's rail mover for rail handling, and Robel's machine for levelling and alignment of the track. Plasser's tamping machine is used for tamping ballast and ballast profiling machine for dressing of the section. Alternatively, plasser and Theurer track-laying machine may be used for mechanising most of the tasks in track construction.
- Track maintenance involves the maintenance of gauge, longitudinal and cross-levels, ballast section, rails and rail joints, sleepers and drainage structures such as drains and cross-drainage works.
- Conventional manual method of maintenance involves the use of equipment and tools such as rail gauge and gauge-cum-level for checking the gauge and track levels, Canne-a-boule for looseness of packing, sleepers and rails tongs for handling sleepers and rails, beater or pick-axe for packing ballast, Jim-crow for bending/breaking of rails, keying hammer for tightening keys, wire claw for taking out ballast and phowrah (shovel) for handling and placing ballast.
- An annual maintenance cycle involves through packing by opening up and packing all sleepers, overhauling through shallow screening of ballast and making up the ballast section, and picking up slacks at places where poor riding is noticed due to looseness of ballast.
- Maintenance of rails involves creep adjustment and buckling adjustment, in addition to restoration of surface. maintenance of rail joints involves rectification of hogging, high

joints, blowing joints, pumping joints—most of which involve the adjustment of ballast packing. Rail joints are lubricated at a predetermined frequency. Work with sleepers involves squaring and picking up slacks; maintenance of level crossings and track-circuited sections need special attention. An annual cyclic maintenance programme aims at the protection of track structure for safe and efficient operation.

- Gang strength required to maintain a standard beat length is determined based on a scientific and rational approach; such a method has been developed by the Indian Railways. Track patrolling is carried out on a regular basis; further, special patrolling during heavy rains and monsoon periods is also carried out.
- Track recording cars are used to identify defects and their location in the track for rectification later.
- Modern method of track maintenance involves a mechanised approach using off-track and on-track tampers.
- Measured shovel packing (MSP) is the process of packing only disturbed and loosened sleepers, without touching well-settled sleepers. MSP needs special implements such as dansometer and fleximeter, as also cann-a-boule for assessing the loosening of the ballast under the sleepers. Based on the result, the ballast feed required for packing is calculated and used.
- Direct Track Maintenance (DTM) is a 'need-based' approach; defects in track are located and rectified. This is the precursor to the scientific method—Modern Track Management system.
- Track standards with respect to the components are evolved and used on Indian Railways. Track rehabilitation involves attending to worn out components to extend the service life pending track renewal. Dedicated freight corridors are being developed for fast and efficient transportation of goods traffic.
- Track renewal may be primary, using new components, or secondary, using released serviceable components. This may be complete track renewal, through rail or sleeper renewal in a given stretch, or casual renewal at stray locations where defects are noticed. Criteria for renewal are fixed on the basis of field research by the RDSO as the values of rail wear and sleeper defects. Manual method of track renewal may be carried under short traffic blocks of two hours duration.
- Mechanised track renewal involves the use of plasser quick relaying system (PQRS), which is semi-mechanised, or Track relaying Train (TRT) which is fully mechanised.
- Diversions are required when engineering construction goes on, which may be temporary (short duration) or semi-permanent (longer duration). work is carried out under speed restriction, coupled with appropriate temporary signals.

## QUESTIONS

1. Define 'plate laying' and explain the telescopic method used on Indian Railways.
2. Describe briefly the various steps involved in the construction of a new railway line.
3. Write brief notes on:

   (a) Doubling of a railway line

   (b) Gauge conversion work

4. Write short notes on:
   (a) Ballast trains
   (b) American method of plate laying
   (c) Different gangs in track construction
5. Write brief critical notes on:
   (a) Fixing and packing sleepers on ballasted bed
   (b) Gaps in fish-plated joints
   (c) Materials per kilometre of track
6. Why does a railway track need more frequent attention regarding maintenance than other civil engineering structures? Explain briefly.
7. What are the advantages of good track maintenance?
8. Differentiate between:
   (a) Blowing and pumping joint
   (b) Rail tong and sleeper tong
   (c) Off-track tamper and on-track tamper
9. Write notes on:
   (a) Squaring of sleepers
   (b) Longitudinal alignment of track
   (c) Lateral alignment of track
10. Write brief notes on:
    (a) Track rehabilitation
    (b) Track patrolling
    (c) Deep screening of ballast
11. Write a note on 'through packing' of a railway track.
12. Write a note on annual track maintenance programme by the conventional manual method on Indian Railways.
13. Differentiate between:
    (a) High joint and hogged joint
    (b) Insulated joint and fish-plated joints
    (c) Lifting and lowering of track
14. Write a note on maintenance of track-circulated sections.
15. Write a note on the method of fixing the gang strength.
16. Write a note on Amsler Track recording Car.
17. Write a note on measured shovel packing and its advantages in track maintenance.

18. Write a note on 'diversions,' differentiating between temporary and semi-permanent diversions.
19. Write notes on:
    (a) Classification of track renewals
    (b) Speed restrictions
    (c) Track relaying train
20. Explain the telescopic method of laying a new track. (Kalinga Institute of Technology, Orissa)

Chapter 9

# MODERNISATION OF RAILWAYS AND URBAN RAILWAY SYSTEMS

**CHAPTER OBJECTIVES**

This chapter analyses:

- The methods of track modernisation for high-speed trains such as improving the track structure and track geometry
- The use of tilting train technology, tracked air cushion vehicle (TACV) and MAGLEV trains for super-high speeds on ballastless tracks.
- The use of dedicated freight corridors for fast movement of goods
- Urban railway systems such as mass rapid transit (MRT) systems, their principles and operation using case studies of the operation of a few such systems in Indian Metros.
- The significance of tunnelling for railways and the methods used to bore through hard rock and soft ground, including modern methods using the shield and tunnel boring machines
- The various aspects of tunnel maintenance such as ventilation, lighting and drainage.

## 9.1 INTRODUCTION

The objective of modernisation of railways is to facilitate the operation of heavier trains at relatively higher speeds safely, in addition to providing better customer service to all categories of users.

Modernisation consists of not only upgrading the track structure, but also improving the design of rolling stock, use of a superior mode of traction, better signalling and interlocking systems, and the adoption of modern techniques in all the other areas of operation of railways.

Modernisation of track structure and maintenance involves the use of:

- Heavier rail sections
- Long welded rails and switch-expansion joints
- Pre-stressed concrete sleepers
- Elastic fastening of superior design
- Curved switches of 1 in 16 and 1 in 20 size at points and crossings

- Modern and mechanised methods of track maintenance (MSP, DTM and TMS)
- Modern track monitoring systems (Amsler track recording car) to aid track maintenance.

Modernisation also includes better-designed, all-coiled, anti-telescopic coaches with better braking systems, diesel and electric traction, modern signalling and interlocking systems to ensure safety, and computerisation of all the aspects of operation including customer service pre-travel, during travel, and post-travel.

Urban railway systems include Mass Rapid Transport Systems, Suburban Railway Systems, Underground Railways, and Metro Railways—both elevated systems and tube Railways.

Recent developments in railway networking are container transport and dedicated freight corridors for faster movement of goods traffic, increase in axle loads and ballastless track under special locations.

Tunnels may be required for railways, depending upon the terrain for short distances, or for longer stretches for tube railways; this needs the development of modern techniques of tunnelling with specific reference to railways.

## 9.2 TRACK MODERNISATION FOR HIGH SPEEDS

Reduction in journey time is not the only advantage of high-speed trains; the following are the additional ones:

- Greater frequency of service
- Operation of trains at predetermined regular intervals
- Complementing other modes of transport for maximum efficiency in the transportation sector as a whole
- Better passenger service both in the stations and in the trains
- A superior level of comfort and safety
- Environmental friendliness and reduction in noise pollution.

### Effect of High Speeds on Track Structure

On Indian Railways, permissible speeds of 120 km/h or more are considered to be high speeds; the present speeds on the BG trunk routes on which Rajdhani Expresses are operated are 120 to 140 km/h.

Investigations carried out on these routes have revealed that increased speeds do not necessarily cause increases in the stress and deformation of the track components which necessitates the use of a heavier track structure. However, such increase in stress and deformation is caused by the incongruous movement of vehicles, including pitching, bouncing, rolling, and lateral oscillations caused by repeated applications of wheel loads. It is well known that these phenomena are a result of poor maintenance of track.

It is, therefore, logical to understand that the same vehicles can be operated at a higher speed with a given track structure without any increase in track stresses, provided the standards of maintenance of the track and the rolling stock are improved so as to minimise the distorted movements referred to above.

Thus, in order to achieve speeds of the order of 160 to 200 km/h, the standard of track maintenance required is much higher, as to keep the track parameters to very close tolerances. If this is to be achieved with the existing track structure, it may prove to be uneconomical. A better option would be to adopt an improved track structure, which can be maintained at closer tolerance limits at a relatively low cost.

The modern track structure, consisting of long-welded rails on concrete sleepers with elastic fastenings, and ballastless tracks fulfils this condition; although the initial cost of such a track is comparatively high, its maintenance cost is low.

## Technologies for High-speed Train Operation

The following are the two distinct options for the operation of high-speed trains:

**(a) Upgradation or improvement:** The existing system of operation and infrastructure can be upgraded to make them suitable for speeds of the order of 200 km/h or more. A well-designed and heavier track structure, well-designed and properly maintained rolling stock, and appropriate signalling, interlocking and train control systems are elements of upgradation.

**(b) Construction of dedicated high-speed corridors:** On high-speed corridors, appropriate high-tech infrastructure and appropriately selected track construction and maintenance parameters to the desired standards are the prerequisites for safe and efficient operation of trains at designated speeds of 250 km/h and above. The International Union of Railways (UIC) defines a high-speed train as one that runs at 200 km/h on conventional upgraded tracks or at over 250 km/h on dedicated high-speed corridors for a significant part of the journey.

### Upgradation of the Existing Railway System and Infrastructure for High-speed Lines

A *sound infrastructure for high-speed lines* involves the following components:

**(i) Track structure:** Specially designed for high-speed with consolidated subgrade and appropriate track geometry.

**(ii) Rolling stock:** Diesel and electric locomotives designed to attain high speeds, and specially designed coaches and wagons.

**(iii) Signalling and interlocking:** Modern and advanced techniques of signalling and interlocking.

**(iv) Major cross-drainage works such as bridges:** Properly-designed bridges for high speeds.

**(v) Grade-separations and fencing:** All level crossings must be replaced by grade separations—underbridges/overbridges.

**(vi) Environmental protection measures:** Adequate precautions to ensure that the environment is not degraded during the laying of high-speed lines.

**(vii) Track structure and track geometry for high-speed lines**

*Rails*: 60 kg/m rails; LWR/CWR is used to improve riding quality and reduce vibrations and noise.

*Sleepers*:Pre-stressed concrete sleepers; sleeper density of 1660.

*Fastenings*: Double elastic fastenings; rubber pads between the rail and sleepers, fastened by leaf spring/wire spring (TGV Nabla /ICE Vossloh fittings are used on high speed lines of advanced countries such as France and Japan.)

*Curves*: Flat curves must be adopted. Minimum radius varies from 4000 m to 7000 m.

*Cant*: Maximum 180 mm

***Cant*** deficiency: Maximum 100 mm
***Grade separations***: Overbridges and underbridges in place of level crossings.
***Fencing***: Required to enhance safety and avoid trespassing.

### Tilting Train Technology

The maximum permissible speed on railways has all along been governed by the maximum cant (or superelevation) and the cant deficiency values permissible for the track.

With the advent of bogie tilting technology developed by a number of firms, higher speeds than those possible until now on horizontal curves could be permitted for a given cant without reducing the comfort to the passengers. This has facilitated the operation of high-speed trains without the need for speed reduction on curves.

Tilting of bogies is achieved by the use of hydraulic/pneumatic/electrical power; the vehicles get tilted as they negotiate a curve depending upon the degree of curvature. With a cant of 180 mm and a cant deficiency of 275 mm for Indian Railways, a significant reduction in the radius of curves is possible along with an increase of about 30% in the maximum permissible speed when compared with the conventional non-tilting trains.(Tilting technology is normally used for speeds over 350 km/h.)

### Dedicated High-speed Corridors

Exclusive corridors have been designed and constructed for operating high-speed trains. The construction parameters which need special attention for such corridors, in order to ensure smooth, safe and efficient operation at the designated speed, are given below:

**(i) Horizontal curves:** Easiest possible curves in terms of their radius, cant and cant deficiency are to be provided.

**(ii) Vertical curves:** Gentler vertical curves are preferred on high-speed lines than on conventional lines in order to ensure better comfort to passengers.

**(iii) Ruling gradient:** steeper gradients can be provided on high-speed lines in view of greater tractive power and smaller loads of high-speed trains.

**(iv) Spacing of tracks:** Greater centre-to-centre spacing between lines on double-line sections is preferred to deal with the higher air-pressure generated when trains cross each other.

**(v) Track structure:** The subgrade must be more stable and unyielding. Embankments and cuttings must be carefully designed with suitably graded strong granular material for ballasted tracks. Track drainage becomes very important. *Ballastless tacks* are now being preferred on high-speed lines (Refer Chapter 3).

## Track Structure for High-speed Lines on Indian Railways

Since ballastless track structure has several advantages over the conventional ballasted tracks, the adoption of the former on high-speed lines is inevitable. The track structure and track geometry recommended for adoption on Indian Railways for speeds in the range of 300 to 350 km/h, for dedicated high-speed corridors are given in Table 9.1.

**Table 9.1** Recommended track structure for high-speed lines on Indian Railways

| **Track geometry** | |
|---|---|
| Radius of horizontal curves | 10,000 m |
| Maximum cant | 180 mm |
| Cant deficiency | 100 mm |
| Minimum radius of vertical curves | 24,000 m |
| Maximum longitudinal gradient | 1 in 1000 |
| Spacing of tracks on double lines | 5.35 m |
| **Track structure** | |
| ***Rails***: 60 kg UIC 90 UTS rails, long-welded to 480 m lengths which are converted to continuously welded rails (CWR), for a safe and durable track structure | |
| ***Sleepers***: Pre-stressed concrete sleepers at 60 cm c/c spacing (around 1660 sleepers per km). | |
| ***Points and crossings***: High-speed turnouts of 1 in 16/1 in 20 size, with switch expansion joints | |
| ***Fastenings***: Elastic fastenings with rubber pads and leaf springs. Joints, if found inevitable at certain locations, shall be glued and insulated. | |

### 'Shatabdi' Trains on Indian Railways

These are high-speed intercity trains which run at a speed of 140 km/h. The *main features* are:

- Vestibuled air-conditioned coaches, mainly chair cars
- Hauled by electric locomotives with twin-pipe air-brake system.
- Track structure as on Rajdhani routes except that 90R rails and CST-9 sleepers are not permitted
- Tolerances for track maintenance are the same as those for Rajdhani routes.
- Tolerances regarding lateral and vertical oscillations are same as those for Rajdhani routes.
- Total visibility distance of 1800 m between the distant and home signals; if such distance is not available, a speed restriction of 120 km/h is imposed.
- Automatic warning signals (AWS) located 800 m ahead of all the distant signals.
- All level crossings are manned and provided with telecommunication facilities.

## Superhigh-speed Tracks and Trains

The *limitations* of conventional railway tracks in developing high and super high-speeds for the vehicles with steel wheels plying over them are:

(i) Air resistance through wave formation
(ii) Adhesion and friction between the wheels and the rails
(iii) Oscillations and vibrations of the railway vehicles at high speeds (including 'hunting' or self-excited vibrations of the vehicle), jeopardising riding comfort and even safety of the trains.
(iv) Centrifugal effects and cant requirements on a curved track which increase with the square of the speed and the degree of curvature (cant cannot be increased to offset the centrifugal force beyond a limit, as it affects the stability of the vehicle.)

### Concepts for the Development of Super-high Speeds

Two of the more important concepts evolved for the development of super-high speeds for trains are:

1. Tracked air-cushion vehicle—(TACV)
2. Magnetic levitation—(MAGLEV)

These two are briefly described here.

**1. Tracked air-cushion vehicle (TACV):** This is a type of high-speed train that replaces conventional steel wheels with *hovercraft* lift pads, and the conventional railway bed with a paved road-like surface known as the *guideway*. This aims to eliminate rolling resistance and provide superior performance. In the UK, they are known as *tracked hovercraft* or *hovertrains;* in France, they are called *aerotrains* and in the US they are *tracked air-cushion vehicles* (TACV).

At high speeds, trains running on conventional tracks suffer from a form of instability known as *hunting oscillation*, forcing the wheel flanges to hit the sides of the rails as on a sharp curve. At speeds of over 200 km/h, these hits form a major drag, increasing the rolling resistance and potentially causing a derailment. This means that above such speeds, a hovercraft with air cushion would be more efficient. The load in this case will be spread out on the underside of the vehicle, the pressure being greatly reduced compared to that on a train wheel. Hovertrains could be supported on surfaces similar to light-duty roadways, instead of the much more complex and expensive conventional railway tracks.

Although gas turbines and jet engines were tried, *linear induction motors (LIM)* were found to be much more efficient. TACV envisioned a linear induction motor-powered hovertrain with a speed of 480 km/h

LIM, in its simplest form, consists of an active portion on the vehicle corresponding to the windings on a conventional motor, and a metal plate on the tracks acting as a stator. When the windings are energised, the magnetic field they produce causes an opposite field to be induced in the plate. There is a small delay between the original field and the induced field due to *hysteresis*; thus, the fields in the windings and the 'reaction rail' will be slightly offset due to the hysteresis. This offset results in a net thrust along the reaction rail, allowing the LIM to pull itself along the rail without any physical contact; thus, with no moving parts and no physical contact, this system needs little maintenance and facilitates the development of super-high speeds.

Different shapes of guideways such as the inverted 'T', 'L', 'U', and box-shaped have been tried in different countries

*Aerotrains* offer many *advantages*:

- Increased safety
- Better comfort
- No noise
- No vibrations
- Excellent braking
- Savings in track structure and maintenance.

**2. Magnetic levitation trains (MAGLEV):** The technology of MAGLEV trains is as follows: When an electric current is passed through propulsion coils on the ground, a magnetic field is produced. The train is propelled forward by the attraction of unlike poles and the repulsion of like poles of the magnetic field acting between the ground coils and the superconducting magnets built into the vehicles. This is based on the linear motor concept. The guideways and the energisation system of a MAGLEV train are shown in Fig. 9.1.

A simple schematic/ general view of a MAGLEV train is shown in Fig. 9.2.

MAGLEV trains are similar to aeroplanes; they are provided with retractable landing wheels and horizontal guide wheels which are necessary when the speed comes below 200 km/h, the minimum speed for levitation. The three phase coils forming the starter of the linear motor are installed in the side walls of the guideways. The engine design is aerodynamic.

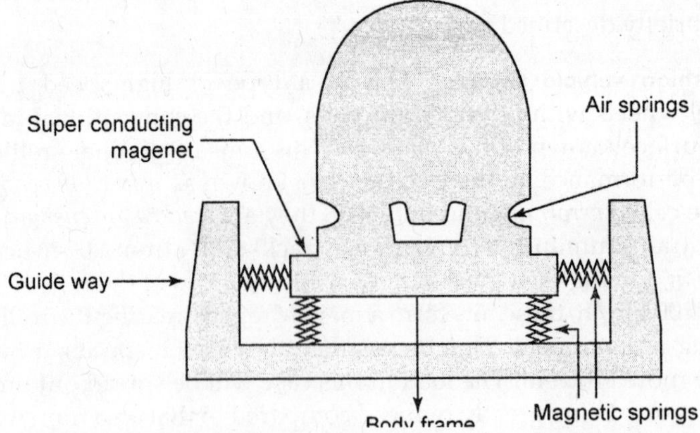


**FIG. 9.1** *Schematic of a MAGLEV train and energisation system*

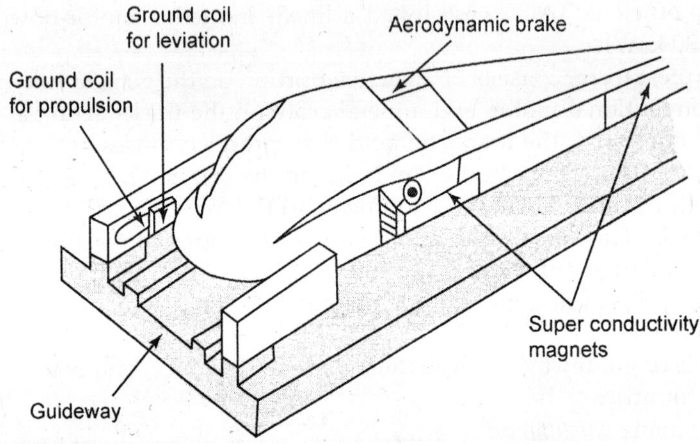


**FIG. 9.2** *General view of a MAGLEV train on guideways*

The speed is controlled by varying the frequency of power passing through the coils. Overhead platforms are built for passengers to board the train, the entry being from the top. The speeds range from 400 to 500 km/h. there are two primary types of MAGLEV technology:

***Electromagnetic suspension (EMS)***, which uses the attraction force of a magnet beneath a rail to lift the train up.

***Electrodynamic suspension (EDS)***, which uses the repulsion force between two magnetic fields to push the train away from the rail.

A German monorail system – *TRANSRAPID* – is based on EMS technology; in this, no wheels are necessary. [1]*Japanese MAGLEV system between Tokyo and Osaka* is based on EDS system; the vehicles require wheels for travel at speeds less than 200 km/h. *china* has adopted transrapid MAGLEV trains in Shanghai to its Pudong international airport. The 30-km stretch is covered in 7 m 20s at an average speed of 250 km/h and a top speed of 431 km/h.

[1] The height of levitation is normally about 100 mm.

### Super MAGLEV Trains

The concept envisages a MAGLEV train running in a vacuum tube without any aerodynamic resistance. The power to levitate as well as to propel the train is provided by superconducting magnets; this technology minimises the power requirements.

The combination of the concepts of superconductivity and vacuum tube is expected to propel MAGLEV trains at as high a speed as 3700 km/h. The dedicated tube corridors for this could be assembled at a depth of 50 m.

Thus, in about three to four decades, supersonic speeds for MAGLEV trains appear probable.

### High-speed Rail (HSR) in India

The Government of India has proposed to build the 'Diamond Quadrilateral project,' which would connect via high-speed rail, the cities of Chennai, Delhi, Kolkata and Mumbai.

Mumbai Rajdhani Express will be upgraded to run at 200 km/h; 'Vision 2020' envisages the implementation of regional high-speed rail corridors for services at 250–350 km/h, High-speed Rail Corporation of India (HSRC), a subsidiary of Rail Vikas Nigam Ltd (RVNL), will be entrusted the job of implementation of these corridors on Design, Build, Finance, Operate and Transfer (DBFOT) basis as Public–Private Partership (PPP).

The Government of India has recently an MOU with Japan for the implementation of high-speed lines to run MAGLEV trains.

## Ballastless Track

With the conventional ballasted track structure, the speed is normally limited to 200–250 km/h.

For higher speeds than this, the construction of a *ballastless track*, in which the rails are directly fastened to the concrete slab using elastic fastenings, is considered necessary. Ballastless track[2] is being adopted for metro rail systems as well as high-speed tracks all over the world. In India, Delhi Metro and Calcutta (Kolkata) Metro systems are built on ballastless tracks.

The initial cost of a ballastless track is very high but it may not need any maintenance during its life time.

The *advantages* of adopting a *ballastless track* for high-speed lines are:

- Practically no maintenance requirements
- Long service life
- High resistance to lateral loads, permitting higher speeds with the adoption of tilting train technology
- Environment friendly due to the absence of dust
- Economy in construction, especially in tunnels and on viaducts.
- Monsoon maintenance which is difficult in ballasted track is avoided.

There are several types of ballastless track assemblies that are used on high-speed lines and at special locations. Two types of assemblies, which are used in the Kolkata Metro and Delhi Metro, are described here:

### Plinth-type Ballastless Track Assembly

This assembly consists of:

- Elastic rail pad, 6 mm-thick

---

[2] Ballastless track is used on girder bridges, washable aprons and near washing platforms; even through sleepers are avoided in such locations.

- Fittings consisting of elastic rail clips, high tensile-steel bolts screwed into polythene inserts, triple coil spring washer, insulating washers—eccentric for lateral adjustment.
- Cast-iron base plate for load distribution
- An elastomeric pad, 12 mm thick, to function as ballast
- Steel plates for vertical adjustment.

This indigenous assembly has been used in the Kolkata Metro with satisfactory results.

**Vossloh 336 Ballastless Track Assembly**

This assembly consists of:

- Elastic rail pad, 6 mm-thick
- 10mm-thick elastic base plate pad and 5 mm-thick plastic pad.
- Malleable CI plate for load distribution
- Fittings consisting of Vossloh rail clip, HT steel anchors set in concrete, triple coil spring washers and eccentric insulating washers.

This assembly has been used in Delhi Metro, with excellent results.

An improved version is the *Logwell ballastless track assembly (LM-1)* consisting of base plate pad, 10 mm thick and plastic pad, 5 mm thick, fastened by six clips, and anchored strongly to concrete.

## 9.3 RECENT DEVELOPMENT IN INDIAN RAILWAYS AND NETWORKING

Recent developments in Indian Railways have been towards modernisation of the services in respect of passenger as well as goods traffic. Further, certain services and projects are planned for networking with other modes of transport such as road transport and sea ports for ships.

### Dedicated Freight Corridors (DFC)

In view of increasing demand of freight traffic, Indian Railways has planned to construct new double lines dedicated to operate goods trains with higher axle load and special type of wagons to carry the increased freight traffic. Since these lines are meant only for goods trains with no interference from passenger traffic, higher speeds of operation are possible, resulting in quicker movement and better utilisation of the track capacity.

Dedicated freight corridors are planned for the entire Golden Quadrilateral and its two diagonals, thus making the existing system a passenger corridor. (The approximate route kilometerage is 11,500 km or 23,000 km of lines.)

The axle loads are to be increased to 30-t as against the existing 22.9 t.

In the first phase, the Eastern and the Western corridors – 1805 km and 1490 km, respectively –, has been taken up and is expected to be completed by 2016 and 2017, respectively.

**Advantages of Dedicated Freight Corridors**

- Creation of infrastructure for carrying higher quantum of freight traffic
- Higher productivity by speeding up freight traffic and better utilisation of track capacity
- Increased throughput of freight movement by higher axle loads and improved pay load/tare ratio of wagons
- Reduced cost of goods transport resulting in greater customer satisfaction
- Relief for the existing routes from operating additional passenger traffic.

**Track structure and track geometry for DFC's:** The details of the structure and geometry of the track for DFCs have been covered at length in Chapter 8.

### New or Improved Goods Wagons

The existing goods wagons mostly have a value of 1:2.7 for the tare-to-payload ratio; this is sought to be improved to 1:4 by using higher steel alloys to increase the carrying capacity of wagons for a given axle load.

### Standards of Track Maintenance on DFC

(a) Fully mechanised maintenance of the track structure and geometry to minimise manual labour
(b) Through tamping of track by Plasser UNIMAT machines
(c) Rectification of isolated track defects using portable off-track tampers
(d) Use of Ballast Cleaning Machine (Plasser's) at least once in ten years.

In addition, mobile maintenance gangs for picking slacks and monitoring gangs to identify defects are also deployed.

### Basic Design Features of DFC

| S. No | Parameter | Specification |
|---|---|---|
| 1 | Length of train | 1500 m (max.) |
| 2 | Moving dimensions (H × B) | 7.10 m × 3.66 m (Western corridor) |
| | | 5.10 m × 3.66 m (Eastern corridor) |
| 3 | Train load | 15,000 t |
| 4 | Axle load | 32.5 t |
| 5 | Traction | Electrical (2 × 25 kV) |
| 6 | Maximum speed | 100 km/h |
| 7 | Gradient | 1 in 200 |
| 8 | Curvature | Up to 2.5° |
| 9 | Signalling | Automatic with 2 km spacing |

## Container Transport Service

Rail-cum-road transport services using containers for loading the freight, have been introduced on Indian Railways to attract high-rated freight traffic. The containers in the form of open or closed boxes of different sizes and designs are made of steel, aluminium or stainless steel. They can be handled by means of cranes and transferred from rail wagons to road trucks, ships or aircraft, and vice-versa. In the case of highway, door delivery and collection services are possible, and prove economical for long hauls, when compared to road transport by trucks.

The present capacity of containers on Indian Railways is only 5 t; thus, there is a need to increase it to at least 20 t for export and import to be compatible with international standards.

## Konkan Railway

Konkan railway line runs along the west coast of India from Mangalore in Karnataka to Roha in Maharashtra through Goa and the Western ghats. It was built by the Konkan Railway

Corporation. The 760 km line connects Maharashtra, Goa and Karnataka. In terms of the magnitude, technical challenges and the variety of structures in the hilly terrain, the railway line has no parallel.

The Konkan Railway line is a single-line track, and is not electrified. It has been designed for a maximum speed of 160 km/h, but presently the fastest train runs only at 130 km/h. The major challenge was to limit the ruling gradient to 1 in 150 and to avoid sharp curves. Hundreds of bridges and tunnels had to be constructed.

The project took about 8 years to build and became operational on 26 January 1998. The salient features of the Konkan Railway have already been given in Chapter 1.

## Kashmir Rail Link Project

This is a 326 km-long railway line joining the Kashmir Valley with the network of Indian Railways. This is considered to be a project of national importance. It joins Jammu and Baramulla through Udhampur, Katra and Qazigund. It is the biggest mountain railway line, taken up in 1983. Although it was scheduled for completion in 2007, it was pushed back to 2017 owing to several unforeseen complications.

The line crosses earthquake zones, inhospitable terrain exposed to extreme heat and cold, Himalayan ranges with many peaks over 4600 m high, hundreds of bridges and over 100 km of tunnels (the longest being about 11 km).

The first leg from Jammu to Udhampur (53 km), the second leg between Udhampur and Katra (25 km), and the last leg from Qazigund to Baramullah (119 km) have been completed. The third leg between Katra and Qazigund (129 km) is expected be completed in 2017.

The bridge over Chenab River, 1.3 km long, is to be 359 m above the riverbed, is expected to be highest railway bridge in the world (35 m higher than the Eiffel Tower).

The maximum line speed will be 100 km/h. Three-aspect colour-light signals are being installed.

The *advantages* of this rail link are:

- Connectivity of the Kashmir Valley with the rest of the country
- Construction of access roads
- Employment generation for local population.

## Increase in Axle Load

Indian Railways decided to increase its axle loads for better utilisation of its track capacity and other assets based on a joint study of the Railway staff college, Vadodara and IIM Ahmedabad.

BOXN wagons were permitted to be loaded up to 22.90 t, and on selected routes up to 25 t of axle loads. This meant a significant increase in the capacity of the wagons, which led to an increase in income form freight traffic.

The following measures were taken by the different departments of Railways to allow higher axle loads:

- All new lines, renewals, and gauge conversions have been designed for 25 t axle loads.
- Thick web switches and CMS crossings are used.
- Lubrication of rails is done by mechanised methods.
- Mechanised ultrasonic flaw detectors have been introduced for track maintenance.
- Rail grinding is planned to be introduced progressively to get better service life of 90 UTS rails.

### Rail Grinding

Irregularities of rail surface due to spalling, scabbing and corrugations can be removed by *rail grinding*, which leads to reduction in track maintenance costs and improvement of riding quality.

Rail grinding is done by modern rail grinding machines to produce a good longitudinal profile of the rails, and to control the effect of rolling fatigue in order to maximise rail life.

**Rail grinding trains:** These are provided with a number of rotating grinding wheels, which can smoothen out corrugations up to a depth of about 0.4 mm in a single pass on the track. Five to six passes may be needed to completely grind off the corrugations. These trains are equipped for complete re-profiling of the rail table, which is usually needed for the worn-out rails on curves.

All the grinding movements – horizontal, vertical and angular – of the grinding unit are computer-controlled; a set of rail grinding patterns are programmed by the operator, one of which is chosen and accomplished.

[It must be noted that rail grinding is not a permanent solution to prevent rail corrugation, as the fine grooves and ridges reappear after some time and need grinding again!]

Rail grinding may be for any the following reasons:

**Initial grinding:** When rails are laid during new construction or rail renewals, initial grinding is aimed at correction of construction damage and removal of decarbonised surface area of the rail head.

**Preventive grinding:** This is done when the damage is at the initial stage; grinding has to be done cyclically from time to time in this approach.

**Corrective grinding:** This approach is symptom-related. Short pitch corrugations are directed to be removed once they reach a depth of 0.05 mm.

Indian Railways has procured a few rail grinding machines from M/s Loram of the USA and has put them to use on heavy-haul routes.

The *salient features* of these machines are:

| | |
|---|---|
| Size | 12.142 m × 3.351 m × 3.810 m |
| Travel speed | 64.4 km/h |
| Speed while grindig | 6.4 km/h |
| Water system | Spray bars at each end. |
| Hourly output | Removal of defects of 0.4mm size in one pass on the track |

In view of the increase of axle loads on Indian Railways, rail grinding has become a necessity in order to extend rail life.

It is planned to progressively introduce rail grinding on different routes so as to bring the entire track under rail grinding regime by the year 2020.

## 9.4 URBAN RAILWAY SYSTEMS

The term *urban* refers to a city or a town as against *rural*, which refers to villages. When a city grows further into one of a much bigger size, it becomes a *metropolis* or a *metropolitan city* which is often a major or an important city. Indian cities have often grown in an unplanned manner, with an unbalanced distribution of population in the colonies or neighbourhood. our capital city, Delhi, is a typical example.

The objective of urban transport is to facilitate the movement of people from their residences to their workplaces, educational institutions and shopping areas.

Road transport, which caters to the needs of self-propelled vehicles such as cars, buses, vans and bikes, has been the primary mode of intra-city transport. However in metropolitan cities, this mode of transport has been utterly inadequate to meet the needs of the travelling public for several reasons:

- Severe congestion of roads with low capacity, particularly during peak hours.
- Very low average speed of traffic, 5 to 15 km/h, causing a lot of inconvenience to the people.
- Inordinate increase in the number of vehicles (Bengaluru, Hyderabad and Delhi are notorious in this respect with traffic jams)
- Traffic hazards such as accidents, causing injury and loss of life.
- Environmental pollution caused by the large number of motor vehicles.

A good solution to all these problems is to establish a Mass Rapid Transit System (MRTS).

## Railway Systems for Urban Transport

The best solution is to provide railway systems with *electric traction* in view of the following *advantages*:

(i) Environment-friendly, as it does not cause atmospheric pollution
(ii) Acceleration is faster than other modes
(iii) Facilitates modern automatic signalling
(iv) Same efficiency in forward as well as backward directions of movement
(v) No need for reversal of engine
(vi) Special coaches, called *Electric Multiple Units (EMU)*, are deployed for mass transit

The different railway systems which come under the category MRTS are the following:

1. Surface railways/suburban railway networks
2. Underground railways/Tube railways
3. Elevated railways
4. Monorail

## Surface Railways/Suburban Railways

Surface railways are the cheapest and hence are extensively used; in the case of urban transport, they are referred to as 'suburban' railways, for which electric traction is the preferred mode.

In addition, the railway line may have to cross different roads, some of which may be busy with traffic of their own. At such points, the options are the provision of a level crossing, road overbridge (the road passing over the railway line), or an underbridge (the road passing under the railway line). An underbridge generally costs less, but may present construction difficulties from high ground water table and drainage problems.

### Mumbai Suburban Railway

This covers the Mumbai Metropolitan region and consists of six lines totalling 427.5 km on BG. Electrification is on 1500 V DC/25 kV AC overhead category. The average speed is 50 km/h and the top speed is 100 km/h. The system operates over 2300 train services and carries more than 70 million commuters daily; it is considered to be the busiest MRTS in the world. Trains run from 4 AM until 1 AM and they are generally overcrowded.

This system is an offshoot of the first railway built in India. The Central Railway operates the Central line, the Harbour line and the Trans-harbour line, as also Panvel-Diva-Vasai line. The Western railway operates the Western line only. Electrical Multiple Units (EMU) ply on these lines, with 9, 12 and 15-coach rakes.

The Mumbai Suburban Railway is administered by the Mumbai Rail Vikas Corporation (MRVC), a public sector unit under the Ministry of Railways, instituted in 1999. This is also responsible for implementing the rail component of an integrated rail-cum-road project, called Mumbai Urban Transport Project (MUTP).

### Chennai Suburban Railway

This is operated by the Southern Railway in Chennai city, the capital of Tamil Nadu. There are four lines—the North line, the West line, the South line, and the MRTS line. This began operation in 1931. There are 73 stations, with 303 km of suburban routes on BG. Electrification is on 25 kV AC overhead catenary system. The average speed is 50 km/h, the top speed being 100 km/h. EMU's are used to operate the system with 9/12-coach rakes. The Beach–Tambaram section and the Central–Tiruvallur sections are the busier ones. About 1.46 million commuters travel daily.

### Delhi Suburban Railway

This serves the National Capital Region of India, operated by the Northern Railway on BG. This railway service covers Delhi, along with the districts of Faridabad and Ghaziabad, and other adjoining places in Haryana and UP. These services are mostly run using EMU and MEMU rakes. It includes passenger trains and Diesel Multiple Unit (DMU) services up to Rewari in Haryana, a part of the National Capital Region.

Delhi Suburban Railway uses the same tracks which are used for long distances trains. Ladies special trains were introduced between New Delhi and Palwal, and from New Delhi to Ghaziabad and Panipat. over 110 suburban trains are run using EMU's with 12 coaches.

**Delhi ring railway:** This is part of Delhi suburban railway services and is a circular rail network, which runs parallel to the Ring Road. It started in 1975 to service goods, and later upgraded to serve passengers during the 1982 Asian Games. Its circular route is 35 km long and runs in both clockwise and anti-clockwise directions from Hazrat Nizamuddin Railway station, from 8 AM to 7 PM. There are 12 electric trains on the ring railways.

This is not popular and is an example of a failed mass transit system. The network is now utilised as a freight corridor, though limited EMU passenger services are available during peak hours.

An integrated rail-bus transit system (IRBT) is being promoted by the Government to connect Delhi with the satellite towns of Gurgaon, Ghaziabad, and Sahibabad. The IRBT will have two dedicated tracks.

## Underground Railways

In this case, the railway line is constructed below the ground level. The construction work is done mostly by the *cut-and-cover method*.

This is a simple method for the construction of shallow tunnels where a trench is excavated with ground support as necessary, and a tunnel of the required size is constructed. The track is laid on a prepared formation and the necessary overhead structures are provided. Finally, the trench is back-filled and the ground surface is restored.

Shallow tunnels are often of the cut-and-cover type, while deep tunnels are bored or driven using a tunnel shield. A major drawback of the cut-and-cover method is the widespread disruption at the surface during construction.

An underground railway uses mostly electric traction to avoid pollution caused by other modes. The following are the primary advantages of underground railways

- Trains can run fast without obstruction or interference from other modes.
- Land-use patterns on the surface are preserved with no wastage of land for these railways.
- Safety from aerial attacks during war.

The *disadvantages* are:

- High degree of inconvenience to the public during construction.
- Special attention and arrangements are needed for drainage and ventilation.
- More expensive than surface railways.
- Public utility services have to be diverted or suspended temporarily—water supply, sewerage, electricity cables and telephone cables come under these services.

A typical cross-section of an underground railway constructed by the cut-and-cover method is shown in Fig. 9.3:

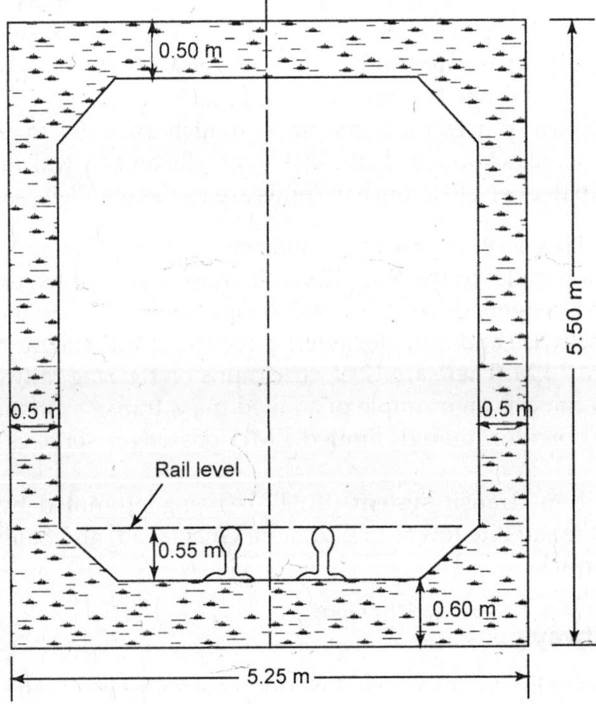


**FIG. 9.3** *Typical cross-section of an underground railway (cut-and-cover method)*

### Kolkata Metro

This is in Kolkata (formerly Calcutta), the capital city of West Bengal. This is the Metro Rapid Transit system for the metropolitan area of Kolkata.

One line is under operation and five are under construction. There are presently 24 stations serving over 6.5 lakh commuters daily. It is operated by the Kolkata Metro Rail corporation. It is 28.14 km long and is mostly on BG. Electrification is 750 V DC through third rail.

This metro serves the city of Kolkata and the districts of 24 paraganas (South and North). It is the first underground metro railway in India, which began operation in October, 1984. It is considered to be the 17th zone of the Indian Railways.

It was taken up as the Metropolitan Transport Project of the Railways in 1969. The first line from Dum Dum to Tollygunge was started in 1973, but could be opened for operation only in October, 1984 (16.45 km). Extension to this line is built as elevated corridors.

**Salient features of Kolkata metro:** The following are the essential features of Kolkata Metro.

- Use of cut-and-cover method with diaphragm walls and sheet piles
- Use of extensive decking to keep the traffic flowing during the construction
- Shield tunnelling using compressed air for the new lines in some stretches
- Ballastless track using elastic fastenings, rubber pads and epoxy mortar and nylon inserts
- AC and ventilation systems for tunnels and stations
- Third rail current collection system for traction
- VHF-radio communications system
- Microprocessor-based train control
- Automatic ticket vending and checking systems.

The first line has a total length of 28 km, with a combination of underground, ground-level, and elevated lines. The maximum speed is 80 km/h and the average operating speed is 35 km/h. Eight-coach rakes are in operation.

Route Relay Interlocking system and Train protection and warning system (TPWS) are provided for ensuring safety. The hours of operation are 0700 to 2145 hours with 10-minute frequency (5-minute frequency during peak hours). It operates 270 services daily.

## Tube Railways

Tube railways are underground railways constructed at a depth of more than 25 m. This needs the construction of a tunnel which may be circular or tubular in cross-section. The reason for the construction of a tunnel at such depths is to avoid interference with water supply mains or sewerage pipe lines, which are normally located within 10 m below the ground level.

*Salient features* of a tube railway are:

- The stations are generally cylindrical in shape.
- Only electric traction is used in order to avoid environmental pollution.
- Escalators are provided at stations for the convenience of passengers.
- Automatic signalling is installed.
- Automatic closing of the coach doors is activated by means of a switch in the driver's cabin before the train starts
- Automatic ticket-vending machines are installed for the convenience of passengers.
- Adequate information boards are provided to give the necessary indications to passengers.

The first tube railway of the world is the London underground railway or the London Tube constructed in the year 1863. Then the tube railways of New York, Paris, berlin, Tokyo, Seoul, Shanghai, Beijing, Madrid, and a number of others followed.

(The term *subway* is used in the USA.)

## Elevated Railways

These are railways provided at an elevation above the ground level. The track is laid on a deck supported by RCC columns. The stations are also provided at an elevation for easy access and convenience of passengers.

The primary *advantage* of elevated railways is that it avoids interference with road traffic.

The *disadvantages* are:

- Prohibitive cost
- Noise pollution
- Heavy loss of life and property in case of accidents.

## Monorail

This is a form of elevated railway with only one rail on which the train runs. This is recommended only in exceptional cases. The trains are mounted on pylons in Tokyo monorail, while these are suspended on the monorail in Montreal, Canada.

## Delhi Metro

This is world-class Metro, operated by the Delhi Metro Rail Corporation (DMRC), and serves Gurgaon, Noida, and Ghaziabad, besides Delhi city.

The *metro corridor* is underground, while the *rail corridor* is elevated or at-grade (at ground level); the former is in core city area.

The first phase—consisting of 12.1 km underground, 11.8 km at-grade, and 38.2 km elevated—became operational in December 2005.

The second phase consists of 125 km (operational from 2011); the third phase, 103 km (expected to be ready by 2016); and the fourth phase, 108 km (expected to be completed by 2020).

| Technical details | |
|---|---|
| Gauge | 1.676 m (BG)* |
| Frequency of trains | 5 min (3 min. during peak hours) |
| Speeds | 30 km/h (maximum 80 km/h) |
| Traction | 25 kV AC |
| Rakes | 8-coach (air-cooled) |
| Signalling and interlocking | Cab signalling and solid-state interlocking (continuous automatic train control) |
| Telecommunication | Fibre optic system (Train mobile radio and CCTV-system) |
| Platform | Height 1.08 m; Length 185 m |

Cont'd

**Table Cont'd**

| Technical details | |
|---|---|
| Ruling gradient | Main line – 4% |
| Radius of curves | Main line–300 m |
| Spacing of two tracks-4.1 m (For underground tunnels-5.4 dia. At 13m c/c | |

*From Connaught Place to Indira Gandhi International Airport, the line is likely to be on Standard Gauge of 1.435 m.

Escalators are installed at the metro stations. The ultimate aim of Delhi Metro is to integrate it with other modes of public transport so as to enable commuters to interchange from one mode to another.

**Rolling Stock**

The special features are:

- Automatic door closing mechanism in 2.5 to 3.5 seconds
- Train cannot move until all doors are closed and locked
- If a door opens during running, the train can be halted by applying emergency brakes
- Provision of emergency exits
- Emergency illumination in case of power failure.

**Other Special Features**

- Automatic train control and automatic operation
- Automatic train protection by cab signalling
- Optic fibre telecommunication
- Computer-controlled passenger information system
- CCTV systems at all critical locations
- 60-kg UIC rails on concrete sleepers, fitted with elastic fastenings
- Sleeper density – 1660 sleepers per km
- Hard stone ballast with a cushion of 300 mm
- Ballastless track on elevated sections, viaducts and tunnels; 60 kg-rails fixed with Vossloh fastenings
- Use of LWR.
- Bored tunnel construction in metro corridor at many places; tunnels lined with precast RC segments.

## Mumbai Metro

This serves the capital city of Maharashtra, Mumbai, the financial capital of India, and is owned and operated by the *Mumbai Metropolitan Region Development Authority (MMRDA).* This is a Mass Rapid Transit system (MRTS) designed to reduce the heavy burden on the overloaded Mumbai Suburban Railway system. One line began operation on 8 June 2014, and three more lines are planned.

A total of over 15 lakh commuters are expected to use the first line daily. The trains consist of 4- and 6-coach sets, running on the 11.4 km-long route. The total planned length of the route is 161 km. The track is on Standard Gauge of 1.435 m. The electrification system is on 25 kV,

50 Hz AC through overhead catenary. The average speed is 33 km/h and top speed, 80 km/h. The entire work is planned in three phases over a 15-year period, expected to be completed by 2021.

Mumbai has the advantage of a high modal share of 88% of a public mass transport system. The *Brihan Mumbai Electric Supply and Transport (BEST)* bus system provides the feeder services to the commuters of the existing Mumbai suburban Railway system and the Mumbai Metro under construction and planning. The main objective of Mumbai Metro is to provide mass rapid service to people within 1 to 2 km of approach distance. The first line serves the Versova–Andheri–Ghatkopar corridor. The second line, 33.5 km long, is planned to run on the colaba–Bandra–Chatrapathi Shivaji International Airport sector and will be the city's first *underground metro* railway line, with 27 stations. The stretch from Mahlakshmi to Bandra is going to be on an *elevated track*.

The Dahisar–Bandra–Mankhurd line will be 40.2 km long with 37 underground stations. The rolling stock is being supplied by CSR Nanjing of China; the coaches are fire retardant, air-conditioned, and designed to reduce noise and vibration. A number of features for safety and convenience are fitted – each coach will have a black box to assist in accident investigations. A four-car unit can carry 1100 passengers. The Mumbai Metro will feature an advanced signalling system, including the Automatic Train Protection system (ATPS).

## Jaipur Metro

This is expected to serve Jaipur city, the capital of Rajasthan; *Jaipur Metro Rail Corporation (JMRC)* is the operator. The system is initially planned to be 9.25 km long and eventually reach up to 32.5 km. It is being constructed to the standard gauge (1.435m) and shall be using the same rolling stock as that of Delhi Metro's standard gauge stock.

Trains on the metro operate at a maximum speed of 80 km/h and an average speed of 32 km/h.

The JMRC is a PSU under the Government of Rajasthan. The trial run was conducted on 18 September 2013, in a record time of two years and nine months.

The operation is a Public–Private Partnership (PPP). Jaipur Metro phase-1 has become operational on 03 June 2015.

There have been controversies regarding court litigations, accidents, and allegations about perceived risks to World Heritage sites such as Hawa Mahal, Jantar Mantar, and so on.

The Jaipur Metro will be built in two phases—the first phase is the *Green Line* and the second, *Orange Line*. The first phase is being managed by DMRC.

## Bengaluru Metro

This Metro, called *Namma-Metro* in Kannada, is planned to serve Begaluru city, the capital of Karnataka. It is owned and operated by *Bengaluru Metro Rail corporation (BMRCL)* as the Mass Rapid Transit system for this city.

This railway system is planned to consist of 8 lines – four lines in the first phase and the remaining four in the second phase. Two lines of the first phase are now operational; the remaining two lines are targeted to be operational by June, 2016 and the second phase lines by December, 2019.

Over 40,000 commuters are currently using the lines being operated daily, according to one estimate. Trains of 3-coach sets with headway of 10 to 15 minutes are being run. Phase I is planned to be 42.3 km long and Phase II, 114.4 km. Two tracks of standard gauge are being built. Electrification is by 750 V-DC- third rail system. The average speed is 40 km/h, the top speed being 80 km/h.

BMRCL is a joint venture of the Government of India and the Government of Karnataka. The first stretch between Baiyappanhalli and MG Road was inaugurated on 20 October 2011, and the stretch between sampige Road and peenya Industrial Estate on 1 March 2014.

Each corridor consists of two tunnels, the first underground railway tunnels in South India. They are located 18 m below the ground with a diameter of 5.5 m and are 5 m apart. The two tunnels are ready for track laying. Some portions of this metro rail are on elevated tracks.

### MRTS for Thane

The Thane Municipal Corporation will develop an elevated mass rapid transit system to reduce traffic congestions at thane Railway station. This is planned as a 16.2 km line, with options such as *light rail*, monorail and sky bus being considered. 'Light rail' or light rail transit uses rolling stock similar to a tramway. It operates at a higher capacity often with an exclusive right of way. The MRTS will have 11 stations. At Thane college and along the creek, it will be at the ground level. It will be a kind of Ring Railway. The *Maharashtra state Road Development Corporation (MSRDC)* is the nodal agency, which has submitted the project report. The project is currently awaiting enviornmental clearance and approval of the Government of Maharashtra.

#### Salient Features of the Project

Phase I of this MRTS is proposed as a Green Corridor – an area or habitat connecting wild life populations separated by structures such as railway embankments, river banks and road side grass verges. In general, it is a corridor with vegetation. In Thane, the corridor is expected to be 16.05 km with an elevated stretch of up to 14.65 km and the remaining 1.4 km at the surface level.

For the first phase, it is proposed to have a light rail system with elevated dual track MRT system on Standard Gauge. Of the 11 stations, one will be at-grade and the remaining 10 will be elevated. The expected period of construction is 4 years. The necessary clearances from the Government of India and the Government of Maharashtra are awaited.

### Other Metros in Progress in Indian Cities

Metro Railway projects are in progress in several other Indian cities such as Chennai, Hyderabad, Gurgaon, Ahmedabad, and Kozhikode. It is in the planning stage in several more cities such as Lucknow, Chandigarh, Kochi, Bhopal, Nagpur, Patna, Pune, Kanpur, Indore, Surat and Vadodara. The first phases of Chennai, Hyderabad and Ahmedabad are expected to be completed in 2016.

Metro railways are sanctioned in principle by the Government of India in the AP cities of Vijayawada and Visakhapatinam.

Thus, it can be observed that the scenario of urban Railway system in India is truly encouraging, in spite of financial and several other constraints.

## 9.5 TUNNELLING FOR RAILWAYS

A 'tunnel' may be defined as an underground passage for the transport of passengers, goods, water, sewage, oil, gas, and so on.

Tunnels have been constructed from time immemorial; it dates back to 2000 BC in Babylon. Tunnels have been constructed for the construction of highways as well as for railways.

The first tunnel for a railway was constructed towards the end of the nineteenth century to connect Switzerland and Italy—the Simplon tunnel, about 20 km long. It was the longest railway tunnel, until the Seikan tunnel of Japanese Railways under the Pacific Ocean, 54 km long, was opened in 1988.

The first tunnel on Indian Railways was built near thane, known as the 'Parsik Tunnel', 1.317 km long. Presently, there are hundreds of tunnels on Indian Railways; a large number of tunnels have been constructed for the Konkan Railway, and several tunnels are under construction on the Udhampur–Quazikund sector of Kashmir Rail Link project.

(**Note:** Tunnelling is a vast subject and a somewhat detailed treatment is planned for the last part of this volume. Hence, only a brief treatment of tunnelling, especially for railways, is presented here.)

**Advantages of a Tunnel**

The following are the advantages of a tunnel when compared to an open cut:

- The length of the railway in a hilly terrain is considerably reduced by avoiding a detour along a contour around of a hill.
- A tunnel is more economical than an open cut in a rocky terrain.
- A tunnel may be more economical under a large water-body like a river, than building a bridge over it.
- A tunnel constructed to carry an underground railway provides a fast and unobstructed means of transport for a metropolitan city.

However, tunnel construction needs high technical skills and specialised machinery and equipment; also, it is laborious and time-consuming, and may prove to be expensive.

**Horizontal and Vertical Alignment of Tunnels**

Besides reconnaissance and preliminary surveys, a precise location survey is necessary for setting out the horizontal alignment of a tunnel. The centre line and the ends have to be marked accurately on the ground.

The centre line and the levels are transferred to their correct underground positions.

The horizontal alignment should normally be straight, as it would be the shortest route between the ends.

The vertical alignment of the tunnel, the gradient of the tunnel and its approaches should be as flat as possible.

Proper drainage of tunnel requires the provision of side drains in it with a minimum gradient of 1 in 500 for effective outflow. In the case of long tunnels, the gradient should be from the centre towards either end.

## Shape of a Tunnel Cross-Section

The shape of the cross-section should be such that it is stable under the pressures exerted over the roof and on the sides by the unsupported walls.

If the tunnel is to be driven in hard rock, there would be choice with regard to its shape. A circular shape is best suited irrespective of the purpose of the tunnel; theoretically speaking, it provides the largest cross-sectional area for a given perimeter and also provides the greatest resistance to external pressure. This shape is more popular if the tunnel is to carry water, sewage, and other fluids.

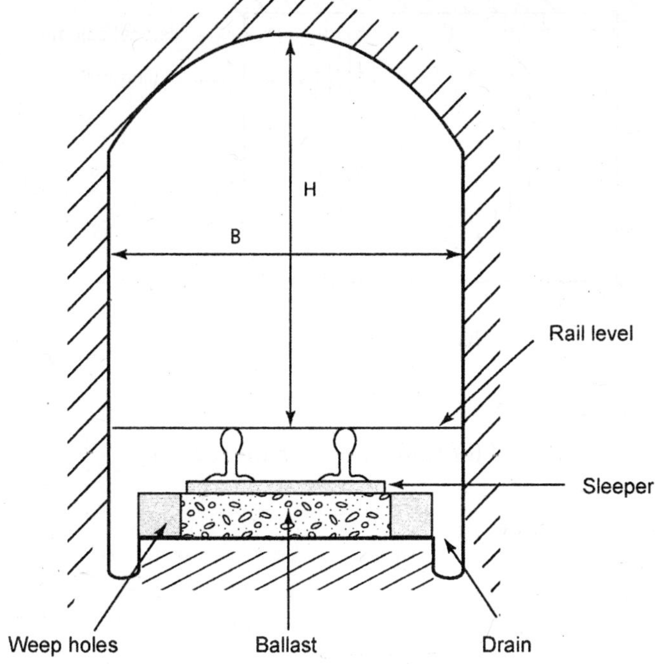


**FIG. 9.4** *A typical cross-section of a railway tunnel*

For a railway tunnel, the bottom portion has to be levelled for laying the track and placing the concrete base.

Circular and elliptical shapes are common for carrying water and sewage, while *horseshoe* and *arched roof with vertical walls* are commonly adopted for carrying roads and railways.

A typical cross-section of a railway tunnel is shown in Fig. 9.4.

### Size of a Tunnel

The size of a railway tunnel depends upon the gauge and the number of lines. Typical dimensions for BG are given below:

| | | |
|---|---|---|
| Single line | Breadth (B) | 4880–5490 mm |
| | Height (H) | 6700–7320 mm |
| Double line | Breadth (B) | 8530–9140 mm |
| | Height (H) | 6700–7320 mm |

## Methods of Tunnelling

There are various methods of tunnelling; the selection of the method depends upon the type of soil/rock, the size and shape of the tunnel, the extent to which timbering is required, and the equipment available for construction of the tunnel.

Tunnelling is broadly classified into two groups:

1. Tunnelling in hard rock
2. Tunnelling in soft rock/soft soils

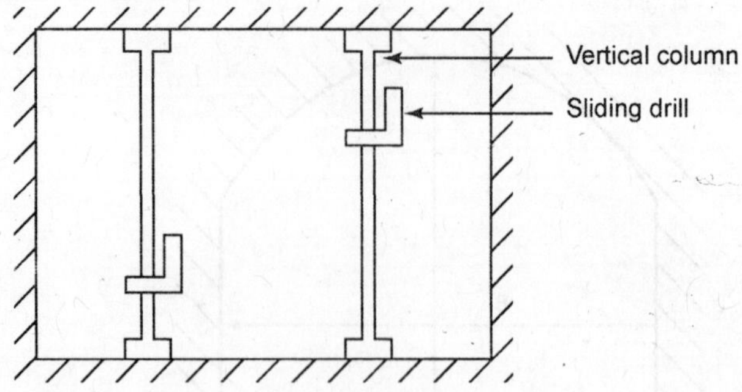


**FIG. 9.5** *Full face method*

## Tunnelling in Hard Rock

The different methods employed for tunnelling in hard rock are:

(a) Full face method
(b) Heading and bench method
(c) Drift method
(d) Pilot tunnel method

### Full Face Method

This is used for small-sized tunnels not exceeding 3 m. The full face or the entire facade is dealt with at the same time. Sliding drills attached to vertical columns and erected near the tunnel face, are used to drill holes of 10 to 40 mm diameter at about 1200 mm c/c in the rock face, preferably in two rows. These holes are filled with explosives and ignited. After blasting, the muck or waste material is removed before proceeding with further repetition at the new face (Fig. 9.5).

### Heading and Bench Method

In this method, the heading or top half of the tunnel is bored first and then the bench or bottom half is tackled. The heading portion is made to progress about 4 m ahead of the bench portion. The drill holes for the bench can be made as the muck is being removed. The roof can stay without supports as it is in hard rock (Fig. 9.6):

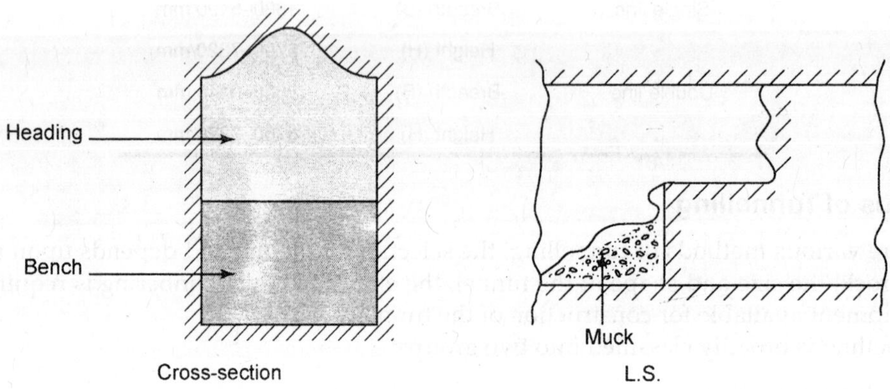


**FIG. 9.6** *Heading and bench method*

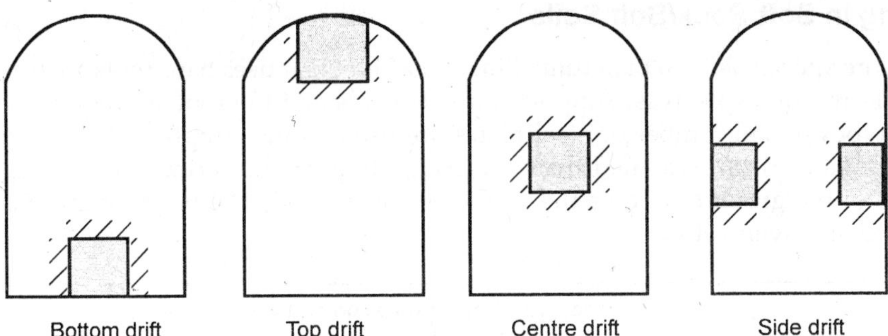


**FIG. 9.7** *Drift method*

### Drift Method

A *drift* is a small tunnel, 3 m × 3 m, driven into the rock; the section of this drift is widened in the subsequent stages until the desired full section of the tunnel is obtained. Drill holes are made all around the drift, filled with explosives and blasted.

The position of the drift in relation to the full cross-section of the tunnel may in the centre, top, bottom, or side (Fig. 9.7). In general, the centre drift is considered the best choice.

### Pilot Tunnel Method

This method involves digging of two tunnels – a 'pilot' tunnel and the 'main' tunnel; the former measures about 2.4 m × 2.4 m. The pilot tunnel is driven parallel to the centre line of the main tunnel and connected to it with cross cuts at several points. The main tunnel is then excavated from these points. The pilot tunnel facilitates quick removal of muck from the main tunnel and also makes a provision of good lighting and ventilation to the main tunnel (Fig. 9.8). This tunnel is usually retained as a refuge for maintenance operations after the main tunnel work and also as additional access during emergencies.

Since an additional tunnel is to be drilled, this is expensive and time-consuming.

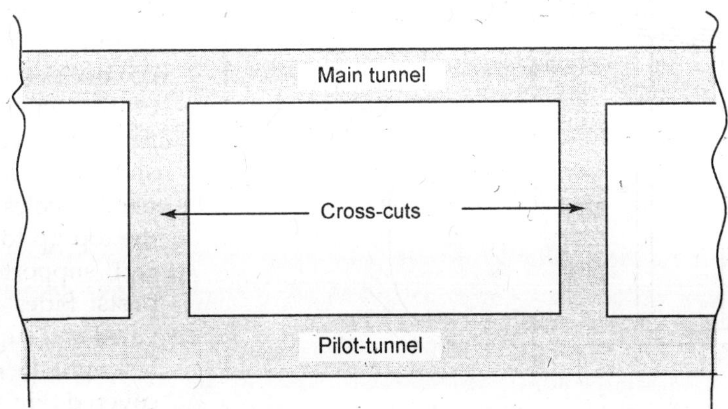


**FIG. 9.8** *Pilot tunnel method*

## Tunnelling in Soft Rock/Soft Soils

This is a more specialised job than tunnelling in hard rock. It uses hard tools such as pick-axes and shovels and no explosives. Supports for side walls and the roof are required depending upon the types of soil; timber or steel plates are used as the supports. The operations are: excavation, removal of excavated material, scaffolding or shuttering, and lining of tunnel. The nature of the ground governs the choice of the method. The types of ground generally encountered are given below:

| | |
|---|---|
| | **Dry sand, gravel, slit, mud, sand and water** |
| Running ground | Requires instant support on excavation |
| | **Damp sand, soft soil** |
| Soft ground | Requires quick support on excavation, but can stand for a few minutes. |
| | **Firm clay, gravel and dry earth** |
| Firm ground | Sidewalls and the tunnel face can stand for one or two hours, but the roof can last only a few minutes |
| | **Sand stone, hard clay** |
| Self-supporting ground | Tunnel work can proceed without support for small lengths of 2 to 5 m. |

Besides the nature of the ground, the size of the tunnel, equipment available and the sequence of operations also govern the choice of the method.

The following are the important methods:

(a) Forepoling method
(b) Linear plate method
(c) Needle beam method
(d) American method.

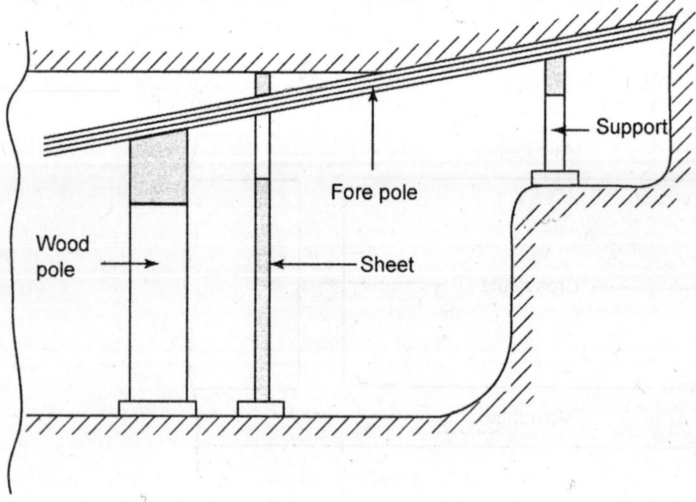


**FIG. 9.9** *Forepoling method*

### Forepoling Method

This is one of the earliest methods of tunnelling in soft ground. A frame in the shape of the letter 'A' is placed near the face of the tunnel and covered with suitable planks. Poles are inserted at the top up to a viable depth and supported by vertical posts. Sides are then excavated and supported by timber, until the entire section is covered (Fig. 9.9).

This is a slow and tedious method.

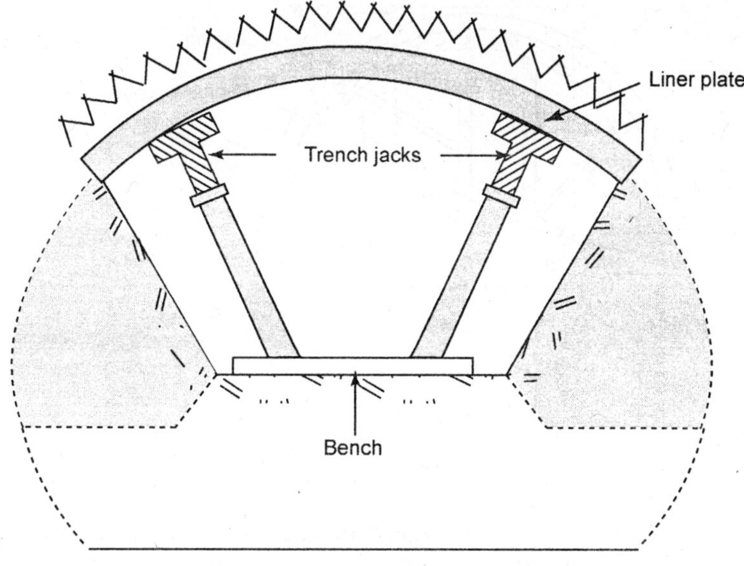


**FIG. 9.10** *Liner plate method*

### Liner Plate Method

This is a recent method in which instead of timber plates, pressed steel plates are used (Fig. 9.10).

### Needle Beam Method

This method is used in sections where the soil permits the roof to stand without support for a few minutes. A small drift is driven for inserting a needle beam, which consists of two rail steel joints or I-sections, bolted together with a wooden block in the centre. The roof is supported on laggings on wooden beam. Jacks are fixed on the needle beam. This method is considered to be economical (Fig. 9.11).

### American Method

A drift is driven into the top of the tunnel and is supported by laggings, caps, and two vertical posts. The sides are then widened and additional supports are given by means of timber

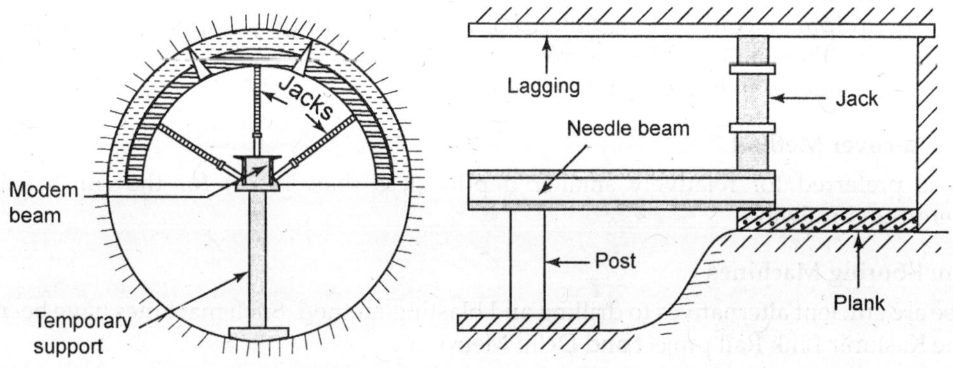


**FIG. 9.11** *Needle beam method*

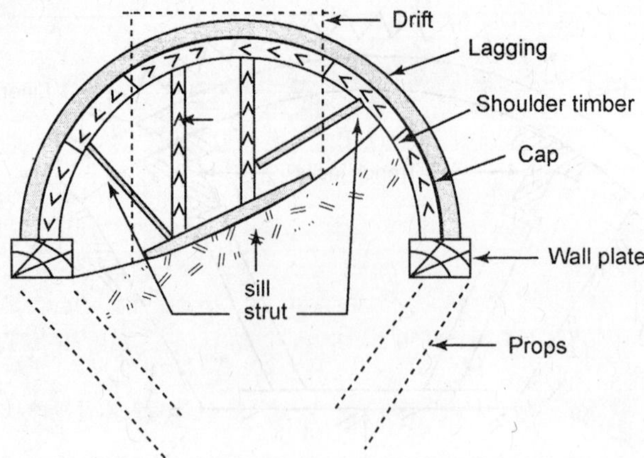


**FIG. 9.12** *American method*

planks and struts. The process of widening is continued up to the springing level. Wall plates are fixed at this level and supported by vertical posts. The supports for the drift can now be removed and the work is continued further in a similar manner (Fig. 9.12).

Other methods such as the English, Austrian, and Belgian methods are modifications, and are not presented here.

## Modern Methods of Tunnelling

### Shield Tunnelling

This is one of the modern approaches, particularly suited for tunnelling through water-bearing strata. A *shield* is a movable frame used to support the face of a tunnel; the excavation and lining of the tunnel proceeds under the protection of the shield.

A shield consists of a cutting edge, a skin plate in the form of a shell, and hood of jacks, ring girders, and stiffeners consisting of steel plates. A circular section is preferred for shield tunnelling.

### Compressed Air Method of Tunnelling

This is perhaps the most modern approach. Compressed air at a pressure of about 1 kg/cm$^2$ is forced into the enclosed space within the tunnel to prevent the collapse of the sides and the tops. The equipment consists of a bulk head, which is an airtight diaphragm with an airlock and doors at either end opening inwards.

This is considered to be a difficult process.

### Cut-and-cover Method

This is preferred for relatively smaller depths (less than 10 m) for the construction of underground railways.

### Tunnel-boring Machines

These are efficient alternatives to drilling and blasting method. Such machines have been used in the Kashmir Link Rail project and Delhi Metro.

## Safety Aspects of Tunnels

### Ventilation

Ventilation is important for tunnels during as well as after construction in order to provide fresh air and to remove dust and other gases due to the use of explosives.

Natural ventilation by means of drifts driven from portal to portal may not be adequate; mechanical ventilation by blow-in or blow-out method, or a combination of both are used.

### Lighting of Tunnels

Good lighting is necessary both during construction and later for maintenance of the track and the tunnel.

### Drainage of Tunnels

Good drainage is essential both during the construction and later, for safe operation of the facility.

### Tunnel Shafts

Shafts are vertical passages driven along the length at a few points. Working shafts facilitate the construction from a number of points, the minimum size being 4.30 m.

Ventilation shafts are generally inclined, with a minimum size of 1.2 m.

### Lining of Tunnels

Lining is necessary for tunnels in soft rock or soft soils to strengthen the roof and sides in order to prevent possible collapse, as also to obtain a good finished shape and to check possible leakage of water into the tunnel. Concrete lining is preferable.

### Maintenance of Tunnels

This is perhaps the most important post-construction aspect and encompasses the various components of a tunnel.

## SUMMARY

- Track modernisation for high speed may be achieved through upgradation or improvement of an existing track by improvement of track structure and track geometry. Alternatively, dedicated high-speed corridors may be constructed with higher standards for track structure, track geometry, track maintenance, and improved methods of signalling and interlocking.
- Tilting train technology may be used on high-speed routes to maintain higher speeds even on sharp curves.
- Track structure for high-speed tracks shall consist of heavier and stronger rails, provision of LWR/CWR, PSC sleepers with high sleeper density, stronger points and crossings and elastic fastenings.
- For achieving supper-high speeds, friction at rail-wheel contact should be eliminated. Curves should be made flatter.

  Tracked Air cushion vehicles (TACV) or Aerotrains run on Linear Induction Motor (LIM) concept. MAGLEV trains use the magnetic levitation concept coupled with LIM and move with a levitation of about 100 mm. Electromagnetic suspension (EMS) or Electrodynamic

Suspension (EDS) is also used for achieving high track speeds. Ballastless track assembles are used for these tracks.

- Dedicated Freight Corridors are used for the fast movement of goods. container transport service, increase in axial loads, and rail grinding for enhancing the service life of rails are a few of the modern developments on Indian Railways.
- Konkan Railway and Kashmir Link Railway are two challenging railway projects; the former is in operation while the latter is under construction. Major bridges and tunnels are a part of these big railway projects.
- Urban railway systems primarily consist of Mass Rapid Transit System (MRTS). They may be surface railways consisting of Suburban railway systems with electric traction and EMU as rolling stock. The suburban railway systems of Mumbai, Chennai and Delhi are excellent examples. Underground railways (at less than 10 m depth below the ground) and tube railways (at depths ranging from 10 m to 30 m), which involve the use of tunnels are excellent Mass Rapid Transit Systems.

  Elevated railway systems are supported on bridge structures so that they do not hinder the highway traffic. Kolkata Metro, Delhi Metro and Mumbai Metro are all combinations of these systems, depending upon the site conditions. Delhi Metro is considered to be a world-class facility.

  Metros are either under construction or are in different stages of planning in several Indian cities such as Bengaluru, Thane, Jaipur, Chennai, Hyderabad, Gurgaon, Ahmedabad and Kozhikode.
- Tunnelling is required for railways depending on the terrain – mountainous or hilly, or for underground/tube railways.

  Various methods are available for tunnelling in hard rock and in soft ground. Shield tunnelling and the use of Tunnel Boring Machines are modern developments used under special circumstances.

  Ventilation, lighting, drainage, lining, and maintenance aspects of tunnels need special attention. Ballastless tracks and electric traction are preferred on railways in tunnels.

## QUESTIONS

1. Write brief notes on:

   (a) Tube railways

   (b) Kolkata Metro
2. Write a note on underground railways and its advantages.
3. Differentiate between an underground railway and a tube railway.
4. Write a note on the special features of Delhi Metro.
5. What are the various improvements needed to achieve higher speeds on an existing railway line.
6. Write short notes on:

   (a) Dedicated Freight Corridor

   (b) Container Transport Services
7. Write a note on elevated railway systems.

8. Differentiate between high speeds and super-high speeds on railways.
9. Write a note on the advantages of electric traction on railways.
10. Write a note on MAGLEV trains and the technology involved in their operation.
11. What is 'rail grinding' and what are its objectives?
12. What are the advantages of increasing the 'axle load'? What are the factors to be considered if the axle load has to be increased?
13. Write a note on 'Ballastless Track' and its uses.
14. Write a note on 'Tilting Train Technology' and explain where it is used.
15. Define a tunnel and mention its uses in railways.
16. What are the advantages and disadvantages of tunnels?
17. What are methods available for tunnelling in hard rock?
18. Differentiate between:
    (a) Main tunnel and pilot tunnel
    (b) Heading portion and bench portion
    (c) Central drift and side drift
    (d) Working shaft and ventilation shaft.

# PART II

## Airport Engineering

# Chapter 10

# GENERAL CONSIDERATIONS AND AIRPORT PLANNING

**CHAPTER OBJECTIVES**

This chapter enables the student to understand the following:
- The definition and purpose of air transport and airport engineering
- The historical aspects of the development of air transport and the relevant organisations
- The characteristics of aircraft
- ICAO classification of airports
- Obstructions around airports and restrictions on the heights of objects
- Zones around airports and zoning laws to ensure safety of operations
- Airport surveys and their role in planning and design of airports.

## 10.1 INTRODUCTION

As seen in the earlier section, the progress of civilisation was marked by the development of various modes of transport to move people and goods from one place to another.

The primary modes by which transportation could take place are:

- on land
- through water
- through air

Transportation on land primary consists of movement by road and on rail. (Highway Engineering related to road transport and railway engineering related to rail transport have already been considered in detail)

Transportation through water primarily consists of movement using ships, which requires docks and harbours or ports.

(Dock and Harbour Engineering related to shipping will be studied in a later part of this volume.)

Transportation through air is effected by using different types of aircraft such as aeroplanes and helicopters.

Airport Engineering is associated with the engineering aspects of air transport. An *airport* is defined as a place where aircraft can take off and land for operating commercial services. Facilities for passengers as well as the maintenance of aircraft are provided at airports.

## Historical Aspects of Air Transport

Roger Bacon, a 13th century British Philosopher who is well known for several inventions including the magnifying glass, was the first who envisioned flight in air based on his studies on the flight of birds.

Leonardo da Vinci, perhaps the greatest inventor of all time, made drawings depicting the mechanical principles of wing movement of birds, air resistance and airflow around flying bodies in early sixteenth century. These studies later formed the basis of the modern principles of flying. This Italian intellectual is credited to have invented the parachute.

Going up in hot air balloons started in the eighteenth century, but these flights lacked the flier's control.

Wright brothers of the USA are said to be the pioneers who invented the heavier-than-air flying aircraft. These first flying machines were devised in 1903 and could fly a distance of 35 m.

Zeppelin airships, devised by Zeppelin, a German army officer, were used during the First World War by Germany.

In 1918, the first international air service was operated in France, and long distance air mail service was introduced between Washington and New York in the USA.

The first trans-Atlantic flight was by Charles Lindberg in 1927 between New York and Paris in a monoplane.

With the advent of radio-communication, air travel became relatively safe.

The world's first aircraft to fly under turbojet power was manufactured by Heinkel Aircraft Company; it was the first practical jet aircraft and its maiden flight was on Aug 27, 1939.

Air transport made rapid strides after the Second World War. International Civil Aviation Organisation (ICAO) was established in 1947 as a specialised agency of the UNO, with headquarters in Montreal, Canada. ICAO functions in close collaboration with World Meteorological Organisation. Another international organisation in the field of civil aviation (flying for public purpose) is the International Air Transport Association (IATA), a trade association of 250 major airlines, founded in 1945, and headquartered in Montreal, Canada.

## Characteristics of Air Transport

The *positive attributes* of air transport are:

- Rapidity or high speeds
- Continuous travel for long distances.
- Accessibility–even to remote locations, which are normally inaccessible by other modes of transport.

The *disadvantages* are:

- Need for high skill and sophisticated equipment for ensuring safety in operation
- Difficulty in operating under unfavourable weather conditions
- Prohibitive cost of flying.

## Air Transport Scenario in India

The development of air transport in India was initially slower when compared to that in the Western and the European countries.

The pioneer of civil aviation in India was J R D Tata, whose 'Tata Airways Limited', started domestic services in 1932. Indian Transcontinental Airways Ltd was instituted in 1933 for international services.

Air transport in India received an impetus during the Second World War and immediately thereafter. Post independence, the Tata Airlines Limited became Air India Limited, and the Government of India established Air India International Limited for its international flights.

In the year 1953, the airlines were nationalised and two corporations – Indian Airlines for domestic services and Air India International for international services –were established. Air India International had been operating international flights from the four international airports at Mumbai, Kolkata, Delhi and Chennai ever since. The first airport to be upgraded to international status after independence is the Trivandrum (Thiruvananthapuram) airport in January 1991.

Later on, Bengaluru, Hyderabad and Ahmedabad got their international airports. At present, India's airline industry is a fast-growing one.

### Development of National Organisations for Civil Aviation

The following is the sequence of development of National Organisations for controlling civil aviation in India:

**International Airport Authority of India (IAAI):** This organisation was set up in April 1972 for the operation, planning and development of international airports in the country. It has carried out modernisation of the four international airports in the cities of Mumbai, Kolkata, Delhi and Chennai. Further, it has provided consultancy and construction services for several developing nations of the world.

**Civil Aviation Development and National Airport Authority (NAA):** The NAA was established on June 1, 1986, through an Act of Parliament and it was managed by broad-based board.

The functions of this body were to ensure safety in civil aviation, manage all civil airports, and provide all essential facilities for the smooth operation of aircraft and improved passenger amenity.

All regulatory functions for the aircraft and their operation were managed by the Director General of Civil Aviation (DGCA).

**Airport Authority of India (AAI):** The AAI was constituted on April 1, 1995, by merging the IAAI and the NAA, under the administrative control of the Ministry of Civil Aviation of the Government of India. This organisation has been vested with the responsibility of managing both domestic and international airports.

### Open Sky Policy

The Air Transport Corporation Act of 1953 was repealed on January 29, 1994, to introduce the *Open Sky Policy* in civil aviation. According to this policy, private airlines are permitted to operate their aircraft under the supervision of the Director General of Civil Aviation (DGCA).

This step encouraged competition between national airlines and private airlines in providing the best possible services to air passengers.

## Components of Airports and their Functions

The two main components of an airport are:

- Terminal area
- Landing area

**Terminal area:** This is the focal point of an airport, with several elements, each of which has its own functions. They are:

(i) Apron: Paved area parking of the aircraft for passengers to emplane and deplane and for cargo to be loaded and unloaded. This is analogous to platforms in railways.
(ii) Terminal building: This building houses all facilities for the passengers, operational staff, and control towers.
(iii) Motor vehicle parking and circulation area: This is meant to provide the facility of parking for the motor vehicles by which passengers arrive at and depart from the airport.
(iv) Hangars: These are large sheds which house aircraft for servicing, fuelling and repairs. Machine shops and spare parts are necessary adjuncts for hangers.

**Landing area:** This is very critical for the operation of an aircraft. Landing as well as take-off operations are performed in this area with the aid of runways and taxiways. This area includes the approach zone and clear zone, and is governed by zoning laws and restrictions.

## Aeroplane

This is a flying machine with fixed wings, heavier than air, and power-driven. It derives its lift mainly from aerodynamic effects on its exposed surfaces.

The following are the *important parts* of aeroplane:

(i) Engine
(ii) Wings
(iii) Propeller
(iv) Fuselage
(v) Flaps
(vi) Controls
(vii) Tricycle undercarriage

A general view of an aeroplane is shown in Fig.10.1.

### Engine

The engine is the heart of the aeroplane that propels it through the air. It may be a piston engine, jet engine or rocket engine.

**Piston engine:** This is a conventional type of engine suitable for moderate speeds at low attitudes. It consists of a reciprocating engine that drives a propeller. The torque developed by the engine rotates a shaft connected to the propellers, which then rotates and hurls air backward; this propels the plane forward and lifts the wings. The engine runs on gasoline.

**Jet engine:** There are no propellers; this is suitable for high speeds at high altitudes. In the case of turbojets, the hot exhaust gases come out at high velocity, propelling the plane forward. A *ramjet* does not have any moving parts since its fuel flow and combustion are continuous; though the fuel consumption is high, it is suitable for very high speeds and preferred for guided missiles.

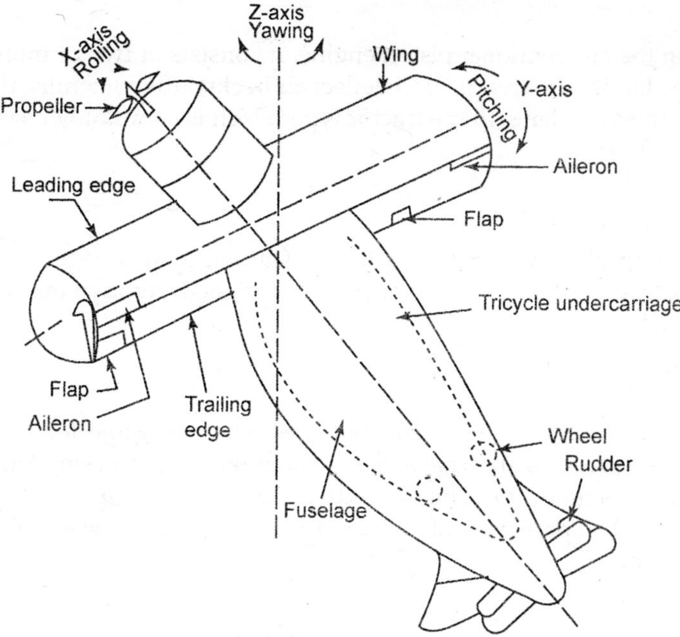


**FIG. 10.1** *General view of an aeroplane and its main parts*

### Wings

The aircraft is supported by the wings when it is airborne. The wings are slightly curved in section and are like cambered aerofoils. This helps the wings to obtain lift from the difference in air pressure between the bottom and the top surfaces (Fig. 10.2)

An aeroplane with a single pair of wings is a *monoplane*; If there are two wings, one above the other, on either side; it is a *biplane*. Although more sets of wings may be used, these are the common ones.

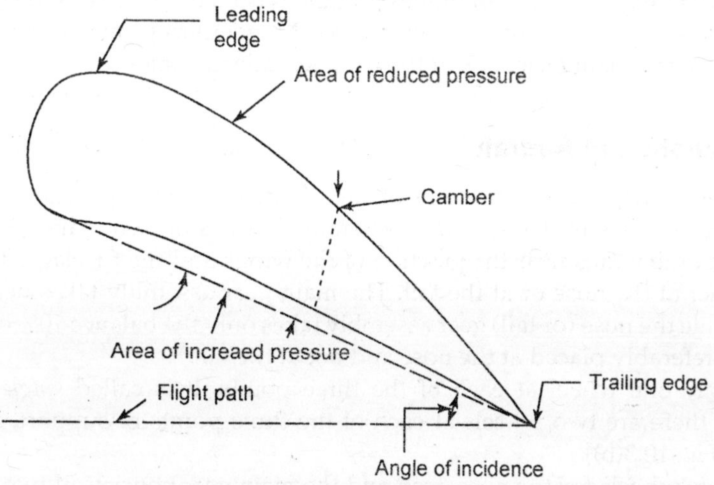


**FIG. 10.2** *Cambered aerofoil*

### Propeller

This is provided in the conventional piston engine. It consists of two or more blades rotating about the shaft; the blades are designed to deflect air backwards forcefully, thus developing a forward thrust on the aeroplane. In the tractor type which is commonly used, the engine and propeller are in the front.

### Fuselage

This is body of the aeroplane which houses the cockpit, cargo, passengers, ovens, bathrooms, and so on. The shape of the fuselage should be designed aerodynamically so that wind resistance is reduced.

### Flaps

A flap is the hinged portion of an aeroplane wing, which when projected into the air, reduces the speed of the aeroplane, somewhat like an air brake by increasing the drag. Flaps are provided on the inner side of the wings on either side of the aeroplane. These are operated together by the pilot. The flaps provide necessary lift at low speeds, and help in controlling the landing operation.

### Controls

An aeroplane can move in the three principal directions in space and also rotate about these axes. As shown in Fig.10.1, the rotations about the x-, y- and z-axes are known as *rolling*, *pitching* and *yawing* respectively.

The *aileron,* fixed to the trailing edge of the wing, serves as the control for movements in the x-direction. The *elevator,* in the form of two flaps, is provided at the extreme rear end of the fuselage, can move up to 60° and control the movements in relation to the y-axis. The *rudder* controls movements in relation to the vertical or z-axis, and it is provided at the tail end of the fuselage.

### Tricycle Undercarriage

This houses the landing-gear system with sets of wheels to support the aeroplane while on the runway. Different types of wheel assemblies – single-wheel, dual wheel and so on – are used, depending upon the type and size of the aircraft. Manoeuvring on ground and absorbing landing shocks are the main functions of the tricycle undercarriage.

## Wheel Configuration of Aircraft

The various possible wheel configurations of aircraft are shown in Fig.10.3.

The landing gear system of an aircraft generally consists of two main gears provided in the fuselage or in the wings near the junction of the wings and the fuselage. The third wheel is provided either at the nose or at the tail. The main gear assembly takes nearly 90% of the aircraft load, while the nose (or tail) gear assembly takes only the balance 10% of the load. (The third wheel is preferably placed at the nose and not at the tail.)

If there is only one wheel at each of the three points, it is called *single-wheel assembly* [Fig. 10.3(a)]. If there are two wheels at each of the three points of support, it is called *dual wheel assembly* [Fig. 10.3(b)].

If there are two wheels for the nose gear, and the main gear consists of two wheels at each point with four points of support, it is known as *twin-tandem gear assembly* [Fig. 10.3(c)]. There

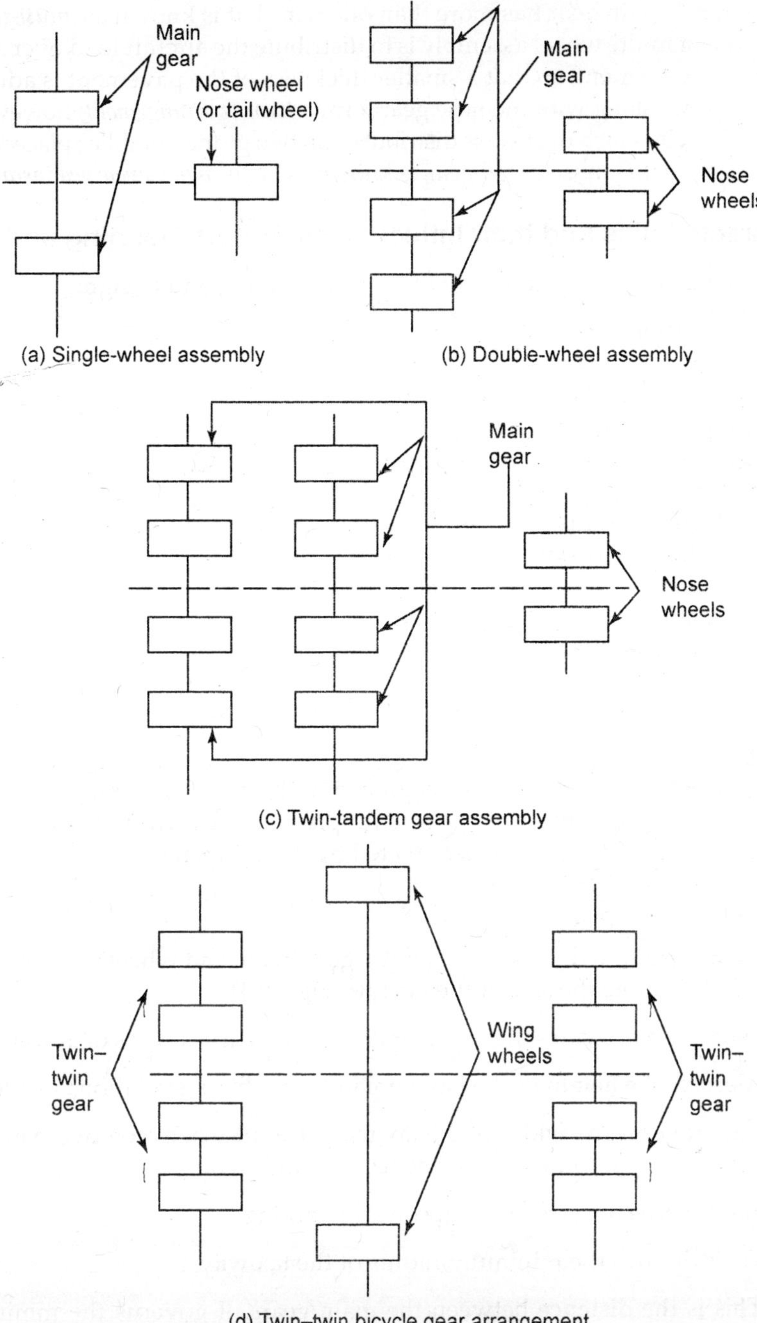


**FIG.10.3** *Aircraft wheel configurations*

are eight main gear wheels arranged in two rows in tandem, with four wheels in each point of support as in the dual wheel assembly.

In general, when a main gear has more than one wheel, it is known as *multi-wheel assembly*. The main purpose of multi-wheel assembly is to distribute the aircraft load over a large contact area of the runway pavement, so that a smaller thickness of the pavement is adequate.

The two main gears along with the nose gear form a *tricycle arrangement*; however, the system shown in Fig.10.3 (d), in which the load is distributed on two gear assemblies placed along the axis of the fuselage, without any nose or tail gear, is known as, *twin-twin bicycle gear arrangement*.

## Aircraft Characteristics and their Influence on Airport Planning and Design

The following characteristics are relevant to airport planning and design:

(i) Type of propulsion
(ii) Size of aircraft
(iii) Aircraft weight and wheel configuration
(iv) Tyre pressure and contact area
(v) Speed of aircraft
(vi) Minimum circling radius
(vii) Minimum turning radius
(viii) Take-off and landing distances
(ix) Range of aircraft
(x) Noise of aircraft
(xi) Fuel spilling and jet blast
(xii) Aircraft capacity

### Type of Propulsion

The type of propulsion – reciprocating engine running with gasoline as fuel and a shaft connected to a propeller or a turbo jet or turbo propulsion – governs the size, weight-carrying capacity, speed, circling radius, and noise created by the aeroplane.

### Size of Aircraft

Length of fuselage, height, wing span, tail width, gear tread, and wheel base are the important constituents that determine the size of the aircraft (Fig. 10.4).

**Length of fuselage:** This decides the width of taxiway on curves, sizes of aprons and hangars.

**Height:** This decides the height of the hangar gate and other installations inside the hangar.

**Wing span:** This governs the width of the taxiway, clearance between two parallel runways, sizes of aprons and hangars, gate size for the hangar, and so on.

**Tail width:** This governs the size of the apron and parking.

**Wheel base:** This decides the minimum radius of the taxiway.

**Gear tread:** This is the distance between the main gears. It governs the minimum turning radius of the aircraft.

### Weight of Aircraft and Wheel Configuration

The weight of the aircraft and wheel configuration play an important role in deciding the runway design, range of aircraft, fuel requirements, and so on. This is composed of the tare or empty weight which is constant for a given aircraft, and the pay load, trip fuel and fuel reserve, which are variable.

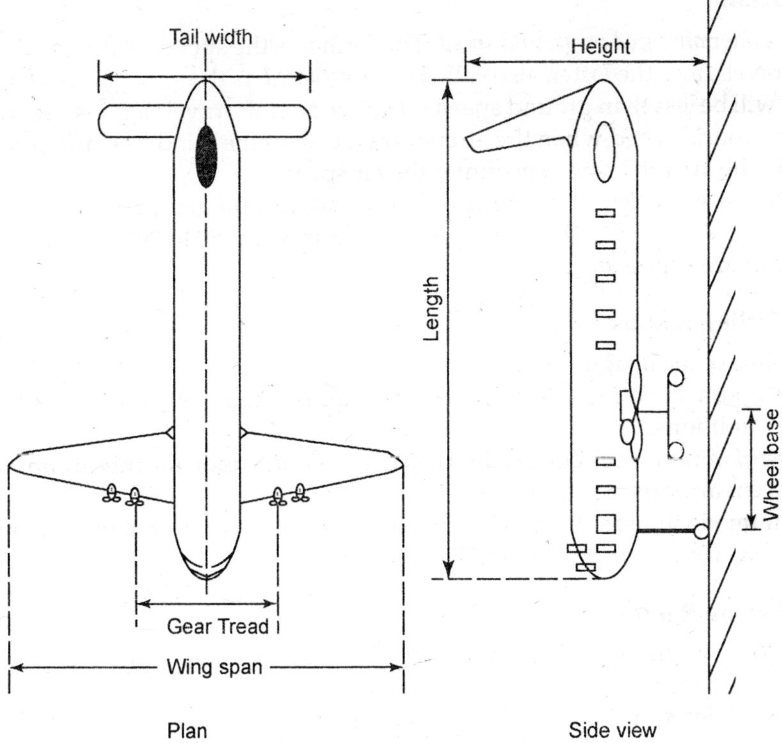


**FIG. 10.4** *Constituents of the size of aircraft*

*Pay load* means the weight of passengers and their baggage, and the mail and cargo, which fetch revenue.

*Maximum gross take-off weight* is the maximum load which the aircraft can carry during take-off; the pavements of the runways and taxiways are designed for this load.

*Maximum structural landing weight* is the difference between the gross take-off weight and the weight of the fuel consumed for the trip.

*Operating empty weight* is the weight of the aircraft including the crew and all the necessary items needed for the flight; it does not include the weights of the pay load and the fuel.

*Zero fuel weight* means the weight of the aircraft excluding only the weight of the fuel.

The landing weight should not exceed the maximum structural weight used for design of the runways.

*Fuel reserve* is the fuel necessary to take care of any contingency like emergency landing at an alternative airport.

*Trip fuel* will depend upon the distance to be travelled, altitude and speed of the aircraft, the pay load, and the atmospheric conditions.

As the range of the aircraft increases, the percentage of pay load decreases and that of the trip fuel increases.

**Tyre Pressure and Contact Area**

These will govern the design of the thickness of the airport pavements – runways and taxiways.

## Speed of Aircraft

The speed is either *air speed* or *ground speed*. The former is the speed relative to the air in which the aircraft travels, and the latter, also called *cruising speed*, is the speed relative to the ground.

Air speed will be less than ground speed when the aircraft travels against the wind, and will be more than ground speed when the aircraft travels with the wind; the difference constitutes the air speed. The aircraft wings encounter the air speed.

When the air speed is less than the speed of sound, it is *subsonic*, and when it is more than that of sound, it is *supersonic*. While most jet aircrafts travel at 80 to 90% of the speed of sound, military aircraft are supersonic.

## Minimum Circling Radius

This is the radius of the imaginary circle in space which should be free of any obstruction if the aircraft has to take a smooth and safe turn when required. It depends upon the type of aircraft and weather conditions.

Knowledge of minimum circling radius helps in deciding upon the minimum clear distances between adjacent airports.

The minimum circle radii required for jet aircraft can be up to 80km, thus restricting the operations in an adjacent airport within such distance.

## Minimum Turning Radius

This governs the design of taxiways and helps to ascertain the positioning of aircraft on the aprons and in the hangars.

The method of determining this is shown in Fig. 10.5.

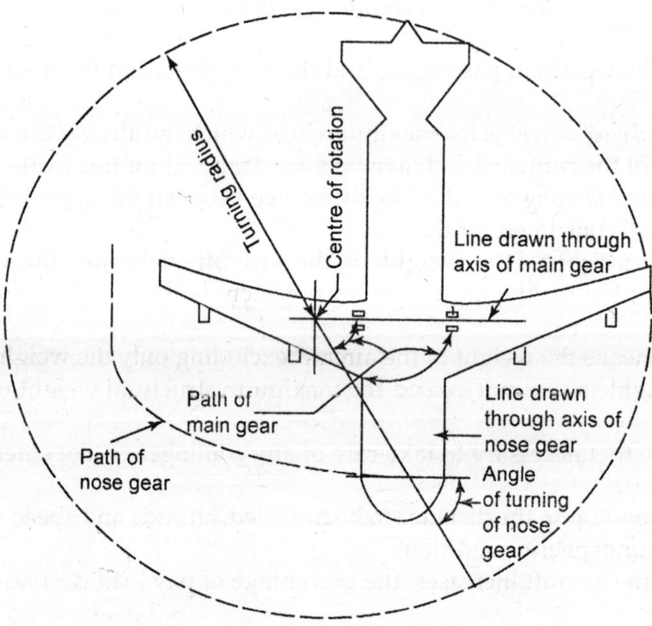


**FIG. 10.5** *Method of determining minimum turning radius*

The steps are:

(a) The line through the axis of the nose gear at its maximum angle of rotation is drawn (between 50° and 60° for a turbo jet).
(b) The line passing through the axis of the two main gears is also drawn.
(c) The intersection of these two lines gives the centre of rotation.
(d) The line joining this centre to the tip of the farther wing is the minimum turning radius

The paths of the main gear and nose gear may also be drawn as shown.

### Take-off and Landing Distances

These will govern the minimum length of runway required. These distances depend on the altitude of the airport, wind speed and direction, gradient of the runway, weight of aircraft at take-off and landing, and atmosphere temperature.

### Range of Aircraft

The maximum distance that an aircraft can fly without refuelling is called the *range;* this is inversely related to the pay load. Other conditions which affect the range are weather conditions, speed, flight altitude and reserve fuel.

### Noise of Aircraft

This is also crucial to minimise noise pollution. The noise level during take-off is more severe than that during landing.

*Perceived noise level* means the measured noise level adjusted by giving greater weightage to annoying decibel levels.

A further refinement is the *effective perceived noise level* when the perceived noise level is corrected for the duration of the sound and adjusted for the presence of pure tones. This is the index used for certification of an aircraft.

Modern technology has been making it possible to produce quieter aircraft.

### Fuel Spilling and Jet Blast

The spilling of fuel and lubricant and the blast from the jet affect bituminous surfaces; therefore, concrete is preferred for aprons and taxiways.

### Aircraft Capacity

This refers to the number of passengers (along with their baggage cargo) that can be accommodated. The terminal facilities are planned to deal with the total number of passengers who depart and arrive in the various flights in a close period of time.

### Civil and Military Aircraft

The functions of the two categories are totally different.

Military aircraft include fighter planes during emergencies and war. They are designed to travel at supersonic speeds and for heavier pay loads. High degree of manoeuvrability is a characteristic of such aircraft.

## Classification of Airports

Airports can be classified, in general, as follows:

- *Military airports*: These are meant to operate military aircraft for emergencies and natural calamities at times of peace, and for defence purposes during wartime.
- *Civil or commercial airports*: These are meant for civil aviation for ferrying passengers and cargo/ mail, both within the country and abroad. Airports which can cater to foreign travel are known as *international airports,* which are generally large in size. Medium and small airports cater to domestic flights, such airports are *domestic airports.*

The ownership or mode of operation may be *Public* or *Government, Private,* or *Private–Public Partnership (PPP or Joint ventures).*

Regular airports with all the facilities to operate flights are known as *Terminal Airports. Intermediate landing ports* may also be provided to facilitate emergency landings to ensure safety.

### ICAO Classification of Airports

The most popular classification of airports is the one given by the International Civil Aviation Organisation (ICAO).

The following are the two criteria used by the ICAO:

(a) Code letters A to E are used to indicate the basic length of the runway, width of runway, and the maximum longitudinal gradient (percent) as given below:

| Code letter | Basic length of runway(m) | | Width of runway (m) | Max. Longitudinal Grade (%) |
|---|---|---|---|---|
| | Maximum | Minimum | | |
| A | Over 2100 | 2100 | 45 | 1.5 |
| B | 2099 | 1500 | 45 | 1.5 |
| C | 1499 | 900 | 30 | 1.5 |
| D | 899 | 750 | 22.5 | 2.0 |
| E | 749 | 600 | 18 | 2.0 |

(b) The numbers 1 to 7 are used to indicate the isolated wheel load and tyre pressure, as shown below:

| Code Number | Isolated wheel load (kg) | Tyre pressure ($kg/cm^2$) |
|---|---|---|
| 1 | 45,000 | 8.5 |
| 2 | 34,000 | 7.0 |
| 3 | 27,000 | 7.0 |
| 4 | 20,000 | 7.0 |
| 5 | 13,000 | 6.0 |
| 6 | 7,000 | 5.0 |
| 7 | 2,000 | 2.5 |

The designations would be A–3, B–2, C–1, etc., depending upon the values of these parameters.

## Different Kinds of Flights

Flights may fall into one of the following categories:

(i) *Scheduled commercial flights*: These are operated to a certain plan at a particular time. The flights may be on domestic or international routes.
(ii) *Non-scheduled commercial flights*: These are not operated according to any plan. The purpose may be scientific research, aerial survey, aerial photogrammetry, or aircraft testing.
(iii) *Military flights*: These may be for specific military purposes such as training, bombing in war, air surveillance, military intelligence, transport of military personnel, medical aid, and so on.
(iv) *Personal flights*: These may be chartered flights or those using personal aircraft.

### Forecasting Air Traffic Potential

This is an important requirement for planning an airport and the facilities needed to handle the expected air traffic in a specified design period.

The data needed for such a forecast are:

- Population
- Topographical and geographic features
- Existing airports in the vicinity
- Air traffic characteristics vis-à-vis the existing modes of transport facilities.

Future estimates of the following are normally required for planning an airport.

1. Annual volume of passenger traffic.
2. Annual volume of aircraft
3. Peak-daily volume and peak-hourly volume of passenger traffic and of aircraft.
4. Air cargo
5. Air mail
6. General purpose aviation

### Annual Volume of Passenger Traffic

The data of the number of emplaned passengers for a particular region, *R*, as well those for the nation as a whole, are collected for the past decade (or as many years as possible).

The percentages of emplaned passengers for the region with respect to the national values are obtained. From the past trends, a reasonable growth potential is decided. (This involves an element of the subjectivity in approach). This growth percentage is applied first to the national value to forecast the annual volume of passengers; then the particular regional forecast value is obtained by applying the same growth factor.

This does not account for the population growth of the region during the forecast period. When the population statistics are available for the nation and the region, the whole procedure can be modified to account for this growth also.

### Annual Volume of Aircraft Traffic

The technique used for the forecast of annual volume of passengers can also be extended for the forecast of annual volume of aircraft departures and the number of passengers emplaned per departure.

**Peak-daily Volume and Peak-hourly Volume of Passenger Traffic and of Aircraft**

Peak-hourly volume of passenger traffic and of aircraft are required to decide upon the extent of the facilities required in the terminal building and facilities, as well as the design of runways, taxiways and aprons.

Although a specific hourly volume – the thirtieth highest hourly volume – is used for the geometric design of highways, such a criterion is not chosen for the design of airports. A reasonable peak-hourly volume is chosen based on the available past records and interaction with the authorities responsible for operation of the airport.

To estimate the peak-hourly volume, peak-monthly and peak-daily volumes of passenger traffic are determined. The procedure for this is the same as that for estimating the annual volume of passenger traffic. The peak-hour aircraft movements are obtained from the service schedules by selecting a suitable percentage for forecasting.

$$\text{Peak-hour aircraft movement} = \frac{\text{Peak month}}{\text{Annual}} \times \frac{\text{Peak day}}{\text{Peak month}} \times \frac{\text{Peak hour}}{\text{Peak day}} \times \text{Annual volume}$$

The value of the peak-hour aircraft movement forecast in this manner is multiplied by the forecast of emplaned passengers per aircraft departure. This gives the forecast peak-hour volume of passenger traffic.

Of course, this does not include the traffic due to general-purpose aviation and / or military operations; these should be estimated separately and added.

**Air Cargo**

Although this industry is of recent origin, the available records and statistics may be used in exactly the same way as for the peak-hourly volume of passenger traffic to forecast the total tonnage of air cargo. From this, the number of aircraft required to carry this cargo is estimated.

**Air Mail**

The responsibility of forecasting the aircraft and other terminal facilities required for dealing with air mail cargo rests with the Department of Posts, a similar procedure as for cargo may be employed for this too, if necessary.

**General-purpose Aviation**

This includes traffic for miscellaneous purposes which are not pre-planned, such as business aerial surveys, aerial photography, aerial spraying, tourism and training.

Each component must be dealt with separately, with appropriate estimates, using the same technique as for the items, in order to obtain the forecast design volumes of aircraft and of traffic in this regard.

## 10.2 IMPORTANT DEFINITIONS

Certain important terms and their definitions according to airport terminology are given here to facilitate the understanding of airport engineering

1. *Aerodrome*: A defined area intended to be used for the arrival and departure of aircraft; facilities for dealing with the passengers and cargo, as well as for shelter and repairs of aircraft may be provided here.
2. *Aeroplane*: A power-driven heavier-than-air flying machine with fixed wings.

3. *Aircraft*: Any machine designed to navigate through atmospheric air; it may or may not be power-driven and may be lighter or heavier than air. A general term which includes a glider, helicopter, aeroplane and rocket.
4. *Airfield*: An area for aircraft landing and take-off; this may not be provided with the facilities for passengers and the aircraft.
5. *Airport*: An aerodrome mainly intended for the operation of commercial flights. It is provided with all the necessary facilities for passengers as well as for aircraft, commensurate with its classification.
6. *Airport elevation*: The elevation of the highest point of the landing area above mean sea level.
7. *Approach zone*: The wide area on either side of a runway up to a certain distance which is kept clear of any obstruction; this area is trapezoidal in shape when projected on to the ground with increasing width from the end of the runway outwards.
8. *Apron*: A defined area to accommodate aircraft for loading of cargo, emplaning and deplaning of passengers, parking and refuelling of aircraft, and so on. It is paved and is located near the terminal building or adjacent to hangars.
9. *Beaufort scale*: A scale to indicate the wind velocity, devised by Admiral Beaufort of the British Navy, and adopted by the World Metrological Committee. It is used as part of the code in communications for aircraft control.
   It is reproduced below:

| Beaufort Number | Wind description | Velocity at a height 10 m above flat open ground (km/h) |
|---|---|---|
| 0 | Calm | 1.60 |
| 1 | Light | 3.20 |
| 2 | Slight breeze | 8 |
| 3 | Gentle breeze | 16 |
| 4 | Moderate breeze | 24 |
| 5 | Fresh breeze | 34 |
| 6 | Strong breeze | 43 |
| 7 | Moderate gale | 56 |
| 8 | Gale | 67 |
| 9 | Strong gale | 80 |
| 10 | Whole gale | 95 |
| 11 | Strom | 110 |
| 12 | Hurricane | More than 120 |

10. *Cargo*: Freight, other than passengers, baggage and mail, carried by an aircraft.
11. *Clearway*: A rectangular area at the end of a strip in the direction of take-off over which an aircraft makes its initial climb.
12. *Control area*: The airspace of specified dimensions within which air traffic control is exercised.
13. *Control tower*: A tower which is normally situated at the top of the terminal building, enclosed in glass, with an unobstructed view of the entire air-field. This is used to control flight operations within the control area.

14. *CTOL*: Conventional take-off and landing.
15. *Flight time*: The total time from the moment the aircraft first moves under its own power for taking off to the moment it stops at the end of the flight.
16. *Flight visibility*: The average range of visibility in the forward direction from the cockpit of an aircraft in flight.
17. *Hangar*: A large shed at the airport for housing, servicing, and repairs of an aircraft.
18. *Holding apron*: The designated portion placed adjacent to the ends of runways for facilitating check of aircraft instruments and engine operation prior to take-off, and for waiting till clearance for take-off is given.
19. *Landing area*: The portion of the airport excluding the buildings, used for landing and take-off of the aircraft.
20. *Landing strip*: A long and narrow area, suitable for landing and take-off operations; it consists of the runway plus the shoulders on either side of it.
21. *Mach number*: The ratio of the velocity of a body to that of sound. Mach 1 indicates the speed equal to that of the sound, which is nearly 1130 km/h.
22. *Runway*: A long a narrow strip, prepared for landing and take-off of aircraft; it is usually paved with cement concrete.
23. *Standard atmosphere*: It is an imaginary content of dry air around the earth and is defined by the ICAO to have the following conditions:

    (i) The air is perfectly dry.
    (ii) The temperature at sea level is 15°C.
    (iii) The pressure at sea level is 760 mm of Hg.
    (iv) The temperature gradient from sea level is –0.0065°C per metre up to the altitude at which the temperature becomes –15.5°, and zero above it.

24. *Stopway*: The rectangular area at the end of the runway in the direction of take-off in which an aircraft can be stopped after an interrupted take-off. Its width is equal to the width of runway and thickness sufficient to bear the weight of the aircraft.
25. *Surveillance radar*: This provides an overall picture of the surrounding atmosphere within a radius of 50 km to 100 km; it moves through 360°, and an aircraft within its range appears as a dot with a luminous tail, indicating its path of movement.
26. *Taxiway*: A path defined for the use of taxiing aircraft to and from the runway and the apron.
27. *Terminal area*: The portion of the airport other than the landing area, which includes the terminal building apron, cargo storage, hangars, car parking area and so on.
28. *Wind rose*: A diagram showing the direction, duration, and intensity of wind over a certain period in a specified region; it resembles the shape of a rose.
29. *Zoning*: Specifying certain areas or zones for specific functions for the operations of flights in the vicinity of airports in order to ensure safety in aviation.

## 10.3 AIRPORT PLANNING

The planning of an airport is a complex process since it encompasses a wide range of activities with different and often conflicting requirements. Airport planning has to take into account, not only the local or regional aviation needs, but also the national needs as a whole and follow the guidelines established by the national and international authorities concerned. Thus, the overall development of civil aviation in the country has to be kept in mind, while planning an airport in a specific place.

**Objectives of Airport Planning:**

- Justification of the need for the airport
- Formulation of the layout plan based on the assessed requirements
- Integration of the planning with regional needs and national needs.

**Steps in Airport Planning**

- Estimation of traffic potential
- Site selection
- Layout plan
- Design of various components
- Preparation of working drawings
- Cost estimation
- Proposals for financing and institutional arrangements
- Environmental impact assessment (EIA)

## Master Plan

A *master plan* may be made for the development of airports in the entire country on the basis of needs as well as priorities, bearing in mind national interests. There can also be a master plan for a particular state or a region based on similar criteria.

In the same way, a master plan has to be prepared for any newly planned airport. This is because, it is sufficient to start an airport with minimum functional requirements, and then continue to develop it further as the demand for air traffic increases. Thus, the primary aim of a master plan is to provide details for future development of an airport to satisfy the gradually increasing air traffic requirements.

If the master plan is prepared taking into consideration only the traffic potential and growth forecast, it is said to be an *unconstrained* forecast; if, however, the constraints at the particular location, such as non-availability of land, restriction on noise levels, etc., are also considered in the preparation, it is said to be a *constrained* forecast.

The ICAO, based on the US-based *Federal Aviation Agency* (FAA) recommendations has given guidelines for the preparation of the master plan.

## Regional planning

This aims at the formation of an effective network of airports in a zone or region, and ultimately in a country, in such a way that the planning of an individual airport smoothly merges into the regional plan of other airports, and the regional plan, in turn, merges into the national plan.

Improper location of airports may lead to airspace conflicts and wastage of resources. The civil aviation authorities must be involved in the preparation of the regional plan in collaboration with the State Government.

In order to make the regional planning of airports sound and effective, data should be collected pertaining to the anticipated air traffic, existing airports in the region, population and their patterns of usage of air travel, in addition to the relevant topographical and geographical features of the proposed airport.

The ICAO has stipulated certain minimum distance between adjacent airports in order to avoid any interference in the working of electronic instruments used by the control tower operator and the aircraft pilot; also, this minimum distance ensures adequate traffic potential to all airports in the region.

These are:

| | |
|---|---|
| Small airports (under visibility flight rules (VFR)) | 3 km |
| Bigger airports (under VFR) | 6 km |
| Airports operating piston engine aircraft | 25 km |
| Airports operating jet aircraft | 160 km |

These are only guidance values for regional planning.

**Improvement of an Existing Airport**

It is essential to consider making improvements to an existing airport to cater to the needs of increased air traffic before deciding to construct an entirely new facility.

For this, an appropriate traffic forecast for a reasonable design period, say 20 years, is the first step.

Next, the capacity of the existing airport should be assessed in respect of the facilities such as aprons, hangars, taxiways, runways, motor vehicle parking areas, passenger facilities in the terminal building, baggage and cargo handling facilities, and so on.

This should be compared to the anticipated short-term and long-term air traffic and the facilities required to deal with them efficiently.

It will then be possible to plan the measures needed to improve the existing capacity – construction of new runways, exit taxiways, traffic control devices, loading apron and terminal facilities.

The economics of such renovation should be studied, and a decision to build a new airport should be taken only if the improvements are not justifiable from the viewpoint of economy.

**Site Selection for an Airport**

Site selection is an important element in airport planning. Several factors influence the location of an airport. The capacity, safety and efficiency of airport operation depend very much on the choice of an appropriate site.

Hence, alternative sites should be studied and the best among them chosen based on all the relevant considerations.

**Factors affecting the site selection for an airport:** The following are the important factors that govern the selection of the site for an airport.

(i) Atmospheric and climatic conditions
(ii) Topography
(iii) Location of other airports in the vicinity
(iv) Accessibility
(v) Availability of land for future needs
(vi) Availability of utility services
(vii) Land use pattern of the surrounding area
(viii) Regional plan
(ix) Soil characteristics
(x) Obstructions surrounding the site

(xi) Economy of construction
(xii) Purpose of the airport.

(i) ***Atmospheric and climatic conditions***: Visibility is affected by fog, haze and smoke, which hamper flight operations and impair its capacity. Fog tends to settle into areas with less wind. Smoke and haze are present near industrial areas. Thus, analysis of weather records including wind velocity and direction will be helpful in the site selection.
(ii) ***Topography***: Topographical features like ground contours, streams, trees and vegetation should be studied with the aid of toposheets for the area. A raised ground with good drainage and good visibility will be an ideal site.
(iii) ***Location of other airports in the vicinity***: According to the recommendations of the ICAO, a certain minimum distance is necessary between adjacent airports to prevent obstructions in the operation of the aircraft and possible interference of signals.
(iv) ***Accessibility***: This is an important criterion, not only for airline users, but also for the other sections of the public. Proper accessibility reduces the overall travel time and hence, it is an attractive feature for selecting an airport site.
(v) ***Availability of land for future needs***: This is necessary for future expansion needs.
(vi) ***Availability of utility services***: Utility services like water supply, electricity and telephones are essential for an airport.
(vii) ***Land use pattern of the surrounding area***: Proximity to residential areas, educational institutions, and hospitals should be avoided. Noise pollution near these areas is objectionable
(viii) ***Regional plan***: The site should fit appropriately into the overall regional plan, and ultimately into the national plan of development.
(ix) ***Soil characteristics***: Good pervious soil is desirable for effective drainage; a soil with good bearing capacity is desirable for the construction of buildings and runways.
(x) ***Obstructions surrounding the site***: For efficient landing operations, the site should be clear of possible obstructions for visibility as well as circling of the aircraft.
(xi) ***Economy of construction***: This is obviously an important criterion for site selection.
(xii) ***Purpose of the airport***: If the airport is for military use, or may be taken over for this purpose during war, protection from possible air attacks also becomes a criterion.

### Size of an Airport

The factors governing the size of an airport are the following:

- Meteorological conditions
- Elevation of the site
- Characteristics of the aircraft
- Volume of traffic

Wind and temperature characteristics, runway length, landing and take-off requirements of the airport, and the volume and intensity of the passenger traffic to which the airport has to cater, are all important factors that govern the required size of the airport. All these factors are embodied in the four main criteria listed above.

### Capacity of an Airport

The number of aircraft movements which an airport can process or handle in a specified period of time such as one hour (or one day) is called *airport capacity*.

For this purpose, a landing or a take-off operation is taken as one movement.

A reasonable delay in the departure is allowed for in reckoning the capacity.

## 10.4 AIRPORT OBSTRUCTIONS

Avoiding obstructions in and around airports and imposing certain restrictions in this regard are essential to ensure safety in aircraft operations.

These obstructions are broadly classified as (1) Imaginary surfaces and (2) Objects with actual heights.

### Imaginary Surfaces

The following are the types of imaginary surfaces.

(a) Approach surface
(b) Conical surface
(c) Horizontal surface
(d) Take-off climb surface
(e) Transitional surface
(f) Baulked landing surface

#### Approach Surface

This is taken to be trapezoidal in shape. Its longitudinal centre line is got by imagining the runway centreline to be extended. It slopes upwards from the end of the runway on landing side, and its sides diverge as they extend away from the end of the runway. It is applicable to each end of the runway.

A rectangular strip of the approach surface lying inside, near the threshold or end of the runway, for a relatively smaller length within the conical surface is called the *inner approach*.

The ICAO standards for approach surface are given below for airport/runway codes A, B and C:

| Approach surface item | Instrument landing | Non-instrument landing |
|---|---|---|
| Width at the landing end | 300 m | 150 m |
| Length of horizontal projection | 15 km | 3 km |
| Longitudinal gradient | 2 % | 2.5 % |
| Divergence of the sides | 15 % | 10 % |

#### Conical Surface

This extends upwards and outwards from the outer circular edge of the horizontal surface to a point which is at a specified height above it. The conical surface has a larger radius at the top than at the bottom, thus resembling a frustum of a cone, sitting on the horizontal surface. The slope of the conical surface is measured in a vertical plane perpendicular to the periphery of the inner horizontal surface.

The ICAO standards for conical surface are given below for A, B, C runways:

| | |
|---|---|
| Height of upper circular edge above the horizontal surface | 100 m (A and B) |
| | 75 m (C) |
| Radius of inner circular edge (with airport reference point as origin) | 4 km |
| Side slope | 5 % |

### Horizontal Surface

This surface extends from the upper edge of the transitional surface to the inner or lower edge of the conical surface; thus, the outer edge of the *inner horizontal surface (IHS)* is the same as the inner edge of the conical surface. The radius of the IHS is measured from the airport reference point (ARP).

The *outer horizontal surface (OHS)* is circular in plan with its height 150 m above ARP elevation. For runways of 900 m to 1500 m length, the OHS extends to 9.9 km from the elevation of ARP. For runways over 1500 m long, the OHS extends to 15 km above ARP. No structure should be allowed to protrude above the OHS.

### Take-off Climb Surface

This surface is trapezoidal in shape and rises from the end of the runway on the take-off side. Its sides diverge as they extend away from the end of the runway.

The ICAO standards for A, B, C runways are:

| | |
|---|---|
| Width near the running end | 180 m |
| Length of the horizontal projection | 15 km |
| Longitudinal upgrade | 2 % |
| Divergence of the sides | 12.5 % |

### Transitional Surface

This is trapezoidal in shape with side slopes, and extends along the landing strip and into a part of the approach surface. It slopes (at 14.5% for A, B & C runways) upwards and outwards to the inner horizontal surface.

### Baulked Landing Surface

This is an inclined surface located at a specified distance from the threshold of the runway, extending up to the inner transitional surface. It is meant to take care of baulked landings in a safe manner.

These surfaces[1] are shown schematically in the layout given in Fig. 10.6.

[1] All of these or only some may be required depending upon the purpose of the runway and operation control systems.

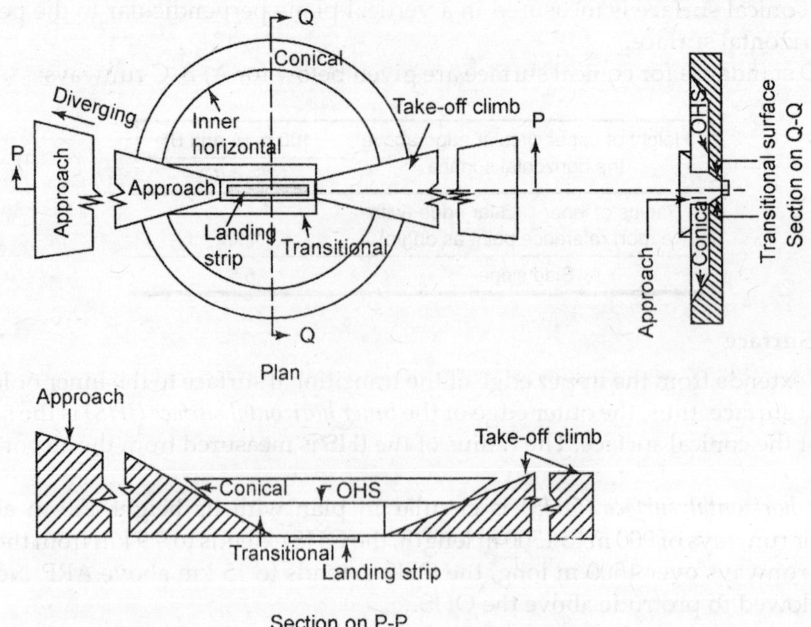


**FIG.10.6** *Schematic layout of imaginary surfaces*

## Objects with Actual Heights

Any object which exceeds a specified height above the ground is treated as an obstruction for operating flights.

The following are the ICAO stipulations for airports equipped with instrument landing facilities.

(i) Any object within 4.5 km distance from the end of the runway, with an actual height of 30 m above the level of the approach end of the runway, is considered an obstruction.

(ii) Any object located beyond a distance of 4.5 km from the end of the runway, with an actual height above 30 m and increasing at the rate of more than 7.5 m for each additional 1.5 km distance, or exceeding 75 m within 15 km distance, is considered an obstruction.

(iii) Any object projecting above the minimum approach flight attitude, or exceeding 150 m above the ground level, is also considered an obstruction.

Keeping in view these stipulations, a profile of the approach zone may be drawn for reference.

## 10.5 AIRPORT ZONES AND ZONING LAWS

A *zone* near an airport is a well-defined and demarcated area with certain common requirements. It is typical of an aircraft to gradually change its altitude in landing and take-off operations. This requires large clearances on both sides of the paths during landing and take-off, which may range up to a distance of 15 km on either side of the runway. It is practically impossible to acquire such large areas by the airport authorities. Therefore, certain restrictions are enforced by the authorities regarding the heights of structures and types of land use in airport zones.

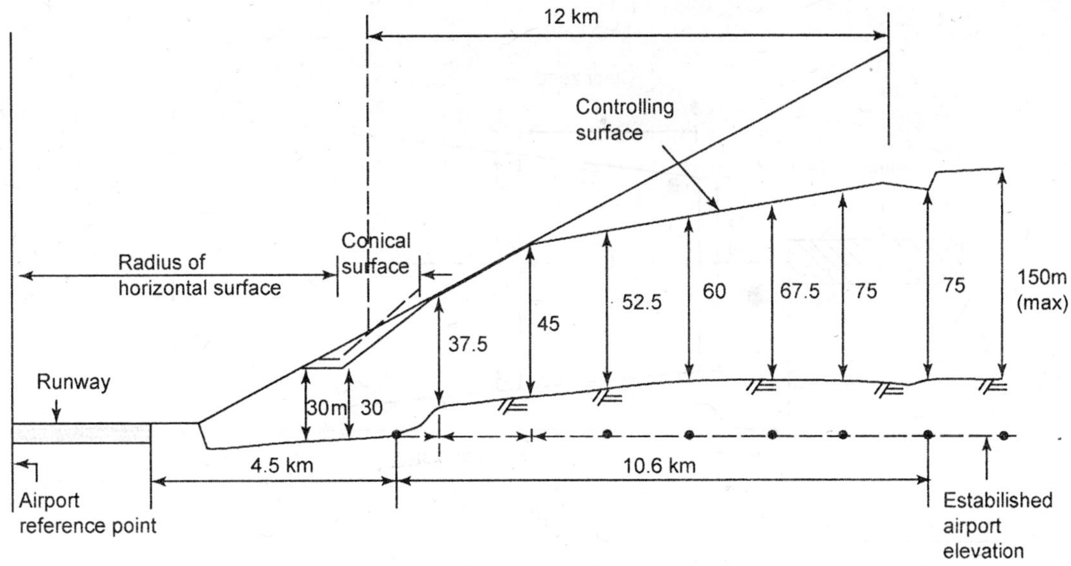


**FIG. 10.7** *Typical profile of approach zone for ILS runway*

These zones are:

(a) Approach zone
(b) Clear zone
(c) Turning zone

## Approach Zone

This is an area through which aircraft approach the airport during landing, or leave the airport during take-off. Since the change in attitude in either case is gradual, a wide clearance is required on either side of the runway in the approach zone. Height restrictions are imposed for structures that come up in this zone as given below:

(i) Structures up to 30 m height are permitted up to a distance of 4.5 km, and 60 m from the edge of runways.
(ii) Beyond 4.5 km and up to 10.5 km, an increase in height of 7.5 m for every 1.5 km distance is permitted.
(iii) Within 15 km from the edges of runways, no development should be allowed to exceed a height of 150 m.

A typical profile of an approach zone for a runway operated by Instrument Landing System (ILS) is shown in Fig. 10.7.

## Clear Zone

This is the innermost portion of the approach zone and is provided at the ends of the runway. This should preferably be a level area that is free of all major obstructions, since it is a critical zone for all aircraft operations.

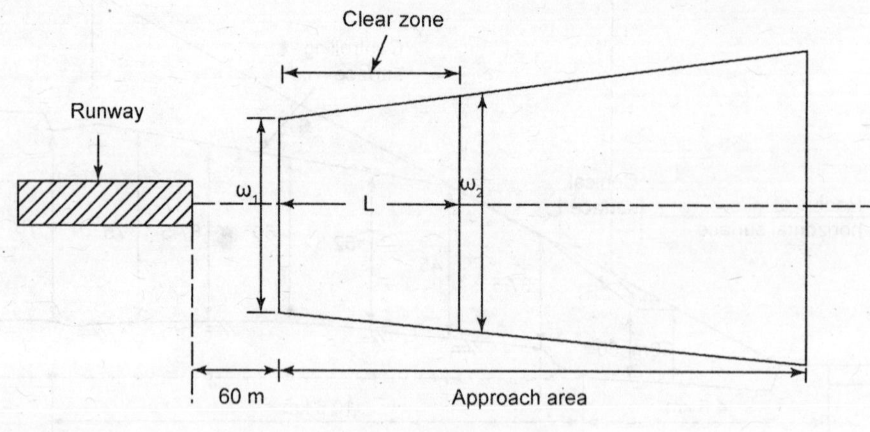


**FIG. 10.8** *Clear zone*

The clear zone is of trapezoidal shape in plan, and the important dimensions are the end widths and the length (Fig. 10.8).

ICAO guidelines for this zone are given below:

| Type of landing | Instrumental Landing | Non-instrumental landing | |
|---|---|---|---|
| | | Large airport | Small airport |
| $w_1$ (m) | 300 | 150 | 75 |
| $w_2$ (m) | 525 | 270 | 135 |
| L (m) | 750 | 600 | 300 |

The length of the clear zone is determined by the distance required to reach a height of 30 m for the approach surface.

## Turning Zone

This is the area of an airport other than the approach area, which is meant for turning operations of the aircraft in case of emergencies such as engine failure of an aircraft at the beginning of take-off. The pilot then takes the turn and comes in the line of the runway before landing. The aircraft, at such instances, operates at a low height in this zone, and it is therefore essential to make this free from obstructions of any kind.

The guidelines for the turning zone profile are:

(i) The height of any object within 4.5 km from the airport reference point (ARP)[2] should not exceed 51 m above the ground or the established airport elevation, whichever is greater.

(II) The height of any object located beyond 4.5 km from the ARP should not exceed 51m plus 30 m for each additional 1.5 km distance, and also should not exceed 150 m within a distance of 15 km from the ARP.

[2]ARP is the centre point of the airport, located at the geometric centre of all the usable runways—the weighted average of the end of runway coordinates.

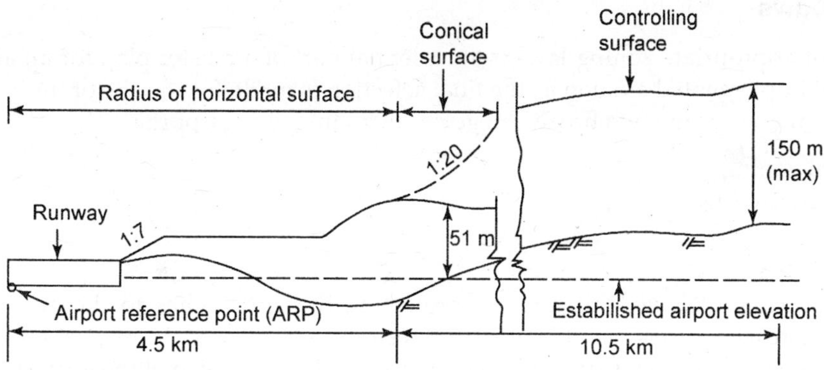


**FIG. 10.9** *Typical profile of airport turning zone for a runway with ILS*

A typical profile of turning zone for a runway with instrument landing system is shown in Fig. 10.9.

## Other Airport Zones

### Buffer Zone

This is the zone in the direction of take-off of aircraft beyond the runway for about 4.5 km, in which no development is permitted. It is meant primarily for noise abatement and minimising noise pollution of developed residential areas.

### Highways and Railways in Clear Zones

Highways and railways can be allowed in clear zones provided the clearance standards are adhered to and the vehicles are always in motion. The essential clearance requirements over a highway or a railway located in the approach area are shown in Fig. 10.10:

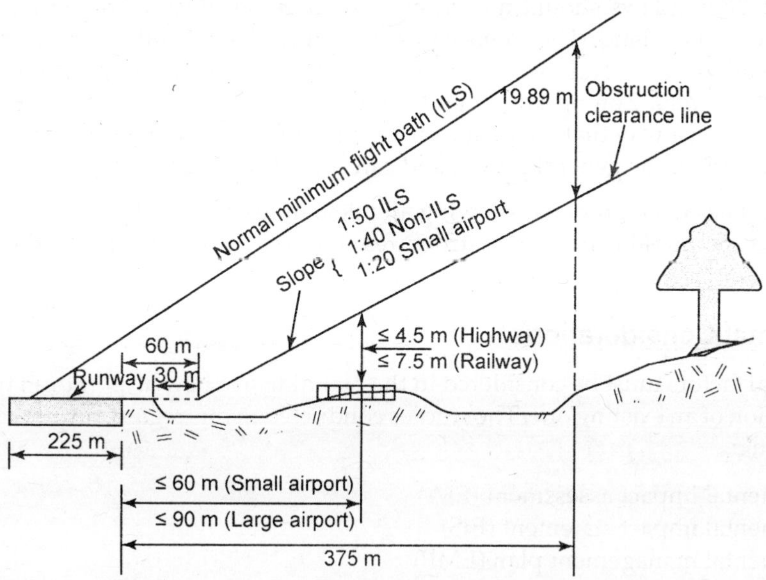


**FIG. 10.10** *Clearance over highways/railways near airports*

## Zoning Laws

Framing of appropriate zoning laws is an essential part of a master plan for an airport; these have to be implemented as soon as the final selection is made for the site of the airport.

The following are the two broad categories of zoning near airports:

(a) Height zoning
(b) Land-use zoning

### Height Zoning

Also called *hazard zoning*, this aims at protecting the approaches to the airports from all kinds of obstructions. The heights of structures in the vicinity of the airport are regulated or restricted. These restrictions depend, magnitude -wise, on the size of the airport and the type of aircraft operated.

### Land-use Zoning

This places restrictions on land-use and development in the vicinity of an airport.

Some types of land-use may be closely related to aviation – for example, terminal buildings, parking, aprons, runways and taxiways, hangars, and other maintenance facilities.

Other types of land use such as for commercial, agricultural, industrial and recreational purposes come under non-aviation category. Zoning laws relating to this must ensure that such types of land use do not interfere with aircraft operations in any manner. Industries which give rise to smoke are not desirable as smoke affects the visibility for aircraft operation.

Agricultural use may be permitted if the particular crops do not attract birds, which are hazardous as they could cause accidents through bird hits. Recreational facilities that do not need the erection of high structures and which do not cause much noise may be permitted; for example, golf courses.

The following *factors* are to be borne in mind while *framing zoning laws*:

- *Legal rights*: Zoning laws should not interfere with the legal rights of land owners on the airspace above their land. The interests of both airport authorities – regarding safety of aircraft operation – and land-owners in respect of their rights should be protected.
- *Nature of zoning laws*: Zoning laws should not be unreasonable or arbitrary. They should be framed in such a way that a judicious balance is achieved between the welfare, comfort, safety and health of the general public, and safe and efficient aircraft operations.

A *zoning map* has to be prepared showing all relevant features such as property lines, land-use regulations, permissible heights of structures , the area of applicability of the zoning laws, and so on.

## Environmental Considerations

Environmental factors must be considered in the planning of a new airport or in the expansion and upgradation of an existing one. The studies conducted in this regard aim at the preparation of the following:

(a) Environmental impact assessment (EIA)
(b) Environmental impact statement (EIS)
(c) Environmental management plan (EMP)

**Environmental Impact Assessment (EIA)**

This is meant to determine the potential effects on the environment around the proposed airport construction/expansion.

When large airports are planned in strategic or ecologically sensitive zones, certain effects of long-term consequence may be involved—for example, disturbance of wild-life habitats, streams and major natural drainage courses, and recreational areas of tourist interest/ potential.

These effects may be studied by comparing alternative sites and assessing the extent of damage possible in each case.

**Environmental Impact Statement (EIS)**

The EIA report is used to prepare this impact statement, which includes the following:

(i) Unavoidable effects
(ii) Avoidable alternative developments.
(iii) Environmental impact of the proposed development.
(iv) Overall view of the merits and demerits of the proposal.
(v) Overall advantage/justification of the proposal.
(vi) Remedial measures to minimise the impact.
(vii) Miscellaneous factors such as land-use, water table, traffic congestion in the vicinity, interference with wild-life, increase in noise pollution levels, air pollution, water pollution, disturbance to the aesthetics and natural beauty of the region and so on.

**Environmental management plan (EMP)**

The EMP contains the measures for implementation of recommendations made in the EIS for controlling the adverse effects on the environment due to the proposed development.

It has to include the plan for mobilisation of financial and other resources, and details of phasing, if any, needed for the implementation of remedial measures to contain the possible damage to the environment as brought out in the EIA, and summarised in the EIS.

## 10.6 AIRPORT SURVEYS AND DRAWINGS

As with highway and railway projects, surveys of different kinds play a major role in the formulation of airport projects. The three major types – reconnaissance survey, preliminary survey, and finally, location survey (paper location and ground location) – are common to all such projects. The objectives, instruments used, data collected, and the conclusions sought to be drawn, are similar in all these projects. These have been studied in detail in both Highway Engineering and Railway Engineering. However, certain aspects which are relevant to airport engineering will be considered here.

### Objectives of Airport Surveys

(i) To demarcate the airport site on the ground selected after considering the data obtained from preliminary survey.
(ii) To collect data relating to the meteorological conditions at the site.

(iii) To collect data necessary for design of the various components of the airport.
(iv) To ascertain the soil characteristics at the site.
(v) To suggest measures for improving the site conditions, if necessary.
(vi) To suggest the use of locally available materials for construction.
(vii) To make provision for future expansion of the airport.
(viii) To prepare the necessary working drawings for the execution of the project.

## Types of Airport Surveys

The following surveys are relevant to airport projects:

1. Topographical survey
2. Meteorological survey
3. Approach zone survey
4. Soil survey
5. Drainage survey
6. Survey of natural resources
7. Traffic survey

These are now briefly considered below:

### Topographical Survey

This involves the study of topo-sheets of the area available from the Survey of India. Surface features such as hillocks, valleys, streams, rivers, and other relevant features which can affect the design of the various components of the airport are studied.

### Meteorological Survey

The study of the weather and the climatic conditions at the site is known as meteorological survey.

Information is collected about the following parameters in this survey:

(i) Atmospheric pressure
(ii) Temperature
(iii) Rainfall intensity and duration
(iv) Wind direction, duration, and intensity
(v) Snow, frost and fog, which affect visibility

Wind data is needed for deciding the orientation of runways, while atmospheric pressure impacts the length of the runways.

Rainfall data will be helpful in the design of airport pavement and drainage systems.

### Approach Zone Survey

This has relevance to the landing and take-off operations and the clearances associated with it.

The heights of objects and ownership of structures in the approach zone are ascertained, and the zoning laws and regulations are imposed based on the data from this survey.

In fact, this may be considered as part of topographical survey, or in other words, as its extension.

### Soil Survey

This has relevance to the design of runways, buildings, and other airport structures.

Soil survey is also helpful in the assessment of drainage needs and the design of the systems needed for efficient drainage of the airport area.

Further, it helps to assess the need for improvement of the soils available at the site in terms of its bearing power and drainage qualities.

### Drainage Survey

Strictly speaking, this may be considered to be a part of soil survey. In fact, a combined consideration of the soil conditions and the rainfall characteristics is necessary to study the drainage characteristics of the airport area. This is the objective of drainage survey.

### Survey of Natural Resources

The aim of this survey is to collect the information about locally available construction materials, the varieties and the quantities available, and the nearby water source to facilitate the construction of the various airport structures.

### Traffic Survey

Data relating to the existing air traffic and an estimation of future traffic are helpful, both in the design of facilities for new airport and for improvement of an existing one; statistical methods are used for such forecasting. The design period could be 10 to 20 years, depending upon the importance of the airport.

## Airport Drawings

The following are the drawings required to be prepared for the execution of an airport project:

- Master plan
- Topographic plan
- Obstruction plan
- Grading plan
- Drainage plan
- Paving plan
- Layout plan
- Lighting plan

All these plans in the form of working drawings that include relevant details of each these aspects are helpful in the systematic execution of an airport project and various components thereof.

### Cost Estimation

This consists of two parts – capital cost or initial cost of construction, and annual maintenance cost of the various components of the airport. Capital cost includes the cost of land acquisition, construction cost of the various structures such as terminal buildings, landing area, taxiways, runways and hangars, the cost of air-traffic control devices and the provision of visual aids.

### Economic Evaluation

This involves the economic justification of the project over the design period; it is a cost–benefit analysis of the airport project.

This evaluation is done in addition to studying the engineering feasibility, environmental justification and social benefits.

**Institutional Arrangements**

These are concerned with the organisation of planning, design, execution and maintenance.

Mobilisation of financial resources for the airport project is based on the participation of the Government and private funding agencies, and on the concepts of Build–Operate–Transfer (BOT), Build–Own–Operate–Transfer (BOOT), and other similar patterns of public–private partnership (PPP) for handling such big projects.

## SUMMARY

- Transportation through air using an aircraft is called *Air Transport*.
- All the engineering aspects of air transport and its operation constitute *Airport Engineering*. An airport is the place where take-off and landing operations of aircraft take place for commercial services.
- The Wright brothers of the USA are credited with the invention of aeroplane in 1903. Air transport made rapid strides during and after the Second World War. The important characteristics of air transport are high speeds, continuous long-distance travel, and providing access to remote locations.
- The International Civil Aviation Organisation, the National Airports Authority and the Airport Authority of India in our country are responsible for laying down standards and regulations for operating international and domestic flights for commercial purposes.
- Several characteristics of aircraft – the type of propulsion, size and capacity among others – influence the planning and design of an airport.
- ICAO classification of airports is based on runway dimensions – length, width and grade – as A, B, C, D, and E, and on the maximum wheel load and tyre contact pressure (1 to 7). Estimation of future air-traffic is necessary for the design of an airport and the facilities to be provided.
- Site selection for an airport is an important part of planning; several factors relevant to the safe and efficient operation of aircraft for commercial services have to be considered.
- Obstructions in and around airports are of two kinds – imaginary surfaces and objects with actual heights. Under imaginary surfaces, approach, conical, horizontal, take-off climb, and transitional types are included. Restrictions on the heights of objects around airports fall in the latter category. The aim of such restrictions is to ensure safe landing and take-off operations in accordance with the stipulations laid down by ICAO.
- There are different zones in and around an airport – approach zone, clear zone, and turning zone, with ICAO stipulations for each.
- Zoning laws are framed with regard to height zoning and land-use zoning. These stipulations and restrictions are meant to ensure safety in aircraft operations. Environmental considerations also have to be borne in mind in airport planning.
- Airport surveys help in the planning and site selection – topographical, meteorological and traffic surveys are the more important among them. Detailed working drawings have to be prepared for the execution of an airport project. Cost estimation and economic evaluation are essential for justification of the project. Institutional arrangements are required for planning, design, construction and maintenance; they are also needed for mobilisation of financial resources. PPP models such as BOT and BOOT are common in the execution of such projects.

## QUESTIONS

1. What is the importance of airport planning?
2. Explain the important factors to be considered in the location of an airport.
3. What are the factors which affect the size of an airport?
4. Explain what is meant by 'regional planning'.
5. Explain the importance of 'Master Plan' for a new airport.
6. Mention airport obstructions and their types
7. Explain the ICAO classification of airports.
8. What are the various types of imaginary surfaces near an airport?
9. What is 'airport capacity'? What are the factors which influence it?
10. Define the following:
    (i) Airport
    (ii) Terminal area
    (iii) Airport elevation
    (iv) Approach surface
11. What is 'clear zone' in airport planning?
12. Explain the importance of 'airport zoning'.
13. Explain the 'turning zone' with a diagram.
14. Differentiate between height zoning and land-use zoning.
15. What are the details gathered during a meteorological survey?
16. Write short notes on:
    (a) Topographical survey
    (b) Survey of natural resources
    (c) Soil survey
17. What are the various drawings needed for the execution of an airport project?
18. What are 'environmental impact assessment' and 'Environmental impact statement'?
19 Explain the important characteristics of an aircraft and their importance in airport planning (JNTU-H)
20. Explain the various factors to be considered for site selection of an international airport as per ICAO guidelines. (JNTU-H)
21. List the various elements of an airport and explain them with a neat sketch (VTU).
22. (a) List and explain the aircraft characteristics which affect the planning and design of airports.
    (b) Write a brief note on airport classification.
    (c) Explain the factors which influence airport site selection (VTU).

23. What do you understand by zoning laws? (Kerala University)
24. (a) Explain imaginary surfaces.

    (b) What do you understand by 'airport capacity'?

    What are the factors which affect it? (Punjab Tech. University)
25. Mention briefly the various surveys required for collection of data for planning of an airport. (UP Tech. University)
26. (a) State the advantages and limitations of air transportation.

    (b) Discuss the types of surveys to be carried out for site selection for an airport.

    (c) Write a note on aircraft characteristics. (Pune University)

Chapter 11

# AIRPORT LAYOUT, GEOMETRICS AND DESIGN

**CHAPTER OBJECTIVES**

From this chapter, the student understands the following:

- Different areas and structures at an airport terminal
- Aprons, gates, and the factors affecting gate capacity
- Hangars, types and ancillary structures like workshops
- Airport capacity and the factors affecting it
- Typical airport layouts and case studies of certain airports
- Runways and runway design – orientation, patterns and the geometric elements
- Wind-rose diagrams for deciding upon the orientation
- Basic runway length and corrections for elevation, temperature and gradient, including runway patterns
- Taxiways, types and the design of their geometric elements.

## 11.1 AIRPORT LAYOUT

An airport terminal, the focal point of the airport, consists of the terminal area with the buildings necessary for its operation and the motor vehicle parking area.

The buildings necessary are the terminal building, the hangars and the control tower along with other miscellaneous structures needed for the airport operations.

In addition, the aprons required for emplaning and deplaning are provided near the terminal building. Also, hangars for stationing the aircraft for the purpose of inspection, servicing and repairs have to be constructed in the terminal area.

Since passengers are expected to arrive at the airport and depart therefrom in a constant repetitive flow, adequate facilities for the parking of motor vehicles have to be provided in front of the terminal building. The circulation pattern has to be properly laid out.

Thus, the layout of an airport has to be designed appropriately to accommodate all these facilities.

The degree of development in the terminal area depends upon the volume of operations, type of aircraft and air traffic using the airport, the number of passengers, employees, and visitors to be served.

It is also necessary to design the airport terminal facilities such that it serves not only the current needs but also the traffic needs anticipated in a reasonably long future or design period such as 20 to 25 years.

The architecture of the terminal building should be aesthetic in order to attract the passengers and the public.

## Terminal Building

The primary portion of the terminal area is the *Terminal building*, which houses the facilities required for the passengers, airport staff and airlines staff.

The location of the terminal building in the overall layout of the airport should be such that it is easily accessible and centrally located; it should have adequate parking facilities for motor vehicles on the front side; it should have a convenient natural drainage, utility services such as water supply, sewerage, and telephones, and finally a favourable orientation with respect to wind and other meteorological factors.

### Passenger Facilities and Services

The following facilities and services are required to be provided in the terminal building for the convenience of the passengers, the staff, and the general public:

(i) Inter-terminal transport
(ii) Arrival hall
(iii) Lounges for seating (Economy class and Executive class)
(iv) Television
(v) Departure hall
(vi) Foreign exchange counters
(vii) Medical services
(viii) Pharmacy
(ix) Post-office
(x) Telephones
(xi) Baggage counters
(xii) Shopping mall
(xiii) Supermarket
(xiv) Flight information counter
(xv) Facilities such as ramps, elevators, escalators, specially designed toilets and telephones for differently abled persons
(xvi) Facilities for babies and children
(xvii) Hotel reservations counter
(xviii) Car rentals counter

### Space Requirements

The level of service desired governs the space requirements for the terminal building. The ICAO requirement based on FAA recommendations for space in the terminal building ranges from 7.43 $m^2$ to 11.15 $m^2$ per one thousand emplaned passengers per annum; of this space, 30% is needed for public space, 40% for airline operations, and 15% each for airport administration and utilities, approximately.

### Planning Concepts for Terminal Building

The following are the concepts in use:

(a) Centralised system
(b) Decentralised system
(c) Unit terminal principle or decentralised–centralised system.

**(a) Centralised system:** In this system, the passengers, baggage and cargo are all dealt with in a central location, and routed to the respective aircraft positions.

This system is convenient and economical when the aircraft parking area is within walking distance (say, 200 m or less). The primary and secondary stages of this system are shown schematically in Fig. 11.1.

**(b) Decentralised system:** In this pattern, passenger facilities are arranged in smaller units and are provided in two or more buildings. Each unit is designed to serve the passengers using one or more gates. This system becomes uneconomical if the number of gates required by an airliner exceeds a certain number, say six. This is shown schematically in Fig. 11.2.

**(c) Unit terminal principle:** This concept is a combination of the centralised and the decentralised systems; it is considered better in serving a large volume of air traffic.

A schematic of this is shown in Fig. 11.3.

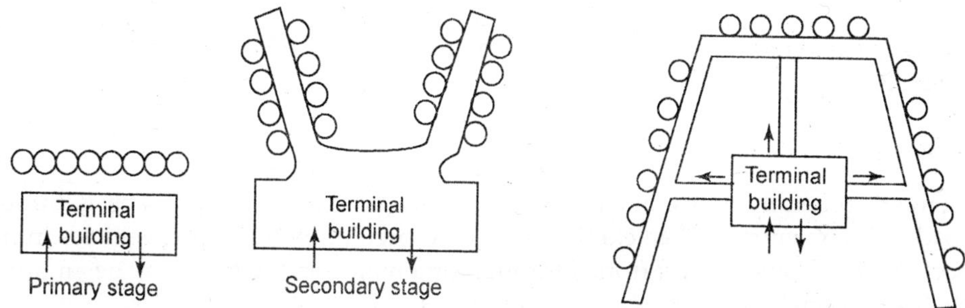


**FIG. 11.1** *Centralised system*

**FIG. 11.2** *Decentralised system*

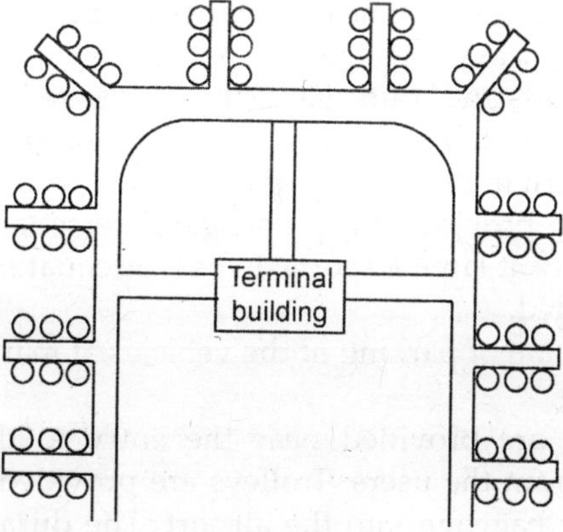


**FIG. 11.3** *Decentralised–centralised system*

**Advantages of the Unit Terminal Principle**

- Centralisation of amenities
- Expansion by adding further units in case of need
- Reduced initial cost
- Suitable for parallel runway pattern at large airports

**Passenger Flow**

The broad principles of passenger flow at a terminal building are:

- Separation of arrival and departure areas
- Efficient baggage delivery system with minimum walking time for passengers
- Provision of escalators, elevators, and vans and buses for transport to the aircraft from sky-walks
- Efficient information systems through CCTV and neon sign boards.

The sequence of activities are:

(i) Arrival
(ii) Check-in
(iii) Waiting in the boarding area
(iv) Security screening
(v) Departure

Similarly, the activities of passengers on deplaning may also be arranged sequentially for efficient passenger flow.

## Parking Area

Air passengers generally arrive at the airport mostly by road. The approach is through an access road, with an entry and an exit in a one-way circulation for motor vehicles in front of the airport. Access roads take off from a through or a main road, which may be an important one such as a National highway or a State highway.

Parking means leaving a motor vehicle in a particular place called a *parking lot*, for a limited duration. Parking areas have to take into account not only the needs of accommodating the motor vehicles of passengers but also that of other visitors who come to the airport to see off or receive them. The parking area has to be near the entry to the airport terminal building since air passengers generally carry heavy luggage.

In a city with congested roads and traffic jams, the time saved by air travel is partly lost in road transport from the airport to the passenger's destination. At least, appropriate approach, parking, and circulation pattern in the vicinity of the airport will mitigate this to some extent.

Separate parking areas are provided for passengers' personal vehicles, taxies/cabs, and employees' vehicles, especially at large airports. In case adequate area is not available, multi-storey parking may be provided.

Depending upon the duration of parking of the vehicles, it may be *short-term parking*, *long-term parking*, or *remote parking*.

Short-term parking areas are provided near the entry to the terminal building, and parking charge is collected from the users. Trolleys are provided very near the parking for the passengers to carry their baggage into the airport. The duration of short-term parking is usually 3 hours or less; this accounts for nearly 75 to 80% of the total number of vehicles. Therefore, accumulation of vehicles due to them is insignificant.

Long-term parking areas are provided a little farther away from the terminal building, with discounted charges for such users. In fact, employees' parking comes in this category.

Remote parking is even farther than long-term parking with further discounted rates and is the most economical; it is meant to be used for a few days at a time by users who need such a facility. Of course, appropriate facilities must be provided for carrying the baggage.

The size of the parking area should be based on 1.5 to 2 cars for each passenger during the peak hour, according to FAA recommendations. The geometrics of the design, the characteristics of the vehicles and the users, and continuity of flow or circulation, among other things, govern the number of vehicles that could be accommodated.

### Access and Circulation Pattern

Easy access is an important criterion in planning a parking area/parking lot. Entry and exit and an appropriate circulation pattern with adequate aisle widths will minimise the delay in the parking and unparking operations.

Car parking bays are usually 2.5 m × 5.0 m in India. Driveways shall have a minimum width of 3.75 m for one-way and 6.25 m for two-way operation. These and turn-around aisles should not have dead-ends.

The maximum length of the aisle should be 100 m for easy parking, especially when the parking lot is close to its capacity. The width of aisle should be adequate to allow for parking and unparking of a vehicle in a single smooth manoeuvre.

### Geometry of Parking Lots

Common patterns of parking are:

- Parallel parking
- Right-angle parking
- Angled parking

*Parallel parking* is preferred when the vehicles are to be placed parallel to the kerb of a foot-path.

*Right-angle parking* is simple, but only a less number of vehicles can be accommodated.

*Angled-parking,* at 45° to 60°, is considered to be the best and the most elegant; it can accommodate the maximum number of vehicles.

Fig. 11.4 shows a schematic of these parking patterns:

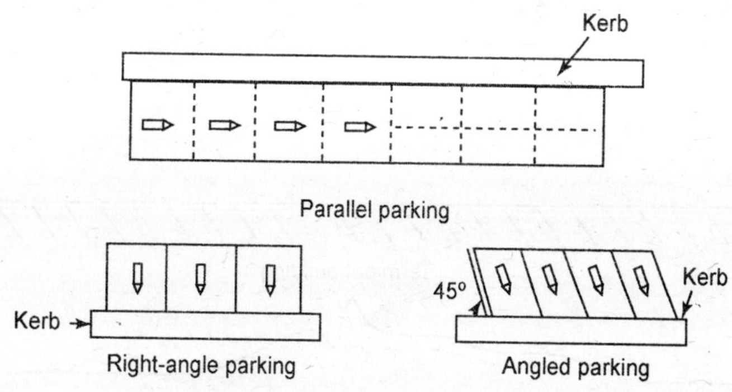


**FIG. 11.4** *Parking patterns for cars*

## Design of Apron

An apron is a paved area located adjacent to the terminal building and is used to hold aircraft for loading and unloading of passengers, cargo and fuel.

The size of the apron required depends on (a) the size of the gate, (b) the number of gates, and (c) the system of aircraft parking.

### Size of the Gate

Gate means parking space for an aircraft adjacent to the terminal building. It is used by one aircraft at a time for loading before take-off and unloading after landing.

The size of a gate depends on the size of the aircraft and the method of parking the aircraft at the gate.

The size of the aircraft determines the space required, not only for parking but also for manoeuvring; this is also governed by the servicing equipment for the aircraft. In addition, the method of parking also influences the space requirement for the aircraft and hence the size of the gate to handle the operations of the particular aircraft.

The following are the different *methods of parking the aircraft at the gate*:

(i) Nose-in
(ii) Nose-out
(iii) Angled nose-in
(iv) Angled nose-out
(v) Parallel

These are shown schematically in Fig. 11.5.

**(i) Nose-in:** The aircraft is parked at right angles to the terminal building as close as possible, with its nose near it. It can manoeuvre into the position by its own power; but it has to be towed out sufficiently for it to proceed by its own power.

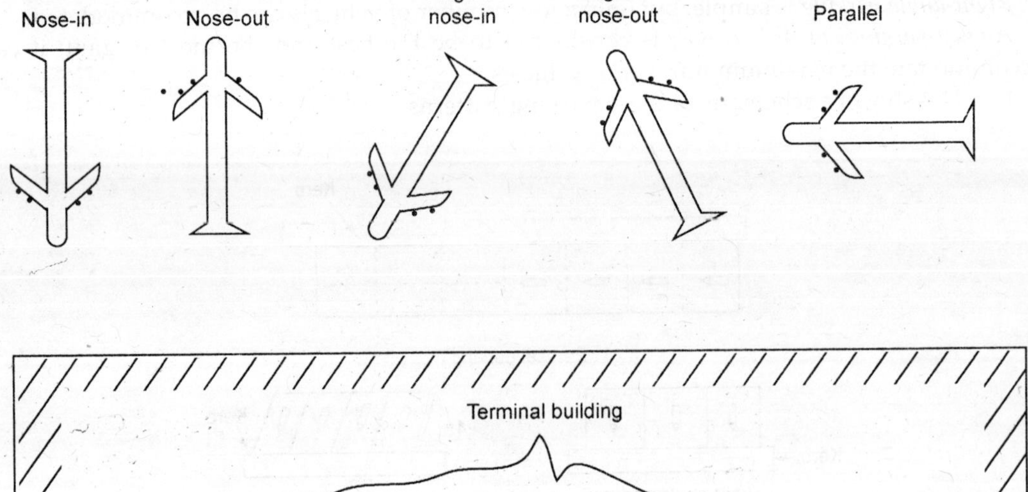


**FIG. 11.5** *Methods of aircraft parking at the gate*

This requires the smallest gate area and does not send the jet blast towards the terminal building. However, it needs greater effort while towing out as the aircraft is loaded.

**(ii) Nose-out:** The aircraft is parked at right angles to the terminal building with its tail close to the terminal building.

This permits the effective use of the rear door for loading, and requires less power while manoeuvring the loaded aircraft out of the gate position.

**(iii) Angled nose-in:** This is similar to the nose-in method except that the aircraft is parked at an angle other than 90° with the terminal building. The aircraft is allowed to manoeuvre in and out of the gate under its power. However, this needs larger area than nose-in method, and also gives rise to higher noise level.

**(iv) Angled nose-out:** This is similar to the nose-out method except that the aircraft is parked at an angle other than 90° with the terminal building. No towing is required either for the inward or for the outward manoeuvre. The main drawback is that the jet blast as well as the noise is directed towards the terminal building.

**(v) Parallel:** This is the easiest with respect to manoeuvring the aircraft in and out of the gate position. Also, both the front and rear doors can be used for loading and unloading. But it needs a larger gate area parallel to the terminal building, and hence longer loading bridges. Also, the jet blast and noise are directed towards the adjacent gate position.

Thus, each method has its own advantages and drawbacks; the choice rests with the airline operating the aircraft depending on their characteristics and the method of operation.

However, the nose-in parking method is generally preferred for saving in area and for directing the jet blast and noise away from the terminal building.

### Number of Gates

Gates are designed in such a way that the expected hourly flow of aircraft can be conveniently handled. This depends upon the *estimated peak-hour volume of aircraft'*, *gate occupancy time*, *gate capacity*, and *gate utilisation factor*.

**Estimated peak-hour volume of aircraft:** The peak-hour volume of aircraft to be handled depends primarily on the capacity and number of runways.

**Gate occupancy time:** This is the time duration for which an aircraft occupies a gate. It depends primarily on the size and characteristics of the aircraft and on the flight type – whether it is a through flight or a turnaround flight. An aircraft on a through flight may require very little servicing; the gate occupancy time in such a case may not exceed 30 minutes. However, for a turnaround flight, an aircraft may require servicing and refuelling, and thus, it may require a gate occupancy time of at least one hour.

**Gate capacity:** The number of arrivals or departures of aircraft in one hour that can be conveniently handled at one gate is the gate capacity. The gate time demanded by an aircraft should be less than the time that can be allocated to it.

Gate capacity depends on whether the pattern of gate-use is *unrestricted* for all the aircraft or *restricted* for certain size of aircraft or airline, for which the gate is specifically designed or intended.

**Gate utilisation factor:** All the gates available at a terminal are unlikely to be utilised to their full capacity; this is because, the time gaps between successive aircraft operations may be inadequate to accommodate other flights, or because the manoeuvres of an aircraft may block those of aircrafts meant for adjacent gates. The percent utilisation of the capacity of a gate is characterised by the *Gate utilisation factor*. This factor may be in the order of 0.60 to 0.80 for unrestricted gate use and 0.50 to 0.60 for restricted use.

**Determination of gate capacity:** Gate capacity and the number of gates are related by the following formula:

$$C = \frac{G}{T} \cdot U \tag{11.1}$$

where

C = Gate capacity

G = Number of gates

T = Weighted average gate occupancy time in hours

U = Gate utilisation factor.

This formula is applicable to the case of unrestricted use of all the gates at the terminal.

For restricted-use, the mix of aircraft and the number of gates may not match; in such a case, the gate capacity of each type is determined separately and the minimum of these values is taken to obtain the overall capacity of the airport.

(The norm is 3 to 5 gates per million passengers per annum.)

The use of these principles will be illustrated in the Numerical Examples Section.

### Systems of Aircraft Parking

Depending on the horizontal terminal concept used in the construction, the aircraft can be parked or stationed adjacent to the terminal building in different ways. These are known as *parking systems*. The following are the common systems used:

(i) Frontal system
(ii) Open-apron system
(iii) Pier system
(iv) Satellite system

**Frontal system:** In this system, the aircraft are parked along the face of the terminal building (Fig. 11.6):

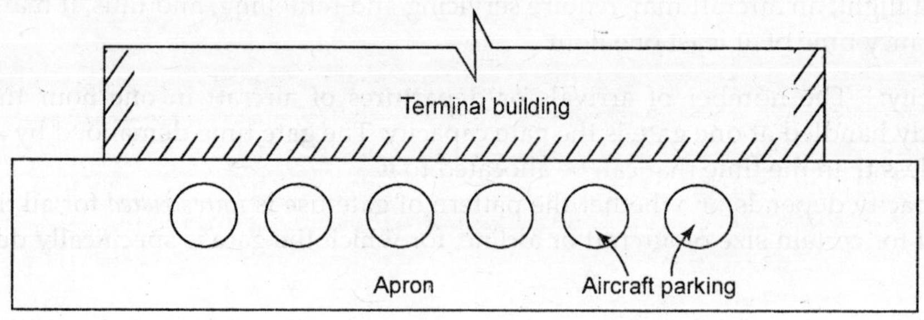


**FIG. 11.6** *Frontal system*

Also called the *'linear'* system, this is the simplest. The apron provides close-in parking for three to six aircraft. It provides easy access with minimum walking distances from the gate, and flexibility for expansion. However, it is considered suitable only for small airports with low air traffic.

**Open-apron system:** Also called the *transporter* system, the aircraft are located in rows as shown in Fig. 11.7.

Since a long distance is involved for the last or outermost rows, vehicular transport is essential for emplaning and deplaning of passengers.

Expansion for future needs is easy and capital cost is relatively low. Greater flexibility and manoeuvrability of the aircraft is an attractive feature of this system.

**Pier system:** Also called the *finger* system, piers are built from the main terminal in the shape of a straight line, 'T' or 'Y'; aircraft gates can be provided on both sides of the piers, as shown in Fig. 11.8.

**Satellite system:** In this system, small buildings, called *satellites*, are built on the apron; gates are provided around these as shown in Fig. 11.9.

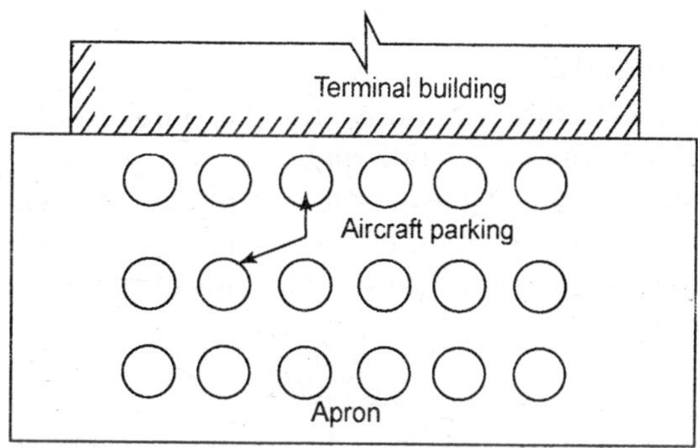


**FIG. 11.7** *Open-apron system*

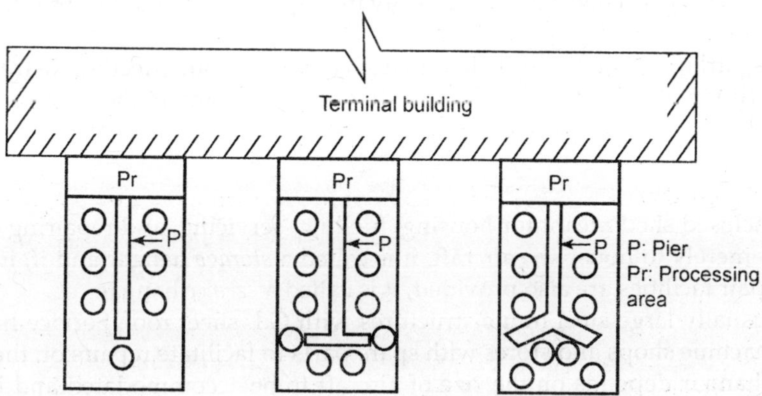


**FIG. 11.8** *Pier system*

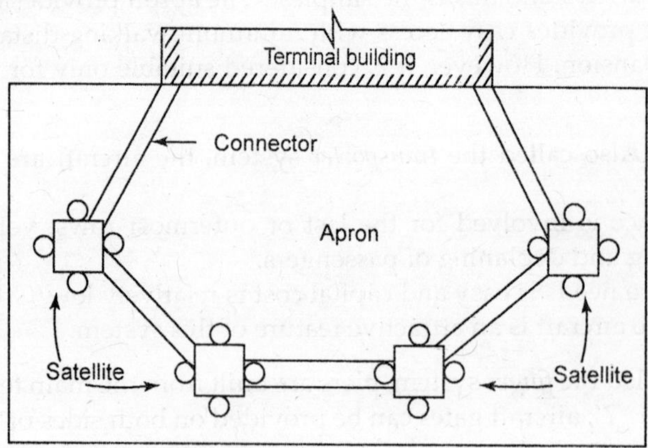


**FIG. 11.9** *Satellite system*

The arrangement of common facilities and aircraft manoeuvring is easy; but this needs mechanical system for transport of the passengers and the baggage. Also, this system entails greater capital cost.

**Loading Aprons and Holding Aprons**

Aprons built adjacent to the terminal building for the purpose of loading and unloading operations of the aircraft are known as *loading aprons*.

In contrast, aprons built at the starting points of runways, which are meant to hold the aircraft just prior to take-off for checking of the instruments and the engines are called *holding aprons*. These are provided only in busy airports. One wide apron for adjacent runways serves the purpose of holding apron for both.

**Apron Turntable**

The pavement of the apron is designed to carry the largest wheel load of the aircraft. In order to limit the area of the thick heavily reinforced apron, an *apron turntable* is sometimes built.

The rotating portion is made flush with the remaining apron. An aircraft parked on this is rotated to come into the desired direction by merely rotating the circular movable portion of the apron after unbolting it. Then it is bolted firmly in that position, and the aircraft is allowed to take-off.

This facilitates parking of the aircraft close to each other without affecting their freedom of movement; in addition, it helps to achieve some degree of economy in the construction of the apron pavement.

**Hangars**

A *hangar* is an enclosed shed meant for housing, fuelling, servicing, and repairing of aircraft. If the purpose is merely to house the aircraft, it is called a *storage* hangar and if, in addition, servicing and repair facilities are also provided, it is called a *service* hangar.

Hangars are usually large steel frame structures with G.I. sheet roof. Service hangars are provided with machine shops and stores with spare parts to facilitate repairs on the aircraft.

The size of a hangar depends on the size of aircraft to be accommodated and its turning radius. The number of storage and service hangars needed at an airport depends upon peak-hour traffic volume as also on the demand from airline agencies on rental basis.

Adequate ventilation and lighting are essential for hangars.

The *criteria for locating hangars are*:

- Proximity to the terminal building and the loading aprons.
- Site with effective drainage
- Scope for future expansion.

The plan shape should be such that the hangar covers minimum land area, yet providing all the necessary facilities.

The *types of hangars* used are:

(a) T-hangars
(b) Nose-hangars
(c) Grouped hangars

**T-hangars:** These are of the shape of 'T' in plan and are considered suitable for small-sized aircraft. The area should be adequate to allow easy manoeuvring into and out of the hangar; the minimum clearance is 300 mm (Fig. 11.10).

**Nose hangars:** This type of hangar encloses only the nose portion of the aircraft; a major portion of the fuselage and the tail are exposed. This design is convenient for servicing and repairs, and also leads to economy in construction. It is considered suitable for large aircraft (Fig. 11.11).

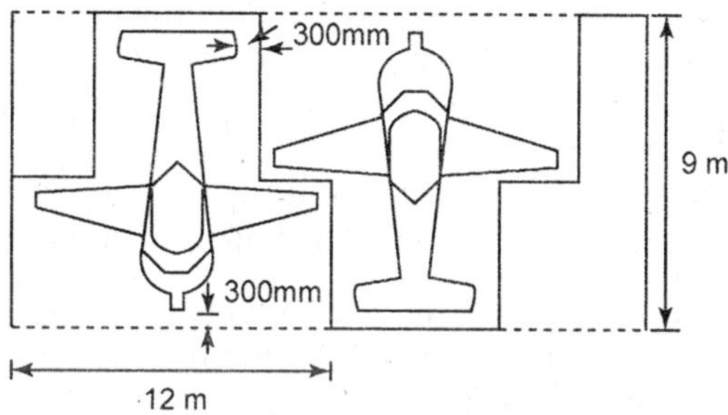


**FIG. 11.10** *T-hangars*

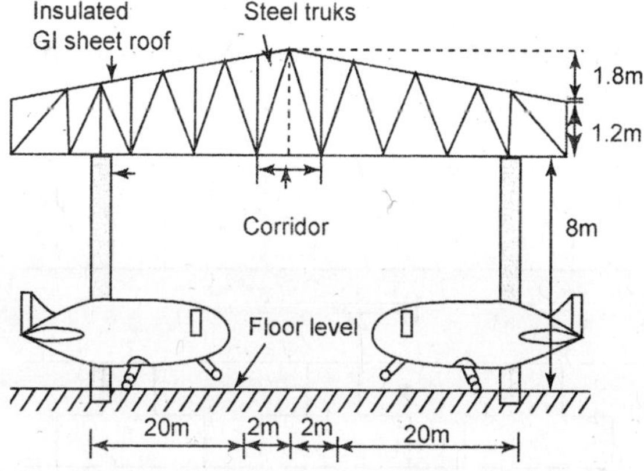


**FIG. 11.11** *Nose hangar*

**Grouped hangars:** These are hangars grouped in a long shed and are meant to serve a large number of aircraft.

Grouped hangars of the T-type are shown in Fig. 11.12.

### Blast Considerations

Blast from jet exhaust can cause discomfort and in extreme cases, injury to the passengers while arriving to board an aircraft. Damage can be caused to servicing equipment, buildings, and the pavements. Holding aprons, ends of runways, terminal area, and taxiways may have to be protected from jet blast.

Blast fences and erosion control methods may be used for protection against jet blast. A *blast fence* may consist of a single curved vane or multiple curved vanes, which diffuse the smoke from the jet exhaust. The height of these vanes may be 2.5 to 3 m.

*Erosion control* from jet blast may be achieved by using bituminous surface, turfing, stone pitching and so on. Shoulders of aprons and taxiways need protection.

### Oil Spilling

Spilling of oil from the engines can damage the surfaces of aprons and taxiways; especially vulnerable are bituminous surfaces.

### Airport Capacity

This is defined as the number of aircraft movements (landing and take-off are taken to be two movements of each aircraft), which an airport can handle within a specified period of time such as one-hour or one day. An acceptable time delay has to be allowed in reckoning airport capacity.

In general, landing is given priority over take-off for obvious safety reasons.

A number of *factors influence airport capacity*. The following are some important ones:

- Size, number of gates and their pattern in the apron area
- Availability of air-space for operations in and around the airport
- Characteristics of aircraft using the facility
- Navigational aids and air-traffic control facilities
- Number and frequency of operations
- Runway and gate occupancy times of different aircraft using the airport
- Wind conditions
- Weather conditions affecting visibility, since the regulations are different from the normal ones in case of bad weather.

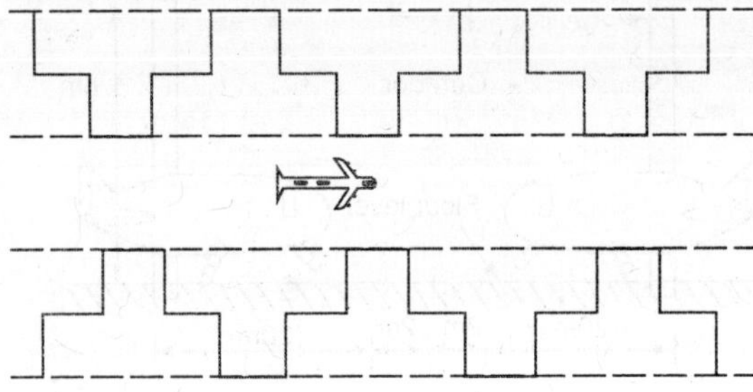

**FIG. 11.12** *Grouped T-hangars*

## 11.2 TYPICAL AIRPORT LAYOUTS

The layout of an airport depends primarily on the configuration of the runways planned for it. The other airport elements such as taxiways, hangars, aprons and the terminal building along with the various facilities required to be provided in it, are all planned in such a manner that a smooth flow of traffic is ensured to produce an integrated layout.

### Characteristics of a Well-planned Airport Layout

- Adequate facilities in the terminal building
- Ample and well-planned parking area for motor vehicles
- Adequate number of loading aprons and gates to handle the anticipated air traffic, with easy access to passengers
- Smooth operations of landing, taxiing and take-off with least interference to each other
- Adequate runway lengths and efficient design for safe operations
- Well-planned layout for the control tower for good visibility
- Provision for future expansion
- Minimum capital cost and maintenance cost for the layout.

It is difficult to fulfil all these conditions in each layout; it is therefore necessary to arrive at a judicious compromise depending upon the requirements and site conditions.

Figs. 11.13 to 11.15 show typical layout patterns:

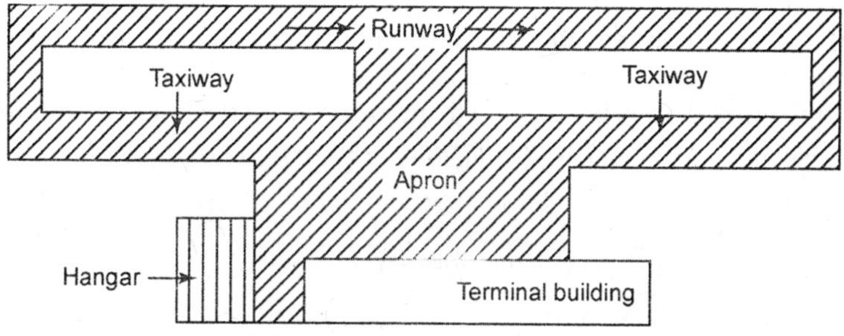


**FIG. 11.13** *Layout of a small airport with a single runway*

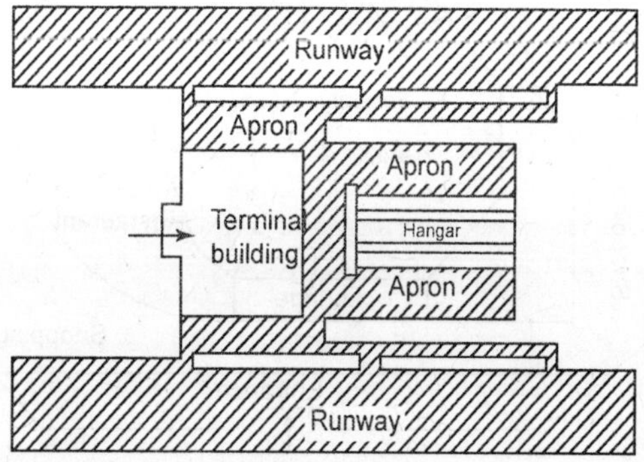


**FIG. 11.14** *Layout of an airport with two parallel runways*

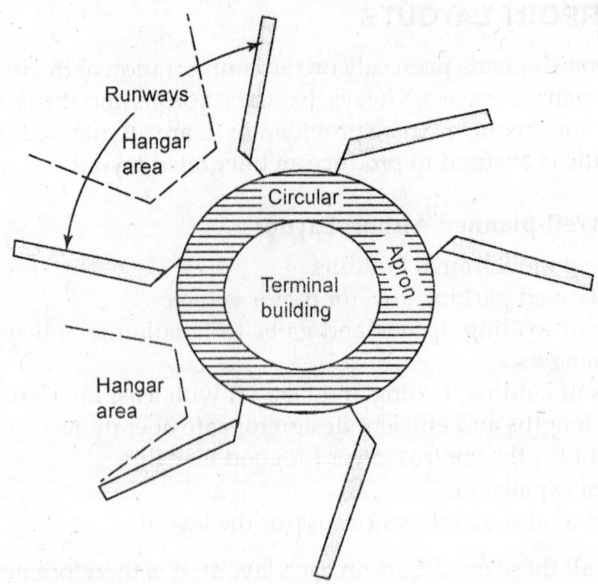


**FIG. 11.15** *Layout of an airport with multiple tangential runways*

Several additional facilities such as multiple check-in counters, baggage claim area, customs check area, immigration areas, transit lounges, departure lounges, restaurants, shopping malls, telephone facilities, foreign exchange counters, bus facilities for inter-terminal transport, and transport vans at the gates in the apron areas for emplaning and deplaning have to be provided at international terminals at large airports. This is in addition to aircraft control facilities for staff responsible for air operations. A typical layout of a domestic terminal is shown in Fig. 11.16:

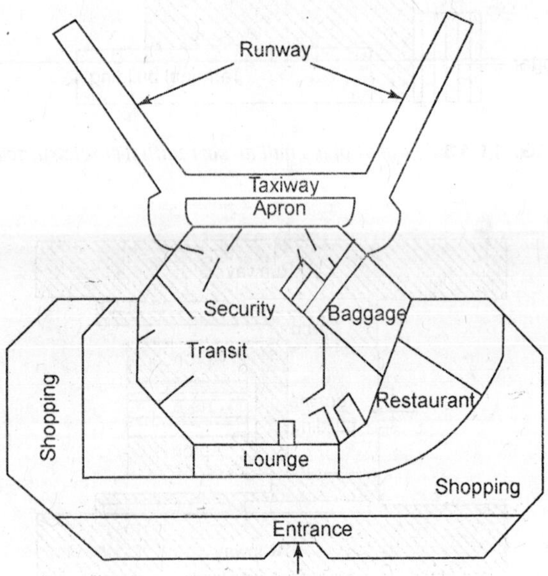


**FIG. 11.16** *Typical layout of a domestic terminal*

## CASE STUDIES OF INDIAN AIRPORTS

### Chennai International Airport

Owned by the Government of India and operated by the Airports Authority of India (AAI), the Chennai international airport is the biggest in south India and the third busiest airport in India in terms of passenger traffic.

It lies just 16 m above main sea level (MSL).

| | | |
|---|---|---|
| Primary runway | 3,658 m long (Asphalt) | |
| Secondary runway | 2,925 m long (Asphalt concrete) | |
| Passenger traffic | 12.896 million | (during 2013–14) |
| Aircraft movements | 1.218 lakh | (during 2013–14) |
| Cargo tonnage | 2.921 lakh | (during 2013–14) |

The domestic terminal (Kamaraj terminal) and the International terminal (Annadurai terminal) are adjacent to each other. The old terminal at Meenambakkam is used for cargo, and the new terminal complex at Tirusulam is used for passenger operations.

The link building between the domestic and international terminals houses the administrative offices and a restaurant. The international terminal has two aerobridges and the domestic terminal has three. In the recent extension of the international terminal, three more aerobridges are added for departures. The international and domestic terminals cover an area of 1.5 sq. km and 1.8 sq. km, respectively. The airport is divided into two circles, with five zones each, for administrative convenience.

The Kamaraj terminal has an area of 19,250 m$^2$ with 48 check-in counters. The Annadurai terminal has an area of 42,870 m$^2$ with 45 check-in counters, 38 immigration counters, and 18 customs counters. There are two entry gates for each terminal.

The airport has 70 parking bays, one of which can accommodate the superjumbo jet Airbus-A 380.

#### Air Traffic Control (ATC) Tower

The Chennai ATC tower is one of the biggest in the country; advanced surface movement guidance and control system has been introduced in the tower. This ATC has Hyderabad, Mangalore, Bangalore, and Tiruvananthapuram under its control. With advanced integrated radar technology, the ATC in Chennai has the entire South Indian region on its radar screens, and co-ordinates flight movements in the range of 26,000 to 46,000 ft flight altitude.

### Delhi International Airport

Also called *Indira Gandhi international airport*, this is owned by AAI and operated by the Delhi International Airport Private Limited (DIAL) and located at an elevation of 237 m above MSL.

| | | |
|---|---|---|
| Runways | 3,810 m<br>2,813 m<br>4,430 m | (Asphalt) |
| Passenger movements | 37 million | (during 2013–14) |
| Aircraft movements | 2.90 lakh | |
| Cargo tonnage | 6.06 lakh | |

This is the busiest airport in India and is spread over an area of 2,500 ha. Delhi airport serves the National Capital Region.

The Safdarjung airport was built in 1930 and was the main airport until 1962. The old terminal (Terminal-1) handles domestic flights, and is divided into three terminals. There is also a separate technical Area for VVIP passengers.

It has three runways, 60 m wide that run nearly parallel to each other.

Terminal T-3 is the eighth largest airport terminal in the world. Delhi Airport Express train run by *Delhi Metro* and the Delhi–Gurgaon Expressway provide access to the airport.

## Mumbai International Airport (MIAL)

(*Chhatrapati Shivaji international airport*)

Owned by the Airports Authority of India, it was originally opened in 1942. It is located at an elevation of 11 m above MSL.

| | | |
|---|---|---|
| Runways | 3,660 m<br>2,990 m | (Asphalt) |
| Passenger movements | 32.222 | (during 2013–14) |
| Aircraft movements | 2.61 lakh | |
| Cargo tonnage | 6.50 lakh | |

It is the second busiest airport in the country and has 5 operating terminals spread over an area of 1,160 ha.

The airport has two intersecting runways, which have been upgraded to accommodate Airbus-A 380 aircraft.

The new Air Traffic Control Tower is the tallest in India at 85 m and has a visibility of 8 km.

At the new Terminal 2, there are 135 check-in counters, and 3600 cars can be accommodated in the parking area.

It was ranked the third-best airport in the world in the 25 to 40 million passengers category by Airport Council International in the years 2011 and 2012.

## Kolkata International Airport

(*Netaji Subhas Chandra Bose international airport*)

Owned and operated by the Airport Authority of India (AAI), it is located at an elevation of 5 m above MSL.

| | | |
|---|---|---|
| Runways | 2,285 m long<br>3,660 m long<br>and 50 m wide. | (Asphalt) |
| Passenger movements | 10.1 million | (during 2013–14) |
| Aircraft movements | 92,870 | |
| Cargo tonnage | 1.30 lakh tonnes | |

This serves the Kolkata metropolitan area and is located 17 km from the city. It was known previously as Dum Dum Airport. Spread over an area of 1000 ha, it is the largest in eastern India and is the fifth busiest airport in India in terms of passenger traffic. It operates flights to North-east India, Bangladesh and South-east Asia.

The Kolkata airport is one of the oldest in the country, through which flights from London to Australia were operated in 1933. It played an important role during the Second World War for the military operations of the Allied Forces.

A new domestic terminal was opened in 1995, and the airport was renamed in honour of Netaji Subhas Chandra Bose. A new international arrival hall was opened in the year 2000. The runways were extended 400 m on the northern side and 300 m on the southern side.

A new integrated terminal was opened in January 2013. It is spread over 2.33 lakh sq. m to handle 25 million passengers annually. It has 128 check-in counters, and 78 immigration counters, and 12 customs counters, and is equipped with 18 aerobridges.

The airport is served by the Kolkata Suburban Railway.

### Bengaluru International Airport

Also called *Kempegowda international airport*, it is located near Devanhalli, 40 km from Bengaluru. Owned and operated by Bangalore International Airport Limited (BIAL) and the GVK Group, the airport was opened on 24 May 2008. It is situated at an elevation of 915 m above MSL and is spread over an area of 2,100 ha. It replaced the HAL airport for commercial aviation.

| | | |
|---|---|---|
| Runway | 4000 m long (Asphalt) and 60 m wide | |
| Passenger movements | 12.87 million | (during 2013–14) |
| Aircraft movements | 1.18 lakh | |
| Cargo tonnage | 2.42 lakh tonnes | |

It is the fourth busiest airport in India in terms of passenger traffic after Delhi, Mumbai and Chennai. It handled over 18 million passengers in 2015.

Stage II expansion of the airport has been cleared recently and includes the construction of a new terminal along with the second runway along with taxiways, aprons, aircraft parking, and modern aviation facilities.

BIAL has been given the rights to build, own and operate, and manage the airport for 30 years in the first instance.

The passenger terminal is a four-level building that accommodates domestic and international operations; the arrival and departure areas are separated vertically. The check-in facilities and baggage reclaim area are located on the ground floor; the total floor area is 170,000 m$^2$. It has 38 gates, 20 aero-bridges, and 45 aircraft stands. It has a car-parking area for 2000 cars in front of the Terminal building. The airport was officially renamed after the city's founder, Kempegowda, on 14 December 2013.

## Hyderabad International Airport

This has been renamed as *Rajiv Gandhi international airport*. Located at Shamshabad, 22 km south of Hyderabad, it has replaced the former airport of Begumpet since 23 March 2008. It is the second PPP-venture after Cochin international airport, and its domestic terminal is named after the late N T Rama Rao, the former CM of united Andhra Pradesh.

This is a joint venture between GMR group, Malaysia Airport Holdings Berhad, the Govt. of Telangana, and the Airport Authority of India. The maximum equity of 63% is held by the GMR group.

| | | |
|---|---|---|
| Runways | 4,260 m long<br>3,800 m long | (Asphalt) |
| Passenger movements | 8.654 million | (during 2013–'14) |
| Aircraft movements | 87,740 | |
| Cargo tonnage | 86,670 | |

It is being developed in three phases, to eventually provide infrastructure for 40 million passengers per annum.

Terminal 1 has 18 contact and 36 remote stands for aircraft parking. Cargo hangars have a capacity of 100,000 tonnes. The parking area in front of Terminal 1 can accommodate 1800 cars.

The ATC Tower is 75 m tall and gives an uninterrupted view of the airfield.

The 11.6 km PV Narasimha Rao Expressway provides dedicated high-speed corridor to the airport and is the second longest flyover in India.

## Visakhapatnam International Airport

Owned and operated by the Indian Navy and the Airports Authority of India, it is located at an elevation of 5 m above MSL and is the biggest and the busiest in the state. It has been operating international flights to Dubai and Singapore since 2012.

| | | |
|---|---|---|
| Runways (60 m wide) | 3,200 m long<br>1,594 m long | (Asphalt) |
| Passenger traffic | 1.013 million | (2013–14) |
| Aircraft movements | 11,008 | |
| Cargo tonnage | 1,823 | |

The airport's instrument landing system (ILS) and the air traffic control system is managed by the Indian Navy. It is also equipped with night landing facilities.

There is also a Flying Club and an Aeronautical Institute started in the city to cater to the needs of frequent fliers who travel to and from the city

Flights are operated in domestic routes to Bengaluru, Hyderabad, Chennai, Tirupati, Vijayawada, Mumbai, Delhi, Kolkata, Bhubaneshwar and Port Blair, by different airlines.

## CASE STUDIES OF FOREIGN AIRPORTS

### New York International Airport *(John F. Kennedy International Airport)*

It is the busiest international air passenger gateway in the USA. Together, JFK International, LaGuardia, and Newark Liberty airports, all operated by the Port Authority of New York and New Jersey, are the largest airport system in the USA, second largest in the world in terms of passenger traffic, and first in the world by total flight operations. Started in 1948 as the New York International Airport and known as the Idlewild Airport until 1963, it was renamed in memory of John F. Kennedy, the 35th President of the United States.

It is situated 4 m above MSL.

| | | |
|---|---|---|
| Runways | 3,800 m | (Asphalt) (61 m wide) |
| | 2,560 m | |
| | 3,448 m | |
| | 4,480 m (Concrete) | |
| Helipads | 4 nos., 18 m long (Asphalt) | |
| Passengers | 50.42 million (2013) | |
| Aircraft operations | 4.06 lakh | |
| Taxi ways | Over 40 km long, 23 m wide | |
| Air-traffic control Tower | 97.5 m high | |
| Terminals | 6 nos. | |
| Gates | 128 | |

Over 90 airlines operate out of JFK airport.

### London Heathrow Airport

Situated 25 m above MSL, it is the primary airport which serves the city of London, the capital of the UK.

| | | |
|---|---|---|
| Runways | 3902 m | (Grooved asphalt) |
| | 3660 m | |
| Passengers | 72.37 million | (2013) |
| Aircraft movements | 4.72 lakh | |

This is the busiest airport in the UK and Europe by passenger traffic, and the third busiest in the world.

There are 5 terminals covering 12.14 sq. km of area.

Started in 1929 as a small airfield near the hamlet of Heathrow, the airport takes its name from this; it was developed after the Second World War. This is used by over 90 airlines and is the primary hub of British Airways.

Both the runways are utilised to 98% of their capacity.

London Tube Railway provides access to the airport, as do several buses and coaches. Inter-terminal transfer is by Heathrow Express trains or bus.

Plans for expansion of the Heathrow Airport are afoot.

### Paris International Airport: (Charles de Gaulle Airport or Roissy Airport)

Located 25 km from Paris at an elevation of 119 m above MSL, it is the largest airport in France and is one of the world's principal aviation centres.

| | | |
|---|---|---|
| Runways | 4,215 m<br>4,200 m | (Asphalt) |
| | 2,700 m<br>2,700 m | (Concrete) |
| Passengers | 62 million | (2013) |
| Aircraft movements | 4.98 lakh | |

Named after Charles de Gaulle, the President of France from 1959 to 1969, it serves as the principal hub for Air France.

It is the second busiest airport in Europe, next to Heathrow of London, and is the world's eighth busiest in terms of passenger traffic. It covers an area of 32.38 sq.km of area.

Terminal 1 is a ten-floor high circular building surrounded by seven satellite buildings, each with six gates.

Terminal 2 was exclusive to Air France initially, but now it is allowed to be used by other airlines also.

Terminal 3 hosts charter and low-cost airlines.

Roissy pole is a complex of offices, shopping areas and hotels within Charles de Gaulle airport. A free automatic shuttle train service is available for inter-terminal transport.

High speed TGV train services connect the airport to Paris.

### Dubai International Airport

Governed by the Govt of Dubai (United Arab Emirates) and operated by Dubai Airports Company, Dubai international airport is a major airline hub in the Middle East, serving mainly Emirates airlines. It is also a secondary hub for the Australian airlines, Qantas. It is spread over an area of 3,400 ha at an elevation of 19 m above MSL.

| | | |
|---|---|---|
| Runways | 4000 m long<br>4450 m long | (Asphalt) |
| Passenger traffic | 66.40 million | (2013) |
| Cargo traffic | 2.44 million tonnes | |

It is the seventh busiest airport in the world in terms of passenger traffic and second busiest in terms of international passenger traffic; it is also the fifth busiest in terms of cargo handling.

It has three passenger terminals and an Executive flight terminal besides a VIP pavilion. The airport marked its 50th anniversary in October 2010. It serves 140 airlines for flights to 239 international destinations.

The passenger terminal building has a floor area of nearly 2 million sq. m and is equipped with 74 aerobridges and an 87 m-high control tower. It has one of the largest areas of retail space and duty-free shopping.

It is served by Dubai Metro and buses on land.

### Layout of Military Airports

Military airports have the following special features:

- Special colours for runways
- Inconspicuousness of the facilities
- Strategic location from defence point of view
- Decentralisation of the facilities
- Scattered location of hangars
- Natural vegetation preserved to the extent possible.

## 11.3 DESIGN OF RUNWAYS

Runway design is concerned with planning the pattern and configuration of runways at an airport; it further includes the geometric elements of runways and their design.

The number of runways at an airport depends on the volume of air traffic expected to use the facility. For a small airport, one or two runways may be adequate, but several runways may be needed to handle the traffic at a large airport.

*Orientation* of a runway is its direction along its length and its positioning with respect to the direction of the wind prevailing at the airport site. Runways are positioned based on the presumption of adequacy for the area of the site available for this purpose. In order to ensure uniformity in the operational facilities at airports in different countries, the International Civil Aviation Organisation (ICAO) has evolved certain common design standards, and their recommendations have to be borne in mind in the design of the runways and taxiways.

### Geometrics of Runways

The important geometric elements of runways are the following:

(a) Length
(b) Width
(c) Longitudinal gradient
(d) Rate of change of gradient
(e) Transverse gradient
(f) Sight distance
(g) Runway shoulders
(h) Safety area

### Orientation of a Runway

The orientation of a runway is its direction at an airport; this is an important aspect in runway design and is primarily dependent on the wind characteristics at the airport site.

**Preliminary information:** The basic information required consists of the topographical details of the site and a record of the wind characteristics such as direction, force and duration of the wind in the area for as many years as possible, at least for the previous 5 to 10 years. Other details such as that related to fog, snow and rainfall which affect the visibility in general are also valuable.

**Head wind:** A runway is usually oriented in the direction of the prevailing winds; *head wind* means wind against the direction of the head or nose of the aircraft, while it is landing or taking off.

The *advantages* of head wind are:

- Braking effect during landing helps the aircraft to come to a stop in a relatively short length of the runway.
- Greater lift on aircraft wings during take-off makes the aircraft to rise above the ground using a relatively shorter length of the runway.

It is considered that the effective length of the runway needed for landing and take-off operations gets reduced by about 10 per cent if such operations take place against the direction of the wind.

This is the primary reason for orienting a runway in the predominant direction of the winds prevailing at the airport site. The wind records data are analysed for this purpose.

Note:

The best orientation from the analysis may not always be practicable due to excessive earth work/grading unexpected obstructions, or noise pollution near residential areas.)

**Cross-wind component:** It is not possible to obtain the head wind along the centre line of the runway on all the days of the year, or everyday throughout the year. Thus, for some period of the year, the wind blows at a certain angle with respect to the centre line of the runway.

In such a case, while the component parallel to the centre line of the runway is helpful in the operations, the component perpendicular to it, which is called the *cross-wind component*, is considered to be undesirable beyond a certain limit; this is because it can affect the safety of the landing and take-off of light- and medium-weight aircraft.

The orientation of the runway should be such that the cross-wind component is kept to a minimum (Fig. 11.17).

In general, the cross-wind component should not exceed 25 km/h.

**Wind coverage:** This is defined as the percentage of time in a year during which the cross-wind component is within the stipulated permissible limit.

For the purpose of calculating the wind coverage in any particular direction, a deviation up to ±11.25° (or 11.25° on either side) is considered permissible, For example, if N–S direction is taken as the orientation of the runway, the wind coverage is obtained by summing up the durations of wind in the directions — NNE, N, NNW, SSE, S and SSW respectively (Fig. 11.18). The desirable minimum wind coverage is taken as 95%.

**Calm period:** This is the period, expressed as percentage in a year, during which velocity is less than a stipulated minimum. This value is considered to be a velocity of 6 km/h, which will have negligible effect.

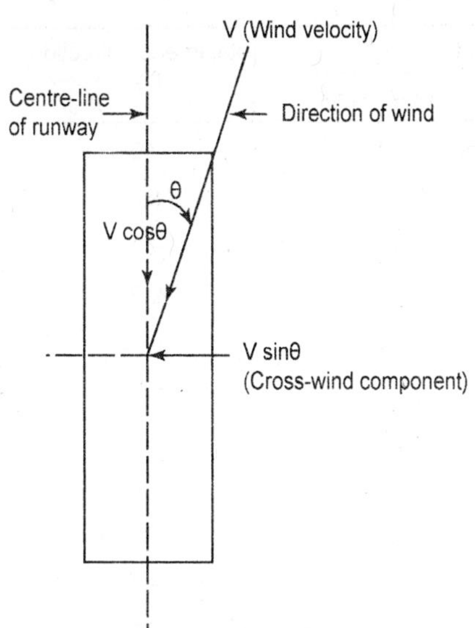


**FIG. 11.17** *Cross-wind component*

**FIG. 11.18** *Deviations for wind coverage*

**Wind-rose**

The average wind data of a design period (5 to 10 years) are represented in a graphical form by means of a chart, known as *wind-rose*. (It is called so because the plotted chart appears in the form of rose petals!)

This diagram helps in the determination of the best orientation of the runway for maximum wind coverage, and consequently for the maximum possible runway capacity. (This will be dealt with in some detail later.)

Wind-rose diagrams may be plotted in two different ways:

Type-I shows the direction and duration of the wind in percent in a year.

Type-II shows the direction, percent duration in a year and the wind velocity or intensity.

For the purpose of reckoning the wind blowing in any particular direction, a deviation of $\pm 11.25°$ is considered. Therefore, the wind blowing in a sector of 22.5° central angle (provided the particular direction chosen is exactly midway in the sector) is taken as the wind data for that direction. The two types of wind-rose, their construction, and interpretation for use in the selection of the most suitable orientation of the runway for maximum wind coverage, are better understood through the following illustrative example:

**Type-I wind-rose:** A typical set of wind data (an average of say, 5 years) for a site for an airport is given as shown below:

| Wind direction | Duration of wind (percent) | | | Total in each direction (%) |
|---|---|---|---|---|
| | 6–25 km/h | 25–50 km/h | 50–75 km/h | |
| N | 7.2 | 2.8 | 0.2 | 10.2 |
| NNE | 5.4 | 2.4 | 0.2 | 8.0 |
| NE | 2.7 | 0.7 | 0.6 | 4.0 |
| ENE | 1.2 | 0.5 | 0.1 | 1.8 |

| Wind direction | Duration of wind (percent) | | | Total in each direction (%) |
|---|---|---|---|---|
| | 6–25 km/h | 25–50 km/h | 50–75 km/h | |
| E | 0.7 | 0.3 | 0 | 1.0 |
| ESE | 0.4 | 0.2 | 0 | 0.6 |
| SE | 4.2 | 2.8 | 0.2 | 7.2 |
| SSE | 5.6 | 3.6 | 0 | 9.2 |
| S | 9.6 | 4.8 | 0 | 14.4 |
| SSW | 6.3 | 3.6 | 0.6 | 10.5 |
| SW | 3.6 | 1.9 | 0.5 | 6.0 |
| WSW | 1.2 | 0.6 | 0.2 | 2.0 |
| W | 0.5 | 0.1 | 0 | 0.6 |
| WNW | 0.3 | 0.1 | 0 | 0.4 |
| NW | 5.4 | 2.0 | 0 | 7.4 |
| NNW | 5.0 | 1.4 | 0.3 | 6.7 |
| | | | Total | 90.0 |

Calm period (percent time during which the wind velocity is less than 6 km/h)

$$= (100 - 90)\ \% = 10\%.$$

This period does not influence the landing or take-off operations because of low wind velocity.

Type-I wind-rose plotted from this set of data is shown in Fig. 11.19:

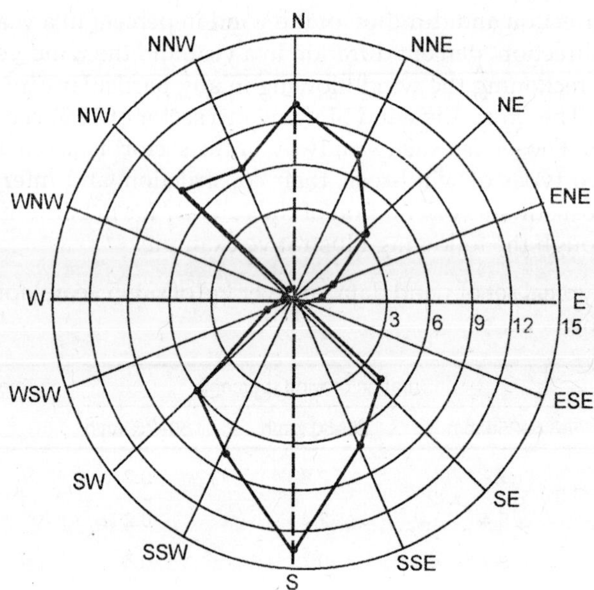


**FIG. 11.19** *Type-I wind-rose*

The following are the steps in the procedure:

(a) The radial lines indicate the wind direction, and each circle represents the duration of the wind to a certain scale.
(b) The percentage of the wind blowing in each direction is marked along the corresponding radial line to obtain a point.
(c) After marking the points along all the directions, these are sequentially joined as shown to get the wind-rose.
(d) The best orientation of the runway is the direction of the longest line of the wind-rose. (In this case, it is the N–S direction).

(The effect of the cross-wind is not considered in this method.)

If deviation of wind direction up to (22.5° + 11.25°) or 33.75° from this direction is permissible, the wind coverage in a year for safe landing and take-off on this runway can be obtained by summing up the percentages of time along NNW, N, NNE, SSE, S, and SSW directions.

This comes to 59% in this case.

Adding the calm period percentage to this value, the total percentage comes to (59 + 10) or 69%. This is not considered to be adequate. This indicates the next best direction. (In this case, NNW–SSE direction, it is with wind coverage of over 40 %.) The wind coverage for any direction should be added only once, while considering second or further runways.

**Type-II wind-rose:** For the same data, Type-II wind-rose is obtained as follows:

(a) Concentric circles corresponding to 6, 25, 50 and 75 km/h of wind velocity (or intensity) are drawn to a convenient scale.
(b) The 16 radial directions are shown on the outermost circle as shown. The mid-points of the respective arcs have to represent the cardinal directions such as N, NNE, NE, and so on.
(c) The respective wind durations (percentages) for each direction are marked in the corresponding sector or part of it – the segments of the sectors.

The wind-rose diagram is obtained as shown in Fig. 11.20.This is used in the following manner to obtain the best orientation of the runway for these conditions:

(i) A transparent paper template, rectangular in shape, is taken such that its length is a little more than the diameter of the outermost circle. The centre line parallel to the length is drawn to represent the runway.
(ii) Two more parallel lines are drawn to this line such that they are at a distance of 25 km/h (the maximum permissible cross-wind component) on either side of the centre line, to the same scale used on the wind-rose.
(iii) This template is placed on the wind-rose such that their centres coincide.
(iv) The template is now turned and oriented in such a way that the sum of all the values of wind duration between the two outer parallel lines is the maximum value.
(v) The centre line of the template in this orientation is the best orientation of the runway.
(vi) The wind coverage percentage is obtained by summing up the percentages marked in the segments encompassed by the template.

The percentage is considered to be the same for the entire segment.

If the outer parallel lines of the strip cross a segment, the percentage for this is assessed from its area by eye judgement with respect to the total area of the segment or sector. The wind coverage should be at least 95% by means of the runways — one or more than one –for effective utilisation of the runway and airport capacities (Fig. 11.20).

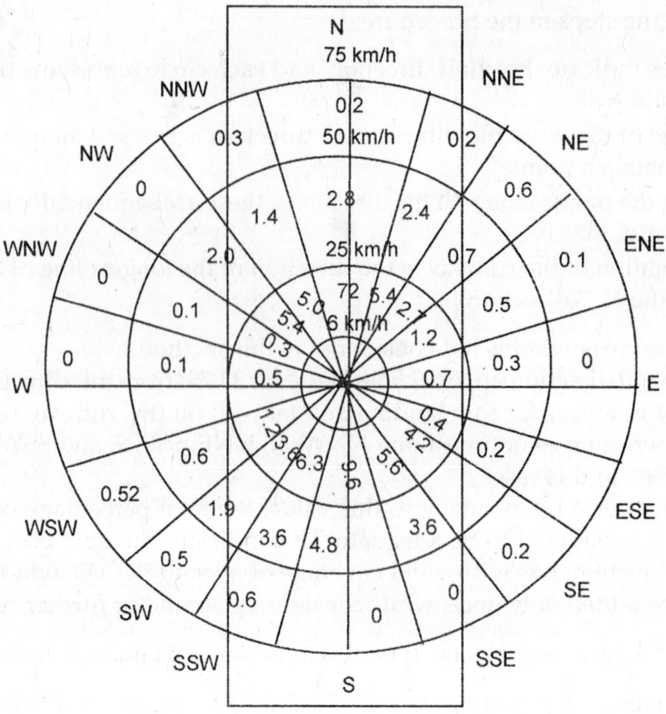


**FIG. 11.20** *Type-II wind-rose*

For N–S, the wind coverage comes to nearly 93%. (This just falls short of the minimum of 95%)

It is to be noted that, even by this method, the best orientation is the North–South direction for the runway.

**Note**: This may be got by trial and error, by orienting the template in all the cardinal directions and determining the wind coverage for each.

If wind data are not available for the site, those for the nearest location may be used; however, in hilly terrain, there may be significant difference in wind speed for different locations even if such locations are not far off.

## Length of Runway

The length of a runway has to be carefully determined for safe and efficient landing and take-off of the various aircraft expected to use the airport. The maximum length needed for the heaviest aircraft has to be provided, although it may be uneconomical.

**Basic runway length:** The length of the runway under standard prevalent conditions is known as the *basic runway length*.
These conditions are:

(i) The airport is situated at sea-level.
(ii) The temperature at the airport is the standard value of 15°C.[1]
(iii) The temperature along the way to the destination is also the standard value.

[1] This is fixed by the ICAO as the standard temperature at MSL to start with and corrected later for the elevation of the airport site.

(iv) The runway is level along its longitudinal direction.
(v) No wind blows on the runway.
(vi) No wind blows even on the way to the destination.
(vii) The aircraft is loaded to its full capacity.

The basic length of the runway is determined based on the performance characteristics of the aircraft during landing and take-off.
The following *three cases* are considered:

(a) Normal landing
(b) Normal take-off
(c) Engine failure or emergency

For jet aircraft, all the three cases are considered; for piston-engine aircraft, the first and the third are considered. Engine failure is not common for jet aircraft.

**(a) *Normal landing*:** The aircraft should come to a stop within 60 per cent of the landing distance. It is assumed that the pilot approaches at the proper speed and crosses the threshold of the runway at a height of 15 m (Fig. 11.21).

The runway pavement should be strong enough to support the weight and speed of the aircraft for the entire landing distance.

**(b) *Normal take-off*:** The *take-off distance* must be 1.15 times the actual distance the aircraft needs to reach a height of 10.5 m; the *lift-off distance* is that required by the aircraft to just lift itself off the runway (Fig. 11.22).

Normal take-off requires a clearway beyond the runway, not less than 150 m wide. The upward slope of the clearway plane beyond the end of the runway should not exceed 1.25%; the clearway should not be more than one-half the difference between the TOD and 1.15 times the LOD, as shown.

**(c) *Engine failure or emergency*:** In case of engine failure, TOD is the actual distance required to reach a height of 10.5 m with no extra factor applied. In general, this does not occur frequently.

In case of engine failure, adequate distance must be available for the aircraft to stop, rather than continue to take-off. This distance is called the *accelerate-stop distance* (Fig. 11.23).

A stopway or a clearway or both may be needed here. A rectangular paved area at the take-off end of the runway in which the aircraft can be stopped in case of engine failure is called the *stopway*. The thickness of the pavement can be less for this, but should be adequate to take the aircraft load.

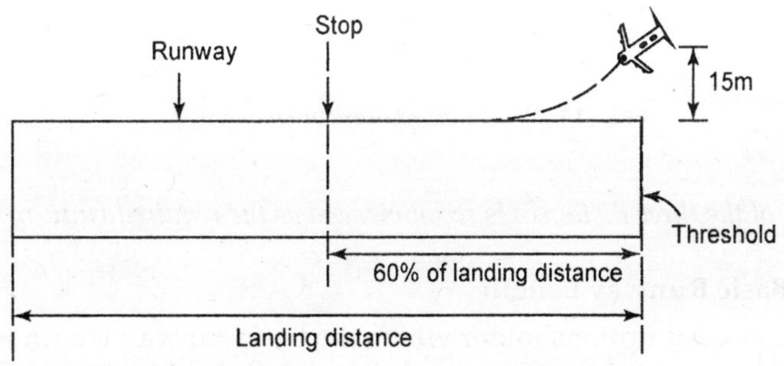


**FIG. 11.21** *Normal landing*

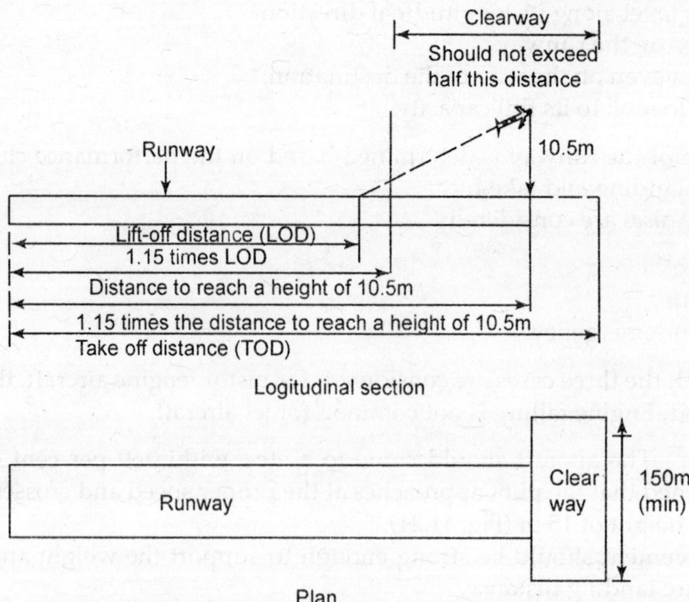


**FIG. 11.22** *Normal take-off*

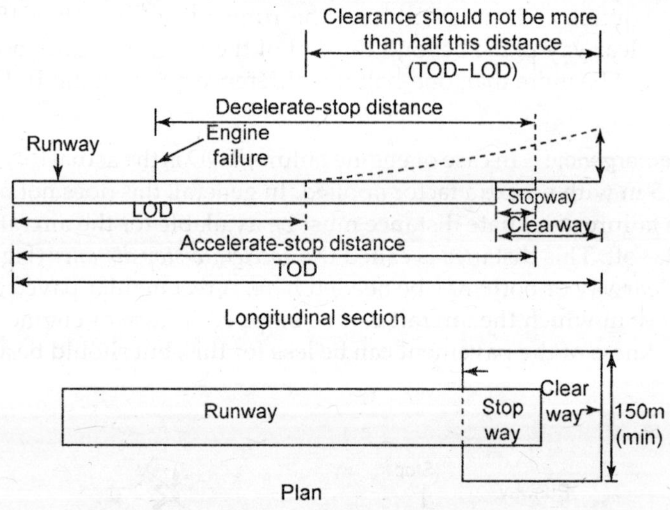


**FIG. 11.23** *Engine failure or emergency*

*The maximum length of the three distances is recommended as the required runway length.*

### Corrections to the Basic Runway Length

Since the conditions or assumptions under which the basic runway length is determined are rarely satisfied, certain corrections are required to be applied to this length in order to arrive at the actual length of the runway.

The following corrections are needed:

1. Correction for elevation
2. Correction for temperature
3. Correction for gradient

**Correction for elevation:** The density of atmospheric air decreases as the elevation above the mean sea level increases; this, in turn, reduces the lift on the wings of the aircraft. Hence, greater ground speed becomes necessary for the aircraft to rise into air and this, in turn, requires greater length of runway.

Therefore, ICAO recommends that the basic runway length be increased at the rate of 7% for every 300 m rise in elevation of the airport above the mean sea level.

**Correction for temperature:** The rise in airport reference temperature has an effect similar to that of the increase in its elevation above mean sea level.

The *airport reference temperature* (ART) is determined as follows:

$$\text{ART} = T_{\text{mean}} + \frac{1}{3}(T_{\text{max}} - T_{\text{mean}}) \tag{11.2}$$

where

$T_{\text{max}}$ = Monthly average of the maximum daily temperature for the hottest month of the year

$T_{\text{mean}}$ = Monthly average of the mean daily temperature for the same month.

ICAO recommends that, after the basic runway length is corrected for the elevation of the airport, the length be further increased at the rate of 1% for every 1°C rise of the airport reference temperature above the standard atmospheric temperature at that elevation.

The temperature gradient of the standard atmosphere from the mean sea level to the altitude at which the temperature becomes –15.6°C is –0.0065°C per metre; thereafter, the temperature gradient is zero.

The standard mean sea level temperature is 15°C.

Therefore, the standard temperature at the airport site can be obtained by reducing the standard temperature at MSL of 15°C at the rate of 0.0065°C per metre (or 6.5°C per thousand metre) rise of the airport elevation above MSL.

It should be obvious how the temperature correction for ART has to be applied.

The ICAO further recommends that if the total correction for elevation and temperature exceeds 35% of the basic runway length, specific studies at the site be conducted by model tests, before finally adopting the runway length.

**Correction for gradient** Steeper gradient results in greater consumption of energy, and needs a longer runway to attain the desired ground speed during take-off.

ICAO does not recommend any specific correction for gradient.

But FAA recommends that the runway length, corrected for elevation and temperature, be further increased at the rate of 20% for every 1% of effective gradient.

[*Effective gradient* is defined as the maximum difference in elevation of the site or the difference in level between the highest and the lowest points of the runway, divided by the total length of the runway.]

The actual runway length is got after applying all the necessary corrections to the basic runway length.

The application of these corrections will be better understood from the problems in the Numerical Examples Section.

**Note**: Corrections for gradient and temperature are not required for length of runway for landing; only elevation correction is needed. All the corrections are necessary for the runway length for take-off.

### Geometric Design for Runway Elements

The geometric design of runway depends upon the characteristics of various aircraft expected to use the airport. Critical aircraft characteristics are published by the ICAO in *Airport Reference Codes* so that uniform standards are practised in the geometric design of the runway elements.

Aircraft reference field length, the wing span and the outer main gear wheel span of the aircraft govern the geometric design of runway elements. (Wing span may range up to 65 m and the outer main gear wheel span may be up to 14 m.)

**Runway length:** This has already been considered; the length of the *primary runway* should be adequate for landing and take-off operations of the various aircraft.

*Secondary runways* are additional runways built to bring the usability factor to 95%.

**Runway Width:** This depends on the classification of the airport. ICAO recommends runway widths ranging from 18 m to 45 m; jet aircraft need a minimum width of 45 m.

**Longitudinal gradient:** ICAO recommends gradients ranging from 1% to 2%.

***Rate of change of longitudinal gradient*:** Abrupt change is undesirable. Rate of change of gradient should be 0.1% to 0.4% per 30 m, depending upon the class of airport.

**Transverse gradient:** This is necessary to drain off rain water. It should also be in the range of 1.5% to 2%. The rate of change of transverse gradient may range from 0.1% to 0.4% per 30 m.

**Sight distance:** Minimum sight distance is required only where two runways intersect, or a runway intersects a taxiway.

Any object that is 1.5 m to 3 m above the runway should be visible within a distance of at least half the length of the runway.

**Runway shoulders:** These are provided on either side when the runway width is less than 60 m, to make the overall width not less than 60 m. Transverse slope of the shoulders should not exceed 2.5%.

**Runway safety area:** This consists of the runway and shoulders, and the area that is cleared and graded for proper drainage. Minimum width is 75 to 150 m for ILS runways and 30 to 75 m for non-ILS runways.

*Fillets* at junctions of runways and at junctions of runways with taxiways are meant to smoothen out acute and obtuse angles, as in the case of highway intersections.

## Configurations of Runways

*Configuration* refers to the shape or arrangement of runways.

Runways may be *independent units* or *interdependent groups*; they may be parallel or intersecting.

Two runways intersecting near their thresholds is the usual pattern at international airports.

### Runway Patterns

These depend on the air traffic volume and the wind characteristics at the site. The basic patterns of runways are:

(i) Single runways
(ii) Parallel runways
(iii) Intersecting runways
(iv) Open-V runways

These are shown schematically in Fig. 11.24.

### Runway Capacity

*Runway capacity* is defined as the number of operations per hour (or per year) which a runway can accommodate. (Landing and take-off are considered as independent operations for this purpose.)

This depends primarily on the wind coverage and the type of aircraft; a reasonable delay is supposed to be allowed.

Mathematical models are available for the computation of *saturation capacity,* at least for simple systems.

**Factors affecting runway capacity:** The following aspects affect runway capacity.

- Air traffic control system
- Characteristics of demand
- Environmental factors
- Layout and design of the runway system

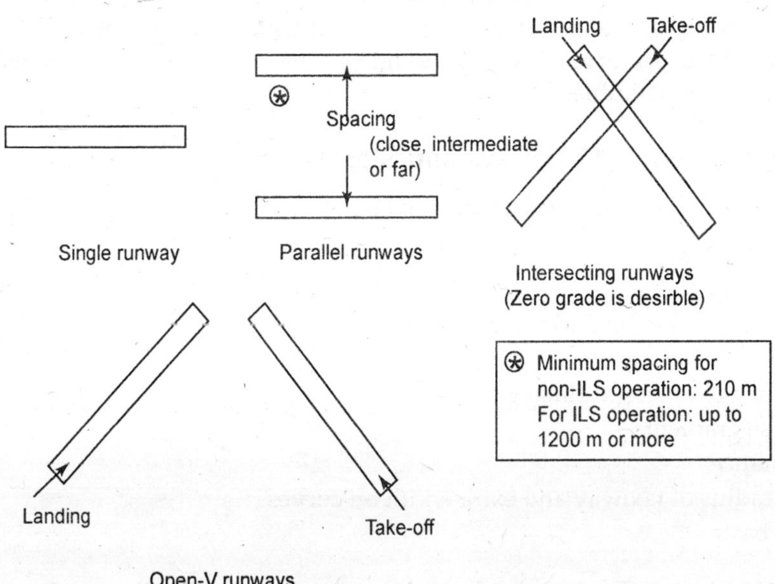


**FIG. 11.24** *Basic runway patterns*

**Practical capacity of a runway:** Practical annual capacity of a runway depends on:

- Practical hourly capacity
- Weather conditions and runway use
- Allowable delay
- Extent of usage of off-peak periods.

Practical capacity of a runway, therefore, is much less than the theoretical capacity.

## 11.4 DESIGN OF TAXIWAYS

The primary function of a *taxiway* is to provide access to the terminal area and service hangars from the runway.

The speed of an aircraft on a taxiway will be much less than that on the runway; as such, the standards of design, construction and strength requirements for taxiways will be somewhat less than those for runways.

*Factors governing taxiway layouts*:

- *Arrangement*: It should be such that aircraft taxiing on landing does not interfere with one taxiing to take-off.
- *Location*: At busy airports, it is desirable to locate taxiways at various points along the runways. Such taxiways are known as *exit taxiways*. The *main taxiway* is the one located around the apron area. It is also referred to as the *apron taxiway*. Usually *dual (or parallel) taxiways* are provided such that aircraft which are landing and are taking off can use them in opposite directions. The portion of the taxiway nearest the gate positions is the *terminal taxilane*.
- *Intersection*: As far as possible, a taxiway should not cross an active runway.
- *Turn-off speeds*[2]: Exit taxiways should be designed for higher turn-off speed so as to reduce the runway occupancy of landing aircraft. This contributes to an increase in airport capacity.
- *Optimal route*: The route for a taxiway should be such that its distance from the terminal area is as small as is practicable.

### Geometric Elements of Taxiways and Standards

The following are the geometric elements of taxiways:

(1) Length of taxiway
(2) Width of taxiway
(3) Width of safety area (including shoulders)
(4) Longitudinal gradient
(5) Rate of change of longitudinal gradient
(6) Transverse gradient
(7) Sight distance
(8) Turning radius of taxiway and extra width on curves
(9) Holding bays

**(1) *Length of taxiway*:** The length of a taxiway depends on the distance between the apron and the end of the runway at its entry or exit. The limiting length is not recommended by any organisation; however, this should be as short as possible in order to save fuel.

[2] Turn-off speed is the speed of the aircraft while turning away from the exit taxiway just before entering the runway.

**(2) *Width of taxiway*:** This need not be as much as for a runway because the speed of aircraft on taxiway is much less, making the manoeuvring relatively easy. ICAO recommends widths ranging from 7.5 m to 22.5 m, depending upon the class of the airport.

**(3) *Width of safety area (including shoulders)*:** This consists of partially paved shoulders on either side of the taxiway, as also the area which is graded and drained. This area may extend even up to a parallel taxiway, apron, or a runway.

Before the advent of jet aircraft, the shoulders consisted of stabilised granular material. But with the advent of jet planes, shoulders are constructed for a minimum width of 7.5 m on either side of the taxiway with light paving material. The maximum transverse gradient of shoulders is 5% for the first 3 m and 2% thereafter (FAA).

**(4) *Longitudinal gradient*:** ICAO recommends longitudinal gradients of 1.5% (max.) for A and B classes of airports and 3% (max.) for C, D and E classes of airports.

**(5) *Rate of change of longitudinal gradient*:** The maximum rates of change of longitudinal gradient recommended by the ICAO are:

1% per 30 m for A, B and C classes of airports (min. radius 3000 m)
1.2% per 30 m for D and E classes of airports (min. radius 2500 m)

**(6) *Transverse gradient*:** This is needed for effective drainage of rain water.

ICAO recommends the following *maximum* values —
1.5% for A, B and C classes of airports
2% for D and E classes of airports
A *minimum* of 0.5% is desirable.

**(7) *Sight distance*:** Smaller values are adequate in view of smaller speeds on taxiways.

ICAO recommends the following values:
For A and B classes, surface should be seen for a distance of 200 m from a height of 2.1 m
For C, D and E classes, surface should be seen for a distance of 300 m from a height of 3 m.

**(8) *Turning radius of taxiway and extra width on curves*:** A change in the direction of a taxiway has to be achieved through a horizontal curve; this should be designed in such a way that an aircraft can negotiate it without significantly reducing its speed. This is possible if the radius of the curve is large.

The radius R may be got from the formula:

$$R = \frac{V^2}{125\,f} \tag{11.3}$$

where

$R$ = Radius of the curve (m)
$V$ = Speed of aircraft (km/h)
$f$ = Coefficient of friction between the tyre and the pavement surface (usually 0.13)

The minimum value of the radius of curvature is taken as 120 m, irrespective of the speed, for an airport serving large subsonic jet planes. For supersonic jet planes, a minimum radius of 180 m is recommended. In this case, the Horonjeff's equation which takes into account the wheel base of the aircraft, is considered applicable but this is considered out of scope of the present treatment.

Extra width for the taxiway is recommended on the inside of the horizontal curve for better safety of the aircraft while negotiating the curve.

***Holding bays:*** Holding bays should be provided when the volume of air traffic is high.

A taxi-holding position is provided at the intersection of a taxiway with a runway or at that between two runways; the location is such that a taxiing aircraft can interfere with the operation of navigating aids. Figure 11.25 shows a typical holding bay.

### Turnaround Taxiway

In the initial stages of an airport, when the air traffic volume does not justify the construction of parallel taxiways, a bypass or turnaround taxiway is provided, as shown in Fig. 11.26, which serves the purpose of a holding apron also.

### Separation Clearance for Parallel Taxiways

Certain standards of separation clearance between taxiways/runways have to be maintained in order to ensure a margin of safety during landing, take-offs, and taxiing operations of aircraft.

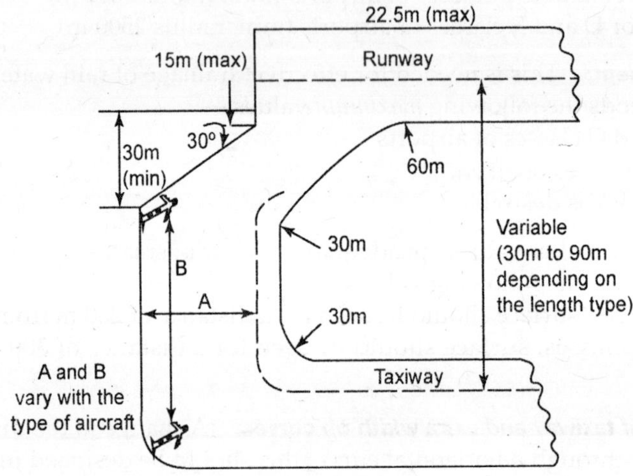


**FIG. 11.25** *Holding bay*

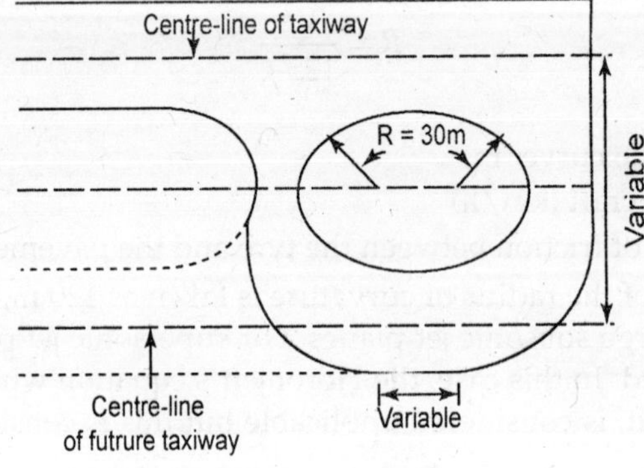


**FIG. 11.26** *Turnaround taxiway*

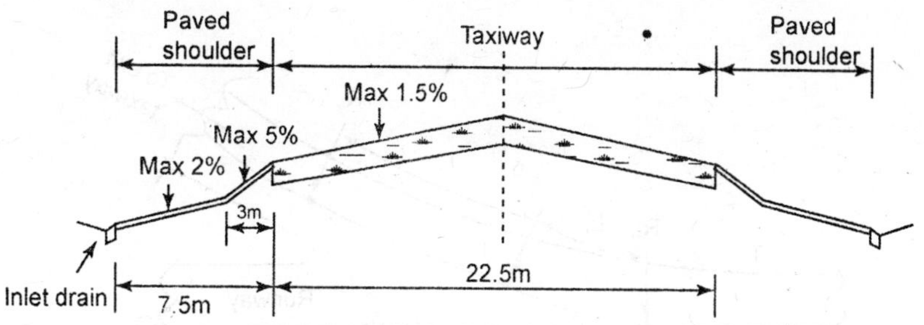


**FIG. 11.27** *A typical cross-section of a taxiway*

This clearance depends on the type of airport, wing span of aircraft, and the navigational aids available at the airport.

ICAO recommends 184 m for ILS and 109 m for non ILS as separation clearances for parallel taxiways for A-class airport. A clearance of 75 m to 150 m is adequate for building lines.

### Cross-section of a Taxiway

A typical cross-section of a taxiway is shown in Fig. 11.27.

### Exit Taxiways

The function of exit taxiways is to minimise runway occupancy by the landing aircraft. Their *location* depends on the *following factors*:

(i) Exit speed
(ii) Location of runways relative to the terminal area
(iii) Type of aircraft and their landing speed
(iv) Number of exit taxiways
(v) Topographical features of the airport
(vi) Weather conditions
(vii) Air traffic control
(viii) Personal factors of the pilot.

The optimum location of exitway is that which yields the highest rate of acceptance for landing. This is where *runway saturation*[3] occurs. It varies with the type of aircraft; FAA recommends a distance of 1800 m to the exit taxiway from the runway threshold for large jet aircraft.

**Design of exit taxiways:** The following are the aspects of design:

***(i) Angle of turn***: An angle of turn of 30° to 45° is desirable for negotiating a curve in a smooth and satisfactory manner.

***(2) Compound curve***: For high turn-off speeds (65 to 95 km/h), a compound curve is provided to minimise tyre wear of the nose gear (Fig. 11.28).

$R_1$: Radius of entrance curve

$R_2$: Radius of central curve

$L_1$: Length of entrance curve

[3] Runway saturation means the situation when the runway cannot take any more landing/take-off operations, or when its capacity has been realised.

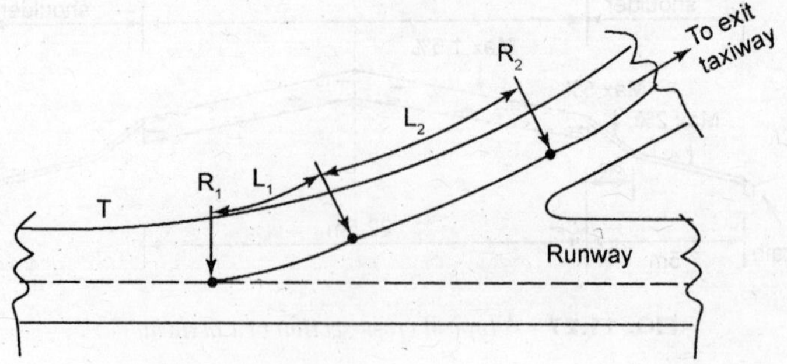


**FIG. 11.28** *Compound curve for exit taxiway*

$L_2$: Length of central curve
The recommendations for $R_1$ are:

| Speed Km/h | $R_1$(m) |
|---|---|
| 65 | 517 |
| 80 | 731 |
| 95 | 941 |

***(3) Exit speed***: This is the most important factor affecting the turning radius, and not the total angle of turn.

***(4) Lengths of entrance and central curves***:

$$L_1 = \frac{V^3}{45.5 C R_2} \tag{11.4}$$

$$L_2 = \frac{\pi R_2 D_2}{180} \tag{11.5}$$

Here, C is a constant taken as 0.39 and $D_2$ is the deflection angle of the central curve.

***(5) Occupancy time***: This is the time before an aircraft enters the taxiway from the runway and consists of the following components:

(i) Time required for the flight from runway threshold to touch-down of the back wheels that make up the main landing gear.
(ii) Time for the nose gear to make contact with the runway after the touchdown of the main gear (usually 3 seconds).
(iii) Time to reach exit velocity from touchdown velocity after the nose gear makes runway contact and brakes are applied.
(iv) Time needed for the aircraft to turn to the taxiway and clear the runway (about 10 seconds).

These factors govern the location of the exit taxiway from the threshold of the runway.

***(6) Shape of the taxiway***: A widened entrance of about 30 m, gradually tapering to the normal width of the taxiway, is preferred for greater latitude in manoeuvring into the taxiway.

***(7) Stopping distance***: This is measured from the edge of the runway along the taxiway. The equation for this is given by

$$\text{SSD} = \frac{V^2}{25.5d} \tag{11.6}$$

Where $V$ = Speed of aircraft (km/h)

and $d$ = Deceleration rate (usually 1 m/sec²)..

***(8) Turning radius***: To ensure a smooth and comfortable turn, the turning radius is calculated by the equation given earlier:

$$R = \frac{V^2}{125\,f} \tag{11.3}$$

Here, $R$ represents the radius of the central curve, $R_2$, depicted in Fig. 11.28.

## 11.5 NUMERICAL EXAMPLES

### EXAMPLE 11.1

An airport has 4 gates which are available for all the aircraft. Three classes of aircraft use them with the following mix and average gate occupancy times during peak hour:

Class-I ··· 35% ··· 60 min.
Class-II ··· 50% ··· 45 min.
Class-III ··· 15% ··· 30 min.

Find the gate capacity if the gate utilisation factor is (i) unity and (ii) 60%.

**Solution**

Gate capacity, $$C = \frac{G}{T}\cdot U \quad \text{(Eq. 11.1)}$$

$$T = \text{Weighted average gate occupancy}$$

$$= (0.35\times 60 + 0.50\times 45 + 0.15\times 30)\times \frac{1}{60}\ \text{hr}$$

$$= 0.80\ \text{hour.}$$

Gate utilisation factor, $U = 1$

$$C = \frac{4}{0.8}\times 1 = 5\ \text{aircraft/hour}$$

Hence, the number of operations is <u>10/hr</u>

If $U = 0.60$

$$C = \frac{4 \times 0.60}{0.8} = 3 \text{ aircraft/hour}$$

Hence, the number of operations—arrival and departure—are 6 per hour.

## EXAMPLE 11.2

Determine the capacity of the 12 gates which are used by three classes of aircraft – I, II and III – if the number of gates in each group, their mix percentages, and average gate occupancy time in minutes for each group are as follows:

| | | | |
|---|---|---|---|
| I ··· | 2 gates ··· | 20% ··· | 25 min |
| II ··· | 4 gates ··· | 30% ··· | 45 min |
| III··· | 6 gates ··· | 50% ··· | 50 min |

Gate utilisation factor is unity.

**Solution**

The capacity of each group of gates has to be got separately.

$$C_1 = \frac{2}{T_1} \cdot U \ (U = 1)$$

$$T_1 = \frac{0.20 \times 25}{60}$$

$$\therefore \quad C_1 = \frac{2 \times 60}{0.2 \times 25} = 24 \text{ aircraft/hour}$$

$$T_2 = \frac{0.30 \times 45}{60}$$

$$\therefore \quad C_2 = \frac{4}{T_2} = \frac{4 \times 60}{0.3 \times 45} = 17.77 \text{ aircraft/hour}$$

$$T_3 = \frac{0.50 \times 50}{60}$$

$$C_3 = \frac{6}{T_3} = \frac{6 \times 60}{0.5 \times 50} = 14.4 \text{ aircraft/hour}$$

The minimum of these is the gate capacity i.e., 14.4 aircraft/hour.

(*Check*: Gate time supplied $= 12 \times 60 = 720$ min

Gate time demanded $= [0.20 \times 25 + 0.30 \times 45 + 0.50 \times 60] \times 14.4$
$= 626.4$ min

Supply is more than demand, and hence the value obtained is okay)

**EXAMPLE 11.3**

An airport has 10 gates with restrictions on the types of aircraft which can be accommodated in each group. The particulars are given below:

| Aircraft type | Gate group | No. of gates | Mix % | Gate occupancy (min.) | Aircraft type |
|---|---|---|---|---|---|
| A | I | 5 | 35 | 60 | A, B, C |
| B | II | 3 | 50 | 45 | B, C |
| C | III | 2 | 15 | 30 | C only |

Determine the gate capacity assuming the gate utilisation factor as unity.

### Solution

The number of gates available for each type of aircraft is:

$$\left.\begin{array}{lcr} A & \cdots & 5 \\ B & \cdots & 8 \\ C & \cdots & 10 \end{array}\right\} U = 1$$

Let the capacity of each gate group be $C_1$, $C_2$ and $C_3$

$$T_1 = \frac{0.35 \times 60}{60}$$

$$\therefore \quad C_1 = \frac{G_1}{T_1}.U = \frac{5 \times 60}{0.35 \times 60} = 14.28$$

$$T_2 = \frac{0.35 \times 60}{60} + \frac{0.50 \times 45}{60} = \frac{43.5}{60}$$

$$\therefore \quad C_2 = \frac{G_2}{T_2} \cdot U = \frac{8 \times 60}{43.5} = 11$$

$$T_3 = \frac{(0.35 \times 60 + 0.50 \times 45 + 0.15 \times 30)}{60} = \frac{48}{60}$$

$$\therefore \quad C_3 = \frac{G_3}{T_3}.U = \frac{10 \times 60}{48} = 12.5$$

The least of these is the great capacity; i.e., 11 aircraft/hour.

(*Check*: Gate time supplied $= 10 \times 60 = 600$ min

Gate time demanded $= (0.35 \times 60 + 0.50 \times 45 + 0.15 \times 30) \times 11$
$= 48 \times 11 = 528$ min

Supply is more than demand; hence, the value is okay)

## EXAMPLE 11.4

The average wind data at an airport site are given below when the wind velocity is more than 6 km/h:

| Direction | N | NNE | NE | ENE | E | ESE | SE | SSE | S | SSW | SW | WSW | W | WNW | NW | NNW |
|---|---|---|---|---|---|---|---|---|---|---|---|---|---|---|---|---|
| % of time | 6 | 11 | 8 | 4 | 2 | 2 | 2 | 4 | 8 | 15 | 10 | 6 | 4 | 2 | 1 | 5 |

Determine the best orientation of the runway and the wind coverage using the wind-rose diagram.

**Solution:**

Total coverage in all the directions = 90%

$\therefore$ Calm period = (100 – 90) = 10%

The wind-rose is plotted below:

The best orientation of runway is NNE–SSW (Bearing N 22°30′E).

Wind coverage → Sum of those in N, NNE, NE, S, SSW, and SW and Calm period %

$$= 6 + 11 + 8 + 8 + 15 + 10 + (10) = 68\%$$

This is not adequate. Hence, the second runway can be in the NE–SW direction.

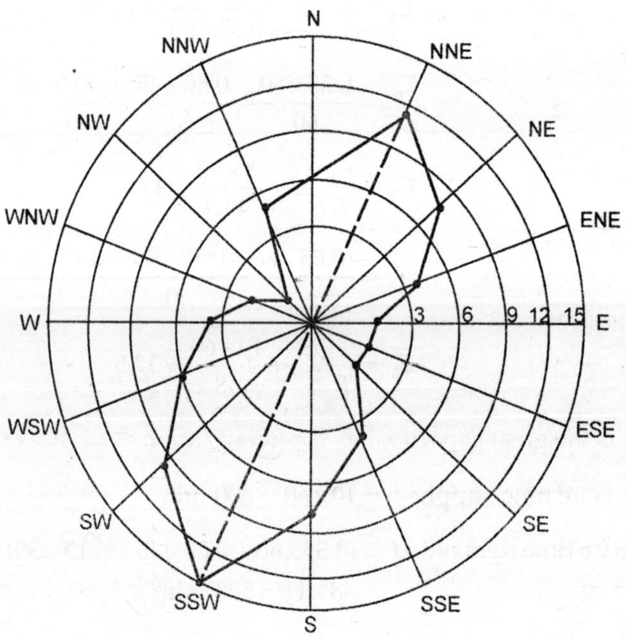


**FIG. 11.29** *Wind-rose (Example 2.4)*

## EXAMPLE 11.5

The average wind data for a 10-year period at a proposed airport site are given below when the wind velocity is greater than 6 km/h:

| Direction | N | NNE | NE | ENE | E | ESE | SE | SSE | S | SSW | SW | WSW | W | WNW | NW | NNW |
|---|---|---|---|---|---|---|---|---|---|---|---|---|---|---|---|---|
| % of time | 12 | 9 | 3 | 3 | 0 | 0 | 9 | 6 | 12 | 9 | 6 | 0 | 0 | 0 | 9 | 6 |

Determine the best orientation of the runway and wind coverage using the wind-rose diagram. Is a second runway needed? If so, which is the best orientation for it? What is the total wind coverage?

**Solution:**

Total coverage in all the directions = 84%

∴ Calm period = (100 – 84) = 16%

The best orientation of the runway is *N-S (Bearing–0°)*.

Wind coverage = Sum of those in NNW, N, NNE, SSE, S AND SSW and calm period %

$$= (6 + 12 + 9 + 6 + 12 + 9) + (16) = 70\%$$

Therefore, a second runway is needed.

This can be in the NNE–SSW direction or NW–SE direction.

The wind coverage in the NW–SE direction is better with an additional 18% wind coverage or a total of 88%.

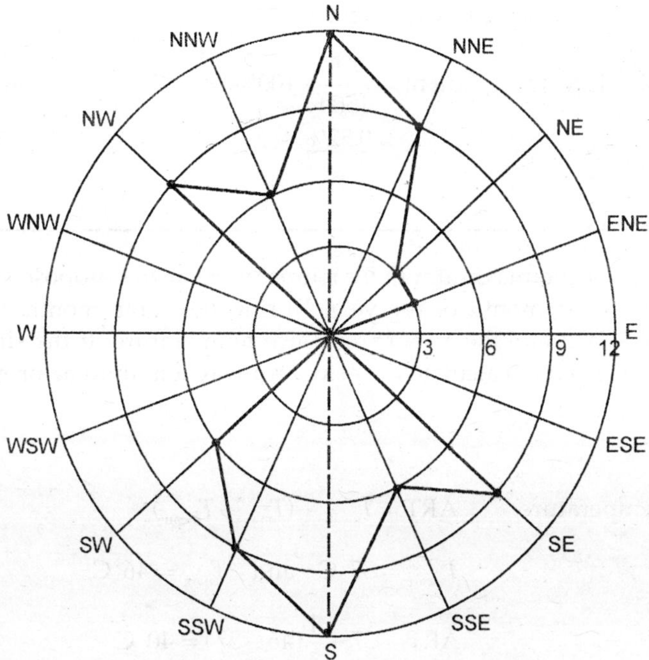


**FIG. 11.30** *Wind-rose (Example 2.5)*

## EXAMPLE 11.6

The gradients along a proposed 1500 m long runway are measured as follows:

| | | | | |
|---|---|---|---|---|
| 0 | to | 300 m | ··· | +1% |
| 300 | to | 600 m | ··· | −1% |
| 600 | to | 1200 m | ··· | +1.2% |
| 1200 | to | 1500 m | ··· | +0.2% |

What is the *effective* gradient?

**Solution:**

Let the reduced level (RL) of the starting end be 100 m.

$$\text{RL of 300m chainage} = 100 + \frac{1}{100} \times 300 = 103.0$$

$$\text{RL of 600m chainage} = 103 - \frac{1}{100} \times 300 = 100.0$$

$$\text{RL of 1200m chainage} = 100 + \frac{1.2}{100} \times 600 = 107.2$$

$$\text{RL of 1500m chainage} = 107.2 + \frac{0.2}{100} \times 300 = 107.8$$

$$\text{Maximum difference in elevation} = (107.8 - 100) = 7.8\,\text{m}$$

$$\text{Length of the proposed runway} = 1500\ \text{m}$$

$$\therefore \quad \text{Effective gradient} = \frac{7.8}{1500} \times 100\% = \underline{0.52\%}$$

## EXAMPLE 11.7

The monthly average of the maximum daily temperature at the proposed site of an airport is 46°C during the hottest month of the year. During the same month, the average daily temperature is 37°C. Calculate the airport reference temperature. If the site is at mean sea-level and is on a level ground, calculate the actual runway length to be provided, if the basic length is 2,000 m.

**Solution**

Airport reference temperature, $\quad \text{ART} = T_{mean} + (T_{max} - T_{mean})$

$$T_{mean} = 37^\circ\text{C} \quad \text{and} \quad T_{max} = 46^\circ\text{C}$$

$$\therefore \quad \text{ART} = 37 + \frac{1}{3}(46 - 37) = 40^\circ\text{C}$$

Standard atmospheric temperature at the mean sea-level = 15°C

Since the proposed site is at MSL and on a level ground, no corrections are needed for elevation and gradient.

$$\text{Correction for temperature} = 1 \times (40 - 15)\% = 25\% \text{ of basic length}$$

$$= \frac{25}{100} \times 2000 = 500\,\text{m}$$

$\therefore$ The actual length of runway to be provided $= (2000 + 500)\,\text{m}$

$= \underline{2{,}500\text{ m}}$

## EXAMPLE 11.8

An airport is proposed at an elevation of 300 m above mean sea-level. The maximum and average daily mean values of temperature for the hottest month of the year at the site are 45°C and 27°C, respectively. The maximum difference in elevation is 7 m along the proposed profile of the runway of 1400 m. Determine the actual length of the runway to be provided.

### Solution

*Correction for elevation*

$$\frac{7}{100} \times \frac{300}{300} \times 1400\,\text{m} = 98\,\text{m}, \quad \text{say } 100\,\text{m}.$$

$$\text{Length of runway corrected for elevation} = (1400 + 100)\,\text{m} = 1500\,\text{m}$$

*Correction for temperature*

$$\text{Airport reference temperature, ART} = T_{mean} + \frac{1}{3}(T_{max} - T_{mean})$$

$$= 27 + \frac{1}{3}(45 - 27) = 33^\circ\text{C}$$

Standard atmospheric temperature at an elevation of 300 m

$$= 15 - \frac{300}{1000} \times 6.5 = 13^\circ\text{C nearly.}$$

$$\text{Rise in temperature} = (33 - 13) = 20^\circ\text{C}$$

$$\text{Correction for temperature} = \frac{20}{100} \times 1500 = 300\,\text{m}$$

$$\text{Corrected length} = 1500 + 300\,\text{m}$$

$$= 1800\,\text{m}$$

Total correction for elevation and temperature = (100 + 300) = <u>400 m</u>

This increase is $\frac{400}{1400} \times 100\%$ or 30% nearly.

Since this is less than 35%, it is okay.

*Correction for gradient*

$$\text{Effective gradient} = \frac{7}{1400} \times 100 = 0.5\%$$

$$\therefore \quad \text{Correction for gradient} = 0.5 \times \frac{20}{100} \times 1800\,\text{m}$$

$$= 180\,\text{m}$$

$\therefore$ Actual length of the runway required = (1800+180) m = 1,980 m

## EXAMPLE 11.9

The basic lengths of runway required for take-off and landing are 2000 m and 2400 m respectively at the proposed site for an airport situated 450 m above MSL. The airport reference temperature is 24°C and the effective gradient along the proposed runway is 0.40%. Calculate the actual length of the runway to be provided.

**Solution**

The following are the corrections in runway length for take-off:

*Correction for elevation*

$$= \frac{7}{100} \times 2000 \times \frac{450}{300}\,\text{m}$$

$$= 210\,\text{m}$$

$$\text{Corrected length} = (2000 + 210)\,\text{m} = 2,210\,\text{m}.$$

*Correction for temperature*

$$\text{Standard atmospheric temperature at 450 m above MSL} = \left(15 - \frac{450}{1000} \times 6.5\right) = 12^{\circ}\text{C, nearly}$$

$$\therefore \quad \text{Correction for temperature} = (24 - 12) \times \frac{2210}{100} = 265\,\text{m}$$

$$\text{Corrected length} = (2210 + 265) = 2,475\,\text{m}$$

Total correction at this stage = 475m

This is 23.75% of the basic length; i.e., < 35%
Hence, it is okay.

*Correction for gradient*

$$\frac{20}{100} \times 0.40 \times 2475 = 200\,\text{m, nearly.}$$

$\therefore$ Corrected runway length for take-off = 2475 + 200 = 2,675 m.

For landing, no corrections are required for temperature and gradient.

*Correction for elevation*

$$\frac{7}{100} \times 2400 \times \frac{450}{300} = 252\,\text{m}$$

$$\therefore \text{Actual runway length required for landing} = (2400 + 252)\,\text{m}$$

$$= \underline{2{,}652 \text{ m.}}$$

Adopting the higher of the two values, the actual runway length to be provided is <u>2675 m</u>.

## EXAMPLE 11.10

Determine the turning radius of a taxiway for subsonic aircraft for a design turning speed of 45 km/h. Assume the coefficient of friction between the tyre and the pavement as 0.13.

**Solution**

$$R = \frac{V^2}{125\,f}$$

Turning radius,

$$= \frac{(45)^2}{125 \times 0.13}\text{m}$$

$$= \underline{125 \text{ m}}$$

## EXAMPLE 11.11

Design an exit taxiway for a total angle of turn of 36° and a maximum turn-off speed of 75 km/h.

Radius of the entrance curve = 720 m

Width of runway = 45 m

Width of taxi-way = 22.5 m

Suggest the necessary separation clearance.

**Solution**

*Radius of central curve*

$$R_2 = \frac{V^2}{125\,f} = \frac{(75)^2}{125 \times 0.13} = 346.15\,\text{m or } 350\,\text{m}$$

*Length of entrance curve*

$$L_1 = \frac{V^3}{45.5\,CR_2} = \frac{(75)^3}{45.5 \times 0.39 \times 350} = 68\,\text{m} \text{ or } 70\,\text{m}$$

*Deflection angle of entrance curve*

$$D_1 = \frac{180L_1}{R_2} = \frac{180}{\pi} \times \frac{70}{350} = 11.5$$

*Deflection angle of the central curve*

$$D_2 = 36^\circ - 11.5^\circ = 24.5^\circ$$

*Length of the central curve*

$$L_2 = \frac{\pi R_2 D_2}{180} = \frac{3.14 \times 350 \times 24.5}{180} = 150 \text{ m}$$

*Stopping distance*

$$\text{SSD} = \frac{V^2}{25.50\,d} = \frac{(75)^2}{25.50 \times 1} = 221 \text{ m.}$$

*Separation Clearance*

Assuming instrumental landing facilities, ICAO recommends a separation clearance

$$\text{of 184 m + Width of taxiway + Half width of taxiway} \begin{cases} = (184 + 22.50 + 11.25) \text{ m} \\ = 217.75 \text{ m, say } \underline{220 \text{ m}} \end{cases}$$

$$\therefore \quad \text{Available length of taxiway} = \left[\frac{220}{\sin 36^\circ} - \frac{(45 + 22.50)}{2 \sin 36^\circ}\right] \text{m}$$

$$= \underline{317 \text{ m}}$$

This is more than the calculated SSD. Thus, the SSD is fully accommodated within the length of the taxiway.

*Fillets*

The fillet radii recommended are 22.5 m and 60 m respectively at the 36° and 144° changes in directions.

## SUMMARY

- An airport terminal is a hub of activity comprising the terminal area and the motor-vehicle parking area. The terminal area consists of the terminal building, aprons, hangars, and the air traffic control tower.
- The Terminal building is the focal point at an airport where several kinds of facilities required for the passengers and airport staff are planned and provided to ensure passenger convenience, comfort and smooth flow of air traffic. The planning concepts of terminal buildings may be classified into centralised, de-centralised, and the unit terminal or decentralized–centralised system.
- Motor-vehicle parking area should be provided near the terminal building, with appropriate circulation pattern. Parking areas may be designed for short-term (3 hours or less), long-term, or remote parking. Patterns of parking may be parallel, right-angled or angled parking – the last one increases the capacity.
- An apron is a paved area adjacent to the terminal building, meant for loading and unloading of passengers and cargo into or from the aircraft.
- Parking of aircraft may be nose-in, nose-out, parallel, angled-nose-in or angled-nose-out. Angled nose-in considered the best option.
- The number of gates at the terminal for each aircraft operation of take-off/landing needed depends on the peak-hour traffic volume, gate occupancy time and gate-utilisation factor. Gate capacity, which is the number of aircraft that can be handled per hour, either for landing or for take-off, also depends on these factors.

- Aircraft parking near terminal building in the apron area may be according to frontal system, open-apron system, pier system, or satellite system.
- Aprons meant for loading operations are called loading aprons; holding aprons are those near the starting end of the runway and are meant to hold aircraft for checks before take-off.
- Hangars enclosed or semi-enclosed sheds to house, refuel, service and repair aircraft; workshops and spare parts are also provided. Hangars may be T-hangars, Nose-hangars or Grouped-T hangars in respect of their layout in plan.
- Jet blast has certain ill-effects on runways, taxiways, aprons, and terminal buildings. Blast fences or erosion control methods are used against jet blast. Oil spilling can affect bituminous surfaces.
- Airport capacity means the number of aircraft that can be handled per hour or per year. This depends upon several factors such as the type of aircraft, the number of gates and air traffic control methods among others.
- Typical airport layouts for domestic as well as international traffic are discussed and case studies of a few prominent airports in India and abroad are presented.
- Runway design pertains to orientation and pattern as well as its geometric elements such as the length, width, gradients-longitudinal and transverse, runway shoulders and safety area.
- Runway orientation is decided based on wind characteristics at the site; operations are preferred in the head-wind (or against the wind), with minimum cross-wind component for safety. Wind-rose diagram, showing wind direction and percent duration of the wind in a year in a particular direction, helps in deciding the best runway orientation based on the wind coverage, (with leeway of 33°.75 on either side) which should be of the order of 95%. If one runway cannot provide the necessary orientation, a second runway may be provided to enhance coverage. Another type of wind-rose which depicts the direction, percentage of time in a year, and the wind velocity may be plotted; this helps in ascertaining the best orientation of the runway, accounting for the cross-wind component also.
- The basic runway length required is got for three cases– normal landing, normal take-off, and engine failure or emergency. The greatest of these values is taken as runway length. Since the basic length is got for standard conditions of temperature and gradient at sea-level, corrections will have to be arrived at to determine the actual length.

  The corrections are those for site elevation above sea-level, temperature rise – airport reference temperature minus the standard temperature at that elevation, and the effective gradient correction, in this order. These corrections may necessitate an increase in runway length. Runway length required for landing need not be corrected for gradient and temperature – the maximum of the corrected lengths for take-off and landing is provided as the actual runway length.

- Runway patterns may consist of single runway, parallel runways, intersecting runways, or open-V (diverging) type of runways. The type to be used depends upon the volume and nature of the air traffic expected.
- Taxiways provide access to the terminal area and hangars. The geometric elements of taxiways are length, width, shoulders and safety area, longitudinal and transverse gradients, sight distance, turning radius, extra width on curves near their intersection with runways or other taxi-ways, and fillets.

  Holding bays near intersection with runways are meant for holding aircraft for checks before take-off.

- Exit-taxiways taking off from runways at intervals along the runway enhance the airport capacity.

Exit taxiways are designed for correct curvatures and changes in direction for safe manoeuvres of aircraft towards the aprons or away from them.

## QUESTIONS AND PROBLEMS

1. What are the functions of a terminal building at an airport?
2. List the facilities required to be provided for passengers and airport staff at an airport terminal building.
3. Explain briefly the following:

   (a) Decentralised system

   (b) Angled parking

   (c) Grouped-T hangars
4. Write shorts notes on:

   (a) Satellite system

   (b) Nose-hangar

   (c) Blast fence
5. Write a note on the design of motor-parking area in front an airport terminal and its circulation pattern.
6. What is a *gate* at an airport terminal? What factors govern the size of a gate?
7. How is the number of gates at an airport terminal decided?
8. What are the types of aircraft parking at a gate? Which one is the best?
9. What are the functions of hangars? Explain a T-hangar with a sketch.
10. How is protection from the ill-effects of a jet blast achieved?
11. Sketch a typical layout of a small domestic terminal with a single runway.
12. Write notes on:

    (a) Passenger flow at a terminal

    (b) Short-term parking and long-term parking

    (c) Nose-in and Nose-out parking of aircraft
13. Write short notes on:

    (a) Unit terminal principle

    (b) Gate occupancy time

    (c) Gate utilisation factor
14. What is a wind-rose? How is the Type-I wind-rose plotted?
15. How is Type-II wind-rose superior to Type-I wind-rose?

16. Define basic runway length. (JNTU-H)
17. Explain the three standard cases for obtaining the basic runway length. (JNTU-H)
18. Explain the corrections required to be applied to the basic runway length to arrive at the actual runway length. (JNTU-H)
19. Write short notes on:
    (a) Head wind
    (b) Cross-wind component
    (c) Wind coverage
20. Write critical notes on:
    (a) Runway patterns
    (b) Sight distance
    (c) Transverse gradient
21. Explain the following:
    (a) Normal landing and normal take-off
    (b) Parallel runways and intersecting runways
    (c) Runway occupancy time
22. Write a note on *runway capacity*.
23. What are the functions of a taxiway? What is an *apron taxiway*?
24. Write short notes on:
    (a) Turning radius
    (b) Safety area of a taxiway
    (c) Separation clearance of parallel taxiways
25. Write a note on the functions of an exit taxiway and its location with respect to a runway. (VTU)
26. List the geometric elements of a taxiway.
27. Write brief notes on:
    (a) Holding bay
    (b) Turnaround taxiway
    (c) Entrance curve and central curve of an exit taxiway
28. Define 'Airport capacity' and explain the factors which govern it.
29. Write short notes on:
    (a) Airport reference temperature
    (b) Effective gradient
    (c) Apron turntable

30. An airport has 6 gates which are available for all the three types of aircraft which it serves. The details are given below.

| Aircraft class | Mix % | Average occupancy time in minutes |
|---|---|---|
| I | 25 | 60 |
| II | 45 | 45 |
| III | 30 | 30 |

Find the gate capacity if the gate utilisation factor is (a) 100% and (b) 75%

31. The length of a runway under standard conditions is 1800 m. The airport site is at an elevation of 110 m above mean sea-level. The airport reference temperature is 22°C. The construction plan includes the following data:

| End to end of runway (m) | Grade (percent) |
|---|---|
| 0 to 300 | + 1.00 |
| 300 to 900 | – 0.25 |
| 900 to 1500 | + 0.50 |
| 1500 to 1800 | + 1.00 |
| 1800 to 2100 | – 0.30 |

Determine the actual length of the runway to be provided. (JNTU-H)

32. The proposed runway length under standard conditions is 1900 m. The airport site has an elevation of 365 m. The monthly mean average daily temperature is 40°C and the monthly mean of the maximum daily temperature is 44°C at the site. Find out the corrected runway length if the effective gradient is 0.16%. (JNTU-H)

33. Determine the corrected length of runway with the following data:

    Basic runway length = 2520 m

    Airport elevation = 500 m

    Airport reference temperature = 21°C

    Runway effective gradient = 0.2% (VTU)

34. Plot the wind-rose (Type-I) for the following data and find the orientation of the runway and the percent wind coverage:

| Direction | Duration % time | Direction | Duration % time |
|---|---|---|---|
| N | 7.3 | S | 8.6 |
| NNE | 12.4 | SSW | 14.5 |
| NE | 9.0 | SW | 10.3 |
| ENE | 3.3 | WSW | 6.8 |

Cont'd

Table Cont'd

| Direction | Duration % time | Direction | Duration % time |
|---|---|---|---|
| E | 1.2 | W | 4.0 |
| ESE | 1.6 | WNW | 1.1 |
| SE | 0.3 | NW | 0.7 |
| SSE | 0.8 | NNW | 4.3 |

35. Determine the effective gradient with the following data:

| Chainage (m) | Gradient (%) |
|---|---|
| 0–300 | + 1.50 |
| 300–900 | – 1.00 |
| 900–1200 | – 1.00 |
| 1200–1800 | + 0.50 |
| 1800–2100 | + 0.50 |

(VTU)

36. Runway length for landing at sea level in standard atmospheric conditions is 3000 m. Similarly, runway length for take-off is 2500 m. Aerodrome reference temperature is 25°C and that of the standard atmosphere at the aerodrome elevation of 150 m is 14.025°C. If the effective runway gradient is 0.5%, determine the runway length to be provided. (UP Tech. University)

37. Calculate the actual length of the runway from the following data:

| | | |
|---|---|---|
| Airport elevation | ··· | RL 100 |
| Airport reference temperature | ··· | 28° C |
| Basic length of runway | ··· | 600 m |
| Highest point along the length | ··· | RL 98.2 |
| Lowest point along the length | ··· | RL 95.2 |

(West Bengal Tech. University)

38. Given the following wind data, draw the wind-rose diagram and show the best runway orientation:

| Wind direction | % of time | Wind direction | % of time |
|---|---|---|---|
| N | 2.9 | S | 2.3 |
| NNE | 5.7 | SSW | 3.2 |
| NE | 8.3 | SW | 2.3 |
| ENE | 11.7 | WSW | 3.2 |
| E | 11.5 | W | 5.4 |

Cont'd

Table Cont'd

| Wind direction | % of time | Wind direction | % of time |
|---|---|---|---|
| ESE | 11.9 | WNW | 4.6 |
| SE | 8.8 | NW | 2.5 |
| SSE | 4.9 | NNW | 2.7 |

Calm wind = 8.1%

Total = 100% (West Bengal Tech. University)

39. The length of runway at sea-level, standard atmospheric condition and zero gradient is 1500 m. The airport site has an elevation of 900 m and reference temperature of 20°C. If the proposed effective gradient is 0.2%, determine the actual length of the runway required. (Kalinga Institute of Technology, Odisha)

40. Determine the effective gradient of the proposed airport runway from the following data:

| End to end of runway | Gradient % |
|---|---|
| 0 to 5 chains | +1.0 |
| 5 to 15 chains | −1.0 |
| 15 to 30 chains | +0.8 |
| 30 to 40 chains | +0.2% |

(Pune University)

41. The length of a runway under standard conditions is 2400 m. The airport is to be provided at an elevation of 380 m above mean sea level. The airport reference temperature is 32°C and the gradient details are given below:

| End to end of runway | Gradient % |
|---|---|
| 0–300 m | +1.0 |
| 300–1200 m | −0.6 |
| 1200–1800 m | +0.5 |
| 1800–2400 m | −0.6 |
| 2400–3000 m | +0.5 |

Apply the corrections for elevation and temperature as per ICAO and the correction for gradient as per FAA specifications, and determine the corrected length of the runway. (Mumbai University)

42. The average wind data collected at a particular site are given. Determine the calm period, orientation of runway, and percentage of time in a year during which the runway can be used with satisfactory wind condition. Use Type-I wind-rose.

| Direction | Percentage of time with velocity in km/h | | | |
|---|---|---|---|---|
| | 6.4–25 | 25–35 | 35–55 | 55–80 |
| N | 7.8 | 2.7 | 1.6 | 0.2 |
| NNE | 5.1 | 2.1 | 0.8 | 0.1 |
| NE | 2.4 | 0.9 | 0.6 | 0.0 |
| ENE | 1.2 | 0.4 | 0.2 | 0.0 |
| E | 0.8 | 0.2 | 0.9 | 0.0 |
| ESE | 0.3 | 0.1 | 0.0 | 0.0 |
| SE | 2.3 | 1.8 | 0.9 | 0.1 |
| SSE | 4.5 | 3.0 | 1.0 | 0.2 |
| S | 6.7 | 5.6 | 2.9 | 0.1 |
| SSW | 4.8 | 2.7 | 1.3 | 0.1 |
| SW | 1.6 | 0.8 | 0.3 | 0.0 |
| WSW | 2.2 | 0.6 | 0.1 | 0.0 |
| W | 0.4 | 0.1 | 0.0 | 0.0 |
| WNW | 0.2 | 0.1 | 0.0 | 0.0 |
| NW | 2.3 | 1.8 | 0.9 | 0.2 |
| NNW | 4.0 | 3.2 | 1.3 | 0.1 |

(Kalinga Institute of Technology, Odisha)

Chapter 12

# DESIGN AND MAINTENANCE OF AIRPORT PAVEMENTS

**CHAPTER OBJECTIVES**

This chapter enables the student to understand the following:

- The factors for design of pavement thickness are similar to those for flexible and rigid highway pavements; ESWL of the aircraft loads is used for design.
- Thickness design methods are similar to those for highway pavements – CBR and Burmister's approaches are for flexible type.
- Load Classification Number (LCN) method is particularly used for airport pavements.
- For structural evaluation of airport pavements, Aircraft Classification Number (ACN) and Pavement Classification Number (PCN) are used.
- Overlay design procedures are evolved for repair and rehabilitation of runway pavements.

## 12.1 INTRODUCTION

The main function of an airport pavement, as in the case of a highway pavement, is to resist the imposed loads from the traffic, withstand their abrasive action, and transmit them to the soil subgrade through the different layers of which the pavement itself is constructed, in such a manner that the stresses in any layer or in the subgrade soil do not cause failure.

The traffic on airport pavements comes from aircraft, which transmit much higher wheel loads on to the pavement than those from motor vehicles on highway pavements. The pavement must withstand adverse weather conditions and other forces tending to cause deterioration, and provide a smooth riding surface.

### Types of Pavements

From the design and construction points of view, airport pavements (as in the case of highway pavements) are classified as *flexible pavements* and *rigid pavements*.

**Flexible pavements:** A typical flexible pavement consists of several layers. It comprises the surface course, base course, sub-base course, and soil sub-grade at the bottom-most layer (the subgrade is usually a well-prepared and compacted earth). This layered system, the strongest being at or near the surface, derives its load-carrying capacity through its own load-distribution characteristics and from the strength of the subgrade. The thickness required for the layered system depends, therefore, on the subgrade strength. However, it cannot resist tensile stresses.

**Rigid pavements:** A rigid pavement typically consists of cement concrete, which can be plain, reinforced or pre-stressed. In view of the rigidity of the concrete slab and its inherent flexural strength, a rigid pavement derives its structural load-bearing capacity from the cement concrete slab itself. However, the provision of a good base course or a sub-base enhances its load-carrying capacity and durability in its service life.

A rigid pavement distributes the load on to a larger area of the underlying layers, including the subgrade. The structural strength of the concrete slab governs the thickness required; the support from the base or subgrade should be as uniform as possible, but minor variations in the subgrade strength do not affect the performance of the pavement.

### Choice of the Type of Pavement

This depends on the following factors:

(i) Soil characteristics
(ii) Nature and volume of the air traffic
(iii) Climatic conditions such as humidity, temperature, and rainfall.
(iv) Wheel loads
(v) Maintenance cost
(vi) Availability of materials, labour and equipment
(vii) Durability or design period
(viii) Overall cost – capital cost and maintenance cost.

### Differences between Airport Pavements and Highway Pavements

The following are the points of difference between the design of airport pavements and highway pavements, although the methodology is conceptually the same for both:

| S. No. | Item | Highway pavement | Airport pavement |
|---|---|---|---|
| 1. | Maximum or design wheel load | 4.1 tonnes | 100 tonnes |
| 2. | Tyre pressure | 5 to 7 kg/cm$^2$ | 17 to 24 kg/cm$^2$ |
| 3. | Load repetition | 1000 to 2000 commercial vehicles per day | 20,000 to 40,000 repetitions in the service life |
| 4. | Placement of load | Within 1 m from the edge | In the central 10 m portion of the runway width |
| 5. | Thickness | Uniform | Greater thickness at runway ends, taxiways and aprons. |

Fuel spillage, heat and blast of engine exhaust, high tyre pressure due to smaller contact areas of tyres with the pavement, longitudinal rutting, wavy surface, and changes of grade caused by repetitive loading in a narrow band, especially in the case of taxiways, are special problems in the case of airport pavements, which have to be taken care of in their design.

## 12.2 DESIGN FACTORS

The following are the important factors to be considered for determining the thickness required for an airport pavement:

(a) Strength of the construction materials in the layers
(b) Bearing capacity of the subgrade soil
(c) Design wheel load of the aircraft.

**Strength of the Construction Materials in the Layers**

The strength of the granular materials used in the different layers has to be evaluated, characterised by the CBR-values or by the plate load test values. These will be utilised in the appropriate design approach selected.

**Bearing Capacity of the Subgrade Soil**

This plays a key role in the design of the thickness of a flexible pavement. In addition, the drainage characteristics of the subgrade also have to be studied.

The strength of the subgrade soil should be obtained under the most adverse conditions anticipated, especially under submergence because of high ground-water table. This is because the strength of the soil or its bearing capacity tends to decrease almost by 50 per cent under complete submergence.

Appropriate laboratory tests, simulating the field conditions, have to be conducted to arrive at a reliable value for the bearing strength of the subgrade soil. (However, this does not influence the design of rigid pavements.)

**Design Wheel Load of the Aircraft**

Airport pavements are designed for the wheel loads imposed by the aircraft on the airport pavement through the contact areas.

In general, an aircraft is supported by a nose gear and two main landing gears located near the wing area on each side. Modern aircraft are supported by a 'tricycle' assembly.

Each landing gear may consist of one, two or four wheels; the first is a single-wheel type, the second is a dual-wheel assembly, and the one with four wheels is a dual-in-tandem assembly. Depending upon the wheel configuration, multiple wheel loads will be imposed on the runway pavement. For convenience in design, these are converted into an *equivalent single wheel load (ESWL)*.

The ESWL-values are given below for a few common types of aircraft, along with tyre pressures:

| S. No. | Type of aircraft | ESWL (tonnes) | Tyre pressure kg/cm² |
|---|---|---|---|
| 1. | A-300 Air Bus | 26.60 | 10.0 |
| 2. | Boeing 707 | 23.40 | 11.6 |
| 3. | Boeing 727 | 25.00 | 9.2 |
| 4. | Boeing 737 | 15.40 | 2.8 |
| 5. | Boeing 747 | 27.40 | 12.6 |
| 6. | DC-8 | 27.20 | 13.5 |
| 7. | Dakota | 6.90 | 3.4 |
| 8. | Viscount | 12.40 | 8.8 |

Besides these primary factors, other factors such as wheel load repetitions, resistance to fatigue, resistance to impact, climatic conditions such as temperature variations and distribution of traffic across the pavement have also to be considered in the design.

## 12.3 DESIGN OF FLEXIBLE PAVEMENTS

The design of flexible pavements is a complex problem since it involves the consideration of a large number of variables. The methods developed initially were empirical or semi-empirical in nature; these were based on empirical laboratory tests such as the *California Bearing Ratio (CBR)* of the earth and granular materials used, or field tests such as the *penetration tests* or the *plate load tests*. The results have been employed in the development of empirical design equations or design charts for determination of the thickness required for the various layers and the total thickness of construction for a flexible pavement.

However, these methods needed correlation with the field performance of the pavements. Hence, of late, analytical and mathematical methods have been developed for accurate calculation of pavement strength through application of *Computer Aided Design (CAD)* software which, in conjunction with digital computers, makes the solution of complex theoretical equations relatively simpler.

Thus, the design of airport pavements has been on a more rational basis in the recent past than it was with the empirical approaches used earlier.

The methods of designing pavement thickness are:

1. CBR method (US corps of Engineers method)
2. McLeod method
3. Boussinesq's (single-layer) method
4. Burmister's (two-layer) method (US Navy method)
5. Analytical methods (US corps of Engineers method using CAD)

Of these, McLeod's and Boussinesq's methods are superseded by the other methods. Therefore, the CBR method, Burmister's method, and the analytical methods are presented here.

### CBR Method (US Corps of Engineers Method)

The method of design using the California Bearing Ratio (CBR) values of the pavement materials was developed by the California Division of Highways for the determination of the thickness of a flexible pavement for highways. The method was adopted by US Corps of Engineers during the Second World War for the thickness design of the pavements of military airports.

The concept of CBR is purely empirical, as this indicates the strength of a soil or other pavement material relative to a standard crushed stone material; the load required for a standard plunger to penetrate a standard depth through the material, compacted and placed in a standard manner in the CBR mould, is expressed as a percentage of the load for the same penetration for the standard crushed stone material. (The material must be kept soaked in the mould for three days.) This is known as the CBR-value of the soil or other pavement material. The details of this test are available in *Transportation Engineering* Volume-I, as also in other standard books on geotechnical engineering.

A set of design curves was developed by the California Division of Highways relating the CBR-value and the total thickness of construction of the base and surfacing for different wheel loads. These curves were satisfactory only for highway vehicle wheel loads.

Since the wheel loads of aircraft are much higher than those of highway vehicles, the US Corps of Engineers developed a different chart that was similar to the highway design curves.

The general procedure of using these charts to determine the thickness of the different layers and that for constructing a flexible pavement is given below:

(i) The total thickness of the pavement—sub-base, base and surface courses—for a given design wheel load is obtained corresponding to the CBR-value of the subgrade soil.

(ii) Similarly, the thickness of the base and surface courses for the particular wheel load is got corresponding to the CBR-value of the sub-base material.

(iii) The thickness of the surface course is got corresponding to the CBR-value of the base material.

The base course thickness should not be less than 150 mm. The recommended ranges of CBR-values for the subgrade, sub-base, and base materials are:

| | |
|---|---|
| Subgrade | 5 to 10% |
| Sub-base | 20 to 50% |
| Base | 80 to 90% |

The surface course of bituminous material can be 50 mm to 100 mm in two layers as the minimum. This satisfies the principle that the strongest material should be at or near the surface with the strength decreasing from top to bottom in the case of a flexible pavement.

The modified design chart developed by the US Corps of Engineers for obtaining the thickness of an airport pavement of the flexible type is shown in Fig. 12.1.

### Determination of Equivalent Single Wheel Load (ESWL)

This step is essential for the use of design charts developed for determination of the thickness required for flexible pavement using the CBR method.

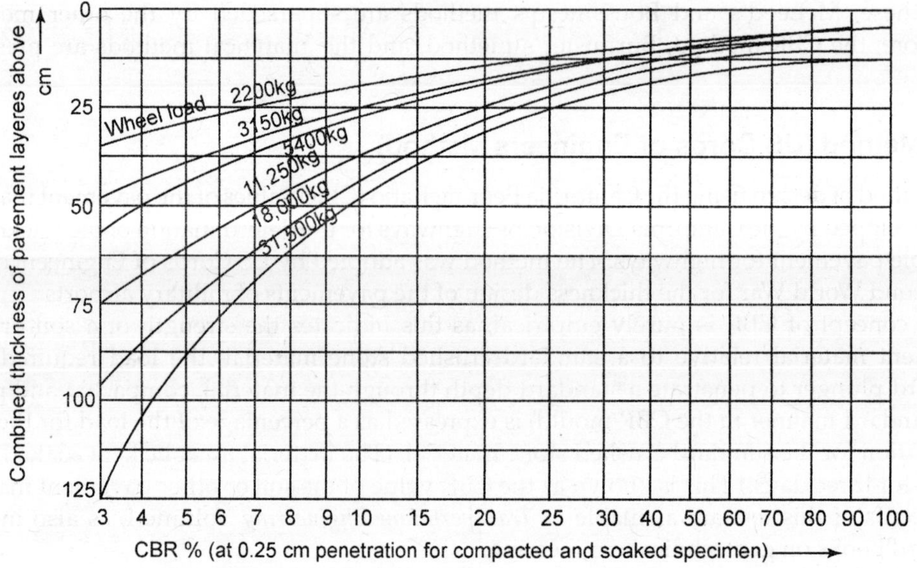


**FIG. 12.1** *Design chart for pavement thickness using CBR*

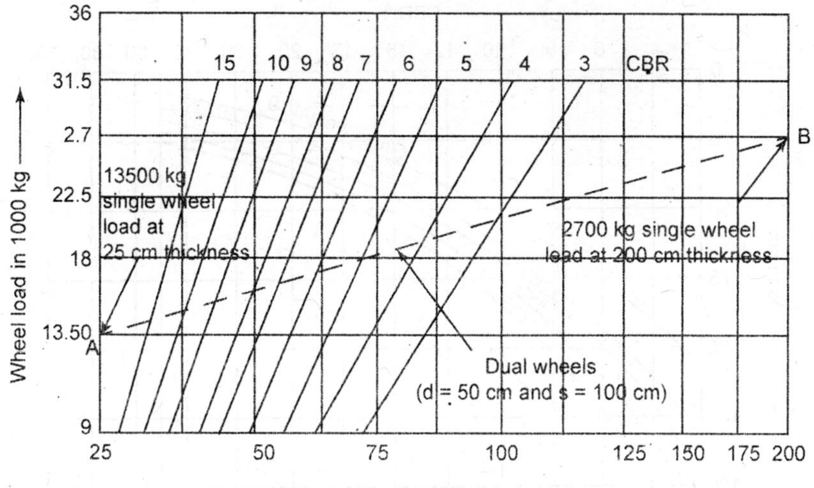


**FIG. 12.2** *Determination of ESWL for a dual-wheel assembly*

The equivalent single wheel load of a dual wheel assembly may be determined using the chart shown in Fig. 12.2 developed by the US Corps of Engineers.

The co-ordinates of A and B are ($d/2$, $P$) and ($2S$, $2P$), respectively, where $d$ is the clear distance between the tyres, $P$ is the wheel load, and $S$ is the centre-to-centre distance between the lines of tyres, as shown in Fig. 12.3.

For twin-tandem landing gear assembly, the co-ordinates of A and B are ($d/2$, $P$) and (2S, 4P), respectively, with the definition of S as shown in Fig. 12.4.

A number of design charts were developed by the US Corps of Engineers for different group wheel loadings and tyre pressures. A typical design chart giving the thickness of construction for different CBR-values and for different gross weights of the aircraft is shown in Fig. 12.5.

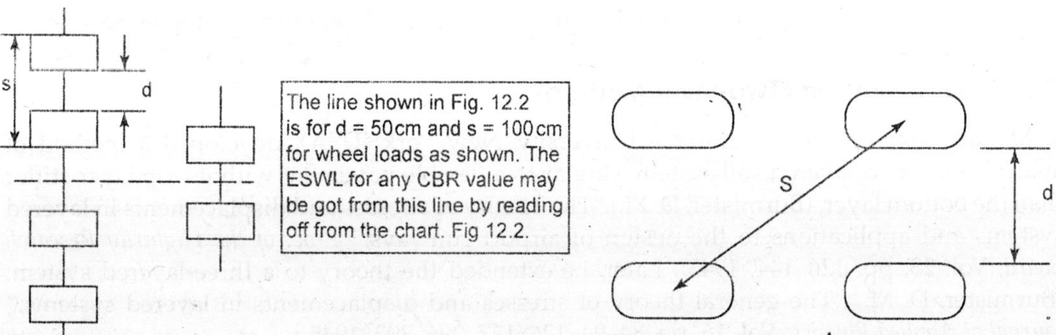


**FIG. 12.3** *Dual wheel assembly definition of d and S*

**FIG. 12.4** *Twin-tandem gear assembly-definition of d and S*

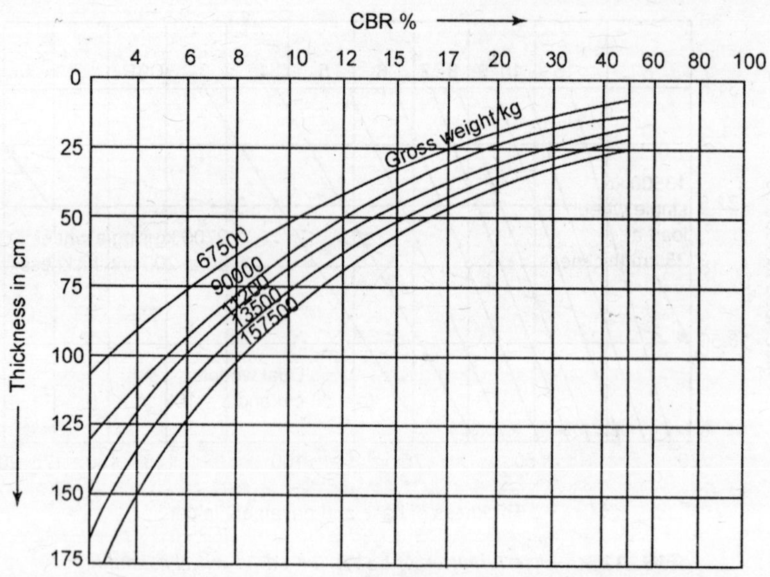


**FIG. 12.5** *A typical design chart for design of flexible pavement (US Corps of Engineers)*

### Effect of Number of Coverages

The number of coverages anticipated by the US Corps of Engineers was about 5000 in the life of the pavement.

If the number of coverages is more than 5,000, it is recommended that the thickness of the pavement be increased as given below:

| No. of coverages | Percent increase in thickness |
|---|---|
| 10,000 | 7 |
| 15,000 | 11 |
| 20,000 | 14 |

The use of these ideas and charts will be illustrated in the Numerical Examples Section.

## Burmister's Method (Two-layer Method)

D M Burmister (1943) of Columbia University, New York (USA), developed a method of analysis for a two-layered soil system, similar to a flexible pavement with its top layer stiffer than the bottom layer. (Burmister, D. M., "The Theory of stresses and displacements in layered systems and applications to the design of airport runways," *Proc. of the Highway Research Board*, Vol. 23, pp. 126–144, 1943.) Later, he extended the theory to a three-layered system. (Burmister, D. M., "The general theory of stresses and displacements in layered systems," *Journal of Applied Physics*, Vol. 16, pp. 84–94, 126, 127, 296–302, 1945.]

The analysis of multi-layered systems based on elastic, non-linear, visco-elastic and finite element methods involves complex mathematical operations, which could be dealt with only by the use of computers.

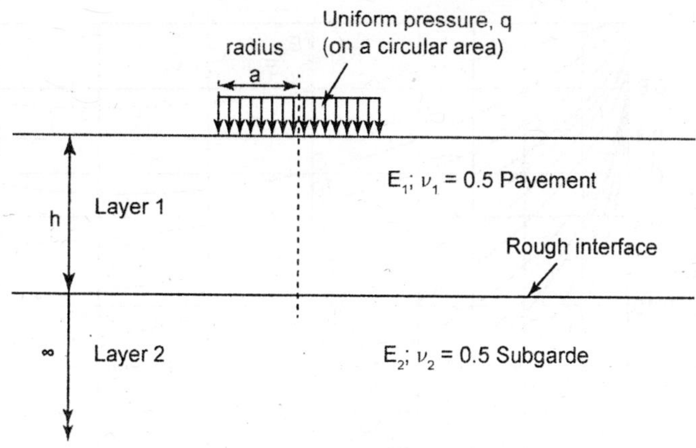


**FIG. 12.6** *Burmister's two-layer system*

Burmister's two-layered system is based on the following assumptions:

(i) Each layer is elastic, homogeneous and isotropic, with unique values of modulus of elasticity and Poisson's ratio.
(ii) The elastic modulus of the top layer, $E_1$, is greater than that of the bottom one, $E_2$.
(iii) The layers extend infinitely in the horizontal direction.
(iv) The interface is rough.
(v) The applied pressure is uniform over a circular area.
(vi) The top layer is of finite thickness, $h$, while the bottom layer is infinitely thick.

The system is shown in Fig. 12.6.

The equations derived by Burmister for surface deflections[1], Δ, under the centre of the loaded plate are:

*For a flexible plate*:

$$\Delta = \frac{1.5qa}{E_2} \cdot F_d \tag{12.1}$$

*For a rigid plate*:

$$\Delta = \frac{1.18qa}{E_2} \cdot F_d \tag{12.2}$$

$F_d$ is the *displacement factor*, which depends on the ratios, $\left(\frac{E_1}{E_2}\right)$ and $\left(\frac{h}{a}\right)$.

Figure 12.7 shows the relationship between $F_d$ and $\frac{h}{a}$ for different values of $\left(\frac{E_1}{E_2}\right)$.

By introducing additional pavement layers over the subgrade, the vertical stress induced in them can be significantly reduced – 30% to 68%. For this to happen, the thickness of the additional layer has to be equal to the radius of the loaded area, and its elastic modulus

[1] Surface deflections are those observed at the surface of the layer just below the plate; those below this level depend upon the elastic properties of the layer.

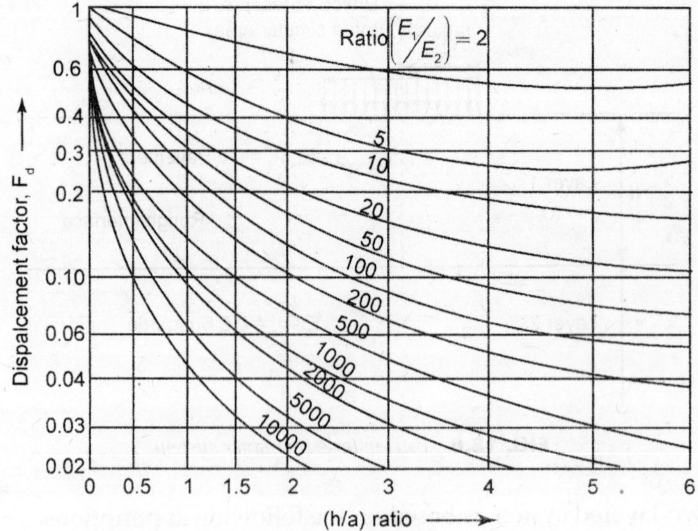


**FIG. 12.7** *Burmister's two-layered system* – $F_d$ *vs* $\frac{h}{a}$

should be approximately ten times that of the subgrade; also, the Poisson's ratio has to be nearly 0.5.

For this reason, multiple layers are preferred to a single thick layer in the construction of flexible pavements.

For multi-layered systems, several software programs have been developed for linear elastic, non-linear, and visco-elastic analyses using finite element method, which gives accurate results. But the computational effort required for this is relatively high.

Burmister's method has been adopted by the *US Navy*; based on their extensive research on trial pavement sections, they recommend that the design thickness be increased by 30 to 40% to account for possible variations of the elastic modulus value of the pavement.

The *US Navy recommends* the following *minimum thickness* values of *bituminous surfacing* for *different wheel loads*:

| Wheel load (kg) | Thickness (cm) |
|---|---|
| Up to 7000 | 3.75 |
| 7000–11250 | 6.25 |
| 11,250–22500 | 7.40 |
| 22,500 or more | 10.00 |

The use of Burmister's (US Navy) method will be illustrated in the Numerical Examples Section.

## Analytical Methods (US Corps of Engineer's Method Using CAD)

Empirical and semi-empirical methods give satisfactory results so long as the conditions under which they were developed do not change significantly. In view of the rapid increase of aircraft loadings in the recent past, extension of these methods to the altered conditions is inappropriate without further field experiments.

Analytical design methods have been developed for the design of airfield pavements; the criterion underlying these methods is that the stresses, strains, and deformations produced by the design wheel loads do not exceed the values that can be safely carried by the pavement materials.

The *essential features* are:

- Selection of a suitable design model
- Criteria for design and performance in terms of stress, strain and deformation
- Solution of the equations for the stresses, strains, and deformations due to applied loads
- Selection of the pavement system to satisfy the stability and the performance criteria.

**Application of Computer Aided Design (CAD)**

Software packages or computer programs for implementing the selected analytical methods have been developed in advanced countries. For example, the US Corps of Engineers have developed packages for implementing the method advocated by Brown and Rice, based on their equation. This takes into account the contact area, tyre pressure, and the CBR-value of the pavement material, and also the load repetition factor based on the air traffic volume assumed for design.

CAD has several advantages since any changes in the conditions can be considered and the new set of results obtained very fast using the digital computer.

## 12.4 DESIGN OF RIGID PAVEMENTS

Rigid pavements are those constructed of cement concrete, which may be plain, reinforced, or pre-stressed. The primary characteristic of such a pavement is that it derives its load-carrying capacity from its own flexural strength. Therefore, it does not rely much on subgrade support, although its durability gets enhanced if a well-compacted granular base is provided below the concrete slab. Even if certain weak pockets exist in the subgrade, they can be easily bridged by the slab. With good subgrade support and a sub-base, the thickness of the pavement required to carry a certain aircraft loading can get reduced.

### Design Criteria and Types of Stresses Induced

Stresses in a concrete pavement may be induced in the following ways:

(i) Due to wheel loads of the aircraft
(ii) Due to warping because of temperature changes and temperature differential between the top and bottom of the slab
(iii) Due to the resistance to slab contractions offered by frictional forces between the slab and the base/sub-base/subgrade.

In general, the methodology to design rigid pavements is to first determine the stresses due to all the above factors and the critical combination of the stresses, and then compare the maximum stresses with the strength values of concrete in tension, compression and shear. The desired factor of safety shall be ensured.

### Methods of Design of Rigid Pavements

The following are the common methods of design adopted.

1. FAA method
2. Westergaard's method
3. PCA method (based on Westergaard's equations)

### FAA Method

As in the case of the flexible pavements, FAA has prepared charts for the determination of thickness of the concrete slab for different conditions of the subgrade soil. The thickness of the base course and the surface course (or the concrete slab) can be obtained directly by using these ready-made charts, by using all the known parameters.

### Westergaard's Method

H.M. Westergaard of Harvard University is credited to have evolved a scientific analysis of stresses in cement concrete pavement slabs.

The following are the assumptions made in the analysis:

(i) The concrete slab is assumed to be homogeneous, isotropic and elastic, with uniform thickness.

(ii) The subgrade reaction to loading is purely vertical, and is proportional to the deflection at every point; in other words, the subgrade acts like a dense fluid, with no shear strength.

(iii) The constant of proportionality between the subgrade reaction ($q$) and the slab deflection ($S'$), is called the Modulus (or coefficient of subgrade reaction (k):

$$k = \frac{q}{S'} \tag{12.3}$$

This is taken to be a constant.

(iv) The load contact area is taken to be circular in shape at the interior and the corner of the slab and semicircular at the edge of the slab; these are considered to be the critical positions for the load.

### Stresses Due to Wheel Load

The stresses based on Westergaard's approach, revised by Teller and Sutherland, was adopted by the Indian Roads Congress and included in their "Guidelines for the Design of Rigid Pavements" (IRC: 58–2002). These are given below:

*Edge stress:*

$$\sigma_e = 0.529\frac{W}{h^2}(1+0.54\nu)\left[4\log_{10}\left(\frac{l}{b}\right)+\log_{10} b - 0.4048\right] \tag{12.4}$$

*Corner stress:*

*Kelley's equation:*

$$\sigma_c = \frac{3W}{h^2}\left[1-\left(\frac{a\sqrt{2}}{l}\right)^{1.2}\right] \tag{12.5}$$

*Interior stress:*

$$\sigma_i = \frac{0.316}{h^2}\left[4\log_{10}\left(\frac{l}{b}\right)+1.069\right] \tag{12.6}$$

Here,

$W$ = Design wheel load

$h$ = Thickness of concrete slab

$a$ = Radius of the circular area of contact of wheel load

$l$ = Radius of relative stiffness

$$l = \left[\frac{Eh^3}{12k(1-\nu^2)}\right]^{0.25} \tag{12.7}$$

Here,

$E$ = Elastic modulus of cement concrete

$k$ = Modulus of subgrade reaction

$\nu$ = Poisson's ratio of cement concrete (= 0.15)

$b$ = Equivalent radius of resisting section:

$$b = a \quad (\text{for } a \geq 1.724h) \tag{12.8}$$

$$b = \left(\sqrt{1.6a^2 + h^2} - 0.675h\right) \quad (\text{for } a < 1.724h) \tag{12.9}$$

For dual wheel loads,

$$a = \left[\left(\frac{0.8521W_1}{p\pi}\right) + \frac{S_{cc}}{\pi}\left(\frac{W_1}{0.5227p}\right)^{0.5}\right]^{0.5} \tag{12.10}$$

where

$W_1$ = Load on one tyre

$S_{cc}$ = Centre-to-centre distance of dual tyres

$p$ = Tyre pressure.

**Warping Stresses Due to Temperature Differential**

These have been derived by Bradbury, and adopted by IRC. Stresses may be calculated in the $x$- and $y$- directions, both at the interior and edge of the slab. Bradbury's coefficients have to be used along with the warping stress chart.

**Stresses Due to Friction Between the Slab and Subgrade**

The maximum frictional tensile stress, $\sigma_{ft}$, is given by

$$\sigma_{ft} = \frac{\gamma_c \cdot L \cdot \mu}{2} \tag{12.11}$$

where

$\gamma_c$ = Unit weight of concrete

$L$ = Length of the slab

$B$ = Width of the slab

$h$ = Height of the slab

$\mu$ = Average coefficient of friction (taken as 0.15)

**Critical Combination of Stresses**

The critical combination of the above-said stresses is obtained and the maximum values of tensile and compressive stresses are determined.

These are ensured to be less than the corresponding strength values of concrete.

In case this is not satisfied, the assumed slab thickness is increased, and the analysis repeated, until the condition of stability is satisfied.

**PCA Method (Based on Westergaard's Equations)**

Influence charts and design charts for the determination of the thickness of rigid pavements, based on Westergaard's equations for critical stresses, have been developed by G.G. Pickett of the University of Wisconsin (through the Portland Cement Association); this obviates the need for solving the complex stress equations of Westergaard for varying conditions of aircraft loadings and wheel configurations.

Portland Cement Association (PCA) has recommended the following factors of safety to be adopted:

| | |
|---|---|
| Aprons, taxiways and ends of runways | 1.7 to 2.0 |
| Central portion of runway width | 1.25 to 1.50 |

These safety factors are related to the fatigue resistance of concrete. The factor of safety needed is more for aprons, taxiways and ends of runways since the traffic is more channelized in these areas than in the central portion of the runway width owing greater stress concentration under repeated wheel loads.

The PCA charts are based on E-and $\nu$-values of $2.8 \times 10^5$ kg/cm$^2$ and 0.15, respectively, for concrete. PCA indicates that the variation in these values has relatively small effect upon the stress values. According to it, a reduction in E-value from $2.8 \times 10^5$ kg/cm$^2$ to $2.1 \times 10^5$ kg/cm$^2$ decreases the stress by 5%. An increase in $\nu$-values from 0.15 to 0.20 increases stress by 4%.

The charts include wheel loads, k-values, thickness of pavement slab, and the stress-values as variables. Since more rational methods of design have been evolved, the design charts are not presented here.

## 12.5 LCN METHOD OF AIRPORT PAVEMENT DESIGN

The *Load Classification Number (LCN)* method of aircraft evaluation and airport pavement design was formulated by the British Air Ministry Directorate of General Works.

In the LCN system, the supporting capacity of a pavement is expressed in terms of a number, called the Load Classification Number. Similarly, the Equivalent Single Wheel Load (ESWL) of an aircraft (which depends on the gear-and-wheel configuration, tyre contact pressure and the thickness and composition of the pavement) can also be expressed in terms of the LCN.

The pavement can be used safely by an aircraft if the LCN of the pavement is greater than that of the aircraft.

In order to express the capacity of a pavement as a single number that characterises the wheel loads and tyre contact areas, the data relating to the aircraft prevailing at that time were used. These are shown below:

| Wheel load (kg) | Tyre pressure (kg/cm²) | Contact area (cm²) (= ESWL/Tyre pressure) | LCN |
|---|---|---|---|
| 45000 | 8.4 | 5357 | 100 |
| 40500 | 8.1 | 5000 | 90 |
| 36000 | 7.7 | 4675 | 80 |
| 31500 | 7.4 | 4257 | 70 |
| 27000 | 7.0 | 3857 | 60 |
| 22500 | 6.7 | 3358 | 50 |
| 18000 | 6.3 | 2857 | 40 |
| 13500 | 6.0 | 2250 | 30 |
| 9000 | 5.6 | 1607 | 20 |
| 4500 | 5.3 | 849 | 10 |

The above data is represented in the form of a standard load classification curve Fig. 12.8.

Extensive field tests – plate bearing tests – were conducted on different types of flexible and rigid pavements with contact areas of the plates ranging from 1250 cm² to 4375 cm², representative of the aircraft in use at that time.

The correlation between failure load and contact area for both flexible and rigid pavements was obtained; this is depicted as shown in Fig. 12.9. It was observed that similar relationships hold good for both flexible and rigid pavements over this range of load contact areas.

The average curve having the following equation was plotted:

$$\left(\frac{W_1}{W_2}\right) = \left(\frac{A_1}{A_2}\right)^{0.44} \qquad (12.12)$$

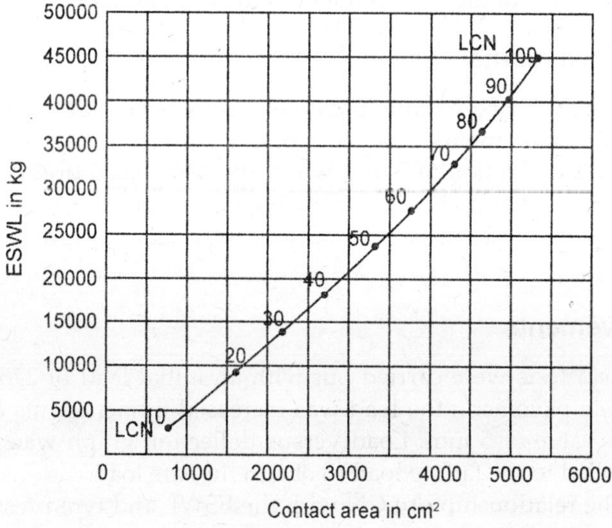


**FIG. 12.8** *Standard load classification curve for aircraft*

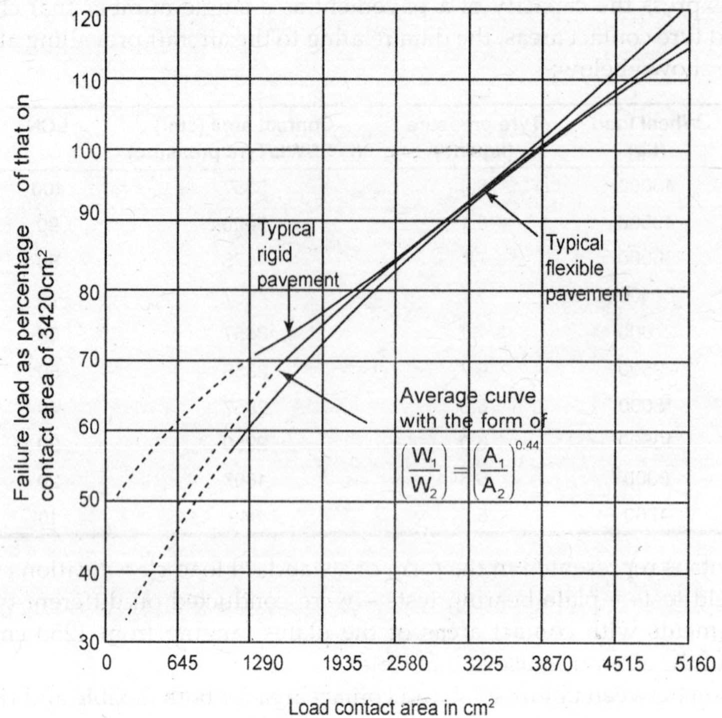


**FIG. 12.9** *Failure load versus contact area relationships*

Here, $W_1$ and $W_2$ are the failure loads corresponding to the contact areas of $A_1$ and $A_2$.

It may be noted that the ordinate of Fig. 12.9 is expressed as a percentage of the load required to produce failure when applied through a plate (3420 cm$^2$) of diameter 66 cm.

This contact area was chosen as the standard for comparison because it was representative of heavy aircraft at the time of the development of the LCN system of pavement design.

## LCN for Flexible Pavements

In order to obtain the LCN of a flexible pavement, repetitive plate load tests were carried out on existing airport pavements. The safe load on the pavement was considered to be that which produced a deflection of 5 mm when applied 10,000 times. The contact area of the plate to be used was taken to be that of the main wheel of the aircraft. A plate of 46 cm diameter with an area of 1652 cm$^2$ is generally used for the design of flexible pavements for civil airports.

## LCN for Rigid Pavements

In this case, plate load tests were carried out with an initial load of 2267 kg at the corners of the cement concrete pavement; the load was increased in increments of 2267 kg until the deflection reached a value of 5 mm. Load versus deflection graph was plotted. A factor of safety of 1.5 was applied to the failure load to obtain the safe load.

Charts showing the relationship of LCN with the ESWL and tyre pressure were prepared based on the results of several tests carried out for this purpose.

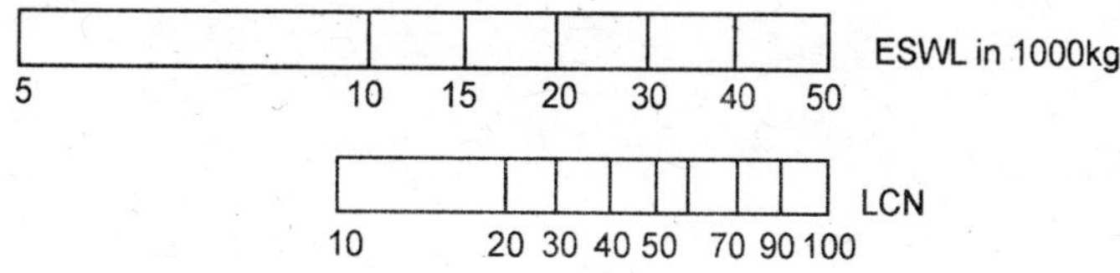


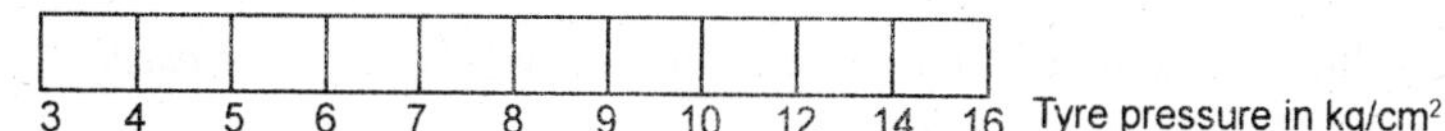


**FIG. 12.10** *Relationship of LCN with ESWL and tyre pressure*

A typical chart is shown in Fig. 12.10.

In view of the large number of variables involved, too much precision in the design and evaluation of a pavement is not warranted.

Aircraft evaluation and pavement classification has been simplified by a broad system of *Load Classification Groups (LCG)* as shown below:

| LCG | I | II | III | IV | V | VI | VII |
|---|---|---|---|---|---|---|---|
| LCN | 101–120 | 76–100 | 51–75 | 31–50 | 16–30 | 11–15 | 10 and less |

The LCG-bands were selected such that aircraft imposing similar stress levels on pavements in reasonable and regular increments of pavement thickness were placed in the same LCG.

New pavement structures, however, are limited to the following three types:

(i) *Rigid pavements* of unreinforced pavement quality Portland cement concrete for runway ends, junctions of runways and taxiways, and aprons where aircraft will stand with the likelihood of jet heat, blast, and fuel spillage. (For other areas, their use is a matter of economic or operational preference).

(ii) *Composite pavements* of continuously reinforced concrete with bituminous surfacing with good riding quality, for stretches of runways and taxiways in which aircraft move at high speeds.

(iii) *Flexible pavements* composed of bound granular materials with bituminous surfacing, though not very economical for heavy duty runways, offer an alternative to composite pavements with similar advantages.

## 12.6 AIRPORT PAVEMENT FAILURES

Airport pavements may be of the flexible type or of the rigid type. Certain defects arise in these pavements during operation, eventually leading to the failure of the pavement, making it unserviceable.

### Flexible Pavement Failures

Some defects arise solely because of *lack of quality in the construction of surface courses*

These are:

- Ravelling
- Stripping
- Cracking
- Plastic deformation
- Disintegration
- Bleeding
- Loss of skid-resistance

The following defects are attributable to the *deficiencies in base/sub-base/subgrade*:

- Pot holes
- Road bumps
- Consolidated settlement
- Wavy surface
- Ruts/Corrugations
- Deformation due to lack of bond between layers
- Edge damage
- Streaking in bituminous surfaces.

**Ravelling:** Progressive dislodging of aggregate due to insufficient binder under the action of traffic.

**Stripping:** Separation of the bitumen coating in the presence of moisture; this leads to loss of binding action and loss of aggregate under the action of traffic.

**Cracking:** Different types of cracks occur.

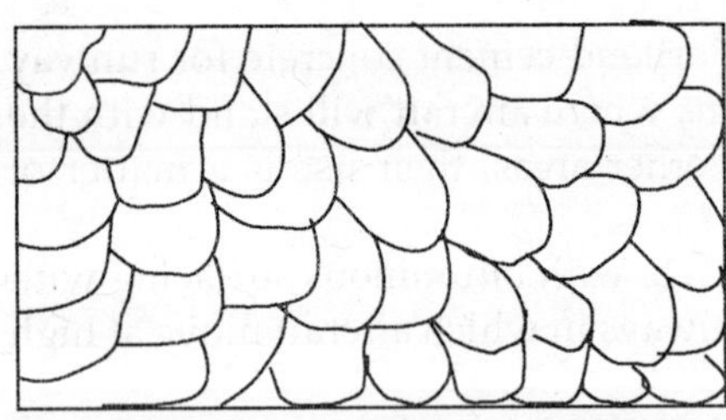

**FIG. 12.11** *Alligator (map) cracking*

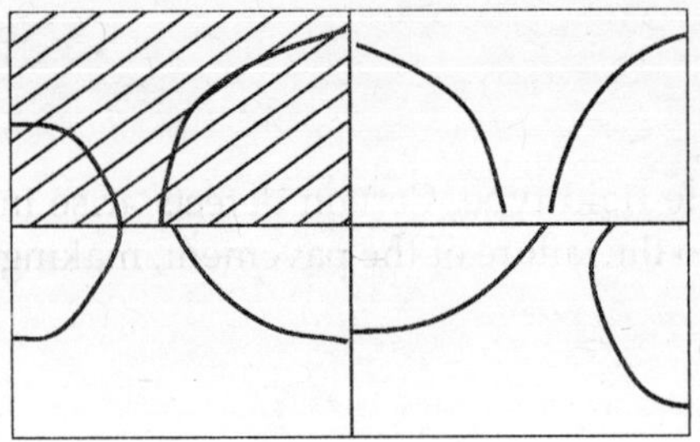

**FIG. 12.12** *Reflection cracking*

***Hair-line cracks*:** These are short and fine cracks at close intervals on the surface due to insufficient binder, excessive filler, or inadequate compaction.

***Alligator cracks (map cracks)*:** Interconnected cracks forming a series of blocks (Fig. 12.11).

***Shrinkage cracks*:** Transverse cracks, usually interconnected to form a series of large blocks.

***Longitudinal cracks*:** Cracks in the longitudinal direction along runways; volume changes due to frost action may cause this.

***Edge cracks*:** Cracks near the edges, probably due to lack of shoulder support and poor drainage.

***Reflection cracking*:** Usually occurs on bituminous overlays in existing cement concrete pavements. The crack pattern from the underlying surface gets reflected on the bituminous surfacing.

These cracks allow percolation of rain water and cause weakening of the subgrade (Fig. 12.12).

**Plastic deformation:** Occurs in wet clay subgrade soil under excessive loading and is not recoverable after removal of the loads. This could lead to surface defects.

**Disintegration:** Causes loss of aggregate in some portions. This could lead to ravelling, stripping, edge damage and pot-holing.

**Bleeding:** Collection of binder on the surface, probably due to excessive binder; the surface appears fatty.

**Loss of skid-resistance:** Polishing of aggregates under traffic or excessive binder may lead to this.

**Pot-holes:** Shallow depressions at certain locations due to inadequate compaction.

***Road bumps*:** Localised upward movements of the pavement due to frost heave or due to swelling of clay on absorption of moisture.

**Consolidation settlement:** Consolidation of poor subgrade containing clay, or inadequate compaction.

**Wavy surface:** Longitudinal/transverse undulations in the surface with crests and troughs. Uneven subsidence of the subgrade may cause this.

**Rutting:** This is caused by longitudinal depressions in the wheel tracks.

***Corrugations*:** Regular undulations, especially in the longitudinal direction. Faulty surface course may cause this.

**Lack of bond between layers:** Formation of crescent shaped cracks pointing in the direction of wheel thrust; it causes slippage. This may be due to lack of bond between the surface course and the underlying courses.

**Edge damage:** Irregular breakage of the edges of the pavement. Poor shoulder support due to percolation of water into it may be the reason.

**Streaking in bituminous surfaces:** Characterised by the presence of alternate lean and heavy lines of bituminous binder. Non-uniform application of the bituminous binder may cause this condition.

## Rigid Pavement Failures

The following defects in rigid pavements or cement concrete pavements may occur due to deficiencies in pavement materials, structural inadequacy or poor workmanship and ultimately lead to failure:

- Scaling of concrete
- Spalling of concrete
- Mud-pumping
- Shrinkage cracking
- Cracking due to warping
- Structural cracking

**Material deficiencies** could be in respect of the requirements for cement, fine aggregate, coarse aggregate, mixing water, joint filler, and sealing compounds.

**Structural inadequacy** could be in respect of inadequate pavement thickness, poor subgrade support and inadequate treatment of joints.

**Poor workmanship** could be in respect of mixing, laying, compaction, screeding/finishing, and curing of the cement concrete; or it could be in respect of the provision of joints including

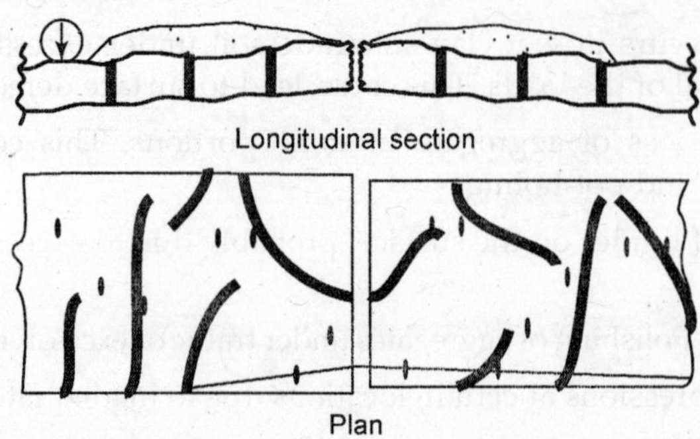


**FIG. 12.13** *Mud-pumping in concrete pavements*

load transfer devices such as dowel bars and tie-bars, and filling and sealing of the joints with appropriate sealants.

***Scaling of concrete:*** Flakes coming off the surface of the road is referred to as *scaling*. Improper mix design and excessive vibration of the mix may cause this. Scaling causes rough surface and affect riding quality.

***Spalling of joints:*** When joint filler is placed in a position ahead of concreting, faulty alignment of the filler may lead to projection of concrete on one side. This projection gets chipped off by the impact of traffic, and lead to cracking and subsidence of the joint.

***Mud-pumping:*** When the pavement slab tends to settle downward under heavy traffic, the sub-base/subgrade material, which forms a slurry in the presence of water, tends to work up through the joints and cracks on to the surface.

This phenomenon is called *mud-pumping*, and is common in rigid pavements. This is observed in rainy season in the case of clayey subgrade.

Pavement damage because of mud-pumping may appear as shown in Fig. 12.13.

Shrinkage cracking: Soon after construction, shrinkage cracks develop in concrete during the curing period of concrete. These may occur both in the longitudinal and transverse directions; however, this may not lead to permanent damage.

***Cracking due to warping:*** Warping cracks may occur if warping joints, meant to control warping stresses, are either not provided, or not properly designed.

***Structural cracking:*** Structural inadequacy, arising out of inadequate thickness of the pavement for the traffic volume and the repetitive loading imposed, leads to the formation of structural cracks. Such cracks generally occur at the edges and corners of pavement slabs.

Cracks in the interior regions of the pavement slabs are due to temperature stresses.

## 12.7 MAINTENANCE AND REHABILITATION OF AIRPORT PAVEMENTS

Preventive maintenance operations may be undertaken periodically on airport pavements in order to ensure that the service life of the pavements is extended as much as possible.

### Maintenance of Flexible Airport Pavements

Depending upon the degree of deterioration of the runway pavement, the following maintenance operations may be needed for bituminous pavements:

(i) Patch repair
(ii) Surface treatment
(iii) Resurfacing

**Patch repair:** This consists of patching up pot-holes and localised depressions, and may be done on up to 25% of the surface area annually. For patch repair, sand premix, open-graded premix, dense-graded premix, or penetration patching may be adopted.

**Surface treatment:** Renewal of the surface course is undertaken when patch repair becomes uneconomical. It may also be done to improve skid resistance when the surface is badly worn out; standard specifications for tack coat, prime coat, and seal coat, along with surface dressing should be used.

**Resurfacing:** This is taken up when the pavement has deteriorated badly and may involve the renewal of not only the surface course but also the base course.

## Rehabilitation of Flexible Airfield Pavements

*Rehabilitation* may involve different steps depending upon the reasons for the damage of the pavement.

If the instability of the underlying course is due to excessive moisture leading to loss of support, suitable subsurface drainage system is necessary to remedy the defect permanently.

If the failure is due to improper compaction of the underlying base/sub-base courses, this can be remedied only by complete renewal of the pavement structure.

If the failure is due to plastic expansive clay in the subgrade soil, the subgrade itself has to be stabilised using an appropriate technique.

Thus, the nature of rehabilitation measures needed would depend upon the identification of the particular layer/stage which was the root cause for deterioration of the flexible pavement used for the airport runways.

## Maintenance of Rigid Airfield Pavements

A cement concrete pavement needs very little maintenance if it is well designed and properly constructed. This is the primary advantage which offsets its initial cost.

However, defects are likely to occur due to ingress of water, especially through ill-maintained joints and cracks, besides inadequate pavement thickness and poor workmanship.

The usual maintenance measures are:

(i) Crack repair
(ii) Joint maintenance
(iii) Patch repair
(iv) Control of mud-pumping
(v) Restoration of anti-skid surface

**Crack repair:** Appearance of cracks is a common phenomenon in cement concrete pavements.

These cracks may be shrinkage cracks or warping cracks due to temperature changes. They may also be structural cracks when they appear at corners and edges of the slab due to excessive loading. This indicates inadequacy of the pavement thickness.

Hair cracks are not harmful, but medium and wide cracks allow water to seep through and cause progressive loss of subgrade support. Such cracks are filled with low-viscosity epoxy grout, after thoroughly cleaning with compressed air, and topped up with sand or fine aggregate chips to prevent disturbance due to traffic.

**Joint maintenance:** This consists of replenishing lost sealant, removing deteriorated joint filler and making a provision of fresh filler material.

The sealant is then poured to an excess height of about 3 mm and sand sprinkled for it to be compressed by the traffic to the level of the pavement surface. This type of treatment is adequate only when the crack depth is not more than one-third of the slab depth.

When the crack extends to a greater depth, cross-stitching with inclined tie-bars may be adopted.

**Patch repair:** Patch repair involves the patching up of scales, spalls, depressions and other irregularities in the slab when they are localised.

Premix bituminous materials are commonly used for this purpose, but they are not satisfactory. Epoxy resin mortars are the best option for such patch repairs. The sides of the slab area to be patched are trimmed; fresh concrete is laid and tamped.

**Control of mud-pumping:** To control the phenomenon of mud-pumping and blowing, defective joints and wide cracks should be refilled and sealed. To prevent further damage and recurrence, the slab is grouted through holes drilled in it. The grout may be of cement mortar, 1:3.5 mix, or of bituminous material; the latter is considered to be more effective in filling the void spaces between the slab and the subgrade and raising the slab to the desired level. This process is known as 'Mud-jacking', and is popular in advanced countries.

**Restoration of anti-skid surface:** When the surface becomes smooth and slippery, anti-skid surface can be restored by cutting grooves with the aid of grooving machines or by grinding machines.

***Rehabilitation of rigid pavements*:** *Emergency repair techniques* have been evolved in advanced countries for the restoration of airports, especially military airports, during war.

## 12.8 EVALUATION AND STRENGTHENING OF AIRPORT PAVEMENTS

*Pavement evaluation* may be defined as a technique of assessing the condition of a pavement, from the point of view of both *functional adequacy* based on surface characteristics and *structural adequacy* based on its load-carrying capacity.

Pavement evaluation serves the following *functions*:

- Assessing the maintenance needs of a pavement
- Assessing the residual life of the pavement
- Determining the pavement surface characteristics existing at the time of testing
- Determining the structural adequacy such as thickness and other details of the pavement
- Assessing the adequacy of the drainage arrangements.

### Functional Evaluation

Surface characteristics of an airfield pavement, such as the riding quality, surface texture, and skid resistance determine passenger comfort, effect of vibration of the aircraft components, and safety during landing and take-off operations.

Instrumentation techniques for the measurement of surface roughness characteristics are being continually improved; expensive and sophisticated instruments have been devised in advanced countries for this purpose.

### Equipment for Measuring Riding Quality

***Simple devices:*** Measurements using these devices are not rigorous and the process is slow.

(i) 1.5 m or 3 m long straight edge measuring the longitudinal unevenness.
(ii) Rod and level for measuring longitudinal and transverse profiles.

***Advanced equipment:*** These are capable of recording continuously the surface unevenness in both the longitudinal and transverse directions of the runway pavement; profiles may be got as a graphic record for analysis in the office.

(i) *Automatic Road Unevenness Recorder* is used for measuring bumps of short wavelengths; electronic recording of the surface unevenness is obtained as the Recorder travels along the pavement at a speed of 30 km/h. The 'Unevenness Index' values may be got in terms of mm/km.
(ii) *High-Speed Profilometers* have been developed in the UK and the USA for obtaining the pavement surface profile and vehicle response to the irregularities. *Bump integrator* has also been developed.

### Aircraft Response to Runway Surface Irregularities

Modern aircraft are equipped with Aircraft Integrated Data System (AIDS) and Digital Flight Data Recording System (DFDRS) to record the vertical acceleration of the aircraft.

The variables considered are:

- Speed
- Take-off and landing speeds
- Aircraft suspension system
- Aircraft load and its distribution
- Natural frequencies
- Nature of the surface unevenness.

### Tolerance Norms of Surface Unevenness

- 3 m straight edge: Maximum permissible deviation of the surface is 3 mm.
- Rod and level method:

| Length of straight edge | Deviation (mm) | |
|---|---|---|
| | Average value | Characteristic* value |
| 1.5 m | 7.6 | 19.0 |
| 3.0 m | 12.4 | 26.0 |

* For 95% confidence level (Average value + 1.245)

- Automatic Road Unevenness Recorder:
  Bituminous surface: 2000 mm/km
  Cement concrete surface: 2250 mm/km

### Skid Resistance Characteristics of Runway Pavement

Skid resistance of a wet pavement surface depends mainly on the macro-texture.

The *main reasons for loss of skid resistance* are:

(i) Waterponding due to poor drainage
(ii) Rubber accumulation in touch-down zones.

(iii) Polishing of surface by the snow-clearing equipment.
(iv) Adverse climatic and weather conditions.

*Methods of restoration of skid resistance:*

(i) Grooving of fresh cement concrete
(ii) Sprinkling of chippings
(iii) Stripping of fresh cement concrete
(iv) Grooving of set concrete

Equipment are devised for the measurement of tyre-pavement friction.

## Structural Evaluation[2]

The objective of structural evaluation is to assess the structural adequacy of pavement vis-a-vis the type of aircraft expected to use the runway. Compatibility between the aircraft and the pavement is necessary to prevent excessive damage to the pavement, and aircraft operations are unrestricted. With the advent of modern aircraft with complex wheel configurations, a uniform standard method had to be developed that is applicable to all types of aircraft.

### Aircraft – Pavement Classification Number System

Developed by the UK Air Ministry and adopted by the ICAO, this system aims at classifying both the aircraft and the pavement for the purpose of assessing their compatibility to each other.

It was observed that the load-carrying capacity of the pavement varied with the tyre pressure (or contact radius) according to the relationship: $\left(\frac{W_1}{W_2}\right) = \left(\frac{A_1}{A_2}\right)^{\alpha}$ where $\alpha = 0.44$ as established from field tests (refer Fig. 12.9).

The aircraft is thus defined by a curve having an exponent, $\alpha$, which passes through the point obtained by the contact area and the all-up weight of the aircraft.

A load classification number is assigned to such sets of load on single wheel and exerted tyre pressure, to give a curve. A family of curves is developed, each curve being allocated a number, called the *load classification Number (LCN)*; this defines the load effect of an aircraft.

The safe load read on such curves for the pavement from plate load test with a plate of a certain contact area gives the LCN of the pavement.

The *Aircraft Classification Number (ACN)* is defined as the load carried by a single wheel at a constant tyre pressure of 12.5 kg/cm$^2$ and exerted on the pavement. It is denoted using the standard CBR method of design for flexible pavement, and the PCA method of design for rigid pavement.

The *Pavement Classification Number (PCN)* is the load-carrying capacity of the pavement and is the ACN of the aircraft which the pavement can carry for unrestricted number of movements.

It may be noted that the ACN/PCN method is for *constant tyre pressure* whereas the LCN is for *constant contact area.* These methods help in computing the permissible load of an aircraft that is compatible with the strength of the pavement.

The International Civil Airport Association (ICAA) has recommended that, in order to define an airport pavement, the following information is necessary to be understood:

(i) PCN is the ACN of the aircraft which is permitted for unrestricted operations.

[2] Refer IRC: 76–1979 and IRC: 85–2015 (listed under references)

(ii) Type of pavement — flexible (F) or rigid (R).

(iii) Subgrade type is classified into four groups:

| | |
|---|---|
| A | CBR > 15 and K ≥ 150 MN/m³ |
| B | CBR – 10 to 15 and K ≥ 80 MN/m³ |
| C | CBR – 6 to 10 and K ≥ 40 MN/m³ |
| D | CBR ≥ 3 and K < 20 MN/m³ |

(iv) Tyre pressures classified into four groups:

| | |
|---|---|
| W | Unlimited |
| X | 1.5 MPa (15 kg/cm²) |
| Y | 1.0 MPa (10 kg/cm²) |
| Z | 0.5 MPa (5 kg/cm²) |

(v) Method of evaluation:

| | |
|---|---|
| Technical | T |
| Experience-based | U |

The classification of a pavement may typically appear as given below:

| |
|---|
| 37/F/B/X/T |
| 40/R/C/W/U |

**Note:** The first number is the PCN

### ACN for Some Types of Aircraft

For flexible pavement, ACN is determined using CBR method; and for rigid pavement, it is determined using PCA method for interior loading condition. These values are specified by the aircraft manufacturers for standard conditions of subgrade strength and tyre pressures. Thus ACN varies with the type of pavement, as also with the subgrade. Hence, characterisation of aircraft-pavement interaction by a single number is not really possible.

### Determination of PCN

Direct measurement is not possible. Methods available presently use the stiffness of the pavement obtained by non-destructive testing. This stiffness multiplied by permissible deflection gives the maximum permissible load. This load, which is for a particular contact area and pressure, is modified for the standard contact pressure of 1.25 MPa. This cycle is repeated for stable values.

The effect of tyre pressure is given below:

$$ACN_2 = ACN_1 \cdot \left(\frac{P_1}{P_2}\right)^{\alpha} \tag{12.13}$$

Here,

$\alpha = 0.50$ for purely granular layers.

$\alpha = 0.68$ for normal flexible pavements with asphalt layers.

$\alpha = 0.73$ for thin rigid pavements ($h < 15$ cm).

$\alpha = 0.885$ for thick rigid pavements.

Similarly, the effects of change in pay load, number of repetitions, and thickness of the pavement on the ACN may be determined.

**Limiting Deformation**

This is an important factor in the determination of PCN and ACN for ensuring smooth ride.

For a flexible pavement, maximum permissible deformation at the surface is taken as 5 mm. Deflections are assumed to vary in logarithmic proportion; therefore the deformation in one cycle is $[5/(\log N + 1)]$ mm, if 5 mm is for *N* cycles.

For a rigid pavement, corner or edge loadings are critical. The limiting deflections are a function of the thickness, failure stress, and the modulus of subgrade reaction.

**Modern Devices for Structural Evaluation**

Deflectometers and Dynamic Deflection Devices have been devised for structural evaluation; however, these are out of the scope of the present treatment.

## Strengthening of Airport Pavements

Strengthening of existing airport pavements may be necessitated for any one of the following reasons:

(i) When the existing pavement has deteriorated to such an extent that patch repair no longer improves its serviceability; and,

(ii) When there is spurt in traffic volume and in wheel loads due to improvements in aircraft construction technology, and the existing pavement is rendered inadequate functionally and structurally. In such a case, it is neither economical nor feasible to replace the existing pavements with an entirely new section.
[In India, most of the military airport pavements were constructed during the Second World War; since then, aircraft gross weight and wheel loads increased considerably, necessitating the strengthening such pavements to make them serviceable for the present day needs.]

In either of these cases, strengthening of the pavement is achieved by providing additional layers of pavement; such additional layers are known as *'Overlays'*.

**Design of Overlays**

Estimation of the design traffic, evaluation of the surface characteristics and structural strength of the existing pavement, and determination of the type and thickness of the overlay are the important steps in overlay design.

Thus the *factors governing the design of overlays* are:

- Bearing capacity of the soil subgrade
- Evaluation of the condition of the existing pavement
- Modified design wheel loads of the aircraft

The three standard types of overlays used for airfield pavements are:

1. *Bituminous overlay*: This consists of bituminous concrete, or a combination of bituminous concrete and penetration macadam.
2. *Flexible overlay*: This consists of a combination of high quality base course and a bituminous surface course.
3. *Rigid overlay*: This consists of a cement concrete slab.

For the purpose of design, overlays are classified into the following broad groups:

(a) Flexible overlays
(b) Rigid overlays

Subgrade strength properties and the actual thickness of the existing pavements are the pre-requisite data for the design of overlays.

### Design of Flexible Overlays

Flexible overlays can be provided over an existing pavement – flexible or rigid.

**Flexible overlay for a flexible pavement:** The basic procedure for the design of a flexible pavement taking into account the subgrade strength is also applicable for the overlay. Any conventional method of design of a flexible pavement may be used. The minimum thickness of the base course recommended by the CAA is 15 cm. The difference between the computed thickness for the new wheel load and the thickness of the existing pavement gives the overlay thickness. US Corps of Engineers recommend a minimum overlay thickness of 10 cm.

**Flexible overlay for a rigid pavement:** First, the total thickness of the rigid pavement required for the new/increased wheel load has to be computed. This value is modified by a factor, $F$, which represents the characteristics of the subgrade or sub-base of the existing rigid pavement.

For this purpose, the subgrade and sub-base are categorised as $R_a$, $R_b$, $R_c$, $R_d$, and $R_e$–types or classes.

The values of $F$ are given below:

| Existing subgrade class | Value of $F$ | | | | |
|---|---|---|---|---|---|
| | Sub-base class | | | | |
| | $R_a$* | $R_b$ | $R_c$ | $R_d$ | $R_e$ |
| $R_a$ | 0.88 | – | – | – | – |
| $R_b$ | 0.90 | 0.80 | – | – | – |
| $R_c$ | 0.94 | 0.90 | 0.80 | – | – |
| $R_d$ | 0.98 | 0.94 | 0.90 | 0.80 | – |
| $R_e$ | 1.00 | 0.98 | 0.94 | 0.90 | 0.80 |

(*Without sub-base)

The thickness of the flexible overlay is calculated using the following equation, given by CAA:

$$h_{fo} = 2.5\,(F \cdot h_d - h_e) \qquad (12.14)$$

Here,

$h_{fo}$ = Flexible overlay thickness

$F$ = Reduction factor

$h_d$ = Required thickness of the rigid pavement over existing sub-base for increased wheel load.

$h_e$ = thickness of existing rigid pavement.

If it is a bituminous overlay, the thickness, $h_{bo}$, may be taken as

$$h_{bo} = \frac{h_{fo}}{1.5} \qquad (12.15)$$

**Design of Rigid Overlays**

A rigid overlay may be provided on an existing flexible pavement or on an existing rigid pavement.

**Rigid overlay for a flexible pavement:** The modulus of subgrade reaction, *k*, is obtained for the existing pavement, and is used to determine the additional pavement thickness (or overlay thickness) by any one of the methods for rigid pavement design—FAA, Westergaard, or PCA method.

**Rigid overlay for a rigid pavement:** The following equation may be used for determining the thickness of the rigid overlay:

$$h_{ro} = (h_d^{1.4} - Ch_e^{1.4})^{(1/1.4)} \qquad (12.16)$$

Here,

$h_{ro}$ = Rigid overlay thickness

$h_d$ = Design rigid pavement thickness required for modified load and subgrade condition

$C$ = Coefficient which depends on the condition of the existing rigid pavement

$h_e$ = Thickness of the existing rigid pavement

[This is the equation recommended by the US Corps of Engineers.]

*Values of C recommended are:*

US Corps of Engineers {C = 1.00 for good condition; C = 0.75 for slightly cracked condition}

FAA, PCA → C = 0.35 for badly cracked condition.

This equation is applicable when the rigid overlay is placed directly over the existing slab. (This is known as the 'partially bonded condition'.)

However, when the rigid overlay is placed on a levelling course, the US Corps of Engineers recommended the following formula for the overlay thickness, $h_{ro}$:

$$h_{ro} = (h_d^2 - Ch_c^2)^{1/2} \tag{12.17}$$

The notation is the same as that for Eq. (12.16). (This is known as 'unbonded condition'.)

## 12.9 NUMERICAL EXAMPLES

### EXAMPLE 12.1

Design a runway pavement section for a gross weight 90,000 kg. The CBR values of the pavement materials were obtained as given below:

Subgrade soil: 4%
Gravel: 17%
Graded gravel: 40%
The runway pavement has to sustain 20,000 load coverages.

**Solution:** Using the design chart shown in Fig. 12.5, the total thickness of construction over the subgrade with CBR of 4% for a gross weight of 90,000 kg is 105 cm.

The thickness of construction required over the sub-base of gravel of CBR of 17% for this gross weight is 30 cm.

Similarly, the surface course thickness required over the base of graded gravel with CBR of 40% is obtained as 10 cm. (The minimum thickness of surface course recommended by the US Corps of Engineers is 7.5 cm).

These thicknesses are for 5000 coverages.

For 20,000 coverages they have to be increased by 14%. The respective increased values are 120 cm, 35 cm, and 12 cm.

By difference, *sub-base thickness is* 85 cm.

*Base thickness is* 23 cm.

*Surface course of bituminous concrete is* 12 cm.

This brings the *total thickness of construction* to 120 cm as designed.

### EXAMPLE 12.2

Design a flexible pavement section using Burmister's two-layered system. Plate load tests were conducted on soil subgrade and 15 cm thick base course using a 75 cm diameter plate. The pressures recorded at 5 mm deflection for the layers were 1.2 kg/cm and 3.6 kg/cm, respectively. The wheel load for design is 27,000 kg with a tyre pressure of 7 kg/cm$^2$.

**Solution:** For a rigid plate,

$$\Delta = \frac{1.18\, q \cdot a}{E_2} \cdot F_d$$

$F_d = 1$ for single layer of soil subgrade.

$$q = 1.2 \text{ kg/cm}^2$$

$$a = \frac{75}{2} = -3.5 \text{ cm}$$

$$h = 15 \text{ cm}$$

$$\Delta = 5 \text{ mm}$$
$$= 0.5 \text{ cm}$$
$$\frac{h}{a} = \frac{15}{37.5} = 0.4$$
$$\therefore \quad E_2 = \frac{1.18 \times 1.2 \times 37.5}{0.5} \text{ kg/cm}^2$$
$$= 106.2 \text{ kg/cm}^2$$

For the base curve,

$$F_d = \frac{106.2 \times 0.5}{1.18 \times 3.6 \times 37.5} = 0.33$$

For $F_d = 0.33$ and $\left(\frac{h}{a}\right) = 0.4$,

$$\frac{E_1}{E_2} = 1000 \quad \text{(from Fig. 12.7)}$$

$\therefore$ $E_1$, elastic modulus of pavement section

$$= 1000 \times 106.200$$
$$= 1{,}06{,}200 \text{ kg/cm}^2$$

Contact area of the wheel load,

$$A = \frac{27000}{7} = 3857 \text{ cm}^2$$
$$a = \sqrt{\frac{3857}{\pi}} = 35 \text{ cm.}$$

For a flexible plate,

$$\Delta = \frac{1.5\, q \cdot a}{E_2} \times F_d$$
$$F_d = \frac{0.5 \times 106.2}{1.5 \times 7 \times 35} = 0.15$$

For $F_d = 0.15$ and $\left(\frac{E_1}{E_2}\right) = 1000$,

$$\left(\frac{h}{a}\right) = 0.7$$

$\therefore$ Thickness of pavement required = 35 × 0.7 cm = 24.5 cm = or say, 25 cm.
US Navy recommends 40% increase on this thickness.
This comes to 35 cm, out of which the top 10 cm can be bituminous concrete and the bottom 25 cm base course.

## EXAMPLE 12.3

An existing runway consists of 15 cm of cement concrete slab resting on subgrade. The subgrade class is $R_b$-type. It is necessary to strengthen the pavement for supporting an equivalent single wheel load of 27,000 kg. For this, the runway thickness required has been got as 25 cm.

Determine the overlay thickness of bituminous concrete.

**Solution:**

$$h_{fo} = 2.5\,(F \cdot h_d - h_e) \qquad \text{(Eq. 12.14)}$$

Value of $F$ for $R_b$-class subgrade is 0.90.

$h_d = 25$ cm and $h_e = 15$ cm

$$\therefore \quad h_{fo} = 2.5(0.90 \times 25 - 15)$$

$$= 18.75 \text{ cm}$$

This is for flexible overlay.

For bituminous overlay,

$$h_{bo} = \frac{h_{fo}}{1.5}$$

$$= \frac{18.75}{1.5} = \underline{12.5 \text{ cm}}$$

## EXAMPLE 12.4

An existing rigid runway pavement, 10 cm thick, has to be provided with a rigid overlay. Due to increase in the aircraft wheel load, it is determined that a rigid pavement of thickness 25 cm is required. The existing runway surface is in good condition. Determine the thickness of the rigid overlay required, if

(i) It is laid directly over the existing runway, and
(ii) It is laid over a levelling course.

**Solution:**

(i) *Overlay directly over the existing runway:*
(*Partially bonded condition*):

$$h_{ro} = (h_d^{1.4} - C\,h_e^{1.4})^{1/1.4}$$

Here, $h_d = 25$ cm $\quad h_e = 10$ cm

and $C = 1$

$$\therefore \quad h_{ro} = (25^{1.4} - 1 \times 10^{1.4})^{1/1.4} = 19.83 \text{ cm, say } \underline{20 \text{ cm}}$$

(ii) *Overlay laid over a levelling course*:
(*Unbonded condition*)

$$h_{ro} = (h_d^2 - C\,h_e^2)^{1/2}$$

Here, $h_d = 25$ cm $\quad h_e = 10$ cm

and $C = 1$

$$h_{ro} = (25^2 - 10^2)^{1/2} = 22.9 \text{ cm or say } \underline{23 \text{ cm}}$$

## SUMMARY

- Airport pavements, as in the case of highway pavements, are broadly classified as *flexible* and *rigid* pavements. A flexible pavement is a layered system consisting of a sub-base course, base course, and a surface course (which may be made of graded granular material, or bituminous material), resting over a well-compacted and prepared soil subgrade. If the surface course is of bituminous material, it is called a *bituminous pavement*. A flexible pavement derives its strength primarily from the subgrade and from the load distribution properties of the overlying layers; the system cannot transmit tensile stresses.

  A rigid pavement usually consists of a cement concrete (plain, reinforced, or pre-stressed) slab laid over a soil subgrade. It derives its load-carrying capacity primarily from its flexural strength, and can bridge any weak spots in the soil subgrade. However, a well-compacted subgrade and additional sub-base/base courses of either granular or bituminous material, enhance its strength and durability.
- The design factors which need consideration in airport pavement thickness design are the strength and drainage characteristics of the subgrade soil, strength of the overlying layers of granular, bituminous, or cement concrete material and their properties, and the design wheel load from the aircraft using the facility.

  Depending upon the wheel and gear configuration, the aircraft loads may be converted into an 'Equivalent Single Wheel Load (ESWL)' for convenience in design.
- For flexible pavement thickness design, the US Corps of Engineers developed the California Bearing Ratio (CBR) method and evolved design charts relating the CBR value of each layer and the thickness of construction required above it for different aircraft wheel loads/gross weight. This is basically an empirical approach.

  Burmister's two-layer theory is considered to be on a more scientific footing involving mathematical equations considering the elastic properties of the soil subgrade and the courses above it. This also involves the use of design charts.

  Analytical methods have been developed and applied by the US Corps of Engineers, utilising the latest Computer Aided Design (CAD) techniques to solve the complex mathematical equations involved in the analytical approach.
- For rigid pavement design, FAA developed design charts relating the subgrade conditions and aircraft wheel loads to the required slab thickness. These are empirical/semi-empirical. Westergaard developed a sound mathematical approach for the computation of critical stresses due to the wheel loads placed at the corner, edge, and interior of the slab; he introduced the concept of 'modulus of subgrade reaction' of the subgrade, which is the pressure required to produce unit deflection/penetration. The load contact area details also play a role. The equivalent radius of the resisting section and the elastic properties of the concrete, characterised by the elastic modulus and the Poisson's ratio are involved in the computations.
- The Load Classification Number (LCN) method of aircraft evaluation and airport pavement design was formulated by the British Air Ministry Directorate of General works. In this system, the supporting capacity of a pavement is expressed in terms of a number, called the Load Classification Number; the equivalent single wheel load (ESWL) of an aircraft can also be expressed in terms of the LCN.

  The pavement can be used safely by an aircraft for an unrestricted number of coverages if the LCN of the pavement is greater than that of the aircraft.

  Besides the ESWL, the tyre pressure and the contact area also play a significant role in the method.

The LCN of both flexible pavements and rigid pavements may be determined by means of plate load tests conducted in a specified manner.

Usually, reinforced concrete (rigid) pavements are used for runway ends and aprons; composite pavements of continuously reinforced cement concrete with bituminous surfacing are used for stretches of runways and taxiways in which aircraft move at high speed; and, flexible pavements composed of bound granular materials with bituminous surfacing, offer an alternative to composite pavements with similar advantages.

- Airport pavements fail during their operation due to certain defects which occur due to repeated aircraft loadings.

  Typical defects in flexible pavements which lead to failure are ravelling, stripping, cracking of different kinds, and disintegration, which result in pot holes, ruts and corrugations, and settlements. Typical defects in rigid pavements leading to their failure are scaling and spalling of concrete, cracking due to shrinkage and warping and structural cracking due to inadequacy of strength, and mud-pumping. Material deficiencies, structural inadequacy, and poor workmanship could be the reasons for these defects and ultimate failure of the pavement.
- Maintenance of flexible pavements involves patch repair, surface treatment and resurfacing, depending on the extent of damage. Rehabilitation measures will be needed to prolong the life of the pavement.

  Maintenance of rigid pavements involves crack repair, joint maintenance, patch repair, control of mud-pumping, and restoration of antiskid surface.
- Evaluation of an airfield pavement involves 'functional' and 'structural' evaluation; the former is concerned with the surface characteristics and riding quality, while the latter with structural adequacy or strength.

  Aircraft Classification Number (ACN) and Pavement Classification Number (PCN) have been defined and developed as part of 'evaluation'.
- When an existing pavement has become unserviceable, or when it is inadequate for operational needs of an immediate future traffic, it is strengthened by the addition of extra layers called *overlays*.
- The overlays can be flexible or rigid, and each type can be used over an existing flexible or rigid pavement. Overlay design procedures have been developed for both these cases.

## QUESTIONS AND PROBLEMS

1. Explain the differences between the two broad categories of airport pavements.
2. Explain briefly the factors governing the design of airport pavements.
3. Describe briefly the CBR method of design of the flexible type of airport pavements.
4. Describe briefly the principle involved in Burmister's two-layer system of design of flexible pavements.
5. What are the advantages of Computer Aided Design used in conjunction with analytical methods of design of airport pavements?
6. What are the design criteria for the rigid type of airfield pavements?
7. Explain briefly Westergaard's approach to the design of rigid pavements.
8. Explain briefly the PCA method of design of airport pavements of the rigid type.
9. What are the characteristics of rigid, composite and flexible pavements used for airports? What are the locations at which these are used?

10. Explain briefly the principles involved in the LCN system of airfield pavement design.
11. Explain briefly the procedures for the determination of LCN of flexible pavements and of rigid pavements.
12. List the typical failures seen in flexible pavements.
13. Enumerate the typical failures seen in rigid pavements.
14. Briefly explain the maintenance operations of flexible pavements.
15. Describe briefly the maintenance aspects of rigid pavements.
16. Write brief notes on:

    (a) Load classification number

    (b) Mud-pumping

    (c) Alligator cracking
17. Differentiate between:

    (a) Longitudinal cracking and reflection cracking

    (b) LCN and ESWL

    (c) PCN and ACN
18. Explain briefly 'functional adequacy' and 'structural adequacy'. How are these evaluated for an existing pavement?
19. Explain the need for overlays and the factors governing their design.
20. Enumerate the different kinds of overlays used for strengthening existing pavements.
21. Design a runway pavement section for a gross weight of 90,000 kg. The CBR-values of subgrade, sub-base, and base course are determined to be 5%, 20% and 44%, respectively. The runway pavement is expected to sustain 20,000 coverages.
22. Design a flexible pavement section using Burmister's two-layered system. Plate load tests were conducted on soil subgrade and a 20 cm-thick base course using a 75 cm-diameter plate. The pressures recorded at 5 mm-deflection for these layers were 1.5 $kg/cm^2$ and 4.5 $kg/cm^2$, respectively. The wheel load for the design is 36,000 kg with a tyre pressure of 9 $kg/cm^2$.
23. An existing runway consists of 18 cm of cement concrete slab resting on a subgrade of $R_c$-type. The thickness required for enhanced wheel load is determined to be 30 cm, and the pavement is to be strengthened by means of a bituminous concrete overlay. Determine the necessary thickness of the overlay.
24. A monolithic slab of cement concrete, 20 cm thick, is required due to increased traffic. Compute the thickness of a rigid overlay when the

    (i) existing cc pavement is 15 cm thick and is in good condition.

    (ii) existing cc pavement is 15 cm thick and is in a slightly cracked condition.

    The interface condition may be taken as partially bonded. Use the US Corps of Engineers method.

Chapter **13**

# AIRPORT DRAINAGE

**CHAPTER OBJECTIVES**

This chapter enables the student to understand:

- The importance of airport drainage for safe operation and durability of the aircraft
- The special characteristics of airport drainage such as large areas, heavy wheel loads and flat gradients involved
- The design of subsurface drainage through drains leading to a suitable outfall at a safe distance
- How surface drainage is channelled through embedded drain pipes of proper size and thickness
- The use of specially designed granular filters to prevent piping and clogging of drain pipes.

## 13.1 INTRODUCTION

A basic requirement of an ideal site for an airport is that it should be free from surface unevenness; it could be gently undulating, or sloping up and down. The area should be properly graded to facilitate quick drainage of rain water and for convenient construction of the various components of the airport.

The two aspects of grading and drainage of an airport are closely related to each other and are therefore planned and executed together.

An efficient drainage system is essential for safe operation of the aircraft, as also for durability of the runway pavement. Ponding of the runway surface can be hazardous to landing or take-off of aircraft.

Thus, it is necessary to provide a well-drained area that is sufficiently firm to permit safe operation of aircraft under all weather conditions. Since it is rare that the natural drainage due to favourable topography of the airport is adequate, it is invariably necessary to provide a well-planned drainage system.

## 13.2 AIRPORT GRADING

Highway grading is essentially a two-dimensional problem as it involves length and height. The width of a highway pavement is relatively small, for which cross-drainage can be easily provided by means of camber in straight reaches and superelevation on curved reaches.

In contrast, airport grading is a bigger problem in view of the significant width of the runway, which warrants efficient drainage in both the longitudinal and lateral directions.

A careful study of the topographic map of the airport site helps to identify the overall flow directions on the natural surface and the existing natural streams or low-lying areas into which the surface runoff flows. This, in turn, helps in the location and orientation of the primary runways and the secondary runways, if any, besides other controlling factors like the wind characteristics.

### Requirements of Airport Grading

(i) The quantity of earthwork should be such that the cut and fill should, as far as possible, balance each other.
(ii) The haulage distance and the quantity of the earth material to be moved should also be borne in mind.
(iii) Proper surface and sub-surface drainage facilities should be provided with the chosen grading plan.
(iv) The grading plan should be prepared with all necessary details of the grades chosen.
(v) At intersections of runways or runways and taxiways, grading should be done in such a way that there are common grade elevations at such points.
(vi) Grading should be done at the areas beyond the runways and landing strips, so that aircraft which overshoot while taking off or undershoot while landing are affected to the least extent.
(vii) The subgrade soils in the area should be preferably of uniform character, so that the design of drainage facilities becomes relatively easy. Soil surveys should be conducted to collect the necessary information.
(viii) The equipment needed for the clearing, grading, and compaction should be chosen considering the nature of the soil material, magnitude of the project, and the time allotted for completion.

An appropriate grading plan is prepared taking into account all these factors.

### Grading Operations

The procedure for establishing a grade line is given below:

(a) Profiles are prepared for each of the proposed alternatives for the runways.
(b) The profile for the runway of the best choice is established carefully.
(c) At the proposed intersections of runways, the grade lines for the other runways are established using the elevation already established as the control point,
(d) More than one trial may be required to evolve the best grading or the best combination of grade lines.

### Computation of Earthwork

The methods used for computing the earthwork involved in grading are:

- Cross-section method
- Mass-haul curve method

**Cross-section Method**

In this method, the cross-sectional areas of the required formation, at suitable intervals such as 30 m, are determined from plotted sections. Areas may be calculated using a graph paper or a planimeter.

The volume of the earth work may now be got using either the trapezoidal formula or the prismoidal formula.

*Trapezoidal formula:*

$$V = d\left(\frac{A_1 + A_n}{2} + A_2 + A_3 + \cdots + A_{n-1}\right) \tag{13.1}$$

*Prismoidal formula:*

$$V = \frac{d}{3}[A_1 + 4A_2 + 2A_3 + 4A_4 + \cdots + 2A_{n-2} + 4A_{n-1} + A_n] \tag{13.2}$$

Here,

$V$ = Volume of the earthwork

$d$ = Distance between consecutive cross-sections

$A_1, A_2, \cdots A_n$ $A_1$, $A_2$, $A_n$ = Areas of the cross-sections at points 1, 2, ......*n*, spaced at equal intervals.

**Mass-haul Curve Method**

The cost of excavation depends on the volume of earth material excavated, as also upon the manner of its disposal. Embankments may be built of material obtained from an excavation during the grading of the area, or from a borrow area.

This distance up to which no charge need be paid for haulage of the excavated material is known as the *free-haul distance*. This distance depends on the means of transport, by trucks or by manual labour; the distance may range from 150 to 300m if it is by manual labour.

There is a limiting distance beyond which it would prove economical to waste the excavated material and to bring material for the fill from a borrow area; this limiting distance is known as the *economic-haul distance*. This is also the distance at which the cost of both the options – haulage of the excavated material and that of the material from the borrow area – would be the same; obviously, this depends upon the relative costs of labour and of the transportation cost from the borrow area.

The economic haul and free haul are related as follows:

$$D_e = D_f + \frac{C_b}{C_o} \tag{13.3}$$

where

$D_e$ = Economic haul

$D_f$ = Free haul

$C_b$ = Cost of borrow per unit material per station

$C_o$ = Cost of overhaul per unit material per station (or that for carrying unit material from one station beyond the free haul distance).

Grading – cuts and fills – should be properly planned so that the cost of earthwork operations is a minimum. To determine the best-planned movement of earth materials to achieve economy, the *mass-haul curve* or the *mass-haul diagram* is utilised.

The *mass-haul diagram is a graphical representation of the amount of earth work in excavation and embankment in a given project, and the programme of earth material haulage.* This curve shows the *cumulative volume* of earthwork at any point. In addition, the following information can be obtained from this diagram:

(i) The volume of earth work
(ii) The direction of haul
(iii) Free haul distance
(iv) Volume of earth required to be borrowed
(v) The economic haul distance.

The *construction of mass-haul diagram,* though somewhat laborious, provides valuable information that helps to compare the economy of using the excavated material with that of using the material hauled from a borrow area, in a given set of conditions.

Mass-haul diagram is a curve plotted with distance as the base. The ordinate at any point is the algebraic sum of the volumes of cut and fill from the starting point (or any other arbitrarily chosen point) up to that point. Cuts are considered positive and fills negative. Thus, the diagram is an integral or cumulative sum of the volumes of the several cuts and fills of the earthwork. When cut and fill occur in the same section as in side-hill section, their difference alone is to be used in the summation, the sign being that of the greater of the two.

A typical mass-haul diagram is shown in Fig. 13.1.

The important *characteristics of the mass-haul diagram* are given below:

1. An upward slope in the forward direction indicates excavation, and a downward slope, a filling.
2. A positive peak occurs at the end of a cut and a negative peak at the end of a fill.
3. The vertical distance between a positive peak and the negative peak represents the whole volume of a fill. Similarly, that between a negative peak and the next positive peak represents the whole volume of a cut.
4. The vertical distance between two points on the curve with no peak between these two chainages represents the volume of earthwork between these points.

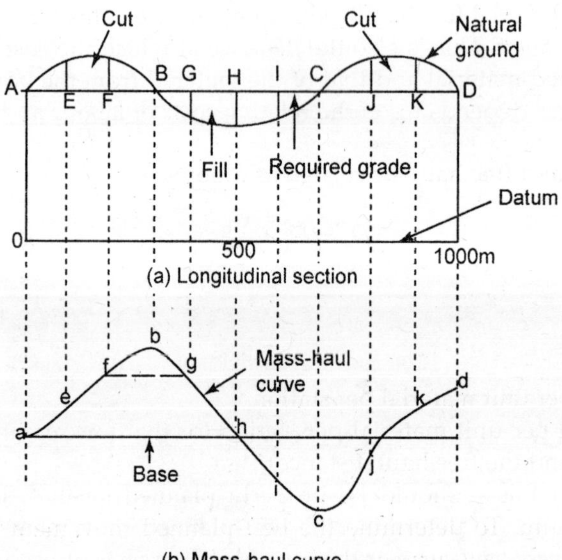


**FIG. 13.1** *A typical mass-haul curve*

5. If the curve cuts the base between two points, the volume of cut equals that of the fill, since the algebraic sum of the earthwork quantities between them is zero.
6. Any horizontal line intersecting the mass curve represents a length over which the cut and fill are equal (fg is the balancing line in Fig. 13.1).
7. When the loop of the mass curve cut off by a balancing line lies above the line of intersection, the excavated material must be hauled forward; when the loop lies below, the haul is backward. The length of the balancing line represents the maximum haul distance in that section.
8. The area bounded by a loop of the mass curve and the balancing line represents the haul volume in that section.
9. The haul over any length is a minimum when the balancing line is so situated that the sum of all the areas cut off by it, without regard to sign, is a minimum.

There could be more than one way to balance the volumes of cut and fill, but the best is that based on haulage and other factors.

The mass-haul diagram is of immense use in the economical planning of major earthwork projects.

## 13.3 AIRPORT DRAINAGE

A well-designed and efficient drainage system is essential for the safety in operation of aircraft and durability of runway pavements. Lack of proper drainage could result in ponding of the runway pavement which is hazardous to the safe landing or take-off of aircraft; it could also lead to loss of subgrade support and failure of the runway pavement. Replacement of ill-designed drainage structures like drains is expensive and time-consuming, especially when they pass below paved areas.

In view of this, a well-designed drainage system which can serve for a well-judged design period is vital in the construction of an airport.

### Factors Affecting Airport Drainage

The following factors are considered significant to ensure effective drainage in an airport:

- Site characteristics such as gradients and soil properties.
- Meteorological and hydrological characteristics of the area such as rainfall, run-off, atmospheric temperature, etc.
- Special needs of drainage of the components of an airport.

### Objectives of Airport Drainage

The following are the objectives of airport drainage:

(i) To avoid or prevent ponding of runways, taxiways and aprons, which could be hazardous to safe operation of aircraft

(ii) To prevent loss of subgrade support due to high ground water table during monsoon leading to structural/functional failure of runway pavements

(iii) To ensure safety in the operation of aircraft

(iv) To reduce the maintenance cost of the airport facilities

(v) To minimise periods of closure of the airport due to possible inundation

(vi) To increase the overall efficiency of the airport during its service life.

### Functions of Airport Drainage

The important functions are as follows:

- To divert the rain water – surface run-off and ground-water flow – to suitable location outside the airport area
- To lower the ground-water level in the airport and restrict it within permissible limit
- To lead the surface run-off to a suitable location outside the airport site.

### Special Characteristics of Airport Drainage

Airport drainage has certain special and unusual characteristics making it somewhat complex. These are given below:

(i) Extensive area to be drained
(ii) Heavy and concentrated wheel loads of aircraft
(iii) Wide runways, taxiways and aprons, which need to be kept relatively dry
(iv) Flat longitudinal and transverse gradients
(v) Absence of surface drains at the sides
(vi) Shallow water courses
(vii) Varying soil conditions, in view of the large area to be drained
(viii) Concentration of the outfall flow
(ix) Rapid drainage requirement for safe operation of aircraft
(x) Combined drain pipes for surface water and subsurface water
(xi) Need of provision for future expansion.

### Basic Requirements of Airport Drainage System

- Capacity of the drain pipes for the needs of both surface and subsurface flows
- Rapid drainage to minimise suspension of airport operation
- Strong drain pipes to withstand heavy wheel loads of the aircraft
- Provision for future extension of runways when such need arises.

### Data for design of Airport Drainage System

(i) Contour map of the airport site and adjacent land, showing all the existing natural features
(ii) Rainfall data for a period of 5 to 10 years, obtained from the Metrological Department
(iii) Centre line profiles of the proposed runways, taxiways and aprons, and their cross-sections
(iv) Temperature data during different seasons for the design of run-ways as well as the drainage facilities
(v) Boring data showing the soil strata and ground-water table conditions at the site
(vi) Run-off records
(vii) Drainage or permeability characteristics of the soil strata for the design of filters
(viii) Hydraulic and structural data for the design of drain pipes
(ix) Working drawings of the drainage system to facilitate the construction of the necessary drainage structures.

## 13.4 COMPONENTS OF AIRPORT DRAINAGE

Airport drainage is made up of the following components:

1. Surface drainage
2. Sub-surface drainage

## Surface Drainage

This consists of the arrangements made for quick and effective draining of rain water which collects on the runway pavement surface and the adjoining areas. Surface drainage should be done in such a way that there is least interruption caused to the air traffic, subsoil is not saturated, and damage to airport facilities is avoided. The system should be designed in such a way that extension for future increase in air traffic can be done with minimal interruption to the existing traffic.

### Time of Concentration

This is defined as the time required for rain water to reach the point under consideration from the farthest point of the catchment area.

It consists of two components – *entry time* and *time of flow*.

*Entry time* is the time taken by the storm water to flow from the farthest point of the catchment area to the drain inlet.

*Time of flow* is the time for the water to flow from the inlet or entry of the drain to the particular point under consideration.

Entry time depends on a number of factors such as the *coefficient of run-off* or the fraction of the precipitation or rainfall which flows on the surface; this depends on the nature of the surface. Other factors are the gradient of the surface and the distance through which the flow occurs from the remotest point of the catchment to the drain inlet.

Charts have been developed between the entry or inlet time and the distance of flow for different surfaces (or coefficients of runoff) and for different gradients or surface slopes (in percent). These may be used for determination of the entry time.

Time of flow, $T$, can be determined as follows:

$$T_f = \frac{D}{\nu} \tag{13.4}$$

where $D$ = Distance of flow

$\nu$ = Velocity of flow in the drain pipe (this depends on the discharge, the area of section of the pipe drain, and the surface roughness of the drain).

### Weighted Coefficient of Run-off

When the catchment area consists of different types of surfaces with different run-off coefficients, the weighted average value of C is obtained as follows:

$$C = \left(\frac{C_1A_1 + C_2A_2 + \ldots}{A_1 + A_2 + \ldots}\right) \tag{13.5}$$

where $A_1$, $A_2$, etc are the areas of surfaces with run-off coefficients $C_1$, $C_2$, . This is to be used in the estimation of the run-off.

The *values of coefficients of run-off for the surfaces commonly encountered in airports*, as recommended by the FAA, are given below:

| S. No. | Nature of surface | Coefficients of run-off (*C*) |
|---|---|---|
| 1. | Watertight roof | 0.75–0.90 |
| 2. | Bituminous runways | 0.80–0.90 |

(Contd)

Table (Contd)

| S. No. | Nature of surface | Coefficients of run-off ($C$) | |
|---|---|---|---|
| 3. | Cement concrete runways | 0.70–0.90 | |
| 4. | WBM or gravel surface | 0.35–0.70 | |
| 5. | Impervious soils | 0.40–0.65 | Slope 1% to 2% |
| 6. | Impervious soils – turfed | 0.30–0.55 | |
| 7. | Slightly pervious soils – turfed | 0.10–0.30 | |
| 8. | Moderately pervious soils | 0.05–0.20 | |
| 9. | Moderately pervious – turfed | 0–0.10 | |

### Intensity–Duration–Frequency Curves

These are provided by the Meteorological Department for a particular area as contours of 1 hour rainfall intensity for frequency periods of 2, 5, 10, 25 and 50 years.

A frequency or return period of 5 years is recommended by the FAA for the design of airport drainage.

These curves are useful for the determination of rainfall intensity, which is required in the estimation of run-off discharge necessary for the design of drain pipes.

**Estimation of run-off:** The following formula is used for the estimation of run-off for the design of drainage facilities:

$$Q = \frac{C \cdot i \cdot A}{36} \approx 0.028 \times C \cdot i \cdot A \qquad (13.6)$$

where $Q$ = Peak run-off ($m^3/s$)

$C$ = Coefficient of run-off (weighted value, if necessary)

$i$ = Critical intensity of rainfall in cm/h, occurring during the time of concentration – for the chosen frequency interval or return period, as it is called.

$A$ = Area of catchment or drainage area which contributes to the run-off at the particular location (hectares)

## Design of Surface Drainage

### General Principles

(i) Each drain pipe is designed to carry the discharge from the inlet at its upstream end plus the contribution from all the preceding inlets.

(ii) The gradient of each drainage line is fixed after careful study of the topographic map of the site. A minimum mean velocity of 0.75 m/s is assumed to prevent silting of the pipe.

(iii) A minimum diameter of 30 cm is assumed to provide adequate cross-section for flow.

(iv) A minimum distance of 22.5 m is ensured for the inlet from the edges of the runway pavement.

(v) In the case of parallel runways or taxiways, inlets are provided in a valley between them.

(vi) On aprons, inlets are placed within the pavement area since large areas are involved.

(vii) Inlets are usually spaced at 60 to 120 m intervals if ponding is allowed; otherwise, they are spaced at 52.5 m intervals.

**Design Procedure**

(i) The coefficients of run-off are selected for all types of surfaces over which the water has to flow.
(ii) The weighted average value of the coefficient of run-off is determined.
(iii) The length of flow over each type of surface is determined.
(iv) The inlet time and times of flow for all the sub-lengths are determined.
(v) The time of concentration for each inlet is determined.
(vi) From the rainfall intensity curves, the intensity of rainfall is obtained.
(vii) The quantity of run-off is got from Eq. (13.6). This amount of flow must be accommodated by the effluent pipe.
(viii) The diameter of the pipe and the slope are calculated using Manning's formula:

$$A = \frac{Q}{V} \qquad (13.7)$$

where $A$ is the area of the section of flow,
$Q$ is the design surface-runoff, and,
$V$ is the permissible velocity for the surface of the drain. (While small velocities can cause silting, large velocities may cause erosion or scouring of the inner surface of the drain).

Manning's formula for the permissible velocity is

$$V = \frac{1}{n} \cdot R^{2/3} \cdot S^{1/2} \qquad (13.8)$$

where $V$ = permissible velocity (m/s)
$n$ = Manning's coefficient of rugosity (depends on the roughness of the surface)
$R$ = Hydraulic mean depth (= $\frac{A}{P}$, where $A$ is the area of flow, $P$ is the wetted perimeter (m) and $S$ is the longitudinal bed slope of the drain (fraction).

[$n$ ranges from 0.012 to 0.200.
$n$ for concrete 0.012 to 0.016
$n$ for bitumen (asphalt) 0.013 to 0.014
$n$ for random stone in mortar 0.017 to 0.200
$n$ for natural earth (without vegetation) 0.035 to 0.040
$V$ ranges from 0.3 m/s to 6.0 m/s
$V$ for concrete 6.0 m/s
for bitumen 5.4 to 6.0 m/s
for rock 6.0 m/s
for well-graded gravel 1.8 to 2.1 m/s]

**Inlet:** It is a concrete box, with a lid of reinforced concrete, cast-iron, or cast steel; it should be strong enough to withstand aircraft wheel loads.

**Ponding**

*Ponding is a means of providing temporary storage of runoff water prior to its entry into the underground drainage system.* Ponding areas would protect the pavements from damage due to storms of unexpected intensities higher than the design value. Ponding is, however, prohibited on runways and taxiways, parking aprons, and shoulders. Ponding is permissible on open and graded areas where it would not damage the subgrade and base courses.

A safe distance of 22.5 m is mandatory for areas of ponding from the edges of the pavements. For the purpose of design, ponded volume is taken to be an inverted frustum of a pyramid, the height of which is equal to the depth of water above the inlet at any stage. The base area is taken as the surface area of the pond.

While the steps in the *design procedure* for drains are applicable, the following are also needed in the *design of ponds*:

(i) The volume of storage in the pond is given by

$$V = \frac{1}{3}h\left(A_t + A_b + \sqrt{A_t . A_b}\right) \tag{13.9}$$

where $V$ = Volume of the pond
$h$ = Height of pond
$A_t$ = Area of the top surface
$A_b$ = Area of the bottom surface

(ii) Rainfall intensities for durations of 5, 10, 15, 20, 30, 60, 90, and 120 minutes are obtained from charts, and the volume of runoff for each case is got as the product of runoff and the duration.

(iii) The cumulative runoff is plotted versus time.

(iv) The diameter of the drain pipe, its slope, and the rugosity coefficient are suitably assumed, and the discharge capacity is calculated from Manning's approach. This is plotted in the cumulative runoff versus time curve.

(v) The largest difference between the cumulative run-off and the discharge capacity of the selected size of the drainpipe is determined, which gives the maximum amount of ponding after the beginning of the storm.

For safety, this should be less than the calculated volume of the pond.

When the airport site is subject to heavy rainfall intensities, the size of the drains required will be large. While the major runways are provided with full complement of drainage facilities, some ponding is tolerated in other areas; in this case, the sizes of the drains required will be based on the difference between the runoff and the ponded volume, effecting economy.

## 13.5 SUBSURFACE DRAINAGE

Moisture changes in the subgrade occur due to percolation of rain water and seepage flow, as also due to the phenomenon of capillary rise. The aim of subsurface drainage is to keep the ground water table (GWT) sufficiently below the level of the subgrade, by at least 1.2 m.

When the water table is almost at the natural ground surface, the best option is to raise the formation of the runway 1.2 m above the ground. If this is not possible due to unfavourable topography, then ground water table may be lowered by means of subsurface drainage arrangements. It must, however, be remembered that only gravitational water in the soil can be drained, but not held water , the film moisture around the grains.

A few drainage arrangements for different situations are discussed below:

**Subsurface Drain for Interception of Ground Water**

This is a drain installed in the pervious layer beside the runway pavement to intercept the ground water before it can reach the subgrade, as shown in Fig. 13.2.

**Longitudinal Drain Trenches and Pipes for Lowering GWT**

This arrangement helps to lower the ground water table below the runway pavement. If the soil is relatively pervious, longitudinal drainage trenches with drain pipes are used, backfilled

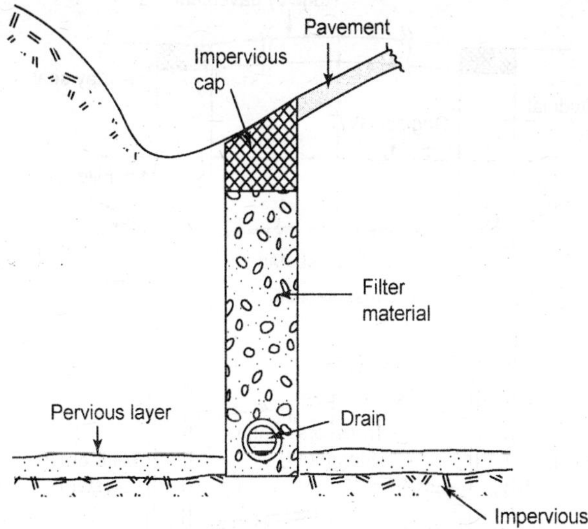


**FIG. 13.2** *Subsurface drain to intercept seepage water*

with filter sand. The depth of the trench depends on the extent of lowering required, soil type, and the distance between the trenches.

A typical arrangement is shown in Fig. 13.3.

### Longitudinal and Transverse Drains for Lowering GWT

If the soil is relatively less permeable, longitudinal as well as transverse drains may be needed to lower the GWT, as shown in Fig. 13.4.

### Subsurface Drains to Control Seepage in Cut Slopes

Sometimes, seepage water renders cut slope unstable; this can be prevented by lowering the seepage line by means of a subsurface longitudinal drain installed to a depth below the pervious layer, as shown in Fig. 13.5.

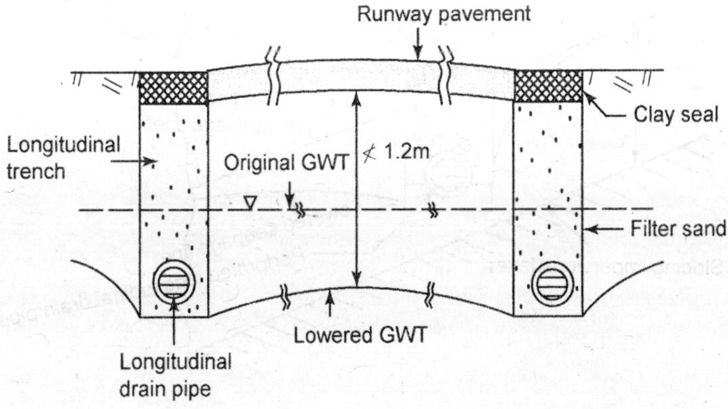


**FIG. 13.3** *Lowering GWT in pervious soil by subsurface drains*

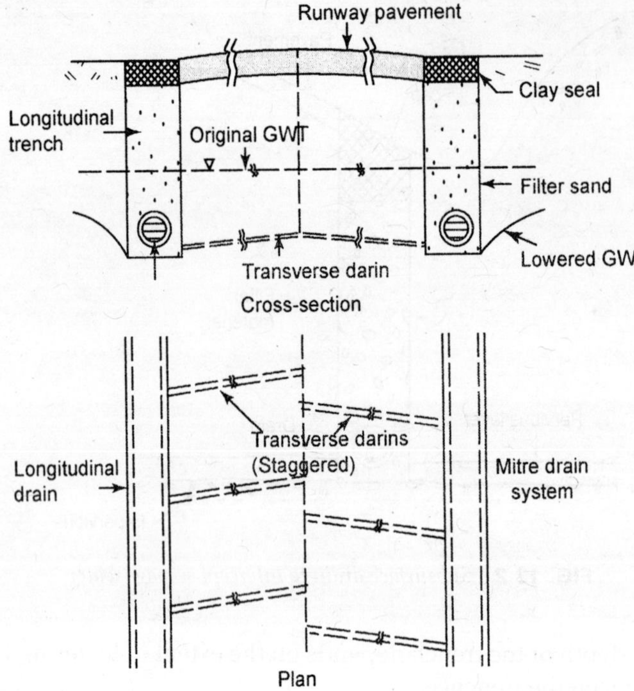


**FIG. 13.4** *Longitudinal and transverse drain system for lowering GWT*

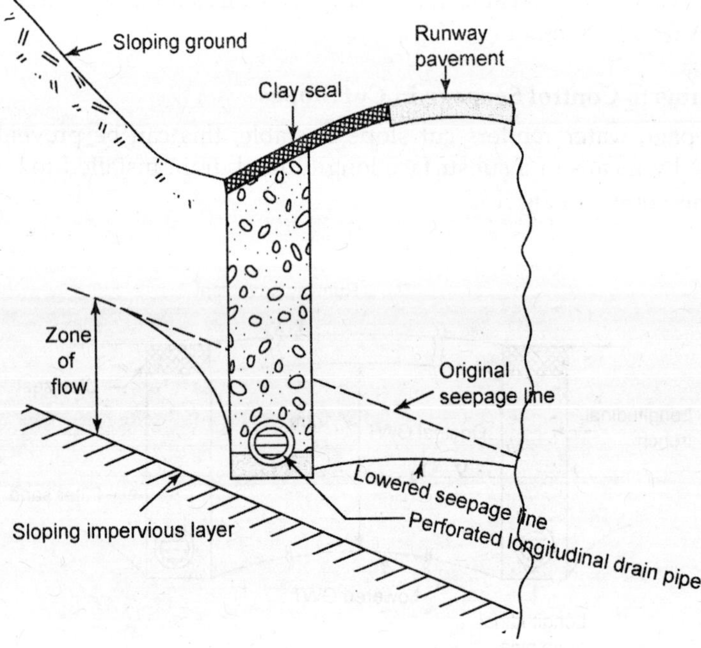


**FIG. 13.5** *Subsurface drains at cut slope to control seepage*

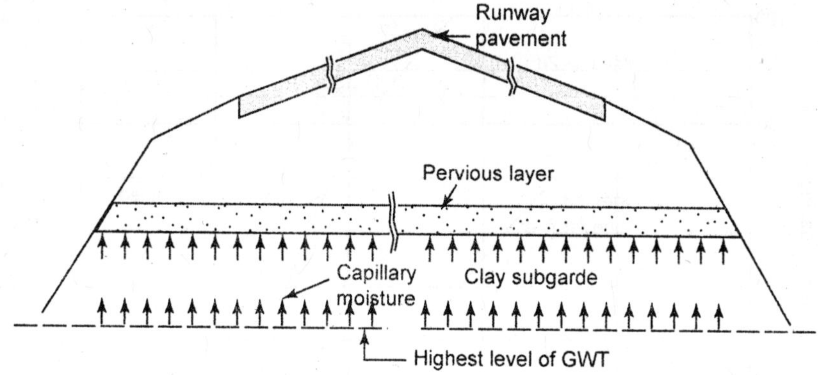


**FIG. 13.6** *Capillary cut-off for a clayey subgrade*

### Capillary Cut-off for Clayey Subgrade

If the subgrade is clayey, the system of subsurface drains on either side will not be effective because of the low permeability of the subgrade. In such a case, the subgrade has to be raised with a free-draining material, or a capillary cut-off has to be applied, as shown in Fig. 13.6.

The capillary cut-off may be even an impermeable bituminous layer.

The location of the cut-off should be above the level of the capillary rise expected for the clayey subgrade.

### Drain Pipes and Filter Media

A subsurface drain may comprise of a perforated pipe, a porous concrete pipe or solid pipe laid with open joints. Alternatively, a trench filled with a free draining material may be used to serve the purpose of a drain.

A perforated pipe, or a porous pipe of no-fines concrete is considered to be a good choice when laid in a trench, backfilled with a granular, free-draining material and covered with an impervious cap at the top. The design of such granular filters with appropriate gradation has to satisfy certain criteria.

**Design of a filter material:** *The gradation requirements of the filter material are based on three criteria:*

(i) Permeability of filter
(ii) Prevention of piping due to high seepage velocity
(iii) Prevention of clogging of the drain pipe.

These requirements are:

(i) Permeability ratio: $\left(\dfrac{D_{15}(\text{filter})}{D_{15}(\text{soil to be drained})}\right) > 5$

(ii) Piping ratio: $\left(\dfrac{D_{15}(\text{filter})}{D_{85}(\text{soil to be drained})}\right) < 5$

(iii) $D_{85}$ (filter) > $2d_p$ ($d_p$ : Diameter of the perforation in the drain pipe)

Figure 13.7 shows an example of the selection of a suitable filter material based on the gradation of the soil to be drained. Based on the gradation of the soil to be drained and its

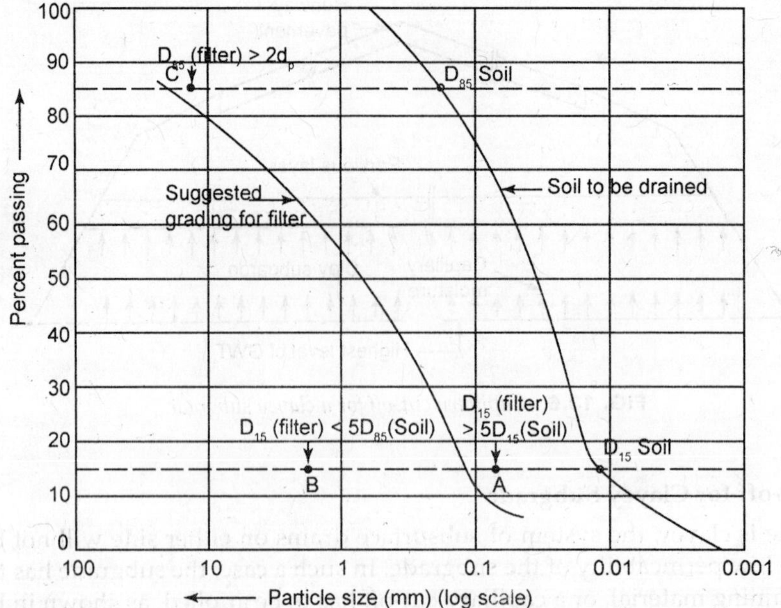


**FIG. 13.7** *Filter design for subsurface drain*

desired filter grading, the grading of the filter layers can be determined on the basis of the criteria given:

Let the grading of the soil to be drained be plotted as shown. $D_{15}$ and $D_{85}$ of the soil are marked.

From the permeability criterion, $D_{15}$ of filter is marked, such that it is greater than $5D_{15}$ of soil (Point A). From piping criterion, $D_{15}$ of filter is marked, such that it is less than $5D_{85}$ of soil (Point B).

On $D_{85}$-line, point C is marked such that $D_{85}$ of the filter is greater than $2d_p$, $d_p$ being the diameter of the perforation of the drain pipe. A suggested grading of the filter may be drawn smoothly such that it lies to the left of C and lies between A and B as shown.

The perforated pipe is usually 100 to 150 mm in diameter with holes in two or more lines towards the bottom side of the pipe.

The collector pipes of porous concrete, metal, or PVC should be laid with a minimum of 100 mm of filter sand around them.

Usually 5 mm diameter holes are considered adequate, if restricted to the lower 60° arc of the pipe. Solid pipes with open joints may be used, but care should be taken to see that silt and fine sand do not enter the pipe.

When the flow of water takes place through porous backfill of graded sand, it is likely to be clogged after some time, thus leading to the maintenance problem of washing of clogged backfill and putting back clean graded sand.

Of late, *geosynthetics* or *geotextiles* are becoming popular as substitutes or alternatives to graded filters. They have high retention of fine particles and permeability similar to graded material in tensile strength. Installation is also easy.

A geosynthetic performs the function of a filter as well as that of a separator. An aggregate drain with a pipe encased in a geosynthetic performs the functions well.

## 13.6 NUMERICAL EXAMPLES

### EXAMPLE 13.1

Determine the weighted coefficient of run-off to be used in the rational equation for the following data relating to the catchment area of a pavement:

| S. No. | Nature of surface | Area (ha) | Coefficient of run-off (C) |
|---|---|---|---|
| 1. | Paved area | 1.00 | 0.90 |
| 2. | Built-up area (60% impervious) | 4.30 | 0.55 |
| 3. | Unpaved area | 0.30 | 0.30 |
| 4. | Loamy area | 0.50 | 0.30 |
| 5. | Sandy area | 0.50 | 0.20 |
| 6. | Lawns | 0.60 | 0.15 |
| Total area | | 7.20 | – |

**Solution:** Weighted coefficient of run-off for use in the rational equation for discharge:

$$C = \left(\frac{1\times 0.9 + 4.3\times 0.55 + 0.3\times 0.3 + 0.5\times 0.3 + 0.5\times 0.3 + 0.6\times 0.15}{7.2}\right)$$

$$= \underline{0.513}$$

### EXAMPLE 13.2

The intensity of rainfall at the proposed site of an airport for designing drainage facilities is determined as 12 cm/h. The weighted coefficient of run-off is 0.9. Determine the run-off to be drained by a surface drainage system, with gutters provided at every 100 m intervals for a runway pavement which is 150 m wide.

**Solution**

$$Q = \frac{C \cdot i \cdot A}{36}$$

Here,

$$C = 0.9$$

$$i = 12 \text{ cm/h}$$

$$A = 100 \times 150 \text{ sq. m}$$

$$= 1.50 \text{ ha}$$

∴ Run-off discharge for design

$$Q = \frac{0.9\times 12\times 1.5}{36} \text{ m}^3/\text{s}$$

$$= \underline{0.45 \text{ m}^3/\text{s}}$$

## EXAMPLE 13.3

The design discharge for draining surface run-off at an airport site is 0.08 m³/s. Using Manning's formula for the velocity of flow in the drain pipe, determine the necessary diameter of a concrete pipe, if the slope of the drain is 1 in 100. Take Manning's $n$ as 0.015.

**Solution:**

Area of pipe section,

$$A = \frac{Q}{V}$$

$$V = \frac{1}{n} \cdot R^{2/3} \cdot S^{1/2}$$

Assuming the pipe diameter as 30 cm,

$$R = \frac{A}{P} = \frac{\pi D^2}{4 \times \pi D}$$

$$= \frac{D}{4} \text{(assuming the pipe to run full)}$$

$$= \frac{30}{4} = 7.5 \text{ cm}$$

$$= 0.075 \text{ m}$$

$$S = \frac{1}{100}; \quad n = 0.015$$

$$\therefore \quad V = \frac{1}{0.015} \times (0.075)^{2/3} \cdot \left(\frac{1}{100}\right)^{1/2}$$

$$= 1.2 \text{ m/s}$$

$$\therefore \quad A = \frac{Q}{V} = \frac{0.08}{1.20}$$

$$= 0.0667 \text{ m}^2$$

$$= 667 \text{ cm}^2$$

Area of section of 30 cm² diameter pipe

$$= \frac{\pi}{4} \times 30^2 = 706 \text{ cm}^2$$

This is more than the section area needed.
Hence, a drain pipe of 30 cm-diameter is adequate.

## EXAMPLE 13.4

A pipe drain with circular perforations of 6 mm diameter is to be designed. The soil gradation is given below. Design the filter material.

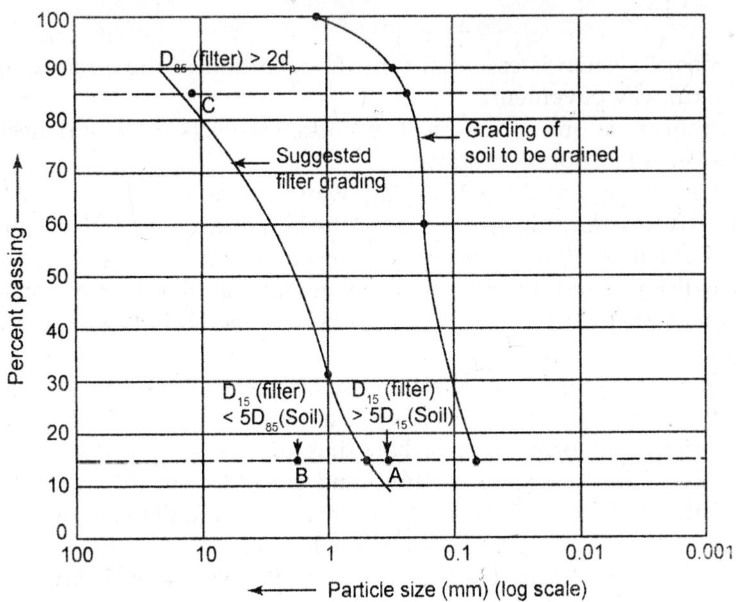


**FIG. 13.8** *Filter design for subsurface drain (Example 13.4)*

| Sieve size | 1.18 mm | 425 μ | 300 μ | 150 μ | 75 μ |
|---|---|---|---|---|---|
| Percent passing | 100 | 90 | 85 | 60 | 15 |

**Solution:** The grading of the soil material to be drained is plotted as shown in Fig. 13.8:

$$D_{15}\text{ (soil)} = 75\alpha = 0.075\text{ mm}$$

∴ From permeability criterion,

$$D_{15}\text{ (filter)} > 5 \times 0.075\text{ mm}$$

or

$$0.375\text{ mm (Point-A)}$$

$$D_{85}\text{ (soil)} = 300\alpha = 0.3\text{ mm}$$

∴ From piping criterion,

$$D_{15}\text{(filter)} < 5 \times 0.3\text{ mm or } 1.5\text{ mm (Point-B)}$$

Size of perforation in the pipe,

$$d_p = 6\text{mm}$$

∴ From prevention of clogging

$$D_{85}\text{ (filter)} > 2 \times 6\text{ mm}$$

or

$$12\text{ mm (Point-C)}$$

The suggested grading for the filter material is sketched smoothly such that it lies to the left of C and in between A and B, as shown.

## SUMMARY

- An efficient drainage system is essential for safe operation of the aircraft, as also for the durability of the runway pavement.
  Thus, airport drainage is one of the most important aspects of airport design to ensure efficiency and safety of airport operation.
- Airport grading is a bigger problem than highway grading since the width of runway is relatively large. This warrants drainage in both the longitudinal and lateral directions.
  The main requirement is that the earthwork in cut-and-fill should balance each other; further, haulage distance and the quantity of earth material to be moved must be optimal.
- The computation of earthwork is done using either the method of cross-sections or that of mass-haul diagram.
  In the method of cross-sections, cross-sectional areas of the earthwork at constant intervals are determined and the volume of earthwork is computed using either the trapezoidal formula or the prismoidal formula. Mass-haul diagram is a graphical representation of the amount of earthwork in excavation (cutting) and embankment (fill), and the program of earth material haulage. Free haul distance, the economic haul distance, as well as a host of other information can be obtained from the mass-haul curve.
- The primary objectives of airport drainage are to prevent ponding of runways to facilitate safe operation of aircraft and to prevent loss of subgrade support.
- The important characteristics of an efficient airport drainage system are that it should take into account the extensive area involved, heavy wheel loads of aircraft, and flat gradients.
  Data relating to the topography of the site and the meteorological details of the area are needed for the design of airport drainage facilities. Broadly speaking, the drainage is divided into surface and sub-surface drainage.
  Draining of the surface water that collects on the runway pavement and adjoining areas is known as surface drainage.
  Time of concentration (the sum of inlet time and time of flow) and rainfall intensity-duration-frequency curves for the area provided by the Meteorological Department are needed for the computation of *run-off*, or the fraction of rainfall that runs on the surface.
  There is also a factor called the *coefficient runoff*, which influences run-off; this coefficient depends on the nature of the surface.
  Run-off is computed using the formula $Q = \frac{C \cdot i \cdot A}{36} (\text{m}^3/\text{s})$,
  C being the coefficient of run-off; *i*, the intensity (critical value) of rainfall; and A being the area drained. The design of drains is based on the formula for area of flow in the drain $\left(A = \frac{Q}{V}\right)$ and Manning's formula for the velocity of flow $\left(V = \frac{1}{n} \cdot R^{2/3} \cdot S^{1/2}\right)$.
- *Ponding* is a means of temporary storage of run-off prior to its entry into the underground drainage system. Occurrence of ponding facilitates economy in allowing smaller pipes for secondary runways for lower discharge values.
- Subsurface drainage aims at preventing water seeping into the ground and weakening the subgrade.
  Intercepting ground water before it reaches the subgrade is achieved by the provision of longitudinal subsurface drains which carry the infiltration to a suitable outfall at a safe distance.

Groundwater lowering is also achieved by means of longitudinal drain trenches and pipes, as also by means of longitudinal and transverse drains.

Subsurface drains can control seepage flow in cut slopes also.

Capillary cut-off, which may be in the pervious stratum or impermeable bitumen layer, can prevent capillary rise in clayey subgrades.

- Drain pipes, which may be porous or perforated, are used for draining off water. Granular filters are designed for permeability of filter, prevention of piping, and prevention of clogging of drain pipes.

## QUESTIONS AND PROBLEMS

1. Explain the importance of proper grading of an airport site, bringing out the differences between highway grading and airport grading.
2. Explain briefly the two methods used for earthwork computations.
3. What is a Mass-haul diagram? What are its characteristics and uses?
4. What are the functions and basic requirements of airport drainage?
5. What are the special characteristics of airport drainage? What are the broad categories of airport drainage?
6. Explain briefly the procedure for estimating runoff.
7. What are the objectives of subsurface drainage?
8. What are the methods of lowering the ground water table with respect to airport drainage? Explain with sketches.
9. Explain the function of capillary cut-off in the protection of the subgrade for an airport pavement from getting weakened in the presence of water.
10. Explain the procedure for the design of drain pipes for the surface drainage of an airport.
11. Explain the method for controlling seepage flow in cut slopes in airport drainage.
12. What are the requirements for the design of a suitable filter material? How is its gradation determined?
13. Write short notes on:
    (a) Time of concentration
    (b) Coefficient of run-off
    (c) Ponding and its role in surface drainage.
14. Differentiate between
    (a) Time of entry and time of flow
    (b) Porous pipes and perforated pipes
    (c) Piping criterion and clogging criterion.
15. The rainfall intensity at an airport site for design of drainage is 90 mm/h. The weighted coefficient of runoff is 0.63. Determine the runoff to be drained by a surface drainage system, with gutters at 120 m intervals for a runway pavement which is 180 m wide.

16. The design discharge for draining surface runoff at an airport site is 0.06 $m^3/s$. Using Manning's formula for the velocity of flow in a pipe, determine the necessary diameter of a concrete pipe. Given the slope of the drain is 1 in 120 and Manning's $n$ is 0.016.

17. With the following data on nature of the surface, area and the runoff coefficient of the catchment area at an airport site, determine the weighted coefficient of runoff:

| | | |
|---|---|---|
| Paved area | 2.00 ha | 0.95 |
| Built-up area | 4.00 ha | 0.60 |
| Unpaved area | 1.00 ha | 0.30 |
| Loamy area | 2.00 ha | 0.30 |
| Sandy area | 0.50 ha | 0.20 |
| Lawns | 0.50 ha | 0.15 |

Chapter 14

# VISUAL AIDS AND AIR-TRAFFIC CONTROL

**CHAPTER OBJECTIVES**

This chapter provides an insight to:

- The use of visual aids such as indicators and markings during the day and lighting during the night to enable operations for safe landing
- The use of Visual Flight Rules and Instrument Flight Rules for safe pilot operations
- Air-traffic control aids such as airway beacons, radio range, and radars that help en-route safety of aircraft
- Landing aids, namely, Instrument Landing System (ILS) and Microwave Landing System (MLS)
- Free flight – a new concept in air-traffic control in which the responsibility of safe flight rests with the pilot, with help from computer systems.

## 14.1 INTRODUCTION

Visual aids are required at an airport to help the pilot of an aircraft land or take-off safely, during both day time and night time. Further, such visual aids are useful for safe operations in unfavourable weather conditions.

Certain critical locations such as runway centre line, runway edges, and runway threshold have to be properly identified by markings and lighting for day and night operations, respectively.

Air-traffic control involves the guidance to be provided to aircraft pilots from the time of take-off to the time of landing at the destination; this is a relatively complex problem. The devices and equipment needed to accomplish air traffic control are usually housed in the Air-Traffic Control tower provided at every airport.

When sophisticated and modern equipment and instruments are used, it is called the *instrument-landing system* (*ILS*); otherwise, it is called *non-instrument landing system* (*Non-ILS*).

## 14.2 VISUAL AIDS

Visual aids are devices, apparatus /equipment, which are meant to help the pilots of aircraft in sighting the various features in and around an airport, so that they land or take-off in a safe manner, either during the day or during the night, and under all weather conditions.

Broadly speaking, visual aids are of the following categories:

1. Indicators or signalling devices
2. Airport markings
3. Airport lighting

While *airport markings* are meant for day-time operations and *airport lighting* for night-time operations, indicators or signalling devices are meant to aid both.

## 14.3 INDICATORS OR SIGNALLING DEVICES

There are two indicators – *wind direction indicator* and *landing direction indicator*. Both are generally meant to aid landing of aircraft and are placed in a large segmented circle near its centre.

### Wind-direction Indicator

This indicates the direction from which the wind blows at the site of an airport. It can be a wind cone, a wind sock, a wind tetrahedral-pyramid. It should be located at a conspicuous place away from the buildings such that it is visible from a height of 300 m.

#### Wind Cone

A typical wind cone (Golland Co.), approved by FAA and ICAO, is briefly described below:

The wind cone facilitates visual identification of direction and speed of wind at airports. When wind speed is at least 6 km/h, the revolving system situated at the top of a pole on bearing will turn in the direction of this movement. The top hole of the wind slope takes in the air stream and the sleeve swings. At a wind speed of at least 28 km/h, the sleeve inflates and indicates visually the wind direction and speed.

The wind cone is made from a mixture of cotton and synthetic material. Colour bands in two different colours are provided for contrast. When inflated by air it takes the shape of a truncated cone.

The shell of the cone keeps the entry hole of the cone open even when there is absolutely no wind. It supports the cone and keeps it in the required shape at a distance of three-eighths of its total length. A structure made of Aluminium tubes prevents the accumulation of water in the cone.

The pole is a structure with support that can be tilted; the top part contains a revolving mechanism that enables rotation of the shell with the cone.

Lighting is provided by light reflectors that facilitate identification during night-time.

The important dimensions are

| | |
|---|---|
| Height | 6500 mm |
| Wind sleeve length | 3.75 m |
| Diameter | 0.90 m |
| Accuracy of wind direction at minimum wind speed | ± 5° |

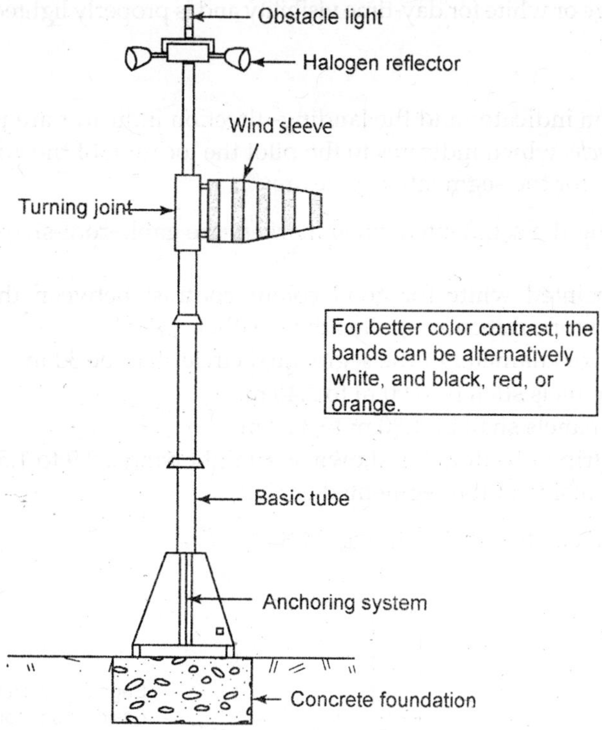


**FIG. 14.1** *Schematic view of a wind cone*

The bottom of the pole is fixed to a concrete foundation by means of metal plates and bolts. A schematic view is shown in Fig. 14.1.

### Wind Sock

It is a conical textile tube, which resembles a giant sock, designed to indicate wind direction and relative wind speed.

Wind direction is the opposite of the direction in which the wind sock is pointing. In low winds, the wind sock drops, and in high winds it flies horizontally.

A 6 km/h wind breeze will cause the wind sock to orient itself in the wind direction, and a 28 km/h wind will fully extend it.

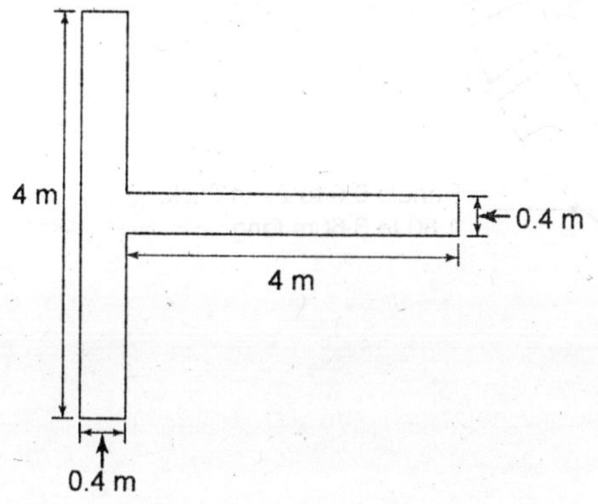


**FIG. 14.2** *Tee-type Landing-direction indicator*

## Landing-direction Indicator

This is in the form of a *Tee* or a *Tetrahedron*. Its purpose is to indicate to the pilot the direction of the active runway of the airport. It is placed at the centre of a segmented circle along with the wind-direction indicator, at a distinct location away from the threshold of the runway.

A typical landing-direction indicator of the Tee-type is shown in Fig. 14.2.

This is painted orange or white for day-time visibility and is properly lighted for night visibility.

## Segmented Circle

Both the wind-direction indicator and the landing-direction indicator are placed at the centre of a large *segmented circle*, which indicates to the pilot the location of the airport.

The *ICAO guidelines* for the segmented circle are:

(i) The panels forming the segmented circle markers are gable-roof-shaped, with a pitch of at least 1 to 1.

(ii) The panels are painted white for good colour contrast between the marker and the surroundings, and for protection against the weather.

(iii) The minimum inside diameter of the segmented circle shall be 30 m.

(iv) The width of the panels shall be 0.90 m to 2.40 m.

(v) The length of the panels shall be 1.80 m to 3.60 m.

(vi) Airport landing strip indicators are shown as straight bands, 0.9 to 1.5 m thick and up to 5 m long, on the outside of the segmented circle.

A typical segmented circle is shown in Fig. 14.3:

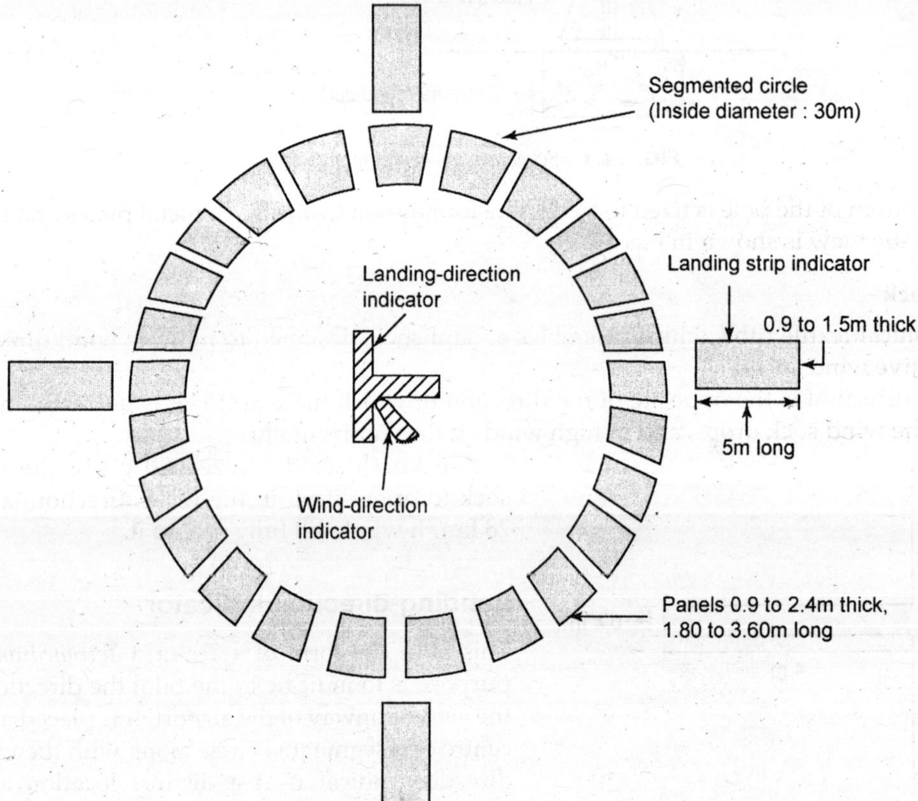


**FIG. 14.3** *Segmented circle with indicators*

## 14.4 AIRPORT MARKINGS

Airport markings are divided into the following groups:

(a) Runway markings
(b) Taxiway markings
(c) Runway and taxiway shoulder markings
(d) Apron markings

### Runway Markings

Runway markings comprise the following:

(i) Threshold marking
(ii) Touch-down or landing zone marking
(iii) Centre-line marking
(iv) Edge-strip marking
(v) Marking of runway designation or numbering.

**Threshold marking:** The threshold of a runway is marked by a series of parallel lines, starting from a distance of 6 m from the end of the runway, as shown in Fig. 14.4.

At some airports, where small obstructions like water courses exist near the runway end, threshold markings are *displaced* by a distance of about 180 m to prevent damage by undershooting while landing. However, the whole length of the runway may be utilised for take-off operations even in this case.

**Touch-down or landing zone marking:** Touch-down or the landing zone is marked by a series of stripes arranged symmetrically with respect to the centre line, with the number of stripes decreasing gradually in the direction of landing. The stripes are 1.80 m wide at 1.50 m clear spacing, and are 22.5 m long. A typical set of markings is shown in Fig. 14.5.

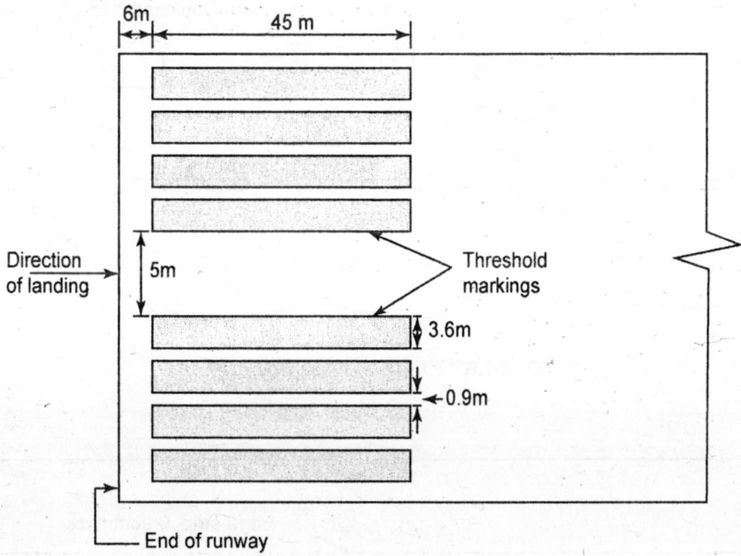


**FIG. 14.4** *Threshold marking of runway*

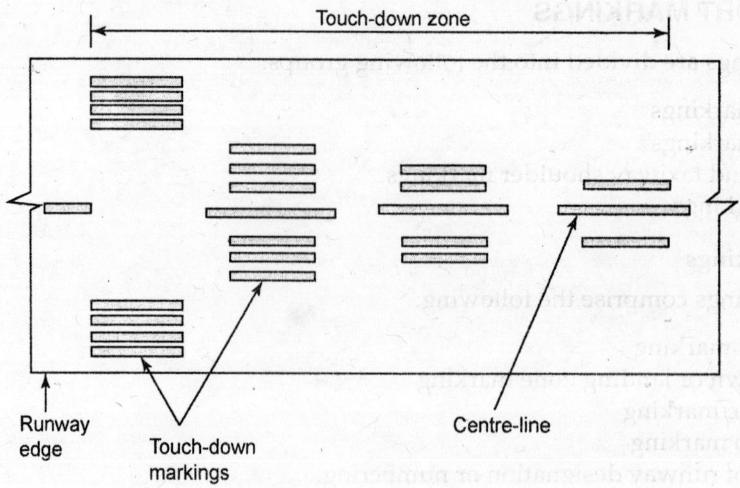


**FIG. 14.5** *Touch-down or landing zone marking*

**Centre-line marking:** The centre-line of a runway is indicated by a broken strip all along the length of the runway. The strip is 0.90 m wide. A typical centre-line marking is shown in Fig. 14.6.

**Edge-strip marking:** This is provided usually when the runway width is 45 m or more; it is a continuous strip marked near the edge of the runway, the thickness being 0.90 m, as shown in Fig. 14.7.

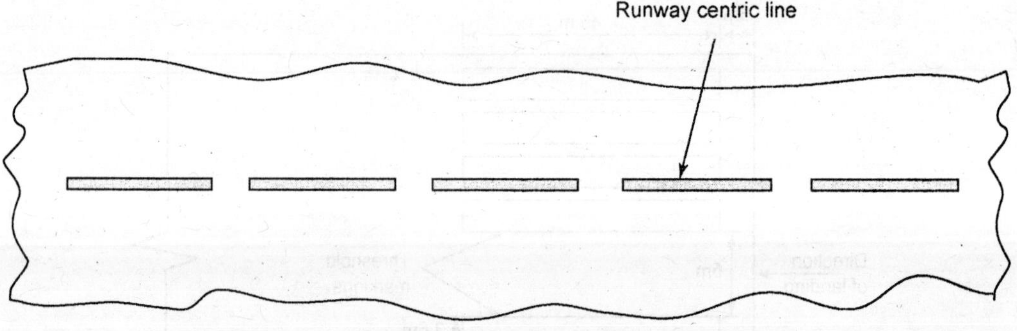


**FIG. 14.6** *Runway centre-line marking*

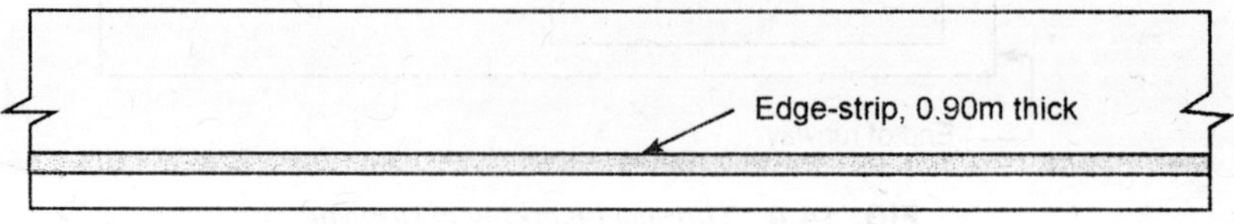


**FIG. 14.7** *Runway edge-strip marking*

**Marking of runway designation or numbering:** Each end of a runway is marked with a number which indicates (in tens of degrees) the magnetic whole-circle bearing or the angle measured clockwise from the north of the runway in the direction of landing; thus, the east end of an East–West runway would be marked 27 (270°) and the west-end 9 (90°), as shown in Fig. 14.8.

This runway is then called 9/27, because its number is 9 while landing or taking-off eastward, and it is 27 while landing or taking off westward. (These numbers are rounded off to the nearest integer depending on the bearing, and they always differ by 18.)

Two-intersecting runways – East–West and North–South – will be numbered as shown in Fig. 14.9.

*Parallel runways* are additionally marked with letters as given below:

| | |
|---|---|
| Two parallel runways | L and R (Left and Right) |
| Three parallel runways | L, C and R (Left, Centre and Right) |
| Four parallel runways | L, LC, RC and R(Left, Left of centre, Right of centre, and Right) |

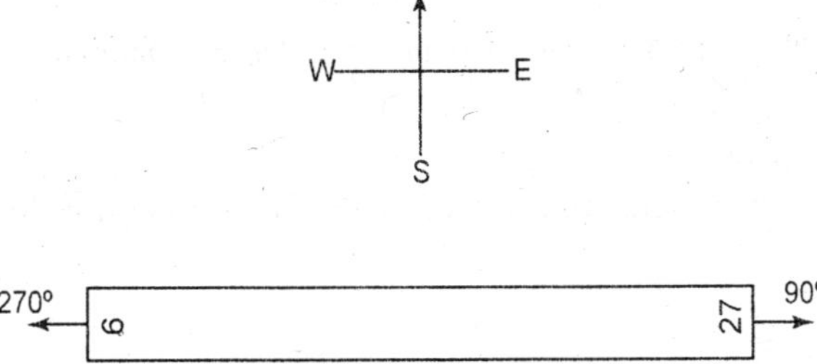


**FIG. 14.8** *Runway numbering of an East–West Runway*

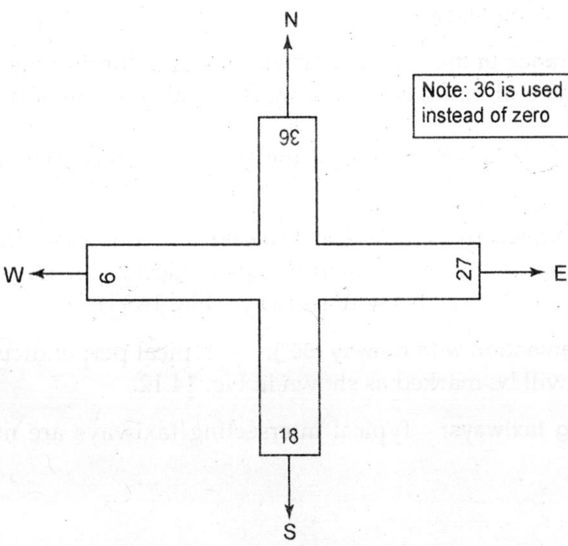


**FIG. 14.9** *Numbering of two intersecting runways*

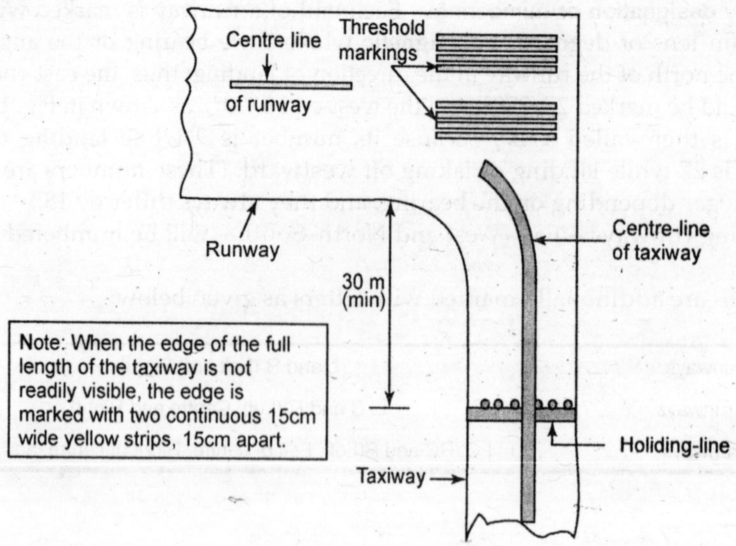


**FIG. 14.10** *Taxiway entrance at the end of a runway*

Sometimes, single digit numbers are prefixed with zero – for example, 09 for 9.)

### Taxiway Markings

These comprise the following:

(i) Markings for taxiway entrance at the end of runway
(ii) Markings for taxiway entrance to runway
(iii) Markings for taxiway intersection with runway (90°)
(iv) Markings for intersecting taxiways

**Markings for taxiway entrance at the end of runway:** A single continuous 15 cm yellow stripe, with a special durable type of paint, is used to mark the centre-line of the taxiway, as shown in Fig. 14.10, above.

It may be observed that the centre-line of the taxiway is terminated at the edge of the runway.

**Markings for taxiway entrance to runway (exit taxiway):** In this case, the centre line of the taxiway is extended to the centre-line of the runway. A holding-line marking is painted at all the intersections of the paved taxiways with runways (Fig. 14.11).

**Markings for taxiway intersection with runway (90°):** A typical perpendicular intersection of a taxiway with a runway will be marked as shown in Fig. 14.12.

**Markings for intersecting taxiways:** Typical intersecting taxiways are marked as shown in Fig. 14.13.

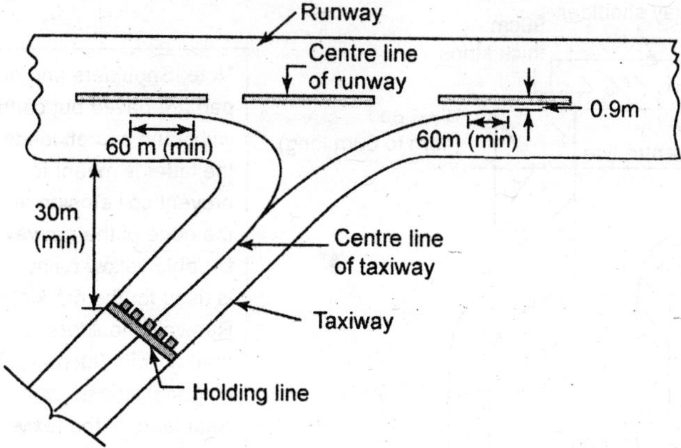


**FIG. 14.11** *Taxiway entrance to runway*

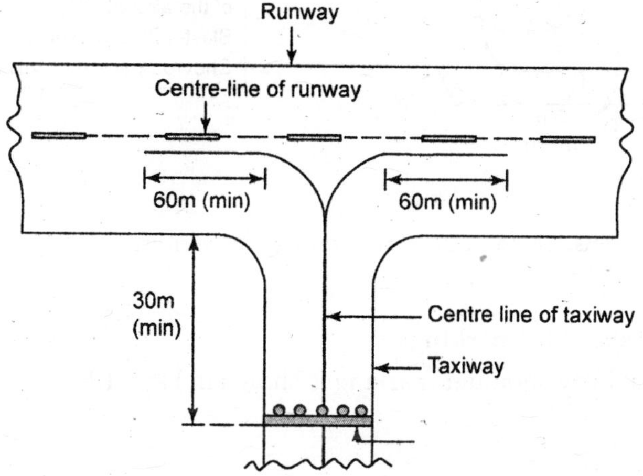


**FIG. 14.12** *Perpendicular intersection of a taxiway with a runway*

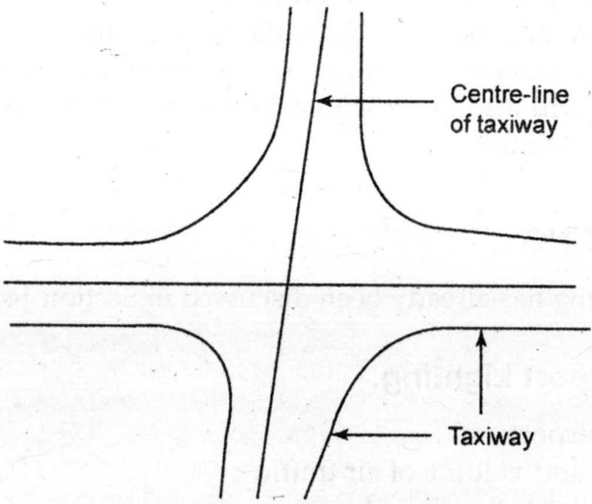


**FIG. 14.13** *Intersecting taxiways*

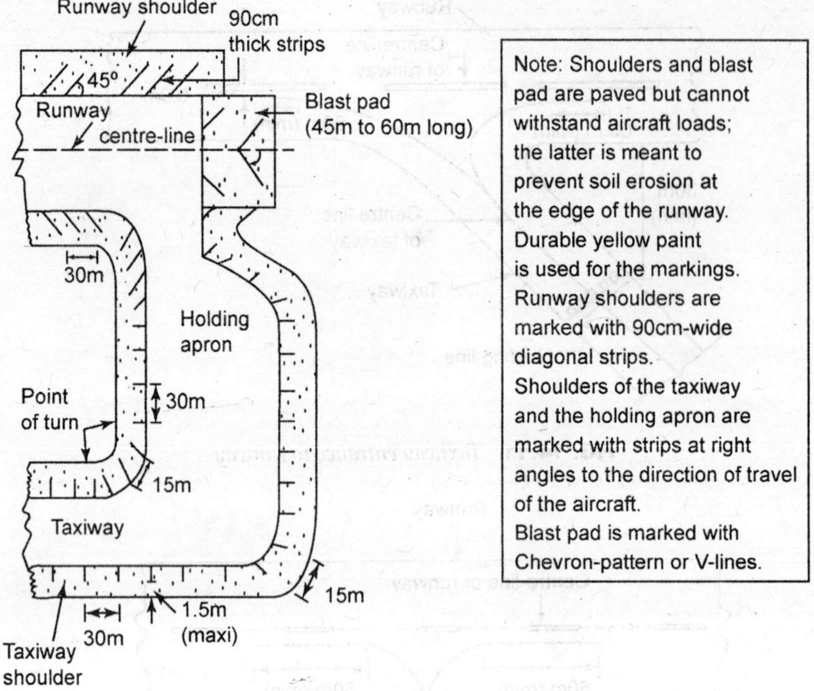


**FIG. 14.14** *Runway and taxiway shoulder markings*

**Runway and taxiway shoulder markings**

Typical runway and taxiway shoulder marking is shown in Fig. 14.14.

**Apron Markings**

Guide-line strips are painted on the apron to help the pilot in manoeuvring the aircraft on the apron and to stand it in the appropriate slot at the gate. Durable and fuel-resistant yellow paint is used for the markings, since there could be fuel spillage on the aprons.

The guide-line strips are painted to indicate the path of the nose gear of critical or large aircraft. Smaller aircraft can be easily manoeuvred along the same path without difficulty.

Appropriate markings will also be provided in the *aircraft standing area* on the apron, in accordance with ICAO guidelines.

## 14.5 AIRPORT LIGHTING

The need for airport lighting has already been discussed in Section 14.1.

### Factors Governing Airport Lighting:

(i) Classification of the airport
(ii) Nature of the aircraft and volume of air traffic
(iii) Type of night operations at the airport.
(iv) Nature of the surfaces of runway, taxiway and aprons

(v) Weather conditions
(vi) Availability of power

For the sake of uniformity of guidance to the pilots, the ICAO guidelines on the arrangement and colour of lighting have to be followed.

Maintenance of airport lighting is another important aspect; it requires a lot of effort to maintain the different kinds of lights, especially at major airports.

Emergency power supply in case of power failure is also an essential feature at major airports with heavy volume of air traffic.

## Components of Airport Lighting

(a) Airport beacon
(b) Code beacon
(c) Lighting of wind-direction indicator
(d) Landing direction indicator lighting
(e) Approach lighting
(f) Boundary lighting
(g) Threshold lighting
(h) Runway lighting
(j) Taxiway lighting
(k) Apron and hangar lighting
(l) Airway beacon
(m) Obstruction lights

**Airport beacon:** A beacon is a strong beam of light which indicates the approximate location of the airfield which is equipped with facilities for night operations. This should be mounted higher than any obstruction, usually over the top of the terminal building; if necessary, it can be mounted on a separate tower. The beacon rotates about the vertical axis, giving out green and white flashes in horizontal directions 180° apart. The speed of rotation is six revolutions per minute.

**Code beacon:** This is mounted at such a height that the beam clears all obstructions. Two 500-watt bulbs are provided with a green colour screen. The code beacon continuously flashes a Morse code signal designating the airport.

**Lighting of wind-direction indicator:** The indicator is illuminated with suitable lighting arrangement such as four 200-watt angle reflectors placed at 1.80 m above the top of the cone. This type of arrangement enables the pilot to discern the wind direction during night time and even in bad weather conditions (Refer Fig. 14.1).

**Lighting of landing-direction indicator:** This indicator is also appropriately lighted so that it is visible to an approaching pilot during night time.

**Approach lighting:** This is a sequence of high-intensity lighting arrangement for a distance of 900 m, before the runway actually begins. These are normally mounted on pedestals such that the lights are at the same level; if they are accidentally hit, they give way without damaging the aircraft.

The ICAO system of approach lighting is shown in Fig. 14.15.

**Boundary lighting:** The entire landing area of the airport is marked by boundary lights spaced 90 m apart. If fencing is also provided all along the boundary of the airfield, the boundary

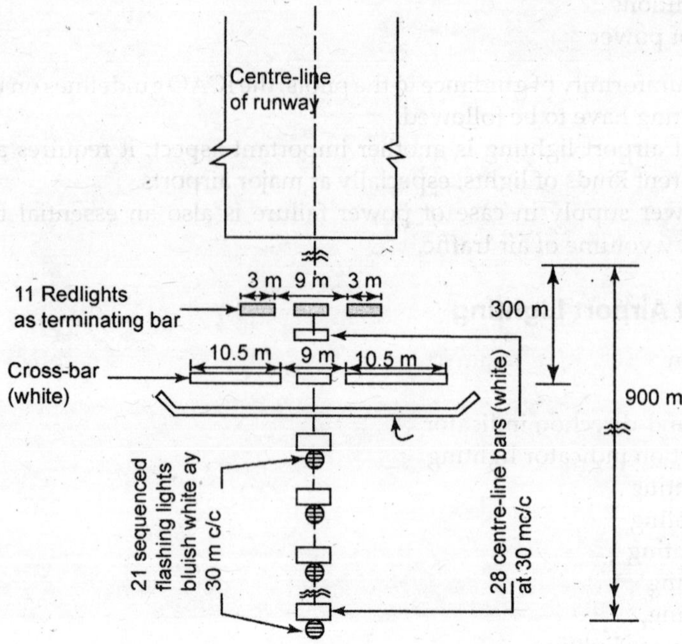


**FIG. 14.15** *ICAO system of approach lighting*

lights should be placed 3 m inside the fencing. For indicating hazardous approach, boundary lights in the region are red in colour.

**Threshold lighting:** Identification of the threshold is important for the pilot to decide whether to land or not. Therefore, special treatment is given to threshold lighting.

The end of approach lighting is marked by a terminating bar of red lights, placed 60 m ahead of the runway threshold.

The threshold wing bars, placed 30 m ahead of the runway, serve as a distance marker. At large airports, the threshold itself is lighted with a continuous bar of green lights for the entire width of the runway; this is placed at a distance not more than 3 m from the runway end.

At small airports, the threshold is indicated by four lights on each side of the runway. The lights are of the semi-flush type, protruding not more than 12 cm above the surface at large airports. However, in small airports, they are of the elevated type and protrude more than 12 cm. A typical treatment of threshold lighting is shown in Fig. 14.16.

**Runway lighting:** Soon after crossing the threshold, the pilot completes the touch-down and rolls the aircraft on the runway. The lighting of the runway should provide guidance to the pilot on alignment, lateral displacement, roll and distance. The visual pattern of the lighting should be easily interpreted by the pilot.

For night landing, the earlier practice was to floodlight the entire landing area. But now floodlighting is restricted to indicate the preferred landing direction.

The runway width is marked by edge lights. As the pilot is about to touch the runway, the central area appears excessively dark. This is known as the *black-hole effect*. Increased intensity of edge lights could not eliminate this effect.

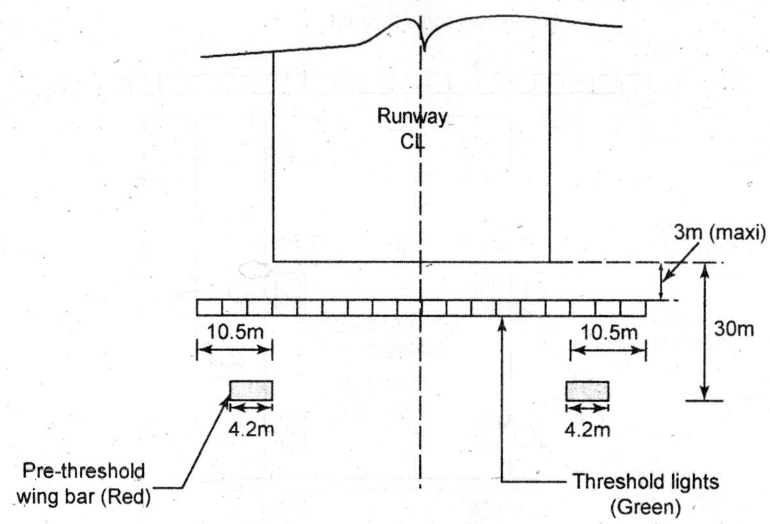


**FIG. 14.16** *Typical threshold lighting*

A more precise design of runway lighting, called *narrow-gauge pattern*, has been devised and is now universally adopted at all major airports. This design lights up the central portion of the runway. A typical pattern of this lighting system is shown in Fig. 14.17.

The narrow-gauge pattern consists of a channel of light, 18 m wide, up to 1140 m from the threshold; beyond this, closely-spaced lights are placed along the centre-line of the runway up to the other end.

All the runway lights are white and of 'flush' type, not protruding more than 1 m above the surface. Runway edge lights are of elevated type and white in colour. In the case of runways at airports using *Instrument Landing System (ILS)*, the lights in the last 600 m facing the pilot are of yellow colour, indicating the caution zone.

**Taxiway lighting:** The design considerations for taxiway lighting are the following:

(i) The lighting of a taxiway should be such that it is distinctly different from that of a runway.
(ii) Adequate guidance should be there all along the taxiway.
(iii) Exit taxiway should be lighted such that the pilot can locate it 360 m to 450 m ahead of the turning point.
(iv) Intersection of taxiway and runway should be clearly identifiable.
(v) While the taxiway centre lights are green, edge lights are blue.

A typical pattern of taxiway lighting is shown in Fig. 14.18.

The spacing of lights is reduced on curves depending upon the sharpness (or the radius) of the curve. On exit taxiway, lights are placed on the centre line in order to identify the point of turn-off.

**Apron and hangar lighting:** These areas are usually flood-lit for convenience in servicing and loading. The height of the mounting should be such that it does not cause glare to the pilots, passengers, and service staff. It is usually not less than 12 m above the pavement.

**Airway beacons:** An air route along which navigation aids such as radio beams and beacon lights are maintained is known as an *airway*. Airway beacons produce narrow horizontal and

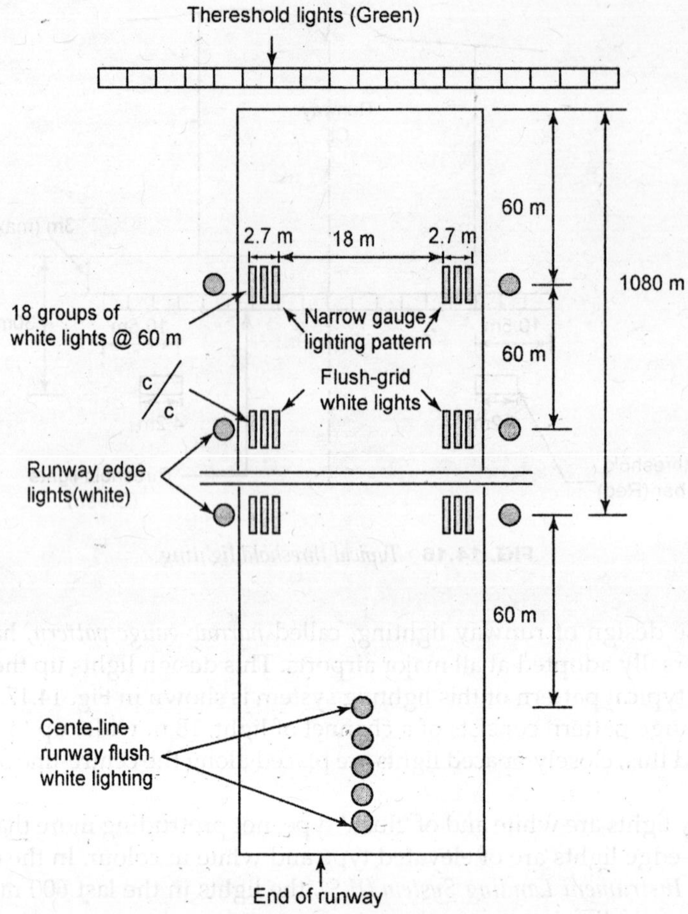


**FIG 14.17** *Narrow gauge pattern of runway lighting*

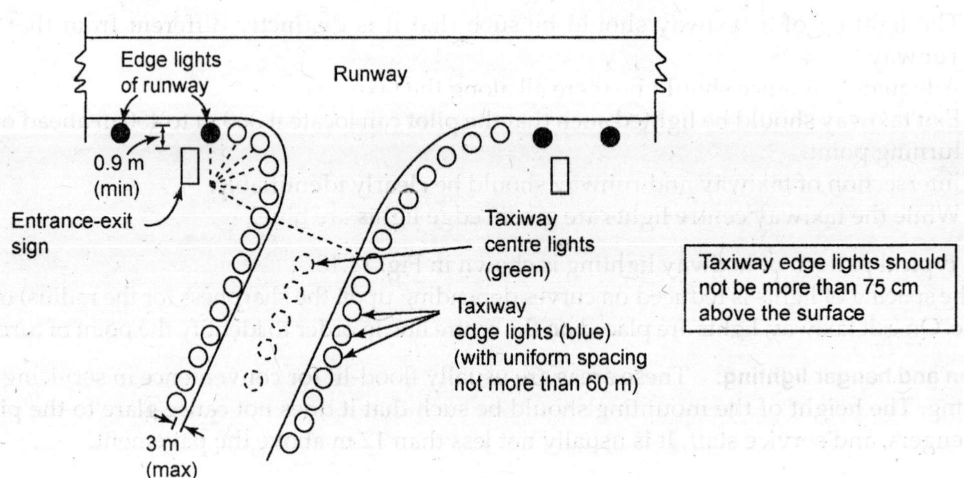


**FIG. 14.18** *Typical taxiway lighting pattern*

vertical beams of high intensity, which rotate at such a speed as to produce 12 flashes per minute for civil airports. The beacons are mounted on a high tower near the airport.

Airway beacons are in addition to ordinary airport beacons.

**Obstruction lights:** All objects/structures which are considered hazardous to aircraft operations in view of their height are marked by means of fixed, flashing, or rotating red lights or beacons. Such lights are known as *obstruction lights.*

They produce a horizontally uniform intensity distribution, and a vertical distribution to give the maximum range at the lower angles of 1.5° to 8°, from which a colliding approach is possible.

Obstruction lights are placed on all obstructions in and around an airport—beacon towers, buildings, flood light masts, wind-cone masts, radio towers, high trees, and any other such objects of hazardous heights.

## Visibility

*Visibility* is defined differently for day and for night in meteorological parlance. During the day, it is the distance from which a black circular target, subtending a visual angle of 1° (at the human eye) can be seen. At night, it is the distance from which a human can see a light of intensity 25 candela.

Visibility, according to meteorology, is associated with the height of the underside of a dense cloud above the ground; this height is called the *ceiling*. To determine the height of a cloud, a *ceiling projector* is used to produce a well-defined illuminated spot on the lower portion of the cloud at night, to serve as the reference mark.

## Runway Visual Range (RVR)

RVR is defined as the distance up to which high-intensity edge lights can be seen by a pilot; this is measured by a *transmissometer* by correlating the calibration with the intensity of the edge lights and the time of the day or night.

## Decision Height

Decision height is defined as the minimum height above the runway at which a pilot can make the decision to continue the landing operation (which is taken to be completed with the touch-down of the aircraft on the runway), or to abort it (or cut it off). In taking this decision, the pilot relies more on visual signals on the ground than on instruments in the cock-pit

Pilots are expected to avoid landing if they are unable to see sufficient number of visual signals from the decision height to ensure safety.

Visibility of the runway is classified primarily into four categories with respect to landing operations, based on the runway visual range (RVR) and the decision height – non-precision and precision categories -I, -II and –III; the last one is again subdivided into III a, III b and III c.

For category-III automatic landing facilities are required but roll-out after landing and taxiing to the terminal are manual.

For category-III b, automatic landing as also automatic roll-out capability is required.

For category III c, all the three operations—landing, roll-out and taxiing—must be automatic.

The recommended minimum values of RVR and decision height for all these categories are given below:

**Visibility**

| ILS-category | | Minimum (m) | |
|---|---|---|---|
| | | RVR | Decision height |
| Non-precision | | 720 | 75 |
| Precision category-I | | 540 | 60 |
| Precision category-II | | 360 | 30 |
| Precision category-III | III a | 210 | 0 |
| | III b | 45 | 0 |
| | III c | 0 | 0 |

## Landing Guidance to Pilots

Guidance and information are necessary for aircraft pilots to ensure safe landing.

The landing operation of an aircraft under good visibility conditions is shown schematically in Fig. 14.19.

The aspects on which the pilot needs guidance and information during the landing operation are the following:

(a) Alignment guidance
(b) Height information
(c) Distance estimation
(d) Roll guidance

### Alignment Guidance

It is essential for the pilot to know the position of the aircraft with respect to the centre-line of the runway. Most runways have lengths many times their width; this makes them appear like long narrow ribbon when viewed from a great distance and height.

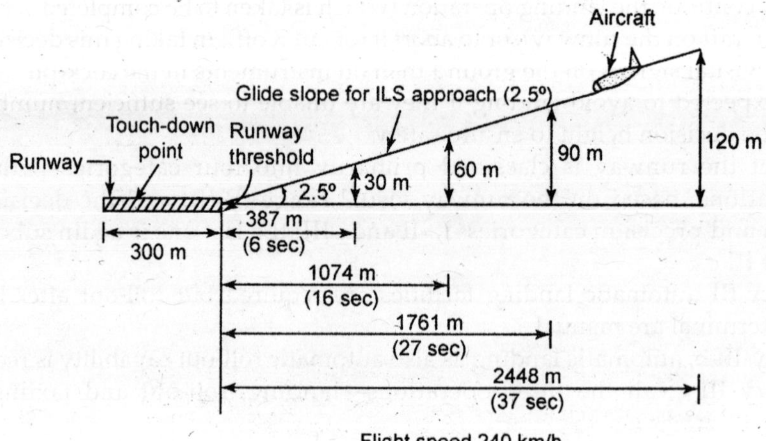


**FIG. 14.19** *Landing operation under good visibility conditions*

Airport marking of the edges and the centre-line of the runway during the day and airport lighting of these elements during the night are designed to develop good contrast with the surroundings and hence, provide the necessary alignment guidance to the pilot.

#### Height Information

A pilot tries to estimate the height of the aircraft while transiting from instrumental guidance to visual guidance. The perspective view of the runway, the ground reference points such as buildings and towers, and tall trees, if any, can be of help in this estimation. Incorrect estimation of the height can lead to accidents.

Well-designed airport markings during the day-time and airport lighting during night-time can provide adequate ground reference data to the pilot to enable him correctly estimate the height of the aircraft.

Of course, the best source of information about the height is the system of cock-pit instrumentation. However, this requires the availability of an Instrument Landing System (ILS) at the airport. Such a system is available only at large airports; in the case of smaller airports, airport markings and lighting are relied upon.

#### Distance Estimation

The pilot needs to estimate the distance of the aircraft from the runway for safe landing. During day time, under clear weather conditions, runway markings and the contrast with the features of the terrain in the vicinity will help the pilot in this regard.

During night time, and under unfavourable weather conditions, the pilot has to rely on the lighting arrangements with colour and intensity contrast. This, once again, emphasises the need for a standardised and well-designed airport markings with different-coloured paints and airport lighting with different types of colours and intensities, for day-operations and night operations, respectively.

#### Roll Guidance

The pilot needs information regarding the aircraft's *roll* or *altitude* to know if the aircraft is banked with respect to the ground surface. Roll is considered to be about the longitudinal axis of the aircraft. If the roll is not properly corrected, the lower wing of the aircraft may hit the runway pavement. Guidance to the pilot on the correct angle of roll may be obtained from the view of the horizon, terrain of the natural ground, and natural water surface, if any.

### Parameters Affecting Visual Range

A number of parameters influence how much distance or area of the ground can be seen by the pilot.

A few important parameters are:

(i) Cockpit cut-off angle, $\alpha$.
(ii) Pitch angle, $\beta$.
(iii) Glide – slope angle, $\Phi$.
(iv) Angle formed by the visual cone with the horizontal, $\theta$.
(v) Height of glide slope above the runway surface, $h$.
(vi) Visual range, $V$.
(vii) Horizontal segment of the visual range, $H$.

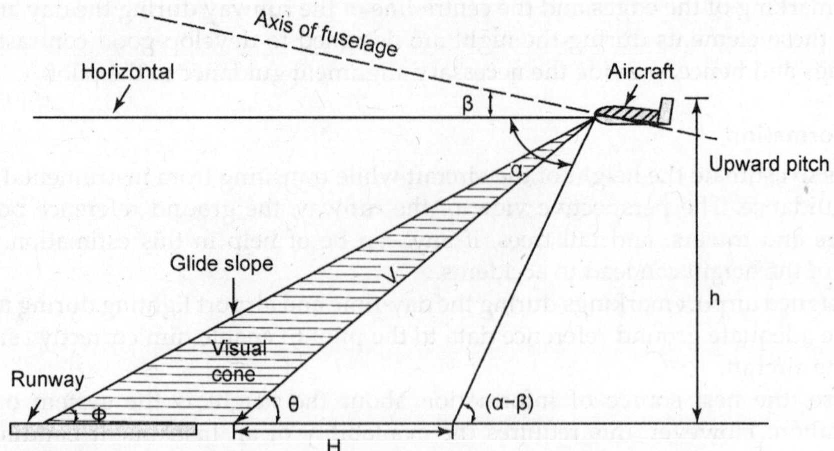


**FIG. 14.20** *Parameters affecting visual range of pilot during landing*

The *cockpit cut-off angle* is the angle between the longitudinal axis of the aircraft fuselage and the plane below which the pilots view is blocked by some part of the aircraft.

*Pitch angle* is the angle made by the longitudinal axis of the fuselage with the horizontal. Usually, the pilot makes the pitch angle negative or positive, as he attempts to land the aircraft. A standard *glide slope angle* is 2.5°.

The relationships among these various parameters can be understood from Fig. 14.20.

From Figure 14.20,

$$\text{Sin}\,\theta = \frac{h}{v} \tag{14.1}$$

$$H = V\cos\theta - h\cot(\alpha - \beta) \tag{14.2}$$

For a particular value of $V$, $H$ increases as $h$ decreases. A pilot usually requires a reaction time of 3 seconds to adjust the aircraft after seeing a visual aid. Thus for an aircraft speed of 240 km/h, the minimum horizontal segment of the visual range should be 200 m.

For $\Phi = 2.5°$ and $(\alpha - \beta) = 12°$, $H$ works out to 60 m when $h = 60$ m. But when $h = 30$ m, $H$ comes to 210 m. It is therefore obvious that the pilot cannot derive visual guidance from the approach lighting until a height of 30 m above the runway is reached – for the assumed conditions in this case.

## 14.6 AIR-TRAFFIC CONTROL

The control of air traffic involves giving guidance to the pilots of aircraft from the time of take-off to the time of landing at the destination. This is more *complex* than the control of road traffic *for the following reasons:*

- An aircraft has to be continuously in flight and cannot afford to stop in base. (Of course, the helicopter is an exception because it can hover in the air at a constant location as long as engine power is available.
- An aircraft lands with a gliding slope, governed by the altitude from which the pilot decides to land. When the visibility is poor and under bad weather conditions, the pilot needs information and guidance regarding alignment, height, and distance from the runway threshold.

In fact, the entire area, both in the air and on the ground, is alive with aircraft operation. Aircraft should follow the correct course in the air space, called the *airway*; this is absolutely necessary to avert the possibility of conflict with other aircraft in flight. No movement of aircraft or ground vehicles in the airport area is permitted without instructions from the Air Traffic Controller.

## Need for Air-traffic Control

The control of air traffic ensures safety, convenience and economy in the movement of aircraft from the airport of origin to that of the destination.

The *basic actions* involved in the flight are:

(i) Take-off from one airport.
(ii) Flight along the airway route at the appropriate altitude.
(iii) Safe navigation through the air route.
(iv) Safe landing at the destination airport.

### Primary Functions of Air-traffic Control Devices

**Traffic control at airport:** This consists of guiding the aircraft for landing or for take-off, and controlling the taxiing operations between the runway and the apron, and vice-versa.

**Airway traffic control:** This consists of the movement of aircraft along the designated air route with adequate separation distances laterally and vertically to avoid collision with other aircraft.

**Airway communication:** This involves the task of communicating the necessary information of flight parameters and weather conditions enroute to the pilot during the whole flight.

**Non-airway traffic control:** When personal flying is also taken into account, the problem becomes more complex and calls for proper regulation to prevent interference with the other flights in the route.

## Flight Rules

In order to handle aircraft movements safely and expeditiously, ICAO has laid down specific flight rules, which have to be followed mandatorily by all the member nations. It is also absolutely necessary for the pilot to remain continuously in touch with one of the controlling stations of the airport of origin or that of destination or with a *flight service station* located along the air route.

The two basic types of flight rules are:

1. Instrument flight rules (IFR)
2. Visual flight rules (VFR)

Differences between IFR and VFR

(i) VFR is applicable during good weather conditions and during the day.
(ii) In VFR conditions, the principle followed is 'see' and be 'seen'. IFR has to be applied only at times of emergency.
(iii) The responsibility for flight safety lies with the pilot in VFR, while it rests with the air-traffic control tower personnel in IFR.
(iv) Under VFR, the control is passive; under IFR, it is positive in insisting on adherence to the route and the altitude assigned.

The airspace prescribed for IFR is called *Positive Control Airspace*.

## 14.7 AIR-TRAFFIC CONTROL NETWORK

The pilot, prior to departure, prepares a flight plan which includes the air route to be followed, the desired flight altitude, the estimated time of departure, and the destination. Once the flight plan is approved by the control authority, no change is allowed without prior approval of the Air Traffic Control Centre.

The network for air-traffic control consists of the following:

1. Control within the terminal area
2. Control over airways
3. Airway communication

### Control within the Terminal Area

The Airport Traffic Control Tower (ATC) is located in the terminal area of the airport; its function is to monitor and control air traffic at the airport. The personnel in the ATC are responsible for issuing clearances to all aircraft for landing/take-off, and for providing information to the pilots about the wind direction, atmospheric temperature and pressure, and all other relevant data.

The ATC is located in such a position from where a view of the entire airfield is obtained; to facilitate this, the walls are fitted with inclined glass panels to minimise glare. It may also be coloured to reduce strain on the eye. Wipers are provided to maintain the panels clear even when it is raining. For a commanding view ATC may be provided over the terminal building atop a tower.

The ATC with approach control facilities directs the aircraft to the airport through a number of points, located within a radius of 40 km from the tower. Until this point, the aircraft are monitored by the *Air Route Traffic Control Centres (ARTCC)*, from which they are taken over by ATC tower of the particular airport.

If the traffic is heavy, the approaching aircraft are detained at this place, known as the *holding fix*; the aircraft are required to keep moving with a vertical separation of 300 m. Then they are released for landing, one by one, by the ATC.

If there are no approach control facilities, the detention and clearing of the aircraft for landing are carried out by the ARTCC, and the ATC takes control of the aircraft from the time they have started the landing approach.

### Control over Airways

Controlling the aircraft along the airways is the responsibility of the *Air Route Traffic Control Centres (ARTCC)*. Each of these centres has control over a definite geographical area; together, all such centres cover the entire country. At the boundary of the control area, the control of the aircraft is released and handed over to the adjacent ARTCC. Although the functions of the ARTCC are different from those of the ATC, they are generally located at major airports in order to eliminate the additional expenditure on communication facilities.

The ARTCC are primarily concerned with the control of the aircraft operating under the Instrument Flight Rules (IFR). Any change in flight plan of an aircraft has to be approved by the ARTCC.

The number of controllers at the ARTCC will be determined based on the volume of air-traffic to be handled, the degree of automation provided, and complexity of the aircraft mix.

In India, ARTCC are located at Delhi, Bombay (Mumbai), Calcutta (Kolkata), Ahmedabad, Allahabad, Madras (Chennai), Nagpur, and Jodhpur.

### Airway Communication

This is facilitated by *Flight Service Stations* (FSS), which are located at the airports and also along the airways. These service stations relay the air traffic control messages between the ARTCC and en-route aircraft. They also provide information to pilots on weather conditions, navigation aids, airports that are out of use, new facilities, and revised procedures.

If radar facility is available, the minimum separation between aircraft is 8 km; if radar coverage is not there, aircraft should be separated by a distance equivalent to a 10-minute run, while flying at the same altitude.

## 14.8 AIR-TRAFFIC CONTROL AIDS

Effective positive aids are required for air-traffic navigation and control. The aids may be external – those located on the ground, or internal – those installed in the cockpit of the aircraft. Aids for special needs such as flying over seas and oceans are also available.

For the sake of convenience, air-traffic control aids are broadly classified as follows:

1. Airway aids or en-route aids
2. Terminal aids or landing aids

### Airway Aids or Enroute Aids

During the flight from one airport to another, the following aids are available to the pilot:

1. Airway beacon
2. Low/Medium frequency (LF/MF) radio range
3. Very high frequency omni-directional range (VOR)
4. Air-route surveillance radar (ARSR)
5. Direction finder
6. Tactical air navigation (TACAN)
7. Distance-measuring equipment (DME)
8. Marker beacon
9. Air-to-ground communication

#### Airway Beacon

Airway beacon lights were the only means of air navigation in the past. They were placed at intervals of 40 km along the air route. They have become obsolete as more effective aids for air navigation and control have become available; however, a few beacons still exist on key locations such as hill-peaks.

#### Low/Medium Frequency (LF/MF) Radio Range

The invention of low/medium frequency (LF/MF) radio range revolutionised air-navigation aids.

The four-way LF/MF radio range emits radio beams along four directions, which may or may not be at right angles. Each beam is directed towards an airway; the facility may be provided at an airport. A pilot flying in an airway along which a radio beam is directed, hears a steady tone when they are on the right course; otherwise, a *dot-dash* or a *dash-dot* signal is heard. The pilot can correct the route so that a steady tone is heard. However, the position of the aircraft with respect to the transmitting station must be known. Another disadvantage is that radio frequency is susceptible to static interference, as a result of which its reception is totally absent during bad weather, such as a thunderstorm, when the need for an effective aid is felt the most.

### Very High Frequency Omni-directional Range (VOR)

Known as *VHF omni-range* or *VOR*, this sends out signals practically in all directions (360 courses or routes at 1°interval, radiating from the VOR station. This station can be located on a hill-top, open field, or any other exposed area; its range is up to 370 km. The VOR receiver in the cockpit of an aircraft is provided with a dial for tuning the desired frequency; it also has a course-line indicator to show whether the aircraft is to the left or right of the correct course. There is also an indicator to show whether the pilot is heading towards or moving away from the VOR station.

This aid is not susceptible to static interference and is effective even during a thunderstorm. Thus, it fully overcomes the disadvantages of LF/MF radio range.

### Air Route Surveillance Radar (ARSR)

*RADAR* is an acronym for *Radio Detection And Ranging*. A pulse of energy in the form of radio frequency is sent out by the transmitter and repeated a certain number of times per second. When the signals strike the object to be detected, a part of the energy is reflected in the form of an echo pulse which travels back to the antenna at the transmitter. The total time for the pulse to travel to the target and back as echo to the transmitter is a measure of the distance to the target. Thus, the transmitter acts as the receiver also in this case.

The air-traffic controller sees the aircraft's picture on the radar screen as a luminous image and gets an idea of its precise position; if necessary, the controller can intervene and convey relevant instructions to the pilots to ensure safe spacing of aircraft travelling at the same altitude. Long-range radars are installed along the airways, and can track the aircraft up to a range of 190 km. Such radars permit closer spacing of the aircraft flying at the same altitude, thereby increasing the capacity of the airways.

### Direction Finder

This may be a *Radio Direction Finder (RDF)* or an *Automatic Direction Finder (ADF)*.

In the case of an RDF, the pilot rotates it to find the direction of the aircraft with respect to the radio transmitter; the correct direction is indicated by a high-pitched tone. A *rull* or *blank* occurs when the RDF is at 90° to the direction of the transmitter.

The Automatic Direction Finder (ADF) is a more sophisticated device which keeps the antenna automatically pointing towards the transmitter, and thus needs no adjustment.

### Tactical Air Navigation (TACAN)

For satisfying the tactical and strategic needs of the military, an enroute navigation aid, known as the *Tactical Air Navigation (TACAN)* was developed by the U.S. Navy in the early 1950's. It operates on ultra-high frequency band and measures the distance and azimuth in one unit instead of two.

### Distance-Measuring Equipment (DME)

This equipment indicates to the pilot the air distance between the aircraft and the particular VOR station at which the ground-based transponder system is installed. This, coupled with the bearing details relative to the airway, gives the exact position of the aircraft to the pilot.

### Marker Beacon

This is used for determining the exact position of the radio range stations on a given airway. Directly above the tower of a radio range station, there is a *cone of silence* in the shape of an inverted cone of a limited area. As the aircraft enters this cone, signals fade out abruptly and remain blank for a few seconds until the aircraft leaves the cone at the opposite point, when the signals again reach high intensity. The cone of silence does not always provide reliable information.

This has led to the development of more precise and effective devices, called *Z-type markers*, which are installed practically at all radio range stations. They operate on ultra-high frequency (75 MHz) and emit an inverted cone-shaped signal directly above the range station. As an aircraft produces a visual indication of maximum brilliance to the pilot when it is directly over the station, the pilot simultaneously hears a steady high-pitched signal.

Both in the cone of silence and the Z-type marker, the duration of marker indication increases with increase in flight altitude. *Fan markers* are used as *position reporting markers* on the beams of certain radio ranges, at a distance of about 30 to 40 km from the range station. They operate on an ultra-high frequency (75 MHz) and send coded identification signals. When an aircraft passes over a fan marker, a small light comes up on the instrument panel in the cockpit, and a steady tone is also heard by the pilot.

### Air-to-ground Communication

*Air-to-ground communication* is necessary in order to provide the pilot information on weather and other flight instructions as the aircraft is flown in its path. Any changes in flight plan warranted by reasons such as bad weather, have to be approved by the Air Route Traffic Control Centre, and communicated to the pilot using this facility.

Similarly, two-way communication between the aircraft pilots and the Flight Service Stations also requires air-to-ground communication facilities. Thus, the importance of these facilities cannot be overemphasised.

## Terminal Aids or Landing Aids

The following aids are available to an aircraft while landing at an airport:

1. Airport-surveillance radar (ASR)
2. Precision approach radar (PAR) or Ground approach control (GAC)
3. Instrument landing system (ILS)
4. Airport surface detention equipment (ASDE)
5. Approach lights

### Airport Surveillance Radar (ASR)

This is an instrument which provides the operator at the control tower an overall picture of all aircraft within the airspace surrounding the terminal. This rotates through 360° and has a range of 50 to 100 km. The relative horizontal positions of the aircraft appear as 'blips' or luminous images on the screen with a luminous trail indicating the direction of movement and speed.

ASR, however, does not indicate the altitude of the aircraft as it simply responds to the reflection of the signal from the skin of the aircraft. ASR is also known as the 'Primary Radar'. In India, ASR is provided at Mumbai and Kolkata airports.

### Precision Approach Radar (PAR) or Ground Approach Control (GAC)

This was developed by the military during the Second World War as a mobile unit independent of airborne navigation equipment. It serves the same purpose as an ILS except that the information is passed on to the pilot through headphones. PAR gives to the air-traffic controller a clear picture of the descending aircraft (both horizontal distance and altitude) on the glide path and whether it is on the correct alignment and glide slope.

The instructions to the pilot are given by voice communication, and therefore the aid of air-borne navigation equipment is not required.

This system is advantageous for military aircraft since the fighter-type of aircraft have very little space for navigation equipment.

Pilots of commercial airlines prefer the ILS as the PAR places too much dependence on the controller in the ATC. At airports where both ILS and PAR are available, commercial pilots request that they be simultaneously monitored by both for extra safety.

### Instrument Landing System (ILS)

This is popularly known as the *blind landing system* and is the most widely used system for aircraft landing operations at all major airports across the world.

This system involves the use of telecommunication aids by the pilot for a successful landing under poor visibility and unfavourable weather conditions, when no ground reference data is useful. Where this facility does not exist, the aircraft has to be diverted to the nearest airport where favourable conditions exist or the ILS facility is available.

According to the ICAO, the visibility conditions are classified into *three* categories for ILS approach:

| | |
|---|---|
| Category-I | Weather conditions in which forward visibility is at least 800 m |
| Category-II | For landing when a forward visibility is at least 400 m |
| Category-III | For landing under zero visibility conditions |

The ILS consists of the following *two* types of signals using which an aircraft can approach for landing:

**Glide slope antenna:** This sends out a radio beam signal from a VOR at a vertical angle of 2° to 3° which is an ideal slope to reach out to all aircraft in the airport's vicinity. It is situated at a distance of 225 to 375 m down the runway from the threshold, and to one side of the runway centre-line at a distance of 120 m to 195 m. The glide slope beam allows a clearance of 15 m from the runway threshold to avoid the very edge of the runway.

**Localiser antenna:** This sends out a radio beam signal (from a VOR) in the vertical plane, and indicates to the pilot whether the aircraft is to the left or right of the correct alignment for approach to the runway.

It is located on the extension of the runway centre-line at about 300 m from the far end. The localiser transmitter is situated in a building located about 90 m to one side of the runway at the same distance from the end of the runway as the antenna.

A typical layout of ILS is shown in Fig. 14.21.

As a further help to pilots using ILS approach, *two low-power fan makers* are installed to enable them track their progress along the runway:

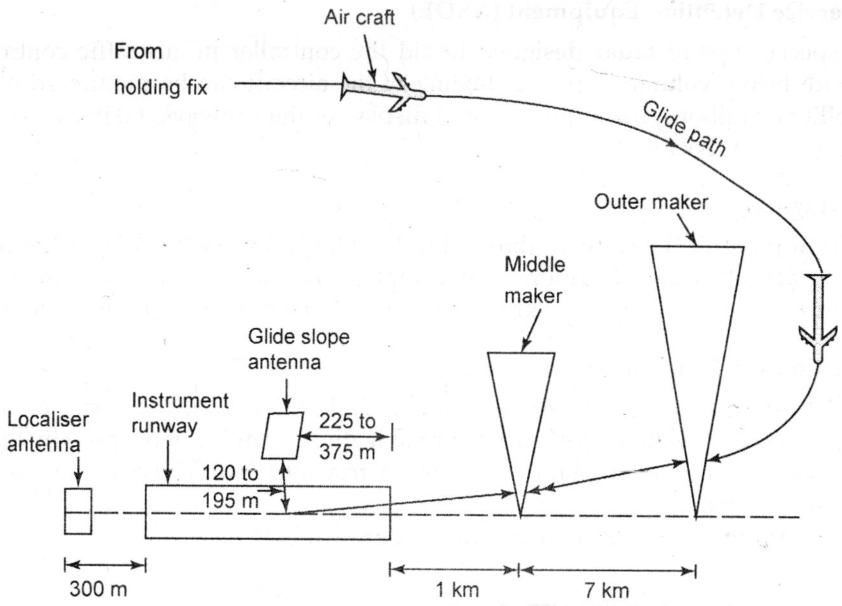


**FIG. 14.21** *A typical layout of an Instrument Landing System*

- *Middle marker*: This is placed 1 km ahead of the runway threshold and produces an amber light on the instrument panel in the cockpit.
- *Outer marker*: This is placed 7 km ahead of the runway threshold-produces a purple light on the instrument panel of the cockpit.

The signal lights are also accompanied by a high pitch tone to indicate to the pilot the location on the approach. The radio beam signals from the glide slope antenna and the localiser antenna show the position of the aircraft. The pilot centralises the aircraft along the localiser beam and descends along the glide slope. When the two beams intersect, it indicates the ideal direction and angle of approach.

When the aircraft crosses the middle marker, the pilot changes from the instrumental to the visual system and uses the visual aids to complete the landing operation safely.

The area adjacent to the antennas must be kept clear of obstructions such as buildings.

The ILS is not suitable to modern aircraft which use a steeper glide angle of about 7°.

**Microwave Landing System**

The limitations of the ILS are overcome by the *Microwave Landing System (MLS)*. The following are the *advantages of MLS*:

- While the ILS provides only one route on the runway, MLS provides for more routes within an area lying between 20° to 60° on either side of the runway centre-line.
- There is significant reduction of noise during landing.
- MLS is less susceptible to interference from surrounding objects.
- Continuous information about the position of the aircraft from the runway end is provided without the need for markers.
- The pilot has reasonable choice regarding the glide slope and the route to the runway.

**Airport Service Detention Equipment (ASDE)**

This is a special type of radar designed to aid the controller in air-traffic control at large airports with heavy volume of traffic. Taxiing of the aircraft can be monitored even under poor visibility conditions, using the pictorial display of the runways, taxiways, the terminal area and the aircraft positions.

**Approach Lights**

At the critical point of the runway threshold, the pilot has to switch over from instrument control to usage of visual conditions and complete the landing operation in the last few seconds with the help of approach lighting system and the runway lighting system.

**Air Traffic Control Through GPS**

The *Global Positioning System (GPS)* is a precise positioning tool, useful in air navigation. Although originally developed for help in air navigation for military purpose by the US Dept of defence, this is being extended for use in commercial aviation. (The details are out of scope of the present treatment).

*Automation* is the modern trend in air traffic control aids.

## 14.9 FREE FLIGHT AIR-TRAFFIC CONTROL

*Free flight* is a new concept in air-traffic control. This eliminates the need for air-traffic control operators, as the responsibility of safe operation of the flight is vested with the pilot of the aircraft. This is possible through the use of advanced techniques of control.

The pilot will be able to make flight path decisions such as the change of trajectory in mid-flight with the aid of computer systems. For the free flight concept to emerge fully, a few more decades of time may be required.

The types of free flight envisaged are:

- *Unrestricted* – very little guidance from the ATC
- *Transition* – slightly restricted, but the pilots are given a little flexibility.
- *Restricted* – significant reduction in the pilots freedom

The following are the different approaches for free flight:

- *Air-borne approach* – full responsibility is with the pilot for safe operation, aided by computer systems.
- *Ground approach* – all data is sent to the ATC and pilot requests for flight path. Communication will be from the aircraft to the ATC.
- *Mixed focus approach* – a combination of both air-borne and ground approaches. An airport operation centre (AOC) coordinates the ATC and the aircraft.

(A more detailed treatment is out of scope here.)

## 14.10 MODERN TECHNOLOGY IN AIR TRANSPORT

Modern technological advances have facilitated improvements in the following aspects of air transport:

- Grooved runways—improved runway surface treatment
- Communication in air navigation and surveillance facilities

- Reciprocal instrument landing system
- Airport lighting
- Cargo handling

The AAI has been introducing improvements in most of these aspects at all major airports in India.

## SUMMARY

- Visual aids are devices or equipment which are meant to help the pilot of an aircraft in sighting the various features in and around an airport for safe landing, either during the day or during the night, and under all kinds of weather conditions.
- Visual aids may be indicators such as wind-direction indicator and landing-direction indicator, which are placed at the centre of a large segmented-circle at the periphery of the airfield. Wind-direction indicator may a wind-cone, wind-sock, or a wind tetrahedral pyramid. Landing-direction indicator may be of the tee-type or tetrahedron-type.
- Airport markings are another important kind of visual aid. Runways are marked at the threshold, touch-down zone, centre-line and edge-strip with standard types and colours for easy identification by the pilot.
- Taxiways are marked at the entrance to the runway and entrance at the end of runway; exit taxiways and taxiway intersections are marked in different standardised patterns for easy identification.
- Shoulders of runways and taxiways are also marked in set patterns at regular intervals.
- Aprons and aircraft standing area near the gates are marked with guideline strips for easy taxiing of aircraft into their respective earmarked positions.
- While visual aids such as markings help the pilot during the day, airport lighting of the different areas and zones is expected to help the pilot during night time.

  Airport beacon, code beacon, approach lighting, boundary lighting, threshold lighting, runway lighting (at the centre-line and the edges), taxiway lighting, lighting of aprons and hangars, airway beacon, and lights to indicate possible obstructions are examples of airport lighting. Each of these has its own standards of lighting with respect to colour, intensity and spacing.
- Visibility at an airport is defined differently during day-time and during night-time.
- Runway visual range (RVR) is the distance up to which high-intensity edge lights can be seen by a pilot. Decision height is the minimum height at which a pilot can decide whether to continue with the landing operation or abort it. Visibility is classified based on these two parameters for different systems of landing.
- Pilots need landing guidance for successful landing. This consists of guidance and information on alignment, height, distance, and roll guidance after touch-down. The parameters which affect the visual range of the runway are the cockpit cut-off angle, pitch angle of the nose of the aircraft, and glide slope angle of the aircraft during landing.
- Air-traffic control is a complex problem due to the need for the aircraft to be in continuous flight, and the difficult task of landing under unfavourable conditions. This consists of control at the airport, traffic control along the airways, and airway communication.
- Flight rules have been laid down by the ICAO to ensure aircraft movements are safe and expeditious; further, these avoid confusion to the pilots of international flights when they fly to countries not known to them.

  Instrument Flight Rules (IFR) are applicable to airports with Instrument Landing System (ILS), while Visual Flight Rules (VFR) are to be followed when ILS facility is not available.

The primary difference between these two sets of rules is that in the former, the responsibility for safe landing lies with the air traffic control tower personnel, whereas in the latter it rests with the pilot.

- Air-traffic control network consists of control within the terminal area through the Airport Traffic Control Tower (ATC), control over airways/air routes through Air Route Traffic Control Centres (ARTCC), and Airway communication, which is facilitated by Flight Service Stations.
- Air-traffic control aids are classified as Airway or en-route aids and terminal or landing aids.
- Airway or en-route aids comprise airway beacons, low/medium frequency (LF/MF) radio range, very high frequency omni-directional range (VOR), air-route surveillance radar (ARSR), direction finders (radio type and automatic type),tactical air navigation (TACAN) for military purpose, distance-measuring equipment (DME), marker beacons to locate radio range stations on an airway, and air-to-ground communication for relay of information and two-way communication between the pilots and flight service stations.
- Landing or terminal aids consist of airport-surveillance radar (ASR), precision approach radar (PAR), Instrument Landing System (ILS) which primarily comprises of glide slope antenna and localiser antenna along with markers, Microwave Landing System (MLS) which is an improvement over ILS, airport service detention equipment (ASDE), and approach lights, all of which are expected to ensure safe and successful landing of aircraft.
- Global Positioning System (GPS) technology, a precise positioning tool, is a useful aid in air navigation.
- *Free flight* is a new concept in air-traffic control, in which the responsibility of safe flight operation is vested with the pilot, with help of computer systems.
- Modern technological advances in air transport comprise runway treatment, communication and surveillance facilities, reciprocal landing system, airport lighting, and cargo handling.

## QUESTIONS

1. What are 'visual aids' at airports? (Punjab Technical University)
2. Write brief notes on:

   (a) Wind-direction indicator

   (b) Landing-direction indicator

   (c) Segmented circle
3. Write a brief note on 'Airport Markings'. (VTU)
4. Write a note on 'Runway markings'. (Mumbai University)
5. Sketch and explain the markings for a taxiway entrance at the end of a runway.
6. Sketch and explain the marking on the shoulders of taxiways and runways.
7. Explain the markings needed on aprons and aircraft standing area.
8. Write a brief note on 'airport lighting' (Kalinga Institute of Technology, Odisha)
9. Write brief notes on:

   (a) Airport beacon

   (b) Code beacon

   (c) Airway beacon

10. Differentiate between visibility during day time and that during night time.
11. Explain 'runway visual range (RVR)' and 'decision height'
12. What are the parameters on which a pilot needs guidance and information during landing?
13. Explain why air traffic control is a complex problem.
14. Differentiate between Instrument Flight Rules (IFR) and Visual Flight Rules (VFR)
15. Explain the three components of air traffic control network.
16. Write a short note on 'Instrument Landing System'. (VTU and Mumbai University.)
17. Write short notes on:
    (a) Very high frequency omni-directional range (VOR)
    (b) Marker beacons
    (c) Air-to-ground communication
18. Write notes on:
    (a) Airport Surveillance Radar
    (b) Cone of silence
    (c) Role of GPS in air traffic control

Chapter

# 15

# HELIPORTS

**CHAPTER OBJECTIVES**

This chapter enables the student to understand that:

- A helicopter is a power-driven machine designed to take off and land almost vertically
- Helicopters are the best form of transport to reach inaccessible areas for military purpose or for relief work during natural calamities
- Planning and design of heliports is similar to that of airports, but on a much smaller scale
- Heliport markings and lighting help during the day and during the night respectively.

## 15.1 INTRODUCTION

A *helicopter* works mainly on the lift generated by one or more power-driven rotors moving about a vertical axis. Its primary characteristic is that it can lift off the ground almost vertically and also land vertically downwards on to the ground. Hence, this is referred to as a vertical take-off and landing (VTOL) machine. A helicopter was successfully devised and developed more than five decades ago.

Such vertical take-off aircraft requires a relatively smaller area for its operation. A specially prepared and designed area for landing and take-off of helicopters is known as a *heliport*. As in the case of airports, fuel, service facilities and air traffic control are available at a heliport.

A *helipad* is a helicopter-landing pad or a landing area or platform for helicopters. While helicopters are able to operate on a variety of relatively flat surfaces, a fabricated helipad provides a clearly marked hard surface away from obstacles where a helicopter can land safely.

Heliports are usually provided as part of airports, since all the facilities including air-traffic control facilities of the airport can be used for helicopter traffic.

## 15.2 HELICOPTERS

Modern helicopters are designed and tested to conform to the specified standards of airworthiness. It is constructed in such a way that there is perfect safety in flying, to avert loss of life and property.

While in steady flight, the speed of the rotating wing and the vertical rate of climb depend on a number of factors. A helicopter is designed to climb at the rate of 60 m/minute, with a gradient of 4 per cent for the flight path. The climbing capacity of 450 m above the level of take-off is another criterion; there should be also some reserve capacity for an additional climb.

The *characteristics of a helicopter* may be classified as:

(a) Physical characteristics
(b) Flight characteristics

## Physical Characteristics

Helicopters may be powered by a single engine or two engines; that is, a single rotor or two rotors in tandem.

The configuration of the landing gear consists of two main wheels and one tail wheel, or four wheels.

The *principal dimensions of a helicopter* are:

(i) Rotor diameter
(ii) Overall length
(iii) Height
(iv) Tread
(v) Wheel base

The other important characteristics are the maximum gross weight and the cruising speed. The principal dimensions are shown schematically in Fig. 15.1 and given in the following table for three kinds of helicopters:

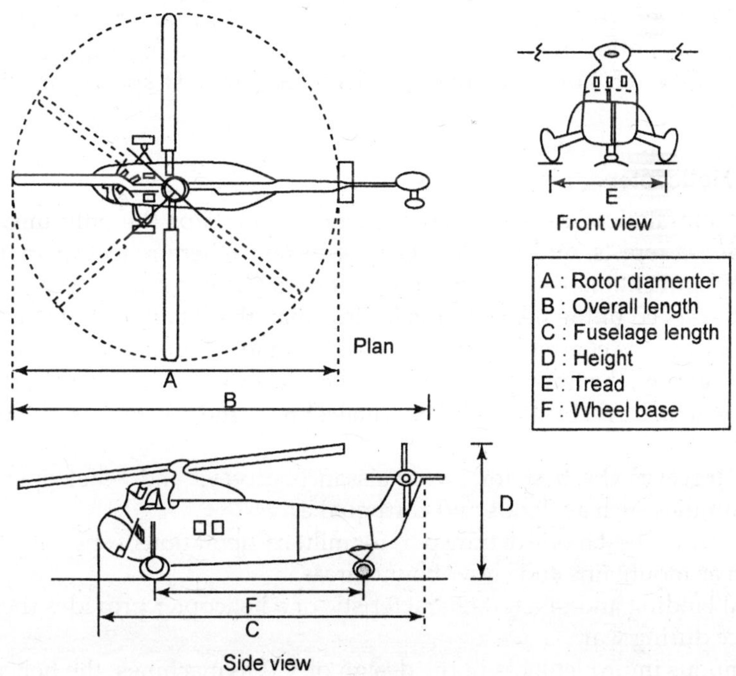


**FIG. 15.1** *Principal dimensions of a single rotor helicopter*

| Make and Model | Dimensions (m) | | | | | | Gross Weight (kg) | No. of engines | Cruising Speed km/h |
|---|---|---|---|---|---|---|---|---|---|
| | A | B | C | D | E | F | | | |
| Bell | 13.41 | 16.15 | 13.00 | 3.42 | 2.53 | 3.28 | 3240 | 1 | 156 |
| Cessna CH-I | 10.66 | 13.00 | 10.01 | 3.55 | 2.53 | ... | 1350 | 1 | 167 |
| Sikorsky S-56 | 21.94 | 25.24 | 19.77 | 6.54 | 6.01 | 11.23 | 13950 | 2 | 185 |

Helicopters may be classified based on the type of power by which they are driven:

- Mechanical
- Jet propulsion

The mechanical drive system is more popular. The gas turbine drive is of recent origin; it is advantageous as the machine becomes lighter, and greater power is available.

The number of rotors is the other important feature. Most helicopters are of the single rotor type. This has fewer parts and has low requirement of floor space. However, twin or tandem rotors give more power and greater cruising speed.

## Flight Characteristics

The primary characteristic of a helicopter is its capacity to hover in air by means of the rotating airfoils. It also has a greater range of cruising speed and flight altitudes than conventional aircraft. In addition, it has an advantage of being able to land on and take off from relatively smaller areas.

The take-off of the machine is initiated by vertical ascent, which is helped by the ground cushion built up by air pressure against the ground by revolving rotors.

After about 2 m of vertical ascent, horizontal acceleration commences. Additional lift for the forward motion is obtained between 8 to 16 km/h, and then the helicopter is accelerated to 65 to 85 km/h, which is considered the best speed for climbing to the cruising altitude.

### Advantages of Helicopters

(i) Helicopter communication is very convenient and perhaps the only means in areas of snowfall, dense forests, and remote, difficult terrain where road transport, or even air transport by conventional aircraft is either not possible or is risky.

(ii) Although the speed of the helicopter is far less than that of aircraft, it saves a lot of time for journeys from the centres of cities by avoiding a lot of ground travel. For distances in the range of 400 km, the time spent on ground travel to and from the airport is much more than that spent on travel by aircraft. Thus, helicopter is the best means of transport in such cases.

(iii) Helicopter travel is the best for reconnaissance surveys and relief operations during natural calamities such as floods and earthquakes.

(iv) Helicopters are the best mode of transport for military operations, especially in inaccessible places such as mountains and snow-bound areas.
The vertical landing and take-off characteristic of a helicopter provides the best strategic convenience during war.

(v) With continuous improvements in the design of VTOL machines, the helicopter is going to be the preferred mode for medium-range travel in future.

## 15.3 HELIPADS

A *helipad* is a helicopter-landing pad or a landing area or platform for helicopters. While helicopters can be operated on a variety of relatively flat surfaces a specially fabricated helipad provides a clearly marked hard surface away from obstacles, where a helicopter can land safely.

Helipads may be located at a heliport or an airport where fuel, service, and air-traffic control facilities are available. However, most helipads are located remotely away from such facilities due to space and cost constraints.

Some office towers maintain a helipad on their roofs to accommodate air-taxi services. Some basic helipads are built on high-rise buildings for evacuation in case of a major fire outbreak. Large ships may have a helicopter deck or a *helideck*. Helipads are also common at hospitals that have air ambulance facility for transfer of patients to trauma units.

Helipads are usually constructed out of concrete and marked with a circle and /or a letter "H", so as to be visible from the air. Landing pads may be constructed even in extreme conditions such as ice.

The world's highest helipad, built in India, is located on the Siachen Glacier at a height of 6400 m above sea level. The world's largest helipad in Morgan city, Louisiana, USA, has a total of 46 helipads.

Portable helipads with telescopic legs may also be used to land a helicopter at unfavourable locations for emergency services.

## 15.4 PLANNING AND DESIGN OF HELIPORTS

A heliport is planned and designed for landing and take-off of helicopters. Facilities for shelter, servicing and repairs, processing of cargo and baggage of passengers, and for parking of motor vehicles are usually provided based on the traffic needs.

The specially prepared area on which a helicopter actually lands and from where it takes off is called the *landing area*; if it is a paved surface, it may be called a helipad. Other facilities may not be available if it is meant only for landing and take-off, as in the case of remote and relatively in accessible areas.

The area of the heliport excluding the landing area is known as the *terminal area*. The following are the *components of planning and design of heliports*:

(a) Site selection and obstruction clearance
(b) Terminal area
(c) Orientation of the landing area
(d) Size of landing area
(e) Heliport marking
(f) Heliport lighting

### Site Selection and Obstruction Clearance

The following are the considerations in the selection of site for a heliport:

(i) It should be located close to areas of potential traffic.
(ii) It should be accessible through a convenient mode of surface transport such as a highway or a railway. (Of course, this cannot be a consideration in the case of snow-clad hills or dense forest areas.)
(iii) The ambient noise level during the operations should be tolerable, depending upon the nature of the area. In the case of areas of industrial activity, greater noise levels may be

tolerated than near residential areas. (Jet engine-powered helicopters produce noise which may be more than that which can be tolerated even in industrial and business areas.)

(iv) Where a heliport is developed at a major airport, a roof site on the terminal building should be preferred to a ground site for reasons of clear visibility and identification from a distance. Further, this can help to save time in handling mail, cargo, and passengers.

(v) Sometimes, helipads are located on the roofs of high-rise buildings for small helicopters; this can be advantageous from the point of view of accessibility from a city centre or a business district.
Visibility is important in this case, since fog is more at high altitudes.

(vi) Although a helicopter is primarily a near vertical landing and take-off machine, it is always better that there is an unobstructed approach – departure path for helicopters at the heliport.

*Obstruction clearance* standards are laid down based on certain assumptions regarding the performance of multi-engine helicopters:

- During take-off, the helicopter rises quickly and achieves a forward speed, so that even if one engine fails, it continues the flight.
- The helicopter reaches the boundary of the heliport before engine failure occurs; with the other engine, the helicopter would be able to manoeuvre a slope of 1 in 20.
- Even with this reduction in power, the helicopter can reach a height of 105 m at a distance of 720 m from the boundary of a heliport. This means that the obstruction clearance line should have a slope of 1in 8.

The obstruction clearance line based on the criteria recommended by FAA is shown in Fig. 15.2.

In the transverse direction, the slope of obstruction clearance plane is 1:2, extended to the minimum en-route altitude of the helicopter.

Zoning is essential to protect the approach–departure path of the helicopter from the development of high-rise buildings and tall structures.

**Terminal Area**

If the volume of traffic is not large, loading and unloading operations can be carried out on the landing and take-off area itself.

For larger volumes of traffic, other facilities are required for the helicopters, passengers and operating staff, on the same lines as for airports, but on a much reduced scale.

The planning considerations for airport terminals are generally applicable to heliports also.

Figure 15.3 shows a typical layout of the terminal area of a heliport for multi-engine helicopters.

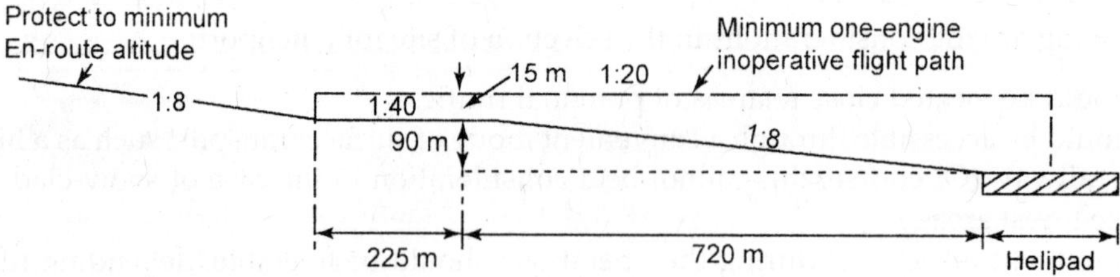


**FIG. 15.2** *Obstruction clearance line in approach area (FAA)*

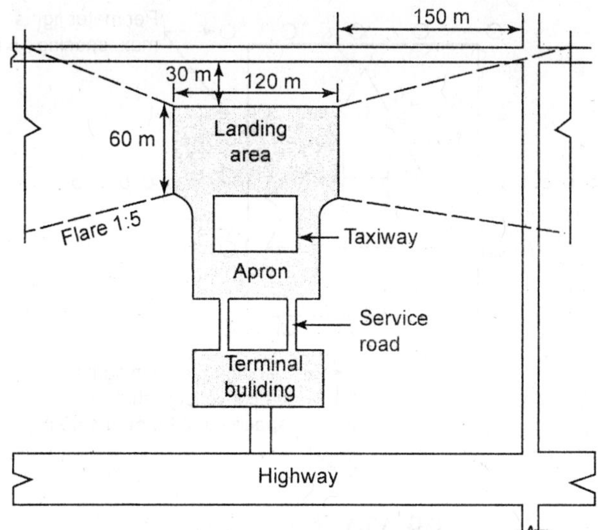


**FIG. 15.3** *Typical layout of the Terminal area of a heliport*

## Orientation of the Landing Area

The permissible cross-wind component for helicopters is much more than that for fixed-wing aircraft. However, it is desirable to orient the landing and take-off area along the wind direction, as nearly as possible.

## Size of Landing Area

The size of a landing and take-off area depends upon the characteristics of the helicopters using the heliport and the permissible weather conditions. While small helicopters are operated under VFR conditions, larger ones operate under IFR conditions, requiring larger space. The recommendations of the FAA with regard to the size of landing area are given below:

| S. No. | Description of helicopter | Size of landing area |
|---|---|---|
| I | *Single-engine* | 22.5 m × 22.5 m or equivalent circular area |
| | (a) 2700 kg Weight (VFR) | |
| | (b) 6800 kg Weight (VFR) | 30 m × 30 m or equivalent circular area |
| II | *Multi-engine* (IFR) | 60 m × 120 m |

A minimum width equal to twice the main rotor diameter of the largest helicopter should be provided. The gradient should not exceed 2 per cent. Adequate area should be available adjacent to the landing area, for emergency landing in case of engine failure.

## Heliport Marking

The landing area is identified by marking the letter 'H' with or without a circle around it. Its size is specified shown in Fig. 15.4.

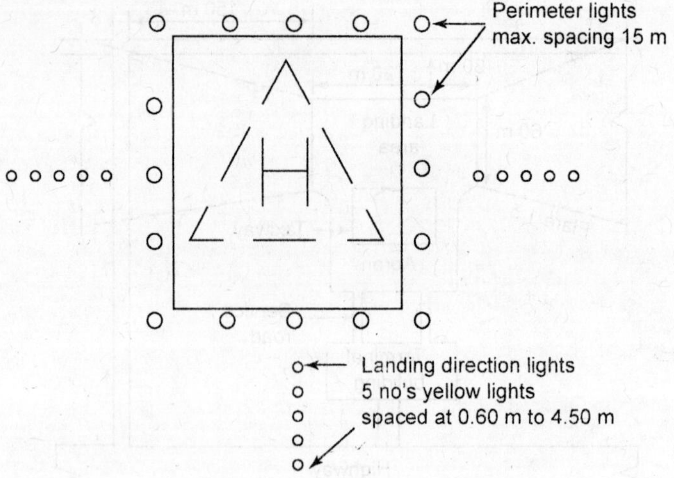


**FIG. 15.4** *Heliport lighting*

Height of letter – 3 to 4.5 m
Width of letter – 2 to 3.5 m
Width of lines – 45 cm

The letter may be coloured in white or yellow paint. A black outline is used around the lines.

Aprons should also be marked appropriately for facilitating the manoeuvring of helicopters.

**Heliport Lighting**

For night operation, heliport lighting is essential.

Lighting is required for the identification of helicopters, wind direction, landing area, and obstruction, if any. The lighting should have special characteristics for differentiating the heliport area from adjacent areas, which may be airport area if the heliport is designed as part of an airport.

FAA recommendations in respect to heliport lighting are given below:

(i) Landing and take-off area: A flood light providing a level of illumination of 15 candela is to be provided.
(ii) The periphery of the landing area should be provided with amber lights of medium intensity, spaced 7.5 m centre-to-centre, as shown in Fig. 15.4.
(iii) Aprons should be lit by flood lights with an illumination level not less than 30 candela over the entire working area.
(iv) Inset lights set into the pavement should be fixed in the touch down-area; they should be white in colour.
(v) Landing direction lights of yellow colour, extending to about 22.5 m distance have to be provided.

## SUMMARY

- A 'helicopter' is a machine, power-driven by one or more rotors, which takes off and lands almost vertically.
- A 'heliport' is a specially prepared area for landing and take-off of helicopters; facilities for loading and unloading of passengers and cargo, fuelling, servicing and repair of helicopters, and traffic control facilities may also be provided at heliports. These may be provided at major airports.
- A 'helipad' is a helicopter-landing pad or a landing area for helicopters; no additional facilities may be available.
- The principal dimensions of a helicopter are the rotor diameter, overall length, height, tread, and wheel base.
- A helicopter may be mechanical-driven or jet-driven; it may be of a single-rotor type or twin-rotor type; the latter has a spare-engine and can cruise at higher attitudes and greater speeds.
- A helicopter can hover in air and keep still as long as engine power is available. In addition, it can land and take off almost vertically and requires smaller area of operation.
- Helicopters are the best mode of transport for medium-range travel (up to 400 km), since it helps to compensate for the time lost in ground travel to and from the airport.
- Helicopters are ideal vehicles that can be used to reach inaccessible areas like mountain peaks, dense forests, and snow-bound areas; it is also convenient for military purposes and for relief work during natural calamities such as floods and earthquakes.
- Helipads may be located on top of high-rise buildings, and in hospital zones for ambulance services in the wake of accidents.
- Considerations for the planning and design of heliports are similar to those for airports, except that the scale on which services and facilities are required is far less than that for airports.
- A 30 m-size square or a 60 m × 120 m rectangular area will be more than adequate for the landing area of a heliport.
- Heliport markings are needed for the landing area (with a large 'H'), and aprons for manoeuvring and standing lines of helicopters.
- Heliport lighting is needed for night operations; landing area, boundary, and obstructions have to be appropriately lit by white/yellow-coloured lights. The lighting for a heliport has to be provided in a manner that differentiates it from that of an adjacent airport.

## QUESTIONS

1. What are the characteristics of a helicopter?
2. List the advantages of a helicopter as a mode of transport over the conventional aircraft.
3. Enumerate the factors to be considered in the selection of site for a heliport.
4. Sketch a typical layout of a heliport for a multi-engine helicopter.
5. Write a short note on the flight characteristics of a helicopter.

6. Write a brief note on 'helipads'.
7. Write short notes on:

   (a) Heliport markings

   (b) Size of landing area

   (c) Physical characteristics of a helicopter.
8. Write a note on heliport lighting. How is it differentiated from airport lighting?
9. Explain, with a sketch, the obstruction clearance requirements at a heliport as recommended by the FAA.

# PART III

# Dock and Harbour Engineering

Chapter **16**

# INTRODUCTION

**CHAPTER OBJECTIVES**

This chapter enables the student to understand:

- The advantages of using ships and steamers on water bodies to transport heavy and bulk cargo
- The definition of harbour as a place of refuge for ships and a port as a harbour having facilities for dealing with cargo and ships
- The historical aspects of ocean waterways and river waterways in India over several centuries and the impetus this has been receiving in the post-independence era
- The evolution of systematic planning procedures using systems approach, data storage and analysis using digital computers, and integrated development with other modes of transport and energy conservation.

## 16.1 INTRODUCTION

Highways and railways are the popular means of transportation on land while aircraft and helicopters are used for transportation through air. *Water transportation* or transportation on water is through the use of ships, steamers and boats.

The *choice of the mode of transportation* depends on the following:

- Haulage distance
- Weight and size of consignment in the case of freight or cargo
- Traffic volume and density
- Nature of route and speed of transport.

Demand for transportation of passengers and freight depends upon the stage of economic development, and the topographical and demographic conditions of the area or country.

## Water Transportation

The *importance of water transportation* can be understood from two facts:

- Nearly two-thirds of the earth's surface is covered by water.
- A vast majority of the world's population lives within 50 to 60 km from the coastal regions; the productivity of a coastal system depends on water transportation.

The vehicles used for transportation on water are those that float on water; they are usually vessels of wood or metal, or a combination of both. These vessels are designed to be hollow and move on water by natural wind power or mechanical power of IC engines; boats fall in the first category while steamers and ships fall in the second. Boats are relatively small while ships are larger in size.

A *waterway* is any navigable water of sufficient depth, which is free of obstructions. Navigable waters consist of rivers, lakes, canals, streams, seas and oceans. While seas and oceans are natural, canals and channels are man-made.

## Types of Water Transportation

Depending on the waterways used, water transportation is broadly classified as follows:

1. Inland waterways
2. Ocean waterways

**Inland waterways:** These are waterways on rivers, lakes, canals and channels within the mainland.

**Ocean waterways:** These are means of transportation of passengers and goods across seas and oceans.

## Advantages and Disadvantages of Water Transportation

### Advantages

(i) This is the cheapest mode of transport because other modes such as railways and highways need special tracks and surfaces, and air transport has special and sophisticated needs. Natural waterways are a gift of nature.
(ii) Heavy freight and cargo can be carried by water transport.
(iii) Waterways are ideal routes to boost international trade and commerce.
(iv) Water transportation needs relatively cheap motive power.
(v) Natural resources such as rivers, seas and oceans can be utilised effectively.
(vi) Countries separated by large water-bodies such as seas and oceans are easily made accessible by means of water transport.
(vii) Water transportation leads to industrialisation and economic progress, as in the case of European countries and the USA.
(viii) The Navy, one of the three primary wings of defence and security of a nation, utilises ocean waterways in the most effective manner.

### Disadvantages

(i) It can be developed only when a water body is available along specific routes only.
(ii) Operations of loading and unloading of cargo are affected during high tides in oceans.

(iii) Mountainous terrain, hill streams and waterfalls hinder smooth operation of waterways.
(iv) Since freight traffic originates on land away from the coast, feeder modes are required in the form of goods transport by means of roads or railways.
(v) It is relatively slow compared to other modes of transportation.
(vi) Ocean storms and hurricanes may lead to accidents on high seas and loss of valuable cargo.
(vii) Fleets of naval ships or war ships often lead to race for power and political domination.

In spite of these disadvantages, the advantages are considered powerful enough to promote water transportation.

## Definitions

**Harbour:** Primarily, a *harbour* is a place of refuge or rest for sea-faring vessels such as ships, which is in the form of a basin in navigable waters, well protected naturally or artificially from the action of wind and waves in seas or oceans. This is usually located along the shore. Provision for supplies, transfer of cargo, refuelling of the engines, and repairs, if necessary, is available at a harbour.

**Port:** A *port* is a harbour where terminal facilities such as piers of wharves where ships may be berthed while loading or unloading of cargo, transit sheds, and storage areas such as warehouses and cargo-handling equipment like cranes are provided. Thus, every port is a harbour, while the converse is not true. A port is a commercial harbour with all infrastructure facilities. In fact, a port is the connecting link between sea traffic and land traffic, and is a kind of gateway to land from sea, and sea from land. Roadways and railways serve as feeder routes for movement of cargo to ports for transportation to other ports, either within a country or to those in other countries.

**Satellite port:** It is a small port with minimum facilities for berthing, loading and unloading of cargo. It depends upon a major port for higher order facilities.

## 16.2 HISTORICAL ASPECTS OF WATER TRANSPORTATION

Waterways are of great historic importance in the development of human civilization and the growth of a nation. Rivers, seas and oceans were the available natural means of mass transport of goods for a long time. Heavy goods such as grain, stone and timber could be carried in bulk only through waterways.

The pyramids of Egypt built around 2500 BCE required large quantities of stone, which could be transported only with the aid of waterways. Canals, supported by locks, were built in Egypt and the Arabian Gulf around 500 BCE. Egyptians were said to have built large ships even around 1200 BCE.

Greek cities thrived on shipping, and the city of Athens had a fine harbour. The lighthouse at Alexandria was built around 279 BCE. Roman trade extended to China by 100 AD through shipping and waterways; Romans could wield power through their shipping and water transportation via the Mediterranean Sea.

The Suez Canal was built in 1869 AD for a length of 112 km. The Panama Canal, 80 km long, connecting the Atlantic and the Pacific Oceans, was built by 1914.

The Europeans used sailing vessels and ships to make sea voyages; Columbus discovered America in 1492 AD in his efforts to establish a sea route to India.

### Inland Waterways in India

Inland waterways could be through natural water courses such as rivers, streams and lakes, or through artificial or man-made canals or channels.

Navigation thrived on the rivers Indus (Sindhu), and the Ganga and its tributaries long ago. Barges weighing 400 tons were used to carry merchandise between Agra and Calcutta in the seventeenth century. Inland navigation thrived in Bengal. Power-driven steamers were used for river navigation between Calcutta and Agra in 1842.

*River transportation*: became popular in India due to the existence of perennial rivers such as the Ganga, Brahmaputra, Godavari, Krishna, and Mahanadi. Ganga is navigable for about 800 km up to Kanpur. River Yamuna is also navigable for some distance from Allahabad.

*'Canal navigation'* has been limited in India with only a few canals:

- Hooghly canals
- Calcutta Eastern canal
- Madras Godavari canal
- Buckingham canal
- Krishna canal
- West Coast canal
- Ganga canal
- Bihar and Odisha canals

Soon after independence, a River Research Institute was set up in Poona. In 1957, the Inland Water Transport Committee was constituted.

### Ocean Transportation in India

Costal shipping is the best means of transporting bulk cargo. This mode helped in establishing cultural and trade contacts with other countries in coastal regions. About three-fourths of international trade is carried out by merchant shipping. Navy shipping is intended for national defence.

Ordinary ships as well as large special ships were used in ancient days in India for trade with Persia, Arabia, Malaya and East Africa.

Excavations in Mohenjodaro and Harappa indicated that the art of ship-building was known as far back as 3000 BCE. Ajanta cave paintings indicate the existence of ship-building during 500 to 600 AD. Mughal rulers encouraged ship-building for war as well as for trade.

During the British-India period, Dacca (presently Dhaka), Machilipatnam, Hooghli, Surat and Goa were ship-building centers.

After independence, shipping and sea-trade received significant boost due to the allocation of finances for this sector during the Five-Year Plans. Berthing and handling facilities were developed at Madras (now Chennai), Cochin (now Kochi), Marmagao, and Visakhapatnam; the last one is the most mechanised port in India, capable of providing shelter for large ships. With a coastline of over 7500 km, India has immense potential for the development of sea-ports.

## 16.3 PLANNING OF WATER TRANSPORTATION

Water transportation planning is a systematic process of developing the infrastructure and services necessary to deal with the needs of water transport anticipated in the near future for a reasonable design period. It involves the formulation of goals and a judicious integration of the economic objectives, technological feasibility and financial constraints to evolve the best alternative of phased development.

The US Army Corps of Engineers developed a planning procedure consisting of three planning stages, each of which involving four tasks:

The *stages* of planning involve development of preliminary plan, intermediate plan and detailed plan.

The four *tasks* in each stage are:

- Identification of problems
- Formulation of alternatives
- Impact assessment
- Evaluation

The important criteria for improvement of navigation are:

(i) Acceptability to the public, users and the government agencies.
(ii) Maintainability of the project during the design period.
(iii) Efficiency of the project in handling the projected capacity.

The last task of evaluation is based on addressing the environmental concerns, is any.

### Elements of Water Transportation

The important elements of water transportation are:

- Waterways
- Terminals or water ports
- Carriers, vessels or ships.

Owing to the increasing complexities of water transportation, several disciplines are involved in the *systems approach* to planning:

(i) Statistics and probability
(ii) Statics
(iii) Model studies
(iv) Gaming and queuing theory.
(v) Linear programming
(vi) Optimisation techniques
(vii) Naval architecture and ship-building
(viii) Marine studies of seas and oceans, and so on.

Thus, the subject of water transportation is a challenging inter-disciplinary task.

## 16.4 MODERN TRENDS IN WATER TRANSPORTATION

Water transportation has gained popularity in the recent past. This surge in activity is attributed to the burgeoning world population and industrial growth of several countries and led to increased tonnage of cargo of raw materials and finished products handled by water transportation. Obviously, this has necessitated the building of larger ocean-liners, innovative and efficient loading, unloading and storage facilities, and creating a provision for modern port facilities at all major ports the world over.

A similar situation exists even in the area of inland water transportation in developing countries like India; grain and agricultural products are invariably transported by means of inland waterways.

Thus, the future of water transportation at both domestic and international levels is highly promising. New ships with larger capacities and greater cruising speeds are being built, and new ports with modern amenities are being developed. India, with a coast line of over 6,600 km, has promising potential for further expansion of water transportation.

The *recent trends* in water transportation are:

- Use of systems approach and optimisation with the aid of operations research techniques in planning
- Use of the digital computer in data processing, storage and analysis
- Integrated development of water transportation, ensuring compatibility with the other modes of land transportation such as the roadways and railways
- Energy conservation techniques through the development of non-conventional energy sources such as the wind energy, wave energy, and solar energy.
- Integration of the system of water transportation with environment, so as to minimise the possible adverse impact.

## SUMMARY

- *Water transportation* means transportation of passengers and goods using ships, steamers and boats moving on water bodies. *Waterway* means any navigable water of sufficient depth which is free of obstructions. Water transportation has great significance as it deals with vast stretches of natural waters in the form of rivers, streams, seas and oceans.
- Broadly speaking, water transportation is classified as inland waterways on rivers, lakes, canals and channels; and ocean waterways on seas and oceans.
- The advantages of water transportation are that it is the cheapest mode using natural water resources and it can carry heavy and bulk cargo. The primary disadvantages are that it is slow and accident-prone during storms, and that feeder modes of transport like roads and railways are essential.
- A *harbour* is defined as a place of refuge for ships, well protected from the action of wind and waves in seas or oceans, in close proximity to the coast.
- A *port* is defined as a harbour with terminal facilities for facilitating loading and unloading of cargo, storage areas, and handling equipment like cranes.
- Historically speaking, water transportation was the first mode to have been developed from millennia. Inland waterways in India were developed on major rivers such as the Ganga, Brahmaputra, Yamuna, Godavari, Krishna and Mahanadi. Ocean waterways existed in India even in the ancient era along with ship-building activity. Significant developments have taken place in India during the British rule, and more so after independence.
- Systematic procedures have been laid down for developing the infrastructure and services necessary to deal with the needs of water transport and integrating the economic objectives, technological feasibility and financial constraints associated with phased development.
- The primary elements of water transportation are waterways, the terminals or water ports, and the sea-faring vessels such as ships.
- Recent trends in water transportation include the use of systems approach and computers for database storage and analysis, planning of integrated development with other modes of transportation and energy conservation techniques.

## QUESTIONS

1. Explain the special features and the importance of water transportation.
2. What are the types of water transportation? Explain briefly.
3. Enumerate two advantages and two disadvantages of water transportation, compared to the other modes.
4. Distinguish between a harbour and a port. (West Bengal University of Technology)
5. "All ports are harbours, but the converse is not true." Justify this statement.
6. Trace the development of inland water transport in India.
7. Write brief notes on
   (a) River transportation
   (b) Canal transportation
8. Trace the development of ocean transport in India from the ancient days till the post-independence era.
9. Explain the three basic elements of water transportation.
10. Enumerate the modern trends in the development of water transportation.

Chapter 17

# PLANNING AND LAYOUT OF HARBOURS AND PORTS

**CHAPTER OBJECTIVES**

This chapter elucidates:

- The difference between a harbour and a port
- The basic types of harbours – natural, semi-natural and artificial – and their classification based on function and on the location
- Site selection and site investigation for a harbour and salient features
- The various facilities required at a port for guiding and protecting vessels, cargo loading and repairs of vessels, for dredging and shipment storage
- The location of major ports in India and outlines their functions
- The importance of Environment Impact Assessment (EIA) in planning water transportation projects
- The regulatory bodies and associations for the promotion and development of water transportation.

## 17.1 DEFINITIONS OF RELEVANT TERMS

A few important terms associated with harbours and ports are defined below:

1. *Harbour*: Partly enclosed area of water to provide safe shelter for sea-faring vessels such as steamers and ships.
2. *Port*: A harbour where marine terminal facilities are provided; the facilities consist of piers and berthing facilities for loading and unloading of cargo, and for storage.
3. *Port of entry*: This is a designated area from where foreign nationals and foreign goods are cleared through customs for entry into a country.
4. *Free port*: It is an enclosed and policed area adjacent to a port of entry, with facilities for loading, unloading and storage of cargo, for supplying fuel and spare parts for ships. This is an area where there is no intervention of customs authorities.
5. *Passenger terminal*: That part of a harbour or a port where only passengers embark and disembark along with their personal luggage or small cargo.

6. *Marine terminal:* That part of a harbour or a port which provides docking facilities in addition to cargo handling and storage facilities.
7. *Cargo terminal:* That part of a port or a harbour where the traffic is mainly cargo, such as cement, grains and petroleum.
8. *Anchorage area:* A place where ships may be anchored for inspection, to await docking space or fair weather conditions, or for quarantine to prevent the spread of infectious diseases.
9. *Offshore moorings:* These are anchorages at places where it is not feasible to build a dock or a protected area like a harbour. A number of such units may be present, each consisting of anchors, sinkers, and buoys, to which the mooring lines of a ship can be attached.
10. *Turning basin:* It is an area of water inside a harbour or an enlargement of a channel to allow the turning of a ship. The area should have a radius of at least twice the length of the ship to permit its free turning. Where the available space is limited, the ship may be turned warping around the end of a pier, with or without the use of line; in this case, the turning basin can be much smaller in area.
11. *Plimsoll mark or load line*: It is the mark or line indicating the level up to which a ship may be loaded in different bodies of water during the various seasons of the year, conforming to the prevalent maritime laws.
12. *Beam*: The width of a vessel.
13. *Draft*: The depth of the keel of a ship below the water level for a particular condition of loading; thus, the ship rests in the water up to this depth when fully loaded.
14. *Fairway*: It is merely the open water of navigable depth.
15. *Low water*: It is the minimum depth of water for a navigation channel between channel lines.
16. *Channel lines*: The limits of navigation channels which are maintained at specified depth, by dredging, if necessary.
17. *Bulkhead line*: This is the farthest line offshore up to which a fill or a solid structure may be constructed without causing interference to navigation.
18. *Ballast*: This is the weight added to hold or to the ballast compartments of a ship, in order to increase the draft after it discharged its cargo. This is necessary to improve the stability of the ship and is usually expressed in tonnes of water. For example, ballast in the form of salt water may be used to replace the amount of petroleum unloaded or discharged, for keeping equilibrium.
19. *Displacement tonnage*: This is the actual weight of the vessel, or that of the water displaced when afloat; this may be either in the loaded or light condition.
20. *Displacement light*: This is the weight of the ship in tonnes, without cargo, fuel and stores.
21. *Dead weight tonnage (DWT)*: This is the weight of cargo and fuel loaded to the Plimsoll mark. This is the carrying capacity of the ship and varies with the season and the latitude. DWT depends on the salinity of the water, which affects the specific gravity, and hence the buoyancy. DWT is taken as the mean of the values in the different seasons.
22. *Gross tonnage*: This is the carrying capacity in terms of the cubic content of the ship, expressed in units of 100 cu. m.
23. *Net tonnage*: This is the gross capacity less the space provided for the crew, machinery, fuel, and engine room.
24. *Cargo tonnage*: This is a commercial term, which forms the basis of freight charge. Ocean freight may be accepted on weight or volume basis at the shipping firm's discretion. It is usually expressed in thousands of tonnage.

## 17.2 PLANNING AND LAYOUT OF HARBOURS

Experience coupled with judgement plays an important role in the planning, layout and design of harbours, as well as in the construction of their component parts or structures. Harbour structures should be built to be simple but robust.

### Requirements of a Harbour

The following are the basic requirements for the proper functioning of a harbour:

- Ship channels or ocean waterways must have sufficient depth of water for adequate draft of vessels.
- Land masses must be available or breakwaters must be constructed for protection against the destructive action of sea or ocean waves.
- The bed of the water body must be strong enough to provide secure anchorage for ships against high wind forces.
- The entrance of the harbour must be wide enough to permit easy passage of ships; at the same time, it must be narrow enough to restrict the transmission of excessive amounts of wave energy in times of stormy weather.

The layout and design of a harbour depends upon the nature of shipping to be provided for. Previously, emphasis was laid on breakwaters, quay walls, and other protection works; of late, the focus has shifted to increasing the depths of water to accommodate to increase in the size of ships.

### Classification of Harbours

Harbours may be classified in different ways based on different criteria, as given below:

(a) Classification based on the nature of protection
(b) Classification based on utility
(c) Classification based on the location.

#### Classification Based on the Nature of Protection

Depending on the nature of protection, harbours may be classified as follows:

(i) Natural harbours
(ii) Semi-natural harbours
(iii) Artificial harbours.

**Natural harbours:** These are inlets or areas of water protected from waves and storms by the natural configuration of land, resulting in creeks or basins. They are located usually in bays, tidal estuaries, and river-mouths. Ships or other vessels can seek refuge from the fury of the sea or ocean, especially during bad weather or storms, within an area of water close to the shore.

When the area within or just outside the harbour provides good anchoring and berthing conditions in the form of a deep navigable channel with a protective natural bank or a shoal, it is referred to as a natural *roadstead*. Sometimes, a confined area naturally enclosed by islands as in a creek may be available to provide shelter to vessels; such an area is referred to as a *circumscribed natural roadstead*. In brief, a roadstead is a partly-sheltered anchorage outside a harbour.

Figure 17.1 shows a simple layout of a natural harbour.

Figures 17.2 (a) and (b) show a natural roadstead with a bank or a shoal and a naturally circumscribed roadstead near a creek, respectively:

New York (USA), San Francisco (USA), Bombay (now Mumbai–India), and Kandla (India) are well-known examples of natural harbours.

## Semi-natural Harbours

This type of harbour has protection on two sides by headlands, requiring artificial protection only at the entrance. Further protection may be obtained by the addition of detached breakwater at the entrance.

A typical layout of a semi-natural harbour is shown in Fig. 17.3.

Visakhapatnam (India) and Plymouth (The UK) are examples of semi-natural harbours.

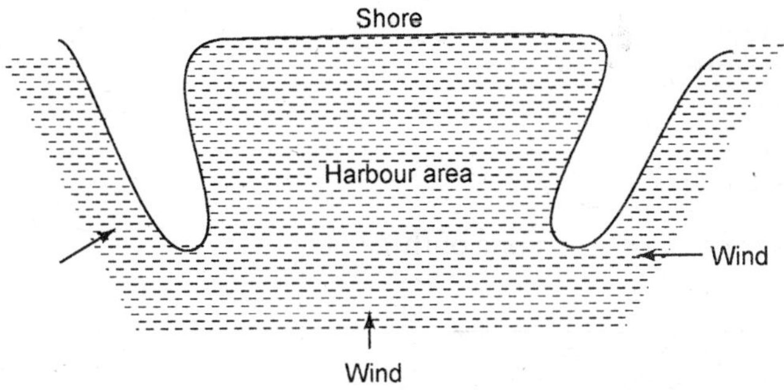


**FIG. 17.1** *Natural harbour*

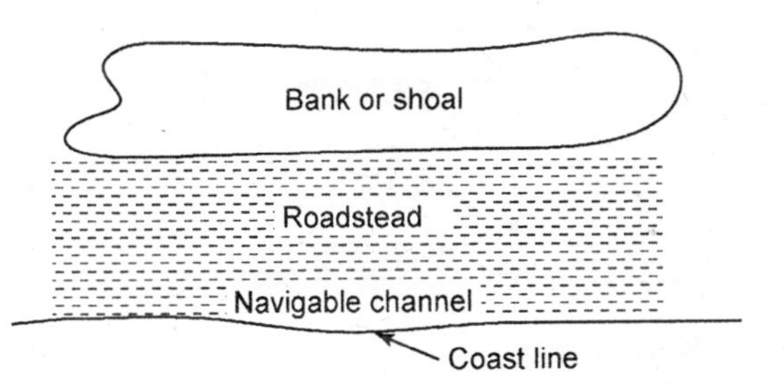


**FIG. 17.2 (a)** *Natural roadstead*

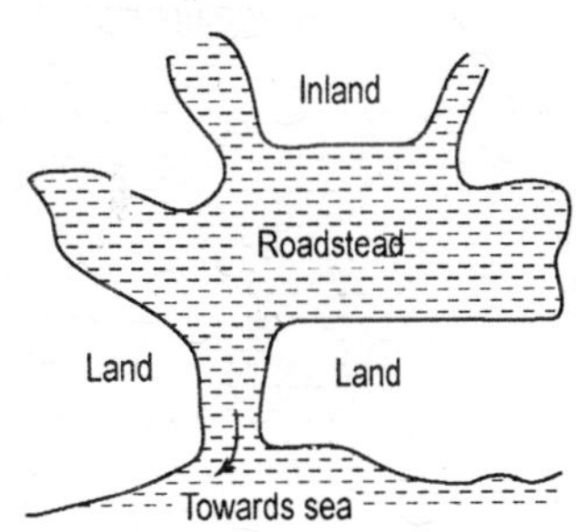


**FIG. 17. 2 (b)** *Naturally circumscribed roadstead*

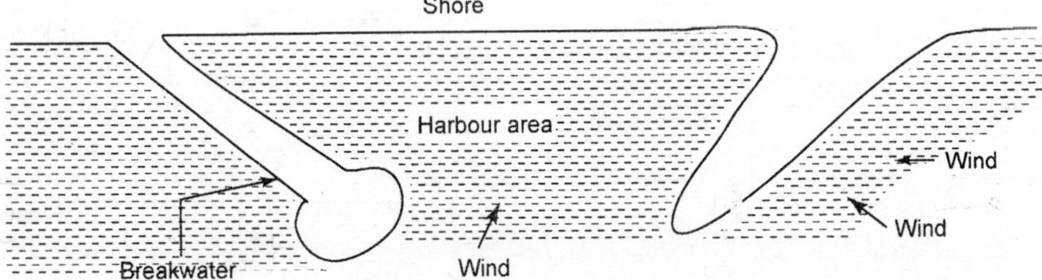


**FIG. 17. 3** *Semi-natural harbour*

**Artificial harbours:** When the facility of a natural harbour or roadstead is not available, artificial formations such as breakwaters are constructed to provide shelter or refuge to ships and vessels during bad weather. Such man-made harbours are known as *artificial harbours*, or *artificial roadsteads*; the latter can be created by constructing a breakwater parallel to the coast or curvilinear from the coast.

An artificial roadstead can also be created by enclosing an area of water by projecting solid walls, known as *jetties*, from the sea-shore. This type is called *artificial circumscribed roadstead*. Simple schematics of an artificial harbour, an artificial roadstead, and an artificial circumscribed roadstead are shown in Figs. 17.4, 17.5 and 17.6 respectively.

Madras (now Chennai) is a typical example of an artificial harbour.

### Classification Based on Utility

Based on utility, harbours can be categorised as follows:

(i) Harbours of refuge
(ii) Commercial harbours
(iii) Fishery harbours
(iv) Military harbours
(v) Marina harbours

Certain harbours serve more than one of these purposes.

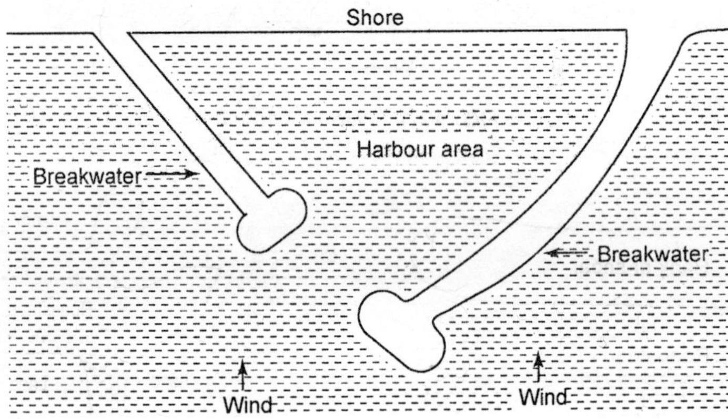


**FIG. 17.4** *Artificial harbour*

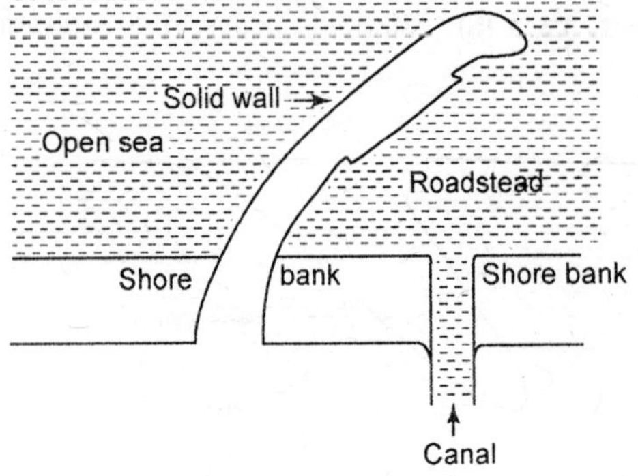


**FIG. 17.5** *Artificial roadstead*

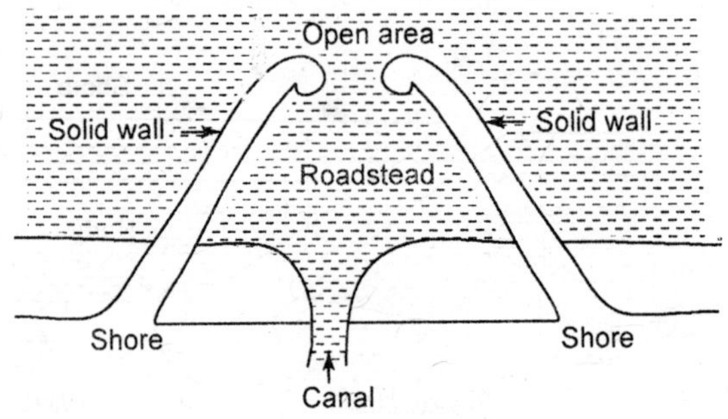


**FIG. 17.6** *Artificially circumscribed roadstead*

**Harbours of refuge:** These are primarily meant to be haven for ships seeking refuge in bad weather and stormy conditions. They should be as spacious as possible to accommodate even large ships.

Ready accessibility, safe anchorage system, and facilities for supplies and repairs are the primary requirements of a harbour of refuge. Damaged or disabled ships or ships in distress due to sudden storms and hurricanes need quick shelter and urgent repairs. Harbours of refuge cater to the needs of such vessels. Outer harbours of Chennai and Visakhapatnam are two examples of such harbours.

**Commercial harbours:** These harbours are meant for commercial activities such as import and export of goods, cargo, or merchandise. They may be located on sea coast, estuaries of big rivers, or even on island river banks.

The requirements of a commercial harbour are plenty of space for marine merchandise (which could be steel, coal, timber, fertiliser, chemicals, drugs and pharmaceuticals, finished products from industries, fruit, and oil and petroleum products), ample quay space for loading, unloading, transporting, and storage facilities, and adequate repair facilities in times of need.

For efficient loading and unloading of cargo, calm water is required, which could be provided by what are known as *docks* or *wet docks; dry docks* facilitate repairs of the vessels.

Commercial harbours do not have emergency demands as in the case of harbours of refuge; practically, the size and number of ships using the facilities are known in advance. Chennai and Mumbai harbours are examples of this type.

**Fishing harbours:** These are meant for the use of fishing vessels.

The requirements for fishing harbours are:

- Constant operation for the arrival and departure of fishing vessels.
- Facilities for quick loading and unloading, and also for the transport of perishable fish catch, such as railway sidings and service roads.
- Refrigerated storage space sufficient for preserving the perishable fish catch. Mangalore harbour is an example of a fishing harbour.

**Military harbours:** These are naval bases meant to accommodate ships and other vessels of a country's navy. They serve as supply depots for ships and the crew.

Bombay (Mumbai), Visakhapatnam, and Cochin (Kochi) are the important naval bases in India.

Hampton roads (Virginia, USA), Guantanamo Bay, and Pearl Harbour (Hawaii, USA) are a few well known naval bases of the USA.

**Marina harbours:** These are harbours which provide facilities for fuel, food, showers, telephones, washing machines and the like for small boats, along with berths on temporary or permanent basis.

Thus, a marina is a small harbour, or boat basin, providing docking facilities, berths, supply, and services for small craft used for pleasure trips associated with tourism. Resort facilities, sport and fishing facilities and pubs/bars are the additional facilities provided at such places.

Large marinas have accommodation for 200 or more berths, while smaller ones cater to the needs of less than 100 crafts. Kochi harbour has a marina.

### Classification of Harbours Based on Location

The layout and the configuration of the entrance of a harbour are influenced by its location. Based on this criterion, harbours are classified as follows:

(i) Sea or ocean harbours
(ii) River or estuary harbours
(iii) Lake harbours
(iv) Canal harbours

**Sea or ocean harbours:** These are harbours which are located on the coast of a sea or an ocean and are meant to serve sea-faring vessels, which may vary significantly in size. Penetration of waves and tides into the harbour, silting or accretion of coastal sand in the area of the harbour and the approach channels, and coastal erosion are the main problems in the layout and design of such harbours.

**River and estuary harbours:** Rivers and estuaries are the natural routes between the sea and the hinterland. Harbours were developed on the banks of these waterways, and were used for small-sized vessels. With increase in the size of vessels, the use of these harbours has been limited to smaller vessels for inland navigation.

Some sea ports have been developed at the mouth of tidal rivers. Continual dredging may be required to maintain such harbours.

**Lake harbours:** These are harbours built along the banks of a lake. The difference between this type and an ocean harbour is that, in the case of a lake, tidal action is absent.

When large lakes are connected by canals, these waterways provide excellent means of transportation of goods. Lake Superior and Lake Michigan in the USA serve such a purpose.

**Canal harbours:** A harbour can be located along the banks of a canal depending on its proximity to the sea; such a harbour can serve both inland navigation and sea-ward navigation. Invariably, minimal dredging is necessary at such harbours to provide the necessary draft for the vessels expected to use the facility.

## Planning of Harbours

The important considerations in the planning of a harbour are:

(i) Need for the harbour and principal use.
(ii) Anticipated volume of seaborne cargo and its nature.
(iii) Availability of inland communications or feeder routes by land as well as water.
(iv) Economic justification.
(v) General location of the harbour.

### Selection of Site for Harbours

The following factors should be considered in the selection of site for a harbour:

(i) Traffic potential due to industrial development.
(ii) Natural protection of wind and waves, if available.
(iii) Sea-bed and subsoil conditions; silting and erosion.
(iv) Marine conditions such as waves and tides.
(v) Meteorological conditions such as winds and cyclones.
(vi) Transport and communication facilities with respect to the feeder routes on land.

(vii) Availability of fresh water and other supplies.
(viii) Availability of electricity.
(ix) Defence and other strategic aspects.

It is of course, next to impossible to arrive at an ideal site that satisfies all the above factors for the location of any harbour. Hence, the best possible option should be chosen under a given set of conditions.

### Site Investigation

When a site is selected for the development of a harbour, all data or particulars necessary for planning, layout and design of the harbour and its component structures are to be collected. Work relating to the site investigation consists of the following:

1. Surveys
2. Meteorological data
3. Oceanographic data
4. Geological conditions
5. Soil investigation
6. Seismic data
7. Model testing
8. Locally available resources

**Surveys:** The surveys required include both topographic and hydrographic surveys.

Topographic details of the terminal area of the harbour on the shore side should be obtained, and the contour map of the area prepared; the location of all important elements such as buildings, utilities, test pits and borings, and any prominent landmarks should be marked on the map.

The contour interval may be 0.50 m to 1.50 m; the higher value may be used in rough terrain and in areas where no important structures are planned. Prominent irregularities on the ground should be examined at 30 m intervals. A miniature map may be drawn to a scale of 1:10,000, and a detailed map to a scale of 1:1000.

Hydrographic surveys consist of locating the shore line at low tide and high tide levels, and the location of possible obstructions along the shore and on the waters.

The sea-bed elevations should be charted by using an appropriate method of sounding; the soundings are taken at 7.50 m intervals along range lines parallel to one another at intervals of 15 m to 30 m, depending on the extent of detail needed. The presence of rocks, reefs, banks and shoals must be detected and marked. Hydrographical charts are prepared to help ship navigators steer clear of any hazards through the approach channels demarcated for the purpose.

The charts so produced are to supplement the charts available from the Indian Naval Hydrographic Office.

**Meteorological data:** Information regarding wind direction, wind velocity and duration should be collected for a period of at least one year, and wind rose diagram prepared. Alignment of the different elements of the harbour will be aided by the wind direction and other related characteristics. Similar data from an adjacent site within reasonable distance, if available, may prove to be a useful guide.

Wind velocities and duration during cyclonic storms and other relevant information should be obtained from the records of the Meteorological Department. Other meteorological data such as rainfall characteristics, temperature, atmospheric pressure, and relative humidity should also be collected and compiled.

**Oceanographic data:** Tidal ranges and details of high-water and low-water in a day and in a monthly cycle are important in the design of the depth of navigation channel, turning basin, and sometimes, even berths.

Characteristics of waves, wave direction, and the height and period should be obtained either by direct observations at the site or by forecasting from the available data from the charts.

Information about the tidal currents at the site will be useful in the design of ancillary structures at the harbour.

Sea-bed characteristics, suspended silt load, and salinity of sea water and temperature should also be collected. This helps in ascertaining silting and erosion characteristics at the site, in order to estimate the dredging needs.

**Geological conditions:** Information regarding the geological conditions of the site, the various strata, and the depth to the underlying rock has to be collected. The strength characteristics, the presence of faults, fissures, folds, and other unconformities of the bed-rock are important in the design of the component structures of a harbour.

**Soil investigation:** Test pits and borings should be made at important locations to obtain information on the sub-soil conditions at harbour structures such as breakwaters, piers, wharves, quay walls, and so on. Spacing of the borings should be determined based on variation of the strata from one boring to the next. Bore-logs should be prepared, giving an idea of the strata with respect to depth.

Main borings should be spaced at 50 m intervals at the edge of the shore, to a minimum depth of 3 m. Bore holes must be drilled to an extra depth of 5 m below the design depth of dredging.

For strength characteristics, undisturbed samples obtained with aid of sampler tubes should be used.

For more detailed information about soil investigation, exploration, sampling, and testing, please refer *Geotechnical Engineering* by C. Venkatramaiah, Universities Press, Hyderabad.

***Seismic data*:** Information on earthquakes that occurred in the past in the vicinity of the harbour site should be gathered. This will be useful in the design of components of the harbour structure with adequate resistance to seismic forces.

***Model testing*:** Testing of a hydraulic model of the harbour is considered to be a necessity to develop and design a good layout for a harbour.

Froude's number and Froude's laws are used in the design and testing of a hydraulic model. Tidal models, wave models, and models to simulate wind forces are considered useful to assess the hydraulic behaviour of water currents at the proposed harbour site. However, there are certain limitations in the interpretation of the results of model testing.

***Locally available resources*:** It would prove economical if locally available resources – materials as well as labour – are effectively utilised in the development of the harbour structures.

Bricks, stones, timber, sand, coarse aggregate or broken stone, available locally, should be preferred. Similarly, skilled as well as unskilled labour available locally should be utilised. This is certain to lead to overall economy.

On the whole, planning and design of a harbour is a challenging task owing to the multi-disciplinary nature of the problem.

### Features of Ships Affecting Harbour Design

Certain characteristics of ships affect the components of harbour design as given below:

(i) The types of vessels likely to utilise the harbour will govern the requirements of land area.
(ii) The width and depth of the approach channel will be governed by the dimensions of the largest ships likely to use the waterway.
(iii) The volume of marine traffic at the proposed harbour will determine whether the approach channel should single line or double line.
(iv) The cruising speed of the ship and the nature of the alignment of the channel will also govern certain aspects of harbour development.

Function-wise, a ship may be a dry-cargo vessel or a tanker.

Dry-cargo ships may be general cargo vessels or bulk-carrier ships; the latter are built to carry grain, ore, coal, and sugar in large quantities. They may be also container ships when cargo loaded in containers is carried. A tanker carries oil or other liquids in bulk. The dimensions of ships depend upon the nature of cargo and the Gross Registered Tonnage (GRT).

The length, beam (breadth), and depth may be as large as 200 m, 25 m, and 12 to 15 m, respectively.

## Salient Features of a Harbour

Salient features and components of a harbour are:

1. Approach channel
2. Anchorages and offshore mooring basin
3. Harbour entrance
4. Turning basin
5. Harbour basin
6. Berths
7. Breakwaters
8. Wharves and quays
9. Jetties and piers
10. Locks and lock basin
11. Harbour docks or wet docks
12. Dry docks
13. Slipways
14. Transit shed and warehouses
15. Ancillaries and miscellaneous facilities
16. Size and shape of harbour
17. Depth of harbour

**Approach channel:** The depth of water naturally available in the entire harbour area including the approach area outside it should be adequate for navigation of the vessels expected at all times.

When such ideal conditions do not prevail, an 'approach channel' of sufficient width and depth must be dredged for the passage of ships between the harbour entrance and the harbour basin, as also from the open sea to the harbour basin.

The position of this channel which lies in the open sea beyond the harbour entrance is called the *outer channel*; and, the remaining portion between the harbour entrance and the harbour basin is called the *inner channel*. The inner channel may be protected from waves and storms by natural configurations or by structures known as breakwaters.

*Channel alignment:* The alignment should be such that difficult manoeuvres and strong cross-currents are avoided. Waves, currents, and littoral drift snould be minimal. Number and angles of curves should be also minimal.

The direction of the prevailing wind should be favourable for navigation through the channel. In general, a minimum radius of 1000 m is desirable for curves; but it can be 400 to 600 m for turning with the assistance of tugs. A minimum stopping distance of five to eight times the length of the ships is desirable.

*Channel depth*: The depth required at harbour entrance is more than that required within the harbour because of exposure to the sea waves and wind at the entrance.
The following factors govern channel depth:

- Size and shape of the design vessel
- Speed of the vessel
- Velocity of the currents in the channel
- Number of lanes in the channel
- Wind and wave action
- Salinity of sea water and the nature of the bed material.

The Permanent International Association of Navigation Congress (PIANC) recommends a minimum design depth of 1.5 to 2.5 m in excess of the static summer salt water draught of the design vessel. An additional 0.5 m is allowed to compensate for oscillating conditions. The total depth may be around 12 m.

*Channel width*: This is measured at the bottom or at the design depth. This depends upon the following:

- Beam and manoeuvrability of the design vessel
- Speed of the design vessel
- Channel alignment.
- Channel depth
- Stability of channel banks
- Winds, waves, currents, and cross-currents in the channel.

PIANC recommends a channel width that is 3 to 4 times the beam of the design vessel for a single-lane channel in favourable conditions.

The total width actually constitutes the following:

(i) Manoeuvring lane (about equal to the beam of the ship).
(ii) Bank clearance lane (around twice the beam).
(iii) Ship clearance lane (equal to the beam of the depth)

(The last component is required for a two-lane channel to provide adequate space for one ship passing another).

**Anchorages and offshore mooring basin:** The area of water required for anchorages and mooring basin depend on the number and size of vessels requiring protection, and the number of mooring berths required for supplementing the handling of bulk cargo.

Large vessels are secured at the sea-bed by anchoring at a minimum of four points along the length of the ship on either side. The mooring system depends on degree of exposure, size of vessel, degree of restraint needed, and the nature of the bed material.

**Harbour entrance:** The depth and width required at the harbour entrance is more than that needed in the approach channel. The entrance width depends upon the volume of traffic, the degree of protection, and the navigational requirements; in order to avoid tidal currents, it should be adequate. If the width is large, it may increase the transmissions of high waves into the harbour basin.

The norm for the entrance width is 0.7 $L$ to $L$, $L$ being the length of the design ship; 100 m to 150 m for medium size vessels and 200 m to 250 m for large vessels, whichever is more. The maximum velocity for navigation is considered to be about 3 knots.

**Turning basin:** This is an area of water used for turning around of the vessel; its size depends on the available area, its depth, and the method of turning employed for the vessel: turn ahead, turn astern and ahead, and warping. Such turning may be with or without tug assistance. The first method of *turn ahead* is the most desirable, as it allows a quick turn-around.

The turning basin is best located at a position that it is tangential to the main channel. In this method, the minimum diameter of the turning basin should be 4 times the length of the design ship without tug assistance, while it can be twice the length of the design ship if tug assistance is taken.

**Harbour basin:** Also called 'sheltered basin', this area is protected by the shore and the breakwaters. The other elements of the harbour including the area for anchorages are located in this basin.

If the harbour receives a wide range of vessels, two zones are provided – one for large vessels and the other for small draught vessels, the draft, being different for the two categories. Anchorage should be clear of the channel, but close to the harbour entrance.

***Depth of basin:*** The minimum depth of the basin should be 0.75 m more than the loaded draft of the design vessel.

***Berthing length and width:*** The length should be 10% to 20% more than the length of the design vessel. The width should not be less than 1.15 times the beam of the design vessel; it should be much more if a number of vessels are to be kept abreast.

***Passing and manoeuvring area:*** This is the space beyond berthing for passage of vehicles and for manoeuvring the vessels to and from the berths. The width required depends upon the angle of the berths with respect to the shore line, the maximum being twice the length of the design vessel for 90° berths.

**Berths:** The form and location of the berths are governed by the considerations of navigability, availability of quays, and bed-soil conditions.

The length of quays should be at least 50 m more than the length of the design vehicle; the space between vehicles should be at least 15 m, if two or more vehicles are to be berthed.

Berths for tankers and general cargo should be separate, and those for explosive cargo should be isolated for safety.

"IS: 4651-Part 4-1980: General requirements for berths" is being followed in India.

**Breakwaters:** The function of 'breakwaters' is to protect the area of the enclosed water from storms and waves, Thus, it makes the harbour basin calm and ensures safety and ease of working with the vessels.

The monolithic structure at the tip of a breakwater is known as the 'pierhead'.

**Wharves and quays:** These are constructed parallel to the breakwater or shore within the harbour for berthing of vessels alongside for loading and unloading of cargo. They have wide platforms built atop earth fills.

**Jetties and piers:** These may be solid or open type structures with a wide platform on top to permit cargo working of vessels. They are built out from the shore in order to reduce silting and the need for dredging.

**Locks and lock basin:** *Lock basin* is an enclosed basin in which a number of vessels are berthed, with an entrance which is controlled by lock gates. The water inside such a basin is generally not affected by changes of water level outside.

**Harbour docks:** These are also known as *wet docks*, and are sheltered basins for berthing ships; water level in the dock may be kept more or less constant to facilitate loading and unloading of cargo. They are surrounded by structures such as quays or piers.

**Dry docks:** Also called *graving docks*, these are places where vessels can be taken in, gates closed, and the water inside pumped out to keep the dock dry. This facilitates repairs to be carried out for the vessel as needed; thus, these may sometimes called *repair docks*. When a dry dock is used to build a ship, it is called a *building lock*.

**Slipways:** These are formed by rails laid on firm ground at a gentle uniform slope. This facilitates a vessel to be hauled up above the water, repairs carried out and pushed back into the water. However, docks are considered to be a better arrangement.

**Transit sheds and warehouses:** These are meant as storage facilities for cargo before loading for export, or for distribution to the destinations on land after import.

**Ancillaries and miscellaneous facilities:** These include anchors, buoys and lights, and miscellaneous facilities for fire protection, service units, and cargo-handling equipment such as cranes.

**Size and shape of harbour:** The 'size' of a harbour depends upon the nature and volume of marine traffic, that is, the number and size of the ships using the facility. The use of tugs to assist the manoeuvring of the ships in docking, as also the conditions at the site affects the size required. The size chosen should be adequate for safe and convenient operations.

The minimum area required is that to accommodate the turning basin and the docking facilities. Since some of the large ships are 300 m long and 30 m wide, the size of the harbour should be such that it can hold the expected number and type of vessels using it at one time.

The *shape* of a harbour is governed by the following considerations:

- One of the pierheads should project beyond the other for affording protection from sea-waves.
- The width should increase rapidly inside the entrance in order to facilitate quick and easy manoeuvring of the vessels.
- The boundary should consist of a series of straight portions; re-entrant angles should be avoided for further safety in operation.

**Depth of harbour:** The depth of the harbour should be adequate for the draft required to accommodate the largest ship expected to use the facility. The vessel should rest in the water in proper balance up to the Plimsoll mark or the legal load line for different waters at different seasons of the year.

The following formula is used as a guideline for the depth, $D$, of a harbour:

$$D = D_1 + \frac{H}{3} + D_S \tag{17.1}$$

where $D_1$ = Draft of the largest ship to be accommodated.

$H$ = Height of the storm waves.

$D_S$ = Allowance for squat of the moving ship. (*Squat* is the maximum variation of level of a moving ship.)

The maximum harbour depth, $D_{max}$, below the lowest level of low water is specified as

$$D_{max} = D_1 + 1.20 \quad \text{(metres)} \quad \text{(for soft bottom of sea-bed)} \tag{17.2 a}$$

$$D_{max} = D_1 + 1.80 \quad \text{(metres)} \tag{17.2 b}$$

where $D_1$ is the draft of the loaded vessel.

**Layout of a Harbour**

There are no specific rules governing the layout of a harbour; ingenuity coupled with experience plays a significant role in producing a successful layout in a given set of circumstances.

There are several instances of excellent harbour layouts – for example, Madras (Chennai) harbour with its strong breakwaters and Visakhapatnam harbour, with an ingenious system of protection for the navigation channel from fouling by silt accumulation and littoral drift, are typically good layouts.

A well-planned harbour layout should provide necessary protection from storm waves and the effect of littoral drift – sand drifts that occur in the proximity of the coast or foreshore, nearly parallel to the coast. The other factors to be considered are the shape and size of the harbour, and the layout of other structures such as wharves and jetties.

Irrespective of the littoral drift, there is bound to be some silting, which needs periodic dredging in order to maintain the necessary draft for the largest vessels expected to use the harbour.

The approach to the harbour entrance should be so designed as to reduce the wave height within the harbour to a safe level.

## Formulae for the Reduction of Wave Height in Harbours

**Stevenson's formula:**

$$h_1 = h.\left[\sqrt{\frac{b}{B}} - 0.027D^{1/4}\left(1 + \sqrt{\frac{b}{B}}\right)\right] \tag{17.3}$$

where

$h_1$ = Reduced wave height within the harbour

$h$ = Wave height in the sea at the entrance

$b$ = Width of harbour entrance

$B$ = Breadth of the harbour at the point of observation

$D$ = Distance from the entrance to the point of observation.

(This formula is applicable only to small harbours and vertical walls.)

**Minikin's formula:**

$$h_1 = h.\left[\left(C_1 - \frac{D}{24,400}\right)\left(\frac{b}{B}\right) + \left(C_2 - \frac{D}{24,400}\right)\right] \quad (17.4)$$

where

$C_1$ = Shape constant (1.0 for medium divergence and 0.9 for curving arm –diverging walls)

$C_2$ = Constant for spending beaches (0.12 where there are no spending beaches, 0.10 for spending beaches 30% of harbour perimeter, and 0.08 for spending beaches 50% of the harbour perimeter.)

Other Symbols are as in Stevenson's Formula

[*Spending beaches* are protective beaches which help in dissipating wave energy without causing adverse effects, affording protection to the upland in the vicinity.]

A schematic diagram of *Madras (Chennai) harbour*, an *excellent artificial harbour*, is shown in Fig. 17.7.

## 17.3 PLANNING AND LAYOUT OF PORTS

A port is a harbour with terminal facilities. Such facilities are needed for the development of commerce and trade, inclusive of international trade. The various facilities required for the development of ports are channels, anchorages, moorings, breakwaters, basins, wharves, piers, docks, cargo-handling equipment, repair plants, slips, and repair plants. All these are necessary requirements for a first class port to afford interchange between marine and land transport.

Port facilities are of utmost importance for efficient operation of the ships.

### Requirements of a Port

(i) The location should be centrally situated to the hinterland, that part of the land area to be served with efficiency by the port.

(ii) It should well-connected to the rest of the country by means of highways and railways.

(iii) The hinterland should be sufficiently developed for generating cargo traffic.

(iv) It should provide the maximum facilities for all the visiting ships.

(v) The passage into the port must have sufficient width and depth of water, and must be properly delineated to aid navigation.

(vi) The soil along the coastline should be sufficiently hard to provide good foundation to all the structures needed for the port.

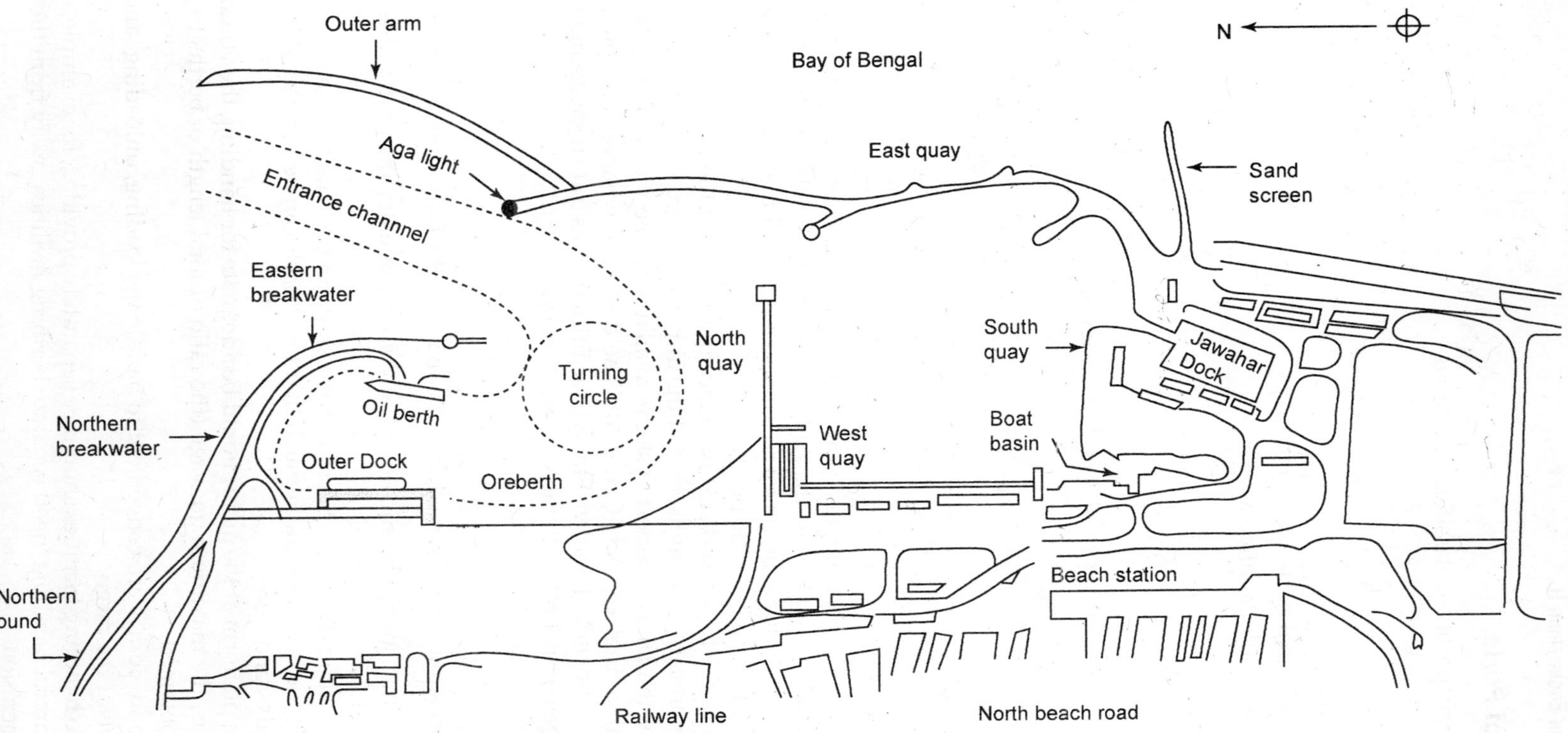


**FIG. 17.7** *Schematic of Madras harbour (Chennai)*

## Classification of Ports

Based on the *location*, ports are classified as follows:

- Sea ports
- River ports
- Canal ports

Of these, river and canal ports are inland ports

Based on the *size*, ports are classified as:

- Major ports
- Intermediate ports
- Minor ports

Based on the *nature of the cargo handled, and the use* they are:

- Coal ports
- Grain ports
- Ports of call
- Trans-shipment ports

## Planning of Ports

The data required for the planning of a port include physical characteristics such as the size, shape, depth, and entrance width of the harbour, facilities available in the proximity or the nearest city, nature and volume of the cargo to be handled, and the seasonal variations of the various factors or components of the port.

Characteristics of the hinterland also influence the design of the port, since it is the point of origin of the export cargo and utilisation of the imported cargo. This naturally is governed, to some extent, by the land transport systems available in the vicinity. Large ships can be operated only with a minimum of 10 m draft. The design of wharves and piers should be suited to the commodity to be handled. Storage facilities such as transit sheds and warehouses provided should be compatible with the volume of marine cargo.

## Port Layout

The layout of a port should be designed, bearing in mind, the following requirements:

- The entrance channel should be such that the vessels can enter the port and leave it in an easy and comfortable manner.
- The turning basin should be of such a size that a ship can turn fully in the basin itself.
- The alignment of the quays should be such that the vessels can be berthed alongside easily even under windy conditions.
- The width of the platform at the quay should be adequate for handling the volume of cargo expected – that is, the maximum quayage (the ratio of total length of berths to the area of basin) must be obtained.
- Railway tracks and sidings are convenient to handle the loading, unloading, and transport to the destination of the cargo.
- Specialised cargo handling facilities should be separately provided; for example, explosives, which require careful handling, need separate berthing facilities, away from the others, for ensuring safe operation.

### Types of Port Layout

(a) *Square layout*: This is the simplest geometrical shape for the port basin.

(b) *Rectangular layout*: This is also simple and affords more quayage than a square layout. Quay walls are built in a direction perpendicular to the longer sides and projecting from them. However, turning operation gets restricted.

(c) *Rectangular layout with a central pier*: The pier projects from the shorter side parallel to the longer sides. Although greater quayage can be got, turning gets restricted by the pier.

(d) *Machicolated layout*: This layout is better than the other rectangular layouts; quays are provided perpendicular to the longer sides in a staggered manner, as shown in Fig. 17.8. This is a layout wherein more quayage is obtained and all amenities are provided for the port. A turning basin is provided at the centre for free turning of the vessels.

(e) *Tridentine layout:* This is a modification of the machicolated layout, in which the longer sides are built to flare out as shown in Fig. 17.9.

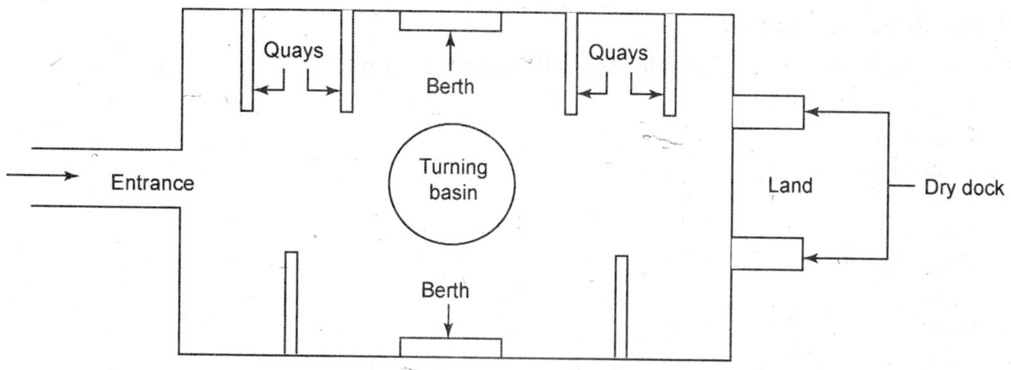


**FIG. 17.8** *Machicolated layout of a port*

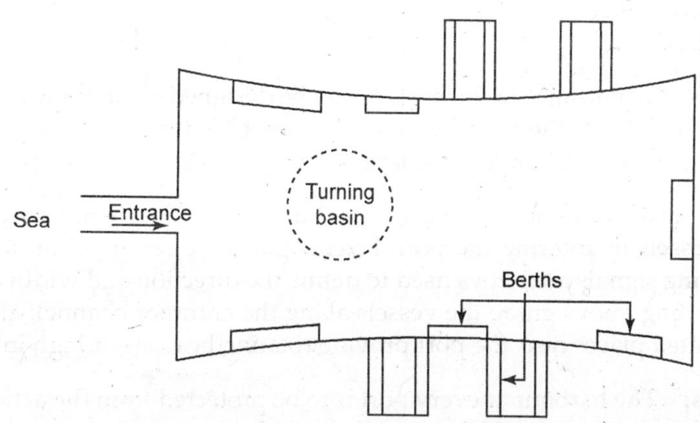


**FIG. 17.9** *Tridentine layout of a port*

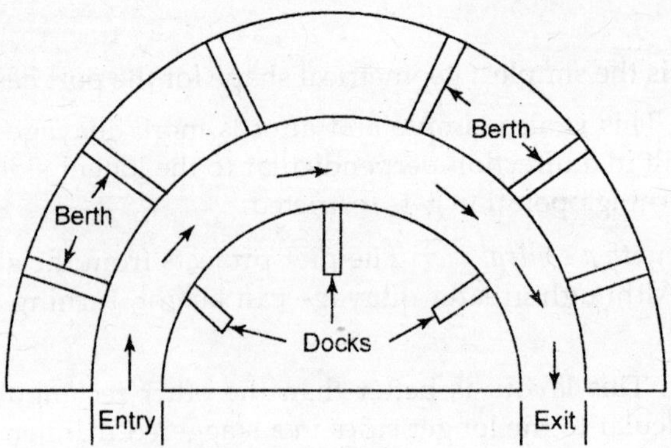


**FIG. 17.10** *Digital layout of a port*

(f) *Digital layout*: In this semi-circular layout, no free turning space is provided, as there are two channels–one for entry and the other for exit. The berths are provided radially, as shown in Fig. 17.10. Large quayage is the good feature of this layout.

## Facilities Required at a Port

A port is a harbour with facilities for the following functions:

1. Entry
2. Guiding
3. Protection
4. Lock
5. Dock
6. Cargo-handling
7. Repair
8. Storage
9. Off-shore anchorage
10. Quarantine inspection
11. Turning
12. Dredging
13. Miscellaneous services
14. Administration

**Entrance facilities:** The entrance to the port should be designed in such a way that the widest ship can be manoeuvred into the harbour easily. The depth of water at the entrance should be marginally more than that required for the largest ship expected to use the port.

**Guiding facilities:** Fixed or floating signals, whistling signals, and light houses are provided for guiding the vessels in entering the port. Fixed signals are stationed on top of piers and breakwaters; floating signals are buoys used to define the direction and width of approach; in bad weather, whistling buoys guide the vessels along the entrance channel; and light houses positioned at elevated places near the port provide rotating beacons of high intensity.

**Protection facilities:** The harbour at every port is to be protected from the action of wind and waves, either by natural formation of the coastal area or by constructing appropriately aligned breakwarters..

**Lock facilities:** If the tidal range is greater than 3 m, a water lock is considered necessary to control the entry and exit of the tidal waters into the harbour basin, in order to provide a relatively calm area.

**Docking facilities:** Wharves, quay walls, or piers are constructed with aprons or platforms to form wet docks, wherein vessels can be berthed in such a way that the oscillations of water due to tidal currents are minimised, to facilitate easy operations.

**Cargo-handling facilities:** These are primarily cranes on permanent track, provided to operate on aprons and supported on quay walls of wet docks. Conveyors and vacuum pumps are also provided for handling certain types of cargo.

**Repair facilities:** Graving or dry docks are provided in which vessels can be berthed and lifted, if necessary, for carrying out the necessary repairs.

**Storage facilities:** Facilities such as transit sheds for temporary storage before export or after import, warehouses for extended storage, and cold storage facilities for perishable goods such as fruit, medicines and chemicals, are also to be provided at major ports.

**Offshore anchorage facilities:** These are necessary when the ships have to be kept in waiting in queue outside the harbour basin, or when the draft inside the port is not sufficient for large visiting vessels. In such cases, alternative procedures are used for loading and unloading of such ships.

**Quarantine inspection facilities:** If passengers from abroad are to be screened medically for infectious diseases, inoculated, and cleared for entry, their vessels are kept outside the port by off-shore anchorages, temporarily.

**Turning facilities:** A turning basin of adequate size has to be necessarily provided within the harbour basin, preferably near the entrance, in order to permit the vessels to turn around for the purpose of entering the docks, or for leaving the port.

**Dredging facilities:** Excavation of the bed in the harbour basin for removing the silt accretion in order to increase the depth of water or draft is known as dredging. This may become necessary if the port has to serve larger vessels than originally planned. Dredgers are used for this purpose. If it is only a one-time operation, this is known as primary dredging; if it is required to be done periodically, it is known as maintenance dredging.

**Miscellaneous services:** These include fire protection measures, service units, and cargo-handling equipment for specialised cargo, as also anchorages and moorings.

**Administrative facilities:** Administrative offices, custom houses, cargo-booking offices, shops, post-offices, banks, security services, communication facilities such as telephones, satellite phones and wireless are necessary in and around a major port.

## Important Ports in India

There are 12 major ports and over 200 notified ports in the minor and intermediate categories. The major ports are in Gujarat, Maharashtra, Karnataka, Tamil Nadu, Kerala, Andhra Pradesh, Odisha, Goa, and West Bengal. The other notified minor ports are also located mostly in these states.

The following are the major ports:

**Nhava Sheva Port**

This is most modern and best equipped harbour, renamed as 'Jawaharlal Nehru Port'. It is built across the eastern shore and Elephant Island in Bombay (Mumbai), and is meant to relieve the congestion at Bombay Port. It is designed to handle dry bulk cargo and container cargo.

**Bombay Port**

This is the biggest port and is a natural harbour, established under the Bombay Port Trust in 1873. It handles foreign trade in petroleum and petroleum products and heavy machine, in addition to dry cargo. It has three Wet Docks – Princess, Victoria and Alexandria Docks, and two Dry Docks – Hughes and Merewether dry docks. The port has its own railway.

**Cochin Port**

The Cochin (Kochi) Port in Kerala is a natural harbour formed at the entrance of a lagoon in the Arabian Sea. It handles petroleum products, fertilisers, and raw materials.

**Kandla Port**

This port lies in the Kutch area of Gujarat, and is an excellent natural harbour. It is a tidal port, developed soon after independence in the wake of the loss of Karachi Port to Pakistan. It handles petroleum products including crude oil, fertilisers, cotton, cement, sugar, and edible oils. It serves the states in its vicinity.

**Madras Port**

It is the oldest artificial port and the third biggest one in India. It is used to import coal, cotton, fertilisers, electrical goods, grain, machinery, and so on.

**Calcutta Port**

It is an inland riverine port, about 130 km from the Bay of Bengal; it is a tidal port, requiring frequent dredging of the Hooghly river. Water from the Farraka Barrage on the river Ganga is used to supply water to ensure navigability.

**Haldia Port**

Developed downstream of the Calcutta Port at Haldia, it supplements the facilities of the latter. It handles petroleum products, fertilisers and mineral oil.

**Visakhapatnam Port**

This is situated in Andhra Pradesh on the coast of the Bay of Bengal. It is a deep, land-locked and protected semi-natural-harbour. The outer harbour exports iron ore, petroleum products and dry cargo.

**Marmugao Port**

This is located in Goa at the mouth of the Zuari River, which joins the Arabian Sea. It has mechanised facilities for handling bulk cargo.

**Paradeep Port**

This is a newly developed port in Odisha on the Bay of Bengal. It exports iron ore, and handles coal and dry cargo.

**Mangalore Port**

It is located in Karnataka on the Arabian Sea. It exports primarily Kudremukh iron ore.

**Tuticorin Port**

This is the latest major port in Tamil Nadu at the south-eastern extremity of India. It handles coal, salt, chemicals, and edible oils.

## 17.4 ENVIRONMENTAL IMPACT ASSESSMENT (EIA)

While planning water transportation projects, *environmental impact assessment* (*EIA*) has to be made in respect of the following *aspects*:

(a) ***Assessment of the existing conditions***

- (i) Flora and fauna
- (ii) Water quality
- (iii) Quantitative and qualitative uses of water
- (iv) Existing effluents
- (v) Area involved in the project.

(b) ***Prediction of future effects***

- (i) Of water quality on aquatic life
- (ii) Of water quality on social and economic factors.

(c) ***Consideration of alternatives*** because of adverse effects, as also for justifying cost benefit.

(d) ***Resource commitments*** for controlling adverse environmental impact.

(e) ***Identification of environmental monitoring processes*** for initiating remedial action.

Items of environmental concern in water transportation planning include moorings, bulk-handling facilities, dredging operations, transportation of petroleum products and toxic substances on inland routes. An *environmental impact statement* (*EIS*) has to accompany such water transportation project reports.

## 17.5 PORT AUTHORITIES AND ASSOCIATED BODIES

These are meant for the administration and promotion of maritime interest such as planning and development of water transportation, promotion of marine traffic, and encouragement of international trade and commerce.

The following are the prominent regulatory authorities and bodies that govern marine traffic:

1. National Association of Port Authorities
2. National Bureau of Customs
3. National Maritime Administration
4. Bureau of Foreign Commerce

5. Inter-state Commerce Commission
6. Department of Agriculture
7. Immigration and Naturalisation Service
8. Quarantine and National Public Health Service
9. Security Association
10. Waterfront Commission
11. National Coast Guard

**National association of port authorities:** Many countries have their own national association; these organisations combine technical and trade associations for the promotion of maritime industry. Separate committees look after different activities such as port planning development and construction, port maintenance and operation, terminal rates, and legislation.

**National bureau of customs:** This is meant to collect tariffs due to the government in accordance with the Tarff Act; prevention of smuggling of undeclared and illegal merchandise is another responsibility.

**National maritime administration:** This is under the Department of Commerce and acts as the administrator of regulation for the control of rates, services, routes, agreement, and such related items in foreign commerce.

**Bureau of foreign commerce:** This also is under the Department of Commerce. It distributes marketing information for manufacturers and exporters.

**Interstate commerce commission:** This is a regulatory body which controls services and rates for coastal and inland water transportation in interstate commerce.

**Department of agriculture:** There is a division called 'Fruit and Vegetable Division' in the Department of Agriculture to control and deal with perishable food such as fruits and vegetables, and to enforce the laws of the Perishable Commodities Act, and other related statutes.

**Immigration and naturalisation services:** This body deals with the entry of foreign citizens into the country. People of foreign origin have to present themselves before the immigration officials with all the necessary documents to determine their eligibility for temporary or permanent residence under the laws.

**Quarantine and national public health service:** Foreign vessels have to be quarantined and the passengers screened for infectious diseases. Such vessels are anchored in a separate area, sufficiently away from the area of regular port operations.

**Security association:** This body provides help in controlling pilferage of goods and cargo, and provide all the necessary security measures.

**Waterfront commission:** This deals with the prevention of unjust practices in the hiring of waterfront labour, and the methods of loading and unloading of cargo. It also ensures that only persons of good service record in their prior work history are hired at the waterfront.

**National coast guard:** This organisation is instituted for the prevention of illegal and unsafe practices on high seas and in all other navigable waters. They also render help in times of bad weather, provide and maintain navigation aids, and also take severable measures to make navigation on high seas safe.

## SUMMARY

- A harbour is a partly enclosed area of water to provide safe shelter for sea-faring vessels. A port is a harbour where several facilities are available for loading and unloading of cargo, storage, and distribution via land transport, and repair facilities for ships.
- Harbours may be natural, semi-natural, or artificial based on how the protection is obtained; they may be for refuge, commercial, fishery or military purposes; ocean, river, lake or canal harbours based on their location.
- Site selection is governed by several factors – availability of natural protection from waves and tides, winds and cyclones, land transport availability for feeder routes, and strategic aspects.
- Site investigation involves surveys and meteorological, oceanographic, geological and soil investigation, all of which govern the design of the components of a harbour.
- The salient features of a harbour are the approach channel, anchorages, size, shape and depth of the harbour, turning basin, harbour basin, breakwaters, docking facilities, and so on.
- Several facilities are required at a port – those for guiding the vessels, protection of vessels, wet docks for cargo-harbour dry docks for repair of vessels, cargo – handling equipment, storage facilities, and dredging facilities for maintaining the necessary draft for the vessel.
- India has 12 major ports – at Nhava Sheva, Mumbai, Kochi, Kandla, Chennai, Kolkata, Haldia, Visakhapatnam, Mormugao, Paradeep, Mangalore and Tuticorin. There are over 200 minor and intermediate ports besides the major ones.
- Environmental Import Assessment (EIA) is necessary in the planning of major water transportation projects.
- Several authorities and bodies are formed in every country for administration and maritime interests, such as the development of water transportation, promotion of marine traffic and encouragement of international trade and commerce.

## QUESTIONS

1. Write a brief note on water transportation.
2. Differentiate between harbour and a port.
3. What are the important requirements of a harbour?
4. Explain the criteria of classifying harbours.
5. Explain briefly the difference between a natural harbour and an artificial harbour. Give one example of each from India.
6. Classify harbours based on their location.
7. What are the considerations in the selection of site for a harbour?
8. Enumerate the salient components of a harbour.
9. What are the requirements of a port?
10. Explain the classification of ports.
11. What are the various kinds of facilities needed at an ocean port?

12. Enumerate the major ports in India.
13. Explain the significance of Environmental Import Assessment in planning a port.
14. What is the role of different Port Authorities in operating commercial ports in our country?
15. Write brief notes on:

    (a) Semi-natural harbour

    (b) Depth of water at a harbour

    (c) Free port.
16. Write a short note on:

    Natural harbour. (VTU)
17. Explain, with neat sketches, the natural classification of harbours. (VTU)
18. Explain briefly, with diagrams, the classification of harbours (West Bengal Tech. University)
19. (a) Define a harbour.

    (b) Name the types of harbour according to the different systems of classification.

    (c) What is the difference between a harbour and a port? (West Bengal Tech. University)
20. What are the requirements of a good harbour? (West Bengal Tech. University)
21. Explain different types classification of harbours. How are they distinguished from ports? (Mumbai University)
22. Explain the various characteristics of ships and their influence on port management and operations. (Mumbai University)
23. Explain the various port facilities with neat sketches. (Mumbai University)
24. Explain the types of cargo-handling equipment at ports. (Mumbai University)
25. Explain the common terms used in connection with the characteristics of ships.
26. Differentiate between:

    (a) Natural roadsteads and artificial roadsteads

    (b) Port and hinterland

    (c) Commercial harbour and fishery harbour.
27. What are the kinds of facilities provided at a marina?

Chapter 18

# NATURAL PHENOMENA AFFECTING HARBOUR DESIGN

**CHAPTER OBJECTIVES**

This chapter explains:

- The effects of natural phenomena such as winds, cyclones and storms, waves and tides on the design of harbours and component structures
- The study of wind velocities and their variations using wind-rose diagrams
- The causes occurrence of sea waves, their characteristics, and their effect on harbour structure
- Tides, Newton's equilibrium theory, effects of attraction forces of the moon and the sun; spring and neap tides, tidal ranges and the prediction of tides
- The effect of tidal forces on harbour structures
- Silting and erosion, the phenomena of littoral drift; and how silting decreases the draft and erosion affects foundation stability.

## 18.1 INTRODUCTION

Certain natural phenomena such as those relating to meteorology and oceanography affect the planning and design of harbours; as such, it is necessary to study such phenomena, and their possible effects on the design of the various harbour components.

Meteorological phenomena include winds, cyclones and storms. Oceanographic phenomena include waves, tides and tidal currents, and silting and erosion. Study of these phenomena enables the harbour engineer to assess the conditions at the proposed site and take them into consideration during the planning and design of the components and structures of the harbour.

This ensures an effective design of the harbour for safe and convenient operation of the marine traffic expected to use the facility.

## 18.2 WIND

*Wind* is defined as air in motion. Wind has to be specified by means of its speed as well as direction. In maritime parlance, wind velocity is expressed in *Knots*. One knot is one *nautical mile* per hour.

A nautical mile is the length of the arc along the meridian of the earth, which subtends a one-minute angle at the centre. Since the equatorial diameter is greater than the polar diameter owing to the earth's spheroidal shape, the arc length equivalent to one minute of central angle varies slightly in different latitudes; an average value accepted in maritime practice is 1852 m (6080 ft).

$$\therefore \quad 1 \text{ Knot} = 1852 \text{ m/h}$$

$$= 1.152 \text{ miles/h}$$

$$= 1.853 \text{ km/h}$$

Air, just like water, flows from a point of higher potential to a point of lower potential. Potential, here, means the sum of elevation head, pressure head and velocity head. It varies from place to place in the atmosphere due to variations of temperature over the earth's surface and consequent changes in pressure.

Land and sea breezes are caused by convection currents. Land has a lower specific heat than water. Thus, land heats up more quickly than water; it also loses its heat more quickly than water. This results in higher temperatures on land during day time than on water. Hot air is pushed upward by the cooler air from the sea. Thus, cool *sea breeze* blows toward the land in the afternoon until the evening.

During night time, the process gets reversed; the land cools more quickly than water, the warm air over the water is forced upward, and the wind blows from the land toward the water, setting up a *land breeze*. This is the general pattern of wind flows during summer; this pattern gets reversed during winter months.

Tropical regions near the equator receive sun's rays more directly than those near the polar region. The heavier cool air from the polar region displaces the lighter warm air in the tropical regions, which tends to move back toward the poles, setting up wind flows. Earth's rotation about its axis form west to east imparts the same angular velocity to all the points on the surface; but the eastward tangential velocity varies from a maximum at the equator to zero at the poles, in proportion to the radial distance to the point. Thus, the equatorial air moves north with a greater tangential velocity than that air at greater latitudes. This process sets up seasonal wind patterns on the globe between the equatorial and the polar region.

### Wind Speed

This is measured in km/h in practice. Surface friction encountered by the winds over water being more than that over land, the wind speed over the sea is about two-thirds of that at a height of 600 m, while over land it is about one-third the speed. This speed is known as the 'gradient speed.'

Prior to the development of precise instruments in meteorology, wind speed was estimated by its effects. In this context, Admiral Beaufort of the British Navy devised a scale for wind strength based on the speed, which ranges from 0 to 12, as the wind speeds and its adverse effects increase. Description of the wind is made using words phrases such as *slight breeze, strong breeze*, and so on.

The estimates from this scale have been found to be satisfactory.

The details of the Beaufort scale, the corresponding wind velocity and its effects on the sea are given below:

## Beaufort Scale

| Wind description | Beaufort no. on the scale | Wind velocity km/h | Effects noticeable on the sea |
|---|---|---|---|
| Calm | 0 | 1.6 | Sea is smooth. |
| Light | 1 | 3.2 | Small wavelet-like scales. |
| Slight breeze | 2 | 8 | Waves are short but more pronounced. |
| Gentle breeze | 3 | 16 | Crests begin to break; foam has glassy look. |
| Moderate breeze | 4 | 24 | Waves are longer with many white horses. |
| Fresh breeze | 5 | 34 | Waves are more pronounced. White foaming crests everywhere. |
| Strong breeze | 6 | 43 | Longer waves foam; foaming crests extensive. |
| Moderate gale | 7 | 56 | Foam begins to flow in streaks. |
| Gale | 8 | 67 | Waves increase; foam is blown in dense streaks. |
| Strong gale | 9 | 80 | Waves increase much more; foam is blown in dense streaks. |
| Whole gale | 10 | 95 | High waves with long over-hanging crests; great foam patches. |
| Storm | 11 | 110 | Waves are high enough to hide ships in their troughs. |
| Hurricane | 12 | More than 120 | Sea covered with streaking foam; air filled with spray. |

## Wind-rose

This is a diagrammatic representation of the wind direction and duration (in percent time) for a significantly long period like one to five years, observed and recorded systematically at the particular site, or at a location in close proximity to the proposed site.

There are two types of wind-rose:

*Type-I*: Shows the direction and percent duration of the wind.

*Type-II*: Shows the direction, duration and wind speed.

Wind roses can be constructed based on monthly data, yearly data, and so on.

This kind of representation of the data on wind characteristics helps in deciding the best orientation of the various components of a harbour.

## Cyclones and Storms

Atmospheric disturbance accompanied by high winds is known as a *storm*. It may be accompanied by low-quantum precipitation and thunder and lightning. A *tropical storm* is one which originates in the tropical regions. A storm is characterised by wind intensity more that of a *gale*, but less than that of a *hurricane*; it is set at No. 11 on the Beaufort scale.

Storms and cyclones can cause severe destruction on the shore by the wind velocities, as also by the violent sea. A cyclone is a violent tropical storm in which the air moves very fast in a circular direction. This is also sometimes referred to as cyclonic storm. Storm and Cyclone are not differentiated in the Beanfort scale.

The meteorological department of the country prepares the *tracks of cyclonic storms*, based on the observed data; this helps in knowing the proneness of a region to cyclones.

Cyclonic storms occur both in the Bay of Bengal and in the Arabian Sea. They are more frequent every year in the Bay of Bengal; however, the few that occur in the Arabian Sea have been known to be more devastating; for example, the one which occurred in October 1975, which crossed the Gujarat coast near Porbandar.

An exceptionally violent hurricane occurred in October 2014 ('Hud hud') in the Bay of Bengal, which devastated Visakhapatnam with wind speeds of over 200 km/h.

## 18.3 WAVES

Waves on the sea are caused by various disturbances – artificial or natural. Artificial sources are moving vessels, especially large ships and ocean-liners, and man-made explosions. Natural sources are winds, tides and earthquakes.

Wave forces are the most powerful forces which affect the design of harbours and related marine structures.

In general, waves produced on bodies of water due to the reasons cited above are called *water waves*, to differentiate them from others such as energy waves, sound waves, light waves, and electromagnetic waves.

### Origin of Waves

Water waves are generated by the transfer of wind energy to the water surface. Tangential stresses between the wind and water surface, and the difference in pressure exerted by the wind at the rear and the front of the waves result in the progression of waves.

Water waves could be either *waves of oscillation* or *waves of translation*. While the former are stationary, the latter start as waves of oscillation and get transformed into waves of translation by the action of wind on the water surface. These are all *gravity waves*. Water waves also arise due to surface tension. These waves are called *ripple* or *capillary waves* and are of small height and short length.

### Factors Governing the Characteristics of Water Waves

The factors which govern the characteristics of water waves are:

(i) Wind velocity
(ii) Wind duration
(iii) Bed or bottom friction
(iv) Depth of water

Water waves attain their typical characteristics for a particular wind velocity only when the wind blows over it continuously for some time.

The wave increases in height rapidly at first but the rate of increase starts decreasing as the wind lasts longer.

Bed or bottom friction and breaking action limit the wave height in shallow waters. A steady state is reached as the wind energy is dissipated by the bottom friction and no energy is left for wave growth.

Waves in shallow water tend to break when the wave height reaches about 0.8 times the still water depth; at this point, the additional wind energy is spent in breaking the crest. Thus, in shallow water, waves break and become translatory; while in deep water, waves remain oscillatory.

When water waves move from deep water towards shallow water, bottom friction reduces the wave height, while 'shoaling effect'[1] tends to increase it; the wave energy will be confined in increasing depth of water.

Thus, the characteristics of the waves depend primarily upon the wind velocity, bottom friction, and the depth of water.

### Wave Decay

This is caused mainly due to air resistance; viscosity and pressure have negligible effect. As waves travel away from the place of their origin, energy is continuously lost, which leads to reduced wave height, increased wave length and increased wave velocity.

## Wave Form

In shallow water with depth of about 1.25 times the wave height, the wave usually breaks. In deep water, each water particle on the wave surface tends to describe a circular path of radius one-half of the wave height; with centre at mid-height. Figure 18.1 shows the wave and the relative positions of the water particles.

Wave height, $H$, is also known as *amplitude* of the wave. Distance between consecutive crests and troughs is the wave length, $L$.

The time for two consecutive crests to pass a particular point is the wave period, $T$. The speed of propagation of the wave is the wave velocity, $V$.

### Determination of Wave Characteristics

The relation between the velocity, $V$, the length, $L$, and the depth of water, $d$, is given by

$$V = \sqrt{\frac{gL}{2\pi} \tan h\left(\frac{2\pi d}{L}\right)} \tag{18.1}$$

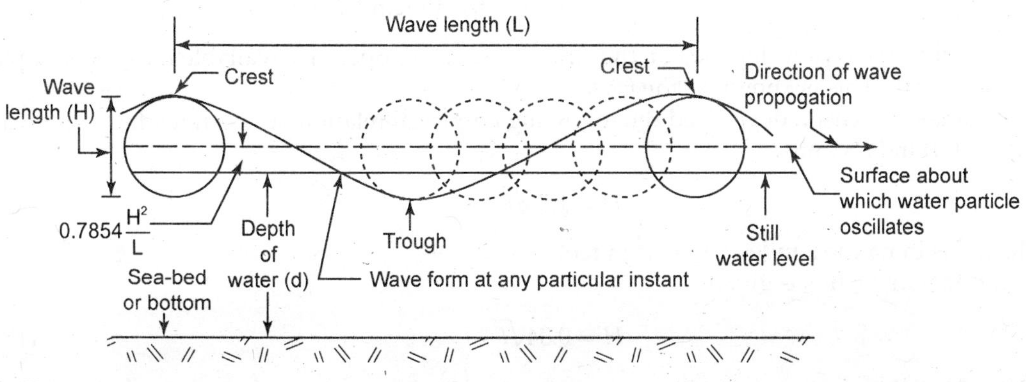


**FIG. 18.1** *Water wave*

[1] Shoaling is the formation of shoals, or raised areas of sand, under the surface of water due to sediment transport and deposition phenomena near the sea-bed.

$g$ is the acceleration due to gravity

If $d = \frac{L}{2}$, this reduces to

$$V = \sqrt{\frac{gL}{2\pi}.\tanh(\pi)} \tag{18.2}$$

$$\tanh(\pi) = 0.9963\text{, or very nearly unity.}$$

For greater values of d, $\tanh\left(\frac{2\pi d}{L}\right)$ approaches unity.

$$\therefore \qquad V = \sqrt{\frac{gL}{2\pi}} \tag{18.3}$$

For $\left(d \geq L/2\right)$

Also, $$V = \frac{L}{T} \tag{18.4}$$

From these two expressions for $V$,

$$L = \frac{gT^2}{2\pi} \tag{18.5}$$

This is known as Bertin's formula.

Or $$T = \sqrt{\frac{2\pi L}{g}} \tag{18.6}$$

With respect to the axis of propagation of the wave, the still water level is lower by a quantity $\rightarrow 0.7854\frac{H^2}{L}$. Substituting for the value of g in $m/s^2$, Bertin's formula reduces to

$$L = 1.56\ T^2\ (L \text{ in m and } T \text{ in s}) \tag{18.7}$$

The height of sea wave depends on the straight stretch of open sea available for uninterrupted wave growth. This is known as the *fetch*.

Thomas Stevenson developed the following empirical relation between fetch ($F$) and wave height ($H$) in FPS units:

$$H = 1.5\sqrt{F} \tag{18.8}$$

Here, $F$ is in nautical miles and $H$ is in feet.

In MKS units, this is given as

$$H = 0.34\sqrt{F} \tag{18.9}$$

Here, $F$ is in kilometres and $H$ is in meters. Fetch, in this context, refers to the extent of sea over which the wind has been blowing in a relatively uniform direction; it does not indicate the free distance. Thus, although there is a stretch of 800 km of open sea near Karachi with great depth, the highest waves do not exceed 4.5 m in height.

The height of storm waves on important waterways, as observed over a long time, is given below:

| Waterway | Wave height |
|---|---|
| Pacific ocean | 15 to 18 m |
| Tropical oceans | 15 to 18 m |
| Atlantic ocean | 9 to 12 m |
| Great lakes | 8 m |
| Black sea | 10 m |
| Mediterranean sea | 4.5 to 6 m |
| North sea | 3.5 to 4.5 m |

Note: Wave length can be 18 to 33 times the wave height depending upon the sea and wind characteristics.

### Significant Wave

Waves are not uniform; they are generated with varying heights, lengths, and periods. When they move over long distances, their characteristics are altered due to wave decay.

In view of this, the concept of *significant wave* has been developed by Munk; the *significant wave height* ($H_s$) is taken to be one-third of the highest wave. The period of such a wave is obtained by observations as the average of 10 to 15 successive prominent waves.

The following relationships are given:

(i) The greatest wave height $H_{max} = 1.87\ H_s$ (18.10)
(ii) Average wave height, $H_a = 0.60\ H_s$ (18.11)
(iii) Average height of 10% of waves, $H_{10} = 1.27\ H_s$ (18.12)
(iv) Average height of 1% of waves, $H_1 = 1.67\ H_s$ (18.13)

*The height of simple sine waves, having the same energy as the actual wave train, is taken to be 0.8 times the significant wave height.*

### Modification of Water Waves

This involves changes in the height and length of waves. Modification may be because of one or more of the following phenomena:

(a) Decay
(b) Breaking
(c) Diffraction
(d) Refraction
(e) Reflection

**Decay of waves:** This occurs because of air resistance; it is a slow process in which the height and length of the wave reduce to zero, if allowed uninterrupted.

**Breaking of waves:** As the depth of water decreases, shallow water waves tend to break when the depth of water equals the wave height.

**Diffraction of waves:** At the entrance to a harbour, waves from the open sea and the approach channel will enter into the harbour through a restricted width, and the energy of the waves gets decreased as the wave front travels inside. The reduced wave height inside the harbour can be determined from Stevenson's formula, as given earlier.

**Refraction of waves:** When water waves travel in zones of decreasing depth, the flow of water nearer to the bottom or bed becomes parallel to the surface at the bottom; this phenomenon is somewhat similar to refraction.

**Reflection of waves:** When water waves strike a vertical wall or a marine structure such as a breakwater, the waves get reflected and impede the crests of the incoming waves; this results in the formation of a standing wave, known as 'Clapotis'. This phenomenon occurs when the depth of water is more than about 1.5 times the wave height. Harbour structures, like piers, should be aligned such that the reflected waves do not impinge on them.

## Dynamic Effects of Waves

The pressure exerted by waves impinging on the surface of a harbour structure can be determined from the principles of dynamics.

This is given by

$$p = \frac{\gamma_w . V^2}{g} \tag{18.14}$$

Here,

$p$ = Pressure exerted by the waves

$\gamma_w$ = Unit weight of water

$V$ = Wave velocity

$g$ = Acceleration due to gravity

In *deep waters*, when the water depth is very large compared to the wave length, velocity is nearly equal to that of a free fall through a height equal to half the radius of the circle, with circumference equal to the wave length.

$$\therefore \quad V = \sqrt{2g . \frac{1}{2} . \frac{L}{2\pi}} = 1.25\sqrt{L} \tag{18.15}$$

Considering the wave as a cycloidal curve, the wave height is $\frac{L}{\pi}$

$$\therefore \quad V = \sqrt{5H} \text{ (approx.)} \tag{18.16}$$

Substituting in Eq. (18.14),

$$p = \frac{\gamma_w . 5H}{g} = \frac{1}{2} \gamma_w . H \text{ (approximate)} \tag{18.17}$$

In *shallow* waters, the velocity is found to be nearly

$$V = 3.16\sqrt{d} \tag{18.18}$$

Substituting in Eq. (18.14),

$$p = \frac{\gamma_w}{g} \times 10 \times d \text{ (approximate)}$$

or,

$$p = \gamma_w \cdot d \text{ (approximate)} \tag{18.19}$$

Therefore, if the depth of water is taken or expressed in terms of wave height, the pressure can be expressed in terms of the latter.

**Wave Action on Marine Structures**

A sea wave exerts different kinds of forces on a marine structure while breaking against it:

(i) A direct horizontal force which causes compression.

(ii) A deflected vertical component of force which tends to cause a shearing effect on the projections on the wall face.

(iii) A downward vertical force on the sea bed due to the wave tends to stress the foundation of the marine structure. This is due to the collapse of the wave.

(iv) The water which returns after striking tends to cause suction effect on the surface of the marine structure.

**Wave Action on Vertical Walls**

Sea walls are subjected to the action of breaking waves; this results in a greater pressure than that exerted by non-breaking waves, since most of the energy of the latter is reflected by the wall. The energy of the breaking waves is destroyed at the wall and so, these waves exert a greater pressure.

Wave pressure against vertical walls comprises hydrostatic pressure as well as dynamic pressure. Hydrostatic pressure is applied as the wave rises and falls along the wall. The dynamic pressure varies from zero at the crest to a maximum at the still water level. The sum of these static and dynamic forces is the resultant force on the wall.

## 18.4 TIDES

### Definition

Tides are periodical variations in the level of the surface of a large body of water such as a sea or an ocean. These variations are mainly caused by the attraction of the sun and the moon. Tides have a bearing on a number of aspects of hydrographic surveying.

### Hydrographic Surveying

*Hydrographic surveys* obtain information about bodies of water such as oceans, seas, rivers, lakes, and streams. The information includes areas, levels, volumes, rates of flow, and the characteristics of underwater surfaces. The basic principles of surveying and levelling are adapted by making use of specialised apparatus and procedure. Horizontal control will be required in *shore-line surveys* and vertical control in *sounding*. Extensive hydrographic surveys are undertaken by government agencies to prepare navigation charts, to determine the depth of water available for shipping, to survey water currents, and to locate hazards, such as shoals, and also anchorages, buoys and lights.

Hydrographic surveying has many applications in the fields of dock and harbour engineering and coast protection; these are in addition to those in other fields of engineering activity related to the use of water.

## Theory of Tide Generation

Although there are different theories of tide generation, *Newton's equilibrium theory*, based on the law of universal gravitation, explains many of the phenomena relating to tides and is popular.

According to this, every body in the universe attracts every other body with a force acting in a straight line between their centres of mass with a magnitude proportional to the products of their masses and inversely proportional to the square of the distance between them.

The earth is assumed to be completely enveloped by an ocean of uniform depth; the inertia and viscosity of the water are considered negligible, so that the ocean is capable of assuming instantaneously the figure of equilibrium required by tide-producing forces.

The moon is the principal tide generator due to its proximity to the earth. Assuming that the earth and the moon are the only two bodies under consideration, the sequence of the forces and the resultant effect are illustrated as shown in Fig. 18.2 and can be explained as follows:

(a) The earth and the moon attract each other, the resultant force acting along the line joining their centres of masses (EM) [Fig. 18.2(a)].

(b) As a result of the monthly rotation of the earth about the common centroid (G) of the earth and the moon, all points on the earth's surface are subjected to a uniform centrifugal force in a direction parallel to the line joining the centres of the earth and the moon (EM). (The daily rotation of the earth about its own axis does not have any tide-producing effect.) [Fig. 18.2(b)].

(c) The moon's attractive forces are not uniform at different points on the earth's surface because of the differences in their distances from the moon. These are shown in Fig. [18.2(c)].

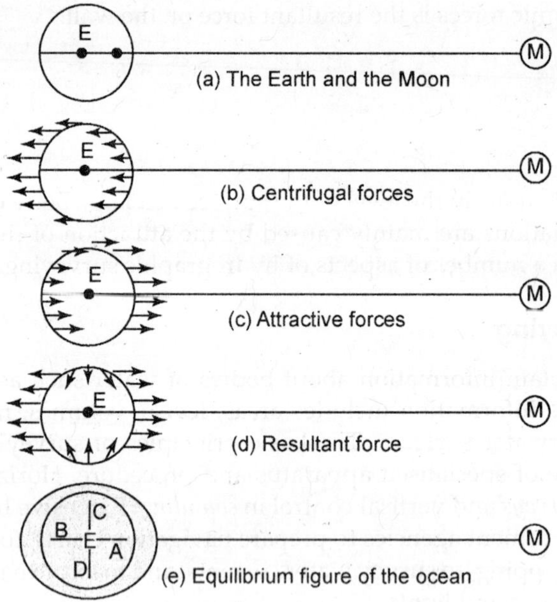


**FIG. 18.2** *Newton's equilibrium theory of tides*

(d)The vector addition of these two systems of forces gives the resultant forces, as shown in Fig. [18.2(d)]. These are the tide-producing forces of the moon.

(e)Since the inertia and viscosity of the water are ignored, the equilibrium figure of the ocean is a prolate ellipsoid of revolution with the major axis directed towards the moon. This is shown in Fig. [18.2(e)].

Two lunar tides are thus simultaneously produced; the *superior lunar tide* at A and the *inferior* or *anti-lunar tide* at B. Low water/tide occurs simultaneously at C and D.

Because of the earth's rotation about its own polar axis, high and low water recur at a given place. If the moon were stationary with respect to the earth, the interval between the high waters would be exactly 12 hours. However, the moon moves around the earth in 29.53 days, which is called the period of *lunation*. The number of transits of the moon during this period is 28.53 days on subtracting the period of the earth's rotation about its own axis, both the motions being in the same sense. That is, the interval between successive transits is, on an average, 24 h 50.5 m, called the *lunar day*. Therefore the average interval between successive high waters at any place is about 12 h 25 min.

The sun, too, generates tides in exactly the same manner. The superior and inferior solar tides occur simultaneously at diametrically opposite points. These times differ by 12 h at a given place. In spite of the sun's large mass, it is a less powerful tide-producing agent than the moon because of its large distance from the earth.

Since both the moon and the sun act simultaneously, the two tides are superimposed, and the ocean assumes an equilibrium figure in accordance with the combined tide-producing forces.

It can be established that the magnitude of the *solar tide is 0.458 times the lunar tide*. Thus, the combined effect of the moon and the sun is to produce a tide which is 1.458 times the lunar tide.

## Spring Tides and Neap Tides

These result from the combined effect of lunar and solar tides.

A *spring tide* is a high tide which is produced when the relative positions of the sun, the moon and the earth are such that the solar and the lunar tides add to each other; high water is higher than normal and low-water is lower than normal.

A *neap tide* is produced when the relative positions of these bodies are such that the tide-producing forces of the moon and the sun are perpendicular to each other; in this case, high water is below the average while low water is above the average.

During a lunar cycle, two spring tides occur, one at new moon and the other at full moon; two neap tides also occur, one at the first quarter and the other at the third quarter of lunation.

The conditions for spring tides and for neap tides are shown in Fig. 18.3.

[It can also be established that the ratio of spring tide to neap tide is 2.8.]

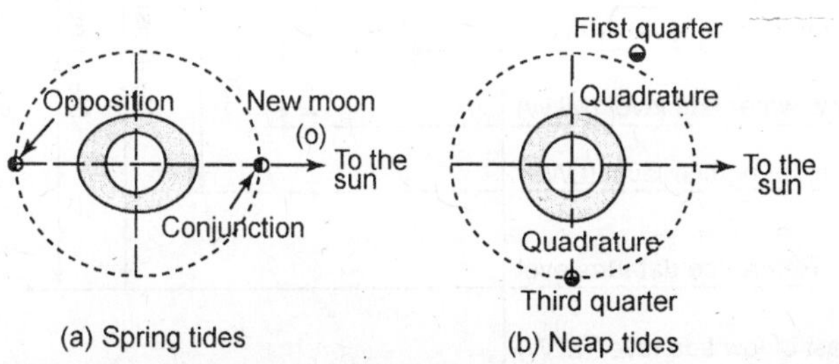


**FIG. 18.3** *Spring tides and neap tides*

### Priming and Lagging of Tides

The varying relative positions of the sun and the moon affect the regularity with which high water recurs at a place, because the tidal day does not have a constant duration of 24 h 50.5 m. Since the moon is the more powerful agent, the crest of the resultant tide lies nearer the meridian of the moon. Hence, it is convenient to refer the time of high water to that of the moon's transit at the meridian of the place.

When high water occurs before the moon's transit, the interval between successive high waters is less than the average, and the tide is said to *prime*. If the high water occurs after the moon's transit, the tide is said *to lag*.

At new moon and full moon, and at the two quadrature points (the first and the third quarters), the crest of the composite tide is under the moon; therefore, the tide is *normal* without priming or lagging. The tide primes during the first and the third quarters of lunation, while it lags during the second and the fourth quadrants of lunation.

### Factors Affecting the Tidal Range and Time Interval Between Successive High Waters at a Place

(i) Varying relative positions of the sun and the moon.
(ii) Ellipticity of the orbit of the moon around the earth and that of the earth around the sun.
(iii) The departure of the earth's shape from a spheroid.
(iv) The effects of land masses in the ocean bed.

### Water Levels During Tides and Tidal Ranges

*Tidal range* is defined as the difference between the water level of high tide and the corresponding low tide. A pictorial representation of sea water levels during different tides is shown in Fig. 18.4:

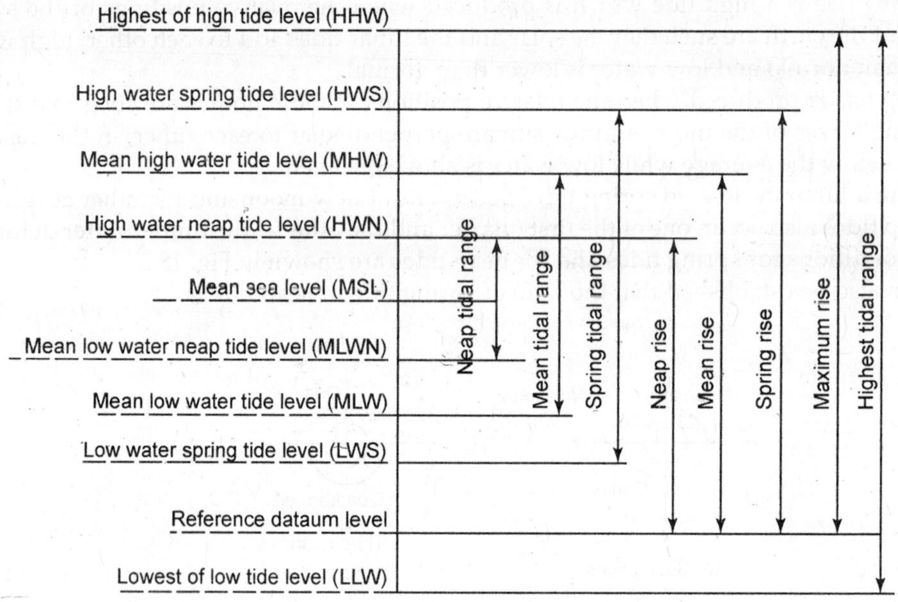


**FIG. 18.4** *Sea water levels during different tides*

### Certain Definitions Relating to Tidal Phenomena

1. *Lunar day*: It is the time taken by the moon to make a complete revolution around the earth – 24 h 50 m, which is 50 min longer than the solar day.
2. *Semi-diurnal tides*: These are tides which occur twice each lunar day.
3. *Diurnal tides*: These are tides that occur only once in a day.
4. *Tidal bores*: These are caused by the rush of flood tides up a river; they occur usually at the mouths of rivers near the sea.
5. *Spring tides*: When the sun, the moon, and the earth are in a straight line, the tide-producing forces of the sun and of the moon are cumulative. These are high tides which occur twice in a lunar-month—at new moon and at full moon.
6. *Neap Tides*: These tides occur twice in a lunar month – at quadratures – the first quarter and the third quarter of the lunar cycle. Since the tide-producing forces of the moon and the sun do not add to each other, neap tides are relatively low water tides.
7. *High water*: This the maximum water level of each rising tide.
8. *Low water*: This the minimum water level of each falling tide.
9. *Mean high water*: This is the average of the high water levels over a period of 19 years.
10. *Mean low water*: This is the average of the low water levels over a period of 19 years.
11. *Tidal range*: This is the difference in level between high water and low water at a tidal station.
12. *Mean range*: This is the average of the differences in level between all high waters and the corresponding low waters; or, the difference in mean high water and mean low water.
13. *Spring range*: This is the difference between the mean high water and mean low water of spring tides.
14. *Neap range*: This is the difference between the mean high water and mean low water of neap tides.
15. *Higher high water*: This is the higher of the two high waters of any two diurnal tidal days.
16. *Mean higher high water*: This is the average of the higher high waters over a 19-year period.
17. *Lower low water*: This is the lower of the two low waters of any two diurnal tidal days.
18. *Mean lower low water*: This is the average of the lower low waters over a 19-year period.
19. *Highest high water:* This is the highest spring tide on record.
20. *Lowest low water*: This is the lowest spring tide on record.
21. *Mean sea level*: This is the mean of the height between mean high water and mean low water; this is the standard datum for all elevations on land.
22. *Diurnal range*: This is the difference in height between mean higher high water and the mean lower low water.
23. *Sieche*: Long period oscillation of water level in a harbour.
24. *Age of tide*: This is the time by which the tide lags with respect to the moon's transit at a place after new or full moon; also called the *establishment*, it can be as high as $2\frac{1}{2}$ days. This is also called the *lunitidal interval*.

### Primary and Secondary Tide Waves

The only part of the ocean which encircles the earth, almost unobstructed, is the Southern Ocean, extending south from about 40°S latitude. The tide waves generated in this zone closely follow the tide-producing forces of the moon and the sun; therefore, these are called the *primary* waves.

From these primary waves, *secondary* waves are propagated into the Pacific, the Atlantic, and the Indian Oceans. The direction of their propagation is influenced by the form of the coast-lines and the intervention of land masses; however, they travel in a general north–south direction in the open oceans.

**Uses of tides:** Energy from the tides, especially those of height more than 5 metres, can be harnessed by means of tidal power stations. Tidal power has the primary advantage of being environment-friendly, and is a perennial source. There is good potential for tidal power in India in Gulfs of Kutch and Cambay.

**Effects of tides:** The following are the important effects of tides:

(i) Static water pressure increases because of the increase in water level due to tides.
(ii) Dynamic force of a tidal wave is not significant.
(iii) Tidal ranges are produced.
(iv) *Tidal prism* is developed in a harbour basin; this is the amount of water displaced in a single tide.
(v) Stability of floating vessels is impaired by tides.
(vi The phenomena of tidal bores and clapotis may occur due to tides, with their consequent effects.

## Prediction of Tides

The sun, the earth and the moon occupy the same relative positions once in a period of nearly 18.6 years. Therefore, tides during this period get repeated cyclically.

Since it is difficult to obtain the records of tides for such long periods, prediction of tides can be made from observations over a relatively shorter period by extrapolation using a system of harmonic analysis. The observations should include at least one set each of spring tides and neap tides. A set of harmonic constants (varying from 10 to 37), typical to the site of observation, is evolved from the available record of observations of the tides. These constants are then used for the prediction of tides.

Tide prediction involves tedious calculations; to overcome this difficulty, tide-predicting machines have been evolved and used. Such machines are designed to produce a graphic representation of the predicted tides, which can be analysed later in an office environment.

With the advent of the digital computer, tide prediction has become relatively easy; if the available data and the relevant harmonic constants are fed into the computer, the results can be obtained in the desired form.

Tide-prediction tables and charts are prepared for particular locations on an annual basis. In some cases, it may be done even on a permanent basis.

## Tidal Currents

These are produced by horizontal movement of water purely due to the tide-producing forces. The velocity and direction of such currents vary with time. Theoretically speaking, the tide-producing forces produce tidal streams first, and the vertical movements of the tides as a secondary effect.

The characteristics of the tidal currents vary between the water surface and the sea bed; it is usual to observe these characteristics up to a depth of about 9 metres, which is considered adequate for most ordinary sea-faring vessels.

*Tidal flows* arise because of the actual movement of water experienced in a given region or zone. These comprise the tidal streams/currents, modified by several other components, such as short-period currents caused by meteorological effects.

Tidal flows or currents are usually expressed in terms of their speeds as well as directions. The speed is usually expressed in knots and the direction as a whole circle bearing, measured clockwise from the north direction.

Knowledge of the tidal flows and tidal currents is useful in the planning and layout of certain harbour components and structures.

## 18.5 SILTING AND EROSION

The coastline and the waters of the sea or ocean near the shore get continuously altered by the action of wind, waves and tides. The material of the bed is removed at some places and it is deposited at some others; the former phenomenon is called *erosion* and the latter, *silting* or *accretion*. It is desirable that harbours are free from both silting and erosion.

Silting causes the reduction of navigable depth of channels and turning basins, requiring constant dredging and maintenance, while erosion causes weakening of the foundations of harbour structures. Periodic scouring is helpful in maintaining the necessary depth of water.

The study of these two phenomena is of interest to a harbour engineer. The zones of the sea bed starting from the shore inward into the sea are schematically shown in Fig. 18.5.

The shallow-water zone AC is known as the *continental shelf*, up to about a depth of 180 m; according to geologists, this had been mostly above the sea level during the Ice Age, but was inundated by the sea due to the melting of ice and settling of the land mass.

It has been observed that the sea-bed material up to a depth of about 10 m in the shallow-water zone is subjected to the action of waves and tidal currents; this zone, AB, is called the *littoral zone* as there is drift or movement of sediments here.

### Sediment Transport

The process of movement of sediments or sea-bed material is known as *sediment transport*. Winds, waves and tidal currents are the three agencies which cause the phenomenon of sediment transport.

The process of movement of drifting bed material is a complex phenomenon. Resistances to the lifting and sliding of the sediment particles govern the characteristics of their motion.

It can be shown that the *force resisting this motion is proportional to the cube of the particle diameter, and the diameter of the largest particle that can be moved is proportional to the square of the velocity of the water currents.*

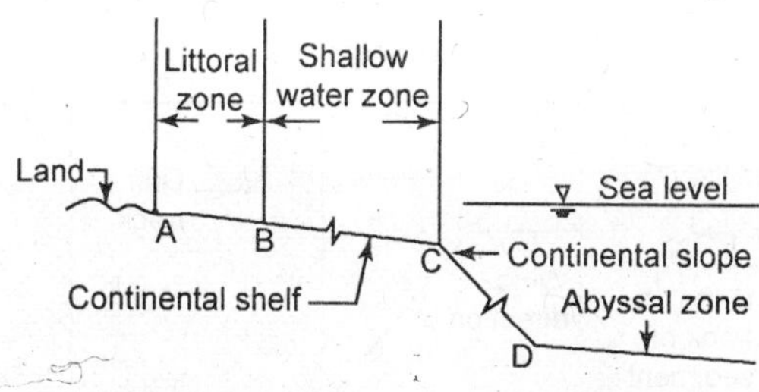


**FIG. 18.5** *Zones of sea bed*

### Littoral Drift

The process of movement and deposition of sand drift in a zigzag manner in the proximity of the foreshore or coastline is referred to as the *littoral drift*.

Sediment transport, and more specifically the littoral drift, is due to the individual as well as the combined influence of winds, waves and tides.

**Effect of winds:** The general effect of the wind, either *offshore* or *onshore* is cause the littoral drift in a zigzag fashion, as shown in Fig. 18.6.

If the coastline is interrupted by a river outlet, the sediment moving along the coast tends to close the outlet, while the river discharge and tidal currents tend to prevent the closure. The current velocities in the area of the river mouth are generally high, thus maintaining a large depth of water in this area; however, the seaward tidal currents are much weaker, and tend to form a *bar*.

**Effect of waves:** Waves strike the coast at a certain angle. In shallow waters, the waves get refracted due to friction at the bed and strike the coast at right angles. From the principles of hydrodynamics, it can be established that the wave form adjusts itself so as to strike the coast at right angles.

However, the waves from deep waters tend to strike the coast at an acute angle to the shore; this contributes a component along the shore. (This can be easily observed by a person standing on the shore.) The water currents generated by the component along the shore cause the drift of sediments parallel to the shore; this is the littoral drift caused by the effect of waves. It is established that the littoral drift, as bed-load, is proportional to the fifth power of the velocity; thus, even a small increase in velocity will cause a significant increase in the littoral drift.

Littoral drift is likely to vary with the season, but may have a predominant effect in a particular period.

**Effect of tides:** Tides or storm waves may not cause much littoral drift when they strike strong tidal currents at nearly right angles, as their effect is mostly reduced by the latter. This results in the formation of sand bars; examples on the east coast are Machilipatnam and Cuddalore ports and Dabbol and Kochi on the west coast of India.

In general, the effects of littoral drift are considered in the design of many harbour structures.

### Estimation of Littoral Drift

Although there are different methods for the *estimation of littoral drift*, they are mostly empirical and have some limitations. Empirical relationships between wave energy and littoral drift, combined with a suitable sediment-transport formula, generally yield satisfactory results.

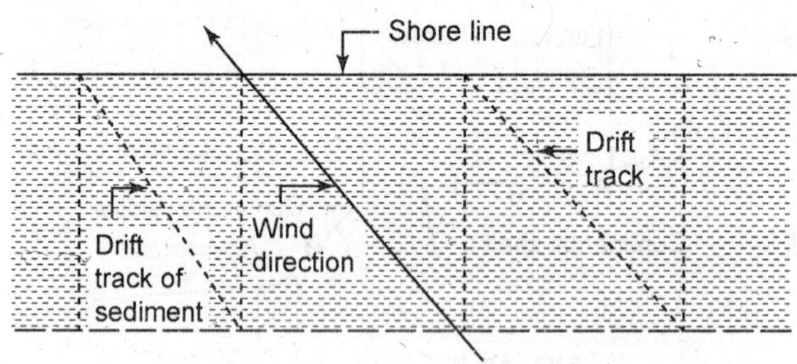


**FIG. 18.6** *Littoral drift due to winds*

Examination of field data near natural or man-made coastal structures is likely to yield valuable information for estimating the littoral drift in the vicinity of such an area.

**Erosion**

Beach profiles vary continually over a long period of time due to the dynamic response of the coastline to the different processes and phenomena to which it is subjected. The shore line is susceptible to major changes due to the effect of severe storms, which cause significant erosion at particular locations.

The bed material and bed slope play an important role in the actual sediment transport, and *building-up* or *erosion* of the beach. The presence of natural or man-made structures near the coast line also affects these processes.

*Beach protection works* are planned in an appropriate manner to tackle the twin effects of silting and erosion; these will be seen in some detail in Chapter 21.

## SUMMARY

- Natural phenomena – meteorological phenomena such as winds, cyclones and storms, and oceanographic phenomena such as waves, tides and tidal currents, and silting and erosion affect the planning and design of harbours and harbour structures.
- Wind is air in motion; wind speed is expressed in 'knots' or 'nautical miles per hour'. One knot is equal to 1.853 km per hour. Besides the speed, wind direction is also important.
- Sea breeze blows towards the land and land breeze blows towards the sea.
- Beaufort scale is used to describe wind characteristics based on the wind speed.
- Wind rose is a diagrammatic representation of the wind direction, duration, as also speed; observed wind data for a long period are used to construct wind rose of either Type-I or Type-II. Cyclones and storms are tracked for a number of years and the data are utilised in the design of harbour structure.
- Waves on the sea are caused by various disturbances, natural or man-made, such as winds, tides and earthquakes, and explosions and large moving vessels. Forces of water waves are powerful, and have to be accounted for in the design of harbour structures. Waves are generated by the transfer of wind energy to the water surface; waves originate as waves of oscillation and get transformed as waves of translation by the action of the wind. These are gravity waves. Surface tension causes ripples or capillary waves, which are short and insignificant when compared to gravity waves. Wind characteristics, water depth and friction of the bed material govern the characteristics of water waves.
- Wave height, wave length and wave velocity are the important characteristics of water waves; wave height depends on the fetch, or the straight stretch of open sea available for uninterrupted wave growth.
  Significant wave height is taken to be one-third of the highest wave height.
  Waves get modified by the phenomena of decay, breaking, diffraction, refraction, and reflection.
  Waves exert pressure on harbour structures and this has to be accounted for in their design.
  There is significant wave action on vertical walls as the waves break upon them and this has to be accounted for in the design of harbour structures.
- Tides are periodical variations in the water level of seas and oceans.
  Hydrographic surveys are conducted as part of shore-line surveys and soundings for the determination of water depth; hence, these surveys have many applications in the field of harbour engineering.

Newton's equilibrium theory is used to explain several phenomena of tides.

The Moon being the nearest celestial body to Earth, as its satellite, has the greatest influence in generating tides in seas and oceans. Two high-water tides and two low-water tides occur simultaneously at four different points on Earth's surface, with two pairs of points being diametrically opposite. The average interval between successive high waters at a place is about 12 h 25 min. The Sun also produces tides; but, owing to its great distance from the earth, solar tides are only about 0.458 times the lunar tides. When the positions of the Moon and the Sun relative to the Earth are in a straight line, the lunar and the solar tides add to each other, producing 'spring tides'. When they are aligned in perpendicular directions, 'neap tides' occur with high water being below the average tide and low-water above the average. Two spring tides occur, one at full moon and the other at new moon in a cycle of lunation, the lunar month. Similarly, two neap tides occur at the first and the third quarter positions of the moon in the cycle of lunation.

Several factors affect the tidal range and the time interval between successive high waters at a place, the chief among them being the variations in the relative positions of the Sun and the Moon and the ellipticity of orbits of the rotation of the Earth around the Sun and the Moon around the Earth.

Tidal range is the difference between the water level of high tide and the respective low tide.

The energy of tides can harnessed to generate tidal power. The stability of floating vessels is impaired by tides; tides have several other effects on harbour components.

Prediction of tides is a complex process requiring harmonic analysis of past data on a long-term basis. Tidal currents are also to be considered in the planning and design of harbours.

- Silting at some points and erosion at some locations near the coast-line and within the harbour area occur due to the phenomenon of sediment transport. This is also a relatively complex process in the shallow zone of the continental shelf of the sea bed near the shore.

The process of movement and deposition of sand drifting in a zigzag manner in the proximity of the coastline is known as littoral drift.

This is caused by the effects of wind, waves, and tides – both individual and combined.

Erosion of the coastline is caused by the forces of nature and sediment transport.

Silting and littoral drift need dredging operations for maintaining the necessary depth for navigation in and around the harbour. Also, erosion has to be considered for preventing the weakening of foundations of harbour structures.

## QUESTIONS

1. Explain the significance of natural phenomena in the planning and design of harbours.
2. How is wind speed stated? Give the relationship between the units of wind speed in the FPS units and the MKS units.
3. Explain briefly the Beaufort scale.
4. What is a wind-rose and how is it constructed? What are its uses?
5. What is the influence of cyclones and storms in the planning of a harbour?
6. What are the causes for the generation of water waves in a sea?
7. What are the important characteristics of water waves? Explain briefly each of them.

8. Distinguish between:
   (a) Waves of oscillation and waves of translation.
   (b) Factors affecting the characteristics of water waves.
   (c) Gravity waves and capillary waves.
9. Write short notes on:
   (a) Significant wave height
   (b) Fetch
   (c) Wave decay
10. Explain 'wave breaking' and 'wave diffraction'.
11. Explain briefly the origin of tides and their generation.
12. Explain the role of hydrographic surveys in the design of harbour structures.
13. Distinguish between lunar tides and solar tides. What is the approximate relationship between them?
14. Write short notes on:
    (a) Spring tides and neap tides
    (b) Superior lunar tide and anti-lunar tide
    (c) Tidal range
15. Distinguish between primary and secondary tide waves.
16. Explain the uses and effects of tides.
17. Write short notes on:
    (a) Tidal currents
    (b) Age of tide
    (c) Tide prediction.
18. Explain the phenomenon of 'littoral drift'. [Anna University]
19. Explain the causes of littoral drift and its effects on harbours.
20. Explain 'sediment transport' and its characteristics.
21. What are the effects of silting and erosion on harbour structures?
22. Explain the term 'tidal currents' and its uses. (West Bengal Tech. University)
23. Explain briefly the terms 'Spring tide' and 'Neap tide'. (West Bengal Tech. University)
24. Explain the terms 'littoral drift' and 'neap tide'. (West Bengal Tech. University)
25. Define 'Significant wave height'. (West Bengal Tech. University)
26. What is the role of the following processes in harbour layout?
    (i) Wind waves
    (ii) Tidal currents
    (iii) Littoral drift
    (iv) Silting and erosion (Anna University)
27. Discuss the wind and wave effects and their action on coastal structures. (Anna University)

Chapter **19**

# FACILITIES FOR PORTS AND HARBOURS

**CHAPTER OBJECTIVES**

This chapter helps the student to gain an insight into:

- The facilities at ports for protection of vessels, cargo handling, docking and navigation
- Breakwaters that are meant for dissipation of wave energy and protection of harbour structures
- Quays, wharves, piers, dolphins, jetties and fenders meant for loading and unloading of cargo
- Aprons and warehouses for storage of cargo
- Buoys, moorings, light houses and beacon lights that are used to aid navigation and anchoring of vessels.

## 19.1 INTRODUCTION

Several kinds of facilities are required to be provided for sea-faring vessels such as ships and ocean-liners in and around a harbour or a port.

The different types of facilities needed are:

(a) *Protection facilities,* or those that provide protection for vessels.
(b) *Loading and unloading facilities,* or those for cargo-handling.
(c) *Cargo-handling and storage facilities* either for transit or for temporary storage of cargo, including cold storage.
(d) *Navigation facilities* meant to facilitate safe navigation of all the vessels which are on the high seas, into the harbour or out of it for further travel to their respective destinations.
(e) *Docks and repairing facilities for vessels.* (This is dealt with in the Chapter 20.)

## 19.2 PROTECTION FACILITIES

Certain places on the sea coast are so situated that they are naturally protected from the fury of the sea or the ocean. Such places serve as protection for vessels, which are berthed with minimal man-made construction; these are therefore called *natural harbours*.

But, if a harbour is to be created on an open coast, certain structures are required to be built for affording protection to the vessels; these primary structures for protection are called *breakwaters*.

### Breakwaters

A *breakwater* is a structure, which, as suggested by its name, is meant to reflect and dissipate the force of wind-generated waves, thereby providing the necessary protection to the vessels inside the harbour area in relatively calm waters.

Thus, an artificial harbour is located behind breakwaters, constructed with a properly planned orientation to protect the harbour area, while providing suitable approach channel and a well-designed entrance.

#### Considerations in the Design of Breakwaters

1. **Alignment**: Straight, converging arms are preferred with an angle of intersection not exceeding 60°. This is necessary to provide a relatively narrow entrance to the harbour in order to provide calm waters inside.
2. **Design data**: The following information has to be collected:
   (i) Nature of coastal currents
   (ii) Wind characteristics
   (iii) Wave characteristics
   (iv) Nature and properties of sea-bed material
   (v) Availability of construction materials.
3. **Design considerations**: The following are the important points to be ensured:
   (i) The maximum wind and wave forces are to be used in the design.
   (ii) The bed material should be protected from scour.
   (iii) The height of the breakwater should be more than the maximum wave height, calculated based on the fetch. The difference between the height of the breakwater and the maximum wave height is the desired *freeboard*.
4. **Forces on breakwaters**: The following are the important forces likely to act on a breakwater:
   (i) Hydrostatic force on the inside
   (ii) Wind and wave forces on the outside
   (iii) Chemical action of seawater on the materials of construction used
   (iv) The effects of marine life and creatures on the material of construction.

#### Classification of Breakwaters

The following is the classification based on the material and the construction:

1. Mound breakwater
2. Composite breakwater
3. Vertical wall breakwater
4. Special types of breakwater.

**Considerations in the Choice of the Type of Breakwater**

(i) Depth of water at the site
(ii) Nature of the sea-bed material
(iii) Availability of the materials of construction
(iv) Equipment available
(v) Resources and the time constraints.

## Mound Breakwater

Also called *heap-type breakwater* or *rubble-mound breakwater*, this is constructed using natural rubble/undressed stone blocks and supplemented, if necessary, by concrete blocks or other artificial blocks of huge size and weight, such as patented stone or concrete blocks. These units are deposited without regard to bond or binding material.

This is the simplest type of breakwater constructed by dumping rubble stone blocks into the sea until the mound or heap rises out of the water. The heap gets consolidated by the impact energy of the waves. This type of construction is called *rip-rap* or *pell-mell* type. Because of the large voids among the stones, there is significant settlement during the first few years; hence, it requires re-dumping of stones to the desired level. No special preparation is made for the foundation of this type of breakwater.

Since the destructive action of the waves is highest between the low and high water levels, necessary protection measures are adopted in this zone at the top and on the front face exposed to the sea.

A mound breakwater usually consists of three zones: the *hearth* or the *core*, *filter* and *armour*. These are briefly described below:

**Hearth or the core:** The function of the core is to increase the bulk by using small-sized quarry waste or washed materials from dredging. These stones weigh from 10 kg to 20 kg each. The core is the least pervious relative to the zone above and encompassing it.

**Filter or secondary armours:** The size of the units in this zone laid around the core is much bigger than the core stones and is much more pervious. They weigh 0.5 to 1.5 tonnes each. The function of the filter is to destroy the energy of the waves penetrating the armour layer above it.

**Armours:** These stone blocks are the primary line of defence against wave energy, resisting most of it and hence the name 'armours' as in the army. They weigh up to 30 tonnes each and consist of dumped stones with maximum voids in the zone.

A mound breakwater is schematically shown in Fig. 19.1:

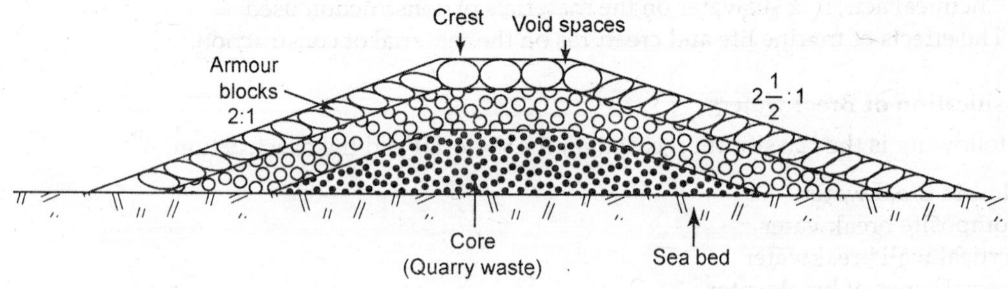


**FIG. 19.1** *A typical section of mound breakwater*

### Energy Dissipation in the Mound-Type Breakwater

The outermost or armour layer dissipates the major portion of the wave energy by the following processes:

- Reflection of some part of the impinging water
- Ingress of some portion of the water into the voids of the layer against hydraulic roughness
- Division of the incoming water into jets which strike against one another and dissipate a portion of the energy
- Allowing the entering water into the next layer and facilitating water cohesion.
- Causing a portion of the impinging water to rise up the slope and fall back into the sea.

All these actions are achieved by large, heavy and rough blocks, irregular in shape with large voids.

The middle layer of secondary armour blocks filters only a small portion of the wave energy. The innermost layer, or the core, allows the portion of the water entering it to move in zigzag paths, the little energy left being dissipated by friction.

The flexibility of the mound breakwater takes care of weak areas through settlement of the inner blocks, and gradual lowering of the top, which can be replenished later. Unskilled labour is adequate for this type of construction.

### Advantage and Disadvantages of Mound Breakwater

#### Advantages

(i) Low initial cost
(ii) High porosity eliminates uplift
(iii) No chance of sliding, slipping or overturning because of the flexibility of construction.
(iv) Because of the flexibility, weak spots and settlement of blocks are transferred to the top, facilitating replenishment of stones or block at the top to the desired level
(v) Preparation of foundation is not required
(vi) Very efficient in the dissipation of wave energy by the very nature of construction
(vii) Sudden and spectacular failures do not occur in the case of mound-type breakwaters.

#### Disadvantages

(i) Requirement of material is huge.
(ii) Rip-rap method of construction allows the projecting surfaces of stone blocks to become rounded by the impact of waves
(iii) The settlement of the breakwater in the first few years is significantly high; this requires frequent replenishing of stone blocks to the desired level
(iv) Maintenance cost is significantly high
(v) High cross-sectional area, especially at the base, occupies a large area of the basin, precluding loading and unloading operations on the inner side of the breakwater
(vi) Design can be only empirical, as rigorous analysis is not possible.

**Note:** *Concrete blocks* of the required irregularity are difficult to be cast and prove to be more expensive than rubble stone blocks. Transportation cost from the factory to the site of construction is also high. Hence, the use of the concrete blocks for breakwaters is not preferred.

**Design of Rubble-Mound Breakwaters**

The important aspects of design are the following:

**Length:** The length of breakwater is governed by the best alignment to cater to the needs of the different types of vessels expected to use the harbour.

**Crest width:** This is chosen to be the height of the highest wave plus the width necessary to move the repairing machinery/equipment.

**Height of breakwater:** The height of construction along the alignment is the difference between the levels of the crest and the sea bed. Since the sea bed level varies, the height of construction along the alignment also varies. The crest level may be given as

Crest level = High-water spring tide level + Height of highest wave + Freeboard (2 to 3 m to take care of possible settlement).

**Side slopes:** The design of side slopes depends upon the stability aspects of the mound.

The following empirical formulae are used:

**Iribarren's Formula**

Although Eduardo de Castro gave an empirical formula, Ramon Iribarren Cavanilles (1938) developed a more popular formula given below:

$$W = \frac{K\gamma_r H^3}{(\gamma_r - 1)^3 \cdot (\cos\alpha - \sin\alpha)^3} \tag{19.1}$$

where

$W$ = Weight of individual stone

$K$ = Coefficient combining certain variables (15 for undressed natural rubble stone and 19 for artificial concrete blocks)

$\gamma_r$ = Unit weight of stone

$H$ = Wave height of the highest wave

$\alpha$ = Angle of slope of the breakwater with horizontal.

Later in 1953,

Castro and Iribarren gave a modified formula for submerged stones in breakwaters.

**Castro–Iribarren's Formula**

$$W = \frac{KH^3\gamma_r\gamma_w}{(\gamma_r - \gamma_w)^3 \cdot (\cos\alpha - \sin\alpha)^3} \tag{19.2}$$

where $\gamma_w$ is the unit weight of sea water and all the other symbols are as given in Iribarren's formula.

### Rubble-Mound Breakwaters Strengthened by Concrete Blocks

The advantages of concrete blocks are that they can be cast in any desired shape and designed with a suitable mix proportion to achieve the desired strength, greater than that for natural rubble, with steel reinforcements.

Concrete blocks are laid around armour stones on the seaward side of the mound to act as extra armour blocks; these are cast in irregular shapes to be as heavy as 25 tonnes. They are laid either in a pell-mell manner or in a compacted manner.

The slope of the layer of concrete blocks can be as steep as $1\frac{1}{2}:1$. In this case, the armour layers are provided with toes on the seaward side; a sufficiently thick crown wall made of concrete block is also built at the crest.

### Rubble-Mound Breakwater Strengthened by Patented Concrete Blocks

The energy dissipation characteristics of concrete blocks in cuboid shape are rather poor. They are also expensive in view of their solid shape.

In view of this, different-shaped concrete blocks, hydraulically rough as also stable, have been developed. These are permeable with large void spaces in between, and have better energy dissipation characteristics.

Several types of concrete armour units have been developed, the chief among them being *tetrapods, tribars,* and *quadripods.*

**Tetrapods:** These are four-legged, truncated cone-shaped, precast concrete units. They can weigh up to 25 tonnes for large breakwaters.

A schematic view is shown in Fig. 19.2, along with a general view.

This type of unit was developed in France. Best results may be got by using two layers, each of different volumes and weights, with different centre-to-centre spacing. Tetrapods are stable on any three of their four legs and when laid properly, they can get interlocked with adjacent units.

They are hydraulically rough with high porosity between adjacent units, and they are economical with high energy dissipation characteristics.

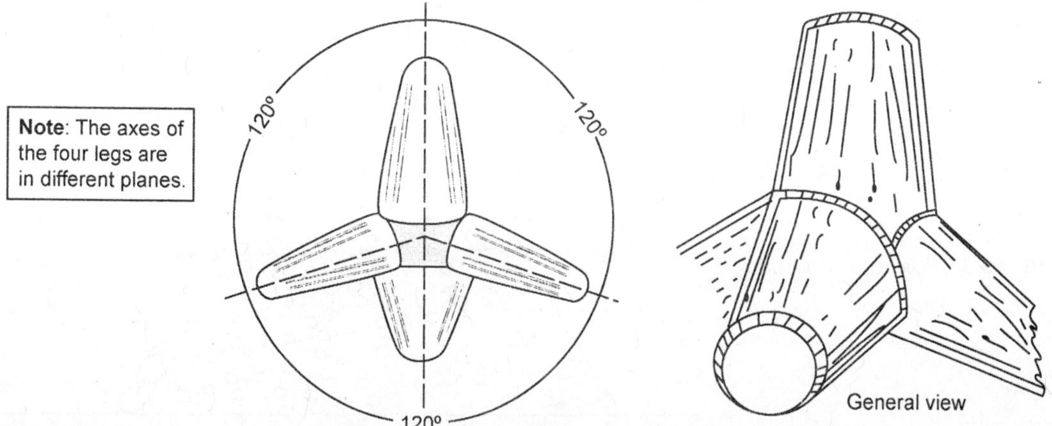


**FIG. 19.2** *A schematic view of a tetrapod*

**Tribars:** These were developed by US Army Engineers; they are three-legged, and the three legs are connected with a plate at mid-height as shown in Fig. 19.3.
The weight of each unit may range from 4 to 40 tonnes. If the units are placed properly they can get interlocked and remain stable.

**Quadripods:** These are similar in shape to tetrapods except that the axes of the three legs forming the base are all in the same plane. [Fig. 19.3 (a)]

[Hexapods, modified cuboids, Dollosse, Akman and Stabit Armours are other patented blocks, which are less popular.]

## Composite Breakwater

A mound breakwater provided with a solid, usually concrete, superstructure at low water level is known as a *composite breakwater*. This provides a platform for handling cargo and also protects the top of the mound. Ships can come to the breakwater wall inside the harbour close enough to facilitate loading and unloading of cargo. The sea face is provided with heavy concrete blocks for withstanding the destructive effect of waves.

A typical composite breakwater is shown in Fig. 19.4.
In deep water, a mound with superstructure founded below the low water level proves more economical than that founded at low water level, since the disturbing effect of the waves is not felt at such depths.

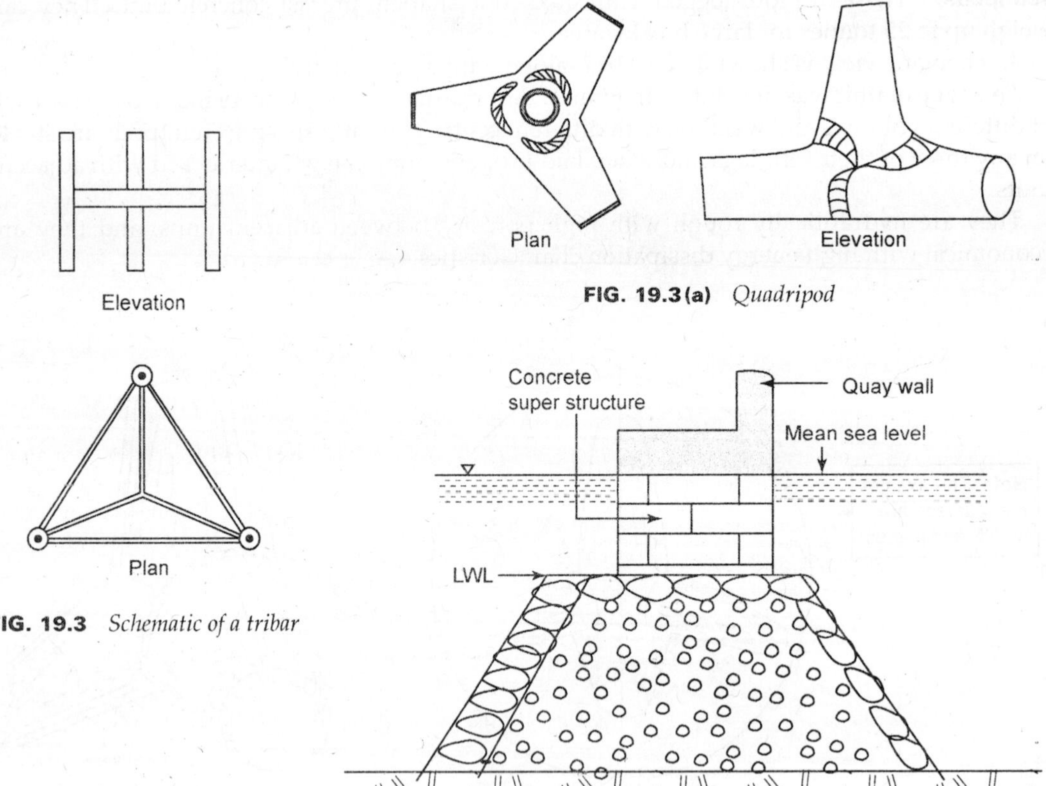


**FIG. 19.3 (a)** *Quadripod*

**FIG. 19.3** *Schematic of a tribar*

**FIG. 19.4** *A typical composite breakwater*

### Methods of Construction of Mound Breakwaters

The following are the methods used:

(i) Barge method
(ii) Staging method
(iii) Low-level method

**Barge method:** In this method, flat-bottom barges with hoppers that have doors fitted at the bottom are used to drop rubble stones into the water. The hoppers are loaded with stones and the barge is aligned in position along the line of construction. The stones are then discharged into sea water by opening the doors at the bottom of the hopper (Fig. 19.5).

The rubble is evenly distributed over the entire base; the width of the breakwater is trimmed and brought to the required section by divers.

The shutters of the door are held in the closed position by means of chains, hooked at the top.

When the mound rises sufficiently and hopper barges cannot be used, *decked barges* are employed. These are loaded with rubble, brought to the site, and slightly canted by flooding the compartments one side, thereby discharging the material (Fig. 19.6).
The rubble can be uniformly deposited over a large area by this method; but this method is not economical and the progress is hampered in unfavourable weather conditions.

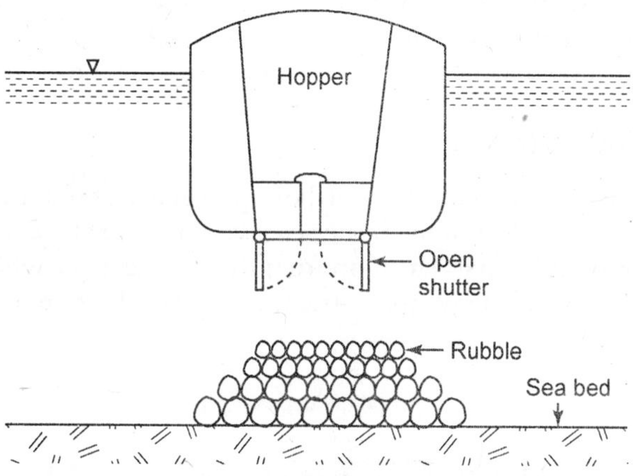


**FIG. 19.5** *Hopper barge*

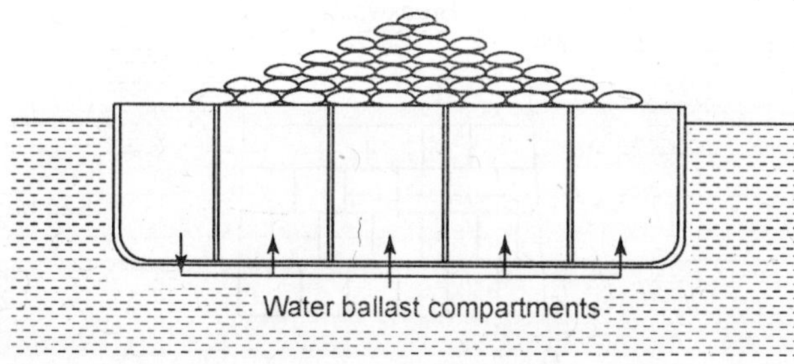


**FIG. 19.6** *Decked barge*

**Staging method:** A staging is erected by driving a series of piles at intervals of 4.5 to 6 m, connecting them by longitudinal runners, struts and braces. A number of parallel tracks can be laid on the staging, and trains of tipping wagons loaded with rubble can be used to form the mound. The staging can be dismantled and reused.

**Low-level method:** This is a slow process in which the length of the mound is formed from the shore well above the high water level of the sea. The mound formed is used for laying tracks to run tipped wagons, and the procedure in the staging method is used to advance the work of construction of the breakwater.

## Vertical Wall Breakwater

This involves the use of concrete block gravity walls, concrete caissons, sheet pile cells with rock-fill/concrete and steel sheet pile walls with a solid concrete wall at the top, and a few other varieties of construction. These are preferred in deep waters with a sea bed resistant to erosion and settlements. If soft materials like fine sand, silt and clay exist near the top of the sea bed, these must be removed by dredging before further work is carried out.

Generally, concrete blocks are preferred for vertical wall breakwaters with slice work; the blocks weigh up to 200 tonnes each and give stability to the wall. Slice blocks weigh only 20 tonnes each.

The typical vertical wall breakwater is shown in Fig. 19.7.

Sudden failure of the breakwater is possible; this is a disadvantage of wall-type breakwater.

Sometimes, the breakwater may be built using a solid concrete wall above the low water level, in a foundation formed by means of concrete bags from the sea bed up to the LWL. Good weather is required, as the construction may be partly lost, especially in caisson and sheet-piling work.

## Special Types of Breakwater

It has been established that the destructive force of the impact of waves in a sea or ocean is maximum only up to about 10 meters depth from the mean sea level. Hence, solid construction is provided for breakwater for a depth of the first 5 to 10 m from the surface, after which it is supported by means of relatively simple and economical construction up to the bottom or the sea-bed.

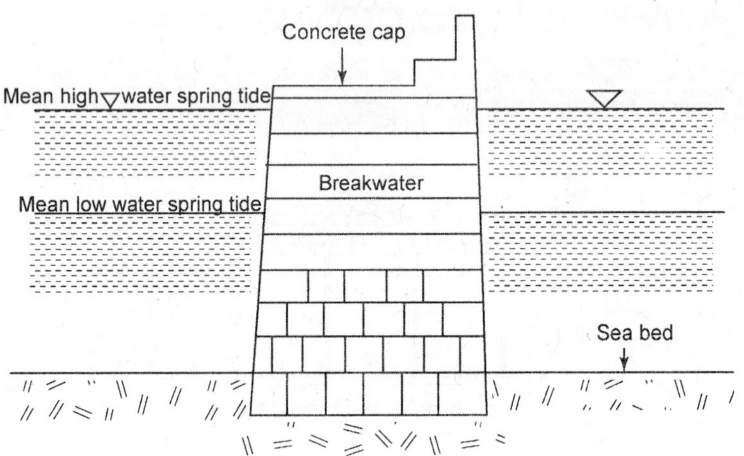


**FIG. 19.7** *Vertical wall breakwater (with concrete blocks)*

If a breakwater is constructed to resist wave forces for the top 5 to 10 m of the depth, with practically no construction below such depth, it is known as a *special-type breakwater*.

The following are the special types of breakwater:

(i) Floating type
(ii) Hydraulic type
(iii) Pneumatic breakwater

**Floating-type breakwater:** This consists of a system of well-connected bulkheads, 5 to 10 m deep, properly anchored, with practically no construction below 10 m. In such cases, harbour-basin water has connection with open-sea water below this depth. Such floating breakwaters are not very strong and may be used only at small harbours with a relatively small basin. The only advantage is the flexibility of movement, facilitating the change of the shape of harbour basin, if required.

A schematic of floating-type breakwater is shown in Fig. 19.8.

**Hydraulic breakwater:** Translatory water waves tend to rotate about their own axes as they move towards the shore or the harbour basin. In this process, they generate a huge amount of energy or destructive force. If, by some means, this force is nullified, the wave gets converted into a silent mass of water. This may be achieved by means of water jets positioned appropriately through nozzles, as shown in Fig. 19.9.

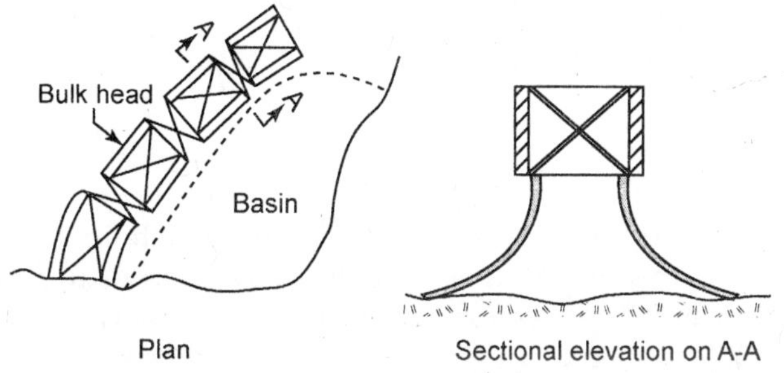


**FIG. 19.8** *Floating-type breakwater*

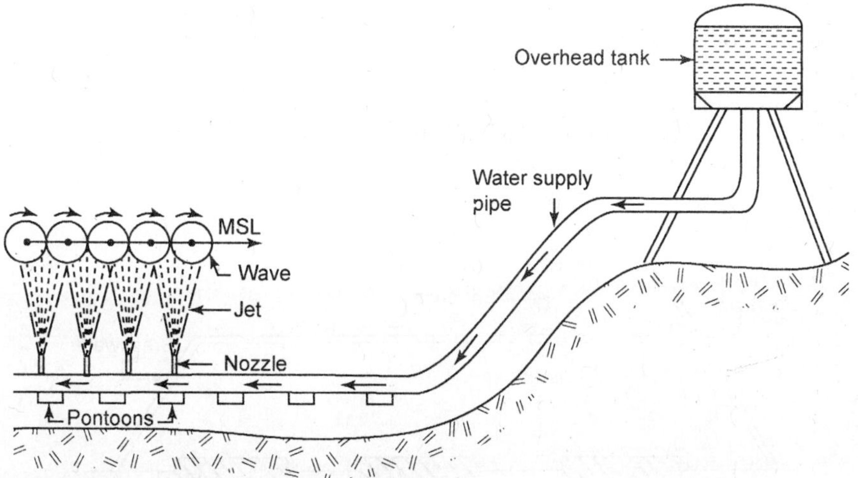


**FIG. 19.9** *Hydraulic breakwater*

**Pneumatic breakwater:** Instead of water jets under pressure, compressed air jets can also serve the same purpose of resisting the wave forces. Such an arrangement is known as pneumatic breakwater.

Since the estimation of the wave pressure is not very reliable, both hydraulic and pneumatic types of breakwater are still in the developmental stage.

## 19.3 LOADING AND UNLOADING FACILITIES

The components at a port, which are required for safe and efficient operations of embarkation and disembarkation of passengers, and loading and unloading of cargo, fall in this category of facilities.

The facilities and components are listed below:

1. Quays and quay walls
2. Wharves
3. Piers
4. Dolphins
5. Jetties
6. Fenders
7. Trestles
8. Moles

These are now considered sequentially.

### Quays and Quay Walls

#### Quays

*Quays* are stationary platforms, built along or parallel to the shore, to serve the purpose of loading and unloading of vessels brought near them. Walls built to protect the quays and provide stability to them are known as *quay walls*.

A schematic of a simple system of quays is shown in Fig. 19.10.

**Quay walls:** These are constructed to retain the filling and embankment.

The following are the *factors governing the design of quay walls*:

(i) Nature of the foundation material
(ii) Dead loads and live loads on the platform
(iii) Uniform surcharge pressure, if any
(iv) Impact force of the vessels berthing near the quay
(v) Earth pressure on the rear, including buoyancy effect
(vi) Self-weight and the construction materials used.

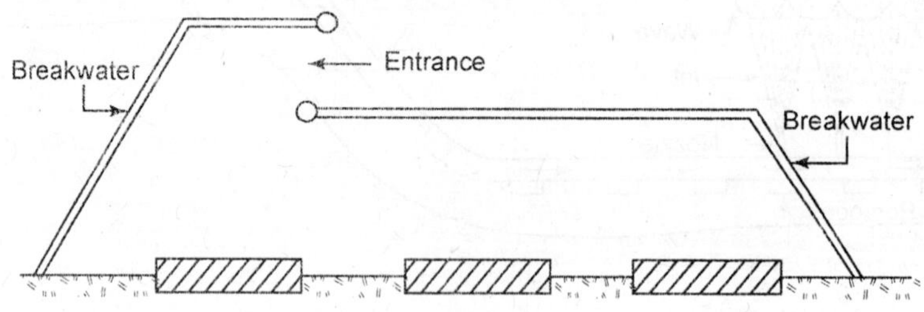


**FIG. 19.10** *A simple system of quays*

**Design of quay walls:** Quay walls are designed in a similar way as retaining walls.

The *primary forces* acting on the quay walls are:

- *Water pressure* (hydrostatic pressure) on the seaward side, acting horizontally.
- *Earth pressure* of the fill on the rear side. (The appropriate coefficients for the earth pressure considering the buoyancy effect of the submerged portion have to be used.)
- *Self-weight* of the wall acting vertically through the center of gravity, (including the surcharge effect.)

All these forces are computed for unit length of the wall, and the stability is checked against overturning and sliding at the base.

### Types of Quay Walls and Their Construction

Quay walls are founded under water and constructed mostly in water.

The following are the different types of quay walls, based on their construction:

(i) Solid quay wall
(ii) Dwarf quay wall
(iii) Quay wall supported on rubble mound
(iv) Quay wall supported on timber crib
(v) Quay wall supported on timber lattice.

**Solid quay wall:** This consists of masonry or large concrete blocks, and is founded on mounds, wells or caissons.
A schematic of a typical solid quay wall is shown in Fig. 19.11.

**Dwarf quay wall:** This is founded on piles and proves economical for small ports and river ports. A schematic of this type is shown in Fig. 19.12.

**Quay walls on other kinds of supports:** Quay walls may also be supported on a rubble mound, timber crib, or timber lattice. Except for the bottom support, these are similar to the other types. [In the case of long quay walls, expansion joints at every 20 m length are provided.]

## Wharves

*Wharves* are platforms in the form of walls near the shore, built for vessels to be berthed; this provides a working platform by the side of the vessel in continuity of the shore.

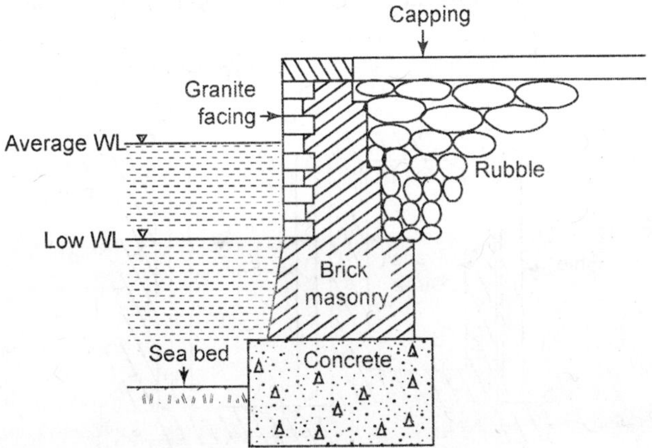


**FIG. 19.11** *Schematic of a solid quay wall*

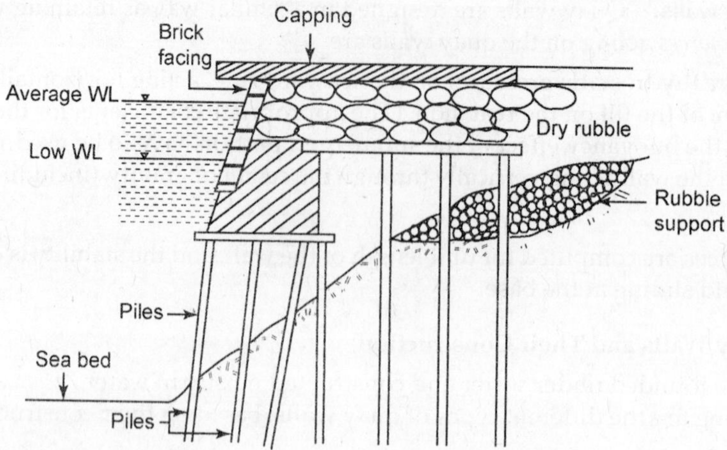


**FIG. 19.12** *Dwarf quay wall supported on piles*

A wharf is similar to a quay, but it is an open or framed structure of posts with bracings, or supported on piling. It may consist of a sheet-pile wall with or without soil retained some distance behind it.

Wharves are built parallel to the shore and abutting it; but the depth of water should be adequate for ships to berth, at least 10 m. The top of the wharf should be kept a little above the high-water level. The bracings must be strong enough to withstand impact from a vessel.

The length of the wharf and the required depth of water depend on the types of vessels to be served. The wharf should be strong enough to withstand cargo-handling equipment.

## Piers

These are structures built perpendicular to the shore, and serve a similar purpose as that of quays and wharves. The depth of water need not be as high as in the approach channel.

Piers may be built as solid structures or structures supported on piles or on columns, struts and braces. Open structures are used when the flow should not be obstructed as in the case of solid structures.

Figs. 19.13 and 19.14 show schematics of two-berth pier and four-berth pier, in plan:

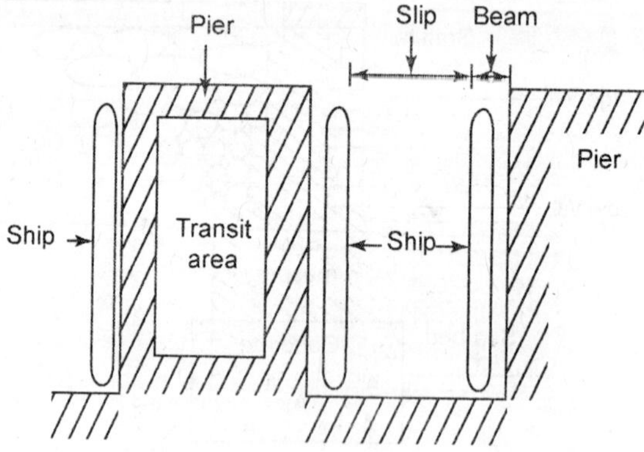


**FIG. 19.13** *Schematic of a two-berth pier*

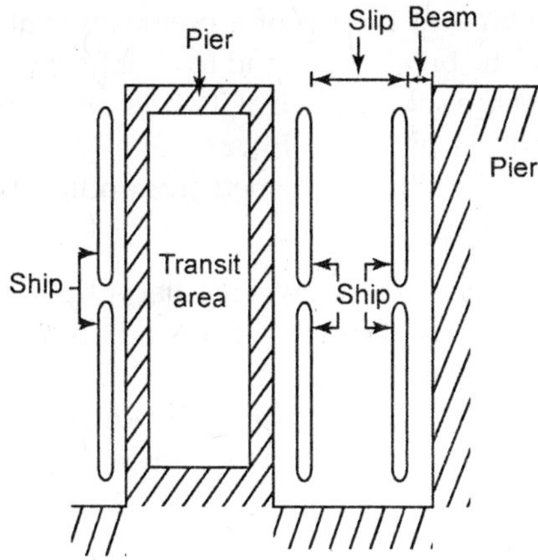


**FIG. 19.14** *Schematic of a four-berth pier*

The length of the pier should be sufficient to accommodate the longest ship, and the width should be adequate for easy unloading of cargo.

The water depth should be such that there is a minimum clearance of 30 cm under the keel of a fully loaded ship, with provision for an abnormally low tide.

Usually, no provision is made for storage of cargo in the pier.

For the design of the pier, allowances are made for live loads from the anticipated cargo and the impact load, taken as one per cent of the maximum weight of the vessel.

While the floor of the pier may be of timber or of RCC, depending upon the loads – whether small or large – the substructure and foundation can be made of piles or frames, or of concrete and masonry.

**Types of piers:** A pier structure may be solid with masonry or concrete, or an open-framed structure made mostly of timber supported on piles, with a central fill or a central mound.

A typical structure of the latter type is shown in Fig. 19.15:

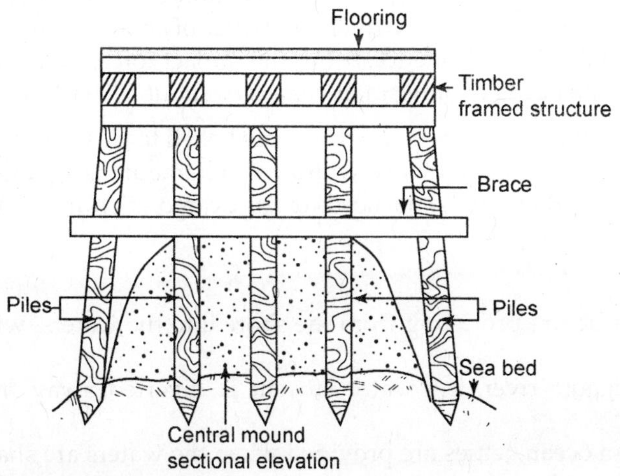


**FIG. 19.15** *Pier with timber-frame structure on piles with a central mound*

**Pierhead:** This is a structure built at the tip of a breakwater near the harbour entrance. Its function is to protect the end of the breakwater and to clearly mark the entrance to the harbour. Since it is exposed to the fury of the sea on three of its sides, it should be designed to withstand the maximum forces, and provided as a monolith of stone or concrete; it is separated from the breakwater by means of a joint. It is distinguished from others by means of a coping and is provided with light for night vision.

**Slip:** This is the space of the water area between adjacent piers where ships are berthed. It should be adequate to accommodate the two vessels adjacent to each of the piers, and for barges to operate for loading and unloading of ships.

The width of the slip should also be adequate for easy entry and exit of the vessels. It should at least be 3 to 4 times the beam of the largest ship to satisfy the above requirements.

## Dolphins

A *dolphin* is a closely spaced cluster of piles, the tops of which are pulled together and tied by means of a cable. Dolphins are used in combination with piles and wharves; they are used for mooring vessels, as also for transferring cargo from one ship to an adjacent vessel, when both the vessels are moored.

These are designed to resist water current forces, impact force of vessels, and wind forces.

**Types of dolphins:** Based on their function, dolphins are either for breasting or for mooring.

***Breasting dolphins*:** These are provided with mooring posts to take the lines of a vessel, move it along the dock or hold it against the water current. They are also provided with what are called *fenders* to take the impact of the vessels and protect them.

***Mooring dolphins*:** These are provided to hold a vessel against its broadside wind force. They are located such that the mooring lines are approximately at right angles to the dock wall. Since they are away from the dock, impact force of the vessels need not be considered in their design. Based on the method of construction, dolphins may be flexible or rigid.

***Flexible dolphins*:** They are a cluster of piles tied with galvanized cable. The number of piles is odd — 3,5,7,9, and so on. The centre pile is made to project above the surrounding ones by at least one metre to provide means of attaching the mooring lines of a vessel. The piles are arranged symmetrically, the outer ones driven to a batter of about 1 in 10. The bearing capacity of the foundation should be adequate to absorb the impact forces of vessels. Small-clustered dolphins can cater to the needs of only small vessels of about 5000 t dead weight.

***Rigid dolphins*:** In this type, the piles in the cluster are bolted to cross members of wood, located just above the low water level, and provided with a wood frame at the top to serve as the platform. This type of dolphin is used to cater to the needs of large cargo ships and oil tankers.

## Jetties

A *jetty* is a narrow structure projecting from the shore into the waters, with berths on one or both sides.

If used in a navigable river, the water current is diverted away from the bank, thus preventing scour.

If it is in a sea or an ocean, jetties are provided when the waters are shallow near the shore for a long distance, or when the harbour entrance is affected by littoral drift. Thus, jetties extend from the shore into deep waters to receive ships. Since jetties are exposed to wave forces, they are designed in a way similar to breakwaters. In plan, a jetty may be curved, or when more than one is used, they may be converging or diverging.

Jetties may serve as berthing facilities either in the initial phase of the development of a harbour, or as additional berthing at a later stage.

The impact caused by berthing ships may be minimised by skilful manoeuvring near the jetty at a moderate speed.

The primary *differences between a jetty and a wharf* are:

- A jetty is perpendicular to the shore, while a wharf is parallel to it.
- A jetty may be provided with berthing facilities on both sides, while a wharf can provide berthing on the water side only.

## Fenders

The walls of a *wet dock* (which is an enclosed area of relatively calm waters of a harbour in which ships may be berthed for loading and unloading operations of cargo) may get damaged by being hit by large vessels during berthing; *buffers* or *cushions* provided permanently on the dock wall to protect the latter from the impact of ill-guided ships are known as *fenders.*

Fenders may be *rubbing strips* of timber, coir padding, or used rubber tyres; these are used for small vessels. They may consist of vertical and horizontal members of wood in the form of a *grill* fixed to the face of the jetty. A *concrete block* of sufficient weight may also be used to absorb the shock of the impact of the ships. After the development of rubber technology, *rubber fenders* have become popular.

**Types of fenders:** The following are the different types of fenders used:

(i) Timber piles with springs
(ii) Rubber fenders
(iii) Pneumatic fenders.

***Timber piles with springs:*** Timber piles are driven at 3 m to 5 m intervals on the harbour basin side of the dock wall, at a distance of about 0.5 m from the wall. At the top, the piles are supported by means of springs attached to the dock or quay wall, as shown in Fig. 19.16:

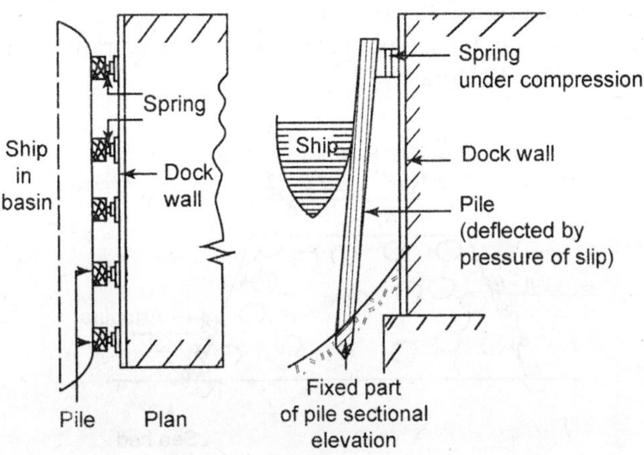


**FIG. 19.16** *Timber pile fender with springs*

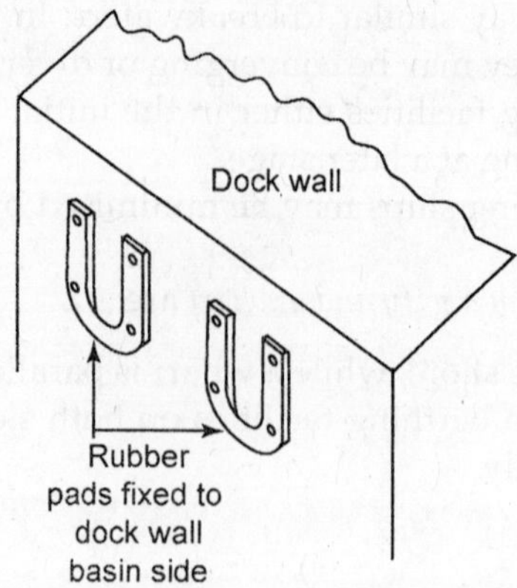


**FIG. 19.17** *Rubber fenders*

***Rubber fenders***: Rubber pads may be fixed on the face of the dock or quay wall at regular intervals to absorb the impact from the ships. They may have to be replaced from time to time if they get damaged under heavy impact (Fig. 19.17).

In place of fixed pads, *loosely anchored* rubber pads may also be used.

***Pneumatic fenders***: These are fenders in which pneumatic rubber tyres or tyres under compressed air pressure are fixed at the top of the wall, projecting a little into the harbour basin, at regular intervals along the length of the wall.

Fenders may be used at the edge of jetties also to protect them from ship impact.

## Trestles

These are timber structures on wood piles and serve the purpose of piers. Being lighter structures, they are economical and are generally used on a temporary basis.

## Moles

Rock-fill structures extending from the shore towards the water to serve the purpose of a pier are called *moles*. These may be preferred to trestles when the depth of water is 3 m or less. The total height of the mole does not usually exceed 8 m.

The top surface of a mole may be used to provide a roadway, railway, conveyors, and side-walk for feeder transport (Fig. 19.18):

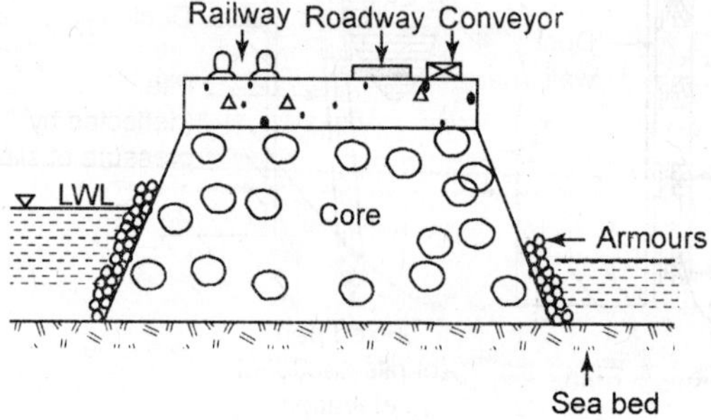


**FIG. 19.18** *Rock mole*

## 19.4 CARGO-HANDLING AND STORAGE FACILITIES

Different types of cargo such as general cargo, bulk cargo and containers will be carried by ships. General cargo comprises units such as automobiles, machinery, and boxes, bags and bales containing different kinds of items, which have to be handled carefully. Bulk cargo consists of loose and unpackaged material which can be poured freely into the ship compartments; this may be grain, ore and coal. Even liquids such as oil, petroleum and petroleum products fall in this category. Containers are large box-like structures designed to hold cargo, which are generally transported by railway wagons or special trucks on land, and special cargo ships. Transportation of the various categories of cargo necessitates the use of specialised cargo-handling equipment.

Further, cargo may require temporary storage facilities such as during transit either before loading into the ship or after unloading from the ship, and before being sent via land transport to the respective destinations. Sometimes, cargo may need to be stored on a short-term basis for a few days or on a semi-permanent basis for a few months or a season.

For such purposes, aprons, transit sheds, warehouses and cold storages are required.

### Cargo-Handling Equipment

Ship-gear for handling cargo, also called *deck furniture*, is necessary for transferring cargo from the ship-hold on to aprons/transit areas, to be transported later to the destinations via surface transport – railways and highways.

#### General-Cargo Equipment

Most general cargo is handled in lifts of 5 tonnes or less. The equipment consists of machines for lifting and lowering of cargo, those for transporting cargo, and for piling and de-piling.

For lifting and lowering of cargo, different types of cranes – fixed crane, mobile crane, and hand crane, are used. Some of these will be stationed on the quays and some on the ship decks.

Lifting a load is called *hoisting*; to move a load by changing the inclination of jib of a crane is referred to as *luffing*, while moving a load radially is termed *slewing*.

Pneumatic-tyred mobile cranes are popular since they can be moved effortlessly to any convenient position.

Hand cranes are used for minor cargo only as they are operated by human power.

Other equipment for general-cargo handling include trucks, tractors and trailers, conveyor belts, and piling and unpiling equipment such as overhead gantry cranes. Heavy machinery can be handled by using heavy duty cranes of capacity 50 t or more.

#### Bulk Cargo Equipment

This is custom made; that is, equipment to handle each type of bulk cargo is specifically designed for the purpose. With such equipment, the rate of loading can be as high as 16,000 tonnes per hour. There can be specific equipment for loading and for discharging dry bulk cargo like grain, ore and coal.

#### Container Cargo Equipment

Packaged cargo containers are limited in size by the clearance requirements of land transport by highways and railways. Containers to be transported by ships are designed for vertical stacking. Container ships need berths of length 300 m or more.

Container terminals require a lot of open space for convenience of operation. Containers may be handled by lifting with gantry cranes, by rolling, or by rolling on or off ships (*Ro-Ro* method, in which the containers are moved on wheels).

The sizes of containers range from 2.44 $m^2$ and 3.048 m long to 2.44 $m^2$ and 12.192 m long; the gross weight of such containers ranges from 10 t to 30 t.

Specialised equipment such as large gantry cranes operating on the wharf side are needed to handle the containers. Generally containers are stored in the open; however, sorting and re-picking is carried out at the terminals.

## Aprons

The open floor space provided in front of a berth is called an *apron;* this is required for expeditious loading and unloading of cargo into and from the vessels. Supporting facilities such as cranes, railway wagons and trucks are also needed.

The width of the apron depends upon the number of rail tracks to be provided, the volume of cargo, and economic considerations, including minimum travel time to the transit sheds. The general norm is 10 m for light traffic and 25 m for heavy cargo traffic.

### Transit Sheds

A *transit shed,* as the name suggests, is meant as a temporary storage space for the cargo unloaded from the ship to be sent to the consignees via land transport, or for that obtained from the consignees for export by the ship to their destinations.

Usually, the cargo is of a mixed variety with different types for different consignees, who are given a free time of 3 to 5 days for clearance; after the free period, 'demurrage' or penalty is levied for storage on 'per day' basis, in order to discourage delay in clearance.

*The general requirements for a transit shed are:*

(i) Proximity to the quay.
(ii) Adequate capacity for the expected volume of cargo.
(iii) Portable cranes for handling the cargo.
(iv) Provision of a roadway and a railway for quick transit of cargo.

*The construction requirements are:*

(i) Doors of a suitable type (say *sliding type* or *rolling type*) and adequate size and number
(ii) Mobile cranes of appropriate capacity
(iii) Fire-resistant light construction
(iv) Fire-fighting arrangements
(v) Good lighting by means of skylight during the day, and electric lights during the night
(vi) Minimum height of 6 to 8 m
(vii) A minimum length of 150 m, or equal to the length of the berth
(viii) A floor area of 8,000 to 10,000 sq. m. to cater to a ship-load of cargo
(ix) A width of around 60 m
(x) Adequate aisle space to facilitate easy and rapid handling of cargo, using mobile equipment
(xi) A uniform live load of 25 kN/$m^2$ for design of the floor
(xii) A transverse slope rising from berth to land side for easy loading into railway wagons
(xiii) Grilled and lock rooms for storing precious cargo items
(xiv) Non-slippery and abrasion-resistant flooring

**Warehouses**

A *warehouse* is a permanent storage godown, provided on shore, behind transit sheds for storing cargo on long-term basis.

If the cargo is under customs authority, until cleared, the warehouse is called *bonded warehouse*. Warehouses are built of R.C.C. with as many floors as are necessary to store the expected volume, as well as the varieties of cargo. Fire-resistant construction, appropriate flooring for abrasion resistance and durability, separate storage spaces for different kinds of cargo, facilities of appropriate handling equipment, and fire-fighting arrangements are the important requirements for warehouse design.

**Cold Storages**

These are specially equipped with refrigeration plants, to store perishable cargo such as fruits, vegetables, fish, meat, and eggs. All the other requirements and facilities for warehouse are also necessary.

**Guard Houses**

These are small cabins specially meant for housing security personnel and guards. The usual places are near the entrance gates of highways and railways, entrances to terminals, piers, warehouses, and bonded warehouses.

## 19.5 NAVIGATION FACILITIES

### Need for Navigation Facilities

*Navigation aids* are essential to ensure safe, efficient and comfortable negotiation of vessels not only in harbours and inland waterways, but also in the open sea.

These aids guide the vessels and the mariners to approach the harbours and ports, especially during nights and bad weather conditions; further, they help the mariner to avoid hazardous areas like sand bars and shallow rocky outcrops, thereby avoiding difficult situations like grounding.

Accidents occur in relatively narrow and congested waters, or even in the open sea, unless the vessels follow two-way sea lanes of about 4 to 5 km width, especially in narrow zones and zones of heavy maritime traffic.

The type of navigation aids required depends upon the nature of waterway, the type of vessels, and size of the harbour.

The need for navigation aids for marine traffic is established; besides these aids, all stipulated maritime regulations in the particular zone of waters have to be mandatorily followed by all those concerned with maritime traffic and international trade.

### Types of Navigation Lights

The following are the two broad categories:

(i) General lights
(ii) Local lights

**General Lights**

These are navigation aids along the coast, which serve the approaches to more than one port/harbour. These are usually *lighthouses* or *lightships* (or moored ships with lights). It is the responsibility of the Union Government to build and maintain lighthouses; the Director General of Lighthouses and Lightships normally looks after these duties.

### Local Lights

These are aids to guide the vessels in the approach channels and harbour entrances and to anchorages and berths. They are lighthouses, channel markers, leading lights on the shore, piers, wharves, and breakwaters. They are meant for a particular port/harbour, and it is the responsibility of the port authorities to install and maintain these aids.

Another classification is based on whether they are of *fixed* or *floating* type.

**Fixed light stations:** These are built on land and are permanent structures called lighthouses. Generally, they are built on the shore, near a harbour, or at some distance from it.

**Floating light stations:** When it is difficult to build a lighthouse, a moored light vessel, or a lightship may be used; Since they float on water, they are called the floating type of light stations. There are also other types of floating-type aids.

Depending upon the type of *service* provided and the *range of visibility*, navigational aids are classified as follows:

(i) Coast approach indicators
(ii) Obstruction indicators
(iii) Approach channel indicators
(iv) Harbour location indicators.

## Fixed Navigation Aids

Fixed navigation aids are –

- Lighthouses
- Beacon lights on breakwaters and shores
- Navigation lights on piers and wharves

### Lighthouses

Lighthouses are tall towers built of masonry or reinforced concrete. A marine beacon light is provided at the top of the tower, which is divided into a number of floors/storeys. The lower floors house services, stores, and living rooms meant for the maintenance and working/operation of the lighthouse.

The lighthouse tower is generally erected at a suitable point on the shore to guide vessels to a port/harbour nearby. Sometimes, they are built on shoals and other hazardous features. The tower should be strong enough to withstand heavy wave forces.

A thick bed of concrete is placed on hard soil or rock as the foundation for the lighthouse. A flight of spiral stairs is provided from the bottom to the top of the tower.

It can be founded on piles or caissons if the bed soil is weak for a significant depth. A typical lighthouse is shown in Fig.19.19.

**Note:** Visible distance of a lighthouse can be computed based on the curvature of the earth and the atmospheric refraction of light rays – factors which govern the distance to the visible horizon.

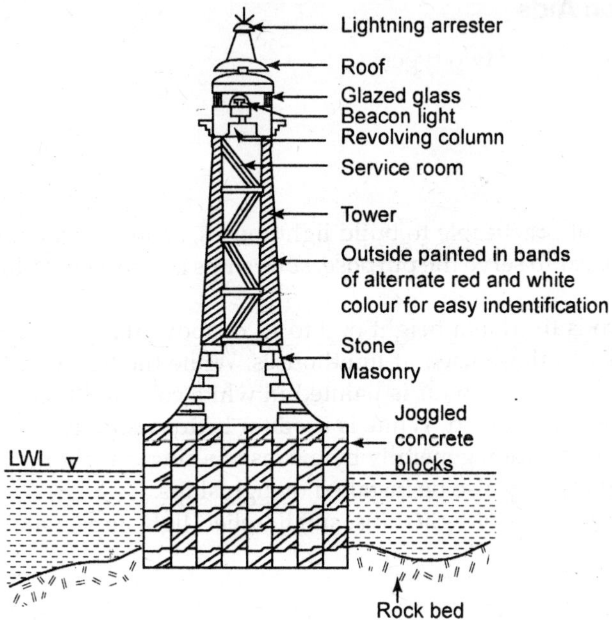


**FIG. 19.19** *Schematic of a lighthouse*

Yet another classification of signals is given below:

(i) Light signals
(ii) Sound signals

Signals should compel attention, with clear meaning and adequate response time. Light signals may be primarily beacon lights, which may be fixed or of the rotating type. Occasionally, sound signals which make audible sounds/whistles are used in foggy weather

## Beacon Lights

These may consists of lights, fixed or flashing, for easy identification during night time. In fact, any easily identifiable natural or artificially constructed feature, object, or tower may serve the purpose of guiding navigation, and may be called 'beacon'. In this sense, lights come in handy for visibility during night time.

The illuminating power of the lights is characterised by its candle power. It is enhanced by refraction through lenses and prisms, and also by reflection by polished hyperbolic concave mirrors. Electric lights are preferred to acetylene gas lights owing to the lower operating cost of the former.

*Flasher mechanisms* are used and regulated to obtain any desired flashing characteristics; these mechanisms are operated by means of a motor-driven circuit interrupter. The flashes may be set to occur at single or multiple intervals and may be a combination of white and red colour.

*Fixed beacon lights* may be used to indicate *breakwaters* and *shores* during night time. These may also be fixed on *piers* and *wharves* to aid navigation.

All these aids help in navigation of vessels to the desired location – into the harbour basin, adjacent to the piers and wharves, and so on.

## Floating Navigation Aids

Floating navigation aids are of two types:

- Lightships
- Buoys

### Lightships

At places where it is not practicable to build lighthouses, small ships with a displacement of about 500 tonnes are used to serve the purpose; such ships are fitted with lights and are known as lightships.

The light or lantern is fixed at a height of 9 to 12 m above the water level and its flashing mechanism are similar to those used at lighthouses. While the hull of a lightship is painted red, the name of the place or harbour is painted in white coloured letters on both sides. The main superstructure is also painted. White is used for better visibility.

A lightship is held in position generally by means of a single anchor.

Storm warning signals may also be installed on lightships.

During overhauling/repair of lightships, small 'relief' lightships, painted red, are used as a temporary measure.

### Buoys

These are floating structures of relatively small size; they are used for demarcation at entrances and approach channel boundaries. Boundaries of the width of the approach channel are defined by two rows of buoys, one on each side.

Buoys are fabricated from cylindrical cans and drums; they are provided with through stays. Buoys are designated 'star-board hand' buoys and 'port-hand' buoys depending on whether they appear to the 'left' or to the 'right' of the approaching navigator.

Near the water line, buoys are provided with wood fendering to protect them from unexpected impact with ships during manoeuvring. They are fabricated as hollow steel-plated structures.

The diameter of a buoy varies from 1.80 to 3.00 m and their spacing from 150 to 300 m, Buoys of different shapes, sizes and colours are used to provide different indications to the approaching mariner.

Buoys may be of the following types:

(i) Mooring buoys
(ii) Wreck buoys
(iii) Buoys for special indications.

**Mooring buoys:** These buoys serve the purpose of moorings for ships inside the harbour basin, and hence the name. Typical mooring buoys are schematically shown in Fig. 19.20.

**Wreck buoys:** These are of special shape and are used to locate wrecks at the exterior of a harbour or on open seas, as also for under-water cable locations. A typical wreck buoy is shown in Fig. 19.21.

**Buoys for special indications:** The purpose of the buoy is generally indicated by the shape and colour.

The practice in India is the same as the British practice in this regard. Fig. 19.22 shows a few special buoys.

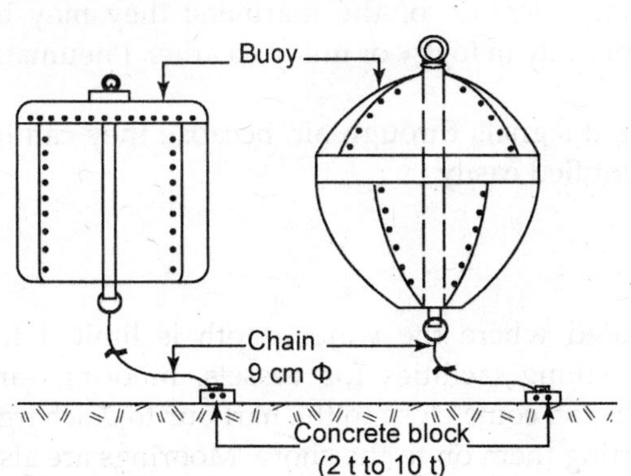


**FIG. 19.20** *Mooring buoys*

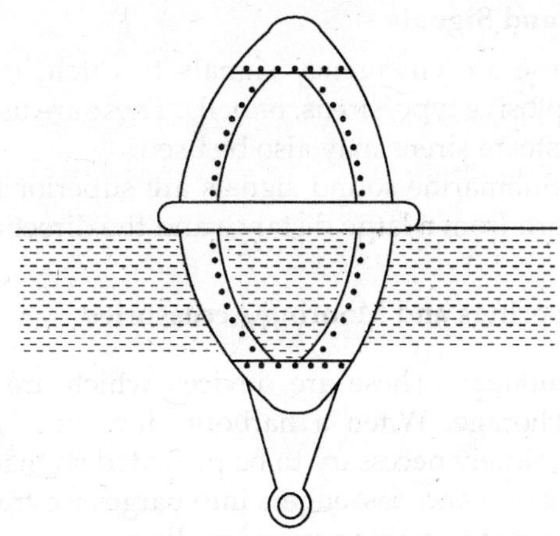

**FIG. 19.21** *Wreck buoy*

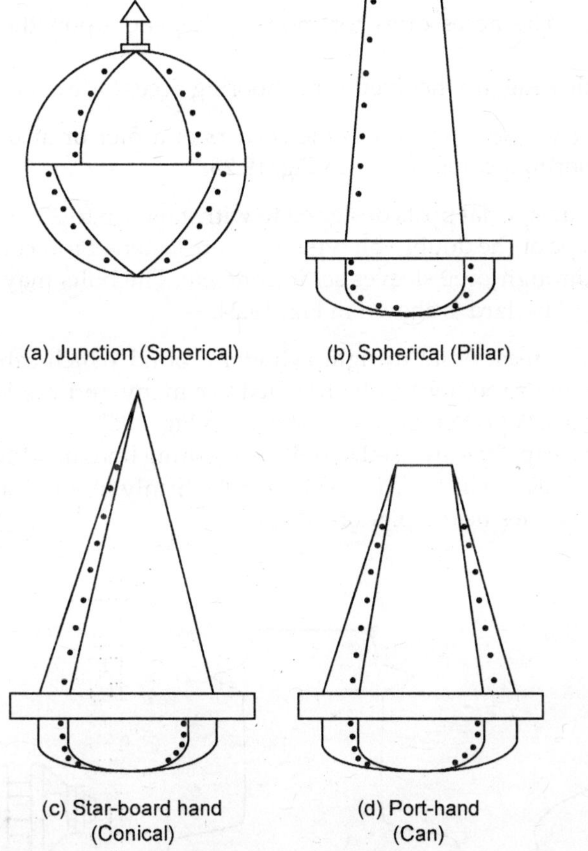


**FIG. 19.22** *Special buoys*

### Sound Signals

These are emergency signals to catch immediate attention of the mariners; they may be explosive type, sirens, or bells. These are used especially in foggy or misty weather. Pneumatic or steam sirens may also be used.

Submarine sound signals are superior to sound signals through air, because they can be heard from a large distance and the direction identified easily.

### Moorings and Mooring Accessories

**Moorings:** These are devices which are provided where the water depth is limited for anchorage. When a harbour does not have berthing facilities for vessels, moorings are absolutely necessary to be provided so that a ship can come close to the harbour to discharge its cargo and passengers into barges for transporting them on to the shore. Moorings are also required to stable a vessel in distress.

Also, when a vessel is made to wait outside the harbour for want of berthing space, moorings are necessary to stable it until the vessel gets the call in its turn to enter the harbour. Moorings may be provided off-shore also, if docks cannot be built.

**Mooring accessories:** The accessories for mooring depend upon the type of moorings – 'fixed' or 'floating'.

*Fixed type*: The following are the fixed-type mooring accessories:

(i) **Mooring posts:** They are attached to the corners of a pier or at the ends of a wharf. A typical corner mooring post is shown in Fig. 19.23.

(ii) **Bollards:** These are special posts designed to withstand up to 35 tonnes pull, and may be of the single-bit type or the double-bit type. They are fastened to a concrete block by means of bolts passing through pipe sleeves set in concrete. The bolts may be easily replaced, if necessary. A typical bollard is shown in Fig. 19.24.

(iii) **Capstans:** These consist of an upright cylinder around which cables are wound. They are helpful when the rope lines to be handled are many and are long, especially when attached to dolphins. A power capstan is shown in Fig. 19.25.

Electrically operated capstans are used to pull the mooring lines into the mooring accessories or fittings when long lines are involved. A ship may be firmly moored using lines at the bow, stern, spring and breast lines near a pier or a dock.

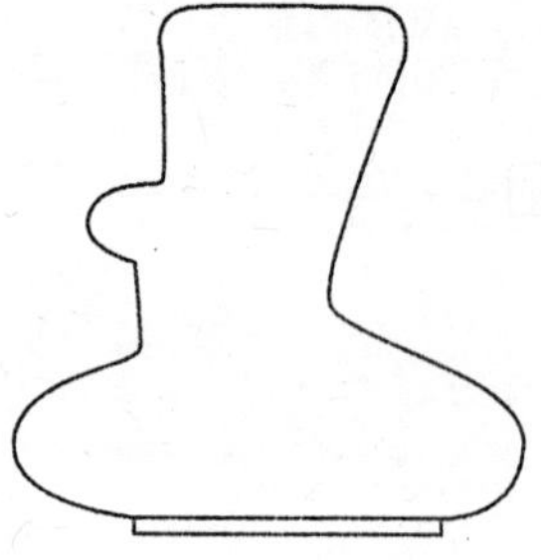

**FIG. 19.23** *Corner mooring post*

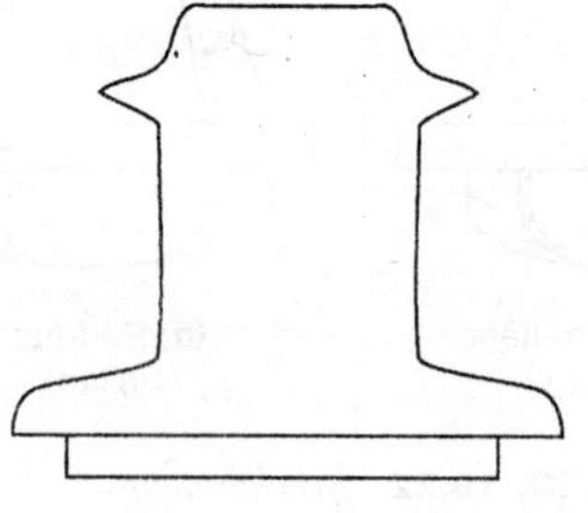

**FIG. 19.24** *Bollard*

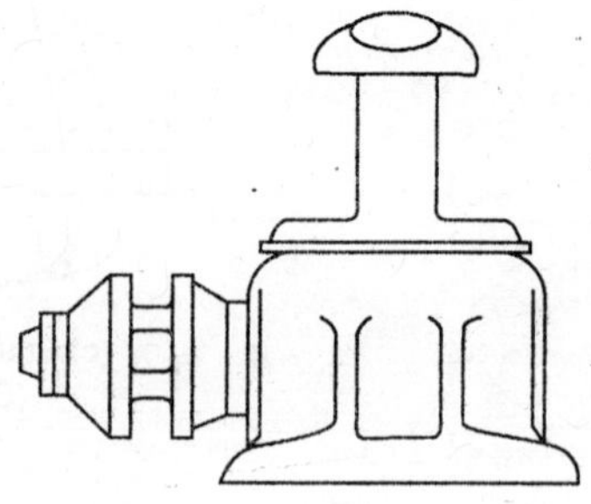

**FIG. 19.25** *Power capstan*

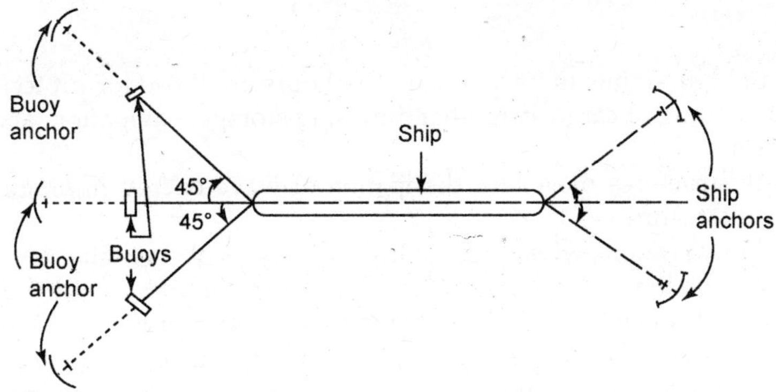


**FIG. 19.26** *Offshore mooring with three buoys*

*Floating-type*: The following are the accessories of the floating type:

**(i) Buoys**: Mooring buoys fall in this category.

**(ii) Cables**: A cable comprises a 'pendant', or a stud or open-link chain, a hanging chain, and a ground chain. The upper portion is fitted to a swivel. The ends of the cable on the sea-bed are suitably anchored. Heavy objects are used as anchors to prevent a vessel from drifting.

**Offshore moorings:** Where a dock or a harbour cannot be provided, vessels are moored offshore with its own anchors and buoys. A vessel moored with its own anchors will not be stable enough for loading and unloading operations; hence, these anchors need to be supplemented with additional mooring buoys. The number of mooring buoys used depends upon the wind, currents, waves, nature of sea-bed, and of course, the size of the vessel; this may range from three to eight. Normally three buoys are adequate (Fig. 19.26).

Two quarter-lines and one stern-line are supplemented by the ships anchors and buoys.

[Super-tankers up to 1,00,000 DWT will require six to eight buoy anchors in addition to their own anchors.]

## Modern Trends in Navigation Control and Safety

Marine radars help in the segregation of sea lanes; the 'Plan Position Indicator (PPI)' is one such radar which helps the mariner to know the position of the vessel.

A more accurate system is the continuous-wave hyperbolic system which uses a chain of four radio-transmitters on land along with receivers on board the ship. This system, which works at frequencies in the order of 100 kHz, is one of the most accurate in the world, and covers several countries including India.

Russia has developed an 'Anti-collision Indicator'. Surveillance satellites are used to obtain position data of obstacles. Partial control of vessels by computers and automation can keep a vessel on a preset course.

Automation of on-board activities such as ballast-trimming, hatch-operation, ventilation and humidity control, and fire-detection, along with centralisation of control room data for all on-board systems of the vessels enhances the safety at sea and in the ports during docking manoeuvres.

The latest development is the use of the 'Global Positioning System (GPS)', which works with the aid of communication satellites, to predict and avert accidents in the navigation of big vessels, ocean liners, and aircraft carriers.

## SUMMARY

- The facilities and requirements for ports and harbours are those for protection of vessels, loading and unloading of cargo, cargo-handling and storage, navigation, and docking and repairs of vessels.
- Breakwaters are structures meant for dissipation of wave energy impinging on harbour components and structures.
- Rubble-mound breakwaters, with stones in the three layers of core, filter and armours have good energy dissipation characteristics. These are sometimes strengthened by concrete blocks and patented concrete blocks such as tetrapods, which prove to be stronger.
- Composite breakwaters consist of solid superstructure of concrete at the top, supported by rubble mounds. Vertical wall breakwaters and special types such as floating and hydraulic breakwaters also serve the purpose. Barge method and staging method are the important methods of construction of rubble-mound breakwaters.
- Loading and unloading facilities for cargo are quays, wharves, piers, dolphins, jetties, and fenders, which are mostly fixed structures.
- Cargo-handling and storage facilities are necessary at ports; the former consist of fixed cranes and mobile cranes. The cargo may be general cargo, bulk cargo, and container cargo, each of which requires specialised handling equipment.
- Aprons, transit sheds for temporary storage, ware-houses for long-term storage, and cold storages for perishable goods are the other facilities needed.
- Navigation facilities are essential for safe and efficient navigation of vessels.

General lights serve more than one port while local lights are those meant only for one particular port or harbour.

Fixed type of aids are lighthouses and beacon lights on breakwaters, piers and wharves. Lightships and buoys are of the floating type. Buoys may be mooring buoys, wreck buoys, and special indicators. Sound signals are used in foggy weather.

Moorings are necessary when water depth is limited. Moorings and mooring accessories may be of the fixed type such as mooring posts and bollards, or of the floating type, of which buoys and cables are examples.

In the absence of a harbour, offshore moorings are useful to stable a vessel near the shore for discharging passengers and cargo into barges for transporting to the shore. Modern trends in navigation include the use of plan position indicator, continuous-wave hyperbolic system and automation of all activities using computers.

The use of Global Positioning System (GPS) utilising satellite data is the latest development used for ensuring safety in navigation.

## QUESTIONS

1. What is a 'breakwater'? Classify breakwaters.
2. Explain the structure of a mound-breakwater and the mechanism of dissipation of wave energy.
3. Explain the important methods of construction of a rubble-mound breakwater.
4. Write short notes on:

   (a) Tetrapods

(b) Staging methods of construction

(c) Composite breakwater.

5. What are the advantages and disadvantages of rubble-mound breakwater?
6. What are the factors governing the height of a breakwater?
7. Write a note on vertical wall breakwater.
8. Explain the action of a floating breakwater.
9. Write short notes on:

   (a) Quay

   (b) Pier

   (d) Jetty
10. Explain the different types of dolphins.
11. Write short notes on:

    (a) Fender

    (b) Wharf

    (c) Pier-head
12. Write a note on the equipment used for handling general cargo.
13. Differentiate between general cargo and bulk cargo, and the manner of handling them.
14. Write a note on container cargo and the manner of handling such cargo.
15. Write short notes on:

    (a) Transit sheds

    (b) Cold storage

    (c) Warehouse
16. What is a lighthouse? What are the salient features of its construction?
17. Differentiate between fixed and floating signals used for navigation near harbours.
18. Write short notes on:

    (a) Lightships

    (b) Flashing beacons

    (c) Mooring buoys
19. Write a note on mooring accessories.
20. Write a note on sound signals and those used in foggy weather.
21. Write a note on the modern trends in navigation aids.
22. What is the purpose of navigation aids? What are the various types of aids used on shore at sea? (Anna University)
23. Explain 'Tetrapod'. Write a short note on vertical breakwater. (Anna University)

24. Explain the different types of breakwaters with sketches. (Anna University)
25. Write a descriptive note on moorings and mooring accessories. (Anna University)
26. Differentiate between quay and pier. (Anna University)
27. Why should a jetty face be provided with a cushion? (Anna University)
28. Distinguish between dolphins and jetties. (Anna University)
29. Write short notes on:
    (a) Breakwater
    (b) Spring fenders (Anna University)
30. What are tetrapods and tribars? (Anna University)
31. Differentiate between a transit shed and a warehouse in respect of
    (i) their use
    (ii) their construction. (Anna University)
32. Briefly explain about navigational aids with suitable sketches. (Anna University)
33. Write short notes on:
    (i) Upright wall breakwater
    (ii) Quays and Jetties
    (iii)Tetrapods (VTU)
34. Compare wall type breakwater with mound type breakwater with suitable sketches. (VTU)
35. What is the purpose of navigation aids? What are the various types of aids used on shore at sea? (UP Tech. University)
36. What is a breakwater? What are the factors required for selection of breakwater? (West Bengal Tech. University)
37. What are quays? (West Bengal Tech. University)
38. (a) Write a short note on Tetrapods.
    (b) What is a lighthouse? (West Bengal Tech. University)
39. (a) Define breakwater.
    (b) Mention its classification.
    (c) Explain the forces acting on a breakwater.
    (d) What are the controlling factors for selection of types of breakwater?
    (e) Draw a section of rubble mound breakwater showing core, filter and armour layers. (West Bengal Tech. University)
40. Write short notes on:
    (a) Bollard
    (b) Buoys

(c) Cables

(d) Anchors (West Bengal Tech. University.)

41. Write short notes on:

    (a) Wharf

    (b) Breakwater

    (c) Jetty (West Bengal Tech. University.)

42. Differentiate between:

    (a) Jetty and fender

    (b) Flexible and rigid dolphins

    (c) Quay and wharf (Mumbai University)

43. (a) Write a note on the facilities needed at a port.

    (b) Explain:

    (i) Jetty

    (ii) Dolphin

    (iii) Wharf

    (iv) Pier (Mumbai University)

44. Differentiate between:

    (i) Jetty and fender

    (ii) Flexible and rigid dolphin. (Mumbai University)

Chapter **20**

# DOCKS AND REPAIR FACILITIES

**CHAPTER OBJECTIVES**

This chapter enables the student to understand the following:

- Docks are sheltered basins for berthing ships either for loading/unloading cargo (wet dock) or for repairs (dry dock).
- Dock walls are designed as gravity retaining walls for the expected forces
- The graving or dry dock is most popular; this is designed for dock empty and dock full conditions.
- Marine railway or slipway is an alternative means of exposing a ship for repairs.
- A floating dock can support a ship above water by its own buoyancy.
- A lock is a device used to overcome differences in water level to maintain continuity in navigation. Lock gates are designed to withstand water pressures. They are similar to river or canal locks, in principle.

## 20.1 INTRODUCTION

Docks are defined as sheltered basins or enclosures for berthing ships. They may be closed basins or open to the surrounding structure such as the quay or pier.

Ships are to be kept afloat at a reasonably uniform level to facilitate loading and unloading of passengers and cargo; docks are necessary since these operations require a few days, during which period, the vessel needs to be relatively stable.

### Classification of Docks

Docks are primarily classified as

(a) Wet docks
(b) Dry docks

Wet docks are also known as harbour docks; vessels are kept afloat in these docks to facilitate loading and unloading operations of passengers and cargo. Dry docks are those in which repairs are carried out on the vessels, if necessary. The vessel is brought into the dock in adequate depth of water, and the water is withdrawn after the vessel is supported appropriately to carry out repairs.

## 20.2 WET DOCKS

Wet docks are useful in harbours where the range and effect of the tides is considerable; they are designed to maintain uniform level of water to facilitate the handling of cargo. The effects of storms in the open sea and also in the harbour basin do not affect the waters in the dock enclosure.

If the water in the harbour is turbid, carrying silt load, the dock can be filled with water from other sources in order to maintain the necessary draft for the vessels.

### Classification of Wet Docks

Wet docks may be classified as

1. Wet docks in tidal basins
2. Wet docks in enclosed basins

#### Wet Docks in Tidal Basins

Tidal basins in the harbours are affected by tides, which cause significant variations in water level. Vessels may be berthed in these basins if there is adequate water depth for the required draft.

In artificial harbours formed by outlying breakwaters, a wet dock may be constructed perpendicular to the shore for permitting vessels to enter it and discharge their cargo by berthing inside the dock, alongside its walls (Fig. 20.1).

**Advantages:** Tidal wet docks have the following advantages:

(i) Unrestricted entry and exit of vessels
(ii) Expensive locks and lock gates can be avoided

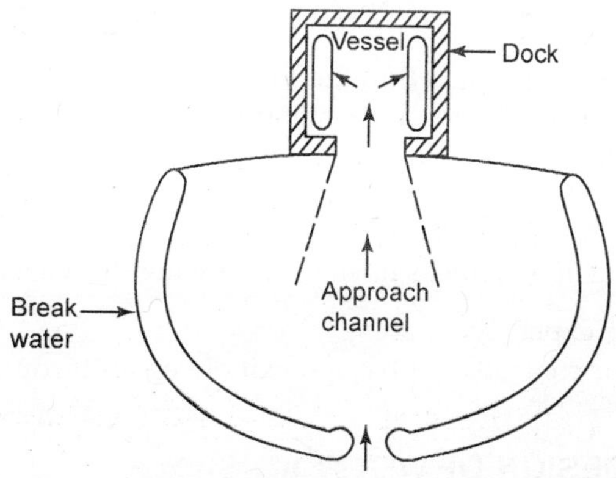


**FIG. 20.1** *Wet dock in tidal basin*

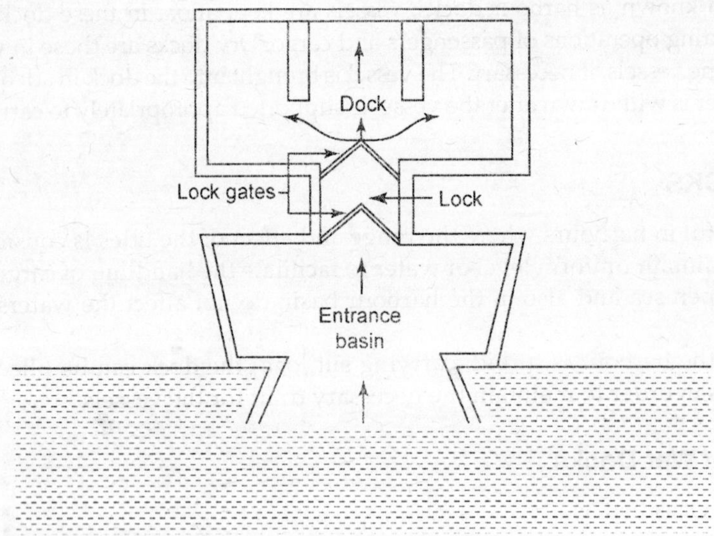


**FIG. 20.2** *Wet dock in an enclosed basin with locks*

**Disadvantages:** The following are the major disadvantages:

(i) When the tidal range is high, loading and unloading of cargo are affected.
(ii) Water level fluctuations can cause the rubbing of the sides of vessels against the berths.

**Wet Docks in Enclosed Basins**

Wet docks in this case are enclosed and are off from the open sea or harbour basin with controlled entry and exit by means of what are called *locks* and *lock gates*. This facilitates the maintenance of uniform water level for convenient loading and unloading of cargo.

Such an arrangement is preferred where the tidal ranges are significantly large, say more than 4.5 m.

A typical wet dock in an enclosed basin with lock arrangement is shown in Fig. 20.2.

**Advantages:** As with wet docks in tidal basins, those in enclosed basins also have a few advantages:

(i) Uniform level of water is maintained through locks.
(ii) It obviates the need for constant attention to mooring.
(iii) The dock enclosure has fully controlled entry and exit.
(iv) The effects of storms are not felt inside.

**Disadvantages:** Enclosed basin wet docks have the following drawbacks:

(i) Locks and lock gates are expensive.
(ii) More time is required for controlled entry and exit of vessels through locks.

## 20.3 PLANNING AND DESIGN OF WET DOCKS

The planning of the layout depends upon the available area at the site; an important criterion is to obtain the maximum length of quay in relation to the area of the basin or dock.

The docks should be rectilinear in shape and not curvilinear, for ease in berthing. The walls of the dock are built of masonry with stone or brick, or of concrete in which case, construction joints are provided.

The shapes of docks and basins commonly used are:

(i) Rectangular
(ii) Diamond
(iii) Inclined quay

### Rectangular Docks

The length-to-breadth ratio should be adjusted to give the maximum quayage. A typical layout of a rectangular-shaped dock basin is shown in Fig. 20.3.

### Diamond-Shaped Dock

This shape gives better quayage as the long sides of the docks can be conveniently extended, as shown in Fig. 20.4.

### Inclined-Quay Docks

The inclined shape of the piers forming the docks and dock basin provide better quayage than the rectangular-shaped basin, as shown in Fig. 20.5.

## Approaches to Docks

Approaches to docks should be sheltered and of sufficient length; the approaches should be kept at navigable depth by means of dredging, if necessary. Otherwise, the docks can be approached only during high tides.

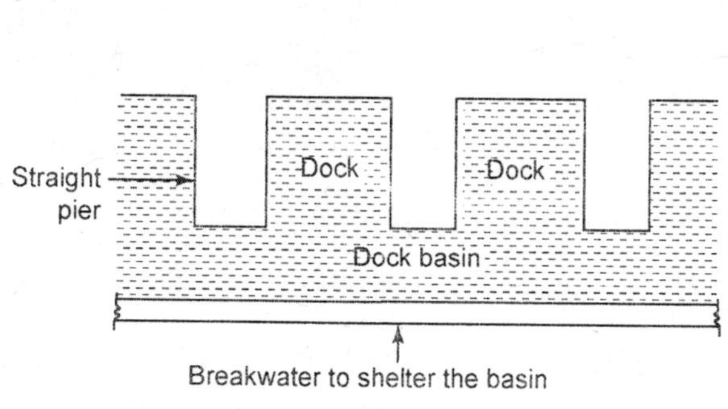


**FIG. 20.3** *Rectangular-shaped docks*

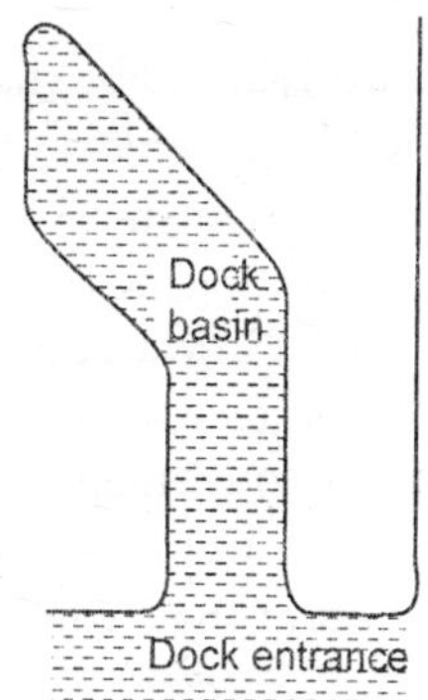


**FIG. 20.4** *Diamond-shaped dock*

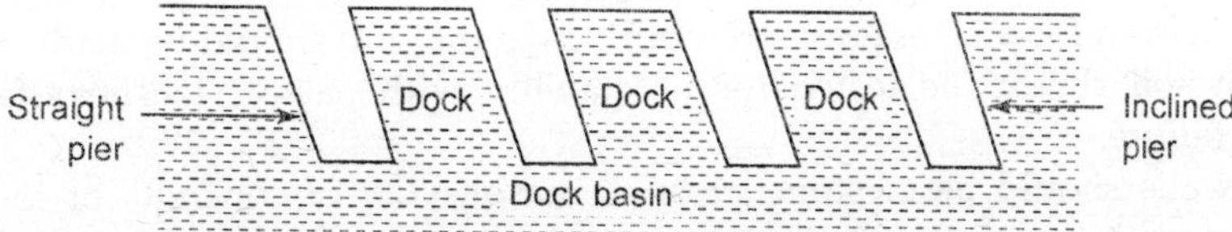


**FIG. 20.5** *Inclined-quay docks*

## Water Depth in Docks and Dock Basins

The depth should be adequate for the draft required by the largest vessel to be accommodated. In case the initial water depth has to be increased as part of future improvements of the port, it can be done by dredging, provided the foundations of the dock walls are not disturbed. The procedure of formation of the dock and basin will depend upon whether the basin is inland, or partially covered by the tide; in the latter case, an embankment properly protected against scour by means of rubble pitching, has to be built to keep the sea out of the basin until the dock construction is completed.

The soil in the bed has to be relatively impervious to keep the water depth adequate for navigation.

## Design of Wet Docks

The following are the loads and pressures which act on a dock wall:

(i) Dead load (self-weight)
(ii) Live load
(iii) Surcharge pressure due to loads of vehicles and trains on quay adjacent to the wall
(iv) Earth pressure from the rear side
(v) Static water pressure
(vi) Dynamic wave pressure
(vii) Ship impact pressure
(viii) Uplift pressure at the base, if any.

A dock wall is designed as a gravity retaining wall, with the standard conditions of stability against bearing and sliding at the base, and overturning about the base, being ensured with a margin of safety. (Unit length of the wall is considered for the analysis.)

The following requirements should also be fulfilled:

(i) The wall should be stable under earth pressure from the backfill when the dock is empty.
(ii) It should be stable under water pressure when the backfill is being placed and compacted.
(iii) It should withstand live loads due to machinery such as cranes, vehicles and trains.
(iv) The height of the basin walls should be more than that of the dock walls, to absorb the variations in water levels due to tides.
(v) The front face should have only a small batter to allow ships to stand close to the wall face, for convenience in operations.
(vi) Fenders should be provided to withstand ship impact.
(vii) The bed of the basin should be relatively impervious in order to retain sufficient water depth.
(viii) The backfill should be pervious in order to make the earth pressure a minimum.
(ix) Drain holes through pipes should be provided at suitable intervals to ensure proper drainage of the backfill.
(x) Suitable provision for anchorages and moorings of the ship should be made.
(xi) Foundation soil should be tested for its stability with respect to shear failure and settlement failure.
(xii) The dock walls should be of stone or brick masonry, or of concrete, or a composite construction with the materials. Construction joints are to be provided in the case of concrete walls.

### Dock Entrances

Vessels can enter the docks directly or through locks. In the either case, the entrances should be provided with gates.

The width of a dock entrance should be more than the beam of the widest ship to be accommodated in the dock. This may be more than 30 m.

**Gates for the dock entrances:** These may be:

(i) Wood gates
(ii) Iron gates
(iii) Caissons

The first two are mostly used for locks. (Refer Section 20.5: Locks and Lock Gates)

Caissons, which are box-like steel structures with bracings, are built and supported on rollers and rails. The entrance is opened by hauling the caisson such that it moves into a recess provided in the side wall of the dock.

The caisson may also serve as a bridge across the dock entrance.

## 20.4 DRY DOCKS AND REPAIR FACILITIES

### Need

Seafaring vessels should be kept in good condition for safe and efficient maritime operations. Therefore, provision of repair facilities for ships and other vessels is necessary at ports and harbours.

Repair involves cleaning, painting and replacement of parts, as required. Since a ship floats on water, these operations can be carried out only by exposing the entire surface of the ship, including its keel, for inspection.

Primarily, a dry dock serves this function. Hence, it is also known as a 'repair dock'.

### Classification of Repair Arrangements

The following are the repair arrangements used:

1. Dry dock, or a graving dock
2. Marine railway or a slipway
3. Lift dock
4. Floating dock

While the first two are of the fixed type, the others are of the floating type.

### Dry Dock or a Graving Dock

This kind of dock appears like a grave; hence, it is also called a graving dock. It is an excavated chamber with side walls, end wall, a solid floor, and a front opening with a gate.

The two side walls are constructed in 2 m steps of masonry or concrete and are covered with granite blocks on the exposed faces. They house pumps on galleries at levels higher than the water level, and culverts and sumps at lower levels to fill or drain out water when required. The outer side of the walls may be battered or stepped in order to increase the self-weight and the base width for stability against the worst combination of the expected loads.

The end wall may be semi-circular or trapezoidal in plan, and no galleries are provided in it. A gate is provided for the front opening.

### Facilities Needed at a Graving Dock

(i) Cranes for placing materials at the required locations and levels
(ii) Salt water supply to fill the dock chamber through culverts
(iii) Fresh water supply for cleaning the surface of the vessel
(iv) Electrical power to run equipment such as pumps and welding sets.
(v) Compressed air for pneumatic tools used for drilling and cutting.
(vi) Capstans, or special electric motors, for pulling any part of a vessel, or the whole vessel into the dock.
(vii) Bollards at 3 to 5 m intervals for mooring or anchoring a vessel; mooring lines or ropes of the ship are wound round the dumb-bell shaped bollards.
(viii) Supports for the ship – in the form of timber shores or struts set inclined with the horizontal against the dock-walls; side shores of steel are used to maintain their spacings. Bottom support for the ship is derived from keel blocks, set longitudinally at intervals, along the middle of the width; and bilge blocks or side blocks are set at intervals along the sides.
(ix) Pumping facilities: The water level in the dock is brought to that in the basin outside, by flooding it with water from elevated tank through pipe culverts which open into the dock; pumps of appropriate power are to be used.

In order to support the ship correctly on the series of keel blocks and bilge blocks, water in the dock is gradually pumped out. The inclined shores help in maintaining the correct position.

### Size of the Graving Dock

The graving dock should be adequate to accommodate the largest docking ship. For big ships, the graving dock has to be 300 m long and 30 m wide.

Sometimes, caisson doors are provided in the middle of the length to facilitate the docking of two small vessels instead one big one, depending on the need.

Figures 20.6 and 20.7 show a schematic of a graving dock and the section of a keel block, respectively:

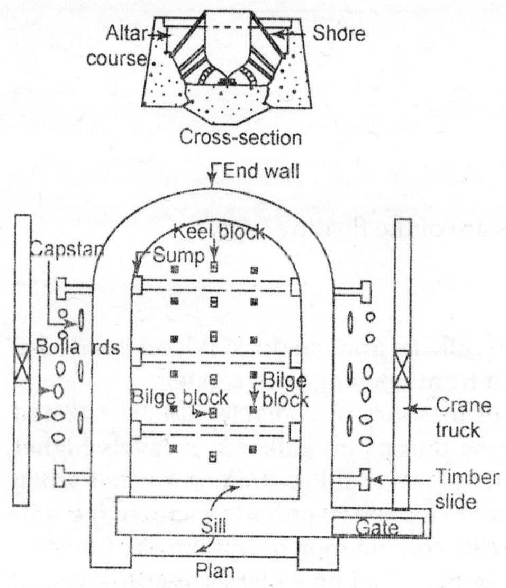


**FIG. 20.6** *Schematic of a graving dock*

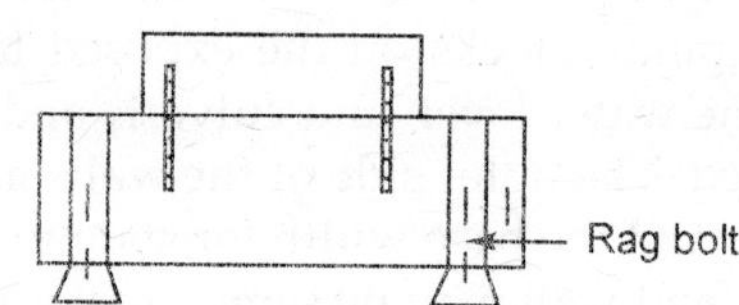


**FIG. 20.7** *Keel block*

### Forces Acting on a Graving Dock

(i) Weight of vessel along the longitudinal centre-line of the dock floor.
(ii) Weight of water when the dock is flooded.
(iii) Earth pressure and hydrostatic pressure on the side walls.
(iv) Pressure exerted by the shores on the side walls.
(v) Surcharge pressure on the side-walls due to heavy equipment such as cranes.
(vi) Uplift pressure on the floor when the chamber is being emptied.

### Design Considerations of Graving Dock

**Dock empty:** In this condition, the dock floor is subjected to uplift, which can exceed the weight of the floor. This excess load is transmitted to the side-walls by inverted arch action (which may be virtual or actual), and has to be resisted by the weight of the side-wall and the horizontal pressure behind it.

The weight of the ship is transmitted to the dock floor through keel blocks and bilge blocks as a number of point loads. It is assumed that nearly 5/8 of the load is borne by the keel blocks and 3/8 by the bilge blocks, equally divided among them.

**Dock flooded with water:** This condition is the harshest as it imposes the greatest load on the dock foundation. The earth pressures on the side walls are mostly balanced by the water pressures from inside. The foundation has to take the full weight of the side walls along with surcharge loads, and the weight of water inside the dock. If the dock is not founded on rock or hard soil, a thick bed of reinforced cement concrete block has to form the dock floor, which has to be designed to withstand the critical stresses. The thickness needed may be as high as 5 m.

### Scheme of Construction of a Graving Dock

The scheme or sequence of construction of a graving dock has to be carefully planned to ensure the stability of the structure during all the stages of construction.

The bottom portion of the side-wall is built first; after excavation, the flooring is constructed in bits, with the outer section being taken up first. The aim is to see that only small lengths of the side-wall are exposed to earth pressures and hydrostatic pressures. The flooring in the middle portion and the upper portions of the side-wall are built last and finally backfilled.

## Marine Railway or Slipway

*Marine railway*, *slipway*, or a *slip dock*, is simply a railway laid at a uniform slope, starting from some distance under water to a point on the shore at which the longest vessel to be accommodated is completely above the range of the tide.

The track should be supported on a firm foundation, or piling taken to rock or a firm stratum.

A *cradle*, or a platform of steel which is mounted on a system of rollers, can move on the inclined track. Keel and bilge blocks of timber are provided on the cradle so that it can receive the ship on to it. The cradle along with the ship is hauled by strong cables attached to the hauling machinery fixed on the shore.

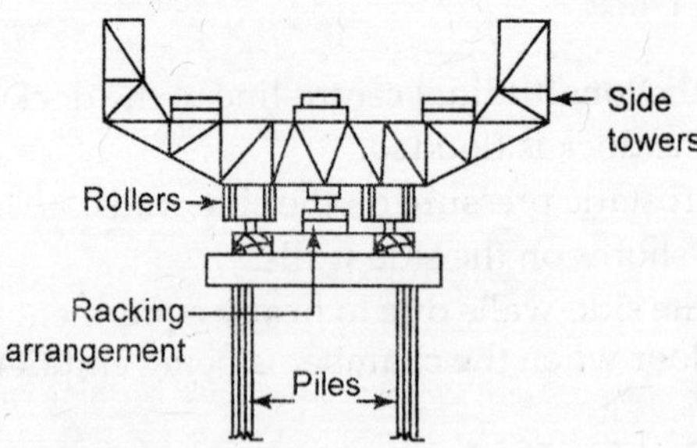


**FIG. 20.8** *Cross-section of a cradle used along with a slipway*

The cross-section of a cradle is shown in Fig. 20.8 while a typical slipway profile is depicted in Fig. 20.9.

The rails are securely fastened to longitudinal sleepers supported on cross-ties. A racking arrangement is provided at the middle of the railway line as a locking device to prevent the cradle from slipping down in the event of breakage of the hauling cable, thus ensuring safety. The longitudinal slope of the slipway may range from 1 in 12 to 1 in 25; a slope of 1 in 15 is generally considered to be suitable.

For dry docking a vessel, the cradle is first lowered into the water; the vessel is towed on to the cradle, positioned to rest on the keel and bilge blocks, and is moored to the towers on either side of the cradle. The cradle along with the vessel is then hauled up along the slipway until it emerges above the high water level, and to the normal docking level and position.

This method of docking is economical but can be used only for small vessels, say up to a displacement of about 2,000 t. Thus, a regular graving dock is preferred for larger vessels.

The slipways used for the construction of ships are specialised and constitute a part of a shipyard and are out of scope of the present treatment.

The length of the slipway can be calculated, with reference to Fig. 20.9, as given below:

Let the ship take the slipway at high water; that is, the length of the ship begins from 0.

Let $l$ be the length of the ship;

$d$ be the draft of the ship; and

$h$ be the height of the cradle from the rail bottom to the keel and bilge blocks.

The length of the slipway, OA, from the low up to high water, $L_1$ given by

$$L_1 = (d + h)\operatorname{cosec}\theta \tag{20.1}$$

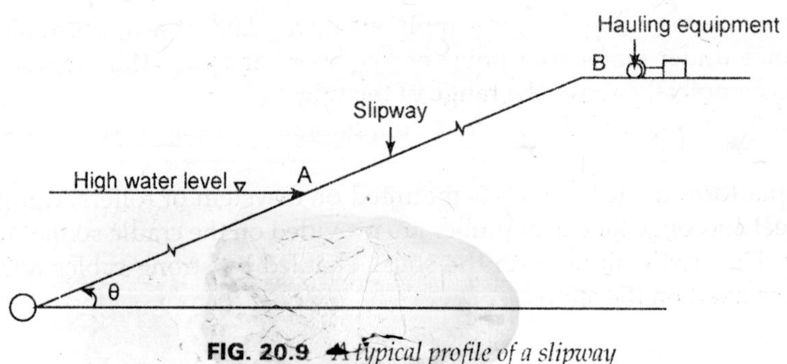


**FIG. 20.9** *A typical profile of a slipway*

$\theta$ being the slope of the slipway with the horizontal. A small margin of 1 to 3 m is added to this.
Let this margin be $k_1$.

$$\therefore \qquad L_1 = (d+h)\operatorname{cosec}\theta + k_1 \qquad (20.2)$$

The length of the slipway, AB, above the high water should be adequate to accommodate the ship- length, with a margin, $k_2$, for any wash due to waves (usually taken as 2 m); it is, say $L_2$.

$$L_2 = l + k_2 \qquad (20.3)$$

The total length, $L$, of the slipway is, therefore given by

$$L = L_1 + L_2 = 2l + (d+h)\operatorname{cosec}\theta + k \qquad (20.4)$$

where $k = k_1 + k_2$ (normally, 2 to 5 m)

For vessels with a pronounced keel, a third rail in between the standard pair of rails is provided.

The pull, $P$, required to haul the ship up the slipway can be got as

$$P = (W_s + W_c)(\sin\theta + \mu) \qquad (20.5)$$

where $W_s$ and $W_c$ are the weights of the ship and cradle respectively, and $\mu$ is the coefficient of friction (taken as 0.05).

## Lift Dock

Also called *shiplift*, ships are bodily lifted out of water after placing them on a platform, which can be lowered into and raised from water. The lifting is accomplished by electric power, hydraulic power, or pneumatic power. This method may be used for launching new ships, or dry docking old ones for repairs.

The most recent development in this regard is the *Syncrolift*, using marine elevators powered by a series of hoists driven by synchronous electric motors.

However, in the wake of the development of significantly large vessels, the method of ship-lift or lift dock has become practically obsolete.

## Floating Dock

A floating dry dock may be defined as a floating vessel which can support a ship on it, and retain it above water (as in the case of a dry chamber) by means of its own buoyancy.

It is a hollow framed steel structure (or an RCC structure) with two side-walls and a floor and with open ends. To receive a ship, the structure is sunk to the required depth by ballasting its chambers with water; the ship is floated into position, and is berthed. The dock gets raised above the water by pumping out water from its chambers, its buoyancy coming into play.

### Advantages of Floating Docks

(i) Mobility and capability of being transported from one port to another.
(ii) It needs much less time for construction than that for a graving dry dock.
(iii) No elaborate entrances and gates

(iv) A damaged ship which cannot be towed into a graving dry dock can be taken in by a floating dry dock for repairs.
(v) It is much less expensive than a graving dry dock.

**Disadvantages**

(i) Regular maintenance by cleaning and painting is needed for a floating dock as it is constantly in water.
(ii) Manoeuvring a floating dry dock needs skill and care.
(iii) Its durability, even with regular up-keep and maintenance, is limited to that of a steel structure, constantly floating in water (about 50 years); but, a well-built graving dry dock is practically indestructible.
(iv) The stability of a floating dry dock in exposed or open sea is much less, which precludes its use in such situations.

## 20.5 LOCKS AND LOCK GATES

**Need**

*Locks* are devices which are employed to overcome differences in water level, and which provide access to enclosed chambers such as docks. These are similar to river or canal locks, which are used in inland waterways.

A lock is an extremely useful device to maintain continuity of navigation between two areas of water at different levels. The entrance into and exit from a lock are controlled by means of lock gates.

Gates of dry docks are similar to those of locks, except that the former are used to exclude water, whereas the latter are used to retain water.

Lock gates are the most important component of a lock since they provide the necessary control on the entry of water and of vessels into the lock chamber.

The size of a lock is determined by the size of the largest vessel utilising the dock. The dimension of the width or beam of the vessel is important in this regard.

Locks are needed in a wet-dock system when it is built in an enclosed basin. However, provisions for such locks are expensive.

**Locks**

**Description:** A lock consists of a chamber which is enclosed by quay walls on two sides; it is provided with an inverted arch flooring at the bottom. Such flooring resists the uplift pressure on it when the water level in the lock is very low; the floor bears against the bottom of the side walls. The lock floor is fully paved with cement mortar.

Locks should be at least 25 m wide, 10 m deep, and 240 m long in order to accommodate large vessels. Longitudinal sluices are provided in the side-walls near the bottom for filling and emptying the lock.

A typical lock is shown schematically is Fig. 20.10.

Foundation for the lock should be provided with special care to prevent settlement and bearing capacity failure, in view of significant fluctuations of the water level in the lock, and the consequent seepage forces on the lock walls and uplift pressures on the underside of the lock floor.

The foundation must be rock or hard pan if available at reasonable depth. Otherwise, the lock walls and floor must be supported by a series of bearing piles, which are protected by sheet piling around the sides.

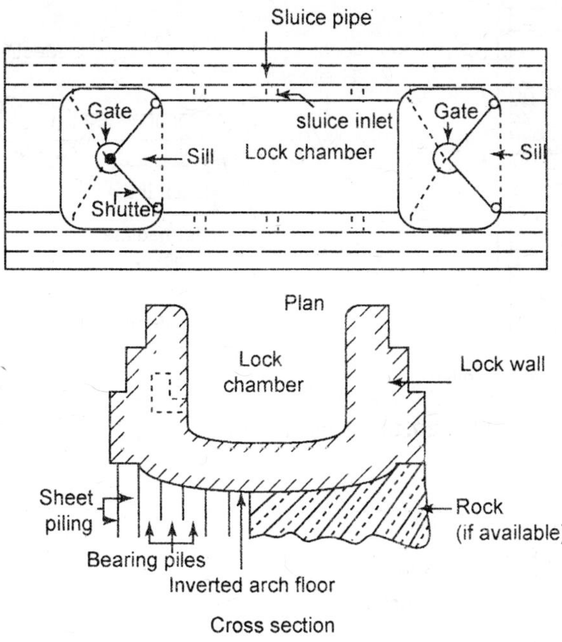


**FIG. 20.10** *Schematic of a typical lock*

## Lock Gates

The water inside a lock chamber is controlled by means of lock gates, which therefore constitute an important component of a lock. Lock gates are made in pairs and point in the same direction – towards the dock.

Generally, two pairs of gates are adequate for a lock. But, if the tides in the outer harbour area are large, a third pair of gates is provided to minimise the tidal effects on the water admitted into the lock chamber. The storm gates point in the opposite direction.

Sometimes, a third pair of gates is provided to conserve water while handling smaller vessels; this pair is referred to as the 'intermediate' gate. Lock gates are made of either timber or of steel.

**Timber or wooden gates:** These consist of a series of horizontal beams, spaced closer near the bottom in order to resist the larger water pressure. The beams are joined together by heel posts at the ends and *meeting posts* at the junction of the gates. There will be upright posts at the middle. Close sheeting boards are fixed to the inner face of the beams. Figure 20.11 shows the plan and sectional elevation of a typical lock gate of wood.

**Steel gates:** These are built or fabricated with iron ribs fixed horizontally and vertically and covered with iron sheeting. Near the bottom of the gate, the thickness of the plate and the ribs is increased and ribs are placed closer, so that the structure withstands the high water pressure near the base of the gate.

A typical lock gate of steel is shown in Fig.20.12.

The compressive force of one gate pressing against the other varies inversely as the rise or curvature. A large curvature decreases the stress, but the length of the gate and the stress due to water pressure increase. A rise of 1 in 4 or 1 in 5 is considered a good compromise between these criteria.

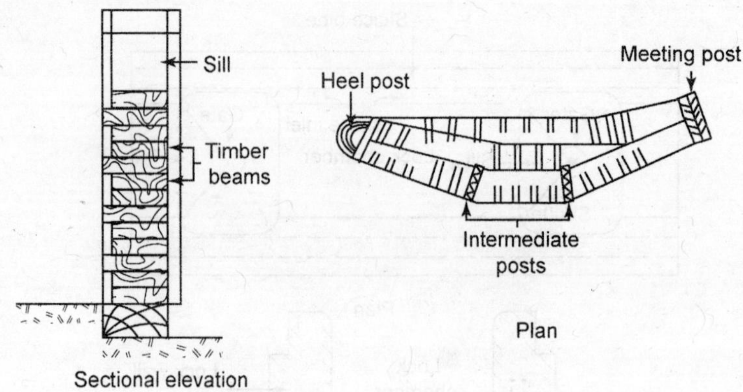


**FIG. 20.11** *A typical wooden lock gate*

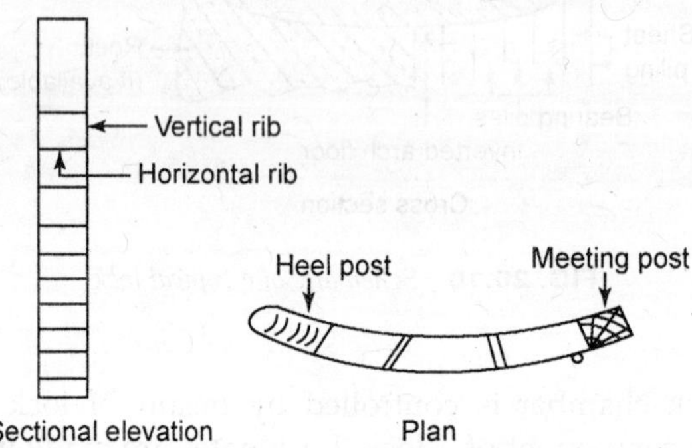


**FIG. 20.12** .*A typical lock gate of steel*

The shape of gates in plan may be straight or curved; the latter is stronger due to arch action.

Gates with a small curvature and a straight sill are preferred.

**Working of gates:** The gates may be operated manually or by means of power. Manual operation is feasible only for small canal locks.

For gates of locks at harbour docks, mechanical or hydraulic power is used for the operation. Electric motors are used for large gates. Strong pivots are needed at the ends, especially at the bottom. Roller supports are used at the bottom for ease of operation.

## SUMMARY

- A 'dock' is defined as a sheltered basin or enclosure for berthing ships; this facilitates maintaining a reasonably uniform level of water for loading and unloading of cargo. Docks are primarily classified as 'wet docks' or harbour docks and 'dry docks' or repair docks.
- Wet docks are useful in harbours where the range and effect of tides is considerable. These may be wet docks in tidal basins or those in enclosed basins; the latter give better convenience for operations.
- The shapes of wet docks in plan may be rectangular, diamond, or inclined quay. The depth of water in docks and dock basins should be adequate for the largest vessel to be operated.

- Several forces act on dock walls – dead loads, live loads, earth pressure including surcharge effect, static and dynamic water pressure, uplift pressure at the base, and so on.
  A dock wall is designed as a gravity retaining wall, against bearing, sliding and overturning, considering unit length of the wall and the worst combination of forces.
- Vessels can enter the docks directly or through locks. Gates for dock entrances can be of wood, iron, or fabricated caissons.
- Dry docks are meant to provide repair facilities for ships by allowing them inside with water, and pumping out water and emptying the dock later. The types of dry docks are graving docks, slipways, lift locks, and floating docks; the first two are of the fixed type, and the other two are of the floating type.
  The graving dock is the most popular. The facilities needed are cranes, water supply, power supply, bollards for mooring, and shoring support for the ships, keel blocks and bilge blocks of timber for bottom and side support, and pumping facilities.
  Design considerations include the dock empty condition and the dock full condition, while design loads take into account the weight of vessel, weight of water, earth pressure, water pressure, pressure exerted by the shores on the side walls, and uplift pressure at the base.
- 'Slipway' is a railway laid at a uniform slope, starting from some distance under water to a point on the shore at which the longest vessel is completely above the range of the tide.
  A platform of steel called the 'cradle', mounted on a system of rollers, moves on the track. The cradle is sunk into water, the ship is brought on to the cradle, and the cradle along with the ship is hauled up the slipway by means of power.
- 'Lift lock' or 'shiplift' is an arrangement by which a ship is lifted bodily out of water, and placed at the desired location for repairs.
- A 'floating dock' is a floating vessel which can support a ship on it, and retain it above water by means of its own buoyancy. The structure is sunk to the required depth by ballasting its chambers with water; the ship is floated into position on to the floating dock and then berthed.
- The dock gets raised above the water by pumping the water out from its chambers, its own buoyancy coming into play.
- A 'lock' is a device employed to overcome differences in water level to maintain continuity in navigation. It consists of a lock chamber, lock gates being used to control the water level inside the lock. Lock gates are generally operated in pairs, and are made of timber or steel. Large lock gates are operated by mechanical means, aided by power.

## QUESTIONS

1. Define a 'dock' and explain the classification of docks.
2. Differentiate between wet docks in tidal basins and those in enclosed basins.
3. Explain, with sketches, the usual plan shapes adopted for docks.
4. What are the various forces acting on dock walls and what are the design considerations?
5. How are dock entrances controlled? Explain the use of sliding caisson for a dock gate.
6. What are the various facilities needed at a dock?
7. What is a dry dock and what is its function?
8. Classify dry docks and explain each type briefly.

9. Explain the features of a graving dry dock and its construction.
10. Enumerate the various forces which act on the side walls of a dry dock.
11. Explain a 'slipway' and its function in hauling a ship above water.
12. Write short notes on:

    (a) Dock gates

    (b) Dock entrance

    (c) Cradle in slipways
13. What is a 'lift dock'? Explain its function.
14. Explain the function of a 'floating dock'.
15. Define a 'lock' and its use in the functioning of docks.
16. Explain, with neat diagrams, the different types of docks. (Anna University)
17. Explain, with a suitable sketch, the working of a dry dock. (VTU)
18. What is a dry dock? Explain the construction and uses of a dry dock. (VTU)
19. What is the name of a chamber in which a ship berthed is lifted by the principles of buoyancy? (West Bengal Tech. University)
20. Differentiate between dry docks and wet docks. (West Bengal Tech. University; Mumbai University)
21. What is the structure that protects the harbour from stormy waves and permits calm in the harbour called? (West Bengal Tech. University)
22. What is the chief function of a dry dock in a harbour? (West Bengal Tech. Unyiversity)
23. Explain a wet dock and its function. (Mumbai University)

Chapter **21**

# MAINTENANCE AND DREDGING OF HARBOURS

**CHAPTER OBJECTIVES**

This chapter enables the student to understand the following:

- Dredging involves the removal of soil material from the bed of a harbour to increase water depth, so as to maintain navigable depth of water.
- Capital or initial dredging is primary and maintenance dredging is secondary.
- Dredging equipment or dredges are primarily classified as mechanical and hydraulic. Grab type, bucket-ladder type and dipper type fall in the first category while rock cutters are in the second.
- Rock-breaker and hopper barge are used for specialised jobs.
- Seawalls, bulkheads, sand dunes and groynes are different types of structures designed for coast or shore protection and control of littoral drift.

## 21.1 INTRODUCTION

### Need for Dredging

The vertical depth of immersion of a vessel floating safely in water is known as its 'draft'. The *navigable depth of water* needed for a vessel is obtained from the sum of the draft and the vertical clearance between the bottom of a ship and the seabed.

A harbour can be fully utilised only if the navigable depth of the class of vessels which it is expected to serve is ensured in the approach area as well as the harbour basin.

The cycle of erosion and silting of seabed is inevitable as a result of natural phenomena. When the natural regime is disturbed by the construction of harbour structures, silting generally takes place in course of time. This reduces the initial water depth, which may lead to the reduction of the navigable depth.

### Definition

*Dredging* is the technique of removal of silt and other materials from the bed to increase the depth of water, so as to restore the navigable depth.

Sometimes, a particular site is favourable for locating a harbour in several respects, but the insufficiency of the depth of water alone is the constraint; in this situation, the best solution is to initially increase the depth of water by dredging and to continue with the operation periodically.

The most important element of maintenance of a harbour/port is, therefore, considered to be dredging; however, maintenance, in a broad sense, includes the upkeep all the facilities and structures at a port/harbour.

The history of the development of dredging dates back to ancient civilisations that were born on the banks of large rivers such as the Nile, Euphrates, Tigris, Indus, Ganges and Yangtze Kiang.

Mastery over water bodies – seas, rivers and canals – was required for the survival of human race for irrigation, navigation and flood control.

Equipment for dredging was envisaged, developed and used for navigation purposes right from the days of Leonardo da Vinci. In the wake of industrial revolution and advancement of civilisation, modern equipment has been developed for dredging and improved continually.

## 21.2 CLASSIFICATION OF DREDGING

The need for dredging may arise under the following circumstances:

(i) To take care of a low-lying country by increasing the capacity of a nearby water body.
(ii) To protect a land area near the coast or shore from the onslaught of waves and tides of the sea or ocean.
(iii) To improve the water body or water course for the purpose of navigation.

Dredging work may be classified as follows:

(a) Capital dredging

  (i) Improvement dredging
  (ii) Sundry dredging

(b) Maintenance dredging

### Capital Dredging

When a particular site from among several considerations is favourable for locating a harbour and the only constraint is inadequacy of water depth, dredging is required initially to increase the water depth to provide enough draft for the vessels expected to utilise the facility. Such an initial removal of soil material is known as *capital dredging*, or *primary dredging*,

The dredging carried out to improve the existing water depth, suitable only for small vessels, in order to cater to the needs of larger vessels, is called *improvement dredging*.

Dredging carried out for reclamation of new areas and for purposes of navigation, and similar jobs, is called as *sundry dredging*.

In a way, improvement and sundry dredging also fall under the category of capital dredging. These dredging jobs may be carried out without interfering with the existing operations.

For assessing the extent and location of capital dredging, hydrographic surveys, with both horizontal control (for location of points/area of dredging) and vertical control (for knowing the depth of excavation), are conducted.

### Maintenance Dredging

Since the forces of nature act continually on large bodies of water such as rivers, seas and oceans, and cause erosion and silting, the water depths created initially in the vicinity of a harbour (in the entrance channels and harbour basins) gets affected regularly. This necessitates periodic dredging to maintain the water depth needed for the vessels; this is known as *maintenance dredging* or *secondary dredging*.

In fact, there will be practically no harbour which does not need maintenance dredging.

If the silting at a particular harbour occurs at regular periodic intervals in a predictable manner, an annual cycle of dredging work may be planned and carried out. This may need a relatively smaller-sized dredging equipment.

If silting recurs at short and unpredictable intervals of time, the dredging job brooks no delay in order to maintain the navigability of the waters, so that the larger vessels which, hitherto, have been utilising the facility, need not be kept out for a significant period. Thus, a proper plan for a maintenance dredging improves the efficiency of a harbour, although larger equipment may be needed.

## 21.3 EQUIPMENT FOR DREDGING

The primary equipment for dredging are known as *dredgers*, or *dredges*.

They can be divided into two categories:

1. Mechanical type
2. Hydraulic type

*Mechanical dredges* use mechanical means to excavate the material, lift it physically, and remove it from the zone of activity.

*Hydraulic dredges* work on the principle of suction; the material is loosened by cutters or jets, if necessary, and the material mixed in water is sucked and transported to a suitable place on the shore.

Special equipment such as *rock breaker* and *Hopper barge* are also used in connection with dredging.

### Mechanical Dredges

Mechanical dredges are of three types:

(a) Grab or clamshell dredger
(b) Bucket-ladder dredger
(c) Dipper dredger

Each of these is useful under certain conditions, depending on the nature of bed material.

#### Grab or Clamshell Dredger

This type consists of a *grab* suspended by a cable from a boom or a crane. When the grab is lowered to the bottom, it excavates the bed material and hoists it to the surface where the dredged material is discharged into a hopper.

This dredger is stationed on a pontoon, on which one or more cranes are mounted and fitted with grabs. The cranes are operated by steam, diesel, or diesel-electric power.

The grabs are specially designed with cutting edges to suit different materials. A grab dredger is extensively used for dredging soft clay or mud. A schematic of a grab dredger is shown in Fig.21.1.

By virtue of its manoeuvrability, this can be used in confined regions in docks, wharves and breakwaters.

**Bucket-Ladder Dredger**

This consists of an endless chain of buckets, mounted on and running round a ladder, which is fitted on the bow of a floating vessel. The buckets dig in the sea-bed, bring up the material to the top of the ladder, and get inverted on descent and discharge the contents into special holds or hoppers.

This dredge is suitable for digging sand, hard clay, coarse gravel, or even broken stone, at depths beyond the reach of other types of equipment.

The bucket chain is driven by electric motors. Now-a-days, this type of dredge is used in mining and in commercial production of gravel or sand. A schematic of a bucket-ladder dredger is shown in Fig.21.2.

Systematic and substantial dredging can be performed by this dredger.

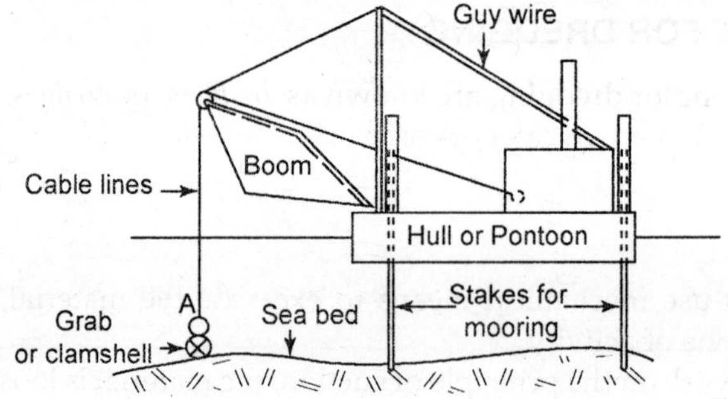


**FIG. 21.1** *Grab or clamshell dredger*

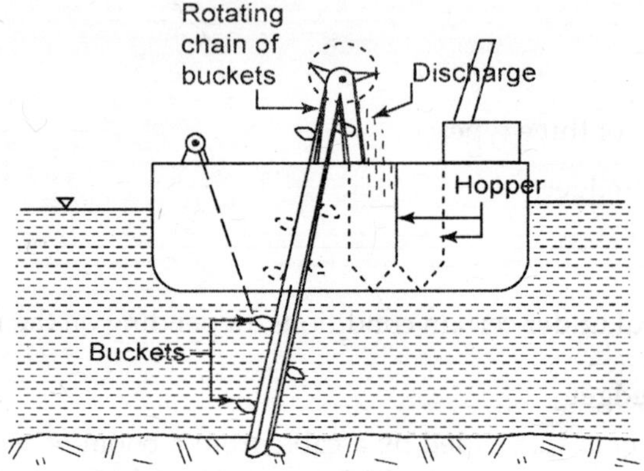


**FIG. 21.2** *Bucket-ladder dredger*

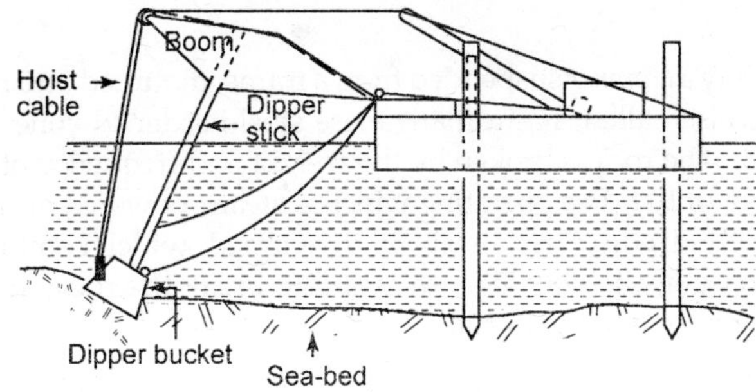


**FIG. 21.3** *Dipper dredger*

### Dipper Dredger

This is a kind of heavy duty equipment; its main part is an arm sliding through a boom, with controlled movements in both the horizontal and vertical planes. The dipper is forced into the bed, lifted when it is filled with the material, and the contents are dumped into hopper barges floating alongside, or on a special spoil area.

Steam-operated dippers are powerful and can excavate hard soil, boulders, and even rock. Easy manoeuvrability makes it suitable for use in confined areas around docks and narrow channels.

A schematic of a dipper dredger is shown in Fig. 21.3.

## Hydraulic or Suction Dredger

This type of dredge combines the operations of excavating and transporting the sea-bed material in a continuous manner. The main component is a *sand pump* which sucks the sludge mixed with water and transports it into a discharging pipe line. Pure suction is adequate for loose material; but for harder material, a rotating cutter is used at the end of the suction pipe, or a system of water jets is used.

Modern hydraulic dredgers can handle soft rock using powerful cutters.

A typical hydraulic dredger is shown in Fig.21.4.

This type is effective in sand, silt, mud, and clay. It is also useful in clearing sand bars.

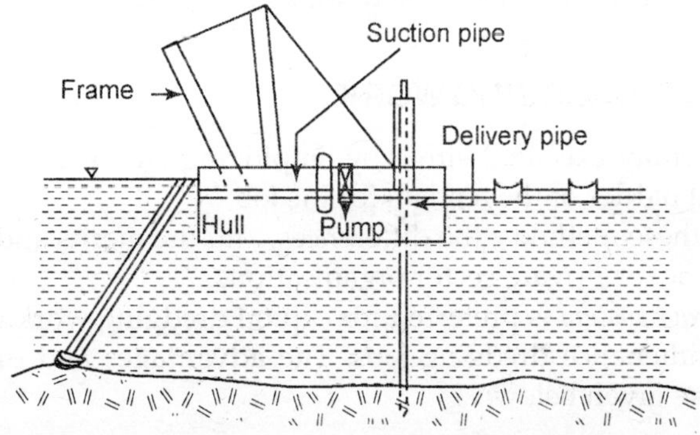


**FIG.21.4** *Hydraulic dredger*

### Rock-Breaker

A rock-breaker is basically a *hammer* suspended from a frame, mounted on a pontoon. The hammer, or the 'needle' as it is also called, has a manganese steel-hardened cone or chisel point which strikes the sea-bed rock. The rock is broken by the weight and frequency of strokes of the needle, which can weigh from 6 to 10 t. The operating winch is steam-powered or runs on diesel-power.

A rock-breaker dipper dredger has also been developed, which combines the operations of the dipper and the dredger; the needle is first used to break the rock, and the dipper is fitted in place of the needle remove the material.

### Hopper Barge

This is a barge which can receive the dredged material and take it to a spoil area some distance away.

It has doors at the bottom which can be opened hydraulically; buoyancy tanks are used to keep the barge afloat.

The hopper barge is brought alongside a dredge so as to receive the dredged material and move it away either by its own power or by means of a tug to the spoil area. The hopper may be made to dump the material in the portion of the sea where the depth is large, or to a spoil area near the shore.

### Choice of Dredger

The primary factor which governs the choice of the dredger is the nature of the sea-bed material.

*The other factors are*:

(i) The volume of dredging needed
(ii) The depth to be attained
(iii) The length of the dredge channels
(iv) Cost and availability of the equipment
(v) Disposal facilities
(vi) Wind, waves and tides
(vii) Restrictions from other waterway users
(viii) Environmental concerns and restrictions imposed, if any
(xi) The time available for the dredging job.

Since many different varieties of modern dredging equipment are now available, the choice of the type appropriate to a particular job should not be difficult.

## 21.4 EXECUTION OF DREDGING WORK

Dredging work is generally executed either by employing labour directly or by awarding a contract to a successful bidder of the tenders for the job.

The comparison of these methods, the differences, and the merits and demerits of each are similar to those of public- or private-sector organisations.

In general, the *contract system* is preferred for capital dredging work while the *direct labour system* is used for maintenance dredging jobs. The advantages and disadvantages of each method of execution are given below:

## Advantages and Disadvantages of the Direct Labour Method

### Advantages

(i) Urgent needs of dredging can be met easily.
(ii) The available resources including the tools and plant can be put to optimal use.
(iii) The overheads and other general costs in the contract system are avoided.
(iv) It obviates the need to check and measure the work executed.
(v) Possible disputes regarding cost escalation and delay in execution are avoided, as also the associated arbitration and possible litigation processes.
(vi) A co-ordinated plan for the dredging work in all the ports can be prepared.

### Disadvantages

(i) Negligence in maintenance is possible.
(ii) Availability of labour and their co-operation could be constraints.
(iii) Skilled supervision and administrative control are necessary for successful completion of the job.
(iv) Lack of flexibility is a constraint.

## Advantages and Disadvantages of Contract System

### Advantages

(i) Competition among the bidders or contractors may result in economy in execution.
(ii) The most suitable plant can be specified in the contract for carrying out the job.
(iii) Expeditious execution of the job can be achieved by stipulations in the contract.
(iv) The cost of the job is known before-hand.
(v) Better flexibility is available in this system.

### Disadvantages

(i) Checking and measuring the executed work is a tedious job.
(ii) The rates for unexpected items of work which may crop up during execution are difficult to decide.
(iii) Price variations during the period of the contract may prove to be irritants.
(iv) Strict supervision by the engineers of the port authority is needed to ensure quality of work.
(v) Disputes with contractors and arbitration may lead to inordinate delay in completion.

## Disposal of Dredged Material

The dredged material may be dumped in deep sea away from the harbour area, especially if suction dredges are employed. Alternatively, the material may be dumped in attendant barges containing hoppers, or hopper barges, and may be carried to a spoil area on shore.

**Pipeline**

A dredge pipeline has the following components:

(i) Floating pipe
(ii) Flexible connection between the floating pipes (usually ball-and-socket joints)
(iii) Shore pipe line.

The floating pipe is connected to the discharged pipe of the dredger and mounted on tubular float-pontoons. Winches are used for bigger-size pipes.

Ball-and-socket joints are popular in connecting different lengths of the floating pipe, because of the flexibility to suit the dredge movements as well as waves.

Shore pipeline can be of steel or RCC. HDPE pipes are used as they are light and unaffected by sea water.

The consistency of the dredged material, improved by water, is important in the use of a pipeline system.

**Reuse of Dredged Material**

The reuse of dredged material for the construction of embankments in large harbours leads to considerable economy.

However, mud is usually dumped in deep sea or in back waters, if available nearby.

Sands are used specially in embankment construction even if they are somewhat muddy. They can be dumped by hopper barges into large ditches; another dredger can be used to pick up this material and use it in the embankment job.

The embankment construction must proceed from the land towards the sea, and not the other way, to avoid the formation of mud pockets in the embankment.

Gravel, pebbles and sand are widely used in embankment and dykes, or for concrete-making. The presence of salt in these materials may accelerate the setting process of concrete, and accelerators should not be used for quick-setting.

Salt may cause corrosion of steel reinforcements; hence, these dredged materials are not generally used in pre-stressed concrete.

## 21.5 COAST PROTECTION

### Need

A coastline continuously changes because of natural phenomena such as winds, waves, currents, and storms and cyclones, which act incessantly. There is either erosion or deposition of earth material on the shore.

If the deposition is excessive at a harbour, silting occurs and reduces the water depth, which thereby calls for dredging. If, on the other hand, erosion is excessive, the neighbouring structures will get affected because of the exposure of their foundations.

Erosion also calls for coast protection in order to protect the land near the coastal zone. Activities such as stabilisation of an existing beach, restoration of an eroded beach, or creation and stabilisation of artificial beaches require shore protection works.

### Shore Protection Works

The following are the common types of shore or coast protection works:

1. Seawalls
2. Revetments

3. Bulkheads
4. Spending beaches
5. Sand dunes
6. Groynes
7. Offshore breakwaters

### Seawalls

Seawalls are heavy structures which derive stability by their own weight. These walls may be vertical, battered, stepped, or retractive (concave towards the sea) for reversing the water jets from waves. The battered wall is subjected to the least thrust. The retractive type of wall is used where the wave action is severe; it is shown schematically in Fig. 21.5.

A concave seawall with a stepped portion at the bottom is considered to be more effective in providing shore protection.

### Revetments

The earth embankment slopes on the seaward side are protected by pitched stone revetments with large rubble stones, as shown in Fig. 21.6.

### Bulkheads

These are constructed along the shore-line to protect the coast from erosion caused by sea waves. These are sheet pile walls or flexible earth retaining structures, which may derive stability by cantilever action, or by the restraining action of an anchorage, as shown in Fig. 21.7.

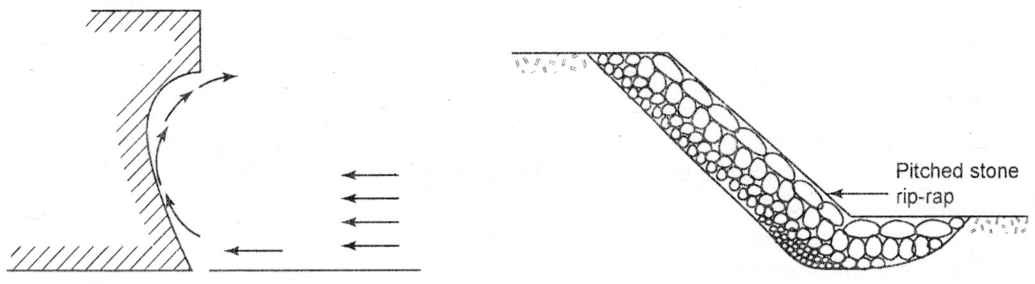


**FIG. 21.5** *Concave seawall*

**FIG. 21.6** *Stone revetment*

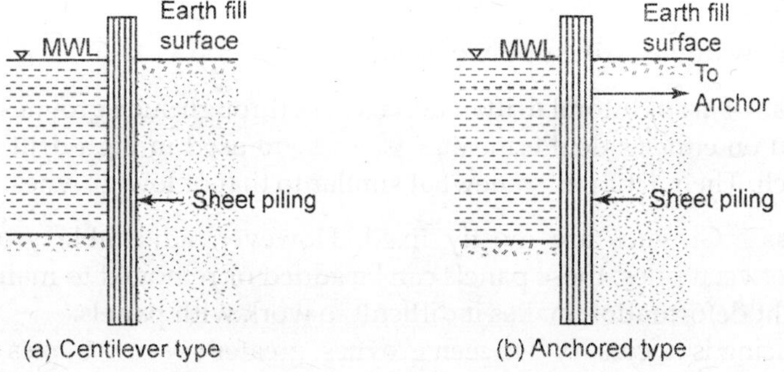


**FIG. 21.7** *Sheet pile walls or bulkheads*

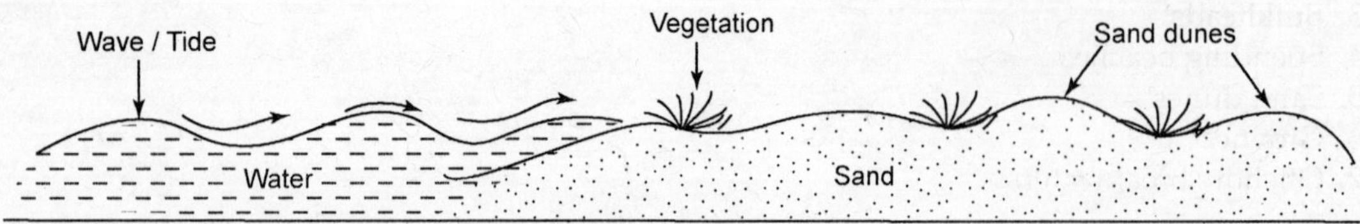


**FIG. 21.8** *Dune action*

### Spending Beaches

Spending beaches are protective beaches, which are effective in dissipating wave energy and providing protection to the adjacent upland. Restoration and stabilisation of a beach involves constant supply of sand at a required rate depending upon the extent of erosion. This periodic replenishment of sand is called *artificial beach nourishment*. It remedies most erosion problems, although it involves some special effort. This aspect has been dealt with briefly in Chapter 3.

### Sand Dunes

*Sand dunes* are formations of sand along the coast, which prevent movement of waves and tides into the areas behind them. Vegetation stabilises dune sand, and prevents its migration into adjacent areas and damage to property, besides protecting the dune itself (Fig 21.8).

In the case of a severe storm, the dune sand falls and replenishes the beach below. On-shore winds transport the beach sand towards the dune, thus building it up progressively. This process is hastened by vegetation or by fencing.

Under the action of severe waves, the dune gets overtopped and the crest gets lowered. This disturbed sand is transported off shore, and is deposited in a bar formation; this is helpful in dissipation of wave energy.

### Groynes

A *groyne* is a coast protection structure which helps in building up a protective beach, or retard the erosion of a restored beach by trapping littoral drift. Groynes also protect the toes of seawalls or bulkheads.

Groynes may be built either perpendicular to the shore or inclined to the shore line. A series of groynes, or a *groyne system*, protects the shore-line.

Groynes may be of different types — permeable or impermeable, high or low, long or short, and fixed or adjustable.

**High and low groynes:** High groynes tend to completely block sand movement; these are also called *terminal groynes*.

Low groynes allow down-drift supply of sand.

**Permeable groynes:** Part of littoral drift materials pass through such groynes and induces the deposition of sand on either side. Permeable groynes are used in high-drift areas to prevent recession of a beach. Their action is somewhat similar to that of low groynes.

**Adjustable groynes:** Groynes are mostly fixed. However, adjustable groynes consist of movable panels between piles; these panels can be added or removed to maintain down-drift beach. Even a slight deformation makes it difficult to work with panels.

An optimal spacing is necessary between groynes; greater spacing makes them ineffective in holding material, while closer spacing will not be advantageous.

Groynes are designed to resist wave action and sand pressure. The shore-end of a groyne should be secured well in a seawall or firm ground. They may be made of timber piles or steel sheet piles.

### Offshore Breakwaters

Breakwaters are often built offshore to protect vulnerable areas from wave action. An offshore breakwater can serve as a shore protection structure, a trap for littoral drift, an aid for navigation, or a combination of these. They are of rubble-mound type and afford protection to harbour entrances. Only small vessels or dredges can seek refuge in an area protected by an offshore breakwater. It is considered to be very effective in intercepting littoral drift.

## SUMMARY

- 'Dredging' is the technique of removal of silt and other materials from the bed of the harbour to increase the water depth.
  Dredging is invariably needed to provide and maintain navigable depth of water in harbours to facilitate the docking of vessels.
- Dredging is classified as capital or primary dredging and maintenance or secondary dredging.
- The initial dredging that is required to provide sufficient depth of water for navigation at a site, which is otherwise suitable for a harbour, is called capital dredging.
- The periodical dredging required to restore the navigable depth of water for the largest vessel expected is known as maintenance dredging.
- The equipment needed for dredging are called 'dredgers'. Dredgers are primarily classified as mechanical and hydraulic types.
- Grab or clamshell type, bucket-ladder type, and dipper type fall under the mechanical category. Suction-type dredger with cutters, if necessary, for dredging soft rock, falls under the hydraulic category.
- Special equipment for dredging like rock breaker and hopper barge are also devised.
- The choice of dredger depends primarily upon the nature of the bed-material and extent of dredging needed, area-wise as well as depth-wise; it may also depend on whether capital- or maintenance-type of dredging is required.
- Dredging jobs may be executed either by direct labour or by awarding contract. Each has its own merits and demerits. Disposal of dredged material may be done by hoppers and hopper barges, or by pipelines and sand pumps. Re-use of dredge material for embankment formation leads to economy.
- Seawalls, bulkheads, sand dunes, groynes, and offshore breakwaters are the different types of coast or shore protection works. These structures can also be used to control littoral drift.

## QUESTIONS

1. Define 'dredging' and bring out its need in the maintenance of a harbour.
2. Classify dredging work and explain the difference between the various categories.
3. Bring out the difference between capital dredging and maintenance dredging.
4. Classify dredging equipment and list the different types of dredgers used.
5. Write short notes on:

(a) Rock breaker

(b) Hopper barge

(c) Pipeline

6. Explain briefly the working of a hydraulic dredger.
7. Explain the differences between the methods of direct labour and contract in the execution of a dredging job.
8. Write a brief note on methods of disposal of dredged material.
9. Write a note on the reuse of dredged materials is harbour works.
10. Why is coast protection needed?
11. Write brief notes on:

    (a) Seawall

    (b) Bulkhead

    (c) Groyne.
12. Write short notes on:

    (a) Revetments

    (b) Sand dunes

    (c) Offshore breakwaters.
13. Write about the characteristics of a 'pipeline'. (Anna University)
14. Write a short note on dredging. (Anna University)
15. Explain the process of dredging. Why is it necessary? (Anna University)
16. Write a short note on coastal structures. (Anna University)
17. Explain about dredging equipment with a neat sketch. (Anna University)
18. What do you understand by dredging of docks? Explain the various methods of dredging. (Mumbai University)

# PART IV

# Bridge Engineering

Chapter 22

# GENERAL CONSIDERATIONS AND CLASSIFICATION

**CHAPTER OBJECTIVES**

This chapter gives an insight into the following aspects of bridges:

- Definitions and components of a bridge
- Criteria for classification of highway and railway bridges based on the primary material function and structural aspects
- Criteria for site selection and surveys needed
- Hydraulic design features such as discharge, waterway afflux and scour depth
- Abutments, piers and wing walls and their types
- Foundation for abutments and piers – shallow and deep foundations
- Cofferdams, their functions and types based on the materials used.

## 22.1 INTRODUCTION

A *bridge* is a structure built to provide passage over an obstacle without closing the way underneath it; the obstacle may be a stream, a river, a roadway, a railway, or just a valley in a mountainous terrain.

The need for a bridge and its importance in the modes of land transportation has been realised long ago; its need to cross a water course such as canal, stream, or a river has been realised even before this.

### Components of a Bridge

The components of a bridge may be primarily grouped into the following broad categories:

1. Superstructure
2. Substructure
3. Foundations

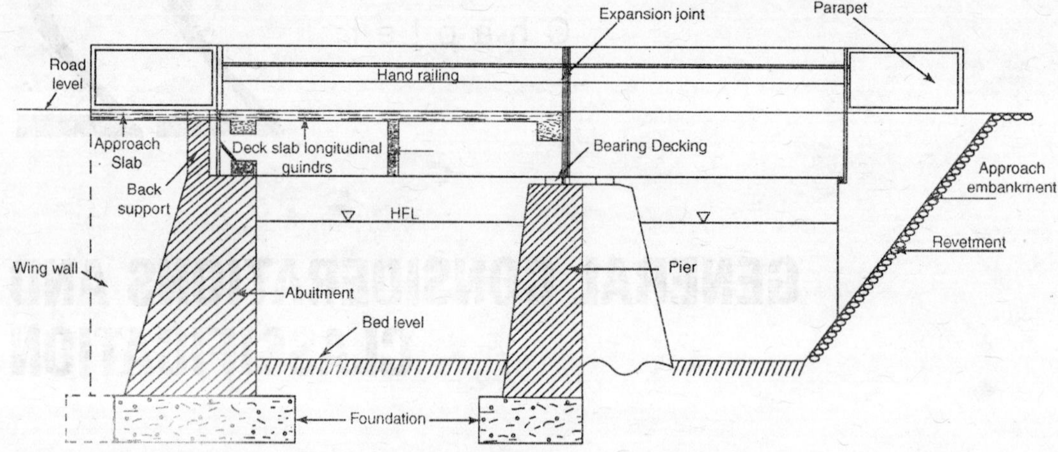


**FIG. 22.1** *Components of a bridge*

**Superstructure:** This consists of the structural system which carries a transportation route such as a roadway or a railway. Decking or flooring supported by load-bearing components like slabs, beams or girders, arches, trusses and cables, and parapets, hand rails, guard stones, and bearings for the decking constitute this system.

**Substructure:** This is the supporting system for the superstructure, comprising abutment, wing walls and piers.

**Foundations:** These are the lowest components of a bridge, and they are meant to support abutments and piers. Different types of foundation may be used depending upon the site conditions.

The *ancillary works* for a bridge are:

(i) *Approaches* to the bridge from either side to connect it to the transportation route.
(ii) *River training works* such as aprons at the bed level of the waterway, and revetments for the *protection of slopes* of the embankments at the abutments.

These components of a bridge are shown schematically in Fig. 22.1.

## Importance of Bridges

A bridge is an important element of a transportation system since its failure or weakening causes serious disruption of traffic. The cost of a bridge per unit length is many times that of a road or railway track; it is therefore necessary to design the bridge with adequate strength and durability, consistent with economy

The mobility of an army at war is very much affected by the availability or otherwise of bridges across rivers, as part of the road or the railway. Hence, military engineers are trained in building temporary bridges quickly while advancing, and destroying bridges while retreating in order to prevent the enemy from advancing.

The capacity of a bridge to allow traffic and handle traffic density governs the capacity of the system or the stretch of which this structure is a part. This adds to the importance of good construction as well as maintenance of bridges.

Bridges have been the visible testimony to the contribution of engineers. Some bridges add to the spirit of the people of a particular city. Famous examples are the Golden Gate

Bridge of San Francisco, Brooklyn bridge of New York, Sydney Harbour bridge, Tower bridge of London, and the Howrah bridge of Kolkata. Such structures enhance the vitality of the cities and characterise the social, cultural and economic progress of the respective places

Thus, the importance of bridges cannot be over-emphasised.

## Classification of Bridges

The criteria for classifying bridges are manifold; but these are not mutually exclusive as one type may overlap with some others. The criteria are listed below:

**Function or purpose:** Highway bridge, railway bridge, aqueduct, viaduct, foot bridge (for pedestrians), or road-cum rail bridge. (An aqueduct carries a canal over a river, while a viaduct bridges a valley.)

**Duration of use or service life:** Permanent (long service life or durability), or temporary (shorter service life), such as military bridges (Bailey or pontoon types)

**Construction material:** Timber, masonry, steel, RCC (reinforced cement concrete), PSC (pre-stressed concrete), or Composite (a combination of two or more materials) bridges.

**Road level relative to high flood level:** For road bridges – non-submersible or high-level, or submersible or low-level bridges

**Types of Superstructure:** Slab, beam, arch, truss, cantilever, balanced cantilever, bow-string girder, cable-stayed or suspension bridges.

**Inter-span relation in structural action:** Simple, cantilever, or continuous bridges

**Position of bridge floor relative to the superstructure:** Deck, through, or semi-through bridges

**Clearance for navigation:** High-level or fixed bridge, or movable types such as swing, lift or bascule bridges

**Span or length:** Culverts (less than 6 m), minor bridges (6 to 30 m), major or long-span bridges (more than 30 m to 100 m)

(This classification based on length is applicable to highway bridges whereas for railways, bridges with a waterway less than 18 m are considered to be minor and those with waterway more than 18 m are major bridges)

**Relative level of railway and highway in grade separations:** Over bridges (railway below highway) and under bridges (highway below railway)

**Alignment relative to the water course:** Straight or rectangular, or skew bridges

**Standard loadings:** Culverts and highway bridges are classified by IRC as Class-AA, Class-A, and Class-B. Similarly, railway standard loadings are given by the Indian Railways.

As already stated, there can be overlaps of more than two types in any particular case.

For example, a railway bridge of steel-truss type may be through-type continuous bridge of permanent nature and a long-span type.

## Historical Aspects of Bridges

Bridges are closely linked to the history of human civilisation. The first bridges were supposed to have been formed by forces of nature. A tree which had fallen accidentally across a stream was the first example of a bridge. The natural rock arch formed by erosion of the soil below, and the aerial roots of banyan tress which allow monkeys to cross a nearby stream are the other examples. These were the forerunners of beams, arches and cables used in modern bridges.

*Rama Sethu*, the bridge across the Indian Ocean from the southern tip of India to the north of Sri Lanka, was supportsed to have been built by Sri Rama and his *Vanara* army, according to the Hindu epic, *Ramayana*; satellite photographs taken from space reveal the existence of limestone shoals in this region set in a unique pattern similar to a bridge. The remnants of *Rama Sethu* are still there, as if in testimony of this belief.

The primitive human was a nomad who wandered in search of food and was in need of land routes to cross streams and water bodies; he imitated nature and constructed the first forms of bridges. As humans gradually settled to community life around 4000 BC, the need for permanent bridges was realised; for example, the timber–trestle construction, the forerunner of timber bridges, was developed by the lake dwellers of Switzerland, around that time. The forerunners of suspension and cantilever bridges were developed in India and China.

Around 2650 BC, a bridge was built over the river Nile in Egypt; Babylonians started building bridges across the river Euphrates. The oldest existing pointed-arch bridge in Babylonia dates back 350 BC.

The Romans were pioneers in the art of bridge-building; they extensively used stone masonry arches with mortar using natural binding materials. The bridges built by them around 200 BC were magnificent structures; the bridge across the Tiber river in Rome, built around 100 BC, is still in use

The Chinese were also active in building stone arch bridges from 250 BC. The Zhao Zhou bridge built around 600 AD with stone arches still stands today, bearing testimony to their skills.

Until the medieval period, bridges were built using stone, while the occasional use of timber was noticed for minor ones.

With the advent of iron and steel and the idea of trusses that developed in the eighteenth and the nineteenth centuries in the wake of the invention of the railway, bridge construction received a fresh impetus. Many major railway bridges of steel trusses were built in India to cross major rivers.

Continuous box girder bridges, cable- stayed bridges and suspension bridges of steel were extensively used for long spans.

As reinforced cement concrete (RCC) came to be used in the late nineteenth century and pre-stressed concrete in the early twentieth century, major bridges were built using these strong materials. The compliance of cement to be moulded into practically any shape has been a valuable feature that helps in obtaining aesthetically pleasing shapes for any construction, including bridges.

The most recent example of an RCC bow-string girder bridge is the third railway bridge across the river Godavari between Kovvur and Rajahmundry in Andhra Pradesh, built in 1996, with 28 spans of 97.5 m each.

Pre-stressed concrete railway bridges on the Assam rail link rear Siliguri were the first such bridges built by the Indian Railways in 1948. The first PSC Highway Bridge was built in 1954 over the Palar river rear Chengalpattu in Tamil Nadu.

Advances made in several areas of civil engineering such as fluid mechanics, soil mechanics and geotechnical engineering, and structural engineering enabled bridge engineers to make similar technological advances in the techniques of bridge construction.

## 22.2 INVESTIGATIONS FOR BRIDGES

### Need for Investigation

The need for building a bridge at a particular site should be established by considering several factors such as the need for a bridge based on traffic needs, characteristics of the stream or river, subsoil conditions, alternative sites, and economic considerations or cost-benefit analysis.

The primary aim of the investigation is to select a good site for the bridge satisfying the cost, safety, and aesthetic consideration.

Technical feasibility and economic considerations should be included in the investigation for a major bridge project to aid the justification of the project. All details of the investigation should be included in the project report

The thoroughness of the investigation ensures the success of the final design and construction of the bridge

### Site Selection

The following factors govern the selection of the site for a proposed bridge across a water course:

(i) Crossing angle of the bridge
(ii) Straight stretch of the stream or river
(iii) Width of the river
(iv) Velocity of the flow of water
(v) Embankments on either side of the bridge
(vi) Foundation soil for the substructure
(vii) Availability of materials and labour
(viii) Connection with the transportation route
(ix) Approaches to the bridge
(x) Need for river training works

An ideal site, satisfying all the requirements is rarely available.

**Crossing angle of the bridge:** It is desirable that the centre line of the bridge is at right angles to the water course; such a crossing is called *right angle, square,* or *normal* crossing.

The reasons for this preference are as follows:

- The bridge length will be shorter
- There will be smooth flow of water under the bridge
- Wing walls and return walls can be oriented to avoid eddies and cross currents.

Skew crossings are not desirable except under unavoidable circumstances, for the following reasons:

- The design and maintenance of skew bridges are difficult.
- The flow of water under the bridge is not smooth and eddy currents are produced
- The piers of a skew bridge have to resist greater water pressures than those encountered in a rectangular crossing.
- The construction of the substructure components as well as the superstructure is rather difficult.
- Greater depth of foundations may be needed owing to scour.

**Straight stretch of the stream or river:** With respect to the proposed site, the stream or river should have a straight stretch on either side. This is to ensure smooth flow of water. A curved stretch creates construction as well as maintenance problems.

**Width of the stream or river:** The width of the river at the site is a measure of the length of the bridge. The smaller it is, the cheaper the bridge is going to cost.

**Velocity of the flow of water:** The velocity of water flow should be within reasonable range; if it is small, silting may occur and if it is large, scouring will occur.

The permissible velocity depends upon the nature of the bed material. Typical values are given in Table 22.1:

**Table 22.1** Permissible velocities of water flow under a bridge

| S. No. | Bed material | Permissible velocity of flow (m/s) |
|---|---|---|
| 1. | Boulders and rock | 2.5 to 5.0 |
| 2. | Coarse gravel | 1.5 to 2.5 |
| 3. | Stiff clay | up to 2.0 |
| 4. | Clay-sand mixture | up to 1.5 |
| 5. | Fine gravel, sandy clay | 1.0 to 1.5 |
| 6. | Coarse sand | 0.5 to 1.0 |
| 7. | Fine sand and soft clay | up to 0.5 |
| 8. | Very fine sand | 0.6 to 0.9 |

**Embankments on either side of the bridge:** The embankments of the river on either side of the bridge should be firm, straight, and of sufficient height. This is necessary for flood protection and prevention of change in the river course.

**Foundation soil for the substructure:** The foundation soil at the bridge site should provide adequate bearing capacity at reasonable depth for the components of the substructure. The scour depth at the site should be small for stability of the foundation.

**Availability of materials and labour:** Materials and labour required for the proposed bridge construction should be available easily with minimum transport charges.

**Connection with the transportation route:** A bridge is constructed to connect the roadway or railway on either side of the water course. The soil and other conditions on either side of the bridge should be favourable for easy connection.

**Approaches to the bridge:** The approaches on either side of the bridge should involve easy gradients that could be negotiated easily by vehicles. They should be laid on firm soil and designed to provide adequate stability, consistent with the expected loadings from the vehicles using the facility.

**Need for river-training works:** Straight stretch of the river on either side of the proposed facility will minimise the need for construction of river-training works, meant to stabilise the regime of flow of water in the stream or the river, with minimal disturbance of the flow.

A site which satisfies the maximum number of the above criteria is chosen.

**Preliminary Data or Information Needed**

The following preliminary information or data has to be collected as part of the investigation for a bridge project:

1. Name of the road/railway, name of the river/ stream, location in km to the centre of the proposed bridge from an intial or a reference point.
2. Identification mark of the crossing
3. Particulars of the great trigonometric survey (GTS) benchmark and its reduced level (RL)
4. Availability and location of quarries for stones for materials for masonry and aggregates for concrete
5. Nearest place where cement, steel and timber are available
6. Details of transport for materials
7. Availability of skilled and unskilled labour required for construction
8. Hydraulic data of the river or stream – highest flood level (HFL), ordinary flood level (OFL), low water level (LWL), catchment details, possible future changes in respect of vegetation such as afforestation and deforestation, and rainfall characteristics
9. Soil profile at locations across the stream, and at approaches on either side
10. Possible need for river training works
11. Navigational requirements of the river, if any
12. Possibility of seismic disturbances
13. Availability of electric power
14. Feasibility of providing public utility services such as water supply, power cables, telephones cables, etc.
15. Facilities required for housing labour during the construction period
16. Details of existing bridges on the same river, if any, within reasonable distance from the proposed bridge.
17. Preliminary drawings:
    - (i) Index map
    - (ii) Catchment area map
    - (iii) Contour plan
    - (iv) Site plan
    - (v) Cross-section of the river at the proposed site
    - (vi) Longitudinal section of the river on either side of the site
    - (vii) Additional cross-sections of the stream or river on to either side of the site
    - (viii) Soil profile.

## Site Investigation and Surveys

Site investigation for major bridges is carried out in the following stages:

(a) Reconnaissance for feasibility, including reconnaissance survey
(b) Preliminary investigation and economic feasibility, including preliminary survey, and traffic projection
(c) Detailed investigation, including detailed survey and working drawings, and project report

Based on the investigation in the first two stages, the best possible site is chosen and detailed investigation is carried out.

**Reconnaissance Study and Reconnaissance Survey**

In this stage, the entire width of the river for a certain length or distance covering the probable crossing sites, is to be surveyed and studied. Based on these studies and surveys, a few alternative sites which satisfy the maximum number of favourable criteria are chosen for the location of the bridge.

The studies in this stage include the following:

(i) Study of the existing traffic
(ii) Study of topographic maps and conducting reconnaissance surveys.
(iii) Spot inspection of possible sites
(iv) Study of data relating to the river at such sites.
(v) Study of possible construction problems.
(vi) Study of the possible benefits that are likely to accrue from the construction of the bridge.

**Preliminary Investigation and Surveys**

In this stage, further studies are carried out with regard to the alternative sites chosen in order to bring out their relative merits and demerits. The estimated costs of the bridge at these alternative sites are worked out to a reasonable degree of accuracy (say, within 15% of the final estimated cost). For this purpose, some more field measurements and detailed surveys are carried out.

Traffic projection for the estimated design period is made using one of the standard methods.

Utilising all these data, the economic feasibility of the alternatives (which should be at least three) is studied, and the final choice is made for the bridge site.

**Detailed Investigation and Surveys**

Once the final selection of the bridge site is made, detailed investigation is made into the nature of the ground, hydraulic data, subsoil characteristics, traffic volumes, structural design of the foundations, substructure components, superstructure components, and the other ancillary and appurtenant works such as approaches, river-training works, embankment protection, and so on.

Based on these investigations and surveys, detailed working drawings are prepared for the benefit of the site engineers responsible for the construction of the bridge.

A project report containing all the relevant details and the techno-economic feasibility of the bridge, construction schedules and funding details, is prepared and submitted to the appropriate sanctioning authority for administrative sanction and technical approval.

In the case of major bridges, model studies are conducted to predict the behaviour of the river after the construction of the bridge. The results of such studies are utilised in the planning and execution of ancillary works such as the river-training works, as well as the design of the substructure components.

## Subsoil Exploration

An accurate assessment of the physical and engineering properties of the soil at the proposed bridge site is essential for the design of the foundations for substructure components — piers and abutments.

The field and laboratory tests required to determine the engineering properties such as shear strength and bearing capacity of the soil at the proposed depth, constitute 'subsoil exploration'. This is a complex problem in view of different soil strata at different depths at any particular site.

A complete soil investigation consists of the following information:

(i) Nature of soil deposits up to an adequate depth.
(ii) Thickness and composition of each stratum, and its depth from the bed level.
(iii) Depth to hard stratum or rock.
(iv) Engineering properties of the soils and rock.
(v) The depth to ground water table and its effect on the engineering properties of the soils.

The following methods are available for soil exploration:

(a) Open, or trial, or test pits.
(b) Borings.
(c) Soundings and penetration tests.
(d) Geophysical methods (seismic refraction and electrical resistivity).

[A more detailed treatment is given in Transportation Engineering, Volume I: Highway Engineering and Section I: Railway Engineering of this Volume.

The importance of site investigation and soil exploration cannot be overemphasised in view of the fact that successful completion of the bridge project depends, to large extent, on these aspects.]

## Alignment of a Bridge

The alignment of a bridge along its centre line is governed by the following factors:

### Rectangular or Square Crossing

The angle of crossing should be as near to 90° as possible; that is, the centre-line of the bridge should be at right angles to the direction of water flow.

### Straight Alignment

The centre-line of the bridge should be straight and not curved; alternatively, it can also be a series of straight lines.

In mountainous or hilly terrain it may become inevitable to align the bridge on a curve. In such cases, an appropriate superstructure design should be adopted and the axis of the piers should be made as nearly parallel to the direction of flow as possible

### Alignment of Approaches

Sometimes, approaches to the bridge may be aligned on curves in order to avoid skew crossing and achieve square crossing for the bridge.

These curves have to be easy for the vehicles to negotiate even if they are reverse curves.

### Silting and Scouring

The alignment should be protected from the effects of silting and scouring on the abutments, piers, and embankments near approaches

### River-training Works

If necessary, appropriate river-training works are adopted to create *nodal points*, where the river regime does not shift significantly. This will ensure stability of the bridge substructure as well as the bridge, as a whole.

Model tests may also be conducted to determine the appropriate location of the river-training works, in the case of major bridge projects.

River-training works are similar to bank-protection works for river banks and embankment, or coast or shore protection works in the case of sea coast.

## Choice of the Type of Bridge

The choice of an appropriate type of bridge for a given set of conditions involves a crucial decision in a major bridge project. The preliminary data collected prior to and after site selection should be carefully considered in choosing the most appropriate type of bridge for the existing site conditions.

The *most important conditions* to be fulfilled by a *completed bridge* structure are:

- Suitability to carry the design traffic
- Efficiency and effectiveness in operation in order to serve the intended purpose
- Minimum overall cost, or initial cost plus maintenance cost, over a design period
- Aesthetically pleasing look for the completed structure.

The important factors influencing the choice of the bridge type and the basic features are the following:

(i) The topography of a site and the soil conditions may limit the choice; for example, the best choice in a rocky area in a ravine is an arch bridge.

(ii) The corrosive atmosphere near a sea coast may preclude the use of some types and indicate the preference for some others; in such conditions, precast pre-stressed concrete is preferred to a cable-stayed steel bridge (Example: Pamban bridge near Rameswaram).

(iii) For highway traffic, deck bridges are preferred to through bridges to afford an uninterrupted view of the scenic beauty of the surroundings.

(iv) A submersible bridge may be preferred to a high-level road bridge when the traffic volume is low, in order to achieve economy.

(v) For a railway bridge, steel truss may be preferable to suspension bridges.

(vi) The need to reduce the number of piers, either because of poor subsoil conditions or navigational requirements, may necessitate a cantilever bridge or a cable-stayed bridge for a National Highway with heavy traffic

(vii) If large navigational clearances are needed, arches, cantilever bridges, or cable-stayed or suspension bridges are preferred (Example: Howrah bridge).

(viii) In order to achieve economy by combining the requirements of both highway and railway traffic, a road-cum-rail bridge may be preferred; the highway traffic and railway traffic are then separated vertically. (Example: The second bridge on Godavari river).

(ix) The specialised expertise of the executing construction company may influence the final choice of the type of bridge; this may happen when freedom is given to the bidder in the case of major bridges of considerable length. Specialised expertise may be available both for the superstructure and for the substructure (Example: Cable-stayed steel or pre-stressed concrete construction for superstructure and pile or well foundations for substructure.)

## 22.3 HYDRAULIC DESIGN

*Hydraulic design* of a bridge involves the following aspects:

1. Design discharge
2. Linear waterway and afflux

3. Economic span
4. Scour depth
5. Depth of foundation
6. Freeboard and clearance

## Design Discharge

The determination of the maximum discharge for the design of the components of a bridge is an important aspect for design. It is usually done by one of the following methods:

(a) Empirical method or formulae
(b) Rational method or formula
(c) Area-velocity method
(d) Unit-hydrograph method.

### Empirical Formula

The general form of the empirical formula is

$$Q = C.A^n \tag{22.1}$$

where $Q$ = Maximum flood discharge (m³/s)

$A$ = Area of the catchment (sq.km)

$C$ = Constant which depends on the nature of the catchment and location

$n$ = a constant.

*Ryve's formula*, which is commonly used for calculating design discharge, is a type of empirical formula and is given as follows:

$$Q = C.A^{2/3} \tag{22.2}$$

Here, $C$ = 6.8 for flat catchment within 25 km from the coast

= 8.5 for areas between 25 and 160 km from the coast

= 10 for areas near hills.

The reliability of empirical formulae is limited because of the oversimplification of the factors involved. Much depends on the value of C assumed; a reliable value can be got only by statistical analysis of a large volume of observed flood data. Further, it is considered valid only for the particular region for which it is developed.

### Rational Formula

This should take into account the intensity, distribution and duration of rainfall, and the characteristics of the catchment such as the area, shape, slope, permeability of the soil and the initial wetness of the basin.

The following formula is recommended by the IRC [IRC: SP: 13–1973 (2004-Revised)-*Guidelines for the design of Small Bridges and Culverts, IRC, New Delhi, 2004.*]:

$$Q = C.i.A \tag{22.3}$$

where $Q$ = Peak runoff or discharge (m³/s)

$C$ = Coefficient which depends on the type of catchment surface and the extent of the catchment area.

$i$ = Critical intensity of rainfall in mm/hr occurring during the time of concentration ($t_c$) and for the chosen frequency interval or return period

$A$ = Catchment area in sq.km.

When the catchment area consists of different types of surface with different run-off coefficients, the weighted average value is taken for computing C since C is dependent on the runoff coefficient.

***Intensity-Duration-Frequency (IDF):*** IDF curves are analysed for any given locality as contours of 1 hour rainfall intensity for frequency periods of 2, 5, 10, 25 and 50 yrs. (Longer frequency periods are chosen for major bridges on national highways, while suitable shorter periods are chosen for minor bridges on secondary and tertiary roads.)

A return period of 10 years and above is considered suitable in many cases.

The design runoff may be got by multiplying the 1-hour rainfall intensity for the specific locality for the chosen frequency period, by a suitable empirical conversion factor.

Time of concentration $t_c$, is given by

$$\underset{\text{(hours)}}{t_c} = \left(0.87 \times \frac{L^3}{H}\right)^{0.385} \tag{22.4}$$

where $L$ = Distance from the critical point to the bridge site in km

$H$ = Difference in elevation between the critical point and bridge site in metres.

C in Eq. (22. 3) is given by

$$C = \frac{0.56\,P \cdot f}{(t_c + 1)} \tag{22.5}$$

where, $P$ = Coefficient of run-off for the particular nature of the catchment

$f$ = A correction factor for the variation of the intensity of rainfall, $i$, over the area of the catchment

$t_c$ = Time of concentration of rainfall (hours).

$P$ is given in Table 22.2 below:

**Table 22.2** Coefficient of run-off for different types of catchment

| Nature of the catchment | P |
|---|---|
| Steep bare rock and pavements | 0.9 |
| Steep rock with thick vegetation | 0.8 |
| Lightly covered plateau | 0.7 |

Conti'd

Table Conti'd

| Nature of the catchment | $P$ |
|---|---|
| Stiff and bare clay | 0.6 |
| Lightly covered clay | 0.5 |
| Lightly covered loam | 0.4 |
| Largely cultivated loam | 0.3 |
| Sandy soil with light growth | 0.2 |
| Sandy soil with heavy bush | 0.1 |

$f$ is given in the following table:

| Area (sq.km) | 20 | 40 | 60 | 80 | 100 | 150 | 200 | 300 | 400 | 2000 |
|---|---|---|---|---|---|---|---|---|---|---|
| $f$ | 0.900 | 0.845 | 0.800 | 0.760 | 0.730 | 0.675 | 0.645 | 0.625 | 0.620 | 0.600 |

**Area–Velocity Method**

This is considered to be the most reliable method as it is based on the hydraulic characteristics of the stream.

The discharge $Q$ is given by

$$Q = A \cdot V \tag{22.6}$$

where $Q$ = Discharge in cum per second

$A$ = Wetted area in sq. m

$V$ = Velocity of flow in metres per second

Manning's formula is used for the velocity of flow:

$$V = \frac{1}{n} \cdot R^{2/3} S^{1/2} \tag{22.7}$$

$S$ = Slope of the stream

$R$ = Hydraulic mean depth in metres $\left(\dfrac{\text{wetted area in m}^2}{\text{wetted perimeter in m}}\right)$

$n$ = Coefficient of rugosity or roughness. $n$ depends on the material of the bed and embankment slopes and their condition. This ranges from 0.025 for clean straight banks to 0.150 for highly weedy reaches in bad condition.

$S$ is the more difficult parameter to determine; usually, it is taken to be the same as the bed slope measured over a sufficiently large distance..

The velocity of flow, $V$, can be obtained by direct measurement also by using surface floats or a device known as the current meter.

Area of flow, $A$, may be obtained by taking soundings for the depth of flow at regular intervals along the surface width of the river.

**Unit-hydrograph method** A *hydrograph* is the graphical representation of discharge in a stream plotted against time due to a rainfall of specified intensity, duration and areal distribution. For a given catchment or drainage basin, the hydrographs of runoff of two rain storms will be similar, as their ordinates are proportional to the rainfall intensity.

A *unit-hydrograph* is defined as the hydrograph representing a unit depth (1 mm) of direct runoff as a result of rainfall excess occurring uniformly over the catchment, at a uniform rate, for a specified duration (say, 6 hrs or 12 hrs). The area under a unit hydrograph represents the volume of rainfall excess due to a rain of 1 mm over the entire basin.

The procedure for calculating the peak discharge is given below:

(i) The hydrograph for the basin for a particular rainfall excess is plotted from the recorded data of runoff $m^3/h$ and time in hours.
(ii) The base flow is separated from the direct runoff.
(iii) The volume of direct runoff is computed from the area under the storm hydrograph. The depth of flow, $d$, in mm, can be obtained by dividing this volume by the area of the basin.
(iv) The ordinates of the unit hydrograph are obtained by dividing the corresponding ordinates of the storm hydrograph by $d$.
(v) Direct runoff for any given storm can then be calculated by multiplying the maximum ordinate of the unit hydrograph by the depth of runoff over the area. (The depth of runoff corresponding to the given storm can be got in proportion to the depth of the storm hydrograph already plotted for deriving the unit hydrograph)
(vi) The maximum runoff rate is obtained by adding the base flow to the maximum direct runoff rate. (The assumptions are not normally satisfactory for large catchments of over 5000 sq.km)

## Linear Waterway and Afflux

The unobstructed area of the river or the stream through which, water flows is the *natural waterway* of the river or the stream.

The area through which water flows under a bridge superstructure is known as the *waterway of the bridge*.

The linear distance or width of this area (the waterway of the bridge) is called the *linear waterway* of the bridge, when measured along the bridge.

The *effective linear waterway* of the bridge is the linear waterway minus the mean submerged width of the piers from the top to the foundation level. (This is nearly equal to the sum of all the clear spans).

The following norms are followed in fixing the effective linear waterway:

### Stream, Channel or Canal with Rigid Boundaries

In the case of artificial channels used for irrigation, drainage or navigation, the boundaries remain the same, irrespective of the flow. The effective linear waterway for artificial channels is defined as *the width of the channel at mid-depth*.

### Waterway for Natural Streams with Well-defined Banks

For natural streams (not wider than 30 m) in alluvial beds with well-defined banks, the effective linear waterway is taken as *the distance between the banks at high flood level water surface*.

### Waterway for Alluvial Streams

For large natural rivers in alluvial beds, with undefined banks, the effective linear waterway may be got from Lacey's formula for regime conditions:

$$L = C\sqrt{Q} \tag{22.8}$$

where $L$ = Linear waterway or the regime surface width of flow (metres)

$Q$ = Design discharge (cu. m/s)

$C$ = Constant (4.8 for regime channel, taken to be 4.5 to 6.3 according to site conditions)

### Waterway for Streams with Shallow Subsections

The waterway required is equal to the actual surface width of the active channel. (Length of a bridge is the overall length from end to end along its centre-line.)

**Afflux:** This is defined as the *heading up of water* on the upstream side of the bridge caused by the constriction of the waterway due to the construction of the bridge.

It is measured by the difference in level of the water surfaces upstream and downstream of the bridge. From the principle of continuity of flow, the velocity of flow through the bridge has to increase in order to pass the discharge in view of the reduction in the waterway at the bridge when compared to the natural waterway upstream of the bridge. This velocity increase is facilitated by the heading up of water or afflux, which creates the necessary head causing flow.

Afflux can be calculated from the following formula:

$$h_a = \frac{V^2}{2g}\left(\frac{L^2}{c^2 L_1^2} - 1\right) \tag{22.9}$$

where $h_a$ = Afflux (m)

$V$ = Velocity of normal flow in the stream (m/s)

$g$ = Acceleration due to gravity (m/s²)

$L$ = Width of stream at HFL (m)

$L_1$ = Linear waterway under the bridge (m)

$c$ = Coefficient of discharge through the bridge (0.7 for sharp entry and 0.9 for bell-mouthed entry).

**Empirical formulae for afflux:** The following empirical formulae are also used to calculate afflux:

(i) **Molesworth's formula**

$$h_a = \left(\frac{V^2}{17.9} + 0.0153\right)\left\{\left(\frac{A}{A_1}\right)^2 - 1\right\} \tag{22.10}$$

Here, $V$ = Velocity of normal flow (m/s)

$A$ = Natural waterway area near the site (m²)

$A_1$ = Contracted waterway area under the bridge (m²)

(ii) **Merriman's formula**

$$h_a = \frac{V^2}{2g}\left\{\left(\frac{A}{cA_1}\right)^2 - \left(\frac{A}{A_e}\right)\right\} \quad (22.11)$$

Here, $A_e$ = Enlarged area of waterway (m²)

$c$ = Coefficient of discharge through the bridge; it is given by

$$c = 0.75 + 0.35\left(\frac{A_1}{A}\right) - 0.10\left(\frac{A_1}{A}\right)^2, \text{ approx} \quad (22.12)$$

$V$, $A$ and $A_1$ are the same as those in Molesworth's formula.

**Increase in velocity due to afflux:** Afflux increases the velocity and this increase is given by

$$v_a = \sqrt{2g \cdot h_a} \quad (22.13)$$

If $g$ is in m/s² and $h_a$ is in m, $v_a$ is obtained in m/s. It is desirable to keep the afflux at as low a value as possible – preferably at 150 mm, and at any rate not more than 300 mm.

**Effects of afflux:** The effects of afflux are to be considered in the design of the bridge with regard to the following aspects:

(i) Increase in velocity of flow on the downstream side of the bridge, increasing scour, and necessitating greater depth of foundation for piers and abutments.
(ii) Protection works such as guide banks are to be raised to maintain the necessary freeboard.
(iii) Afflux should be considered in fixing the bridge deck level to prevent submersion.

## Economic Span

The following definitions are relevant here:

**Clear span:** This is the clear distance between any two consecutive supports (or piers) of a bridge.

**Effective span:** This is the centre-to-centre distance between any two consecutive supports (or piers) of a bridge.

**Economic span:** This is the span for which the total cost of the bridge structure is a minimum.

For a given linear waterway, considering only the items for which the cost varies with span length, the total cost of the superstructure increases and that of the substructure decreases with increase in span length.

The economic span length is that for which the total cost of the bridge is a minimum; this can be shown to occur when the cost of the superstructure equals that of the substructure, as elucidated by the following brief derivation.

Let $C_a$ = Cost of approaches

$C_{ab}$ = Cost of the two abutments, including foundations

$C_p$ = Cost of one pier, including foundation

$s$ = Span length

$n$ = Number of spans

$C$ = Total cost of the bridge

$C_s$ = Cost of superstructure for one span

$C_m$ = Cost of miscellaneous items such as railings, flooring, etc.

$L_b$ = Total length of the bridge

$C_s$ can be shown to vary with the square of $s$ since it depends on the bending moment, or it can be taken as $ks^2$, where $k$ is a constant.

$C_m$ can be shown to vary with $L_b$, or it can be taken as $k_1 L_b$, where $k_1$ is another constant.

Therefore, the total cost, $C$, of the bridge is given by

$$C = C_a + C_{ab} + (n-1)\,C_p + nks^2 + k_1 L_b$$

For minimum cost, $\frac{dC}{ds}$ should be zero. Substituting $n = \frac{L_b}{s}$, differentiating, and equating $\frac{dC}{ds}$ to zero, we eventually get $C_p = ks^2$ on simplifying the resultant expression.

Therefore, for economic span, the cost of superstructure and that of the substructure should be equal for one span.

The economic span, $s_e$, can be computed as

$$s_e = \sqrt{\frac{C_p}{k}} \tag{22.14}$$

$C_p$ and $k$ are to be evaluated as average values over a range of possible span lengths.

For small bridges with open foundations and solid masonry piers and abutments, the economic span may be taken as 1.5 times the height of the pier.

For masonry arch bridges, the economic span may be taken as two times the height of the keystone above the foundation.

For major bridges with more elaborate foundations, the question of economic span has to be examined in greater detail.

As a rule, the number of spans should be kept as low as possible, as piers obstruct the flow of water. An odd number of spans (or an even number of piers) are preferred so as to avoid a pier at the deepest portion of the stream or river.

## Scour Depth

*Scour* may be defined as the removal of bed material and material from the banks of the stream during the passage of a flood discharge; this occurs when the velocity of flow exceeds the limiting velocity that can be withstood by the soil material.

When the construction of the bridge reduces the waterway significantly, severe scour may occur during heavy floods.

The scour is aggravated at the nose of the piers and at bends in the stream. The maximum scour depth should be measured with reference to the existing structures near the proposed bridge site, if possible. For this purpose, soundings are taken during or immediately after a flood.

If the design discharge is more than that for which the sour depth was measured by the soundings, allowance should be made for this; allowance should also be made for the increased velocity of flow expected due to the construction of the proposed bridge. Scour depth is always measured below the highest flood level (HFL).

When this practical method of determining the scour depth is not possible, a theoretical value is obtained by using Lacey's equation for natural streams on alluvial beds:

$$d_s = 0.473\left(\frac{Q}{f}\right)^{1/3} \qquad (22.15)$$

where $d_s$ = Normal scour depth below HFL (metres)

$Q$ = Design discharge (cu m/s)

$f$ = Lacey's silt factor for the bed material

$f$ is given by

$$f = 1.76\sqrt{m} \qquad (22.16)$$

where $m$ = Mean diameter of the bed material particles (mm)

(Note: $Q$ may be increased from 10% to 30% to provide a margin of safety depending upon the catchment area.)

Values of $f$ for representative bed materials are given in Table 22.3:

**Table 22.3** Lacey's silt factor

| S. No | Nature of bed material | Mean dia. of particle (mm) | Value of $f$ |
|---|---|---|---|
| 1. | Very fine silt | 0.052 | 0.40 |
| 2. | Fine silt (as in Godavari western delta) | 0.081 | 0.50 |
| 3. | Fine silt | 0.120 | 0.60 |
| 4. | Fine silt(as in Krishna western delta) | 0.158 | 0.70 |
| 5. | Medium silt | 0.233 | 0.85 |
| 6. | Standard silt | 0.323 | 1.00 |
| 7. | Medium sand | 0.505 | 1.25 |
| 8. | Course sand | 0.725 | 1.50 |
| 9. | Fine sand and gravel | 0.988 | 1.75 |
| 10. | Coarse sand medium gravel | 1.290 | 2.00 |

Eq.(1.15) is applicable when the linear waterway of the bridge is not less than the regime width of the stream; when it is less than the regime width, the normal scour depth under the bridge is given by

$$d_s' = d_s\left(\frac{L_r}{L}\right)^{2/3} \tag{22.17}$$

Here, $d'_s$ = Normal depth under the bridge

$d_s$ = Normal scour depth when $L = L_r$

$L_r$ = Regime width of the stream

$L$ = Waterway provided under the bridge

The maximum scour depth is not uniform even in straight reaches for natural streams The maximum scour depths under different conditions of flow are given in Table 22.4:

**Table 22.4** Maximum scour depth

| S. No | Condition of flow | Maximum Scope depth(in terms of $d_s$) |
|---|---|---|
| 1 | In a straight reach and near abutments | 1.27 $d_s$ |
| 2 | At a moderate bend | 1.50 $d_s$ |
| 3 | At a sharp bend | 1.75 $d_s$ |
| 4 | At a right-angled bend and at noses of piers | 2.00 $d_s$ |
| 5. | At upstream noses of guide banks | 2.75 $d_s$ |

It has been observed that the maximum scour in the case of an abutment occurs at the upstream corner, whereas that in the case of a pier occurs at the downstream end.

## Depth of Foundation

For a preliminary idea, the minimum depth, $h$, of the foundation for abutment and piers is given by

$$h = \frac{q}{\gamma}\left(\frac{1-\sin\phi}{1+\sin\phi}\right)^2 \tag{22.18}$$

where $q$ = Bearing capacity of the soil

$\gamma$ = Unit weight of the soil

$\phi$ = Angle of internal friction of the soil

However, the bearing capacity of a soil depends on a number of factors including the depth of foundation, and cannot be oversimplified. Each type of the foundation needs a specific procedure of determination of the bearing capacity, as it depends not only on the soil characteristics, but also on the size and shape of the foundation, and the ground water conditions.

## Freeboard and Clearance

The term *freeboard* is associated with the approaches to the bridge and the guide bunds or banks of the stream or river. *Freeboard* is defined as the difference between the highest flood level after allowing for afflux, if any, and the formation level of the road embankment on the approaches, or the top level of the guide banks at the point under consideration.

The minimum freeboard recommended for high-level bridges is 1750 mm.

The term *clearance*, in this case *vertical clearance*, is associated with the superstructure of a bridge. *Vertical clearance* is defined as the vertical distance or difference between highest flood level and the lowest point of the superstructure.

Clearance is required for the following reasons:

(i) To allow for any error in the estimation of the design discharge and the HFL, and prevent the possible submergence of the bridge floor
(ii) To allow the floating debris to pass under the bridge without the damaging the bridge structure while passing the maximum discharge including wave action
(iii) To allow for navigation, if required, even at the time of high floods.

For arch bridges, the clearance below the crown of the intrados of the arch should not be less than one-tenth the maximum depth of water plus one-third of the rise of the arch. For structures with metallic bearings, the clearance between the HFL and the base of the bearings should not be less than 500 mm For navigation channels, the clearance requirement may be relaxed at the discretion of the engineer.

It is logical to provide higher values of clearance for greater discharges. The recommendations for vertical clearance is given in Table 22.5:

**Table 22.5** Recommendations for greater discharges

| Discharge ($m^3/s$) | Minimum vertical clearance (mm) |
|---|---|
| Less than 0.3 | 150 |
| 0.3 to 3.0 | 450 |
| 3 to 30 | 600 |
| 30 to 300 | 900 |
| 300 to 3000 | 1200 |
| Over 3000 | 1500 |

## 22.4 CLASSIFICATION OF BRIDGES AND COMPONENTS

The *classification of bridges* was discussed in Section 22.1 with regard to the various criteria applicable.

However, the primary classifications are:

- Road bridges and Railway bridges – based on the function or purpose
- Permanent bridges and Temporary bridges – based on the duration of use

While permanent bridges are designed to serve their function for significantly long periods such as 50 to 75 years, temporary bridges are those that serve for short period, especially for military purpose, for the use of an army during periods of war, and natural calamities such as floods and earthquakes.

Road bridges and railway bridges differ in certain respects with regard to their functional requirements as well as design to suit the respective standard loadings

## Classification of Bridge Components

The primary components of a bridge, as discussed earlier, are the superstructure, substructure and foundations of the substructure components.

### Classification of Superstructure

Based on the *material* used, the superstructure of a bridge can be of the following types:

(i) Timber bridges
(ii) Masonry bridges
(iii) Steel bridges
(iv) Reinforced cement concrete (RCC) bridges
(v) Pre-stressed concrete bridges
(vi) Composite bridges

Based on the *structural action*, the superstructure can be of the following types:

1. Simple or beam bridges
2. Continuous bridges
3. Cantilever bridges
4. Arch bridges
5. Bow-string girder bridges
6. Rigid frame bridges
7. Suspension bridges
8. Cable-stayed bridges

Several combinations are possible among these types.

### Timber Bridges

Timber is used for only minor and temporary bridges, and is meant to carry relatively small loads. It is used only for road bridges and not for railways, as the loadings are high for the latter.

Timber bridges are suited only for small spans of up to 6 m. For bigger spans, timber trusses are used for the superstructure. Tension members of the trusses are reinforced by steel plates. The flooring system consists of wood planks, at least 30 cm wide. The members of the superstructure should be proportioned such that the deflection of the deck does not exceed 1/300 of the span.

The *disadvantages of timber* bridges are:

- Relatively short life
- Not suited to long spans
- Non-uniform strength
- Liable to termite attack
- Liable to fire hazard
- Necessity for constant maintenance
- Strength variation and directional weaknesses of the timber

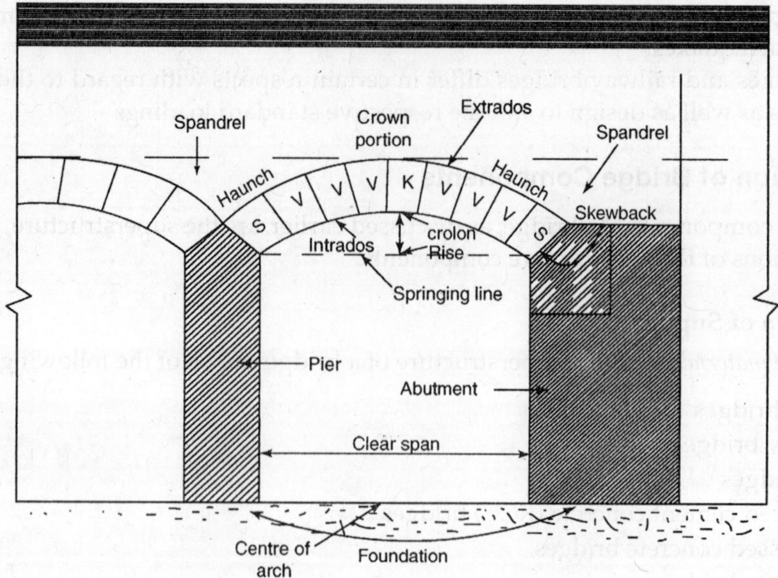


**FIG. 22.2** *A typical masonry segment arch*

**Masonry Bridges**

Masonry bridges may be constructed of brick masonry, stone masonry or precast concrete blocks.

Figure 22.2 shows a typical masonry arch of the segmental type.

The definitions relevant to this are given below:

**Springing points**: These are the points from which an arch springs or rises.

**Springing line**: Line joining the springing points.

**Skewback:** This is the surface of the pier or abutment on which the first block of the arch rests.

**Intrados**: It is the inner surface of the arch; it is also called the 'Soffit'.

**Extrados**: Outer surface of the arch

**Crown**: This is the highest point of the intrados.

**Rise**: It is the vertical distance from the springing points to the crown.

**Voussoirs**: These are the wedge-shaped blocks constituting the arch (*V*)

**Springer**: This is the first voussoir resting on the skewback

**Key**: This is the central voussoir at the crown

**Haunches**: These are the lower flanks of the arch on either side of the crown portion.

**Spandrel**: This is the irregular triangular space enclosed by the extrados and a tangent to the extrados at the summit

The common shapes of the arch are segmental, semi-circular, parabolic, or elliptical. The last two are not generally preferred.

The arch may be pointed at the middle. Semi-circular and pointed arches exert very little thrust on the abutments.

The rise of the arch is kept as large as possible in order to reduce the horizontal thrust and to economise on the design of the substructure and the foundations. It can be in the order of one-third span, but in no case less than one-fifth span. A simple circular shape will facilitate easy setting and construction of the arch.

### Classification of Bridge Arches

The *criteria for classification* of bridge arches are:

- Shape of the arch
- Number of hinges
- Condition of spandrel
- Width of the arch

**Shape of the arch:** The shape of the arch may be segmental, semi-circular or pointed.

If sufficient clearance is available, semi-circular or pointed shapes are preferred. Segmental or parabolic arches are used for medium spans, and pointed arches for long spans.

**Number of hinges:** Bridge arches may be three-hinged, two-hinged, or hingeless or fixed.

A three-hinged arch is hinged at the supports and at the crown, and is statically determinate from the point of view of analysis

A two-hinged arch is hinged at the springings only; it is indeterminate from the point of view of analysis, but is easy from the aspect of construction. Concrete and steel arches and bow-string girder arch of RCC are usually two-hinged

A hinge-less or fixed arch is built-in at the end supports, and is indeterminate from the point of view of analysis. It is usually employed for masonry arches, when hard soil is available for foundation.

**Condition of spandrel:** The spandrel may be filled with earth contained between side walls; or it may be open, in which case the bridge floor is supported by beams and columns, bearing on the arch, as shown in Fig. 22.3

**Width of the arch:** The arch may be of the barrel type, in which the width of the arch is the same as that of the bridge. The portion between the arch and the bridge floor looks like a barrel.

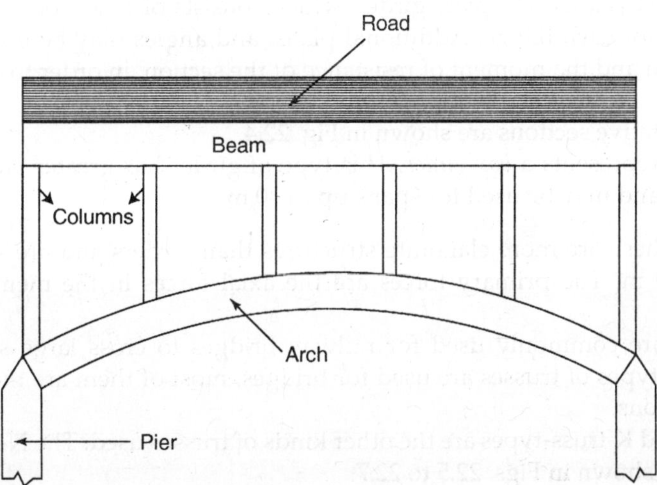


**FIG. 22.3** *Open spandrel arch*

The arch may be of the rib type, in which the arches on either side of the bridge will be of sufficient thickness and bent like ribs. The space in between the ribs may be filled with earth material to support the bridge deck

The *advantages of arch bridges* are:

(i) Aesthetic appeal
(ii) Suitability for medium and long spans
(iii) Easy maintenance
(iv) Better stability under impact and vibratory forces.

**Steel Bridges**

Steel bridges are used for carrying a highway or a railway or for other purposes such as supporting water pipes and gas pipes. Steel is used for a large range of spans when used for railways.

Depending on the method of joint connections, steel bridges are primarily classified as *riveted* and *welded*.

Depending on the *structural action*, steel bridges are classified as given below:

(a) Beam bridges or girder bridges (including plate girder)
(b) Truss bridges
(c) Continuous steel bridges
(d) Cantilever truss bridges
(e) Steel arch bridges
(f) Rigid frame steel bridges
(g) Cable-stayed bridges
(h) Suspension bridges

**Girder bridges:** Rolled steel I-section with or without cover plates are used as main girders. Cross-beams of I-section are used as bracings to support the main bridges.

This type of arrangement is used for culverts or minor bridges.

A *plate girder*, which is essentially a built-up section, is used to carry heavy loads over long spans. The simplest is a riveted plate girder, which consists of a pair of angles connected to a solid web plate for each flange. Additional plates and angles may be used to increase the modulus of section and the moment of resistance of the section, in order to carry heavy loads over long spans.

A few representative sections are shown in Fig. 22.4.

Figure 22.4 (c) represents a *box girder*. This type of girder has greater lateral stability and torsional rigidity, and may be used for spans up to 30 m

**Truss bridges:** These are more elaborate structures than girders and are suitable for spans from 30 m to 300 m. The primary forces are the axial forces in the members – tensile or compressive.

Truss bridges are commonly used for railway bridges to cross large streams or rivers. Although several types of trusses are used for bridges, most of them are the Pratt or *N* type, or their modifications.

Warren truss and K-truss-types are the other kinds of trusses used. The N-type, Warren and K-type trusses are shown in Figs. 22.5 to 22.7.

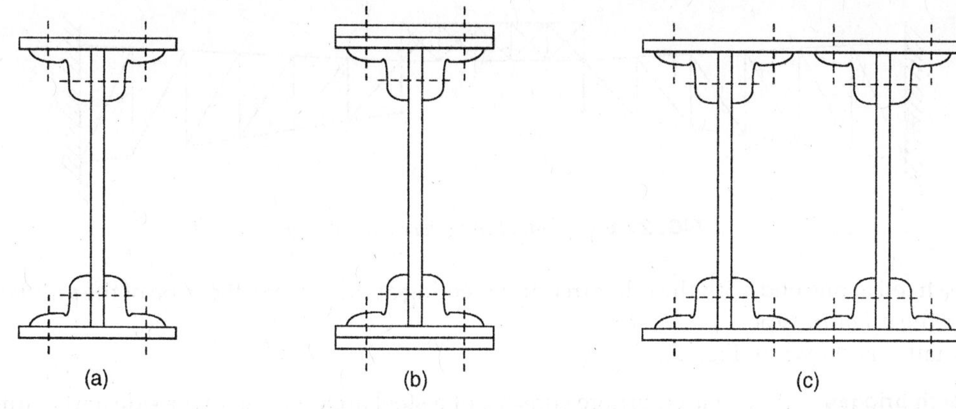


**FIG. 22.4** *Section of plate girders*

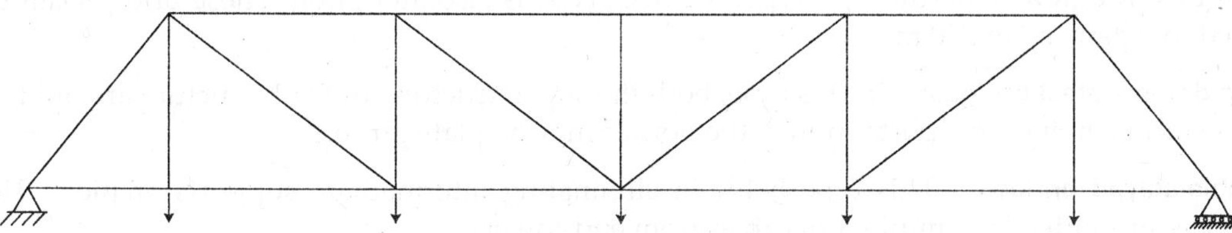

**FIG. 22.5** *Standard Pratt or N-truss*

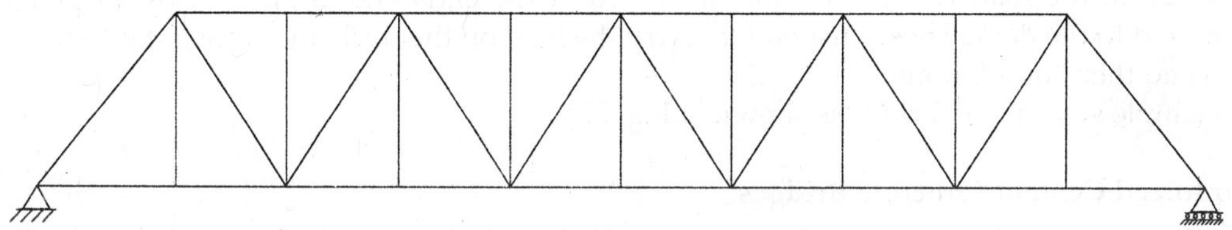

**FIG. 22.6** *Standard Warren truss (with verticals)*

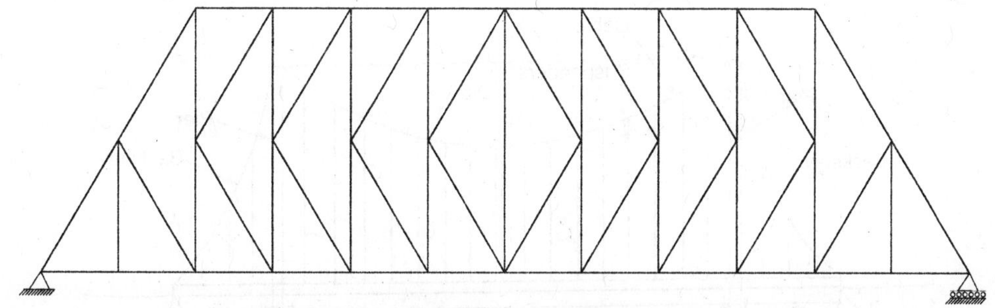

**FIG. 22.7** *K-truss*

### Balanced Cantilever Bridges

Cantilever steel bridges are used to bridge deep valleys, especially in places where a suitable foundation material is not available. The bridge consists of two equal portions of cantilever

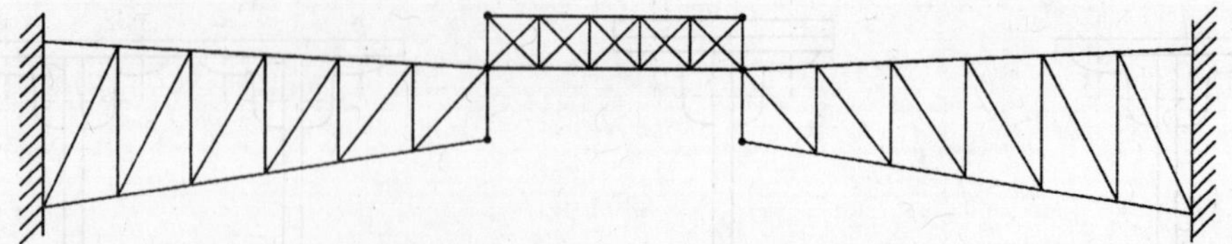

FIG. 22.8 *Balanced cantilever truss bridge*

trusses, fixed at one end with the other free and meeting at the Centre. It can be of the *unbalanced* type or the *balanced* type;

The latter is shown in Fig. 22.8.

**Steel arch bridges:** A steel arch bridge consists of a steel arch on the lower side and columns between the deck and the arch. The arch rib may be built using plate girders or trusses. They can be through or semi-through type; the deck type is also an option. These bridges can be used for spans up to 250 m

**Rigid frame steel bridges:** In this type, both the superstructure and substructure are built so as to form a monolithic portal frame; the portals may be plate girders

**Cable-stayed bridges:** This type helps in eliminating intermediate supports or piers. The deck is supported by a number of cables from end-towers.

**Suspension bridges:** This type of bridge is used for very large spans. Two main cables on each side of the roadway are suitably anchored at the ends and supported by suspenders connected to the deck. These suspenders carry the load on the deck and transmit it to the end piers and their foundations.

A simple suspension bridge is shown in Fig. 22.9.

## Reinforced Cement Concrete Bridges

After its advent in the late nineteenth century, reinforced cement concrete (RCC) became very popular and the most preferred material for bridge construction. Rigidity (as well as flexibility

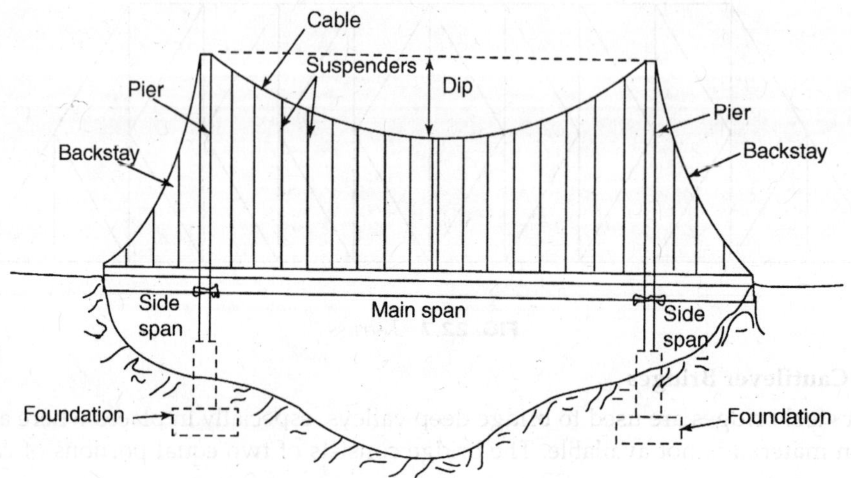


FIG. 22.9 *Simple suspension bridge*

where needed), durability, economy, and the ease with which it can be moulded to achieve an aesthetically pleasing appearance have all contributed to its popularity. Also, the possibility of the use of precast concrete elements for the various bridge components is considered to be another advantage.

The following are the common types of RCC bridges:

(a) Slab type
(b) Girder type (including T-beam type and hollow girder type)
(c) Balanced cantilever type
(d) Continuous beam type
(e) Bow-string girder type

**RCC slab bridges:** This is the simplest type of RCC bridge and the easiest to construct. It is most suited for submersible bridges for spans up to 8 m. The thickness and quantity of steel required for the slab is more than those needed for a girder bridge slab. This type is very economical in view of minimum formwork and labour required

**RCC girder bridges:** Girder bridges are used for spans from 10 to 20 m.

T-beam bridges are very common; the T-beams act as main girders. For wide bridges of two or more lanes, adequate numbers of T-beams are used as longitudinal girders, with or without lateral or transverse beams.

When transverse beams are provided in a T-beam bridge, the stiffness and torsional rigidity are increased. The deck slab is a two-way slab with reinforcements in both directions.

A *four-girder system of T-beam bridge* with foot-paths cantilevered on either side is schematically shown in Fig. 22.10.

*Hollow-girder bridges* consist of closed box girders in place of T-beams, and are used for spans of 25 to 30 m.

**RCC balanced cantilever bridges:** As in the case of steel cantilever bridges, this type consists of spans simply supported over cantilever portions. It is generally used for spans of 35m to 60 m. When the foundation soil is weak, small spans with a large number of piers proves to be uneconomical.

When deep gorges are to be bridged by a single span, a balanced cantilever bridge may be advantageous. The connections between the suspended simple span and the cantilevers

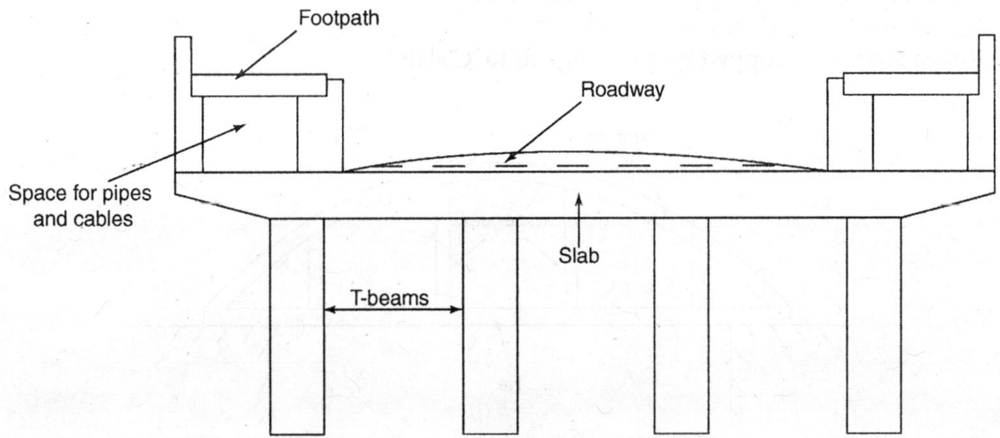


**FIG. 22.10** *Four-girder system of a T-beam bridge*

on either side are called *articulations*. The cantilever spans are usually 20 to 25% each of the suspended spans for which the articulations serve as the supports.

The cross-section of a balanced cantilever bridge may be of the T-beam type or the hollow-girder type.

**RCC continuous beam bridges:** These consist of spans built continuously over the supports. They are suitable for large spans. The deck may be in the form of slab, T-beam, or a box-section. The bending moments at the supports are more than that at mid-span. Hence the thickness at the supports is made 1.3 to 1.8 times that at mid-span; the length of hunched portion will be 20 to 25% of the span. The primary disadvantage of this type is that even small settlements may cause increase in stresses of significant magnitude.

**RCC bow-string girder bridges:** Suitable for spans of 30 m to 45 m, RCC bow-string girder bridges are popular.

A simple bow-string girder bridge is shown in Fig. 22.11.

**Pre-stressed concrete bridges:** Pre-stressed concrete has been widely used in bridge construction due to its inherent advantages; higher load-carrying capacity, fewer expansion joints, light self-weight, and suitability for aesthetic and architectural treatment are some important characteristics of pre-stressed concrete bridges. Its greater capacity to withstand tensile stresses as compared to reinforced cement concrete and its shear capacity as well as torsional resistance are the attractive features of pre-stressed concrete for bridge construction. Another great advantage is the availability of precast pre-stressed concrete girders and other components of bridges

A good example of pre-stressed concrete bridge is the Marmalong bridge across Adyar River in Chennai city. It is 30.5 m wide and consists of 12 spans of 18 m each. The deck is supported by 10 numbers of precast pre-stressed concrete girders of T-section with cast-in-situ slabs.

## Composite Bridges

This type of construction involves the combination of two dissimilar materials into the structural elements needed for bridges. A weak material is reinforced with another with a higher strength.

*Common examples* of composite construction are

(a) Concrete reinforced with steel bars
(b) Rolled steel joists topped by a cast-in-situ RCC slab

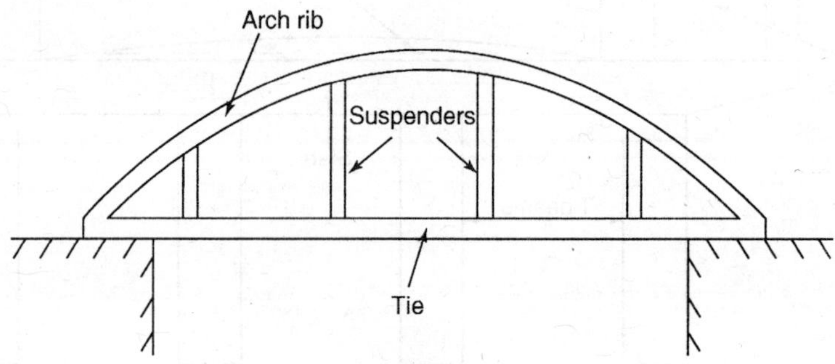


**FIG. 22.11** *RC Bow-string girder bridges*

The *advantages of composite bridges* are:

(i) The depth of the beam and the critical sizes of the various elements are reduced, which leads to economy.
(ii) Speed is achieved in construction
(iii) Deflections and vibrations are reduced for the bridge
(iv) The cost of substructure components and their foundations is reduced by using precast composite construction.
(v) Better quality control is achieved in precast composite construction for bridges.

### Floating Bridges

Also called *Pontoon bridges*, they are used for temporary passage across a river, especially during war. Floating bridges prove to be economical for crossing large rivers with significant depth, where the bottom consists of soft ground, since conventional piers need expensive foundation. (A classic example of a modern pontoon bridge is in Seattle, Washington, USA; this is 2018 m long, supported by hollow concrete pontoons, floating in Lake Washington.)

The concept of a floating bridge is based on the buoyancy of water to support the loads on the bridge. An anchoring system is needed to maintain the longitudinal and transverse alignments of the bridge.

The pontoons consists multi-cellular construction so that if one cell gets damaged for any reason, there is still sufficient buoyancy to keep the bridge afloat.

A floating bridge essentially consists of floating supports or pontoons and the superstructure. The pontoon may be of the rigid type such as decked barge, or of the pneumatic type; the latter consists of three rubberised main tubes which can be inflated prior to use.

The superstructure provides the roadway, and is designed to act as one continuous member when all the parts are connected. In modern floating bridges, aluminium alloy girders of box section are supported on inflated pontoons and connected midway between them to secure continuity. The spacing of the pontoons should be optimal to allow floating debris to pass and to make the superstructure economical.

### Movable Bridges

These are used over a canal or a river, where the requirements of navigation necessitate a high vertical clearance above the water level, making it inconvenient and uneconomical to raise the road or rail levels to the required extent. The choice of a movable bridge depends on this factor and also on the relative amounts of vehicular and waterway traffic and clearance requirements.

(Chicago in the USA has the maximum number of movable bridges.)

### Types of Movable Bridges

(a) Swing bridge
(b) Bascule bridge
(c) Lift bridge
(d) Transporter bridge

**Swing bridge:** This consists of a two-span continuous truss, supported at the centre on rotating gears in the horizontal plane; the gear arrangement is mounted on a pier built in the middle of the waterway. Generally, the two spans are equal.

Some of the width is taken away by the swing deck; the bridge is made to swing by 90° to allow passage, which requires a long time for operation

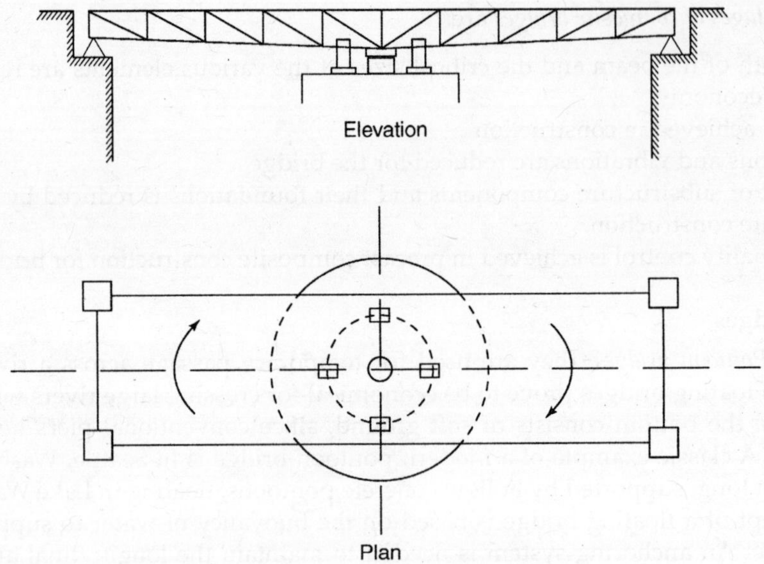


**FIG. 22.12** *Swing bridge*

The machinery required is complicated.

(The largest swing bridge is the railway bridge across the Suez Canal in Egypt, built in 2001, with a main span of 340 m)

A simple schematic of a swing bridge is shown in Fig. 22.12.

**Bascule bridge:** A bascule bridge is one in which the entire superstructure is rotated in the vertical plane up to along about 80° with the horizontal, to create the space for navigation.

For smaller width, a single-bascule bridge will be adequate; while for larger widths, the superstructure is built in two halves, and both the parts can be made to rotate about their respective supports. In the latter case, it is known as a double- bascule bridge. Suitable hinges, rack and pinion arrangements and the necessary counterweights are provided at the bank-end of the bascule, so that minimum power is needed to lift it.

A schematic of a single bascule is shown in Fig. 22.13.

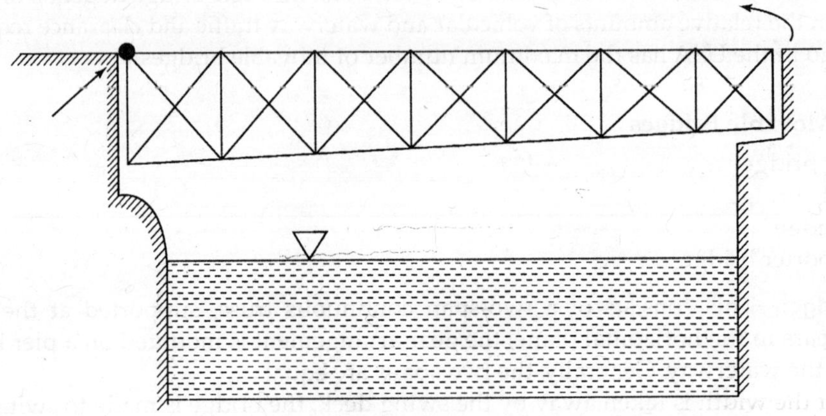

**FIG. 22.13** *Single-bascule bridge*

A double-bascule bridge needs smaller counter weight and can be raised faster than a single large bascule. Pamban bridge in Rameswaram is an example.

A bascule bridge is considered to be more advantageous than a swing bridge as it provides more space for navigation with less of motor power. However, it is difficult to operate this type of bridge under high-velocity winds.

**Lift bridge:** A vertical lift bridge is preferred to provide clearance for navigation across a channel. The whole span of the superstructure can be lifted vertically while the span remains horizontal.

Two towers are provided, one on each bank. The truss of the superstructure is connected to the cables at the end, passing over pulleys mounted at the top of the towers. The cables, in turn, are connected to the counter weight at the other end.

When a vessel is to be allowed, the truss is lifted vertically to the required level. A light overhead truss is provided at the top connecting the two towers for better stiffness and for providing support to a walk-way for maintenance, and space for pipe lines and ducts.

A schematic of a lift bridge is shown in Fig. 22.14.

**Transporter bridge:** This is used within a harbour area to facilitate the movement of persons and materials across a channel. Two towers are provided, one at each end, which are connected by an overhead truss. A cage is hung by wire ropes which are connected to a roller arrangement at the bottom cord of the overhead truss. The cage can be made to move from one end to the other, thus transferring persons end materials; this arrangement is somewhat similar to that of a ferry.

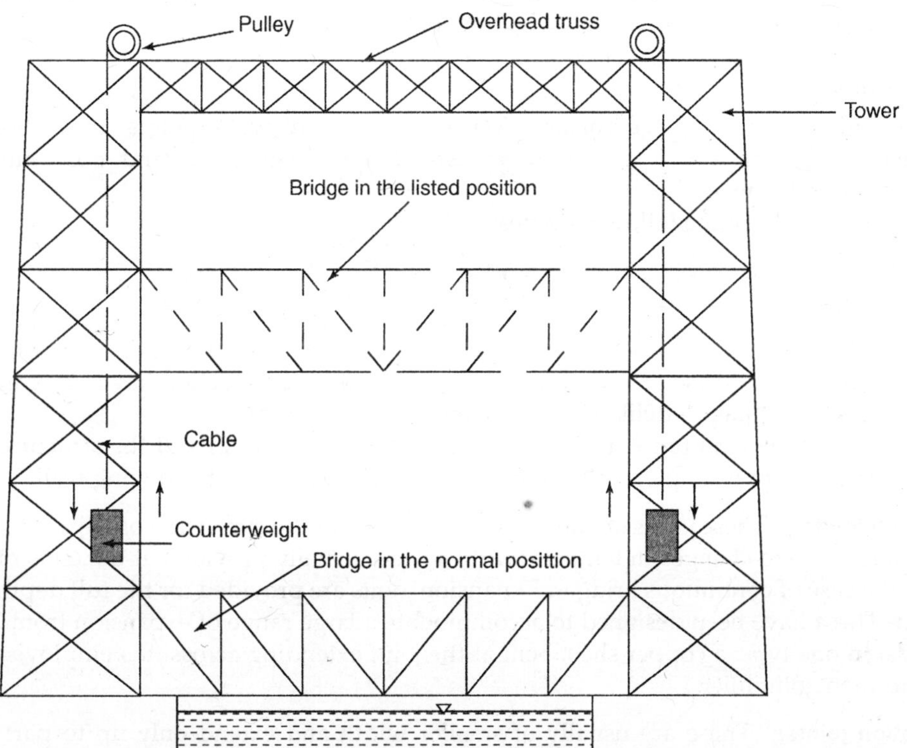


**FIG. 22.14** *Lift bridge*

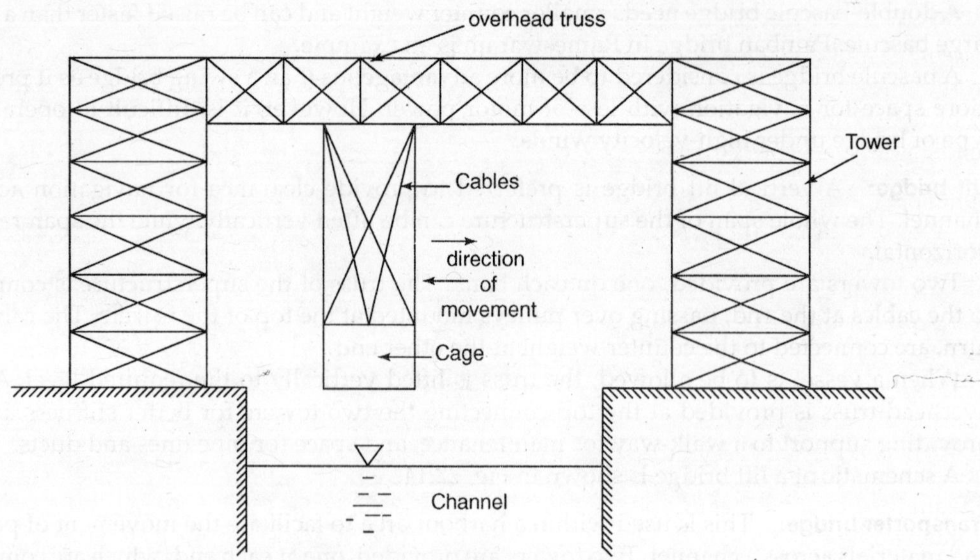


**FIG. 22.15** *Transporter bridge*

A schematic of a transporter bridge is shown in Fig. 22.15.
Movable bridges need more maintenance than fixed bridges, especially because of the machinery and moving components. Aesthetically, they are inferior to fixed bridges. Therefore, they may be used only for essential navigational requirements.

### Bridge Joints

Joints are inevitable in bridge construction; since they are weak points, they should be constructed in such a way that they withstand the design loads and pose minimum maintenance problems

The joints may be of the following types:

1. Construction joints
2. Expansion joints
3. Contraction joints

**Construction joints:** These are needed whenever concrete construction has to be continued after temporary stoppage – deliberate or unexpected.

This may also be required when the foundations of adjacent parts of the structure are at different levels as for example, at the junction between the abutment and wing wall.

**Expansion joints:** These joints are designed to take care of the expansion of adjacent parts of the structure due to changes in temperature. They are usually provided at the ends of every span in the case of continuous bridges. Expansion joints are provided for the full depth of the member. These have been designed to accommodate a large range of expansion from 25 mm upwards. In one type, a copper sheet bent at the joint extending across it is employed along with a resilient joint filler.

**Contraction joints:** These are usually of smaller width and extend only up to part of the depth of the slab. They are spaced closer than expansion joints. Contraction joints are meant to take care of the possible contractions due to changes of temperature and shrinkage of concrete

## Bearings for Bridges

*Bearings* are provided under the supports to accommodate the changes that may take place in the girders due to temperature, deflection, sinking of the supports, shrinkage and creep, in order to keep the bearing stresses within permissible limits by uniformly distributing the loads over the substructure.

**Functions of bearings:** The following are the functions of bearing in bridges.

(i) Permitting movements at the ends of the girder due to settlement of the supports
(ii) Permitting angular movement due to change in temperature
(iii) Absorbing movement due to change in temperature
(iv) Keeping the bearing stresses within permissible limits
(v) Distributing the loads over a large area.

**Importance of bearings:** Improper functioning of bearings could lead even to failure of a bridge. Thus, bearings constitute one of the important components of a bridge; Thus, adequate attention has to be paid to the design, construction and maintenance of bearings. It is interesting to note that the cost of bearings in major bridges could be as much as 10 % of the total cost. Broadly speaking, bearings may allow only sliding or translatory movements of the girder ends, which are then called *free bearings*. Bearings that allow only angular movement of the girder ends but do not allow translatory movement are known as *fixed bearings*.

**Types of bearings:** Several types of bearings have been developed, but only a few of the important types are given here,

These are:

1. Cement mortar pad
2. Expansion bearing
3. Rocker bearing
4. Rocker and roller bearing
5. Neoprene bearing

1. **Cement mortar pad**: This is the simplest type of fixed bearing, and consists of a 30 cm thick rich cement grout (1:1) pad. Dowel bars of size 25 mm are provided to connect the superstructure to the pad
2. **Expansion bearing**: This consists of a bottom shoe and a top shoe. The bottom shoe is shaped in the form of a circular arc to rest on the bed plate, which allows angular movement, while a rocker pin is fixed between the two shoes. Slight angular as well as translatory motion is permitted in this type, which is used only for small spans.
3. **Rocker bearing**: This is a fixed bearing which permits rotation about a hinge or a pin; the hinge is provided between the top and bottom shoes as shown Fig. 22.16.
4. **Rocker and roller bearing**: The bottom shoe rests on rollers that are cylindrical in shape and free to roll on steel bed plate. Rolling and rocking movement are permitted, but not sliding. A typical bearing of this type is shown in Fig. 22.17.
5. **Neoprene bearing**: Neoprene rubber pads are cut to the required sizes from moulded sheets and used as bearings. They are used for steel beams or for precast pre-stressed concrete girders.

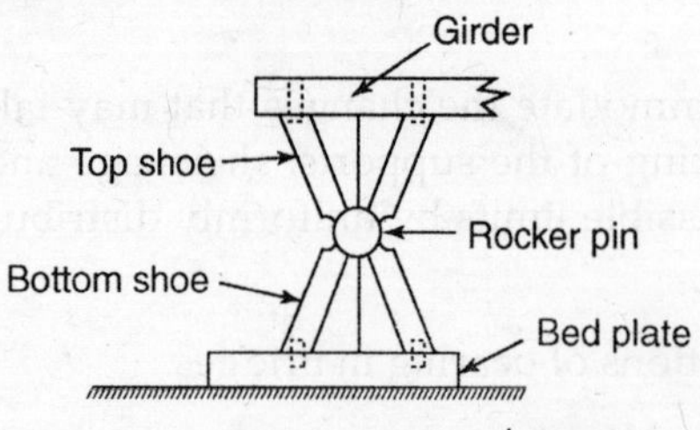


**FIG. 22.16** *Rocker bearing*

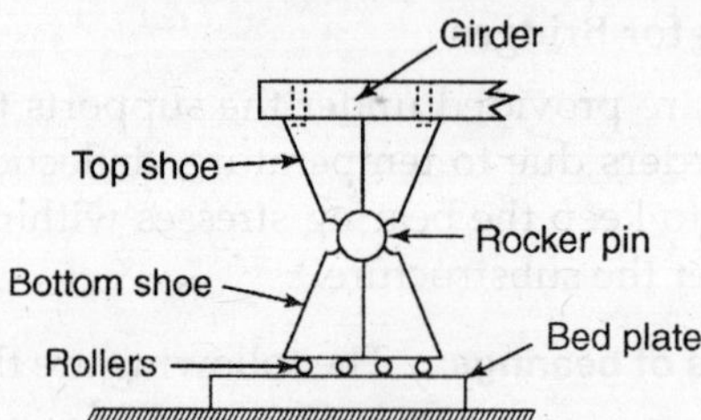


**FIG. 22.17** *Rocker and roller bearing*

They allow movements of the beam including rotation and allow a uniform transfer of loads from the beam to the substructure.

Neoprene bearings serve their function efficiently for very long periods – 75 to 100 years. (*Neoprene* is a synthetic rubber that falls under the category of *elastomers*.)

### Appurtenances for Bridges

These are not major structural components of a bridge; however, they perform useful functions for the benefit and safety of bridge users.

These are:

(a) Parapets and handrails
(b) Footpaths on bridges
(c) Crash barriers
(d) Drainage spouts

*Parapets and handrails* are meant to avert the possibility of the bridge users falling to the sides into the stream or river; parapets are used for culverts and hand rails are used for long bridges. Foot-paths are meant to be utilised by pedestrians.

*Crash barriers* are meant to prevent vehicles from crashing into the streams.

*Drainage spouts* take care of runoff water and lead to pipes through which it is led away.

## Classification of Substructure

The substructure of a bridge is an important component, the design and construction of which will have bearing on the overall economy. The layout of substructure components such as the piers determines the spans, which are one of the most important features in the design of the superstructure; further, this also affects the aesthetics of the bridge structure.

All the possible load combinations should be considered for the design of the substructure in order to produce an effective and efficient design. The following are the components of the substructure:

1. Abutments
2. Wing walls
3. Piers

### Abutments

Abutments are the end supports of the bridge superstructure; they are sometimes called *abutment piers*.

The following are the functions of an abutment:

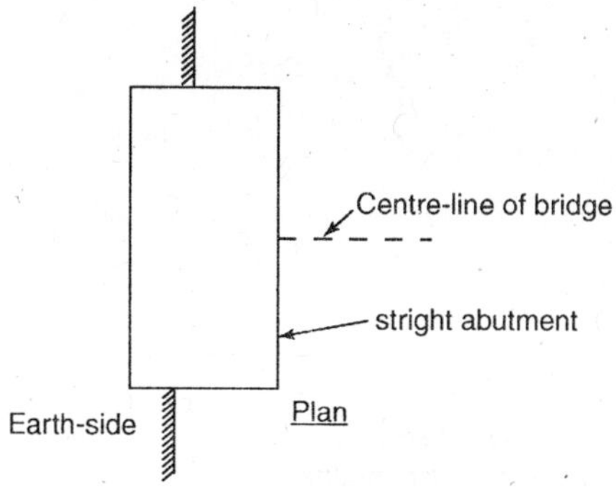


**FIG. 22.18** *Straight abutment (without using walls)*

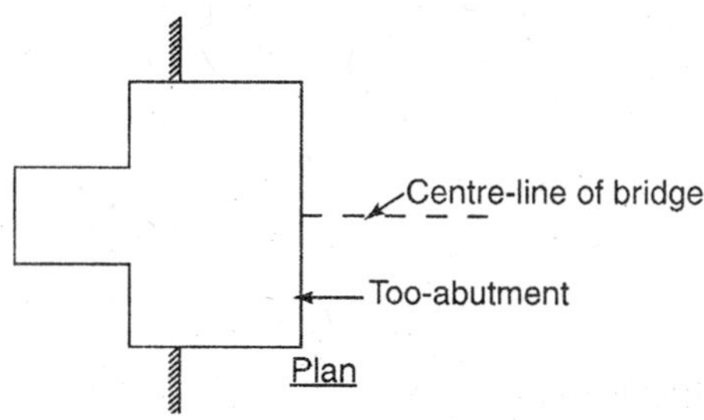


**FIG. 22.19** *T-abutment (without using walls)*

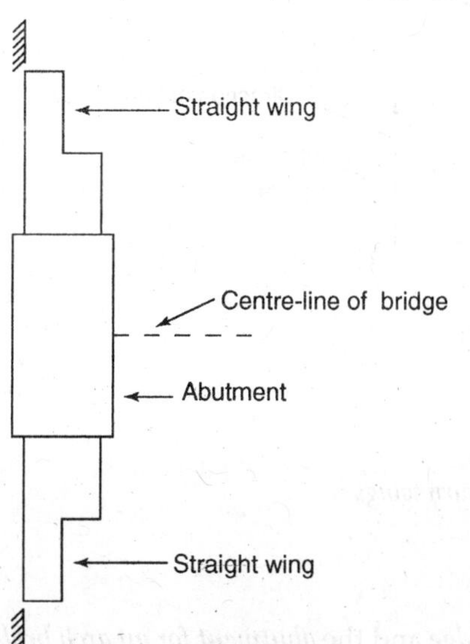


**FIG. 22.20** *Abutment with straight wings*

(i) To transfer loads from the superstructure on to the foundations
(ii) To retain the earth filling at its back to support the roadway or railway at the approaches.
(iii) To provide mandatory end support to the superstructure for effective and safe load transfer.

**Types of abutments:** The criteria for categorisation are the plan layout and the type of superstructure. The layout of an abutment in plan depends on whether it is without wing walls or with wing walls.

***Abutments without wing walls***: When an abutment is constructed without wing walls, it may be a straight or a T-abutment. However, these are not usually adopted becuase of inadequate lateral stability.

***Straight abutment***: A straight abutment without wing walls may be used only when the waterway is negligible. The plan is shown in Fig. 22.18

***T-abutment***: This type of abutment resembles the letter 'T' in plan, and hence the name. The bridge-end is supported by the head of the abutment and the roadway is supported for some distance beyond the embankment by the stem of the abutment (Fig. 22.19).

***Abutments with wing walls***: These may be with straight wings, with splayed wings, or with return wings.

**Abutments with straight wings**: In this type, the wing wall is in line with the abutment. Where there is waterway, this type is not preferred as the flowing water may damage the embankment behind the wing. It may be used for street crossings or railway crossings (Fig. 22.20).

**Abutment with splayed wings**: This type is commonly used for a bridge with significant waterway, because it allows smooth entry and exit of water under the bridge.

The amount of scour below and downstream of the bridge is minimal in this case.

This type of abutment is shown in Fig. 22.21.

**Abutments with return wings**: Also called U-abutment, it is provided with wing walls parallel to the centre-line of the bridge. This type proves to be economical for rivers with steep rocky banks (Fig. 22.22).

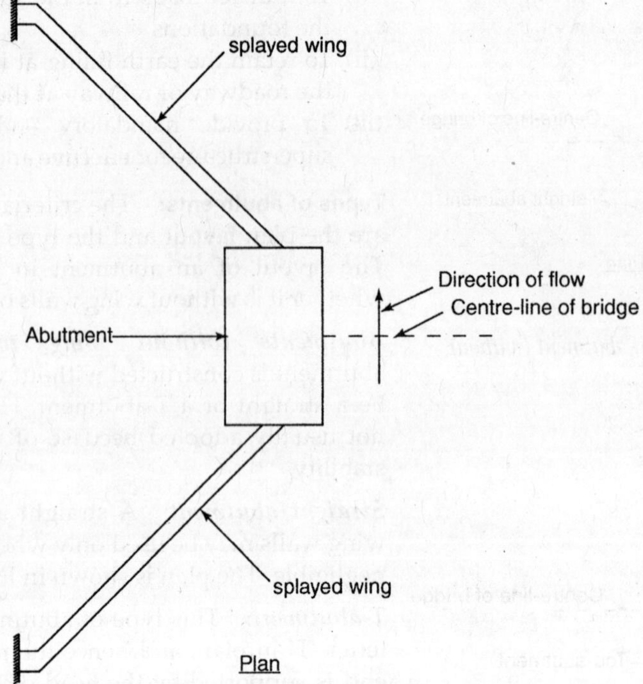


**FIG. 22.21** *Abutment with splayed wings*

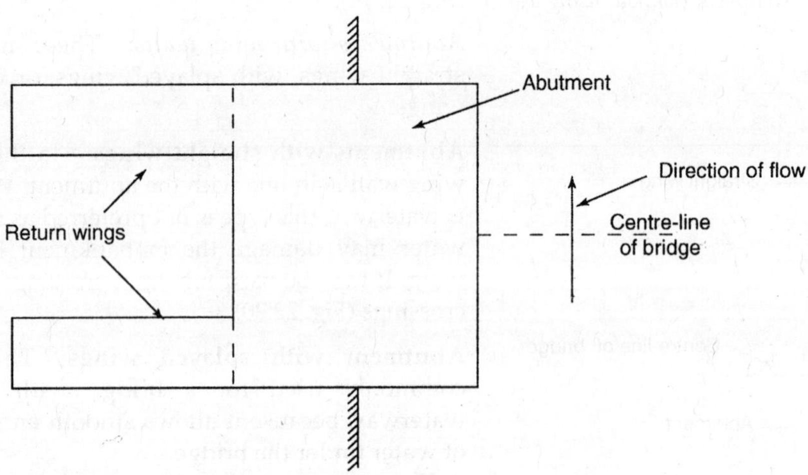


**FIG. 22.22** *Abutment with return wings*

## With Regard to the Superstructure

Figures 22.23 and 22.24 show the *abutment for a girder bridge* and the *abutment for an arch bridge*, respectively.

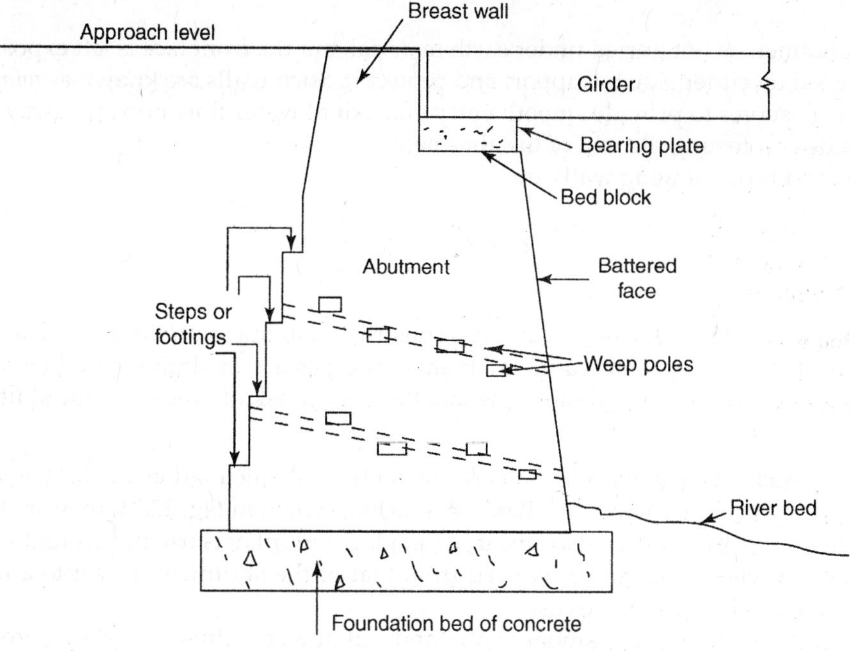


**FIG. 22.23** *Abutment of a girder bridge*

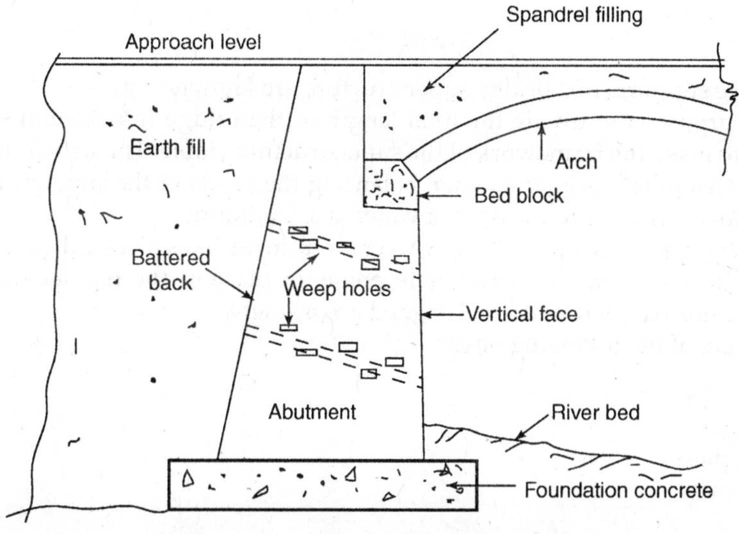


**FIG 22.24** *Abutment for an arch bridge*

[The abutment surface on which a girder rests is horizontal. In the case an arch, this surface is horizontal only for a semi-circular arch, but in segmental arches, it is inclined as shown above]

When the abutment is not buried under earth material, but the front face is left exposed, walls are constructed on either side to support and protect it. Such walls are known as *wing walls*.

A wing wall serves to provide smooth entry and exit of water flow into and away from the bridge, besides protecting the end of the abutment.

The standard types of wing walls are

(i) Straight wing wall
(ii) Splayed wing wall
(iii) Return wing wall

**Straight wing wall:** This type of wing wall is constructed in line with the abutment as shown in Fig. 22.20. This is considered suitable for small bridges across drains with low banks, for street crossings or for railway crossings, where there is no risk of erosion of the abutments or wing walls.

**Splayed wing wall:** When the wing walls are built at an inclination to the length of the abutment in the direction of flow in a flared manner as shown in Fig. 22.21, they are known as splayed wing walls. These may also be curved in plan; the splay is usually around 45°.

The height of wing wall can be reduced from that of the abutment near it to a minimum near the far end. This reduces the cost.

Splayed wing walls facilitate smooth entry and exit of water, thus minimising erosion.

**Return wing wall:** These are constructed perpendicular to the length of the abutment or parallel to the direction of the bridge (or the angle of splay 90°), as shown in Fig. 22.22. They are preferred when the embankments are very high, and they strengthen the approaches.

### Piers

The intermediate supports of a bridge superstructure are known as *piers*

Their only purpose is to divide the total length of the bridge into several spans, usually equal, in order to make the framework of the superstructure viable. Thus, their number should be carefully chosen in an optimal manner, balancing the needs of the superstructure and the need to keep the obstruction to the flow of water at a minimum.

In the case of very long bridges on major rivers, abutment-likes piers, called *'Pier-abutments'* are introduced at one or more intermediate points in place of the regular piers. These are heavier than the normal piers and are designed accordingly.

*Normal piers* are of the following *types*:

(a) Solid piers
(b) Column bents
(c) Cylindrical piers
(d) Dumb-bell piers
(e) Pile bents
(f) Trestle bents

**Solid piers:** This type of pier consists of a solid section of masonry or cement concrete throughout the length of the pier. It is popular because the pier provides good resistance to the impact force of floating debris.

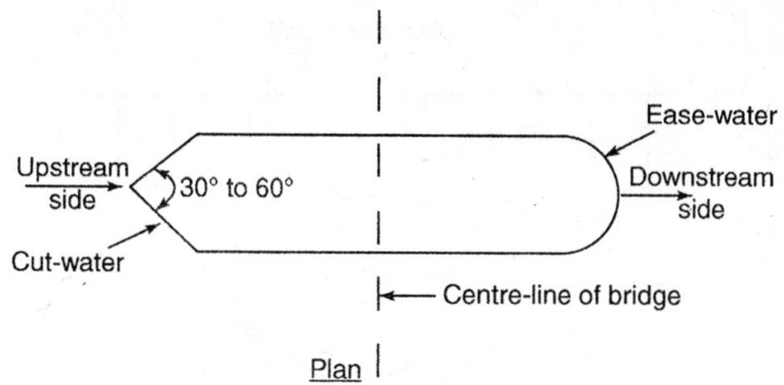


**FIG. 22.25** *Solid pier*

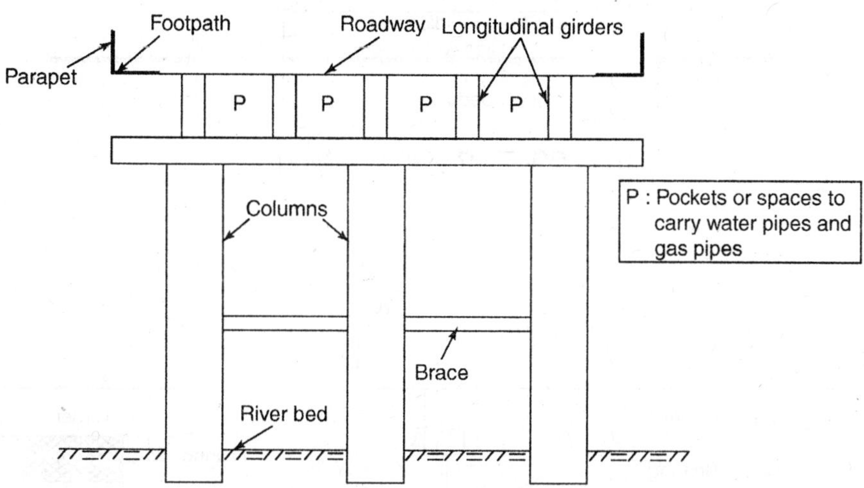


**FIG. 22.26** *Column-bents*

The ends of the pier on the upstream side and the downstream side are called *cutwater* and *ease-water* respectively (Fig. 22.25).

**Column-bents:** This type is preferred for girder bridges in which the girders are closely spaced. 'Bent' is just a supporting frame with vertical members and braces. Longitudinal girders are supported by transverse beams, which, in turn, are supported by columns and braces, as shown in Fig. 22.26.

These are lighter than masonry piers and are used for continuous bridges.

**Cylindrical piers:** These are usually of composite construction – steel cylindrical caissons of hollow section, filled with cement concrete. The steel caissons are first driven to the required depth up to a hard stratum, before filling them with concrete. They are stiffened by steel bracings as shown in Fig. 22.27.

**Dumb-bell piers:** This type of pier appears similar to a dumb-bell with columns supporting the girders; the columns are connected by a thin diaphragm wall (Fig. 22.28).

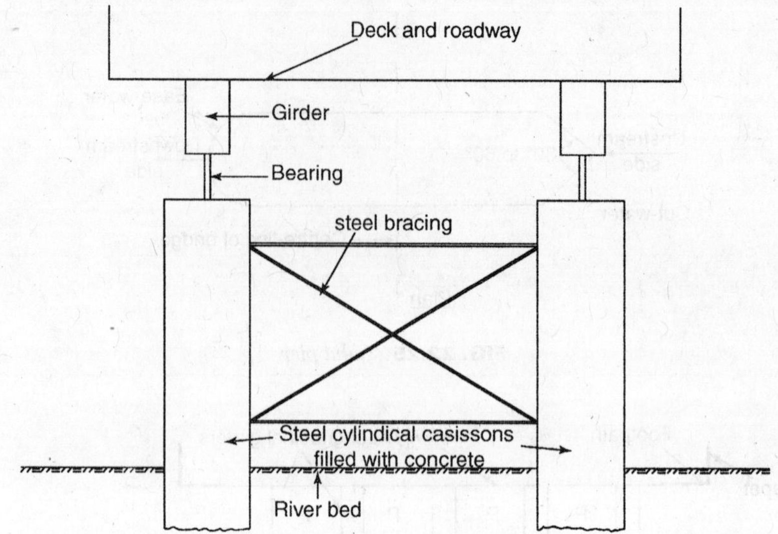


**FIG. 22.27** *Cylindrical pier*

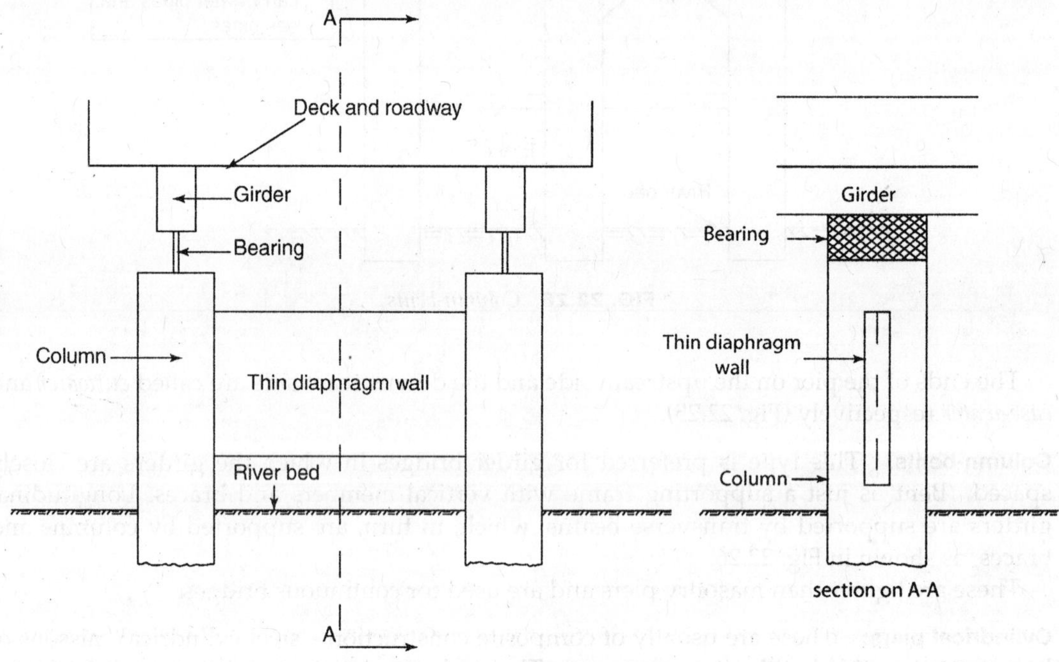


**FIG. 22.28** *Dumb-bell pier*

**Pile bents:** In this type, the superstructure girders are supported on Steel or RCC piles. The piles are connected at the top by means of a pile cap and braced, as shown in Fig. 22.29.

**Trestle bents:** This type consists of a frame of steel; it may be occasionally of concrete. It is useful for viaducts, flyovers, and such structures without waterway (Fig. 22.30).

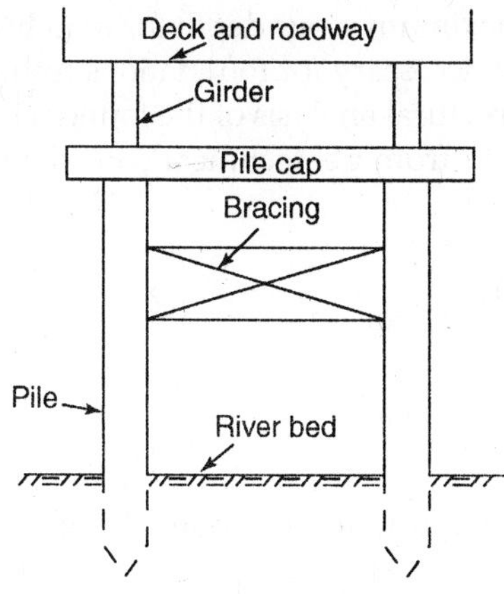


**FIG. 22.29** *Pile bent*

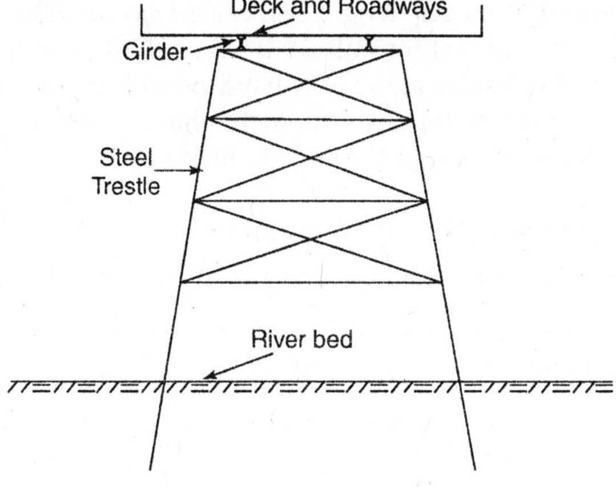


**FIG. 22.30** *Steel trestle*

## Classification of Foundations

The purpose of a foundation is to transfer load from the superstructure to the earth material below in such a manner that stresses in the soil are within the permissible limit with respect to the soil bearing capacity and the resulting deformations are within acceptable limits. For the analysis of a foundation, an adequate knowledge of soil mechanics and foundation engineering (or geotechnical engineering) and hydraulics or fluid mechanics is required; for the design, knowledge of structural analysis and structural design is necessary.

The treatment here will be limited to the basic concepts and types of bridge foundations. For a more detailed treatment, reference may be made to textbooks on foundation engineering listed in References at the end of the book. (For example, *'Foundation Engineering'* by Peek, R.B., Hanson, W.E., and Thornburn, T.H., Asia Publishing House, and *'Geotechnical Engineering'* (Revised Fourth Edition, 2012) by C. Venkatramaiah, New Age International (P) Ltd, New Delhi.)

### Requirements for Stability

The following are the requirements for stability of a bridge foundation, as for any foundation:

(i) The soil pressure at the base does not exceed the safe bearing capacity of the soil.
(ii) The settlement of the foundation is within permissible limits for the type of structure; the differential settlement between different parts of the structure is also within permissible limits.

These criteria prevent failure and ensure the stability of the foundation.
To fulfil these requirements for stability, the following have to be determined:

(i) Maximum scour depth
(ii) Minimum grip length
(iii) Soil pressure at the base
(iv) Total settlement and differential settlement.

The determination of scour depth has already been dealt with in Section 1.3.

**Minimum grip length:** This is the minimum embedment of the foundation required below the maximum scour level for safety and stability of the foundation even under heavy flood conditions.

The minimum grip length is specified as one-third the maximum scour depth for road bridges and one-half of that for railway bridges. (Grip length is not necessary for foundations resting on rock.) Soil pressures at the base have to be obtained from structural analysis of the foundation.

Total and differential settlements have to be determined from geotechnical considerations of the foundation and the soil taken together.

**Broad Classification of Foundations**

Broadly speaking, foundations may be classified as

(a) Shallow foundations (Examples are *footings* and *rafts*).
(b) Deep foundations (Examples are *piles* and *caissons*).

Terzaghi, the father of soil mechanics, defines a shallow foundation as one whose depth is smaller than its width.

For the present purpose, a shallow foundation is one which can be prepared by open excavation, whereas a deep foundation cannot be prepared by open excavation.

**Shallow Foundations**

The important types of shallow foundations for bridges are

(i) Spread footings
(ii) Raft foundations

**Spread footings:** This type is practicable for a depth of about 5 m, especially above the water table. The base of the pier or abutment is widened or enlarged to spread the loading on to a larger area. Generally, a concrete bed is provided at the bottom the footing.

The footing may be constructed either in coursed rubble masonry or concrete. If the bottom layer is of concrete, its thickness can be reduced by introducing reinforcements.

A typical spread footing for a pier is shown in Fig. 22.31:

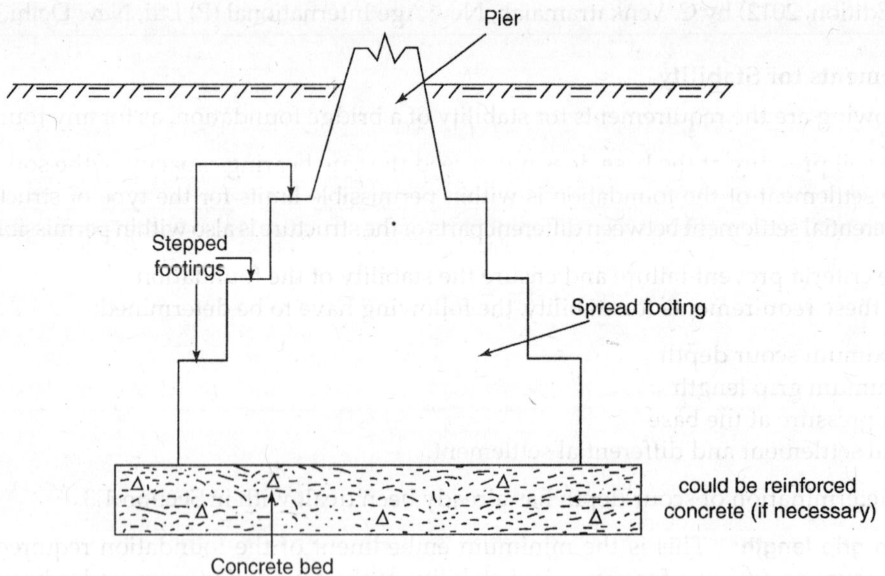


**FIG. 22.31** *Typical spread footing for a pier*

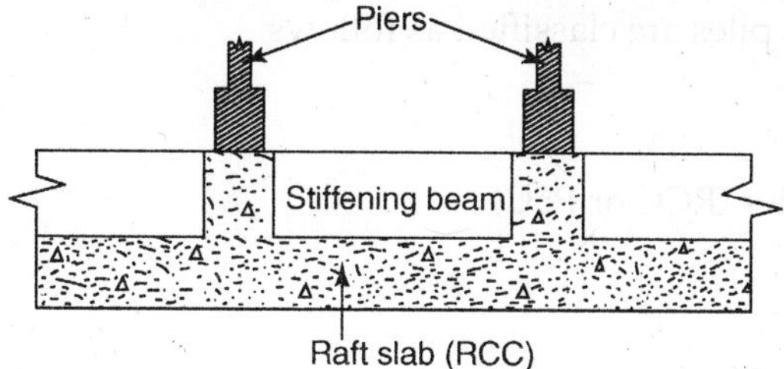


**FIG. 22.32** *Typical raft foundation for bridge piers*

**Raft foundations:** A *raft* foundation is a combined footing which covers the entire area underneath the piers and abutments. When the allowable bearing pressure of the soil is low and the loads on the bridge are heavy, individual spread footings for the piers are likely to cover more than half the area under the bridge. In such a case, a raft foundation (or a *mat*, as it is called) may prove economical.

A raft foundation can also control the differential settlements due to the presence of weak lenses at a few locations, by bridging over the weak spots in view of its flexural strength. Hence even if the total settlements on the raft are relatively large, no damage is caused.

The raft foundation consists of a reinforced concrete slab stiffened by beams parallel to the direction of flow, laid monolithically. A typical foundation of this type is shown in Fig. 22.32.

## Deep Foundations

*Pile foundations* and *caisson foundations* are the common types of deep foundations used for supporting bridge piers and abutments.

**Pile foundations:** These are considered appropriate under the following conditions:

(i) When the foundation soil is soft and under deep standing water
(ii) When a hard stratum is available at a reasonable depth under the soft strata
(iii) When pile foundations are more economical than wells/caissons.

With regard to the nature of load-resistance, piles are primarily classified as

- Bearing piles
- Friction piles

***Bearing piles***: These piles penetrate through the soft soil with their tips resting on a hard stratum. They act like columns and resist load through the resistance generated at the tips or points of the piles; thus, they are also called *point-bearing* pile.

***Friction piles***: A friction pile develops skin friction at its surface of contact through the surrounding soil and thereby resists the load. In practice, depending upon the soil conditions a pile may develop resistance partly through point bearing and partly through skin friction.

Based on their *function*, there can be a few other types of piles also.

***Compaction piles***: The function is to compact loose granular soil in order to increase the bearing capacity

***Uplift piles***: These piles develop uplift resistance and anchor the structure above.

***Batter piles***:They are used to resist large horizontal or inclined loads.

***Sheet piles***: These piles are used for bulkheads.

Based on the *material*, piles are classified as follows:

(i) Timber piles
(ii) Steel piles
(iii) Cement concrete piles (RCC or PSC)
(iv) Sand piles
(v) Composite piles

*Timber piles*: They are made from logs of wood; the section may be square or circular. They have low bearing capacity and need protection against decay.

*Steel piles*: Rolled steel sections are used as H-piles, tube piles, and box piles. They have high bearing capacity and high lateral resistance. Their length can be increased conveniently by welding.

*Cement concrete piles*: They have high resistance to compressive loads.

They may be *precast* to the desired sections and lengths in a factory, transported to the site, and driven to the required depth. Invariably these are made of reinforced cement concrete or pre-stressed concrete.

Alternatively, concrete piles may be *cast-in-situ* by boring through the soil to the desired depth and pouring concrete into the bore hole.

A variety of patented cast in situ piles have been developed – Raymond, Simplex, Franki, and Mac-Aruthur piles are common examples.

Handling stresses and driving stresses are absent, and no reinforcements are needed for cast-in-situ cement concrete piles.

*Under-reamed piles* also fall in the category of cast-in-situ, which are primarily used for stabilising expansive soils such as black cotton soils. One or more bulbs are created along the depth at suitable intervals with the aid of special boring tools.

Every additional bulb is expected to increase the bearing capacity of the pile.

Under-reamed piles have been developed and patented by the Central Building Research Institute, Roorkee.

*Sand piles*: These are also formed similar to the cast-in-situ piles; but here, the holes are filled with sand and capped with concrete. They are also used to stabilise soft soils.

*Composite piles*: These may be driven piles made of two different materials, or a combination of a bored pile and a driven pile. The upper portion of the pile is kept above the water table. Composite piles may prove to be economical under certain conditions.

**Caissons and well foundations:** A 'caisson' is a box-like structure, which is installed in place by appropriate means, and filled with concrete or other materials for making it stable in its final position. Caissons are preferred in sandy soils.

The following are the types of caissons:

(a) Box caissons
(b) Open caissons or wells
(c) Pneumatic caissons

*Box caissons*: This is a water-tight box, open at the top and closed at the bottom. They are generally made of steel or RCC. This type is suitable when the bearing stratum is available at small depths, and the loads are not high. It is floated to the site and sunk to the required position

*Open caissons or wells*: These are open both at the top and at the bottom; they are preferred in sandy or soft soils and where a firm stratum is available at large depths. Such wells may be built of timber, steel, masonry or RCC. They are widely adopted for bridge foundation in India.

The plan shape of a well may be circular, rectangular, twin- hexagonal, dumb-bell, or double-D (with D-shaped ends). The important parts of well are the cutting edge, curb, steining (body), top plug, bottom plug, and well-cap.

***Pneumatic caissons***: These are open at the bottom and closed at the top. These are used for large depths of water, where wells are not convenient.

Compressed air is used to exclude water from the working chamber, and the work of excavation and sinking is carried out when the chamber is dry.

A pneumatic caisson may be made of steel or concrete. Vertical shafts are used for the passage of labour gangs and for materials separately. The depth to which it can be sunk is limited by the pressure which a human being can withstand while working and while excluding water from the working chamber.

### Cofferdams

A *cofferdam* is a temporary structure built to remove water from an area in the river bed, in order to carry out the construction work of the foundation and substructure, in a reasonably dry condition.

Once the construction of the bridge piers, locks, or dams is completed, the cofferdam may be dislodged and salvaged for materials which can be reused.

Pumping out of water from the area enclosed by a cofferdam is part of the job.

**Types of cofferdams:** The following are the common types:

(i) Earth-fill cofferdam
(ii) Rock-fill cofferdam
(iii) Rock-fill crib cofferdam
(iv) Sheet-pile cofferdam

*Sheet-pile cofferdam* is further classified as

- Single-wall type
- Double-wall type
- Cellular type

The *factors governing the selection of the type of cofferdam are:*

- Area from which water is to be excluded
- Depth of water
- Velocity of flow
- Nature of the bed material
- Scour depth
- Availability of the construction materials in the proximity of the site.

## Choice of Materials for Bridge Construction

### For Superstructure

(i) *Stone masonry* arches – For road bridges of moderate spans
*Brick masonry* arches – For culverts
(ii) *Timber structures* – For temporary bridges near forest areas and military bridges of small spans
(iii) *Reinforced cement concrete* – Slabs for small spans up to 8 m

Girders and T-beams – For spans in the range of 10 to 20 m
Hollow grids – For spans from 25 to 30 m
Balanced cantilevers – For spans of 30 to 60 m
Continuous girders – For spans up to 40 m
RCC arches – For spans up to 200 m
RC bow- string girders – For spans 30 to 45 m
RCC rigid frames – For spans up to 20 m
Pre-stressed concrete girders – For major bridges with spans of 30 to 120 m

(iv) *Steel bridges*
Plate girders – For spans up to 50 m
Truss bridges – For long spans in the range of 300 m
Steel arches – For spans from 30 m to 150 m
Steel suspension bridge – For spans from 600 m to 1000 m

**For Substructures**

(i) *Masonry*: This can be of brick or stone.
Brick masonry – For small spans, especially for culverts and minor bridges.
Stone masonry – For large spans, especially for major bridges over large streams and rivers.

(ii) *Steel*: This can be used for piers formed by trestle bents.

(iii) *Cement concrete* – Usually M20 grade (1:1½:3) is adopted for piers and abutments.
Reinforced cement concrete (RCC) – Used when significant tensile stress is expected in addition to compressive stresses
Pre-stressed concrete (PSC) – Used for piers in the case of viaducts with tall piers

**For Foundations**

(i) *Masonry* – Coursed rubble masonry is used for spread footings in the building of piers. Sometimes, it is used for well foundations also.

(ii) *Timber* – For sheet-piling for temporary structures like cofferdams, and bracings for excavations.

(iii) *Steel* – For grillage foundations consisting of steel girders and joists placed in both directions in the form of an open structure. This may be used to support piers of steel columns transmitting large loads from the superstructure. It is also used for H-piles used in pile foundations, cofferdams, caissons and wells.

(iv) *Cement concrete* – Used in spread foundations, either for the footing, as also for the bottom bed, cast-in-situ piles, under-reamed piles
Reinforced cement concrete (RCC) – Used for the bottom layer of spread footings (if the loads are high), for raft foundations, for pile foundations with precast concrete piles, for box caissons, for well foundations, for sheet piles and for cofferdams
Pre-stressed concrete (PSC) – for precast concrete piles used in pile foundations.

In general, the following points govern the choice of material:

- Suitability for construction
- Adequate capacity to bear the expected loads.
- Minimum economic cost – initial cost plus maintenance cost during service life.

- Proximity of the place of availability of the materials.
- Suitability from the point of view of aesthetics of the structure.

## 22.5 NUMERICAL EXAMPLES

### EXAMPLE 22.1

A flat catchment of 150 sq.km area is at an average distance 15 km from the coast. Determine the maximum flood discharge using Ryve's formula.

**Solution**

Ryve's formula: $Q = C \cdot A^{2/3}$

$C$ here is 6.8 and $A = 150$ sq.km

$$Q = 6.8\,(150)^{2/3} = \underline{192\ \text{m}^3/\text{s}}$$

### EXAMPLE 22.2

A catchment area consists of lightly cultivated loam and is 150 sq.km in extent. The distance of the bridge site from the critical point is 18 km and the difference in elevation between these two points is 90 m. The critical intensity of rainfall in the catchment for the chosen conditions is 63 mm/hour. Determine the peak discharge using the Rational Formula recommended by the IRC.

**Solution**

Critical intensity of rainfall, $i = 63$ mm/h

Time of concentration:

$$t_c \text{ (in hours)} = \left(0.87.\frac{L^3}{H}\right)^{0.385}$$

$L = 18$ km $H = 90$ m

$$t_c = \left(\frac{0.87 \times 18^3}{90}\right)^{0.385} = 4.72 \text{ hrs.}$$

Coefficient, $C$, in the Rational Formula:

$$C = \frac{0.56.P.f}{(t_c + 1)}$$

$P = 0.4$ in this case and $f = 0.675$

$$C = \frac{0.56 \times 0.4 \times 0.675}{(4.72 + 1)} = 0.0264$$

Peak discharge:

$$Q = C.\,i.\,A = 0.0314 \times 63 \times 150 \approx \underline{250\ \text{m}^3/\text{s}}$$

## EXAMPLE 22.3

The maximum flood discharge of a regime waterway is 330 cu.m/s.
Using Lacey's formula, determine the linear waterway.

### Solution

*Lacey's formula* for regime waterway:

$$L = C\sqrt{Q}$$

C may be taken as 4.8

Linear waterway, $L = 4.8\sqrt{330} = \underline{87.2\ \text{m}}$

## EXAMPLE 22.4

The area of waterway of a stream is 150 sq.m. Wetted perimeter is 100 m. The slope of the water surface is 1 in 450.

Manning's coefficient of rugosity for the stream condition is 0.028. Determine the discharge by the Area–Velocity method.

### Solution

Area of waterway, $A = 150\ \text{m}^2$

Wetted perimeter, $P = 100\ \text{m}$

Hydraulic mean radius, $R = \dfrac{A}{P} = \dfrac{150}{100} = 1.5$

Slope, $S = \dfrac{1}{450}$

Manning's $n = 0.028$

Velocity of flow, $V = \dfrac{1}{n}.\,R^{2/3}.\,S^{1/2}$

$$= \frac{1}{0.028} \times (1.5)^{2/3} \times \frac{1}{(450)^{1/2}} = 2.206\ \text{m/s}$$

Discharge, $Q = A.\,V = 150 \times 2.206 = \underline{330\ \text{m/s}}$

## EXAMPLE 22.5

Determine the increase in velocity due to an afflux of 200 mm.

### Solution

$$V_a = \sqrt{2gh_a} = \sqrt{2 \times 9.81 \times 0.200} = \underline{2\ \text{m/s}}$$

## EXAMPLE 22.6

The velocity of normal flow in a stream is 2.7 m/s. The width of the stream at HFL is 100 m. The linear waterway under the bridge is 90 m. Assuming the entry to be bell-mouthed, determine the afflux.

**Solution**

Afflux, $h_a$, is given by

$$h_a = \frac{V^2}{2g}\left(\frac{L^2}{cL_1^2} - 1\right)$$

Here,

$$V = 2.7 \text{ m/s}$$

$$L = 100 \text{ m}$$

$$L_1 = 90 \text{ m}$$

$$C = 0.9 \text{ for bell-mouthed entry}$$

Afflux,

$$h_a = \frac{2.7^2}{2 \times 9.81}\left[\frac{100^2}{0.9 \times 90^2} - 1\right]$$

$$= \underline{0.14 \text{ m } (< 300 \text{ mm} \therefore \text{O.K})}$$

## EXAMPLE 22.7

Determine the scour depth for a natural stream in an alluvial bed if the peak discharge is 330 m³/s and Lacey's silt factor for the coarse sand bed material is 1.5. There is a moderate bend at this site.

**Solution**

Lacey's formula for scour depth, $d_s$:

$$d_s = 0.473\left(\frac{Q}{f}\right)^{1/3}$$

$$Q = 330 \text{ m}^3/\text{s and } f = 1.5$$

$$d_s = 0.473\left(\frac{330}{1.5}\right)^{1/3} = 3.27 \text{ m}$$

For a moderate bend,

$$d_s = 1.5 \times 3.27 = \underline{5 \text{ m}}$$

## EXAMPLE 22.8

Normal scour depth when the regime waterway and the waterway under the bridge are equal is calculated as 6 m. The regime waterway in this case is 100 m. If the waterway provided under the bridge is 90 m, determine the modified scour depth. What is the maximum scour depth that can be expected at the nose of the pier?

**Solution**

The normal scour depth, $d_s'$, under the bridge is given by

$$d_s' = d_s\left(\frac{L_r}{L}\right)^{2/3} \quad \text{(when } L < L_r\text{)}$$

Here, $L = 90$ m and $L_r = 100$ m

$d_s$ = normal scour depth (when $L = L_r$)

$= 6$ m

$$d_s' = 6\left(\frac{100}{90}\right)^{2/3}$$

$= 6.44$ m

Maximum scour depth at the nose of the pier $= 2d_s'$

$= 2 \times 6.44 \approx$ 12.9 m

## SUMMARY

- A bridge is a structure built to provide passage over an obstacle such as a river, valley, etc. The main components of a bridge are the superstructure, the substructure and the foundation. Approaches and river training works are ancillary structures to these.
  A bridge is an element of a transportation route such as a roadway or a railway.
- There are several criteria for classification of bridges. They may be highway or railway bridges, steel, RCC or masonry bridges, or basically they may be permanent or temporary. Bridges have been built since prehistoric times.
- Site selection and surveys, including subsoil exploration are important in building a bridge efficiently and economically.
- Hydraulic design aspects include design discharge, linear waterway and afflux, scour depth, foundation depth, and so on.
  Empirical formulae, as well as the Rational Method of IRC are available for the determination of design discharge.
- Different criteria are used to classify superstructures. For example, the classification may be based on the material used – steel, RCC, etc., or the structural action – simple girder bridges, arch bridges, suspension bridges and so on.
  Bearings are to be provided for load transfer at the ends of each of the spans, so that the loads are ultimately transmitted to the foundations through the piers and abutments.
- Abutments at the ends of the bridge, piers for intermediate support, and wing walls to provide additional support to the abutments constitute the substructure. Each of these may be classified depending upon their layout in plan.
- Foundations of piers and abutments are classified as shallow and deep types. Spread footings and rafts are shallow or open type foundations, while pile foundations, caissons and well foundations fall under the category of deep foundations.
- Cofferdams are temporary structures built to exclude water from an enclosed area of the river bed, in order to allow construction in the dry.

They may be built by means of earth or rock-fill, and sheet-piling with single wall, double wall, or cellular type of construction.

Once the construction of the substructure is completed, the cofferdam may be dismantled and the material salvaged for reuse, to the extent possible.

## QUESTIONS AND PROBLEMS

1. Define a 'bridge' and explain briefly its components.
2. Explain the importance of bridges in transportation.
3. Explain the classification of bridges based on their function and the choice of construction material.
4. Explain the factors which govern site selection for a bridge.
5. Enumerate the preliminary information needed as part of the investigation for a bridge project.
6. Write a detailed note on the surveys needed for the investigation of a bridge project including soil exploration.
7. What are the factors to be considered in the alignment of a bridge?
8. List the methods available for the determination of design discharge of a stream, and explain any one method.
9. Explain how the linear waterway of a bridge is determined from the design discharge.
10. Write short notes on:

    (a) Afflux

    (b) Economic span

    (c) Scour depth
11. Write short notes on:

    (a) Depth of foundation

    (b) Freeboard

    (c) Clearance
12. Write a note on classification of bridge superstructure based on the structural action in bearing loads.
13. Write a note on the use of masonry arch bridges.
14. Write short notes on:

    (a) Plate girder bridge

    (b) Balanced cantilever bridge

    (c) Bow-string girder bridge
15. Write short notes on:

    (a) Floating bridge

(b) Swing bridge

(c) Bascule bridge

16. Write short notes on:

    (a) Lift bridge

    (b) Transporter bridge

    (c) Minimum grip length

17. Write a note on bridge joints.

18. Write short notes on:

    (a) Rocker bearing

    (b) Rocker and roller bearing

    (c) Neoprene bearing

19. Write a note on the types of abutments used for bridges.

20. Write short notes on

    (a) Splayed wing wall

    (b) Dumb-bell piers

    (c) Pier-abutments

21. Write short notes on the following, as applicable to bridges:

    (a) Spread foundation

    (b) Raft foundation

22. Write a note on pile foundations as applicable to bridges.

23. Write a note on well foundations as applicable to bridges.

24. Write short notes on

    (a) Pre-cast concrete piles

    (b) Cast-in-situ concrete piles

25. Write a note on the different types of cofferdams used in the construction of bridge piers.

26. Write a note on the choice of materials for bridge superstructures.

27. Write a note on the choice of materials for bridge substructures.

28. Write a note on the appropriate materials for temporary bridges.

29. A bridge has a linear waterway of 150 m constructed across a stream whose natural linear waterway is 220 m. If the average flood discharge is 1200 $m^3/s$ and the average flood depth is 3 m. Calculate the afflux under the bridge. Use Molesworth's formula. 0.20 m (Pune University)

30. The normal velocity of flow in a river is 1.50 m/s. The normal and the artificial waterway under the bridge respectively are 8000 $m^2$ and 7000 $m^2$. Determine the afflux using Molesworth's formula. Also find the increase in velocity due to afflux. Assume $g$ = 9.81 $m/sec^2$ (Pune University)

31. The normal velocity of flow in a river is 2 m/s. The normal waterway, artificial waterway, and the enlarged area upstream of the bridge, respectively, are 8000 $m^2$, 7000 $m^2$ and 9000 $m^2$. Determine the afflux using Merriman's formula. Also find the increase in velocity due to afflux. Assume $g = 9.81$ m/$s^2$. Coefficient of discharge, $c = 0.98$. (Pune University)

33. A bridge is proposed to be constructed across an alluvial stream carrying a discharge of 200 $m^3$/s. Assume Lacey's silt factor as 1.0, and find the maximum scour depth. (Pune University)

33. The approximate costs of one pier and one superstructure span for a multiple-span bridge for various lengths of span are given below. Determine the economic span:

| Span (m) | Cost of one pier (Rs) | Cost of one span of superstructure (Rs) |
|---|---|---|
| 10 | 25000 | 7000 |
| 15 | 28000 | 13815 |
| 20 | 32500 | 31000 |
| 25 | 33700 | 36000 |
| 30 | 34800 | 41400 |

(Pune university)

34. A bridge is constructed across a stream having a linear waterway of 400 $m^2$ and an unobstructed waterway of 500 $m^2$. The flood creates an afflux of 0.3 m. Calculate the flood discharge. (Pune University)

# Chapter 23

# DESIGN ASPECTS OF BRIDGES

**CHAPTER OBJECTIVES**

This chapter describes the following features in the design of bridges:

- Primary and secondary loads to be considered for the design of highway bridges as well as railway bridges
- Live loads on highway bridges as per IRC, Class-AA, Class-70R, Class-A and Class-B loadings; live loads on railway bridges as per 25t loading and DFC loading for BG
- Earth pressures, wind effects and earthquake effects
- Design aspects of bridge superstructure
- Design aspects of bridge substructure – piers, abutments and wing walls
- Design aspects of bridge foundations – shallow and deep – footings, piles and wells
- IRC codes for road bridges and IRC Bridge Rules for railway bridges.

## 23.1 INTRODUCTION

After selecting an appropriate site for a proposed bridge and collecting all the necessary preliminary data through investigation and surveys, the type of bridge and materials of construction are chosen.

The next step is the design of the bridge and its components. The standard procedures for design will depend primarily on whether it is a highway bridge or a railway bridge. Roads and railways are built and controlled by the Central Government in our country; hence, bridges along these paths are to be constructed in accordance with the corresponding Indian Standard Specifications and Codes of Practice, prescribed by the Indian Roads Congress (IRC) and the Indian Railways respectively.

The first set of specifications is collectively known as the IRC Bridge code, and the second set is known as IRS Bridge Rules and IRS Bridge Codes. Separate codes have been evolved for bridges of different materials.

The provisions evolved in these Codes undergo changes from time to time based on the technological developments and practical experience gained from the already built bridges.

## 23.2 DESIGN ASPECTS OF SUPERSTRUCTURES

*The data to be collected for the design* of highway bridges and railway bridges are marginally different because of differences in construction requirements. However, the primary criteria for stability remain the same.

### Highway Bridges – Indian Standards (IRC Standards)

**Design Data for Highway Bridges**

The IRC has developed standard specifications and Codes of Practice for Highway Bridges, called the Bridge Code, which consists of eight sections (See References).

Section I deals with general features of design while Section II discusses loads and stresses. Sections III to VI pertain to bridges of cement concrete, masonry, steel, and composite construction.

Sections VII is concerned with Foundations and Substructure, and Section IX with Bearings.

Section I gives specifications for the preliminary data to be collected, determination of design discharges, clearances, foundations, and so on.

Section II specifies the loadings for which the bridges are to be designed. The other sections set out guidelines in the superstructure design and foundations.

Width of carriageway and clearances for single-lane and multi-lane bridges have been dealt with already in Vol. I Highway Engineering.

### Design Loads

The following loads should be considered in the design of bridges, where applicable: [IRC: 6-2010 (Fifth Revision) – Standard Specifications and Code of Practice for Road Bridges: Section-II – Loads and Stresses.]

(i) Dead load
(ii) Live load
(iii) Snow load
(iv) Impact loads
(v) Wind load
(vi) Longitudinal forces
(vii) Centrifugal forces
(viii) Horizontal forces due to water flow
(ix) Buoyancy effects
(x) Earth pressure, including surcharge
(xi) Temperature effects
(xii) Deformation effects
(xiii) Secondary effects
(xiv) Erection stresses and construction loads.
(xv) Forces due to earthquakes
(xvi) Forces due to collision of vehicles
(xvii) Forces due ship/barge impact
(xviii) Wave pressure

The following are the *criteria* governing the design of bridges:

- Ability to sustain the worst combination of all the design loads with a defined probability during its service life.
- Adequate stability and durability during the service life
- Earthquake-resistant design and construction, depending upon the location of the bridge-site in the standard seismic zones of the country.

## Dead Load

Dead load carried by a member or superstructure consists of fixed loads supported by the member in addition to its own weight.

Dead load can be estimated fairly accurately during design. Typical values of unit weights of a few important materials of construction are given below:

| Material | Unit weight |
|---|---|
| Granite | 27 kN/m³ |
| Stone ballast | 15 kN/m³ |
| Brick work in C.M | 20 kN /m³ |
| Reinforced cement concrete | 25 kN /m³ |
| Compacted earth | 20 kN /m³ |
| Macadam | 22 to 26 kN /m³ |
| Water | 10 kN /m³ |
| Wood | 8 kN /m³ |
| Structural steel | 78 kN /m³ |

Various empirical formulas have been used for the estimation of dead loads.

## Live Load

Live loads are those caused by vehicles which pass over the bridge, and are therefore transient in nature. These loads cannot be estimated precisely, and no control can be exercised once the bridge is opened to traffic.

In view of this difficulty, certain hypothetical loadings which are reasonably realistic have been evolved by the Indian Roads Congress for use in the design of Highway Bridges.

The following four types of IRC standard loadings have been evolved:

1. Class-AA loading
2. Class-70R lading
3. Class-A loading
4. Class-B loading

### Class-AA Loading

This consists of either a tracked vehicle of 700 kN or a wheeled vehicle of 400 kN. The tracked vehicle simulates a combat tank of the Army. The contact length of the track on the road is 3.6 m and the overall nose-to-tail length is 7.2 m. The minimum spacing between two successive vehicles is specified as 90 m. For two-lane bridges, one train of class-AA tracked or wheeled vehicles, whichever produces a more severe effect, shall be considered for every two-lane width. When the train of vehicles is on the bridge, no other live load should be permitted on these two lanes of the bridge.

Class-AA loading is to be adopted for bridges on National Highways and other specified highways and localities.

Bridges designed for Class-AA loading should be checked for Class-A loading, since under certain conditions, stresses more severe than that for Class AA may be produced by the latter.

Class-AA loading is, in general, considered to be heaviest for the purpose of highway bridge design.

Class-AA loading is shown is Fig. 23.1:

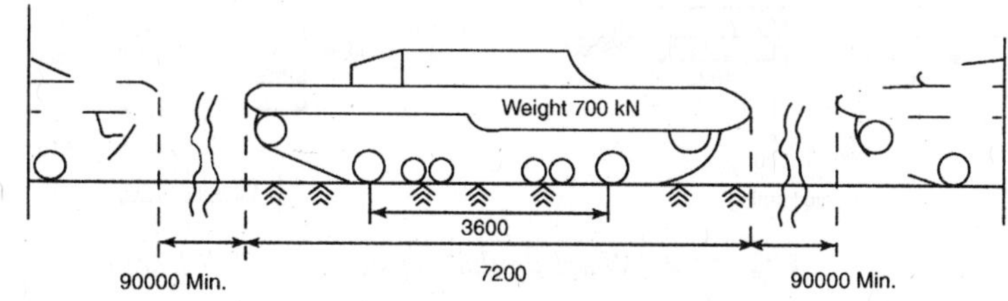


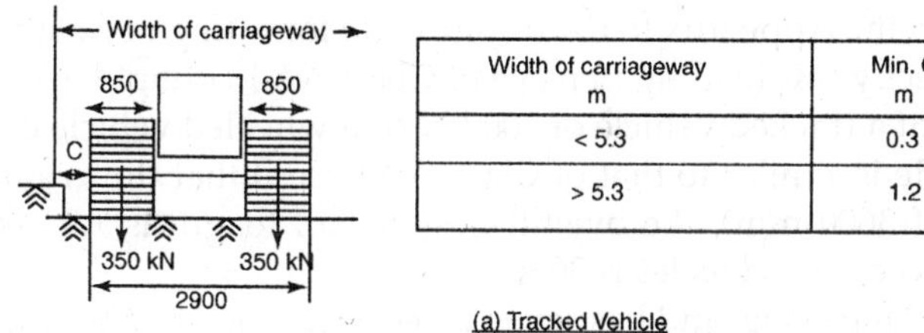


| Width of carriageway m | Min. C m |
|---|---|
| < 5.3 | 0.3 |
| > 5.3 | 1.2 |

(a) Tracked Vehicle

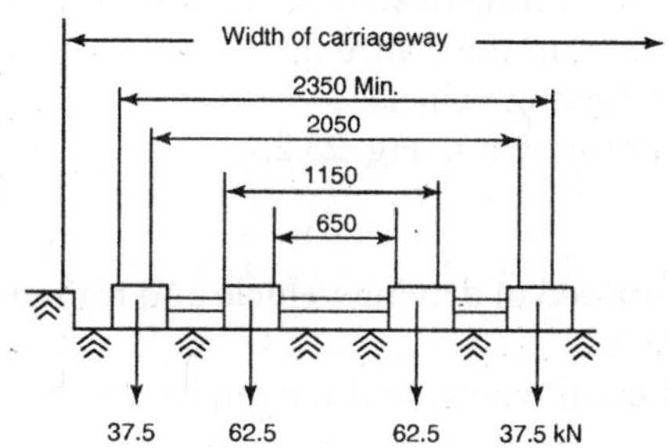


Note : All dimensions are in millimetres.

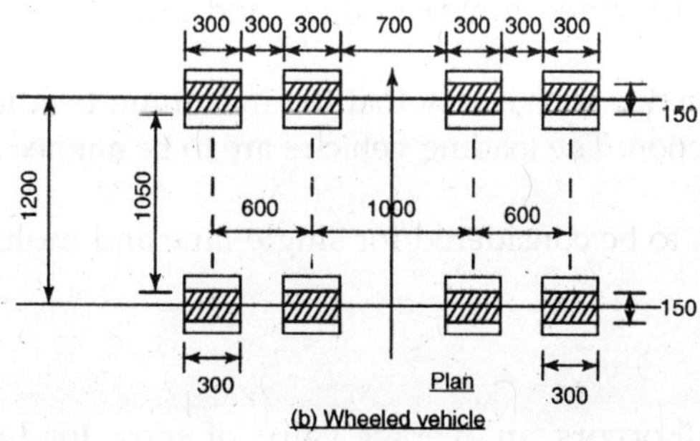


Plan

(b) Wheeled vehicle

**FIG. 23.1** *IRC Class-AA loading*

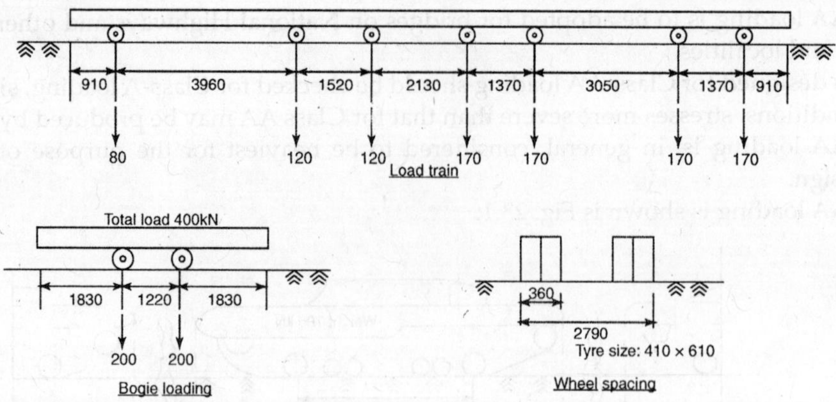


**FIG. 23.2** *Wheeled vehicle of IRC Class-70R loading*

### Class-70R Loading

This was originally included in the Appendix to the bridge code for rating of existing bridges. But the tendency now is to specify this loading in place of Class-AA loading.

Class-70R loading consists of a tracked vehicle of 700 kN or a wheeled vehicle of total load of 1000 kN. The tracked vehicle is similar to that of Class-AA except that the ground contact length is 4570 mm (instead of 3600 mm), the overall nose-to tail length is 7.92 m, and the minimum spacing between successive vehicles is 30 m.

The wheeled vehicle is 15220 mm long, and has seven axles with loads totalling 1000 KN. In addition to this, the effects due to a bogie loading of 400 kN are also to be checked. The spacing between wheeled vehicles is measured from the centre of the rearmost wheel of the leading vehicle to that of the first axle of the following vehicle.

The details of the wheeled vehicle are shown in Fig. 23.2.

### Class-A Loading

This consists of a wheel-load train composed of driving vehicle and trailers with specified axle spacing and loads, as shown in Fig. 23.3.

The ground contact area for the different wheels and the minimum specified clearances are shown in the figure.

Class-A loading is adopted on all roads on which permanent bridges are constructed.

### Class-B Loading

The wheel-load train is similar to that of Class A loading, but with smaller axle loads in Fig. 23.3. This loading is meant for temporary structures, timber bridges, and bridges in areas with low traffic.

The standard loads are to be arranged is such a way that the maximum bending moment or shear is produced at any particular section. The loading vehicles are to be aligned parallel to the bridge length.

The combinations of the live loads to be considered for single-lane and multi-lane bridges are specified in the IRC Bridge Code.

## Snow Load

In the zones in which heavy snowfall occurs, an average value of snow load or pressure is taken as recommended in the Code. It may be based on actual observations or past records

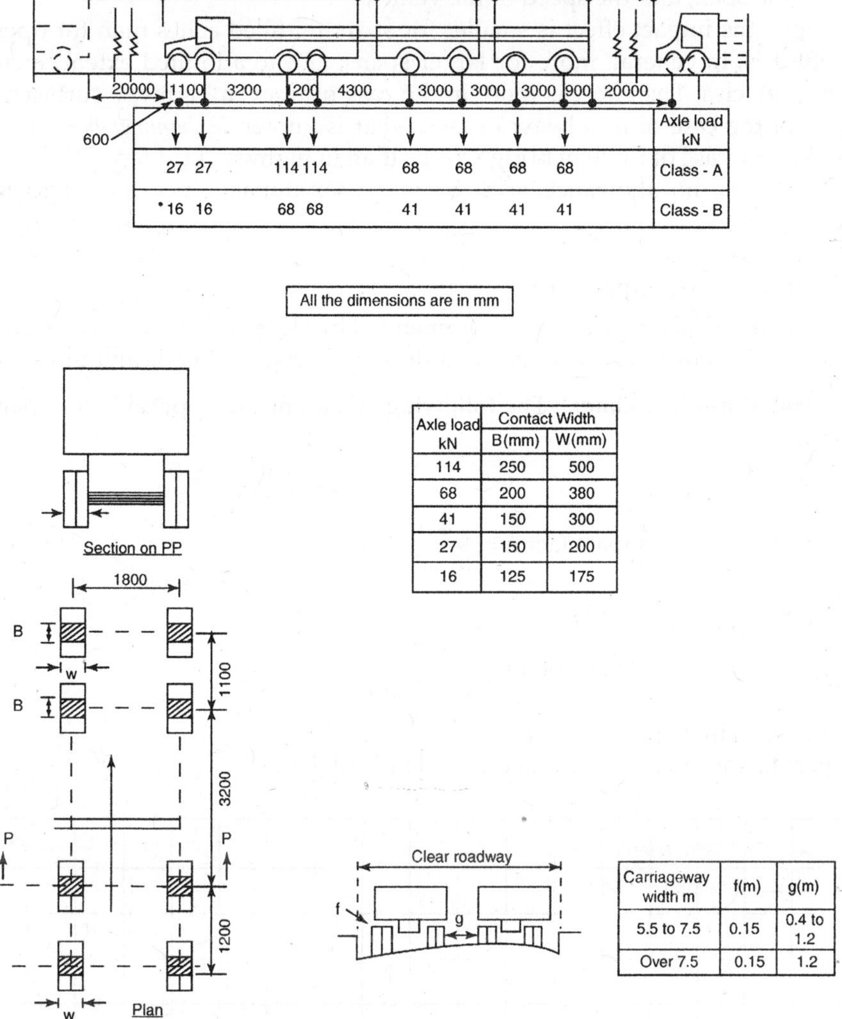


| Axle load kN | Contact Width | |
|---|---|---|
| | B(mm) | W(mm) |
| 114 | 250 | 500 |
| 68 | 200 | 380 |
| 41 | 150 | 300 |
| 27 | 150 | 200 |
| 16 | 125 | 175 |

| Carriageway width m | f(m) | g(m) |
|---|---|---|
| 5.5 to 7.5 | 0.15 | 0.4 to 1.2 |
| Over 7.5 | 0.15 | 1.2 |

**FIG. 23.3** *IRC Class A and Class B loadings*

in the particular area, if available. A snow accumulation of 0.25 m to 0.5 m on the deck is considered for design

## Impact Loads

A series of live loads moving at design speeds over bridges produce higher stresses than those which would be caused if the vehicles causing these are stationary. This is considered to be due to the *dynamic effect* of the loads and is taken into effect by making an allowance for *impact* so as to proceed with a simpler static analysis. The allowance for impact is in the form of an *impact factor,* which is a fraction or a percentage of the applied live load by which the latter is increased.

The impact effect depends upon a few factors, some of which may be different for railway bridges than those for highway bridges.

Factors that affect impact include the type of bridge super-structure and its details of construction, the span, and the speed of the vehicle.

For example, the impact effect is smaller for spandrel-filled aches than for open-spandrel ones; the effect is, in general, more for larger spans; and to a limited extent, increases with speed of the vehicles. The effect is more in the case of uneven highway surfaces or uneven rail surfaces in the case of a railway; in fact, what is known as *hammer-blow action* is more pronounced in the case of railway bridge decks than in highway bridges.

For foot bridges, no allowance need be made for impact since the stress involved is negligible.

### Empirical Formulae for Impact Factor

The following empirical formulae are recommended by "IRC: 6–2010 (fifth revision): Standard Specifications and Code of Practice for Road Bridges, Section-II-Loads and Stresses"

**For Class-A and Class-B loading:** The following equations are applicable for spans between 3 m and 45 m;

For RC bridges,

$$\text{Impact factor} = \frac{4.5}{(6+L)} \tag{23.1}$$

For steel bridges,

$$\text{Impact factor} = \frac{9}{(13.5+L)} \tag{23.2}$$

Here, $L$ is the span in metres

The impact factor can be determined from Fig. 23.4:

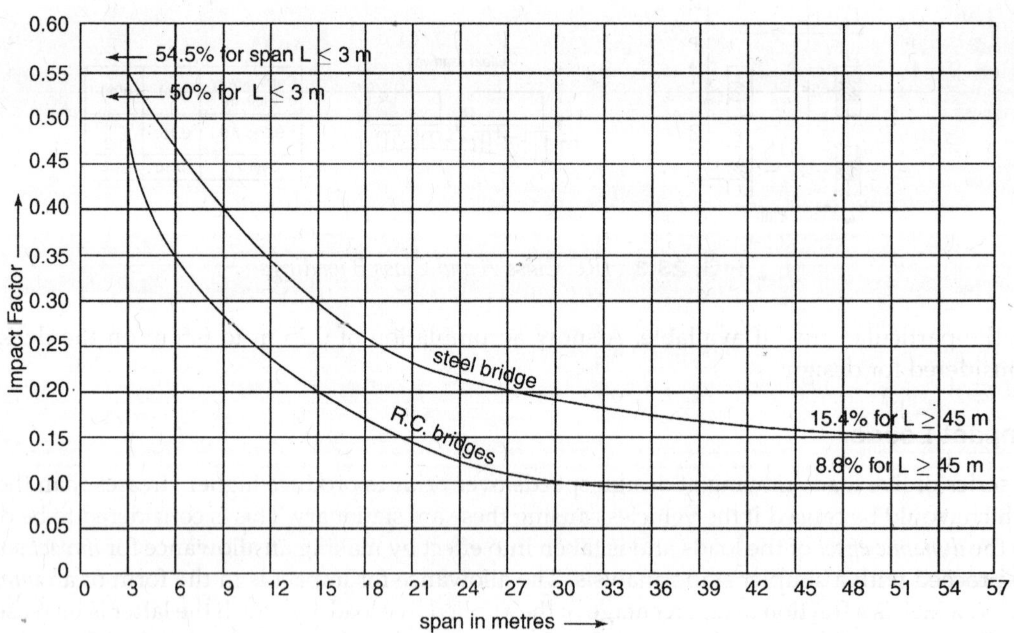


**FIG. 23.4** *Impact factor for highway bridges for Class A and Class B loading*

**For Class-AA and Class-70R loadings:** The value of impact factor shall be taken as follows:

(a) ***For spans less than 9 m:***

(i) Tracked vehicles: 0.25 for spans up to 5 m linearly reducing to 0.10 for spans of 9 m
(ii) Wheeled vehicles: 0.25

(b) ***For spans 9 m or more:***

1. Reinforced concrete bridges:
   (i) Tracked vehicles: 0.10 up to a span of 40 m and in accordance with the curve in Fig. 23.4 for spans in excess of 40 m.
   (ii) Wheeled vehicle: 0.25 for spans up to 12 m and in accordance with the curve in Fig. 23.4 for spans in excess of 12 m.
2. Steel bridges
   (i) Tracked vehicles: 0.10 for all spans
   (ii) Wheeled vehicles: 0.25 for spans up to 23 m and in accordance with the curve in Fig. 23.4 for spans in excess of 23 m.

[If there is a filling of not less than 600 mm including the road crust, the impact factor shall be assumed to be one-half of what is given above.]

**Impact factor for design of bearing:** Full value of the appropriate impact factor is allowed.

**Impact factor for design of piers and abutments:** It varies from half of the appropriate value at the bottom of the bed block at the top in linear manner, decreasing uniformly to zero at a point 3 m below the top.

Thus, for the portion of the structure more than 3 m below the bed block, the impact allowance is zero.

Since a theoretical analysis of impact effect is complex, Bridge odes recommend the use of empirical formulae, which are different for bridges for highways and those for railways.

## Wind Load

The following provisions may be used for spans up to 150 m or for a bridge with pier height up to 100 m.

The wind pressure depends on the geographic location, the terrain, the height of bridge above the ground, the horizontal and lateral dimensions of the bridge.

In addition to considering the prescribed loads in the design of bridge elements, stability against overturning, uplift and sliding due to wind shall be considered

The intensity of wind speed shall be based on hourly mean wind speed. The values given in Table 23.1 correspond to a basic wind speed of 33 m/s (120 km/h) and a return period of 100 years, for bridges situated in plain terrain and for terrain with obstructions with a flat topography.

The hourly mean wind pressure shall be appropriately modified depending on the location of the bridge for other basic wind speeds and used for design

**Table 23.1** Hourly mean wind speed and wind pressure (for a basic wind speed of 33 m/s)

| H(m) | Bridge situated in | | | |
|---|---|---|---|---|
| | Plain terrain | | Terrain with obstructions | |
| | $V_z$(m/s) | $P_z$(N/m²) | $V_z$(N/m²) | $P_z$(n/m²) |
| Upto 10 m | 27.8 | 463.7 | 17.8 | 190.5 |
| 15 | 29.2 | 512.5 | 19.6 | 230.5 |
| 20 | 30.3 | 550.6 | 21.0 | 265.3 |
| 30 | 31.4 | 590.2 | 22.8 | 312.2 |
| 50 | 33.1 | 659.2 | 24.9 | 378.4 |
| 60 | 33.6 | 676.3 | 25.6 | 392.9 |
| 70 | 34.0 | 693.6 | 26.2 | 412.8 |
| 80 | 34.4 | 711.2 | 26.9 | 433.3 |
| 90 | 34.9 | 729.0 | 27.5 | 454.2 |
| 100 | 35.3 | 747.0 | 28.2 | 475.6 |

$H$ = Average height in meters of exposed surface above the mean retarding surface(ground or bed or water level)

$V_z$ = Hourly mean speed of wind in m/s at height $H$

$p_z$ = Horizontal wind pressure in N/m² at height $H$

Note:

1. Intermediate values may be obtained by interpolation.
2. Plain terrain is an open terrain with no obstruction or with well-scattered obstructions having height up to 10 m. Terrain with obstructions is one with numerous closely-spaced structures, forests or trees up to 10 m in height with few isolated tall structures, or with a large number of high closely-spaced obstructions like structures, trees, forests, etc.
3. For other values of basic wind speed as indicated in the Wind Map of India [IS: 875 (Part 3) -1987], the hourly mean wind speed shall be obtained by multiplying the corresponding wind speed value by the ratio of the basic wind speed at the location of the bridge to the value corresponding to Table 23.1 (33 m/s).
4. The hourly mean wind pressure at an appropriate height and terrain shall be obtained by multiplying the corresponding pressure value for base wind speed given in this table by the ratios of the square of the basic wind speed at the location of the bridge to the square of the base wind speed corresponding to this Table 23.1 (33 m/s).
5. If the topography at the site (hill or cliff) can cause acceleration or funnelling of wind, the wind pressure shall be further increased by 20%.
6. For construction stages, the hourly mean wind pressure shall be taken as 70% of the value calculated as given above.

The superstructure shall be designed for wind-induced horizontal forces (acting in the transverse and longitudinal direction) and vertical loads acting simultaneously.

The transverse wind force on the superstructure shall be taken to act on the area calculated as follows:

***For a deck structure***: The area as seen in elevation including the floor system and railing, less area of perforations in hand railing or parapets, shall be considered.

***For a through or a half-through structure***: The area of the elevation of the windward truss, plus half the area of elevation above the deck level of all other trusses of girders.

**Longitudinal Wind Force on Bridge Superstructure**

This shall be taken as 25% of the transverse wind load for beam/box/plate girder bridges; it shall be 50% of the transverse load for truss bridges.

The longitudinal wind load on live load shall be taken as 25% of the transverse wind load; both loads shall be applied simultaneously acting at 1.5 m above the roadway.

[Note: When the wind velocity at deck level exceeds 36 m/s (130 km/h), no live load need be considered to be acting on the bridge.]

A wind pressure of 2.4 kN/m$^2$ on the unloaded structure shall be considered if it produces greater stresses than those produced by the combined wind forces as given above.

**Longitudinal Forces**

These are caused due to one or more of the following:

(i) *Tractive effort* caused through acceleration of the driving wheels.
(ii) *Braking effect* due to the application of brakes.
(iii) *Frictional resistance* offered to the movement of free bearings due to changes in temperature or any other cause.

Braking effect, which is invariably greater than the tractive effort, is computed as follows:

(a) ***Single-lane or two-lane bridge***: For single- or two-lane bridges, braking effect is taken as 20% of the first train load plus 10% of the load in succeeding trains or parts thereof on any one lane only. If the entire train is not on the full span, the braking force is taken as 20% of the loads actually on the span.

(b) ***Multi-lane bridge***: As in (a) above for the first two lanes plus 5% of the loads on the lanes in excess of two.

[**Note**: No impact allowance is included for this computation.]

The force due to braking effect shall be assumed to act parallel to the roadway and 1.2 m above it. While the force is transferred to the bearings, the change in the vertical reaction at the bearings should be considered.

The longitudinal force due to friction at a free bearing is to be limited to the sum of the dead and live load reactions at the bearing multiplied by the appropriate coefficient of friction.

The longitudinal force at a fixed bearing shall be taken as the algebraic sum of the longitudinal forces at the free bearing and the force due to the braking effect on the wheels.

The coefficient of friction is taken as given in Table 23.2.

For bridge structures without bearing such as arches and rigid frames, the effect of braking force shall be calculated in accordance with standard methods of analysis of indeterminate structures.

**Table 23.2** Friction in bearings

| Type of bearing | Coefficient of friction |
|---|---|
| Steel roller bearings | 0.03 |
| Concrete roller | 0.05 |
| *Sliding bearings* | 0.40 |
| Steel on cast iron or steel | |
| Concrete over concrete | 0.50 |
| Teflon on stainless steel | 0.05 |

## Centrifugal Forces

When a road bridge is located on a horizontal curve, the effect of centrifugal forces on vehicles moving at a constant speed along the curve shall be taken into account.

This is given by the equation

$$C = \frac{WV^2}{12.95R} \quad (23.3)$$

where $C$ = Centrifugal force in kN, acting normally to the vehicle at the point of action of the wheel load, or uniformly distributed over every metre length on which the uniform loading acts.

$W$ = Live load in kN for wheel loads, and in kN/m for uniform live load.

$V$ = Design speed in km/h

$R$ = Radius of the curve in metres.

This force is taken to act at a height of 1.2 m above the level of the carriageway. No impact allowance need be given for this force.

### Horizontal Force due to Water Currents

Any part of a bridge structure which may be submerged in running water should be designed to sustain safely the horizontal pressure due to the force of the water current.

The intensity of pressure on piers parallel to the direction of water flow is given by

$$P = 52\,KV^2 \quad (23.4)$$

where $P$ = Intensity of pressure due to water current in kg/m$^2$

$V$ = The velocity of flow in m/s

$K$ = A constant which has the following values:

| | |
|---|---|
| Square-ended piers (and for the superstructure) | 1.50 |
| Circular piers or piers with semi-circular ends | 0.66 |
| Piers with triangular cut waters, the angle included between the faces being 30°or less | 0.50 |

Cont'd

Table Cont'd

| | |
|---|---|
| For angle greater the 30°but less than 60° | 0.50 to 0.70 |
| For angle between 60° and 90° | 0.70 to 0.90 |
| Piers with cut waters of equilateral arcs of circles | 0.45 |
| Piers with arcs of cutwaters intersecting at 90° | 0.50 |

The value of $V^2$ in Eq (23.4) is assumed to vary linearly from zero at the point of deepest scour to the square of the maximum velocity at the free surface of water. The maximum velocity at the water surface is to be taken as $\sqrt{2}$ times the maximum mean velocity of the current. (In other words, the value of $V^2$ at the free surface is assumed to be twice that of the square of the maximum mean velocity

When the current strikes the pier at an angle, the velocity is resolved into two components, parallel and normal to the pier. The pressures are then computed using values of the components in Eq.(23.2), in both directions, except that the normal pressure is calculated with the constant $K$ as 1.5. However, in the case of circular piers it is taken as 0.66.

To provide for the possible variation of direction of the current from the assumed direction, allowance shall be made for a deviation of 20° in the direction of the current.

## Buoyancy Effects

In the design of abutments, especially those of submersible bridges, the effects of buoyancy shall also be considered assuming that the fill behind the abutments has been removed by scour.

To allow for full buoyancy a reduction is made in the gross weight of the member affected, in the following manner:

(a) When the member displaces water only, e.g., a shallow pier or abutment pier founded at or near the bed level, the reduction in weight shall be equal to that of the volume of the displaced water.

(b) When the member under consideration displaces water and also silt or sand as in the case of deep piers, the upward pressure causing the reduction in weight shall be considered as made up of two components:

  (i) Full hydrostatic pressure due to a depth of water from the surface to the foundation of the member; and

  (ii) Upward pressure due to the submerged weight of the silt or sand calculated in accordance with Rankine's theory for appropriate angle of internal friction.

In the design of submerged masonry or concrete structures, the buoyancy effect through pore pressure may be limited to 15% of full buoyancy.

In case of submersible bridges, the full buoyancy effect on the superstructure shall be taken into consideration.

## Earth Pressure

Structures designed to retain earth fills shall be proportioned to withstand pressure calculated in accordance with any rational theory. Coulomb's theory shall be acceptable, subject to the modification that the centre of pressure exerted by a dry backfill is located at 0.42 of the height of the wall above the base instead of 0.33 of the height. The horizontal pressure taken shall not be less than that exerted by a fluid weighing 480 kg/m$^3$. All abutments and return walls shall be designed for a live load surcharge equivalent to 1.2 m earthfill.

Reinforced concrete approach slab shall be provided with 12 mm dia. bars 1150 mm c/c in each direction both at the top and bottom as reinforcements in M30 grade concrete covering the entire width of the roadway, with one end resting on the structure designed to retain earth, and extending for a length not less than 3.5 m into the approach.

Thorough drainage of backfill shall be provided by means of weep holes and crushed rock, gravel drains, pipe drains, or perforated drains.

The pressure of submerged soils (not provided with drainage) shall be considered as made up of two components:

(a) Pressure due to the earth, the unit weight being reduced by buoyancy, and
(b) Full hydrostatic pressure of water.

## Temperature Effects

Daily and seasonal fluctuations in shade air temperature and solar radiation cause changes in the bridge temperature, resulting in a range of effective bridge temperature, with a maximum and a minimum over a prescribed period.

This causes stresses in the bridge due to the restraint offered for the associated expansion/ contraction of the form of construction (portal frame, arch, etc.), and also the friction at roller or sliding bearings.

Another factor is the difference in temperature between the top surface and other levels through the depth of the superstructure, which results in associated loads within the structure.

Provision shall be made for stresses or movements resulting from variations in the temperature.

The maximum and minimum temperature contours for India are available from the India Meteorological Department.

For the purpose of calculating temperature effects, the coefficient of thermal expansion for RCC, PSC and steel structures may be taken as $12 \times 10^{-6}/°C$.

Specified increases in stress may be permitted in the design of bridge superstructures while considering temperature effects.

However, such increase in stress is not applicable for the Limit State Design of Bridges.

## Deformation Effects

These are relevant only to steel bridges. Deformation stress is defined as the bending stress in any member of an open-web girder caused by the vertical deflection of the girder combined with the rigidity of the joints.

The design, manufacture and erection of steel bridges should be carried out so as to keep the deformation stresses to a minimum. In the absence of computations, the deformation stresses should be assumed to be not less than 16% of the dead and live load stresses

### Secondary Effects

In steel structures, secondary stresses are caused due to eccentricity of connections, floor beam loads applied at intermediate points in a panel, cross girders connected away from panel points, lateral wind forces on the end posts of through girders, and movement of supports.

In reinforced concrete structures, secondary stresses may occur due to movements of supports and deformations in the geometrical shape of the structure (which may be caused due to the rigidity of end connections, loads applied at intermediate points of panels, or restrictive shrinkage of concrete floor beams.)

The shrinkage coefficient of RCC may be taken as $2 \times 10^{-4}$ for purposes of design.

All bridges should be designed and constructed in such a manner that the secondary stresses are reduced to a minimum which should be allowed for in the design.

### Erection Stresses and Construction Loads

The effects of erection and the stresses produced during the construction programme shall be accounted for. Construction loads are those which are incident on a structure or any of its components during the construction of the structure. Typical construction loads include plant and equipment, storage of construction material on a partially completed bridge bridge deck, thermal effects during construction due to temporary restraints, loading due to subgrade soil settlement and so on.

## Forces Due to Earthquakes

### Seismic Zones

For the purpose of determining the seismic force India is divided into four types of zones, each zone being associated with a zone factor as given below:

| Zone | V | IV | III | II |
|---|---|---|---|---|
| Zone factor(Z)[1] | 0.36 | 0.24 | 0.16 | 0.10 |

[IS: 1893 (Part-I): 2002][2]

Forces due to earthquakes are also known as *seismic forces*. All bridges supported on piers and arches, directly or through bearings, are to be designed for horizontal and vertical forces due to earthquakes.

However, the following types need not be checked for seismic effects:

(a) Culverts and minor bridges up to 10 m span in all seismic zones.
(b) Bridges in seismic Zones II and III satisfying both limits of total length not exceeding 60 m and spans not exceeding 15 m.

Special investigations should be carried out for bridges of the following description:

(a) Bridges with more than 150 m span
(b) Bridges with piers taller than 30 m in Zones IV and V
(c) Cable-stayed and suspension bridges
(d) Arch bridges with more than 50 m span
(e) Bridges having any of the special seismic resistant features such as dampers and seismic isolators.

Masonry and plain concrete bridges with more than 10 m span shall be avoided in Zones IV and V.

### Components of Seismic Motion

The characteristics of ground motion at any location depend upon the magnitude of the earthquake, depth and distance of epicentre and characteristics of the path through which the seismic wave travels. The ground motion can be resolved into components in three mutually perpendicular directions.

---

[1] Z is used in Eq 23.7 for computing the horizontal seismic coefficient.

[2] IS:1893–2002 Indian Standard Criteria for earth quake-resistant design of structures, BIS, New Delhi, 2002.

The components are considered to act simultaneously, but independently, and the method of recommended combination assumes two horizontal components of equal magnitude, and the vertical component as two-thirds of horizontal component.

The effects of vertical components shall be considered for all elements of the bridge in Zones IV and V; however, these can be omitted for all elements in Zones III and II, except for the following:

- Pre-stressed concrete decks
- Bearings
- Horizontal cantilever structural elements
- Bridges located in the 'Near field regions' (Regions within 10 km of known fault zones).

If $r_1$, $r_2$ and $r_3$ are the full force resultants due to the design seismic forces in three mutually perpendicular directions, the resultant seismic force components for any bridge component are given by

$$\left.\begin{aligned} F1 &= \pm r_1 \pm 0.3r_2 \pm 0.3r_3 \\ F2 &= \pm 0.3r_1 \pm r_2 \pm 0.3r_3 \\ F3 &= \pm 0.3r_1 \pm 0.3r_2 \pm r_3 \end{aligned}\right\} \tag{23.5}$$

A similar approach is adopted for combining the moments about the axes; the largest force and the moments are used for design.

**Computation of Seismic Response**

For most of the bridges, 'Elastic Seismic Acceleration Method' is adequate.

In this method, the first fundamental mode of vibration is calculated and the corresponding acceleration is obtained from Fig. 23.5.

**Horizontal Seismic Force**

The horizontal seismic forces which act at the centre of mass, are to be resisted by the structure as a whole, and are given by

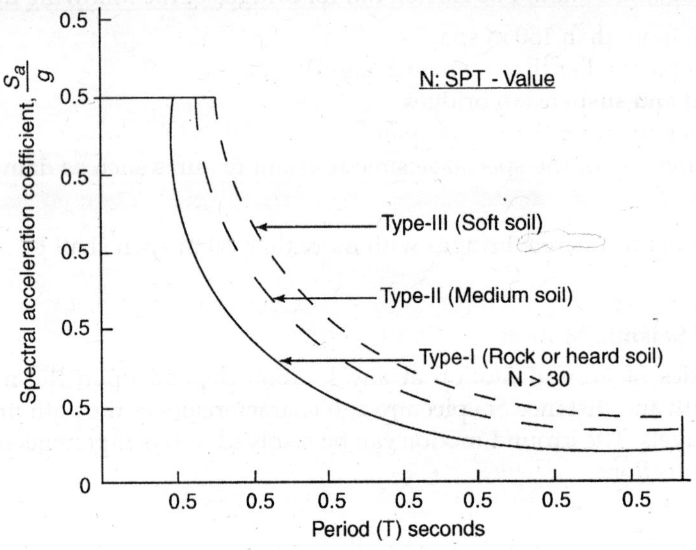


**FIG. 23.5** *Response spectra*

$$F_{eq} = A_h \text{ (dead load + appropriate live load)} \tag{23.6}$$

where $F_{eq}$ = Seismic force to be resisted

$A_h$ = Horizontal seismic coefficient

$A_h$ is given by

$$A_h = \left(\frac{Z}{2}\right)(I)\left(\frac{S_a}{g}\right) \tag{23.7}$$

Here $Z$ = Zone factor for the particular seismic zone

$I$ = Seismic importance factor of the bridge

$T$ = Fundamental period of the bridge for horizontal vibrations (seconds)

$\left(\frac{S_a}{g}\right)$ = Average response acceleration coefficient for 5% damping of load-resisting elements. It depends upon the fundamental period of vibration $T$, as given in Fig. 23.5.

Fundamental time period of the bridge member is to be calculated by any rational method of analysis adopting the Modulus of Elasticity of concrete (IRC: 21) and taking the gross uncracked section for moment of inertia. Alternatively, $T$ (in seconds) of the pier/abutment of the bridge along a horizontal axis may be estimated by the following equation:

$$T = 2.0\sqrt{\frac{D}{1000F}} \tag{23.8}$$

where $D$ = Appropriate dead load of the superstructure and live load (in kN)

$F$ = Horizontal force (in kN) required to be applied at the centre of mass of the superstructure for 1 mm horizontal deflection at the top of the pier/abutment for the earthquake in the transverse direction; and the force to be applied at the top of the bearing for the earthquake in the longitudinal direction.

Figure 23.5 is based on the following equations:

For sites with rocky or hard soil (Type-I soil with N > 30):

$$\frac{S_a}{g}\begin{cases} 2.50 & 0 \le T \le 0.40 \\ 1.00/T & 0.40 \le T \le 4.00 \end{cases} \tag{23.9}$$

For sites with medium soil (Type-II soil with 10 < N ≤ 30):

$$\left(\frac{S_a}{g}\right)\begin{cases} 2.50 & 0 \le T \le 0.55 \\ 1.36/T & 0.55 \le T \le 4.00 \end{cases} \tag{23.10}$$

For sites with soft soil (Type-III soil with $N < 10$):

$$\left(\frac{S_a}{g}\right)\begin{cases} 2.50 & 0 \le T \le 0.67 \\ 1.67/T & 0.67 \le T \le 4.00 \end{cases} \tag{23.11}$$

In the absence of calculations for $T$ for small bridges, the value of $\left(\frac{S_a}{g}\right)$ may be taken as 2.5.

For damping other than 5% offered by load-resisting elements, the following multiplying factors shall be used:

| Damping % | 2 | 5 | 10 |
|---|---|---|---|
| Multiplying factor | 1.4 | 1.0 | 0.8 |
| Application | Pre-stressed concrete, steel and composite steel elements | Reinforced concrete elements | Retrofitting of old bridges with RC piers |

**Seismic Importance Factor of Bridges**

Bridges are usually designed to resist *design basis earthquake* level of forces; alternatively, the design may be made for higher magnitudes of forces, depending on the consequences of their partial or complete non-availability, due to damage or failure from seismic tremors. The factor with which $\left(\frac{Z}{2}\right)$ is multiplied to get the level of design force required (Eq. 23.7) is called the seismic importance factor of the bridge.

Several factors are considered in assessing the consequences of this factor, and hence the choice of factor ($I$).

Some of these are:

(i) Extent of disturbance to traffic
(ii) Possibility of providing temporary diversion
(iii) Availability of alternative routes
(iv) Cost of repairs and time needed in case of damage
(v) Cost of replacement and time for reconstruction in case of failure.
(vi) Indirect economic loss due to partial or full non-availability of the bridge.

The factor $I$, for bridges, is given below:

| | |
|---|---|
| Normal bridges | 1.0 |
| Important bridges(Bridges on NH,SH, crossing rivers or railway lines) | 1.2 |
| Large critical bridges in all seismic zones (long bridges across perennial rivers) | 1.5 |

**Live Load Components**

(i) The seismic force due to live load shall not be considered when it acts in the direction of the traffic, but it shall be considered when it acts in the direction perpendicular to the traffic.
(ii) The horizontal seismic force in the direction perpendicular to the traffic shall be calculated using 20% of live load (excluding impact factor)
(iii) Vertical seismic force shall be calculated using 20% of live load (excluding impact factor).

(**Note**: This is only for calculating the seismic design forces and is based on the assumption that only 20% of the live load is present over the bridge at the time of earthquake)

**Water Current and Depth of Scour**

The depth of scour under seismic condition to be considered for design shall be 0.9 times the maximum scour depth. The flood level, for calculating hydrodynamic force and water current force, is to be taken as the average of yearly maximum design flood for at least 7 years.

### Hydrodynamic and Earth Pressure Forces Under Seismic Condition

Hydrodynamic forces act on the submerged part of the structure and are transmitted to the foundations. Also, additional earth pressures due to earthquake act on the retaining portions of the abutments (IS: 1893). These forces shall be considered for the design of bridges in Zones IV and V.

These additional earth pressure forces need not be considered for other components such as wing walls and return walls since these elements are easily repairable at low cost.

### Response Reduction Factor for Design Forces of Elements

The forces on various members obtained from the elastic analysis of bridge structure are to be divided by *response reduction factor* given in Table 23.3:

**Table 23.3** Response reduction factors

| Bridge component | *R* with ductile detailing | *R* without ductile detailing |
|---|---|---|
| Superstructure | Not applicable | 2.0 |
| Substructure | | |
| Masonry/PCC piers, abutments | – | 1.0 |
| RCC abutments | 3.0 | 2.5 |
| RCC long piers (where plastic hinges can develop) | 4.0 | 3.3 |
| Column | 4.0 | 3.3 |
| Beams of RCC portal frames | 1.0 | 1.0 |
| Bearings | 2.0 | 2.0 |
| Connectors and stoppers (Reaction blocks) | When these are designed primarily to withstand seismic forces, R shall be taken as 1.0 | |
| | When these are designed as additional safety devices in the event of bearing failure, R-value specified in this table shall be used. | |

Response reduction factor is not to be applied for the calculation of displacement of the bridge elements.

### Liquefaction

In loose sands and poorly graded sands, seismic vibrations may cause liquefaction. Appropriate measures have to be taken to prevent this.

### Foundation Design

Seismic loads should be taken as 1.25 times the forces transmitted to it by substructure, so as to provide sufficient safety margin.

## Forces due to Collision of Vehicles

This possibility exists in the case of bridge piers of flyovers built in the median or in the vicinity of the carriageway supporting the superstructure. Such piers shall be designed for vehicle collision loads. The effect of collision load shall also be considered on the supporting elements, such as foundations and bearings.

The effect of collision load shall not be considered on abutments or on structures separated from the edge of the carriageway by a minimum distance of 4.5 m, and shall also not be combined with principal live loads on the carriageway supported by the structural members subjected to such collision loads, as well as wind, or seismic load.

The nominal loads given in Table 23.4 shall be considered to act horizontally as vehicle collision loads. Supports shall be capable of resisting the main and residual loads acting simultaneously. Loads acting normal and parallel to the carriageway below shall be considered to act separately and not be combined.

**Table 23.4** Nominal vehicle collision loads on bridge supports

| Load component | Load normal to the carriageway below (t) | Load parallel to the carriageway below (t) | Point of application on bridge support |
|---|---|---|---|
| Main load component | 50 | 100 | At the most severe point between 0.75 and 1.5 m above carriageway level |
| Residual load component | 25 | 50 | At the most severe point between 1 m and 3 m above carriageway level |

These loads are assumed for vehicle plying with velocity 60 km/h; for smaller velocities, the loads may be reduced in proportion to the square of the velocity, but not to less than 50%.

If protected by a fencing system, the bridge supports shall be designed for the residual load component only.

## Forces due to Ship/Barge Impact

When a bridge is across navigable water, bridge supports/piers shall be designed for ship/barge impact. The ship impact forces and their points of application depend upon the design vessels and their speeds. For larger ships in navigable waterways piers shall be protected by building fenders, sacrificial caissons, etc.

The design impact force shall at least be 100 t acting at a height of 1 m above High Flood Level (HFL), in spite of the provision of fenders.

[The permissible stresses in both steel and concrete shall be increased by 50% and the safe bearing capacity of the foundation soil increased by 25% when considering the effect of collision loads.]

## Wave Pressure

Wave pressure has to be accounted for in the design of bridge components such as piers/abutments when the bridge crosses a portion of a large body of water like a sea. The maximum pressure at 6 m depth for a wind velocity of 160 km/h is 108 kN/m$^2$ (approx)[3]

[**Note:** 1. It has been established from experience that wind forces and seismic forces do not occur simultaneously.

2. All the forces listed and considered do not occur in any particular situation. The designer has to choose the particular set of forces relevant to the bridge project under consideration.]

[3] Paul Anderson: Substructure analysis and design, 2nd edition, Ronald Press, New York, 1956.

### Indian Standards for Railway Bridges

Indian railways have been pioneers in the construction of bridges. At present, there are over 1,33,000 bridges of all types, of which 48 are major bridges of length more than 500 m (as at the end of 2013-14)

Bridges on Indian Railways are built to conform to the Indian Railway Standards (IRS) laid down by the Ministry of Railways (Govt. of India):

The following are the IRS codes relevant to bridges:

(a) IRS Bridge rules
(b) IRS Steel Bridge code
(c) IRS Concrete Bridge code
(d) IRS Arch Bridge code
(e) IRS Bridge substructure code

Railway tracks are classified, according to the gauge, as

1. Broad Gauge(BG)...1676 mm
2. Metre Gauge(MG)...1000 mm
3. Narrow Gauge(NG)...762 mm

Based on the importance and intensity of traffic, the tracks are classified as

(i) Main line
(ii) Branch line

## 23.3 DESIGN ASPECTS OF SUBSTRUCTURES

The components of the substructure of a bridge are the abutments, piers, and the foundations.

The design aspects of these components are covered by the IRC for highway bridges and the Ministry of Railways for railway bridges, The relevant codes of practice are:

(a) IRC: 78-2000 (Second Revision): Standard specifications and code of practice for Road Bridges–Section VII–Foundations and substructure, IRC, 2000.

(b) *Indian Railway Standard Code of Practice* for the design of substructures for Bridges, Government of India, Ministry of Railways-1964 (Revised-2003, with Correction Slips up to 22).

### Abutments

The salient dimensions of a bridge abutment are the following:

**Height**: The height of an abutment is fixed by the difference between the bed level of the stream or river and the formation level of the roadway or the railway.

This will be fixed in such a manner that there is adequate freeboard and clearance with respect to the HFL in order to prevent overtopping of the roadway or the railway track by flood waters.

**Top width**: The top width of the abutment should provide enough space for the bridge seat and bearings, if necessary. It must also accommodate the breast wall to retain earth up to the approach level of the road or the railway track. A minimum of 500 mm is recommended.

**Side batter**: The water side of the abutment may be kept vertical, or a mild batter of 1 in 24 may be given. The face retaining earth is given a batter of 1 in 6 to 1 in 4, or it may be stepped.

**Bottom width**: This is decided by the height, top width and batter, or by the horizontal steps adopted. It should be adequate to provide stability for the abutment against sliding at the base and overturning about the toe under the worst conditions of loading. The maximum pressure at the base calculated from the stability analysis of the adopted section should be less than the bearing capacity of the foundation with adequate margin of safety.

**Length**: The length of the abutment should be at least equal to the width of the roadway including footpaths for a road bridge (a minimum of two lanes shall be used for important bridges on State Highways and National Highways) and the width of the formation or deck for the railway tracks (single or double, as the case may be).

A *footway* may be provided even for a railway bridge for convenience in maintenance operations.

A *cap* of cut-stone or of cement concrete is provided at the top of the abutment for proper bearing and efficient load transfer from the superstructure.

**Loads on Abutments**

The following are the loads and pressures which are to be considered in the design of the abutments:

(i) Own weight or dead load
(ii) Reaction from superstructure
(iii) Impact effect
(iv) Tractive effort and braking force
(v) Earth pressure
(vi) Pressure due to live load surcharge
(vii) Water pressure
(viii) Wind pressure
(ix) Seismic load.

**Own weight or Dead load**: This is from the self weight of the structure or abutment including any permanent fixtures.

**Reaction from superstructure**: This may be vertical or inclined as from superstructure girders or arches, which can be calculated.

**Impact effect**: This has to be allowed for in the case of loads from the superstructure.

**Tractive effort and braking force**: These forces are transmitted from the superstructure.

**Earth pressure**: This is the lateral earth pressure from the backfill which may be normal to the earth face or inclined to the direction of the earth face at an angle called *wall friction*.

**Pressure due to live load surcharge**: This is the additional earth pressure due to live load acting as the surcharge at the top of the fill.

**Water pressure**: This is not to be considered for producing maximum load.

**Wind load**: This also is not considered appreciable.

**Seismic load**: This is considered if the bridge is located in seismic zones.

In a given situation, the particular combination of the loads and pressures applicable which produces the maximum stresses should be considered in the design.

### Design Aspects

An abutment is basically designed as an earth retaining structure – a retaining wall subjected primarily to lateral earth pressure due to the fill material and the pressure due to the surcharge of live load. The maximum stresses occur when there is no water in the river. The backfill must be granular and proper drainage arrangements should be provided.

The *conditions for stability* to be checked are:

(a) **Bearing or crushing**: The maximum compressive stress at any section, including the base section, should be less than the compressive strength of the material of the abutment. In the case of a masonry abutment, *no tension* condition has to be satisfied since masonry is weak in tension; for a rectangular section, the resultant of all the forces should pass within the middle-third of the section. This is called the *middle-third rule*

(b) **Overturning**: The factor of safety against overturning – the ratio of resisting moment to the overturning moment – should be adequate, at least 1.50.

(c) **Sliding**: The ratio of the resisting force against sliding to the force tending to cause sliding at the base is called the factor of safety against sliding; this also should be at least 1.50.

The *footing* for the abutment should be designed based on the bearing capacity of the foundation soil.

## Piers

The salient features of bridge piers are the following:

**Height**: The height of pier is measured up to the support level of the deck girders, or the springing of the arch. The top of the pier is kept such that the freeboard is 1.0 m to 1.5 m above the HFL of the stream or river. This prevents any obstruction to the floating debris, and also any overtopping of the roadway or railway during floods.

**Top width**: This should be adequate to accommodate the bearings on either side with a clearance of at least 150 mm; the top width should be at least 750 mm.

Empirical formulae are also used for a preliminary section of the pier:

**Bligh**: Top width, $b = \sqrt{L}$ ($L$ being the span) (23.12)

**Rankine**: $b = \frac{1}{6}$ to $\frac{1}{7}$ of $L$ (23.13)

The top surface is flat for girders or semi-circular arches, but they will have a skewback for segmental arches.

**Batter**: The sides of a masonry or a concrete pier may be vertical or battered. The batter varies from 1 in 24 to 1 in 12. Short piers are usually constructed with vertical faces or with mild batter; high piers are invariably battered.

**Bottom width**: This depends upon the top width, the batter and the height. This should be adequate from stability considerations. For masonry piers, the general norm for the bottom width is 0.4 times the height.

**Length**: The length is usually kept such that the edges are $1\frac{1}{2}$ times the top width beyond the centre line of the outer girders or trusses plus the length of cut-water and ease water. This is to prevent diagonal shearing at the ends.

**Pier cap**: This is of cutstone or concrete (M150); it may be 450 to 600 mm thick over the straight portion.

**Cutwaters and ease waters**: These facilitate easy passage of water in the upstream side and downstream side respectively. Several different shapes (in plan) have been used.

**Loads on Piers**

The following loads are to be considered in the design of piers:

(i) Dead load or self-weight
(ii) Support reactions from superstructure; these will be eccentric (only from one side) when only one of the adjacent spans in loaded.
(iii) Impact forces due to live load and floating debris for the top 3 m of the pier only
(iv) Longitudinal forces due to tractive effort and braking forces transmitted from the superstructure.
(v) Water pressure—dynamic and impact forces due to cross-currents.
(vi) Buoyancy effect of the submerged part of the substructure.
(vii) Wind pressure on the moving loads, on the superstructure, and on the pier.
(viii) Seismic forces when the bridge is located in the seismic zones, especially zones IV and V.

**Design Aspects of Piers**

The stability considerations for piers are similar to those for abutments–bearing, overturning and sliding. *No-tension* condition is relevant for masonry piers.

The footing for pier should be decided based on bearing capacity criteria of the foundation soil.

## Wing Walls

These are built together with abutments to retain the earth fill of the approaches on either side of the bridge. The wing walls may be built of masonry or RCC, and are built of the same material as the abutments.

Wing walls may be built straight in continuation of the abutment, splayed with respect to the abutments at 45°, or as return wing walls perpendicular to the abutments.

**Salient Dimension of Wing Walls:**

**Height**: The height will be the same as that of the abutments at their junction, and will be made to reduce uniformly, or in steps, towards their ends, to suit the formation of the same height as the abutments depending upon the requirements at the site. Certain minimum height (1.0 to 1.5 m) is kept at the ends.

**Top width**: This may be from 0.5 m to 1.5 m.

**Batter**: The face may be vertical or battered for return wing walls and splayed wings, respectively.

The back batter may range from 1 in 6 to 1 in 4, or may be stepped in case the back of the abutment is stepped. If the wing wall is of RCC, if may be cantilever or counterfort type.

**Bottom width**: This depends on the top width, batter and the height; it must satisfy stability considerations.

**Length**: This should be adequate to provide to stability to the approaches, as also to provide smooth entry and exit in the case of splayed wings.

### Design Aspects of Wing Walls

Wing walls are also designed as retaining walls as in the case of abutments, with the main loads and pressures being lateral earth pressure (including live load surcharge in the case of return wings) besides self weight. The stability considerations are the same as those for abutments.

## Recommendations of IRC on Substructures for Highway Bridges

[IRC: 78-2000 (second Revision): Standard specifications and Code of Practice for Road Bridges –Section VII – Foundations and substructures, IRC, New Delhi, 2000.]

### Substructure

In the case of plain concrete substructure, surface reinforcement at the rate of 2.5 kg/m$^2$ shall be provided in each direction, i.e., both horizontally and vertically. Spacing of such bars shall not exceed 200 mm. If the atmosphere is highly corrosive, surface reinforcement can be dispensed with, but the substructure should be so proportioned as to keep the stresses only up to 90% of the allowable value.

For the substructure below the level of the top of bed block, the live load impact shall be modified as given below.

| |
|---|
| For pressure at the bottom of the pier/abutment cap – 0.5 |
| For pressure on the top 3 m substructure below cap – Decrease 0.5 uniformly to zero. |
| For pressure on the portion of substructure more than 3 m below the cap – zero |

Structures designed to retain earthfill shall be proportioned to withstand pressure calculated in accordance with any rational theory. No structure shall, however, be designed to withstand pressure less than that exerted by a fluid weighing 480 kg/m$^3$, in addition to the live load surcharge, if any.

The backfill behind the wing and return walls shall conform to the specifications in Appendix-VI with provision for proper drainage.

### Piers

Piers should be so located as to meet navigational clearance requirements and give a minimum interference to flood flow.

In general, they should be parallel to the stream current.

Piers shall be provided with suitably shaped cut-waters and ease-waters as given in IRC: 6-2010(Fifth Revision).

Piers may be in PSC, RCC, PCC, or masonry. Only solid section should be adopted for masonry piers.

The thickness of the walls of *hollow concrete piers* should not be less than 300 mm.

*Multi-column piers* of bridges across rivers carrying floating debris should be braced throughout the height of the piers by a diaphragm wall of minimum 200 mm thickness.

Piers shall be designed to withstand the forces transferred from the superstructure and those on the pier itself in addition to its self-weight.

In general, piers may be solid, hollow or framed structures. In the case of piers consisting of two or more columns, the horizontal forces at the bearing can be distributed on all the columns in proportion to their relative rigidities, if the thickness of the pier cap is at least one and a half times that of the column.

If the piers consist of *multiple piles* spaced closer than three times the width of piles across the direction of flow, the group shall be treated as a solid pier of the same overall width. If such piles are braced, the group should be considered a solid pier irrespective of the spacing of the columns.

*Hollow piers* should be provided with suitably located weep-holes of 75 to 100 mm diameter for allowing free flow of water to equalise the water levels on the inside and outside.

The lateral reinforcement of the walls of hollow circular RCC pier shall not be less than 0.3% of the sectional area of the walls of the pier. This reinforcement should be distributed 60% on the outer face and 40% on the inner face.

**Wall Piers**

When the length of a solid pier is more than four times its thickness, it shall also be checked as a wall. The reinforced wall should have a minimum vertical reinforcement of 0.3% of sectional area.

For eccentric axial load, the wall should be designed for axial load with moment.

When walls are fixed with superstructure, the design moment and axial load should be worked out by elastic analysis of the whole structure.

**Abutments**

They carry superstructure from one side and retain earth from the approach embankment.

Abutments should be designed to withstand the loads transferred from the superstructure, the earth pressure, and any load acting on the abutments, including their self-weight.

The live load surcharge equivalent to 1.2 m height of earth should be considered; water current forces during *scour all round* condition should also be considered.

The abutment may be of masonry, or plain or reinforced concrete; it may be either solid, buttressed, counterfort or spill-through type. For the spill-through type, column or wall type analysis may be carried out as for piers. Counterfort-type abutment may be treated as *T*- or *L*-type as the case may be and the slab may be designed as continuous over counterforts.

The saturated weight of fill should be considered for earth pressure during HFL condition. The weight of earth fill on the heel should be considered.

In the counterfort type, the thickness of the front wall should not be less than 200 mm and that of the counterfort 250 mm.

**Wing Walls and Return Walls**

The length of wing walls should be sufficient to retain the roadway to the required extent and to furnish protection against erosion.

Wing walls may be of solid type; the return walls may be of solid or counterfort type.

These may be built of masonry, or plain or reinforced concrete. Drainage arrangement for these walls is similar to that of abutments. If the returns are of cantilever type, their maximum length is 4 m.

Return walls shall be designed for a live load surcharge of 1.2 m height of earth fill.

**Cap for Abutments and Piers**

The minimum thickness of masonry cap is 500 mm. Top width of the cap ranges from 0.40 m to 1.60 m for spans ranging from 3 m to 24 m.

## 23.4 DESIGN ASPECTS OF FOUNDATIONS

The design of foundations is an important part of bridge design, affecting the safety and economy of the bridge. The purpose of a foundation is to transfer the load from the superstructure, through the substructure to the earth or foundation soil in such a manner that the stresses in the soil do not lead to shear failure, and the resulting settlements are within acceptable limits. Foundation design involves the knowledge of hydraulics, geotechnical engineering, and structural analysis and design.

The soil particulars and the waterway characteristics at the site are to be obtained. In addition, the superstructure and substructure details are to be assessed in order to make a choice of the foundation type that is most appropriate for the conditions, consistent with the overall economy of the bridge structure. This involves considerable exercise and judgment.

Thus, foundation engineering is considered as much an art as it is a science.

### Types of Foundations

The foundations used for bridge structures may be broadly classified as:

1. Shallow foundations
2. Deep foundations

*Shallow foundation*, according to Karl Terzaghi, is one with its depth equal to or less than its width. Such a foundation, invariably, is constructed by open excavation.

*Footings* and *raft foundations* are examples of shallow foundations. The load transfer, in this case, is by bearing at the bottom of the foundation.

Deep foundations are further classified as:

(a) Pile foundations
(b) Caisson foundations

*Piles* are relatively long and slender members, which may be precast or prefabricated and driven into the ground to the depth at which the foundation is to be placed for load transfer to the soil below, or cast-in-situ into pre-bored holes and allowed to set. A pile foundation consists of a series of piles – not less than three – which are placed below the substructure with a pile cap in between. In general, the load transfer is partly by point-bearing at the bottom and partly by skin friction with the soil around along the depth of embedment in the soil.

Caisson foundations are of two types:

(a) Open caissons (well foundations, as they are known in India)
(b) Pneumatic caissons

An *open caisson* or a *well* is one with no top or bottom during its sinking. A *pneumatic caisson* is one which is open at the bottom and closed at the top to facilitate the process of sinking it by labour working in the compressed air trapped inside the resulting chamber; the purpose of the compressed air is to keep water out of the work area near the well curb. However, there is a limitation on the depth – about 30 m – up to which it can be used because compressed air pressure increases with the depth of water and exceeds the limit of human endurance beyond this point.

A variety of caisson, called *box caisson*, with a closed bottom and an open top during sinking, which may be filled partly or fully with concrete in its final position, has also been used as a foundation for bridge substructures in some cases.

## Shallow Foundations

A shallow foundation usually consists of spread footing in concrete or coursed rubble masonry. The bottom-most footing over a levelling course (which is about 150 mm thick lean concrete 1:4:8) is of plain concrete 1:3:6 or of reinforced concrete of appropriate thickness. The depth of foundation is chosen such that the bearing capacity of the soil at that depth is adequate. The maximum pressure should be checked to ensure adequate factor of safety for different combinations of loads, as specified in the relevant Code of Practice.

### Spread Footings

These are the most widely used type among shallow foundations. The common types are:

(i) Continuous footings (wall footings)
(ii) Isolated footings (individual footings)
(iii) Combined footings
(iv) Inverted arch footings
(v) Grillage footings

The first one is used primarily in building foundations and the others are used for supporting bridge substructure.

**Isolated footings:** These can be simple, sloped or stepped as shown in Fig. 23.6.

**Combined footings:** A combined footing is used when two columns are so closely spaced that the area of footings require overlap. It is so proportioned that the centroid of the area in contact with the soil lies on the line of action of the resultant of the loads applied to the footing; this is in order to see that the distribution of soil pressure is reasonably uniform. Further, the dimensions of the footing are so chosen that the allowable soil pressure is not exceeded.

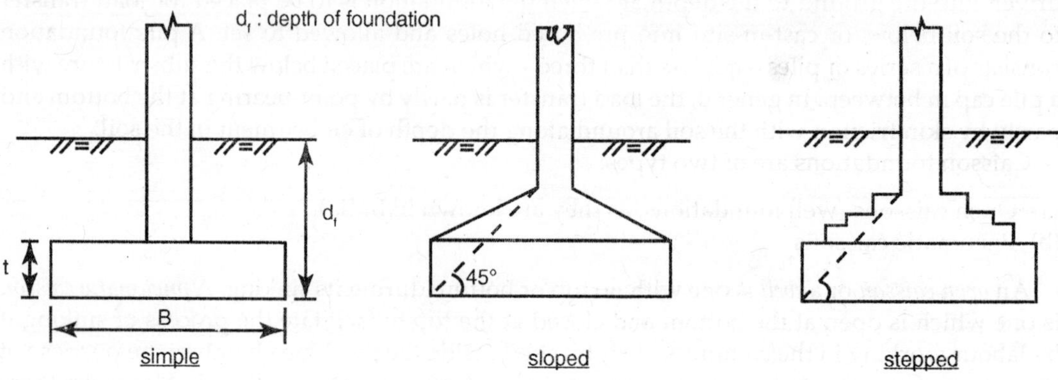


**FIG. 23.6** *Isolated footings*

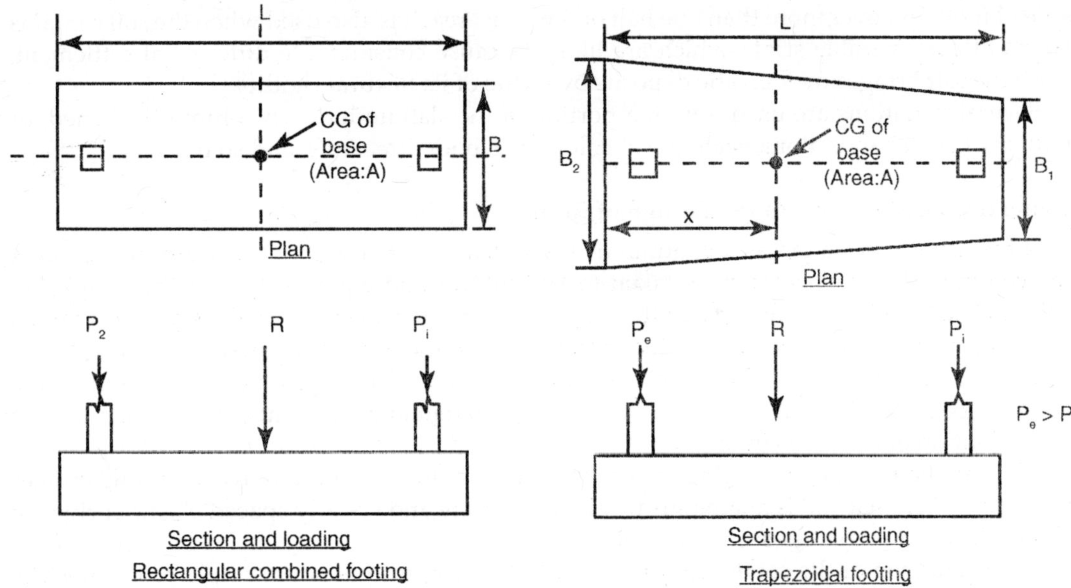


**FIG. 23.7** *Combined footings*

A combined footing may be rectangular or trapezoidal in plan. These are usually constructed in reinforced concrete. Figure 23.7 shows combined footings.

**Inverted arch footings:** These are used to support concrete piers; the footing is shaped like an inverted arch. This causes outward pressure at the ends, which should be resisted by the piers. Inverted arches are used in soft soils to reduce the foundation depth.

**Grillage footings:** This type is used to transmit heavy loads from steel columns through rolled steel joists in perpendicular directions in consecutive tiers: the whole arrangement is encased in concrete. These are used in soft soils.

**Depth of footings:** The following criteria have to be followed while deciding on the depth of footings:

(i) Footings should be taken below the top soil or fill; they should also be taken below the depth of frost action
(ii) They should be taken below the possible depth of erosion due to surface runoff.
(iii) The soil at the required location and depth should have adequate bearing capacity to transmit the loads safely, without shear failure and deleterious settlements.

The minimum depth may range from 300 mm to 900 mm.

**Bearing area of the footings:** This is governed by the total load from the substructure and the safe bearing capacity of the foundation soil. The plan shape depends upon the type of footing and its structural action.

### Raft Foundations

Also called a *Mat*, it covers the entire area beneath a structure and supports all columns/piers. This type is considered suitable when the soil pressure is low, or the loading heavy, and when spread footings cover more than one half of the plan area. It is also used when the soil contains layers of compressible strata which are likely to cause considerable differential settlement, since the raft bridges over erratic deposits by virtue of its flexural rigidity.

Several variations are used – with a portion of the slab under each column thickened, or with thickened bands along each line of columns in both directions, and so on.

### General Considerations in the Design of Rafts

A raft is usually designed as a continuous flat-slab supported without upward deflection at the columns. The soil pressure is taken to be uniform and equal to that of the total of all column loads divided by the area of the raft. The moments and shears in the slab are obtained by using appropriate coefficients listed in the codes for the design of flat-slabs.

The flat-slab analogy is likely to lead to uneconomical design unless the columns are equally spaced and equally loaded. Otherwise, differential settlements, if any, may lead to redistribution of moments in the slab.

Rafts are better designed using the concept of modulus of subgrade reaction; this implies the assumption that soil is analogous to a bed of closely and equally spaced elastic springs of equal stiffness in its stress-strain behaviour.

The evaluation of the coefficient of subgrade reaction itself is not a simple problem, since it is known to vary with a number of factors such as the size, shape and depth of the footing, besides the nature of the soil.

The procedure of design is available in the *IS Code of Practice-"IS: 1950 (Part-I)-1992 (Second Revision) Code of Practice for design and Construction of Raft Foundation-Part-I Design."*

## Deep Foundations

In the case of major bridges across rivers, abutments and piers are invariably supported by means of deep foundations such as pile foundations and caissons or wells; this is in view of the heavy loads expected to be transmitted through the superstructure and the substructure.

### Pile Foundations

Piles can be of steel, concrete (plain or reinforced), or composite nature.

Timber, steel or precast concrete piles are driven into position by using pile-driving equipment.

Only concrete piles can be cast-in-situ by pouring concrete into pre-bored holes.

Piles are used to carry vertical compressive loads, uplift or tensile forces, or resist horizontal or inclined loads.

Bearing piles may carry loads by point-bearing or skin friction, or by both.

### Pile Capacity

This is the maximum load the pile can carry without shear failure or excessive settlement. The allowable load on a pile is the load which it can carry with adequate margin of safety, or the load at which the settlement reaches the allowable value, whichever is lower.

A pile in sand or clay can be designed as if it were a short column as buckling is rare owing to the presence of soil around it.

The ultimate load on the pile, $Q_{up}$, is the sum of the point-bearing component, $Q_{eb}$ and the skin-friction component, $Q_{sf}$

$$Q_{eb} = q_b \cdot A_b \tag{23.14}$$

$$Q_{sf} = f_s \cdot A_s \tag{23.15}$$

$q_b$ is the bearing capacity of the soil at the tip of the pile,

$A_b$ is the bearing area at the tip,

$f_s$ is the unit skin friction for the pile-soil system, and $A_s$ is the surface area of the pile in contact with the soil.

The expression for $q_b$ is given by the Terzaghi Formula:

$$q_b = cN_c + 0.5\gamma bN_r + qN_q = CN_c + o.5\gamma bN_r + \gamma d_f.N_q \tag{23.16}$$

where $\gamma$ is the unit weight of the soil

C: Unit cohesion

*b*: Size of pile

$d_f$: Depth of foundation up to the pile tip

$N_c$, $N_r$ and $N_q$ are the bearing capacity factors of Terzaghi, which depend on the value of the angle of internal friction, $\phi$.

(The coefficient 0.5 in the second term is for square or rectangular pile section; it is 0.3 for circular pile section).

Compared to the first and third terms, the contribution of the second term is considered insignificant.

This is known as *static analysis*.

For piles driven into the soil by pile-driving equipment, the capacity of the pile is determined by dynamic analysis based on the *wave equation method*. Among the dynamic pile driving formulae developed, the Engineering News Formula and the Hiley formula are the more popular ones; in fact, the Hiley formula is considered to be the most complete formula.

However, the *Engineering News Formula*, which is simpler, is given here:

$$Q_{up} = \frac{W_h \cdot H}{(s+C)} \tag{23.17}$$

where $W_h$ = Weight of hammer

$s$ = Final penetration or set per blow (the average value of the last 5 blows for a drop hammer, and for the last 20 blows for a steam hammer.)

C = An empirical constant (represents the temporary elastic compression of the helmet, pile and soil).

*H, s, C* are in the same units; therefore, $Q_{up}$ will be in the same units as $W_h$.

The allowable load, $Q_{ap}$, is got by using a high factor of safety – 6.

C = 25 for drop hammer

= 2.5 for steam hammer

Thus, for double-acting steam hammers,

$$Q_{ap} = \frac{(W_h + a.p)H}{6(s+2.5)} \tag{23.18}$$

where $W_h$ is in newtons

*a* is effective area of piston in mm$^2$

$p$ is the mean effective pressure in N/mm$^2$

$H$ is height of fall of hammer in m

$s$ is the final penetration of pile per blow in mm.

**Load test on a pile:** This is considered to be one of the best methods to determine the load-capacity of a pile. It can be conducted on a test pile, on a single pile, or on a group of piles. The load-settlement curve is plotted, and the point of intersection of the straight-line portions of a log-log plot is considered to be the failure point and the load corresponding to this point is taken as the ultimate load.

**Group capacity of piles:** The capacity of a pile group is not necessarily the capacity of the individual pile multiplied by the number of piles in the group. Disturbance of the soil during installation of the piles and overlapping of stresses between adjacent piles may cause the group capacity to be less than the sum of the capacities of individual piles.

On the other hand, when piles are driven in cohesionless soils, densification occurs during driving, and the group may tend to behave as a unit or an equivalent single large pile. This may lead to the group capacity being larger than the sum of individual pile capacities. When the *block action* is considered, the bearing capacity on the plan area of the entire block and skin friction resistance around the embedded perimeter area for the depth of embedment, put together, will be the capacity of the pile group.

**Pile group efficiency:** This is defined as the ratio of the group capacity to the sum of the individual pile capacities in a group of piles.

## Design Aspects of Pile Foundations

**Length of piles:** This depends upon the soil conditions. End-bearing piles must reach a stratum which has adequate bearing capacity. Friction piles must be long enough to mobilise adequate resistance to carry the load.

**Type and material of piles:** These are decided based on the loads, soil strata, ground-water conditions, availability of equipment, and so on. The characteristics of the flow of water and scour must also be considered in the case of bridge piers.

**Allowable load of pile and pile group:** The load which the pile can carry with adequate margin of safety is considered here.

**Pile spacing:** An optimum spacing is that for which the group efficiency is unity. Slight shifts and tilts are considered permissible for piles in a pile foundation.

## Well Foundations

The elements of a well foundation are:

(i) Cutting edge
(ii) Curb
(iii) Bottom plug (concrete seal)
(iv) Steining
(v) Top plug
(vi) Well cap

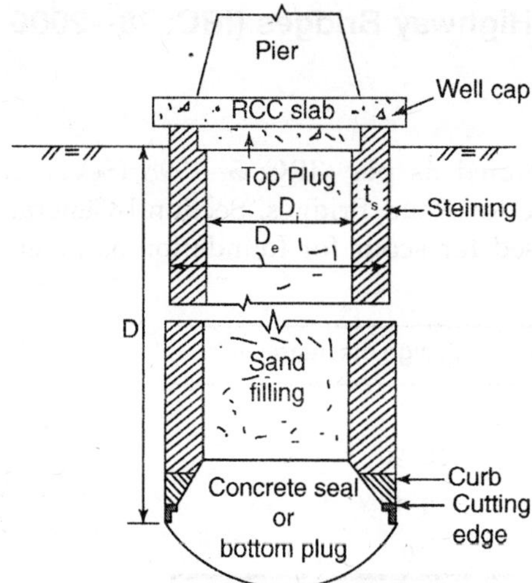


**FIG. 23.8** *Elements of a well foundation*

These are shown schematically in Fig. 23.8.

**Grip length:** This is the depth of the base of the well below the lowest scour level.

If a hard stratum like rock is available at a reasonable depth, the foundation is securely anchored to it for at least 0.3 m depth.

**Bottom plug:** Also called *concrete seal,* this is designed for an upward pressure equal to the hydrostatic pressure at the bottom minus the pressure due to self-weight. It is usually designed as a thick plate; the thickness of the bottom plug, *t*, may be got from the following equations:

***For circular wells*:** For simply supported condition—

$$t = \sqrt{\frac{3W(3+\nu)}{8\pi\sigma_c}} \tag{23.19}$$

where $W$ is the total uniform load on the plug and is equal to $\frac{\pi}{4}D_i^2 \cdot q$, $q$ being uniform pressure on the plug

$\sigma_c$ = Allowable flexural stress in concrete

$\nu$ = Poisson's ratio of concrete

***For rectangular wells*:** For simply supported condition,

$$t = \sqrt{\frac{3q \cdot B_i^2}{4\sigma_c(1+1.61\alpha)}} \tag{23.20}$$

where $B_i$ = Width or smaller dimension of the well

$\alpha = \left(\frac{B_i}{L_i}\right)$, $L_i$ being the longer dimension

**Steining:** The thickness of steining should be adequate to withstand the stresses developed during sinking and after installation.

If the self-weight is to be adequate for sinking without additional dead weight, called *kentledge,* the steining thickness, $t_s$ may be shown to be

$$t_s = \frac{D_e}{2}\left[1 - \sqrt{1 - \frac{4f_s}{D_e\gamma_c}}\right] \tag{23.21}$$

where $f_s$ = Unit skin friction resistance of the soil.

If kentledge is available, the thickness can be reduced. Empirically, the thickness of steining is taken as one-eighth the external diameter for highway bridges and one-fourth for railway bridges.

## Recommendations of IRC on Foundations for Highway Bridges [IRC: 78–2000 (Second Revision)]

### Design Discharge for Foundation

A larger discharge over the design discharge determined as per "IRC: 5–1998 (Seventh Revision) standard specifications and code of Practice for Road Bridges, Section-I-General features of design, IRC, New-Delhi, 1998" shall be used for scour for foundation as given below:

| Catchment area in km² | Increase over design discharge in % |
|---|---|
| 0–3000 | 30 |
| 3000–10000 | 30–20 |
| 10000–40000 | 20–10 |
| Above 40000 | 10 |

Linear interpolation may be used for intermediate values of catchment area.

### Maximum Scour Depth for Design of Foundation

#### Flood without seismic combination

For piers – 2 times the mean scour depth

For abutments – 1.27 times mean scour depth with approach retained or lowest bed level, whichever is deeper; or 2 times the mean scour depth with scour all round.

#### Flood with Seismic Combination

The maximum scour depth given above may be reduced by a multiplying factor of 0.9.

For design of *floor protection works for raft or open foundations,* the following values of maximum scour depth may be used:

| | |
|---|---|
| In a straight reach | 1.27 times mean scour depth |
| In a bend | 1.50 times mean scour depth |

The length of apron upstream may be 0.7 times that on the downstream.

### Depth of Foundation

#### Open Foundations

***In soil***: The minimum depth of open foundations shall be up to a stratum having safe bearing capacity, but not less than 2 m below the scour level or the protected bed level.

***In rocks***: For hard rock with a crushing strength of 10 MPa or above: 0.6 m

All other cases: 1.5 m

#### Pile Foundations

***In soil***: The minimum length required for developing full fixity as calculated by any rational formula.

***In rock***: The foundation should be below erodible rock strata and anchored into sound rock.

**Well Foundations**

***In soil***: A minimum grip of one-third the maximum depth of scour below the design scour level.

***In rock***: It is advisable to make a shear key of 300 mm in hard rock or 600 mm in soft rock inside the well, and seat it evenly all around the periphery.

**Factor of Safety**

It is 2.5 for safe bearing capacity on the ultimate value for the soil.

Factors of safety against overturning and sliding for open foundations:

| | Without seismic case | With seismic case |
|---|---|---|
| Against Overturning | 2 | 1.5 |
| Against sliding | 1.5 | 1.25 |

**Friction Coefficient**

Between soil and concrete : 0.5

Between rock and concrete: 0.8 for good rock; 0.7 for fissured rock

**Open Foundations**

The thickness of footings shall not be less than 300 mm

**Pile Foundations**

Minimum diameter of piles

| | Bridges on land | Bridges on river |
|---|---|---|
| Precast piles | 0.35 m | 1.0 m |
| Bored piles | 1.0 m | 1.2 m |

Grade of concrete, water–cement ratio and slump shall be as follows, both for precast and cast-in-situ piles:

| Item | Tremic concrete cast-in-situ | Driven cast-in-situ | Precast concrete |
|---|---|---|---|
| Grade of concrete | M 35 | M35 | M35 |
| Max. W/c-ratio | 0.4 | 0.4 | 0.4 |
| Slump (mm) | 150–200 | 100–130 | 50 |

**Well foundations**

***Steining thickness***: Minimum thickness–500 mm

Also, the following relationship shall be satisfied:

$$h = Kd\sqrt{l}$$

where $h$ = Minimum thickness of steining in m

$d$ = External diameter of circular well in m

$l$ = Depth of well below top of well cap or LWL, whichever is more

$K$ = A constant

***Value of* K**/Well in cement concrete – 0.03

Well in brick masonry – 0.05

Twin D wells – 0.039

***Concrete mix***: For plain concrete – not leaner than M15

Vertical reinforcements – not less than 0.12% of gross-sectional area, equally distributed on both faces.

For reinforced concrete, vertical reinforcements shall not be less than 0.20%.

***Well curb***: RC of mix not leaner than M25—with minimum reinforcement of 72 kg/m$^3$.

***Plug over the fill***: A 300 mm thick plug of M15 cement concrete.

## SUMMARY

- The loads to be considered for the design of bridges and bridge superstructure are dead load (including self-weight), live load, impact effects, wind effects, longitudinal forces (due to tractive effort and braking forces), forces due to water currents, earth pressure (including live load surcharge), temperature effects, secondary stresses, erection stresses and seismic effect.
- Live loads on highway bridges standardised by IRC are Class-AA, Class-70R, Class-A and Class-B loadings. While the first two are for tracked and wheeled vehicles, the next two are concerned with a train of driving vehicle followed by trailers. Class-AA and Class-70R loadings are used for bridges on National and State Highways, and Class-A and Class-B loadings for bridges on other roads.

  The latest live roads for railway bridges are the 25 t-loading-2008 and DFC loading (32.5 t axle load) for Broad Gauge and the Modified Metre Gauge Loading-1988 (16 t axle load) for Metre Gauge, as recommended by the IRS Bridge Rules of the Ministry of Railways.
- Impact allowance is considered by an *impact factor*, a fraction by which the live load is to be multiplied to arrive at the additional load to be added to the live load.
- Earth pressure due to the earth-fill is to be obtained from any rational earth pressure theory; live load surcharge effect shall also be added, if necessary.
- Wind effects are to be considered based on the wind speed and the corresponding wind pressure on exposed areas.

  Seismic forces depend on the horizontal seismic acceleration; this in turn, depends on the seismic zone in which the bridge is located.
- The design aspects of substructure—piers, abutments and wing walls—are the height, top width, batter and length. The loads to be considered are the dead loads, reactions from superstructure, impact allowance, and earth pressure including surcharge, water pressure, wind and seismic forces.

  Abutments and wing walls are designed similar to retaining walls.
- Foundations for bridges are either shallow or deep in construction. Shallow foundations include footings and rafts; footings can be isolated footings and combined footings.

  Deep foundations include pile foundations, caissons and well foundations.
- The design considerations for foundations are the depth of foundation, bearing capacity of the soil at the required depth, and the expected total and differential settlements.

  In the case of pile foundations, the considerations are the pile capacity to carry loads, group action of pile groups, and group efficiency, besides settlements.

  In the case of well foundations, the consideration is the load capacity which depends on the allowable bearing pressure of the soil at the bottom of the well where it is founded; shifts and tilts of wells which are kept within allowable limits are also to be considered.

  The salient features of a well include the cutting edge, well curb, bottom plug (or concrete seal), well steining, top plug and well cap, all of which have to be designed appropriately for the successful performance of the well foundation.

- The IRC standard specifications and Code of Practice for Road Bridges (Sections I to VIII) are to be adopted for the design of Highway Bridges.
  Similarly, the IRS Bridge Rules and Bridge Codes issued by the Ministry of Railways and the RDSO, Lucknow, have to be followed for the design of Railway Bridges, and the substructures and foundations of these bridges.

## QUESTIONS

1. Enumerate the loads to be considered for the design of bridge superstructures.
2. Explain 'impact allowance' and 'impact factor' in the design of bridge superstructures.
3. Explain the salient features of IRC Class AA loading.
4. Differentiate between IRC Class AA loading and Class 70R loading.
5. What are the circumstances under which bridges are designed for IRC Class A loading?
6. What are the empirical formulae used for the impact factor for RCC bridges and steel bridges?
7. How are wind loads accounted for in the design of highway bridges?
8. Differentiate between basic wind speed and design wind speed for a bridge site.
9. Explain how horizontal seismic force on a bridge superstructure is calculated.
10. Write short notes on:
    (a) Seismic zones of India
    (b) Zone factor
    (c) Importance factor
11. Write short notes on:
    (a) Coefficient of dynamic augment
    (b) Racking force
    (c) Derailment loads
12. Explain Equivalent Uniformly Distributed Loads (EUDL) for bending moment and shear force for railway bridges.
13. Enumerate the forces to be considered for the design of piers and abutments for a road bridge.
14. Enumerate the loads to be considered in the design of abutments and piers for a BG railway bridge across a river.
15. Explain how the following dimensions are decided in the case of abutments and piers: Height, Top width, Batter, and Length.
16. Differentiate between open foundations and deep foundations for bridge piers.
17. Differentiate between bearing piles and friction piles.
18. Explain group action of piles and group efficiency of pile groups.
19. Explain the design features of spread footings for isolated columns and combined footings.
20. Explain the salient design features of raft foundations.
21. Explain the salient design features of a well foundation for a pier of a major bridge across a river.

Chapter 24

# CONSTRUCTION OF FOUNDATIONS AND SUBSTRUCTURES

**CHAPTER OBJECTIVES**

This chapter deals at length with:

- Construction aspects of shallow foundations such as spread footings (including the precautions needed during excavation), and of raft foundations
- Construction aspects of deep foundations such as pile foundations, drilled piers, caissons including pneumatic caissons, and well foundations
- Construction aspects of abutments, piers and wing walls of different types and materials
- Functions of cofferdams and the construction aspects of different types of materials
- Dewatering techniques for facilitating the construction of foundations and substructures.

## 24.1 INTRODUCTION

As discussed in Chapter 23, bridge foundations may be shallow foundations or deep foundations.

When a soil of the required bearing capacity is available to transmit loads from the superstructure through the substructure without shear failure or with settlements (total and differential) within permissible limits, at a reasonably shallow level below the bed or ground at the bridge site, foundations can be laid using open excavation by allowing natural slopes on all sides. Such foundations are known as shallow foundations.

These are normally convenient above the water table, practicable up to a depth of about 5 m. For greater depths or for excavation under water, shoring with sheet piles or timbering will be needed to support the open excavation; alternatively, what are known as *cofferdams* have to be constructed on a temporary basis to exclude water from the work area and allow the foundation to be placed in the dry.

Spread footings and rafts are shallow foundations. On the other hand, if a hard stratum is not available except at a large depth, foundations using piles and caissons or wells are adopted. These are known as deep foundations.

## 24.2 CONSTRUCTION OF OPEN FOUNDATIONS

Spread footings and rafts fall in the category of open foundations; they may be constructed at the required depth by making excavations in the soil and opening the ground to place the foundation.

### Construction of Spread Footings

When excavation reaches the foundation level, the exposed area of the bottom of the pit is levelled and compacted by ramming. A levelling course of about 150 mm thickness is laid with lean concrete (typical cement : sand : coarse aggregate ratio being 1 : 3 : 6 or 1 : 4 : 8).

The spread footing may be constructed of cement concrete or coursed rubble masonry. The bottom-most footing over the levelling course is usually of plain concrete 1:3:6, or of reinforced concrete of suitable thickness and reinforcements in accordance with the design requirements.

Even compacted sand or sand and gravel may be used in place of lean concrete. Spread footings are considered to be relatively easy to construct. Dewatering of excavation when necessary, and bracing for the sides of the excavation, depending on the nature of the soil, require special attention.

Adjacent footings should be constructed such that the lower footing does not affect the stability of the higher one. The difference in levels should not be more than half the clear distance between them. The lower footing is constructed first, so that the level of the upper one may be adjusted if necessary. The area around the footing, left due to excavation, should be backfilled with riprap stone.

### Bracing for Excavations

Excavation of soil is required for several purposes including the placing of foundations for buildings and for bridges. Since open excavations without side support are of limited use, deeper excavations, with lateral support for their stability, are used for open foundations for bridges. Lateral supports in the form of *braced sheeting* along with a number of other structural members, are used at least temporarily, until the main structure or foundation is constructed. This process is also known as *shoring*.

A variety of materials, methods and procedures have been devised for providing lateral support to the sides of excavations. The procedures for design and construction of bracings for excavation have been evolved based on the experience derived from failures, coupled with in-situ experimental observations in subway constructions in New York, Chicago and Berlin.

In an open excavation in clay, tension cracks tend to develop near the top; in such a case, the simplest type of bracing – a *strut* is used to support the soil at the sides of the excavation near about the top, as shown in Fig. 24.1.

When the soil is soft and the depth of excavation is large, a more extensive bracing system is needed. One method is to employ vertical sheeting (either wood or steel sheet piling), as shown in Fig. 24.2.

The vertical sheeting is driven along the line of excavation before the soil is removed. As the excavation proceeds, horizontal members, called *wales*, are placed on the side of the sheeting

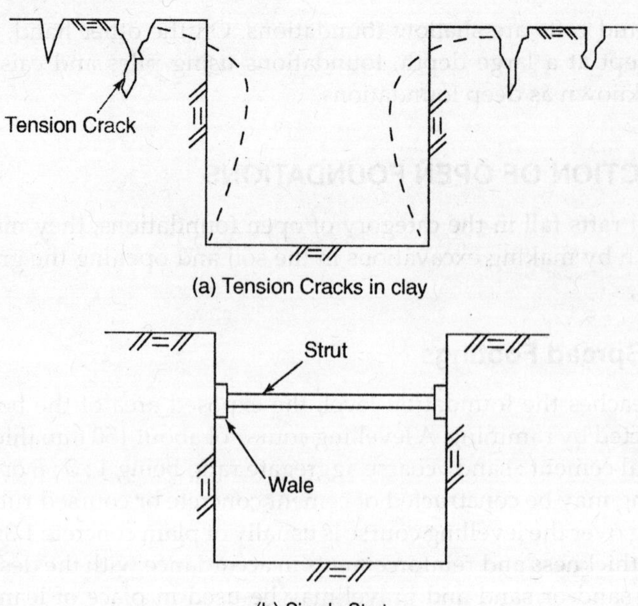


**FIG. 24.1** *Simple strut for shallow excavations*

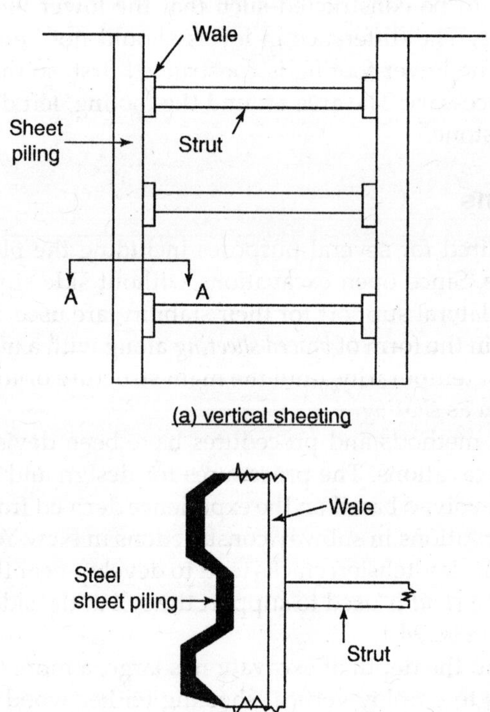


**FIG. 24.2** *Bracing in soft soil with vertical sheeting*

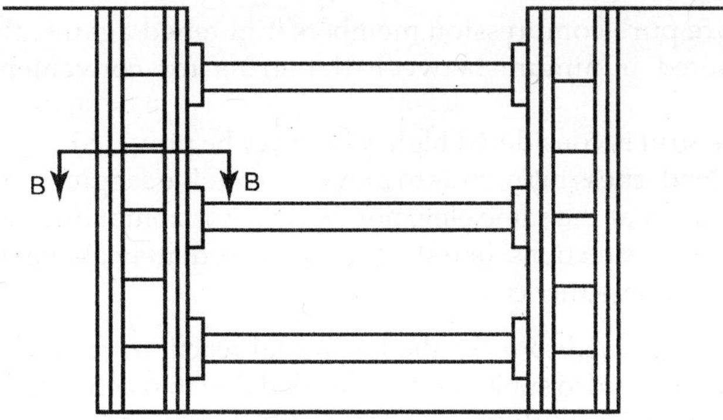


(a) Horizontal lagging with vertical solider beams

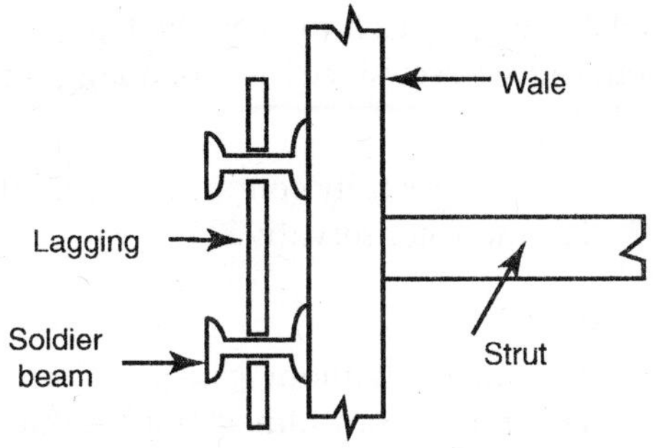


(b) Sectional plan on BB

**FIG. 24.3** *Bracing in medium soil with horizontal lagging and vertical soldier beams*

and braced with struts. In large excavations, complex systems of struts and cross bracings are employed.

When the soil is not very soft, horizontal *lagging* may be used. If the cut can stand without bracing for a few hours, it is excavated without supports and horizontal boards known as *lagging* are placed against the soil. These are held in place by vertical *solider beams* supported by struts, as shown in Fig. 24.3.

The earth pressure on a bracing system depends on the type of soil and the amount of deformation or yield of the bracing. Unlike a retaining wall, which is a structurally rigid unit against which the earth is placed after construction is completed, the bracing system is rather flexible and is called upon to support earth pressure from the soil while it is being constructed. Consequently, the deformation is irregular and the variation in earth pressure with depth is erratic, and cannot be computed by classical earth pressure theories.

### Components of Bracings

**(a) Sheet piling:** This is designed as a continuous member supported at the strut levels. The piling is imagined to be hinged at the bottom which is a fictitious support. The struts may be arranged in such a way that either the maximum bending moments are equal in all spans of the piling, or the reactions at all the struts are equal.

**(b) Struts:** These are pure compression members, tightened against the wales. A spacing of 2.5 m is considered minimum between the struts for convenience of construction operations.

If the length of the strut is considered high, posts may be provided at intermediate points in order to reduce the slenderness ratio, so as to prevent excessive sagging, and possible buckling. The struts may be located at the same elevation in both horizontal directions; or, they may be made continuous in both directions, one strut being immediately above the other. The design of wales will be different in either case.

**(c) Wales:** These are designed to resist the horizontal reactions from sheet piling or soldier beams. In addition, they are also subjected to axial load due to reaction from the perpendicular wales at the corner of the excavation. Usually, moment splices are provided; in such cases, they are designed as continuous beams subjected to transverse and axial loads.

**(d) Lagging:** This is usually made of planks, 5 to 8 cm thick. Soft clay and loose sand must be planked in short depths. Wedges are used to keep the lagging against the soil. Lagging is seldom subjected to high bending stresses since the wood lagging deflects considerably, thus providing some degree of relief.

**(e) Soldier beams:** These are designed as fully continuous over all the supports or as several simple beams, making the design more conservative.

### Construction of Raft Foundations

A raft may undergo large settlements without causing harmful differential settlement. Therefore, permissible settlements for rafts are almost double that for footings.

A raft foundation is adopted when a hard stratum is not available within 2.5 to 3.5 m, and the soil within this depth is very poor, with a low bearing capacity. This is also used when the foundation soil is loose sand extending to great depth, as spread footings may settle excessively.

Raft foundations are invariably constructed of reinforced concrete. Concrete is poured in relatively small areas such as 10 m × 10 m to avoid excessive shrinkage cracking.

Construction joints are located at points of low shear stress, such as the centre lines between columns. Reinforcements should be made continuous across joints. If reinforcing bars have to be spliced, adequate overlap should be provided.

Shear keys should be provided along joints so that the shear stress is safely transmitted across the joint.

In order to provide the necessary strength at the joint, the raft may be thickened at such locations. Depending on the design, the raft may consist of two-way slabs, as also two-way beams. The thickness of the slab is generally increased under the columns.

## Recommendations of IRC for the Construction of Open Foundations (IRC: 78–2000)

"The protective works shall be completed before the floods so that the foundation does not get undermined. Excavation on open foundations shall be done after taking necessary safety precautions from "IS; 3764: 1992 (Revised 2002): Code of Safety for Excavation Work."

Where blasting is required to be done for excavation work in rock, which is likely to endanger adjoining foundations or other structures, necessary precautions such as controlled blasting, providing suitable neat cover to prevent flying of debris, etc., shall be taken to prevent damage.

**Condition for Laying of Foundations**

Normally, open foundations should be laid in the dry and dewatering by pumping or depression of water by well point, or any other method may be used.

A levelling course of 100 mm thickness in M10 (1:3:6) shall be provided below the foundation.

If the foundation cannot be laid in the dry, the concrete for the foundation may be laid under water only by the tremie method, through a tremie pipe.

In the case of flowing water or artesian springs, the flow shall be stopped or reduced as far as possible at the time of placing the concrete. No pumping of water shall be permitted from the time of placing of concrete up to 24 hours after placement.

All spaces excavated and not occupied by abutments, pier, or other permanent works shall be refilled with earth up to the surface of the surrounding ground, with sufficient allowance for settlement. All backfill shall be thoroughly compacted, and its top surface shall be neatly graded.

In the case of excavation in rock, the trenches around the footing shall be filled up with concrete of M15 grade up to the top of the rock.

If the depth of fill required is more than 1.5 m in soft rock or 0.6 m in hard rock above the foundation level, M15 concrete may be used to fill up to this level and the portion above may be filled by concrete or by boulders grouted with cement. This weight of filling need not be considered in stability calculations."

## 24.3 CONSTRUCTION OF DEEP FOUNDATIONS

The following fall under the category of deep foundations:

1. Pile foundations
2. Drilled Piers
3. Caissons
4. Well foundations (a type of caisson)

### Pile Foundations

These have been used from time immemorial in one form or the other. Specifically, pile foundations have been used from Roman times. They are more common than any other type of deep foundation, primarily because of the multifarious uses of piles (the elements which make a pile foundation), in addition to their capacity to bear very heavy loads.

In view of this latter advantage of high load-carrying capacity, a pile foundation is considered to be a preferred type for supporting bridge substructure and superstructure.

**Classification of Piles**

Piles may be classified in a number of ways based on different criteria:

(a) Function or action
(b) Composition and material
(c) Method of installation

Based on *function or action*, piles may be classified as follows:

(i) End-bearing piles
(ii) Friction piles
(iii) Tension or uplift piles

(iv) Anchor piles
(v) Batter piles
(vi) Laterally loaded piles
(vii) Sheet piles
(viii) Fender piles
(ix) Compaction piles

While (i) to (vi) are load-bearing type, the last three are basically non-load-bearing type.

**End-bearing piles:** Used to transfer load through the pile tip to a suitable bearing stratum, passing soft soil/water.

**Friction piles:** Used to transfer load to a particular depth in a soil material by means of skin friction along the surface area of the pile embedded in the soil.

**Tension or uplift piles:** Used to anchor structures subjected to uplift due to hydrostatic pressure or overturning moment due to horizontal forces.

**Anchor piles:** Used to provide anchorage against horizontal pull from sheet-piling or water.

**Batter piles:** Used to resist horizontal and inclined forces, especially in water-front structures.

**Laterally-loaded piles:** Used to support retaining walls, bridges, dams, and wharves, and as fenders for harbour construction.

**Sheet piles:** These are commonly used as bulkheads or cut-offs to reduce seepage and uplift in hydraulic structures, and also in the construction of cofferdams.

**Fender piles:** Used to protect water-front structures from ships or other floating debris.

**Compaction piles:** Used to compact loose granular soil in order to increase the bearing capacity. Since they are not required to carry any load, the material may be sand itself. A pile tube is driven to compact the soil, and is gradually taken out and sand is filled in its place, thus forming a *sand pile.*

Based on *material and composition*, piles may be classified as follows:

(i) Timber piles
(ii) Steel piles
(iii) Concrete piles
(iv) Composite piles

**Timber piles:** Timber of length around 8 m and sound quality is used. Splicing may be adopted for greater lengths. Diameter of the pile lengths may be from 300 to 400 mm. Timber piles perform well either in dry condition or in submerged condition.

Creosoting is used to increase the durability. Its maximum design load is about 250 KN.

**Steel piles:** Usually H-piles, pipe piles, or rolled sections of regular shapes are used. Loads up to 1000 kN may be applied to steel piles.

**Concrete piles:** These may be precast or cast-in-situ. Precast piles are reinforced to withstand handling stresses. Curing time is required and heavy equipment is required for handling and driving. Cast-in-situ piles are installed by pre-excavation. This eliminates vibration due to driving and handling stresses. Patented types such as Raymond, MacArthur and Franki piles are popular.

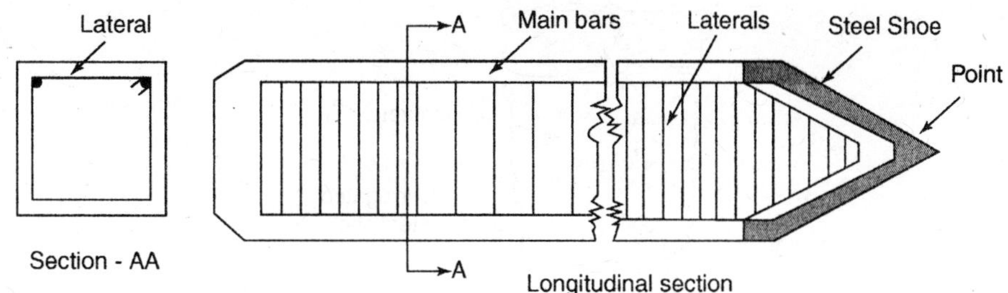


**FIG. 24.4** *Square precast RCC pile*

**Composite piles:** Concrete and timber or concrete and steel may be combined and used, especially when the upper portion of the pile is above the water table. The lower portion may be of steel and the upper one concrete.

Based on *method* of *installation*, piles may be:

(i) Driven piles
(ii) Cast-in-situ piles
(iii) Driven and Cast-in situ piles

**Driven piles:** Timber, steel, or precast concrete piles may be driven into position, either vertically or at an inclination. If inclined, they are termed *batter* or *raking* piles.

Pile driving equipment and pile hammers are used for driving piles. Hammers can be drop hammer, single- or double-acting steam hammer, diesel hammer or vibratory hammer.

**Cast-in-situ piles:** Only concrete piles can be cast-in-situ. Holes are drilled and filled with concrete. These may be *straight-bored* or *under-reamed* with one or more bulbs at intervals along the depth. Reinforcements may be used according to the requirements of load-carrying capacity.

**Driven and cast-in-situ piles:** This is a combination of both types; casing or shell may be used. Franki piles fall in this category.

A typical *precast reinforced cement concrete pile* is shown in Fig. 24.4.

### Raymond Standard Cast-in-situ Concrete Pile

The construction of this type of pile consists in driving into the ground a thin corrugated steel shell, closed at the bottom, with a collapsible steel mandrel or core in it. When the desired depth is reached, the mandrel is collapsed and withdrawn. The concrete is then poured in the shell to complete the pile, as shown in Fig. 24.5.

Spirally-wound wires at a pitch of around 80 mm are used as the reinforcement. Finally, the shell is pulled out after the concrete has fully set.

### Under-reamed Piles

An under-reamed pile is one with an enlarged base or a bulb, which is called an *under-ream*. If there is one under-ream near the bottom, it is called a *single-under-reamed* pile; if there are more than one under-ream, it is called a *multi-under-reamed* pile.

This type of pile was developed by the Central Building Research Institute (CBRI), Roorkee, and standardised by the Bureau of Indian Standards, New-Delhi. [IS: 2911 Part III-1980 (first revision): Code of Practice for Design and Construction of Pile Foundations, Part-III, Under-

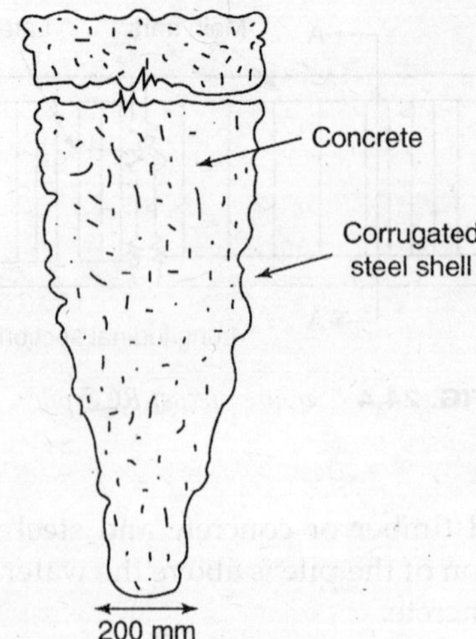


**FIG. 24.5** *A typical Raymond standard pile*

reamed Piles (First Revision)–1980, BIS, New Delhi.] Under-reamed piles were developed as foundations in black-cotton soils or expansive clays.

Under-reamed piles are of the bored cast-in-situ type with one or more bulbs formed by suitably enlarging the bore-hole for the pile stem with a specially developed under-reaming equipment. The sides may be stabilised, if necessary, by using bentonite slurry, also called *drilling mud*. A typical bored cast-in-situ under-reamed pile is shown in Fig. 24.6:

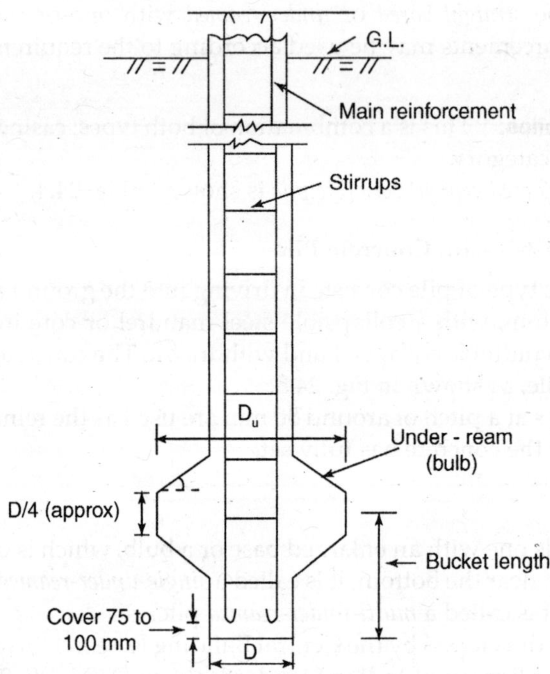


**FIG. 24.6** *Single under-reamed Pile [IS: 2911 (Part-III-1980)]*

Under-reamed piles with more than two under-reams are not advisable in strata needing stabilisation of bore holes with drilling mud.

The following are the uses of under-reamed piles:

(i) To avoid the undesirable effects of seasonal moisture changes, as in expansive soils.
(ii) To reach firm strata for adequate bearing capacity
(iii) To obtain adequate capacity to resist vertical and lateral loads.
(iv) To take the foundation below the scour level.

The ultimate load capacity of an under-reamed pile can be calculated from the soil properties – unit weight, cohesion, angle of internal friction, and the dimensions of the pile and the under-reams.

A factor of safety of 2.5 in compression and 3 in uplift are recommended to obtain the safe load from the ultimate value.

[Appendix-B of IS: 2911 (Part-III) – 1980 lists the safe loads for vertical bored cast-in-situ under-reamed piles in sandy and clayey soils for specific sizes and reinforcements – under compression, under uplift, as also under lateral thrust.]

Another type of under-reamed pile is the *bored compaction pile.* While in the case of the bored and cast-in-situ type, a hole is made by a spiral earth auger and the under-ream created at the desired depth by means of an under-reamer and the bore is filled with concrete around a reinforcement cage, in the case of a bored compaction pile, the concrete is compacted by driving a suitable core directly after placement. Sometimes the reinforcement cage itself is driven through the concrete to effect the necessary compaction. The number of bulbs is limited to two in this case. Bored compaction piles are said to have higher load capacity than the normal under-reamed piles of the same size. However, these are known to have certain drawbacks, the primary one being disturbance of the concrete and the side soil.

### Sheet Piles

These are generally not designed to take vertical loads. They are relatively thinner and are hence, known as sheet piles.

Sheet piles function as:

(i) Cut-off walls under dams to impede sewage.
(ii) Bracings for the sides of foundation trenches.
(iii) The main elements of cofferdams which are used to keep the work area in a stream or a river in a reasonably dry condition to enable the work of placing the foundation for a bridge substructure, or a dam.
(iv) Fenders for water-front structures to take the impact of vessels and floating debris.
(v) Retaining the structures in docks, wharves, and other marine works.
(vi) Walls to isolate the foundations to the adjacent soils and structures from the vibrations of structures.
(vii) Protection for river banks against erosion.
(viii) Confinement walls for soils to increase the bearing capacity.

Sheet piles may be of timber, steel or concrete; they are, in general, made with *interlocking* sections.

**Timber sheet piles:** These are used commonly for temporary works such as cofferdams. They consist of wooden planks 75 to 150 mm thick, 200 to 300 mm wide, and 2 to 4 m in height. They are chamfered at the bottom to act as a cutting edge. They can be fitted with steel shoes at the bottom as well as the top for protection during driving.

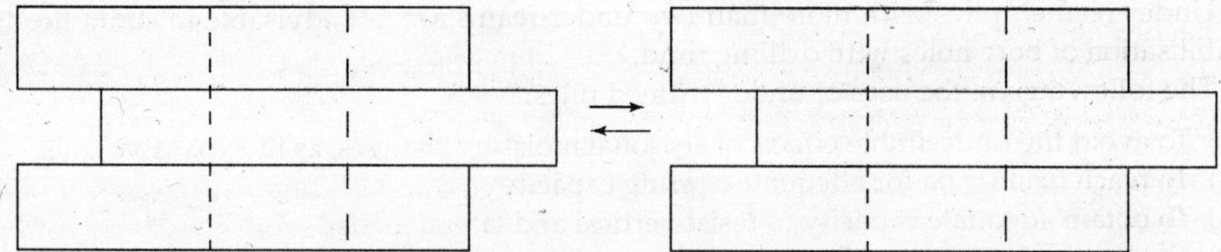

**FIG. 24.7** *Wakefield sheet piling of timber*

For interlocking, the ends of timber sheet piles are made to fit as the *tongue-and-groove* or the *dovetail* for adjacent sheet piles.

*Wakefield* type timber sheet piles are very popular (Fig. 24.7).

**Steel sheet piles:** These are very commonly used. Various patented forms of interlocking steel sheet piling have been rolled and used. Steel sheets, 200 to 300 mm wide and 4 to 5 m high, with suitable interlocking arrangements are produced in order to form fairly water-tight joints; these are very useful in cofferdams and cut-off walls. Two types are shown in Fig. 24.8.

**Concrete sheet piles:** These are always precast and are invariably provided with reinforcements. They are square or rectangular in section and are driven side by side to form a continuous wall.

One popular type is shown in Fig. 24.9.

The circular holes formed by adjacent piles are filled with rich cement mortar (1:3) grout under pressure for water tightness.

**Pulling of Piles**

Pulling out of piles from their position may be required for the following reasons:

(a) Replacement of piles damaged during driving.
(b) Salvage for reuse if the pile foundation is not required for any reason.
(c) Salvage for reuse of the sheet piles in a temporary structure such as a cofferdam.
(d) Exploratory purposes of the soil strata by conducting pull-out tests.

Water jets or compressed air may be used to reduce the skin friction around the piles during the process of pulling out.

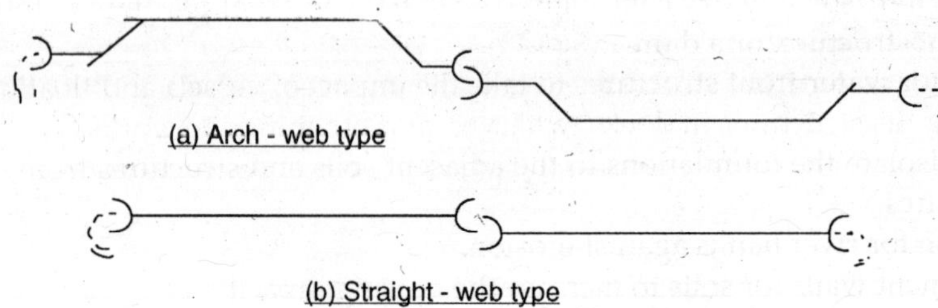


**FIG. 24.8** *Interlocking steel sheet piling*

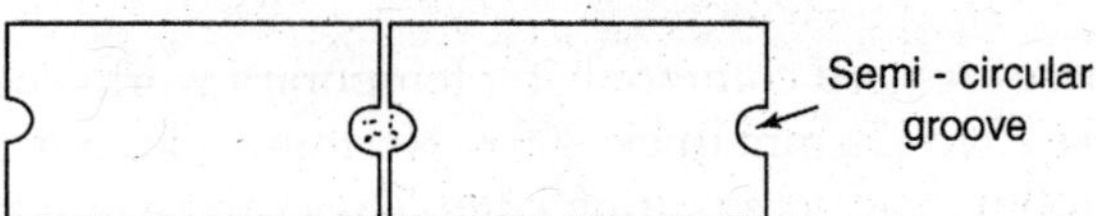


**FIG. 24.9** *Precast concrete sheet pile*

Steel piles can be pulled out relatively easily and, therefore, have the maximum salvage value. Concrete piles cannot be pulled out easily and are liable to damage; hence they have the least salvage value.

Specially designed pile extractors, double-acting steam hammers that can be worked in the reverse direction and vibrators may be successfully used for extracting steel piles from their position, with minimum damage.

### Construction of Pile Foundations

Construction of a pile foundation involves two steps – the installation of piles and the making of pile caps.

Installation of piles depends upon whether the piles are driven or are cast in-place. Pile driving by means of a pile driver and a hammer is the standard procedure for placing timber piles, steel piles, and precast concrete piles in their final position.

Damage due to driving may be minimised if driving is stopped when the penetration offers the desired resistance. Some degree of tolerance in alignment and position has to be permitted since piles can never be driven absolutely vertical and true to position as planned.

A pile may be considered defective if it is damaged by driving or is driven out of position beyond the limits of tolerance, or is bent along its length. Such defective piles must be withdrawn and replaced by another pile; alternatively, they may be left in place and another pile driven adjacent to each of the defective piles firms.

Pile driving may induce subsidence, heave, compaction, and disturbance of the surrounding soil. These effects are to be carefully studied so as to understand their bearing on the capacity of the pile.

Cast-in-place piles are mostly concrete piles of patented type such as Raymond piles, Franki piles and Simplex piles, so called after the piling firms.

Once the group of piles is placed in position and the cap is also made, construction of the substructure above can be commenced.

### Drilled Piers

These are also called *drilled caissons* or drilled foundations. They are typically larger in section than piles. A drilled pier is typically a member designed to take heavy compressive loads.

An opening is drilled to the desired depth and concrete is poured to the required shape. Generally, a drilled pier consists of a bell at the bottom, a shaft in the middle, and a cap at the top. A typical drilled pier is shown in Fig. 24.10:

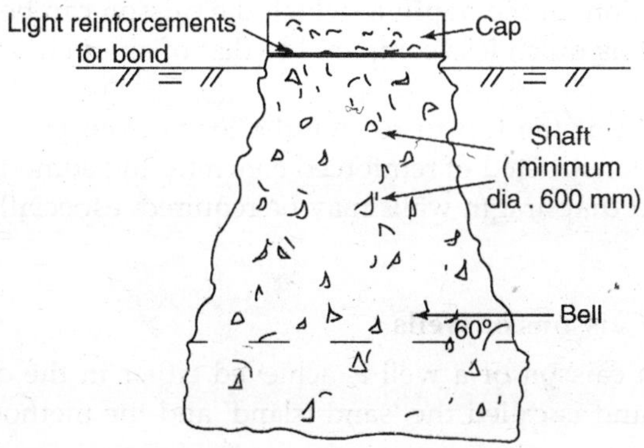

**FIG. 24.10** *Drilled pier*

Usually, pier foundations are used for bridge substructures as heavy loads are expected on to abutments and piers.

## Caissons

Caissons may be of the following types: open caissons or well foundations, pneumatic caissons, or box or floating caissons.

### Construction of Pneumatic Caissons

This type of caisson may be constructed at the site or floated to the site and lowered from barges. The cutting edge is carefully positioned before sinking. Compressed air is introduced into the chamber to keep off mud and water. After dewatering the working chamber and keeping it dry, the working crew descends into the chamber through airshafts or via airlocks. As the excavation proceeds in the dry, the caisson gradually sinks. The air pressure is increased to equalise the pressure due to the head of water outside, as the sinking goes on. The excavated material can be removed by buckets through the air shafts. In granular soils, the excavated material can be removed by the blow-out method through a blow-out pipe. When the valve in the blow-out pipe is opened, the granular material is blown out by high air pressure in the working chamber.

After the caisson has reached the desired depth, the working chamber is filled with concrete. The air pressure in the chamber is kept constant until the concrete has hardened. A rich mix of concrete is then packed into the working chamber up to the 'roof level'. Cement grout is used to pack any space left between the concrete and the underside of the roof of the working chamber. No space should be left since it may lead to settlement when the caisson is loaded.

The shaft tubes are then dismantled, and finally, the shaft itself is filled with lean concrete.

The depth to which a pneumatic caisson may be used is limited by the physiological endurance of humans under compressed air pressure; this is considered to be about 40 m.

### Construction of Floating Caissons

Floating caissons are constructed or cast on land (depending on whether they are of masonry or concrete), floated to the desired location, and sunk to the desired depth in already excavated space. The sinking is aided by a ballast of sand or gravel. Unlike other types, a floating caisson does not penetrate the soil, but is made to rest on a levelled surface of hard stratum/ rock. Only a small depth of anchoring is needed. Thus, it is preparation of the base which involves more work.

There is also a limitation on the depth to which the caisson can be taken, as also the load-carrying capacity, which is much less compared to that of an open or a pneumatic caisson of comparable size.

A concrete cap is cast at the top to receive the loads from the superstructure. Floating caissons are invariably constructed of reinforced concrete, and sometimes with steel.

Internal strutting and diaphragm walls may be required, especially if it is to be floated in rough waters.

### Construction of Open Caissons or Wells

The sinking of an open caisson or a well is achieved either in the dry or from an artificial island. The artificial island is called the 'sand island' and the method itself is known as the 'sand island method'.

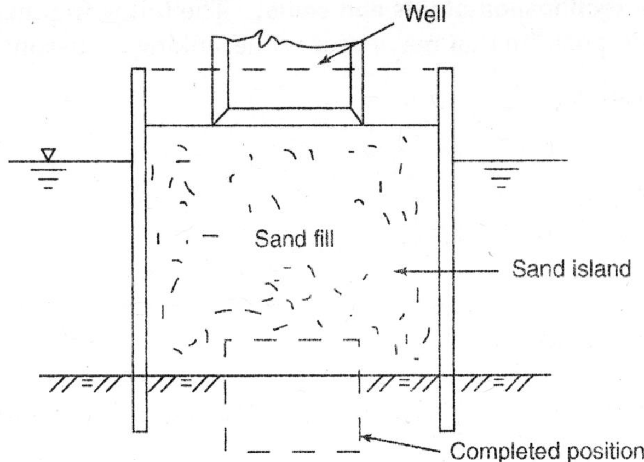


**FIG. 24.11** *Sand island method of placing a well foundation*

An artificial island of sand is made for raising the ground surface temporarily above the water level. Thus a relatively dry area is obtained for sinking the caisson. (Fig. 24.11).

The size of the sand island should be sufficient to provide adequate working space around the caisson.

For the construction of the sand island a woven willow mattress is first sunk to the river bed to protect against scour. A timber staging is constructed around the periphery of the proposed island. Sheet piles are driven to enclose the area and the area is filled with sand up to the proposed level.

Sinking of the caisson is done through water and then it is penetrated into the soil. As the soil is excavated from the well, the self-weight aids the sinking process. During the casting of a lift and curing, the sinking is necessarily stopped. Water jets are occasionally used to aid sinking.

When the well has reached the final location, the bottom is plugged with a concrete seal, using the Tremie method. The water inside the well is pumped out before further concreting is done.

If the depth of water is large and the velocity is high, wells can be fabricated on the banks of the river and floated into the desired position and sunk to the final location. Alternatively, a well can be constructed in lifts, each lift being sunk to its final position.

Kentledge (extra dead weight in the form of sand bags), along with water jets, may be used in the sinking process.

The steps in the entire process are:

(i) Construction of the well curb
(ii) Construction of well steining
(iii) Sinking process

**Shifts and tilts:** It is not easy to sink the well exactly vertically and at the designed location.

The following precautions may be taken in this regard:

- The outer surface of the well curb and steining should be smooth.
- The curb diameter is made 40 to 80 mm larger than the outer diameter of the steining.
- The cutting edge should be uniformly thick and sharp.
- Dredging should be done uniformly on all sides.

**Remedial measures for rectification of tilts and shifts:** The following measures may be taken to rectify the changes in position that may occur while sinking a caisson:

1. Regulation of excavation
2. Eccentric loading
3. Water jetting
4. Excavation under the cutting edge
5. Insertion of wood sleeper under the cutting edge
6. Pulling the well
7. Strutting the well
8. Pushing the well with jacks.

## Recommendations of IRC for the Construction of Deep Foundations for Road Bridges (IRC: 78–2000)

### Pile Foundations

**Driven cast-in-situ piles:** While concreting uncased piles, sufficient head of concrete is to be maintained to prevent inflow of soil into the concrete. Uncased cast-in-situ piles shall not be allowed where mud flow conditions exist.

Before commencement of pouring of concrete, it should be ensured that there is no ingress of water in the casing tube from the bottom. Sufficient head of concrete shall be maintained inside the casing tube at all stages of withdrawal.

Concrete shall be cast up to a minimum height of 600 mm above the top level of pile, which shall be stripped off to obtain sound concrete either before final set or after three days.

**Bored cast-in-situ piles:** Drilling mud, such as bentonite suspension shall be maintained at a level sufficiently above the surrounding ground water level to ensure the stability of the strata being penetrated until the pile has been concreted.

Tremies of 200 mm diameter shall be used for concreting. The tremie shall be withdrawn slowly ensuring adequate height of concrete outside the tremie pipe at all stages of withdrawal.

The tremie pipe should be lowered to the bottom of the borehole, allowing ground water or drilling mud to rise inside it before pouring concrete; the pipe shall always be kept full of concrete and penetrate well into the concrete in the bore hole with adequate margin of safety against accidental withdrawal.

**Driven precast concrete piles:** This type may be adopted when the length of pile is known with reasonable accuracy. An extra length may be cast to avoid lengthening by splicing. During installation, the final set or penetration of pile per blow shall be calculated taking the average of the last 100 blows.

### Well Foundations

**Tilts and shifts:** A tilt of 1 in 80 and a shift of 150 mm due to translation (both additive) in a direction which cause the most severe effect shall be considered in the design.

If the actual tilts and shifts exceed these limits, remedial measures shall be taken to restore the well within the limits

**Cutting edge:** This shall be of mild steel and should be strong enough to withstand the resistance of the strata being penetrated.

**Well curb:** This should offer minimum resistance during sinking, but strong enough to transmit the loads from the steining to the bottom plug. It should be at least of M25 concrete with a reinforcement of 72 kg/m$^3$.

**Bottom plug** It shall be provided in all wells, and the top shall be kept not lower than 300 mm in the centre above the top of the curb. The concrete used shall have a minimum cement content of 330 kg/m$^3$ and a slump of about 150 mm to permit easy flow of concrete through the tremie to fill up all cavities. The concrete shall be laid in one continuous operation.

For under-water concreting, the cement content shall be increased by 10%. If any dewatering is required it shall be done after 7 days have elapsed after bottom plugging.

**Filling the well:** Sand or excavated material shall be used for filling the well above the bottom plug.

**Plug over filling:** A 300 mm thick plug of M15 concrete shall be provided over the filling.

**Well cap:** As many longitudinal bars as possible coming from the well steining shall be anchored into the well cap.

**Sinking of well:** This shall not be started until the well steining has been cured at least for 48 hours.

Blasting may be employed to help sinking of the well for breaking boulders but with due care not to disturb the structures in the vicinity

## 24.4 CONSTRUCTION OF SUBSTRUCTURES

### Abutments and Piers

#### Setting Out

The positions of the abutments and piers have to be set out correctly on the levelling course of lean concrete that is laid on the prepared foundation.

The following are the guidelines for setting out:

**Benchmarks:** It is always desirable to use Great Trigonometrical Survey (GTS) benchmarks, if available nearby; if they are not available, suitable benchmarks should be established near the site, which can then be connected to a GTS benchmark.

Reference points, at least three, should be chosen with respect to which the benchmark at the site can be re-established if it is disturbed. These should be available until at least the work on the substructure is completed.

**Principal reference lines:** The longitudinal and transverse centre lines of the bridge are the principal reference lines, which may be used for setting out the centre lines of the abutments and piers.

**Periodical checks:** These have to be made during the construction to ensure the correct positions and alignment of the abutments and piers.

Unless these precautions and periodical checks are observed, the final position of the tops of abutments and piers may not match the designed levels of the decking and the superstructure. Such a contingency must be averted to see that avoidable corrective measures are not needed.

## Materials for Piers and Abutments

The common materials are masonry, plain or reinforced cement concrete, or in some cases, steel. Steel is used primarily for the trestle bent type of construction.

Masonry may be brick masonry or stone masonry; the former is used for culverts, minor bridges and for small spans. Stone masonry (preferably granite) in cement mortar is to be used. First class coursed rubble masonry in cement mortar 1:4 may be used. Composite construction with stone masonry facing and mass concrete hearting can also be used.

In recent practice, concrete of grade M20 with nominal surface reinforcement is being used. For reinforced concrete piers, concrete of superior grade such as M25 to M35 may be used.

### Caps for Piers and Abutments

The cap at the top of a pier or an abutment, also known as the *bridge seat* or *bed block* is a block, which provides the immediate bearing surface for the support of the superstructure at the pier or at the abutment. It facilitates even dispersion of the superstructure loads through the bearings to the substructure.

The pier cap should cover the entire area of the top of the pier and also project at least 75 mm beyond it. The cap should be of M20 concrete with a minimum thickness of 225 mm up to a span of 25 m and 300 mm for longer spans (IRC: 78–2000). Nominal reinforcement of not less than 1% steel distributed equally at top and bottom is provided; this is provided in two directions, both at the top and at the bottom. The thickness of the cap should not be less than 500 mm over masonry piers.

**Piers:** Piers are meant to transfer the vertical loads to the foundation and to resist all transverse forces acting on the bridge. The appearance of piers is important from the point of view of the aesthetic appearance of the bridge.

Piers can be solid, cellular, or trestle-type; if built of reinforced concrete, piers can be of framed type. These are shown schematically in Fig. 24.12.

Solid piers can be of mass concrete or of masonry for heights up to 6 m and spans up to about 20 m. Composite construction with stone masonry for the outer portion and concrete for the inner enhances the appearance of the pier.

Single-column type R. C. piers are popularly used in flyovers in urban areas, since these allow for free flow of traffic below on either side of the columns.

All the types other than solid masonry piers are suitable for larger heights and greater spans. The thickness of the cellular type of piers should not be less than 300 mm; this leads to saving in concrete but increases the labour cost for formwork. *Trestle-type R. C. piers* consist of circular or octagonal columns with a bent cap at the top. For tall trestles used in flyovers, diaphragms connecting the columns of the trestle may be used for increasing the strength and stiffness.

**Abutments:** An abutment is the substructure provided at the two ends of a bridge for building superstructure at the bridge ends and also to retain and support the embankments needed for approaches to the bridge. In the case of river bridges, an abutment also protects the embankment from scour.

The structural elements of an abutment are:

(i) the *breast wall*, which supports the superstructure loads and also retains the embankment fill in its rear.

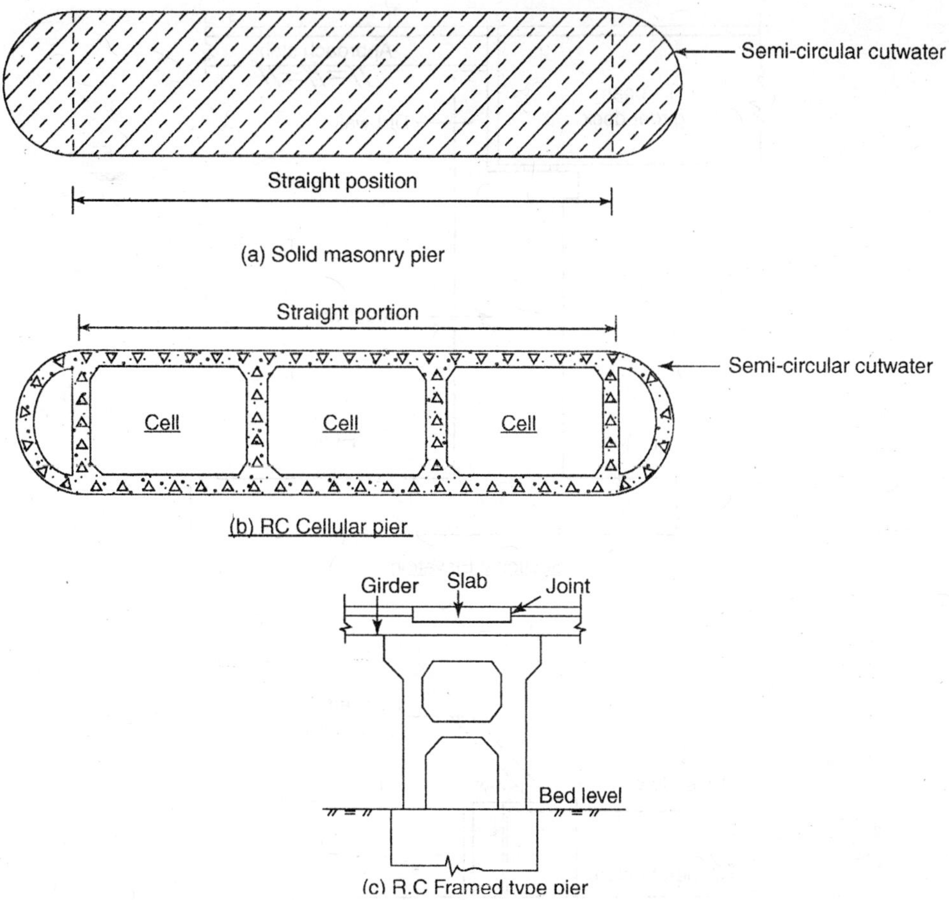


**FIG. 24.12** *Typical piers for bridges*

(ii) the *wing walls* which are the extensions of the breast wall in retaining the backfill.

(iii) the *dirt wall* (or the black wall), which is a small retaining wall just behind the bridge seat, preventing the fill material from falling on the bridge seat. Weep holes are provided at regular intervals along the height and length for effective drainage of the backfill.

A typical abutment for a girder bridge with splayed wing walls is shown in Fig. 24.13.

## Wing Walls and Return Walls

Wing walls can be straight with the abutment or parallel to the stream flow or splayed away from the stream at about 45°, or at right angles to the stream flow at either end of the abutment. In the last case they are called *return wing walls*. Invariably, the first two types are provided with short lengths of walls at right angles to the stream flow, away from the stream. Such short walls are called *returns* or *return walls*. The primary function of the wing walls is to provide a smooth entry into the bridge site for the stream flow and support and protect the embankment near the abutments for the approaches to the bridge.

The short returns serve the purpose of providing anchorage to the wing walls into the earth fill.

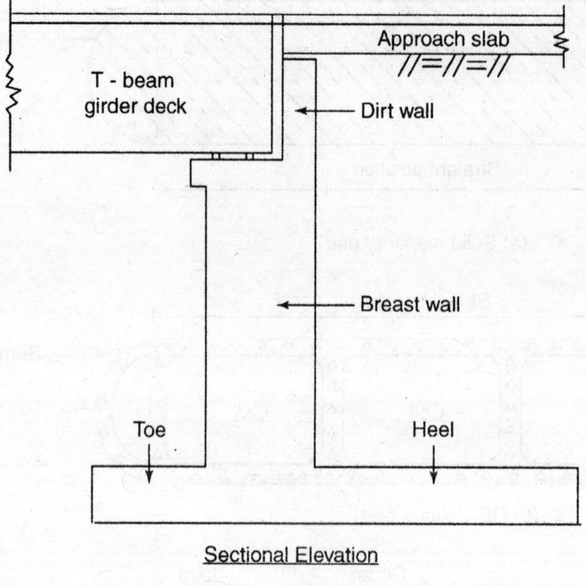


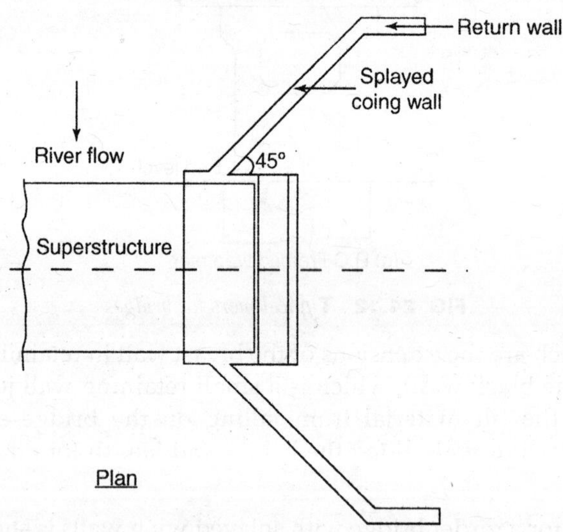


**FIG. 24.13** *A typical abutment with splayed wing walls*

Wing walls and returns are usually constructed of masonry or cement concrete along with steel reinforcements as required, which can be nominal since they need not support vertical compressive loads.

The height of the wing wall is progressively reduced as the height of the fill it has to support decreases along the length of the wall, moving away from the ends of the abutments.

The returns, which are small in length, are usually constructed with a constant height equal to that of the wing wall at their junction.

Weep holes are provided in these substructure elements also to provide effective drainage of the backfill. The backfill material is carefully chosen from suitable granular soils.

## Bridge Approaches

The short lengths of the communication route – the highway or the railway – at either end of the bridge, leading to the bridge proper, are known as the *bridge approaches*. These are an important and integral part of a bridge project, since effective utilisation of the bridge depends, to a certain extent, on the efficient design and construction of the approaches.

The following are the *important considerations* in the *construction of bridge approaches*:

(i) For high-level bridges, the approaches are provided in embankment, while they are in cutting for causeway and submersible bridges.

(ii) It is, in general, preferable to provide approaches in embankment. The material required for the fill has to be obtained from borrow pits, which should be far away from the embankment. This precaution is necessary to ensure the safety of the embankment.

(iii) The fill material should be chosen such that it is mostly granular with minimum clay content; this is to ensure that the fill is free-draining for its stability.

(iv) The approaches should normally be wider than the bridge, or at least, the widths shall be equal.

(v) The approaches should have a minimum straight length of 15 m on either side. This may be increased to match the sight distance requirements for the design speed, if necessary.

(vi) A reinforced concrete approach slab is usually provided on either side of a concrete bridge to function as a smooth transition between the paved roadway and the siding surface of the bridge. The slab should cover the full width of the roadway and extend to a distance not less than 3.5 m into the approach. The top of the approach slab should conform to the cross profile of the top of the deck slab. The minimum slab thickness at the ends should be 300 mm. In general, the approach slab is heavily reinforced with steel bars in both ways at the top as well as the bottom.

(vii) The bridge, as such, may be extended into the banks for some distance in order to provide better substructure to the approaches; the extended portion may or may not have the same type of superstructure as the bridge proper.

(viii) Beyond the minimum straight length, the approaches may be on horizontal curves, if necessary; however, the curvature should be smooth and easy.

(ix) The approaches should have gradients less than the ruling gradient, that is, in the range of 1 in 30 to 1 in 20. It is desirable that the approaches are level for some distance beyond the bridge.

(x) In urban areas, where the land is costly, the approaches are made with near vertical faces of the fill, with retaining walls at either end; this would limit the width of acquisition of land for the approaches.

(xi) The approaches should be carefully treated at their junction with the bridge deck for a smooth transition.

(xii) The importance of bridge approaches cannot be overemphasised; hence, adequate maintenance is required, especially when such approaches are newly constructed.

## Recommendations of IRC for the Construction of Bridge Substructure for Highway Bridges [IRC: 78–2000]

### Piers

Piers may be in PSC, RCC, PCC, or masonry. Only solid section shall be adopted for masonry piers. The thickness of the walls for hollow-concrete piers shall not be less than 300 mm.

Multi-column piers of bridges across rivers carrying floating debris should be braced throughout the height by a diaphragm wall of minimum 200 mm thickness.

Hollow piers shall be provided with suitably spaced weep holes of 75 mm to 100 mm diameter to enable free flow of water to equalise water levels on either side.

**Abutments**

The abutments may be of plain or reinforced concrete or of masonry; they may be of the solid type, buttressed type, counterfort type, or spill-through type.

In the case of spill-through type abutment, it should be ensured that the slope in front of the abutment is well protected by means of stone pitching and aprons.

**Dirt Walls, Wing Walls and Return Walls**

Wing walls shall be of sufficient length to retain the roadway to the required extent and to provide protection against erosion. They may be of solid type.

A dirt wall shall be provided to prevent the earth from approaches spilling on the bearings; the minimum thickness shall be 300 mm.

Return walls may be of solid or counterfort type. The material used may be plain or reinforced concrete, or masonry. The top of wing/return walls shall be carried above the top of the fill by at least 100 mm.

Cantilever returns, where adopted, shall not be more than 4 m long. Wing and return walls shall be provided with separate foundations at their junction; they may be suitably stepped up towards the approaches, if necessary, for the site conditions.

## 24.5 COFFERDAMS

**Definition:** A *cofferdam* is a temporary structure which is built to exclude water from an area in a river/stream bed in order to make it possible to carry on construction work under reasonably dry conditions.

**Uses:** Cofferdams are useful for the following jobs for underwater construction

(i) Pile driving operations
(ii) Raft foundations
(iii) Dams
(iv) Locks in harbour and waterfront structures
(v) Bridge piers and foundations.

**Requirements:** The following conditions are to be satisfied by a cofferdam

(i) Reasonable degree of water tightness
(ii) Stability against bursting, sliding and overturning under floods and anticipated forces.
(iii) Feasibility to construct in-situ.
(iv) Minimum cost for construction, maintenance and pumping out water.
(v) Ease of dismantling to facilitate reuse of salvaged material.

### Types of Cofferdams

Cofferdams may be of the following types, depending on the nature of their construction.

1. Earth-fill
2. Rock-fill

3. Rock-fill crib
4. Single-wall
5. Double wall
6. Cellular

**Choice of type:** The factors which govern the choice of type are

(a) Extent of area to be kept dry.
(b) Depth of water
(c) Velocity of flow
(d) Nature of bed material
(e) Possibility of scour due to reduction in waterway
(f) Availability of construction materials in the vicinity of work.

## Earthfill Cofferdam

This is the simplest type. This may be used in shallow waters with low velocity of flow, if impervious soil is available in the vicinity.

This type is unsuitable where there is the risk of overtopping of the cofferdam by water. (Fig. 24.14).

## Rockfill Cofferdam

This consists of rockfill and can be used for depth of water up to 3 m and even when the velocity of flow is high. It can be preferred where rock is available nearby. An impervious soil layer is laid on the water face in order to prevent water from entering the cofferdam. This type can withstand overtopping without significant damage. A typical rock-fill dam is shown in Fig. 24.15.

## Rockfill Crib Cofferdam

This type consists of timber crib with members stacked in perpendicular directions; the spaces in between are packed with rockfill material, which gives stability to the timber crib.

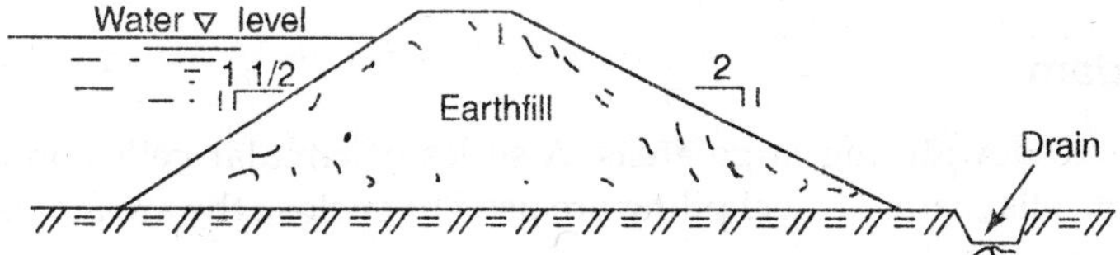


**FIG. 24.14** *Earth fill cofferdam*

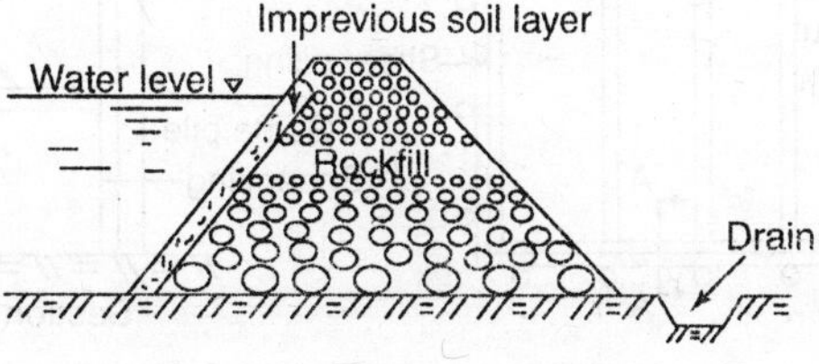


**FIG. 24.15** *Rockfill cofferdam*

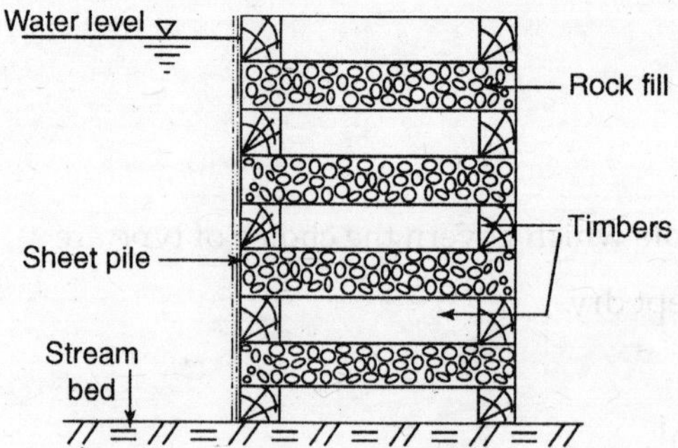


**FIG. 24.16** *Rockfill-crib cofferdam*

This type is suitable under the following conditions:

(i) The stream bed is stiff.
(ii) Depth of water is large and velocity of flow high.
(iii) Overtopping is possible.
(iv) Timber is readily available at low cost.
(v) Working space is limited.

A typical rockfill crib cofferdam is shown in Fig. 24.16.

## Single-Wall Cofferdam

Single-wall cofferdam is preferred when the working space is limited and area to be enclosed is small. It may be used up to a depth of 25 m of water.

The cofferdam wall consists of steel sheet piling; RCC and PSC sheet piles also can be used. Guide piles are used outside the sheet piling and struts are used to give support in the horizontal direction for the cofferdam (Fig. 24.17).

## Double-Wall Cofferdam

Double-wall cofferdams are preferred when a large area is to be enclosed. Since there are two rows of sheet piling at close interval, the stability is high.

## Cellular Cofferdam

This type is suitable dewatering large areas. A series of circular cells connected by arcs or diaphragm type of cells made of steel piling are used to enclose the area to be dewatered. The

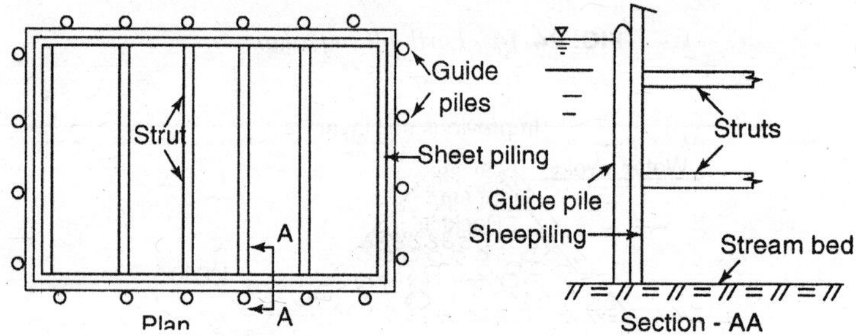


**FIG. 24.17** *Single-wall cofferdam*

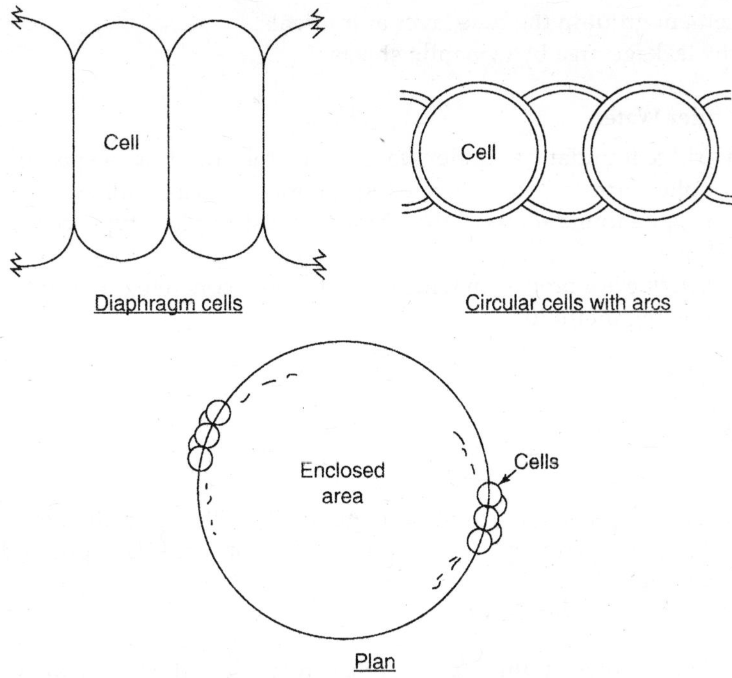


**FIG. 24.18** *Cellular cofferdam*

space inside the cells is uniformly filled with earth material or sand. As the construction is expensive, they are used for bridge substructures across large rivers.

These are schematically shown in Fig. 24.18.

### Cofferdams for Bridge Piers

Medium-sized cofferdams of 10 m to 50 m diameter have been used for the construction of bridge piers in river waters up to 20 m depth. Sheet piling with internal bracings have been used to withstand the external pressures. Concrete cofferdams with pre-cast concrete piles have also been used similar to sheet piling.

Cellular cofferdams made of cells filled with earth can be used for very large bridge piers and for suspension bridge anchorages. A concrete seal is usually provided at the bottom of the cofferdam by *tremie* method to facilitate dewatering by pumping. The seal is provided with emergency relief pipes to prevent failure of the seal in case the dewatering system malfunctions for any reason.

### Measures to Prevent Leakage in Cofferdams

Water may leak through a cofferdam either through the sheet piling or by flow underneath through the base layer. When the leakage is significant, the following measures may be adopted to reduce it:

(i) Clay soil may be laid to a small height all round the cofferdam.
(ii) Sawdust can be laid near the points of leakage, which tends to clog the openings permitting leakage.
(iii) Driving small diameter piles through the fill between the two sheet pile walls of a double wall sheet pile cofferdam, to increase the density of the fill soil.

(iv) Pumping cement grout in the base layer at intervals.
(v) Covering the leakage area by tarpaulin sheets.

**Construction Under Water**

Construction of bridge foundations under water needs the adoption of special techniques and materials. This applies primarily to concrete foundations such as shallow footings of concrete or raft foundations, pile foundations of the precast or cast-in-place type, cofferdams, caissons and well foundations.

*Under water concreting* is a process in which the prepared concrete is poured below the water surface by the following methods:

(a) Tremie method
(b) Pump method
(c) Toggle bags
(d) Bag work

**Tremie method:** A *tremie* is a watertight pipe, generally 250 mm in diameter. It has funnel-shaped hopper at its upper end and a loose plug at the bottom. It is supported on a working platform above water level.

The principle is shown in Fig. 24.19.

**Pump method:** This involves pumping concrete directly into the final position and includes both horizontal and vertical delivery of concrete. Pumping concrete has the advantage of operational efficiency with potential savings of time and labour.

This method is considered unsuitable for massive under-water concrete construction of navigation structures.

**Toggle bags:** These are ideal for placement of small quantities of concrete. Toggle bags are bags filled in the dry with wet concrete. They are then lowered under water to the desired depth at the location of the work and squeezed out by a driver.

This method is used for repair jobs.

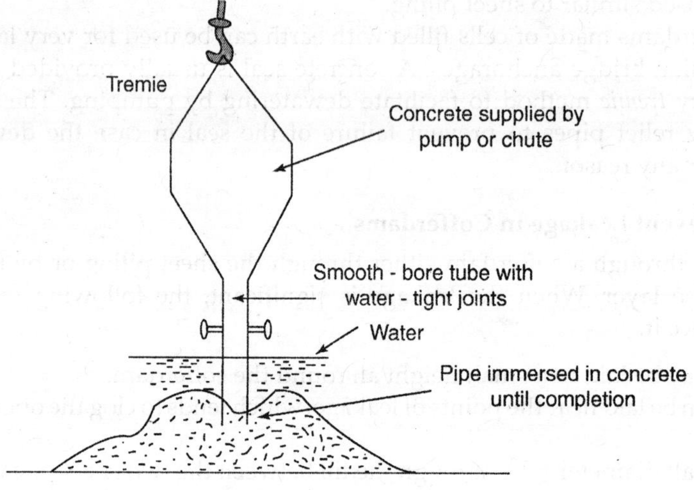


**FIG. 24.19** *Tremie method of underwater concreting*

**Bag work:** Bags, made of open weave material, are filled with wet concrete and lowered to the work spot under water, either by divers or by means of cranes. Diver-handled bags are usually of 10 to 20 litres capacity, while 1 cu m bags can be used for placement with aid of cranes.

Usually, cement for under-water concreting is of the *quick-setting* quality.

### Shotcrete

Shotcrete refers to a process in which *sprayed concrete* is used for several types of construction jobs. Some of the jobs where shotcrete is used are

**New construction**

- Retaining walls
- Diaphragm walls
- Caissons

**Underground construction**

- Tunnel lining
- Lining for storage reservoirs
- New Austrian tunnelling method

**Water retaining structures**

- Sea and river walls
- Reservoirs and dams
- Aqueducts
- Swimming pools
- Water towers
- Canal linings
- Irrigation and drainage channels

**Advantages of shotcrete:** The following are the advantages of shotcrete:

- Reduction in formwork costs
- Capability to form practically any shape
- Ease of access to restricted areas, often at a considerable distance from the point of access.

## Construction Below Ground Water Table

This involves the application of dewatering techniques or methods of lowering ground water table.

The following are the important methods of dewatering of deep excavations:

(i) Pumping out of sumps
(ii) Well point system
(iii) Deep well system
(iv) Vacuum method
(v) Cement grouting method

**Pumping out of sumps:** This method is used in shallow excavations in coarse grained soils with permeability greater than $10^{-3}$ cm/s. Shallow pits, known as 'sumps', are dug along the periphery of the area. Seepage water emerging from the toes of the slopes is guided through drains into one or more sumps; the water so collected is pumped out.

In the case of fine sand and coarse silt, *boils* may occur, which may lead to the phenomenon of *piping* in the bottom of the sump and backward erosion; this is likely to lead to subsidence of the surrounding soil or slides of the slopes. To counteract this, an inverted filter, consisting of layers of successively coarser material from the bottom of the sump upwards, may be employed.

**Well point system:** This method is based on gravity flow obtained by installing well points. These are perforated pipes of 50 to 80 mm diameter and length of about 1 m, covered by a strainer (a cylindrical wire gauge screen), which are inserted into the ground 1 to 2 m apart. The upper ends of the wells are connected to a horizontal header pipe at about the original ground water level. Self-priming pumps are fixed to the header pipe at intervals. By pumping from these wells, the ground water table is lowered sufficiently so that the seepage line does not intercept the slopes of the excavation. The strainer prevents the coarser soil grains from being pumped out.

**Deep well system:** This system is adopted when the depth of excavation is more than 16 m. A hole of 150 mm to 600 mm diameter is bored and a casing is provided. Pumping is carried out from the deep wells by means of pumps which can be lowered into the casing to the desired level. This method is more effective than the ordinary well point system.

**Vacuum method:** This method is considered suitable for soils with permeability between $10^{-5}$ cm/s and $10^{-3}$ cm/s. In this method, a hole of 300 mm diameter is bored around the well point and the riser pipe by water jetting. Meanwhile, coarse sand is shovelled into the hole to fill it up to about 0.75 to 1.00 m from the top. The top of the hole is then sealed by means of bentonite clay.

The header pipe is connected to vacuum pumps as well as to ordinary pumps. The vacuum pumps run continuously, while the ordinary pumps run as and when required.

Vacuum is created inside the sand surrounding the well points, so that the atmospheric pressure squeezes the water from the surrounding sand until the effective pressure is increased by an equivalent amount. This process aids consolidation of the ground. The soil becomes sufficiently stiff for the excavation to be carried out safely.

A schematic of the method is shown in Fig. 24.20:

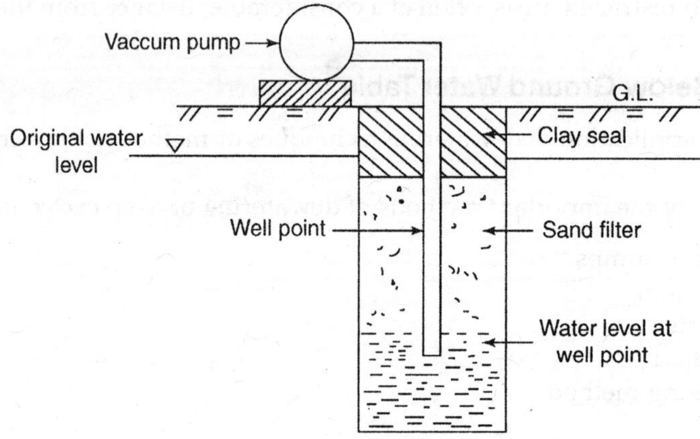


**FIG. 24.20** **A** *schematic of vacuum method of dewatering*

**Cement grouting method:** Grouting strengthens the formation by solidifying it, thereby increasing its bearing capacity. It automatically reduces the flow of water through the formation; further, the hydrostatic uplift pressure gets reduced.

Several combinations of grout materials have been used with cement, sand, clay, rock flour and asphalt, along with water. However, the simplest and best combination is a rich mix of cement and water (1:2 by volume).

Cement grouting is considered very effective in nullifying the deleterious effect of water on the bearing capacity of a soil formation.

### Construction Above Water

This includes the placement of open foundations such as spread footings and rafts above the water table. It is a relatively simple process as long as the sides of the excavation are stiff and stable. Otherwise, the need for shoring and bracing systems may arise.

In the case of deep foundations such as piles and caissons/wells, the construction will be carried out in the open on these elements when they reach or extend above the water level, without any special difficulty.

If some suitable method of lowering the ground water table is employed, what would otherwise have been under-water construction gets transformed into construction above water.

In the case of bridge substructures such as abutments, piers, wing walls, and return walls, the construction becomes relatively simple and easy, once it reaches the water level and extends above it. Here, once again, no special procedures will be needed.

## SUMMARY

- Spread footings are laid on a levelling course of lean concrete about 150 mm thick; they may be constructed of coursed rubble masonry over a plain concrete footing, or made wholly of reinforced concrete.
- Deep excavations in loose/soft soils need lateral supports at the sides, called 'shoring' or 'braced sheeting'. Depending upon the depth and nature of the soil, the bracing may be constructed with simple strut, vertical sheetings and struts, or horizontal lagging with vertical soldier beams, and so on. Sheet piling and/or lagging may be used for the side support.
- Raft foundations are usually constructed of reinforced concrete; concrete is poured in small panels of size about 10 m × 10 m to reduce shrinkage cracking. Reinforcements are made continuous across joints; rafts may be thickened across joints and under column locations.
- Pile foundations are the most popular form of deep foundations. Piles may be classified in different ways – based on function, material, and method of installation. Steel or cement concrete piles are used for bridge foundations. Steel and precast concrete piles are driven to the desired depth by means of pile-driving equipment.
- Cast-in-situ concrete piles are standardised and patented. The common types are Raymond, MacArthur, Simplex and Franki piles. Under-reamed piles developed at CBRI, Roorkee, and standardised by the BIS, serve a variety of functions.
- Sheet piles are non-load bearing as far as vertical compression is concerned; they may be of timber, steel or cement concrete. Their main use is in braced sheeting, cofferdams, and flexible earth-retaining structures such as sheet pile walls.
- Drilled piers or drilled caissons are similar to piles, but typically have much larger section area. They are used for supporting bridge piers.

- In pneumatic caissons, compressed air pressure is used to keep off water inside the caisson near the bottom to provide a relatively dry working area and to facilitate sinking by excavating near the cutting edge.
- An open caisson, called a 'well' in India, is sunk into its final position by means of an artificial sand island created using steel containers that help to retain the sand at the required level. The sand island method is a popular method of sinking wells. Remedial measures are needed to control shifts and tilts that invariably occur during the sinking process.
- Coursed rubble masonry in cement mortar or plain or reinforced concrete is used for the construction of bridge piers and abutments; caps or bed blocks are typically provided at the top.
- Piers may be solid, cellular or framed type; the last two are invariably built using reinforced concrete.
- The structural elements of an abutment are the breast wall, wing walls on either side, and dirt wall near the bridge seat.
- Wing walls may be straight or splayed at about 45° for smooth entry and exit of stream flow. If the splay is 90°, it is called a return wing wall. Short returns or return walls are built at the end of wing walls for proper anchorage into the embankment fill.
- Bridge approaches are the short lengths of roadway/railway on either side of the bridge. These are important for the efficient utilisation of the bridge. Various factors govern the construction of bridge approaches.
- A 'cofferdam' is a temporary structure meant to exclude water from the construction area and provide a reasonably dry working space for the construction of bridge foundations and substructures.
- Depending on their nature of construction, cofferdams may be classified as earth-fill, rock-fill, rock-fill crib, single-wall, double-wall, and cellular types; the last three involve the use of steel sheet piling. Cellular cofferdams with the cells filled with earth or sand have been used for placing large bridge piers on major rivers.
- 'Tremie' method is a popular method of under-water concreting. 'Shotcrete' or sprayed concrete is used for several types of construction jobs, including repair jobs.
- Popular techniques for dewatering for construction below the water table include pumping out water from sumps, well-point system, vacuum method, and cement grouting. Construction becomes relatively simple when the bridge substructure or its elements reach the water level and extend above it.

## QUESTIONS

1. Explain the construction of spread foundations for bridges.
2. Describe the construction aspects of raft foundations.
3. Describe the procedure of casting a pre-cast concrete pile with the aid of a neat sketch.
4. Explain, with a neat sketch, how a Raymond standard pile is cast in place.
5. Describe how a single under-reamed pile is constructed. How are more than one under-ream obtained for such a pile?
6. Explain the differences between the methods of placing steel sheet piles and concrete sheet piles in the foundation for a cofferdam.

7. Differentiate between the construction methods for cast-in-situ concrete piles and drilled piers.
8. Explain the method of placing a pneumatic caisson in its final position.
9. Explain with a neat sketch the sand-island method of placing a well foundation in position.
10. How are the positions of a pier set out on the levelling course at its desired location?
11. Differentiate between the caps for piers and abutments with respect to their construction aspects. Give neat sketches to support your answer.
12. Show with neat sketches the breast wall and the dirt wall of an abutment.
13. Differentiate between straight wings and splayed wings for abutments, and explain their construction features.
14. How are return walls constructed with separate foundations for wing walls?
15. Explain the important considerations in the construction of bridge approaches.
16. Explain the construction features of the following with neat sketches:
    (a) Earth fill cofferdam
    (b) Single-wall sheet pile cofferdam
17. Explain the construction features of a cellular cofferdam with diaphragm types of cells.
18. Explain the 'tremie method' of underwater concreting for bridge piers.
19. Explain 'shotcrete' and its use for repair of damaged piles below water.
20. Explain the dewatering of a formation using the vacuum method.
21. Write short notes on:
    (a) Well point system of dewatering
    (b) Cement grouting.

Chapter 25

# CONSTRUCTION OF SUPERSTRUCTURES

**CHAPTER OBJECTIVES**

This chapter delves into the following aspects of bridge construction:

- The factors governing the choice of superstructures and flooring system for a bridge
- Different types of bridge superstructures such as arches of masonry and concrete; of steel girders (cast in-situ or pre-cast)
- Erection of suspension bridges
- Methods of building superstructure – the Incremental Launching Method and the Span-by-Span Method
- Construction of joints in bridge decks – various types.

## 25.1 INTRODUCTION

The following are the different kinds of superstructures used for bridges:

1. Girder bridges or Simple bridges
2. Continuous bridges
3. Cantilever bridges
4. Arch bridges
5. Bow-string girder bridges
6. Rigid frame bridges
7. Cable-stayed bridges
8. Suspension bridges

(1) to (6) may be constructed of concrete or steel. Masonry may also be used for arch-type bridges in the case of culverts and minor bridges.

In the case of cable-stayed bridges the cables are of steel and the bridge deck may be RCC or PSC girders, or steel box girders.

For suspension bridges, the cables are of steel and bridge deck consists of the roadway or railway; therefore, the deck is much lighter in this case than in the other types. The methods of erection of a few of these, are discussed in Sections 25.4 to 25.7.

## 25.2 CHOICE OF SUPERSTRUCTURE

The following are *the factors which govern the choice of superstructure* for a bridge:

(a) Site conditions
- (i) Approaches
- (ii) Climate conditions
- (iii) Foundation soil

(b) Availability of materials and labour
(c) Availability of equipment
(d) Special requirements
(e) Availability of financial resources
(f) Economy in construction
(g) Traffic needs
(h) Aesthetic considerations

### Site Conditions

The following site conditions influence the choice of the superstructure:

**Approaches:** The formation level and the length of the approaches with respect to the HFL – these will influence the type of bridge, as also whether it has to be deck, through, or semi-through type.

**Climatic conditions:** If the rainfall is heavy, timber components are to be limited since they are likely to be less durable. Similarly, in coastal areas where the effects of corrosion are significant, steel bridges or steel trusses should be avoided to the extent possible.

**Foundation soil:** Poor soil conditions call for restriction on the number of spans and the substructure components such as piers. This may eliminate some choices, and necessitate the adoption of cable-stayed and suspension bridges.

**Availability of materials and labour:** In order to achieve economy in construction, the materials available in the vicinity have to be utilised; the availability of labour should be ensured, if the superstructure construction is labour-intensive.

**Availability of equipment:** The equipment necessary for erection of the chosen bridge superstructure must be available; hence, this may be an important factor influencing the choice.

**Special requirements:** Requirements such as for navigation in major rivers and navigable streams may dictate the choice of the type of bridge superstructure. Suspension bridges, arch bridges and truss bridges are suitable for lift or swing motions needed to facilitate navigation.

**Availability of financial resources:** In the case of a minor bridge, simple masonry arch, or steel or concrete girder bridge may suit in case the financial resources are limited.

However, in case the bridge is a major one across a river, financial resources should not be a constraint.

**Economy in construction:** With certain specialised types of superstructure, such as pre-stressed concrete girders and arches, economy in construction may be difficult to achieve. Consistent with the needs, a simple superstructure may be preferred.

**Traffic needs:** The nature and volume of traffic may govern the choice of superstructure. The number of lanes for a highway bridge and the number of tracks to be carried in the case of a railway bridge are decided based on the traffic requirements extrapolated for the design period or service life. Generally speaking, steel truss bridges are preferred for railway bridges instead of suspension bridges.

**Aesthetic consideration:** For major bridges across rivers, this is an important consideration. Bowstring girders and suspension bridges give a good appearance.

**Choice of the method of construction/erection:** Some of the factors listed above – site conditions, equipment available, availability of labour, and special requirements, if any – govern the choice of the method of construction/erection. In addition, the stipulated period of construction for the bridge project, and the sequence of construction dictated by site conditions and other related factors also may govern the choice of the method of erection. The trend is to use precast and prefabricated members and to adopt heavy machinery such as launching girders and cranes, in order to achieve speed in construction.

## 25.3 FLOORS AND FLOORING SYSTEMS FOR BRIDGES

The *flooring system* for a bridge primarily consists of the following *components*:

1. Cross girders or floor beams
2. Stringers
3. Main flooring
4. Wearing surface

### Cross Girders or Floor Beams

Rolled steel joists or plate girders are used as cross girders, or *floor beams* as they are called, and are placed perpendicular to the longitudinal centre line of the bridge. They are connected to or supported by the main girders or trusses, depending upon the type of the bridge.

In plate girder bridges, the floor beams are connected to the webs of the plate girders at either end. In truss bridges, they are connected either to the lower chord or to the vertical post above it.

### Stringers

Stringers are usually rolled steel joists supported on the floor beams and run perpendicular to them (that is, parallel to the longitudinal centre line of the bridge). They are fixed on top of the floor beams (Fig. 25.1).

### Main Flooring

There are, broadly speaking, two types of main flooring:

(a) Open floor
(b) Solid floor

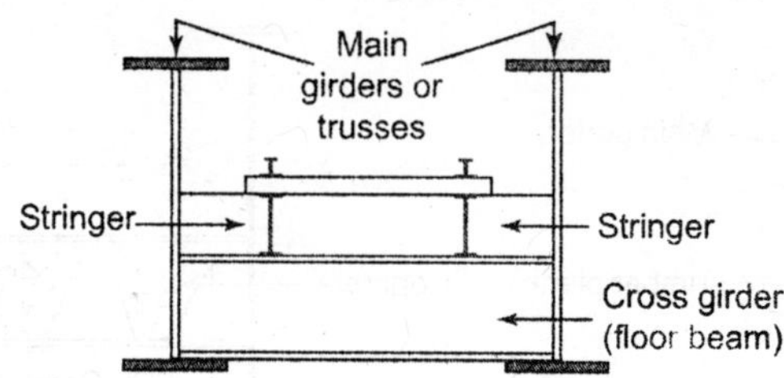


**FIG. 25.1** *Components of a bridge flooring system*

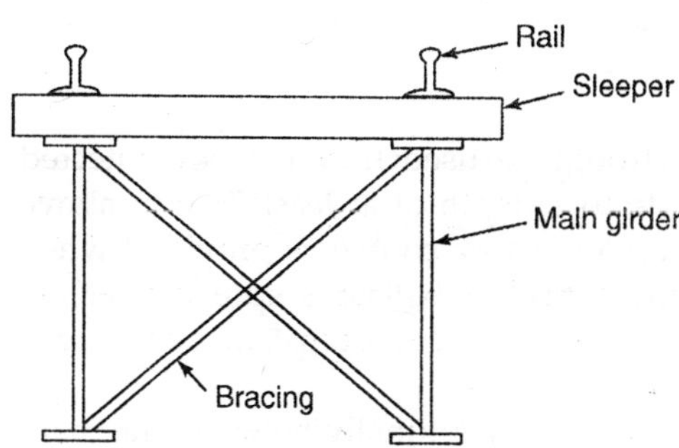


**FIG. 25.2** *Simple open floor for a railway bridge*

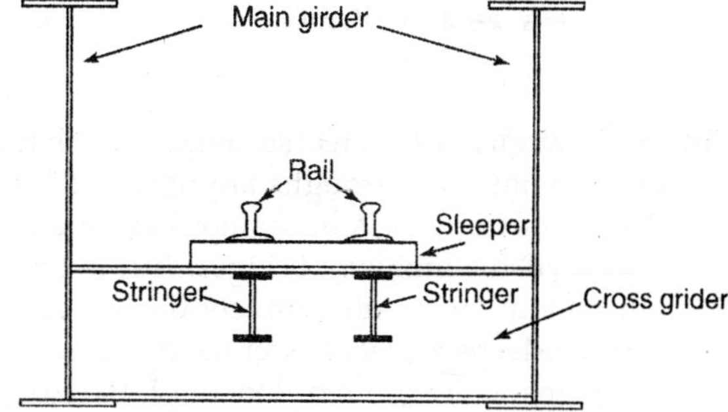


**FIG. 25.3** *Open floor for a through bridge*

**Open floor:** Open floor is used only for railway bridges. In the case of deck-type girder bridges, open floor consists of timber sleepers, placed directly over the main girders, as shown in Fig. 25.2.

In the case of through plate girder bridges, the rails rest on sleepers, which are connected to longitudinally placed stringers; these stringers, in turn, are connected to cross girders transverse to the bridge at designed spacing and connected to the main girders, as shown in Fig. 25.3.

Open floors are easily inspected and maintained; further, they don't need any drainage arrangements. They are, therefore, considered economical.

**Solid floor:** Solid floors cover the space between the girders fully; they are preferred for highway bridges.

The following types of solid floors are commonly used:

(i) Timber floor
(ii) RCC slab floor
(iii) Steel trough floor
(iv) Steel buckle plate floor
(v) RCC Jack arch floor

(i) Timber floor: Wooden planks of suitable thickness are used; this can be used for light traffic as in the case of minor bridges and culverts. The planks are placed between main girders, in a slightly inclined position (in plan), as shown in Fig. 25.4.

(ii) R.C.C. slab floor: Reinforced concrete slabs of designed thickness and reinforcements are used for bridge floors. While they are popular for highway bridges, they are not preferred for railway bridges.

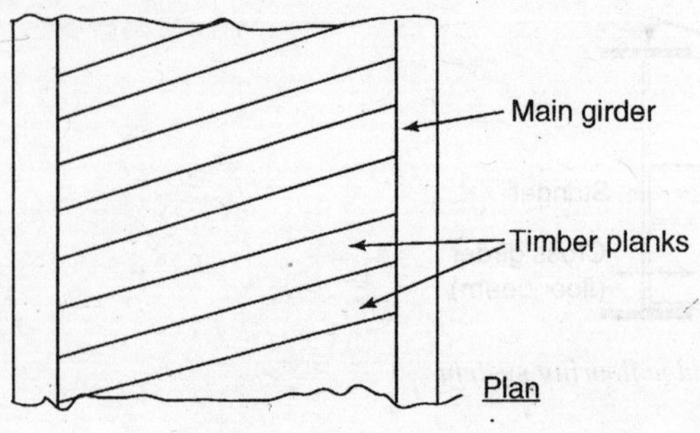


FIG. 25.4 *Timber floor*

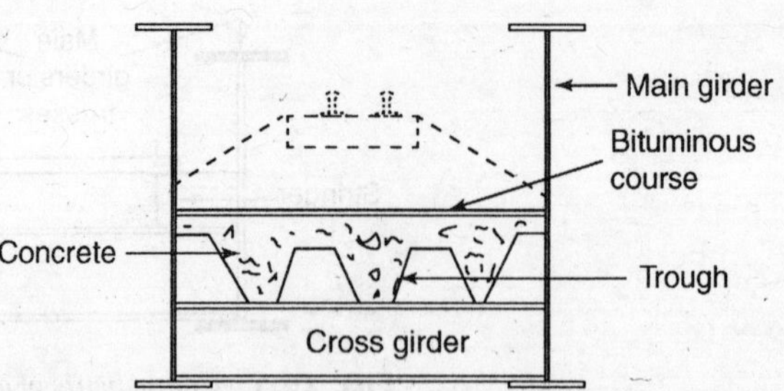


FIG. 25.5 *Steel trough floor (troughs placed longitudinally*

(iii) Steel trough floor: Steel sections of the shape of a trough are used; they may be connected by lap joints. The troughs are filled with concrete to a depth of at least 75 mm above the top of troughs. A bituminous layer, along with a wearing course, is provided when it is used for highway bridges. If it is for a railway bridge, ballast is spread over the bituminous layer. Alternatively, the troughs may be filled with ballast instead of concrete, and the sleepers placed over the troughs.
A schematic of a trough floor, with the troughs placed longitudinally between the main girders, is shown in Fig. 25.5. (The troughs span between the cross-girders in this case.) There is no need for stringers in this case. Alternatively, the troughs may be placed transversely and fixed between the main girders; in this case, cross-girders and stringers are not needed.

(iv) Steel buckle plate floor: Steel buckle plates, bent in the form of an arch, are used in this case. The buckle plates are 6 to 11 mm thick; rise of the arch is from 60 to 90 mm. Their use and placement is similar to that of steel troughs. Figure 25.6 shows a steel buckle plate floor, with buckle plate arches placed longitudinally between the main girders, and spanning between the cross-girders:
There is no need for stringers in this case. As an alternative, the buckle plates can also be placed transversely.

(v) RCC jack arch floor: This is similar to the steel buckle plate floor, except that jack arches in cement concrete are used, as shown in Fig. 25.7.
All the comments about the placement of jack arches are applicable as in the case of steel troughs and buckle plates.

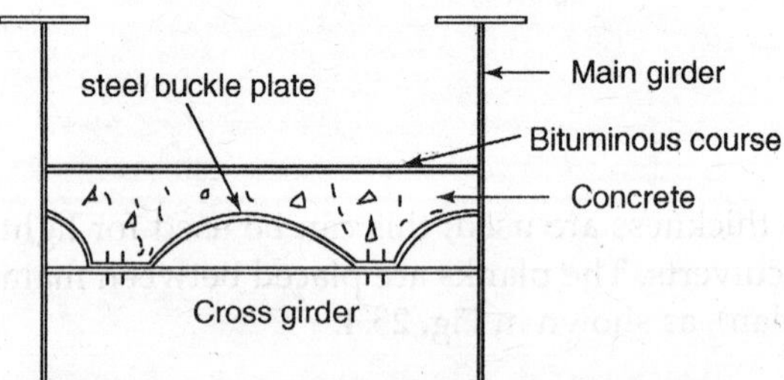


FIG. 25.6 *Steel buckle plate floor (buckle plates placed longitudinally)*

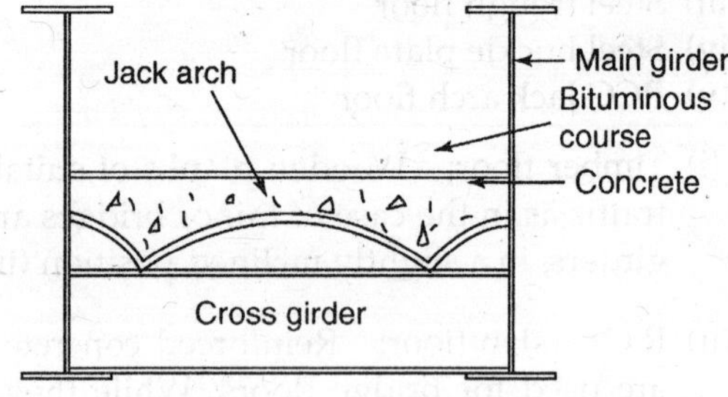


FIG. 25.7 *Concrete jack arch floor*

**Choice of flooring:** This depends upon the nature of traffic, climatic conditions, and cost considerations including maintenance.

**Wearing Surface**

This provides a smooth riding surface to the traffic. A minimum thickness of 75 mm is needed. Camber should be provided for drainage.

**Requirements of a Good Flooring System**

(i) Durability
(ii) Impact resistance
(iii) Silent riding
(iv) Good drainage
(v) Fire-resistance
(vi) Minimum economic cost.

## 25.4 CONSTRUCTION OF SUPERSTRUCTURES

As mentioned in Section 25.1, bridge superstructures are of several different types both with regard to material – steel, RCC, PSC and masonry, and their structural action – simple and continuous girders, arches, trusses, cable-stayed structures, suspension bridges, and so on. In addition, they may be of deck type, through type or semi-through type, based on the position of the bridge floor.

The methods and techniques of construction vary in their details in each case, although the general procedure and approach could be similar.

In view of this, the methods of erection of only a few of the important types – arches, girders (steel and concrete), steel trusses and suspension bridges – are presented here.

### Erection of Arch Bridges

*Centering* or *falsework* is an essential requirement for the construction of arches for bridges. This is a temporary structure to support the permanent structure such as an arch until the latter derives enough strength to support itself.

The *requirements of centering* for arches are:

(i) Adequate strength and stiffness to keep the line, level and shape during construction.
(ii) Maximum salvage value for reuse in other similar jobs.
(iii) Minimum total cost of materials and labour.

The *factors governing the choice of the type and design of the centering* are:

(i) Availability of the materials required and labour
(ii) Depth of water and stream velocity of flow.
(iii) Foundation soil conditions
(iv) Type and dimensions of the arches.

**Types of Centering for Arch Construction**

The common types of centering for arch construction are as follows:

(a) Timber truss centering
(b) Steel truss centering
(c) Trestle type of centering

**Timber Truss Centering**

Timber trusses of shape suitable for the proposed arch are constructed and used as centering.

Timber truss centering may be used if the arch is designed to bridge a deep gorge or valley; it may be used even when the supports are to rest in deep water.

The *disadvantages* of *timber truss centering* are:

- Labour involved in bringing the curvature for the top members is more; hence, the cost may increase.
- Stiffness is low and liability for deformation due to the load of the construction above is more.
- Salvage value is less and continued reuse is not possible.

**Steel Truss Centering**

Steel trusses of suitable shape for the proposed arch are constructed and used. The design of these trusses depends on the type of arch, material of construction, and duration of the work.

The circumstances in which steel truss centering is preferred:

(i) The river on which the bridge is to be built is susceptible to receive flash floods; (steel centering is generally strong enough to resist damage.)
(ii) Depth of water is high.
(iii) Foundation soil offers sufficient skin friction.

The handling of steel truss centering needs machinery such as cranes because of their weight.

A schematic of truss centering is shown in Fig. 25.8.

**Trestle Type of Centering**

In this type of centering, the arch is supported by vertical or inclined posts anchored into the ground.

There are different forms of trestle centering with steel posts:

(i) Vertical posts in a single stage
(ii) Vertical posts in two stages or tiers
(iii) Inclined posts anchored at different points

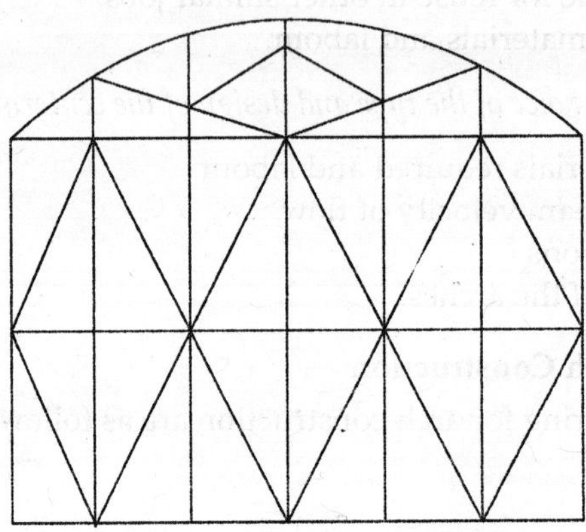

**FIG. 25.8** *Truss-type false work for arch construction*

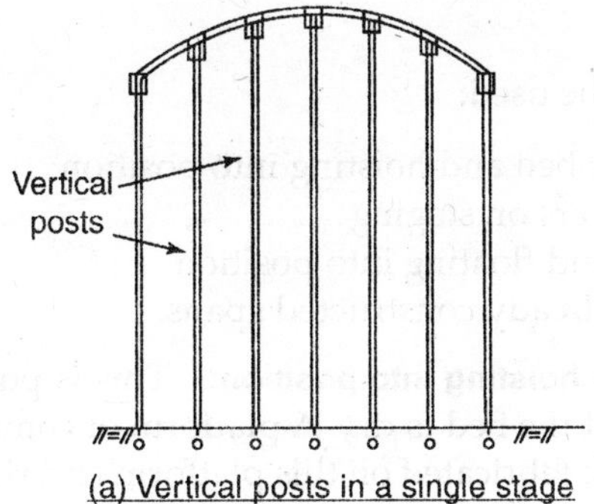


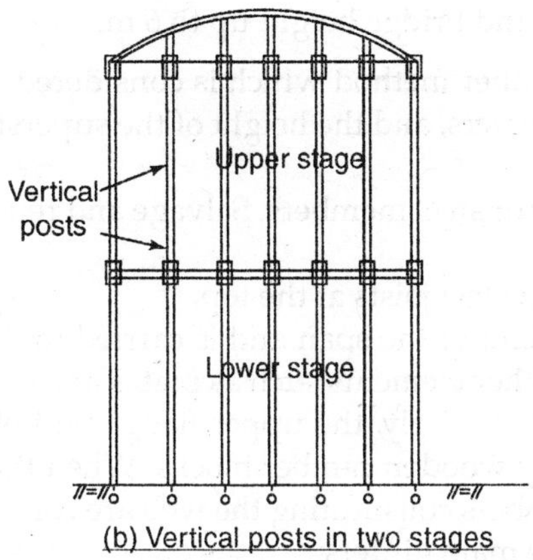


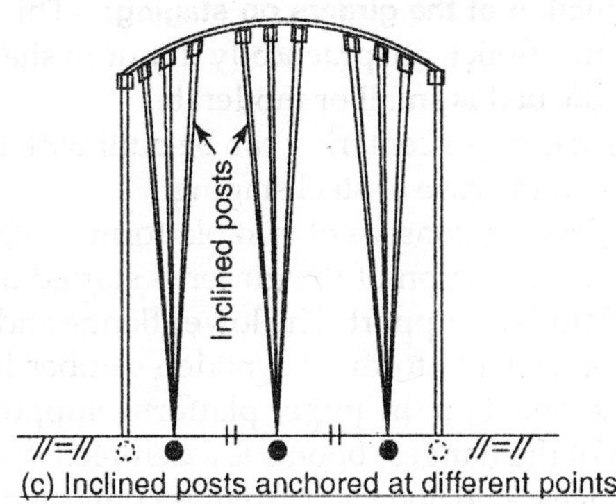


**FIG. 25.9** *Trestle-type of centering for arches*

The first one is used when the height of the arches is low, while the second and the third types are used when the bridge for which the arches are required is relatively high.

Figures 25.9 (a) (b) and (c) show these three types of trestle-type of centering schematically.

Advantages of trestle centering:

- Least deformation
- High salvage value
- It is not expensive compared to the other types

**Formwork for RCC Arches**

For reinforced concrete arches, specially designed formwork are used, over which the reinforcements are placed in position before laying the cement concrete.

## Erection of Girder Bridges

The method of erecting girder bridges depends on whether the girders are of steel, RCC or PSC.

Steel girders are usually prefabricated, while reinforced concrete girders may be precast or cast-in-situ, as also pre-stressed concrete girders. The method of construction or erection may vary in each case.

### Steel Girders

The following methods can be used:

(i) Assembling on the river bed and hoisting into position
(ii) Construction of the girders on staging
(iii) Assembling on barges and floating into position
(iv) Rolling out girders on already constructed spans.

**Assembling on river bed and hoisting into position:** This is possible when the rivers are dry for most part of the year and the bed is dry. A platform of compacted earth fill is formed and levelled. The steel girders are fabricated on this platform and riveted. The girders are lifted by means of hoisting machinery such as cranes, crab winches, and derrick cranes and placed in position. The bracings and the flooring are then connected to the girders.

This is considered suitable for spans up to 30 m and bridge height up to 6 m.

**Construction of the girders on staging:** This is another method which is considered suitable in streams which are practically dry or in shallow waters, and the height of the superstructure above the bed is small or moderate.

The staging or centering can be built with timber or steel members. Salvage and reuse value is more in the case of steel staging.

The staging consists of two platforms with rolled steel joists at the top.

The construction of the girder is started at one end of the span and is carried to the other end resting on support. The lower flange and the other elements such as plates are assembled on the lower platform on wooden camber blocks. Similarly, the upper flange and elements are assembled on the upper platform, supported by wooden camber blocks. When the whole length of the flanges/booms is assembled, the members constituting the web are connected to the flanges. The flooring system is connected to the main girders.

Finally, the bridge girders are slowly lowered on to the bearings on the piers after removing the camber blocks with the aid of jacks.

The girders are appropriately connected to the bearings after ensuring that they are in the correct position.

**Assembling on barges and floating into position:** This method is adopted when the depth of water of the river or stream is large. The bridge superstructure is assembled on barges; when the fabrication is completed, the barges are moved and positioned such that the structure is over the final alignment of the span, as shown in Fig. 25.10:

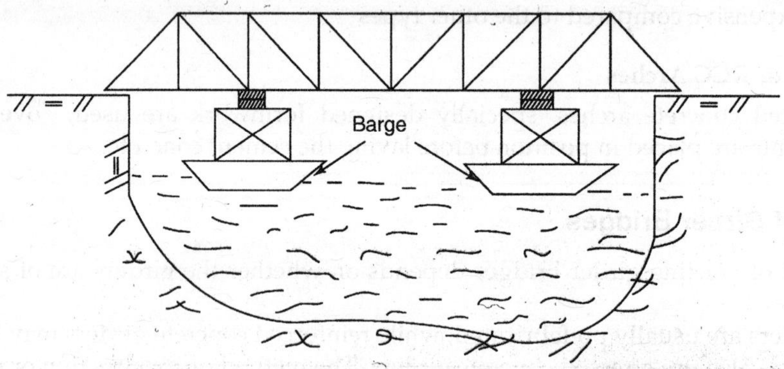


**FIG. 25.10** *Assembling on barges and floating into position*

Finally, the structure is lowered on to the bearings with the aid of jacks; then the barges are sunk and taken away from the superstructure,

Alternative procedure: The girder is assembled on the banks in alignment with the centre line of the bridge. The front end of the girder is supported on rollers and the rear end on a trolley driven by a small engine.

The front end is slightly drawn ahead from the bank and is supported on a pontoon with a staging floated under it. This floating pontoon carries the girder across the stream so that the girder can be lifted and slowly lowered on to the bearings by means of jacks. When the job of erecting the girder is complete, the pontoon can be lowered by weighting it down and withdrawn.

**Rolling out girders on already constructed spans:** This method is considered suitable for the erection of girders for a continuous bridge. The first span from the abutment is erected by one of the methods discussed above.

The girder of the next span is brought to the required position by rolling it on the already erected span from the abutment; as this span is cantilevered from the erected span, the next span, connected to it, acts as the counterweight for stability. This procedure is repeated until all the spans are erected to complete the entire bridge.

This method is advantageous especially when deep gorges are to be bridged.

### Reinforced Concrete and Pre-stressed Concrete Girders

In the case of reinforced concrete girders, the girders may be either precast or cast-in-situ.

When the girders are precast in a casting yard, they are transported up to the abutments making use of a temporary rail track, if the bridge to be built is a highway bridge. If it is for a railway bridge, the railway track is conveniently used for this purpose. The girders are placed on two trolleys, one at the front end and the other at the rear, and are moved slowly on the track up to the abutment, so that the stability of the girders on the trolleys is not disturbed.

After the girder reaches the abutment, a *launching girder* is used to carry it further on to the span. A launching girder is a kind of temporary bridge which is put on a span and is used for moving the trolleys on which the beams are carried.

The following are the steps in this procedure:

(i) The launching girder is moved to the span to be laid in the working position.
(ii) The bridge girders are transferred from the trolleys on to the launching girder.
(iii) The girders are moved along the launching girder.
(iv) The girders are slowly lowered on to the piers of the span.
(v) The girders are connected to the bearings on the piers and the flooring system is completed.

In the case of *pre-stressed concrete girders*, the girders are precast in a casting yard and transported to the abutments on a rail track as in the case of RCC girders. Post-tensioning method is usually adopted. Concrete is poured into the formwork in which corrugate conduits are placed for passing the pre-stressing cables. After the concrete has set, these cables are stretched and anchored at the ends. Only a few cables are pre-stressed before launching, just sufficient to take care of handling stresses. The remaining cables are pre-stressed after the girders are placed in the final position.

As in the case of RCC girders, PSC girders also may be wholly cast-in-situ with suitable staging support. Precautions are necessary in handling precast girders such that the handling stresses are as low as possible.

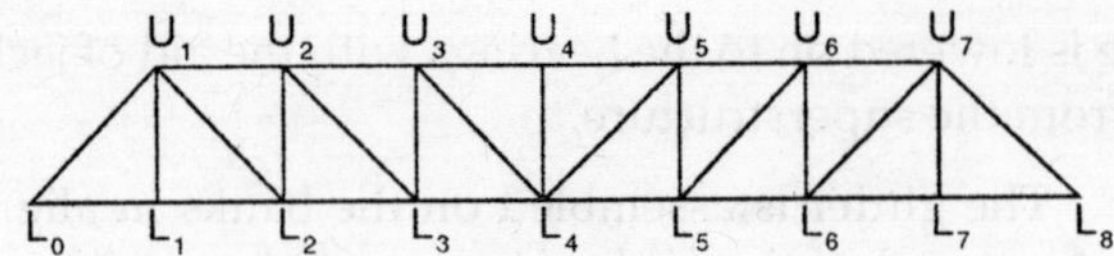


**FIG. 25.11** *Steel truss superstructure*

**Erection of Steel Truss Bridges**

Steel truss superstructures are erected with the help of a stiff derrick crane and two falsework bents. (A 'falsework bent' is a temporary structure built on the stream or river to provide support to part of the superstructure in the process of erection.) Depending on the number of panels of the truss, it is built in viable parts and erected one by one, by making use of the portions erected already.

Let the bridge truss to be erected be as given below (Fig. 25.11).

The erection procedure may be given in a few steps as follows:

***First step***

(i) The first falsework bent $F_1$ is erected in the stream.

(ii) The portions consisting of the members $L_0 L_1$, $L_1L_2$, $L_2U_1$, $U_1L_1$, and $L_0U_1$ are assembled. [Fig. 25.12(a)].

***Second step***

(i) The derrick traveller is moved to $L_2$.

(ii) The second falsework bent $F_2$ is erected.

(iii) The next portion consisting of the members – $L_2 L_3$, $U_1U_2$, $L_2U_2$, $L_3U_2$, $L_3 U_3$, $U_2U_3$, $L_3 L_4$, and $U_3 L_4$ – is then assembled [Fig. 25.12(b)].

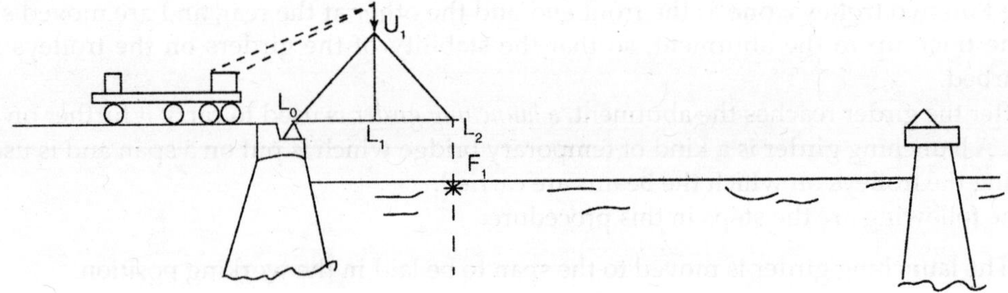


**FIG. 25.12(a)** *First step in erection of truss*

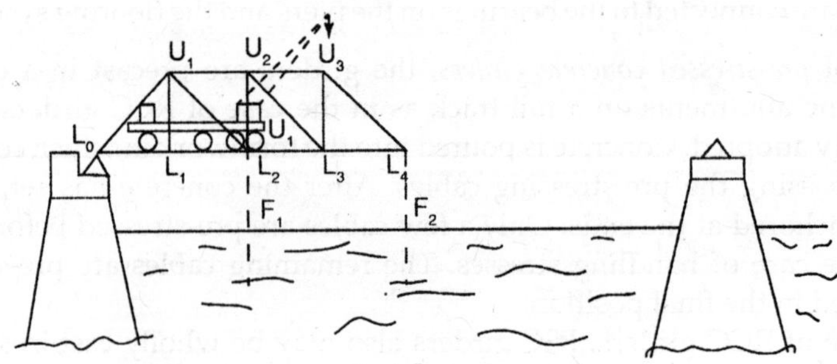


**FIG. 25.12(b)** *Second step in erection truss*

(iv) The level of falsework bent $F_2$ is adjusted to close the joint at $U_3$.
(v) Making use of the support at the falsework bent $F_2$, the truss is jacked down at $F_1$ and released.

This procedure has to be continued for two more steps to complete the erection of the whole truss for the span.

A multiple-span truss bridge which is continuous over the supports can be erected from either end until the centre is reached.

The joining of the two halves of the bridge is achieved by a suitable jacking procedure.

**Erection of Suspension Bridges**

The main elements of a suspension bridge are:

(a) Towers at either end
(b) Suspension cables
(c) Suspenders
(d) Stiffening trusses

**Construction of towers:** The first tower is erected with the help of a *floating crane* (a ship with a crane); the other towers are erected with tower cranes.

**Erection of catwalks:** *Catwalks* are working platforms in the form of foot-bridges over ropes placed parallel to the main cables. A tramway is installed from end to end along the catwalk for pulling the bridge cables.

**Erection of suspension cables:** A spinning wheel is attached at each anchorage; the cable-ends are fastened to the anchorage and looped around the spinning wheels. These wheels are pulled along the catwalks and over the tower to the anchorage on the other side; the cables are then attached to anchorage.

This procedure is repeated if there are multiple towers.

**Erection of suspenders:** Suspenders are vertical members connecting the bridge deck to the suspension cables, which, in turn support the deck. Steel castings, called *cable bands* hold the suspenders in place over the cable. At the lower end, the suspenders are connected to the bridge deck.

**Erections of stiffening trusses:** *Stiffening girders* are truss structures which are connected to the deck; their function is to stiffen the deck with a view to reducing the oscillations under traffic.

Stiffening trusses are fabricated, lifted, and hung off the suspenders before being placed in position, to be connected to the deck. They can be erected from either end near each tower, until they meet at the centre and are connected to the anchorages at the ends. Finally, the flooring system is completed and the road surface is placed over the floor.

## Recent Methods of Bridge Construction

### Incremental Launching Method (ILM)

In modern bridge construction, the *incremental launching method (ILM)* has been used for erection of bridges over deep valleys, deep rivers, steep slopes or regions with poor soil conditions that make access difficult for equipment, with minimal disturbance to the surroundings including environmentally sensitive areas.

In this method, the bridge superstructure is assembled on one side of the obstacle to be crossed; it is then pushed longitudinally and launched into its final position. The launching is typically performed in a series of increments so that additional sections can be added to the rear of the superstructure unit prior to subsequent launches.

This method has the decisive advantage of construction at ground level with a stable work platform. Safety is also enhanced in the construction.

### Span-by-span Construction Method

This is a relatively new technique, which is considered to be rapid and the most economic method of bridge construction for long bridges and viaducts.

Decks are commenced at one abutment and constructed continuously by placing segments to the other end of the bridge. The segments can be positioned either by a staying mast system or an erection girder in the form of a truss assembly.

The segments are placed on sliding pads and slid into position.

The main advantage of this system is that the already constructed portion of the bridge serves as the launching pad for successive segments.

## 25.5 JOINTS IN BRIDGES

Joints are weak spots in bridges. Unless these are properly designed, serious maintenance problems may arise, in view of the high traffic density occurring on most major bridges.

Three types of joints are provided on a bridge deck:

(a) Construction joints
(b) Expansion joints
(c) Contraction joints

### Construction Joints

These are necessary whenever concreting work has to be stopped temporarily for any reason, as at the end of a day's schedule of work, or for any unexpected contingency including failure of machinery such as concrete mixer, vibrator and so on.

Another situation where a construction joint is necessarily provided is when the foundations of different parts of a structure are at different levels; this is possible at the junction of an abutment and its wing walls. This is to minimise the effects of discontinuity on the durability and appearance of a structure due to possible differential settlements.

In general, joints should be located away from the regions of maximum stress, particularly shear and bond stresses.

### Expansion and Contraction Joints

*Expansion joints* and *contraction joints* are meant to take care of the effects of temperature changes on the bridge structure. The difference between these two types of joints is in respect of the depth and width of the joint.

**Contraction joints:** Joints to facilitate contraction, where provided, extend only for a part of the depth of the bridge slab and also will be of smaller width, while expansion joints extend to the full depth and, in general, are of larger width.

**Expansion joints:** These are considered to be an important structural element in a highway bridge. Their main function is to accommodate all movements of the bridge elements, that is, translation and rotation. The movements could be due to shrinkage of concrete, temperature changes, and long-term creep.

Expansion joints are subjected to the impact of wheel loads and the effects of water, dust, oils, and salt solutions. Leakage of water from the joints through the deck could lead to deleterious effects on the bearings and the substructure. Therefore an expansion joint should be watertight, durable, replaceable, and provide good riding quality.

The performance and durability of expansion joint systems directly affect the performance of the bridge itself.

In view of this, appropriate design, good fabrication, systematic inspection and proper maintenance of expansion joints are of utmost importance for bridges.

The Special Publication of the IRC, *"IRC: SP-69-2011 Guidelines and specifications for Expansion Joints (First Revision)"*, IRC, New Delhi, provides the necessary details to be followed in this regard.

Relevant portions of this Special Publication are given here:

In the design of an expansion joint, earthquake and dynamic effects need not be considered since it is a replaceable item with a short design life.

The prime functional requirement of a joint is to accommodate the maximum movement due to contraction and expansion, which is normally the longitudinal movement of the deck. The maximum opening of the joint is the sum of the maximum movement requirement and the minimum opening in the extreme closed condition.

## Joints for Small Openings (Movement up to 25 mm)

The following types fall in this category:

- Buried joint (movement up to 10 mm)
- Filler joint (movement up to 25 mm)
- Asphaltic plug joint (movement up to 25 mm)

### Buried Joint

This type consists of continuously laid bituminous/asphaltic surfacing over the joint gap bridged by a steel plate resting freely over the top surface of the deck slab. The joint is suitable for short-span bridges. A typical buried joint is shown in Fig. 25.13.

Limitations of buried joint are:

- Prone to deterioration due to cracking and squeezing of the bituminous surfacing and consequent bad riding quality.
- The surfacing at the joint has to be renewed frequently.
- Deterioration of steel plate due to rusting since it is inaccessible for maintenance.

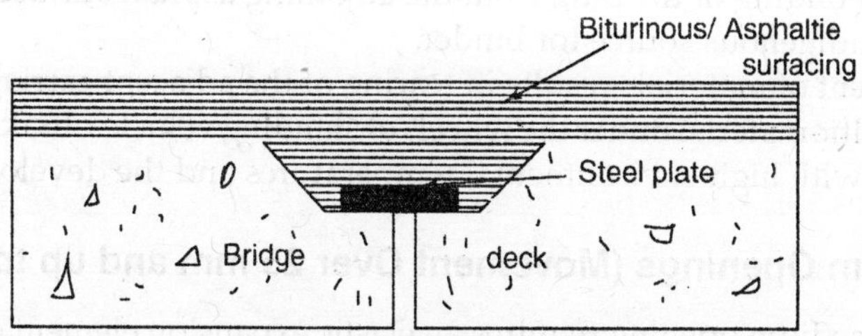


**FIG. 25.13** *Buried joint*

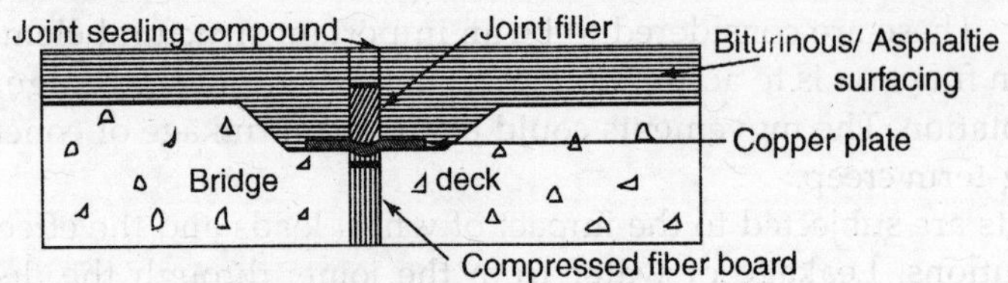


**FIG. 25.14** *Filler joint*

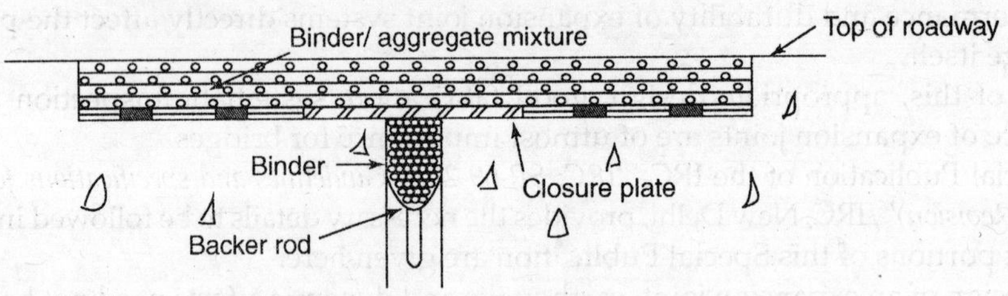


**FIG. 25.15** *Asphaltic plug joint*

### Filler Joint

The components of this type of joint are corrugated copper plate (minimum thickness 2 mm) placed slightly below the wearing coat, a 20- *mm* thick compressible fibre board to protect the edges, and a 20- mm thick pre-moulded joint filler, sealed with a joint sealing compound, filling the gap up to the top level of the wearing coat.

A typical filler joint is shown in Fig. 25.14.

The limitations of filler joints are:

- Short life
- Need to be replaced often, as compressibility of filler gets reduced with age

### Asphaltic Plug Joint

This type of joint consists of a modified bitumen binder, carefully selected single-size aggregate, closure/bridging metallic plate, and heat-resistant foam caulking/backer rod. This joint is especially suitable for rehabilitation jobs. A typical asphaltic plug joint is shown in Fig. 25.15.

The *limitations of asphaltic plug joint* are:

- Prone to deterioration due to flow of material in wheel path in high ambient temperature
- The aggregate may get debonded from the bitumen matrix in winter with subsequent pot-holing under traffic.
- Possibility of de-bonding of the plug from the adjoining asphalt surface.
- Lack of reliable indigenous source for binder.
- Incorrect placement of materials results in tearing of the adjacent carriageway pavement.
- Yielding of asphaltic material under the wheels of standing vehicles, braking and acceleration forces combined with high environmental temperatures and the development of rutting.

## Joints for Medium Openings (Movement Over 25 mm and up to 80 mm)

The absorption of medium opening requires an elastic expansion element or an expansion gap across the carriageway surface. For traffic safety, gaps in deck slab below 5 mm or over 65 mm

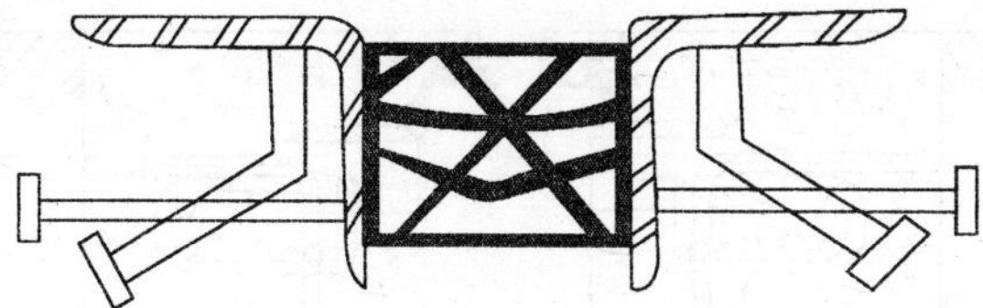

**FIG. 25.16** *Compression seal joint*

are not recommended, Thus, the expansion movement of a simple gap construction is limited to 60 mm. Expansion joints for medium openings consist of a sealing element, edge elements and fixing elements.

The following types fall in this category:

- Compression seal joint (movement up to 40 mm)
- Single strip/box seal joint (movement up to 80 mm)
- Reinforced elastomeric joints (movement up to 80 mm)

### Compression Seal Joint (Movement up to 40 mm)

This type of joint consists of steel armoured nosing at the two edges of the joint gap suitably anchored to the deck concrete and a preformed chloroprene elastomer closed cell foam joint sealer, compressed and fixed into the joint gap with a special adhesive binder.

The seal is supposed to remain in continual compression due to pressing of the seal wall against joint faces throughout the service life to ensure that the joint remains water tight and capable of repelling debris. A typical compression seal joint is shown in Fig. 25.16.

The *limitations of compression seal joint* are:

- Contact pressure of seal with steel nosing, which is the key to effective sealing mechanism, decays considerably with age. Hence more care is required during installation for the desired level of compression in the seal.
- Susceptible to detachment of seals from nosing due to accumulation of debris and the dynamic vehicular impact on the nosing.
- Susceptible to vandalism.

### Single Strip/Box Seal Joint (Movement up to 80 mm)

This type of joint consists of two edge beams with anchorages and an elastomeric sealing element held firmly in the housings of edge beams that guarantee the water-tightness of the joint. The maximum gap between the edge beams at road surface when the joint fully opens due to maximum contraction of deck shall be limited to 80 mm for comfortable passage of traffic. The edge beams protect the adjacent bridge/deck concrete from damage due to vehicular impact and also transfer the vehicular load to the deck structure through a robust anchorage system.

A typical single strip seal joint is shown in Fig. 25.17 (a), along with seals of expansion gaps.

Seals of expansion gaps can be constructed as V-shaped sections [(Fig. 25.17 (b)] or hollow box seals [(Fig. 25.17 (c)]. Movements are absorbed by these elements. Both shear-headed stud type and loop-type anchorages may be used:

The *limitations of single-strip seal joint* are:

Open surface gap tends to accumulate too much of debris on the strip seal element, particularly near the footpath and the kerb.

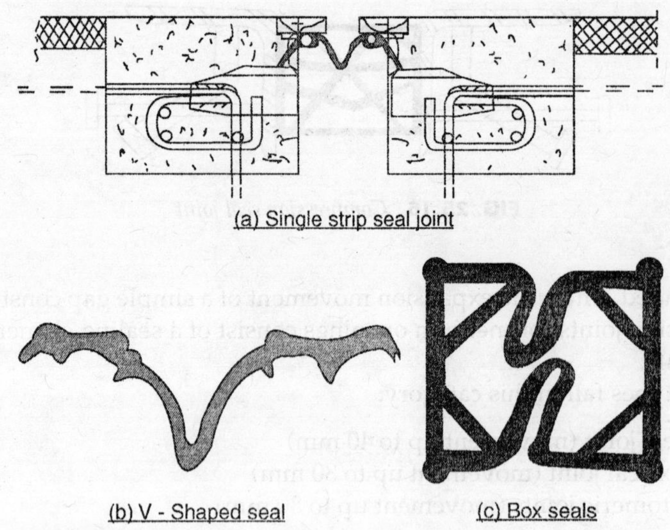


**FIG. 25.17** *Expansion joint of single-strip seal type*

Noise is generated during the passage of vehicles over the open surface gap.

Blocking of open surface gap by entrapped stones or hard pieces may generate locked- in condition restraining free movement of the structure.

### Reinforced Elastomeric Joints (Movement up to 80 mm)

This type of joint consists of reinforced elastomeric pads connected to a supporting steel structure which itself is cast into the concrete deck of the bridge. The joint accommodates movement by shear deformation of elastomer and opening and closing of grooves at the upper and lower surface. The maximum movement is limited by the gap width.

A typical reinforced elastomeric joint is shown in Fig. 25.18.

The following are the limitations of reinforced elastomeric joint:

- In high-volume and high-speed traffic areas, the rubber surface of the bearing plate is susceptible to excessive wear, which lowers the skid resistance.
- Since the reinforced elastomeric pads of most of the elastomeric joints are manufactured in 1 m to 1.5 m sections, there is a potential for water leakage at the tongue-and-groove splicing of the elastomeric pads in case they are not installed properly.
- Additional water seal below the elastomeric seal is required for making the joint water-proof.
- Horizontal force offered by the joint and transmitted to the bridge structure may be considerable due to high resistance of the elastomeric elements.

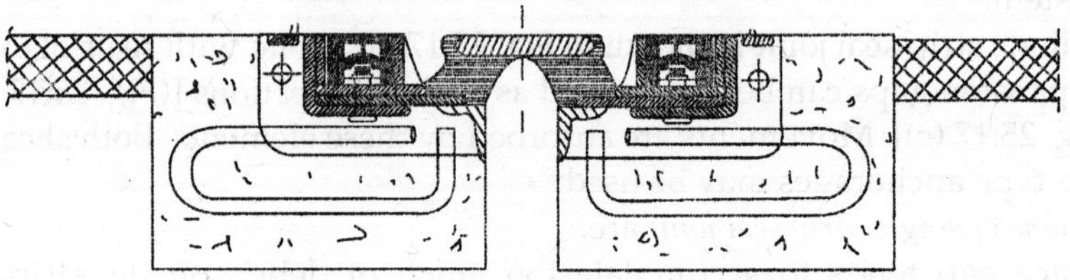

**FIG. 25.18** *Reinforced elastomeric joint*

### Joints for Large Openings (Movement over 80 mm) and Joints for Specific Purposes

Joints for large opening have also been developed. The following are the common types used:

- Modular strip/box seal joint
- Finger joints
- Reinforced coupled elastomeric joint

Likewise, there are *special types of joints* designed for *specific purposes* such as to accommodate very high longitudinal movement and high transverse movements; joints with anti-skid surface, and joints for high seismic zones are also available

The details of these joints along with their limitations, and the criteria for selection of the type of joint to be used, as also the specifications of the material required installation, testing, inspection and maintenance of expansion joints are available in the IRC Special Publication "IRC: SP-69-2011 Guidelines and Specifications for Expansion Joints (Second Revision)".

## SUMMARY

- The factors governing the choice of the type of superstructure are site conditions, materials labour and equipment available, traffic needs, and aesthetic considerations.
  The factors governing the choice of flooring system for a bridge are durability, drainage and silent riding. The important components are floor beams or cross girders, stringers, main flooring, and wearing courses. The basic types of bridge floors are the open floors used for railway bridges, and solid floors used for highway bridges.The types of floorings are RCC or steel troughs, steel buckle plates and RCC jack arch floorings.
- Construction of arch superstructures needs well-designed centering.Centering can be of timber truss, steel truss, and trestle-type; the last one can be with vertical posts and/or inclined posts
  Girder bridge superstructures can be steel girders or RCC girders. Steel girders can be launched by means of construction on staging. Assembling on river bed and hoisting into position, and rolling out girders on already constructed spans are the methods used. RCC girders may be precast or cast-in-situ. Precast girders are carried on rail track trolleys and launching girders are used for erection. Steel trusses are erected by using falsework bents in streams.Suspension bridges are erected making use of catwalks; towers, cables, suspenders, and stiffening girders are the main components to be erected.
- Recent methods of superstructure erection are the Incremental Launching Method and the Span-by-Span Method.
- Joints in bridge decks are weak spots which need good design and maintenance. Construction joints and expansion joints/contraction joints are the types of joints. Contraction joints are similar to expansion joints, but they are only for part of the depth and also are of smaller width. Construction joints are needed when the work is stopped temporarily for any reason. Expansion joints are important in bridge decks. The treatment depends on the expansion gaps needed. Buried and filler type of joints are for movements upto 10 mm; Asphaltic plug joint is for movement from 10 mm to 25 mm; compression seal joint is for movement up to 40 mm.Single strip/box seal joints and Reinforced concrete elastomeric joints are for movements up to 80 mm.

## QUESTIONS

1. What are the factors governing the choice of superstructure for a bridge? Explain each of them briefly.
2. What are factors which govern the method of construction/erection of bridge superstructure?
3. What are the components of the flooring system for a bridge? Explain briefly.
4. Differentiate between open floor and solid floor for bridge decks. When is each of them preferred?
5. Enumerate the types of solid floor for bridge decks.
6. Write short notes on:
   (a) RCC slab floor
   (b) Steel trough floor
   (c) Steel buckle plate floor
7. Write short notes on:
   (a) Wearing surface for a bridge deck
   (b) Timber floor
   (c) RCC jack arch floor
8. Write a note on centering for the erection of arch bridges.
9. Write short notes on:
   (a) Timber truss centering for an arch
   (b) Steel truss centering for an arch
10. Write a detailed note on trestle-type centering for the construction of an arch with suitable sketches.
11. Enumerate the methods of erection of steel girders for bridges and explain any one method.
12. Explain the use of launching girder for the erection of RCC girders for a bridge.
13. Write a note on the erection of suspension bridges.
14. Explain briefly the incremental launching method of bridge construction. What are the advantages of this method?
15. Write brief notes on:
    (a) Construction joints
    (b) Contraction joints
16. Write short notes on:
    (a) Buried joint
    (b) Filler joint
17. Write short notes on:
    (a) Asphaltic plug joint
    (b) Compression seal joint
18. Write a note on box seal type of expansion joint for a bridge deck.
19. Write a brief note on reinforced elastomeric joint for a highway bridge deck.

Chapter 26

# MAINTENANCE, TESTING AND STRENGTHENING OF BRIDGES

**CHAPTER OBJECTIVES**

This chapter enables the student to understand the following:

- Maintenance, testing, assessment of load capacity and strengthening of structure are important to keep a bridge in operating condition.
- Cracking of concrete and corrosion of steel are primary effects that develop.
- Periodical inspection and maintenance of a bridge register are essential.
- Posting of bridges involves speed and load restrictions on weak bridges.
- Bridge testing may involve different kinds of tests – non-destructive tests are preferred.
- Rating of bridges is based on the assessment of load capacity through testing.
- Strengthening and rehabilitation is necessary until repairs or replacement are undertaken
- Causes of bridge failure are multifarious, and include design and construction deficiencies; these should be prevented to avert critical bridge failures.

## 26.1 INTRODUCTION

Maintenance of bridges is an important part of a Bridge Management System to keep a bridge in operating condition. Potential deterioration in the future either due to natural causes such as atmospheric changes and unexpected high floods or seismic activity, or due to unexpected spurt in traffic volume during the service/design life, can also be prevented or minimised through a systematic plan of maintenance. Thus, the maintenance of a bridge includes the identification and repair of the defects and damage that have occurred during the operation, as well as preventive maintenance aimed at enhancing its service life and ensuring trouble-free operation. Maintenance should be based on adequate inspection and monitoring procedures, which are evolved through experience.

Testing of a bridge and assessment of the safe carrying capacity may be required prior to its being thrown open to traffic soon after its construction, or as a part of its maintenance and management to ensure safe operation.

Strengthening of a bridge may be required in case it has deteriorated under a spurt in traffic, in order to enhance its durability and safe operation. This may involve strengthening of the substructure, the superstructure or both.

Study of bridge failures provides valuable information which will be helpful in evolving construction and maintenance procedures to avert potential failures.

## 26.2 IDENTIFICATION OF BRIDGES

All culverts and bridges on a highway should be assigned separate numbers to enable easy identification by the inspecting and maintenance personnel.

In view of the need for a uniform system of numbering for the cross-drainage structures, the IRC has laid down the standard practice as given in *IRC:7-1971 Recommended Practice for Numbering Bridges and Culverts (First Revision).*

The important provisions of this code are given below:

### Method of Numbering

Cross-drainage works on a road shall be numbered in serial order, in each kilometre separately. The number shall be in the form of a fraction, the numerator denoting the number of kilometre in which the structure is situated and the denominator, the kilometre-wise serial number of the structure. For instance, the 5th cross-drainage structure in the fourth kilometre (i.e., between kilometre stones 3 and 4) shall be designated as $\frac{4}{5}$, and the 8th structure in 25th kilometre as $\frac{25}{8}$.

If any new culverts/bridges are built subsequently, say between the 3rd and the 4th structures in kilometre 375, they shall be designated as $\frac{375}{3/1}$, $\frac{375}{3/2}$, etc.

### Manner of Inscription

The number of the structure shall be inscribed near the top of the left hand side parapet wall as seen by traffic in the end elevation when approaching the structure from each direction, as illustrated in Figs. 26.1 and 26.2.

If the structure is provided with railings instead of parapet walls, but without end pillars, the number of the structure shall be indicated by means of a separate numbering plate of size 300 mm × 300 mm. There shall be two such numbering plates, one for each direction of travel. The plates shall be welded or fixed securely to the railing on the left hand side facing the carriageway as close to the entrance of the structure as possible.

In the case of buried culverts such as pipe culverts, where there are usually no parapet walls or railings at road level, two stone or RCC posts having a cross-section of 150 mm × 150 mm, and exposed height 300 mm, shall be set up, one on each side to mark the position of the culvert. The marker posts shall be fully outside the roadway width. The culvert number shall not appear on the marker posts but be either engraved or painted at their base. In addition, the number might also be inscribed at a suitable location on the headwall of the structure, above the HWL.

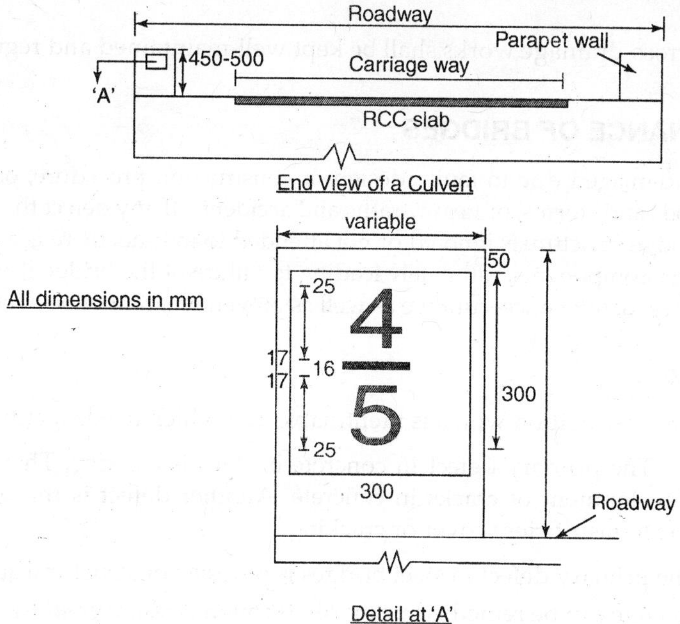


**FIG. 26.1** *Numbering of a culvert*

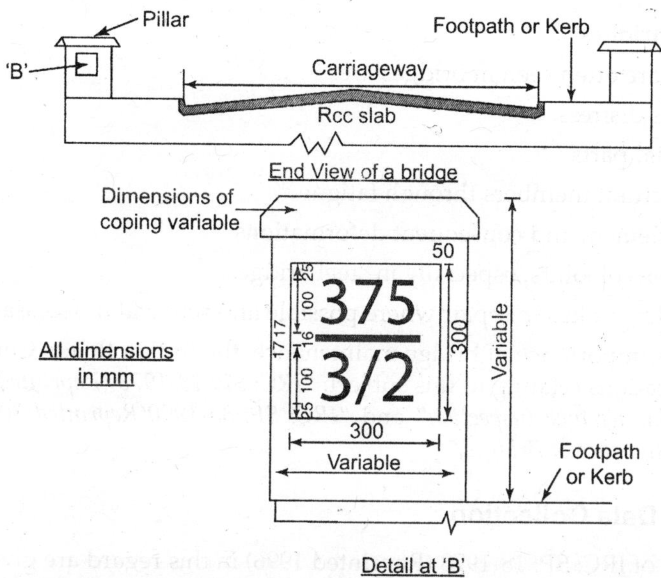


**FIG. 26.2** *Numbering of a bridge*

The numerals shall be 100 mm high and of international form conforming to the IRC specifications — "IRC: 30–1968 Standard Letters and Numerals of Different Heights for use on Highway Signs". These shall be painted in black on a smooth panel of size 300 mm × 300 mm. The colour of the background shall be canary yellow, ISI Shade No. 309.

**Maintenance**

The numbers on cross-drainage works shall be kept well-maintained and regularly painted."

## 26.3 MAINTENANCE OF BRIDGES

A bridge may get damaged due to faulty design or construction procedure, or due to external causes such as floods and storms, or heavy traffic and accidents. If any defect that develops in any component of a bridge structure is ignored or not attended to in time, there is a possibility that it may extend to other components, ultimately leading to failure of the bridge itself. Therefore, the importance of bridge maintenance, curative as well as preventive, cannot be over-emphasised.

### Defects in Bridges

*Defect* is an unwanted condition which is identifiable and which needs rectification.

***Concrete bridges***: The primary defect in concrete bridges is *cracking*. There can be several reasons for the development of cracks in concrete. Another defect is the corrosion of steel reinforcements due to insufficient cover or cracking.

***Steel bridges***: The primary defect in steel bridges is *corrosion* of structural steel components.

Cracking in concrete can be remedied primarily through *pressure grouting*.

Corrosion in steel can be remedied through *periodic cleaning and painting*, and *efficient maintenance of drainage provisions.*

#### Other Defects in Bridges

Other defects that are often seen in bridges are:

Expansion joints distress

Cracking of metal parts

Damage to structural members through fatigue

Foundation settlement and consequent deformations

Loose connections of joints, especially in steel bridges.

Remedy to these defects lies in repair where possible and renewal if necessary.

Recognising the importance of bridge maintenance, the Indian Roads Congress has issued two Special Publications relating to this subject: *"IRC: SP: 18-1978 (Reprinted 1996) Manual for Highway Bridge Maintenance Inspection"* and *"IRC: SP: 35-1990(Reprinted 2000) Guidelines for inspection and maintenance of Bridges,"*

### Inspection and Data Collection

Salient provisions of IRC: SP: 18-1978 (Reprinted 1996) in this regard are given below:

Bridge inspection is meant to ascertain the physical condition of the structure and recommend suitable remedial measures for portions adversely affected. This manual embraces all vital aspects like status of inspecting staff, periodicity of inspection, records to be maintained, complete list of items to be inspected, etc. The inspection report facilitates timely repairs/strengthening measures, where necessary.

**Records:** The prime need for any bridge maintenance operation is a complete, accurate and up-to-date record of the bridge, right from the stages of construction.

**Bridge register:** For maintenance records, a bridge register shall be used. For each bridge, there must be two reports:

(i) Original Bridge Report
(ii) Inspection Report

## Original Bridge Report

For newly constructed bridges, this shall be prepared on the basis of completion drawings supplemented by site inspection. In the case of old bridges where such data are not available, an investigation shall be undertaken for compiling them.

The following items shall be included in the original bridge report:

1. General particulars
2. Details of span
3. Hydraulic particulars
4. Loadings
5. Soil particulars
6. Foundations
7. Substructure
8. Bearings
9. Superstructure
10. Protection works
11. Position of surface utilities
12. Design details and drawings along with references
13. Diagrammatic sketch (LS of bridge showing span arrangement, as also soil profile)[1]
14. Special features (construction or design problems for future guidance)[1]
15. Date of completion and opening to traffic[1]
16. Cost of the project
17. Name of the constructing agency
18. Date of the report
19. Signature of the reporting officer with date.

### Inspection

The time of inspection, status of the inspecting officer and the frequency of inspection shall be recorded.

## Inspection Report

The following items shall be covered in the inspection report

1 Approaches
2. Protection works
3. Waterway
4. Foundations
5. Substructure (Piers, abutments and Wing Walls)
6. Bearings (Metallic bearings and Elastomeric bearings)

[1] These items are not required for minor bridges.

7. Superstructure
   (i) RC members
   (ii) PSC members
   (iii) Structural steel members
   (iv) Masonry arches
   (v) Timber components
   (vi) Suspension bridges
8. Expansion joints
9. Wearing coat
10. Drainage spouts
11. Hand rails
12. Footpath
13. Utilities (sewage pipes, telephone and electric cables, and lighting facilities)
14. Bridge numbering
15. Aesthetics

**Follow-up Action on Inspection Report**

Remedial measures immediately necessary shall be pointed out to the competent authority.

**Proforma for inspection report:** A detailed proforma is provided for the preparation of the inspection report. The final recommendations are given at the end of the Report for necessary action."

Salient provisions of IRC: SP: 35-1990 (Reprinted 2000) with regard to maintenance of bridges are given below:

**Scope**

(a) Inspection
(b) Maintenance
   (i) Defining the goals of bridge management and maintenance policy indicating the means and strategies.
   (ii) Evolving a methodology of utilising data provided by bridge inspection.
   (iii) Assessment of present situation as regard to bridge maintenance.
   (iv) Making a brief inventory of the different methods and techniques of bridge maintenance.
   (v) Defining the research areas to improve the techniques of bridge maintenance.

These guidelines cover only normal types of bridges; special types such as suspension bridges and cable-stayed bridges are excluded as they are very few in number and their maintenance needs a more specialised approach.

The guidelines are to be looked upon as suggested "good practices" and not as mandatory specifications.

**Bridge Management System**

This embraces the following aspects:

1. **Maintenance**: It is the work needed to preserve the intended load-carrying capacity of the bridge. This excludes improvement of the structure, either by widening or by strengthening

to carry heavier loads. It also excludes repairs of any damage caused in exceptional circumstances such as landslides, earthquake, cyclones, fire, etc., but includes preventive maintenance.

2. **Repairs and rehabilitation**: These activities are larger in scope and in cost than maintenance. Rehabilitation work aims at restoring the bridge to the service level it once had and has now lost.
3. **Improvements**: Strengthening, widening, raising, etc. aim at upgrading the level of service of a structure.
   The basic parameters are load-carrying capacity and geometric design elements.
4. **Replacement or reconstruction**: Involve the works carried out when the whole structure or at least its major components such as the whole superstructure are removed and replaced.
5. **A scientific bridge management system (BMS):** This is a collection of organisational elements, regulations, standards and procedures to monitor, analyse and implement the activities following bridge construction. Bridge management system is useful to achieve optimum use of funds, and to adapt existing bridges to traffic needs and public safety. This is a rational and systematic approach to organising and carrying out all activities concerning bridges, such as predicting bridge needs, allocating funds for maintenance, rehabilitation, strengthening, replacement and reconstruction and rating of bridges, and maintaining information data base.
   The major modules of BMS are:
   (i) Database module
   (ii) Major maintenance, rehabilitation and replacement (MR & R) selection module.
   (iii) Maintenance module
   (iv) Historical data analysis module
   (v) Project level Interface Module
   (vi) Reporting Module

   Each module will have a few associated sub-modules.

**Maintenance Policy**

This consists of several aspects:

- Safety aspects
- Economic aspects
- Traffic aspects
- Engineering aspects

**Training**

Bridge management requires extensive team work involving various levels of responsibility and skills. Training programmes need to be organised in order to develop such expertise and team work.

**Inspection of Bridges and Documentation**

Major bridges: Length up to 60 m or more
Minor bridges: Length less than 60 m.

**Documents:** In accordance with the provisions contained in "IRC: SP: 18-1978 (Reprinted 1996) Manual for Highway Bridge Maintenance Inspection", it is necessary to compile an original bridge report on the basis of as-built drawings along with the details of reinforcements. The information contained in the original Bridge Report will be utilised to compile the bridge inventory record, which will form a part of the data base module.

There should be three classes of maintenance inspection as given below:

**Routine inspection:** This shall be carried out periodically to report the fairly obvious deficiencies which might lead to traffic accidents or cause high maintenance or repair cost, if not attended to promptly. Routine inspection will have to rely mainly on visual assessment using conventional standard tools and method. The frequency shall be preferably twice a year, before and after monsoon.

**Principal inspection:** This is a more intensive and detailed inspection and will involve close examination of the elements of the structure. It will be close visual assessment supplemented by standard instrumental aids. Detailed inspection of the foundation, which may involve underwater inspection, should also be included. This should be done against a comprehensive checklist of items related to the material, condition and situation of the structure. The first principal inspection shall be done not later than 6 months after opening the bridge to traffic, and later at intervals of maximum three years. More frequent inspections shall be essential if the routine inspection reveals any distress in the bridge structure.

**Special inspection:** This shall be undertaken in the event of unusual occurrences such as strong earthquake, accidents, passage of unusual loads or floods, major weaknesses noticed during routine or principal inspection, unusual settlement of foundations, and substantial changes of traffic pattern. Such inspections may require a good deal of supplementary testing and structural analysis.

**Inspection procedure:** Effectiveness of an inspection depends on proper recording of the actual stage of each component, in a systematic manner to ensure that no component is overlooked. A typical pattern may be as follows:

(i) Foundations
(ii) Abutments
(iii) Wing walls/returns
(iv) Piers
(v) Columns and bearings
(vi) Soffits of the deck including beams
(vii) Details under the deck
(viii) Condition of road surface, drainage, parapets
(ix) Expansion joints
(x) Condition of approaches
(xi) Condition of protection works
(xii) Underwater inspection, where necessary.

**Means of access:** Appropriate means of access are a prerequisite for any inspection. Safety, ease and convenience are the criteria. Possible means of access can be of the following types:

(i) Built-in access
(ii) Access ladders
(iii) Hand-rails

(iv) Catwalks
(v) Platforms
(vi) Scaffolding
(vii) Manholes and inspection pits
(viii) Equipment operating from under the bridge–boat/barges
(ix) Equipment operating from the bridge deck (Snoopers)

**Assessment techniques and equipment:** Assessment is usually done by the following techniques.

(i) Routine inspection: Visual assessment using conventional standard tools
(ii) Principal inspection: Primarily visual assessment, supported by more advanced tools.
(iii) Special inspection: Advanced techniques and equipment apart from visual assessment.

**Inspection report:** This shall be in the prescribed format. Sketches and photographs may be used to illustrate the defects observed. Any recommendations for repairs and strengthening of the affected components shall be included.

**Follow-up action:** Necessary follow-up action shall be taken in time. If the assessment shows the bridge to be inadequate for the design loads, one of the following actions should be taken:

(i) The load should be restricted to the assessed capacity.
(ii) The bridge should be closed for all traffic where the rated capacity is lower than the lowest level of traffic load expected to ply on it.
(iii) Replacement or strengthening of the bridge or the affected part should be undertaken

### Maintenance Techniques

Guidelines on methods and techniques for repairs for strengthening and rehabilitation are covered under a separate document.

**Maintenance operations:** These operations can be of two types – ordinary maintenance operations and specialised maintenance operations.

Operations that are repetitive or periodical in nature are, in general, technically simple and could be classified as ordinary maintenance operations. Specialised maintenance operations are those which are unpredictable regarding their necessity and are relatively more complex.

Specialised maintenance operations are listed separately.

### Maintenance: Research and Development

A great deal of research is needed and is going on in India on the structural aspects of maintenance, especially of concrete bridge decks.

### Appendices

Appendix-1: Inventory of road bridges; Details of the bridge.
Appendix-2: Specimen inventory for one bridge
Appendix-3: Simplified bridge inventory proforma – General data and Technical Data
Appendix-4: Inspection proforma (for routine inspection)
Appendix-5: Checklist of items for inspection report
Appendix-6: Standard tools
Appendix-7: Possible assessment methods (only indicative):
Concrete – strength, quality and cover

| | |
|---|---|
| | Steel – Cracks, corrosion, movements, pressures. |
| Appendix-8: | Non-destructive and destructive tests for concrete strength, cover, corrosion of reinforcing steel, voids and cement content in concrete. Cracking, corrosion, pressures, and fatigue tests for structural steel components. |
| Appendix-9: | Suggested Action Plan for Organisations in charge of bridges and for IRC. |

## Posting Bridges

The term 'Posting a bridge' is used to indicate restrictions on the use of a bridge to ensure safety. Posting may be for load limits or for speed limits.

### Load Limit Postings

The primary purpose of load restriction is to ensure safety of the bridge and of the traffic; it may also indicate rehabilitation or replacement needs. Until such time when measures are taken in this direction, load limit postings are necessary.

The restrictions may be necessitated when there are changes in dead or live loads, physical condition of the bridge, or in the laws and specifications governing the bridges.

### Rating Methods

Load and resistance factor rating (LRFR) for new and major bridges

Provide uniform reliability

These are probabilistic methods to derive load and resistance factors

Strength-based load factor rating for existing bridges and minor repairs

Allowable stress ratings are given for timber bridges

### Posting Signs

Advance warning signs indicating the load limit on the bridge ahead are placed about 200 m from the bridge.

Load restriction signs, mentioning the maximum axle load or gross load, are placed 60 m in advance.

### Speed Limit Postings

These are used to reduce the impact effect on a bridge which is considered weaker than when it was constructed. Heavier loads may be carried at reduced speeds.

In general, speed restrictions are not preferred, except under extraordinary circumstances.

## 26.4 TESTING OF BRIDGES

Testing of bridges becomes necessary if old bridges that have been in operation for some time are affected to such an extent that the deficiencies are seen even on visual inspection. Testing is also essential in the case of newly constructed bridges in which new materials and/or new techniques of construction have been used.

The ultimate aim of testing a bridge is to assess its safe carrying-capacity under the existing conditions, and rate it in relation to other similar structures.

IRC has issued two special publications relevant to the testing and assessment of load capacity of bridges:

1. IRC: SP: 51-2015 Guidelines for Load Testing of Bridges (First Revision).
2. IRC: SP: 37-2010 Evaluation for Load-carrying Capacity of Bridges (First Revision).

These two publications have to go hand-in-hand and should be studied in conjunction with each other for a good understanding of the two aspects—testing and assessment of load capacity of bridges.

Salient provisions of *IRC: SP: 51-2015 Guidelines for Load Testing of Bridges* (First Revision) are presented below:

**Types of Tests**

The load tests on bridges could be classified under five heads:

(a) Behaviour tests
(b) Proof load tests
(c) Stress history tests
(d) Ultimate load tests
(e) Diagnostic Tests

*Behaviour tests* are carried out to verify the results of any method of analysis or design. The test load could be equal to or lower than the design load.

*Proof load tests* are commonly adopted for operational rating of bridges. The tests are done on new structures which had design or construction problems, or for the rating of an existing bridge.

*Stress history tests* are carried out to establish the distribution of stress ranges in fatigue-prone areas of the bridge. The data obtained from the passage of regular traffic is used to assess the fatigue life.

*Ultimate load tests* are performed to understand the global behaviour when sufficient theoretical knowledge is not available to predict the structural performance. The tests provide valuable information regarding the sequence and mode of failure.

*Diagnostic tests* are meant to monitor the behaviour of a component of a bridge either to establish the cause of damage or its share in transfer of loads when sufficient theoretical analysis is not developed.

The type, magnitude, application and duration of different tests are decided on the basis of an objective evaluation procedure. Each methodology depends on exigencies and site conditions.

Thus, unless otherwise specified, the term 'Load Test' used without qualification commonly denotes *Proof Load Test* only.

**Scope**: These guidelines deal with Proof Load Test. They cover testing of superstructures, excluding arches, for evaluating their flexural rigidity and capacity. This test is not intended to assess ultimate load carrying capacity of bridge superstructure.

**Test Procedures**

**Selection of span:** In the case of new multi-span bridges, at least one span is chosen for load test when the number of spans is up to 15; if it exceeds 15, two spans are chosen for testing.

**Method of loading:** Loading should be such that it either simulates the specific class of vehicle or induces in the member(s) the calculated forces, viz., the bending moments at critical sections.

The test loads may be in the form of:

- mobile test vehicles;
- static loads on wheel/track imprints of the specific class of vehicle for which the bridge is to be rated;
- static loads on a configuration which produces the calculated force in the member under test.

**Test vehicles:** The usage of commercial vehicles has been recommended in IRC: SP: 37. The vehicles and/or their trailers are uniformly loaded with pre-weighed units such as concrete cubes or sand soil containers and the axle loads determined on weigh bridges. However, commercial vehicles do not produce the expected load effects of standard loadings. Use of special test vehicles would be ideal. The advantage of such vehicles is that they can be moved on and off the structure quickly.

### Static Loads

Simulation of the specific IRC vehicle can be achieved by building up pre-weighed units on loading imprints spaced as per codal provisions; the imprints are built either with brick masonry or concrete and rolled steel sections placed across pairs of imprints, so that platforms could be built on a group of four imprints for placement of pre-weighed units. Large concrete blocks or steel ingots could be used if mechanical handling facilities are available. The loads are placed eccentrically on the carriageway in such a way that maximum bending moment is produced in any longitudinal section.

**Loading and unloading sequence:** The test load shall be applied in stages. In most cases, the design live load effect would be equal to or less than that due to dead load. With the dead load already there, the test load is added as some specified multiple of live load more than one. The suggested stages of test load are 30%, 50 %, 70%, 90%, and 100%. Unloading should also be in the same stages. Each load increment shall be added only after the deflections under the previous load have stabilised and all the stipulated observations are completed.

It is advisable to monitor the appearance and widening of flexural cracking at every stage of loading. The load-deflection relation is expected to be linear at every increment of load. If the observed deflection exceeds the limit prescribed in the code, further loading is stopped. Spalling of delaminated concrete is possible during load tests.

**Precautions:** The following precautions have to be taken.

- Staging should be stable and safe.
- Staging for instruments and that for the observer should be independent and rigid.
- Contact with the spindle of the dial gauge should be ensured when the superstructure moves due to temperature change.

**Observations:** The following should be recorded at hourly intervals over a period of 24 hours:

- deflections at critical sections
- appearance of cracks – length, width, location, and orientation with respect to load.
- deformation of bearings.
- ambient temperature and related temperature in the body of the structure.

**Measurement of deflections:** The following devices could be used to measure deflections:

(a) Dial gauges
(b) Ruler and cursor
(c) Deflectometers
(d) Precision level
(e) Water level

While (a) to (c) could be used wherever dry bed is available under the span (d) and (e) could be used with a reference station near the abutment. The debris at the expansion gaps should be cleared and the bearings lubricated to permit translation and rotation freely.

**Temperature correction:** The temperature of the bridge deck is monitored either by means of thermo couples or thermometers. The superstructure tends to hog or sag due to variation in ambient temperature; this needs correction to the observed deflections because the duration of loading/unloading operation is as much as 4 to 5 hours.

The thermal response deflections of the deck should be measured prior to load test for 2 to 3 days at hourly intervals, and a basic curve for temperature correction is plotted for each gauge station. The deflections at any location due to load and temperature during load test are superimposed on the basic curve. The difference between the two values gives the true deflection for the location under reference corresponding to the same temperature.

**Percentage recovery of deflection:** This is calculated at 24 hours after removal of loads after applying temperature correction to deflection data:

$$\left.\begin{array}{l}\text{Value of deflection after}\\ \text{24 hours of placement of}\\ \text{test load}\end{array}\right\} R_i$$

$$\left.\begin{array}{l}\text{Value of deflection after}\\ \text{24 hours of removal of}\\ \text{test load}\end{array}\right\} R_f$$

$$\left.\begin{array}{l}\text{Percentage recovery of deflection}\\ \text{after 24 hours after removal of test load}\end{array}\right\} \frac{(R_i - R_f)}{R_i} \times 100$$

**Acceptance criteria:** These are based on the recovery of deflection after removal of test load. It is necessary to specify the quantum of the applied load, the duration of the load on the span, and the percentage recovery of deflection on removal of load.

For bridges designed for IRC standard of loadings, the acceptance criteria are given below:

| Type of bridge | Minimum percentage recovery of deflection |
|---|---|
| RCC bridge | 75 |
| PSC bridge | 85 |
| Steel bridge | 85 |
| Composite bridge | 75 |

(Live load shall be 1.0 L plus corresponding impact as per IRC)

A general acceptance criterion for the behaviour of the structure under applied load is that no 'visible evidence' of failure such as cracks of width more than 0.30 mm, spalling or deflections which are excessive with respect to safety requirements, shall occur."

**Rating of Bridges**

Assessment of the safe load-carrying capacity of a bridge with a view to comparing it with that of similar structures is known *rating*. The purpose of rating of a bridge is to impose only the necessary restrictions relating to the axle load or gross load of vehicles using the bridge, with a view to ensuring safety to the users of the bridge as well as to the bridge itself.

This involves, primarily thorough inspection, collection of the necessary data for assessing the existing condition of the bridge, conducting tests on the components which appear to have been affected, and finally assessing the safe load-carrying capacity of the bridge.

The tests that may be conducted are comprehensively covered in the Appendix-8 of "IRC: SP: 35-1990 (Reprinted 2000) Guidelines for inspection and maintenance of bridges."

The salient features are given below:

"A large number of destructive and non-destructive test procedures are available for evaluating the strength of the components of bridges. However, the limitations, especially of non-destructive testing and the reliability of the results should be closely examined.

**Non-destructive Tests**

1. ***Rebound and penetration tests***: These tests measure the hardness of concrete and are used to predict the strength of concrete.
   The relative strength of concrete can be assessed by the Windsor probe.
2. ***Sonic methods***: Electronic and sonic methods have been used for assessing the strength of concrete. The travel time of a sonic pulse between monitoring points is related to the modulus of elasticity and hence the strength.
3. ***Ultrasonic techniques***: These are used to assess the maturity of concrete and its elastic properties by measuring the velocity of pulse in concrete. The correlation between the pulse velocity and compressive strength is reasonably good with proper calibration.
4. ***Magnetic methods***: These are used for determining the position of reinforcement as well as the assessment of the concrete cover. Portable battery-operated cover meters are available, which can operate with an accuracy of $\pm$ 1 mm in the range of up to 75 mm.
5. ***Electrical methods***: Measurements of electrical resistance and potential are used to detect corrosion of reinforcement.
6. ***Nuclear methods***: These are used to measure the moisture content in concrete by neutron absorption techniques; this gives an idea of the possibility of corrosion of reinforcement.
7. ***Radiography***: Gamma radiation techniques have been used for assessing voids in concrete as well as in grouted tendon ducts
8. ***X-ray diffraction and differential thermal analysis***: These tests are useful in determining hydration characteristics. The results indicate extent of attack due to sulphates or sea water.
9. ***Electrochemical potentials***: The procedure is as per ASTM C 876. This is useful in finding the condition of embedded steel.
10. ***Electrical Resistivity Management***: This is also useful in indicating the condition of embedded steel.

11. *Microscopic examination using optical and polarized light microscopy*: Useful in identifying aggregate source, presence of slag, etc.
12. *Dye penetration test and optical microscopy*: These tests give qualitative evaluation of incidence of voids in concrete.
13. *Deflection monitoring*: This is done with precision levels, theodolites and laser aligners.
14. *Strain measurements*: These are done with ER strain gauges and strain-measuring bridges.
15. *Crack meters*: These are used to measure crack width.

A combination of various non-destructive tests may be required assess the structure.

### Destructive Tests

1. *Concrete strength*: Testing core samples from concrete components yields the actual strength of concrete. Core drilling equipment is used for taking core samples.
2. *Carbonation*: Carbonation of concrete in the cover region results in loss of protection of steel against corrosion. The depth of carbonation is measured by painting a fresh concrete surface with 2% solution of phenolphthalein. The colour of concrete surface after painting may be compared with standard charts to indicate the areas of carbonation.
3. *Endorscope*: Endorscope consists of usually flexible viewing tubes that can be inserted into holes drilled into the concrete bridge components. A light is provided by optic fibres from an external source. Endorscopes are used for detailed examination of parts of the bridge structure which could not otherwise be assessed. These may also be used with attachments for a camera or TV monitor.
4. *Pull-out test*: This measures the force required to pull out a steel rod with enlarged head cast in concrete or introduced by drilling. It is possible to estimate the compressive and tensile strengths of concrete by correlating it with the pull out force.
5. *Pull-off test:* This is used to evaluate the tensile strength between two different concrete or mortar layers. Tensile force is applied to a circular steel disk of 50 mm diameter glued to a concrete surface with an epoxy resin. The pull-off force ranges from 500 N to 50000 N depending upon the type of equipment. This test is easy to operate on site and can be used specially for the quality control of repaired concrete surfaces.
6. *Internal fracture test*: In this test, the force required to pull out an anchor bolt is measured, and it is used to estimate the tensile strength of concrete through a correlation with the pull-out force.
7. *Break-off test*: This is mainly used to estimate the flexural tensile strength of concrete. A core, 55 mm dia. and 70 mm deep, is prepared, and the force required at the top and at right angles to the axis to break off the core at the bottom is measured. The core is drilled in the hardened concrete. The test can also be used to evaluate the bond between two concretes or mortars of different ages.
8. *Water penetration test*: In this, the flow rate of water through concrete is measured. This indicates the degree of protection of reinforcement offered by the concrete cover. The specimen is placed in a permeability test rig and water pressure is applied on one face, while the flow rate is measured.
9. *Rapid chloride test and rapid sulphate test*: Results in terms of chloride content and also depth-wise penetration are possible.

10. ***Analysis of hardened concrete***: Determination of cement content in hardened concrete is possible to be obtained.

11. ***Optical microscopy and expansion test on concrete cores***: This test is useful to detect the alkali-aggregate reaction."

## Assessment of Load-capacity of Bridges

This is the natural follow-up activity to testing of bridges. IRC Special Publication-"IRC: SP: 37-1991 (First Revision 2010) – Guidelines for evaluation of load-carrying capacity of bridges. Salient features of these guidelines are given here:

### Objective

The objective of these guidelines is to establish a common procedure for assessing the strength, evaluating the safe load-carrying capacity and provide information to the users about rating and posting of the bridge.

The guidelines also include recommendations for dealing with over-dimensioned and overweight vehicles; these terms have the following meaning:

(a) Rating of a bridge: The safe permissible load-carrying capacity of the bridge in terms of standard IRC loadings.
(b) Posting of a bridge: The limitations of dimensions and weight of vehicles commonly used in the country, which can be permitted to ply without requiring specific permission or escorted by traffic controllers of the authorities concerned.
(c) Over-dimensioned/overweight vehicles: Vehicles carrying consignments, resulting in any of the dimensions – height, width or length – of the loaded vehicles exceeding the legal limits are termed *over-dimensioned consignment* (ODC). Overweight consignments (OWC) are those carrying total loads beyond 100 tonnes. Special multi-axle vehicles are also included in this category.

Rating and posting of a bridge are desirable for all old and new bridges. It becomes essential to do so when:

(a) The design live load is less than that of the heaviest statutory commercial vehicle plying or likely to ply on the bridge.
(b) The design live load is not known.
(c) Records and drawings are not available.
(d) The bridge indicates distress of a serious nature leading to questions about its structural and/or functional adequacy.

### Assessment of Condition of Bridge

Inspection and maintenance: Routine inspection of newly constructed bridges will lead to normal maintenance, of which full record should be maintained.

Old bridges shall be inspected in detail and the survey and investigation of the condition shall be carried out. This may lead to minor repair/major repair/rehabilitation.Rating and posting activities shall take cognisance of the existing conditions.

Assessment of condition: Detailed guidelines and an exhaustive data base are essential for assessing the condition of a bridge.

The following IRC Special Publications are useful for this purpose:

IRC: SP: 35-1990 – Guidelines for inspection and maintenance of bridges

IRC: SP: 40-1993 – Guidelines on techniques for strengthening and rehabilitation of bridges

IRC: SP: 51-2015 – Guidelines for load testing of bridges

IRC: SP: 52-1999 – Bridge inspector's reference manual

IRC: SP: 60-2002 – An approach document for assessment of remaining life of concrete bridges.

IRC: SP: 74-2007 – Specifications for mix seal surfacing (MSS), close-graded premix surfacing (CGPS)

IRC: SP: 80-2008 – Guidelines for corrosion protection, monitoring and remedial measures for concrete bridges.

For the bridges identified for detailed investigation, field and laboratory testing may be required, depending on the degree of deterioration of the structure. A separate study may be required for deficiencies in the hydraulic parameters.

## Data Needed for Assessment

**Collection of data:** Drawings to reflect as built details, design details, construction of records, soil investigation details, material tests and load test data, post construction inspection and maintenance reports, details of repair work, commercial vehicular loads plying on the bridge and traffic data.

**Structural condition:** Cracking, spalling and loss of material of concrete members, corrosion and cracking of structural steel members, condition of connections in steel bridges, condition of expansion joints and bearings, settlement and rotation characteristics resulting in redistribution of stress, any movements of substructure components and hydraulic data such as scour, HFL, afflux, GWT, and so on.

### Preliminary Assessment

This can be based on the observation of visible deterioration such as cracking, spalling of concrete, and large deflection, that are likely to affect the load-carrying capacity.

The decision to proceed with detailed investigation depends on the results of the preliminary assessment.

### Detailed Assessment

*Structural inadequacy* arises from faulty design, material or workmanship and increase in traffic loading and environmental influences.

Mapping of cracking pattern of all cracks of 0.20 mm width or more is necessary.

### Estimation of Traffic Loads

Inspection and testing shall be in accordance with the IRC guidelines-IRC: SP: 35, IRC: SP: 74 and IRC: SP: 80.

**Assessment of strength of the materials:** In-situ testing at selected locations is undertaken to derive strength values indirectly through related parameters. Upper and lower bounds for these values shall be fixed with the aid of statistical techniques. Sectional areas of structural members and location of reinforcements, settlement, deformation, or rotation of structural members shall be ascertained by means of appropriate equipment, which are generally simple but effective.

## Full-scale Load Testing of Bridges

Full-scale load testing of bridges has been covered in detail in IRC: SP: 51 : 2015 (Second Revision)

### Traffic Factors

Review of axle and vehicle weights:

This guideline covers only commercial vehicles with gross vehicle weight (GVW) more than 16 tons.

The *maximum safe axle load* ranges (according to the Motor Vehicles Act, 1988) from 3 tonnes for single axle fitted with 1 tyre to 24 tonnes for tandem axle fitted with 12 tyres. Multi-axle fitted with 4 tyres on each axle can carry 10.2 tonnes on each side; this, however, is a non-standard vehicle.

The *laden weight of the vehicle* must not be more than the sum of the safe axle loads for all the axles fitted to be vehicle.

The GVW of commercial vehicles presently plying in India ranges from 12 t to 49 t; the permitted overload factor is 1.4. A multi-axle heavy vehicle with 15 axles can carry 147.4 t, without any overload factor.

The vehicle dimensions – height, width and length – must not exceed 4.75 m (for a double-decker vehicle), 2.5 m, and 11 m (rigid truck) or 18 m (trucks with trailer combinations), respectively.

Speed limit ranges from 40 km/h to 65 km/h. Special traffic surveys need to be conducted using automatic type of equipment for a realistic assessment of traffic on major bridges.

### Rating Methods

The following three methods may be followed for rating of bridge structures:

1. *Analytical method*: This is applicable when the as-built drawings and specifications followed are available.
2. *Load testing method*: This is suitable when no construction drawings and specifications followed are available, and the loss of strength cannot be assessed during condition survey.
3. *Correlation method*: This method can be applied if the sectional details of the structure can be correlated with those of bridges having identical specifications and sectional details, since the load-carrying capacity of the other bridges are supposed to be known.

For *recently built and undeteriorated bridges*, the highest class of design live load as envisaged in the original design shall be taken as the load for rating of the bridge.

*Rating and re-rating of old bridges* requires careful and detailed evaluation of several factors and conditions with respect to loads, allowable stresses, materials, fatigue, and the design philosophy –working stress or limit state design.

## Analytical Method

The following factors are relevant to this method:

(i) Carriageway width
(ii) Loads and load combinations for rating
(iii) Design method
(iv) Allowable stresses for the materials used.

**Logical Sequence for Rating by Analysis**

**Stage-I: Establishing need for re-rating**

The need to re-rate should be based on:

1. Field observations, inspection reports, change in use, change in design codes, etc.
2. Inspection of all components to establish the defects and their severity in order to plan the order of priority for repair.
3. Identification of defects which are likely to affect the load-carrying capacity.
4. Identification of critical elements for each group; foundation/pier/bearing/superstructure, the repair of which is likely to raise the overall rating significantly.

**Stage-II: Initial desk studies**

These include the following:

1. Collection of all design drawings and construction data available.
2. Collection of reports of the bridge, if any.
3. Preparation of a 'design basis report' including applicable design codes, design loads, materials, etc.
4. Establishing the strengths of the materials as per the original data.

**Stage-III: Field investigations**

The following are needed:

1. Estimation of corrosion by observing the 'in-place' loss of section, and collection of samples for lab tests.
2. Assessment of the quality of the concrete by non-destructive testing (NDT).
3. Repair of concrete broken for inspection of steel.

**Stage-IV: Desk study for fresh design assessment**

This includes the following:

1. Completion of the design basis report
2. Identification of structural parameters to be studied.
3. Calculation of actual strength parameters.
4. Analysis of the desirable strength for the desired class of loading. If the actual strength is 90% of the desired strength, the bridge is satisfactory for that class. Otherwise the above steps have to be repeated for the next lower class of loading

**Capacity of Superstructure**

(a) *Detailing of steel in existing bridges*: Fully detailed drawings are used if available. Non-destructive testing of critical sections is carried out.
(b) *Use of analytical methods*: The methods available at the time of construction for calculation of the strength of the superstructure shall be used.

(c) *Slab bridges*: The following table can be used to arrive at the safe axle loads at primary evaluation:

**Table 26.1** Safe axle load for RCC slab bridges

| Effective span(m) | Slab thickness (mm) | Safe axle load (t) | Effective span (m) | Slab thickness (mm) | Safe axle load (t) |
|---|---|---|---|---|---|
| 2 | 150 | 9.5 | 6 | 300 | 10.0 |
| | | | | 325 | 13.0 |
| | 175 | 14.5 | | | |
| | 200 | 21.0 | | 350 | 16.0 |
| 3 | 200 | 11.5 | | 375 | 19.0 |
| | 225 | 15.5 | | 400 | 24.0 |
| | 250 | 20.0 | 7 | 325 | 9.0 |
| | 275 | 25.5 | | 350 | 11.5 |
| 4 | 225 | 9.5 | | 375 | 14.0 |
| | 250 | 13.0 | | 400 | 17.5 |
| | 275 | 17.0 | | 425 | 21.0 |
| | 300 | 21.5 | 8 | 375 | 9.0 |
| 5 | 250 | 9.0 | | 400 | 12.0 |
| | 275 | 11.5 | | 425 | 15.0 |
| | 300 | 15.0 | | 450 | 18.0 |
| | 325 | 19.0 | | 475 | 21.5 |
| | 350 | 23.0 | | | |

[Notes:
Slab thickness includes a cover of 25 mm
A 75 mm-thick wearing coat is assumed over the slab.
No separate impact allowance need be made as it has already been accounted for.]

(d) *Masonry and plain concrete arch bridges*: Only the methods used in the relevant period of construction have to be used. Often, the design has been based on the rule of thumb, or use of 'standard' section or nomogram. The maximum provisional axle load (before applying various factors) may be got by any of these methods. From this, the allowable axle load may be got by multiplying the provisional value by appropriate profile factors, material factors, support factors, etc., as given in the Annexure to this Special Publication of the IRC; this is used for rating. For plain concrete arches, a value of 1.5 shall be used for the material factor.

**Substructure Including Bearings**

The overall rating of the bridge is not expected to be affected by bearings, which can be inspected and repaired

(a) Condition survey of piers and pier caps is important, because the bridge rating can be affected. This is especially true for old masonry/plain concrete piers. Local repairs may turn out to be cheaper than to "derate" the bridge.
Otherwise, re-rating will have to be done as governed by the strength of substructure.

(b) If development of deep scour is the cause of distress, it can be tackled by filling the scour holes by stone/gabions and general bed protection, so that re-rating of the bridge can be avoided. If, however, the general bed level has also changed, the resultant changes in hydraulic aspects may call for re-rating.

**Foundation Condition**

This is the most difficult element to inspect, or strengthen.

(a) If excessive scour is the source of danger, it can be tackled as given in b) above.
(b) If excessive settlements have been observed, the foundation might have stabilized after settlement as in open foundations in sand or rocky strata. In such a case, resetting the bearings will solve the problem. If it is clay in which long-term settlements are expected, it is better to derate the bridge.

**Analysis for Posting Based on Actually Plying Vehicle Population**

The rating of the bridge based on hypothetical trains of live load cannot be directly used for placing restrictions on traffic actually plying on the bridge. For this purpose, the maximum effects of loads plying on bridges need to be calculated instead of IRC live loads.

Bridges should be posted for one of the nominal GVW classes, except over-dimensioned vehicles. For over-dimensioned and over-weight consignments carried on multi-axle vehicles, special studies are needed. After selecting the reference GVW class of vehicle, distance between successive vehicles may be taken as 4 m without impact factor but with appropriate overload factor; or, as 20 m with impact factor as also with appropriate overload factor. Load combinations with other loads of IRC: 6: GVW loading replaces only the IRC: 6 live loads. As such, other load and load combinations should be used as per IRC: 6 and the relevant design codes.

## Load Testing for Rating and Posting

**Load test for rating:** This is done when the essential details are not available for determining the rated capacity of the bridge. For rating of masonry arches, load testing is recommended.

**Load test for posting:** This is done when the strength of all elements of the existing structure cannot be verified by analytical method due to lack of reliable data.

**Test vehicle for rating:** One of the standard IRC loadings, with the aid of mobile test vehicles, shall be used.

**Test vehicles for posting:** The test vehicles will be from amongst those which are commercially available. The next heavier vehicle than the predominant heavy vehicles is used as the test vehicle, allowing for the appropriate overload factor, based on actual traffic data. The procedure for load testing shall be as per IRC: SP: 51 as given earlier.

**Bridge Posting**

**Temporary traffic restriction:** Necessary traffic restrictions on the bridge are decided based on safety considerations.

(i) *Speed restriction*: To be effective until the detailed investigations and strengthening or rehabilitation on the bridge is completed.

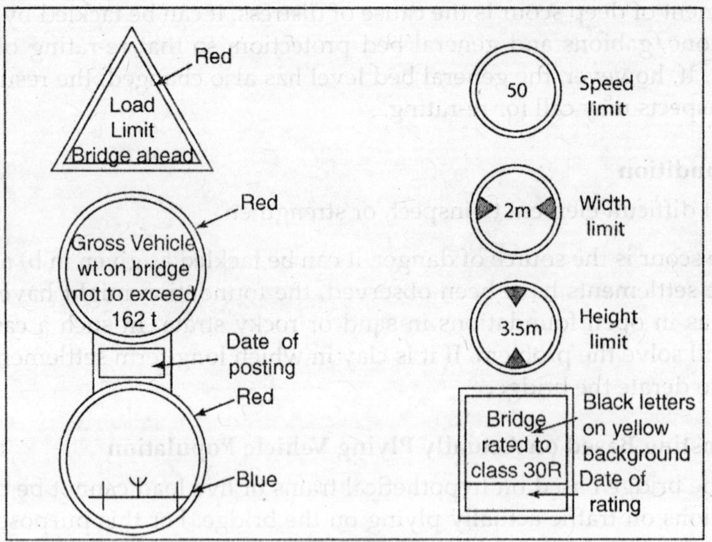


**FIG. 26.3** *Bridge posting Signs*

(ii) ***Geometrical restriction***: This would involve curtailing the carriageway width to ensure less magnitude of live load on the bridge, as also installing height barriers on either side to restrict passage of over-sized and overloaded commercial vehicles.

(iii) ***Footpath loading***: Depending upon the condition of the footpath slab, restriction of load on the footpath may be imposed until the distressed part is rehabilitated.

### Posting Signs

Postings for bridges shall be made as shown in Fig. 26.3, on both sides of the bridge. Load regulatory and advanced warning signs shall the installed on both sides of the bridge on the approaches and at all road junctions leading to the posted bridge. Load regulatory signs shall be at least 100 m ahead of the bridge, while advance warning signs shall be placed at least 200 m ahead.

## Repair, Strengthening and Rehabilitation of Bridges

After detailed inspection, testing and assessment of the load-carrying capacity of an existing distressed bridge, the options available and the follow-up action would be carefully evaluated.

One of these options will be to undertake immediate repair, strengthening and rehabilitation of the bridge. The procedure for repair and strengthening depends upon the nature and extent of the distress in the bridge superstructure, substructure and foundations. It is not always necessary to restore the original condition; it is sufficient, both economically and technically, to provide proper strengthening while at the same time derating the safe load-carrying capacity of the bridge. The more economical of the two methods in the long-run would be chosen

These aspects are covered in detail in IRC guidelines SP: 40 "Guidelines on Techniques for Strengthening and Rehabilitation of Bridges", IRC: SP: 75 "Guidelines for Retrofitting of steel Bridges by Pre-stressing," and IRC: SP: 80 " Guidelines for corrosion Prevention, Monitoring and Remedial Measures for Concrete Bridge Structures".

**Permissible Stresses in Different Materials**

For working stress method of analysis, the permissible stresses shall be as under:

(i) In structural steel and mild steel, 45% extra shall be allowed over the values specified in the relevant IRC Codes for Road Bridges.

(ii) In concrete and masonry, 33.3% extra shall be allowed over the values specified in the relevant IRC codes for Road Bridges and Design Criteria.

Several factors are used for rating of masonry arch bridges. These are:

(a) Profile factors
   (i) Span/Rise factor – depends on span/rise ratio
   (ii) Shape factor – depends on rise at quarter point/rise at the centre

(b) Material factors – depend on the material and the fill – Ring factor

(c) Joint factors
   (i) Depth factor – depends on width and depth of joint.
   (ii) Mortar factor – depends on the condition of the joint

(d) Support factor – depends on condition of supports

(e) Cracks factor – depends on crack characteristics

(f) Deformation factors – depend on deformation of the arch

(g) Abutment fault factors – depend on the movements of the abutment

The Annexure contains comparison between live load forces (maximum B.M. at mid-span and maximum shear at support) generated by IRC live load and GVW class loads for simply supported spans for different ranges of spans for concrete bridges and steel bridges.

For complete details, IRC: SP: 37-2010 (First Revision) may be consulted.

## 26.5 STRENGTHENING OF BRIDGES

Strengthening and rehabilitation of bridges is an important aspect in prolonging their service life, without total renewal or replacement by constructing a new one in place of a damaged structure.

The IRC have given guidelines for this purpose in their Special Publication "IRC: SP: 40-1993 –Guidelines on Techniques for strengthening and Rehabilitation of Bridges".

Important provisions of these guidelines are presented here:

**Policy for Decisions**

The individual decision considerations will vary in accordance with the size and importance of the bridge in question. For rehabilitating/strengthening major bridges, elaborate analytical techniques for evaluation of various solutions may be applied, but such work of minor bridges may be carried out in accordance with the general principles.

The decisions may vary from no action, some degree of temporary action, full rehabilitation or strengthening, to a replacement; a cost-benefit analysis may help in decision-making. However, replacement may prove to be laborious and unjustified in majority of the cases. A simple operational framework for decision-making is advisable as given below:

A bridge may be divided into elements with relatively short life-spans such as pavements, water-proofings, expansion joints, paint, etc., and those with long ife-spans such as structural elements in bridge decks, columns, piers, foundations, etc. This division may be based on difference in technology and economy, or on traffic considerations between bridges on rural roads, state highways or national highways, as also on environmental considerations.

The general policy for rehabilitation/strengthening of bridges must, therefore, be based on the following factors:

(i) Short-life elements
(ii) Long-life elements – important bridges and those on Expressways, National Highways, State Highways and Arterial roads.
(iii) Long-life elements – other bridges
(iv) Type of bridge, viz., ordinary, major, and important.

Apart from these, the following factors are also considered:

(i) Future in functional demands of traffic.
(ii) Environmental impact
(iii) Technical limitations, e.g., for strengthening of foundations

**Technical Approach for Rehabilitation/Strengthening Solutions**

The steps involved could be as below:

(a) Evaluation of the structure from documented database and inspections (IRC: SP: 35)
(b) Locating defects/distresses through visual inspection and testing techniques such as non-destructive and semi-destructive methods.
(c) Analysis of causes of defects/distresses – this is to design the repair measures in such a way that the risk of repetition is minimised.
(d) Evaluation of the results of structural assessment. It may be necessary to evaluate the load-carrying capacity of the bridge (IRC: SP: 37).
(e) Design of repairs for rehabilitation/strengthening works – generally the structural design for repairs shall conform to the relevant IRC codes.
(f) Proposals and estimation of costs – There may be several options for restoration of the functionality of a damaged structure: total replacement, partial replacement and repair, and extensive rehabilitation/strengthening measures. However, reduction of live loads and speeds may be sometimes the best option. (IRC: SP: 37).

**Normally Observed Distresses**

**Arch bridges:** The most common defects are –

(a) Changes in profile of the arch
(b) Loosening of mortar – ageing effect
(c) Arch-ring deformation or partial failure of the ring.
(d) Movement of the abutment/pier.
(e) Longitudinal cracks of varying susbsidence along the length of the abutment/pier
(f) Lateral and diagonal cracks – these indicate a dangerous state
(g) Cracks between the arch ring, spandrel or parapet wall
(h) Old cracks no longer widening – these would probably have occurred immediately after the bridge was built.
(i) A vertical crack in the return wall, where foundations on yielding soil are stepped.
(j) Bulging of wall – due to absence or clogging of weep holes.

**RCC bridges:** The most common distresses are –

(a) Cracking due to plastic shrinkage, drying shrinkage settlements, structural deficiency, reactive aggregates, corrosion of reinforcements, frost damage, sulphate attack and physical salt weathering.
The significance of the cracks depends upon the structure type, location of the cracks, their origin, and whether the width and length increase with time and load.
(b) Scaling – This is characterised by loss of concrete in patches on the surface (kerbs and parapet walls are particularly susceptible to scaling).
(c) Delamination – Separations along a plane parallel to the surface of the concrete due to corrosion of reinforcements.
(d) Spalling – generally considered to be a serious defect as it can cause structural failure ultimately. Spall is a depression caused by separation and removal of surface concrete.
(e) Leaching – it is the accumulation of salt or lime on the concrete surface, indicating porous or cracked concrete; migration of moisture may initiate severe early deterioration.
(f) Stains – most significant are those due to rust which indicate corrosion. But absence of rust is not necessarily indicative of the absence of corrosion.
(g) Hollow or dead sound – if tapping with a hammer or rod produces a 'dead' sound, this is indication of delamination.
(h) Deformations – effects of distress which may show up in the form of deflection, spalling, delamination, scaling, cracks, etc. Twisting of substructure or superstructure units may be evidence of a settlement of foundation.
(i) Excessive deflections – due to passage of abnormal loads. Rheological effects such as creep also can be the cause.
(j) Holes in deck slab – could be due to local weakness in concrete.

**Pre-stressed concrete bridges:** Most of the forms of distress in pre-stressed concrete are similar to those in RCC.

However certain special features are given below:

(a) Cracking – In pre-stressed concrete, cracking is a serious problem. The reasons could be deficiency of reinforcing steel, overstressing or loss of pre-stress, restricted movement in bearings and mishandling in transportation and erection.
(b) Leaching – associated moisture movements will aggravate corrosion risk.
(c) Stains – rust stains indicate corrosion of pre-stressing cables, posing a serious threat to structural stability. However, the absence of rust stain does not necessarily indicate lack of corrosion.
(d) Spalling – a serious problem which can result in loss of pre-stress
(e) Excessive deformations – These could occur due to loss of pre-stress with time.
(f) Abnormal vibrations – Such vibrations could be due to slender members or a combination of various reasons.

**Steel bridges:** The defects in steel bridges include –

(i) Corrosion
(ii) Excessive vibrations
(iii) Excessive deformations like buckling, kinking, warping and waving
(iv) Fractures
(v) Distress in connections and
(vi) Fatigue cracking.

Overloading, fatigue, cracking in welds, and loosening of bolts and rivets could be the reasons leading to failure, if not attended to in time.

**Composite construction:** The distresses are normally the same as those for the concrete and/or steel bridges. However, the distresses are more common at the interface between two materials due to horizontal shear, the shear connectors being absent or of insufficient capacity.

**Timber bridges:** The normally observed distresses are –

(i) Cracking or splitting of members due to overload, ageing, or under-designing of members.
(ii) Abnormal deflections due to overloads or imperfect joints.
(iii) Infestations and decay due to environmental aggressiveness.
(iv) Looseness of joints due to lack of good workmanship.

**Miscellaneous:** Distresses in bearings and expansion joints manifest themselves in various forms.

### Testing and Diagnosis

Most bridges built in India in the recent past are of RCC or pre-stressed concrete. Therefore, the testing techniques and diagnosis of the defects developed relate mostly to this type of bridges.

Testing methods could be non-destructive, partially destructive, or a full-scale load testing of bridges, besides the simple process of visual inspection. Such testing and diagnosis is necessary not only for maintenance and evaluation of the load-carrying capacity of bridges, but also for strengthening/rehabilitation of bridges.

These have already been covered by the IRC Publications-SP: 35 and SP: 37, which have been presented in Sections 26.3 and 26.4, respectively.

### Repairs and Strengthening Techniques – General

The criteria for selection of materials and techniques for repairs and strengthening could be:

(i) Causes of distresses
(ii) Efficacies of the materials and techniques in the preservation and/or enhancing the load-carrying capacity of the structure
(iii) Availability of materials and equipment
(iv) Importance of the bridge
(v) Time available
(vi) Life expectancy
(vii) Feasibility of traffic diversion

### Repairs of Foundations

Most repair works for foundations are in the category of protection and strengthening. Some examples are given below:

- Scour and erosion protection and repair of washed away/damaged protection works.
- Repair of foundations built on soft ground subjected to erosion.
- Strengthening of foundations affected by extra load on existing piles due to settlement of soft ground under the foundation
- Remedying the effects of horizontal movements of abutments built on soft ground.
- Strengthening of foundations due to widening of a channel or a road

- Repair of underwater bridge structures.
- Repair of serrations on the surface caused by high velocity of particles in the water.

**Note**: Foundation movements can substantially increase the loads and moments in some parts of superstructures by redistribution. This must always be checked for.

Some of the repair works carried out for foundations are given below *briefly* for information and guidance. IRC: 89-1985 "Guidelines for Design and Construction of River training and control works for Road Bridges" may be consulted for further information on this.

**Erosion problems:** Stone rip-rap is placed on a mattress at or beneath the bed level. The weight of the mattress should be a minimum of 150 kg/m$^2$. The slope of the rip-rap should be 1 in 3 to 1 in 3.5. Heavier stones should be used for rip-rap, for steeper slopes.

**Protection against scour:** Spur dykes, jetties, deflectors and other devices may be constructed to direct water away from the fill, bridge pier, or abutment. At sites where soil erosion has occurred because of stream or tidal action, rock or rip-rap in the voids or bagged concrete rip-rap are used. As an alternative, the affected area is grouted or repaired by placing sheet piling to keep material in place or to prevent further scour. Sheet piling may also be used to prevent scour near piers and abutments. If piles are exposed due to scour, supplemental piles may be driven. To restrict scour around piers, very heavy concrete blocks or stones of designed weight are placed around the pier foundations below the bed level by excavation. This is known as the *garlanding technique.*

**Foundations on soft rock** subject to erosion can be protected by reinforced concrete curtain walls enclosing the footing or piles

**Increasing the bearing capacity of the soil** by injecting cement or chemical grout taking care to ensure that grout pressure does not exceed the overburden pressure.

**Rock and ground anchors** are used for abutments where protective slope had to be removed for, say, widening of navigable channel, road, etc. Now-a-days, pre-stressed anchors are also used.

**Extension of existing foundations**: This would be necessary while widening an existing bridge.

**Liquefaction of foundation soil**: There are two approaches of retrofitting that will mitigate this type of failure:

(i) Eliminate or improve soil conditions that tend to be responsible for seismic liquefaction; and,

(ii) Increase the ability of the structure to withstand large relative displacements similar to those caused by liquefaction or large soil movements.

Some methods available for stabilisation of soil in-situ are:

- Lowering of ground water table
- Consolidation of soil by vibroflotation or sand compaction
- Placement of permeable overburden
- Soil grouting or chemical injection.

**Underwater work**: Normally, inspection of under-water components of a bridge is carried out with the aid of divers. Underwater photographic techniques are also available where in damages are detected by divers who can then take photos of affected areas. Similarly, underwater cameras (mounted on diver's head gears) can be used to scan the various components of submerged portions of the structure continuously and signals can be read on a TV monitor kept on the bridge deck.

Placing of concrete under water can be carried out with the help of conventional underwater bucket or tremie concrete, although under certain conditions placing of pre-packed concrete or bagged concrete or pumped concrete may be more suitable. Piles or columns which are substantially disintegrated or deteriorated through corrosion can be provided with integral jackets which may or may not be reinforced depending on the thickness of the jacket. A cofferdam may be used, if necessary, to facilitate repairs. Grouting with quick-setting cement of epoxy can also be carried out, if necessary.

Now-a-days, devices called habitats are used for under-water repair work. Habitat is a multi-cell metal unit open at the bottom with water-tight joints, installed around the member to be repaired; it is kept dry with compressed air, so that the divers can undertake repairs.

**Repairs to Masonry Structures**

Strengthening and widening of existing masonry bridges are done only for preserving and maintaining the existing appearances as historical landmarks.

The general defects and remedial measures for arch bridges in stone or brick masonry are given below briefly:

(i) **Loss of bond for the crown stone**: Low pressure cement grouting for the old mortar and the use of flat jacks for pushing the crown stone back to its original position are commonly employed.

(ii) **Longitudinal cracks along the direction of traffic**: Raking and refilling the mortar joints with fine cement grouting is used.

(iii) **Transverse cracks**: Injection of cement provides good bond between stones and brick masonry.

iv) **Strengthening of arch rings**: Material can be added either to the intrados or the extrados. Extra material may be placed by shuttering and pumping concrete, or by fixing a mesh to the intrados and spraying concrete.

Sprayed concrete will, in any case, change the appearance of an arch constructed of stone, brick or a combination of both.

A more effective, but more expensive, treatment is to remove the fill and cast the extra thickness on the extrados of the arch. Usually a full ring is cast, but occasionally only the end quarters are strengthened to reduce the effective span. Replacement of backfill may be with normal or light-weight concrete.

Another remedial measure for small-span bridges is to cast a slab at road level to act as an auxiliary deck in order to increase the load-carrying capacity. For cracks in arches, grouting with cement at pressure 4 to 6 $kg/cm^2$ is sometimes quite effective.

**Repairs to Steel Structures**

**Deck replacement of older steel bridges**: Many of the old steel truss or arch bridges have either warped steel plates with bituminous surfacing or a concrete deck. The steel plates often get rusted due to insufficient water proofing.

Bridge decks can be replaced by new concrete decks. Bolting is the preferred method of connecting the new deck system to the existing structural members.

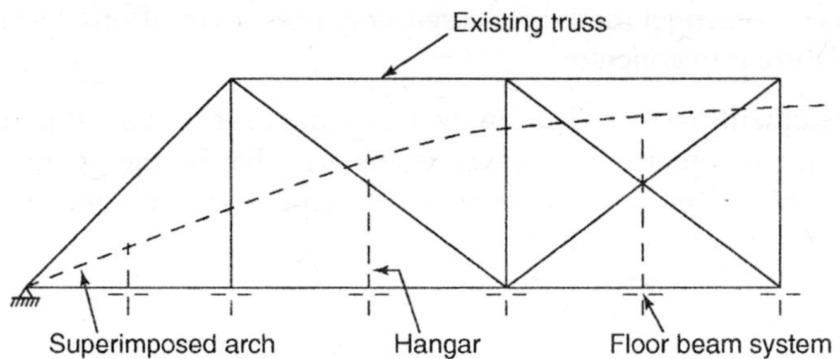


**FIG. 26.4** *Steel arch superimposition to strengthen old truss bridge*

**Strengthening of structural members**: Strengthening usually involves conventional techniques such as installing new diaphragms to existing double compression members, and strengthening or replacement of diagonals. Plate girders may be strengthened by external pre-stressing cables, anchored and fixed on the web in the required parabolic curvature acting in a similar way as in pre-stressed concrete. Strengthening by the addition of stiffeners to flanges, webs and diaphragms helps to prevent compression failure.

**Repair of cracks**: Crack repair methods depend on the root cause of crack initiated. Repairs can be made by techniques such as drilling holes at the crack tip, cutting out the cracked material and bolting plates in place, cutting out the crack and re-welding with a higher class weld.

Arc welding is an accepted procedure in underwater construction, salvage and repair operation.

Use of steel arch superposition scheme can be used to strengthen old truss bridges made of superimposed arches, hangers and additional floor beams.

This is illustrated in Fig. 26.4.

This method is an overall strengthening measure which increases the load-carrying capacity of the entire structure. There is no need for temporary shoring of jacking for the installation of the super-positioned elements. The dead load increases by about 15% to 20%.

**Repairs and Strengthening Techniques for Concrete Bridges**

Repair work not requiring strengthening:

This may be categorized as bellow –

(i) Repair of the concrete surface
(ii) Repair of cracks in concrete
(iii) Repair of corroded steel reinforcement and repairs related to high tensile steel in PSC
(iv) Repair of porous concrete and holes

**Repair of the concrete surface:** The damaged and fractured concrete must be removed to a sound surface for proper adhesion of the treatment material

Mechanical methods used for this purpose are milling, chipping, sand blasting, water or steam blasting, and compressed air cleaning.

Hydraulic methods include water-jetting at different pressure ranges–10 to 40 MPa for removal of scaling and vegetation coatings, 40 to 120 MPa (high pressure water-jet) for removal of soft areas of the concrete surface, and 140 to 240 MPa (hydro-jet) for deep penetration or even cutting grooves. Hydro-jet method is essentially free of vibrations, but there will be deep penetration of moisture into concrete.

**Bonding agents:** Cement paste with a low water-cement ratio, cement mortar (1:1) in slurry form, polymer-modified cement, and epoxy resins with hardening agents (with or without fillers) are used to improve the bond between old concrete and new concrete. The choice of the bonding agent must be made carefully.

The *surface protection methods* used are

- hydrophobation
- painting
- impregnation
- sealers
- coatings,

in the increasing order of protection.

The *materials* used are:

Silicon organic impregnation materials such as siliconates and silicon resins, oils such as linseed oil; and sealers such as epoxy resins for coatings

Repair involving a substantial depth of concrete section is made through the use of polymer-modified cement-bond systems.

*Shotcrete* (*gunite*) is considered suitable for the strengthening of structural elements

**Repair of cracks in concrete:** The methods used depend upon the nature of the cracks and the cause of formation. The techniques used are caulking, jacketing, stitching, injection, grouting, coating, thin resurfacing, and concrete replacement which may be partial or total. The materials used for crack repair are epoxy and polyurethane resins used along with appropriate injection equipment.

For *PSC members*, most of the above crack repair methods are applicable.

### Corrosion Protection of Steel Reinforcement

The corrosion protection system of steel reinforcement is restored by adding concrete cover or cement mortar with polymers after removing the chloride-contaminated concrete cover, cleaning of the rusted steel .

**Repair of pre-stressing steel** should be by adding concrete cover and cement grout in the ducts. Corrosion protection for external tendons is by means of an envelope of plastic pipe or painted steel pipe, the internal void being filled with cement grout or grease. The relative merits of rehabilitation methods shall be weighed before finally selecting the method to be used in any particular case.

### Strengthening of Concrete Structures

Strengthening of structural members can be achieved by:

- replacing defective material by better quality material
- providing additional load-bearing material
- redistribution of the loads through imposed deformation of the structural system.

The new load-bearing material will usually be

- high-quality concrete
- reinforcing steel bars

- thin steel plates and straps
- post-tensioning tendons (for PSC members)
- various combinations of these materials.

The main problem in strengthening is to achieve compatibility and continuity of structure and behaviour between the original material/structure and the new material/repaired structure. Two possibilities exist in the load-sharing pattern between these:

(a) The strengthening part of the structure participates only under live load.
(b) The strengthening part of the structure participates under live and dead load (or a part of it).

These strengthening measures improve the strength, but not necessarily the durability of the original structure.

Strengthening should be designed and constructed in accordance with appropriate codes. Typical problems occur in the transfer of shear forces between old concrete and new concrete. Satisfactory interaction between existing concrete and new concrete is generally required in strengthening and repair. For homogeneous action, the joint between old and new concretes must be capable of transferring shear stresses without significant relative movements. Differences in creep and shrinkage properties between old and new structural elements will require careful evaluation; the strength and modulus of elasticity of the old and the new material must match as closely as possible. Traffic may have to be suspended for 3 to 14 hours after strengthening.

**Strengthening of the reinforcement:** The strengthening of reinforcement subject to tensile forces can be achieved by:

- replacement of reinforcement damaged by corrosion
- additional reinforcement placed in the old cross-section or placed in an additional concrete layer
- pre-stressing
- epoxy-bonded steel plates.

In special cases, severely damaged reinforcing bars must be replaced. The new bars must be joined to the ends of the old ones by lapped splices, welding, or coupling devices. Strengthening by means of epoxy-bonded steel plates is a popularly used technology in many countries.

The load-carrying capacity depends on the strength of the reinforcement, the concrete and the adhesive.

The question of long-term behaviour is of particular importance for these materials, the properties of which are time-dependent.

Creep of epoxy resins is considerably greater than that for concrete. Change of properties due to ageing causes a loss of nearly 50% of the strength on a long-term basis. The same applies for fatigue strength also.

In correctly designed structures with bonded-plate reinforcement, a ductile failure can be expected.

### Strengthening of Pre-stressed Concrete Members

This is achieved through supplementary pre-stressing, which is considered effective, both for reinforced concrete and pre-stressed concrete structures. So far, only post-tensioning systems have been developed for this purpose. Both unbonded and bonded tendons can be used. If short pre-stressing elements are required, a post-tensioning system with minimum anchorage slip should be chosen.

**Strengthening with prefabricated RCC or PSC elements:** This method requires de-stressing of the original cross-section. The composite cross-section of precast elements and the original cross-section are then restressed/loaded.

**Note**: A *concrete overlay* is the best for most cases of deck rehabilitation and strengthening jobs.

## Repair to Expansion Joints, Bearings, Footpaths and Railings

Operational life of expansion joints, bearings, foot paths and railings is usually shorter than that of the bridge. These elements need special attention for repairs, replacements/renewals. The strength and efficiency of bearings may be a limiting factor in situations where it is required to increase the load-carrying capacity of a bridge.

**Expansion joints:** Since these are not expected to last for the entire life of the bridge, the joints must be replaced on a regular cycle. Elastomeric joints perform better than the conventional type of joints.

The joints must be kept watertight to prevent corrosion on the beam ends, bearing shelves and substructure. Leaks must not be tolerated. Any replacement joint should be watertight across the full width of the deck, kerb, footpath, etc.

If complete watertightness is not possible, adequate drainage arrangements should be provided. Regular maintenance of the bearings and pedestals will prevent the water from damaging the concrete. Joints may be filled with a sealant or filler, which must be designed for watertightness.

All damaged joints should be replaced. The sealant filler shall be replaced periodically. Cracked concrete in the zone of anchoring the joints shall be replaced. Periodic cleaning and removal of debris is a must.

**Bearings:** Most bearings need replacement as soon as they show signs of distress/damage. However, careful inspection and periodic maintenance of bearings can extend their service life.

Several types of defects are possible in bearings:

- Corrosion, defective seal in neoprene bearings, broken guides, cracked rollers and plates, cracks, splits and tears in neoprene pads, accumulation of debris at the bearing point, failure of anchorage system, creep of parts out of position, partial contact of bearing plates, and excessive tilt/shift.

Timely corrective action shall be taken for any of these defects. Repair or replacement of bearings requires traffic restrictions, or even temporary suspension of traffic. Lifting of the superstructure may be achieved by using flat jacks.

**Footpaths:** These have been constructed either in-situ or using precast slabs with a gap between the top of the deck slab and soffit of the footpath slab. The common distress noticed is the formation of cracks. Also precast slab components may be found displaced or missing. The kerb line and the joint area with deck slab are particularly prone to deterioration.

The cracks shall be repaired by injecting either cement grout or epoxy resin. The broken/missing precast panels shall be replaced with stronger panels. Mastic topping may also be provided over the cracked precast slabs. Where major replacements of precast elements are needed, it may be better to provide a solid in-situ foot path slab.

**Railings/Parapets:** Corrosion of holding-down bolts on railway is a common problem. A decision to repair or replace these elements shall be based on the economics and importance of the appearance of the bridge.

**Miscellaneous Aspects**

The following measures, apart from technical, have to be taken:

- Safety of the public using the bridge until the repairs are satisfactorily completed
- Effective traffic control and regulation in terms of speed as well as loads until the repairs are completed. (Radar meters and weigh-bridges are used respectively for monitoring speeds and loads)
- Contingency plans for diversion of traffic
- Providing proper information to the public about the restrictions.
- The morale of the engineers in charge of the strengthening job has to be kept high in view of the risks involved.

## 26.6 BRIDGE FAILURE

Although absolute safety is not attainable for any structure including a bridge, every possible care should be taken in the design and construction of a bridge structure. This is because the failure of bridges, especially of major ones, entails interruption to traffic and failure of vital communication links, and loss of human life and property.

In view of the importance of preventing failure of bridges, a critical study of any bridge failure that has unfortunately occurred, would prove to be very valuable in that the lessons learnt from such failure could be of immense help for future construction activity. This may even lead to revision of the standard codes of practice governing bridge design.

In the wake of increasingly long spans for bridges, innovative designs and construction methods are being tried by bridge engineers; under these circumstances, the bridges tend to become more vulnerable to failure.

Bridge failure may be total or partial; total failure means 'collapse' of the bridge and partial failure results in deficiencies in meeting the intended requirements. This necessitates reduced load limit, decreased speed, and substantial repairs and rehabilitation – rating and posting of bridges, pending repair and rehabilitation.

Usually, collapse of a bridge attracts greater attention; but study and rectification of partial failure provide valuable information to avoid recurrence, besides affording economical solutions to enhance their service life.

### Causes of Bridge Failure

The failure of a bridge is normally due to a combination of several causes, defects and deficiencies. However, a few major causes contribute to the failure of one or more of the components – foundations, substructure and the superstructure, ultimately leading to the failure of the bridge.

Based on a study of several cases of bridge failure all over the world, Smith categorised the causes of failure as given below (Smith, D.W., 'Bridge failures', Proc. Institution of Civil Engineers, London, Vol. 60, pt.1, August., 1976, pp 367-382.):

1. Floods, scour and foundation movements
2. Unsuitable or defective material or poor workmanship

3. Overloading and accidents
4. Inadequate temporary works during construction or improper erection procedure
5. Earthquake effects
6 Inadequate design procedures
7. Wind forces
8. Fatigue under traffic
9. Corrosion of steel components

In general, Factors 4) and 6) cause failure during construction only, while the others may cause failure either during construction, during service, or both. Earthquakes, fatigue and corrosion have caused failure of bridges in service.

It has been observed in this study that about sixty percent of failures have been due to natural phenomena such as floods, earthquakes and wind forces. Other major causes are faulty design and unsuitable material, overloading and accidents and erection errors.

**Floods, Scour and Foundation Movements**

Scouring of the foundation of a bridge during floods beyond the limits envisaged during design has been a common cause of damage to the bridge. The reduction of the waterway and obstruction to free flow caused by the construction of the piers and abutments could modify the hydraulic conditions of the river flow, resulting in new-patterns of silting and erosion.

Thus, special attention is to be devoted to the correct determination of the design flood discharge, the maximum flood level, maximum scour depth, and the characteristics of subsoil below the river bed.

The foundation should be taken to a level well below the computed maximum scour depth in order to ensure that the foundation is not exposed to the river water flow even during high floods.

Another measure is to undertake scour protection works near piers and abutments by providing rip-rap fill. The reduction in waterway due to the construction of the bridge should be relatively inconsiderable, so that increase in velocity of flow is not unduly high

Quarrying of sand in around the bridge substructure should not be allowed, as this could lead to further erosion by scour during floods.

There have been examples of bridge failure due to scour; a well known example is the Schoharie Creek Bridge in New York State in 1987.

If unexpected settlements have occurred resulting in excessive movements for the pier, failure may be caused unless remedial measures are taken in time.

**Unsuitable or Defective Material or Poor Workmanship**

An example for this is the use of steel with inadequate ductility, which may result in brittle fracture with sudden failure of a steel member. This kind of difficulty used to arise when welding of steel members was adopted in place of riveting during the initial phase of the development of welding of steel trusses; defective welding led to brittle fracture failures especially when the steel used was not ductile enough. Later on, the situation changed for the better will the availability of weldable ductile steels and improvements in welding techniques.

**Overloading and Accidents**

Overloading of heavy commercial vehicles plying on a road bridge can cause failure, if allowed over a period of time. This is possible due to consequent fatigue.

Severe accidents involving heavy commercial vehicles on road bridges may damage the bridge superstructure and lead to failure. Even in the case of railway bridges, derailment and capsizing of wagons and coaches can damage the bridge superstructure.

In the case of bridge over navigable waters, collision of ships and barges with the bridge substructure/superstructure has been the cause of failure in some cases. An example of failure due to ship impact is that of Sunshine Skyway Bridge in Florida, USA, in 1980. Another example of failure due to barge impact is that of Arkansas River Bridge in Oklahoma, USA, in 2002.

Of course, the forces due to accidents can be taken into account in the design to some extent and provision of suitable fenders supported on heavy caisson foundations.

### Inadequate Falsework and Improper Erection Procedure

Failure of temporary arrangements such as 'falsework', which facilitate construction and operation of the bridge in the initial stages, can cause loss of material and human life, besides interruption to traffic. Falsework failure can cause excessive deflection and catastrophic failure of the bridge superstructure. Such failures are attributable to human error as reuse of material is common for economic reasons. Falsework design has become complicated with increasing spans in modern bridge construction.

A thorough check of all the components of a falsework system just before concreting and constant vigilance go a long way in avoiding falsework failures. Similarly, erection of a major bridge invariably involves risks, which could lead to superstructure failures and consequent loss of material and lives. Underestimation of construction loads during erection and handling stresses in major components have led to bridge failures due to improper erection procedures.

An example of failure of this kind is that of the Injaka Bridge in South Africa; the prestressed concrete bridge collapsed during erection by incremental launching method, due to wrong placement of temporary bearings.

The importance of correct falsework design and the adoption of proper erection procedures appropriate to the type of bridge and the site conditions has now been realised.

### Earthquake Effects

Several bridge failures have been caused by severe earthquakes. Span shortening through violent shaking and movements of piers and abutments during severe earthquakes have been the causes of bridge failure.

Bridges with simply supported spans and foundations located in soft soil are vulnerable during earthquakes. Differential settlements of piers and abutments are caused by seismic vibrations. Liquefaction of foundation soils or fill materials has led to substructure failure during earthquakes.

Concrete columns or steel columns in the substructure have failed due to inadequate reinforcements or torsional shear strength of concrete and local buckling of steel elements, respectively.

Several preventive measures can be taken in the design of the RCC or steel columns to withstand the seismic forces in what is known as earthquake-resistant design.

Expansion joints are designed with adequate restraint against large movements due to earthquakes, while permitting smaller ones due to temperature changes.

The Niigata earthquake (1964) and the Kobe earthquake (1995) in Japan caused bridge failures; the collapsed bridges had to be rebuilt with greater care.

### Inadequate Design Procedures

In general, bridge failures due to inadequate design procedures are relatively few, although there have been some examples. In the quest for maximum strength and economy, sometimes bridge engineers have been faced with catastrophic failures.

The Quebec railway bridge (1907) in Canada failed due to buckling of the main lower chord. The Tacoma Narrows Bridge (1940) in Washington State, USA, a suspension bridge with the longest span at that time, failed due to aero-dynamic instability in high winds in the Narrows at the site.

### Wind Forces

If the effects of wind forces on the bridge superstructure are not adequately considered in the design, it could lead to failure of the bridge due to aerodynamic instability.

A well-known example of such failure is that of the Tay Bridge (1879) over the Tay River near Dundee, Scotland. It was a railway bridge built of steel lattice girders, which failed due to aerodynamic instability during a violent cyclonic storm.

Another example was the Chester bridge (1944), which was destroyed by a tornado. It was a continuous truss bridge built on the Mississippi river in the USA. The Tacoma bridge disaster also falls in this category.

### Fatigue Under Traffic

Fatigue is the gradual weakening of a structure under repetitive loading; it has the effect of spreading of a crack. The crack progresses fast, especially if the steel near it is corroded.

An example of failure of this type is that of the Silver Bridge at Point Pleasant in West Virginia, USA, in 1967. It was an eye-bar-chain suspension bridge built in 1928. The reason was established as the failure of a single eye-bar in a suspension chain, due to a small defect, 25 mm deep. Unfortunately, this had gone unnoticed in its 39-years of service, due to poor maintenance strategies.

This failure led to strict specifications for bridge inspection, maintenance and bridge management system.

### Corrosion of Steel Components

Corrosion of the steel reinforcements in RCC and PSC bridges, especially in the latter, has led to bridge failures.

In pre-tensioned structures, high-performance concrete in which corrosion inhibiting admixtures are added, tends to prevent corrosion. In post-tensioned structures, grouting of the pre-stressed cables soon after pre-stressing is an essential prerequisite for preventing failure due to corrosion

An example of this type of failure is that of the Mandovi first bridge, Panjim, Goa in 1986; corrosion of the pre-stressed steel was cited as the reason for failure. Another example is the failure of Ynys-y-Gwas Bridge, Wales, UK (1985); it failed because of grouting deficiencies which led to the corrosion of the tendons at transverse joints due to the penetration of chloride-contaminated water.

Partial failures of PSC bridges due to loss of pre-stress due to corrosion can be dealt with through rehabilitation techniques.

## SUMMARY

- Maintenance, testing, assessment of load-carrying capacity, and strengthening, if necessary, are all important aspects of bridge management.
- Identification of culverts and bridges is facilitated by a method of numbering, recommended by the IRC in the relevant specification/guidelines.
- The defects that develop in bridges are primarily cracking of concrete and corrosion of steel components and steel reinforcements in RCC.
- Defects may develop also in expansion joints; distress in any component may be caused due to fatigue under traffic loads, or due to foundation settlements.
- IRC recommends periodical inspections as well as special inspections and maintenance of a bridge register; this register should contain the original report on the bridge immediately after construction and also the inspection reports from time to time after the bridge is opened to traffic. All these are part of an efficient bridge management system.
- Posting of bridges means placing restrictions on the traffic using the bridge both with regard to load limit and speed limit, and giving advance warning ahead of the approach to the bridge.
- Bridge testing involves the use of non-destructive tests as well as destructive tests on core samples. The tests may be behaviour-, proof load-, stress history-, ultimate load-, or diagnostic tests. However, only proof-load tests are employed for rating of bridges.
- Rating of bridges is based on the assessment of the load-carrying capacity by conducting appropriate tests; preliminary assessment is primarily based on visual inspection, while detailed assessment is based on the analysis of test results.
- The rating methods employed may fall under the analytical, load testing and correlation methods. Load testing becomes mandatory in the absence of original drawings, specifications, and other relevant information about the bridge.
- Strengthening and rehabilitation of bridges becomes necessary when the bridge needs to be operated until the major repairs/replacements are undertaken as found necessary. Strengthening may be required for the substructure and/or the superstructure.
- Bridge failures are costly and have to be averted as far as possible.
- The primary causes for bridge failure are floods and scour, overloading and accidents, falsework failure and erection errors, earthquakes, wind forces, fatigue under traffic and corrosion of steel components. Preventives measures are preferably adopted to avert bridge failure.

## QUESTIONS

1. State how a bridge on a highway may be numbered as per IRC Specifications.
2. List and briefly explain the various defects that can occur in a bridge.
3. Explain the importance of bridge inspection in bridge maintenance.
4. List the equipment needed for routine periodical inspection of a bridge.
5. Differentiate between periodic inspection and special inspection of a bridge.
6. Enumerate the critical aspects to be checked during bridge inspection.
7. Discuss the important features of Bridge Management System.
8. Write short notes on:

(a) Numbering of culverts

(b) Effect of foundation settlements on bridge stability.

(c) Bridge register

9. Write brief critical notes on:

(a) Posting of bridges

(b) Non-destructive tests on bridge components

(c) Proof-load tests on bridges

10. Write short notes on:

(a) Behaviour tests

(b) Ultimate load tests

11. Write a note on analytical methods of bridge rating.
12. Write a note on the assessment of the load-carrying capacity of an existing bridge
13. Write a note on load testing for bridge rating. When does load testing become mandatory?
14. Write a note on strengthening of the superstructure of a steel girder bridge
15. Discuss the important causes of bridge failure.
16. Discuss the precautions to be taken in the construction of a bridge to prevent possible failure.
17. Explain how accidents and ship-impact can cause bridge failure. What precautions may be taken to prevent such a failure?
18. Explain how wind forces can cause bridge failure. Give one well-known example of such a failure.

## APPENDIX

### List of some major bridges of India and other countries

| India * | | | | |
|---|---|---|---|---|
| **Years** | **Name of Bridge** | **Location** | **Type of bridge particulars** | **Main Span (m)** |
| 1900 | Sone bridge (Nehru Sethu) (Railway bridge) | Dehri-on-Sone (Bihar) | Prestressed concrete girder bridge | 3060 |
| 1914 | Pamban Rly Sea- bridge on Palk strait | Pamban near Rameswaram (Tamil Nadu) | Cantilever bridge (Vertical lift central span of steel trusses) | 2065 |
| 1943 | Howrah bridge (over Hooghly river) | Kolkata (West Bengal) | Cantilever type Suspension bridge (with suspended span) | 457 |
| 1974 | Godavari Rail-cum-Road bridge | Kovvur–Rajahmundry (Andhra Pradesh) | Steel truss bridge (Road deck above rail track) | 2700 Rail 4200 Road |
| 1985 | Krishna Rail bridge | Vijayawada (Andhra Pradesh) | Steel truss bridge | 1100 |
| 1992 | Vidyasagar Sethu | Kolkata (West Bengali) | Cable-stayed bridge (composite type) | 457 |
| 1997 | Godavari Rail Bridge | Kovvur–Rajahmundry (Andhra Pradesh) | Bow-string Girder Bridge (Twin- arch type) | 2745 |
| 1998 | Sharavathy Rail Bridge | Hornavor (Karnataka) (Konkan Railway) | PSC box girder and PSC girder bridge | 2060 |
| 2001 | Bassein creek Bridge | Ghodbunder (Maharashtra) | Pre-stressed Concrete Girder bridge | 556 |
| 2011 | Vembanad Rail Bridge | Over Vembanad Lake (Kerala) | Pre-stressed Concrete Girder bridge | 4620 |

The first bridge on the Godavari river was the Havelock Bridge built in 1897 for the rail link between Madras and Calcutta. It was a deck type steel bridge, which was decommissioned in 1997, after a century of service.

*This list is only indicative and by no means exhaustive; There are many more Road as well as Railway bridges in Operation and more under construction.

| Other Countries** | | | | |
|---|---|---|---|---|
| **Years** | **Name of Bridge** | **Location** | **Type of bridge particulars** | **Main Span (m)** |
| 1883 | Brooklyn Bridge | New York City, USA | Suspension bridge | 486 |
| 1930 | Harbour Bridge | Montreal, Canada | Cantilever bridge of steel with concrete deck | 334 |
| 1932 | Sydney harbour Bridge | Sydney, Australia | Suspension bridge | 503 |
| 1936 | Trans bay Bridge | San Francisco, California, USA | Cantilever bridge | 427 |
| 1937 | Golden Gate Bridge | San Francisco, California, USA | Suspension bridge | 1280 |
| 1950 | Tacoma Narrows —II | Puget Sound, Washington State, USA | Suspension bridge | 853 |
| 1955 | Nagasaki Bridge | Nagasaki, Japan | Steel arch bridge | 317 |
| 1959 | Arthur Kill Bridge | Elizabeth, New Jersey, USA | Movable bridge, vertical lift type, steel truss | 170 |
| 1964 | Verrazano Narrows Bridge | New York City USA | Suspension bridge | 1298 |
| 1966 | Astoria Bridge | Columbia River, Oregon, USA | Continuous steel Truss bridge | 376 |
| 1973 | Bosphorus Bridge | Istanbul, Turkey | Suspension bridge | 1074 |
| 1987 | Sunshine Skyway Bridge | Florida, USA | Cable stayed bridge with concrete deck | 366 |
| 1994 | Normandy bridge | Over Seine, Normandy France | Cable stayed bridge with concrete deck | 856 |
| 1996 | Wanxiang bridge | Over Yangzi River, China | Concrete arch bridge | 420 |
| 1998 | Akashi kaikyo Bridge | Kobe–Naruto Japan | Suspension bridge | 1991 |
| 2001 | Suez Canal Bridge | El Ferdan, Egypt | Movable bridge of steel swing-span type | 340 |
| 2004 | Lupu Bridge | Shanghai, China | Steel arch bridge | 550 |

****This list is only indicative and by no means exhaustive;**

Bridge construction is a continuing activity everywhere in the world.

# PART V

## Tunnel Engineering

Chapter 27

# GENERAL CONSIDERATIONS IN TUNNELLING

**CHAPTER OBJECTIVES**

This chapter enables the student to understand the following:

- The definition and uses of a tunnel for a roadway for roadways, railways and several other purposes
- The advantages of a tunnel over other alternatives
- Historical aspects of tunnels from the time of Egyptians and Romans
- Shapes of the cross-section and their relative merits
- Classification of tunnels based on different criteria
- Stability of a tunnel – arch action and pressure relief phenomenon
- Salient features of the tunnel built in India and several other countries around the world.

## 27.1 INTRODUCTION

Roadways and railways are the general means of land transportation, besides waterways. In the alignment of these, the engineer may encounter different types of obstacles such as a hill or a hill range, a vast stretch of marshy land, complex road intersections, crossing between a roadway and a railway at the same level or at different levels, a lake or a river.

In order to overcome these obstacles, a *tunnel* may be necessitated in order to keep the length of the roadway or the railway to a minimum, especially in the case of a hill. An alternative solution in this case could be an *open cut*, or a *detour* around the hill.

In the case of other obstacles, the alternative could be a bridge or a flyover.

**Definition:** A *tunnel* is defined as a horizontal passage made in an obstacle such as a hill, mountain, or underground between two points on either side, in order to take the transportation route forward in as short a distance as possible.

A tunnel can be used for a roadway, railway, or water for drainage, water supply, navigation, irrigation, or for the generation of hydroelectric power.

The *alternatives* to tunnelling through a hill are:

- To make an open cut (a cut in the rock or earth material, which is open to the sky) in the hill to accommodate the roadway, the railway, or the canal.
- To take a detour around the hill, usually in a contour or on allowable gradients.

The second option of detour involves a much greater distance than the first, the tunnel being the one with minimum length.

## Tunnel Versus Open Cut – Advantages and Disadvantages

An open cut is found to be cheaper up to about 18 to 20 m depth, beyond which a tunnel would be economical. However, this depth varies with the nature of the strata; in rock, a deeper cut may be economical, but in loose or soft soil, a tunnel would be cheaper than an open cut even for relatively less depth.

### Advantages of Tunnels

(i) The excavation for a tunnel is less than that for an open cut.
(ii) The land acquisition costs for a tunnel are less than that for an open cut; some land is needed only for the shafts and spoil banks.
(iii) Tunnel avoids a division of the land through which it passes, which is inevitable in the case of an open cut.
(iv) There is minimal obstruction to the traffic above during the construction of a tunnel.
(v) Tunnels are not vulnerable to bombing in war unlike bridges and open cuts.
(vi) A tunnel is advisable in cities as it causes the least interference with traffic and avoids tearing up of expensive pavements.
(vii) A tunnel is the cheapest means of carrying water from one watershed to another, especially when crossing a narrow but long ridge.

### Disadvantages of Tunnels

- High initial cost
- Slow progress due to the mandatory safety precautions during construction.
- Need for skilled supervision and specialised equipment.
- Need for illumination both during and after construction.
- Limited working space.
- Need to acquire underground rights, if any.

In contrast, open cuts have the following advantages and disadvantages.

### Advantages of Open Cuts

(i) Initial cost is less than that for a tunnel.
(ii) The work can proceed at a greater pace than for a tunnel.
(iii) Ordinary earth-moving equipment is adequate without the need for highly specialised equipment.
(iv) Larger working space provides convenience.
(v) Minimum illumination is adequate during construction.
(vi) Availability of natural ventilation.

**Disadvantages of Open Cuts**

- Greater lengths are needed due to approaches.
- Higher tractive force is needed for the vehicles due to steeper gradients.
- Roadways/railways in open cuts are subjected to weathering action of the atmosphere leading to greater maintenance costs.
- Lack of safety in case of bombing in times of war.

## Site Selection for a Tunnel

For the selection of an appropriate site for a tunnel, the following are the criteria:

- Convenience of ingress and egress.
- High hill with a large circumference at the base indicates the need for a tunnel. (Study of topographical sheets is necessary).
- The cost of acquisition of private property with unfettered right of way and a convenient access often governs the choice.
- When the obstacle is a hill range, the minimum economic cost of a tunnel indicates it as an attractive choice.

## Historical Aspects of Tunnelling

Tunnelling is a very old art. The oldest tunnels in recorded history were constructed in Egypt and Babylon around 2000 BCE, basically for carrying water through canals. Later, in the first few centuries of the Christian era, Romans improved the techniques of tunnelling using fire and vinegar for disintegrating rocks and by working at two or more points by introducing what are now known as *shafts*.

With the advent of gunpowder or dynamite, blasting has become a popular technique in tunnelling through rock. Tunnels for carrying water through canals were constructed in the 17th century in France. The first tunnel for a roadway was built in England in 1766, and the first railway tunnel was constructed in 1826 in France. Tunnelling activity began in the USA in the early nineteenth century for carrying a canal. A large number of tunnels were constructed through the Alps from 1857 to 1913, primarily for railways.

In India, the Ajantha and Ellora caves near Aurangabad in Maharashtra date back from 2$^{nd}$ century BCE to 7$^{th}$ century CE and 6$^{th}$ to 10$^{th}$ century CE, respectively; the Elephanta caves near Mumbai were supposed to have been built from 5$^{th}$ to 8$^{th}$ century CE. Railway tunnels have been built for Indian Railways from the early twentieth century. Several tunnels have been built in the recent past.

The World's longest tunnel, the Gotthard Base Tunnel under the Swiss Alps for the Swiss Railways, is the latest and is 57 km long. The next is the Seikan Tunnel, 53.85 km long with a 23.3 km long portion under the ocean bed, built for the Japanese Railways in 1988. The longest underwater tunnel, 50.5 km long, has opened in 1994 for the Railways, passing under the English Channel. This is considered to be an excellent feat of twentieth century tunnel engineering.

## Developments in Tunnelling Technique

Compared to the manual method of tunnel construction used in olden days, tunnelling techniques in the modern era are far more advanced.

With the advent of explosives such as dynamite, steam power and compressed air pressure, rapid strides have been made in the construction of long tunnels through all kinds of strata, especially with regard to the speed of tunnelling. Safety in tunnelling has also been enhanced through appropriate measures. Special devices have been developed and used for rock blasting and drilling.

Tunnelling in soft soil has always been a challenge; excavation in small stretches with quick provision for temporary supports along with good foundations has been used successfully. What is known as *shield tunnelling* is a very successful approach in soft soil

Most of the devices and equipment for successful tunnelling have been developed during the nineteenth century; modern tunnel boring machines have been manufactured and used in European Countries such as France, England, Germany, Austria, Norway and Belgium, as also in the USA, Canada, Japan and Australia.

## 27.2 SIZE AND SHAPE OF A TUNNEL

Various shapes have been adopted for a tunnel-cross section. The choice of the shape is governed by the nature of the soil through which the tunnel is to be driven and certain practical considerations such as the purpose of the tunnel, the available equipment and the method of construction.

The most common shapes used for tunnels are given below:

(a) D-section
(b) Horse-shoe section
(c) Egg-shaped or oval section
(d) Circular section
(e) Oval section with the major axis horizontal
(f) Rectangular section

**D-section:** This consists of an arch roof supported by straight sides. The arch may be segmental or semi-circular and the sides may be vertical or slightly inclined (Fig. 27.1). The section of the tunnel appears like the letter-'D' when rotated clockwise by 90º, and hence the name.

This section is commonly used for tunnels in rock since the risk of collapse due to external pressure from unstable soil on the tunnel roof is practically absent. Under these conditions, it is convenient to have an arch roof, which transfers the external loads to the side walls and then to the ground at the base.

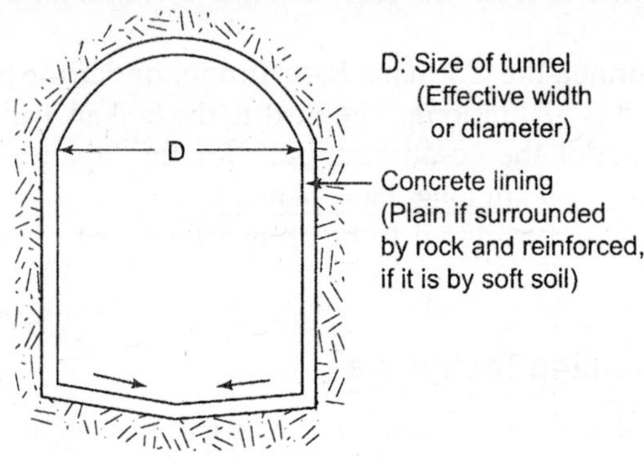


**FIG. 27.1** *D-section*

**Horse-shoe section:** This is a popular section in the shape of a horse-shoe – -a semi-circular roof with sides which are either slightly arched or straight (Fig. 27.2).

When lined, this section can withstand the external pressure well. This shape is best suited for traffic tunnels with a nearly flat floor along with adequate working space. This section is the most suitable one for soft rock or soft soil, and is commonly used for highway and railway tunnels. It can also be used for carrying water or sewage in view of the fact that its wetted perimeter is not much greater than that of a circular section of comparable size .

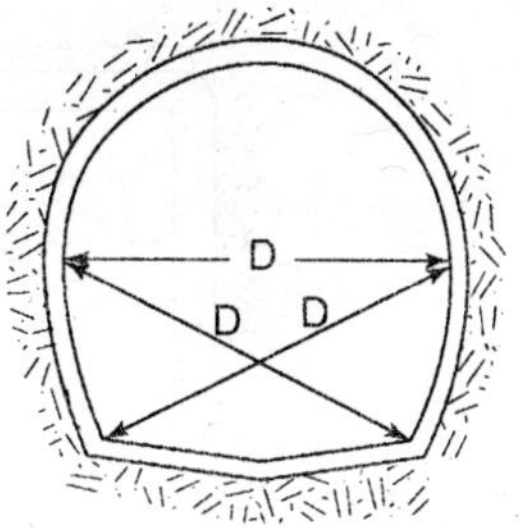


**FIG. 27.2** *Horse-shoe section*

**Egg-shaped or oval section:** This section is considered to be suitable in firm soil, but it is also effective in resisting external pressures. Therefore, it is commonly used for carrying sewage. Another advantage of choosing this shape is that it can maintain self-cleansing velocity even in dry-weather flow (Fig. 27.3).

**Circular section:** This cross-section is theoretically the best for two reasons: First, it is the best form to resist both external and internal pressures, and can be used in firm soil as well as in soft soils with a cement concrete lining; second, it provides the maximum cross-sectional area for the least perimeter.

Thus, it is considered most suitable for carrying water or sewage. But it is not suitable for carrying vehicular traffic since the flat base needed at the bottom can be created only by losing some valuable space (Fig. 27.4).

However, it is not convenient for providing concrete lining in view of the difficulty in providing form work.

**Oval section with the major axis horizontal:** This kind of section is sometimes used when the ground under the tunnel met with is water-bearing (Fig. 27.5).

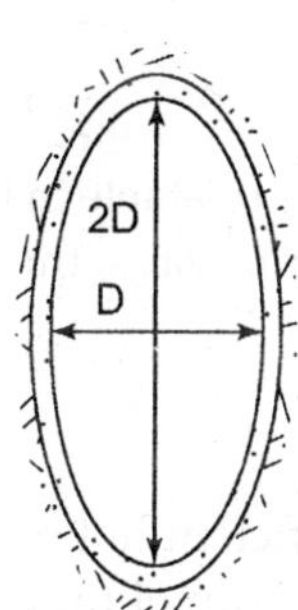


**FIG. 27.3** *Egg-shaped or oval section*

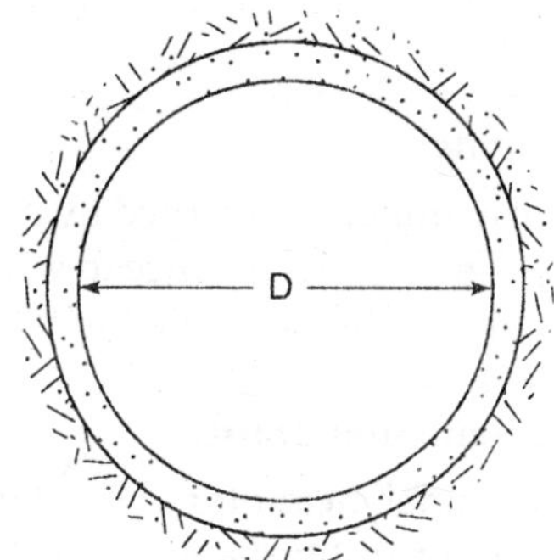


**FIG. 27.4** *Circular section*

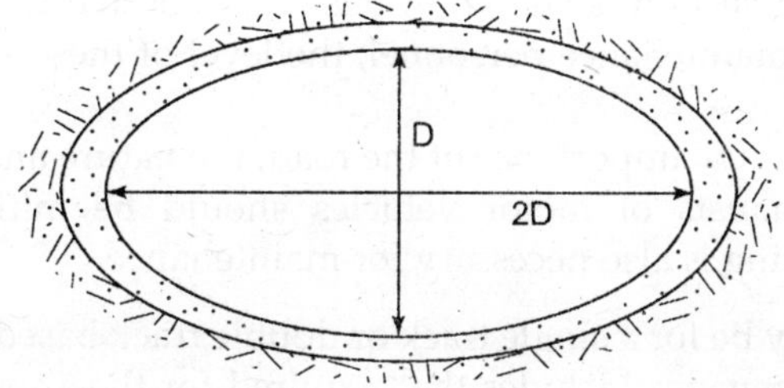


**FIG. 27.5** *Oval section with major axis horizontal*

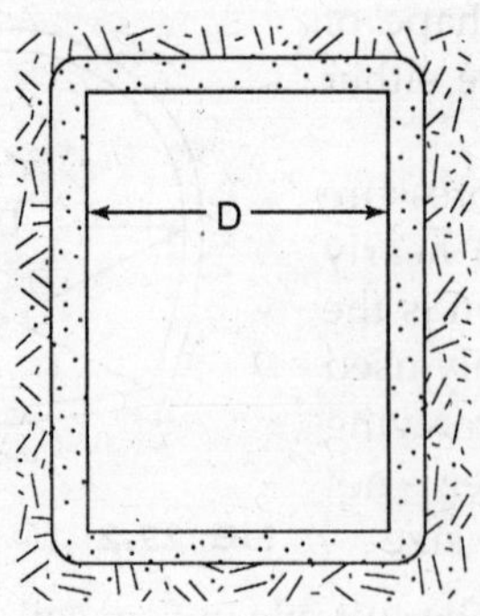

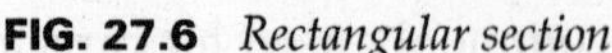

FIG. 27.6 *Rectangular section*

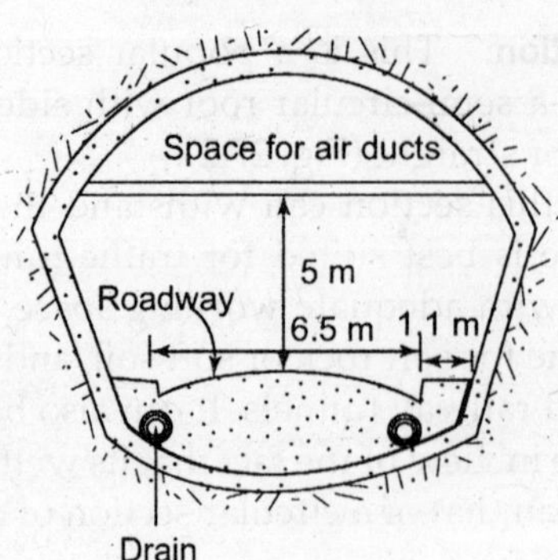


FIG. 27.7 *Highway tunnel*

**Rectangular section:** This kind of section is suitable only in hard rock. An RCC box structure is used; it may be used only for pedestrian traffic (Fig. 27.6). It is inconvenient for construction, and so is not generally preferred

## Size of Tunnel

The size of a tunnel is will be determined by its function. For highway tunnels, the size will be determined by the number of traffic lanes; for railway tunnels it depends on the gauge of the railway and the number of tracks to be accommodated. A clear opening that provides for the volume and the nature of the traffic must be made available. The thickness of the lining, the space required for utilities needed for the operation of the tunnel like ventilation, lighting and communications should also be accounted for.

The size of tunnels required to carry water/sewage is governed by the capacity for which they are designed. Footway tunnels must provide the necessary headroom and lateral clearances.

### Length of Tunnel

The length of a tunnel is governed by the nature and size of the obstacle in the case of highways and railways and on the distance to which the underground railways, the water supply mains or sewers may have to be carried in the respective cases.

### Tunnels for Vehicular Traffic

The size of a tunnel meant for vehicular traffic should be sufficient not only for the passage of traffic but also for utilities and maintenance operations. The tunnel may be for a highway or a railway.

**Highway tunnel:** A typical section is shown in Fig. 27.7. Footpaths should be provided for pedestrians as well as for maintenance personnel; the level of these should be at least 30 cm higher than the road level.

The width depends upon the importance of the road, the nature and volume of traffic.

Pollution from the exhausts of motor vehicles should be mitigated through proper ventilation. Adequate lighting is also necessary for maintenance.

**Railway tunnel:** These may be for a single track or double track based on the demands of the traffic. The space for maintenance includes that required for the appurtenances and that for free movement of the maintenance personnel. The minimum width needed depends upon the

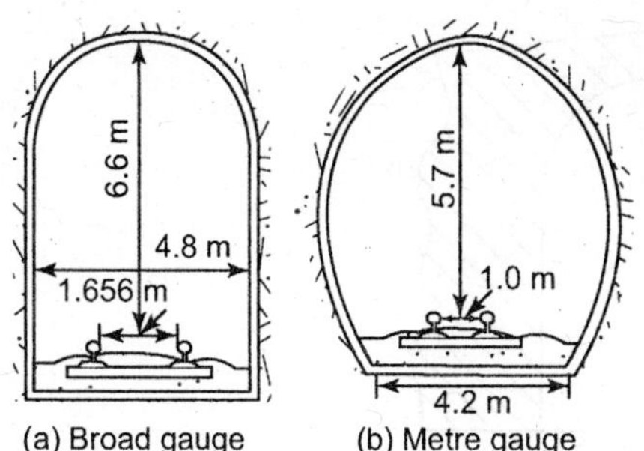


**FIG. 27.8** *Railway tunnels for single track*

gauge and the number of tracks to be accommodated. At stations, the section is widened to accommodate passengers on platforms.

Typical railway tunnels for single track for broad gauge and metre gauge is shown in Fig. 27.8.

Railway tunnels are shown for double track in Fig. 27.9.

### Tunnels for Pedestrians

Pedestrian tunnels are also called footway tunnels. In busy cities, they are constructed for pedestrian traffic to cross the roads. The size need not be large for these tunnels. The width may be 1.0 to 3.5 m and the height, about 3 m. A typical footway tunnel is shown in Fig. 27.10.

## Tunnel Approaches and Portals

Tunnel approaches are open cuts at either end leading to the tunnel proper. The tunnel entrance (or exit) is called the 'portal'. Thus, there will be two approaches and two portals on either side of a hill.

The approach is relatively short for steep slopes and long for flat slopes of the hill.

Both these are shown schematically in Fig. 27.11.

The portal should be massive and aesthetically designed. A retaining wall with wing walls and parapet should be constructed as support for the portal.

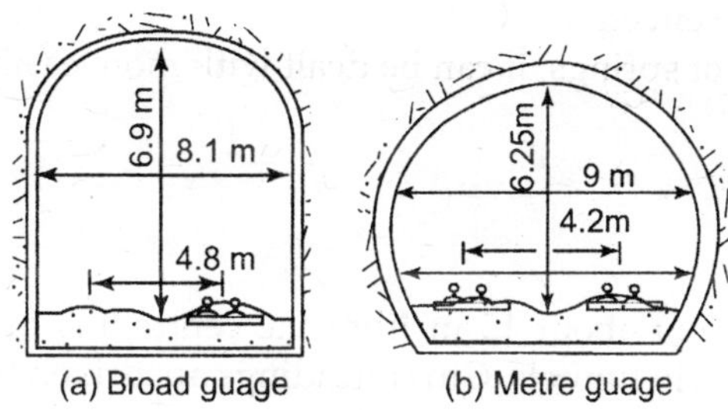


**FIG. 27.9** *Railway tunnels for double track*

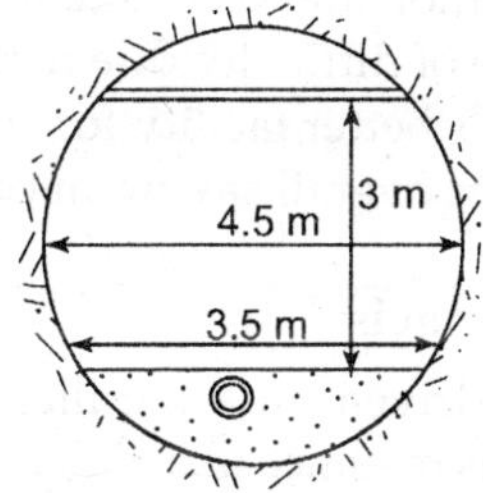


**FIG. 27.10** *Footway tunnel for heavy traffic*

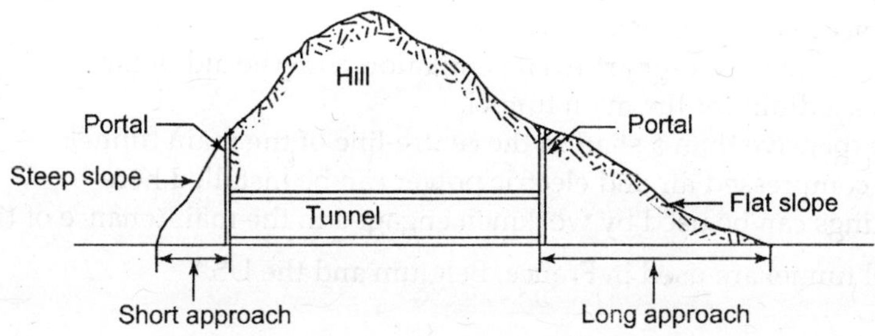


**FIG. 27.11** *Tunnel approaches and portals*

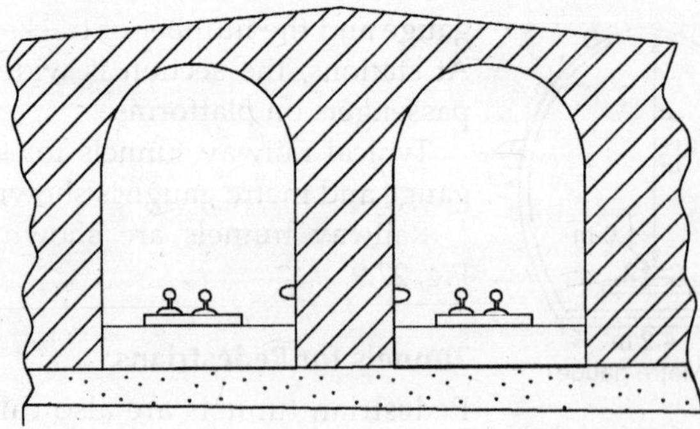

**FIG. 27.12** *A typical twin tunnel for railways*

## Twin Tunnels

For a railway with double track, it is more advantageous to have two independent small tunnels than a single large tunnel, especially in the case of long tunnels.

Each tunnel has to be excavated separately and constructed with sufficient space between them such that the work of one tunnel does not cause disturbance to that of the other. The spacing between the two tunnels is governed by the geological conditions. The size need not be the same for both the tunnels.

A typical twin tunnel is shown in Fig. 27.12.

Also called the *double- bore method*, the *advantages of twin tunnels* are:

(i) If one tunnel is blocked for any reason, the other tunnel may be used for traffic in both directions.
(ii) Ventilation and drainage may be made more effective.
(iii) In case of difficulty due to the existence of hot springs, it can be dealt with more easily.
(iv) There is better facility for transportation
(v) There is overall saving in cost.

### Parallel Tunnels

These are driven on a parallel line at a distance of about 15 m from the centre-line of the tunnel. This distance is usually much smaller in twin tunnels. Cross-headings are driven from these to the main tunnel. Setting the centre line, in this case, needs special care.

The advantages claimed for parallel tunnels are:

(i) There is less danger of material falling during construction.
(ii) Cross-headings are convenient for storing the tools and other minor equipment of the maintenance and construction workers.
(iii) The shafts can be used for artificial ventilation with the aid of fans.
(iv) It serves as a drain for the main tunnel.
(v) It is less expensive than a shaft in the centre-line of the main tunnel.
(vi) Pipes for compressed air and electric power can be installed here
(vii) The headings can be used by workmen engaged in the maintenance of the tunnel.

Such parallel tunnel are used in France, Belgium and the USA

## Classification of Tunnels

Tunnels can be classified in different ways based on the following criteria:

(a) Purpose of the tunnel
(b) Nature of material met with during construction
(c) Alignment

**Classification based on the purpose of the tunnel**

(i) Conveyance tunnels for carrying drinking water, water for hydro-electric power generation, and sewage.
(ii) Traffic tunnels – highway and railway tunnels, subways or pedestrian tunnels and tunnels for navigation.

**Classification based on the nature of material met with during construction**

(i) Tunnel in hard rock/stiff soil
(ii) Tunnel in soft rock/ soft soil
(iii) Tunnel in water bearing soils or under riverbeds

**Classification based on the alignment of the tunnel**

(i) Short tunnels or 'off-spur' tunnels to negotiate minor obstacles such as hillocks.
(ii) Base tunnels – Constructed in valleys along the natural slope up to the point where the slope does not exceed the ruling gradient.
(iii) Slope tunnels – Constructed in steep hills for economic and safe operation of roads and railways.
(iv) Spiral tunnels – Provided in narrow valleys in the form of loops in the interior of the hill, so as to increase the length of the tunnel to avoid steep slopes.

## 27.3 STABILITY OF TUNNELS

In general, tunnels derive their stability through bridge action as well as pressure relief phenomenon.

## Bridge Action

Bridge action is the mechanism by which the imposed external loads get transmitted to the supports at the ends of a span, which, in turn, transmit them to the elements below and ultimately to the ground below.

A tunnel may also derive stability by transmitting the loads due to overburden imposed on its roof to the side walls of the tunnel, and ultimately to the ground, similar to bridge action.

When the tunnel is lined, the side walls, which are of concrete and reinforced, if necessary, transmit the loads safely.

Since the overburden pressure occurs directly at the invert of the tunnel, this is usually strengthened by lining with additional material, and by means of structural steel reinforcements, if necessary.

This, coupled with the mechanism of bridge action, contributes to the stability and the safety of the tunnel.

### Pressure Relief Phenomenon

Rocks, especially those at depth, are affected by the weight of the overburden, resulting in a build up of stress in them. If this stress is not dissipated or released to cause the corresponding strain, it remains in the rocks as *residual stress*. While the rocks remain in a confined condition, the stresses accumulate and may reach high values. If the confining condition is removed, as in tunnelling, the residual stress can cause displacement, depending on the magnitude of the residual stress.

The *pressure relief*, which is characterised by a decrease in residual stress, may be instantaneous or slow in character, and is accompanied by movement of the rock mass with variable degrees of violence.

In tunnels driven at great depths below the surface, rock may suddenly break from the sides of the excavation. This phenomenon is referred to as *rock bursting*. Rock bursts are due to the dispersion of residual stresses which may exceed the strength of the ground around the excavation; this usually happens at depths over 600 m.

*Popping* is a similar but less violent form of failure; the sides of an excavation bulgê before exfoliating. Spalling tends to occur in cleaved rocks.

Another pressure relief phenomenon is *bumping ground*. Bumps are sudden and cause somewhat violent earth tremors that may dislodge rock from the sides as a consequence of the newly created stress conditions.

Pressure relief can be accomplished if the tunnels have large free cross-sectional area and are connected to vertical shafts constructed alongside the tunnel and connected by a short horizontal shaft. The area of cross-section of the pressure-relief shaft may be as much as 15% of that of the tunnel. This acts as an enhanced safety measure.

## 27.4 TUNNELS IN INDIA AND IN OTHER COUNTRIES

A select group of tunnels in operation for various purposes both in India and in other countries are presented in following tables:

**Tunnels in India**

| S. No | Tunnel | | Location | Purpose | Year | Remarks |
|---|---|---|---|---|---|---|
| | Name | Length (km) | | | | |
| 1 | Jawahar tunnel | 2.50 | Banihal, J&K | Roadway | 1956 | |
| 2 | Bhatan tunnel | 1.05 | Mumbai–Pune Expressway | Roadway | 2000 | |
| 3 | Malabar Tunnel | 3.60 | Mumbai | Roadway | 2014 | |

Cont'd

**Table Cont'd**

| S. No | Tunnel | | Location | Purpose | Year | Remarks |
|---|---|---|---|---|---|---|
| | Name | Length (km) | | | | |
| 4 | Kamshet tunnel | 1.00 | Mumbai–Pune Expressway | Roadway | 2000 | |
| 5 | Banihal Tunnel | 8.45 | Jammu & Kashmir | Roadway | 2015 | Between Banihal and Qazikund |
| 6 | Patnitop Tunnel | 9.20 | Jammu & Kashmir | Roadway | 2016 (expected) | 90 km- long network of tunnel in this project |
| 7 | Rohtang Tunnel | 8.80 | Himachal Pradesh | Roadway | 2017 (expected) | Manali-Leh bhighway |
| 8 | Parsik tunnel | 1.60 | Maharashtra | Railway | 1916 | Between Mumbai and Kalyan |
| 9 | Saranda (T-1) tunnel | 2.00 | Goilkera–Jharkhand | Railway | 1900 | |
| 10 | Monkey Hill Tunnel | 2.16 | Karjal–Maharashtra | Railway | 1982 | |
| 11 | Karbude tunnel | 6.50 | Ukshi–Maharashtra | Railway | 1997 | |
| 12 | Tike (T-39) tunnel | 4.30 | Natuwadi | Railway | 1997 | Konkan Railway |
| 13 | Ratnagiri tunnel | 4.08 | Ratnagiri, Maharashtra | Railway | 1997 | Konkan Railway |
| 14 | Panvel tunnel | 2.83 | Chowk-T-3 Maharashtra | Railway | 2006 | Konkan Railway |
| 15 | Khowai (T.2) Tunnel | 2.48 | Munigiabam, Tripura | Railway | 2008 | |
| 16 | Pir Paigal tunnel | 11.22 | Banihal–Jammu & kashmir | Railway | 2013 | |
| 17 | Metro tunnel | 6.50 | Delhi Metro | Railway | 2005 | Metro Railway |
| 18 | Netaji Subhash Place tunnel | 1.25 | Delhi Metro | Railway | 2013 | Metro Railway |

## Tunnels in Other Countries

| S. No | Tunnel Name | Tunnel Length (km) | location | Purpose | Year | Remarks |
|---|---|---|---|---|---|---|
| 1 | Laerdal | 24.51 | Norway | Roadway | 2000 | Longest road tunnel |
| 2 | Zhongnau-Shan | 18.04 | China | Roadway | 2007 | |
| 3 | St.Gotthard | 16.94 | Switzerland | Roadway | 1980 | Alps |
| 4 | Ariberg | 15.52 | Austria | Roadway | 1978 | |
| 5 | Mount Ovit | 14.70 | Turkey | Roadway | 2015 | |
| 6 | Frejus | 12.90 | France–Italy | Roadway | 1980 | |
| 7 | Mont Blanc | 11.60 | France–Italy | Roadway | 1965 | Alps |
| 8 | Kan-etsu | 11.00 | Japan | Roadway | 1991 | |
| 9 | Duplex A-86 | 10.00 | France | Roadway | 2011 | |
| 10 | Seikan Tunnel | 53.85 | Japan | Railway (23.3 km below sea bed) | 1988 | Located 100 m below ocean bed |
| 11 | Channel tunnel | 50.50 | UK–France | Railway | 1994 | 38 km under the English Channel |
| 12 | Lötschberg Base tunnel | 34.60 | Switzerland | Railway | 2007 | Alps |
| 13 | Guadarrama | 28.40 | Spain | Railway | 2007 | Twin tunnels 30 m apart |
| 14 | Taihong | 27.80 | China | Railway | 2007 | |
| 15 | Hakkoda | 26.45 | Japan | Railway | 2005 | |
| 16 | Geumjeong | 20.30 | South Korea | Railway | 2009 | |
| 17 | Guangzhou | 60.40 | China | Railway | 2010 | Metro Railway |
| 18 | Seoul Subway | 47.60 | South Korea | Railway | 1995 | Metro Railway |
| 19 | Singapore Metro tunnel | 35.70 | Singapore | Railway | 2012 | Metro Railway |
| 20 | Delaware Aqueduct | 137.00 | New York, USA | Water supply | 1945 | For drinking water |
| 21 | Päijänne | 120.00 | Finland | Water supply | 1982 | For drinking water |
| 22 | Dahuofang | 85.32 | China | Water supply | 2009 | For drinking water |
| 23 | Orange-Fish River tunnel | 82.80 | South Africa | Water supply | 1972 | For drinking water |
| 24 | Bolmen Water tunnel | 82.00 | Sweden | Water supply | 1987 | For drinking water |

Cont'd

**Table Cont'd**

| S. No | Tunnel | | location | Purpose | Year | Remarks |
|---|---|---|---|---|---|---|
| | Name | Length (km) | | | | |
| 25 | Karahnjukar | 39.70 | Iceland | Water supply for hydel power | 2007 | Water for hydel generation |
| 26 | Emisor Oriente | 62.50 | Mexico city | Sewage | 2012 | Sewerage tunnel |
| 27 | Sanliurfa | 26.40 | Turkey | Irrigation | 2005 | Water supply for irrigation. |

## SUMMARY

- A tunnel is a horizontal passage made through an obstacle such as a hill or underground earth between two points on either side to take the route forward in a relatively short distance. It can be used for a roadway, railway, or water for various purposes such as drinking, hydroelectric generation, drainage, irrigation, navigation, or for sewerage.

  The alternatives for a tunnel are an open cut (up to a certain depth) or a detour around the obstacle.
- There are several advantages of a tunnel -- there is least interference to traffic and it is secure against bombing during times of war. It is the cheapest means of carrying water through long distances.

  However, the disadvantages are high initial cost, slow progress in construction and need for special equipment and skilled supervision.
- Tunnelling is a very old art practised by Egyptians and Babylonians around 4000 years ago. Tunnels were driven for canals in France as early as in the 17th century,; tunnels for traffic were built in England and France later.

  Modern tunnels are very long and run up to a distance of 50 km or more
- The shape of the cross-section of a tunnel is usually a D-section or a horse-shoe section for traffic tunnels and circular for tunnels carrying water/sewage.

  Footway tunnels are smaller in size and are usually egg-shaped in section
- Tunnel approaches are open cuts leading to the tunnel, while the actual entrances proper are called 'portals.'

  Twin tunnels are considered to be advantageous in many ways as compared to single large tunnels, especially for traffic tunnels.
- Tunnels are classified based on their purpose, material met with during construction and the nature of the tunnel alignment.
- A tunnel derives its stability through bridge or arch action by transmitting the loads from above to the side walls.

  Pressure relief phenomenon is also an important factor to be considered for ensuring stability of a tunnel during its construction and later operation.
- In India, the earliest tunnels were built for the railways, e.g., Saranda and Parsik tunnels in Maharashtra. The Karbude tunnel in Maharashtra is 6.5 km long; a few tunnels on the Konkan Railway are over 4 km long

  The Jawahar tunnel in Jammu & Kashmir is the oldest roadway tunnel in India built in 1956.

The Malabar hill tunnel and the tunnels in the Mumbai–Pune Expressway are of more recent origin, built in the 21th century.

The tunnels in the ongoing Jammu & Kashmir National Highway and the Rohtang tunnel in the Manali–Leh Highway in Himachal Pradesh are long tunnels and are challenging engineering feats.

The Delhi-Metro tunnel is 6.50 km long and was built in 2005.

- The Monblanc and St. Gotthard tunnels in the Alps in France and Switzerland, respectively, are long road tunnels built in 1965 and 1980, respectively. Among the railway tunnels, the channel tunnel between England and France under the England Channel is 50.5 km long, and was built in 1994.

  The Seikan tunnel in Japan is nearly 54 km long and was built under the Pacific Ocean for most part. It was an engineering feat achieved in 1988. The water supply tunnel for New York City is 137 km long, and was built in 1945.

## QUESTIONS

1. What is a tunnel? Mention its advantages. (VTU)
2. Explain the various shapes of tunnels with neat sketches (VTU)
3. Differentiate between tunnel and open cut. (KIIT University, Odisha)
4. What are the disadvantages of a tunnel when compared to an open cut?
5. What are the limitations of an open cut? Under what circumstances is a tunnel considered a better option?
6. Write a brief note on the historical aspects of tunnelling.
7. Write short notes on:
   (a) Tunnels for carrying water
   (b) D-section
   (c) Horse-shoe section
8. Differentiate between tunnel approaches and portals.
9. Classify tunnels according to their purpose.
10. Classify tunnels according to the material met with during construction.
11. Write a brief note on the pressure relief phenomenon in tunnels.
12. Explain the salient features of any one railway tunnel in India and an under-water tunnel outside the country.

Chapter 28

# ALIGNMENT AND SURVEYING OF TUNNELS

**CHAPTER OBJECTIVES**

This chapter focuses on the following aspects of tunnels:

- The investigations needed for tunnel construction – soil exploration, and geological, hydrological and meteorological data; the role of holdings and drifts
- The alignment, surface surveys, pegging the centre-line using surveying techniques
- Transfer of the alignment to the underground, Weisbach triangle
- Role of shafts and pilot tunnels in construction and maintenance
- Different aspects of shafts, sinking of shafts and their functions.

## 28.1 INVESTIGATIONS

Investigations are necessarily made to collect information on several aspects of tunnel construction even during the planning stage.

Studies are primarily made on the following aspects:

(i) Soil exploration
(ii) Geological conditions of the hill strata
(iii) Atmospheric conditions
(iv) Hydrological particulars at the site

**Soil exploration:** Determination of the nature of the soil and engineering properties such as permeability, shear strength and bearing capacity, and settlement characteristics have to be ascertained.

Appropriate sampling techniques have to be used. Undisturbed samples are required to be taken for determination of the engineering properties such as shear strength and bearing capacity, using samplers. In the case of rocks, core samples should be obtained using percussion or rotary drilling methods; strength tests, especially for compressive strength are to be conducted. Permeability tests are to be conducted even in the case of rocks, as for soils at the proposed route.

**Geological conditions of the hill strata:** These include mineral composition of the soils and rocks, defects in the rock strata such as folds and faults, and petrology of the rocks. A study of the hill strata aids to determine the stability of the strata, which is a crucial factor on which the safety of the tunnel is dependent.

**Atmospheric conditions:** Temperature, pressure and humidity are important in ascertaining the requirements for ventilation and lighting of the tunnel.

**Hydrological particulars at the site:** Rainfall and run-off characteristics of the site chosen for the tunnel have to be collected and analysed based on sound principles. The data will be particularly useful in the provision of appropriately designed drainage arrangements for the tunnel.

All these investigations should be completed even during the *planning stage*.

## Investigations During the Construction Stage

During the construction stage, just before embarking on a full-scale tunnel-driving, *heading* or a *drift* may be driven to obtain valuable information on the rock/soil strata for visual inspection.

**Heading:** This is the excavation of a part of the tunnel cross-section for a *small length*. A heading may be driven at the top, bottom, or the side of the proposed cross-section to be excavated for the tunnel.

Valuable information can be obtained by the driving characteristics and the material met with, by appropriate observations and testing. This will help in deciding on the method to be followed in the construction of the tunnel, the equipment needed and the precautions to be taken for safety of the operations.

Since this forms part of the final cross-section, no additional cost is involved for the heading.

**Drift:** This is part of the tunnel cross-section, excavated for the entire length of the tunnel. Such excavation is done only when the length of the tunnel is relatively small. The size of the resulting drift hole can be 3 to 5 sq.m. in cross-section, so as obtain valuable information through visual inspection and conducting minor tests.

The drift hole can be used for service lines also.

A drift gives complete information on the strata for the entire length of the tunnel, whereas a heading gives information only for a short length, as the driving of the tunnel proceeds in short stretches.

### Information From Heading and Drift

The aim or function of both heading and drift are primarily the same; information may be got on the following aspects:

- Rock stratification
- Layer thicknesses
- Rock texture
- Hardness of rock
- Ground water particulars to help decide on the use of compressed air to keep out water.
- Possibility of land slides
- Blow condition to decide on the safety measures to be adopted during construction
- Constituent minerals
- Presence of foul gases such as $CO_2$, $H_2S$, CO, $CH_4$ and $SO_2$.

The nature of the rock and its condition may be ascertained by geological examination – as intact, stratified, moderated jointed, squeezing, swelling, or popping rock.

## 28.2 ALIGNMENT OF TUNNELS

Alignment of a tunnel is an important aspect, which has bearing on many crucial items involved in the construction of a tunnel.

Ordinarily, the alignment is made straight from one end to the other. However, the final alignment will be fixed taking into account the constraints in any particular case. If curves are considered to be unavoidable, they should be made as flat as possible.

The primary criteria of the topography of the obstacle/hill, favourable geological conditions, and site constraints such as legal difficulties in securing underground rights of a private property have already been considered in Chapter 27.

### Constraints in the Alignment of a Tunnel

- Heterogeneous strata with widely varying properties.
- Water table problems
- High temperature zones posing construction problems.
- Fractured rocks
- Filled-up ground, lacking stability
- Presence of service lines for electricity and communication facilities.

A thorough evaluation of these constraints should be made so that the best alignment and the most appropriate construction procedure may be chosen.

### Environmental Considerations

The tunnel site as well as the alignment should be chosen in such a way that minimum difficulty is experienced in tackling the disposal of ground-water, muck and foul gases.

In addition, tunnel alignment shall be avoided in environmentally protected zones notified by the Government.

### Gradients in Tunnels

The minimum longitudinal gradient recommended for a tunnel is 0.25 to 0.50%. For long railway tunnels, the gradient should be 25% less than the ruling gradient. Gradients in the ascending direction tend to increase the atmospheric resistance; also, the friction of the wheels on damp rails gets reduced, posing traction problems.

Hence, it is recommended that, unless it is unavoidable, only the minimum gradient needed from drainage considerations be adopted for tunnels. If the tunnel is short, the gradient is given on to one side only; however, if the tunnel is long, it can be made to slope down on to either side from the mid-point of the tunnel.

## 28.3 SURVEYING FOR TUNNELS

The setting out of a tunnel or marking its centre line underground consists of the following:

(a) Surface survey
(b) Surface alignment

(c) Transfer of surface survey to underground
(d) Setting out underground
(e) Levelling underground in the tunnel

### Surface Survey

A preliminary study should be carried out by means of a transit theodolite fitted with a tacheometer (for stadia tacheometry) for at least two kilometres on either side of the proposed alignment and plotted to a small scale (say, 1: 20,000) with contours at 5 m intervals. On such a plan, the final alignment is selected and a detailed survey conducted along it as closely as possible on the surface of the hill.

A detailed study of the geological structure of the site is also recommended at this stage.

Once the route is decided upon, the following steps are taken:

(i) Alignment of the centre-line of the tunnel
(ii) Gradient of the tunnel
(iii) Determination of the exact length of the tunnel
(iv) Establishment of permanent stations marking the centre line.

Instead of a transit theodolite, a mining transit may be used, if available. A 30-m steel tape along with two tripods to support the tape is also needed.

### Surface Alignment

It is necessary to set out the centre-line of the tunnel, on the surface accurately, from one end to the other. If there is only a single peak or ridge from which both ends of the tunnel are visible, the method of *balancing-in* is adopted repeatedly to fix the intermediate points until a straight line between the two ends is obtained. Theodolites with a least count of one second are used for angular measurements.

A point roughly in the middle near the peak may be first established with the aid of two theodolites set up simultaneously at either end of the centre line. The length and bearing of the proposed alignment may be obtained by running an open traverse. A *laser pole unit* is used to define a thin visible vertical plane up to a large height; this facilitates the viewing from a theodolite beyond the peak (Fig. 28.1):

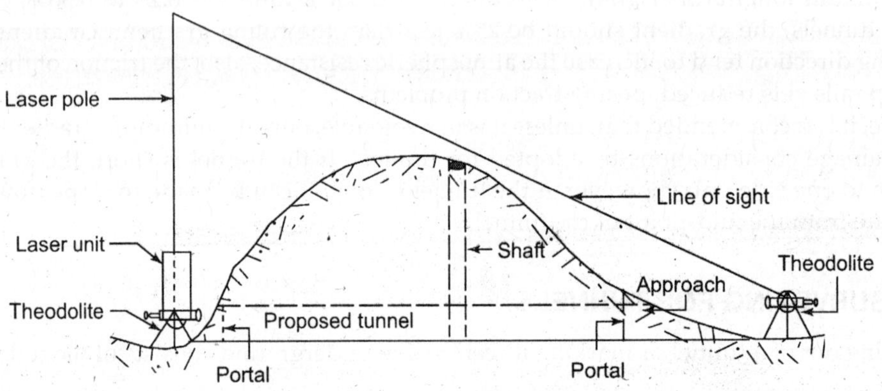


**FIG. 28.1** *Surface alignments of a tunnel*

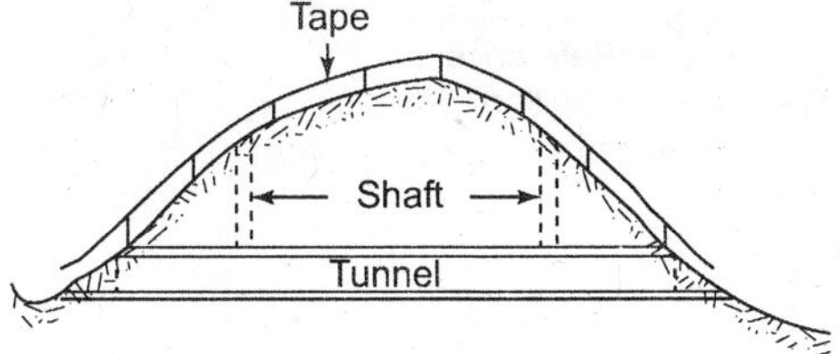


**FIG. 28.2** *Direct measurement of the length of tunnel*

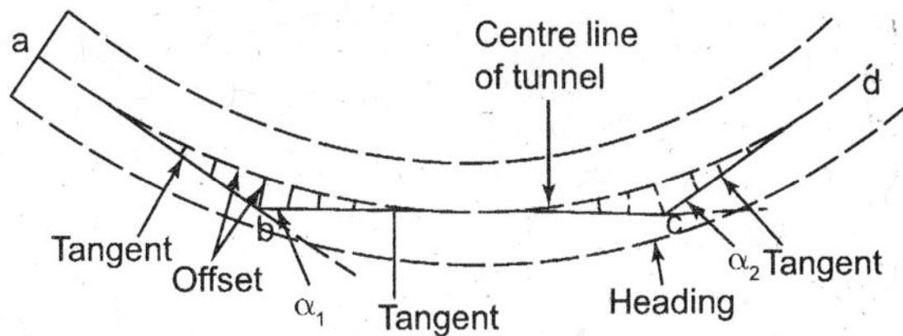


**FIG. 28.3** *Setting out a curved alignment for a tunnel*

Permanent stations are located at salient points with a high degree of accuracy on the centre-line. Observatory pillars are erected at either end of the tunnel beyond the portals.

The exact horizontal distance between the two terminals may be obtained by direct measurement similar to the procedure adopted in the case of a base line for a triangulation network. All the corrections necessary or applicable to the observed/measured length should be applied in order to obtain the true horizontal length. The corrections are for standard, temperature, pull, sag (when measured with tape supported above ground on tripod supports at the ends), gradient or slope, and finally, reduction to mean sea level. Of these, only those corrections which are applicable for the particular case need be applied. (Fig. 28.2).

If direct measurement is considered difficult due to rough terrain, the length and the co-ordinates of the end points are obtained by indirect methods such as traversing or triangulation, the former being simpler. Shafts may be driven at intervals for checking the accuracy of surface alignment and levels. Shafts are further useful to provide additional working faces, as also to provide ventilation. (Shafts are discussed in more detail in Section 28.4)

The centre-line may be marked by nails driven on wooden plugs fixed in holes drilled to the roof of the tunnel.

### Curved Alignment

A straight alignment is always preferred for a tunnel; however, a curved one may be adopted if it is necessary at a particular site because of the nature of the terrain and other considerations. It is desirable to set out the curve on the surface. The curve is usually set out underground by means of offsets from short tangents to the curved alignment; in fact, these tangents form the headings for the driving of the tunnel and the procedure of setting out the curve is repeated after each blasting (Fig. 28.3).

An alternative method of setting out the curve is by the method of tangential angles; a theodolite and a tape may be used after calculating and tabulating the tangential angles for the different chords into which the curve is to be divided.

If two theodolites are available, the use of a tape may be avoided. [Refer any standard book on Surveying.]

## Transfer of Surface Survey to Underground

The transfer of the surface alignment of the tunnel to underground is a difficult task and requires skill and care. This is done through vertical shafts which are excavated at regular intervals of around 500 m for long tunnels, and at closer intervals for short tunnels. Since a short line is to be

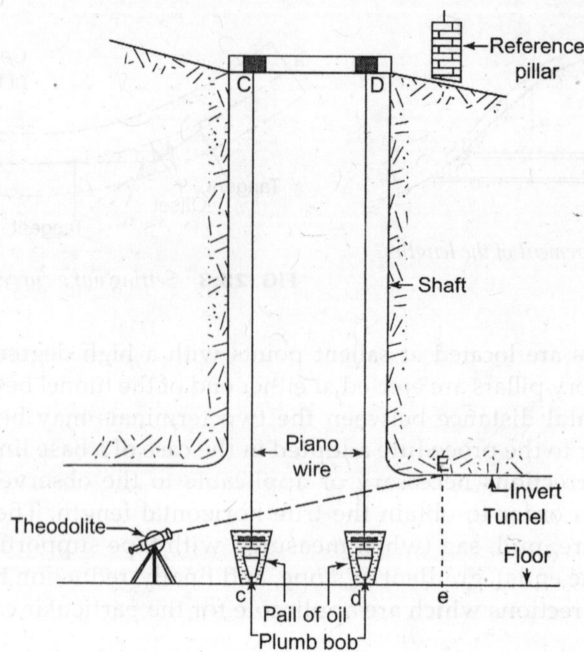


**FIG. 28.4** *Transfer of alignment from surface to underground*

transferred underground accurately and produced without error through shafts of 2.5 m diameter or less, great care is to be exercised. The procedure for the transfer is given below:

Two timber baulks are fixed across the top of the shaft close to the edges at right angles to the direction of the tunnel, as far apart as possible. A theodolite is set up over one of the previously established stations on the centre-line and accurately sighted on the other. The points on the centre line are set carefully on the beams by the method of *lining-in,* taking a number of observations on both faces of the theodolite and averaging the results. From these points, two long piano wires (one wire preferably thicker than the other) are suspended into the tunnel through the shaft, with heavy plumb bobs fixed to their lower ends. The plumb bobs are at least 10 to 12 kg in weight and are suspended into a pail of thick oil in order to damp their oscillations. They are fitted with projecting vanes to prevent rotation.

Once the plumb bobs are in equilibrium and the wires hang vertically in a stable manner, the line joining the two wires gives the alignment of the tunnel underground. For a check, the perpendicular distance between the wires should be precisely the same both at the top and at the bottom (Fig. 28.4).

## Prolonging the Alignment Underground

In order to prolong the centre-line at the bottom or floor of the tunnel, the theodolite is transferred underground through the shaft, and set up exactly in line with both the wires by trial and error. To aid the observation, a white card is placed between the two wires, the farther wire being made a little thicker. The line of sight should cut both the wires and appear as shown in Fig. 28.5.

After the theodolite is set up such that its horizontal axis is exactly in the vertical plane containing the two wires, the alignment is marked using nails or dog spikes set in wooden

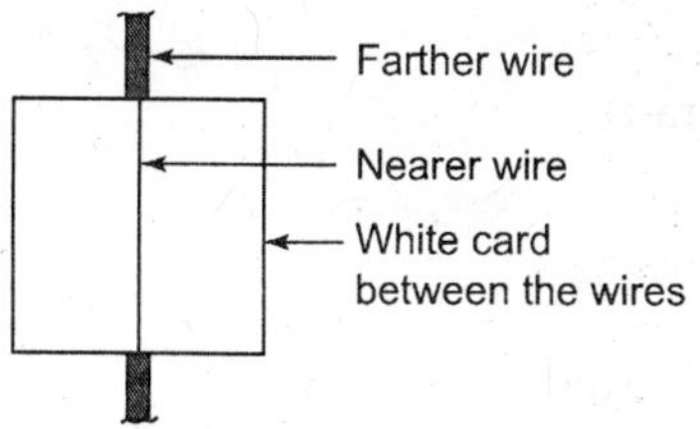


**FIG. 28.5** *White card between wires*

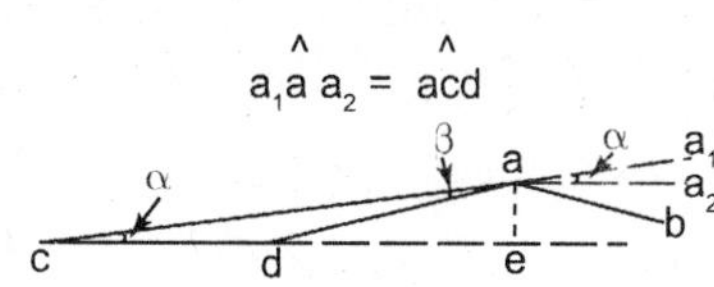


**FIG. 28.6** *Weisbach triangle*

plugs driven into holes drilled into the roof or invert of the tunnel. The marks may be easily transferred to the tunnel floor with the aid of the theodolite by sighting the nails on the roof, clamping the horizontal movement and moving the telescope down about the horizontal axis in a vertical plane to cut the floor.

(Refer Fig. 28.4. 'cd' is extended to 'e' vertically below E on the roof; C and D from the surface correspond to e and d on the floor.)

**Alternative method:** This method, popularly used for prolonging a short stretch of the centre-line in the tunnel, is known as the *Weisbach Triangle* (Fig. 28.6).

The theodolite is set up at 'a', with the two wires nearly in line such that the angle $c\hat{a}d$ is just a few minutes. It is accurately measured on both faces of the instrument; and the distances $\overline{ca}$ and $\overline{da}$ are also measured accurately. As a check, distance 'ed' is also measured.

The angles $c\hat{d}a$ and $d\hat{a}b$, which *ca* and *da* make in any other random direction, *ab*, are also measured in order to check the angle, $c\hat{a}d$, which is equal to the difference between these two angles.

The deviation $\overline{ea}$ of *a* from the plane of *cd* is then calculated as below:

The angles $d\hat{c}a$ and $c\hat{a}d$ being small, $\overline{cd}$ may be taken as equal to $(\overline{ca}-\overline{da})$.

$$\overline{ea}=\overline{ca}\,.\,\text{Sin}\ d\hat{c}a=\overline{ca}\times d\hat{c}a\ \text{(in radians)}=\overline{ca}\times\alpha\ \text{(rad.)} \tag{28.1}$$

Also, from the sine rule,

$$\text{Sin}\,d\hat{c}a=\frac{\overline{da}}{\overline{cd}}\,.\,\text{Sin}\ d\hat{a}c\,,\ \text{or Sin}\,\alpha=\frac{\overline{da}}{\overline{cd}}\cdot\sin\beta \tag{28.2}$$

(Very small angles, in radian measure, are almost equal to the Sine of the angle.)

**Example:** In the application of the Weisbach triangle method, the following are observed:

Distance $\overline{cd}=4\text{m}$

Distance $\overline{da}=5\text{m}$

Angle $\alpha=0°20'00''=0.005818\,\text{rad.}$

Angle $\beta=0°16'00''=0.004654\,\text{rad.}$

From the Sine rule applied to the Weisbach triangle, $c\hat{d}a$,

$$\text{Sin}\,\alpha=\frac{\overline{da}}{\overline{cd}}\cdot\text{Sin}\beta$$

$$\alpha(\text{rad}) = \frac{\overline{da}}{\overline{cd}} \cdot \beta\ (\text{rad})$$

$$\therefore \quad \alpha = \frac{5}{4} \times 0.004654$$

$$= 0.0058175\,\text{rad.}$$

This is very nearly the same as the observed value of α, which is 20′ or 0.005818 rad. Now the shift $\overline{ae}$, perpendicular to cde, required to locate the true position of e on the line cd produced, is given by

$$\overline{ae} = \overline{ca} \cdot \alpha(\text{rad})$$

$$= \overline{ce} \cdot \alpha(\text{rad})$$

$$= (4+5) \times 0.005818\,\text{m}$$

$$= 0.05236 \text{ m or } 5.236 \text{ cm}$$

This is because $\overline{ca}$ is, for all practical purposes, equal to $\overline{ce}$.

[**Note**: the angles α and β and the distances $\overline{cd}$ and $\overline{da}$ have to be measured very accurately with a theodolite and a steel tape, respectively, applying all the necessary corrections.]

A line $aa_2$, parallel to the centre line cd, is set out by taking a backsight on c, then plunging the telescope and finally turning off the deflection angle $\hat{dca}$ from ca produced (or $aa_1$).

By shifting the mark 'a' in the direction perpendicular to $aa_2$ by a distance equal to the calculated value $\overline{ea}$, the correct position of the point 'e' on $\overline{cd}$ produced is obtained.

## Setting out Underground

Extending or prolonging the centre-line underground, or in the tunnel, may also be carried out in short stretches of 10 metres by using a theodolite which is suitable for precise measurements. This procedure is the same as that used for such a job on the surface.

The procedure, in general, is to set the theodolite on the forward point on the centre line, backsight on the previous point and transit the theodolite to turn the line of sight by 180°. A point on this line is marked at about a distance of 10 m. The work is repeated on the other face of the instrument, and another point is marked similarly. If these two points are the same, the point lies on the centre-line extended; otherwise, a point midway between these two points is considered to be the required point on the extension of the centre-line (Fig. 28.7).

In Figure 28.7, $f$ is midway between $f_1$ and $f_2$.

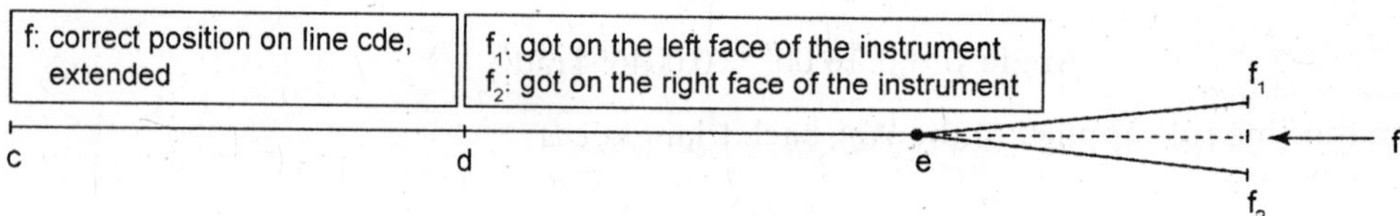


**FIG. 28.7** *Extension of centre line of the tunnel*

## Levelling Underground in the Tunnel

Levelling along the surface alignment is done in the usual way. Benchmarks are established at each shaft and at the ends of a tunnel. The difference in level between the ends of the tunnel may be determined by trigonometric levelling while conducting triangulation for horizontal control.

The levels are transferred underground from the nearest benchmark previously established on the surface. Steel band, chain, or rod is used to transfer the levels down a vertical shaft.

A mark is made near the top of one of the guides in which the cage travels, and its elevation determined by levelling from the benchmark. An assistant holds the zero end of the steel tape to the mark and the engineer descends the shaft in the cage and they mark the position of the tape on the guide, stretching the tape vertically. The assistant then descends the shaft and holds the end on the mark made by the engineer. Then, the engineer makes another mark on the guide at the other end of the tape. The process is repeated until the bottom of the tunnel is reached.

In the case of very deep shafts, the vertical measurement is changed to the horizontal one by the arrangement shown in Fig. 28.8.

A fine steel wire loaded with a weight of 5 to 12 kg passes over a pulley at the top of the shaft from a windlass or a winch mechanism. It is then lowered down the shaft, care being taken to see that the wire is in contact with the horizontal threads AA and BB, stretched at the top and bottom of the shaft respectively. The points of contact are then marked on the wire with a chalk piece, the exact point being marked with a red pencil. The wire, which is still loaded, is wound up and stretched on the surface of the horizontal planks on the surface. The distance between the marks on the wire is than accurately measured with a steel tape. The method is rapid, convenient and accurate. Since the wire is under constant tension throughout the operation, no correction for pull is necessary.

### Underground Benchmarks

The levels established near the bottom of the shaft are taken as benchmarks and fixed on the top of stakes driven firmly into the ground or on the side of the tunnel. If the tunnel is in rock, permanent bench marks are established on flat projections of the rock, or on a short piece of

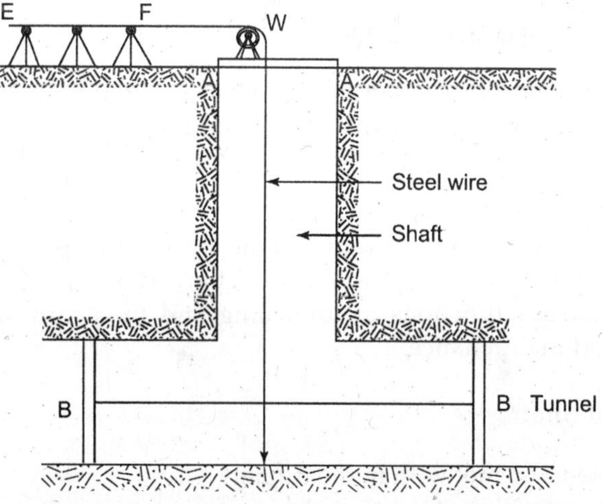


**FIG. 28.8** *Transferring levels through deep shafts*

drill steel grouted in a hole drilled in the sidewall. In brick-lined tunnels, bench marks are established on iron spikes or wedges driven into the joints if the brickwork is in the side of the tunnel at a convenient height. In the case of steel-lined tunnels, they are established on the flanges of the segments of the lining.

A line of levels is run through the tunnel from end to end of the tunnel for checking after the construction is completed.

It is desirable to connect the levels to the nearest GTS benchmark if possible.

**Special Facilities for Underground Sights**

- Plumb line against a white background of a sheet of oiled paper, illuminated by a lamp behind, for short sights.
- Carriage candles in a weighted frame,for long sights.
- Argand oil lamp of 40 to 50 CP in a metal frame, for very long sights
- A plummet lamp, or a plumb line illuminated from behind, or a vertical illuminated slit for sighting on floor stations.

Thus, in general, illuminated signals are necessary for underground sights.

### Accuracy of Tunnel Surveying

A high degree of precision is necessary in tunnel surveying, as there is no way to check the work until the tunnel is driven through. The permissible error in alignment of railway tunnels is 25 mm. If the headings do not meet, the error being appreciable, it will be necessary to introduce a very flat curve (sometimes a reverse curve) into the alignment at their junction, and also to widen the tunnel to accommodate the curve; this results in increased cost and permanent annoyance.

The range of error of alignment in the past tunnel work was 1 mm to 120 mm, while that in level was 0.3 mm to 55 mm, depending upon the importance of the tunnel.

However, in modern tunnelling jobs, the tolerance limits imposed are even less in view of improved equipment and techniques developed for tunnel driving.

## 28.4 ADDITIONAL WORKING FACES

The methods of obtaining additional working faces for facilitating the construction of tunnels are by driving or sinking what are known as *shafts* and *pilot tunnels*.

### Shafts

A shaft is a vertical bore on the centre-line of a tunnel or on a line parallel, but close to it. They are sunk from the ground surface down to the tunnel invert.

Shafts serve multifarious functions in tunnelling and thus form an inseparable part of tunnel construction and maintenance.

**Functions and Uses of Shafts**

(i) Inspection of strata to be met with in the tunnel
(ii) Alignment of the centre-line of the tunnel

(iii) Transfer of levels from the surface to the underground.
(iv) Assessment of the volume of water that is likely to be met with, which will help in deciding upon the lining to be adopted.
(v) Provision of extra faces for working and driving headings for the tunnel in order to avoid long lead for removal of muck.
(vi) Provision of additional ventilation for the tunnel.

**Location of shafts:** It is desirable to locate a shaft where its depth would be a minimum; from this criterion, valleys are better than ridges, but the former present drainage difficulties.

In general, shafts are located directly over the centre line of the tunnel, as it is convenient to remove materials from the tunnel and to transfer levels underground. In some cases, side shafts or shafts shifted to one side of the centre-line of the tunnel are sunk; in such cases, a transverse passage, called an *adit* has to be bored to connect the shaft to the tunnel. When machinery has to be lowered into the tunnel through the shaft, the passage may prove to be a bottle neck for turning.

**Size of the shaft:** The shaft size depends upon the size of the cage to be used for the removal of muck, or excavated material, the quantity of muck to be handled, the type of equipment used for tunnelling, and space required to carry ancillary items such as pipes and wires.

The minimum size required for a single cage is 3 m × 4 m and that for a double cage is 3 m × 6 m. Usually, a width of 1.5 m is separately used as a sideway for ladders, pumping equipment, etc., which is included in the overall size.

**Shape of shaft section:** This may be rectangular for temporary shafts to be filled up later. Permanent shafts, which are lined with cement concrete or pre-stressed steel liner plates, are invariably of circular section.

**Number of shafts:** This depends upon the depth of the tunnel, the intended uses of the tunnel, the nature of difficulties met with during the construction and the time available for completion.

## Classification of Shafts

There are three types of shafts based on their use:

(i) Temporary shafts
(ii) Working shafts
(iii) Permanent shafts

**Temporary shafts:** These are located either at the centre-line of the tunnel or to one side of it; in the latter case, it has to be connected to the tunnel through a horizontal bore. Their main purpose is to ascertain the nature of the strata to be met with in the tunnel. A shaft gives a better idea of the earth material than a bore hole, which is small and provides the material as a highly disturbed sample.

The common size of a temporary shaft is 2 m square.

**Working shafts:** The functions of the working shafts are;

1. For carrying the material to and from the heading.
2. For pumping water when there is inflow.
3. Serving and augmenting the ventilation needs.
4. Helping in alignment and transfer of levels.

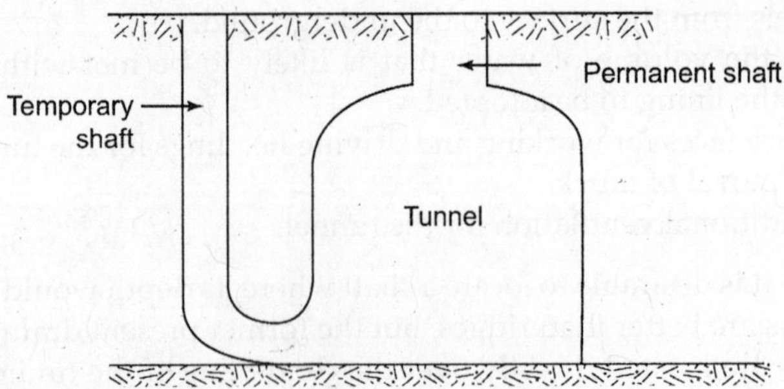


**FIG. 28.9** *Permanent and temporary shafts*

**Permanent shafts:** These shafts are permanently left even after the completion of the construction. They are raised about 2.5 m above the ground in order to avoid accidents.

Usually a circular section with diameter 2.5 m to 4.5 m is used.

A combination of a temporary and permanent shaft is shown in Fig. 28.9.

### Methods of Sinking Shafts

The method of sinking a shaft depends upon the nature of the material met with.

**Sinking or construction of shafts in rock:** The steps involved in the construction are-

1. Drilling – a convenient pattern of drill-holes which are charged with explosives
2. Blasting – this results in splitting of the rock into pieces of several sizes.
3. Mucking – removal of the loose material/pieces for dumping outside.
4. Timbering – for carrying guides for the cages and support lagging for stability
5. Pumping – for removal of water, if any, by a submersible pump.

These are the same steps in tunnelling in rock. The only difference is in the scale, which is much smaller in the case of shafts than that for tunnels.

If the rock is sound, shafts may be *raised* or constructed upwards from the tunnel heading, instead of being sunk from above. The advantage is the muck drops into the tunnel and pumping will not be needed. The initial size, in this case, is around 1.50 m, which can be subsequently enlarged to the desired size.

**Sinking of shafts in soft soil** There are two distinct approaches to sinking a shaft in soft soil:

1. Drop shaft method
2. Underpinning method

***Drop shaft method*** A curb of steel or timber, with a cutting edge, is placed firmly on the ground and the lining is built above it up to about 2 m above ground level. It is then gradually lowered bodily by excavating under the bottom. This is known as the *drop shaft method*.

***Underpinning method***: In this method, excavation is made to the required section and up to a depth where the sides do not fall, or are self-supporting. At the bottom of the excavation, a circular curb is placed, and the masonry lining is built on this curb. The excavation is continued in the centre for a smaller area to a further depth of 2 to 2.5 m, leaving a berm to carry brickwork. Ribs are fixed, and the material between the top curb and the bottom curb is removed to the full area of the top curb.

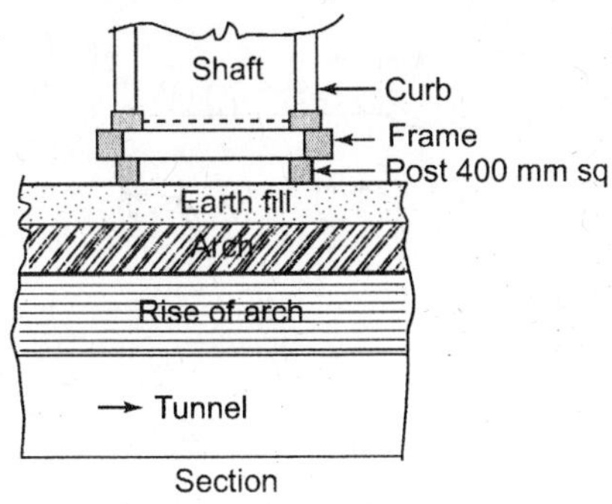


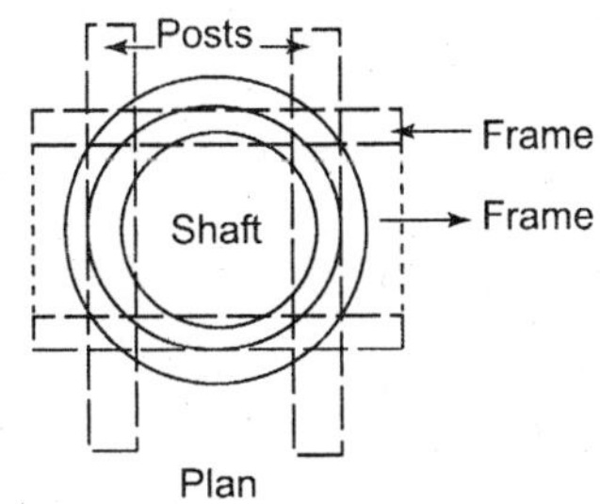


**FIG. 28.10** *Supporting the shaft from bottom (by means of an arch)*

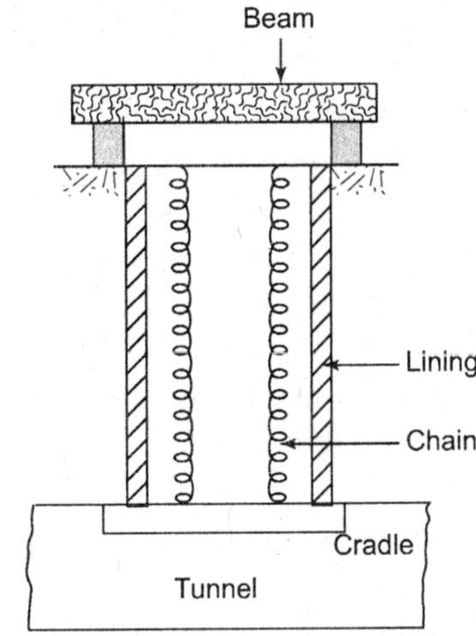


**FIG. 28.11** *Supporting the shaft from top*

Another curb as big as the top curb is introduced at the level of the bottom curb, and the masonry lining is built up to the bottom of the first curb. The ribs are then removed and the process is further repeated until the shaft touches the roof of the tunnel.

This method is known as the *underpinning method*. The shaft may be supported from the bottom by an arch or a wooden frame, if the soil is hard (Fig. 28.10).

If the soil is loose, the lining of the shaft is suspended from above the surface. This may be done by using wooden baulks, iron bars, or iron chains (Fig. 28.11).

Drop caissons and pneumatic caissons may also be used for very important shafts, but they are expensive.

**Pressure on Shaft Supports**

The pressure on the supporting members of the shaft in soft soil is given by the lateral earth pressure, $P_a$, which in turn, is governed by Rankine's theory as below:

$$P_a = K_a.r.H \tag{28.3}$$

where $K_a$ = Coefficient of active earth pressure

$r$ = Unit weight of surrounding soil

H = Height of earth-fill above the bottom of the shaft

$K_a$ is given by

$$K_a = \left(\frac{1-\sin\phi}{1+\sin\phi}\right) \tag{28.4}$$

Here, $\phi$ is the angle of internal friction.

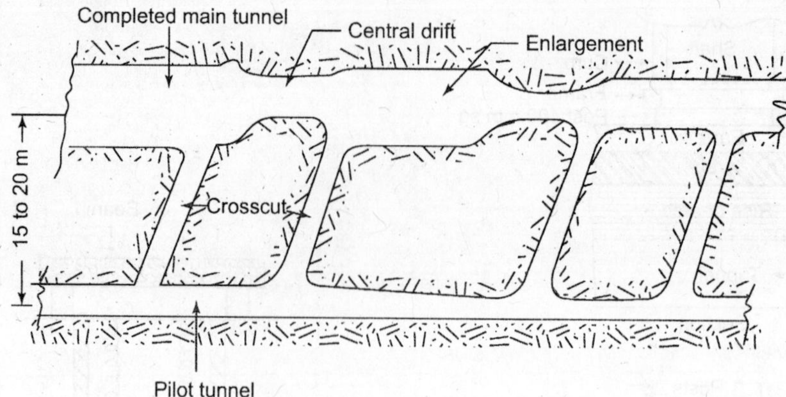


**FIG. 28.12** *Pilot tunnel system*

The total horizontal force,

$$P_a = \frac{1}{2} K_a \cdot r H^2 \tag{28.5}$$

The total force on any member may thus be got at its location.

## Pilot Tunnels

Pilot tunnels are considered to be a better alternative to deep shafts since the latter may be difficult to sink and are expensive. In such cases, a tunnel of small size, called a *pilot tunnel*, is driven parallel to the main tunnel, sufficiently close to the proposed main tunnel. Short cross-cuts or connecting tunnels are driven from the pilot tunnel to the main tunnel, thus forming a system of pilot tunnels. (Fig. 28.12).

In the main tunnel, a central drift is driven from one portal; crosscuts from the pilot tunnel to the main tunnel are driven at about 500 m intervals in the case of long tunnels. At the junction of these two, headings are formed, which facilitates expeditious driving of the main tunnel.

The *advantages* claimed *for pilot tunnels* are:

- Additional headings open up to expedite the work by simultaneous driving of the main tunnel from these faces.
- Cross cuts may be used for storing tools and material during construction.
- Cross cuts may also be used for the movement of maintenance personnel.
- Less expensive than deep shafts.

In fact, this is considered to be one of the methods of tunnelling in rock.

## SUMMARY

- The investigations required for a tunnelling job are soil exploration, geological investigations, and collection of the relevant hydrological and meteorological data.
  A *heading* is driven in a part of the tunnel section for a short length, while a *drift* is one that is driven for the entire length of the tunnel.
- The alignment of a tunnel should be preferably straight, but in exceptional cases or because of alignment errors, it can be on a flat curve.

- Surface survey involves fixing the centre-line, determining the exact length of the tunnel, and fixing permanent stakes on the alignment. The procedure of 'balancing-in' is used for prolonging the centre-line using a precision theodolite. The length of the tunnel may be obtained by direct measurements as in the case of a base-line in triangulation, or by the indirect methods of traversing and triangulation.
- Transfer of surface survey and alignment to underground is accomplished through vertical shafts on the centre-line of the proposed tunnel. Wires and heavy plumb bobs kept in pails of oil (to damp the oscillations) are used. Prolonging the centre-line underground is accomplished by means of a precision theodolite. The Weisbach triangle is a popular method used for this purpose.
- Transfer of levels from the surface to the underground is also done through shafts. Steel wires suspended by heavy weights are used. Benchmarks are established underground, either on the roof or on the side of a tunnel. It is desirable to connect all benchmarks to the nearest GTS benchmark. The accuracy desired in alignment is 25 mm and in levels, it is 0.5 mm.
- Additional working faces are obtained through the sinking of shafts and pilot tunnels. Shafts as well as pilot tunnels have other uses also for exploration and surveys and for maintenance operations.
- Shafts may be temporary, permanent or working shafts. Sinking of shafts through rock involves drilling, blasting and mucking, as in the case of tunnelling through rock. Sinking of shafts through soft soil may be by the drop shaft method or underpinning method. Necessary timbering is used.
- A pilot tunnel is a parallel tunnel close to the main tunnel but of smaller size. Cross-cut connections are provided between the pilot tunnel and the main tunnel. Apart from facilitating quick progress through additional working faces, there are other uses for pilot tunnels and cross-cuts, as in the case of shafts.

## QUESTIONS AND PROBLEMS

1. Describe the usefulness of shafts during the construction of tunnel. (KIIT University, Odisha)
2. Explain the term 'pilot tunnel'. (KIIT University, Odisha)
3. Differentiate between main tunnel and pilot tunnel. (KIIT University, Odisha)
4. Write short notes on:

   (a) Transfer of surface alignment

   (b) Use of shafts (KIIT University, Odisha)
5. Write a note on driving 'drift' (KIIT University, Odisha)
6. Write a note on the investigations required for a tunnelling job.
7. How is the length of a tunnel determined directly or indirectly?
8. Explain the use of 'Weisbach triangle'.
9. In the application of the Weisbach triangle method, the following were observed:

   Length of existing line = 5.4 m

Length of extension needed = 6 m

Angle $\alpha = 0°21'00''$ Angle $\beta = 0°18'54''$

Verify the relation between $\alpha$ and $\beta$.

Also determine the shift needed from the trial station to the true station to fix the correct point on the extension of the existing line.

10. Explain how surface levels are transferred underground.
11. Explain what is meant by a 'shaft' and classify shafts.
12. How are shafts sunk through rock?
13. Explain the drop shaft method of sinking shafts through soft soil.
14. What are the various uses of shafts in tunnels?
15. Explain how a system of pilot tunnelling is more advantageous than deep shafts.

Chapter 29

# TUNNELLING IN ROCK

**CHAPTER OBJECTIVES**

This chapter discusses the techniques used for tunnelling in rock with emphasis on:

- Special features and essential steps in the procedure for tunnel construction
- Ring drilling and bench drilling of blast holes of different cuts and patterns
- Rotary, percussion and drag type drilling
- Explosives, detonators and delays for firing; quantity of rock mass released
- Temporary and permanent supports for the walls and roof of the tunnel
- New Austrian Tunnelling Method and its advantages for large tunnels
- Mucking and haulage of debris – the equipment and appliances used
- Precautions necessary for safe operations in the process of tunnelling in rock.

## 29.1 INTRODUCTION

Tunnelling in rock is advantageous because rock is self-supporting and does not need timbering to support the excavation. When the alignment and centre-line are at large depths below the natural ground, tunnelling is more economical than open cuts apart from being safe, with the development of efficient techniques.

However, compared to tunnelling through soft soil, tunnelling through rock is expensive. In view of the self- supporting nature of rock, work can go on in several sections along the length of the tunnel, which makes the pace of work relatively fast.

The advent of specialised machinery has made rock tunnelling safe, easy and economical. Power requirement is high for rock tunnelling because of heavy drilling equipment; a separate power plant near the tunnel is convenient as it reduces the distance through which power is to be transmitted.

Thus, the following are the *essential features* of rock tunnelling:

- Practically, no timbering will be needed.
- Work can go on simultaneously at several faces.
- Power requirement is high.
- Specialised machinery has to be employed.
- Care is to be taken to avoid over-cutting, for otherwise, it becomes difficult to get the required size.
- It is more expensive, but safe, when compared to tunnelling in soft ground.

## 29.2 GENERAL PROCEDURE OF TUNNELLING

The following are the steps in the general procedure of tunnelling:

(i) Shaft construction
(ii) Excavation
  - Drilling
  - Blasting
(iii) Ventilation and dust removal
(iv) Mucking or removal of excavated material
(v) Pumping out water, if necessary.
(vi) Timbering and trimming
(vii) Temporary and permanent supports
(viii) Lining
(ix) Throwing open for use

The first step of shaft construction has already been dealt with in Chapter 28. We shall now consider the remaining steps of tunnel excavation in detail.

The excavation process and the equipment to be used depend upon the nature of the material to be excavated.

For tunnelling in rock, the two major steps for excavation are drilling of holes and blasting.

### Drilling

Holes are to be drilled in the rock, cleaned and charged (or filled up) with explosives. Mechanised drilling using compressed air pressure has become common.

A wide variety of drills have been developed, the important types being rotary drills, percussion drills and abrasion drills.

The *rotary drill* achieves cutting by rotation under pressure at the top of the drill; the *percussion drill* breaks the rock by impact of blows; and, the *abrasion drill* does so by abrasion under pressure.

Among the percussion drills, jack hammer and drifter types are popularly used. The *jack hammer* drill weighs from 10 to 40 kg and works on compressed air; it is used to drill holes up to 5 m depth.

The *drifter* drill weighs 30 to 110 kg; holes up to 120 mm diameter can be drilled in any direction using compressed air.

Among the abrasion drills are *shot* drills which can drill holes up to 750 mm diameter and a depth of about 200 metres.

*Diamond* drills can be used for drilling holes up to 300 mm diameter and a depth of about 800 m.

Selection of drilling equipment: The following factors govern the selection:

(i) Purpose of the hole—for blasting or for grouting
(ii) Diameter and depth of hole
(iii) Nature of rock and its hardness
(iv) Extent or area of rock to be blasted.

In rough terrain, jack hammer drills may be used up to a diameter of 120 mm for drill holes. For larger diameter holes, shot drills may be used.

### Mountings for Drilling Equipment

In view of the weight and vibrations, the drilling equipment needs to be supported on mountings, which may be fixed or movable.

In the fixed category, *horizontal bar mounting* is used when the width of the tunnel is less than its height.

A *column mounting* is used when the width is greater than the height

In the movable category, a *jumbo drill carriage on wheels* is used. Continuous penetrator drills have been devised and used in recent times since they do not need changing of drill bits. These are unlike conventional drills in which the drill bits become blunt in about 20 to 100 hours (diamond drill bit lasts longer) and have to be replaced frequently resulting in loss of valuable man-hours.

### Methods of Drilling Holes

Different methods/patterns of drilling holes are used.

1. American method
2. English method
3. Ring-drilling method
4. Bench-drilling method

**American method:** This consists of drilling a large number of deep holes, loaded fully with a small quantity of explosive powder.

**English method:** In this, the holes are shallow and straight. The drill is mounted on a horizontal bar. This method is suitable only for a narrow cutting.

**Ring-drilling method:** This is used in conjunction with the centre-heading method, to enlarge the section to the full size of the tunnel (Fig. 29.1):

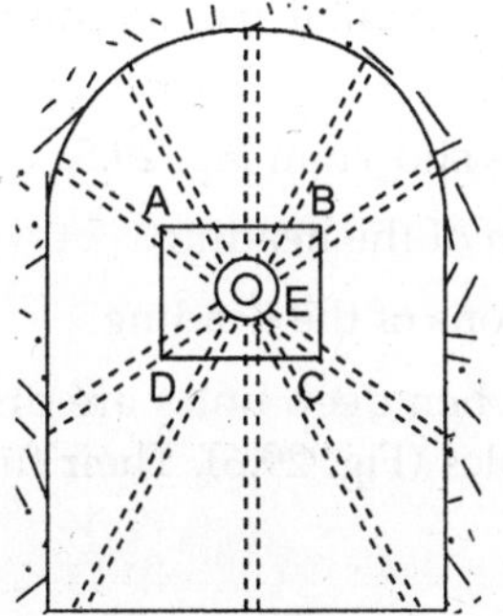


**FIG. 29.1** *Ring-drilling method*

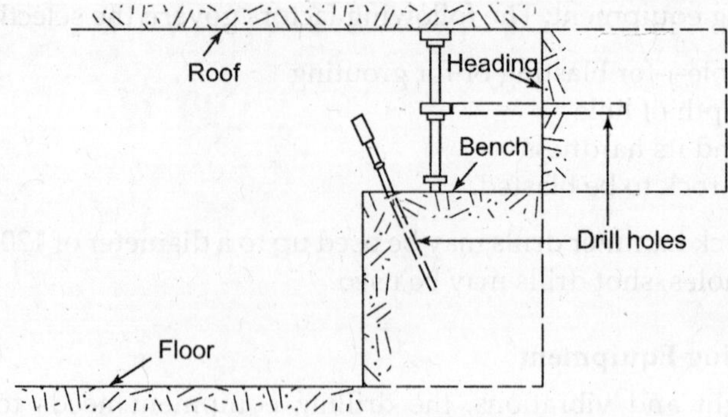


**Fig. 29.2** *Bench-drilling method*

**Bench-drilling method:** Jack hammers are used to drill vertical holes in this method; the holes are about 100 mm apart. A bench[1] is also drilled horizontally (Fig. 29.2).

## Types of Cuts

The following are the types of cuts:

(i) Wedge cut
(ii) Pyramid cut
(iii) Bottom cut
(iv) Top cut
(v) Side cut
(vi) Buster cut
(vii) Burn cut

**Wedge cut:** This is also called *vent* cut or *V-cut*, and is most commonly used. From the opposite sides of the heading, several pairs of holes are drilled so that when properly charged and blasted, a wedge–shaped cone of rock is removed from the face. Two patterns of holes are shown in Fig. 29.3.

**Pyramid cut:** This is a modification of the wedge cut. In this, the holes slope both in the horizontal and vertical planes. Cut holes are drilled in such a manner that they meet near the axis of the tunnel. When this cut is blasted through holes, a pyramid-shaped rock gets removed from the rock face.

This type of cut is used in circular or horse-shaped tunnels along with vertical column–mounted drills.

Two typical patterns of holes — one with horizontal bar mounting and the other with vertical mounting are shown in Fig. 29.4.

**Bottom cut:** This is also called 'draw cut' and is shown in Fig. 29.5.

**Top cut:** These are cuts made in the top portion of the heading.

**Side cut:** These are cuts made in the side portions of the heading.

**Buster cut:** Also called *baby cut*, this is used when deep holes are drilled. The depth of the buster cut holes is half of that of the regular holes (Fig. 29.6). Their function is to relieve the burden of the main holes.

[1] Bench is a horizontal portion on which workers can stand and drilling equipment can be placed. This helps to drive the portion of the heading above the given level.

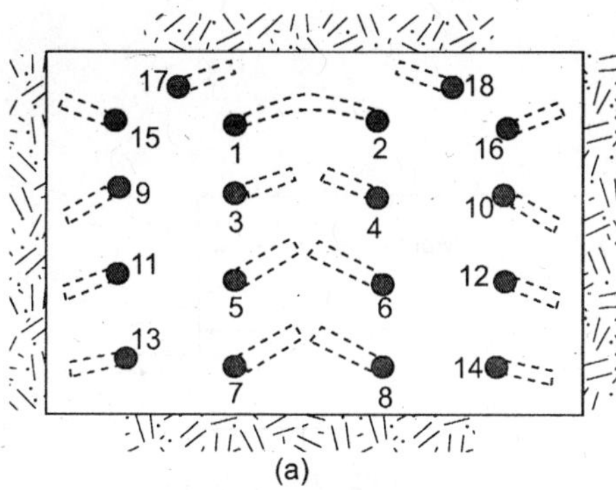


(a)

Holes 1 to 8 wedge cut-first
9 to 14 side holes-next
15 to 18 back holes-last

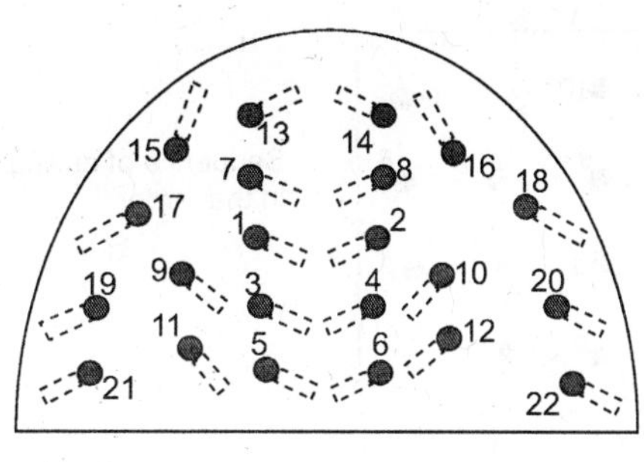


(b)

Holes 1 to 6 wedge cut-first
7 to 12 relievers-next
13 to 22 trimmers-last

Elevation of the heading

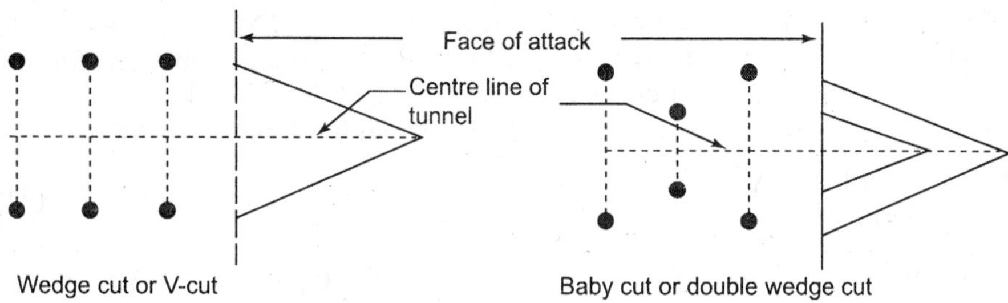


(c) Longitudinal section along the central vertical axis

**FIG. 29.3** *Patterns of wedge-cut holes*

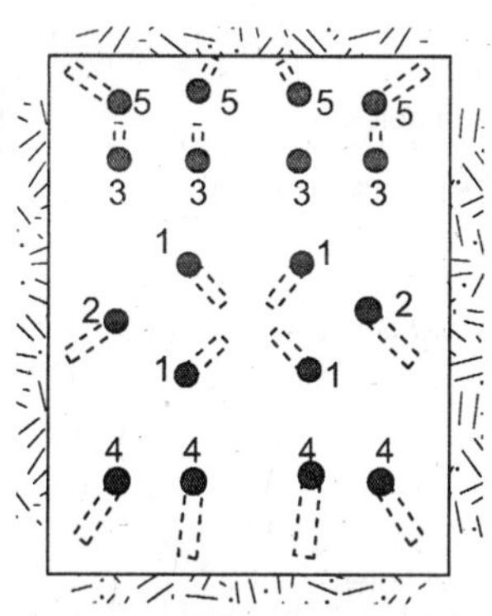


(a) With horizontal bar
Holes marked 1-first
Holes marked 2-next
Holes marked 3-next
Holes marked 4-next
Holes marked 5-last

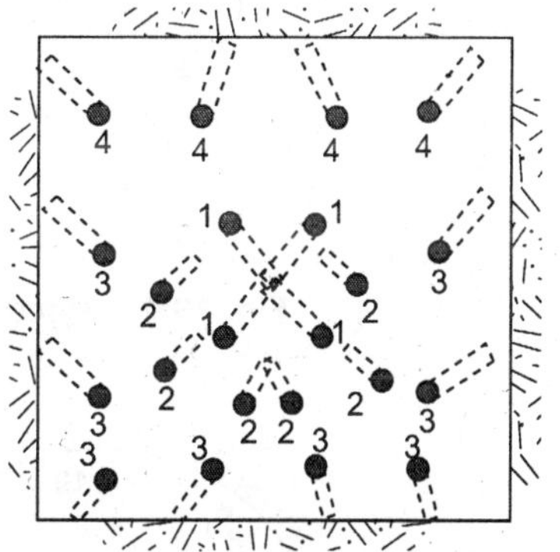


(b) With vertical column
Holes marked 1-first
Holes marked 2-next
Holes marked 3-next
Holes marked 4-last

**FIG. 29.4** *Patterns of pyramid cut holes*

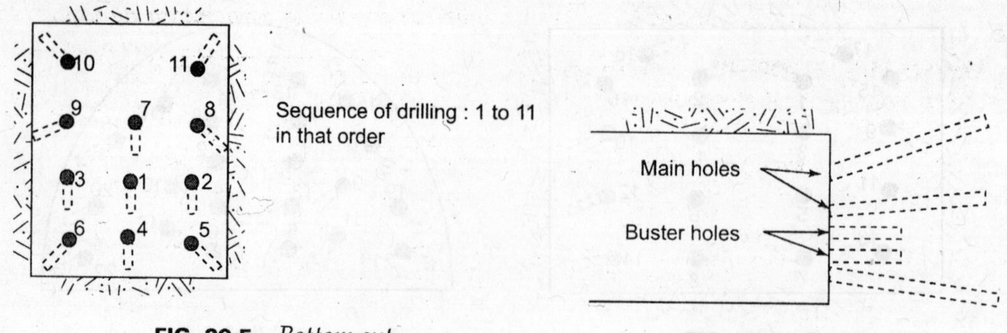


**FIG. 29.5** *Bottom cut*

**FIG. 29.6** *Buster cut*

**Burn cut:** In this, the holes are drilled parallel to the axis of the tunnel at close spacing. However, some of these holes are of larger diameter and not loaded with explosives. These are meant to provide extra faces to the loaded holes. Holes without explosive charge are called *open* or *relief holes*. The number and diameter of relief holes should be 10 to 15% of total volume of the cut. Blast holes are located in the shadow of relief holes and fired with a time lag of about half-a-second.

Two typical patterns of burn cut holes are shown in Fig. 29.7.

Based on the *function* and *sequence of firing*, the holes can be classified as follows:

*Cut holes*: These are located near the centre of the tunnel section. They are the main blast holes with maximum charge.

*Easers*: These are located around the cut holes and are blasted after the cut holes.

*Side holes or lifters*: These are drilled at the side and bottom, and are blasted after the easers.

*Top holes*: These are located near the top of the section and are blasted last.

The various classes of holes are shown in Fig. 29.8.

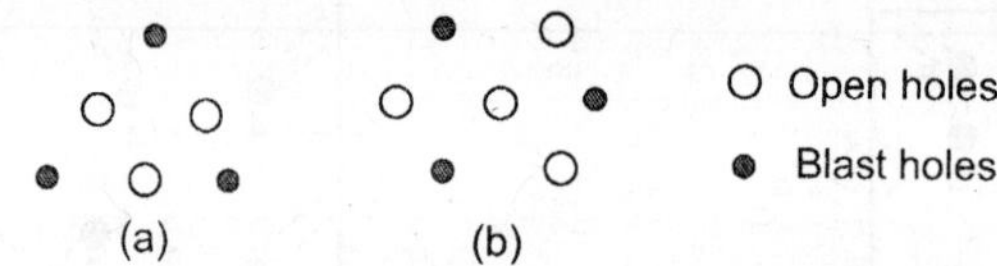


**Fig. 29.7** *Typical patterns of burn cut holes*

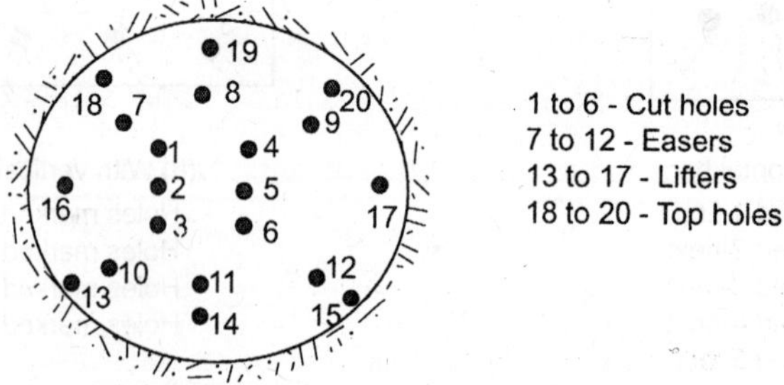


**FIG. 29.8** *Types of holes based on function*

Usually, the depth of the holes is not more than the width of the tunnel. Holes that are too deep are not advantageous. The economical depth is around 1.5 to 2.5 m. The drilling speed is such that it takes 3 to 9 minutes per metre run of the hole.

Drill bits are forged and are about 20 to 25 mm hexagonal hollow steel and about 1 m long. They need to be sharpened often.

### Excavation in Rock Using Machinery

Many kinds of machines have been developed for tunnel excavation in rock. The equipment fall into the following categories based on their action on rock breakage:

(a) Percussion type
(b) Ripper type
(c) Drag type
(d) Cutter type

**Percussion type:** This uses pneumatic/hydraulic hammers and is used for weathered rock and boulders.

**Ripper type:** This type of machine is used for weathered and jointed rock; it may be used in the heading and bench method

**Drag type:** Machines of this kind are fitted with conically shaped self-sharpening drag bits.

**Rolling/rotating type:** Machines of this type use disc cutters or tungsten carbide drill bits.

Modern tunnelling machines use water jets to assist in the excavation and are considered to be very efficient. Miscellaneous methods using cannons and lasers have not proved to be popular.

## Blasting

Great care is to be exercised in the storage and handling of explosives used for blasting in the construction of tunnels. They should be stored in a dry, well-ventilated and fire-proof place.

### Requirements of a Good Explosive

A good explosive should have the characteristics listed below:

(i) Poisonous gases should not evolve on explosion.
(ii) It should be sufficiently stable in composition.
(iii) It should not attract moisture.
(iv) It should not be sensitive to shock/vibration for ease and safety in transport.

### Explosives Used for Rock Blasting

**Ammonia dynamites:** High velocity high density (HVHD) as well as low velocity low density (LVLD) types are used; HVHD type are considered to have better fume characteristics while LVLD are sensitive to water.

**Ammonia gelatine:** This is considered to be better than ammonium dynamites since it can be fired even in the presence of water.

**Ammonium nitrate ($NH_4NO_3$):** Popularly used as a mixture of 94.5% $NH_4NO_3$ and 5.5% fuel oil, it is very effective and not expensive. However, it is highly sensitive to water.

**Water gels:** High percentage of ammonium nitrate in aqueous solution has resistance to water and is popularly used.

**Liquid oxygen:** Also called *Oxyliquit* or *liquid air explosive,* it is a mixture of liquid oxygen with a suitable fuel such as carbon (lampblack) or an organic chemical (a mixture of soot and naphthalene), aluminium powder or sponge. These fuels have the ability to absorb liquid oxygen in quantities several times their own weight and can be ignited by a safety fuse.

However, the disadvantage of using this fuel is that, once mixed, it is sensitive to shock, heat and sparks.

**Firing:** Initially, acetylene lamp or a spitter fuse was used. However, electric detonators have become popular in view of its water-proof quality, safety and certainty. Smoke and dangerous gases do not evolve; also, they permit successive firing of holes or groups of holes using *delay* mechanism.

Misfires should be checked and the safety of the work-persons should be ensured.

## Quantity of Rock Released on Blasting

This depends upon various factors such as

(i) Rock characteristics
(ii) Explosive power of the charge
(iii) Number, depth, inclination, and free faces of the holes
(iv) Stages of firing the holes
(v) Patterns of the drilled holes.

**Rock Characteristics:** The volume of rock released on blasting will be minimal if the rock is hard and sound, without any defects such as stratification and faults. In contrast, if the rock is soft, weathered, stratified, and is with structural defects, the quantity of rock released on blasting is much more.

**Explosive power of the charge:** The nature of the explosive chosen governs the explosive power of the charge. While gunpowder is a low-power explosive, dynamite, ammonia gelatine, nitroglycerine, trinitrotoluene, and oxyliquit are high-power explosives.

For detonating the blast, a small quantity of the explosive, called *primer,* is connected to a fuse via blasting cap/detonator. When the fuse is ignited, it burns at a constant rate producing a blast of the primer, which, in turn, blasts the charge. An exothermic chemical reaction produces a large amount of heat and pressure energy which travels in the form of shock waves; this flow of energy from high pressure to low pressure crushes the intervening rock mass.

**Number, depth, inclination and free faces of the holes:** It is always better to fire a number of holes in *stages* so that wastage of energy, which might occur if the holes are fired one at a time, is avoided, and the maximum destructive effect is achieved in dislodging the rock.

Holes *inclined* to the surface have proved to be more effective than those normal to the surface; the chances of misfiring are much less in the case of inclined holes.

The *depth* of hole is an important factor which governs the power of the blast up to a certain limit; when the charge in a hole is blasted, a cone of rock mass is released as shown in Fig. 29.9.

Volume, $V$, of the rock released:

$$V = \frac{1}{3}\pi R^2 D$$

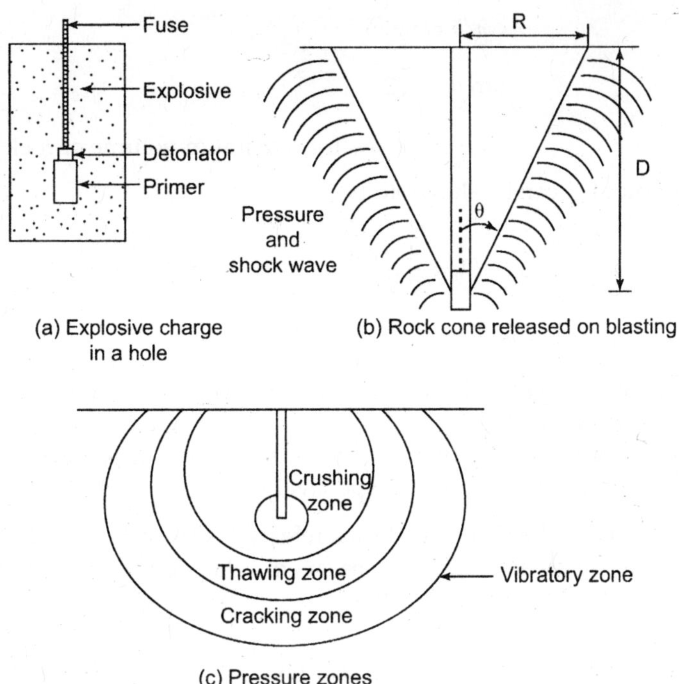


**FIG. 29.9** *Effect of depth of hole on the rock mass released*

Here, $R = D \tan\theta$

$$V = \frac{1}{3}\pi D^3 \tan^2\theta$$

Lebrun's (2000) relation between the weight of the charge *W* and the depth of cut *D* is given by

$$W = C\,\frac{D^3}{\cos^3\theta} \tag{29.1}$$

where C = a coefficient which depends on strength of rock and the depth of cut, *D*.

$$D^3 = \frac{W}{C}.\cos^3\theta$$

$$V = \frac{1}{3}\pi\frac{W}{C}.\cos^3\theta\tan^2\theta$$

$$\frac{dV}{d\theta} = \frac{\pi}{3}.\frac{W}{C}(2\sin\theta\cos^2\theta - \sin^3\theta)$$

Equating $\frac{dV}{d\theta}$ to zero for *V* to be maximum,

$$2\cos^2\theta = \sin^2\theta$$

$$\tan\theta = \sqrt{2}$$

$$\theta = 54°54' \text{(for maximum volume of rock released)}$$

The optimum depth, $D_0$, can be shown to be

$$D_0 = \frac{4}{7}\sqrt[3]{\frac{W}{C}} \tag{29.2}$$

An appropriate value of $C$ has to be used.

**Number of free faces of the hole:** The quantity of rock mass released depends on the number of free faces of the hole (Fig. 29.10).

The quantity of rock released by two and three free faces compared to that with one free face, $V$, is shown in Figs. 29.10 (b) and (c) respectively.

**Stages of firing the holes:** The quantity of rock released is largest when the holes are fired in stages, in the following sequence: cut holes, easers or reliever holes, side holes or lifters, and top holes. These have been shown in Fig. 29.8.

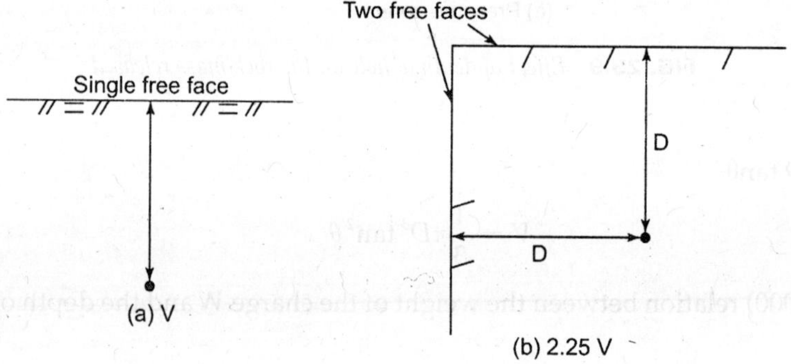


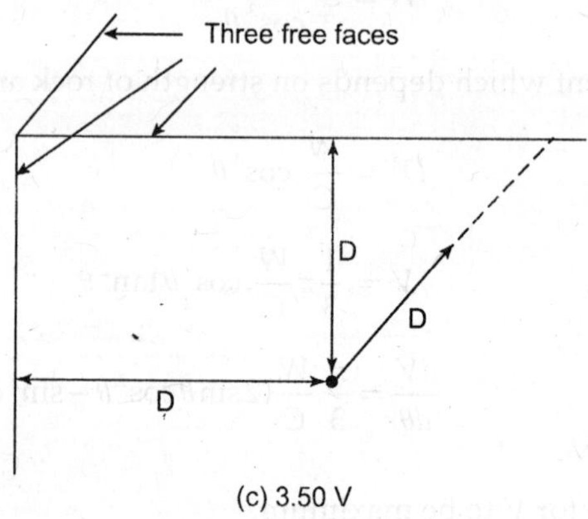


**FIG. 29.10** *Number of free faces of a hole*

The time interval between the stages of firing is crucial. It should only be a fraction of a second between the different stages of firing for making the best use of the pressure waves generated by each stage of blasting. This is called *millisecond firing,* and is considered to be the best approach.

The detonators can be connected in series, in parallel, or in series and parallel.

**Patterns of the drilled holes:** These have already been shown in Figs. 29.3 to 29.8. The quantity of rock released on blasting depends also on the pattern of the drill holes to a certain extent.

## Supports for a Tunnel

The sides and roof of a tunnel need supports, especially when freshly excavated or blasted. They may remain stable only for some time after excavation due to what is known as *bridge action*. Unless supports are provided, the sides/roof of the tunnel may cave in due to stress concentration. This depends on, among other things, the nature of the material through which the tunnel is being excavated.

The *temporary* supports may be independent of the *permanent* support to be erected, or can be used as a framework for laying or casting of the permanent support.

Permanent support may be provided by grouting or by construction of a lining for the tunnel.

### Temporary Support Systems

*Temporary support systems* may consist of units for roof or for sides. The connections may be rigid or flexible, depending upon the need. The support systems may be of timber or steel. Three temporary support systems of timber are given below:

1. For small tunnels of about 2.5 m size, the *cap and post* method is used. The cap is supported on temporary beams and the posts are then fixed and wedged on foot-blocks at the bottom. The sets are braced. This method is also called the *square set* method (Fig. 29.11).
2. For medium-size tunnels, the *three-piece arch* method is used. The foot of the arch is supported on a small bench, cut in the sidewall (Fig. 29.12).
3. In the case of larger tunnels, the *five-piece arch* method is used. The longitudinal spacing of the sets may be about 2 m (Fig. 29.13).

Support systems of *steel* may consist of *ribs and posts* or *ribs and wall beams*. Rolled steel sections of I-, H-, T- or inverted T-sections may be used (Fig. 29.14).

Hinges, rollers or bolts may be used for connecting steel elements.

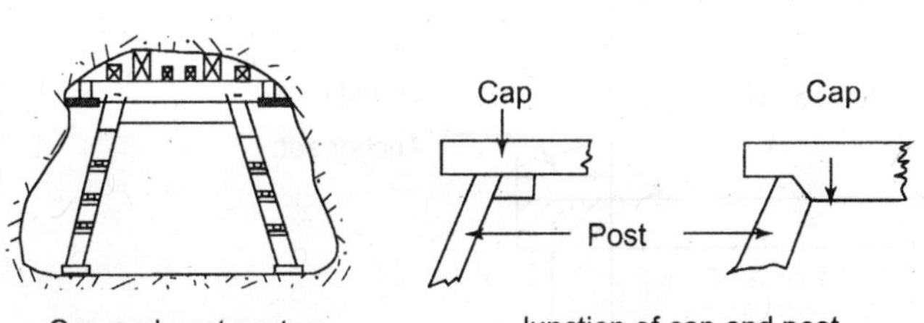


**FIG. 29.11** *Square set method of timbering*

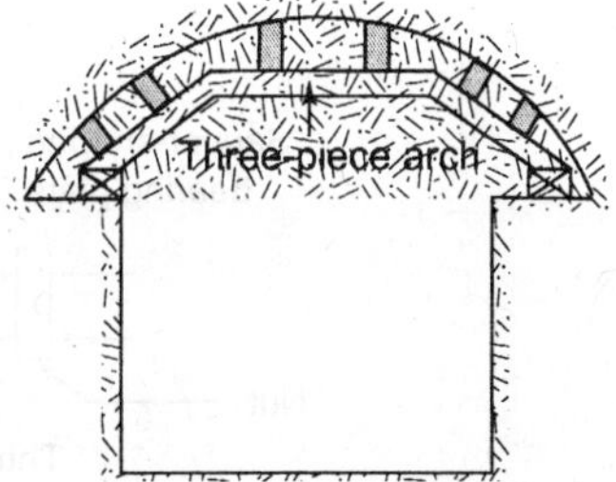


**FIG. 29.12** *Three-piece arch method*

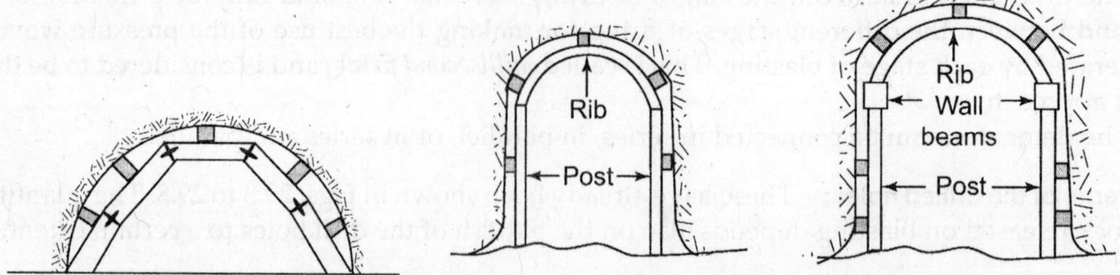


**FIG. 29.13** *Five-piece arch method*

**FIG. 29.14** *Supporting systems of steel*

### Permanent Supports

Permanent supports may consist of rock bolts and bearing plates.

**Rock bolts:** Pre-tensioned rock bolts are effective in providing supports for a tunnel; they take the load immediately on installation and prevent deformation and fracturing of rock.

Round steel bolts of 16 to 18 mm diameter are threaded at both ends along with expansion shells.

Bolts of stainless steel and chromium/nickel steel are also used.

**Bearing plates:** These may be circular, triangular, rectangular, or pressed spherical plates in plan shape (Fig. 29.15).

Rock bolts are installed usually with bearing plates. *Pre-stressing* a bolt means stretching it between an anchor set in sound rock and the bearing plate on the outer surface of the rock. The strength of the bearing plate should be at least equal to that of the rock bolt. A typical set-up is shown in Fig. 29.16.

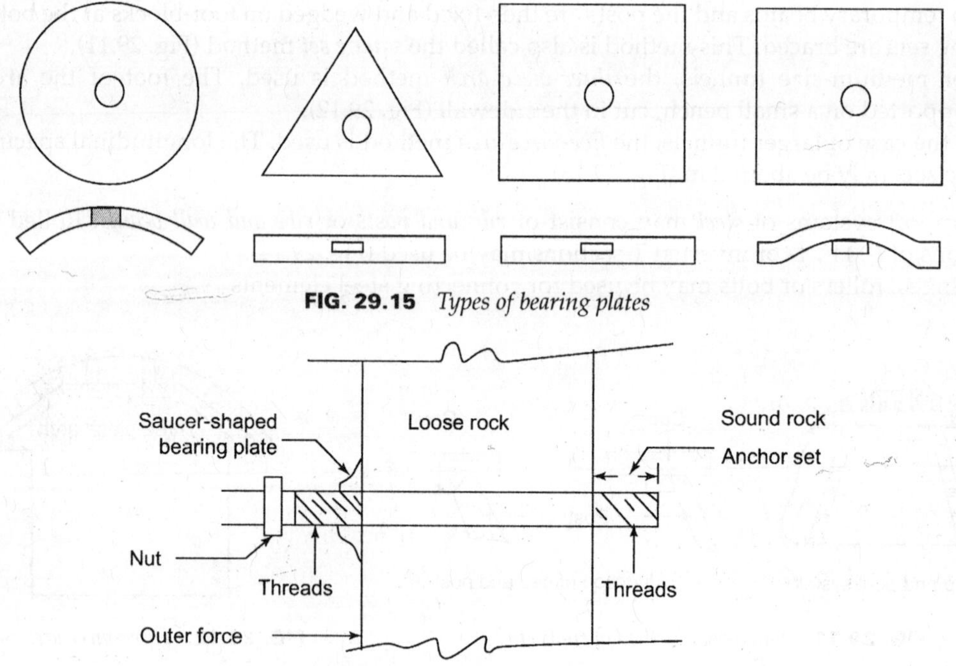


**FIG. 29.15** *Types of bearing plates*

**FIG. 29.16** *Rock bolt with bearing plate*

### Dust Removal and Ventilation

Dust and smoke should be removed expeditiously before the waste material can be taken away. Exhaust fans are used along with water sprays to quench the smoke and dust. About 30 minutes to one hour should be allowed before entering the tunnel after blasting.

### Muck Removal

Removal of the blasted material is called *mucking*. This consists of loading the muck into trucks and trolleys and moving it outside. For faster disposal, the blast debris is loaded by mechanical shovels into power-driven trucks.

### Pumping Out Water/Drainage

If water accumulates in the tunnel for any reason, it should be effectively pumped out to the keep the tunnel dry for further progress of the work.

### Lining of Tunnel

*Lining* is necessary to provide a finishing to the tunnel surface. In the case of tunnels in rock, lining serves the purpose of covering the irregularities and providing an aesthetic appearance to the surface. In this case, the thickness of the lining is only nominal.

However, in the case of tunnels in soft soil, lining is expected to serve as a permanent support to the roof and sides of the tunnel to prevent the possibility of spalling, or failure in the extreme.

[**Note**: While *mucking* is considered in more detail in Section 29.5, the aspects of ventilation and drainage are covered in Chapter 31. Lining is discussed in Chapter 30.]

## 29.3 METHODS OF TUNNELLING IN ROCK

The following are the methods used for tunnelling in rock:

(a) Top heading method
   - (i) Top pilot-heading or drift method
   - (ii) Heading and bench method

(b) Centre heading method
(c) Full face method
(d) Pilot tunnel method

### Top Heading Method

There are two different approaches to this method, namely, the *top pilot-heading* or *drift* method and the *heading and bench method*.

### Top Pilot-heading (Drift) Method

A top heading of about 2.5 m square is first driven at the middle portion, making its top to conform to the final shape of the tunnel; this is driven from end to end of the tunnel, making it a drift.

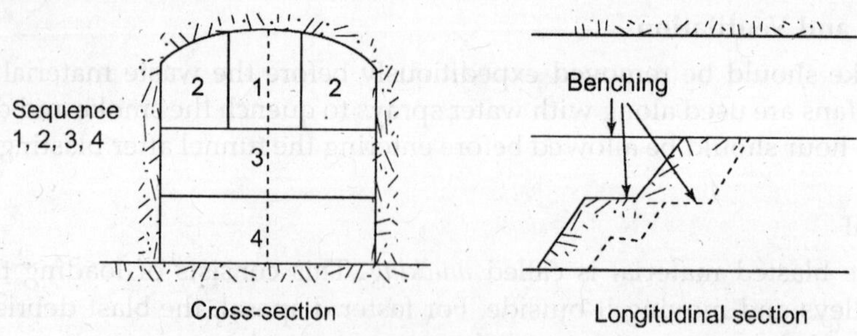


**FIG. 29.17** *Top pilot-heading method*

Then, the sides are enlarged for the whole length and later, the bottom portion or benching is driven. This is also known as the *multiple-drift* method since more than one drift is involved. The sequence of excavation is shown in Fig. 29.17.

The bottom portion is excavated by benching as shown. The disadvantage of this method is that the mucking tracks have to be shifted frequently during enlargement and benching.

**Heading and Bench Method**

A top heading of about 2.5 m × 2 m is first driven for a length of about 3 to 4 metres. This is widened later for the full width and then the lower portion or the bench is removed. The top heading is made to advance one round ahead of the bottom, so that the heading and benching processes follow each other. If the rock is hard and sound, no support is needed; however, if it unsound with defects, the top heading needs support, and the bench will provide a platform for the timbering (Fig. 29.18).

Bench holes are fired first, followed by the heading holes. This is done for the convenience of mucking.

## Centre-heading Method

In this method, a central drift of about 3 m square is first driven and then, it is enlarged to the required dimensions. Blast holes, in this case, are ring-drilled, with the rings at a spacing of about 1 m. (Fig. 29.19). Good ventilation during construction and economy in the use of explosives are the advantages. However, the drilling and the mucking jobs clash with each other, causing delay and inconvenience.

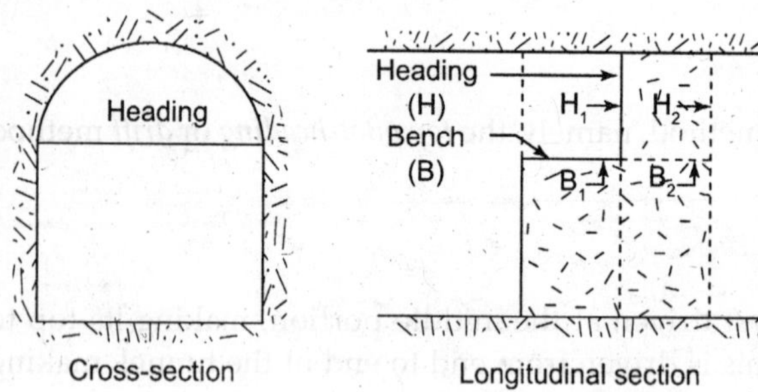


**FIG. 29.18** *Heading and bench method*

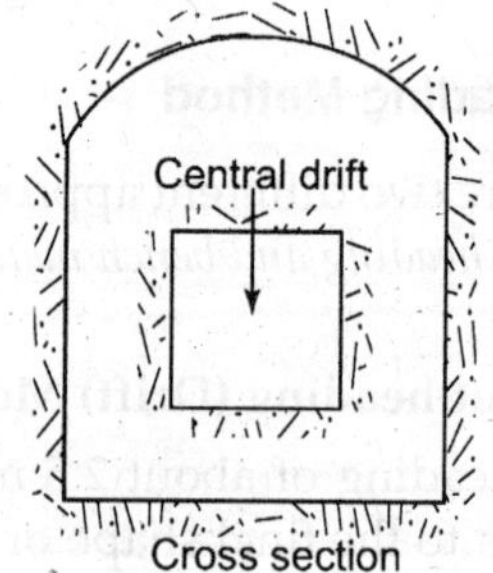


**FIG. 29.19** *Centre-heading method*

### Full Face Method

Initially, this method of tackling the entire section of the tunnel in a single stretch was used for tunnels of small size. However, with the advent of modern tunnel boring machines and shield tunnelling, the full face method has become popular. Simultaneous drilling of several holes on the entire face has become possible by means of a jumbo-drill carriage mounted on wheels with a battery of drills on its front face.

The advantages claimed are

- Speedy execution of the job.
- Mucking can be carried out from the tunnel floor and progressively extended.
- Advantage of prevention of instability in multi-phase excavation, especially in weak strata.
- Minimising ground disturbance.

In spite of these advantages, the full face method cannot be used for large tunnels in view of equipment costs and problems of support. An optimum upper limit for the size of the tunnel is considered to be 6 m.

### Pilot Tunnel Method

This has already been discussed in the previous chapter in Section 28.4.

A smaller-size tunnel is first driven parallel and close to the main tunnel and the two are inter-connected, as the job progresses, through several cross cuts. This provides several working faces for the main tunnel simultaneously, facilitating speedy progress and is considered less expensive than sinking several vertical shafts. The cross tunnels connect to a central drift in the main tunnel and have several uses during and after the completion of the tunnel.

## 29.4 NEW AUSTRIAN TUNNELLING METHOD (NATM)

The *New Austrian Tunnelling Method (NATM)*, also known as the *Sequential Excavation Method (SEM)* is a popular method of modern tunnel design and construction. This technique was developed by Austrian Tunnel Engineers in the 1960's, and has been widely used since 1964 in the construction of many famous modern tunnels. The term 'NATM' distinguishes it from the old Austrian tunnelling approach.

NATM/SEM is considered to have helped revolutionise the modern tunnelling industry.

### Elements of NATM

- Exploitation and preservation of the strength of the rock mass to be used as the main component of tunnel support.
- 'Shotcrete' protection – loosening and deformation of rock must be minimised by applying a thin layer of shotcrete immediately after the face advances.
- Measurement and monitoring – Potential deformations of the excavation must be carefully monitored by installing sophisticated instrumentation, embedded in lining, ground and boreholes. Thus, additional supports are installed only when needed leading to overall economy.
- Flexible support – The primary lining is thin and reflects recent strata conditions. The tunnel is strengthened by a flexible combination of rock bolts, wire mesh and steel ribs, and not by a thicker concrete lining.

- Closing of the invert – Especially crucial in soft ground, the quick closing of the bottom portion of the tunnel is important because it creates a load-bearing ring, exploiting the inherent strength of the rock mass surrounding the tunnel.
- Since NATM is based on monitoring measurements, changes in supports and construction method are possible; hence the contractual system must be tailored to enable this.
- Rock mass classification, ranging from very hard to very soft, determines the minimum support measures required, leading to economy. Only a thin shotcrete protection is necessary. It is applied immediately behind the tunnel boring machine to create a natural load-bearing ring and minimise rock deformation. Reinforcement is provided in the form of wired concrete that can be combined with steel ribs or lug bolts. Geotechnical instruments are installed to measure the later deformation of the excavated portion. This method makes monitoring of stress distribution possible.

NATM can be used successfully for tunnelling in soft ground also.

The Austrian Society of Engineers and Architects defines NATM as "a method where the surrounding rock or soil formations of a tunnel are integrated into an overall ring-like support structure. Thus the supporting formations will themselves be part of this supporting structure." NATM can refer both to the design philosophy and to the construction method. It is desirable to excavate the complete section; it is also better to avoid sharp corners as they develop stress concentrations. Structurally, the tunnel is considered to be a thick-walled pipe.

### Key Features of NATM as a Design Philosophy

- The strength of the ground around a tunnel is deliberately mobilised to the maximum extent possible.
- Mobilisation of ground strength is achieved by allowing controlled deformation of the ground.
- Initial primary support is installed appropriate to the load-deformation characteristics of the existing ground conditions, and the installation is timed with respect to ground deformations.
- Instrumentation is installed to monitor deformations in the initial support system, as well as to form the basis of varying initial support design and the sequence of excavation.

### Key Features of NATM as a Construction Method

- The tunnel is sequentially excavated and supported and the excavation sequence can be varied to address the specific rock conditions being encountered.
- The initial ground support is provided by shotcrete in combination with fibre or welded-wire fabric reinforcement, steel arches (usually lattice girders) and sometimes ground reinforcement (e.g., soil nails)
- The permanent support is typically a cast-in-place concrete lining placed over a water-proofing membrane. The rock is reinforced by systematic rock-bolting for preserving its original strength.

NATM was originally developed for use in the Alps, where tunnels are excavated at great depths and in high in-situ stress conditions.

Most city tunnels are built at a shallow depth and need not control the release of in-situ stress seen with the original NATM in the Alps. Here, different support methods are used, the aim being the control of settlement. NATM is a scientifically sound concept where a new secondary state of equilibrium is reached for the rock mass surrounding the hole by controlled pressure release, and by improving the carrying capacity of the rock mass surrounding the tunnel hole.

## 29.5 MUCKING AND HAULING

*Mucking* is the removal of debris from the blasting/excavation site in the tunnel to places outside. This is a major element of tunnel construction, especially in the case of large tunnels. Hence, efficient and quick removal of muck leads to overall economy.

Initially, muck removal was done manually. Later on, various types of machines that run on compressed air or electric power were developed for this purpose.

*Eimco loader* and *Conway shovel* are two such machines, which are fitted with a bucket/shovel at the leading end and are actuated by a jib of a crane. The machines move forward into the muck to fill the bucket; the jib raises it and throws the muck into hoppers, which are moved out of the tunnel when they are full.

The mucking machines depend on an efficient disposal method such as a light railway of 600 mm to 900 mm gauge. Muck cars of about 2 to 4 $m^3$ capacity are used. The disposal technique allows the replacement of a loaded car by an empty one without much loss of time. This is achieved either by the use of loops, sidings, and shunting operations on the light railway tracks, or by using mobile appliances. The important types of appliances used are:

(a) California crossing
(b) Grass hopper
(c) Cherry picker

**California crossing:** This consists of passing loops parallel to the normal track on a light railway used for muck removal operations in the construction of tunnel. It facilitates quick exchange of loaded muck cars with empty ones for efficient muck removal (Fig. 29.20).

**Grass hopper:** This consists of a gantry with hinged ramps at either end. Empty cars are kept on rails on the gantry and are moved down the forward ramp to replace the loaded car moved away from the face on the normal track below the gantry (Fig. 29.21).

**Cherry picker:** This consists of an overhead gantry girder with a track. An empty car can be lifted bodily and moved sideways in order to allow a loaded car to run back with the muck. The empty car can then be lowered on to track near the working face for loading.

The California crossing has been used on a 600 mm track in 3 m wide tunnels; the other two appliances need a minimum of 4 m headroom for operation. The muck cars are to be moved out by an electric locomotive to avoid ventilation problems

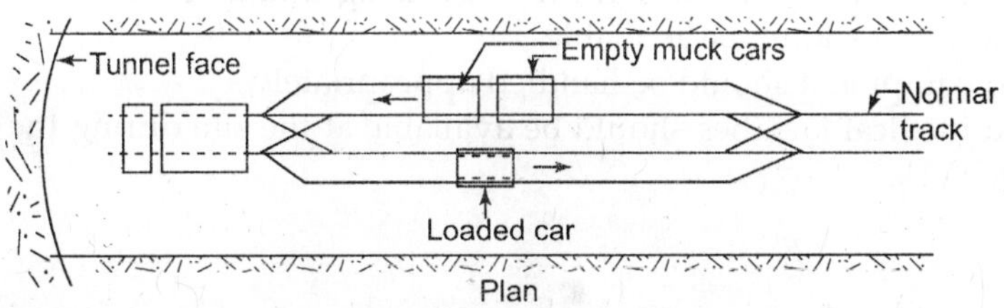


**FIG.29.20** *California crossing*

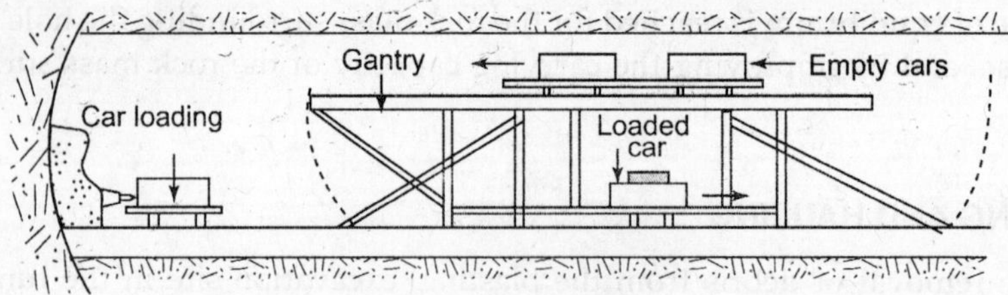


**FIG.29.21** *Grass hopper*

## Hauling

In the early days, animals such as horses and mules were employed for hauling carts loaded with muck. Wheel barrows were used later. Now-a-days electric-powered locos are used for hauling muck cars; this would ensure that additional ventilation arrangements, needed for steam and diesel locos, are avoided.

The muck cars should be hauled by locos with adequate tractive power. The tracks and loops should be laid true to the gauge and maintained well. The movement of trains of muck cars should be controlled and regulated for ensuring smooth, safe and efficient transportation of the muck. The spoil dump spaces should be located not too far away from the entrance of the tunnel for saving transit time.

## 29.6 PRECAUTIONS FOR TUNNELLING IN ROCK

Tunnelling is a hazardous task often performed underground and it may endanger the lives of the personnel working on the job through rock falls and earth slides, besides exposing them to the risks of blasts and fire accidents.

The following safety precautions are to be observed while working on tunnels:

1. Scaling of rock results in loose material on blasting, which should be checked and removed.
2. Dangerously perched rock pieces should be carefully removed
3. Quick provision of temporary supports should be made to avoid collapse of roof or the sides.
4. Power lines should be insulated properly.
5. Water should not be allowed on the tunnel floor as far as possible.
6. The personnel should be provided with safety gear for the head and the body
7. Explosives should be handled with great care; fire-fighting equipment should kept within easy reach.
8. Good illumination should be provided inside the tunnel.
9. Dust should be brought down to safe limits by using exhaust fans.
10. Standby power unit should be available.
11. All tools and equipment should be handled appropriately.
12. First aid and medical facilities should be available at the site during the entire period of construction work.

## SUMMARY

- The important features of tunnelling in rock are that it requires specialised machinery and high power supply. However, it usually needs no timbering.
- The essential steps in the general procedure of tunnelling in rock are: shaft construction, excavation by drilling holes, charging with explosives and blasting, dust control and ventilation, temporary and permanent support erection, muck removal, and finally lining.
- Ring drilling and bench drilling of holes are used for effective blasting, based on the rock material and the method of construction. The holes are classified as wedge, pyramid, buster, and burn cuts based on the shape of the material likely to be removed; and as top, centre, bottom, or side cuts based on their location in the tunnel section.
- The machinery for drilling could be of the rotary type, percussion type, or drag type, based on the action of the drilling machine.
- Ammonia gelatine is a popular explosive used for tunnelling. The firing is accomplished by using detonators and delays, if a series of holes are to be blasted with time lag.
- The quantity of rock released on blasting depends upon the rock characteristics, explosive power and the characteristics of the holes and their pattern in the section.
- Temporary supports for the tunnel could be of timber or steel; permanent supports are by means of rock bolts and bearing plates. Pre-tensioned rock bolts, used in conjunction with bearing plates give good support to the roof and sides of the tunnel.
- New Austrian Tunnelling Method (NATM) is the most recent technique which is popularly used for large tunnels. This method exploits the original strength of the rock mass, protects by using shotcrete measurement, and monitors deformations during and after construction, so that additional support is provided only when needed.
- Mucking and haulage of the debris or loose material obtained after blasting is an important component of tunnel construction. The use of Eimco loader and Conway shovel expedites the work. Appliances such as the California Crossing, Grass Hopper and Cherry Picker are used along with tracks of 600 mm to 900 mm gauge for running muck cars. These appliances help in quick placement of empty cars immediately after withdrawing loaded cars from the face.
- The haulage should be by means of electric power-driven cars for avoiding pollution of the inside of the tunnel. The site for dumping the muck outside should be close to the tunnel's entrance so that the haulage cost is reduced.
- Precautions are necessary in tunnel construction as it is a hazardous underground work. Careful handling of explosives, personal protective equipment, and timely identification and removal of loose material are essential in order to ensure the safety of the work personnel involved in the tunnelling work.

## QUESTIONS

1. Write the operations of tunnel construction in rocks. (KIIT, Odisha)
2. Define 'mucking'. Mention the different methods of mucking process and explain any one method. (VTU)
3. What are the important features of tunnelling in rock?
4. Explain sequentially the steps in tunnelling through rock.

5. Explain the different types of mountings used for drilling equipment.
6. Explain 'ring-drilling' and 'bench-drilling' of blast holes in rock.
7. Write short notes on:
   (a) Wedge cut
   (b) Pyramid cut
   (c) Bottom cut
8. Explain briefly the following:
   (a) Cut holes
   (b) Easers
   (c) Lifters
9. Explain the percussion and ripper types of machinery for excavation in rock
10. Differentiate between low-power explosives and high-power explosives. Give one example for each type.
11. What are the important factors which govern the quantity of rock mass released on blasting?
12. Differentiate between temporary and permanent supports for a tunnel. Give one example for each type.
13. Explain the use of pre-tensioned rock bolts along with bearing plates.
14. Explain what is meant by 'mucking' with reference to tunnelling
15. What are the desirable characteristics of equipment for haulage of muck?
16. Write a note on 'California crossing'.
17. Differentiate between 'Grass Hopper' and 'Cherry Picker' in mucking operations.
18. Enumerate the precautions necessary to ensure safety in rock tunnelling.

Chapter 30

# TUNNELLING IN SOFT SOIL

**CHAPTER OBJECTIVES**

This chapter enables the student to understand the following:

- Classification of soil based on the degree of support needed during excavation
- Methods of tunnelling in soft soil – forepoling, needle-beam and classical methods such as American, Belgian, and Austrian Methods
- Timbering methods for tunnels in soft soil
- Liner plate method and shield tunnelling using modern Tunnel Boring Machines
- Tunnelling in water-bearing soil using compressed air to preclude water
- Tunnel lining, functions and uses – cement concrete lining for water-proofing
- Precautions necessary for safe operations during tunnelling in soft soil.

## 30.1 INTRODUCTION

The method chosen for tunnelling in soft soil depends primarily on the nature of the soil.

### Classification of Soft Soils for Tunnelling

For the purpose of tunnelling, soft soils are classified as follows:

(i) *Ordinary type of soil,* which is firm enough for the tunnel roof to stand unsupported for a short period; in fact, the sides can stand for a little longer period. Dry earth material and firm clay are examples of this type.

(ii) *Loose or soft type,* which needs instant support for roof; soft clay is an example.

(iii) *Treacherous soil or running ground,* which needs instant support all around. Loose cohesionless soils, water-bearing granular soil and very soft clay soil are examples.

In the case of the second and the third categories, excavation and timbering should go hand in hand.

### Choice of Method for Tunnelling in Soft Soil

The following factors govern the choice of the method:

- Nature of the soil and the extent of support needed during and after construction.
- Sequence of excavation
- Size of the tunnel
- Equipment available.

## 30.2 METHODS OF TUNNELLING IN SOFT SOILS

The following are the methods used for tunnelling in soft soils:

1. Forepoling method
2. Needle beam method
3. Five-piece set method
4. Classical methods
   (a) American method
   (b) English method
   (c) Belgian method
   (d) German method
   (e) Austrian method
5. Army or case method
6. Cut-and-cover method

### Forepoling Method

This is perhaps the only method advised for tunnelling in running ground.

The salient features are:

- Old, slow and tedious method.
- Small-sized tunnels for sewers or gas pipes are driven by this method.
- The correct sequence of operations is to be strictly followed.

The *sequence of operations* is given below:

(i) A shaft is sunk from the surface and protected by timber sheeting.
(ii) A wooden bent is set up covering the first face of attack.
(iii) A line of sheeting above the cap is cut along the top line of holes and the soil from behind the sheeting is permitted to come out.
(iv) Small holes at close intervals are drilled through the sheeting.
(v) *Forepoles* or *Spiles* – with wedge ends – are entered and driven through the cut into the ground with an upward inclination of 170 mm per metre.
(vi) After all the roof and part of the side spiles are driven to half their lengths, timber is laid across the back ends of the spiles, wedging them down; the front ends of spiles are cantilevered up.
(vii) The face sheeting is cut across the lower line of holes and loose soil is run into the tunnel until the face assumes a natural slope.
(viii) A *horse head* is set up about 600 mm from the sheeting and earth beneath the forward end is scooped out for a length of 500 mm, and; the face supported by a 'breast board' is placed underneath the forward end of the spile.

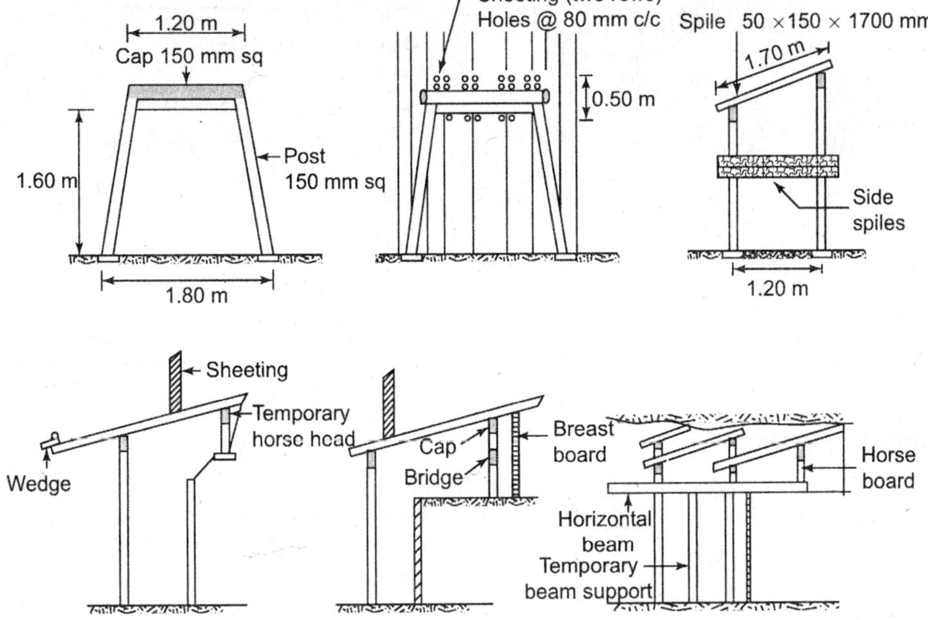


**FIG. 30.1** *Stages or sequence in forepoling method*

(ix) The next cap, supported on a bridge, is set and temporarily supported on a single post.
(x) Side spiles are driven to their full lengths.
(xi) A heavy horizontal beam is pushed forward to support the forward cap.
(xii) The breast board is extended to grade level by adding new sheets.

This *sequence of operations* is shown in Fig. 30.1.

## Needle Beam Method

This method is used in soils in which the tunnel roof can stand for some time without any support. In this method, 5 to 6 m long R.S. joists or timber beams are required in addition to other timber boards and struts. Apart from these, this method requires a large number of jacks, which are likely to cause obstruction of work space. However, it is more economical than other methods for tunnelling in soft soils.

The *sequence of steps* in this method are given below:

(i) First, a small drift of about 1 m is prepared on the working face of the tunnel.
(ii) The needle beam, which consists of two I-girders bolted together with a wooden block at the centre, is inserted in the drift and its roof is supported on lagging carried on the wooden segment [Fig. 30.2(a)].

These segments are supported by jacks resting on the needle beam.

(iii) The needle beam is placed horizontally with its front-end resting on the drift itself and the rear end supported on a vertical stout post, resting on the lining of the tunnel. [Fig. 30.2(b)]
(iv) The jack is placed on top of the needle beam to support the roof with lagging and then the drift is widened sideways before the whole section is excavated.
(v) After the excavation is completed, a lining may be provided for the tunnel.

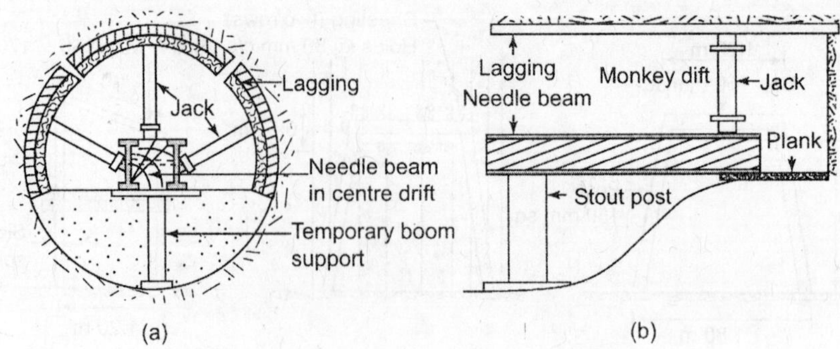


**FIG. 30.2** *Needle beam method*

### Five-piece Set Method

A small drift of about 1.5 m × 2 m is driven in the face. This is supported by caps and the heading is widened in the face using the required timbering (Fig. 30.3). This method is used in soft clay and other soft soils.

## Classical Methods

Among the other methods of tunnelling in soft soils, the following are also adopted:

### American Method

The *salient features* of this method are:

- Suitable for large-sized railway or highway tunnels.
- Top drift is driven and supported by laggings, segment and two posts.
- Sides of the drift are widened and supported on the shoulder segment using timber and struts.
- Wall plates of 5 m length are introduced at springing, supporting arch set.
- Wall plates are underpinned by props.
- Timber arch segment and roof load are transferred to wall plate and posts for support.
- The sides and benching are fully cleared and then the tunnel is lined (Fig. 30.4).

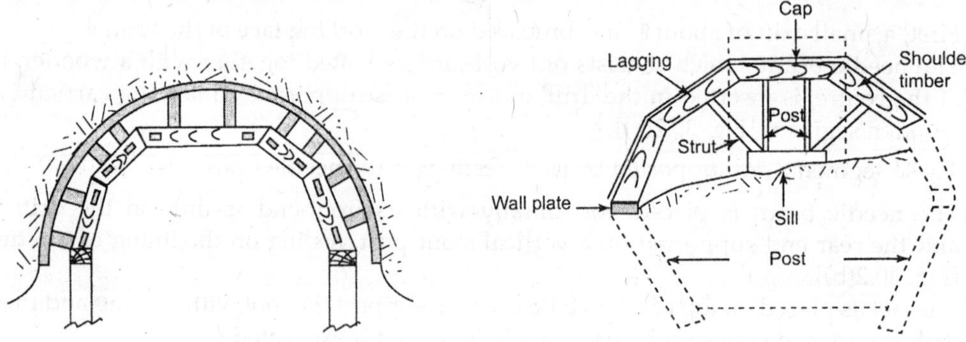


**FIG. 30.3** *Five-piece set method*

**FIG. 30.4** *American method*

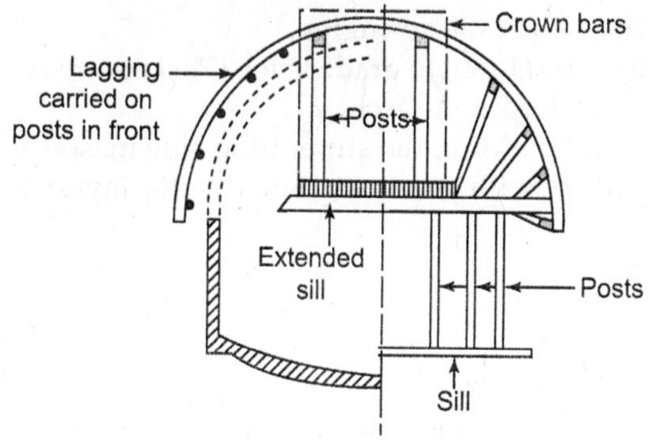


**FIG. 30.5** *English method*

## English Method

The *sequence of steps* in this method is as follows:

- Driving of a central top heading about 5 m ahead of the existing arch.
- Supporting by crown bars, which, in turn, are supported on posts in the front, and blocked by the face of the completed arch in the rear.
- Widening of the heading as in the American method, the sill piece being extended right across the tunnel.
- Underpinning the extended sill on supports, the entire arch now being carried on the longitudinal crown bars (Fig. 30.5).

The *disadvantages* of this method are:

(i) Use of a lot of timber and frequent shifting of the timber as the job progresses.
(ii) The masons and excavators have to work alternately, thus causing more time to be taken for job completion.
(iii) In unstable soil, excavation of the full section of the tunnel entails the danger of caving.

The *advantages* claimed are that the hauling is simple and the construction of the masonry lining in one piece results in a strong and homogenous layer.

## Belgian Method

This is considered suitable for moderately firm soils .The *sequence* of *steps* can be understood from Fig. 30.6.

- Driving a top centre heading (abcd) for the full rise of the arch, and supporting it by crown bars, posts and laggings as in the case of the English method. (The posts are supported on sills.)
- Widening of the heading sideways and providing support by using additional crown bars and posts on the same sills.

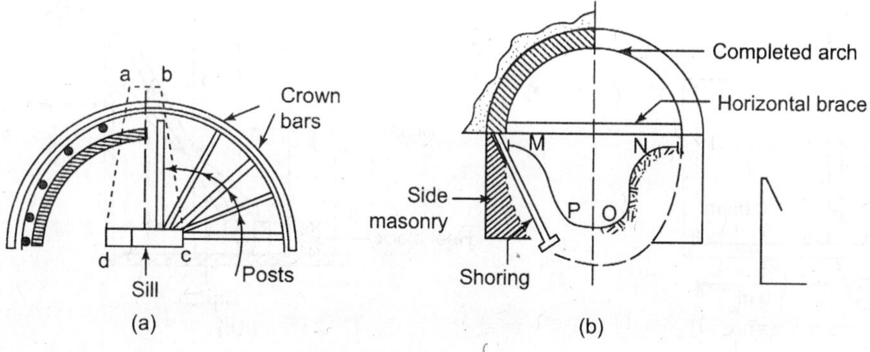


**FIG. 30.6** *Belgian method*

- Fixing a horizontal brace between the ends of the arch at the springing.
- Excavating a trench MNOP to clear the benching right down to grade level. (Pockets are cut at intervals in the trench sides to insert shores to underpin the arch).
- Clearing the alternating spaces between shores and building the supporting side masonry. Then the shoring is removed and the space filled with masonry. Finally, the invert is constructed.

The *advantages* claimed for this method are:

(i) Lighter sections of timbers are used.
(ii) Starting excavation from different points simultaneously.
(iii) Constructing the roof of the arch first, which is crucial, and then the side walls and the invert.

The *disadvantage* is said to be the underpinning of the built arch, which may cause settlement and cracks in the arch masonry. RCC arches may be used to overcome this contingency.

### German Method

In this method, three drifts are employed – one at the crown and two at the bottom, close to the side walls. The roof and the sides are thus supported appropriately. The section is then widened at the locations where necessary to complete the section.

### Austrian Method

A centre cut is made for the full height of the tunnel and it is then widened to the full face to allow for the completion of short sections of the masonry.

The New Austrian Tunnelling Method (NATM) method may be employed for soft soils also as for rock, since the principles are equally applicable for both.

### Army or Case Method

This method was devised by the US Army engineers and is considered good for small tunnels at shallow depths. Thus it is mainly used for underground sewers (Fig. 30.7):

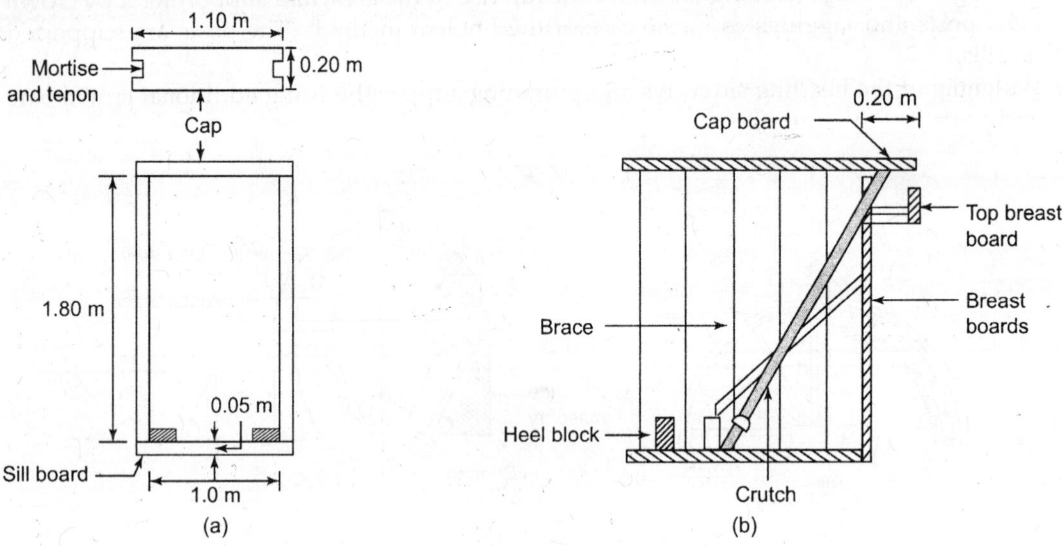


**FIG. 30.7** *Army or Case method*

Steps: (i) The top breast board is removed and the excavation is advanced for about 250 mm ahead. The breast board is then reset in the new-face.

(ii) The next cap board is set and held in position by a crutch or trench jack.

(iii) The breast boards are removed one at a time and reset in a line below the top board after removing the earth behind each board.

(iv) The sill boards are now advanced after checking the grade level.

(v) The side posts are fixed between the top cap and the sill board.

The *advantage* of this method is that it can be done using only simple timbers and trench jack. However, the *disadvantage* is that it is good only for short shallow tunnels.

### Cut-and-Cover method

Shallow tunnels such as rapid transit lines under city streets, underpasses, land sections of underwater tunnels, and end sections of tunnels through hills, are built by the *'cut-and-cover'* method.

A trench is excavated from the surface and a concrete tunnel is constructed within it. With *bottom-up* construction, the completed tunnel is covered-up, and the surface reinstated.

With *top-down* construction, the walls are constructed first, perhaps using bentonite slurry in narrow trenches. The roof is constructed next, backfilled and the surface reinstated. Excavation and construction of the floors below the roof level then follows, using access from the ends, or glory holes.

Both these types of construction almost always use cast-in-place concrete. Depth of invert on subways or underpasses does not exceed 12 m.

The excavation should be carried out in such a way that all surface encumbrances and underground utility installations such as sewers, electrical and telephone conduits, and water pipes are protected. The safety of workers also should be ensured.

Groundwater may be lowered, as needed, by well points. In confined areas, the sides of the excavation are necessarily steep, and so they may need support through timbering.

## Timbering Methods for Tunnels in Soft Soils

The *general principles* to be followed in the design of timbering are

(i) All the members should be in compression and flexural stresses avoided as far as possible.
(ii) The effective length of each member should be reduced by the provision of bracings.
(iii) The details of the joints should be made simple.
(iv) The system should be able to resist the pressure from the roof, sides and the floor.

The *kinds* of *timbering* commonly used during construction are:

**One-piece set:** This is supported in the recess of the side wall, and is used for low side pressures (Fig. 30.8):

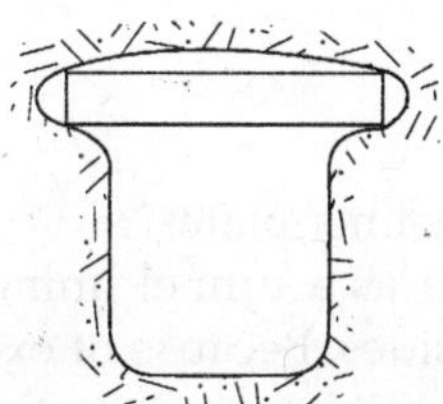

**FIG. 30.8** *One-piece set*

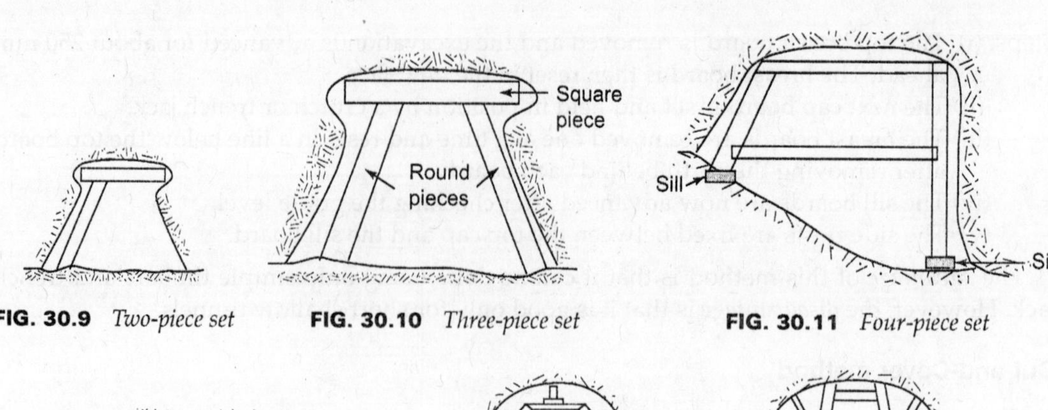


**FIG. 30.9** *Two-piece set* **FIG. 30.10** *Three-piece set* **FIG. 30.11** *Four-piece set*

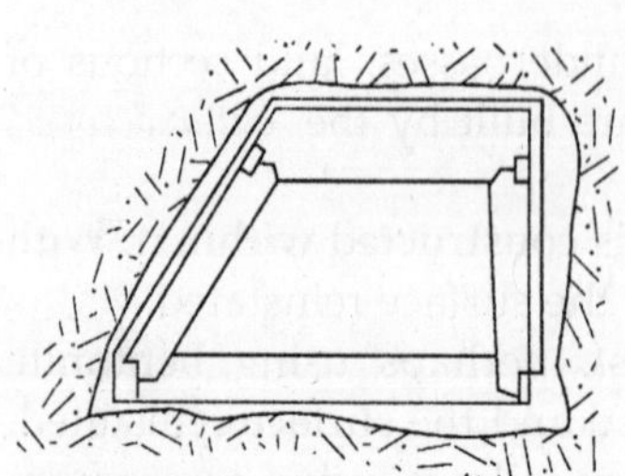

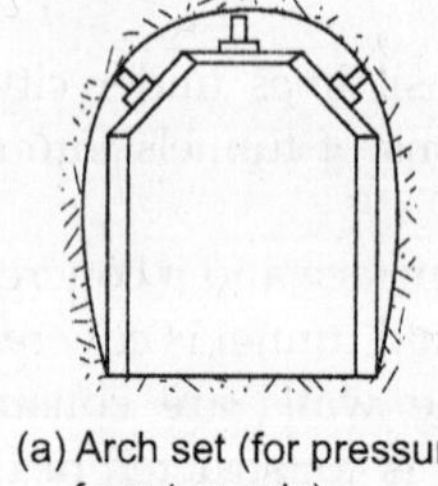

(a) Arch set (for pressure from top only)

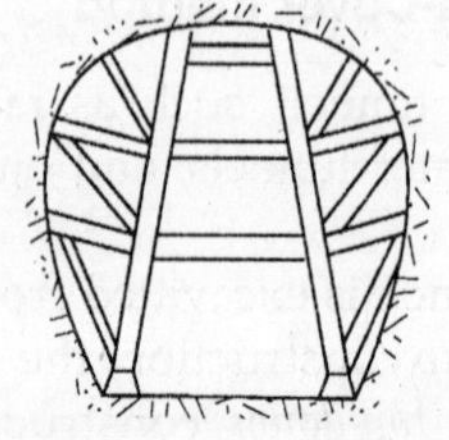

(b) Segmental arch rib system (for loads from top and sides)

**FIG. 30.12** *Normal four-piece set* **FIG. 30.13** *Typical timbering used after the construction of tunnel*

**Two-piece set:** This is used when there is vertical pressure and side pressure from one side only (Fig. 30.9).

**Three-piece set:** This is used when there is pressure from both sides and from the top (Fig. 30.10).

**Four-piece set:** If the floor cannot bear the load from the posts, they are made to rest on sills. Wooden pieces are used for timbering (Fig. 30.11).

**Normal four-piece set:** This is a modification of the four-piece set and is preferred when the trouble due to the presence of water is high (Fig. 30.12).

### Timbering Used After the Construction of the Tunnel

Two typical arrangements are shown in Fig. 30.13.

## 30.3 SPECIAL METHODS

The following two methods fall in this category:

(a) Liner plate method
(b) Shield method

### Liner Plate method

Of late, timbering is being replaced by liner plates.

*Liner plates,* which were developed as a tunnel lining material, are corrugated thin steel sheets with flanges attached on four sides. Because of excellent features, liner plates are used widely for shafts and tunnel linings during construction. The flanges carry holes through which bolts and nuts are passed to connect the plates longitudinally and transversely.

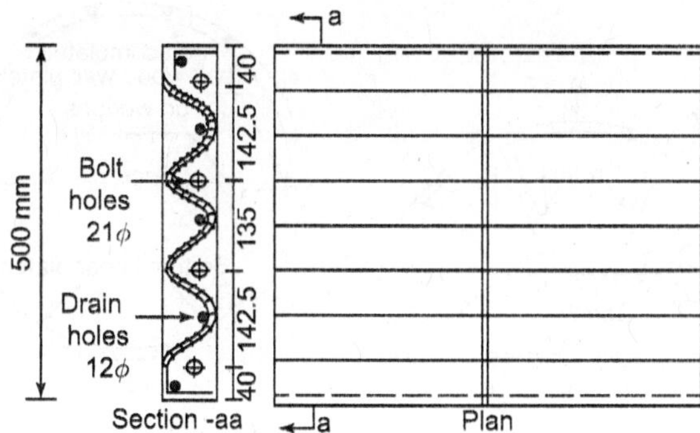


**FIG. 30.14** *A simple liner plate*

The following are the *advantages of liner plates*:

(i) They are of light weight and easy to handle.

(ii) They have high strength and rigidity due to corrugations; they have high durability due to hot-dip galvanisation on both the faces.

(iii) They are easy to transport and store.

(iv) They have very few joints because of their relatively large size compared to timber components.

(v) Because the bolts can be fastened from the inside, liner plates can be installed close to the ground. This saves time and cost of construction.

(vi) Unskilled labour is adequate for erection.

(vii) They are fire-proof and so the work can be carried out even under compressed air pressure.

(viii) Running ground and water can be effectively controlled by means of liner plate construction.

The primary disadvantage is its relatively high cost.

Figure 30.14 shows the details of a simple liner plate.

The liners are stiffened by means of ribs consisting of rolled steel joists, to the web of which the flanges of the liner plates are bolted. The depth of such joists may vary from 25 to 40 mm per diameter of the tunnel. Stiffened liner plates are preferred for tunnels larger than 3 m diameter. The ribs are spliced near the crown and on the flanges as the web connects the flanges of the liner plate. The stiffener beams are usually spaced at 400 mm c/c. The thickness of the liner plate may be from 3 to 6 mm, depending upon the size of the tunnel.

Details of a 2-flange liner plate are shown in Fig.30.15:

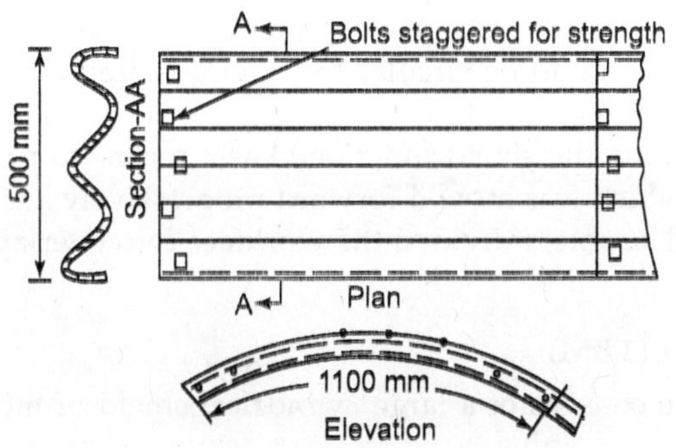


**FIG. 30.15** *A 2-flange liner plate*

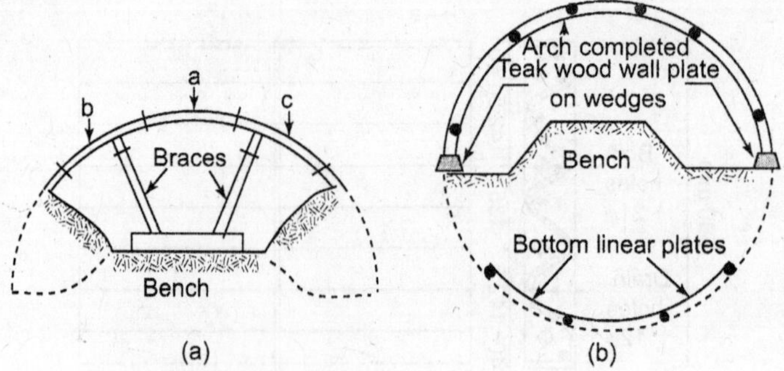


**FIG. 30.16** *Tunnelling with linear plates*

The *sequence of operations* in tunnelling with liner plates is given below with reference to Fig. 30.16.

(i) First, a crown linear plate, 'a', is inserted and set correctly [30.16 (a)].
(ii) The excavation is widened sideways and the adjacent liner plates, 'b' and 'c', are inserted and set correctly. Trench braces may be used for support, if necessary.
(iii) The top portion is then extended for a total length of three liner plates; then, it is widened sideways as earlier up to the springing line [30.16 (b)].
(iv) Wall plates along with wedges are spiked under the last plates at either end.
(v) The trench braces are then removed; the liner plate arch then supports the roof.
(vi) The bottom liner plate is then placed, which may be used as sill for supporting the braces. The wall plate is thus underpinned. The liner plates in the bottom half are then inserted symmetrically from either side of the invert plate.
(vii) Finally, after the plates are securely tightened, the braces are removed.

## Shield Method

*Shield tunnelling* consists of using a protective metallic shield in the excavation of tunnels through soil that is too soft to remain stable during the time it takes to line the tunnel. with an appropriate support structure of concrete, cast iron or steel. In effect, the shield serves as a temporary support structure for the tunnel while it is being excavated.

A successful tunnelling shield was developed by M.I. Brunel, and patented by him and Lord Cochrane in 1818. Brunel is said to have been inspired in his design by the shell of the shipworm which bores through submerged timber.

Brunel's design was substantially improved upon by P.W. Barlow, who suggested that the cross-section of the shield should be circular to provide better support to the weight of the surrounding soil.

In early shield tunnelling, the shield functioned as a means to protect workers who were engaged in excavation, which was moved forward progressively and replaced with pre-built sections of tunnel wall. The shield divided the workface into overlapping portions that each worker could excavate.

### Tunnel Boring Machines (TBM)

A tunnel boring machine consists of a large cylindrical shield of metal and trailing support mechanisms.

A rotating cutting wheel is located at the front end; behind this wheel there is a chamber where, depending on the type of the TBM, the excavated soil is either mixed with *slurry*, or left as it is (in an earth pressure balance or EPB shield). The choice of the type of TBM depends upon the soil conditions. Systems for removal of the soil are available. There are hydraulic jacks supported by the finished part of the tunnel which are used to push the TBM forward. When a distance of 1.5 to 2.0 m has been excavated, a new tunnel ring is built. This whole process is repeated until the job is completed.

### Tunnelling With Shield

The shield does not actually displace the work in the tunnel, but only expedites it with less timbering. It should be strong enough to withstand earth pressure from outside.

Introduction of compressed air pressure into the tunnel corresponding to the pressure head of water acting on the tunnel would hold back the water, so that the work may be done in the dry. An air pressure corresponding to the average pressure of water at the top and bottom of the tunnel should be used, whereby working conditions are achieved with only some water entering at the bottom, and some air escaping at the top.

Tunnelling has become relatively cheap and safe in recent times with the advent of the use of shield and compressed air.

A typical tunnel shield is shown in Fig. 30.17:

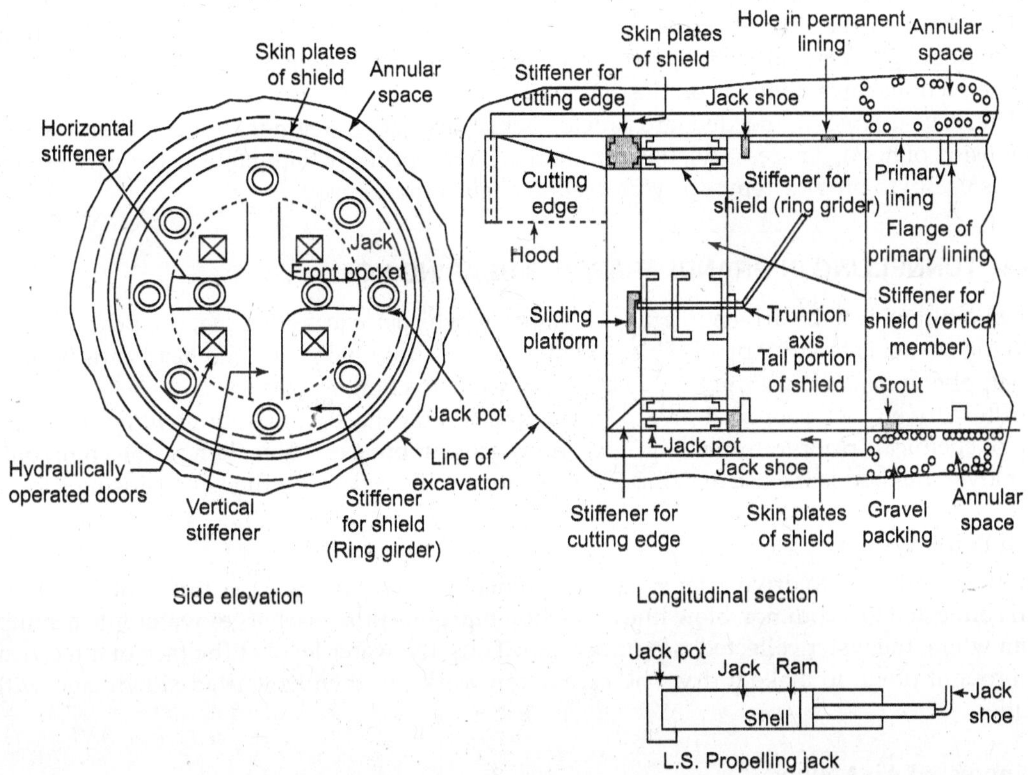


**FIG. 30.17** *A typical tunnel shield*

The main parts of a shield are the cutting edge (or leading edge), central portion (or trunk) and tail (or rear portion). With the cutting edge, a hood or a detachable roof is provided inat the top portion of the shield to provide shelter or protection for the workers in the case of loose soils. Forward movement of the shield is called the 'shove'. The parts or doors in the front portion are kept open (in the case of granular soils), or partially kept open (in the case of compressible soils, depending upon their compressibility); in unstable soils such as quick sand, they are closed. The three situations are called *open shove, partially blind shove* and *blind shove* respectively.

The pressures applied may range from 20 to 30 MPa.

The *brief steps in the procedure* of using the shield are as follows:

- The forward end of the cutting edge is made to project into the earth to be excavated and the rear end backwards for some distance a over the completed lining.
- Hydraulic jacks attached to the shield are made to react against the lining and they shove the shield forward as the earth is excavated.
- As soon as the shield advances a short distance, the jacks are withdrawn, and a ring of the lining is built.
- The operation is repeated until the job is completed.

If the ground is very soft, the shield may often be propelled forward without any excavation, the silt or loose soil being forced out before the front end is closed.

**Size of the shield:** The length of the shield depends upon the storage space required.

The shape of the shield section, more popularly used, is often circular, although square, rectangular or hexagonal sections have also been employed once in a while.

The sectional area depends on the diameter.

The *main equipment* needed are shield jacks, face jacks, segment erectors, and visors (extension of hood, if necessary). It is common to have a primary lining fabricated in the shield as it is useful to provide temporary protection against the soft soil.

## 30.4 TUNNELLING IN GRANULAR/WATER BEARING SOIL

When a tunnel is to be driven in granular deposits which are *water-bearing*, the flow of water into the tunnel is to be stopped. This is achieved by counterbalancing the head causing flow by supplying *compressed air* pressure in the opposite direction.

Alternatively, the proposed area of the tunnel should be *stabilised* by *grouting* or by the use of chemicals prior to tunnelling. This approach is more expensive; hence, the commonly employed method is the use of compressed air for counterbalancing the head causing flow.

### Well Point System

Another method is to drive a series of well points with riser pipes set at different elevations and connected to a common blow line or suction pipe; this pipe discharges water into a sump, from where the water collected is pumped out. Thus, the water level at the face of attack can be brought down to make it dry. The excavation work can then be started slowly and with caution.

### Compressed Air Method

The principle of this method is to create a dry working space for the excavation of the tunnel by means of a chamber near the working face which is kept under compressed air pressure of sufficient magnitude to counterbalance the pressure caused by the hydraulic head of percolation near the face. This pressure depends upon the depth below the water table, which can be calculated easily.

The *primary equipment* needed for this method are:

(a) Bulkhead
(b) Air-locks
  (i) Air-lock for workers
  (ii) Air-lock for materials
  (iii) Emergency air-lock
(c) Pipe and conduits
(d) Compressors
(e) Pumps.

**Bulkhead:** This is a wall constructed across the tunnel section in order to separate the two zones of pressure – high air pressure zone at the front end and the atmospheric pressure zone at the rear end. Hence, this wall must be designed to withstand this pressure difference. The bulkhead is built of plain or reinforced concrete, steel, or masonry.

**Air-locks:** An air-lock is a reserved space in the bulkhead for transferring workmen/materials from atmospheric pressure to the high pressure zone. The lock is 2 m to 3 m in diameter, and 25 m long. There will be an entry door from the atmospheric pressure side and an exit door into the high pressure side.

If the air- pressure is atmospheric in the air- lock, the door from outside will be open. When this door is closed and the air pressure is raised to that in the work chamber, and the door leading to the work chamber will be open.

When the workmen have to come out, the procedure is simply reversed.

Facilities for sitting/sleeping are provided in the air lock. Both recompression from low to high pressure and the decompression from high to low pressure must be done very gradually to avoid health problems for the workers.

***Air-locks for workers*** may be 1.8 m to 2.0 m with doors of 0.75 m × 1.20 m size.

***Air-locks for materials*** may vary in size from 1.8 m to 3 m with increasing size of the doors.

***Emergency air-lock***: This is meant to provide escape route for workers when fire/flood conditions develop due to what is known as a *blow-out* condition. This condition occurs when compressed air in the working chamber finds an escape route to the surface through loose soil. When the air pressure in the work chamber falls suddenly, water flows into the tunnel and causes flooding. Alternatively, a foggy condition may also develop due to such sudden fall of air pressure. Since compressed air has high amount of oxygen, any accidental triggering of a matchstick might cause sudden fire resulting in the formation of smoke.

Under such conditions, workmen would be advised to escape via the emergency air lock in which loose-rope stair cases are provided to climb to safety.

The *blow-out condition* will then be fought from inside as well as outside to control the flood/fire.

**Pipes and conduits:** Separate pipes/conduits are provided for low pressure air supply/high pressure air supply/below line or exhaust.

**Compressors:** These units are placed outside the tunnel. At least two units of the required capacity are set up, so that , one acts as a stand-by.

Similarly, two *generator units* are also set up for electric power supply.

**Pumps:** Two units of the required capacity are set up, one of them being a standby.

*Fire-fighting* units and *alarms* also are provided to meet any contingency.

Compressed air work in tunnelling is similar to pneumatic caissons for bridge foundations, except that the former is horizontal while the latter is vertical in direction.

## 30.5 TUNNEL LINING

*Tunnel lining* is the final finishing given to the inner surface of the tunnel.

The lining may be temporary as in the construction of a tunnel for temporarily supporting the roof and sides.

Permanent lining, on the other hand, is that provided in the case of soft ground that is liable to disintegrate with time. In other words, it prevents possible failure of the roof or sides of the tunnel.

Even in the case of tunnels in sound rock, lining may be provided for some distance from the portals for smooth and aesthetic appearance. The remaining portion of the tunnel may be left unlined, as long as it does not jeopardise the stability.

### Uses or Functions of Tunnel Lining

(i) Suitable lining provided behind the tunnel surface averts the possibility of falling of the crumbled material.

(ii) Even in rock, lining helps to supports large blocks that might otherwise be loosened due to blasting.

(iii) Lining prevents the chemical action of water on certain types or rock that could lead to their disintegration in the long run.

(iv) It reduces the maintenance cost of the tunnel by preventing the falling of debris, which might necessitate a provision for track and muck cars, and also by reducing the pumping cost by preventing percolation of water.

(v) Lining provides a neat appearance and aesthetic look for the tunnel.

### Material for Lining

1. *Brick masonry* or *stone masonry* may be used if the pressures expected are not high. Brick masonry is considered suitable for tunnels carrying sewage since it is more resistant to acid attack.

   Stone masonry is used to line the sides of the tunnel than the roof in view of its weight and the difficulty to pack the space on the extrados of the arch.

2. *Cement concrete* is the most common material used for tunnel lining because of the several *advantages* it offers:

   - Suitability for rock as well as all types of earth materials.
   - Ease of packing and moulding to any shape.
   - Convenience for pre-casting as well as casting-in-situ.
   - Excellent water proof qualities.
   - Amenability for rational thickness design for unusual soil pressures.

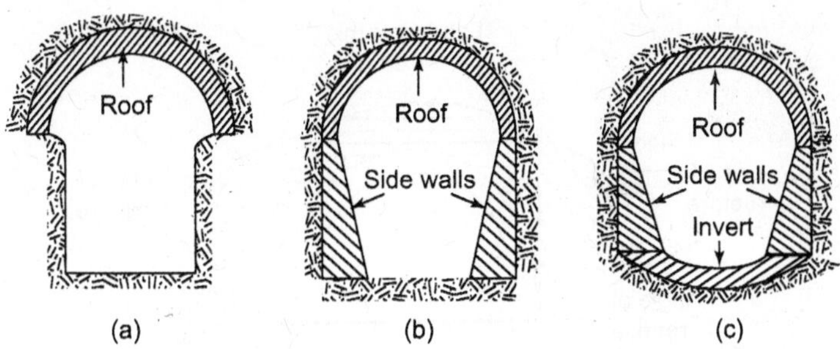


**FIG. 30.18** *Methods of lining*

- Possibility to increase the strength by means of reinforcements, including thin mesh.
- Finishing with a smooth surface, both for appearance and for reducing friction for water and sewage tunnels.
- Least maintenance cost.

3. *Cast iron* or *steel*, as in the case of shield tunnelling, offers the advantages of high strength even immediately on being placed in position.
4. *Timber* may be used both as temporary and permanent lining because of its low cost. However, the disadvantages are that it is liable to decay, vulnerable to fire and lacks water-proofing quality.

**Methods of Lining**

- In firm soil/rock, only the roof is provided with lining.
- If the soil is moderately soft or if the rock is not sound, side walls are also lined (preferably with masonry)
- In soft or treacherous soil, lining is provided on all faces including an invert; this helps in resisting the upward hydrostatic pressure from below the tunnel floor.

These are shown pictorially in Fig. 30.18.

The back of the lining is made to conform to the surface of the cut soil/rock and designed to be slightly greater in size than that of the finished tunnel so as to avoid trimming.

Reinforcements are generally avoided in concrete lining owing to the inconvenience of cramped space. Since the lining has to resist considerable head of water and be water-tight, grouting is considered necessary. Pressure grouting at a pressure of 0.20 to 0.40 MPa is considered adequate.

*Guniting* is the process of providing a coating of thin cement mortar over a surface in order to protect it from seepage and weathering. This is a form of pressure grouting and used not only for tunnel lining, but also for several other applications such as for strengthening and water-proofing.

*Formwork* for placing concrete lining is similar to timbering for temporary support of roof/side walls. Wooden timbering or steel form work may be used, depending on the size of the tunnel and the pressures expected.

The formwork used for the lining of sidewalls is called the *'leading frame'* and that used for floor lining or invert concreting is known as the *'ground mould'*:. In the case of invert, the formwork has to be arranged to suit the curved shape of the desired invert.

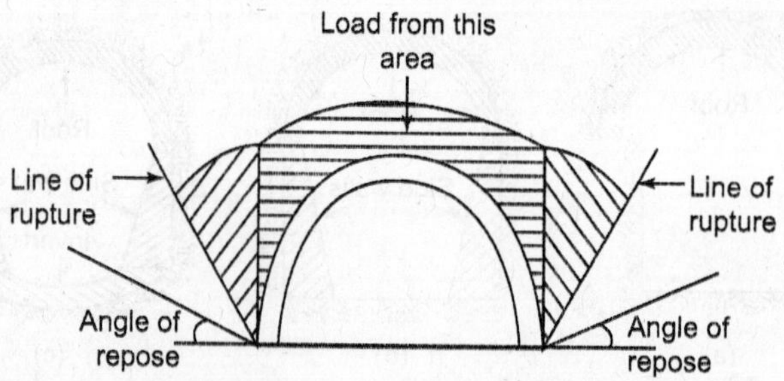


**FIG. 30. 19** *Pressure of tunnel lining*

### Thickness of Lining

The loads will be from the over-lying masses and the earth material on the sides. The wedge-shaped earth shown in Fig. 30.19 causes the pressure on the lining.

This gives a parabolic section of the roof.

A thumb-rule is that the thickness of the lining should be $\frac{1}{12}$th of the tunnel diameter. Empirical formulae have been used, but more rational methods of thickness design based on numerical methods such as the *finite element analysis* are preferred for large tunnels.

### Sequence of Operations for Lining

Different methods are used:

Placing the entire lining in one operation; invert first and the rest of the lining next; invert, sides and the roof in that order; walls, roof, and invert in that order; and, curbs at the sides, side walls, roof, and finally invert — all these approaches are possible, depending upon the size of the tunnel and other considerations.

## 30.6 SAFETY MEASURES AND PRECAUTIONS IN TUNNELLING

The following are the safety measures and precautions-:

1. Stability of working platforms should be ensured.
2. All machines and equipment should be maintained well so that they are in perfect working condition.
3. The equipment and machines should be used only up to just below their capacity and rating.
4. Freshly excavated surfaces should be immediately provided with suitable temporary supports.
5. Proper ventilation and lighting should be provided.
6. Protective gear – clothing, helmet, gloves, and goggles – should be provided for all the workers.
7. Fire-fighting arrangements should be ready at hand.
8. Telephonic communication facilities among the various units are a must.
9. Good publicity and awareness of the precautions and safety measures are necessary for the workers, engineers and the staff involved in the making of the tunnel.

In general, occupational hazards such as *silicosis* due to continued inhalation of silica particles in the dust inside the tunnel causing lung disease, and *caisson disease* due to continued exposure and/or fast recompression and decompression in the course of compressed air work, leading to haematological (blood) disorders should be checked to the extent possible. Strict implementation of the safety precautions is thus mandatory for tunnelling, which is a hazardous underground construction job.

## SUMMARY

- Soft soils for tunnelling are categorized as ordinary type which can stand unsupported for some time, soft soils which need instant support for roof only, and treacherous soils or running ground which need instant support all-round. Except for the first type, excavation and timbering should go hand in hand in the case of other soft soils.
- The methods of tunnelling in soft soils include the forepoling method, needle beam method, and classical methods such as the American, English, Belgian, German and Austrian methods, army or case method, and cut-and- cover method.
- The forepoling method is useful in running ground; the needle beam is useful when the roof can stand for some time; the American method for large highway and railway tunnels; the German method involves three drifts – one near the crown and the other two at the bottom; the Belgian method is useful for moderately firm soil; the Austrian and the New Austrian tunnelling methods are good for any kind of soft soil; the Army or case method is good for small tunnells at shallow depths as for underground sewers; and, the cut-and-cover method is suited for the construction of shallow tunnels for underpasses on city streets.
- The timbering methods for tunnels in soft soils depend on the nature of the soil and the pressures to be resisted; they may range from simple two — and three — piece methods to complex arrangements of several pieces.
- The liner plate method involves the use of corrugated thin steel sheets with flanges to support the tunnel section around it.
- Shield tunnelling involves the use of a metal shield and is the most popular modern technique for tunnelling in soft soils. Hydraulic jacks are used to take the pressure and push or shove the shield forward.

  Modern Tunnel Boring Machines (TMB's) are based on the use of shield as the basic component.
- Tunnelling in water-bearing soils is accomplished by the use of compressed air pressure for keeping off water from the working chamber for excavation. The air pressure counterbalances the hydraulic head of water in the soil outside and prevents the flow of water into the tunnel. The work relating to this technique is similar to that of a pneumatic caisson, which is used for placing a bridge foundation for piers and abutments, except for the direction of advancement, which is vertical for bridges, while it is horizontal in the case of tunnelling.
- Tunnel lining may be temporary or permanent. Permanent lining serves several functions such as providing water-tightness and an aesthetic and smooth appearance. While masonry, steel and cement concrete are the usual materials used for lining tunnels, cement concrete is the most popular one for its strength, durability and waterproofing qualities.
- Since tunnelling is a hazardous work carried out underground, several safety measures and precautions have to be meticulously followed to avert accidents and loss of life and property. The important precautions include immediate provision of temporary supports for tunnel roof, proper ventilation and lighting, adequate drainage to prevent flooding, fire-fighting preparedness, availability of personal protective gear for workers, careful control of work involving compressed air to prevent caisson disease and protect the worker's health, providing good communication facilities and educating all concerned about the right job safety measures.

## QUESTIONS

1. Describe the various methods of tunnelling in soft soil. Discuss elaborately one of the methods of tunnelling in soft soil. (KIIT, Odisha)
2. Explain with a neat sketch the needle beam method of tunnelling. (VTU)
3. Mention the objects of tunnel lining. List the materials used for lining. (VTU)
4. Explain, with neat sketches, the operations involved in the needle beam method of constructing a tunnel. (VTU)
5. With the help of a neat sketch, explain forepoling method of tunnelling in soft soils. (VTU)
6. What are the factors governing the choice of method for tunnelling in soft soil?
7. Write short notes on the following methods of tunnelling in soft soil:

   (a) American method

   (b) Belgian method

   (c) Austrian method

   Mention the nature of soil for which each is suited.
8. Write short notes on:

   (a) Cut-and-cover method

   (b) Army or case method

   Mention the circumstances under which each is suited.
9. Explain the timbering needed for a shallow tunnel of small size with a neat sketch.
10. Explain the liner plate method of tunnelling in soft soil with neat sketches.
11. What is 'Shield Tunnelling'? What are its advantages when used in soft soil?
12. Explain the compressed air method of tunnelling. For which kind of soil is it best suited?
13. What are the precautions to be taken in the compressed air method to ensure the safety of workers?
14. Write short notes on:

    (a) Bulkhead

    (b) Air-locks
15. What are the functions of tunnel lining? What are the materials used?
16. Write a note on cement concrete lining for a tunnel.
17. Enumerate the safety measures and precautions to be taken in tunnelling work.

Chapter 31

# VENTILATION, LIGHTING AND DRAINAGE OF TUNNELS

**CHAPTER OBJECTIVES**

This chapter enables the student to understand the following:

- Ventilation and its purpose in a tunnel – removal of foul gases and dust
- Natural and mechanical ventilation methods; use of reversible fans for exhaust and blowing
- Ventilation systems – longitudinal, transverse and semi-transverse systems
- Equipment for ventilation and dust control methods
- Lighting in tunnels – during construction, operation and maintenance; types of tunnel light for day-time and night-time lighting
- Drainage of tunnels – during construction and during operation
- Site ground water rating for predicting water in flow
- Drainage structures – weep holes, drains, sumps, pumps, and their maintenance.

## 31.1 INTRODUCTION

*Ventilation* means the provision or supply of adequate quantity of fresh air so as to dilute any foul gases that may emanate in the course of construction, operation or use of the tunnel (for the purpose for which it is intended) such that all the people concerned are comfortable and safe.

### Need for Ventilation in Tunnels

The atmosphere in a tunnel may be foul either during construction or during operation due to various reasons such as dust from drilling, fumes from blasting, dust from falling debris, exhaust from vehicles or cars for carrying equipment and mucking, burning of lamps, crowding of persons in a restricted space, and presence of gases due to chemical reactions during decomposition as in the case of tunnels carrying sewage, oils or gases—to name a few.

During operation after construction, exhausts from vehicles and engines in the case of highway and railway tunnels, and decomposition/chemical reactions in tunnels carrying sewage, oils, etc. could be the causes of fouling.

In order to keep the tunnel environment safe for working/maintenance, it is imperative to provide appropriate ventilation systems because even 5 to 8% of carbon dioxide may make humans feel uncomfortable and even lead to their death.

Ventilation may be natural or mechanical (or artificial or engineered). There are different methods and equipment for providing mechanical or artificial ventilation in tunnels.

*Lighting* of tunnels means to provide sufficient illumination at all critical locations of the tunnel to facilitate the various operations, both during its construction and during its working and maintenance, at a later point. The critical locations during construction are the entry, working faces/headings, drilling and mucking areas, shafts, storage areas, repair workshops and pumping stations. During the operation and maintenance, proper illumination is required at suitable intervals along the length of the tunnel.

*Drainage* of tunnels is another important aspect. Water may accumulate in the tunnel during its construction due to percolation from the earth material all round; in fact, even a sudden rush of water may occur under certain conditions. All such water has to be drained away from the work places to a suitable location outside the tunnel area so as to facilitate the construction operations.

Drainage will be required on a permanent basis during the operation of the tunnel, except in the case of those which are meant for transportation of water and sewage.

## 31.2 VENTILATION OF TUNNELS

The *objectives* of ventilation of tunnels are:

- Removal of foul gases in the tunnel which primarily evolve on blasting of explosives.
- Removal of dust caused by drilling, blasting and mucking operations.
- Supply of fresh air to the tunnel workers who may be engaged in different construction activities.

The *requirement* of any ventilation system is to satisfy these objectives.

### Methods of Ventilation

Ventilation may be classified as *natural ventilation* and *artificial* or *mechanical ventilation.*

#### Natural Ventilation

This is possible when the tunnel is short and is of large diameter. (Say, less than 300 metres long and about 3 metres or more in diameter.) Even a drift from end to end provides reasonable ventilation.

The minimum volume of fresh air required per person per minute is about 6.5 cu. m; however, up to 15 cu. m is provided for a comfortable working ambience.

#### Artificial or Mechanical Ventilation

(a) Ventilation by mechanical means consists of *exhausting foul air* by means of exhaust fans which suck air and send it out through pipes or ducts provided near the working face. This causes a vacuum into which fresh air from outside flows in, naturally through the tunnel entrance or portal.

This method has the advantage of quick removal of dust, smoke and gases from in and around the working face. However, it also has the disadvantage that fresh air has to flow in from the entrance or portal which may be some distance away. This would mean that workers near the heading may be momentarily deprived of fresh air and feel inconvenienced. This method of ventilation is also called the *exhaust* or *vacuum* process.

(b) An alternative approach is *blowing* fresh air into the tunnel by means of blowers. As this air flows towards the open entrance or portal of the tunnel, dust and gases are pushed out. Although this method has the advantage of having a supply of fresh air near the working face, the disadvantage is that the dust and foul air move out slowly, fogging the atmosphere inside and reducing visibility, especially in long tunnels.
This method is called the *blowing* or *plenum* process.

(c) *A combination of exhausting and blowing*: This system combines the processes of exhausting and blowing alternately and exploits the advantages of both the processes. The exhaust and blower systems are both operated by means of a reversible type of fan, which can be made to act as an exhaust as well as a blower.

The reversal of operations is accomplished by a valve and duct arrangement, shown in Fig. 31.1.

The reversible fan can rotate either in the clockwise direction or in the counter-clockwise direction. Both directions can be activated by the operation of a single switch. When the switch is pressed once, the fan rotates in the counter-clockwise direction and when it is pressed for the second time, clockwise rotation occurs.

The flow of air occurs through a duct, which is a flexible hose connected always to the fan at one end, the other end being moved to the working face as required.

The duct is bifurcated into two portions near the fan. One branch is connected to the inlet side and the other to the outlet side. Airtight valves A and B are set in the bifurcated branches of the hose, and valves C and D are set on the outlet side and inlet side of the duct, respectively.

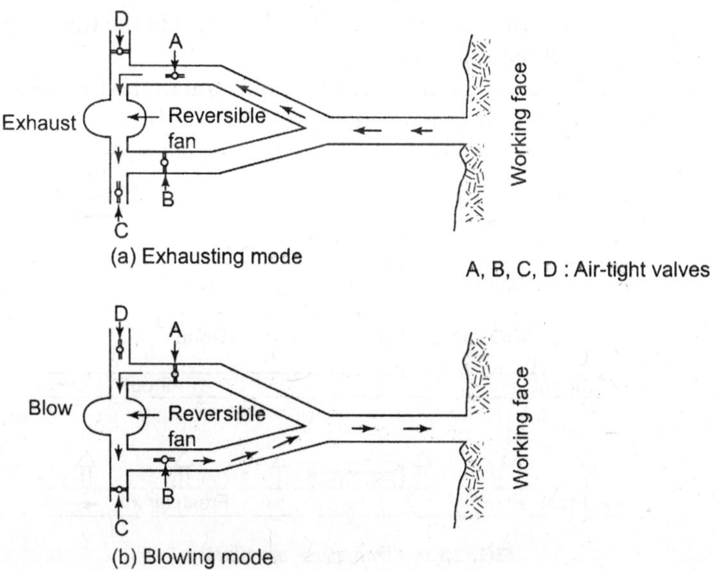


**FIG. 31.1** *Combination of exhaust and blow process*

For making the system act as an exhaust, valves B and D are closed, while A and C are kept open as in Fig.31.1 (a). When it acts as a blower, valves A and C are closed, while valves B and D are kept open as in Fig. 31.1 (b).

The fan is run in exhaust mode for about 20 minutes immediately after the blasting; then the system is reversed to make the fan run in the blower mode for 30 to 60 minutes, depending on the need.

Since this system is portable, it can be moved as required towards the direction of the new working face.

## Types of Ventilation Systems

The following are the types of ventilation systems:

1. Longitudinal ventilation
2. Transverse ventilation
3. Semi-transverse ventilation

### Longitudinal Ventilation

This is the simplest form in which fresh air is introduced into the entry portal and exhaust air is expelled out of the exit portal (Fig. 31.2).

The volume of fresh air coming into the portal entry is controlled to dilute the pollutants. This flow is also induced by moving vehicles, by what is called the *piston effect*.[1]

Since piston effect alone is not adequate to dilute the pollutants to the required degree in the case of long tunnels, this is supplemented by ventilation fans along the length of the tunnel.

### Transverse Ventilation

This works on the same principle of dilution and removal as in the case of longitudinal ventilation; however, these processes occur across the tunnel, i.e., transversely. This system requires two ducts along the length of the tunnel, one for the supply of fresh air and the other for removing the polluted air (Fig. 31.3).

These ducts can be located either at high level or at low level in the tunnel; alternatively, one can be at high level and the other at low level.

Transverse ventilation is used when the extent of pollution in the tunnel is higher than that which could be controlled by longitudinal ventilation.

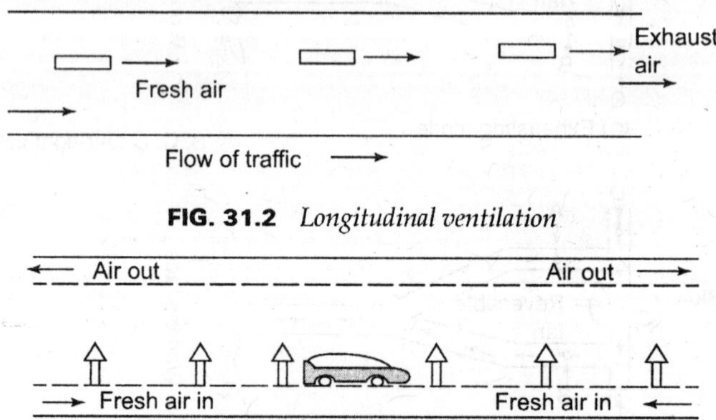


**FIG. 31.2** *Longitudinal ventilation*



**FIG. 31.3** *Transverse ventilation*

[1] This is the flow of air/gas behind a pistion moving in a cylinder

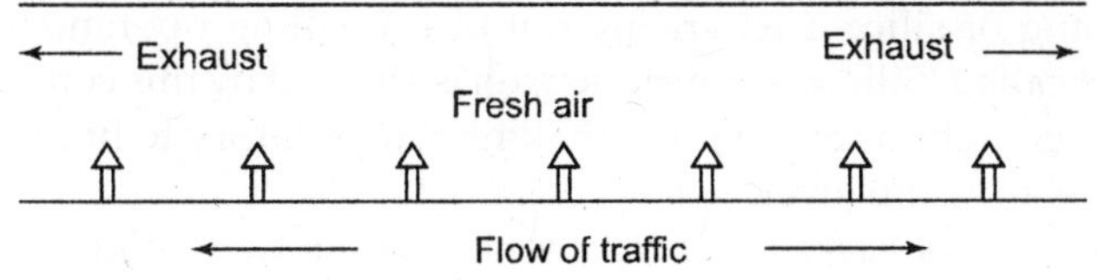


**FIG. 31.4** *Semi-transverse ventilation*

Transverse ventilation is also considered suitable in bi-directional tunnels in which traffic flows in both directions; this cancels the 'piston effect,' thereby increasing the pollutant levels along the length of the tunnel.

### Semi-transverse Ventilation

This is a combination of both longitudinal and transverse ventilation. Fresh air can be supplied from the portals and be continuously exhausted along the tunnel through a duct. Alternatively, fresh air can be continuously supplied along the tunnel via a duct, and exhausted out of the tunnel via the portals.

This is shown in Fig. 31.4.

## Design of Tunnel Ventilating Systems

The design of tunnel ventilation depends upon the requirements of air-quality within the tunnel, ambient air quality criteria just outside the tunnel and any other restrictions specific to the country.

The system capacity can be computed as the quantity or rate of fresh air flow required to bring down or dilute the pollutant levels to meet the air-quality standards specified. Automatic control systems, monitored by operators, are used to ensure continuous ventilation in the tunnel.

Tunnel ventilation is the component which consumes maximum energy in the operation and maintenance of tunnels.

### Criteria governing the capacity of a ventilation system

(i) Length and sectional area of the tunnel.
(ii) Explosives and frequency of blasting.
(iii) Gases expected to be released.
(iv) Carbon dioxide content likely from all sources.
(v) Heat generated by lighting and workers.
(vi) Temperature and humidity inside the tunnel.

### Equipment required for ventilation

Fans and blowers driven by electric power and ducts or pipes which may be of flexible cloth fabric or wire-reinforced fabric are the common equipment used for tunnel ventilation.

## 31.3 DUST CONTROL AND PREVENTION

The operations involved in tunnelling such as drilling, blasting, excavation, and muck removal cause dust inside the tunnel, thereby leading to heavy air pollution. This poses a serious health hazard for those who are involved in the tunnelling work.

Continual breathing of silica dust arising out of tunnelling operations in rock leads to a serious lung disease, called 'Silicosis'; once a person is affected by this condition, it is practically irreversible. Hence, regulations are in place making it mandatory to limit the concentration of dust inside the tunnel to a safe level.

The following *dust control methods* are commonly used in tunnelling:

(a) Wet drilling
(b) Use of vacuum hood
(c) Use of respirators by workers.

### Wet Drilling

Drilling machines have in-built arrangements by means of which water jets are directed into the drill holes while working with the drills. This prevents dust particles from flying into the air to a considerable extent. The quantity of water and speed of the jet are determined by the class of drill and drilling speed. However, this is not fool-proof because microscopic dust particles escape with water.

### Use of Vacuum Hood

A hood is fitted around the drill bit working at the heading or working face, along with a hose pipe and used to collect the rock dust produced on drilling by sucking it. The particles so collected are led to a receiver containing separators and filters. Thus, the dust is removed safely out of the tunnel. This system, along with wet drilling, is considered to be satisfactory in dust control.

### Use of Respirators by Workers

Respirators have been developed for workers to wear while working inside the tunnel. These offer them protection against dust inhalation.

This is considered to be the best personal practice to protect the workers and avert the possibility of their being afflicted by the deadly hazard of silicosis.

Among the dust prevention practices during mucking operation, one effective method is to sprinkle copious amounts of water on piles of muck, allow it to settle, and then load the muck into cars for transporting it outside.

**Noise pollution:** Continuous noise of more than 100 decibels produces discomfort and may prove to be health hazard also. To mitigate this, *resonators* (plastic boxes with spouts, mounted on aluminium cribs) are fixed at the crown portion of the tunnel.

## 31.4 LIGHTING IN TUNNELS

### Need and Function

*Lighting* is needed in tunnels both during construction and later during operation and maintenance during its service life.

During construction, all the areas which are critical, such as working faces, bottoms of shafts, storage points, mucking zones, and other areas where construction-related activities take place, have to be illuminated by means of appropriate lighting arrangements.

After the tunnel is thrown open for use, lighting arrangements are required along the length of the tunnel for safe passage of traffic, especially in the case of long tunnels. Lighting will also be an important aid during the routine maintenance of the tunnel as well as on occasions of any special need.

Another important aspect is that the level of lighting and illumination required in tunnels are different during the day from that which are required in the night. All vehicles which travel through the tunnel should have their own lighting systems, so as to ensure their safe passage through the tunnel, even in the event of unexpected and sudden failure of the lighting system provided for the tunnel.

Lights may be provided in the roof, on the side walls at a height or at a short height above the invert level.

## Recommendations of the International Commission on Illumination

### Definition of the Problem

At night, the problem is relatively simple and consists in providing luminance levels on lit routes inside the tunnel at least equal to those outside the tunnel.

The design of the lighting during day time is particularly critical because of the human visual system. The driver outside the tunnel cannot simultaneously perceive details on the road/floor of the tunnel under lighting levels existing in a highly illuminated exterior and a relatively dark interior (i.e., transient adaptation).

While the visual system can adapt to rapid reduction in ambient illumination such as that produced when passing from daylight into the darkness of a tunnel, these adjustments are not instantaneous. The adaptation process takes a certain time, depending on the amplitude of the reduction – the greater the difference, the longer the adaptation time.

For a given speed, this means that the greater the difference between the lighting level outside and that inside the tunnel, the longer will be the distance over which the visual system of the driver has to adapt.

### Traffic and Visual Tasks

The visual task comprises the detection of the presence and movement of objects on the road in front of the driver. This depends on the type of the tunnel – urban tunnel, road tunnel – and the type of traffic, i.e., motor vehicles, cyclists, pedestrians. As the problem is complicated, a reference task used is the visibility of an object of dimension 0.2 m × 0.2 m having a specified reflection factor such as 0.2. The distance at which this should be seen should correspond to the design speed in the tunnel.

This is essential in spite of the *black hole effect*[2] experienced by the driver while entering an insufficiently illuminated tunnel.

A higher design speed requires an increased lighting requirement. For continuity of flow over a traffic route, it is considered desirable to have the same design speed in the tunnel as in the adjacent open road.

## Distinction Between Long and Short Tunnels

This depends on the degree to which a motorist can see through the tunnel; this ability depends primarily on the length of the tunnel; but in addition, it also depends on other design parameters such as width, height, curvature of the tunnel, and so on.

A tunnel where the driver cannot see the exit from the entry into the tunnel needs to be illuminated like a long tunnel, even if its length would seem to make it a *short* tunnel. Such a

[2] This makes the driver feel that they are entering a black hole while entering a tunnel during day time as the eyes take time to adapt to the sudden reduction of light.

Tunnel lighting is designed to mitigate this effect by gradual reduction of lighting from the entry.

tunnel is designated as an *optically long tunnel* while a tunnel which can be seen through on entering is said to be an *optically short tunnel.*

With regard to lighting, tunnels are subdivided into *three classes*:

1. Geometrically long tunnels
2. Optically long tunnels
3. Short tunnels.

*Day time lighting* of tunnels for different lengths is governed by the following *factors*:

- Length of tunnel
- Visibility of the exit from stopping distance
- Daylight penetration into the tunnel
- Wall reflectance (low < 0.2 and high > 0.4)
- Traffic density (heavy or light)

**Tunnel-related Zones**

The following are the various tunnel-related zones:
- Access zone
- Threshold zone
- Transition zone
- Interior zone
- Exit zone
- Parting zone

**Daytime Lighting for Long Tunnels**

This depends on the zone of the tunnel being considered. Daylight screens are sometimes provided over the tunnel entrance for reflecting daylight into the tunnel.

**Glare restriction:** As glare reduces visibility, it is important to minimise it.

**Night-time Lighting**

The quality of the lighting inside the tunnel should be at least equal to that outside the tunnel.

**Emergency Lighting**

In the event of failure of the normal power source, an emergency system, equal to the normal night-time level, should be provided and activated throughout the tunnel."[3]

## Types of Tunnel Lights

The common types are —

(a) Lanterns or oil lamps: These are used during reconnaissance and surveying
(b) Coal gas lights: The light is bright and steady; but it is vulnerable to accidental explosion.
(c) Acetylene gas lamps: These are also not used much now-a-days
(d) Electric lights: This is the most popular type because of the advantages it offers, primarily the possibility of designing the lighting systems in a scientific manner.

[3] More complete information may be found in the Technical Report of the Committee of the CIE titled "Guide for the Lighting of Road Tunnels and Underpasses" – (CI8 88: 2004 – 2nd edition)

## 31.5 DRAINAGE OF TUNNELS

**Importance**: Tunnels are equipped with a drainage system to deal with surface water from portals, groundwater infiltration through the lining, wall washing water, spillages from tankers, and fire-fighting water.[4]

Where the transport of dangerous goods is permitted, the drainage of flammable and toxic liquids is an important concern. Drainage is important to minimise the size of pools that may otherwise form in the event of a spillage from a road tanker. If there is spillage of a flammable liquid, the drainage system may have a major effect on the size of a resulting fire.

Drainage consists of preventing excessive quantities of water collecting in the tunnel and hindering work, and removing the collected water from the tunnel by suitable means. Water from all the sources other than that from groundwater leakage can be easily determined. But prediction of the quantity of water from ground water leakage is a complex problem, which requires careful investigation.

Sometimes, immediately after a blasting operation, a sudden gush of water may occur. This may indicate that an underground pocket of water has been opened; such water may be allowed to drain away through the portal until the source is fully depleted. The problem will be more serious if the leakage is from a large source such as a lake or a river; in such a case, a bulkhead may have to be erected urgently, and pressure grouting of the area of leakage may have to be done to control the problem fully.

### Water Inflow Into Tunnels

There is a lack of simple and realistic equations or models that can be applied for estimating water inflow into tunnels. However, a few equations have been developed for inflow through fractured rock [Refer: "Drainage systems", Ed.: Prof. Muhammad Salik Javaid, In Tech., Print Edition & On-line Edition, March, 2012 – Ch. 4

**Common Approximations to the Water inflow into Tunnels**

Homayoon Katibeh and Ali Aalianvari.]

The following are the *equations*:

1. Goodman (1965): $$Q = 2\pi K \cdot \frac{H_o}{\ln\left(\frac{2z}{r}\right)} \tag{31.1}$$

2. Heuer (1995, 2005): $$Q = \frac{2\pi K H_o}{\ln\left(\frac{2z}{r}\right)} \times \frac{1}{8} \tag{31.2}$$

3. Lei (1999): $$Q = 2\pi K h \Big/ \left[\ln\left(\frac{h}{r} + \sqrt{\left(\frac{h}{r}\right)^2 - 1}\right)\right] \tag{31.3}$$

4. Karlsrud (2001): $$Q = \frac{2\pi K H_o}{\ln\left(\frac{2H_o}{r}\right)} \tag{31.4}$$

---

[4] Permanent International Association of Road Congresses (PIARC) or World Road Association, Paris – Road Tunnels Manual

5. El Tani (2003):
$$Q = 2\pi K \left[ \frac{1-3\left(\frac{r}{2h}\right)^2}{\left[\left\{1-\left(\frac{r}{2h}\right)^2\right\}\ln\left\{\frac{2h}{r}-\left(\frac{r}{2h}\right)^2\right\}\right]} \right] \quad (31.5)$$

In these equations,

$K$ = Hydraulic conductivity/Equivalent permeability

$H_o$ = Initial head of water above the centre of tunnel

$h$ = Head of water above the centre of tunnel

$z$ = Distance to tunnel from the water source (lake, etc.)

$r$ = Radius of the tunnel.

**Tunnel site rating from ground water hazard point of view**

Katibeh and Aalianvari (2009) have proposed a new way for rating tunnel site from ground water hazard point of view, called *Site Groundwater Rating (SGR)*[5] method.

In this method, the tunnel site is categorized into six ratings, based on engineering, geological and hydro-geological properties: No risk, Low risk, Moderate risk, Risky, High risk, and Critical.

The parameters considered are: Joint frequency, joint aperture, karstification, crashed zone, schistosity, head of water above the tunnel, soil permeability and annual rainfall.

The advantage of this method is that it facilitates the design of suitable drainage systems and choosing suitable drilling techniques according to the potential of groundwater flow, calculated in SGR.

The equation for SGR is given by:

$$SGR = [S_1 + S_2 + S_3 + S_4) + S_5]\, S_6\, S_7 \quad (31..6)$$

Here, the factors $S_1$ to $S_7$ depend on the following parameters:

$S_1$: Frequency and aperture of joints

$S_2$: Schistosity (based on schist planes formed during tectonic processes)

$S_3$: Crashed zone of rock

$S_4$: Karstification (geologic process of chemical and mechanical erosion by water on soluble bodies of rock)

$S_5$: Soil permeability (If the tunnel is excavated in soil, the parameters $S_1$ to $S_4$ are automatically zero as they relate to rock). It is calculated as $K.c$ where $K$ is the soil permeability (m/day) and c is the unit factor (day/m), making $S_5$ dimensionless.

$S_6$: Water head above the tunnel

$$S_6 = \frac{H}{\ln(H.d)} \times d \quad (31.7)$$

$H$ is the head of water above the tunnel in metres and $d$ is the unit factor $\left(\frac{1}{\text{metre}}\right)$

[5] Katibeh, H., Aalianvari, A. (2009): "Development of a new method for tunnel site rating from groundwater hazard point of view," *Journal of Applied Sciences*, Vol. 9, pp. 1496 – 1502.

$S_6$ is unity when the tunnel is excavated above the GWT.
$S_7$: Annual rainfall

$$S_7 = \frac{P_y}{5000} \tag{31.8}$$

$P_y$ = annual rainfall (mm)
$S_7 \leq 1$ ($S_7 = 1$ when the tunnel is in saturated zone of soil.)

**SGR factor:** This is computed by means of Eq.(31.6)
Based on the SGR-value, the tunnel rating is as given in Table 31.1:

**Table 31.1** SGR-rating of tunnel

| SGR | Tunnel rating | Class | Probable ground water inflow into tunnel (l/s/m) |
|---|---|---|---|
| 0–100 | No danger | I | 0 – 0.04 |
| 100–300 | Low danger | II | 0.04 – 0.10 |
| 300–500 | Relatively dangerous | III | 0.10 – 0.16 |
| 500–700 | Dangerous | IV | 0.16 – 0.28 |
| 700–1000 | Highly dangerous | V | Q > 0.28 Inflow of water and mud is possible |
| > 1000 | Critical | VI | Inflow of ground water and mud highly probable |

**Component Structures for Effecting Drainage**

The following are the component structures commonly used for effective drainage in a tunnel:

1. Weep holes in lining of tunnel
2. Concealed conduits
3. Longitudinal drains (Side drains/bottom central drain)
4. Gulleys, channels, pipes
5. Sumps and pumps
6. Oil/water separators
7. Control systems for storage, separation and disposal of effluent.
8. Covered drains under the floor/invert of the tunnel.

Some or all of these structures may be used in tunnel drainage depending on the need in any given case.

To some extent, the method of tunnelling determines the method of handling inflow of water in tunnels. For example, if the heading is driven level or down-slope, water is led away to the portal or a shaft; if the heading is driven up-slope, gravitation aids drainage towards a portal; and, if the heading is driven from a shaft, water may be collected in a sump near the shaft, and pumped out through the shaft to the surface and led away from the tunnel.

## Drainage in Tunnels – A Classification

(a) Drainage prior to the construction of tunnel
(b) Drainage during the construction of tunnel
(c) Drainage during the operation of the tunnel.

### Drainage prior to the construction of tunnel

If the drainage expected is only for a small leakage from the roof of the tunnel, such leakage is made to fall on a temporary pitched roof of corrugated sheets, and into longitudinal drains on both sides, and led outside the tunnel.

If the width of the tunnel is small, drain on one side is considered adequate.

Exploratory holes are drilled into the face to investigate the nature of the material, and estimate the quantity of ground water. If the cost associated with this is expected to be high, preventive grouting of the vulnerable or weak zones is carried out before the commencement of the tunnelling work.

A blanket of clay may be applied to the face if heavy inflow of water is expected from behind.

### Drainage during construction of the tunnel

Open drains on both sides in the tunnel for collecting water is a simple system, useful in hard rock or stiff clay which does not allow water to percolate. However, open drains system suffers from the following disadvantages:

- It is difficult to obtain self-cleansing velocity because of mild grades.
- Valuable working space is reduced.
- Debris from muck can enter the open drains and clog them.
- Open drains affect the integrity of the tunnel section.

Hence, the sump system is generally preferred to open drains. In short tunnels one sump at the middle of the tunnel may be adequate. In long tunnels, sumps are provided at intervals of about 300 to 500 m.

Water collected in the sumps is pumped out from individual sumps with the aid of small-capacity pumps. A system of pumps leads the water into a pipe at different points, and finally to a place outside the tunnel. An added advantage of this method is that dust and debris tend to settle at the bottom of the sump.

Usually centrifugal pumps are employed; the pumps are made to run on electric power, for minimum pollution in the tunnel.

### Drainage during the operation of the tunnel

A permanent drainage arrangement for the tunnel is necessary during its operation throughout its service life.

A simple method is a central drain which is covered under the tunnel invert (Fig. 31.5).

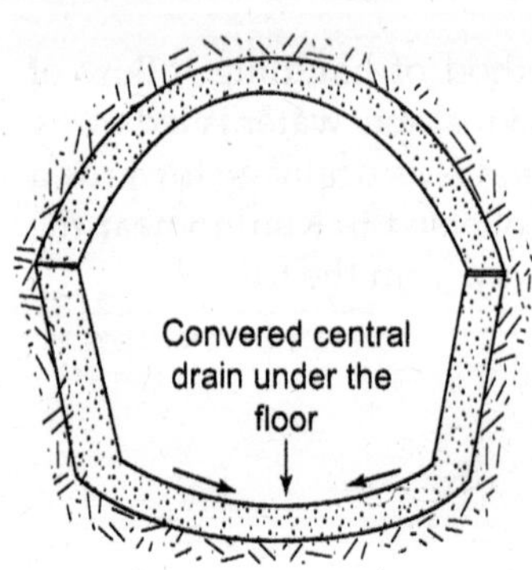


**FIG. 31.5** *Central drain/duct*

Adequate facility should be provided for inspection of drains along the length of the tunnel. Usually, removable covers in manholes provided at intervals of about 50 m are considered adequate.

The type of permanent drainage arrangements needed depends upon various factors such as the nature of the material through which the tunnel is driven, purpose of the tunnel, lining material used and so on. If the leakage is small, grouting or sealing the cracks at the points of leakage is adequate. The aim should not be to make the tunnel perfectly dry, as otherwise, periodic maintenance might prove uneconomical.

## SUMMARY

- Ventilation means the provision of adequate quantity of fresh air to workers and users of the tunnels for their safety and comfort. The need for ventilation arises because of the accumulation of dust and foul gases from the various construction operations of a tunnel.
- The objectives of ventilation are to remove foul gases and dust, and to supply fresh air to workers in the tunnel and for users later.
- Ventilation may be broadly categorised as natural ventilation and artificial/mechanical ventilation; the former is possible only in short tunnels.
- Mechanical ventilation involves exhausting foul air or blowing fresh air, or a combination of both these using a reversible fan.
- The types of ventilating systems are:

  Longitudinal ventilation – blowing fresh air into the entry portal and allowing foul air to pass through the exit portal.

  Transverse ventilation – providing two ducts along the length of the tunnel – one for supply of fresh air transversely and the other for exhausting foul air longitudinally via the portals, or vice-versa.

  Semi-transverse ventilation – a combination of longitudinal and transverse systems using ducts and portals for exhausting foul air and supplying fresh air.

  Design of a ventilation system depends primarily on the specified standards of air quality to be maintained inside the tunnel.
- The primary equipment for ventilation are fans and blowers, driven by electric power.

  The ducts or pipes used for ventilation are usually of flexible, thick cloth fabrics.
- Dust control methods include wet drilling, use of vacuum hood around drills and sucking away the dust, and the use of respirators by workers for their safety. Control of noise pollution is also necessary.
- Lighting is needed in tunnels both during construction, operation and maintenance. This is for ensuring safety of the persons and vehicles.

  Day-time lighting and night-time lighting differ in their standards and characteristics.

  The types of tunnel lights used are oil lamps, coal gas lights and electric lights. However, electric lights are superior as they can be scientifically designed.
- Drainage is very important in tunnels both during construction and operation. Calculation of inflow of water through fractured rock is complex; analytical equations have therefore been developed.

  Site Ground Water Rating (SGR) has been suggested as an effective method for predicting water inflow into a tunnel through rock/soil.
- The component structures used to effect drainage are weep holes in lining, drains, sumps and pumps, and concealed conduits.

  Tunnel drainage may be classified as drainage required prior to the commencement of construction, during construction, and later, during the operation and maintenance of the tunnel. The main factors which govern the drainage arrangements are the size and length of tunnel, purpose of the tunnel, and the characteristics of the surrounding strata.

## QUESTIONS

1. Define 'ventilation' of tunnels and bring out its need.
2. Enumerate the objectives of ventilation. Give a broad classification of types of ventilation.
3. Explain the methods of mechanical ventilation of tunnels. What is a 'reversible fan'?
4. Differentiate between longitudinal and transverse ventilation systems.
5. What are the factors governing the design of a ventilation system?
6. Write a note on 'semi-transverse' ventilation system.
7. What are different types of equipment needed for ventilation?
8. Write a note on dust control methods during the construction of a tunnel.
9. Write short notes on:
   (a) Natural ventilation
   (b) Ducts for exhaust
   (c) Noise pollution control in tunnels.
10. Bring out the need for lighting in tunnels.
11. Differentiate between the standards for day-time lighting and night-time lighting in tunnels.
12. Write short notes on:
    (a) Piston effect in tunnel
    (b) Advantages of electric lights over other kinds of lights.
13. Bring out the importance of drainage in tunnels.
14. Write a note on Site Groundwater Rating (SGR) concept for predicting groundwater inflow into tunnels.
15. Enumerate the component structures used to effect drainage in tunnels during construction.
16. Explain the salient features of permanent drainage after the construction of a tunnel.
17. Explain how drainage requirements may be predicted prior to the commencement of construction of a tunnel.
18. Enumerate the main factors which govern the drainage arrangements needed for a tunnel.

Chapter 32

# OPERATION AND MAINTENANCE OF TUNNELS

**CHAPTER OBJECTIVES**

This chapter enables the student to understand the following:

- Importance of maintenance of a tunnel – maintenance of auxiliary civil works, shafts, tunnel lining, drainage components, ventilation and lighting components
- Recommendations of World Road Association (PIARC) for tunnel maintenance
- OECD and PIARC recommendations for safety in tunnels during transport of hazardous goods
- Groupings A to E of hazardous goods based on the damage likely to be caused in case of accidents
- Quantitative Risk Assessment and Decision Support Models; risk reduction measures for enhancing safety in tunnels.

## 32.1 INTRODUCTION

Operation and maintenance of tunnels is an important aspect of tunnelling. The purpose for which a tunnel is constructed is served only if it is maintained properly and systematically during its service life.

Different organisations are concerned with the operation and maintenance of road tunnels, railway tunnels, water supply tunnels, sewage tunnels, underpasses or subway tunnels for pedestrian traffic in city streets, and so on. These organisations may have their own practices, standards, and strategies for the operation and maintenance of the tunnels for which they are responsible.

The maintenance of tunnels pertains to the shafts, lining, drainage arrangements, ventilation and dust control, and lighting arrangements for night time and day time. Some of these aspects are not relevant for tunnels other than those for traffic use.

Thus, the importance of maintenance in the operation of tunnels cannot be overemphasised.

## 32.2 COMPONENTS AND ASPECTS NEEDING MAINTENANCE

The various components and aspects needing maintenance of tunnels are:

- Ancillary civil engineering works for shelter and maintenance
- Shafts and adits
- Tunnel lining
- Tunnel drainage
- Ventilation and dust control
- Tunnel lighting

## 32.3 METHODS OF MAINTENANCE OF TUNNELS

The following are excerpts from the PIARC[1] Road Tunnels Manual[2] about *Operation and Maintenance of Tunnels*:

"Operation and maintenance activities can be divided into three main streams of activity:

1. **Daily management** This includes all activities to monitor the traffic and to cater to the efficient functioning of all equipment during normal operation, and in the case of an emergency, to ensure the proper functioning of all structural and electro-mechanical installations.
2. **Training of the staff** This is normally a multi-organisation task, considering not only the operator, but also the traffic police, fire brigade, and other emergency services which cooperate to deliver an acceptable level of safety in road tunnels. It includes the basic training and exercises given to the staff and service personnel.
3. **Continuous improvement of safety** This includes all activities of study and planning aiming at a continuous improvement of safety (emergency planning, feedback of experience of accidents, replacement of tunnel equipment, and so on).

### Operational Tasks

Generally speaking, tunnels are considered as part of the road network, which have to be maintained at a suitable or even higher level of safety. This is because the consequences of specific incidents such as breakdown, accident or fire may be far more serious in tunnels than in the open air. Moreover, tunnels may very often be obligatory crossing points and therefore cause bottlenecks in the network, and each total or partial closure can lead to major disturbances of traffic. Hence, the operational continuity and safety of road tunnels has to be ensured, guaranteeing a level of service quality and safety complying strictly with the regulatory requirements in force.

Tunnel operators and traffic police have to manage traffic in the tunnels ensuring safety of the tunnel users as well as the operating staff. The operator is responsible for maintenance, operation of equipment and traffic surveillance, while the traffic police has the overall charge of traffic management.

*Typical tasks of tunnel operators are*:

(i) **Traffic surveillance and operation of tunnel equipment**: Major tunnels are managed by a traffic control centre equipped with surveillance systems such as closed circuit television

[1] PIARC is the acronym for *Permanent International Association of Road Congresses*, established in 1909. This is also called the *World Road Association*.

[2] Permanent International Association of Road Congresses (PIARC), "Operation and Maintenance of Tunnels", *Road Tunnels Manual*, 2011. Reproduced with permission.

and automatic incident detection facilities, which can remote control certain equipment for signalling, ventilation and closure of the tunnel.

(ii) **Technical patrolling inside the tunnel**: Periodic patrolling of the tunnel's interiors for preventive and corrective maintenance helps to increase the service life of the tunnel. This includes:

(a) Management of civil engineering works of the tunnel by conducting regular inspection and surveying; these works also involve—drainage systems, gutters, and all secondary structures inside the tunnel.

(b) Management of equipment, facilitated by computerised tools, for periodic maintenance.

(iii) **Management of emergency situations**: The tunnel operator has to intervene in case of problems related to traffic–accident, fire, etc., or to equipment such as loss of power supply or malfunctioning of data transmission network and activate the pertinent service.

(iv) **Technical and administrative management**: The operator also provides for upgrading the design of any equipment and for the investment and operational budgets for the proper functioning of the tunnel. Periodic reports are also prepared on the operation of the tunnel.

**Players in tunnel operation and their co-operation**

The main stakeholders that need to co-operate are—

- The tunnel operator
- The operators of different parts of the road network
- National and local administrative authorities
- The tunnel owner, if other than the operator
- Public services such as fire and rescue services, traffic police and medical services
- Subcontractors, if any, for cleaning, maintenance, etc.

### Organisation of Operations

Even though there could be variations from one country to another, very generally, the structure of the operation is organized into three principal groups—

- Operating staff, (including emergency rescue services)
- Technical staff
- Administrative personnel

### Operating Instructions and Emergency Plans

These instructions are passed on to the operators at the time of training and also laid down in writing.

### Operating Costs

A kilometre of tunnel is always costlier than a kilometre of road in an open area. This is because for an underground structure, several systems and equipment are deployed which increase the operating costs, which could be considerable for ensuring safety of all tunnel users.

### Maintenance of Equipment

The operator has to carry out the maintenance of both engineering works and the tunnel equipment. For the latter, preventive measures as well as corrective actions might be needed. Preventive maintenance could be valuable in minimising the need for corrective actions.

**Operation during Maintenance and Rehabilitation**

In the case of tunnels with dense traffic, even partial closures may be difficult and a certain operational level may have to be maintained in the event of such closure.

**Short and/or Very Low-trafficked Tunnels**

Most of the above guidelines may be difficult to implement in the case of short low-trafficked tunnels. For each such tunnel, a detailed analysis may have to be made considering geographical and climatic conditions, local and regional resources available, economic context, and exposure to risk and level of risk.

Such an analysis will make it possible to organise and implement the most suitable operating system, according to the specific conditions of these tunnels."

**Ancillary Civil Engineering Works for Shelter and Maintenance**

Ancillary civil engineering structures will be constructed for providing shelter for maintenance staff, especially in the case of long tunnels. These have to be inspected periodically and defects, if any, should be rectified in time.

**Shafts**

Shafts and adits during the construction of a tunnel will continue to be used normally for later maintenance operations. The equipment in the shafts meant to move workers and materials for maintenance have to be kept in working condition by periodic preventive attention.

**Lining**

The lining of a tunnel is usually of cement mortar or cement concrete, depending upon the strata and the loads to be borne from above and from the sides. Although the thickness of the lining is designed considering all the forces expected, sometimes attention may be needed for the upkeep of the lining. This may arise when unexpected and sudden gush of water occurs from the roof or from the sides of the tunnel, or when the lining is disturbed by an adverse environment created by chemical reactions of the salts in the surrounding rock and soil.

Irrespective of whether or not there is water seepage, the lining thickness may have to be restored by shotcreting or by guniting whichever method is appropriate to the situation.

**Ventilation**

Maintenance of the equipment installed for ventilation of the tunnel is imperative for ensuring good ventilation and dust control, and control of air pollution due to the exhausts of the vehicles that use the tunnel.

Depending on the system of ventilation used, the equipment may be fans or blowers, or a combination of these.

**Lighting**

The system of lights installed for night lighting and day-time lighting have to be periodically inspected, and the defective lights have to be promptly replaced with new ones.

Lighting is an important component for the convenience of the users of a tunnel, and its efficient maintenance is essential for their safety as well as that of the operating staff.

## 32.4 SAFETY IN TUNNELS

The following *guidelines and recommendations* for safety in tunnels are based on the findings of a joint research project sponsored by the following entities:

- Road Transport and Intermodal Linkages Research programme of the Organisation for Economic Co-operation and Development (OECD)
- PIARC's Committee on Road Tunnels
- European Commission (by way of financial support).

The aims of the research project were:

(a) To develop a system of regulations for international use as a common reference for harmonised tunnel regulations.
(b) To examine the risk assessment and decision processes in current use and develop tools to improve these processes.
(c) To review risk reduction measures and to evaluate their effectiveness to improve safety according to tunnel and traffic characteristics.

The following excerpts are from the executive summary and recommendations of the Reports:[3]

"In view of the fact that the safety in the operation of tunnels is endangered through the transport of dangerous goods by means of heavy commercial vehicles and multi-axle commercial vehicles, common *lists* or *groupings* of dangerous good have been proposed, along with common regulations. The adoption of the proposed system would improve safety because such harmonised regulations would be easier to comply with and to enforce. It would also facilitate international transport and improve international trade.

All dangerous goods loadings would be split into a small number of groupings. This should be done in such a way that all loadings referred to in the grouping could be accepted together in the same tunnel. The number of groupings must remain reasonably low for the system to be practicable.

The *proposed grouping system* is based on the assumption that there are three *major hazards* in tunnel, which may cause numerous casualties and possibly serious damage to the structure:

- Explosion
- Release of toxic gas or volatile toxic liquid
- Fire

The *main consequences* of these hazards, and the efficiency of possible mitigating measures, are roughly as follows:

- *Large explosions* – two levels of large explosions can be distinguished:
  — *Very large* explosion such as the explosion of a full load of LPG in bulk heated by a fire (Boling Liquid Expanding Vapour Explosion-BLEVE – followed by a fireball, referred to as *hot BLEVE*).
  — *Large* explosion, typically the explosion of a full loading of a non-inflammable compressed gas in bulk heated by a fire (BLEVE with no fireball, referred to as *cold BLEVE*).

[3] OECD/PIARC (2001): Safety in Tunnels – Transport of dangerous goods through Road Tunnels, OECD Publishing, Paris [http://dx.doi.org/10.1787/9789264195691–en]

A very large explosion will kill all the people present in the whole tunnel and cause serious damage to the tunnel equipment and possibly its structure.

A large explosion will cause more limited damage, especially to the tunnel structure.

There is generally no possibility to mitigate the consequences in these cases.

- *Large toxic gas releases*—due to leakage from a tank containing toxic gas (compressed, liquefied, dissolved) or a volatile toxic liquid. It will kill all the people near the release and in the zone where the ventilation will push the gas; only a part of the tunnel may be protected.
- *Large fires*—these may cause a few to several dozens of deaths and serious damage to the tunnel.

The hazards mentioned above are laid out in the order of decreasing damage. From the assumptions given against each hazard on the magnitude of damage caused, a system with *five groupings*, A to E, in the order of increasing restrictions has been devised concerning goods permitted in tunnels[4]:

A – All dangerous goods loadings authorised on open roads.
B – All loadings in A, except those which may lead to a very large explosion
C – All loadings in grouping B, except those which may lead to a large explosion
D – All loadings in C, except those which may lead to a large fire.
E – No dangerous goods (except those which require no special marking on the vehicle) most restrictive category.

*Differentiated time regulations* may be placed on the movement of hazardous goods as per the model scheme given below:

- Grouping C (06.00 to 22.00 hours) – may be transported any time.
  Grouping A (22.00 to 06.00 hours) – authorised only during these hours.
- Grouping C (free passage)
  Grouping B (under escort) } Loadings in Grouping A are forbidden.

For mixed loadings of dangerous goods on the same transport unit, the grouping for each type of dangerous goods is identified. For the whole loading, the first alphabetical loading is used.

## Quantitative Risk Assessment Model (QRAM)

Quantification of risk is difficult because there are numerous factors and variables which influence probabilities that affect the consequences of incidents involving dangerous goods both within and outside tunnels. Even with expert knowledge, it is difficult to assess risk for all circumstances and weather conditions. Nowadays, computers form an indispensable tool for risk assessment and are used to simulate circumstances and weather conditions to develop a sound and rational approach to the problem.

In order to evaluate the risks and set regulations, a comprehensive model is needed to deal with both tunnels and the open road. The resulting model called *Quantitative Risk Assessment Model (QRAM)* is a unique tool.

The model currently considers 13 accident scenarios which are representative of the groupings described in the regulations. The QRAM can produce risk indicators for the various

[4] All loadings in Grouping E are included in Grouping D, all loadings in Grouping D are included in Grouping C, and so on.

groupings and provide a scientific basis for the regulations. The scenarios considered in QRAM are:

Scenarios representative of each grouping in QRAM

| Grouping | Scenario Description |
|---|---|
| E | Heavy goods vehicle fire with no dangerous goods (20 MW)<br>Heavy goods vehicle fire with no dangerous goods (100 MW) |
| D | In addition to scenarios for Grouping E:<br>BLEVE* of Liquid Petroleum Gas (LPG) in cylinders<br>Release of acrolein in cylinders |
| C | In addition to scenarios for Grouping D:<br>Pool fire of motor spirit in bulk<br>Vapour cloud explosion (VCE)of motor spirit in bulk |
| B | In addition to scenarios for grouping C:<br>Release of ammonia in bulk<br>Release of chlorine** in bulk<br>Release of acrolein in bulk<br>BLEVE of carbon dioxide in bulk (not including toxic effects) |
| A | In addition to scenarios for grouping B:<br>BLEVE of Liquid Petroleum Gas (VCE) in bulk<br>Vapour cloud Explosion (VCE) of LPG in bulk<br>Torch fire of LPG in bulk |
| * Boiling Liquid Expanding Vapour Explosion<br>** Chlorine is considered in countries where its transport is allowed in appreciable quantities. | |

## Decision Support Model (DSM)

Decision support methodologies have been studied theoretically and are applied in various fields. A survey and evaluation of proven state-of-the-art decision support tools was carried out. The various, potentially conflicting, objectives are subjected to a mutual weighting—no matter how delicate it may seem to quantify these objectives and weights. In cases where no formalized decision support tool is used, the weighting is made instinctively.

While deciding about which groupings are to be permitted in tunnels, it must be borne in mind that the goods prohibited in the tunnel must be transported on an alternative route. The risk and inconvenience on the alternative route will directly influence which grouping is the best from a societal point of view. This implies that it might not be rational to give the same grouping to two identical tunnels carrying the same traffic if the alternative routes differ significantly—in terms if length and population density along the route.

One of the primary objectives of choosing a given grouping to be permitted in a tunnel is to minimise the risk to human life. Apart from this, there are several other factors that need to be considered while deciding on the routing of dangerous goods. The decision process is a, complex problem and a *decision support model (DSM)* is therefore required to assist a rational decision-making.

The *attributes that are evaluated and weighted* by the DSM are:

- Injury and fatality risks to road-users and the local population using the indicators from the QRAM. The DSM helps the decision makers to weight his concerns. (For example, a risk-averse decision maker considers one incident with 100 fatalities less acceptable than 100 incidents with one fatality each.)
- Material damage due to possible incidents in the tunnel or detour route.
- Environmental impact due to an incident in the tunnel or detour route. QRAM gives only approximate indicators for environmental risk. The DSM can be expanded to accept more detailed environmental information.
- Direct expenses (investment and operational cost of tunnel risk reduction measures, as well as possible additional costs in the transportation of dangerous goods).
- Inconvenience to road users due to a possible incident (time lost during repair works after an incident in the tunnel).
- Nuisance to local population (environmental impact of dangerous goods traffic, with the exclusion of possible incident consequences, but possibly including psychological impact).

In order to make a decision, the decision-maker must determine which attributes are relevant and how these should be weighed against each other. These choices must reflect the preferences of the decision maker.

A computerised tool has been developed, making it possible to take into account the relevant attributes in a rational manner. The DSM includes the option of choosing between the classical Bayesian decision methodology and multi-attribute methodologies. The DSM utilises the QRAM output directly. Other technical data is used as input – for example, reparation costs following an accident, or additional costs for transporting dangerous goods by a longer route. The decision-maker, thus, has all technical inputs and must provide only policy-based preferences.

## Risk-reduction Measures

There are several measures that can be implemented in tunnels which will reduce either the probability or the consequences of an incident in a tunnel. These are included in the QRAM and will influence the regulations governing the restriction on the transport of dangerous goods through a tunnel. Extensive studies were carried out to determine the effectiveness of these measures before they were included in the QRAM.

The QRAM model can thus be used to examine the effects of introducing these measures to enhance the safety of tunnel transport. A number of other measures which were not there in the original model were also examined and included in the model.

Both qualitative and quantitative methods for the analysis of the effects of risk reduction measures are presented. Using QRAM together with these methods, it is possible to assess the effects of these measures for a given tunnel.

The effects of measures are unique for each tunnel, depending on the traffic characteristics and local circumstances. A general effect of the measures applicable to all tunnels could not therefore be generated. Likewise, the costs of the measures vary with the type of tunnel considered. Costs will also differ considerably if the measures are incorporated during the initial design and construction stage than if they are retrofitted. The costs are therefore best estimated for each tunnel on a case-by-case basis, so that the efficiency or cost-effectiveness ratio of the measures can be properly evaluated for the specific case under consideration. The list of risk reduction measures are given here:

**List of Risk Reduction Measures Classified According to their Main Purpose**

| MEASURES TO REDUCE THE PROBABILITY OF AN ACCIDENT | | |
|---|---|---|
| **Related to tunnel design and maintenance** | | |
| Tunnel cross-section and visual design | Alignment<br>Lighting (normal) | Maintenance of road surface (friction) |
| **Related to traffic and vehicles** | | |
| Speed limit<br>Prohibition to overtake | Escort<br>Distance between vehicles | Vehicle checks |
| **MEASURES TO REDUCE THE CONSEQUENES OF AN ACCIDENT** | | |
| **Alarm, information, communication of operator and rescue services** | | |
| Closed-circuit television<br>Automatic incident detection | Automatic fire detection<br>Radio communication (services) | Automatic vehicle identification<br>Emergency telephone |
| **Communication with users** | | |
| Emergency telephones<br>Radio communication (users) | Alarm signs/signals | Loudspeakers |
| **Evacuation or protection of users** | | |
| Emergency exits<br>Smoke control | Lighting (emergency)<br>Fire-resistant equipment | Failure management |
| **Reduction of accident importance** | | |
| Fire-fighting equipment<br>Rescue teams | Drainage<br>Road surface (non-porous) | Emergency action plan<br>Escort |
| **Reduction of the consequences on the tunnel** | | |
| Fire-resistant structure | Explosion-resistant structure | |

## Recommendations

### Implementation of a Consistent Regulatory and Technical Framework

The results of this study are applicable in all countries with tunnels. The analysis of risks and the development of decision support tools achieved through this project provide road administrations with options to improve the transport of dangerous goods through tunnels:

- It is strongly recommended that the 'groupings of dangerous loadings' system be implemented as the basis of regulations. This system should be implemented both at national and international levels.
- It is recommended that, through the adoption of these regulations, all tunnels are assigned a category. This will require new 'sign-posting', both at the tunnel approach and alternative routing signs.

- The adoption of the 'groupings' system requires a systematic and scientific basis for decision-making. To this end, QRAM and DSM are currently state-of-the-art in the field, and are recommended for use in all countries since these models support the adoption of the proposed 'grouping' system.

## International Regulatory Framework

As a global body, the United Nations Committee of Experts for the Transport of Dangerous Goods is the most appropriate body to act as the promoter and developer of this system of groupings.

It is recommended that the system be included in the UN's Model Regulations and promoted in all the regions of the world. This is an important mechanism to promote global transport efficiency through the implementation of a consistent and harmonised regulatory framework.

It is also recommended that the relevant United Nations Committee should be charged with developing the signs necessary for implementation and enforcement of the regulations.

### Expected Benefits

- Reduction in the cost of damage to road infrastructure arising from possible incidents in tunnels or detour routes.
- Reduction in the environmental impact due to an incident in tunnels or on detour routes.
- Improvement in network efficiency by implementing consistent and harmonised regulations for the transport of dangerous goods through tunnels.
- Improvement in overall transport efficiency through reduction in the time costs associated with a possible incident, to road users (time lost arising from the incident itself, detour routes, and due to repair works after an incident in the tunnel).
- Increased efficiency in the deployment of funds invested in upgrading/constructing tunnel infrastructure, management systems and risk reduction measures.
- Increased efficiency of road transport operations arising from compliance with regulations and correct routing of vehicles."

## SUMMARY

- Maintenance of a tunnel is very important for its successful operation.
- The components and aspects needing maintenance are: auxiliary civil engineering works inside the tunnel, shafts and adits, tunnel lining, drainage works such as drains and ducts, ventilation and dust control, and tunnel lighting.
- The recommendations contained in PIARC Tunnel Manual regarding the operation and maintenance of tunnels are useful in tunnel maintenance.
  The main streams of activity in operation and maintenance of tunnels are—daily management, staff training and continuous improvement of safety.
- Operational tasks are set for the purpose of traffic surveillance, operation and upkeep of tunnel equipment, inspection and periodical maintenance of civil engineering works inside the tunnel.
- The staff associated with tunnel management are grouped as Operating staff (including emergency staff), Technical staff and Administrative personnel.

- Shafts need periodic inspection and strengthening when needed as they are useful components in the operation and maintenance of a tunnel.
- Tunnel lining needs periodical strengthening by shotcreting or guniting depending on the specific need.
- Maintenance of ventilation equipment such as fans, blowers and ducts is essential for efficient management of ventilation.
- Lighting in tunnels has to be maintained effectively by timely replacement of defective lights with new ones—both for night-lighting and day-lighting.
- Safety in tunnels is very important in tunnel operation. The 'Grouping' system of dangerous goods concept recommended on the basis of a research project jointly sponsored by OECD and PIARC is considered good for management of tunnels with the desired degree of safety in transporting hazardous goods through tunnels.
- The major hazards are those due to explosives, release of toxic gases and fires.
- Five groupings, A to E, are recommended based on decreasing damage due to possible accidents while transporting hazardous goods through tunnels.
- Quantitative Risk Assessment Model (QRAM) has been evolved on a rational basis of the 'five groupings' system of dangerous goods.
- Decision Support Model (DSM) has been developed based on the evaluation and weighting of relevant attributes.
- Risk-reduction measures have been proposed based on the application of the principles of QRAM and DSM. These are recommended to be followed uniformly in all countries in which tunnels are operated, for enhancing safety in tunnels and boosting international trade and commerce.

## QUESTIONS

1. Bring out the importance of maintenance of a tunnel in its successful operation.
2. Enumerate the components and aspects of a tunnel which need maintenance.
3. Write a brief note on the recommendations of the PIARC in road tunnel operation and maintenance.
4. List the operational tasks for maintenance of tunnels as recommended by PIARC.
5. Explain the maintenance aspects of
   (a) Shafts and adits
   (b) Civil engineering structures inside the tunnel.
6. Explain the maintenance needs of
   (a) Tunnel lining
   (b) Tunnel ventilation
7. What are the maintenance aspects of the following in tunnels?
   (a) Drainage structures
   (b) Lighting arrangements.
8. Write a brief note on safety in tunnels.

9. Explain the 'Grouping' system of classification of dangerous goods transported through road tunnels, as recommended by the OECD and PIARC, based on their joint research study.
10. With reference to transportation of dangerous goods through tunnels, write brief notes on:
    (a) QRAM
    (b) DSM
11. Write a brief note on the risk-reduction measures recommended, based on QRAM and DSM, while transporting dangerous goods through road tunnels.

# APPENDICES

## APPENDIX I

## Answers to Numerical Problems

## Part I Railway Engineering

**Chapter 3**

(3.31) 2.16 t (3.32) 1 in 183 (3.33) 81 t 32 km/h (3.34) 1260 t and 41 km/h (3.35) 1in 75
(3.36) 1 in 400 (3.37) 175 m (3.38) 322 cm.t 1.46 t/cm$^2$ (C); 1.33 t/cm$^2$ (T)
(3.43) (b) 17 (3.48) (b) Rails 2028 t; Sleepers 28530; Ballast 26815 m$^3$ (c) 1369 t; 41 km/h; 50 km/h (3.49) (c) 40 (3.51) 180 m; 360 m (3.56) (b) 1740 t (3.57) 890
(3.58) (a) 1265 t; (b) (N + 3) (3.59) 1350 t; 32 km/h

**Chapter 4**

(4.28) 1 in 230 (4.29) 150 mm; 100 km/h (4.30) 69 km/h (4.31) 85 km/h (4.32) 96 m
(4.33) 320 mm (inside) and 25 mm (outside) (4.34) 36 mm (but only 10 mm is allowed on BG)
(4.35) old – 915 m; new – 120 m (4.36) (b) 170 km/h (4.37) (c) 70 mm (4.38) 103 mm; 73 km/h
(4.39) 65 km/h (4.40) 60 mm (4.41) 50 km/h (4.42) 60 km/h (4.43) 100 mm; 110 km/h
(4.44) 60 m; 0.34 m (4.45) 240 m

**Chapter 5**

(5.26) 1°32′24″; 5023 mm
(5.27) 2° 51′ 51″; 2° 51′ 54″; 2° 51′ 45″; the lowest value is got by the right-angled method, and it is recommended.
(5.31) (c) CL = 28.6 m; R = 245 m; SL = 7.5 m; L = 21.1 m

## Part II Airport Engineering

**Chapter 11**

(11.30) 8 and 6 aircraft/hr (11.31) 2115 m
(11.32) 2720 m; field tests are needed to ascertain the corrections.
(11.33) 3200 m (11.34) NNE-SSW (bearing N 22°30′E); 76 per cent (11.35) 0.43 per cent
(11.36) 3160 m for take-off; 3105 m for landing; hence runway length = 3160 m
(11.37) 755 m (11.38) E-W (270° bearing); 56.4 per cent (11.39) 2100 m
(11.40) 0.35 per cent for a chain length of 30 m (11.41) 3300 m
(11.42) 13.4 per cent; N-S (180° bearing); 75.3%

**Chapter 12**

(12.21) Total thickness = 125 cm (10 cm bituminous surfacing; 25 cm granular base, and 90 cm granular sub-base) (12.22) Total thickness = 75 cm (10 cm bituminous concrete, and 65 cm granular base course) (12.23) 17 cm (12.24) 10 cm; 15 cm

**Chapter 13**

(13.15) 0.34 $m^3/s$ (13.16) 30 cm (13.17) 0.5375

## Part IV Bridge Engineering

**Chapter 22**

(22.29) 0.23 m (22.30) 0.043 m; 0.92 m/s (22.31) 0.10 m; 1.40 m/s (22.32) 2.77 m
(22.33) 22.2 m (22.34) 1304 $m^3/s$

## Part V Tunnel Engineering

**Chapter 28**

(28.9) 6.964 cm

# APPENDIX II

## Notations

## Part I Railway Engineering

### ENGLISH

| S. No. | Symbol | Meaning |
|---|---|---|
| 1. | $A$ | Exposed area of vehicle; Area of section of rail |
| 2. | $B$ | Breadth of loading gauge |
| 3. | $H$ | Height of loading gauge |
| 4. | $D$ | Degree of curve (Chapter 3) |
| | | Distance between parallel tracks (Chapter 5) |
| 5. | $E$ | Modulus of elasticity (Chapter 3); Amount of Super elevation (Chapter 4) |
| 6. | $F$ | Sliding frictional force |
| 7. | $G$ | Gauge of railway (Chapter 3) |
| | | Algebraic difference of grades (Chapter 4) |
| 8. | $L$ | Length of transition/Length of vertical curve (Chapter 3) |
| | | Lead of crossing (Chapter 5) |
| | | Running track kilometre (Chapter 8) |
| 9. | $N$ | Crossing number (Chapter 5) |
| | | Length of rail in metres (Chapter 3) |
| | | Number of workers per km (Chapter 8) |
| 10. | $M$ | Bending moment in rail (Chapter 3) |
| | | Mass of vehicle (Chapter 4) |
| | | Mean power factor (Chapter 8) |
| 11. | $P$ | Force; wheel load |
| 12. | $R$ | Mean radius of curve |
| 13. | $R_a$ | Resistance due to acceleration |
| 14. | $R_c$ | Curve resistance |
| 15. | $R_f$ | Resistance due to friction |
| 16. | $R_g$ | Resistance due to gradient |
| 17. | $R_s$ | Resistance due to starting |
| 18. | $R_{at}$ | Atmospheric wind resistance |
| 19. | $R_{wa}$ | Resistance due to wave action |
| 20. | $S$ | Spacing of sleepers(c/c) |
| 21. | $T$ | Tangent length of a curve |
| 22. | $T_e$ | Tractive effect of a locomotive |
| 23. | $U$ | Track modulus (Chapter 3); Track density factor (Chapter 8) |

| S. No. | Symbol | Meaning |
|---|---|---|
| 24. | $V$ | Speed of train (km/h) |
| 25. | $W$ | Weight of train |
| 26. | $W_a$ | Axle load |
| 27. | $W_l$ | Weight of locomotive |
| 28. | $W_v$ | Weight of trailing vehicles |
| 29. | $Z$ | Section modulus |
| 30. | $L_a$ | Apex distance of a curve |
| 31. | $L_m$ | Versed-sine or mid-ordinate of a curve |
| 32. | $E_d$ | Permissible cant deficiency |
| 33. | $G_e$ | Extra width of gauge on a curve |
| 34. | $L_c$ | Long chord of a curve |
| 35. | CL | Curve lead |
| 36. | SF | Speed factor |
| 37. | SL | Switch lead |
| 38. | EI | Flexural rigidity |
| 39. | ST | Horizontal projection of straight portion of a turnover |
| 40. | CMS | Cast-manganese steel |
| 41. | $w$ | Weight of rail (kg/m) |
| 42. | $t$ | Change in temperature |
| 43. | $l$ | Breathing length of rail (Chapter 3) |
| | | Characteristic length (Chapter 3) |
| | | Length of curve (Chapter 4) |
| 44. | $s$ | Slip of wheel (Chapter 3) |
| | | Shift of circular curve due to transition (Chapter 4) |
| 45. | $w$ | Width of sleeper |
| 46. | $d_b$ | Depth of ballast section |
| 47. | $g$ | Gradient per cent; acceleration due to gravity |
| 48. | $a$ | Acceleration (Chapter 3); radial acceleration (Chapter 4) |
| 49. | $f_c/f_t$ | Compressive/tensile stress |
| 50. | $e$ | Rate of super elevation |
| 51. | $h$ | Depth of flange below top of rail |
| 52. | $d$ | Heel divergence |
| 53. | $p$ | Perpendicular distance between parallel tangents |

## GREEK

| S. No. | Symbol | Meaning |
|---|---|---|
| 1. | $\alpha$ | Coefficient of linear expansion (Chapter 3) |
| | | Polar angle (Chapter 4) |

| S. No. | Symbol | Meaning |
|---|---|---|
| | | Crossing angle (Chapter 5) |
| 2. | $\beta$ | Angle of heel (Chapter 5) |
| 3. | $\rho$ | Length of polar ray (Chapter 4) |

## Part II Airport Engineering

### ENGLISH

| S. No. | Symbol | Meaning |
|---|---|---|
| 1. | $C$ | Gate capacity |
| 2. | $G$ | Number of gates |
| 3. | $U$ | Gate utilisation factor |
| 4. | $T$ | Gate occupancy time (weighted average) |
| 5. | $V$ | Wind velocity (Chapter 11); permissible velocity (Chapter 13); |
| | | Volume of earthwork (Chapter 13) |
| | | Visual range (Chapter 14) |
| 6. | $F_d$ | Displacement factor |
| 7. | $W$ | Wheel load |
| 8. | $E$ | Elastic modulus of concrete |
| 9. | $S_{cc}$ | c/c distance of dual tyres |
| 10. | $D_f$ | Free haul distance |
| 11. | $D_e$ | Economic haul distance |
| 12. | $T_f$ | Time of flow |
| 13. | $Q$ | Peak run-off |
| 14. | $C$ | Coefficient of run-off |
| 15. | $A$ | Area of catchment |
| 16. | $R$ | Hydraulic mean depth |
| 17. | $S$ | Longitudinal bed slope |
| 18. | $H$ | Horizontal segment of visual range |
| 19. | $d$ | Deceleration rate |
| 20. | $f$ | Coefficient of friction between tyre and pavement |
| 21. | $a$ | Radius of load contact circular area |
| 22. | $q$ | Uniform pressure |
| 23. | $h$ | Thickness of concrete slab (Chapter 11); height of glide slope above runway surface |
| 24. | $l$ | Radius of relative stiffness |
| 25. | $b$ | Equivalent radius of resisting section |
| 26. | $k$ | Modulus of subgrade reaction |
| 27. | $p$ | Tyre pressure |
| 28. | $v$ | Velocity of flow |

| S. No. | Symbol | Meaning |
|---|---|---|
| 29. | $i$ | Critical rainfall intensity |
| 30. | $n$ | Coefficient of rugosity |

GREEK

| S. No. | Symbol | Meaning |
|---|---|---|
| 1. | $\alpha$ | Cock-pit cut-off angle |
| 2. | $\beta$ | Pitch-angle |
| 3. | $\gamma_c$ | Unit weight of concrete |
| 4. | $\Delta$ | deflection |
| 5. | $\phi$ | Glide-slope angle |
| 6. | $\theta$ | Angle formed by visual cone with horizontal |

## Part III Dock and Harbour Engineering

ENGLISH

| S. No. | Symbol | Meaning |
|---|---|---|
| 1. | $B$ | Breadth of harbour at the point of consideration |
| 2. | $D$ | Depth of harbour; distance to the point of consideration from the harbour entrance |
| 3. | $D_l$ | Draft of loaded vessel |
| 4. | $F$ | Fetch |
| 5. | $H$ | Height of storm waves |
| 6. | $L$ | Length of design ship (Chapter 17); wave length (Chapter 18); length of slipway (Chapter 20) |
| 7. | $H_{10}$ | Average wave height of 10 percent of waves |
| 8. | $P$ | Pull |
| 9. | $T$ | Wave period |
| 10. | $V$ | Wave velocity |
| 11. | $W$ | Weight of individual stone |
| 12. | $W_s$ | Weight of ship |
| 13. | $W_c$ | Weight of cradle |
| 14. | $b$ | Width of harbour entrance |
| 15. | $d$ | Depth of water (Chapter 18); draft of ship (Chapter 20) |
| 16. | $h$ | Wave height in the sea (Chapter 17); height of cradle from rail bottom to the keel and bilge blocks (Chapter 20) |
| 17. | $h_1$ | Reduced wave height within the harbour |
| 18. | $p$ | Wave pressure |

## GREEK

| S. No. | Symbol | Meaning |
|---|---|---|
| 1. | $\alpha$ | Angle of slope of breakwater with horizontal |
| 2. | $\gamma_r$ | Unit weight of stone |
| 3. | $\mu$ | Coefficient of friction |

# Part IV Bridge Engineering

## ENGLISH

| S. No. | Symbol | Meaning |
|---|---|---|
| 1. | $A$ | Area of catchment/wetted area of flow |
| 2. | $A_h$ | Horizontal seismic coefficient |
| 3. | $C$ | Coefficient which depends on the nature of the catchment and its location |
| 4. | $F$ | Horizontal force to be applied at the centre of mass of superstructure for 1 mm horizontal deflection at the top of pier |
| 5. | $F_{eq}$ | Seismic force |
| 6. | $H$ | Difference in elevation between critical point and bridge site (Chapter 22); ht of fall of hammer (Chapter 23) |
| 7. | $L$ | Distance from the critical point to the bridge site (Chapter 22);<br>Effective span (Chapter 23); linear waterway;<br>width of stream at HFL (Chapter 22) |
| 8. | $P$ | Coefficient of run-off |
| 9. | $R$ | Hydraulic mean depth |
| 10. | $S$ | Bed slope of stream |
| 11. | $L_r$ | Regime width of stream |
| 12. | $L_1$ | Linear waterway under the bridge |
| 13. | $D$ | Appropriate load |
| 14. | $V$ | Velocity of flow; design speed of vehicle |
| 15. | $Q$ | Maximum flood discharge/Peak run-off |
| 16. | $W$ | Total live road |
| 17. | $V_z$ | Hourly mean speed of wind at height H |
| 18. | $P_z$ | Horizontal wind pressure at height H |
| 19. | $I$ | Seismic importance factor of the bridge |
| 20. | $Z$ | Zone factor for the particular seismic zone |
| 21. | $Q_{up}$ | Ultimate load on pile |
| 22. | $h$ | Depth of foundation; min-thickness of well steining |
| 23. | $l$ | Depth of well below the top |
| 24. | $m$ | Mean diameter of particles of bed material |
| 25. | $r$ | Number of spans; coefficient of rugosity; constant index in empirical formula for Q |
| 26. | $d_s$ | Scour depth |

| S. No. | Symbol | Meaning |
|---|---|---|
| 27. | $h_a$ | Afflux |
| 28. | $q$ | Bearing capacity of soil |
| 29. | $f$ | Lacey's silt factor; factor to correct for the variation of the intensity of rainfall over catchment. |
| 30. | $i$ | Critical intensity of rainfall |
| 31. | $s$ | Span (Chapter 22); final penetration of pile or set per blow (Chapter 23) |
| 32. | $s_e$ | Economic span |
| 33. | $t_c$ | Time of concentration |
| 34. | $V_a$ | Increase in velocity due to afflux |
| 35. | $t$ | Thickness of bottom plug of pile |

## GREEK

| S. No. | Symbol | Meaning |
|---|---|---|
| 1. | $\gamma$ | Unit weight of soil |
| 2. | $\phi$ | Angle of internal friction of soil |

# Part V Tunnel Engineering

## ENGLISH

| S. No. | Symbol | Meaning |
|---|---|---|
| 1. | $D$ | Size of tunnel (effective width or diameter) (Chapter 27); |
| | | Depth of rock cone (Chapter 29) |
| 2. | $D_0$ | Optimum depth of rock cone |
| 3. | $H$ | Height of earth fill above the bottom |
| 4. | $K$ | Hydraulic conductivity/Equivalent permeability |
| 5. | $H_O$ | Initial head of water above centre of tunnel |
| 6. | $K_a$ | Coefficient of active earth pressure |
| 7. | $R$ | Top radius of the rock cone |
| 8. | $V$ | Volume of rock cone |
| 9. | $W$ | Weight of charge |
| 10. | $S_1$ to $S_7$ | Factors in the equation for site ground water rating (SGR) |
| 11. | $h$ | Head of water above the centre of tunnel |
| 12. | $p_a$ | Active earth pressure |
| 13. | $r$ | Radius of tunnel |
| 14. | $z$ | Distance of tunnel from the water source |

**GREEK**

| S. No. | Symbol | Meaning |
|---|---|---|
| 1. | $\alpha, \beta$ | Angles in Weisbach triangle |
| 2. | $\alpha_1, \alpha_2$ | Deflection angles |
| 3. | $\gamma$ | Unit weight of soil |
| 4. | $\phi$ | Angle of internal friction of soil |
| 5. | $\theta$ | Semi-apex angle of rock cone |

# APPENDIX III

## Abbrevations and Acronyms

## Part I Railway Engineering

| | | |
|---|---|---|
| 1. | RDSO | Research, Design and Standards Organisation, Lucknow |
| 2. | AREA | American Railway Engineers Association |
| 3. | CR | Central Railway |
| 4. | ER | Eastern Railway |
| 5. | ECR | East Central Railway |
| 6. | ECoR | East Coast Railway |
| 7. | NR | Northern Railway |
| 8. | NCR | North Central Railway |
| 9. | NER | North Eastern Railway |
| 10. | NFR | Northeast Frontier Railway |
| 11. | NWR | North Western Railway |
| 12. | SR | Southern Railway |
| 13. | SCR | South Central Railway |
| 14. | SER | South Eastern Railway |
| 15. | SECR | South East Central Railway |
| 16. | SWR | South Western Railway |
| 17. | WR | Western Railway |
| 18. | WCR | West Central Railway |
| 19. | KRCL | Konkan Railway Corporation Limited |
| 20. | COFMOW | Central Organisation For modernisation of Workshops |
| 21. | RITES | Rail India Technical and Economic Services Limited |
| 22. | IRCON | Indian Railways Construction Company Limited |
| 23. | DFCCIL | Dedicated Freight Corridor Corporation of India Limited |
| 24. | IRFC | Indian Railway Financial Corporation Limited |
| 25. | CONCOR | Container Corporation of India Limited |
| 26. | IRCTC | Indian Railway Catering and Tourism Corporation Limited |

| | | |
|---|---|---|
| 27. | BMRCL | Bangalore Metro Rail Corporation Limited |
| 28. | JMRC | Jaipur Metro Railway Corporation |
| 29. | Rail Tel | Railtel Corporation of India Limited |
| 30. | OFC | Optical Fibre Cable |
| 31. | CRIS | Centre for Railway Information System |
| 32. | RLDA | Rail Land Development Authority |
| 33. | IRWO | Indian Railway Welfare Organisation |
| 34. | CTI | Centralised Training Institute |
| 35. | TACV | Tracked Air-Cushioned Vehicle |
| 36. | MAGLEV | Magnetic Levitation Vehicle |
| 37. | EDM | Electronic Distance Measurement |
| 38. | TS | Total Station |
| 39. | GIS | Geographic Information System |
| 40. | GPS | Global Positioning System |
| 41. | DTM | Digital Terrain Modelling |
| 42. | EIA | Environmental Impact assessment |
| 43. | BG | Broad Gauge |
| 44. | SG | Standard Gauge |
| 45. | CG | Cape Gauge |
| 46. | MG | Metre Gauge |
| 47. | NG | Narrow Gauge |
| 48. | SAIL | Steel Authority of India Limited |
| 49. | FF | Flat-Footed (rail) |
| 50. | DH | Double-Headed (rail) |
| 51. | BH | Bull-Headed (rail) |
| 52. | UIC | International Union of Railways |
| 53. | IRS | Indian Railway Standard |
| 54. | RBS | Revised British Standard |
| 55. | UTS | Ultimate Tensile Strength |
| 56. | LWR | Long-Welded Rail |
| 57. | SWR | Short-Welded Rail |
| 58. | CWR | Continuous Welded Rail |
| 59. | SEJ | Switch-Expansion Joint |
| 60. | CSI | Complete Sleeper Index |
| 61. | CST | Central Standard Trial (Sleeper) |
| 62. | CF | Conventional Fastening |
| 63. | EF | Elastic Fastening |
| 64. | TNC | Theoretical Nose of Crossing |
| 65. | ANC | Actual Nose of Crossing |
| 66. | TVU | Train Vehicle Units |
| 67. | CTC | Centralised Train Control |

| | | |
|---|---|---|
| 68. | TPWS | Train Protection and Warning System |
| 69. | AWS | Automatic Warning System |
| 70. | ACD | Anti-Collision Device |
| 71. | ETKM | Equated Track Kilometre |
| 72. | PWI | Permanent Way Inspector |
| 73. | LVDT | Linear Voltage Differential Transducer |
| 74. | CTR | Combined Track Recording Index |
| 75. | MSP | Measured Shovel Packing |
| 76. | DTM | Directed Track Maintenance |
| 77. | TMS | Track Management System |
| 78. | PQRS | Plasser Quick Relaying System |
| 79. | TRT | Track Relaying Train |
| 80. | LIM | Linear Induction Motor |
| 81. | EMS | Electromagnetic Suspension |
| 82. | EDS | Electrodynamic Suspension |
| 83. | HSR | High Speed Rail |
| 84. | HSRC | High Speed Rail Corporation of India |
| 85. | RVNL | Rail Vikas Nigam Limited |
| 86. | EMU | Electric Multiple Unit |
| 87. | DMU | Diesel Multiple Unit |
| 88. | MRTS | Mass Rapid Transit System |
| 89. | BEST | Brihan Mumbai Electric Supply and Transport |

## PART II Airport Engineering

| | | |
|---|---|---|
| 1. | IAAI | International Airport Authority of India |
| 2. | NAA | National Airport Authority |
| 3. | AAI | Airport Authority of India |
| 4. | DGCA | Director-General of Civil Aviation |
| 5. | ICAO | International Civil Aviation Organisation |
| 6. | VFR | Visual Flight Rules |
| 7. | ILS | Instrument Landing System |
| 8. | EMP | Environmental Management Plan |
| 9. | ATC | Air Traffic Control |
| 10. | DIAL | Delhi International Airport Pvt. Ltd |
| 11. | MIAL | Mumbai International Airport Pvt. Ltd |
| 12. | BIAL | Bengaluru International Airport Pvt. Ltd |
| 13. | LOD | Lift-off Distance |
| 14. | TOD | Take-off Distance |
| 15. | ART | Airport Reference Temperature |
| 16. | FAA | Federal Aviation Administration |

| | | |
|---|---|---|
| 17. | PCA | Portland Cement Aviation |
| 18. | LCN | Load Classification Number |
| 19. | ESWL | Equivalent Single Wheel Load |
| 20. | PCN | Pavement Classification Number |
| 21. | ACN | Aircraft Classification Number |
| 22. | RVR | Runway Visual Range |
| 23. | IFR | Instrument Flight Rules |
| 24. | ARTCC | Air Route Traffic Control Centre |
| 25. | FSS | Flight Service Station |
| 26. | DVOR | Doppler Very High Frequency Omni Range |
| 27. | VOR | Very High Frequency Omnidirectional Range |
| 28. | ARSR | Air Route Surveillance Radar |
| 29. | TACAN | Tactical Air Navigation |
| 30. | DME | Distance-Measuring Equipment |
| 31. | RDF | Radio Direction Finder |
| 32. | ADF | Automatic Direction Finder |
| 33. | PAR | Precision Approach Radar |
| 34. | GAC | Ground Approach Control |
| 35. | ASDE | Airport Surface Detection Equipment |
| 36. | ALS | Airport Lighting System |
| 37. | ARP | Airport Reference Point |
| 38. | CTOL | Conventional Take-off and Landing |
| 39. | GCA | Ground Control Approach |
| 40. | LF/MF/HF | Low Frequency/Medium Frequency/High Frequency |
| 41. | WMO | World Meteorological Organisation |
| 42. | DFDRS | Digital Flight Data Recording System |
| 43. | AIDS | Aircraft Integrated Data System |
| 44. | WHO | World Health Organisation |

## Part III Dock and Harbour Engineering

| | | |
|---|---|---|
| 1. | GRT | Gross Registered Tonnage |
| 2. | PIANC | Permanent International Association of Navigation Congress |
| 3. | MSL | Mean Sea Level |
| 4. | HHW | Highest of High Tide Level |
| 5. | LLW | Lowest of Low Tide Level |
| 6. | PPI | Plan Position Indicator |
| 7. | DWT | Dead Weight Tonnage |

## Part IV Bridge Engineering

| | | |
|---|---|---|
| 1. | IDF | Intensity-Duration-Frequency (curves) |
| 2. | IRSDO | Indian Railway Standards and Design Organisation, Lucknow |
| 3. | CDA | Coefficient of Dynamic Augment |
| 4. | ILM | Incremental Launching Method (of bridge superstructure) |

## Part V Tunnel Engineering

| | | |
|---|---|---|
| 1. | NATM | New Austrian Tunnelling Method |
| 2. | SEM | Sequential Excavation Method |
| 3. | TBM | Tunnel Boring Machine |
| 4. | EPB | Earth Pressure Balance |
| 5. | WRA | World Road Association |
| 6. | PIARC | Permanent International Association of Road Congresses (WRA) |
| 7. | SGR | Site Ground Water Rating |
| 8. | QRAM | Quantitative Risk Assessment Model |
| 9. | DSM | Decision Support Model |
| 10. | OECD | Organisation for Economic Co-operation and Development |
| 11. | HVHD | High Velocity High Density (explosive) |
| 12. | LVLD | Low Velocity Low Density (explosive) |

# APPENDIX IV

## References

## Part I Railway Engineering


| S. No. | Author(s) | Title and Description |
|---|---|---|
| 1. | Agarwal MM | *Indian Railway Track*, 19th ed., Ruby-Jubilee Edition, Standard Publishers Distributers, New Delhi, 2013. |
| 2. | Antia KF | *Railway Track*, The New Book Co. Pvt Ltd, Bombay, 1960. |
| 3. | Cowi | "Environmental Impact Assessment of Railway Projects", Consultancy Services, Lyngby, Denmark (www.cowi.com). |
| 4. | Hay WW | *Railroad Engineering*, 2nd ed., Wiley, 1982. |
| 5. | Indian Rail Board | Rules for the preparation of railway project |
| 6. | Inglis RA | *An Introduction to Railway Engineering*, Chapman and Hall Ltd, London, 1953. |
| 7. | Mundrey JS | *Railway Track Engineering*, 4th ed., Tata McGraw–Hill Education Pvt Ltd, New Delhi, 2010. |
| 8. | Satish Chandra and Agarwal MM | *Railway Engineering*, 2nd ed., Oxford University Press, New Delhi, 2013. |
| 9. | Srinivasan M | *Modern Permanent Way*, Somaiya Publications Pvt Ltd, Bombay, 1969. |

| S. No. | Author(s) | Title and Description |
|---|---|---|
| 10. | Talbot AN (1920) | "Stresses in railway track", *AREA Manual*, 1991. |
| 11. | Webb WL | *Railroad Construction – Theory and Practice*, John Wiley and Sons, New York, 1917. |


## Part II Airport Engineering


| S. No. | Author(s) | Title and Description |
|---|---|---|
| 1. | Ashford NJ and Wright PH | *Airport Engineering*, 3rd ed., John Wiley and Sons, New York, USA, 1992. |
| 2. | Burmister DM | "The theory of stresses and displacements in layered systems and application to the design of airport runways," *Proceedings of Highway Research Board*, Vol.23, 1943. |
| 3. | Burmister DM | "The general theory of stresses and displacements in layered systems", *Journal of Applied Physics*, Vol. 16, pp. 84–94, 1945. |
| 4. | Directorate of Civil Engineering Development, Dept of Environment, U.K. | "Design and Evaluation of Airfield Pavements", 1971. |
| 5. | Khannna SK; Arora MG, and Jain SS | "Airport Planning and Design (Sixth Edition)", Nem Chand & Bros., Roorkee, 2009. |
| 6. | Ministry of Environment and Forests, Dept. of Environment, Forests and Wildlife, Govt of India | "Environmental Guidelines for Airport Projects" (Report prepared by a Working Group), 1989. |
| 7. | Walter GE and Roggeveen WJ | "Airport Approach, Runway and Taxiway Lighting Systems", *Journal of Air-transport Division*, *ASCE*, Vol.84, No.1669, 1958. |
| 8. | Westergaard HM | "New formulas for stresses in Concrete Airfield Pavement", *Proc. ASCE*, 1947. |


## Part III Dock and Harbour Enginering


| S. No. | Author(s) | Title and description |
|---|---|---|
| 1. | Doodson AT and Warburg HD | "Admiralty Manual of Tides", Her Majesty's stationery office, London, U.K., 1941. |
| 2. | Eduardo Castro and Ramon Iribarren | "Formula for the Construction of Rockfill dykes", 1953 |
| 3. | Indian Meteorological Department | "Winds, weather and currents on the coast of India and the Laws of Storms, IMD, 2008 |
| 4. | Johnson DW | "Shore Processes and shore line development", John Wiley & Sons, New York, USA, 1919 |
| 5. | Minikin RR | "Winds, waves and maritime structures", Charles Griffin and Company Ltd, London, U.K., 1950 |

| S. No. | Author(s) | Title and description |
|---|---|---|
| 6. | Oza HP and Oza GH | *Dock and Harbour Engineering*, 7th ed., Charotar Publishing House Pvt Ltd, Anand, Gujarat, India, 2013. |
| 7. | Ramon Iribarren Cavanilles | A formula for the construction of Rockfill Dykes" (IER Series), 1938. |
| 8. | Srinivasan R | *Dock, Harbour and Tunnel Engineering*, 7th ed., Charotar Publishing House Pvt. Ltd, Anand, Gujarat, India, 2013. |
| 9. | Survey of India | "The Indian Tide Tables", 1998 |
| 10. | US Environmental Protection Agency | "Indicators of the Environmental Impacts of Transportation", *USEPA*, 1996 |


## Part IV Bridge Engineering


| S. No. | Author(s) | Title and description |
|---|---|---|
| 1. | Anderson, Paul | *Substructure Analysis and Design*, 2nd ed., Ronal Press, New York, USA, 1956. |
| 2. | Jacoby HS and Davis RP | *Foundations of Bridges and Buildings*, 3rd ed., McGraw Hill Book Company, NY, USA, 1941. |
| 3. | Mike La Violette | "Incremental Launching Method" (for AASHTO), 2007 |
| 4. | Ratay RT (Ed.) | *Handbook of temporary structures in construction*, 2nd ed. McGraw Hill Book Co, NY, USA, 1996. |
| 5. | Smith DW | "Bridge Failures", *Proc. Institution of Civil Engineers*, London, U. K., Vol. 61, No. 3, pp. 367–382, 1976. |
| 6. | Steinman DB | *A Practical Treatise on Suspension Bridges*, John Wiley, New York, USA. |
| 7. | Steinman DB and Watson SR | "Bridges and their Builders", Dover Publications, New York, USA, 1957. |
| 8. | Victor DJ | *Essentials of Bridge Engineering*, 6th ed., Oxford and IBH Co., New Delhi, 2007. |
| 9. | Wai Fah Chen and Lian Duan | *Bridge Engineering Handbook*, 2nd ed., Construction and Maintenance, 2014. |
| 10. | Yihong Gao | "Bridge Load Ratings," MnDOT Bridge Office, *LRFD Workshop*, June, 2012 |


## Part-V Tunnel Engineering


| S. No. | Author(s) | Title and description |
|---|---|---|
| 1. | COWI | "Decision Support Model (DSM)" (for OECD), ERS 2 Group, Report No.42069-006, Issue no.2, April, 2000. |
| 2. | COWI | "Tunnel Engineering", Cowi Consultancy Services, Denmark, 2009. |
| 3. | El Tani M | "Circular tunnel in a semi-infinite aquifer", *Journal of Tunnelling and Underground Space Technology*, Vol. 18, pp 49–55, 2003. |
| 4. | Goodman et al | "Groundwater inflow during tunnel driving", *Engineering Geology*, Vol. 1, pp 150–162, 1965 |
| 5. | Heuer RE | "Estimating Rock Tunnel Water inflow", Proc of the Rapid Excavation and Tunnelling Conference, 1995 and 2009. |

| S. No. | Author(s) | Title and description |
|---|---|---|
| 6. | Karlsrud K | "Water control when tunnelling under urban areas in the Oslo region", *NFF Publication*, Vol. 12, No. 4, pp. 27–35, 2001. |
| 7. | Katibeh H and Aalianvari A | "Development of a new method for tunnel site rating from ground water rating point of view", *Journal of Applied Sciences*, Vol.9, pp. 1496–1502, 2009. |
| 8. | Katibeh H and Aalianvari A | "Common approximations to the water inflow into tunnels", Chapter 4 of *Drainage Systems* (Ed), Prof. Muhammad Salik Javid, InTech , 2012. |
| 9. | Lei | "Groundwater inflow into hard rock tunnels", Common approximations to the water inflow into tunnels, Chapter 4 of Drainage Systems, InTech, 1999. |
| 10. | OECD and PIARC | Safety in Tunnels: "Transport of dangerous goods through road tunnels", OECD Publishing, Paris, 2001 [http://dx.doi.org/10.1787/9789264195691-en] |
| 11. | PIARC | "Fire and smoke control in road tunnels", *Tunnels Committee*, PIARC Paris, 1999. |
| 12. | World Road Association (PIARC) | Chapter 4, "Operation and maintenance of tunnels", *Road Tunnels Manual*, 2011 (electronic version). [http://tunnels.piarc.org/en] |

The following are the IRS, IRC and IRS Special Publication (SP) codes of relevance, along with their title and description:

## Indian Railways Standard Codes

## (Issued by RDSO, Ministry of Railways, Govt of India)

| S. No. | IRS Code | Title and description |
|---|---|---|
| 1. | IRS Bridge Rules<br>(Second Revision 2008)<br>[First Adopted: 1941<br>First Revision: 1964] | Indian Railway Standard Bridge Rules – Second Revision, 2008 (with correction slips up to 39) |
| 2. | IRS Steel Bridge Code<br>[Adopted: 1941<br>2003 – Incorporating A & C Slip No 17] | Indian Railway Standard Code of Practice – For the design of steel or wrought from bridges carrying rail, road, or pedestrian traffic. |
| 3. | IRS Concrete Bridge Code<br>[Adopted: 1936<br>2003 – Incorporating A & C Slip No.7] | Indian Railway Standard Code of Practice – For plain, reinforced and prestressed concrete for general bridge construction. |
| 4. | IRS Bridge Rules (in SI Units)<br>Adopted:1941<br>First Revision – 1964<br>Second Reprint – 2008<br>(Incorporating correction slips up to 39) | Rules specifying the loads for design of superstructure and substructure of bridges and for assessment of the strength of existing bridges. |

| S. No. | IRS Code | Title and description |
|---|---|---|
| 5. | IRS Arch Bridge Code | |
| 6. | IRS Bridge Substructures and Foundation Code<br>Adopted: 1936<br>Revised: 1985<br>Reprinted: 2009 (A & C Slips up to No.29) | Indian Railway Code of Practice – For the design of substructure and foundations of bridges. |
| 7. | IRS Manual on the Design and Construction of Well and Pile Foundations<br>Adopted: 1941; Revised:1985<br>Revised :2010 (A & C Slips up to No.2) | Indian Railway Standard Manual on the Design and Construction of Well and Pile Foundations (Well and Pile Foundation Code) |

## IRC Codes

| S. No. | Code No. and Year | Title and description |
|---|---|---|
| 1. | 5-2015 | Standard specifications and code of practice for road bridges, Section I: General features of design (Eighth revision) |
| 2. | 6-2014 | Standard specifications and code of practice for road bridges, Section II: Loads and stresses (Sixth revision) |
| 3. | 7-1971 | Recommended practice for numbering bridges and culverts (First revision) |
| 4. | 18-2000 | Design criteria for prestressed concrete road bridges (Post-tensioned concrete) (Third revision) |
| 5. | 21-2000 | Standard specifications and code of practice for road bridges, Section III: Cement concrete (Plain and reinforced)(Third revision) |
| 6. | 22-2015 | Standard specifications and code of practice for road bridges, Section VI: Composite construction (Limit state) |
| 7. | 24-2010 | Standard specifications and code of practice for road bridges, Section V: Steel road bridges (Limit state method) (Third revision) |
| 8. | 40-2002 | Standard specifications and code of practice for road bridges, Section IV: Brick, stone and black masonry (Second revision) |
| 9. | 76-1979 | Tentative guidelines for structural strength evaluation of rigid airfield pavements |
| 10. | 78-2014 | Standard specifications and code of practice for road bridges, Section VII: Foundations and substructures (Revised edition) |
| 11. | 83-2015(Part I) | Standard specifications and code of practice for road bridges, Section IX: Bearings (Part I): Roller and booker bearings (Second revision) |
| 12. | 83-2015(Part II) | Standard specifications and code of practice for road bridges, Section IX: Bearings (Part II): Elastomeric bearings (First revision) |
| 13. | 85-2015 | Recommended practice for accelerated strength testing and evaluation of concrete (First revision) |
| 14. | 87-2011 | Guidelines for formwork, false work and temporary structures (First revision) |
| 15. | 89-1997 | Guidelines for design and construction of river training and control works for road bridges (First revision) |

## IRC Special Publications (SP)

| S. No. | SP No. and Year | Title and description |
|---|---|---|
| 1. | SP:18 (1978) | Manual for highway bridge maintenance inspection |
| 2. | SP:35 (1990) | Guidelines for inspection and maintenance of bridges |
| 3. | SP:37 (2010) | Guidelines for evaluation of load-carrying capacity of bridges (First revision) |
| 4. | SP:40 (1993) | Guidelines on techniques for strengthening and rehabilitation of bridges |
| 5. | SP:51 (2015) | Guidelines for load testing of bridges (First revision) |
| 6. | SP:52 (1999) | Bridge inspector's reference manual |
| 7. | SP: 60 (2002) | An approach document for assessment of remaining life of concrete bridges |
| 8. | SP: 69 (2011) | Guidelines and specification for expansion joints (First revision) |
| 9. | SP:74 (2007) | Guidelines for repair and rehabilitation of steel bridges |
| 10. | SP:75 (2008) | Guidelines for retrofitting of steel bridges by prestressing |
| 11. | SP:80 (2008) | Guidelines for corrosion prevention, monitoring and remedial measures for concrete bridge structures |

The following are the relevant Indian Standard Codes of Practice along with their title:

## Indian Standard Codes of Practice

## IS Codes

| S. No. | Code No. and Year | Title |
|---|---|---|
| 1. | 7314 – 1974 | Glossary of terms relating to Port and Harbour Engineering |
| 2. | 4651(Parts I–V) (1974–1989) | Code of practice for planning and design of ports and harbours |
| 3. | 9527 (Parts I–VI) (1980–1989) | Code of practice for design and construction of port and harbour structures |
| 4. | 875 (Part III) – 1987 | Wind map of India |
| 5. | 1893(Part I) – 2002 | Indian standard criteria for earthquake resistant design of structures |
| 6. | 2911 (Part III) –1980 (First revision) | Code of practice for construction of pile foundations: Under-ream piles |
| 7. | 2950 (Part I) –1992 (Second revision) | Code of practice for design and construction of raft foundation, Part I (Design) |
| 8. | 3764 – 2002 | Code of safety for excavation work |
| 9. | 5878–6 (1975) | Code of practice for construction of tunnels conveying water – Part 6: Steel lining |
| 10. | 15026 (2002) | Guidelines for tunnelling methods in rock masses |

# INDEX

## A

## B

## C

## D

## E

## F

## G



## H

## L



## M

## N



## O

## P

## Q



## R

## S

## T

## U



## V



## W

**X**



**Z**